Some Physical Constants

Quantity	Symbol	Value[a]
Atomic mass unit	u	$1.660\ 538\ 782\ (83) \times 10^{-27}$ kg $931.494\ 028\ (23)$ MeV/c^2
Avogadro's number	N_A	$6.022\ 141\ 79\ (30) \times 10^{23}$ particles/mol
Bohr magneton	$\mu_B = \dfrac{e\hbar}{2m_e}$	$9.274\ 009\ 15\ (23) \times 10^{-24}$ J/T
Bohr radius	$a_0 = \dfrac{\hbar^2}{m_e e^2 k_e}$	$5.291\ 772\ 085\ 9\ (36) \times 10^{-11}$ m
Boltzmann's constant	$k_B = \dfrac{R}{N_A}$	$1.380\ 650\ 4\ (24) \times 10^{-23}$ J/K
Compton wavelength	$\lambda_C = \dfrac{h}{m_e c}$	$2.426\ 310\ 217\ 5\ (33) \times 10^{-12}$ m
Coulomb constant	$k_e = \dfrac{1}{4\pi\epsilon_0}$	$8.987\ 551\ 788 \ldots \times 10^9$ N·m²/C² (exact)
Deuteron mass	m_d	$3.343\ 583\ 20\ (17) \times 10^{-27}$ kg $2.013\ 553\ 212\ 724\ (78)$ u
Electron mass	m_e	$9.109\ 382\ 15\ (45) \times 10^{-31}$ kg $5.485\ 799\ 094\ 3\ (23) \times 10^{-4}$ u $0.510\ 998\ 910\ (13)$ MeV/c^2
Electron volt	eV	$1.602\ 176\ 487\ (40) \times 10^{-19}$ J
Elementary charge	e	$1.602\ 176\ 487\ (40) \times 10^{-19}$ C
Gas constant	R	$8.314\ 472\ (15)$ J/mol·K
Gravitational constant	G	$6.674\ 28\ (67) \times 10^{-11}$ N·m²/kg²
Neutron mass	m_n	$1.674\ 927\ 211\ (84) \times 10^{-27}$ kg $1.008\ 664\ 915\ 97\ (43)$ u $939.565\ 346\ (23)$ MeV/c^2
Nuclear magneton	$\mu_n = \dfrac{e\hbar}{2m_p}$	$5.050\ 783\ 24\ (13) \times 10^{-27}$ J/T
Permeability of free space	μ_0	$4\pi \times 10^{-7}$ T·m/A (exact)
Permittivity of free space	$\epsilon_0 = \dfrac{1}{\mu_0 c^2}$	$8.854\ 187\ 817 \ldots \times 10^{-12}$ C²/N·m² (exact)
Planck's constant	h	$6.626\ 068\ 96\ (33) \times 10^{-34}$ J·s
	$\hbar = \dfrac{h}{2\pi}$	$1.054\ 571\ 628\ (53) \times 10^{-34}$ J·s
Proton mass	m_p	$1.672\ 621\ 637\ (83) \times 10^{-27}$ kg $1.007\ 276\ 466\ 77\ (10)$ u $938.272\ 013\ (23)$ MeV/c^2
Rydberg constant	R_H	$1.097\ 373\ 156\ 852\ 7\ (73) \times 10^7$ m^{-1}
Speed of light in vacuum	c	$2.997\ 924\ 58 \times 10^8$ m/s (exact)

Note: These constants are the values recommended in 2006 by CODATA, based on a least-squares adjustment of data from different measurements. For a more complete list, see P. J. Mohr, B. N. Taylor, and D. B. Newell, "CODATA Recommended Values of the Fundamental Physical Constants: 2006." *Rev. Mod. Phys.* **80:**2, 633–730, 2008.

[a]The numbers in parentheses for the values represent the uncertainties of the last two digits.

Solar System Data

Body	Mass (kg)	Mean Radius (m)	Period (s)	Mean Distance from the Sun (m)
Mercury	3.30×10^{23}	2.44×10^{6}	7.60×10^{6}	5.79×10^{10}
Venus	4.87×10^{24}	6.05×10^{6}	1.94×10^{7}	1.08×10^{11}
Earth	5.97×10^{24}	6.37×10^{6}	3.156×10^{7}	1.496×10^{11}
Mars	6.42×10^{23}	3.39×10^{6}	5.94×10^{7}	2.28×10^{11}
Jupiter	1.90×10^{27}	6.99×10^{7}	3.74×10^{8}	7.78×10^{11}
Saturn	5.68×10^{26}	5.82×10^{7}	9.29×10^{8}	1.43×10^{12}
Uranus	8.68×10^{25}	2.54×10^{7}	2.65×10^{9}	2.87×10^{12}
Neptune	1.02×10^{26}	2.46×10^{7}	5.18×10^{9}	4.50×10^{12}
Pluto[a]	1.25×10^{22}	1.20×10^{6}	7.82×10^{9}	5.91×10^{12}
Moon	7.35×10^{22}	1.74×10^{6}	—	—
Sun	1.989×10^{30}	6.96×10^{8}	—	—

[a]In August 2006, the International Astronomical Union adopted a definition of a planet that separates Pluto from the other eight planets. Pluto is now defined as a "dwarf planet" (like the asteroid Ceres).

Physical Data Often Used

Average Earth–Moon distance	3.84×10^{8} m
Average Earth–Sun distance	1.496×10^{11} m
Average radius of the Earth	6.37×10^{6} m
Density of air (20°C and 1 atm)	1.20 kg/m^3
Density of air (0°C and 1 atm)	1.29 kg/m^3
Density of water (20°C and 1 atm)	1.00×10^{3} kg/m^3
Free-fall acceleration	9.80 m/s^2
Mass of the Earth	5.97×10^{24} kg
Mass of the Moon	7.35×10^{22} kg
Mass of the Sun	1.99×10^{30} kg
Standard atmospheric pressure	1.013×10^{5} Pa

Note: These values are the ones used in the text.

Some Prefixes for Powers of Ten

Power	Prefix	Abbreviation	Power	Prefix	Abbreviation
10^{-24}	yocto	y	10^{1}	deka	da
10^{-21}	zepto	z	10^{2}	hecto	h
10^{-18}	atto	a	10^{3}	kilo	k
10^{-15}	femto	f	10^{6}	mega	M
10^{-12}	pico	p	10^{9}	giga	G
10^{-9}	nano	n	10^{12}	tera	T
10^{-6}	micro	μ	10^{15}	peta	P
10^{-3}	milli	m	10^{18}	exa	E
10^{-2}	centi	c	10^{21}	zetta	Z
10^{-1}	deci	d	10^{24}	yotta	Y

Physics

for Scientists and Engineers

Physics
for Scientists and Engineers

Volume 1

eighth edition

Raymond A. Serway
Emeritus, James Madison University

John W. Jewett, Jr.
Emeritus, California State Polytechnic University, Pomona

With contributions from Vahé Peroomian, *University of California at Los Angeles*

BROOKS/COLE
CENGAGE Learning™

Australia • Brazil • Japan • Korea • Mexico • Singapore • Spain • United Kingdom • United States

BROOKS/COLE
CENGAGE Learning™

Physics for Scientists and Engineers, Volume 1, Eighth Edition
Raymond A. Serway and John W. Jewett, Jr.

Executive Vice President and Publisher, Arts and Sciences: Sean Wakely
Vice President, Editor-in-Chief, Sciences: Michelle Julet
Publisher: Mary Finch
Development Editor: Ed Dodd
Associate Developmental Editor: Brandi Kirksey
Editorial Assistant: Joshua Duncan
Senior Media Editor: Rebecca Berardy Schwartz
Marketing Manager: Nicole Mollica
Marketing Communications Manager: Belinda Krohmer
Marketing Coordinator: Kevin Carroll
Project Manager, Editorial Production: Cathy L. Brooks
Creative Director: Bruce Bond
Senior Art Directors: Jill Haber Atkins and Cate Rickard Barr
Print Buyer: Diane Gibbons
Permissions Editor: Timothy Sisler
Senior Permissions Account Manager, Images: Robyn Young
Text Designer: Brian Salisbury
Copyeditors: Kathleen Lafferty, Sara Black
Illustrators: Dartmouth Publishing Inc., Lachina Publishing Services
Cover Designer: Brian Salisbury
Cover Image: © Stirling Energy Systems, SunCatcher™
Compositor: Lachina Publishing Services

Library of Congress Control Number: 2009923972

ISBN-13: 978-1-439-04838-2

ISBN-10: 1-439-04838-X

Brooks/Cole
10 Davis Drive
Belmont, CA 94002-3098
USA

Printed in China
3 4 5 6 7 13 12 11 10

We dedicate this book to our wives Elizabeth and Lisa and all our children and grandchildren for their loving understanding when we spent time on writing instead of being with them.

brief contents

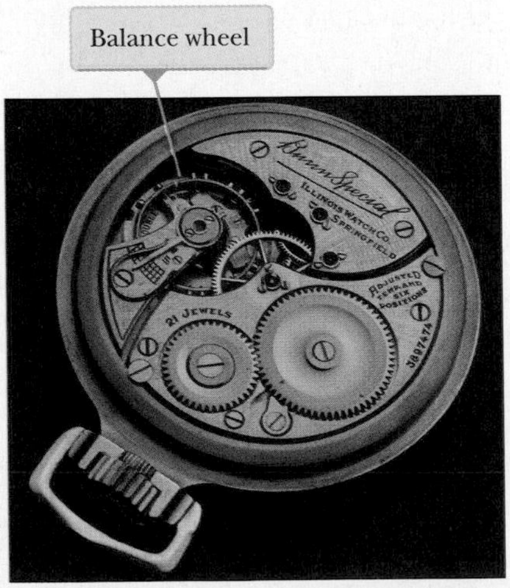

contents

The velocity of the projectile (red arrows) changes in direction and magnitude, but its acceleration (purple arrows) remains constant.

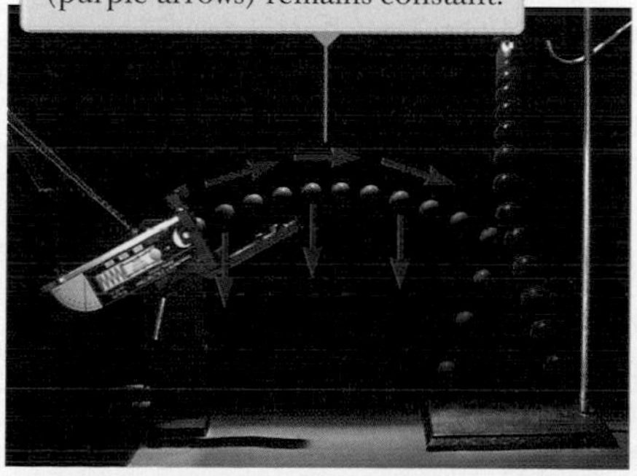

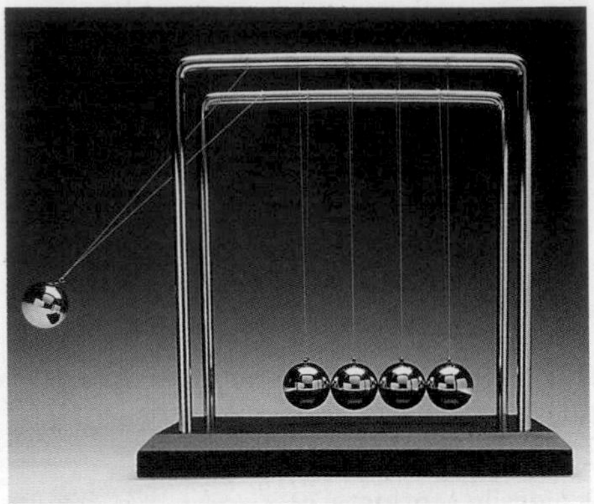

© Cengage Learning/Charles D. Winters

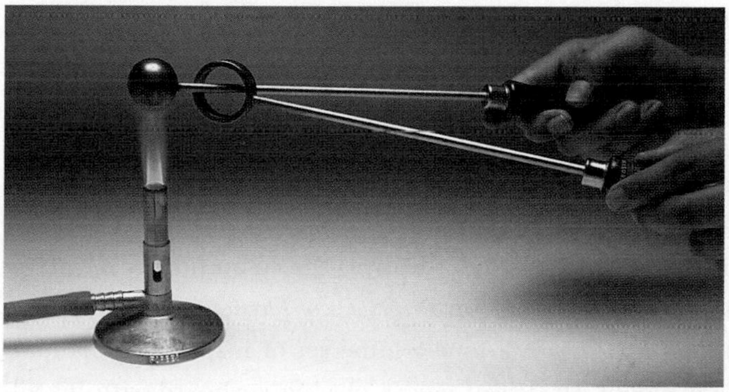

about the authors

Raymond A. Serway received his doctorate at Illinois Institute of Technology and is Professor Emeritus at James Madison University. In 1990, he received the Madison Scholar Award at James Madison University, where he taught for 17 years. Dr. Serway began his teaching career at Clarkson University, where he conducted research and taught from 1967 to 1980. He was the recipient of the Distinguished Teaching Award at Clarkson University in 1977 and the Alumni Achievement Award from Utica College in 1985. As Guest Scientist at the IBM Research Laboratory in Zurich, Switzerland, he worked with K. Alex Müller, 1987 Nobel Prize recipient. Dr. Serway also was a visiting scientist at Argonne National Laboratory, where he collaborated with his mentor and friend, the late Dr. Sam Marshall. Dr. Serway is the coauthor of *College Physics,* eighth edition; *Principles of Physics: A Calculus-Based Text,* fourth edition, *Essentials of College Physics, Modern Physics,* third edition, and the high school textbook *Physics,* published by Holt McDougal. In addition, Dr. Serway has published more than 40 research papers in the field of condensed matter physics and has given more than 60 presentations at professional meetings. Dr. Serway and his wife Elizabeth enjoy traveling, playing golf, fishing, gardening, singing in the church choir, and especially spending quality time with their four children and nine grandchildren.

John W. Jewett, Jr. earned his undergraduate degree in physics at Drexel University and his doctorate at Ohio State University, specializing in optical and magnetic properties of condensed matter. Dr. Jewett began his academic career at Richard Stockton College of New Jersey, where he taught from 1974 to 1984. He is currently Emeritus Professor of Physics at California State Polytechnic University, Pomona. Through his teaching career, Dr. Jewett has been active in promoting science education. In addition to receiving four National Science Foundation grants, he helped found and direct the Southern California Area Modern Physics Institute (SCAMPI) and Science IMPACT (Institute for Modern Pedagogy and Creative Teaching), both of which work with teachers and schools to develop effective science curricula. Dr. Jewett's honors include four Meritorious Performance and Professional Promise awards, the Stockton Merit Award at Richard Stockton College in 1980, selection as Outstanding Professor at California State Polytechnic University for 1991/1992, and the Excellence in Undergraduate Physics Teaching Award from the American Association of Physics Teachers (AAPT) in 1998. He has given over 90 presentations both domestically and abroad, including multiple presentations at national meetings of the AAPT. Dr. Jewett is the author of *The World of Physics: Mysteries, Magic, and Myth,* which provides many connections between physics and everyday experiences. In addition to his work as the co-author for *Physics for Scientists and Engineers* he is also the co-author on *Principles of Physics: A Calculus-Based Text,* fourth edition, as well as *Global Issues,* a four-volume set of instruction manuals in integrated science for high school. Dr. Jewett enjoys playing keyboard with his all-physicist band, traveling, underwater photography, running, and collecting antique quack medical devices that can be used as demonstration apparatus in physics lectures. Most importantly, he relishes spending time with his wife Lisa and their children and grandchildren.

preface

In writing this eighth edition of *Physics for Scientists and Engineers*, we continue our ongoing efforts to improve the clarity of presentation and include new pedagogical features that help support the learning and teaching processes. Drawing on positive feedback from users of the seventh edition, data gathered from both professors and students who use Enhanced WebAssign, as well as reviewers' suggestions, we have refined the text to better meet the needs of students and teachers.

This textbook is intended for a course in introductory physics for students majoring in science or engineering. The entire contents of the book in its extended version could be covered in a three-semester course, but it is possible to use the material in shorter sequences with the omission of selected chapters and sections. The mathematical background of the student taking this course should ideally include one semester of calculus. If that is not possible, the student should be enrolled in a concurrent course in introductory calculus.

Objectives

This introductory physics textbook has two main objectives: to provide the student with a clear and logical presentation of the basic concepts and principles of physics and to strengthen an understanding of the concepts and principles through a broad range of interesting real-world applications. To meet these objectives, we emphasize sound physical arguments and problem-solving methodology. At the same time, we attempt to motivate the student through practical examples that demonstrate the role of physics in other disciplines, including engineering, chemistry, and medicine.

Changes in the Eighth Edition

A large number of changes and improvements were made for the Eighth Edition of this text. Some of the new features are based on our experiences and on current trends in science education. Other changes were incorporated in response to comments and suggestions offered by users of the seventh edition and by reviewers of the manuscript. The features listed here represent the major changes in the Eighth Edition.

Line-by-Line Revision of the Questions and Problems Set. For the Eighth Edition, the authors reviewed each Question and Problem and incorporated revisions designed to improve both readability and assignability. To make problems clearer to both students and instructors, this extensive process involved editing problems for clarity, editing for length, adding figures where appropriate, and introducing better problem architecture by breaking up problems into clearly defined parts.

Data from Enhanced WebAssign Used to Improve Questions and Problems. As part of the full-scale analysis and revision of the questions and problems sets, the authors utilized extensive user data gathered by WebAssign, from both instructors who assigned and students who worked on problems from previous editions of *Physics for Scientists and Engineers*. This data helped tremendously, indicating when the phrasing in problems could be clearer, thus providing guidance on how to revise problems so that they are more easily understandable for students and more easily assignable by instructors in Enhanced WebAssign. Finally, the data was used to ensure that the problems most often assigned were retained for this new edition. In each chapter's problems set, the top quartile of problems assigned in Enhanced WebAssign have blue-shaded problem numbers for easy identification, allowing professors to quickly and easily find the most popular problems assigned in Enhanced WebAssign.

To provide an idea of the types of improvements that were made to the problems, here are problems from the seventh edition, followed by the problem as it now appears in the eighth edition, with explanations of how the problems were improved.

Problem from the Seventh Edition . . .

38. (a) Consider an extended object whose different portions have different elevations. Assume the free-fall acceleration is uniform over the object. Prove that the gravitational potential energy of the object–Earth system is given by $U_g = Mgy_{CM}$, where M is the total mass of the object and y_{CM} is the elevation of its center of mass above the chosen reference level. (b) Calculate the gravitational potential energy associated with a ramp constructed on level ground with stone with density 3 800 kg/m^3 and everywhere 3.60 m wide. In a side view, the ramp appears as a right triangle with height 15.7 m at the top end and base 64.8 m (Fig. P9.38).

Figure P9.38

. . . As revised for the Eighth Edition:

39. Explorers in the jungle find an ancient monument in the shape of a large isosceles triangle as shown in Figure P9.39. The monument is made from tens of thousands of small stone blocks of density 3 800 kg/m^3. The monument is 15.7 m high and 64.8 m wide at its base and is everywhere 3.60 m thick from front to back. Before the monument was built many years ago, all the stone blocks lay on the ground. How much work did laborers do on the blocks to put them in position while building the entire monument? *Note*: The gravitational potential energy of an object–Earth system is given by $U_g = Mgy_{CM}$, where M is the total mass of the object and y_{CM} is the elevation of its center of mass above the chosen reference level.

A storyline for the problem is provided.

The requested quantity is made more personal by asking for work done by humans rather than asking for the gravitational potential energy.

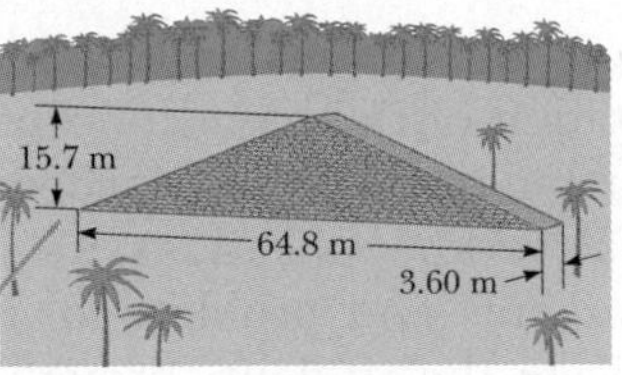

Figure P9.39

The expression for the gravitational potential energy is provided, whereas it was requested to be proven in the original. This allows the problem to work better in Enhanced WebAssign.

The figure has been revised and dimensions added.

Problem from the Seventh Edition . . .

67. A bicycle is turned upside down while its owner repairs a flat tire. A friend spins the other wheel, of radius 0.381 m, and observes that drops of water fly off tangentially. She measures the height reached by drops moving vertically (Fig. P10.67). A drop that breaks loose from the tire on one turn rises h = 54.0 cm above the tangent point. A drop that breaks loose on the next turn rises 51.0 cm above the tangent point. The height to which the drops rise decreases because the angular speed of the wheel decreases. From this information, determine the magnitude of the average angular acceleration of the wheel.

Figure P10.67 Problems 67 and 68.

. . . As revised for the Eighth Edition:

68. A bicycle is turned upside down while its owner repairs a flat tire on the rear wheel. A friend spins the front wheel, of radius 0.381 m, and observes that drops of water fly off tangentially in an upward direction when the drops are at the same level as the center of the wheel. She measures the height reached by drops moving vertically (Fig. P10.68). A drop that breaks loose from the tire on one turn rises h = 54.0 cm above the tangent point. A drop that breaks loose on the next turn rises 51.0 cm above the tangent point. The height to which the drops rise decreases because the angular speed of the wheel decreases. From this information, determine the magnitude of the average angular acceleration of the wheel.

Information about drops leaving the wheel is clarified.

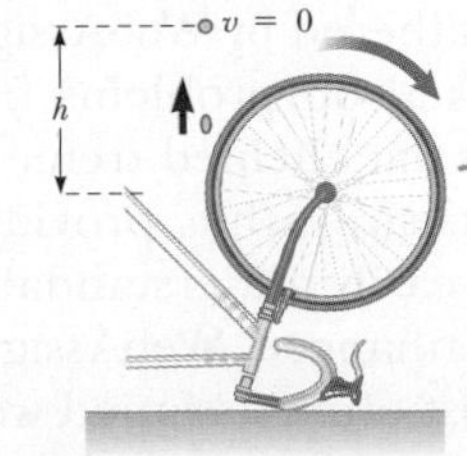

Figure P10.68 Problems 68 and 69.

The figure accompanying the problem has been redrawn to show the front wheel rather than the back wheel, to remove the complicating features of the pedals, chain, and derailleur gear.

Revised Questions and Problems Set Organization. We reorganized the end-of-chapter questions and problems sets for this new edition. The previous edition's Questions section is now divided into two sections: Objective Questions and Conceptual Questions.

Objective Questions are multiple-choice, true/false, ranking, or other multiple guess-type questions. Some require calculations designed to facilitate students' familiarity with the equations, the variables used, the concepts the variables represent, and the relationships between the concepts. Others are more conceptual in nature and are designed to encourage conceptual thinking. Objective Questions are also written with the personal response system user in mind, and most of the questions could easily be used in these systems.

Conceptual Questions are more traditional short-answer and essay-type questions that require students to think conceptually about a physical situation.

The first part of the *Problems* set is organized by the sections in each chapter, but within each section the problems now "platform" students to higher-order thinking by presenting all the straightforward problems in the section first, followed by the intermediate problems. (The problem numbers for straightforward problems are printed in **black**; intermediate-level problems are in **blue.**) The *Additional Problems* section remains in its usual place, but at the end of each chapter there is a new section, *Challenge Problems,* that gathers the most difficult problems for a given chapter in one place. (Challenge problems have problem numbers marked in **red**.)

New Types of Problems. We have introduced four new problem types for this edition:

QC *Quantitative/Conceptual* problems contain parts that ask students to think both quantitatively and conceptually. An example of a Quantitative/Conceptual problem appears here:

53. **QC** A horizontal spring attached to a wall has a force constant of $k = 850$ N/m. A block of mass $m = 1.00$ kg is attached to the spring and rests on a frictionless, horizontal surface as in Figure P8.53. (a) The block is pulled to a position $x_i = 6.00$ cm from equilibrium and released. Find the elastic potential energy stored in the spring when the block is 6.00 cm from equilibrium and when the block passes through equilibrium. (b) Find the speed of the block as it passes through the equilibrium point. (c) What is the speed of the block when it is at a position $x_i/2 = 3.00$ cm? (d) Why isn't the answer to part (c) half the answer to part (b)?

The problem is identified with a **QC** icon.

Parts (a)–(c) of the problem ask for quantitative calculations.

Part (d) asks a conceptual question about the situation.

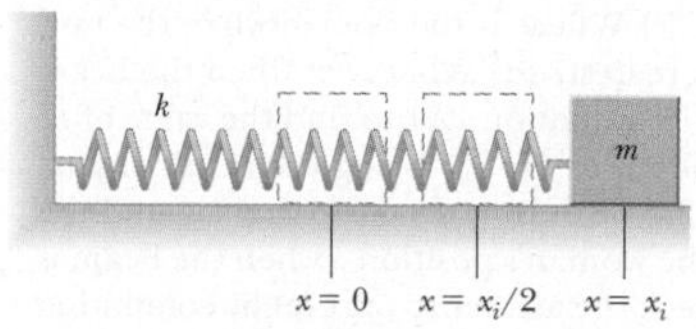

Figure P8.53

S *Symbolic* problems ask students to solve a problem using only symbolic manipulation. Reviewers of the seventh edition (as well as the majority of respondents to a large survey) asked specifically for an increase in the number of symbolic problems found in the text, as this better reflects the way instructors want their students to think when solving physics problems. An example of a Symbolic problem appears here:

The problem is identified with a **S** icon.

No numbers appear in the problem statement.

S A truck is moving with constant acceleration a up a hill that makes an angle ϕ with the horizontal as in Figure P6.51. A small sphere of mass m is suspended from the ceiling of the truck by a light cord. If the pendulum makes a constant angle θ with the perpendicular to the ceiling, what is a?

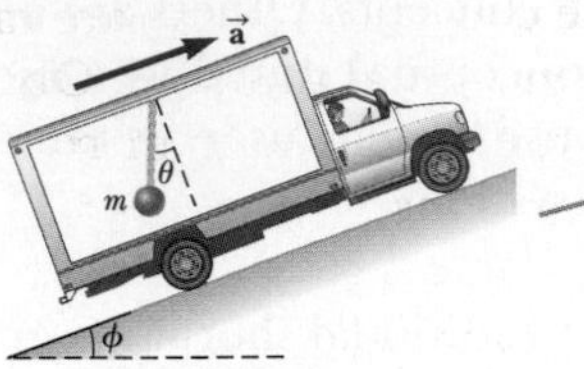

Figure P6.51

The figure shows only symbolic quantities.

51. $g(\cos\phi\tan\theta - \sin\phi)$

The answer to the problem is purely symbolic.

GP *Guided Problems* help students break problems into steps. A physics problem typically asks for one physical quantity in a given context. Often, however, several concepts must be used and a number of calculations are required to obtain that final answer. Many students are not accustomed to this level of complexity and often don't know where to start. A Guided Problem breaks a standard problem into smaller steps, enabling students to grasp all the concepts and strategies required to arrive at a correct solution. Unlike standard physics problems, guidance is often built into the problem statement. Guided Problems are reminiscent of how a student might interact with a professor in an office visit. These problems (there is one in most chapters of the text) help train students to break down complex problems into a series of simpler problems, an essential problem-solving skill. An example of a Guided Problem appears here:

The problem is identified with a **GP** icon.

38. **GP** A uniform beam resting on two pivots has a length $L = 6.00$ m and mass $M = 90.0$ kg. The pivot under the left end exerts a normal force n_1 on the beam, and the second pivot located a distance $\ell = 4.00$ m from the left end exerts a normal force n_2. A woman of mass $m = 55.0$ kg steps onto the left end of the beam and begins walking to the right as in Figure P12.38. The goal is to find the woman's position when the beam begins to tip. (a) What is the appropriate analysis model for the beam before it begins to tip? (b) Sketch a force diagram for the beam, labeling the gravitational and normal forces acting on the beam and placing the woman a distance x to the right of the first pivot, which is the origin. (c) Where is the woman when the normal force n_1 is the greatest? (d) What is n_1 when the beam is about to tip? (e) Use Equation 12.1 to find the value of n_2 when the beam is about to tip. (f) Using the result of part (d) and Equation 12.2, with torques computed around the second pivot, find the woman's position x when the beam is about to tip. (g) Check the answer to part (e) by computing torques around the first pivot point.

The goal of the problem is identified.

Analysis begins by identifying the appropriate analysis model.

Students are provided with suggestions for steps to solve the problem.

The calculation associated with the goal is requested.

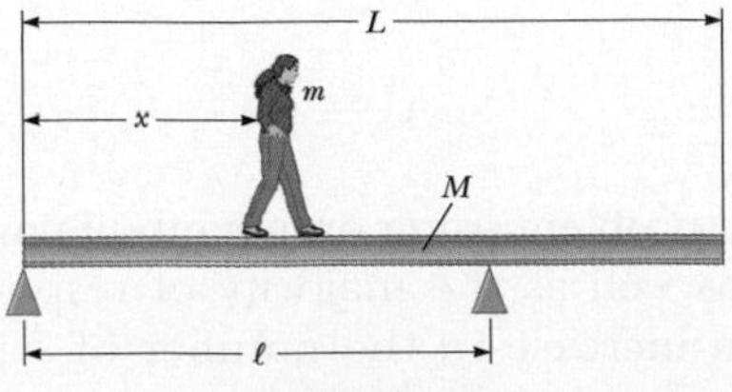

Figure P12.38

Impossibility problems. Physics education research has focused heavily on the problem-solving skills of students. While most problems in this text are structured in the form of providing data and asking for a result of computation, an average of two problems in each chapter are structured as impossibility problems. These begin with the phrase *Why is the following situation impossible?* This is followed by the description of a situation. The striking aspect of these problems is that no question is asked of the students, other than that in the initial italics. The student must determine what questions need to be asked and what calculations need to be performed. Based on the results of these calculations, the student must determine why the situation described is not possible. This determination may require information from personal experience, common sense, web or print research, measurement, mathematical skills, knowledge of human norms, or scientific thinking.

These problems can be assigned to build critical thinking skills in students. They are also fun, having the aspect of physics "mysteries" to be solved by students individually or in groups. An example of an impossibility problem appears here:

> The initial phrase in italics signals an impossibility problem.

> A situation is described.

53. *Why is the following situation impossible?* Manny Ramírez hits a home run so that the baseball just clears the top row of bleachers, 24.0 m high, located 130 m from home plate. The ball is hit at 41.7 m/s at an angle of 35.0° to the horizontal, and air resistance is negligible.

> No question is asked. The student must determine what needs to be calculated and why the situation is impossible.

Increased Number of Paired Problems. Based on the positive feedback we received in a survey of the market, we have increased the number of paired problems in this edition. These are otherwise identical problems, one asking for a numerical solution and one asking for a symbolic derivation. There are now three pairs of these problems in most chapters, indicated by tan shading in the end-of-chapter problems set.

Integration with Enhanced WebAssign. The textbook's tight integration with Enhanced WebAssign content facilitates an online learning environment that helps students improve their problem solving skills and gives them a variety of tools to meet their individual learning styles. New to this edition, Master It tutorials help students solve problems by having them work through a stepped-out solution. Problems with Master It tutorials are indicated in each chapter's problem set with a **M** icon. In addition, Watch It solution videos explain fundamental problem-solving strategies to help students step through the problem. The problems most often assigned in Enhanced WebAssign (shaded in blue) include either a Master It tutorial or a Watch It solution video to support students. In addition, these problems also have feedback to address student misconceptions, helping students avoid common pitfalls.

Thorough Revision of Artwork. Every piece of artwork in the Eighth Edition was revised in a new and modern style that helps express the physics principles at work in a clearer and more precise fashion. Every piece of art was also revised to make certain that the physical situations presented correspond exactly to the text discussion at hand.

Also added for this edition is a new feature for many pieces of art: "focus pointers" that either point out important aspects of a figure or guide students through a process illustrated by the artwork or photo. This format helps those students who are more visual learners. Examples of figures with focus pointers appear below.

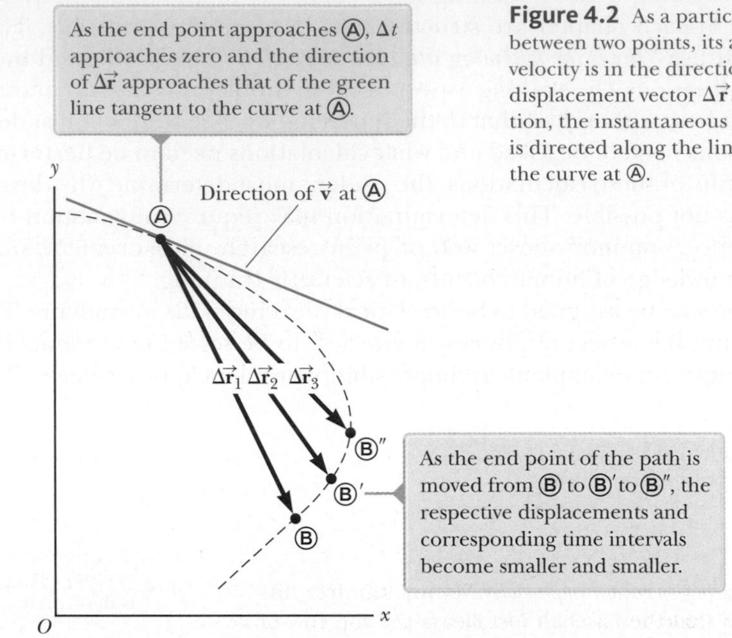

Figure 4.2 As a particle moves between two points, its average velocity is in the direction of the displacement vector $\Delta\vec{r}$. By definition, the instantaneous velocity at Ⓐ is directed along the line tangent to the curve at Ⓐ.

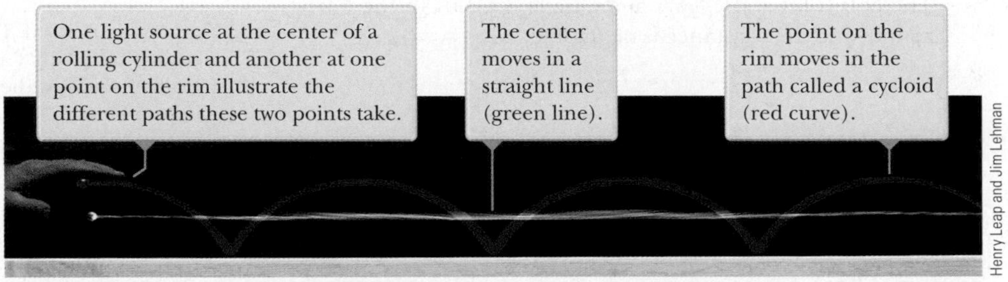

Figure 10.23 Two points on a rolling object take different paths through space.

Expansion of the Analysis Model Approach. Students are faced with hundreds of problems during their physics courses. Instructors realize that a relatively small number of fundamental principles form the basis of these problems. When faced with a new problem, a physicist forms a *model* of the problem that can be solved in a simple way by identifying the fundamental principle that is applicable in the problem. For example, many problems involve conservation of energy, Newton's second law, or kinematic equations. Because the physicist has studied these principles extensively and understands the associated applications, he or she can apply this knowledge as a model for solving a new problem.

While it would be ideal for students to follow this same process, most students have difficulty becoming familiar with the entire palette of fundamental principles that are available. It is easier for students to identify a *situation* rather than a fundamental principle. The *Analysis Model* approach we focus on in this revision lays out a standard set of situations that appear in most physics problems. These situations are based on an entity in one of four simplification models: particle, system, rigid object, and wave.

Once the simplification model is identified, the student thinks about what the entity is doing or how it interacts with its environment; this leads the student to identify a particular analysis model for the problem. For example, if an object is falling, the object is modeled as a particle. What it is doing is undergoing a constant acceleration, due to gravity. The student has learned that this situation is described by the analysis model of a particle under constant acceleration. Furthermore, this model has a small number of equations associated with it for use in starting problems—the kinematic equations in Chapter 2. Therefore, an understanding of the situation has led to an analysis model, which then identifies a very small number of equations to start the problem, rather than the myriad of equations that students see in the chapter. In this way, the use of analysis models leads the student to the fundamental principle that the physicist would identify. As the student gains more experience, he or she will lean less on the analysis model approach and begin to identify fundamental principles directly, more like the physicist does. This approach is further reinforced in the end-of-chapter summary under the heading *Analysis Models for Problem Solving.*

Revision of Worked Examples. Based on reviewer feedback on the last edition, we have made careful revisions to the worked examples so that the solutions are presented symbolically as far as possible and that numbers are substituted at the end. This approach will help students think symbolically when they solve problems, instead of automatically looking to insert numbers into an equation to solve a problem.

Content changes. The content and organization of the textbook are essentially the same as in the seventh edition. Several sections in various chapters have been streamlined, deleted, or combined with other sections to allow for a more balanced presentation. Updates have been added to reflect the current status of several areas of research and application of physics, including a new section on dark matter and information on discoveries of new Kuiper Belt Objects (Chapter 13), developments at the Laser Interferometer Gravitational-Wave Observatory (Chapter 37), progress in using grating light valves for optical applications (Chapter 38), continued plans for building the ITER international fusion reactor (Chapter 45), and the status of the Large Hadron Collider (Chapter 46).

Content

The material in this book covers fundamental topics in classical physics and provides an introduction to modern physics. The book is divided into six parts. Part 1 (Chapters 1 to 14) deals with the fundamentals of Newtonian mechanics and the physics of fluids; Part 2 (Chapters 15 to 18) covers oscillations, mechanical waves, and sound; Part 3 (Chapters 19 to 22) addresses heat and thermodynamics; Part 4 (Chapters 23 to 34) treats electricity and magnetism; Part 5 (Chapters 35 to 38) covers light and optics; and Part 6 (Chapters 39 to 46) deals with relativity and modern physics.

Text Features

Most instructors believe that the textbook selected for a course should be the student's primary guide for understanding and learning the subject matter. Furthermore, the textbook should be easily accessible and should be styled and written to facilitate instruction and learning. With these points in mind, we have included many pedagogical features, listed below, that are intended to enhance its usefulness to both students and instructors.

Problem Solving and Conceptual Understanding

General Problem-Solving Strategy. A general strategy outlined at the end of Chapter 2 (pages 43–44) provides students with a structured process for solving problems. In all remaining chapters, the strategy is employed explicitly in every example so that students learn how it is applied. Students are encouraged to follow this strategy when working end-of-chapter problems.

Worked examples. All in-text worked examples are presented in a two-column format to better reinforce physical concepts. The left column shows textual information that describes the steps for solving the problem. The right column shows the mathematical manipulations and results of taking these steps. This layout facilitates matching the concept with its mathematical execution and helps students organize their work. The examples closely follow the General Problem-Solving Strategy introduced in Chapter 2 to reinforce effective problem-solving habits. All worked examples in the text may be assigned for homework in Enhanced WebAssign. A sample of a worked example can be found on the opposite page.

Examples consist of two types. The first (and most common) example type presents a problem and numerical answer. The second type of example is conceptual in nature. To accommodate increased emphasis on understanding physical concepts, the many conceptual examples are labeled as such and are designed to help students focus on the physical situation in the problem.

What If? Approximately one-third of the worked examples in the text contain a **What If?** feature. At the completion of the example solution, a **What If?** question offers a variation on the situation posed in the text of the example. This feature encourages students to think about the results of the example, and it also assists in conceptual understanding of the principles. **What If?** questions also prepare students to encounter novel problems that may be included on exams. Some of the end-of-chapter problems also include this feature.

Quick Quizzes provide students an opportunity to test their understanding of the physical concepts presented. The questions require students to make decisions on the basis of sound reasoning, and some of the questions have been written to help students overcome common misconceptions. Quick Quizzes have been cast in an objective format, including multiple-choice, true–false, and ranking. Answers to all Quick Quiz questions are found at the end of the text, with additional explanations in the Instructor's Manual. Many instructors choose to use such questions in a "peer instruction" teaching style or with the use of personal response system "clickers," but they can be used in standard quiz format as well. An example of a Quick Quiz follows below.

***Quick Quiz* 7.5** A dart is loaded into a spring-loaded toy dart gun by pushing the spring in by a distance x. For the next loading, the spring is compressed a distance $2x$. How much faster does the second dart leave the gun compared with the first? **(a)** four times as fast **(b)** two times as fast **(c)** the same **(d)** half as fast **(e)** one-fourth as fast

Pitfall Prevention 16.2

Two Kinds of Speed/Velocity

Do not confuse v, the speed of the wave as it propagates along the string, with v_y, the transverse velocity of a point on the string. The speed v is constant for a uniform medium, whereas v_y varies sinusoidally.

Pitfall Preventions. More than two hundred Pitfall Preventions (such as the one to the left) are provided to help students avoid common mistakes and misunderstandings. These features, which are placed in the margins of the text, address both common student misconceptions and situations in which students often follow unproductive paths.

Summaries. Each chapter contains a summary that reviews the important concepts and equations discussed in that chapter. The summary is divided into three sections: Definitions, Concepts and Principles, and Analysis Models for Problem Solving. In each section, flashcard-type boxes focus on each separate definition, concept, principle, or analysis model.

Questions. As mentioned previously, the previous edition's Questions section is now divided into two sections: *Objective Questions* and *Conceptual Questions*. The instructor may select items to assign as homework or use in the classroom, possi-

ENHANCED WebAssign All worked examples are also available to be assigned as interactive examples in the Enhanced WebAssign homework management system.

Example 3.2 A Vacation Trip

A car travels 20.0 km due north and then 35.0 km in a direction 60.0° west of north as shown in Figure 3.11a. Find the magnitude and direction of the car's resultant displacement.

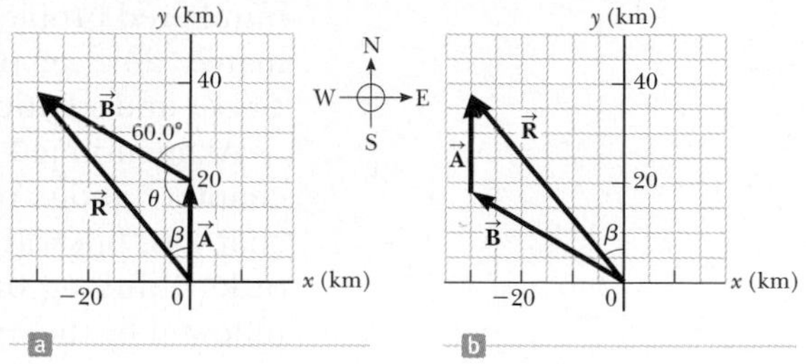

Figure 3.11 (Example 3.2) (a) Graphical method for finding the resultant displacement vector $\vec{\mathbf{R}} = \vec{\mathbf{A}} + \vec{\mathbf{B}}$. (b) Adding the vectors in reverse order $(\vec{\mathbf{B}} + \vec{\mathbf{A}})$ gives the same result for $\vec{\mathbf{R}}$.

Each solution has been written to closely follow the General Problem-Solving Strategy as outlined on pages 43–44 in Chapter 2, so as to reinforce good problem solving habits.

SOLUTION

Conceptualize The vectors $\vec{\mathbf{A}}$ and $\vec{\mathbf{B}}$ drawn in Figure 3.11a help us conceptualize the problem.

Categorize We can categorize this example as a simple analysis problem in vector addition. The displacement $\vec{\mathbf{R}}$ is the resultant when the two individual displacements $\vec{\mathbf{A}}$ and $\vec{\mathbf{B}}$ are added. We can further categorize it as a problem about the analysis of triangles, so we appeal to our expertise in geometry and trigonometry.

Analyze In this example, we show two ways to analyze the problem of finding the resultant of two vectors. The first way is to solve the problem geometrically, using graph paper and a protractor to measure the magnitude of $\vec{\mathbf{R}}$ and its direction in Figure 3.11a. (In fact, even when you know you are going to be carrying out a calculation, you should sketch the vectors to check your results.) With an ordinary ruler and protractor, a large diagram typically gives answers to two-digit but not to three-digit precision. Try using these tools on $\vec{\mathbf{R}}$ in Figure 3.11a!

The second way to solve the problem is to analyze it algebraically. The magnitude of $\vec{\mathbf{R}}$ can be obtained from the law of cosines as applied to the triangle in Figure 3.11a (see Appendix B.4).

Each step of the solution is detailed in a two-column format. The left column provides an explanation for each mathematical step in the right column, to better reinforce the physical concepts.

Use $R^2 = A^2 + B^2 - 2AB\cos\theta$ from the law of cosines to find R:

$$R = \sqrt{A^2 + B^2 - 2AB\cos\theta}$$

Substitute numerical values, noting that $\theta = 180° - 60° = 120°$:

$$R = \sqrt{(20.0\text{ km})^2 + (35.0\text{ km})^2 - 2(20.0\text{ km})(35.0\text{ km})\cos 120°}$$

$$= \boxed{48.2\text{ km}}$$

Use the law of sines (Appendix B.4) to find the direction of $\vec{\mathbf{R}}$ measured from the northerly direction:

$$\frac{\sin\beta}{B} = \frac{\sin\theta}{R}$$

$$\sin\beta = \frac{B}{R}\sin\theta = \frac{35.0\text{ km}}{48.2\text{ km}}\sin 120° = 0.629$$

$$\beta = \boxed{38.9°}$$

The resultant displacement of the car is 48.2 km in a direction 38.9° west of north.

Finalize Does the angle β that we calculated agree with an estimate made by looking at Figure 3.11a or with an actual angle measured from the diagram using the graphical method? Is it reasonable that the magnitude of $\vec{\mathbf{R}}$ is larger than that of both $\vec{\mathbf{A}}$ and $\vec{\mathbf{B}}$? Are the units of $\vec{\mathbf{R}}$ correct?

Although the head to tail method of adding vectors works well, it suffers from two disadvantages. First, some people find using the laws of cosines and sines to be awkward. Second, a triangle only results if you are adding two vectors. If you are adding three or more vectors, the resulting geometric shape is usually not a triangle. In Section 3.4, we explore a new method of adding vectors that will address both of these disadvantages.

WHAT IF? Suppose the trip were taken with the two vectors in reverse order: 35.0 km at 60.0° west of north first and then 20.0 km due north. How would the magnitude and the direction of the resultant vector change?

Answer They would not change. The commutative law for vector addition tells us that the order of vectors in an addition is irrelevant. Graphically, Figure 3.11b shows that the vectors added in the reverse order give us the same resultant vector.

What If? statements appear in about 1/3 of the worked examples and offer a variation on the situation posed in the text of the example. For instance, this feature might explore the effects of changing the conditions of the situation, determine what happens when a quantity is taken to a particular limiting value, or question whether additional information can be determined about the problem situation. This feature encourages students to think about the results of the example and assists in conceptual understanding of the principles.

bly with "peer instruction" methods and possibly with "personal response" systems. More than five hundred objective and conceptual questions are included in Volume 1. Answers for selected questions are included in the *Student Solutions Manual/Study Guide,* and answers for all questions are found in the *Instructor's Solutions Manual.*

Problems. An extensive set of problems is included at the end of each chapter; in all, Volume 1 contains approximately sixteen hundred problems. Answers for odd-numbered problems are provided at the end of the book. Full solutions for approximately 20% of the problems are included in the *Student Solutions Manual/Study Guide,* and solutions for all problems are found in the *Instructor's Solutions Manual.*

As mentioned previously, the *Problems* set is organized by the sections in each chapter (about two-thirds of the problems are keyed to specific sections of the chapter), but within each section the problems now "platform" students to higher-order thinking by presenting all the straightforward problems in the section first, followed by the intermediate problems. (The problem numbers for straightforward problems are printed in **black**; intermediate-level problems are in **blue**.) The *Additional Problems* section remains in its usual place, but at the end of each chapter there is a new section, *Challenge Problems,* that gathers the most difficult problems for a given chapter in one place. (Challenge problems have problem numbers marked in **red**.)

In addition to the new problem types mentioned previously, there are several other kinds of problems featured in this text:

- **Review problems.** Many chapters include review problems requiring the student to combine concepts covered in the chapter with those discussed in previous chapters. These problems (marked **Review**) reflect the cohesive nature of the principles in the text and verify that physics is not a scattered set of ideas. When facing a real-world issue such as global warming or nuclear weapons, it may be necessary to call on ideas in physics from several parts of a textbook such as this one.
- **"Fermi problems."** One or more problems in most chapters ask the student to reason in order-of-magnitude terms.
- **Design problems.** Several chapters contain problems that ask the student to determine design parameters for a practical device so that it can function as required.
- **Calculus-based problems.** Every chapter contains at least one problem applying ideas and methods from differential calculus and one problem using integral calculus.
- **Biomedical problems.** We added a number of problems related to biomedical situations in this edition, to highlight the relevance of physics principles to those students taking this course who are majoring in one of the life sciences.

The instructor's Web site, **www.cengage.com/physics/serway,** provides lists of all the various problem types, including problems most often assigned in Enhanced WebAssign, symbolic problems, quantitative/conceptual problems, Master It tutorials, Watch It solution videos, impossibility problems, paired problems, problems using calculus, problems encouraging or requiring computer use, problems with "**What If?**" parts, problems referred to in the chapter text, problems based on experimental data, order-of-magnitude problems, problems about biological applications, design problems, review problems, problems reflecting historical reasoning, and ranking questions.

Math Appendix. The math appendix (Appendix B), a valuable tool for students, shows the math tools in a physics context. This resource is ideal for students who need a quick review on topics such as algebra, trigonometry, and calculus.

Helpful Features

Style. To facilitate rapid comprehension, we have written the book in a clear, logical, and engaging style. We have chosen a writing style that is somewhat informal and relaxed so that students will find the text appealing and enjoyable to read. New terms are carefully defined, and we have avoided the use of jargon.

Important definitions and equations. Most important definitions are set in **boldface** or are highlighted with a background screen for added emphasis and ease of review. Similarly, important equations are also highlighted with a background screen to facilitate location.

Marginal notes. Comments and notes appearing in the margin with a ▶ icon can be used to locate important statements, equations, and concepts in the text.

Pedagogical use of color. Readers should consult the **pedagogical color chart** (inside the front cover) for a listing of the color-coded symbols used in the text diagrams. This system is followed consistently throughout the text.

Mathematical level. We have introduced calculus gradually, keeping in mind that students often take introductory courses in calculus and physics concurrently. Most steps are shown when basic equations are developed, and reference is often made to mathematical appendices near the end of the textbook. While vectors are discussed in detail in Chapter 3, vector products are introduced later in the text, where they are needed in physical applications. The dot product is introduced in Chapter 7, which addresses energy of a system; the cross product is introduced in Chapter 11, which deals with angular momentum.

Significant figures in both worked examples and end-of-chapter problems have been handled with care. Most numerical examples are worked to either two or three significant figures, depending on the precision of the data provided. End-of-chapter problems regularly state data and answers to three-digit precision. When carrying out estimation calculations, we shall typically work with a single significant figure. (More discussion of significant figures can be found in Chapter 1, pages 11–13.)

Units. The international system of units (SI) is used throughout the text. The U.S. customary system of units is used only to a limited extent in the chapters on mechanics and thermodynamics.

Appendices and endpapers. Several appendices are provided near the end of the textbook. Most of the appendix material represents a review of mathematical concepts and techniques used in the text, including scientific notation, algebra, geometry, trigonometry, differential calculus, and integral calculus. Reference to these appendices is made throughout the text. Most mathematical review sections in the appendices include worked examples and exercises with answers. In addition to the mathematical reviews, the appendices contain tables of physical data, conversion factors, and the SI units of physical quantities as well as a periodic table of the elements. Other useful information—fundamental constants and physical data, planetary data, a list of standard prefixes, mathematical symbols, the Greek alphabet, and standard abbreviations of units of measure—appears on the endpapers.

TextChoice Custom Options

Create a text to match your syllabus. Realizing that not all instructors cover all material from the text, we have included this book in our custom publishing program, TextChoice (**www.textchoice.com**). This extensive digital library lets you customize learning materials on your own computer by previewing and assembling content from a growing list of Cengage Learning titles including *Physics for Scientists and Engineers,* eighth edition. Search for content by course name, keyword, author, title, ISBN, and other categories. You can add your own course notes, supplements, lecture outlines, and other materials to the beginning or end of any chapter as well as arrange text chapters in any order or eliminate chapters that you don't cover in the course. Within 48 hours after you save your project and submit your order, a consultant will call you with a quote and answer any questions you may have. Once your project is finalized, Cengage Learning Custom Solutions will print the product and ship it to your bookstore.

Course Solutions That Fit Your Teaching Goals and Your Students' Learning Needs

Recent advances in educational technology have made homework management systems and audience response systems powerful and affordable tools to enhance the way you teach your course. Whether you offer a more traditional text-based course, are interested in using or are currently using an online homework management system such as Enhanced WebAssign, or are ready to turn your lecture into an interactive learning environment with JoinIn on TurningPoint, you can be confident that the text's proven content provides the foundation for each and every component of our technology and ancillary package.

Homework Management Systems

Enhanced WebAssign. Online homework has never been easier! Whether you're an experienced veteran or a beginner, WebAssign is the market leader in online homework solutions and the perfect solution to fit your homework management needs. Designed by physicists for physicists, this system is a reliable and user-friendly teaching companion. **Enhanced WebAssign** is available for *Physics for Scientists and Engineers,* giving you the freedom to assign

- every end-of-chapter Problem and Question
- the Problems most often assigned by your colleagues in Enhanced WebAssign (the blue-shaded problems in each chapter's problems set), enhanced with targeted feedback and either a Master It tutorial or a Watch It solution video. An example of targeted feedback appears below:

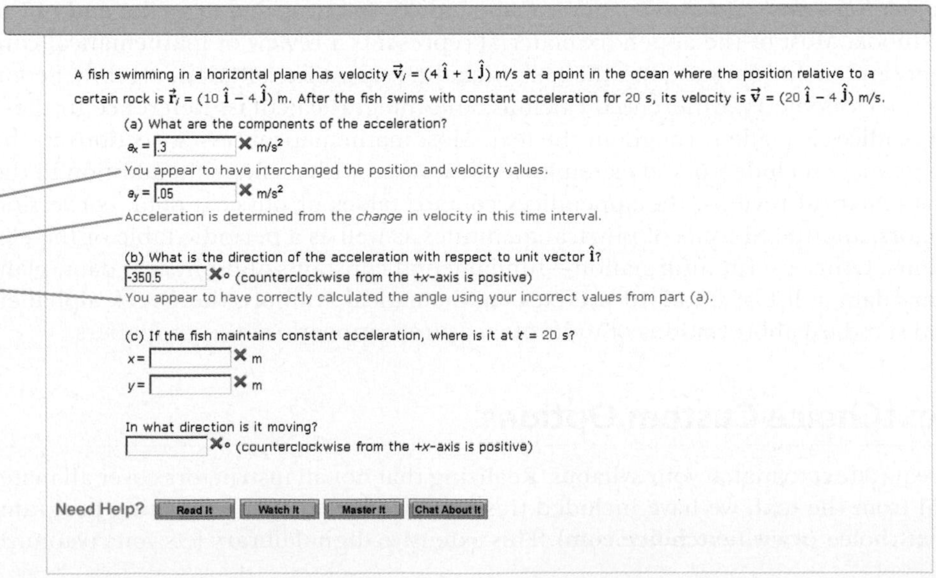

The most widely assigned problems in Enhanced WebAssign include feedback to address common mistakes that students make. This feedback was developed by professors with years of classroom experience.

- Master It tutorials, to help students work through the problem one step at a time. An example of a Master It tutorial appears below.

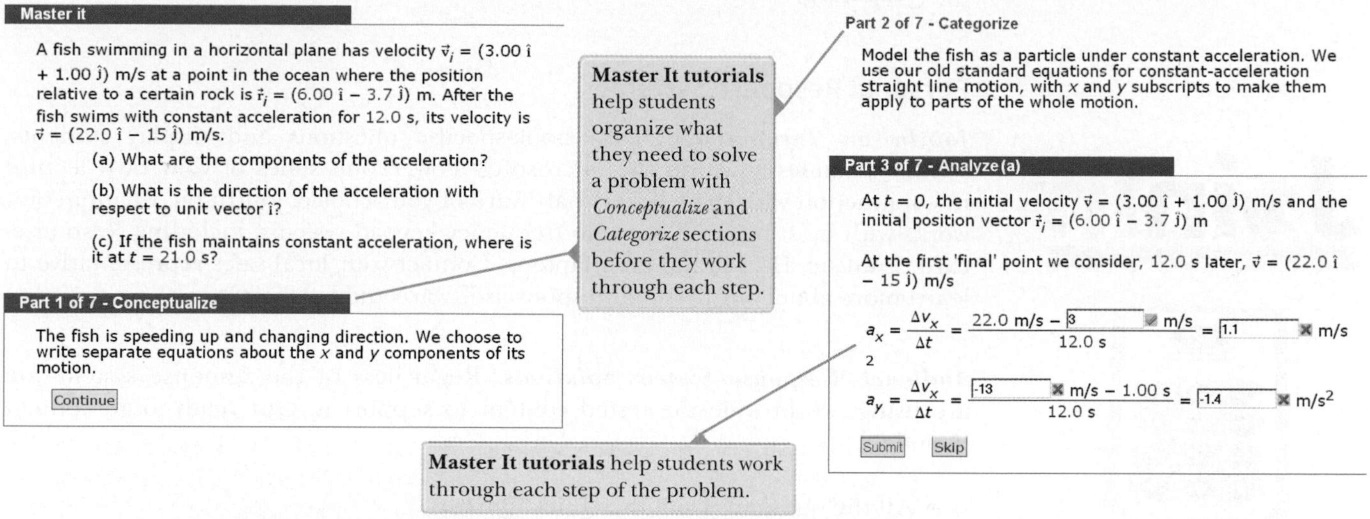

- Watch It solution videos that explain fundamental problem-solving strategies, to help students step through the problem. In addition, instructors can choose to include video hints of problem-solving strategies. A screen shot from a Watch It solution video appears below:

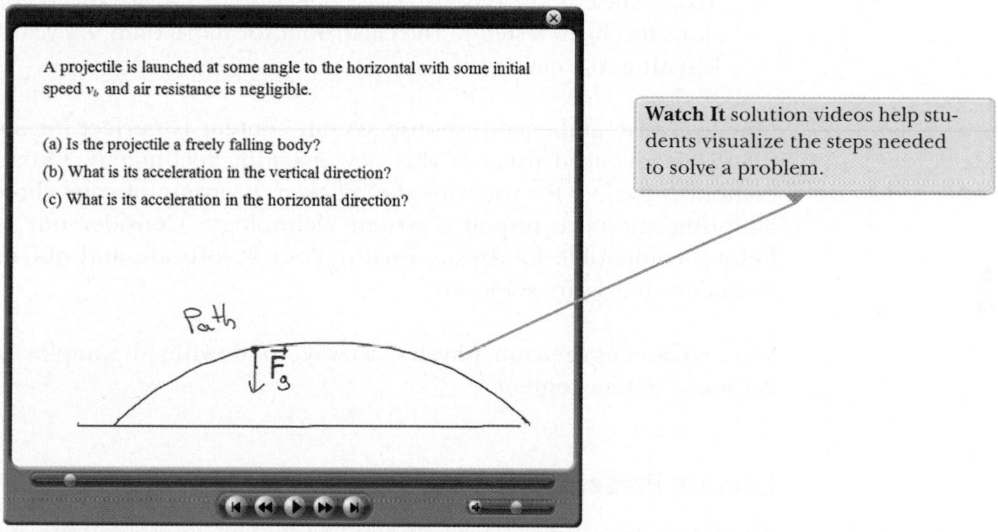

- every worked example, enhanced with hints and feedback, to help strengthen students' problem-solving skills
- every Quick Quiz, giving your students ample opportunity to test their conceptual understanding.

Also available in Enhanced WebAssign are:

- animated Active Figures, enhanced with hints and feedback, to help students develop their visualization skills
- a math review to help students brush up on key quantitative concepts in algebra, trigonometry, and calculus.

Please visit **www.webassign.net/brookscole** to view a live demonstration of Enhanced WebAssign.

The text also supports the following Homework Management System:

LON-CAPA: A Computer-Assisted Personalized Approach **http://www.lon-capa.org/**

Personal Response Systems

JoinIn on TurningPoint. Pose book-specific questions and display students' answers seamlessly within the Microsoft® PowerPoint slides of your own lecture in conjunction with the "clicker" hardware of your choice. JoinIn on TurningPoint works with most infrared or radio frequency keypad systems, including Responsecard, EduCue, H-ITT, and even laptops. Contact your local sales representative to learn more about our personal response software and hardware.

Audience Response System Solutions. Regardless of the response system you are using, we provide the tested content to support it. Our ready-to-go content includes:

- All the questions from the Quick Quizzes
- All end-of-chapter objective questions to provide helpful conceptual checkpoints to drop into your lecture
- Animated Active Figures enhanced with multiple-choice questions to help test students' observational skills.
- *Assessing to Learn in the Classroom* questions developed at the University of Massachusetts at Amherst. This collection of 250 advanced conceptual questions has been tested in the classroom for more than ten years and takes peer learning to a new level.

Our exclusive audience response system content is perfect for amateur, intermediate, or advanced users of this new learning technology. Our platform-neutral content is perfect for use with the "clicker" program of your choice. Interested in adopting audience response system technology? Consider our Microsoft PowerPoint®-compatible JoinIn on TurningPoint® software and our infrared or radio frequency hardware solutions.

Visit **www.cengage.com/physics/serway** to download samples of our personal response system content.

Lecture Presentation Resources

The following resources provide support for your presentations in lecture.

PowerLecture™ DVD-ROM is an easy-to-use multimedia tool allowing instructors to assemble art, animations, and digital video to create fluid lectures quickly. The two-volume DVD-ROM (Volume 1: Chapters 1–22; Volume 2: Chapters 23–46) includes prepared PowerPoint® lectures and files for all of the line art from the text as well as editable solutions available through Solution Builder. The DVD-ROM also includes the ExamView® Computerized Test Bank, giving you the ability to build tests featuring an unlimited number of new questions or any of the existing questions from the preloaded *Test Bank*. Finally, the DVD-ROM includes audience response system content specific to the textbook. Contact your local sales representative to find out about our audience response software and hardware.

Assessment and Course Preparation Resources

A number of resources listed below will assist with your assessment and preparation processes.

Instructor's Solutions Manual by Mike Ziegler (The Ohio State University). This two-volume manual, thoroughly revised for this edition, contains complete worked solutions to all end-of-chapter problems in the textbook as well as answers to the even-numbered problems and all the questions. The solutions to problems new to the Eighth Edition are marked for easy identification. Volume 1 contains Chapters 1 through 22, and Volume 2 contains Chapters 23 through 46. Electronic files of the Instructor's Solutions are available on the *PowerLecture*™ DVD-ROM as well.

Test Bank by Ed Oberhofer (University of North Carolina at Charlotte and Lake-Sumter Community College). The test bank is available on the two-volume *PowerLecture*™ DVD-ROM via the ExamView® test software. This two-volume test bank contains approximately 2 200 multiple-choice questions. Instructors may print and duplicate pages for distribution to students. Volume 1 contains Chapters 1 through 22, and Volume 2 contains Chapters 23 through 46. WebCT and Blackboard versions of the test bank are available on the instructor's companion site at **www.cengage/physics/serway.**

Instructor's Companion Web Site. Consult the instructor's site by pointing your browser to **www.cengage.com/physics/serway** for a problem correlation guide, PowerPoint lectures, and JoinIn audience response content. Instructors adopting the eighth edition of *Physics for Scientists and Engineers* may download these materials after securing the appropriate password from their local sales representative.

Supporting Materials for the Instructor

Supporting instructor materials are available to qualified adopters. Please consult your local Cengage Learning, Brooks/Cole representative for details. Visit **www.cengage.com/physics/serway** to:

- Request a desk copy
- Locate your local representative
- Download electronic files of select support materials

Student Resources

Visit the *Physics for Scientists and Engineers* website at **www.cengage.com/physics/serway** to see samples of select student supplements. Students can purchase any Cengage Learning product at your local college store or at our preferred online store **www.ichapters.com.**

Student Solutions Manual/Study Guide by John R. Gordon, Ralph McGrew, Raymond Serway, and John W. Jewett, Jr. This two-volume manual features detailed solutions to 20% of the end-of-chapter problems from the text. The manual also features a list of important equations, concepts, and notes from key sections of the text in addition to answers to selected end-of-chapter questions. Volume 1 contains Chapters 1 through 22, and Volume 2 contains Chapters 23 through 46.

Premium eBook. This rich interactive eBook includes links to animated Active Figures and allows students to highlight the text, add their own notes, and bookmark pages. Students can access the eBook through Enhanced WebAssign.

Teaching Options

The topics in this textbook are presented in the following sequence: classical mechanics, oscillations and mechanical waves, and heat and thermodynamics followed by electricity and magnetism, electromagnetic waves, optics, relativity, and modern physics. This presentation represents a traditional sequence, with the subject of mechanical waves being presented before electricity and magnetism. Some instructors may prefer to discuss both mechanical and electromagnetic waves together after completing electricity and magnetism. In this case, Chapters 16 through 18 could be covered along with Chapter 34. The chapter on relativity is placed near the end of the text because this topic often is treated as an introduction to the era of "modern physics." If time permits, instructors may choose to cover Chapter 39 after completing Chapter 13 as a conclusion to the material on Newtonian mechanics. For those instructors teaching a two-semester sequence, some sections and chapters in Volume 1 could be deleted without any loss of continuity. The following sections can be considered optional for this purpose:

2.8 Kinematic Equations Derived from Calculus
4.6 Relative Velocity and Relative Acceleration
6.3 Motion in Accelerated Frames
6.4 Motion in the Presence of Resistive Forces
7.9 Energy Diagrams and Equilibrium of a System
9.9 Rocket Propulsion
11.5 The Motion of Gyroscopes and Tops
14.7 Other Applications of Fluid Dynamics
15.6 Damped Oscillations
15.7 Forced Oscillations
18.6 Standing Waves in Rods and Membranes
18.8 Nonsinusoidal Wave Patterns
22.8 Entropy on a Microscopic Scale

Acknowledgments

This eighth edition of *Physics for Scientists and Engineers* was prepared with the guidance and assistance of many professors who reviewed selections of the manuscript, the prerevision text, or both. We wish to acknowledge the following scholars and express our sincere appreciation for their suggestions, criticisms, and encouragement:

Jennifer Blue, *Miami University of Ohio*
Norbert Chencinski, *College of Staten Island/The City University of New York*
Jeffrey Christafferson, *Ferris State University*
Brent A. Corbin, *University of California at Los Angeles*
Michael Dennin, *University of California, Irvine*
Elena S. Flitsiyan, *University of Central Florida*
Chris Littler, *University of North Texas*
Steven Morris, *Los Angeles Harbor College*
Vahé Peroomian, *University of California at Los Angeles*
Alexander L. Rudolph, *California State Polytechnic University, Pomona*
Marllin L. Simon, *Auburn University*
Edward A. Whittaker, *Stevens Institute of Technology*

Prior to our work on this revision, we conducted a survey of professors to gauge how they used end-of-chapter questions and problems in their classes. We were overwhelmed not only by the number of professors who wanted to take part in the survey, but also by their insightful comments. Their feedback and suggestions helped shape the revision of the end-of-chapter questions and problems in this edition, and so we would like to thank the survey participants:

Wagih Abdel-Kader, *South Carolina State University;* Elise Adamson, *Wayland Baptist University;* Shireen Adenwalla, *University of Nebraska-Lincoln;* Rhett Allain, *Southeastern Louisiana University;* David S. Armstrong, *College of William & Mary;* Robert Astalos, *Adams State College;* Abdel Bachri, *Southern Arkansas University;* Colley Baldwin, *Medgar Evers College;* Steve Barnes, *California State University, San Bernardino;* Robert Bass, *Gardner-Webb University;* Chris Berven, *University of Idaho;* Andrew Blauch, *Charleston Southern University;* Paul Bloom, *North Central College;* Carolyn Boesse, *McLennan Community College;* Mary Boleware, *Jones County Junior College;* Catalina Boudreaux, *University of Texas at San Antonio;* John Carzoli, *Oakton Community College;* Ken Caviness, *Southern Adventist University;* Eugene Chaffin, *Bob Jones University;* Robert Chavez, *College of Marin;* Norbert Chencinski, *College of Staten Island, The City University of New York;* Kelvin Chu, *University of Vermont;* Sr. Marie Anselm Cooper, *Immaculata University;* Brent Corbin, *University of California at Los Angeles;* Kevin Cornelius, *Ouachita Baptist University;* Sarah Crowe, *University of Kentucky;* Linda S. Dake, *Utica College;* Ethan Deneault, *University of Tampa;* Gregory Derry, *Loyola College;* Joseph Di Rienzi, *College of Notre Dame of Maryland;* Ryan Droste, *Trident Technical College;* Gintaras Duda, *Creighton University;* Mike Durren, *Lake Michigan College;* John Edwards, *Piedmont Technical College;* Mark Edwards, *Hofstra University;* Efremfon F. Ekpo, *Bethune-Cookman University;* Michael Fauerbach, *Florida Gulf Coast University;* Nail Fazleev, *University of Texas, Arlington;* Terrence F. Flower, *College of Saint Catherine;* Marco Fornari, *Central Michigan University;* Tom French, *Montgomery County Community College;* Richard Gelderman, *Western Kentucky University;* Anthony Gerig, *Viterbo University;* Mikhail Goloubev, *Bowie State University;* Joshua Guttman, *Bergen Community College;* Dean Hamden, *Montclair State University;* Mark Hardies, *St. Petersburg College;* Kathleen Harper, *The Ohio State University;* Wayne Hayes, *Greenville Technical College;* Paul Henriksen, *James Madison University;* David Heskett, *University of Rhode Island;* Scott Hildreth, *Chabot College;* Tracy Hodge, *Berea College;* Dawn Hollenbeck, *Rochester Institute of Technology;* William Hollerman, *University of Louisiana at Lafayette;* George K. Horton, *Rutgers University;* David C. Ingram, *Ohio University;* Shawn Jackson, *University of Arizona;* Mario Klaric, *Midlands Technical College;* Burair Kothari, *Indiana University;* Thomas Krause, *Towson University;* Fred Kuttner, *University of California, Santa Cruz;* Douglas Kurtze, *Saint Joseph's University;* Dan Lawrence, *Northwest Nazarene University;* Lynne Lawson, *Providence College;* David Locke, *College of San Mateo;* Thomas Lockhart, *University of Wisconsin-Eau Claire;* Virginia Long, *Colby College;* Igor Makasyuk, *San Francisco State University;* Jimmy McCoy, *Tarleton State University;* Kenneth W. McLaughlin, *Loras College;* Rahul Mehta, *University of Central Arkansas;* Dominick Misciascio, *Mercer County Community College;* Sudipa Mitra-Kirtley, *Rose-Hulman Institute of Technology;* Poovan Murugesan, *San Diego City College;* Robert Napora, *Purdue University-Calumet;* Joseph L. Nothnagel, *McHenry Community College;* Lauren Novatne-Harris, *Reedley College;* Terry F. O'Dwyer, *Nassau Community College;* Adebanjo Oriade, *Bethany College;* Michael Panunto, *Butte College;* John Phillips, *Capital University;* Robert Pompi, *Binghamton University, State University of New York;* Dale Powers, *Elmira College;* Richard Powers, *Los Angeles Trade Technical College;* Stanley Radford, *The College at Brockport, State University of New York;* Beatrice Rasmussen, *University of Texas at Dallas;* Cameron Reed, *Alma College;* Richard Rees, *Westfield State College;* Ken Reyzer, *Cuyamaca College;* Thomas R. Roose, *Trinity Christian College;* Nanjundiah Sadanand, *Central Connecticut State University;* Joshua Sasmor, *Seton Hill University;* Andria Schwortz, *Quinsigamond Community College;* Mariana Sendova, *New College of Florida;* Henry R. Setze, *Pearl River Community College;* Anwar Shiekh, *Diné College;* Gurbax Singh, *University of Maryland Eastern Shore;* Xiangning Song, *Richland College;* Lawrence P. Staunton, *Drake University;* Glenn B. Stracher, *East Georgia College;* Jeff Sundquist, *Palm Beach Community College;* Gerald Taylor, *James Madison University;* Valentina Tobos, *Lawrence Tech University;* John Vassiliou, *Villanova University;* Jogindra Wadehra, *Wayne State University;* Bill Warren, *Lord Fairfax Community College;* Michael

Weber, *Brigham Young University-Hawaii;* Zodiac Webster, *Columbus State University;* Margaret Wessling, *Pierce College;* Joseph O. West, *Indiana State University;* Dennis P. Weygand, *Thomas Nelson Community College;* Tom Wilbur, *Anne Arundel Community College;* Weldon Wilson, *University of Central Oklahoma;* Stephen Wimpenny, *University of California, Riverside;* Frederick Wolf, *Keene State College;* Alexander Wurm, *Western New England College;* Robert Zbikowski, *Hibbing Community College*

This title was carefully checked for accuracy by Grant Hart, *Brigham Young University;* Michael Kotlarchyk, *Rochester Institute of Technology;* Brian A. Raue, *Florida International University;* James E. Rutledge, *University of California at Irvine;* Greg Severn, *University of San Diego;* Harry W. K. Tom, *University of California at Riverside;* and Som Tyagi, *Drexel University.* We thank them for their diligent efforts under schedule pressure.

Vahé Peroomian reviewed the end-of-chapter questions and problems sets, and offered valuable suggestions for improving them; we are very thankful for his help. We are also grateful to Ralph McGrew for writing some new problems, and suggesting improvements in the content of the textbook. Belal Abas, Zinoviy Akkerman, Eric Boyd, Hal Falk, Melanie Martin, Steve McCauley, and Glenn Stracher made corrections to problems taken from previous editions. We are grateful to authors John R. Gordon and Ralph McGrew for preparing the *Student Solutions Manual/Study Guide.* Author Mike Ziegler has prepared an excellent *Instructor's Solutions Manual.* Ed Oberhofer has carefully edited and improved the test bank.

Special thanks and recognition go to the professional staff at Brooks/Cole Cengage Learning—in particular, Mary Finch, Ed Dodd, Brandi Kirksey (who managed the ancillary program and so much more), Cathy Brooks, Robyn Young, Joshua Duncan, Rebecca Berardy Schwartz, Sam Subity, Nicole Mollica, and Michelle Julet—for their fine work during the development, production, and promotion of this textbook. We recognize the skilled production service and excellent artwork provided by the staff at Lachina Publishing Services and Dartmouth Publishing, Inc., and the dedicated photo research efforts of Michelle Vitiello at the Bill Smith Group.

Finally, we are deeply indebted to our wives, children, and grandchildren for their love, support, and long-term sacrifices.

Raymond A. Serway
St. Petersburg, Florida

John W. Jewett, Jr.
Anaheim, California

to the student

It is appropriate to offer some words of advice that should be of benefit to you, the student. Before doing so, we assume you have read the Preface, which describes the various features of the text and support materials that will help you through the course.

How to Study

Instructors are often asked, "How should I study physics and prepare for examinations?" There is no simple answer to this question, but we can offer some suggestions based on our own experiences in learning and teaching over the years.

First and foremost, maintain a positive attitude toward the subject matter, keeping in mind that physics is the most fundamental of all natural sciences. Other science courses that follow will use the same physical principles, so it is important that you understand and are able to apply the various concepts and theories discussed in the text.

Concepts and Principles

It is essential that you understand the basic concepts and principles before attempting to solve assigned problems. You can best accomplish this goal by carefully reading the textbook before you attend your lecture on the covered material. When reading the text, you should jot down those points that are not clear to you. Also be sure to make a diligent attempt at answering the questions in the Quick Quizzes as you come to them in your reading. We have worked hard to prepare questions that help you judge for yourself how well you understand the material. Study the **What If?** features that appear in many of the worked examples carefully. They will help you extend your understanding beyond the simple act of arriving at a numerical result. The Pitfall Preventions will also help guide you away from common misunderstandings about physics. During class, take careful notes and ask questions about those ideas that are unclear to you. Keep in mind that few people are able to absorb the full meaning of scientific material after only one reading; several readings of the text and your notes may be necessary. Your lectures and laboratory work supplement the textbook and should clarify some of the more difficult material. You should minimize your memorization of material. Successful memorization of passages from the text, equations, and derivations does not necessarily indicate that you understand the material. Your understanding of the material will be enhanced through a combination of efficient study habits, discussions with other students and with instructors, and your ability to solve the problems presented in the textbook. Ask questions whenever you believe that clarification of a concept is necessary.

Study Schedule

It is important that you set up a regular study schedule, preferably a daily one. Make sure that you read the syllabus for the course and adhere to the schedule set by your instructor. The lectures will make much more sense if you read the corresponding text material *before* attending them. As a general rule, you should devote about two hours of study time for each hour you are in class. If you are having trouble with the course, seek the advice of the instructor or other students who have taken the course. You may find it necessary to seek further instruction from experienced students. Very often, instructors offer review sessions in addition to regular class periods. Avoid the practice of delaying study until a day or two before an exam. More often than not, this approach has disastrous results. Rather than undertake an all-night study session before a test, briefly review the basic concepts and equations, and then get a good night's rest. If you believe that you need additional help in understanding the concepts, in preparing for exams, or in problem solving, we suggest that you acquire a copy of the *Student Solutions Manual/Study Guide* that accompanies this textbook.

Visit the *Physics for Scientists and Engineers* website at **www.cengage.com/physics/serway** to see samples of select student supplements. Students can purchase any Cengage Learning product at your local college store or at our preferred online store **www.ichapters.com.**

Use the Features

You should make full use of the various features of the text discussed in the Preface. For example, marginal notes are useful for locating and describing important equations and concepts, and **boldface** indicates important definitions. Many useful tables are contained in the appendices, but most are incorporated in the text where they are most often referenced. Appendix B is a convenient review of mathematical tools used in the text.

Answers to Quick Quizzes and odd-numbered problems are given at the end of the textbook, and solutions to selected end-of-chapter questions and problems are provided in the *Student Solutions Manual/Study Guide.* The table of contents provides an overview of the entire text, and the index enables you to locate specific material quickly. Footnotes are sometimes used to supplement the text or to cite other references on the subject discussed.

After reading a chapter, you should be able to define any new quantities introduced in that chapter and discuss the principles and assumptions that were used to arrive at certain key relations. The chapter summaries and the review sections of the *Student Solutions Manual/Study Guide* should help you in this regard. In some cases, you may find it necessary to refer to the textbook's index to locate certain topics. You should be able to associate with each physical quantity the correct symbol used to represent that quantity and the unit in which the quantity is specified. Furthermore, you should be able to express each important equation in concise and accurate prose.

Problem Solving

R. P. Feynman, Nobel laureate in physics, once said, "You do not know anything until you have practiced." In keeping with this statement, we strongly advise you to develop the skills necessary to solve a wide range of problems. Your ability to solve problems will be one of the main tests of your knowledge of physics; therefore, you should try to solve as many problems as possible. It is essential that you understand basic concepts and principles before attempting to solve problems. It is good practice to try to find alternate solutions to the same problem. For example, you can solve problems in mechanics using Newton's laws, but very often an alternative method that draws on energy considerations is more direct. You should not deceive yourself into thinking that you understand a problem merely because you have seen it solved in class. You must be able to solve the problem and similar problems on your own.

The approach to solving problems should be carefully planned. A systematic plan is especially important when a problem involves several concepts. First, read the problem several times until you are confident you understand what is being asked. Look for any key words that will help you interpret the problem and perhaps allow you to make certain assumptions. Your ability to interpret a question properly is an integral part of problem solving. Second, you should acquire the habit of writing down the information given in a problem and those quantities that need to be found; for example, you might construct a table listing both the quantities given and the quantities to be found. This procedure is sometimes used in the worked examples of the textbook. Finally, after you have decided on the method you believe is appropriate for a given problem, proceed with your solution. The General Problem-Solving Strategy will guide you through complex problems. If you follow the steps of this procedure *(Conceptualize, Categorize, Analyze, Finalize),* you will find it easier to come up with a solution and gain more from your efforts. This Strategy, located at the end of Chapter 2 (pages 43–44), is used in all worked examples in the remaining chapters so that you can learn how to apply it. Specific problem-solving strategies for certain types of situations are included in the text and appear with a special heading. These specific strategies follow the outline of the General Problem-Solving Strategy.

Often, students fail to recognize the limitations of certain equations or physical laws in a particular situation. It is very important that you understand and remember the assumptions that underlie a particular theory or formalism. For example, certain equations in kinematics apply only to a particle moving with constant acceleration. These equations are not valid for describing motion whose acceleration is not constant such as the motion of an object connected to a spring or the motion of an object through a fluid. Study the Analysis Models for Problem Solving in the chapter summaries carefully so that you know how each model can be applied to a specific situation. The analysis models provide you with a logical structure for solving problems and help you to develop your thinking skills to become more like those of a physicist. Use the analysis model approach to save you hours of looking for the correct equation and to make you a faster and more efficient problem solver.

Experiments

Physics is a science based on experimental observations. Therefore, we recommend that you try to supplement the text by performing various types of "hands-on" experiments either at home or in the laboratory. These experiments can be used to test ideas and models discussed in class or in the textbook. For example, the common Slinky toy is excellent for studying traveling waves, a ball swinging on the end of a long string can be used to investigate pendulum motion, various masses attached to the end of a vertical spring or rubber band can be used to determine their elastic nature, an old pair of polarized sunglasses and some discarded lenses and a magnifying glass are the components of various experiments in optics, and an approximate measure of the free-fall acceleration can be determined simply by measuring with a stopwatch the time it takes for a ball to drop from a known height. The list of such experiments is endless. When physical models are not available, be imaginative and try to develop models of your own.

New Media

If available, we strongly encourage you to use the **Enhanced WebAssign** product that is available with this textbook. It is far easier to understand physics if you see it in action, and the materials available in Enhanced WebAsign will enable you to become a part of that action.

It is our sincere hope that you will find physics an exciting and enjoyable experience and that you will benefit from this experience, regardless of your chosen profession. Welcome to the exciting world of physics!

The scientist does not study nature because it is useful; he studies it because he delights in it, and he delights in it because it is beautiful. If nature were not beautiful, it would not be worth knowing, and if nature were not worth knowing, life would not be worth living.

—Henri Poincaré

Mechanics

part 1

A Smart car, imported from Germany, moves along an American roadway. Gasoline-powered automobiles such as the Smart car, as well as hybrid and purely electric vehicles, use many of the concepts and principles of mechanics that we will study in this first part of the book. Quantities that we can use to describe the operation of vehicles include position, velocity, acceleration, force, energy, and momentum. *(Lee Waters/TRANSTOCK/Jupiter Images)*

Physics, the most fundamental physical science, is concerned with the fundamental principles of the Universe. It is the foundation upon which the other sciences—astronomy, biology, chemistry, and geology—are based. The beauty of physics lies in the simplicity of its fundamental principles and in the manner in which just a small number of concepts and models can alter and expand our view of the world around us.

The study of physics can be divided into six main areas:

1. *classical mechanics,* concerning the motion of objects that are large relative to atoms and move at speeds much slower than the speed of light
2. *relativity,* a theory describing objects moving at any speed, even speeds approaching the speed of light
3. *thermodynamics,* dealing with heat, work, temperature, and the statistical behavior of systems with large numbers of particles
4. *electromagnetism,* concerning electricity, magnetism, and electromagnetic fields
5. *optics,* the study of the behavior of light and its interaction with materials
6. *quantum mechanics,* a collection of theories connecting the behavior of matter at the submicroscopic level to macroscopic observations

The disciplines of mechanics and electromagnetism are basic to all other branches of classical physics (developed before 1900) and modern physics (c. 1900–present). The first part of this textbook deals with classical mechanics, sometimes referred to as *Newtonian mechanics* or simply *mechanics*. Many principles and models used to understand mechanical systems retain their importance in the theories of other areas of physics and can later be used to describe many natural phenomena. Therefore, classical mechanics is of vital importance to students from all disciplines. ■

chapter **1**

Physics and Measurement

A rolling ball clock, invented in the early nineteenth century as an alternative to the pendulum clock. A ball rolls back and forth through grooves in the tilted platform near the bottom. When the ball reaches the end of the grooved track, it trips the escapement mechanism of the clock and the platform tilts the other way, sending the ball on a return journey. Time is one of the basic quantities that we use in studying the motion of objects. *(Ray Bates, The British Clockmaker Inc.)*

Like all other sciences, physics is based on experimental observations and quantitative measurements. The main objectives of physics are to identify a limited number of fundamental laws that govern natural phenomena and use them to develop theories that can predict the results of future experiments. The fundamental laws used in developing theories are expressed in the language of mathematics, the tool that provides a bridge between theory and experiment.

When there is a discrepancy between the prediction of a theory and experimental results, new or modified theories must be formulated to remove the discrepancy. Many times a theory is satisfactory only under limited conditions; a more general theory might be satisfactory without such limitations. For example, the laws of motion discovered by Isaac Newton (1642–1727) accurately describe the motion of objects moving at normal speeds but do not apply to objects moving at speeds comparable to the speed of light. In contrast, the special theory of relativity developed later by Albert Einstein (1879–1955) gives the same results as Newton's laws at low speeds but also correctly describes the motion of objects at speeds approaching the speed of light. Hence, Einstein's special theory of relativity is a more general theory of motion than that formed from Newton's laws.

Classical physics includes the principles of classical mechanics, thermodynamics, optics, and electromagnetism developed before 1900. Important contributions to classical physics were provided by Newton, who was also one of the originators of calculus as a mathematical tool. Major developments in mechanics continued in the 18th century, but the fields of thermodynamics and electromagnetism were not developed until the latter part of the 19th century, principally because before that time the apparatus for controlled experiments in these disciplines was either too crude or unavailable.

A major revolution in physics, usually referred to as *modern physics,* began near the end of the 19th century. Modern physics developed mainly because many physical phenomena could not be explained by classical physics. The two most important developments in this modern era were the theories of relativity and quantum mechanics. Einstein's special theory of relativity not only correctly describes the motion of objects moving at speeds comparable to the speed of light; it also completely modifies the traditional concepts of space, time, and energy. The theory also shows that the speed of light is the upper limit of the speed of an object and that mass and energy are related. Quantum mechanics was formulated by a number of distinguished scientists to provide descriptions of physical phenomena at the atomic level. Many practical devices have been developed using the principles of quantum mechanics.

Scientists continually work at improving our understanding of fundamental laws. Numerous technological advances in recent times are the result of the efforts of many scientists, engineers, and technicians, such as unmanned planetary explorations and manned moon landings, microcircuitry and high-speed computers, sophisticated imaging techniques used in scientific research and medicine, and several remarkable results in genetic engineering. The effects of such developments and discoveries on our society have indeed been great, and it is very likely that future discoveries and developments will be exciting, challenging, and of great benefit to humanity.

1.1 Standards of Length, Mass, and Time

To describe natural phenomena, we must make measurements of various aspects of nature. Each measurement is associated with a physical quantity, such as the length of an object. The laws of physics are expressed as mathematical relationships among physical quantities that we will introduce and discuss throughout the book. In mechanics, the three fundamental quantities are length, mass, and time. All other quantities in mechanics can be expressed in terms of these three.

If we are to report the results of a measurement to someone who wishes to reproduce this measurement, a *standard* must be defined. It would be meaningless if a visitor from another planet were to talk to us about a length of 8 "glitches" if we do not know the meaning of the unit glitch. On the other hand, if someone familiar with our system of measurement reports that a wall is 2 meters high and our unit of length is defined to be 1 meter, we know that the height of the wall is twice our basic length unit. Whatever is chosen as a standard must be readily accessible and must possess some property that can be measured reliably. Measurement standards used by different people in different places—throughout the Universe—must yield the same result. In addition, standards used for measurements must not change with time.

In 1960, an international committee established a set of standards for the fundamental quantities of science. It is called the **SI** (Système International), and its fundamental units of length, mass, and time are the *meter, kilogram,* and *second,* respectively. Other standards for SI fundamental units established by the committee are those for temperature (the *kelvin*), electric current (the *ampere*), luminous intensity (the *candela*), and the amount of substance (the *mole*).

Length

We can identify **length** as the distance between two points in space. In 1120, the king of England decreed that the standard of length in his country would be named the *yard* and would be precisely equal to the distance from the tip of his nose to the end of his outstretched arm. Similarly, the original standard for the foot adopted by the French was the length of the royal foot of King Louis XIV. Neither of these standards is constant in time; when a new king took the throne, length measurements changed! The French standard prevailed until 1799, when the legal standard of length in France became the **meter** (m), defined as one ten-millionth of the distance from the equator to the North Pole along one particular longitudinal line that passes through Paris. Notice that this value is an Earth-based standard that does not satisfy the requirement that it can be used throughout the Universe.

As recently as 1960, the length of the meter was defined as the distance between two lines on a specific platinum–iridium bar stored under controlled conditions in France. Current requirements of science and technology, however, necessitate more accuracy than that with which the separation between the lines on the bar can be determined. In the 1960s and 1970s, the meter was defined as 1 650 763.73 wavelengths[1] of orange-red light emitted from a krypton-86 lamp. In October 1983, however, the meter was redefined as **the distance traveled by light in vacuum during a time of 1/299 792 458 second.** In effect, this latest definition establishes that the speed of light in vacuum is precisely 299 792 458 meters per second. This definition of the meter is valid throughout the Universe based on our assumption that light is the same everywhere.

Pitfall Prevention 1.1

Reasonable Values

Generating intuition about typical values of quantities when solving problems is important because you must think about your end result and determine if it seems reasonable. For example, if you are calculating the mass of a housefly and arrive at a value of 100 kg, this answer is *unreasonable* and there is an error somewhere.

Table 1.1 lists approximate values of some measured lengths. You should study this table as well as the next two tables and begin to generate an intuition for what is meant by, for example, a length of 20 centimeters, a mass of 100 kilograms, or a time interval of 3.2×10^7 seconds.

Mass

The SI fundamental unit of **mass,** the **kilogram** (kg), is defined as **the mass of a specific platinum–iridium alloy cylinder kept at the International Bureau of Weights and Measures at Sèvres, France.** This mass standard was established in 1887 and has not been changed since that time because platinum–iridium is an unusually

[1]We will use the standard international notation for numbers with more than three digits, in which groups of three digits are separated by spaces rather than commas. Therefore, 10 000 is the same as the common American notation of 10,000. Similarly, $\pi = 3.14159265$ is written as 3.141 592 65.

TABLE 1.1 *Approximate Values of Some Measured Lengths*

	Length (m)
Distance from the Earth to the most remote known quasar	1.4×10^{26}
Distance from the Earth to the most remote normal galaxies	9×10^{25}
Distance from the Earth to the nearest large galaxy (Andromeda)	2×10^{22}
Distance from the Sun to the nearest star (Proxima Centauri)	4×10^{16}
One light-year	9.46×10^{15}
Mean orbit radius of the Earth about the Sun	1.50×10^{11}
Mean distance from the Earth to the Moon	3.84×10^{8}
Distance from the equator to the North Pole	1.00×10^{7}
Mean radius of the Earth	6.37×10^{6}
Typical altitude (above the surface) of a satellite orbiting the Earth	2×10^{5}
Length of a football field	9.1×10^{1}
Length of a housefly	5×10^{-3}
Size of smallest dust particles	$\sim 10^{-4}$
Size of cells of most living organisms	$\sim 10^{-5}$
Diameter of a hydrogen atom	$\sim 10^{-10}$
Diameter of an atomic nucleus	$\sim 10^{-14}$
Diameter of a proton	$\sim 10^{-15}$

stable alloy. A duplicate of the Sèvres cylinder is kept at the National Institute of Standards and Technology (NIST) in Gaithersburg, Maryland (Fig. 1.1a). Table 1.2 lists approximate values of the masses of various objects.

Time

Before 1967, the standard of **time** was defined in terms of the *mean solar day.* (A solar day is the time interval between successive appearances of the Sun at the highest point it reaches in the sky each day.) The fundamental unit of a **second** (s) was defined as $(\frac{1}{60})(\frac{1}{60})(\frac{1}{24})$ of a mean solar day. This definition is based on the rotation of one planet, the Earth. Therefore, this motion does not provide a time standard that is universal.

In 1967, the second was redefined to take advantage of the high precision attainable in a device known as an *atomic clock* (Fig. 1.1b), which measures vibrations of cesium atoms. One second is now defined as **9 192 631 770 times the period of vibration of radiation from the cesium-133 atom.**[2] Approximate values of time intervals are presented in Table 1.3.

TABLE 1.2 *Approximate Masses of Various Objects*

	Mass (kg)
Observable Universe	$\sim 10^{52}$
Milky Way galaxy	$\sim 10^{42}$
Sun	1.99×10^{30}
Earth	5.98×10^{24}
Moon	7.36×10^{22}
Shark	$\sim 10^{3}$
Human	$\sim 10^{2}$
Frog	$\sim 10^{-1}$
Mosquito	$\sim 10^{-5}$
Bacterium	$\sim 1 \times 10^{-15}$
Hydrogen atom	1.67×10^{-27}
Electron	9.11×10^{-31}

TABLE 1.3 *Approximate Values of Some Time Intervals*

	Time Interval (s)
Age of the Universe	4×10^{17}
Age of the Earth	1.3×10^{17}
Average age of a college student	6.3×10^{8}
One year	3.2×10^{7}
One day	8.6×10^{4}
One class period	3.0×10^{3}
Time interval between normal heartbeats	8×10^{-1}
Period of audible sound waves	$\sim 10^{-3}$
Period of typical radio waves	$\sim 10^{-6}$
Period of vibration of an atom in a solid	$\sim 10^{-13}$
Period of visible light waves	$\sim 10^{-15}$
Duration of a nuclear collision	$\sim 10^{-22}$
Time interval for light to cross a proton	$\sim 10^{-24}$

[2]Period is defined as the time interval needed for one complete vibration.

Figure 1.1 (a) The National Standard Kilogram No. 20, an accurate copy of the International Standard Kilogram kept at Sèvres, France, is housed under a double bell jar in a vault at the National Institute of Standards and Technology. (b) A cesium fountain atomic clock. The clock will neither gain nor lose a second in 20 million years.

TABLE 1.4 *Prefixes for Powers of Ten*

Power	Prefix	Abbreviation	Power	Prefix	Abbreviation
10^{-24}	yocto	y	10^{3}	kilo	k
10^{-21}	zepto	z	10^{6}	mega	M
10^{-18}	atto	a	10^{9}	giga	G
10^{-15}	femto	f	10^{12}	tera	T
10^{-12}	pico	p	10^{15}	peta	P
10^{-9}	nano	n	10^{18}	exa	E
10^{-6}	micro	μ	10^{21}	zetta	Z
10^{-3}	milli	m	10^{24}	yotta	Y
10^{-2}	centi	c			
10^{-1}	deci	d			

In addition to SI, another system of units, the *U.S. customary system,* is still used in the United States despite acceptance of SI by the rest of the world. In this system, the units of length, mass, and time are the foot (ft), slug, and second, respectively. In this book, we shall use SI units because they are almost universally accepted in science and industry. We shall make some limited use of U.S. customary units in the study of classical mechanics.

In addition to the fundamental SI units of meter, kilogram, and second, we can also use other units, such as millimeters and nanoseconds, where the prefixes *milli-* and *nano-* denote multipliers of the basic units based on various powers of ten. Prefixes for the various powers of ten and their abbreviations are listed in Table 1.4. For example, 10^{-3} m is equivalent to 1 millimeter (mm), and 10^{3} m corresponds to 1 kilometer (km). Likewise, 1 kilogram (kg) is 10^{3} grams (g), and 1 mega volt (MV) is 10^{6} volts (V).

The variables length, time, and mass are examples of *fundamental quantities.* Most other variables are *derived quantities,* those that can be expressed as a mathematical combination of fundamental quantities. Common examples are *area* (a product of two lengths) and *speed* (a ratio of a length to a time interval).

A table of the letters in the ▶ Greek alphabet is provided on the back endpaper of this book.

Another example of a derived quantity is **density.** The density ρ (Greek letter rho) of any substance is defined as its *mass per unit volume:*

$$\rho \equiv \frac{m}{V} \quad \textbf{(1.1)}$$

In terms of fundamental quantities, density is a ratio of a mass to a product of three lengths. Aluminum, for example, has a density of 2.70×10^{3} kg/m^3, and iron has a density of 7.86×10^{3} kg/m^3. An extreme difference in density can be imagined by thinking about holding a 10-centimeter (cm) cube of Styrofoam in one hand and a 10-cm cube of lead in the other. See Table 14.1 in Chapter 14 for densities of several materials.

***Quick Quiz* 1.1** In a machine shop, two cams are produced, one of aluminum and one of iron. Both cams have the same mass. Which cam is larger? **(a)** The aluminum cam is larger. **(b)** The iron cam is larger. **(c)** Both cams have the same size.

1.2 Matter and Model Building

If physicists cannot interact with some phenomenon directly, they often imagine a **model** for a physical system that is related to the phenomenon. For example, we cannot interact directly with atoms because they are too small. Therefore, we build a mental model of an atom based on a system of a nucleus and one or more electrons outside the nucleus. Once we have identified the physical components of the model, we make predictions about its behavior based on the interactions among the components of the system or the interaction between the system and the environment outside the system.

As an example, consider the behavior of *matter.* A sample of solid gold is shown at the top of Figure 1.2. Is this sample nothing but wall-to-wall gold, with no empty space? If the sample is cut in half, the two pieces still retain their chemical identity as solid gold. What if the pieces are cut again and again, indefinitely? Will the smaller and smaller pieces always be gold? Such questions can be traced to early Greek philosophers. Two of them—Leucippus and his student Democritus—could not accept the idea that such cuttings could go on forever. They developed a model for matter by speculating that the process ultimately must end when it produces a particle that can no longer be cut. In Greek, *atomos* means "not sliceable." From this Greek term comes our English word *atom.*

The Greek model of the structure of matter was that all ordinary matter consists of atoms, as suggested in the middle of Figure 1.2. Beyond that, no additional structure was specified in the model; atoms acted as small particles that interacted with one another, but internal structure of the atom was not a part of the model.

In 1897, J. J. Thomson identified the electron as a charged particle and as a constituent of the atom. This led to the first atomic model that contained internal structure. We shall discuss this model in Chapter 42.

Following the discovery of the nucleus in 1911, an atomic model was developed in which each atom is made up of electrons surrounding a central nucleus. A nucleus of gold is shown in Figure 1.2. This model leads, however, to a new question: Does the nucleus have structure? That is, is the nucleus a single particle or a collection of particles? By the early 1930s, a model evolved that described two basic entities in the nucleus: protons and neutrons. The proton carries a positive electric charge, and a specific chemical element is identified by the number of protons in its nucleus. This number is called the **atomic number** of the element. For instance, the nucleus of a hydrogen atom contains one proton (so the atomic number of hydrogen is 1), the nucleus of a helium atom contains two protons (atomic number 2), and the nucleus of a uranium atom contains 92 protons (atomic number 92). In addition to atomic number, a second number—**mass number,** defined as the number of protons plus neutrons in a nucleus—characterizes atoms. The atomic number of a specific element never varies (i.e., the number of protons does not vary), but the mass number can vary (i.e., the number of neutrons varies).

Is that, however, where the process of breaking down stops? Protons, neutrons, and a host of other exotic particles are now known to be composed of six different varieties of particles called **quarks,** which have been given the names of *up, down, strange, charmed, bottom,* and *top.* The up, charmed, and top quarks have electric charges of $+\frac{2}{3}$ that of the proton, whereas the down, strange, and bottom quarks have charges of $-\frac{1}{3}$ that of the proton. The proton consists of two up quarks and one down quark as shown at the bottom of Figure 1.2 and labeled u and d. This structure predicts the correct charge for the proton. Likewise, the neutron consists of two down quarks and one up quark, giving a net charge of zero.

You should develop a process of building models as you study physics. In this study, you will be challenged with many mathematical problems to solve. One of the most important problem-solving techniques is to build a model for the problem: identify a system of physical components for the problem and make predictions of the behavior of the system based on the interactions among its components or the interaction between the system and its surrounding environment.

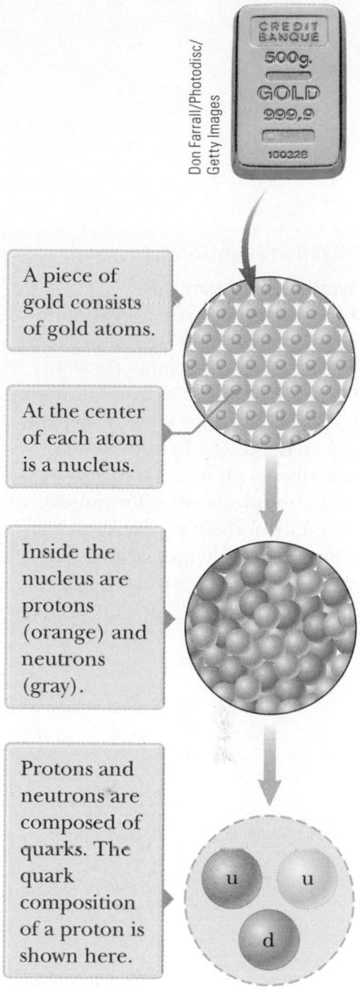

Figure 1.2 Levels of organization in matter.

1.3 Dimensional Analysis

In physics, the word *dimension* denotes the physical nature of a quantity. The distance between two points, for example, can be measured in feet, meters, or furlongs, which are all different ways of expressing the dimension of length.

TABLE 1.5 *Dimensions and Units of Four Derived Quantities*

Quantity	Area (A)	Volume (V)	Speed (v)	Acceleration (a)
Dimensions	L^2	L^3	L/T	L/T^2
SI units	m^2	m^3	m/s	m/s^2
U.S. customary units	ft^2	ft^3	ft/s	ft/s^2

Pitfall Prevention 1.2

Symbols for Quantities

Some quantities have a small number of symbols that represent them. For example, the symbol for time is almost always t. Other quantities might have various symbols depending on the usage. Length may be described with symbols such as x, y, and z (for position); r (for radius); a, b, and c (for the legs of a right triangle); ℓ (for the length of an object); d (for a distance); h (for a height); and so forth.

The symbols we use in this book to specify the dimensions of length, mass, and time are L, M, and T, respectively.[3] We shall often use brackets [ ] to denote the dimensions of a physical quantity. For example, the symbol we use for speed in this book is v, and in our notation, the dimensions of speed are written $[v] = L/T$. As another example, the dimensions of area A are $[A] = L^2$. The dimensions and units of area, volume, speed, and acceleration are listed in Table 1.5. The dimensions of other quantities, such as force and energy, will be described as they are introduced in the text.

In many situations, you may have to check a specific equation to see if it matches your expectations. A useful procedure for doing that, called **dimensional analysis,** can be used because dimensions can be treated as algebraic quantities. For example, quantities can be added or subtracted only if they have the same dimensions. Furthermore, the terms on both sides of an equation must have the same dimensions. By following these simple rules, you can use dimensional analysis to determine whether an expression has the correct form. Any relationship can be correct only if the dimensions on both sides of the equation are the same.

To illustrate this procedure, suppose you are interested in an equation for the position x of a car at a time t if the car starts from rest at $x = 0$ and moves with constant acceleration a. The correct expression for this situation is $x = \frac{1}{2}at^2$ as we show in Chapter 2. The quantity x on the left side has the dimension of length. For the equation to be dimensionally correct, the quantity on the right side must also have the dimension of length. We can perform a dimensional check by substituting the dimensions for acceleration, L/T^2 (Table 1.5), and time, T, into the equation. That is, the dimensional form of the equation $x = \frac{1}{2}at^2$ is

$$L = \frac{L}{\cancel{T^2}} \cdot \cancel{T^2} = L$$

The dimensions of time cancel as shown, leaving the dimension of length on the right-hand side to match that on the left.

A more general procedure using dimensional analysis is to set up an expression of the form

$$x \propto a^n t^m$$

where n and m are exponents that must be determined and the symbol $\propto$ indicates a proportionality. This relationship is correct only if the dimensions of both sides are the same. Because the dimension of the left side is length, the dimension of the right side must also be length. That is,

$$[a^n t^m] = L = L^1T^0$$

Because the dimensions of acceleration are L/T^2 and the dimension of time is T, we have

$$(L/T^2)^n T^m = L^1T^0 \quad \rightarrow \quad (L^n T^{m-2n}) = L^1T^0$$

The exponents of L and T must be the same on both sides of the equation. From the exponents of L, we see immediately that $n = 1$. From the exponents of T, we see that $m - 2n = 0$, which, once we substitute for n, gives us $m = 2$. Returning to our original expression $x \propto a^n t^m$, we conclude that $x \propto at^2$.

[3]The *dimensions* of a quantity will be symbolized by a capitalized, nonitalic letter such as L or T. The *algebraic symbol* for the quantity itself will be an italicized letter such as L for the length of an object or t for time.

Quick Quiz **1.2** True or False: Dimensional analysis can give you the numerical value of constants of proportionality that may appear in an algebraic expression.

Example 1.1 Analysis of an Equation

Show that the expression $v = at$, where v represents speed, a acceleration, and t an instant of time, is dimensionally correct.

SOLUTION

Identify the dimensions of v from Table 1.5: $[v] = \frac{L}{T}$

Identify the dimensions of a from Table 1.5 and multiply by the dimensions of t: $[at] = \frac{L}{T^{\not 2}} \not T = \frac{L}{T}$

Therefore, $v = at$ is dimensionally correct because we have the same dimensions on both sides. (If the expression were given as $v = at^2$, it would be dimensionally *incorrect*. Try it and see!)

Example 1.2 Analysis of a Power Law

Suppose we are told that the acceleration a of a particle moving with uniform speed v in a circle of radius r is proportional to some power of r, say r^n, and some power of v, say v^m. Determine the values of n and m and write the simplest form of an equation for the acceleration.

SOLUTION

Write an expression for a with a dimensionless constant of proportionality k: $a = kr^n v^m$

Substitute the dimensions of a, r, and v: $\frac{L}{T^2} = L^n \left(\frac{L}{T}\right)^m = \frac{L^{n+m}}{T^m}$

Equate the exponents of L and T so that the dimensional equation is balanced: $n + m = 1$ and $m = 2$

Solve the two equations for n: $n = -1$

Write the acceleration expression: $a = kr^{-1} v^2 = k\frac{v^2}{r}$

In Section 4.4 on uniform circular motion, we show that $k = 1$ if a consistent set of units is used. The constant k would not equal 1 if, for example, v were in km/h and you wanted a in m/s^2.

1.4 Conversion of Units

Sometimes it is necessary to convert units from one measurement system to another or convert within a system (for example, from kilometers to meters). Conversion factors between SI and U.S. customary units of length are as follows:

1 mile	= 1 609 m	= 1.609 km	1 ft	=	0.304 8 m	=	30.48 cm
1 m	= 39.37 in.	= 3.281 ft	1 in.	=	0.025 4 m	=	2.54 cm (exactly)

A more complete list of conversion factors can be found in Appendix A.

Pitfall Prevention 1.3

Always Include Units

When performing calculations, include the units for every quantity and carry the units through the entire calculation. Avoid the temptation to drop the units early and then attach the expected units once you have an answer. By including the units in every step, you can detect errors if the units for the answer turn out to be incorrect.

Like dimensions, units can be treated as algebraic quantities that can cancel each other. For example, suppose we wish to convert 15.0 in. to centimeters. Because 1 in. is defined as exactly 2.54 cm, we find that

$$15.0 \text{ in.} = (15.0 \cancel{\text{in.}})\left(\frac{2.54 \text{ cm}}{1 \cancel{\text{in.}}}\right) = 38.1 \text{ cm}$$

where the ratio in parentheses is equal to 1. We express 1 as 2.54 cm/1 in. (rather than 1 in./2.54 cm) so that the unit "inch" in the denominator cancels with the unit in the original quantity. The remaining unit is the centimeter, our desired result.

***Quick Quiz* 1.3** The distance between two cities is 100 mi. What is the number of kilometers between the two cities? **(a)** smaller than 100 **(b)** larger than 100 **(c)** equal to 100

Example 1.3 Is He Speeding?

On an interstate highway in a rural region of Wyoming, a car is traveling at a speed of 38.0 m/s. Is the driver exceeding the speed limit of 75.0 mi/h?

SOLUTION

Convert meters in the speed to miles:

$$(38.0 \cancel{\text{m}}/\text{s})\left(\frac{1 \text{ mi}}{1\,609 \cancel{\text{m}}}\right) = 2.36 \times 10^{-2} \text{ mi/s}$$

Convert seconds to hours:

$$(2.36 \times 10^{-2} \text{ mi}/\cancel{\text{s}})\left(\frac{60 \cancel{\text{s}}}{1 \cancel{\text{min}}}\right)\left(\frac{60 \cancel{\text{min}}}{1 \text{ h}}\right) = 85.0 \text{ mi/h}$$

The driver is indeed exceeding the speed limit and should slow down.

WHAT IF? What if the driver were from outside the United States and is familiar with speeds measured in kilometers per hour? What is the speed of the car in km/h?

Answer We can convert our final answer to the appropriate units:

$$(85.0 \cancel{\text{mi}}/\text{h})\left(\frac{1.609 \text{ km}}{1 \cancel{\text{mi}}}\right) = 137 \text{ km/h}$$

Figure 1.3 shows an automobile speedometer displaying speeds in both mi/h and km/h. Can you check the conversion we just performed using this photograph?

Figure 1.3 The speedometer of a vehicle that shows speeds in both miles per hour and kilometers per hour.

1.5 Estimates and Order-of-Magnitude Calculations

Suppose someone asks you the number of bits of data on a typical musical compact disc. In response, it is not generally expected that you would provide the exact number but rather an estimate, which may be expressed in scientific notation. The estimate may be made even more approximate by expressing it as an *order of magnitude,* which is a power of ten determined as follows:

1. Express the number in scientific notation, with the multiplier of the power of ten between 1 and 10 and a unit.

2. If the multiplier is less than 3.162 (the square root of ten), the order of magnitude of the number is the power of ten in the scientific notation. If the multiplier is greater than 3.162, the order of magnitude is one larger than the power of ten in the scientific notation.

We use the symbol ~ for "is on the order of." Use the procedure above to verify the orders of magnitude for the following lengths:

$$0.008\,6\text{ m} \sim 10^{-2}\text{ m} \qquad 0.002\,1\text{ m} \sim 10^{-3}\text{ m} \qquad 720\text{ m} \sim 10^{3}\text{ m}$$

Usually, when an order-of-magnitude estimate is made, the results are reliable to within about a factor of ten. If a quantity increases in value by three orders of magnitude, its value increases by a factor of about $10^3 = 1\,000$.

Inaccuracies caused by guessing too low for one number are often canceled by other guesses that are too high. You will find that with practice your guesstimates become better and better. Estimation problems can be fun to work because you freely drop digits, venture reasonable approximations for unknown numbers, make simplifying assumptions, and turn the question around into something you can answer in your head or with minimal mathematical manipulation on paper. Because of the simplicity of these types of calculations, they can be performed on a *small* scrap of paper and are often called "back-of-the-envelope calculations."

Example 1.4 Breaths in a Lifetime

Estimate the number of breaths taken during an average human lifetime.

SOLUTION

We start by guessing that the typical human lifetime is about 70 years. Think about the average number of breaths that a person takes in 1 min. This number varies depending on whether the person is exercising, sleeping, angry, serene, and so forth. To the nearest order of magnitude, we shall choose 10 breaths per minute as our estimate. (This estimate is certainly closer to the true average value than an estimate of 1 breath per minute or 100 breaths per minute.)

Find the approximate number of minutes in a year:

$$1\,\cancel{\text{yr}}\left(\frac{400\,\cancel{\text{days}}}{1\,\cancel{\text{yr}}}\right)\left(\frac{25\,\cancel{\text{h}}}{1\,\cancel{\text{day}}}\right)\left(\frac{60\text{ min}}{1\,\cancel{\text{h}}}\right) = 6 \times 10^5\text{ min}$$

Find the approximate number of minutes in a 70-year lifetime:

$$\text{number of minutes} = (70\text{ yr})(6 \times 10^5\text{ min/yr}) = 4 \times 10^7\text{ min}$$

Find the approximate number of breaths in a lifetime:

$$\text{number of breaths} = (10\text{ breaths/min})(4 \times 10^7\text{ min}) = \boxed{4 \times 10^8\text{ breaths}}$$

Therefore, a person takes on the order of 10^9 breaths in a lifetime. Notice how much simpler it is in the first calculation above to multiply 400 × 25 than it is to work with the more accurate 365 × 24.

WHAT IF? What if the average lifetime were estimated as 80 years instead of 70? Would that change our final estimate?

Answer We could claim that $(80\text{ yr})(6 \times 10^5\text{ min/yr}) = 5 \times 10^7$ min, so our final estimate should be 5×10^8 breaths. This answer is still on the order of 10^9 breaths, so an order-of-magnitude estimate would be unchanged.

1.6 Significant Figures

When certain quantities are measured, the measured values are known only to within the limits of the experimental uncertainty. The value of this uncertainty can depend on various factors, such as the quality of the apparatus, the skill of the experimenter, and the number of measurements performed. The number of **significant figures** in a measurement can be used to express something about the uncertainty. The number of significant figures is related to the number of numerical digits used to express the measurement, as we discuss below.

As an example of significant figures, suppose we are asked to measure the radius of a compact disc using a meterstick as a measuring instrument. Let us assume the accuracy to which we can measure the radius of the disc is ± 0.1 cm. Because of the uncertainty of ± 0.1 cm, if the radius is measured to be 6.0 cm, we can claim only that its radius lies somewhere between 5.9 cm and 6.1 cm. In this case, we say that the measured value of 6.0 cm has two significant figures. Note that *the significant figures include the first estimated digit.* Therefore, we could write the radius as (6.0 ± 0.1) cm.

Zeros may or may not be significant figures. Those used to position the decimal point in such numbers as 0.03 and 0.007 5 are not significant. Therefore, there are one and two significant figures, respectively, in these two values. When the zeros come after other digits, however, there is the possibility of misinterpretation. For example, suppose the mass of an object is given as 1 500 g. This value is ambiguous because we do not know whether the last two zeros are being used to locate the decimal point or whether they represent significant figures in the measurement. To remove this ambiguity, it is common to use scientific notation to indicate the number of significant figures. In this case, we would express the mass as 1.5×10^3 g if there are two significant figures in the measured value, 1.50×10^3 g if there are three significant figures, and 1.500×10^3 g if there are four. The same rule holds for numbers less than 1, so 2.3×10^{-4} has two significant figures (and therefore could be written 0.000 23) and 2.30×10^{-4} has three significant figures (also written as 0.000 230).

In problem solving, we often combine quantities mathematically through multiplication, division, addition, subtraction, and so forth. When doing so, you must make sure that the result has the appropriate number of significant figures. A good rule of thumb to use in determining the number of significant figures that can be claimed in a multiplication or a division is as follows:

> When multiplying several quantities, the number of significant figures in the final answer is the same as the number of significant figures in the quantity having the smallest number of significant figures. The same rule applies to division.

Let's apply this rule to find the area of the compact disc whose radius we measured above. Using the equation for the area of a circle,

$$A = \pi r^2 = \pi(6.0 \text{ cm})^2 = 1.1 \times 10^2 \text{ cm}^2$$

If you perform this calculation on your calculator, you will likely see 113.097 335 5. It should be clear that you don't want to keep all of these digits, but you might be tempted to report the result as 113 cm^2. This result is not justified because it has three significant figures, whereas the radius only has two. Therefore, we must report the result with only two significant figures as shown above.

For addition and subtraction, you must consider the number of decimal places when you are determining how many significant figures to report:

> When numbers are added or subtracted, the number of decimal places in the result should equal the smallest number of decimal places of any term in the sum or difference.

Pitfall Prevention 1.4

Read Carefully

Notice that the rule for addition and subtraction is different from that for multiplication and division. For addition and subtraction, the important consideration is the number of *decimal places,* not the number of *significant figures.*

As an example of this rule, consider the sum

$$23.2 + 5.174 = 28.4$$

Notice that we do not report the answer as 28.374 because the lowest number of decimal places is one, for 23.2. Therefore, our answer must have only one decimal place.

The rules for addition and subtraction can often result in answers that have a different number of significant figures than the quantities with which you start. For example, consider these operations that satisfy the rule:

$$1.000\,1 + 0.000\,3 = 1.000\,4$$

$$1.002 - 0.998 = 0.004$$

In the first example, the result has five significant figures even though one of the terms, 0.000 3, has only one significant figure. Similarly, in the second calculation, the result has only one significant figure even though the numbers being subtracted have four and three, respectively.

In this book, most of the numerical examples and end-of-chapter problems will yield answers having three significant figures. When carrying out estimation calculations, we shall typically work with a single significant figure.

◀ **Significant figure guidelines used in this book**

If the number of significant figures in the result of a calculation must be reduced, there is a general rule for rounding numbers: the last digit retained is increased by 1 if the last digit dropped is greater than 5. (For example, 1.346 becomes 1.35.) If the last digit dropped is less than 5, the last digit retained remains as it is. (For example, 1.343 becomes 1.34.) If the last digit dropped is equal to 5, the remaining digit should be rounded to the nearest even number. (This rule helps avoid accumulation of errors in long arithmetic processes.)

A technique for avoiding error accumulation is to delay the rounding of numbers in a long calculation until you have the final result. Wait until you are ready to copy the final answer from your calculator before rounding to the correct number of significant figures. In this book, we display numerical values rounded off to two or three significant figures. This occasionally makes some mathematical manipulations look odd or incorrect. For instance, looking ahead to Example 3.5 on page 65, you will see the operation −17.7 km + 34.6 km = 17.0 km. This looks like an incorrect subtraction, but that is only because we have rounded the numbers 17.7 km and 34.6 km for display. If all digits in these two intermediate numbers are retained and the rounding is only performed on the final number, the correct three-digit result of 17.0 km is obtained.

Pitfall Prevention 1.5

Symbolic Solutions

When solving problems, it is very useful to perform the solution completely in algebraic form and wait until the very end to enter numerical values into the final symbolic expression. This method will save many calculator keystrokes, especially if some quantities cancel so that you never have to enter their values into your calculator! In addition, you will only need to round once, on the final result.

Example 1.5 Installing a Carpet

A carpet is to be installed in a rectangular room whose length is measured to be 12.71 m and whose width is measured to be 3.46 m. Find the area of the room.

SOLUTION

If you multiply 12.71 m by 3.46 m on your calculator, you will see an answer of 43.976 6 m^2. How many of these numbers should you claim? Our rule of thumb for multiplication tells us that you can claim only the number of significant figures in your answer as are present in the measured quantity having the lowest number of significant figures. In this example, the lowest number of significant figures is three in 3.46 m, so we should express our final answer as 44.0 m^2.

Summary

Definitions

The three fundamental physical quantities of mechanics are **length, mass,** and **time,** which in the SI system have the units **meter** (m), **kilogram** (kg), and **second** (s), respectively. These fundamental quantities cannot be defined in terms of more basic quantities.

The **density** of a substance is defined as its *mass per unit volume:*

$$\rho \equiv \frac{m}{V} \qquad \textbf{(1.1)}$$

continued

Concepts and Principles

The method of **dimensional analysis** is very powerful in solving physics problems. Dimensions can be treated as algebraic quantities. By making estimates and performing order-of-magnitude calculations, you should be able to approximate the answer to a problem when there is not enough information available to specify an exact solution completely.

When you compute a result from several measured numbers, each of which has a certain accuracy, you should give the result with the correct number of **significant figures.**

When **multiplying** several quantities, the number of significant figures in the final answer is the same as the number of significant figures in the quantity having the smallest number of significant figures. The same rule applies to **division.**

When numbers are **added** or **subtracted,** the number of decimal places in the result should equal the smallest number of decimal places of any term in the sum or difference.

Objective Questions

☐ denotes answer available in Student Solutions Manual/Study Guide

1. Rank the following five quantities in order from the largest to the smallest. If two of the quantities are equal, give them equal rank in your list. (a) 0.032 kg (b) 15 g (c) 2.7×10^5 mg (d) 4.1×10^{-8} Gg (e) 2.7×10^8 μg

2. (a) If an equation is dimensionally correct, does that mean that the equation must be true? (b) If an equation is not dimensionally correct, does that mean that the equation cannot be true?

3. Answer each question yes or no. Must two quantities have the same dimensions (a) if you are adding them? (b) If you are multiplying them? (c) If you are subtracting them? (d) If you are dividing them? (e) If you are equating them?

4. The price of gasoline at a particular station is 1.5 euros per liter. An American student can use 33 euros to buy gasoline. Knowing that 4 quarts make a gallon and that 1 liter is close to 1 quart, she quickly reasons that she can buy how many gallons of gasoline? (a) less than 1 gallon (b) about 5 gallons (c) about 8 gallons (d) more than 10 gallons

5. One student uses a meterstick to measure the thickness of a textbook and obtains 4.3 cm ± 0.1 cm. Other students measure the thickness with vernier calipers and obtain four different measurements: (a) 4.32 cm ± 0.01 cm, (b) 4.31 cm ± 0.01 cm, (c) 4.24 cm ± 0.01 cm, and (d) 4.43 cm ± 0.01 cm. Which of these four measurements, if any, agree with that obtained by the first student?

6. A calculator displays a result as $1.365\,248\,0 \times 10^7$ kg. The estimated uncertainty in the result is ±2%. How many digits should be included as significant when the result is written down? (a) zero (b) one (c) two (d) three (e) four

7. Newton's second law of motion (Chapter 5) says that the mass of an object times its acceleration is equal to the net force on the object. Which of the following gives the correct units for force? (a) kg · m/s² (b) kg · m²/s² (c) kg/m · s² (d) kg · m²/s (e) none of those answers

8. A house is advertised as having 1 420 square feet under its roof. What is its area in square meters? (a) 4 660 m² (b) 432 m² (c) 158 m² (d) 132 m² (e) 40.2 m²

9. Which of the following is the best estimate for the mass of all the people living on the Earth? (a) 2×10^8 kg (b) 1×10^9 kg (c) 2×10^{10} kg (d) 3×10^{11} kg (e) 4×10^{12} kg

10. What is the sum of the measured values 21.4 s + 15 s + 17.17 s + 4.00 3 s? (a) 57.573 s (b) 57.57 s (c) 57.6 s (d) 58 s (e) 60 s

Conceptual Questions

☐ denotes answer available in Student Solutions Manual/Study Guide

1. Suppose the three fundamental standards of the metric system were length, *density,* and time rather than length, *mass,* and time. The standard of density in this system is to be defined as that of water. What considerations about water would you need to address to make sure that the standard of density is as accurate as possible?

2. Express the following quantities using the prefixes given in Table 1.4. (a) 3×10^{-4} m (b) 5×10^{-5} s (c) 72×10^2 g

3. What natural phenomena could serve as alternative time standards?

4. Why is the metric system of units considered superior to most other systems of units?

Problems

ENHANCED WebAssign The problems found in this chapter may be assigned online in Enhanced WebAssign

1. denotes straightforward problem; **2.** denotes intermediate problem;
3. denotes challenging problem

1. full solution available in the Student Solutions Manual/Study Guide

1. denotes problems most often assigned in Enhanced WebAssign; these provide students with targeted feedback and either a Master It tutorial or a Watch It solution video.

QC denotes asking for quantitative and conceptual reasoning

S denotes symbolic reasoning problem

M denotes Master It tutorial available in Enhanced WebAssign

GP denotes guided problem

shaded denotes "paired problems" that develop reasoning with symbols and numerical values

Section 1.1 Standards of Length, Mass, and Time

Note: Consult the endpapers, appendices, and tables in the text whenever necessary in solving problems. For this chapter, Table 14.1 and Appendix B.3 may be particularly useful. Answers to odd-numbered problems appear in the back of the book.

1. QC (a) Use information on the endpapers of this book to calculate the average density of the Earth. (b) Where does the value fit among those listed in Table 14.1 in Chapter 14? Look up the density of a typical surface rock like granite in another source and compare it with the density of the Earth.

2. The standard kilogram (Fig. 1.1a) is a platinum–iridium cylinder 39.0 mm in height and 39.0 mm in diameter. What is the density of the material?

3. An automobile company displays a die-cast model of its first car, made from 9.35 kg of iron. To celebrate its hundredth year in business, a worker will recast the model in solid gold from the original dies. What mass of gold is needed to make the new model?

4. QC A proton, which is the nucleus of a hydrogen atom, can be modeled as a sphere with a diameter of 2.4 fm and a mass of 1.67×10^{-27} kg. (a) Determine the density of the proton. (b) State how your answer to part (a) compares with the density of osmium, given in Table 14.1 in Chapter 14.

5. Two spheres are cut from a certain uniform rock. One has radius 4.50 cm. The mass of the other is five times greater. Find its radius.

Section 1.2 Matter and Model Building

6. The mass of a copper atom is 1.06×10^{-25} kg, and the density of copper is 8 920 kg/m^3. (a) Determine the number of atoms in 1 cm^3 of copper. (b) Visualize the one cubic centimeter as formed by stacking up identical cubes, with one copper atom at the center of each. Determine the volume of each cube. (c) Find the edge dimension of each cube, which represents an estimate for the spacing between atoms.

7. A crystalline solid consists of atoms stacked up in a repeating lattice structure. Consider a crystal as shown in Figure P1.7a. The atoms reside at the corners of cubes of side L = 0.200 nm. One piece of evidence for the regular arrangement of atoms comes from the flat surfaces along which a crystal separates, or cleaves, when it is broken. Suppose this crystal cleaves along a face diagonal as shown in Figure P1.7b. Calculate the spacing d between two adjacent atomic planes that separate when the crystal cleaves.

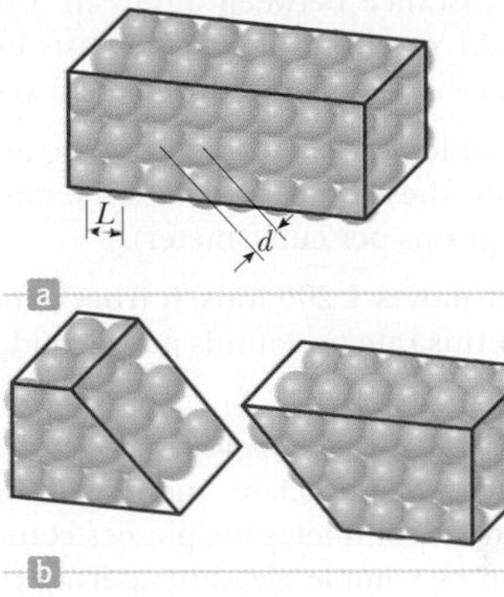

Figure P1.7

Section 1.3 Dimensional Analysis

8. Figure P1.8 shows a *frustum of a cone.* Match each of the expressions (a) $\pi(r_1 + r_2)[h^2 + (r_2 - r_1)^2]^{1/2}$, (b) $2\pi(r_1 + r_2)$, and (c) $\pi h(r_1^2 + r_1r_2 + r_2^2)/3$ with the quantity it describes: (d) the total circumference of the flat circular faces, (e) the volume, or (f) the area of the curved surface.

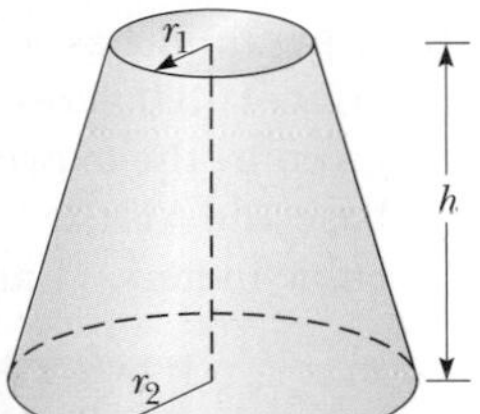

Figure P1.8

9. **Which of the following equations are dimensionally correct? (a) $v_f = v_i + ax$ (b) $y = (2\text{ m}) \cos (kx)$, where $k = 2\text{ m}^{-1}$**

10. Newton's law of universal gravitation is represented by

$$F = \frac{GMm}{r^2}$$

where F is the magnitude of the gravitational force exerted by one small object on another, M and m are the masses of the objects, and r is a distance. Force has the SI units kg · m/s^2. What are the SI units of the proportionality constant G?

11. Kinetic energy K (Chapter 7) has dimensions kg · m^2/s^2. It can be written in terms of the momentum p (Chapter 9) and mass m as

$$K = \frac{p^2}{2m}$$

(a) Determine the proper units for momentum using dimensional analysis. (b) The unit of force is the newton N, where 1 N = 1 kg · m/s^2. What are the units of momentum p in terms of a newton and another fundamental SI unit?

12. (a) Assume the equation $x = At^3 + Bt$ describes the motion of a particular object, with x having the dimension of length and t having the dimension of time. Determine the dimensions of the constants A and B. (b) Determine the dimensions of the derivative $dx/dt = 3At^2 + B$.

Section 1.4 Conversion of Units

13. A rectangular building lot has a width of 75.0 ft and a length of 125 ft. Determine the area of this lot in square meters.

14. Suppose your hair grows at the rate 1/32 in. per day. Find the rate at which it grows in nanometers per second. Because the distance between atoms in a molecule is on the order of 0.1 nm, your answer suggests how rapidly layers of atoms are assembled in this protein synthesis.

15. A solid piece of lead has a mass of 23.94 g and a volume of 2.10 cm^3. From these data, calculate the density of lead in SI units (kilograms per cubic meter).

16. An ore loader moves 1 200 tons/h from a mine to the surface. Convert this rate to pounds per second, using 1 ton = 2 000 lb.

17. *Why is the following situation impossible?* A student's dormitory room measures 3.8 m by 3.6 m, and its ceiling is 2.5 m high. After the student completes his physics course, he displays his dedication by completely wallpapering the walls of the room with the pages from his copy of volume 1 (Chapters 1–22) of this textbook. He even covers the door and window.

18. A pyramid has a height of 481 ft, and its base covers an area of 13.0 acres (Fig. P1.18). The volume of a pyramid is given by the expression $V = \frac{1}{3}Bh$, where B is the area of the base and h is the height. Find the volume of this pyramid in cubic meters. (1 acre = 43 560 ft^2)

Adam Sylvester/Photo Researchers, Inc.

Figure P1.18 Problems 18 and 19.

19. The pyramid described in Problem 18 contains approximately 2 million stone blocks that average 2.50 tons each. Find the weight of this pyramid in pounds.

20. Assume it takes 7.00 min to fill a 30.0-gal gasoline tank. (a) Calculate the rate at which the tank is filled in gallons per second. (b) Calculate the rate at which the tank is filled in cubic meters per second. (c) Determine the time interval, in hours, required to fill a 1.00-m^3 volume at the same rate. (1 U.S. gal = 231 in.^{3})

21. M One cubic meter (1.00 m^3) of aluminum has a mass of 2.70×10^3 kg, and the same volume of iron has a mass of 7.86×10^3 kg. Find the radius of a solid aluminum sphere that will balance a solid iron sphere of radius 2.00 cm on an equal-arm balance.

22. S Let ρ_{Al} represent the density of aluminum and ρ_{Fe} that of iron. Find the radius of a solid aluminum sphere that balances a solid iron sphere of radius r_{Fe} on an equal-arm balance.

23. One gallon of paint (volume = 3.78×10^{-3} m^3) covers an area of 25.0 m^2. What is the thickness of the fresh paint on the wall?

24. An auditorium measures 40.0 m × 20.0 m × 12.0 m. The density of air is 1.20 kg/m^3. What are (a) the volume of the room in cubic feet and (b) the weight of air in the room in pounds?

25. (a) At the time of this book's printing, the U.S. national debt is about \$10 trillion. If payments were made at the rate of \$1 000 per second, how many years would it take to pay off the debt, assuming no interest were charged? (b) A dollar bill is about 15.5 cm long. How many dollar bills attached end to end would it take to reach the Moon? The front endpapers give the Earth–Moon distance. *Note:* Before doing these calculations, try to guess at the answers. You may be very surprised.

26. A hydrogen atom has a diameter of 1.06×10^{-10} m. The nucleus of the hydrogen atom has a diameter of approximately 2.40×10^{-15} m. (a) For a scale model, represent the diameter of the hydrogen atom by the playing length of an American football field (100 yards = 300 ft) and determine the diameter of the nucleus in millimeters. (b) Find the ratio of the volume of the hydrogen atom to the volume of its nucleus.

Section 1.5 Estimates and Order-of-Magnitude Calculations

Note: In your solutions to Problems 27 through 30, state the quantities you measure or estimate and the values you take for them.

27. Find the order of magnitude of the number of table-tennis balls that would fit into a typical-size room (without being crushed).

28. (a) Compute the order of magnitude of the mass of a bathtub half full of water. (b) Compute the order of magnitude of the mass of a bathtub half full of copper coins.

29. To an order of magnitude, how many piano tuners reside in New York City? The physicist Enrico Fermi was famous for asking questions like this one on oral Ph.D. qualifying examinations.

30. An automobile tire is rated to last for 50 000 miles. To an order of magnitude, through how many revolutions will it turn over its lifetime?

Section 1.6 Significant Figures

Note: Appendix B.8 on propagation of uncertainty may be useful in solving some problems in this section.

31. The *tropical year,* the time interval from one vernal equinox to the next vernal equinox, is the basis for our calendar. It contains 365.242 199 days. Find the number of seconds in a tropical year.

32. How many significant figures are in the following numbers? (a) 78.9 ± 0.2 (b) 3.788×10^9 (c) 2.46×10^{-6} (d) 0.005 3

33. A rectangular plate has a length of (21.3 ± 0.2) cm and a width of (9.8 ± 0.1) cm. Calculate the area of the plate, including its uncertainty.

34. Carry out the arithmetic operations (a) the sum of the measured values 756, 37.2, 0.83, and 2; (b) the product $0.003\,2 \times 356.3$; and (c) the product $5.620 \times \pi$.

Note: The next 12 problems call on mathematical skills from your prior education that will be useful throughout this course.

35. Review. A child is surprised that because of sales tax she must pay \$1.36 for a toy marked \$1.25. What is the effective tax rate on this purchase, expressed as a percentage?

36. Review. The average density of the planet Uranus is $1.27 \times 10^3\ \text{kg/m}^3$. The ratio of the mass of Neptune to that of Uranus is 1.19. The ratio of the radius of Neptune to that of Uranus is 0.969. Find the average density of Neptune.

37. Review. In a community college parking lot, the number of ordinary cars is larger than the number of sport utility vehicles by 94.7%. The difference between the number of cars and the number of SUVs is 18. Find the number of SUVs in the lot.

38. Review. While you are on a trip to Europe, you must purchase hazelnut chocolate bars for your grandmother. Eating just one square each day, she makes each large bar last for one and one-third months. How many bars will constitute a year's supply for her?

39. Review. The ratio of the number of sparrows visiting a bird feeder to the number of more interesting birds is 2.25. On a morning when altogether 91 birds visit the feeder, what is the number of sparrows?

40. Review. Find every angle θ between 0 and 360° for which the ratio of $\sin\theta$ to $\cos\theta$ is -3.00.

41. Review. Prove that one solution of the equation

$$2.00x^4 - 3.00x^3 + 5.00x = 70.0$$

is $x = -2.22$.

42. Review. A highway curve forms a section of a circle. A car goes around the curve as shown in the helicopter view of Figure P1.42. Its dashboard compass shows that the car is initially heading due east. After it travels $d = 840$ m, it is heading $\theta = 35.0°$ south of east. Find the radius of curvature of its path. *Suggestion:* You may find it useful to learn a geometric theorem stated in Appendix B.3.

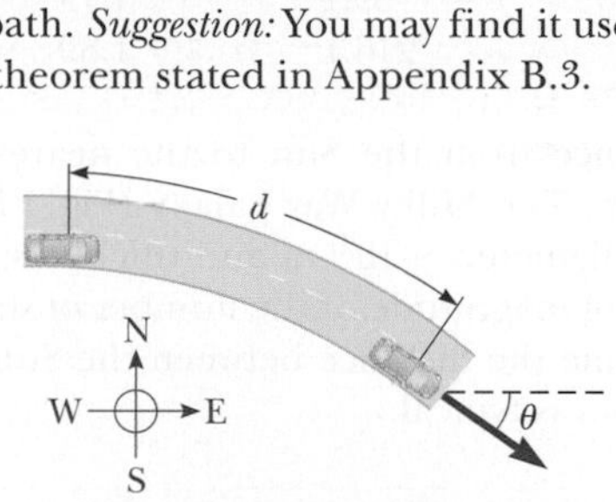

Figure P1.42

43. **M** **Review.** A pet lamb grows rapidly, with its mass proportional to the cube of its length. When the lamb's length changes by 15.8%, its mass increases by 17.3 kg. Find the lamb's mass at the end of this process.

44. **S** **Review.** From the set of equations

$$p = 3q$$
$$pr = qs$$
$$\tfrac{1}{2}pr^2 + \tfrac{1}{2}qs^2 = \tfrac{1}{2}qt^2$$

involving the unknowns p, q, r, s, and t, find the value of the ratio of t to r.

45. **Q|C** **Review.** A student is supplied with a stack of copy paper, ruler, compass, scissors, and a sensitive balance. He cuts out various shapes in various sizes, calculates their areas, measures their masses, and prepares the graph of Figure P1.45. (a) Consider the fourth experimental point from the top. How far is it from the best-fit straight line? Express your answer as a difference in vertical-axis coordinate. (b) Express your answer as a percentage. (c) Calculate the slope of the line. (d) State what the graph demonstrates, referring to the shape of the graph and the results of parts (b) and (c). (e) Describe whether this result should be expected theoretically. (f) Describe the physical meaning of the slope.

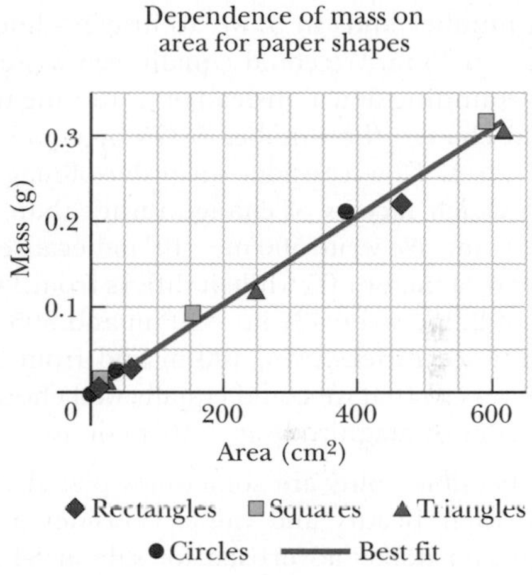

Figure P1.45

46. **Q|C** **S** **Review.** Figure P1.46 shows students studying the thermal conduction of energy into cylindrical blocks of ice. As we will see in Chapter 20, this process is described by the equation

$$\frac{Q}{\Delta t} = \frac{k\pi d^2(T_h - T_c)}{4L}$$

For experimental control, in one set of trials all quantities except d and Δt are constant. (a) If d is made three times larger, does the equation predict that Δt will get larger or get smaller? By what factor? (b) What pattern of proportionality of Δt to d does the equation predict? (c) To display this proportionality as a straight line on a graph, what quantities should you plot on the horizontal and vertical axes? (d) What expression represents the theoretical slope of this graph?

Figure P1.46

47. The radius of a uniform solid sphere is measured to be (6.50 ± 0.20) cm, and its mass is measured to be (1.85 ± 0.02) kg. Determine the density of the sphere in kilograms per cubic meter and the uncertainty in the density.

Additional Problems

48. **Q|C** (a) What is the order of magnitude of the number of microorganisms in the human intestinal tract? A typical bacterial length scale is 10^{-6} m. Estimate the intestinal volume and assume 1% of it is occupied by bacteria. (b) Does the number of bacteria suggest whether the bacteria are beneficial, dangerous, or neutral for the human body? What functions could they serve?

49. In a situation in which data are known to three significant digits, we write 6.379 m = 6.38 m and 6.374 m = 6.37 m. When a number ends in 5, we arbitrarily choose to write 6.375 m = 6.38 m. We could equally well write 6.375 m = 6.37 m, "rounding down" instead of "rounding up," because we would change the number 6.375 by equal increments in both cases. Now consider an order-of-magnitude estimate, in which factors of change rather than increments are important. We write 500 m $\sim 10^3$ m because 500 differs from 100 by a factor of 5 while it differs from 1 000 by only a factor of 2. We write 437 m $\sim 10^3$ m and 305 m $\sim 10^2$ m. What distance differs from 100 m and from 1 000 m by equal factors so that we could equally well choose to represent its order of magnitude as $\sim 10^2$ m or as $\sim 10^3$ m?

50. **Q|C** Collectible coins are sometimes plated with gold to enhance their beauty and value. Consider a commemorative quarter-dollar advertised for sale at \$4.98. It has a diameter of 24.1 mm and a thickness of 1.78 mm, and it is completely covered with a layer of pure gold 0.180 μm thick. The volume of the plating is equal to the thickness of the layer multiplied by the area to which it is applied. The patterns on the faces of the coin and the grooves on its edge have a negligible effect on its area. Assume the price of gold is \$25.0 per gram. (a) Find the cost of the gold added to the coin. (b) Does the cost of the gold significantly enhance the value of the coin? Explain your answer.

51. The diameter of our disk-shaped galaxy, the Milky Way, is about 1.0×10^5 light-years (ly). The distance to the Andromeda galaxy (Fig. P1.51), which is the spiral galaxy nearest to the Milky Way, is about 2.0 million ly. If a scale model represents the Milky Way and Andromeda galaxies as dinner plates 25 cm in diameter, determine the distance between the centers of the two plates.

Robert Gendler/NASA

Figure P1.51 The Andromeda galaxy

52. *Why is the following situation impossible?* In an effort to boost interest in a television game show, each weekly winner is offered an additional \$1 million bonus prize if he or she can personally count out that exact amount from a supply of one-dollar bills. The winner must do this task under supervision by television show executives and within one 40-hour work week. To the dismay of the show's producers, most contestants succeed at the challenge.

53. A high fountain of water is located at the center of a circular pool as shown in Figure P1.53. A student walks around the pool and measures its circumference to be 15.0 m. Next, the student stands at the edge of the pool and uses a protractor to gauge the angle of elevation of the top of the fountain to be $\phi = 55.0°$. How high is the fountain?

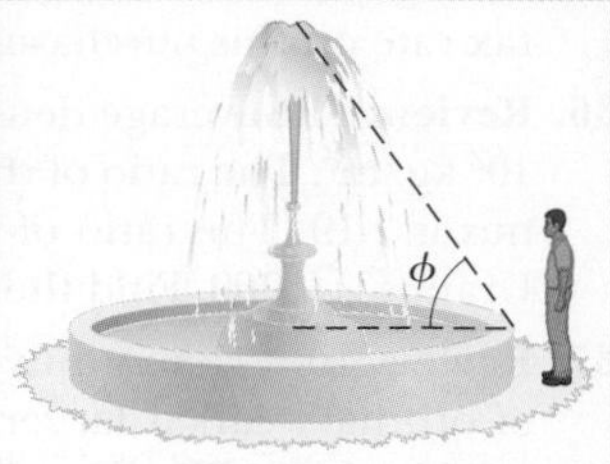

Figure P1.53
Problems 53 and 54.

54. **S** A water fountain is at the center of a circular pool as shown in Figure P1.53. A student walks around the pool and measures its circumference C. Next, he stands at the edge of the pool and uses a protractor to measure the angle of elevation ϕ of his sightline to the top of the water jet. How high is the fountain?

55. **Q|C** The data in the following table represent measurements of the masses and dimensions of solid cylinders of aluminum, copper, brass, tin, and iron. (a) Use these data to calculate the densities of these substances. (b) State how your results compare with those given in Table 14.1.

Substance	Mass (g)	Diameter (cm)	Length (cm)
Aluminum	51.5	2.52	3.75
Copper	56.3	1.23	5.06
Brass	94.4	1.54	5.69
Tin	69.1	1.75	3.74
Iron	216.1	1.89	9.77

56. The distance from the Sun to the nearest star is about 4×10^{16} m. The Milky Way galaxy (Fig. P1.56) is roughly a disk of diameter $\sim 10^{21}$ m and thickness $\sim 10^{19}$ m. Find the order of magnitude of the number of stars in the Milky Way. Assume the distance between the Sun and our nearest neighbor is typical.

Richard Payne/NASA

Figure P1.56 The Milky Way galaxy

57. M Assume there are 100 million passenger cars in the United States and the average fuel consumption is 20 mi/gal of gasoline. If the average distance traveled by each car is 10 000 mi/yr, how much gasoline would be saved per year if average fuel consumption could be increased to 25 mi/gal?

58. Q|C A spherical shell has an outside radius of 2.60 cm and an inside radius of a. The shell wall has uniform thickness and is made of a material with density 4.70 g/cm^3. The space inside the shell is filled with a liquid having a density of 1.23 g/cm^3. (a) Find the mass m of the sphere, including its contents, as a function of a. (b) For what value of the variable a does m have its maximum possible value? (c) What is this maximum mass? (d) Explain whether the value from part (c) agrees with the result of a direct calculation of the mass of a solid sphere of uniform density made of the same material as the shell. (e) **What If?** Would the answer to part (a) change if the inner wall were not concentric with the outer wall?

59. Bacteria and other prokaryotes are found deep underground, in water, and in the air. One micron (10^{-6} m) is a typical length scale associated with these microbes. (a) Estimate the total number of bacteria and other prokaryotes on the Earth. (b) Estimate the total mass of all such microbes.

60. Q|C Air is blown into a spherical balloon so that, when its radius is 6.50 cm, its radius is increasing at the rate 0.900 cm/s. (a) Find the rate at which the volume of the balloon is increasing. (b) If this volume flow rate of air entering the balloon is constant, at what rate will the radius be increasing when the radius is 13.0 cm? (c) Explain physically why the answer to part (b) is larger or smaller than 0.9 cm/s, if it is different.

61. A rod extending between $x = 0$ and $x = 14.0$ cm has uniform cross-sectional area $A = 9.00$ cm^2. Its density increases steadily between its ends from 2.70 g/cm^3 to 19.3 g/cm^3. (a) Identify the constants B and C required in the expression $\rho = B + Cx$ to describe the variable density. (b) The mass of the rod is given by

$$m = \int_{\text{all material}} \rho\, dV = \int_{\text{all } x} \rho A\, dx = \int_0^{14.0\text{ cm}} (B + Cx)(9.00\text{ cm}^2)\, dx$$

Carry out the integration to find the mass of the rod.

62. In physics, it is important to use mathematical approximations. (a) Demonstrate that for small angles ($< 20°$)

$$\tan \alpha \approx \sin \alpha \approx \alpha = \frac{\pi \alpha'}{180°}$$

where α is in radians and α' is in degrees. (b) Use a calculator to find the largest angle for which $\tan \alpha$ may be approximated by α with an error less than 10.0%.

63. M The consumption of natural gas by a company satisfies the empirical equation $V = 1.50t + 0.008\,00t^2$, where V is the volume of gas in millions of cubic feet and t is the time in months. Express this equation in units of cubic feet and seconds. Assume a month is 30.0 days.

64. GP A woman wishing to know the height of a mountain measures the angle of elevation of the mountaintop as 12.0°. After walking 1.00 km closer to the mountain on level ground, she finds the angle to be 14.0°. (a) Draw a picture of the problem, neglecting the height of the woman's eyes above the ground. *Hint:* Use two triangles. (b) Using the symbol y to represent the mountain height and the symbol x to represent the woman's original distance from the mountain, label the picture. (c) Using the labeled picture, write two trigonometric equations relating the two selected variables. (d) Find the height y.

65. A child loves to watch as you fill a transparent plastic bottle with shampoo (Fig P1.65). Every horizontal cross section of the bottle is circular, but the diameters of the circles have different values. You pour the brightly colored shampoo into the bottle at a constant rate of 16.5 cm^3/s. At what rate is its level in the bottle rising (a) at a point where the diameter of the bottle is 6.30 cm and (b) at a point where the diameter is 1.35 cm?

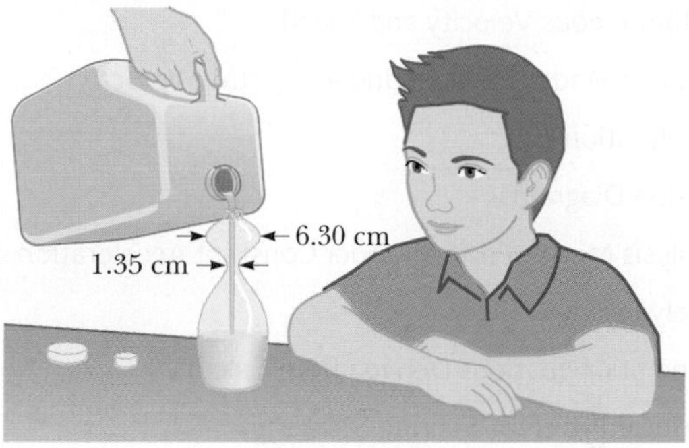

Figure P1.65

Challenge Problems

66. S A woman stands at a horizontal distance x from a mountain and measures the angle of elevation of the mountaintop above the horizontal as θ. After walking a distance d closer to the mountain on level ground, she finds the angle to be ϕ. Find a general equation for the height y of the mountain in terms of d, ϕ, and θ, neglecting the height of her eyes above the ground.

67. You stand in a flat meadow and observe two cows (Fig. P1.67). Cow A is due north of you and 15.0 m from your position. Cow B is 25.0 m from your position. From your point of view, the angle between cow A and cow B is 20.0°, with cow B appearing to the right of cow A. (a) How far apart are cow A and cow B? (b) Consider the view seen by cow A. According to this cow, what is the angle between you and cow B? (c) Consider the view seen by cow B. According to this cow, what is the angle between you and cow A? *Hint:* What does the situation look like to a hummingbird hovering above the meadow? (d) Two stars in the sky appear to be 20.0° apart. Star A is 15.0 ly from the Earth, and star B, appearing to the right of star A, is 25.0 ly from the Earth. To an inhabitant of a planet orbiting star A, what is the angle in the sky between star B and our Sun?

Figure P1.67 Your view of two cows in a meadow. Cow A is due north of you. You must rotate your eyes through an angle of 20.0° to look from cow A to cow B.

chapter 2

Motion in One Dimension

In drag racing, a driver wants as large an acceleration as possible. In a distance of one-quarter mile, a vehicle reaches speeds of more than 320 mi/h, covering the entire distance in under 5 s. *(George Lepp/Stone/Getty Images)*

As a first step in studying classical mechanics, we describe the motion of an object while ignoring the interactions with external agents that might be causing or modifying that motion. This portion of classical mechanics is called *kinematics*. (The word *kinematics* has the same root as *cinema*.) In this chapter, we consider only motion in one dimension, that is, motion of an object along a straight line.

From everyday experience, we recognize that motion of an object represents a continuous change in the object's position. In physics, we can categorize motion into three types: translational, rotational, and vibrational. A car traveling on a highway is an example of translational motion, the Earth's spin on its axis is an example of rotational motion, and the back-and-forth movement of a pendulum is an example of vibrational motion. In this and the next few chapters, we are concerned only with translational motion. (Later in the book we shall discuss rotational and vibrational motions.)

In our study of translational motion, we use what is called the **particle model** and describe the moving object as a *particle* regardless of its size. Remember our discussion of making models for physical situations in Section 1.2. In general, **a particle is a point-like object, that is, an object that has mass but is of infinitesimal size.** For example, if we wish to describe the motion of the Earth around the Sun, we can treat the Earth as a particle and obtain reasonably accurate data about its orbit. This approximation is justified because the radius of the Earth's orbit is large compared with the dimensions of the Earth and the Sun. As an example on a much smaller scale, it is possible to explain the pressure exerted by a gas on the walls of a container by treating the gas molecules as particles, without regard for the internal structure of the molecules.

2.1 Position, Velocity, and Speed

◀ **Position**

A particle's **position** x is the location of the particle with respect to a chosen reference point that we can consider to be the origin of a coordinate system. The motion of a particle is completely known if the particle's position in space is known at all times.

Consider a car moving back and forth along the x axis as in Active Figure 2.1a. When we begin collecting position data, the car is 30 m to the right of the reference position $x = 0$. We will use the particle model by identifying some point on the car, perhaps the front door handle, as a particle representing the entire car.

We start our clock, and once every 10 s we note the car's position. As you can see from Table 2.1, the car moves to the right (which we have defined as the positive direction) during the first 10 s of motion, from position Ⓐ to position Ⓑ. After Ⓑ, the position values begin to decrease, suggesting the car is backing up from position Ⓑ through position Ⓕ. In fact, at Ⓓ, 30 s after we start measuring, the car is at the origin of coordinates (see Active Fig. 2.1a). It continues moving to the left and is more than 50 m to the left of $x = 0$ when we stop recording information after our sixth data point. A graphical representation of this information is presented in Active Figure 2.1b. Such a plot is called a *position–time graph.*

Notice the *alternative representations* of information that we have used for the motion of the car. Active Figure 2.1a is a *pictorial representation,* whereas Active Figure 2.1b is a *graphical representation.* Table 2.1 is a *tabular representation* of the

TABLE 2.1

Position of the Car at Various Times

Position	t (s)	x (m)
Ⓐ	0	30
Ⓑ	10	52
Ⓒ	20	38
Ⓓ	30	0
Ⓔ	40	−37
Ⓕ	50	−53

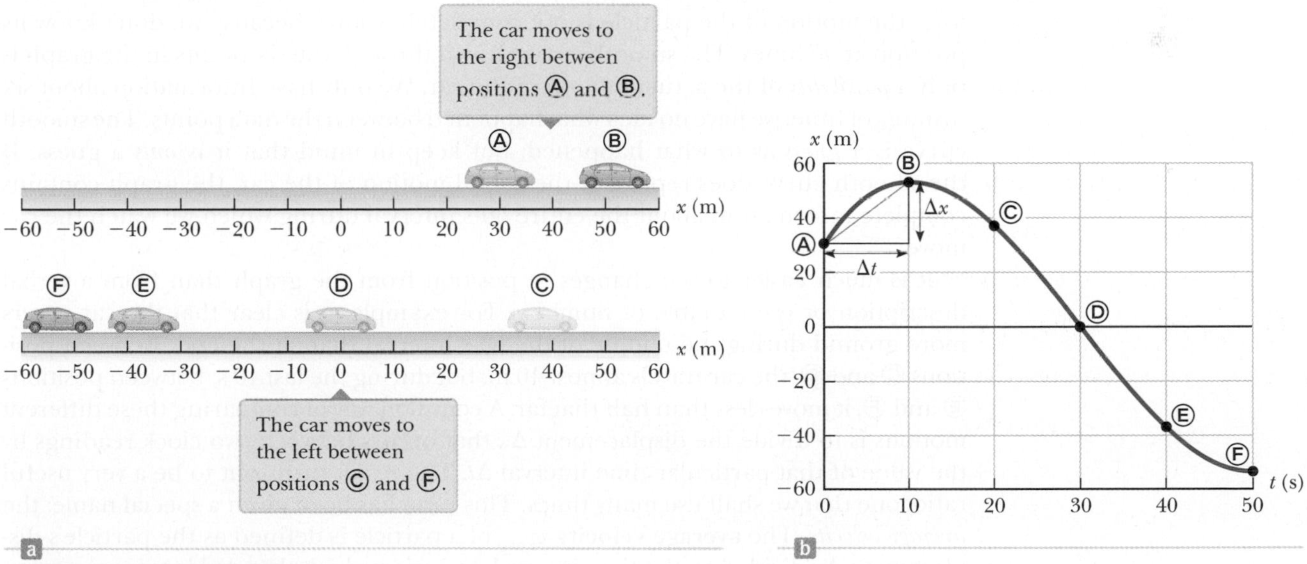

ACTIVE FIGURE 2.1

A car moves back and forth along a straight line. Because we are interested only in the car's translational motion, we can model it as a particle. Several representations of the information about the motion of the car can be used. Table 2.1 is a tabular representation of the information. (a) A pictorial representation of the motion of the car. (b) A graphical representation (position–time graph) of the motion of the car.

same information. Using an alternative representation is often an excellent strategy for understanding the situation in a given problem. The ultimate goal in many problems is a *mathematical representation,* which can be analyzed to solve for some requested piece of information.

Given the data in Table 2.1, we can easily determine the change in position of the car for various time intervals. The **displacement** Δx of a particle is defined as its change in position in some time interval. As the particle moves from an initial position x_i to a final position x_f, its displacement is given by

Displacement ▶

$$\Delta x \equiv x_f - x_i \tag{2.1}$$

We use the capital Greek letter delta (Δ) to denote the *change* in a quantity. From this definition, we see that Δx is positive if x_f is greater than x_i and negative if x_f is less than x_i.

It is very important to recognize the difference between displacement and distance traveled. **Distance** is the length of a path followed by a particle. Consider, for example, the basketball players in Figure 2.2. If a player runs from his own team's basket down the court to the other team's basket and then returns to his own basket, the *displacement* of the player during this time interval is zero because he ended up at the same point as he started: $x_f = x_i$, so $\Delta x = 0$. During this time interval, however, he moved through a *distance* of twice the length of the basketball court. Distance is always represented as a positive number, whereas displacement can be either positive or negative.

Brian Drake/Time Life Pictures/Getty Images

Figure 2.2 On this basketball court, players run back and forth for the entire game. The distance that the players run over the duration of the game is nonzero. The displacement of the players over the duration of the game is approximately zero because they keep returning to the same point over and over again.

Displacement is an example of a vector quantity. Many other physical quantities, including position, velocity, and acceleration, also are vectors. In general, a **vector quantity** requires the specification of both direction and magnitude. By contrast, a **scalar quantity** has a numerical value and no direction. In this chapter, we use positive (+) and negative (−) signs to indicate vector direction. For example, for horizontal motion let us arbitrarily specify to the right as being the positive direction. It follows that any object always moving to the right undergoes a positive displacement $\Delta x > 0$, and any object moving to the left undergoes a negative displacement so that $\Delta x < 0$. We shall treat vector quantities in greater detail in Chapter 3.

One very important point has not yet been mentioned. Notice that the data in Table 2.1 result only in the six data points in the graph in Active Figure 2.1b. Therefore, the motion of the particle is not completely known because we don't know its position at *all* times. The smooth curve drawn through the six points in the graph is only a *possibility* of the actual motion of the car. We only have information about six instants of time; we have no idea what happened between the data points. The smooth curve is a *guess* as to what happened, but keep in mind that it is *only* a guess. If the smooth curve does represent the actual motion of the car, the graph contains complete information about the entire 50-s interval during which we watch the car move.

It is much easier to see changes in position from the graph than from a verbal description or even a table of numbers. For example, it is clear that the car covers more ground during the middle of the 50-s interval than at the end. Between positions Ⓒ and Ⓓ, the car travels almost 40 m, but during the last 10 s, between positions Ⓔ and Ⓕ, it moves less than half that far. A common way of comparing these different motions is to divide the displacement Δx that occurs between two clock readings by the value of that particular time interval Δt. The result turns out to be a very useful ratio, one that we shall use many times. This ratio has been given a special name: the *average velocity*. The **average velocity** $v_{x,\text{avg}}$ of a particle is defined as the particle's displacement Δx divided by the time interval Δt during which that displacement occurs:

Average velocity ▶

$$v_{x,\text{avg}} \equiv \frac{\Delta x}{\Delta t} \tag{2.2}$$

where the subscript x indicates motion along the x axis. From this definition we see that average velocity has dimensions of length divided by time (L/T), or meters per second in SI units.

The average velocity of a particle moving in one dimension can be positive or negative, depending on the sign of the displacement. (The time interval Δt is always positive.) If the coordinate of the particle increases in time (that is, if $x_f > x_i$), Δx is positive and $v_{x,\text{avg}} = \Delta x/\Delta t$ is positive. This case corresponds to a particle moving in the positive x direction, that is, toward larger values of x. If the coordinate decreases in time (that is, if $x_f < x_i$), Δx is negative and hence $v_{x,\text{avg}}$ is negative. This case corresponds to a particle moving in the negative x direction.

We can interpret average velocity geometrically by drawing a straight line between any two points on the position–time graph in Active Figure 2.1b. This line forms the hypotenuse of a right triangle of height Δx and base Δt. The slope of this line is the ratio $\Delta x/\Delta t$, which is what we have defined as average velocity in Equation 2.2. For example, the line between positions Ⓐ and Ⓑ in Active Figure 2.1b has a slope equal to the average velocity of the car between those two times, $(52\text{ m} - 30\text{ m})/(10\text{ s} - 0) = 2.2\text{ m/s}$.

In everyday usage, the terms *speed* and *velocity* are interchangeable. In physics, however, there is a clear distinction between these two quantities. Consider a marathon runner who runs a distance d of more than 40 km and yet ends up at her starting point. Her total displacement is zero, so her average velocity is zero! Nonetheless, we need to be able to quantify how fast she was running. A slightly different ratio accomplishes that for us. The **average speed** v_{avg} of a particle, a scalar quantity, is defined as the total distance d traveled divided by the total time interval required to travel that distance:

$$v_{\text{avg}} = \frac{d}{\Delta t} \qquad (2.3)$$

◀ **Average speed**

The SI unit of average speed is the same as the unit of average velocity: meters per second. Unlike average velocity, however, average speed has no direction and is always expressed as a positive number. Notice the clear distinction between the definitions of average velocity and average speed: average velocity (Eq. 2.2) is the *displacement* divided by the time interval, whereas average speed (Eq. 2.3) is the *distance* divided by the time interval.

> **Pitfall Prevention 2.1**
>
> **Average Speed and Average Velocity**
> The magnitude of the average velocity is *not* the average speed. For example, consider the marathon runner discussed before Equation 2.3. The magnitude of her average velocity is zero, but her average speed is clearly not zero.

Knowledge of the average velocity or average speed of a particle does not provide information about the details of the trip. For example, suppose it takes you 45.0 s to travel 100 m down a long, straight hallway toward your departure gate at an airport. At the 100-m mark, you realize you missed the restroom, and you return back 25.0 m along the same hallway, taking 10.0 s to make the return trip. The magnitude of your average *velocity* is $+75.0\text{ m}/55.0\text{ s} = +1.36\text{ m/s}$. The average *speed* for your trip is $125\text{ m}/55.0\text{ s} = 2.27\text{ m/s}$. You may have traveled at various speeds during the walk and, of course, you changed direction. Neither average velocity nor average speed provides information about these details.

***Quick Quiz* 2.1** Under which of the following conditions is the magnitude of the average velocity of a particle moving in one dimension smaller than the average speed over some time interval? **(a)** A particle moves in the $+x$ direction without reversing. **(b)** A particle moves in the $-x$ direction without reversing. **(c)** A particle moves in the $+x$ direction and then reverses the direction of its motion. **(d)** There are no conditions for which this is true.

Example 2.1 Calculating the Average Velocity and Speed

Find the displacement, average velocity, and average speed of the car in Active Figure 2.1a between positions Ⓐ and Ⓕ.

continued

2.1 cont.

SOLUTION

Consult Active Figure 2.1 to form a mental image of the car and its motion. We model the car as a particle. From the position–time graph given in Active Figure 2.1b, notice that $x_{Ⓐ} = 30$ m at $t_{Ⓐ} = 0$ s and that $x_{Ⓕ} = -53$ m at $t_{Ⓕ} = 50$ s.

Use Equation 2.1 to find the displacement of the car:

$$\Delta x = x_{Ⓕ} - x_{Ⓐ} = -53 \text{ m} - 30 \text{ m} = \boxed{-83 \text{ m}}$$

This result means that the car ends up 83 m in the negative direction (to the left, in this case) from where it started. This number has the correct units and is of the same order of magnitude as the supplied data. A quick look at Active Figure 2.1a indicates that it is the correct answer.

Use Equation 2.2 to find the car's average velocity:

$$v_{x,\text{avg}} = \frac{x_{Ⓕ} - x_{Ⓐ}}{t_{Ⓕ} - t_{Ⓐ}}$$

$$= \frac{-53 \text{ m} - 30 \text{ m}}{50 \text{ s} - 0 \text{ s}} = \frac{-83 \text{ m}}{50 \text{ s}} = \boxed{-1.7 \text{ m/s}}$$

We cannot unambiguously find the average speed of the car from the data in Table 2.1 because we do not have information about the positions of the car between the data points. If we adopt the assumption that the details of the car's position are described by the curve in Active Figure 2.1b, the distance traveled is 22 m (from Ⓐ to Ⓑ) plus 105 m (from Ⓑ to Ⓕ), for a total of 127 m.

Use Equation 2.3 to find the car's average speed:

$$v_{\text{avg}} = \frac{127 \text{ m}}{50 \text{ s}} = \boxed{2.5 \text{ m/s}}$$

Notice that the average speed is positive, as it must be. Suppose the red-brown curve in Active Figure 2.1b were different so that between 0 s and 10 s it went from Ⓐ up to 100 m and then came back down to Ⓑ. The average speed of the car would change because the distance is different, but the average velocity would not change.

2.2 Instantaneous Velocity and Speed

Often we need to know the velocity of a particle at a particular instant in time rather than the average velocity over a finite time interval. In other words, you would like to be able to specify your velocity just as precisely as you can specify your position by noting what is happening at a specific clock reading, that is, at some specific instant. What does it mean to talk about how quickly something is moving if we "freeze time" and talk only about an individual instant? In the late 1600s, with the invention of calculus, scientists began to understand how to describe an object's motion at any moment in time.

To see how that is done, consider Active Figure 2.3a, which is a reproduction of the graph in Active Figure 2.1b. We have already discussed the average velocity for the interval during which the car moved from position Ⓐ to position Ⓑ (given by the slope of the blue line) and for the interval during which it moved from Ⓐ to Ⓕ (represented by the slope of the longer blue line and calculated in Example 2.1). The car starts out by moving to the right, which we defined to be the positive direction. Therefore, being positive, the value of the average velocity during the interval from Ⓐ to Ⓑ is more representative of the initial velocity than is the value of the average velocity during the interval from Ⓐ to Ⓕ, which we determined to be negative in Example 2.1. Now let us focus on the short blue line and slide point Ⓑ to the left along the curve, toward point Ⓐ, as in Active Figure 2.3b. The line between the points becomes steeper and steeper, and as the two points become extremely close together, the line becomes a tangent line to the curve, indicated by the green line in Active Figure 2.3b. The slope of this tangent line represents the velocity of the car at point Ⓐ. What we have done is determine the *instantaneous velocity* at that

Pitfall Prevention 2.2

Slopes of Graphs

In any graph of physical data, the *slope* represents the ratio of the change in the quantity represented on the vertical axis to the change in the quantity represented on the horizontal axis. Remember that *a slope has units* (unless both axes have the same units). The units of slope in Active Figures 2.1b and 2.3 are meters per second, the units of velocity.

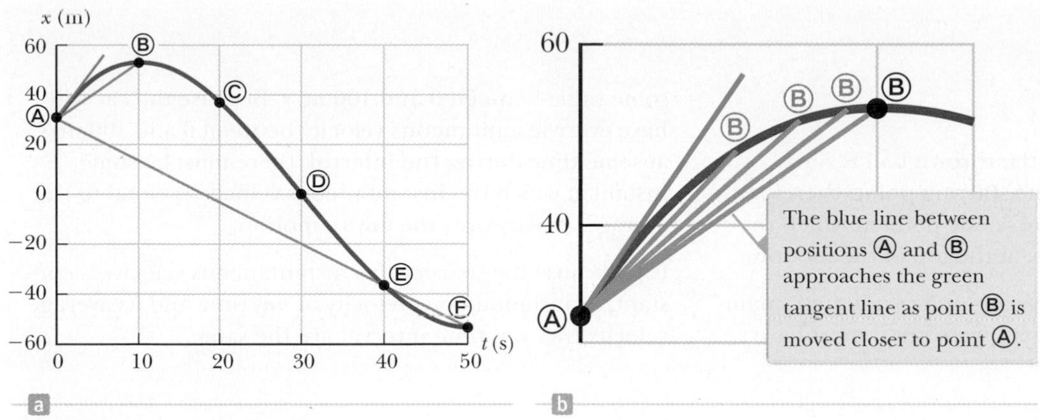

ACTIVE FIGURE 2.3

(a) Graph representing the motion of the car in Active Figure 2.1. (b) An enlargement of the upper-left-hand corner of the graph.

moment. In other words, the **instantaneous velocity** v_x equals the limiting value of the ratio $\Delta x/\Delta t$ as Δt approaches zero:[1]

$$v_x \equiv \lim_{\Delta t \to 0} \frac{\Delta x}{\Delta t} \quad (2.4)$$

In calculus notation, this limit is called the *derivative* of x with respect to t, written dx/dt:

$$v_x \equiv \lim_{\Delta t \to 0} \frac{\Delta x}{\Delta t} = \frac{dx}{dt} \quad (2.5)$$

◀ **Instantaneous velocity**

The instantaneous velocity can be positive, negative, or zero. When the slope of the position–time graph is positive, such as at any time during the first 10 s in Active Figure 2.3, v_x is positive and the car is moving toward larger values of x. After point Ⓑ, v_x is negative because the slope is negative and the car is moving toward smaller values of x. At point Ⓑ, the slope and the instantaneous velocity are zero and the car is momentarily at rest.

From here on, we use the word *velocity* to designate instantaneous velocity. When we are interested in *average velocity,* we shall always use the adjective *average*.

The **instantaneous speed** of a particle is defined as the magnitude of its instantaneous velocity. As with average speed, instantaneous speed has no direction associated with it. For example, if one particle has an instantaneous velocity of +25 m/s along a given line and another particle has an instantaneous velocity of −25 m/s along the same line, both have a speed[2] of 25 m/s.

Pitfall Prevention 2.3

Instantaneous Speed and Instantaneous Velocity

In Pitfall Prevention 2.1, we argued that the magnitude of the average velocity is not the average speed. The magnitude of the instantaneous velocity, however, *is* the instantaneous speed. In an infinitesimal time interval, the magnitude of the displacement is equal to the distance traveled by the particle.

***Quick Quiz* 2.2** Are members of the highway patrol more interested in **(a)** your average speed or **(b)** your instantaneous speed as you drive?

Conceptual Example 2.2 The Velocity of Different Objects

Consider the following one-dimensional motions: **(A)** a ball thrown directly upward rises to a highest point and falls back into the thrower's hand; **(B)** a race car starts from rest and speeds up to 100 m/s; and **(C)** a spacecraft drifts through space at constant velocity. Are there any points in the motion of these objects at which the instantaneous velocity has the same value as the average velocity over the entire motion? If so, identify the point(s).

continued

[1]Notice that the displacement Δx also approaches zero as Δt approaches zero, so the ratio looks like 0/0. As Δx and Δt become smaller and smaller, the ratio $\Delta x/\Delta t$ approaches a value equal to the slope of the line tangent to the x-versus-t curve.

[2]As with velocity, we drop the adjective for instantaneous speed. "Speed" means instantaneous speed.

2.2 *cont.*

SOLUTION

(A) The average velocity for the thrown ball is zero because the ball returns to the starting point; therefore, its displacement is zero. There is one point at which the instantaneous velocity is zero: at the top of the motion.

(B) The car's average velocity cannot be evaluated unambiguously with the information given, but it must have some value between 0 and 100 m/s. Because the car will have every instantaneous velocity between 0 and 100 m/s at some time during the interval, there must be some instant at which the instantaneous velocity is equal to the average velocity over the entire motion.

(C) Because the spacecraft's instantaneous velocity is constant, its instantaneous velocity at *any* time and its average velocity over *any* time interval are the same.

Example 2.3 Average and Instantaneous Velocity

A particle moves along the x axis. Its position varies with time according to the expression $x = -4t + 2t^2$, where x is in meters and t is in seconds.[3] The position–time graph for this motion is shown in Figure 2.4a. Because the position of the particle is given by a mathematical function, the motion of the particle is completely known, unlike that of the car in Active Figure 2.1. Notice that the particle moves in the negative x direction for the first second of motion, is momentarily at rest at the moment $t = 1$ s, and moves in the positive x direction at times $t > 1$ s.

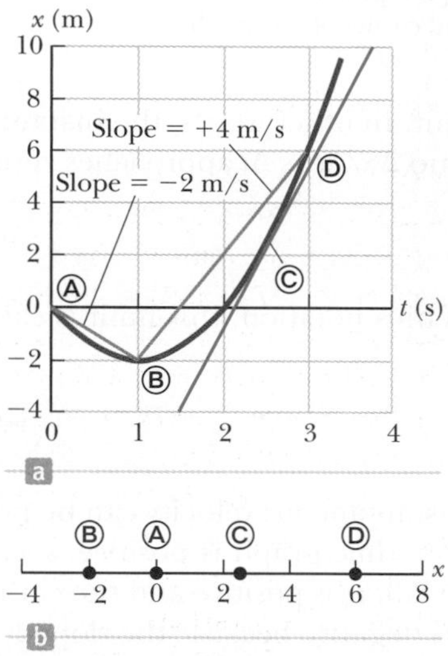

Figure 2.4 (Example 2.3) (a) Position–time graph for a particle having an x coordinate that varies in time according to the expression $x = -4t + 2t^2$. (b) The particle moves in one dimension along the x axis.

(A) Determine the displacement of the particle in the time intervals $t = 0$ to $t = 1$ s and $t = 1$ s to $t = 3$ s.

SOLUTION

From the graph in Figure 2.4a, form a mental representation of the particle's motion. Keep in mind that the particle does not move in a curved path in space such as that shown by the red-brown curve in the graphical representation. The particle moves only along the x axis in one dimension as shown in Figure 2.4b. At $t = 0$, is it moving to the right or to the left?

During the first time interval, the slope is negative and hence the average velocity is negative. Therefore, we know that the displacement between Ⓐ and Ⓑ must be a negative number having units of meters. Similarly, we expect the displacement between Ⓑ and Ⓓ to be positive.

In the first time interval, set $t_i = t_Ⓐ = 0$ and $t_f = t_Ⓑ = 1$ s and use Equation 2.1 to find the displacement:

$$\Delta x_{Ⓐ\to Ⓑ} = x_f - x_i = x_Ⓑ - x_Ⓐ$$
$$= [-4(1) + 2(1)^2] - [-4(0) + 2(0)^2] = \boxed{-2 \text{ m}}$$

For the second time interval ($t = 1$ s to $t = 3$ s), set $t_i = t_Ⓑ = 1$ s and $t_f = t_Ⓓ = 3$ s:

$$\Delta x_{Ⓑ\to Ⓓ} = x_f - x_i = x_Ⓓ - x_Ⓑ$$
$$= [-4(3) + 2(3)^2] - [-4(1) + 2(1)^2] = \boxed{+8 \text{ m}}$$

These displacements can also be read directly from the position–time graph.

(B) Calculate the average velocity during these two time intervals.

SOLUTION

In the first time interval, use Equation 2.2 with $\Delta t = t_f - t_i = t_Ⓑ - t_Ⓐ = 1$ s:

$$v_{x,\text{avg}\,(Ⓐ\to Ⓑ)} = \frac{\Delta x_{Ⓐ\to Ⓑ}}{\Delta t} = \frac{-2 \text{ m}}{1 \text{ s}} = \boxed{-2 \text{ m/s}}$$

[3]Simply to make it easier to read, we write the expression as $x = -4t + 2t^2$ rather than as $x = (-4.00 \text{ m/s})t + (2.00 \text{ m/s}^2)t^{2.00}$. When an equation summarizes measurements, consider its coefficients and exponents to have as many significant figures as other data quoted in a problem. Consider its coefficients to have the units required for dimensional consistency. When we start our clocks at $t = 0$, we usually do not mean to limit the precision to a single digit. Consider any zero value in this book to have as many significant figures as you need.

2.3 *cont.*

In the second time interval, $\Delta t = 2$ s: $v_{x,\text{avg}\,(Ⓑ\to Ⓓ)} = \dfrac{\Delta x_{Ⓑ\to Ⓓ}}{\Delta t} = \dfrac{8\text{ m}}{2\text{ s}} = \boxed{+4\text{ m/s}}$

These values are the same as the slopes of the blue lines joining these points in Figure 2.4a.

(C) Find the instantaneous velocity of the particle at $t = 2.5$ s.

SOLUTION

Measure the slope of the green line at $t = 2.5$ s (point Ⓒ) in Figure 2.4a:

$$v_x = \frac{10\text{ m} - (-4\text{ m})}{3.8\text{ s} - 1.5\text{ s}} = \boxed{+6\text{ m/s}}$$

Notice that this instantaneous velocity is on the same order of magnitude as our previous results, that is, a few meters per second. Is that what you would have expected?

2.3 Analysis Model: Particle Under Constant Velocity

In Section 1.2 we discussed the importance of making models. A particularly important model used in the solution to physics problems is an *analysis model.* Analysis models help us analyze common situations in physics problems and guide us toward a solution. An **analysis model** is a description of either (1) the behavior of some physical entity or (2) the interaction between that entity and the environment. When you encounter a new problem, you should identify the fundamental details of the problem and attempt to recognize which of the types of problems you have already solved might be used as a model for the new problem. For example, suppose an automobile is moving along a straight freeway at a constant speed. Is it important that it is an automobile? Is it important that it is a freeway? If the answers to both questions are no, we model the automobile as a *particle under constant velocity,* which we will discuss in this section.

This method is somewhat similar to the common practice in the legal profession of finding "legal precedents." If a previously resolved case can be found that is very similar legally to the current one, it is used as a model and an argument is made in court to link them logically. The finding in the previous case can then be used to sway the finding in the current case. We will do something similar in physics. For a given problem, we search for a "physics precedent," a model with which we are already familiar and that can be applied to the current problem.

We shall generate analysis models based on four fundamental simplification models. The first of the four is the particle model discussed in the introduction to this chapter. We will look at a particle under various behaviors and environmental interactions. Further analysis models are introduced in later chapters based on simplification models of a *system,* a *rigid object,* and a *wave.* Once we have introduced these analysis models, we shall see that they appear again and again in different problem situations.

When solving a problem, you should avoid browsing through the chapter looking for an equation that contains the unknown variable that is requested in the problem. In many cases, the equation you find may have nothing to do with the problem you are attempting to solve. It is *much* better to take this first step: **Identify the analysis model that is appropriate for the problem.** To do so, think carefully about what is going on in the problem and match it to a situation you have seen before. Once the analysis model is identified, there are a small number of equations from which to choose that are appropriate for that model. Therefore, **the model tells you which equation(s) to use for the mathematical representation.**

Let us use Equation 2.2 to build our first analysis model for solving problems. We imagine a particle moving with a constant velocity. The model of a **particle**

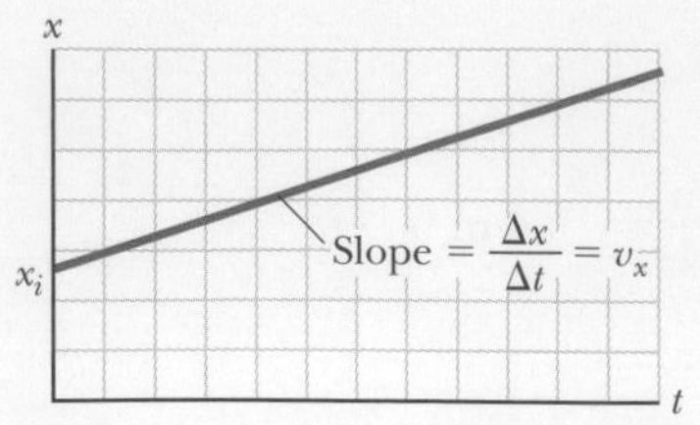

Figure 2.5 Position–time graph for a particle under constant velocity. The value of the constant velocity is the slope of the line.

under constant velocity can be applied in *any* situation in which an entity that can be modeled as a particle is moving with constant velocity. This situation occurs frequently, so this model is important.

If the velocity of a particle is constant, its instantaneous velocity at any instant during a time interval is the same as the average velocity over the interval. That is, $v_x = v_{x,\text{avg}}$. Therefore, Equation 2.2 gives us an equation to be used in the mathematical representation of this situation:

$$v_x = \frac{\Delta x}{\Delta t} \tag{2.6}$$

Remembering that $\Delta x = x_f - x_i$, we see that $v_x = (x_f - x_i)/\Delta t$, or

$$x_f = x_i + v_x \Delta t$$

This equation tells us that the position of the particle is given by the sum of its original position x_i at time $t = 0$ plus the displacement $v_x\,\Delta t$ that occurs during the time interval Δt. In practice, we usually choose the time at the beginning of the interval to be $t_i = 0$ and the time at the end of the interval to be $t_f = t$, so our equation becomes

Position as a function of time for the particle under constant velocity model ▶

$$x_f = x_i + v_x t \quad \text{(for constant } v_x) \tag{2.7}$$

Equations 2.6 and 2.7 are the primary equations used in the model of a particle under constant velocity. Whenever you have identified the analysis model in a problem to be the particle under constant velocity, you can immediately turn to these equations.

Figure 2.5 is a graphical representation of the particle under constant velocity. On this position–time graph, the slope of the line representing the motion is constant and equal to the magnitude of the velocity. Equation 2.7, which is the equation of a straight line, is the mathematical representation of the particle-under-constant-velocity model. The slope of the straight line is v_x and the y intercept is x_i in both representations.

Example 2.4 Modeling a Runner as a Particle

A scientist is studying the biomechanics of the human body. She determines the velocity of an experimental subject while he runs along a straight line at a constant rate. The scientist starts the stopwatch at the moment the runner passes a given point and stops it after the runner has passed another point 20 m away. The time interval indicated on the stopwatch is 4.0 s.

(A) What is the runner's velocity?

SOLUTION

We model the moving runner as a particle because the size of the runner and the movement of arms and legs are unnecessary details. Because the problem states that the subject runs at a constant rate, we can model him as a particle under constant velocity.

Having identified the model, we can use Equation 2.6 to find the constant velocity of the runner:

$$v_x = \frac{\Delta x}{\Delta t} = \frac{x_f - x_i}{\Delta t} = \frac{20\text{ m} - 0}{4.0\text{ s}} = \boxed{5.0\text{ m/s}}$$

(B) If the runner continues his motion after the stopwatch is stopped, what is his position after 10 s has passed?

SOLUTION

Use Equation 2.7 and the velocity found in part (A) to find the position of the particle at time $t = 10$ s:

$$x_f = x_i + v_x t = 0 + (5.0\text{ m/s})(10\text{ s}) = \boxed{50\text{ m}}$$

Notice that this value is more than twice that of the 20-m position at which the stopwatch was stopped. Is this value consistent with the time of 10 s being more than twice the time of 4.0 s?

The mathematical manipulations for the particle under constant velocity stem from Equation 2.6 and its descendent, Equation 2.7. These equations can be used to solve for any variable in the equations that happens to be unknown if the other variables are known. For example, in part (B) of Example 2.4, we find the position when the velocity and the time are known. Similarly, if we know the velocity and the final position, we could use Equation 2.7 to find the time at which the runner is at this position.

A particle under constant velocity moves with a constant speed along a straight line. Now consider a particle moving with a constant speed along a curved path. This situation can be represented with the model of a **particle under constant speed.** The primary equation for this model is Equation 2.3, with the average speed v_{avg} replaced by the constant speed v:

$$v = \frac{d}{\Delta t} \tag{2.8}$$

As an example, imagine a particle moving at a constant speed in a circular path. If the speed is 5.00 m/s and the radius of the path is 10.0 m, we can calculate the time interval required to complete one trip around the circle:

$$v = \frac{d}{\Delta t} \rightarrow \Delta t = \frac{d}{v} = \frac{2\pi r}{v} = \frac{2\pi(10.0 \text{ m})}{5.00 \text{ m/s}} = 12.6 \text{ s}$$

2.4 Acceleration

In Example 2.3, we worked with a common situation in which the velocity of a particle changes while the particle is moving. When the velocity of a particle changes with time, the particle is said to be *accelerating*. For example, the magnitude of a car's velocity increases when you step on the gas and decreases when you apply the brakes. Let us see how to quantify acceleration.

Suppose an object that can be modeled as a particle moving along the x axis has an initial velocity v_{xi} at time t_i at position Ⓐ and a final velocity v_{xf} at time t_f at position Ⓑ as in Figure 2.6a (page 30). The **average acceleration** $a_{x,\text{avg}}$ of the particle is defined as the *change* in velocity Δv_x divided by the time interval Δt during which that change occurs:

$$a_{x,\text{avg}} \equiv \frac{\Delta v_x}{\Delta t} = \frac{v_{xf} - v_{xi}}{t_f - t_i} \tag{2.9}$$

◀ Average acceleration

As with velocity, when the motion being analyzed is one dimensional, we can use positive and negative signs to indicate the direction of the acceleration. Because the dimensions of velocity are L/T and the dimension of time is T, acceleration has dimensions of length divided by time squared, or L/T^2. The SI unit of acceleration is meters per second squared (m/s^2). It might be easier to interpret these units if you think of them as meters per second per second. For example, suppose an object has an acceleration of $+2$ m/s^2. You should form a mental image of the object having a velocity that is along a straight line and is increasing by 2 m/s during every time interval of 1 s. If the object starts from rest, you should be able to picture it moving at a velocity of $+2$ m/s after 1 s, at $+4$ m/s after 2 s, and so on.

In some situations, the value of the average acceleration may be different over different time intervals. It is therefore useful to define the **instantaneous acceleration** as the limit of the average acceleration as Δt approaches zero. This concept is analogous to the definition of instantaneous velocity discussed in Section 2.2. If we imagine that point Ⓐ is brought closer and closer to point Ⓑ in Figure 2.6a and we take the limit of $\Delta v_x/\Delta t$ as Δt approaches zero, we obtain the instantaneous acceleration at point Ⓑ:

$$a_x \equiv \lim_{\Delta t \to 0} \frac{\Delta v_x}{\Delta t} = \frac{dv_x}{dt} \tag{2.10}$$

◀ Instantaneous acceleration

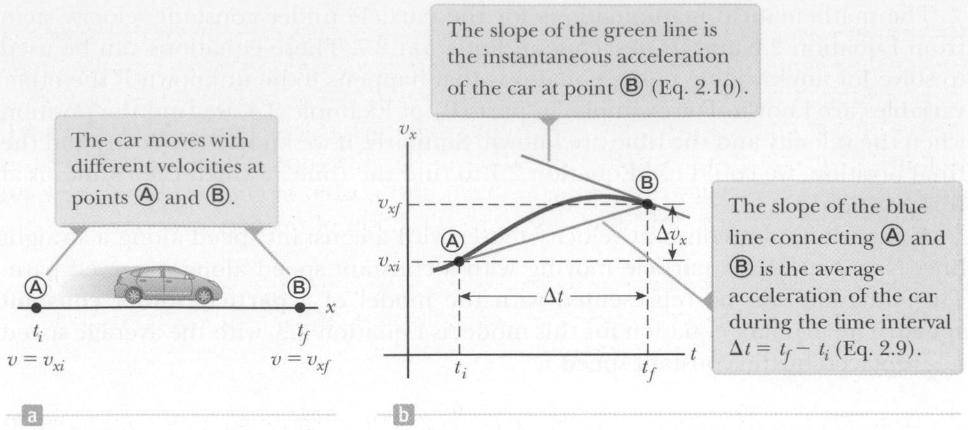

Figure 2.6 (a) A car, modeled as a particle, moving along the x axis from Ⓐ to Ⓑ, has velocity v_{xi} at $t = t_i$ and velocity v_{xf} at $t = t_f$. (b) Velocity–time graph (red-brown) for the particle moving in a straight line.

Pitfall Prevention 2.4

Negative Acceleration

Keep in mind that *negative acceleration does not necessarily mean that an object is slowing down.* If the acceleration is negative and the velocity is negative, the object is speeding up!

Pitfall Prevention 2.5

Deceleration

The word *deceleration* has the common popular connotation of *slowing down.* We will not use this word in this book because it confuses the definition we have given for negative acceleration.

That is, the instantaneous acceleration equals the derivative of the velocity with respect to time, which by definition is the slope of the velocity–time graph. The slope of the green line in Figure 2.6b is equal to the instantaneous acceleration at point Ⓑ. Notice that Figure 2.6b is a *velocity–time* graph, not a *position–time* graph like Active Figures 2.1b and 2.3 and Figures 2.4 and 2.5. Therefore, we see that just as the velocity of a moving particle is the slope at a point on the particle's x–t graph, the acceleration of a particle is the slope at a point on the particle's v_x–t graph. One can interpret the derivative of the velocity with respect to time as the time rate of change of velocity. If a_x is positive, the acceleration is in the positive x direction; if a_x is negative, the acceleration is in the negative x direction.

Figure 2.7 illustrates how an acceleration–time graph is related to a velocity–time graph. The acceleration at any time is the slope of the velocity–time graph at that time. Positive values of acceleration correspond to those points in Figure 2.7a where the velocity is increasing in the positive x direction. The acceleration reaches a maximum at time $t_Ⓐ$, when the slope of the velocity–time graph is a maximum. The acceleration then goes to zero at time $t_Ⓑ$, when the velocity is a maximum (that is, when the slope of the v_x–t graph is zero). The acceleration is negative when the velocity is decreasing in the positive x direction, and it reaches its most negative value at time $t_Ⓒ$.

Quick Quiz **2.3** Make a velocity–time graph for the car in Active Figure 2.1a. Suppose the speed limit for the road on which the car is driving is 30 km/h. True or False? The car exceeds the speed limit at some time within the time interval 0 – 50 s.

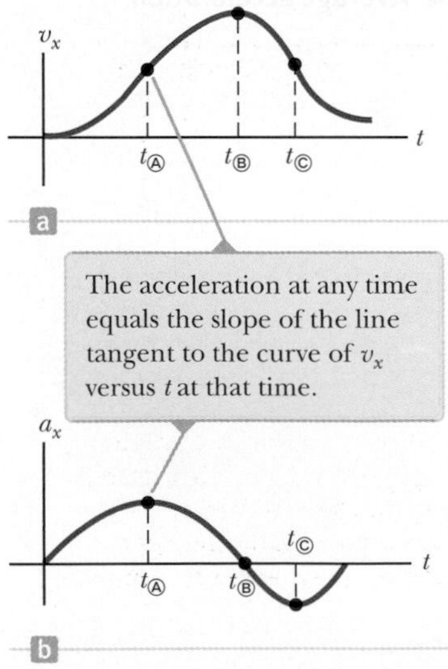

Figure 2.7 (a) The velocity–time graph for a particle moving along the x axis. (b) The instantaneous acceleration can be obtained from the velocity–time graph.

For the case of motion in a straight line, the direction of the velocity of an object and the direction of its acceleration are related as follows. When the object's velocity and acceleration are in the same direction, the object is speeding up. On the other hand, when the object's velocity and acceleration are in opposite directions, the object is slowing down.

To help with this discussion of the signs of velocity and acceleration, we can relate the acceleration of an object to the total *force* exerted on the object. In Chapter 5, we formally establish that **the force on an object is proportional to the acceleration of the object:**

$$F_x \propto a_x \quad \textbf{(2.11)}$$

This proportionality indicates that acceleration is caused by force. Furthermore, force and acceleration are both vectors, and the vectors act in the same direction. Therefore, let us think about the signs of velocity and acceleration by imagining a force applied to an object and causing it to accelerate. Let us assume the velocity and acceleration are in the same direction. This situation corresponds to an object that experiences a force acting in the same direction as its velocity. In this case, the object speeds up! Now suppose the velocity and acceleration are in opposite directions. In this situation, the object moves in some direction and experiences a force acting in the opposite direction. Therefore, the object slows down! It is very useful to equate the direction of the acceleration to the direction of a force because it is easier from our everyday experience to think about what effect a force will have on an object than to think only in terms of the direction of the acceleration.

***Quick Quiz* 2.4** If a car is traveling eastward and slowing down, what is the direction of the force on the car that causes it to slow down? **(a)** eastward **(b)** westward **(c)** neither eastward nor westward

From now on, we shall use the term *acceleration* to mean instantaneous acceleration. When we mean average acceleration, we shall always use the adjective *average*. Because $v_x = dx/dt$, the acceleration can also be written as

$$a_x = \frac{dv_x}{dt} = \frac{d}{dt}\left(\frac{dx}{dt}\right) = \frac{d^2x}{dt^2} \qquad \textbf{(2.12)}$$

That is, in one-dimensional motion, the acceleration equals the *second derivative* of x with respect to time.

Conceptual Example 2.5 Graphical Relationships Between x, v_x, and a_x

The position of an object moving along the x axis varies with time as in Figure 2.8a. Graph the velocity versus time and the acceleration versus time for the object.

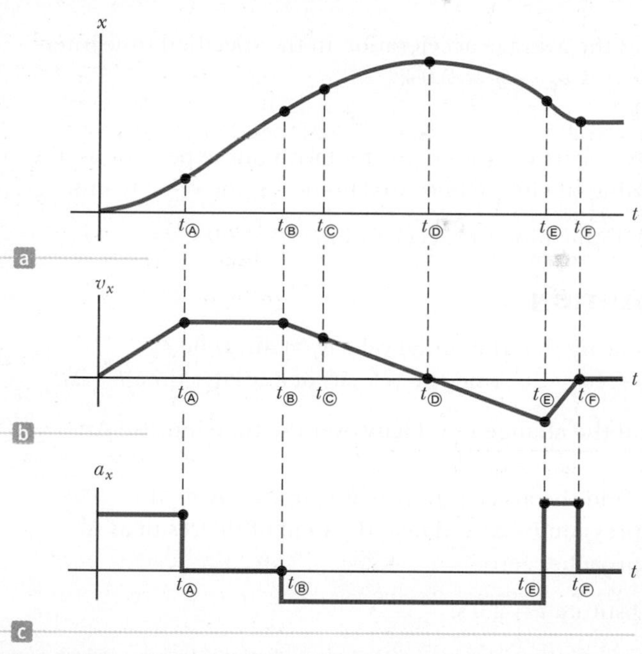

Figure 2.8 (Conceptual Example 2.5) (a) Position–time graph for an object moving along the x axis. (b) The velocity–time graph for the object is obtained by measuring the slope of the position–time graph at each instant. (c) The acceleration–time graph for the object is obtained by measuring the slope of the velocity–time graph at each instant.

SOLUTION

The velocity at any instant is the slope of the tangent to the x–t graph at that instant. Between $t = 0$ and $t = t_Ⓐ$, the slope of the x–t graph increases uniformly, so the velocity increases linearly as shown in Figure 2.8b. Between $t_Ⓐ$ and $t_Ⓑ$, the slope of the x–t graph is constant, so the velocity remains constant. Between $t_Ⓑ$ and $t_Ⓓ$, the slope of the x–t graph decreases, so the value of the velocity in the v_x–t graph decreases. At $t_Ⓓ$, the slope of the x–t graph is zero, so the velocity is zero at that instant. Between $t_Ⓓ$ and $t_Ⓔ$, the slope of the x–t graph and therefore the velocity are negative and decrease uniformly in this interval. In the interval $t_Ⓔ$ to $t_Ⓕ$, the slope of the x–t graph is still negative, and at $t_Ⓕ$ it goes to zero. Finally, after $t_Ⓕ$, the slope of the x–t graph is zero, meaning that the object is at rest for $t > t_Ⓕ$.

The acceleration at any instant is the slope of the tangent to the v_x–t graph at that instant. The graph of acceleration versus time for this object is shown in Figure 2.8c. The acceleration is constant and positive between 0 and $t_Ⓐ$, where the slope of the v_x–t graph is positive. It is zero between $t_Ⓐ$ and $t_Ⓑ$ and for $t > t_Ⓕ$ because the slope of the v_x–t graph is zero at these times. It is negative between $t_Ⓑ$ and $t_Ⓔ$ because the slope of the v_x–t graph is negative during this interval. Between $t_Ⓔ$ and $t_Ⓕ$, the acceleration is positive like it is between 0 and $t_Ⓐ$, but higher in value because the slope of the v_x–t graph is steeper.

Notice that the sudden changes in acceleration shown in Figure 2.8c are unphysical. Such instantaneous changes cannot occur in reality.

Example 2.6 Average and Instantaneous Acceleration

The velocity of a particle moving along the x axis varies according to the expression $v_x = 40 - 5t^2$, where v_x is in meters per second and t is in seconds.

(A) Find the average acceleration in the time interval $t = 0$ to $t = 2.0$ s.

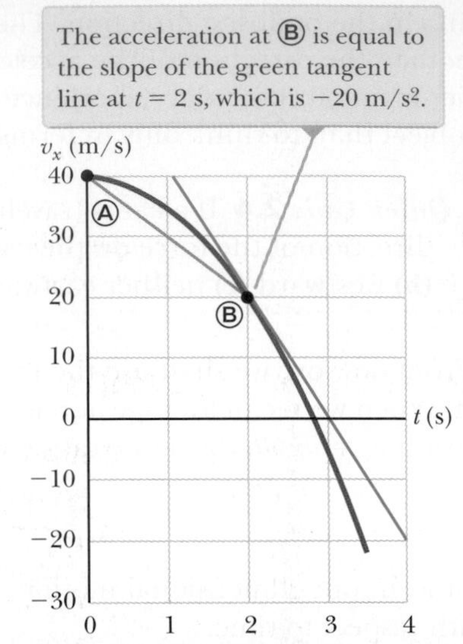

Figure 2.9 (Example 2.6) The velocity–time graph for a particle moving along the x axis according to the expression $v_x = 40 - 5t^2$.

SOLUTION

Think about what the particle is doing from the mathematical representation. Is it moving at $t = 0$? In which direction? Does it speed up or slow down? Figure 2.9 is a v_x–t graph that was created from the velocity versus time expression given in the problem statement. Because the slope of the entire v_x–t curve is negative, we expect the acceleration to be negative.

Find the velocities at $t_i = t_Ⓐ = 0$ and $t_f = t_Ⓑ = 2.0$ s by substituting these values of t into the expression for the velocity:

$$v_{xⒶ} = 40 - 5t_Ⓐ^2 = 40 - 5(0)^2 = +40 \text{ m/s}$$

$$v_{xⒷ} = 40 - 5t_Ⓑ^2 = 40 - 5(2.0)^2 = +20 \text{ m/s}$$

Find the average acceleration in the specified time interval $\Delta t = t_Ⓑ - t_Ⓐ = 2.0$ s:

$$a_{x,\text{avg}} = \frac{v_{xf} - v_{xi}}{t_f - t_i} = \frac{v_{xⒷ} - v_{xⒶ}}{t_Ⓑ - t_Ⓐ} = \frac{20 \text{ m/s} - 40 \text{ m/s}}{2.0 \text{ s} - 0 \text{ s}}$$

$$= \boxed{-10 \text{ m/s}^2}$$

The negative sign is consistent with our expectations: the average acceleration, represented by the slope of the blue line joining the initial and final points on the velocity–time graph, is negative.

(B) Determine the acceleration at $t = 2.0$ s.

SOLUTION

Knowing that the initial velocity at any time t is $v_{xi} = 40 - 5t^2$, find the velocity at any later time $t + \Delta t$:

$$v_{xf} = 40 - 5(t + \Delta t)^2 = 40 - 5t^2 - 10t\,\Delta t - 5(\Delta t)^2$$

Find the change in velocity over the time interval Δt:

$$\Delta v_x = v_{xf} - v_{xi} = -10t\,\Delta t - 5(\Delta t)^2$$

To find the acceleration at any time t, divide this expression by Δt and take the limit of the result as Δt approaches zero:

$$a_x = \lim_{\Delta t \to 0} \frac{\Delta v_x}{\Delta t} = \lim_{\Delta t \to 0} (-10t - 5\,\Delta t) = -10t$$

Substitute $t = 2.0$ s:

$$a_x = (-10)(2.0) \text{ m/s}^2 = \boxed{-20 \text{ m/s}^2}$$

Because the velocity of the particle is positive and the acceleration is negative at this instant, the particle is slowing down.

Notice that the answers to parts (A) and (B) are different. The average acceleration in part (A) is the slope of the blue line in Figure 2.9 connecting points Ⓐ and Ⓑ. The instantaneous acceleration in part (B) is the slope of the green line tangent to the curve at point Ⓑ. Notice also that the acceleration is *not* constant in this example. Situations involving constant acceleration are treated in Section 2.6.

So far, we have evaluated the derivatives of a function by starting with the definition of the function and then taking the limit of a specific ratio. If you are familiar with calculus, you should recognize that there are specific rules for taking derivatives. These rules, which are listed in Appendix B.6, enable us to evaluate derivatives quickly. For instance, one rule tells us that the derivative of any constant is zero. As another example, suppose x is proportional to some power of t such as in the expression

$$x = At^n$$

where A and n are constants. (This expression is a very common functional form.) The derivative of x with respect to t is

$$\frac{dx}{dt} = nAt^{n-1}$$

Applying this rule to Example 2.6, in which $v_x = 40 - 5t^2$, we quickly find that the acceleration is $a_x = dv_x/dt = -10t$.

2.5 Motion Diagrams

The concepts of velocity and acceleration are often confused with each other, but in fact they are quite different quantities. In forming a mental representation of a moving object, a pictorial representation called a *motion diagram* is sometimes useful to describe the velocity and acceleration while an object is in motion.

A motion diagram can be formed by imagining a *stroboscopic* photograph of a moving object, which shows several images of the object taken as the strobe light flashes at a constant rate. Active Figure 2.1a is a motion diagram for the car studied in Section 2.1. Active Figure 2.10 represents three sets of strobe photographs of cars moving along a straight roadway in a single direction, from left to right. The time intervals between flashes of the stroboscope are equal in each part of the diagram. So as to not confuse the two vector quantities, we use red arrows for velocity and purple arrows for acceleration in Active Figure 2.10. The arrows are shown at several instants during the motion of the object. Let us describe the motion of the car in each diagram.

In Active Figure 2.10a, the images of the car are equally spaced, showing us that the car moves through the same displacement in each time interval. This equal spacing is consistent with the car moving with *constant positive velocity* and *zero acceleration*. We could model the car as a particle and describe it with the particle under constant velocity model.

In Active Figure 2.10b, the images become farther apart as time progresses. In this case, the velocity arrow increases in length with time because the car's displacement between adjacent positions increases in time. These features suggest the car is moving with a *positive velocity* and a *positive acceleration*. The velocity and acceleration are in the same direction. In terms of our earlier force discussion, imagine a force pulling on the car in the same direction it is moving: it speeds up.

In Active Figure 2.10c, we can tell that the car slows as it moves to the right because its displacement between adjacent images decreases with time. This case suggests the car moves to the right with a negative acceleration. The length of the

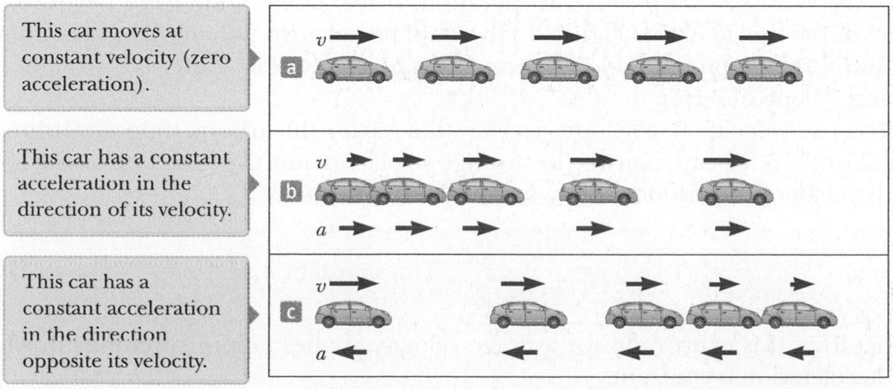

ACTIVE FIGURE 2.10

Motion diagrams of a car moving along a straight roadway in a single direction. The velocity at each instant is indicated by a red arrow, and the constant acceleration is indicated by a purple arrow.

velocity arrow decreases in time and eventually reaches zero. From this diagram, we see that the acceleration and velocity arrows are *not* in the same direction. The car is moving with a *positive velocity,* but with a *negative acceleration.* (This type of motion is exhibited by a car that skids to a stop after its brakes are applied.) The velocity and acceleration are in opposite directions. In terms of our earlier force discussion, imagine a force pulling on the car opposite to the direction it is moving: it slows down.

Each purple acceleration arrow in parts (b) and (c) of Active Figure 2.10 is the same length. Therefore, these diagrams represent motion of a *particle under constant acceleration.* This important analysis model will be discussed in the next section.

***Quick Quiz* 2.5** Which one of the following statements is true? **(a)** If a car is traveling eastward, its acceleration must be eastward. **(b)** If a car is slowing down, its acceleration must be negative. **(c)** A particle with constant acceleration can never stop and stay stopped.

2.6 Analysis Model: Particle Under Constant Acceleration

If the acceleration of a particle varies in time, its motion can be complex and difficult to analyze. A very common and simple type of one-dimensional motion, however, is that in which the acceleration is constant. In such a case, the average acceleration $a_{x,\text{avg}}$ over any time interval is numerically equal to the instantaneous acceleration a_x at any instant within the interval, and the velocity changes at the same rate throughout the motion. This situation occurs often enough that we identify it as an analysis model: the **particle under constant acceleration.** In the discussion that follows, we generate several equations that describe the motion of a particle for this model.

If we replace $a_{x,\text{avg}}$ by a_x in Equation 2.9 and take $t_i = 0$ and t_f to be any later time t, we find that

$$a_x = \frac{v_{xf} - v_{xi}}{t - 0}$$

or

$$v_{xf} = v_{xi} + a_x t \quad (\text{for constant } a_x) \qquad \textbf{(2.13)}$$

This powerful expression enables us to determine an object's velocity at *any* time t if we know the object's initial velocity v_{xi} and its (constant) acceleration a_x. A velocity–time graph for this constant-acceleration motion is shown in Active Figure 2.11b. The graph is a straight line, the slope of which is the acceleration a_x; the (constant) slope is consistent with $a_x = dv_x/dt$ being a constant. Notice that the slope is positive, which indicates a positive acceleration. If the acceleration were negative, the slope of the line in Active Figure 2.11b would be negative. When the acceleration is constant, the graph of acceleration versus time (Active Fig. 2.11c) is a straight line having a slope of zero.

Because velocity at constant acceleration varies linearly in time according to Equation 2.13, we can express the average velocity in any time interval as the arithmetic mean of the initial velocity v_{xi} and the final velocity v_{xf}:

$$v_{x,\text{avg}} = \frac{v_{xi} + v_{xf}}{2} \quad (\text{for constant } a_x) \qquad \textbf{(2.14)}$$

Notice that this expression for average velocity applies *only* in situations in which the acceleration is constant.

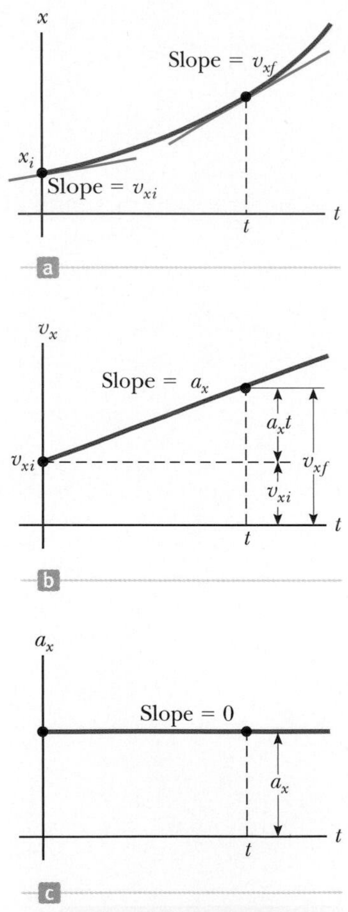

ACTIVE FIGURE 2.11

A particle under constant acceleration a_x moving along the x axis: (a) the position–time graph, (b) the velocity–time graph, and (c) the acceleration–time graph.

We can now use Equations 2.1, 2.2, and 2.14 to obtain the position of an object as a function of time. Recalling that Δx in Equation 2.2 represents $x_f - x_i$ and recognizing that $\Delta t = t_f - t_i = t - 0 = t$, we find that

$$x_f - x_i = v_{x,\text{avg}}\, t = \tfrac{1}{2}(v_{xi} + v_{xf})t$$

$$x_f = x_i + \tfrac{1}{2}(v_{xi} + v_{xf})t \quad \text{(for constant } a_x) \tag{2.15}$$

◀ **Position as a function of velocity and time for the particle under constant acceleration model**

This equation provides the final position of the particle at time t in terms of the initial and final velocities.

We can obtain another useful expression for the position of a particle under constant acceleration by substituting Equation 2.13 into Equation 2.15:

$$x_f = x_i + \tfrac{1}{2}[v_{xi} + (v_{xi} + a_x t)]t$$

$$x_f = x_i + v_{xi}t + \tfrac{1}{2}a_x t^2 \quad \text{(for constant } a_x) \tag{2.16}$$

◀ **Position as a function of time for the particle under constant acceleration model**

This equation provides the final position of the particle at time t in terms of the initial position, the initial velocity, and the constant acceleration.

The position–time graph for motion at constant (positive) acceleration shown in Active Figure 2.11a is obtained from Equation 2.16. Notice that the curve is a parabola. The slope of the tangent line to this curve at $t = 0$ equals the initial velocity v_{xi}, and the slope of the tangent line at any later time t equals the velocity v_{xf} at that time.

Finally, we can obtain an expression for the final velocity that does not contain time as a variable by substituting the value of t from Equation 2.13 into Equation 2.15:

$$x_f = x_i + \tfrac{1}{2}(v_{xi} + v_{xf})\left(\frac{v_{xf} - v_{xi}}{a_x}\right) = x_i + \frac{v_{xf}^{\,2} - v_{xi}^{\,2}}{2a_x}$$

$$v_{xf}^{\,2} = v_{xi}^{\,2} + 2a_x(x_f - x_i) \quad \text{(for constant } a_x) \tag{2.17}$$

◀ **Velocity as a function of position for the particle under constant acceleration model**

This equation provides the final velocity in terms of the initial velocity, the constant acceleration, and the position of the particle.

For motion at *zero* acceleration, we see from Equations 2.13 and 2.16 that

$$\left.\begin{aligned} v_{xf} &= v_{xi} = v_x \\ x_f &= x_i + v_x t \end{aligned}\right\} \text{ when } a_x = 0$$

That is, when the acceleration of a particle is zero, its velocity is constant and its position changes linearly with time. In terms of models, when the acceleration of a particle is zero, the particle under constant acceleration model reduces to the particle under constant velocity model (Section 2.3).

***Quick Quiz* 2.6** In Active Figure 2.12, match each v_x–t graph on the top with the a_x–t graph on the bottom that best describes the motion.

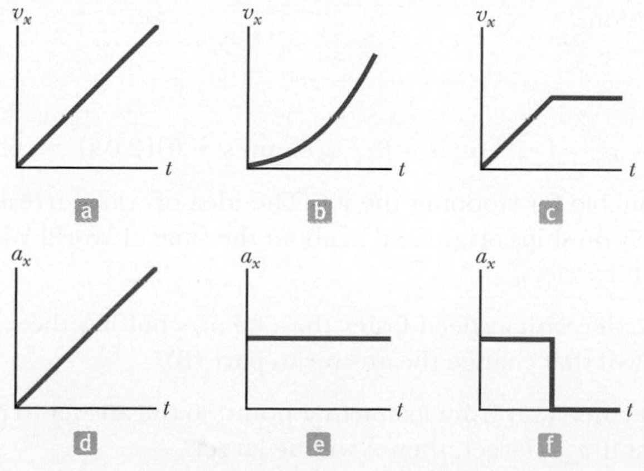

ACTIVE FIGURE 2.12

(Quick Quiz 2.6) Parts (a), (b), and (c) are v_x–t graphs of objects in one-dimensional motion. The possible accelerations of each object as a function of time are shown in scrambled order in (d), (e), and (f).

TABLE 2.2 *Kinematic Equations for Motion of a Particle Under Constant Acceleration*

Equation Number	Equation	Information Given by Equation
2.13	$v_{xf} = v_{xi} + a_x t$	Velocity as a function of time
2.15	$x_f = x_i + \frac{1}{2}(v_{xi} + v_{xf})t$	Position as a function of velocity and time
2.16	$x_f = x_i + v_{xi}t + \frac{1}{2}a_x t^2$	Position as a function of time
2.17	$v_{xf}^{\ 2} = v_{xi}^{\ 2} + 2a_x(x_f - x_i)$	Velocity as a function of position

Note: Motion is along the x axis.

Equations 2.13 through 2.17 are **kinematic equations** that may be used to solve any problem involving a particle under constant acceleration in one dimension. The four kinematic equations used most often are listed for convenience in Table 2.2. The choice of which equation you use in a given situation depends on what you know beforehand. Sometimes it is necessary to use two of these equations to solve for two unknowns. You should recognize that the quantities that vary during the motion are position x_f, velocity v_{xf}, and time t.

You will gain a great deal of experience in the use of these equations by solving a number of exercises and problems. Many times you will discover that more than one method can be used to obtain a solution. Remember that these equations of kinematics *cannot* be used in a situation in which the acceleration varies with time. They can be used only when the acceleration is constant.

Example 2.7 Carrier Landing

A jet lands on an aircraft carrier at a speed of 140 mi/h ($\approx$ 63 m/s).

(A) What is its acceleration (assumed constant) if it stops in 2.0 s due to an arresting cable that snags the jet and brings it to a stop?

SOLUTION

You might have seen movies or television shows in which a jet lands on an aircraft carrier and is brought to rest surprisingly fast by an arresting cable. A careful reading of the problem reveals that in addition to being given the initial speed of 63 m/s, we also know that the final speed is zero. Because the acceleration of the jet is assumed constant, we model it as a particle under constant acceleration. We define our x axis as the direction of motion of the jet. Notice that we have no information about the change in position of the jet while it is slowing down.

Equation 2.13 is the only equation in Table 2.2 that does not involve position, so we use it to find the acceleration of the jet, modeled as a particle:

$$a_x = \frac{v_{xf} - v_{xi}}{t} \approx \frac{0 - 63 \text{ m/s}}{2.0 \text{ s}}$$

$$= \boxed{-32 \text{ m/s}^2}$$

(B) If the jet touches down at position $x_i = 0$, what is its final position?

SOLUTION

Use Equation 2.15 to solve for the final position:

$$x_f = x_i + \tfrac{1}{2}(v_{xi} + v_{xf})t = 0 + \tfrac{1}{2}(63 \text{ m/s} + 0)(2.0 \text{ s}) = \boxed{63 \text{ m}}$$

Given the size of aircraft carriers, a length of 63 m seems reasonable for stopping the jet. The idea of using arresting cables to slow down landing aircraft and enable them to land safely on ships originated at about the time of World War I. The cables are still a vital part of the operation of modern aircraft carriers.

WHAT IF? Suppose the jet lands on the deck of the aircraft carrier with a speed faster than 63 m/s but has the same acceleration due to the cable as that calculated in part (A). How will that change the answer to part (B)?

Answer If the jet is traveling faster at the beginning, it will stop farther away from its starting point, so the answer to part (B) should be larger. Mathematically, we see in Equation 2.15 that if v_{xi} is larger, then x_f will be larger.

Example 2.8 Watch Out for the Speed Limit!

A car traveling at a constant speed of 45.0 m/s passes a trooper on a motorcycle hidden behind a billboard. One second after the speeding car passes the billboard, the trooper sets out from the billboard to catch the car, accelerating at a constant rate of 3.00 m/s². How long does it take her to overtake the car?

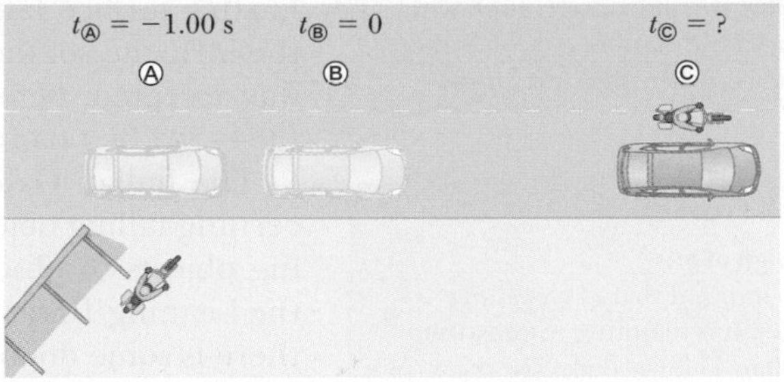

Figure 2.13 (Example 2.8) A speeding car passes a hidden trooper.

SOLUTION

A pictorial representation (Fig. 2.13) helps clarify the sequence of events. The car is modeled as a particle under constant velocity, and the trooper is modeled as a particle under constant acceleration.

First, we write expressions for the position of each vehicle as a function of time. It is convenient to choose the position of the billboard as the origin and to set $t_{Ⓑ} = 0$ as the time the trooper begins moving. At that instant, the car has already traveled a distance of 45.0 m from the billboard because it has traveled at a constant speed of $v_x = 45.0$ m/s for 1 s. Therefore, the initial position of the speeding car is $x_{Ⓑ} = 45.0$ m.

Using the particle under constant velocity model, apply Equation 2.7 to give the car's position at any time t:

$$x_{\text{car}} = x_{Ⓑ} + v_{x\,\text{car}}t$$

A quick check shows that at $t = 0$, this expression gives the car's correct initial position when the trooper begins to move: $x_{\text{car}} = x_{Ⓑ} = 45.0$ m.

The trooper starts from rest at $t_{Ⓑ} = 0$ and accelerates at $a_x = 3.00$ m/s² away from the origin. Use Equation 2.16 to give her position at any time t:

$$x_f = x_i + v_{xi}t + \tfrac{1}{2}a_xt^2$$

$$x_{\text{trooper}} = 0 + (0)t + \tfrac{1}{2}a_xt^2 = \tfrac{1}{2}a_xt^2$$

Set the positions of the car and trooper equal to represent the trooper overtaking the car at position Ⓒ:

$$x_{\text{trooper}} = x_{\text{car}}$$

$$\tfrac{1}{2}a_xt^2 = x_{Ⓑ} + v_{x\,\text{car}}t$$

Rearrange to give a quadratic equation:

$$\tfrac{1}{2}a_xt^2 - v_{x\,\text{car}}t - x_{Ⓑ} = 0$$

Solve the quadratic equation for the time at which the trooper catches the car (for help in solving quadratic equations, see Appendix B.2.):

$$t = \frac{v_{x\,\text{car}} \pm \sqrt{v_{x\,\text{car}}^2 + 2a_xx_{Ⓑ}}}{a_x}$$

$$(1)\quad t = \frac{v_{x\,\text{car}}}{a_x} \pm \sqrt{\frac{v_{x\,\text{car}}^2}{a_x^2} + \frac{2x_{Ⓑ}}{a_x}}$$

Evaluate the solution, choosing the positive root because that is the only choice consistent with a time $t > 0$:

$$t = \frac{45.0\text{ m/s}}{3.00\text{ m/s}^2} + \sqrt{\frac{(45.0\text{ m/s})^2}{(3.00\text{ m/s}^2)^2} + \frac{2(45.0\text{ m})}{3.00\text{ m/s}^2}} = \boxed{31.0\text{ s}}$$

Why didn't we choose $t = 0$ as the time at which the car passes the trooper? If we did so, we would not be able to use the particle under constant acceleration model for the trooper. Her acceleration would be zero for the first second and then 3.00 m/s² for the remaining time. By defining the time $t = 0$ as when the trooper begins moving, we can use the particle under constant acceleration model for her movement for all positive times.

WHAT IF? What if the trooper had a more powerful motorcycle with a larger acceleration? How would that change the time at which the trooper catches the car?

Answer If the motorcycle has a larger acceleration, the trooper should catch up to the car sooner, so the answer for the time should be less than 31 s. Because all terms on the right side of Equation (1) have the acceleration a_x in the denominator, we see symbolically that increasing the acceleration will decrease the time at which the trooper catches the car.

Pitfall Prevention 2.6

g and g

Be sure not to confuse the italic symbol g for free-fall acceleration with the nonitalic symbol g used as the abbreviation for the unit gram.

Pitfall Prevention 2.7

The Sign of *g*

Keep in mind that g is a *positive number.* It is tempting to substitute $-9.80\ \text{m/s}^2$ for g, but resist the temptation. Downward gravitational acceleration is indicated explicitly by stating the acceleration as $a_y = -g$.

North Wind Picture Archives

Galileo Galilei

Italian physicist and astronomer (1564–1642)

Galileo formulated the laws that govern the motion of objects in free fall and made many other significant discoveries in physics and astronomy. Galileo publicly defended Nicolaus Copernicus's assertion that the Sun is at the center of the Universe (the heliocentric system). He published *Dialogue Concerning Two New World Systems* to support the Copernican model, a view that the Catholic Church declared to be heretical.

Pitfall Prevention 2.8

Acceleration at the Top of the Motion

A common misconception is that the acceleration of a projectile at the top of its trajectory is zero. Although the velocity at the top of the motion of an object thrown upward momentarily goes to zero, *the acceleration is still that due to gravity* at this point. If the velocity and acceleration were both zero, the projectile would stay at the top.

2.7 Freely Falling Objects

It is well known that, in the absence of air resistance, all objects dropped near the Earth's surface fall toward the Earth with the same constant acceleration under the influence of the Earth's gravity. It was not until about 1600 that this conclusion was accepted. Before that time, the teachings of the Greek philosopher Aristotle (384–322 BC) had held that heavier objects fall faster than lighter ones.

The Italian Galileo Galilei (1564–1642) originated our present-day ideas concerning falling objects. There is a legend that he demonstrated the behavior of falling objects by observing that two different weights dropped simultaneously from the Leaning Tower of Pisa hit the ground at approximately the same time. Although there is some doubt that he carried out this particular experiment, it is well established that Galileo performed many experiments on objects moving on inclined planes. In his experiments, he rolled balls down a slight incline and measured the distances they covered in successive time intervals. The purpose of the incline was to reduce the acceleration, which made it possible for him to make accurate measurements of the time intervals. By gradually increasing the slope of the incline, he was finally able to draw conclusions about freely falling objects because a freely falling ball is equivalent to a ball moving down a vertical incline.

You might want to try the following experiment. Simultaneously drop a coin and a crumpled-up piece of paper from the same height. If the effects of air resistance are negligible, both will have the same motion and will hit the floor at the same time. In the idealized case, in which air resistance is absent, such motion is referred to as *free-fall* motion. If this same experiment could be conducted in a vacuum, in which air resistance is truly negligible, the paper and the coin would fall with the same acceleration even when the paper is not crumpled. On August 2, 1971, astronaut David Scott conducted such a demonstration on the Moon. He simultaneously released a hammer and a feather, and the two objects fell together to the lunar surface. This simple demonstration surely would have pleased Galileo!

When we use the expression *freely falling object,* we do not necessarily refer to an object dropped from rest. A freely falling object is any object moving freely under the influence of gravity alone, regardless of its initial motion. Objects thrown upward or downward and those released from rest are all falling freely once they are released. Any freely falling object experiences an acceleration directed *downward,* regardless of its initial motion.

We shall denote the magnitude of the *free-fall acceleration* by the symbol g. The value of g decreases with increasing altitude above the Earth's surface. Furthermore, slight variations in g occur with changes in latitude. At the Earth's surface, the value of g is approximately $9.80\ \text{m/s}^2$. Unless stated otherwise, we shall use this value for g when performing calculations. For making quick estimates, use $g = 10\ \text{m/s}^2$.

If we neglect air resistance and assume the free-fall acceleration does not vary with altitude over short vertical distances, the motion of a freely falling object moving vertically is equivalent to the motion of a particle under constant acceleration in one dimension. Therefore, the equations developed in Section 2.6 for the particle under constant acceleration model can be applied. The only modification for freely falling objects that we need to make in these equations is to note that the motion is in the vertical direction (the y direction) rather than in the horizontal direction (x) and that the acceleration is downward and has a magnitude of $9.80\ \text{m/s}^2$. Therefore, we choose $a_y = -g = -9.80\ \text{m/s}^2$, where the negative sign means that the acceleration of a freely falling object is downward. In Chapter 13, we shall study how to deal with variations in g with altitude.

***Quick Quiz* 2.7** Consider the following choices: (a) increases, (b) decreases, (c) increases and then decreases, (d) decreases and then increases, (e) remains the same. From these choices, select what happens to **(i)** the acceleration and **(ii)** the speed of a ball after it is thrown upward into the air.

Conceptual Example 2.9 The Daring Skydivers

A skydiver jumps out of a hovering helicopter. A few seconds later, another skydiver jumps out, and they both fall along the same vertical line. Ignore air resistance so that both skydivers fall with the same acceleration. Does the difference in their speeds stay the same throughout the fall? Does the vertical distance between them stay the same throughout the fall?

SOLUTION

At any given instant, the speeds of the skydivers are different because one had a head start. In any time interval Δt after this instant, however, the two skydivers increase their speeds by the same amount because they have the same acceleration. Therefore, the difference in their speeds remains the same throughout the fall.

The first jumper always has a greater speed than the second. Therefore, in a given time interval, the first skydiver covers a greater distance than the second. Consequently, the separation distance between them increases.

Example 2.10 Not a Bad Throw for a Rookie!

A stone thrown from the top of a building is given an initial velocity of 20.0 m/s straight upward. The stone is launched 50.0 m above the ground, and the stone just misses the edge of the roof on its way down as shown in Figure 2.14.

(A) Using $t_{Ⓐ} = 0$ as the time the stone leaves the thrower's hand at position Ⓐ, determine the time at which the stone reaches its maximum height.

SOLUTION

You most likely have experience with dropping objects or throwing them upward and watching them fall, so this problem should describe a familiar experience. To simulate this situation, toss a small object upward and notice the time interval required for it to fall to the floor. Now imagine throwing that object upward from the roof of a building. Because the stone is in free fall, it is modeled as a particle under constant acceleration due to gravity.

Recognize that the initial velocity is positive because the stone is launched upward. The velocity will change sign after the stone reaches its highest point, but the acceleration of the stone will *always* be downward. Choose an initial point just after the stone leaves the person's hand and a final point at the top of its flight.

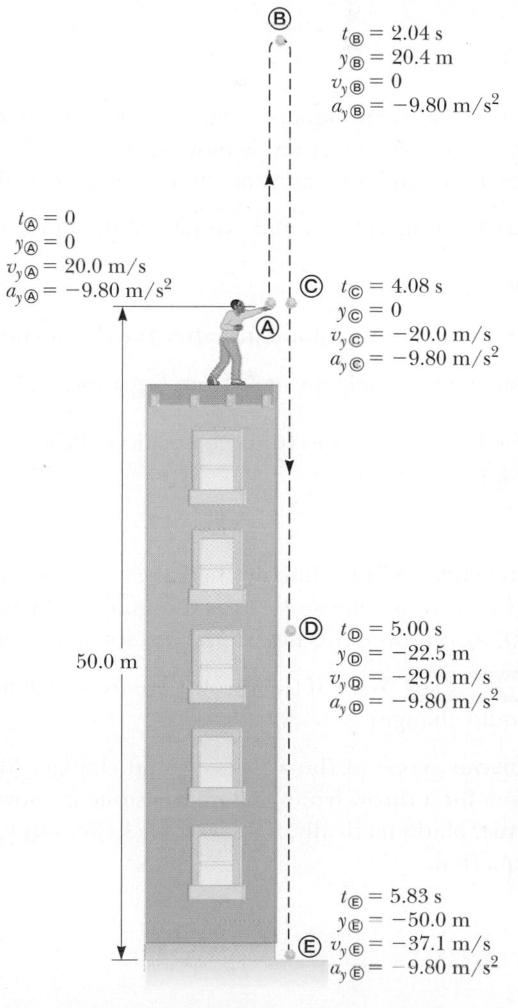

Figure 2.14 (Example 2.10) Position and velocity versus time for a freely falling stone thrown initially upward with a velocity $v_{yi} = 20.0$ m/s. Many of the quantities in the labels for points in the motion of the stone are calculated in the example. Can you verify the other values that are not?

continued

2.10 cont.

Use Equation 2.13 to calculate the time at which the stone reaches its maximum height:

$$v_{yf} = v_{yi} + a_y t \rightarrow t = \frac{v_{yf} - v_{yi}}{a_y}$$

Substitute numerical values:

$$t = t_Ⓑ = \frac{0 - 20.0 \text{ m/s}}{-9.80 \text{ m/s}^2} = \boxed{2.04 \text{ s}}$$

(B) Find the maximum height of the stone.

SOLUTION

As in part (A), choose the initial and final points at the beginning and the end of the upward flight.

Set $y_Ⓐ = 0$ and substitute the time from part (A) into Equation 2.16 to find the maximum height:

$$y_{\text{max}} = y_Ⓑ = y_Ⓐ + v_{xⒶ}t + \tfrac{1}{2}a_y t^2$$

$$y_Ⓑ = 0 + (20.0 \text{ m/s})(2.04 \text{ s}) + \tfrac{1}{2}(-9.80 \text{ m/s}^2)(2.04 \text{ s})^2 = \boxed{20.4 \text{ m}}$$

(C) Determine the velocity of the stone when it returns to the height from which it was thrown.

SOLUTION

Choose the initial point where the stone is launched and the final point when it passes this position coming down.

Substitute known values into Equation 2.17:

$$v_{yⒸ}^2 = v_{yⒶ}^2 + 2a_y(y_Ⓒ - y_Ⓐ)$$

$$v_{yⒸ}^2 = (20.0 \text{ m/s})^2 + 2(-9.80 \text{ m/s}^2)(0 - 0) = 400 \text{ m}^2/\text{s}^2$$

$$v_{yⒸ} = \boxed{-20.0 \text{ m/s}}$$

When taking the square root, we could choose either a positive or a negative root. We choose the negative root because we know that the stone is moving downward at point Ⓒ. The velocity of the stone when it arrives back at its original height is equal in magnitude to its initial velocity but is opposite in direction.

(D) Find the velocity and position of the stone at $t = 5.00$ s.

SOLUTION

Choose the initial point just after the throw and the final point 5.00 s later.

Calculate the velocity at Ⓓ from Equation 2.13:

$$v_{yⒹ} = v_{yⒶ} + a_y t = 20.0 \text{ m/s} + (-9.80 \text{ m/s}^2)(5.00 \text{ s}) = \boxed{-29.0 \text{ m/s}}$$

Use Equation 2.16 to find the position of the stone at $t_Ⓓ = 5.00$ s:

$$y_Ⓓ = y_Ⓐ + v_{yⒶ}t + \tfrac{1}{2}a_y t^2$$

$$= 0 + (20.0 \text{ m/s})(5.00 \text{ s}) + \tfrac{1}{2}(-9.80 \text{ m/s}^2)(5.00 \text{ s})^2$$

$$= \boxed{-22.5 \text{ m}}$$

The choice of the time defined as $t = 0$ is arbitrary and up to you to select as the problem solver. As an example of this arbitrariness, choose $t = 0$ as the time at which the stone is at the highest point in its motion. Then solve parts (C) and (D) again using this new initial instant and notice that your answers are the same as those above.

WHAT IF? What if the throw were from 30.0 m above the ground instead of 50.0 m? Which answers in parts (A) to (D) would change?

Answer None of the answers would change. All the motion takes place in the air during the first 5.00 s. (Notice that even for a throw from 30.0 m, the stone is above the ground at $t = 5.00$ s.) Therefore, the height of the throw is not an issue. Mathematically, if we look back over our calculations, we see that we never entered the height of the throw into any equation.

2.8 Kinematic Equations Derived from Calculus

This section assumes the reader is familiar with the techniques of integral calculus. If you have not yet studied integration in your calculus course, you should skip this section or cover it after you become familiar with integration.

The velocity of a particle moving in a straight line can be obtained if its position as a function of time is known. Mathematically, the velocity equals the derivative of the position with respect to time. It is also possible to find the position of a particle if its velocity is known as a function of time. In calculus, the procedure used to perform this task is referred to either as *integration* or as finding the *antiderivative.* Graphically, it is equivalent to finding the area under a curve.

Suppose the v_x–t graph for a particle moving along the x axis is as shown in Figure 2.15. Let us divide the time interval $t_f - t_i$ into many small intervals, each of duration Δt_n. From the definition of average velocity, we see that the displacement of the particle during any small interval, such as the one shaded in Figure 2.15, is given by $\Delta x_n = v_{xn,\text{avg}}\,\Delta t_n$, where $v_{xn,\text{avg}}$ is the average velocity in that interval. Therefore, the displacement during this small interval is simply the area of the shaded rectangle in Figure 2.15. The total displacement for the interval $t_f - t_i$ is the sum of the areas of all the rectangles from t_i to t_f:

$$\Delta x = \sum_n v_{xn,\text{avg}}\,\Delta t_n$$

where the symbol Σ (uppercase Greek sigma) signifies a sum over all terms, that is, over all values of n. Now, as the intervals are made smaller and smaller, the number of terms in the sum increases and the sum approaches a value equal to the area under the curve in the velocity–time graph. Therefore, in the limit $n \to \infty$, or $\Delta t_n \to 0$, the displacement is

$$\Delta x = \lim_{\Delta t_n \to 0} \sum_n v_{xn}\,\Delta t_n \tag{2.18}$$

Notice that we have replaced the average velocity $v_{xn,\text{avg}}$ with the instantaneous velocity v_{xn} in the sum because the stepwise velocity $v_{xn,\text{avg}}$ approaches a continuous function v_{xn} as the time intervals shrink to zero. As you can see from Figure 2.15, this approximation is valid in the limit of very small intervals. Therefore, if we know the v_x–t graph for motion along a straight line, we can obtain the displacement during any time interval by measuring the area under the curve corresponding to that time interval.

The limit of the sum shown in Equation 2.18 is called a **definite integral** and is written

◀ **Definite integral**

$$\lim_{\Delta t_n \to 0} \sum_n v_{xn}\,\Delta t_n = \int_{t_i}^{t_f} v_x(t)\,dt \tag{2.19}$$

where $v_x(t)$ denotes the velocity at any time t. If the explicit functional form of $v_x(t)$ is known and the limits are given, the integral can be evaluated. Sometimes the

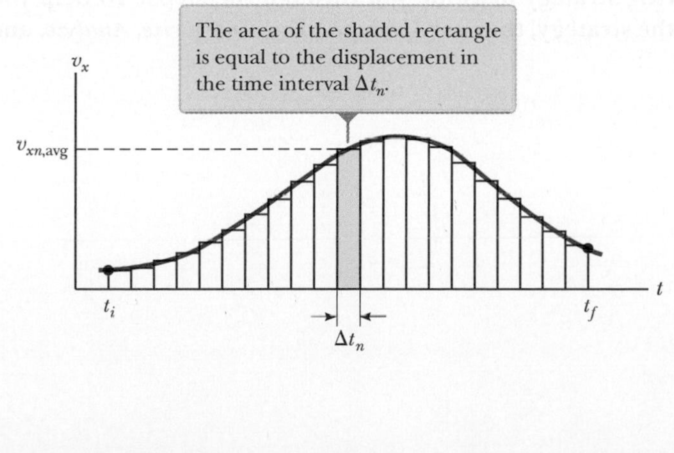

Figure 2.15 Velocity versus time for a particle moving along the x axis. The total area under the curve is the total displacement of the particle.

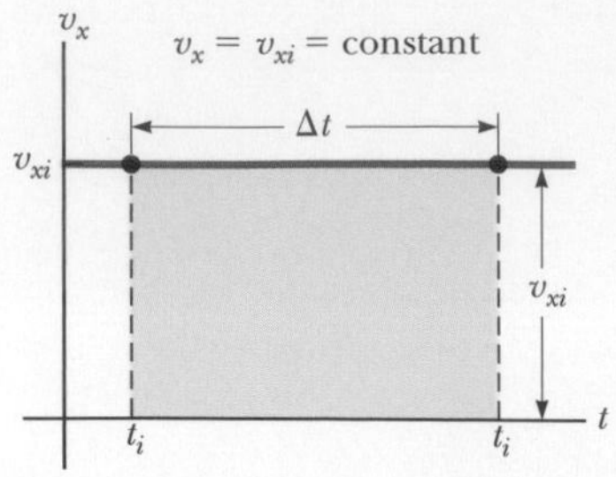

Figure 2.16 The velocity–time curve for a particle moving with constant velocity v_{xi}. The displacement of the particle during the time interval $t_f - t_i$ is equal to the area of the shaded rectangle.

v_x–t graph for a moving particle has a shape much simpler than that shown in Figure 2.15. For example, suppose a particle moves at a constant velocity v_{xi}. In this case, the v_x–t graph is a horizontal line as in Figure 2.16 and the displacement of the particle during the time interval Δt is simply the area of the shaded rectangle:

$$\Delta x = v_{xi}\,\Delta t \text{ (when } v_x = v_{xi} = \text{constant)}$$

Kinematic Equations

We now use the defining equations for acceleration and velocity to derive two of our kinematic equations, Equations 2.13 and 2.16.

The defining equation for acceleration (Eq. 2.10),

$$a_x = \frac{dv_x}{dt}$$

may be written as $dv_x = a_x\,dt$ or, in terms of an integral (or antiderivative), as

$$v_{xf} - v_{xi} = \int_0^t a_x\,dt$$

For the special case in which the acceleration is constant, a_x can be removed from the integral to give

$$v_{xf} - v_{xi} = a_x\int_0^t dt = a_x(t - 0) = a_x t \qquad \textbf{(2.20)}$$

which is Equation 2.13.

Now let us consider the defining equation for velocity (Eq. 2.5):

$$v_x = \frac{dx}{dt}$$

We can write this equation as $dx = v_x\,dt$ or in integral form as

$$x_f - x_i = \int_0^t v_x\,dt$$

Because $v_x = v_{xf} = v_{xi} + a_x t$, this expression becomes

$$x_f - x_i = \int_0^t (v_{xi} + a_x t)\,dt = \int_0^t v_{xi}\,dt + a_x\int_0^t t\,dt = v_{xi}(t - 0) + a_x\left(\frac{t^2}{2} - 0\right)$$

$$x_f - x_i = v_{xi}t + \tfrac{1}{2}a_x t^2$$

which is Equation 2.16.

Besides what you might expect to learn about physics concepts, a very valuable skill you should hope to take away from your physics course is the ability to solve complicated problems. The way physicists approach complex situations and break them into manageable pieces is extremely useful. The following is a general problem-solving strategy to guide you through the steps. To help you remember the steps of the strategy, they are *Conceptualize, Categorize, Analyze,* and *Finalize.*

GENERAL PROBLEM-SOLVING STRATEGY

Conceptualize

- The first things to do when approaching a problem are to *think about* and *understand* the situation. Study carefully any representations of the information (for example, diagrams, graphs, tables, or photographs) that accompany the problem. Imagine a movie, running in your mind, of what happens in the problem.
- If a pictorial representation is not provided, you should almost always make a quick drawing of the situation. Indicate any known values, perhaps in a table or directly on your sketch.
- Now focus on what algebraic or numerical information is given in the problem. Carefully read the problem statement, looking for key phrases such as "starts from rest" ($v_i = 0$), "stops" ($v_f = 0$), or "falls freely" ($a_y = -g = -9.80\text{ m/s}^2$).
- Now focus on the expected result of solving the problem. Exactly what is the question asking? Will the final result be numerical or algebraic? Do you know what units to expect?
- Don't forget to incorporate information from your own experiences and common sense. What should a reasonable answer look like? For example, you wouldn't expect to calculate the speed of an automobile to be 5×10^6 m/s.

Categorize

- Once you have a good idea of what the problem is about, you need to *simplify* the problem. Remove the details that are not important to the solution. For example, model a moving object as a particle. If appropriate, ignore air resistance or friction between a sliding object and a surface.
- Once the problem is simplified, it is important to *categorize* the problem. Is it a simple *substitution problem* such that numbers can be substituted into an equation? If so, the problem is likely to be finished when this substitution is done. If not, you face what we call an *analysis problem:* the situation must be analyzed more deeply to reach a solution.
- If it is an analysis problem, it needs to be categorized further. Have you seen this type of problem before? Does it fall into the growing list of types of problems that you have solved previously? If so, identify any analysis model(s) appropriate for the problem to prepare for the Analyze step below. We saw the first three analysis models in this chapter: the particle under constant velocity, the particle under constant speed, and the particle under constant acceleration. Being able to classify a problem with an analysis model can make it much easier to lay out a plan to solve it. For example, if your simplification shows that the problem can be treated as a particle under constant acceleration and you have already solved such a problem (such as the examples in Section 2.6), the solution to the present problem follows a similar pattern.

Analyze

- Now you must analyze the problem and strive for a mathematical solution. Because you have already categorized the problem and identified an analysis model, it should not be too difficult to select relevant equations that apply to the type of situation in the problem. For example, if the problem involves a particle under constant acceleration, Equations 2.13 to 2.17 are relevant.
- Use algebra (and calculus, if necessary) to solve symbolically for the unknown variable in terms of what is given. Substitute in the appropriate numbers, calculate the result, and round it to the proper number of significant figures.

Finalize

- Examine your numerical answer. Does it have the correct units? Does it meet your expectations from your conceptualization of the problem? What about the algebraic form of the result? Does it make sense? Examine the variables in the problem to see whether the answer would change in a physically meaningful way if the variables were drastically increased or decreased or even became zero. Looking at limiting cases to see whether they yield expected values is a very useful way to make sure that you are obtaining reasonable results.
- Think about how this problem compared with others you have solved. How was it similar? In what critical ways did it differ? Why was this problem assigned? Can you figure out what you have learned by doing it? If it is a new category of problem, be sure you understand it so that you can use it as a model for solving similar problems in the future.

When solving complex problems, you may need to identify a series of subproblems and apply the problem-solving strategy to each. For simple problems, you probably don't need this strategy. When you are trying to solve a problem and you don't know what to do next, however, remember the steps in the strategy and use them as a guide.

For practice, it would be useful for you to revisit the worked examples in this chapter and identify the *Conceptualize, Categorize, Analyze,* and *Finalize* steps. In the rest of this book, we will label these steps explicitly in the worked examples. Many chapters in this book include a section labeled Problem-Solving Strategy that should help you through the rough spots. These sections are organized according to the General Problem-Solving Strategy outlined above and are tailored to the specific types of problems addressed in that chapter.

To clarify how this Strategy works, we repeat Example 2.7 on the next page with the particular steps of the Strategy identified.

When you **Conceptualize** a problem, try to understand the situation that is presented in the problem statement. Study carefully any representations of the information (for example, diagrams, graphs, tables, or photographs) that accompany the problem. Imagine a movie, running in your mind, of what happens in the problem.

Simplify the problem. Remove the details that are not important to the solution. Then **Categorize** the problem. Is it a simple substitution problem such that numbers can be substituted into an equation? If not, you face an analysis problem. In this case, identify the appropriate analysis model.

Example 2.7 Carrier Landing

A jet lands on an aircraft carrier at a speed of 140 mi/h ($\approx$ 63 m/s).

(A) What is its acceleration (assumed constant) if it stops in 2.0 s due to an arresting cable that snags the jet and brings it to a stop?

SOLUTION

Conceptualize

You might have seen movies or television shows in which a jet lands on an aircraft carrier and is brought to rest surprisingly fast by an arresting cable. A careful reading of the problem reveals that in addition to being given the initial speed of 63 m/s, we also know that the final speed is zero.

Categorize

Because the acceleration of the jet is assumed constant, we model it as a particle under constant acceleration.

Analyze

We define our x axis as the direction of motion of the jet. Notice that we have no information about the change in position of the jet while it is slowing down.

Equation 2.13 is the only equation in Table 2.2 that does not involve position, so we use it to find the acceleration of the jet, modeled as a particle:

$$a_x = \frac{v_{xf} - v_{xi}}{t} = \frac{0 - 63 \text{ m/s}}{2.0 \text{ s}}$$

$$= -32 \text{ m/s}^2$$

(B) If the jet touches down at position $x_i = 0$, what is its final position?

SOLUTION

Use Equation 2.15 to solve for the final position:

$$x_f = x_i + \tfrac{1}{2}(v_{xi} + v_{xf})t = 0 + \tfrac{1}{2}(63 \text{ m/s} + 0)(2.0 \text{ s}) = 63 \text{ m}$$

Finalize

Given the size of aircraft carriers, a length of 63 m seems reasonable for stopping the jet. The idea of using arresting cables to slow down landing aircraft and enable them to land safely on ships originated at about the time of World War I. The cables are still a vital part of the operation of modern aircraft carriers.

WHAT IF? Suppose the jet lands on the deck of the aircraft carrier with a speed higher than 63 m/s but has the same acceleration due to the cable as that calculated in part (A). How will that change the answer to part (B)?

Answer If the jet is traveling faster at the beginning, it will stop farther away from its starting point, so the answer to part (B) should be larger. Mathematically, we see in Equation 2.15 that if v_{xi} is larger, x_f will be larger.

Now **Analyze** the problem. Select relevant equations from the analysis model. Solve symbolically for the unknown variable in terms of what is given. Substitute in the appropriate numbers, calculate the result, and round it to the proper number of significant figures.

Finalize the problem. Examine the numerical answer. Does it have the correct units? Does it meet your expectations from your conceptualization of the problem? Does the answer make sense? What about the algebraic form of the result? Examine the variables in the problem to see whether the answer would change in a physically meaningful way if the variables were drastically increased or decreased or even became zero.

What If? questions will appear in many examples in the text, and offer a variation on the situation just explored. This feature encourages you to think about the results of the example and assists in conceptual understanding of the principles.

Summary

Definitions

When a particle moves along the x axis from some initial position x_i to some final position x_f, its **displacement** is

$$\Delta x \equiv x_f - x_i \quad (2.1)$$

The **average velocity** of a particle during some time interval is the displacement Δx divided by the time interval Δt during which that displacement occurs:

$$v_{x,\text{avg}} \equiv \frac{\Delta x}{\Delta t} \quad (2.2)$$

The **average speed** of a particle is equal to the ratio of the total distance it travels to the total time interval during which it travels that distance:

$$v_{\text{avg}} \equiv \frac{d}{\Delta t} \quad (2.3)$$

The **instantaneous velocity** of a particle is defined as the limit of the ratio $\Delta x/\Delta t$ as Δt approaches zero. By definition, this limit equals the derivative of x with respect to t, or the time rate of change of the position:

$$v_x \equiv \lim_{\Delta t \to 0} \frac{\Delta x}{\Delta t} = \frac{dx}{dt} \quad (2.5)$$

The **instantaneous speed** of a particle is equal to the magnitude of its instantaneous velocity.

The **average acceleration** of a particle is defined as the ratio of the change in its velocity Δv_x divided by the time interval Δt during which that change occurs:

$$a_{x,\text{avg}} \equiv \frac{\Delta v_x}{\Delta t} = \frac{v_{xf} - v_{xi}}{t_f - t_i} \quad (2.9)$$

The **instantaneous acceleration** is equal to the limit of the ratio $\Delta v_x/\Delta t$ as Δt approaches 0. By definition, this limit equals the derivative of v_x with respect to t, or the time rate of change of the velocity:

$$a_x \equiv \lim_{\Delta t \to 0} \frac{\Delta v_x}{\Delta t} = \frac{dv_x}{dt} \quad (2.10)$$

Concepts and Principles

When an object's velocity and acceleration are in the same direction, the object is speeding up. On the other hand, when the object's velocity and acceleration are in opposite directions, the object is slowing down. Remembering that $F_x \propto a_x$ is a useful way to identify the direction of the acceleration by associating it with a force.

An object falling freely in the presence of the Earth's gravity experiences free-fall acceleration directed toward the center of the Earth. If air resistance is neglected, if the motion occurs near the surface of the Earth, and if the range of the motion is small compared with the Earth's radius, the free-fall acceleration $a_y = -g$ is constant over the range of motion, where g is equal to 9.80 m/s^2.

Complicated problems are best approached in an organized manner. Recall and apply the *Conceptualize, Categorize, Analyze,* and *Finalize* steps of the **General Problem-Solving Strategy** when you need them.

An important aid to problem solving is the use of **analysis models.** Analysis models are situations that we have seen in previous problems. Each analysis model has one or more equations associated with it. When solving a new problem, identify the analysis model that corresponds to the problem. The model will tell you which equations to use. The first three analysis models introduced in this chapter are summarized on the next page.

continued

Analysis Models for Problem-Solving

Particle Under Constant Velocity. If a particle moves in a straight line with a constant speed v_x, its constant velocity is given by

$$v_x = \frac{\Delta x}{\Delta t} \qquad (2.6)$$

and its position is given by

$$x_f = x_i + v_x t \qquad (2.7)$$

Particle Under Constant Speed. If a particle moves a distance d along a curved or straight path with a constant speed, its constant speed is given by

$$v = \frac{d}{\Delta t} \qquad (2.8)$$

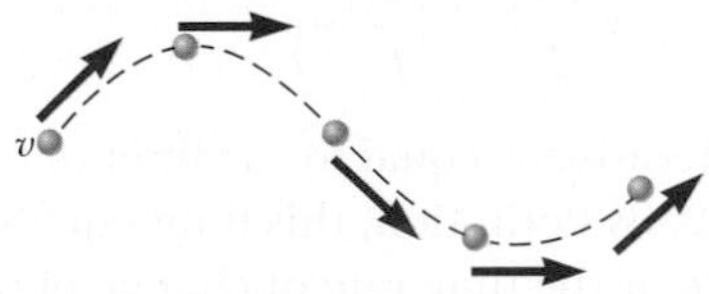

Particle Under Constant Acceleration. If a particle moves in a straight line with a constant acceleration a_x, its motion is described by the kinematic equations:

$$v_{xf} = v_{xi} + a_x t \qquad (2.13)$$

$$v_{x,\text{avg}} = \frac{v_{xi} + v_{xf}}{2} \qquad (2.14)$$

$$x_f = x_i + \tfrac{1}{2}(v_{xi} + v_{xf})t \qquad (2.15)$$

$$x_f = x_i + v_{xi}t + \tfrac{1}{2}a_x t^2 \qquad (2.16)$$

$$v_{xf}^{\,2} = v_{xi}^{\,2} + 2a_x(x_f - x_i) \qquad (2.17)$$

Objective Questions

☐ denotes answer available in Student Solutions Manual/Study Guide

1. One drop of oil falls straight down onto the road from the engine of a moving car every 5 s. Figure OQ2.1 shows the pattern of the drops left behind on the pavement. What is the average speed of the car over this section of its motion? (a) 20 m/s (b) 24 m/s (c) 30 m/s (d) 100 m/s (e) 120 m/s

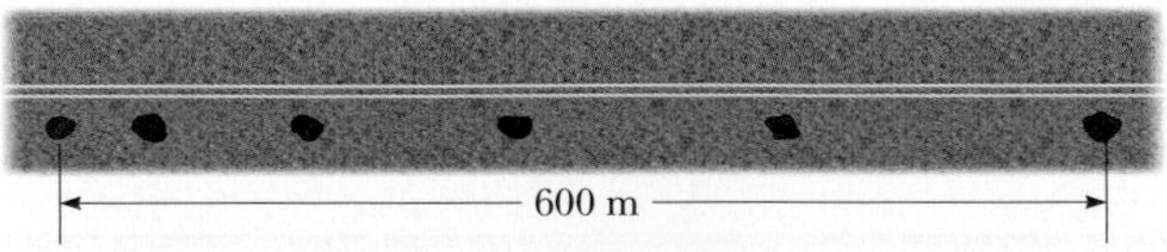

Figure OQ2.1

2. An arrow is shot straight up in the air at an initial speed of 15.0 m/s. After how much time is the arrow moving downward at a speed of 8.00 m/s? (a) 0.714 s (b) 1.24 s (c) 1.87 s (d) 2.35 s (e) 3.22 s

3. A cannon shell is fired straight up from the ground at an initial speed of 225 m/s. After how much time is the shell at a height of 6.20×10^2 m above the ground and moving downward? (a) 2.96 s (b) 17.3 s (c) 25.4 s (d) 33.6 s (e) 43.0 s

4. When applying the equations of kinematics for an object moving in one dimension, which of the following statements *must* be true? (a) The velocity of the object must remain constant. (b) The acceleration of the object must remain constant. (c) The velocity of the object must increase with time. (d) The position of the object must increase with time. (e) The velocity of the object must always be in the same direction as its acceleration.

5. A juggler throws a bowling pin straight up in the air. After the pin leaves his hand and while it is in the air, which statement is true? (a) The velocity of the pin is always in the same direction as its acceleration. (b) The velocity of the pin is never in the same direction as its acceleration. (c) The acceleration of the pin is zero. (d) The velocity of the pin is opposite its acceleration on the way up. (e) The velocity of the pin is in the same direction as its acceleration on the way up.

6. A racing car starts from rest at $t = 0$ and reaches a final speed v at time t. If the acceleration of the car is constant during this time, which of the following statements are true? (a) The car travels a distance vt. (b) The average speed of the car is $v/2$. (c) The magnitude of the acceleration of the car is v/t. (d) The velocity of the car remains constant. (e) None of statements (a) through (d) is true.

7. When the pilot reverses the propeller in a boat moving north, the boat moves with an acceleration directed south. Assume the acceleration of the boat remains constant in magnitude and direction. What happens to the boat? (a) It eventually stops and remains stopped. (b) It eventually stops and then speeds up in the forward direction. (c) It eventually stops and then speeds up in the reverse direction. (d) It never stops but loses speed more and more slowly forever. (e) It never stops but continues to speed up in the forward direction.

8. A rock is thrown downward from the top of a 40.0-m-tall tower with an initial speed of 12 m/s. Assuming negligible air resistance, what is the speed of the rock just before hitting the ground? (a) 28 m/s (b) 30 m/s (c) 56 m/s (d) 784 m/s (e) More information is needed.

9. A skateboarder starts from rest and moves down a hill with constant acceleration in a straight line, traveling for 6 s. In a second trial, he starts from rest and moves along the same straight line with the same acceleration for only 2 s. How does his displacement from his starting point in this second trial compare with that from the first trial? (a) one-third as large (b) three times larger (c) one-ninth as large (d) nine times larger (e) $1/\sqrt{3}$ times as large

10. A pebble is dropped from rest from the top of a tall cliff and falls 4.9 m after 1.0 s has elapsed. How much farther does it drop in the next 2.0 s? (a) 9.8 m (b) 19.6 m (c) 39 m (d) 44 m (e) none of the above

11. As an object moves along the x axis, many measurements are made of its position, enough to generate a smooth, accurate graph of x versus t. Which of the following quantities for the object *cannot* be obtained from this graph *alone*? (a) the velocity at any instant (b) the acceleration at any instant (c) the displacement during some time interval (d) the average velocity during some time interval (e) the speed at any instant

12. On another planet, a marble is released from rest at the top of a high cliff. It falls 4.00 m in the first 1 s of its motion. Through what additional distance does it fall in the next 1 s? (a) 4.00 m (b) 8.00 m (c) 12.0 m (d) 16.0 m (e) 20.0 m

13. A student at the top of a building of height h throws one ball upward with a speed of v_i and then throws a second ball downward with the same initial speed v_i. Just before it reaches the ground, is the final speed of the ball thrown upward (a) larger, (b) smaller, or (c) the same in magnitude, compared with the final speed of the ball thrown downward?

14. A ball is thrown straight up in the air. For which situation are both the instantaneous velocity and the acceleration zero? (a) on the way up (b) at the top of its flight path (c) on the way down (d) halfway up and halfway down (e) none of the above

15. A pebble is released from rest at a certain height and falls freely, reaching an impact speed of 4 m/s at the floor. Next, the pebble is thrown down with an initial speed of 3 m/s from the same height. What is its speed at the floor? (a) 4 m/s (b) 5 m/s (c) 6 m/s (d) 7 m/s (e) 8 m/s

16. You drop a ball from a window located on an upper floor of a building. It strikes the ground with speed v. You now repeat the drop, but your friend down on the ground throws another ball upward at the same speed v, releasing her ball at the same moment that you drop yours from the window. At some location, the balls pass each other. Is this location (a) *at* the halfway point between window and ground, (b) *above* this point, or (c) *below* this point?

17. A hard rubber ball, not affected by air resistance in its motion, is tossed upward from shoulder height, falls to the sidewalk, rebounds to a smaller maximum height, and is caught on its way down again. This motion is represented in Figure OQ2.17, where the successive positions of the ball Ⓐ through Ⓔ are not equally spaced in time. At point Ⓓ the center of the ball is at its lowest point in the motion. The motion of the ball is along a straight, vertical line, but the diagram shows successive positions offset to the right to avoid overlapping. Choose the positive y direction to be upward. (a) Rank the situations Ⓐ through Ⓔ according to the speed of the ball $|v_y|$ at each point, with the largest speed first. (b) Rank the same situations according to the acceleration a_y of the ball at each point. (In both rankings, remember that zero is greater than a negative value. If two values are equal, show that they are equal in your ranking.)

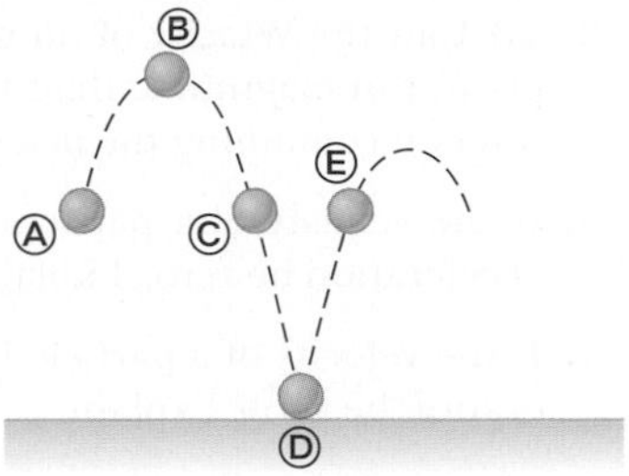

Figure OQ2.17

18. Each of the strobe photographs (a), (b), and (c) in Figure OQ2.18 was taken of a single disk moving toward the right, which we take as the positive direction. Within each photograph, the time interval between images is constant. **(i)** Which photograph shows motion with zero acceleration? **(ii)** Which photograph shows motion with positive acceleration? **(iii)** Which photograph shows motion with negative acceleration?

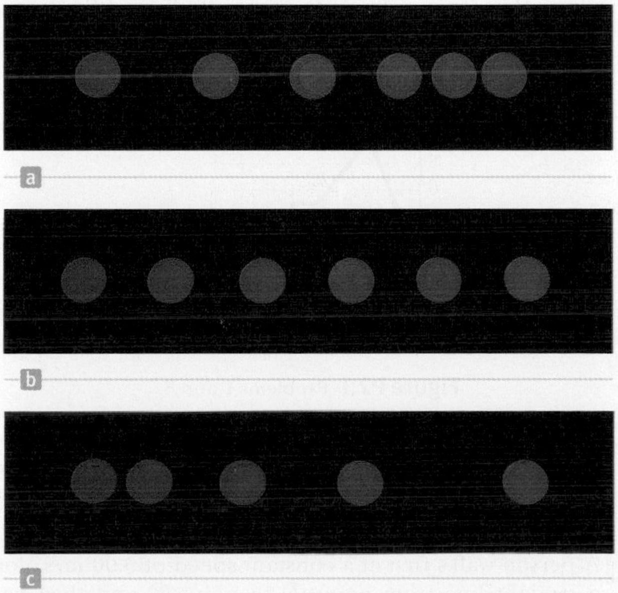

Figure OQ2.18 Objective Question 18 and Problem 19.

Conceptual Questions

□ denotes answer available in Student Solutions Manual/Study Guide

1. If the average velocity of an object is zero in some time interval, what can you say about the displacement of the object for that interval?

2. (a) Can the velocity of an object at an instant of time be greater in magnitude than the average velocity over a time interval containing the instant? (b) Can it be less?

3. If the velocity of a particle is nonzero, can the particle's acceleration be zero? Explain.

4. If the velocity of a particle is zero, can the particle's acceleration be zero? Explain.

5. If a car is traveling eastward, can its acceleration be westward? Explain.

6. You throw a ball vertically upward so that it leaves the ground with velocity +5.00 m/s. (a) What is its velocity when it reaches its maximum altitude? (b) What is its acceleration at this point? (c) What is the velocity with which it returns to ground level? (d) What is its acceleration at this point?

7. (a) Can the equations of kinematics (Eqs. 2.13–2.17) be used in a situation in which the acceleration varies in time? (b) Can they be used when the acceleration is zero?

8. Try the following experiment away from traffic where you can do it safely. With the car you are driving moving slowly on a straight, level road, shift the transmission into neutral and let the car coast. At the moment the car comes to a complete stop, step hard on the brake and notice what you feel. Now repeat the same experiment on a fairly gentle, uphill slope. Explain the difference in what a person riding in the car feels in the two cases. (Brian Popp suggested the idea for this question.)

9. Two cars are moving in the same direction in parallel lanes along a highway. At some instant, the velocity of car A exceeds the velocity of car B. Does that mean that the acceleration of car A is greater than that of car B? Explain.

Problems

ENHANCED WebAssign The problems found in this chapter may be assigned online in Enhanced WebAssign

1. denotes straightforward problem; **2.** denotes intermediate problem; **3.** denotes challenging problem

□1. full solution available in the Student Solutions Manual/Study Guide

1. denotes problems most often assigned in Enhanced WebAssign; these provide students with targeted feedback and either a Master It tutorial or a Watch It solution video.

QC denotes asking for quantitative and conceptual reasoning

S denotes symbolic reasoning problem

M denotes Master It tutorial available in Enhanced WebAssign

GP denotes guided problem

shaded denotes "paired problems" that develop reasoning with symbols and numerical values

Section 2.1 Position, Velocity, and Speed

1. The position versus time for a certain particle moving along the x axis is shown in Figure P2.1. Find the average velocity in the time intervals (a) 0 to 2 s, (b) 0 to 4 s, (c) 2 s to 4 s, (d) 4 s to 7 s, and (e) 0 to 8 s.

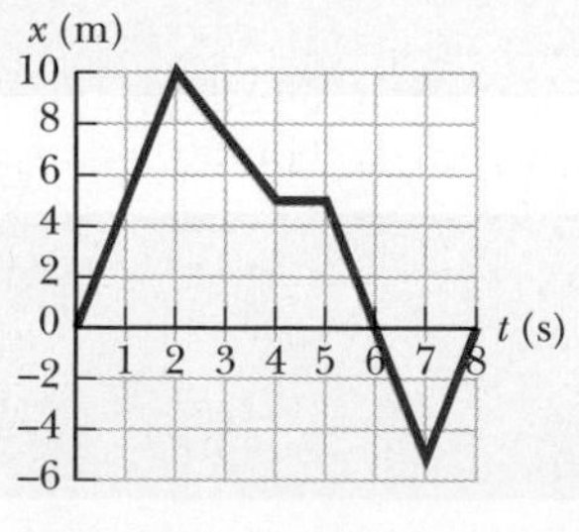

Figure P2.1 Problems 1 and 8.

2. The speed of a nerve impulse in the human body is about 100 m/s. If you accidentally stub your toe in the dark, estimate the time it takes the nerve impulse to travel to your brain.

3. A person walks first at a constant speed of 5.00 m/s along a straight line from point Ⓐ to point Ⓑ and then back along the line from Ⓑ to Ⓐ at a constant speed of 3.00 m/s. (a) What is her average speed over the entire trip? (b) What is her average velocity over the entire trip?

4. A particle moves according to the equation $x = 10t^2$, where x is in meters and t is in seconds. (a) Find the average velocity for the time interval from 2.00 s to 3.00 s. (b) Find the average velocity for the time interval from 2.00 to 2.10 s.

Section 2.2 Instantaneous Velocity and Speed

5. A position–time graph for a particle moving along the x axis is shown in Figure P2.5. (a) Find the average velocity in the time interval $t = 1.50$ s to $t = 4.00$ s. (b) Determine

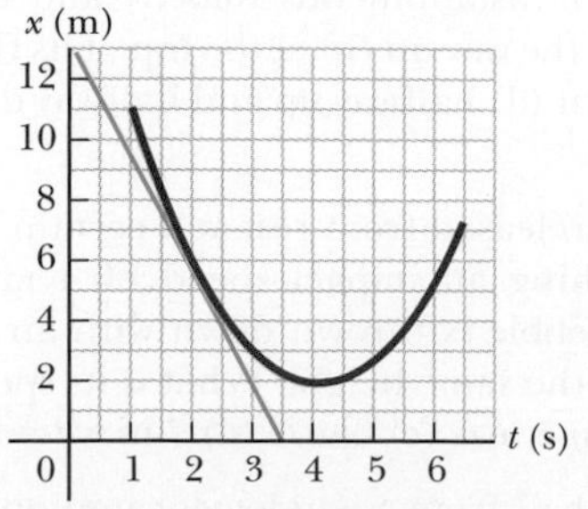

Figure P2.5

the instantaneous velocity at $t = 2.00$ s by measuring the slope of the tangent line shown in the graph. (c) At what value of t is the velocity zero?

6. The position of a particle moving along the x axis varies in time according to the expression $x = 3t^2$, where x is in meters and t is in seconds. Evaluate its position (a) at $t = 3.00$ s and (b) at $3.00\text{ s} + \Delta t$. (c) Evaluate the limit of $\Delta x/\Delta t$ as Δt approaches zero to find the velocity at $t = 3.00$ s.

7. **S** An athlete swims the length L of a pool in a time t_1 and makes the return trip to the starting position in a time t_2. If she is swimming initially in the positive x direction, determine her average velocities symbolically in (a) the first half of the swim, (b) the second half of the swim, and (c) the round trip. (d) What is her average speed for the round trip?

8. Find the instantaneous velocity of the particle described in Figure P2.1 at the following times: (a) $t = 1.0$ s, (b) $t = 3.0$ s, (c) $t = 4.5$ s, and (d) $t = 7.5$ s.

Section 2.3 Analysis Model: Particle Under Constant Velocity

9. **Review.** The North American and European plates of the Earth's crust are drifting apart with a relative speed of about 25 mm/yr. Take the speed as constant and find when the rift between them started to open, to reach a current width of 2.9×10^3 mi.

10. A car travels along a straight line at a constant speed of 60.0 mi/h for a distance d and then another distance d in the same direction at another constant speed. The average velocity for the entire trip is 30.0 mi/h. (a) What is the constant speed with which the car moved during the second distance d? (b) **What If?** Suppose the second distance d were traveled in the opposite direction; you forgot something and had to return home at the same constant speed as found in part (a). What is the average velocity for this trip? (c) What is the average speed for this new trip?

11. A hare and a tortoise compete in a race over a straight course 1.00 km long. The tortoise crawls at a speed of 0.200 m/s toward the finish line. The hare runs at a speed of 8.00 m/s toward the finish line for 0.800 km and then stops to tease the slow-moving tortoise as the tortoise eventually passes by. The hare waits for a while after the tortoise passes and then runs toward the finish line again at 8.00 m/s. Both the hare and the tortoise cross the finish line at the exact same instant. Assume both animals, when moving, move steadily at their respective speeds. (a) How far is the tortoise from the finish line when the hare resumes the race? (b) For how long in time was the hare stationary?

Section 2.4 Acceleration

12. **Review.** A 50.0-g Super Ball traveling at 25.0 m/s bounces off a brick wall and rebounds at 22.0 m/s. A high-speed camera records this event. If the ball is in contact with the wall for 3.50 ms, what is the magnitude of the average acceleration of the ball during this time interval?

13. A velocity–time graph for an object moving along the x axis is shown in Figure P2.13. (a) Plot a graph of the acceleration versus time. Determine the average acceleration of the object (b) in the time interval $t = 5.00$ s to $t = 15.0$ s and (c) in the time interval $t = 0$ to $t = 20.0$ s.

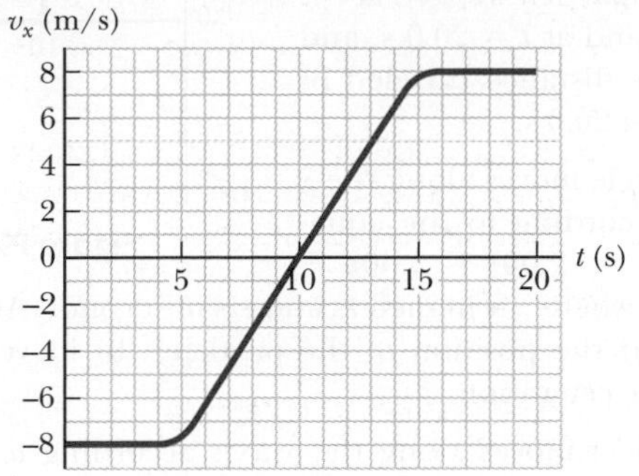

Figure P2.13

14. A child rolls a marble on a bent track that is 100 cm long as shown in Figure P2.14. We use x to represent the position of the marble along the track. On the horizontal sections from $x = 0$ to $x = 20$ cm and from $x = 40$ cm to $x = 60$ cm, the marble rolls with constant speed. On the sloping sections, the marble's speed changes steadily. At the places where the slope changes, the marble stays on the track and does not undergo any sudden changes in speed. The child gives the marble some initial speed at $x = 0$ and $t = 0$ and then watches it roll to $x = 90$ cm, where it turns around, eventually returning to $x = 0$ with the same speed with which the child released it. Prepare graphs of x versus t, v_x versus t, and a_x versus t, vertically aligned with their time axes identical, to show the motion of the marble. You will not be able to place numbers other than zero on the horizontal axis or on the velocity or acceleration axes, but show the correct graph shapes.

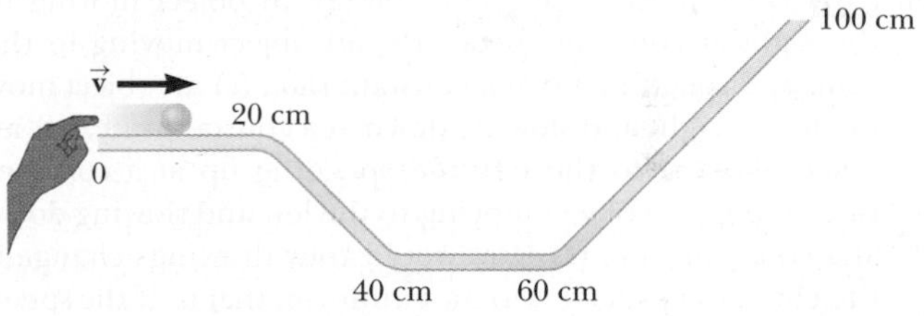

Figure P2.14

15. Figure P2.15 shows a graph of v_x versus t for the motion of a motorcyclist as he starts from rest and moves along the road in a straight line. (a) Find the average acceleration for the time interval $t = 0$ to $t = 6.00$ s. (b) Estimate the time at which the acceleration has its greatest positive value and the value of the acceleration at that instant. (c) When is the acceleration zero? (d) Estimate the maximum negative value of the acceleration and the time at which it occurs.

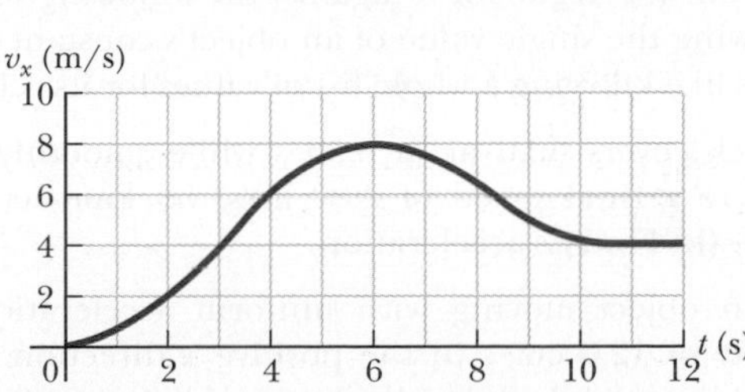

Figure P2.15

16. A particle starts from rest and accelerates as shown in Figure P2.16. Determine (a) the particle's speed at $t = 10.0$ s and at $t = 20.0$ s, and (b) the distance traveled in the first 20.0 s.

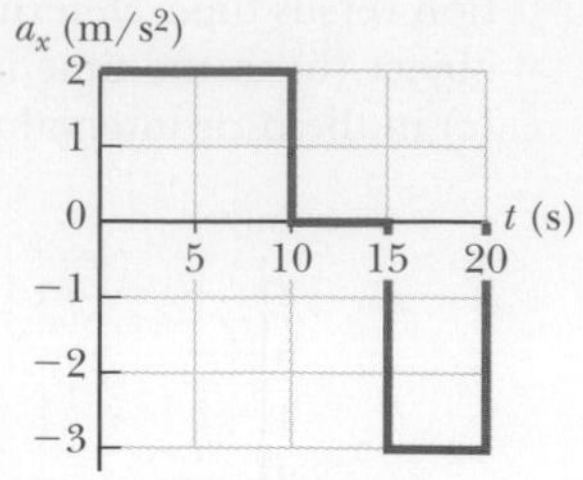

Figure P2.16

17. A particle moves along the x axis according to the equation $x = 2.00 + 3.00t - 1.00t^2$, where x is in meters and t is in seconds. At $t = 3.00$ s, find (a) the position of the particle, (b) its velocity, and (c) its acceleration.

18. An object moves along the x axis according to the equation $x = 3.00t^2 - 2.00t + 3.00$, where x is in meters and t is in seconds. Determine (a) the average speed between $t = 2.00$ s and $t = 3.00$ s, (b) the instantaneous speed at $t = 2.00$ s and at $t = 3.00$ s, (c) the average acceleration between $t = 2.00$ s and $t = 3.00$ s, and (d) the instantaneous acceleration at $t = 2.00$ s and $t = 3.00$ s. (e) At what time is the object at rest?

Section 2.5 Motion Diagrams

19. Q|C Each of the strobe photographs (a), (b), and (c) in Figure OQ2.18 was taken of a single disk moving toward the right, which we take as the positive direction. Within each photograph the time interval between images is constant. For each photograph, prepare graphs of x versus t, v_x versus t, and a_x versus t, vertically aligned with their time axes identical, to show the motion of the disk. You will not be able to place numbers other than zero on the axes, but show the correct shapes for the graph lines.

20. Q|C Draw motion diagrams for (a) an object moving to the right at constant speed, (b) an object moving to the right and speeding up at a constant rate, (c) an object moving to the right and slowing down at a constant rate, (d) an object moving to the left and speeding up at a constant rate, and (e) an object moving to the left and slowing down at a constant rate. (f) How would your drawings change if the changes in speed were not uniform, that is, if the speed were not changing at a constant rate?

Section 2.6 Analysis Model: Particle Under Constant Acceleration

21. Q|C A parcel of air moving in a straight tube with a constant acceleration of -4.00 m/s² has a velocity of 13.0 m/s at 10:05:00 a.m. (a) What is its velocity at 10:05:01 a.m.? (b) At 10:05:04 a.m.? (c) At 10:04:59 a.m.? (d) Describe the shape of a graph of velocity versus time for this parcel of air. (e) Argue for or against the following statement: "Knowing the single value of an object's constant acceleration is like knowing a whole list of values for its velocity."

22. A truck covers 40.0 m in 8.50 s while smoothly slowing down to a final speed of 2.80 m/s. (a) Find its original speed. (b) Find its acceleration.

23. M An object moving with uniform acceleration has a velocity of 12.0 cm/s in the positive x direction when its x coordinate is 3.00 cm. If its x coordinate 2.00 s later is -5.00 cm, what is its acceleration?

24. Q|C In Example 2.7, we investigated a jet landing on an aircraft carrier. In a later maneuver, the jet comes in for a landing on solid ground with a speed of 100 m/s, and its acceleration can have a maximum magnitude of 5.00 m/s² as it comes to rest. (a) From the instant the jet touches the runway, what is the minimum time interval needed before it can come to rest? (b) Can this jet land at a small tropical island airport where the runway is 0.800 km long? (c) Explain your answer.

25. M **Review.** Colonel John P. Stapp, USAF, participated in studying whether a jet pilot could survive emergency ejection. On March 19, 1954, he rode a rocket-propelled sled that moved down a track at a speed of 632 mi/h. He and the sled were safely brought to rest in 1.40 s (Fig. P2.25). Determine (a) the negative acceleration he experienced and (b) the distance he traveled during this negative acceleration.

left, Courtesy U.S. Air Force; *right*, Photri, Inc.

Figure P2.25 (*left*) Col. John Stapp and his rocket sled are brought to rest in a very short time interval. (*right*) Stapp's face is contorted by the stress of rapid negative acceleration.

26. Solve Example 2.8 by a graphical method. On the same graph, plot position versus time for the car and the trooper. From the intersection of the two curves, read the time at which the trooper overtakes the car.

27. GP A speedboat travels in a straight line and increases in speed uniformly from $v_i = 20.0$ m/s to $v_f = 30.0$ m/s in a displacement Δx of 200 m. We wish to find the time interval required for the boat to move through this displacement. (a) Draw a coordinate system for this situation. (b) What analysis model is most appropriate for describing this situation? (c) From the analysis model, what equation is most appropriate for finding the acceleration of the speedboat? (d) Solve the equation selected in part (c) symbolically for the boat's acceleration in terms of v_i, v_f, and Δx. (e) Substitute numerical values to obtain the acceleration numerically. (f) Find the time interval mentioned above.

28. *Why is the following situation impossible?* Starting from rest, a charging rhinoceros moves 50.0 m in a straight line in 10.0 s. Her acceleration is constant during the entire motion, and her final speed is 8.00 m/s.

29. The driver of a car slams on the brakes when he sees a tree blocking the road. The car slows uniformly with an acceleration of -5.60 m/s² for 4.20 s, making straight skid marks 62.4 m long, all the way to the tree. With what speed does the car then strike the tree?

30. S In the particle under constant acceleration model, we identify the variables and parameters v_{xi}, v_{xf}, a_x, t, and $x_f - x_i$. Of the equations in Table 2.2, the first does not involve $x_f - x_i$, the second does not contain a_x, the third omits v_{xf}, and the last leaves out t. So, to complete the set, there should be an equation *not* involving v_{xi}. (a) Derive it from the others. (b) Use the equation in part (a) to solve Problem 29 in one step.

31. **Q|C** A glider of length ℓ moves through a stationary photogate on an air track. A photogate (Fig. P2.31) is a device that measures the time interval Δt_d during which the glider blocks a beam of infrared light passing across the photogate. The ratio $v_d = \ell/\Delta t_d$ is the average velocity of the glider over this part of its motion. Suppose the glider moves with constant acceleration. (a) Argue for or against the idea that v_d is equal to the instantaneous velocity of the glider when it is halfway through the photogate in space. (b) Argue for or against the idea that v_d is equal to the instantaneous velocity of the glider when it is halfway through the photogate in time.

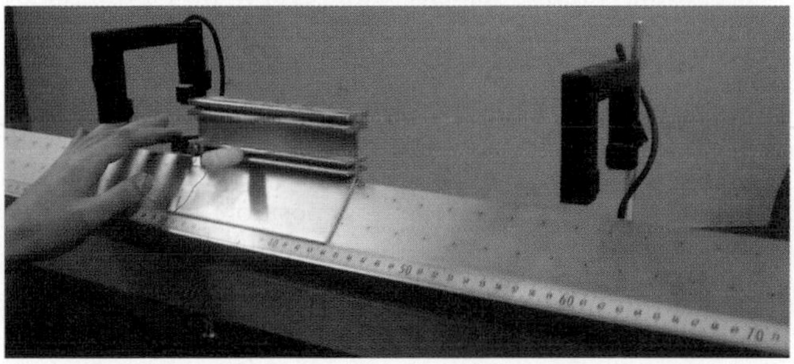

Ralph McGrew

Figure P2.31 Problems 31 and 34.

32. A particle moves along the x axis. Its position is given by the equation $x = 2 + 3t - 4t^2$, with x in meters and t in seconds. Determine (a) its position when it changes direction and (b) its velocity when it returns to the position it had at $t = 0$.

33. An object moves with constant acceleration 4.00 m/s^2 and over a time interval reaches a final velocity of 12.0 m/s. (a) If its initial velocity is 6.00 m/s, what is its displacement during the time interval? (b) What is the distance it travels during this interval? (c) If its initial velocity is -6.00 m/s, what is its displacement during the time interval? (d) What is the total distance it travels during the interval in part (c) ?

34. **Q|C** A glider of length 12.4 cm moves on an air track with constant acceleration (Fig P2.31). A time interval of 0.628 s elapses between the moment when its front end passes a fixed point Ⓐ along the track and the moment when its back end passes this point. Next, a time interval of 1.39 s elapses between the moment when the back end of the glider passes the point Ⓐ and the moment when the front end of the glider passes a second point Ⓑ farther down the track. After that, an additional 0.431 s elapses until the back end of the glider passes point Ⓑ. (a) Find the average speed of the glider as it passes point Ⓐ. (b) Find the acceleration of the glider. (c) Explain how you can compute the acceleration without knowing the distance between points Ⓐ and Ⓑ.

35. Figure P2.35 represents part of the performance data of a car owned by a proud physics student. (a) Calculate the total distance traveled by computing the area under the red-brown graph line. (b) What distance does the car travel between the times $t = 10$ s and $t = 40$ s? (c) Draw a graph of its acceleration versus time between $t = 0$ and $t = 50$ s. (d) Write an equation for x as a function of time for each phase of the motion, represented by the segments $0a$, ab, and bc. (e) What is the average velocity of the car between $t = 0$ and $t = 50$ s?

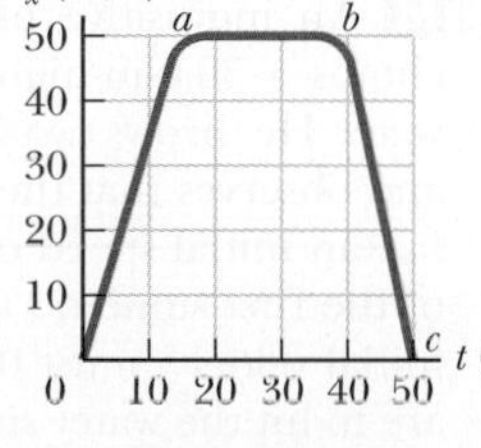

Figure P2.35

36. **Q|C** At $t = 0$, one toy car is set rolling on a straight track with initial position 15.0 cm, initial velocity -3.50 cm/s, and constant acceleration 2.40 cm/s^2. At the same moment, another toy car is set rolling on an adjacent track with initial position 10.0 cm, initial velocity $+5.50$ cm/s, and constant acceleration zero. (a) At what time, if any, do the two cars have equal speeds? (b) What are their speeds at that time? (c) At what time(s), if any, do the cars pass each other? (d) What are their locations at that time? (e) Explain the difference between question (a) and question (c) as clearly as possible.

Section 2.7 Freely Falling Objects

Note: In all problems in this section, ignore the effects of air resistance.

37. In Chapter 9, we will define the center of mass of an object and prove that its motion is described by the particle under constant acceleration model when constant forces act on the object. A gymnast jumps straight up, with her center of mass moving at 2.80 m/s as she leaves the ground. How high above this point is her center of mass (a) 0.100 s, (b) 0.200 s, (c) 0.300 s, and (d) 0.500 s thereafter?

38. **Q|C** An attacker at the base of a castle wall 3.65 m high throws a rock straight up with speed 7.40 m/s from a height of 1.55 m above the ground. (a) Will the rock reach the top of the wall? (b) If so, what is its speed at the top? If not, what initial speed must it have to reach the top? (c) Find the change in speed of a rock thrown straight down from the top of the wall at an initial speed of 7.40 m/s and moving between the same two points. (d) Does the change in speed of the downward-moving rock agree with the magnitude of the speed change of the rock moving upward between the same elevations? (e) Explain physically why it does or does not agree.

39. *Why is the following situation impossible?* Emily challenges her friend David to catch a \$1 bill as follows. She holds the bill vertically as shown in Figure P2.39, with the center of the bill between but not touching David's index finger and thumb. Without warning, Emily releases the bill. David catches the bill without moving his hand downward. David's reaction time is equal to the average human reaction time.

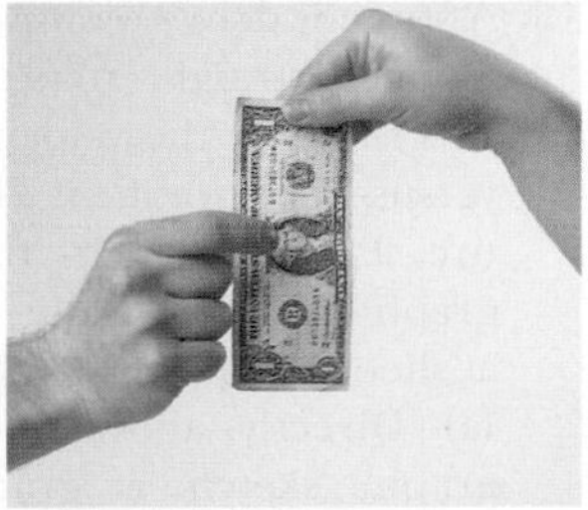

© Cengage Learning/George Semple

Figure P2.39

40. A baseball is hit so that it travels straight upward after being struck by the bat. A fan observes that it takes 3.00 s for the ball to reach its maximum height. Find (a) the ball's initial velocity and (b) the height it reaches.

41. A ball is thrown directly downward with an initial speed of 8.00 m/s from a height of 30.0 m. After what time interval does it strike the ground?

42. The height of a helicopter above the ground is given by $h = 3.00t^3$, where h is in meters and t is in seconds. At $t = 2.00$ s, the helicopter releases a small mailbag. How long after its release does the mailbag reach the ground?

43. M A student throws a set of keys vertically upward to her sorority sister, who is in a window 4.00 m above. The second student catches the keys 1.50 s later. (a) With what initial velocity were the keys thrown? (b) What was the velocity of the keys just before they were caught?

44. S At time $t = 0$, a student throws a set of keys vertically upward to her sorority sister, who is in a window at distance h above. The second student catches the keys at time t. (a) With what initial velocity were the keys thrown? (b) What was the velocity of the keys just before they were caught?

45. A daring ranch hand sitting on a tree limb wishes to drop vertically onto a horse galloping under the tree. The constant speed of the horse is 10.0 m/s, and the distance from the limb to the level of the saddle is 3.00 m. (a) What must be the horizontal distance between the saddle and limb when the ranch hand makes his move? (b) For what time interval is he in the air?

46. S A package is dropped at time $t = 0$ from a helicopter that is descending steadily at a speed v_i. (a) What is the speed of the package in terms of v_i, g, and t? (b) What vertical distance d is it from the helicopter in terms of g and t? (c) What are the answers to parts (a) and (b) if the helicopter is rising steadily at the same speed?

Section 2.8 Kinematic Equations Derived from Calculus

47. S Automotive engineers refer to the time rate of change of acceleration as the "jerk." Assume an object moves in one dimension such that its jerk J is constant. (a) Determine expressions for its acceleration $a_x(t)$, velocity $v_x(t)$, and position $x(t)$, given that its initial acceleration, velocity, and position are a_{xi}, v_{xi}, and x_i, respectively. (b) Show that $a_x^{\,2} = a_{xi}^{\,2} + 2J(v_x - v_{xi})$.

48. A student drives a moped along a straight road as described by the velocity-versus-time graph in Figure P2.48. Sketch this graph in the middle of a sheet of graph paper. (a) Directly above your graph, sketch a graph of the position versus time, aligning the time coordinates of the two graphs. (b) Sketch a graph of the acceleration versus time directly below the velocity-versus-time graph, again aligning the time coordinates. On each graph, show the numerical values of x and a_x for all points of inflection. (c) What is the acceleration at $t = 6.00$ s? (d) Find the position (relative to the starting point) at $t = 6.00$ s. (e) What is the moped's final position at $t = 9.00$ s?

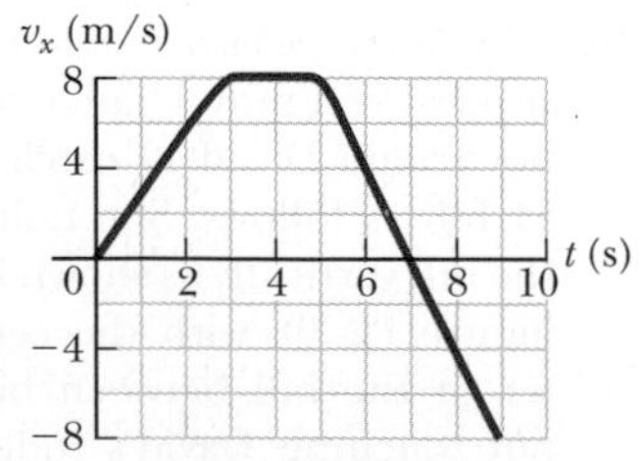

Figure P2.48

49. The speed of a bullet as it travels down the barrel of a rifle toward the opening is given by

$$v = (-5.00 \times 10^7)t^2 + (3.00 \times 10^5)t$$

where v is in meters per second and t is in seconds. The acceleration of the bullet just as it leaves the barrel is zero. (a) Determine the acceleration and position of the bullet as functions of time when the bullet is in the barrel. (b) Determine the time interval over which the bullet is accelerated. (c) Find the speed at which the bullet leaves the barrel. (d) What is the length of the barrel?

Additional Problems

50. An object is at $x = 0$ at $t = 0$ and moves along the x axis according to the velocity–time graph in Figure P2.50. (a) What is the object's acceleration between 0 and 4.0 s? (b) What is the object's acceleration between 4.0 s and 9.0 s? (c) What is the object's acceleration between 13.0 s and 18.0 s? (d) At what time(s) is the object moving with the lowest speed? (e) At what time is the object farthest from $x = 0$? (f) What is the final position x of the object at $t = 18.0$ s? (g) Through what total distance has the object moved between $t = 0$ and $t = 18.0$ s?

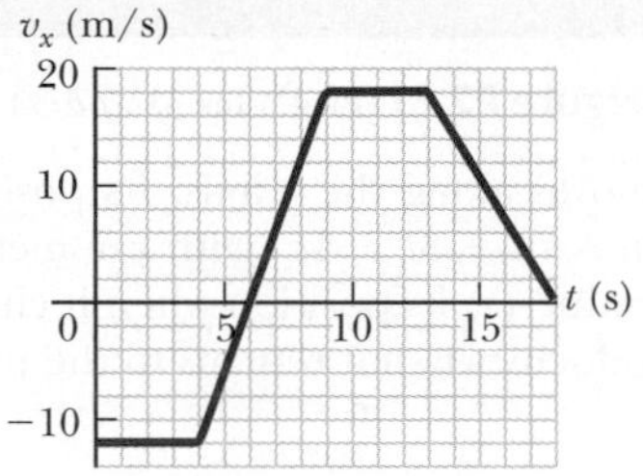

Figure P2.50

51. The froghopper *Philaenus spumarius* is supposedly the best jumper in the animal kingdom. To start a jump, this insect can accelerate at 4.00 km/s^2 over a distance of 2.00 mm as it straightens its specially adapted "jumping legs." Assume the acceleration is constant. (a) Find the upward velocity with which the insect takes off. (b) In what time interval does it reach this velocity? (c) How high would the insect jump if air resistance were negligible? The actual height it reaches is about 70 cm, so air resistance must be a noticeable force on the leaping froghopper.

52. QC S In Active Figure 2.11b, the area under the velocity-versus-time graph and between the vertical axis and time t (vertical dashed line) represents the displacement. As shown, this area consists of a rectangle and a triangle. (a) Compute their areas. (b) Explain how the sum of the two areas compares with the expression on the right-hand side of Equation 2.16.

53. M An inquisitive physics student and mountain climber climbs a 50.0-m-high cliff that overhangs a calm pool of water. He throws two stones vertically downward, 1.00 s apart, and observes that they cause a single splash. The first stone has an initial speed of 2.00 m/s. (a) How long after release of the first stone do the two stones hit the water? (b) What initial velocity must the second stone have if the two stones are to hit the water simultaneously? (c) What is the speed of each stone at the instant the two stones hit the water?

54. *Why is the following situation impossible?* A freight train is lumbering along at a constant speed of 16.0 m/s. Behind the freight train on the same track is a passenger train traveling in the same direction at 40.0 m/s. When the front of the passenger train is 58.5 m from the back of the freight train, the engineer on the passenger train recognizes the danger and hits the brakes of his train, causing the train to move with acceleration $-3.00\ \text{m/s}^2$. Because of the engineer's action, the trains do not collide.

55. **Q|C** An elevator moves downward in a tall building at a constant speed of 5.00 m/s. Exactly 5.00 s after the top of the elevator car passes a bolt loosely attached to the wall of the elevator shaft, the bolt falls from rest. (a) At what time does the bolt hit the top of the still-descending elevator? (b) In what way is this problem similar to Example 2.8? (c) Estimate the highest floor from which the bolt can fall if the elevator reaches the ground floor before the bolt hits the top of the elevator.

56. Two objects move with initial velocity -8.00 m/s, final velocity 16.0 m/s, and constant accelerations. (a) The first object has displacement 20.0 m. Find its acceleration. (b) The second object travels a total distance of 22.0 m. Find its acceleration.

57. **Q|C** The Acela is an electric train on the Washington–New York–Boston run, carrying passengers at 170 mi/h. A velocity–time graph for the Acela is shown in Figure P2.57. (a) Describe the train's motion in each successive time interval. (b) Find the train's peak positive acceleration in the motion graphed. (c) Find the train's displacement in miles between $t = 0$ and $t = 200$ s.

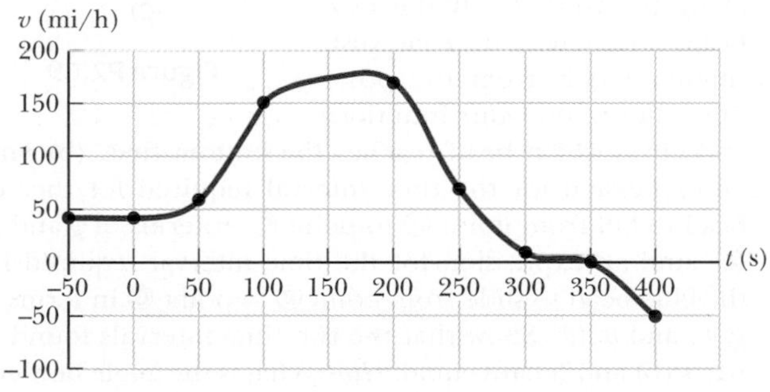

Figure P2.57 Velocity-versus-time graph for the Acela.

58. A catapult launches a test rocket vertically upward from a well, giving the rocket an initial speed of 80.0 m/s at ground level. The engines then fire, and the rocket accelerates upward at $4.00\ \text{m/s}^2$ until it reaches an altitude of 1 000 m. At that point, its engines fail and the rocket goes into free fall, with an acceleration of $-9.80\ \text{m/s}^2$. (a) For what time interval is the rocket in motion above the ground? (b) What is its maximum altitude? (c) What is its velocity just before it hits the ground? (You will need to consider the motion while the engine is operating and the free fall motion separately.)

59. **Q|C** **S** At $t = 0$, one athlete in a race running on a long, straight track with a constant speed v_1 is a distance d_1 behind a second athlete running with a constant speed v_2. (a) Under what circumstances is the first athlete able to overtake the second athlete? (b) Find the time t at which the first athlete overtakes the second athlete, in terms of d_1, v_1, and v_2. (c) At what minimum distance d_2 from the leading athlete must the finish line be located so that the trailing athlete can at least tie for first place? Express d_2 in terms of d_1, v_1, and v_2 by using the result of part (b).

60. **S** Two students are on a balcony a distance h above the street. One student throws a ball vertically downward at a speed v_i; at the same time, the other student throws a ball vertically upward at the same speed. Answer the following symbolically in terms of v_i, g, h, and t. (a) What is the time interval between when the first ball strikes the ground and the second ball strikes the ground? (b) Find the velocity of each ball as it strikes the ground. (c) How far apart are the balls at a time t after they are thrown and before they strike the ground?

61. **M** Kathy tests her new sports car by racing with Stan, an experienced racer. Both start from rest, but Kathy leaves the starting line 1.00 s after Stan does. Stan moves with a constant acceleration of $3.50\ \text{m/s}^2$, while Kathy maintains an acceleration of $4.90\ \text{m/s}^2$. Find (a) the time at which Kathy overtakes Stan, (b) the distance she travels before she catches him, and (c) the speeds of both cars at the instant Kathy overtakes Stan.

62. **Q|C** Astronauts on a distant planet toss a rock into the air. With the aid of a camera that takes pictures at a steady rate, they record the rock's height as a function of time as given in the following table. (a) Find the rock's average velocity in the time interval between each measurement and the next. (b) Using these average velocities to approximate instantaneous velocities at the midpoints of the time intervals, make a graph of velocity as a function of time. (c) Does the rock move with constant acceleration? If so, plot a straight line of best fit on the graph and calculate its slope to find the acceleration.

Time (s)	Height (m)	Time (s)	Height (m)
0.00	5.00	2.75	7.62
0.25	5.75	3.00	7.25
0.50	6.40	3.25	6.77
0.75	6.94	3.50	6.20
1.00	7.38	3.75	5.52
1.25	7.72	4.00	4.73
1.50	7.96	4.25	3.85
1.75	8.10	4.50	2.86
2.00	8.13	4.75	1.77
2.25	8.07	5.00	0.58
2.50	7.90		

63. **Q|C** **S** Two objects, A and B, are connected by hinges to a rigid rod that has a length L. The objects slide along perpendicular guide rails as shown in Figure P2.63. Assume object A slides to the left with a constant speed v. (a) Find the velocity v_B of object B as a function of the angle θ. (b) Describe v_B relative to v. Is v_B always smaller than v, larger than v, or the same as v, or does it have some other relationship?

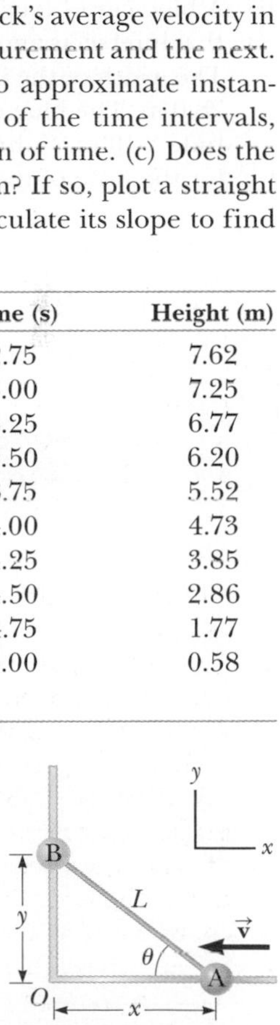

Figure P2.63

64. **Q|C** A hard rubber ball, released at chest height, falls to the pavement and bounces back to nearly the same height. When it is in contact with the pavement, the lower side of the ball is temporarily flattened. Suppose the maximum depth of the dent is on the order of 1 cm. Find the order of magnitude of the maximum acceleration of the ball while it is in contact with the pavement. State your assumptions, the quantities you estimate, and the values you estimate for them.

Challenge Problems

65. In a women's 100-m race, accelerating uniformly, Laura takes 2.00 s and Healan 3.00 s to attain their maximum speeds, which they each maintain for the rest of the race. They cross the finish line simultaneously, both setting a world record of 10.4 s. (a) What is the acceleration of each sprinter? (b) What are their respective maximum speeds? (c) Which sprinter is ahead at the 6.00-s mark, and by how much? (d) What is the maximum distance by which Healan is behind Laura, and at what time does that occur?

66. A man drops a rock into a well. (a) The man hears the sound of the splash 2.40 s after he releases the rock from rest. The speed of sound in air (at the ambient temperature) is 336 m/s. How far below the top of the well is the surface of the water? (b) **What If?** If the travel time for the sound is ignored, what percentage error is introduced when the depth of the well is calculated?

67. A blue car of length 4.52 m is moving north on a roadway that intersects another perpendicular roadway (Fig. P2.67). The width of the intersection from near edge to far edge is 28.0 m. The blue car has a constant acceleration of magnitude 2.10 m/s^2 directed south. The time interval required for the nose of the blue car to move from the near (south) edge of the intersection to the north edge of the intersection is 3.10 s. (a) How far is the nose of the blue car from the south edge of the intersection when it stops? (b) For what time interval is *any* part of the blue car within the boundaries of the intersection? (c) A red car is at rest on the perpendicular intersecting roadway. As the nose of the blue car enters the intersection, the red car starts from rest and accelerates east at 5.60 m/s^2. What is the minimum distance from the near (west) edge of the intersection at which the nose of the red car can begin its motion if it is to enter the intersection after the blue car has entirely left the intersection? (d) If the red car begins its motion at the position given by the answer to part (c), with what speed does it enter the intersection?

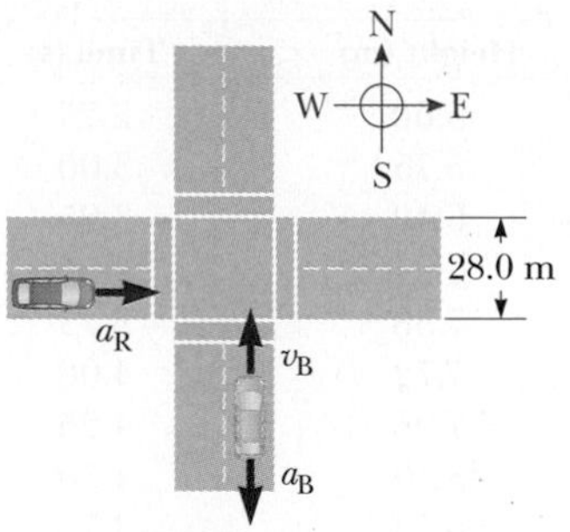

Figure P2.67

68. **Review.** As soon as a traffic light turns green, a car speeds up from rest to 50.0 mi/h with constant acceleration 9.00 mi/h/s. In the adjoining bicycle lane, a cyclist speeds up from rest to 20.0 mi/h with constant acceleration 13.0 mi/h/s. Each vehicle maintains constant velocity after reaching its cruising speed. (a) For what time interval is the bicycle ahead of the car? (b) By what maximum distance does the bicycle lead the car?

69. Two thin rods are fastened to the inside of a circular ring as shown in Figure P2.69. One rod of length D is vertical, and the other of length L makes an angle θ with the horizontal. The two rods and the ring lie in a vertical plane. Two small beads are free to slide without friction along the rods. (a) If the two beads are released from rest simultaneously from the positions shown, use your intuition and guess which bead reaches the bottom first. (b) Find an expression for the time interval required for the red bead to fall from point Ⓐ to point Ⓒ in terms of g and D. (c) Find an expression for the time interval required for the blue bead to slide from point Ⓑ to point Ⓒ in terms of g, L, and θ. (d) Show that the two time intervals found in parts (b) and (c) are equal. *Hint:* What is the angle between the chords of the circle Ⓐ Ⓑ and Ⓑ Ⓒ? (e) Do these results surprise you? Was your intuitive guess in part (a) correct? This problem was inspired by an article by Thomas B. Greenslade, Jr., "Galileo's Paradox," *Phys. Teach.* **46**, 294 (May 2008).

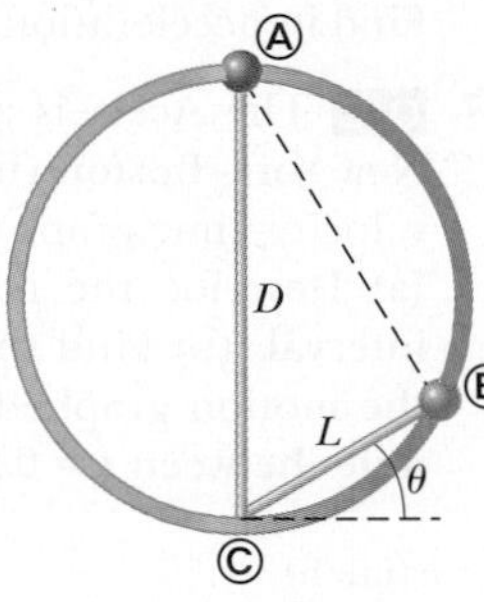

Figure P2.69

chapter 3

Vectors

A signpost in Westport, New Zealand shows the distance and direction to several cities. Quantities that are defined by both a magnitude and a direction are called *vector quantities.* (© Alan Chant)

In our study of physics, we often need to work with physical quantities that have both numerical and directional properties. As noted in Section 2.1, quantities of this nature are vector quantities. This chapter is primarily concerned with general properties of vector quantities. We discuss the addition and subtraction of vector quantities, together with some common applications to physical situations.

Vector quantities are used throughout this text. Therefore, it is imperative that you master the techniques discussed in this chapter.

3.1 Coordinate Systems

Many aspects of physics involve a description of a location in space. In Chapter 2, for example, we saw that the mathematical description of an object's motion requires a method for describing the object's position at various times. In two dimensions, this description is accomplished with the use of the Cartesian coordinate system, in which perpendicular axes intersect at a point defined as the origin (Fig. 3.1). Cartesian coordinates are also called *rectangular coordinates.*

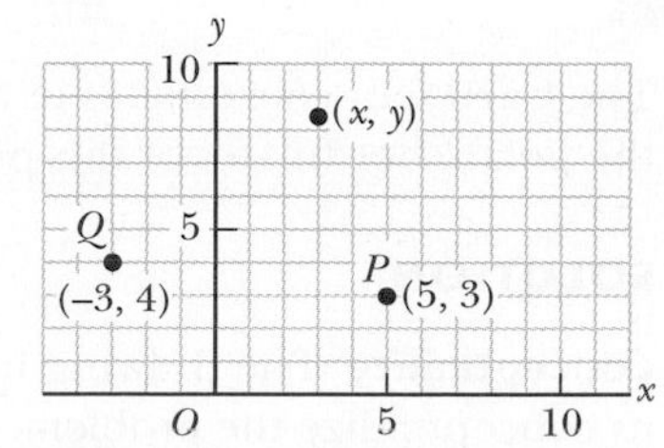

Figure 3.1 Designation of points in a Cartesian coordinate system. Every point is labeled with coordinates (x, y).

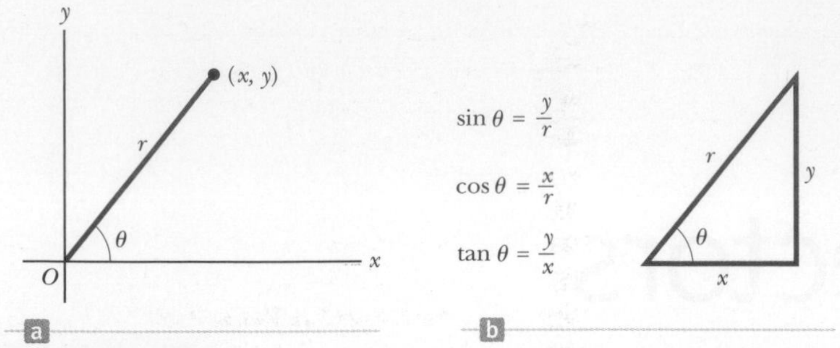

ACTIVE FIGURE 3.2

(a) The plane polar coordinates of a point are represented by the distance r and the angle θ, where θ is measured counterclockwise from the positive x axis. (b) The right triangle used to relate (x, y) to (r, θ).

Sometimes it is more convenient to represent a point in a plane by its *plane polar coordinates* (r, θ) as shown in Active Figure 3.2a. In this *polar coordinate system,* r is the distance from the origin to the point having Cartesian coordinates (x, y) and θ is the angle between a fixed axis and a line drawn from the origin to the point. The fixed axis is often the positive x axis, and θ is usually measured counterclockwise from it. From the right triangle in Active Figure 3.2b, we find that $\sin\theta = y/r$ and that $\cos\theta = x/r$. (A review of trigonometric functions is given in Appendix B.4.) Therefore, starting with the plane polar coordinates of any point, we can obtain the Cartesian coordinates by using the equations

$$x = r\cos\theta \qquad \textbf{(3.1)}$$

$$y = r\sin\theta \qquad \textbf{(3.2)}$$

Furthermore, if we know the Cartesian coordinates, the definitions of trigonometry tell us that

$$\tan\theta = \frac{y}{x} \qquad \textbf{(3.3)}$$

$$r = \sqrt{x^2 + y^2} \qquad \textbf{(3.4)}$$

Equation 3.4 is the familiar Pythagorean theorem.

These four expressions relating the coordinates (x, y) to the coordinates (r, θ) apply only when θ is defined as shown in Active Figure 3.2a—in other words, when positive θ is an angle measured counterclockwise from the positive x axis. (Some scientific calculators perform conversions between Cartesian and polar coordinates based on these standard conventions.) If the reference axis for the polar angle θ is chosen to be one other than the positive x axis or if the sense of increasing θ is chosen differently, the expressions relating the two sets of coordinates will change.

Example 3.1 Polar Coordinates

The Cartesian coordinates of a point in the xy plane are $(x, y) = (-3.50, -2.50)$ m as shown in Active Figure 3.3. Find the polar coordinates of this point.

SOLUTION

Conceptualize The drawing in Active Figure 3.3 helps us conceptualize the problem.

Categorize Based on the statement of the problem and the Conceptualize step, we recognize that we are simply converting from Cartesian coordinates to polar coordinates. We therefore categorize this example as a sub-

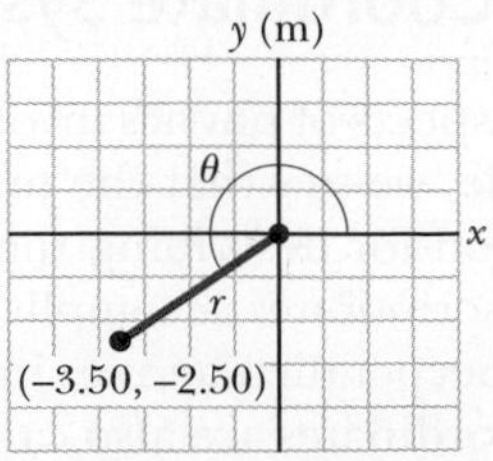

ACTIVE FIGURE 3.3

(Example 3.1) Finding polar coordinates when Cartesian coordinates are given.

stitution problem. Substitution problems generally do not have an extensive Analyze step other than the substitution of numbers into a given equation. Similarly, the Finalize step consists primarily of checking the units and making sure that the answer is reasonable. Therefore, for substitution problems, we will not label Analyze or Finalize steps.

Use Equation 3.4 to find r: $$r = \sqrt{x^2 + y^2} = \sqrt{(-3.50\text{ m})^2 + (-2.50\text{ m})^2} = \boxed{4.30\text{ m}}$$

Use Equation 3.3 to find θ: $$\tan\theta = \frac{y}{x} = \frac{-2.50\text{ m}}{-3.50\text{ m}} = 0.714$$

$$\theta = \boxed{216°}$$

Notice that you must use the signs of x and y to find that the point lies in the third quadrant of the coordinate system. That is, $\theta = 216°$, not 35.5°, whose tangent is also 0.714.

3.2 Vector and Scalar Quantities

We now formally describe the difference between scalar quantities and vector quantities. When you want to know the temperature outside so that you will know how to dress, the only information you need is a number and the unit "degrees C" or "degrees F." Temperature is therefore an example of a *scalar quantity:*

A **scalar quantity** is completely specified by a single value with an appropriate unit and has no direction.

Other examples of scalar quantities are volume, mass, speed, and time intervals. Some scalars are always positive, such as mass and speed. Others, such as temperature, can have either positive or negative values. The rules of ordinary arithmetic are used to manipulate scalar quantities.

If you are preparing to pilot a small plane and need to know the wind velocity, you must know both the speed of the wind and its direction. Because direction is important for its complete specification, velocity is a *vector quantity:*

A **vector quantity** is completely specified by a number with an appropriate unit plus a direction.

Another example of a vector quantity is displacement, as you know from Chapter 2. Suppose a particle moves from some point Ⓐ to some point Ⓑ along a straight path as shown in Figure 3.4. We represent this displacement by drawing an arrow from Ⓐ to Ⓑ, with the tip of the arrow pointing away from the starting point. The direction of the arrowhead represents the direction of the displacement, and the length of the arrow represents the magnitude of the displacement. If the particle travels along some other path from Ⓐ to Ⓑ such as shown by the broken line in Figure 3.4, its displacement is still the arrow drawn from Ⓐ to Ⓑ. Displacement depends only on the initial and final positions, so the displacement vector is independent of the path taken by the particle between these two points.

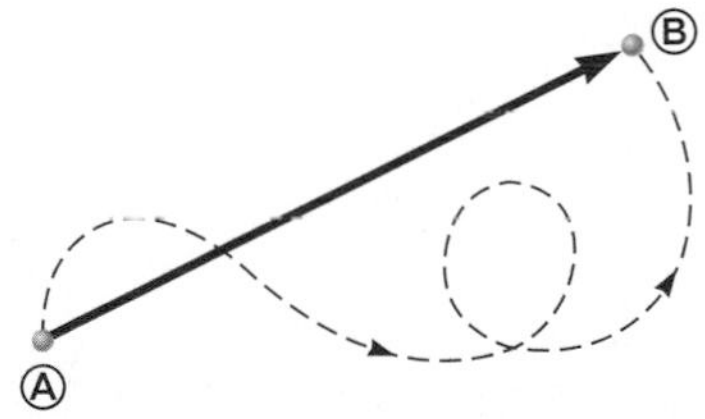

Figure 3.4 As a particle moves from Ⓐ to Ⓑ along an arbitrary path represented by the broken line, its displacement is a vector quantity shown by the arrow drawn from Ⓐ to Ⓑ.

In this text, we use a boldface letter with an arrow over the letter, such as $\vec{\mathbf{A}}$, to represent a vector. Another common notation for vectors with which you should be familiar is a simple boldface character: **A**. The magnitude of the vector $\vec{\mathbf{A}}$ is written either A or $|\vec{\mathbf{A}}|$. The magnitude of a vector has physical units, such as meters for displacement or meters per second for velocity. The magnitude of a vector is *always* a positive number.

***Quick Quiz* 3.1** Which of the following are vector quantities and which are scalar quantities? **(a)** your age **(b)** acceleration **(c)** velocity **(d)** speed **(e)** mass

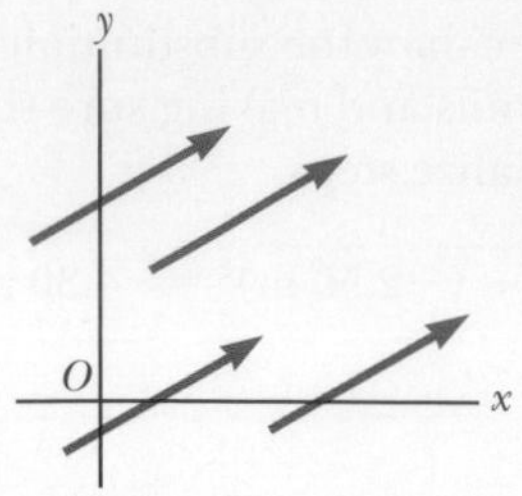

Figure 3.5 These four vectors are equal because they have equal lengths and point in the same direction.

3.3 Some Properties of Vectors

In this section, we shall investigate general properties of vectors representing physical quantities. We also discuss how to add and subtract vectors using both algebraic and geometric methods.

Equality of Two Vectors

For many purposes, two vectors $\vec{\mathbf{A}}$ and $\vec{\mathbf{B}}$ may be defined to be equal if they have the same magnitude and if they point in the same direction. That is, $\vec{\mathbf{A}} = \vec{\mathbf{B}}$ only if $A = B$ and if $\vec{\mathbf{A}}$ and $\vec{\mathbf{B}}$ point in the same direction along parallel lines. For example, all the vectors in Figure 3.5 are equal even though they have different starting points. This property allows us to move a vector to a position parallel to itself in a diagram without affecting the vector.

Adding Vectors

Pitfall Prevention 3.1

Vector Addition Versus Scalar Addition

Notice that $\vec{\mathbf{A}} + \vec{\mathbf{B}} = \vec{\mathbf{C}}$ is very different from $A + B = C$. The first equation is a vector sum, which must be handled carefully, such as with the graphical method. The second equation is a simple algebraic addition of numbers that is handled with the normal rules of arithmetic.

The rules for adding vectors are conveniently described by a graphical method. To add vector $\vec{\mathbf{B}}$ to vector $\vec{\mathbf{A}}$, first draw vector $\vec{\mathbf{A}}$ on graph paper, with its magnitude represented by a convenient length scale, and then draw vector $\vec{\mathbf{B}}$ to the same scale, with its tail starting from the tip of $\vec{\mathbf{A}}$, as shown in Active Figure 3.6. The **resultant vector** $\vec{\mathbf{R}} = \vec{\mathbf{A}} + \vec{\mathbf{B}}$ is the vector drawn from the tail of $\vec{\mathbf{A}}$ to the tip of $\vec{\mathbf{B}}$.

A geometric construction can also be used to add more than two vectors as shown in Figure 3.7 for the case of four vectors. The resultant vector $\vec{\mathbf{R}} = \vec{\mathbf{A}} + \vec{\mathbf{B}} + \vec{\mathbf{C}} + \vec{\mathbf{D}}$ is the vector that completes the polygon. In other words, $\vec{\mathbf{R}}$ is the vector drawn from the tail of the first vector to the tip of the last vector. This technique for adding vectors is often called the "head to tail method."

When two vectors are added, the sum is independent of the order of the addition. (This fact may seem trivial, but as you will see in Chapter 11, the order is important when vectors are multiplied. Procedures for multiplying vectors are discussed in Chapters 7 and 11.) This property, which can be seen from the geometric construction in Figure 3.8, is known as the **commutative law of addition:**

Commutative law of addition ▶

$$\vec{\mathbf{A}} + \vec{\mathbf{B}} = \vec{\mathbf{B}} + \vec{\mathbf{A}} \tag{3.5}$$

When three or more vectors are added, their sum is independent of the way in which the individual vectors are grouped together. A geometric proof of this rule

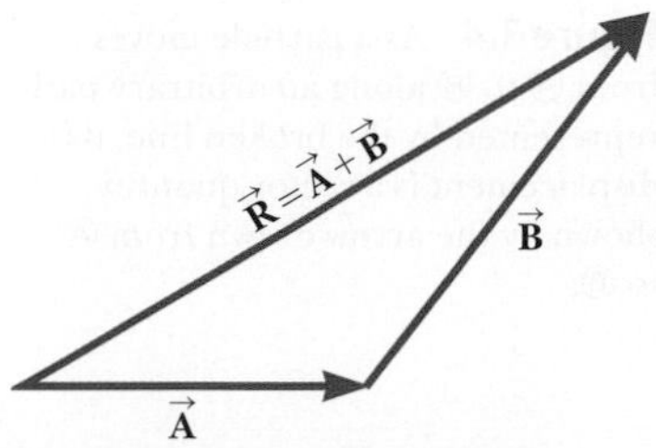

ACTIVE FIGURE 3.6

When vector $\vec{\mathbf{B}}$ is added to vector $\vec{\mathbf{A}}$, the resultant $\vec{\mathbf{R}}$ is the vector that runs from the tail of $\vec{\mathbf{A}}$ to the tip of $\vec{\mathbf{B}}$.

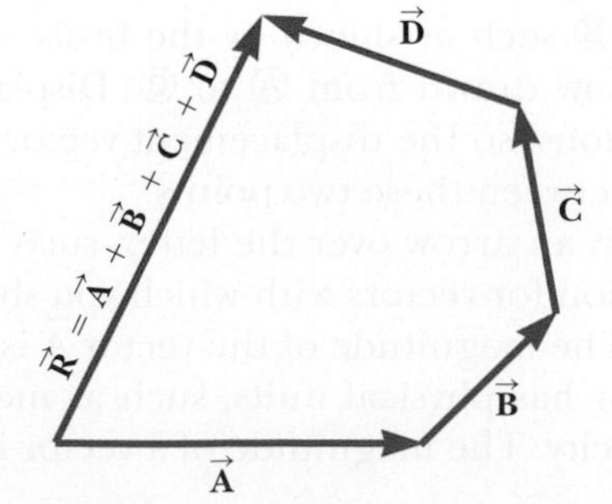

Figure 3.7 Geometric construction for summing four vectors. The resultant vector $\vec{\mathbf{R}}$ is by definition the one that completes the polygon.

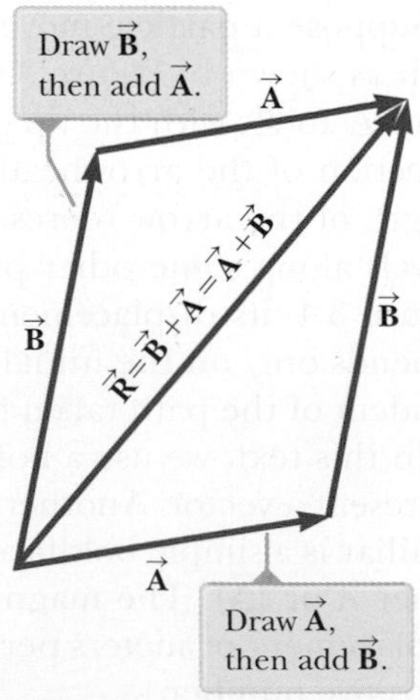

Figure 3.8 This construction shows that $\vec{\mathbf{A}} + \vec{\mathbf{B}} = \vec{\mathbf{B}} + \vec{\mathbf{A}}$ or, in other words, that vector addition is commutative.

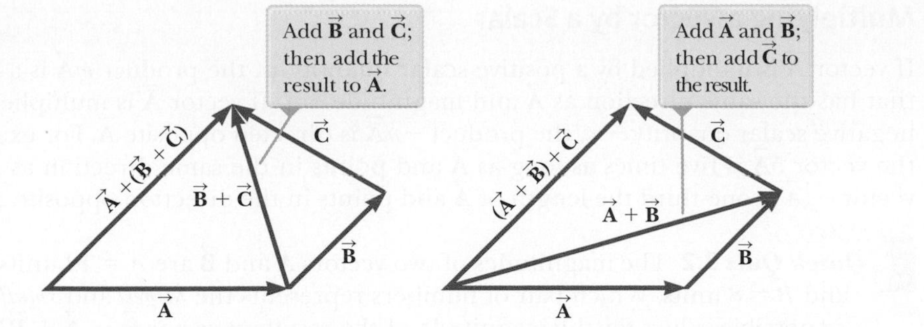

Figure 3.9 Geometric constructions for verifying the associative law of addition.

for three vectors is given in Figure 3.9. This property is called the **associative law of addition:**

$$\vec{A} + (\vec{B} + \vec{C}) = (\vec{A} + \vec{B}) + \vec{C} \quad \textbf{(3.6)}$$

◀ **Associative law of addition**

In summary, **a vector quantity has both magnitude and direction and also obeys the laws of vector addition** as described in Active Figure 3.6 and Figures 3.7 to 3.9. When two or more vectors are added together, they must all have the same units and they must all be the same type of quantity. It would be meaningless to add a velocity vector (for example, 60 km/h to the east) to a displacement vector (for example, 200 km to the north) because these vectors represent different physical quantities. The same rule also applies to scalars. For example, it would be meaningless to add time intervals to temperatures.

Negative of a Vector

The negative of the vector $\vec{A}$ is defined as the vector that when added to $\vec{A}$ gives zero for the vector sum. That is, $\vec{A} + (-\vec{A}) = 0$. The vectors $\vec{A}$ and $-\vec{A}$ have the same magnitude but point in opposite directions.

Subtracting Vectors

The operation of vector subtraction makes use of the definition of the negative of a vector. We define the operation $\vec{A} - \vec{B}$ as vector $-\vec{B}$ added to vector $\vec{A}$:

$$\vec{A} - \vec{B} = \vec{A} + (-\vec{B}) \quad \textbf{(3.7)}$$

The geometric construction for subtracting two vectors in this way is illustrated in Figure 3.10a.

Another way of looking at vector subtraction is to notice that the difference $\vec{A} - \vec{B}$ between two vectors $\vec{A}$ and $\vec{B}$ is what you have to add to the second vector to obtain the first. In this case, as Figure 3.10b shows, the vector $\vec{A} - \vec{B}$ points from the tip of the second vector to the tip of the first.

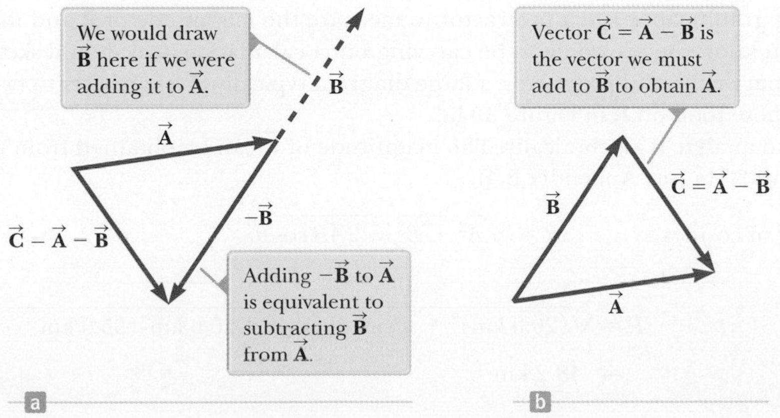

Figure 3.10 (a) Subtracting vector $\vec{B}$ from vector $\vec{A}$. The vector $-\vec{B}$ is equal in magnitude to vector $\vec{B}$ and points in the opposite direction. (b) A second way of looking at vector subtraction.

Multiplying a Vector by a Scalar

If vector $\vec{\mathbf{A}}$ is multiplied by a positive scalar quantity m, the product $m\vec{\mathbf{A}}$ is a vector that has the same direction as $\vec{\mathbf{A}}$ and magnitude mA. If vector $\vec{\mathbf{A}}$ is multiplied by a negative scalar quantity $-m$, the product $-m\vec{\mathbf{A}}$ is directed opposite $\vec{\mathbf{A}}$. For example, the vector $5\vec{\mathbf{A}}$ is five times as long as $\vec{\mathbf{A}}$ and points in the same direction as $\vec{\mathbf{A}}$; the vector $-\frac{1}{3}\vec{\mathbf{A}}$ is one-third the length of $\vec{\mathbf{A}}$ and points in the direction opposite $\vec{\mathbf{A}}$.

***Quick Quiz* 3.2** The magnitudes of two vectors $\vec{\mathbf{A}}$ and $\vec{\mathbf{B}}$ are $A = 12$ units and $B = 8$ units. Which pair of numbers represents the *largest* and *smallest* possible values for the magnitude of the resultant vector $\vec{\mathbf{R}} = \vec{\mathbf{A}} + \vec{\mathbf{B}}$? **(a)** 14.4 units, 4 units **(b)** 12 units, 8 units **(c)** 20 units, 4 units **(d)** none of these answers

***Quick Quiz* 3.3** If vector $\vec{\mathbf{B}}$ is added to vector $\vec{\mathbf{A}}$, which *two* of the following choices must be true for the resultant vector to be equal to zero? **(a)** $\vec{\mathbf{A}}$ and $\vec{\mathbf{B}}$ are parallel and in the same direction. **(b)** $\vec{\mathbf{A}}$ and $\vec{\mathbf{B}}$ are parallel and in opposite directions. **(c)** $\vec{\mathbf{A}}$ and $\vec{\mathbf{B}}$ have the same magnitude. **(d)** $\vec{\mathbf{A}}$ and $\vec{\mathbf{B}}$ are perpendicular.

Example 3.2 A Vacation Trip

A car travels 20.0 km due north and then 35.0 km in a direction 60.0° west of north as shown in Figure 3.11a. Find the magnitude and direction of the car's resultant displacement.

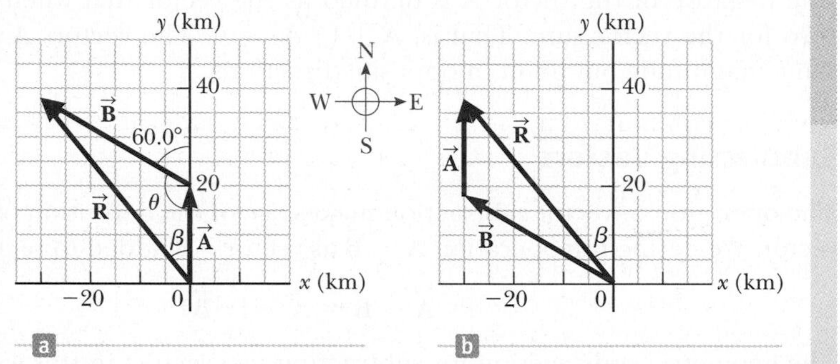

Figure 3.11 (Example 3.2) (a) Graphical method for finding the resultant displacement vector $\vec{\mathbf{R}} = \vec{\mathbf{A}} + \vec{\mathbf{B}}$. (b) Adding the vectors in reverse order $(\vec{\mathbf{B}} + \vec{\mathbf{A}})$ gives the same result for $\vec{\mathbf{R}}$.

SOLUTION

Conceptualize The vectors $\vec{\mathbf{A}}$ and $\vec{\mathbf{B}}$ drawn in Figure 3.11a help us conceptualize the problem.

Categorize We can categorize this example as a simple analysis problem in vector addition. The displacement $\vec{\mathbf{R}}$ is the resultant when the two individual displacements $\vec{\mathbf{A}}$ and $\vec{\mathbf{B}}$ are added. We can further categorize it as a problem about the analysis of triangles, so we appeal to our expertise in geometry and trigonometry.

Analyze In this example, we show two ways to analyze the problem of finding the resultant of two vectors. The first way is to solve the problem geometrically, using graph paper and a protractor to measure the magnitude of $\vec{\mathbf{R}}$ and its direction in Figure 3.11a. (In fact, even when you know you are going to be carrying out a calculation, you should sketch the vectors to check your results.) With an ordinary ruler and protractor, a large diagram typically gives answers to two-digit but not to three-digit precision. Try using these tools on $\vec{\mathbf{R}}$ in Figure 3.11a!

The second way to solve the problem is to analyze it algebraically. The magnitude of $\vec{\mathbf{R}}$ can be obtained from the law of cosines as applied to the triangle in Figure 3.11a (see Appendix B.4).

Use $R^2 = A^2 + B^2 - 2AB\cos\theta$ from the law of cosines to find R:

$$R = \sqrt{A^2 + B^2 - 2AB\cos\theta}$$

Substitute numerical values, noting that $\theta = 180° - 60° = 120°$:

$$R = \sqrt{(20.0\text{ km})^2 + (35.0\text{ km})^2 - 2(20.0\text{ km})(35.0\text{ km})\cos 120°}$$

$$= \boxed{48.2\text{ km}}$$

Use the law of sines (Appendix B.4) to find the direction of $\vec{\mathbf{R}}$ measured from the northerly direction:

$$\frac{\sin\beta}{B} = \frac{\sin\theta}{R}$$

$$\sin\beta = \frac{B}{R}\sin\theta = \frac{35.0\text{ km}}{48.2\text{ km}}\sin 120° = 0.629$$

$$\beta = \boxed{38.9°}$$

The resultant displacement of the car is 48.2 km in a direction 38.9° west of north.

Finalize Does the angle β that we calculated agree with an estimate made by looking at Figure 3.11a or with an actual angle measured from the diagram using the graphical method? Is it reasonable that the magnitude of $\vec{\mathbf{R}}$ is larger than that of both $\vec{\mathbf{A}}$ and $\vec{\mathbf{B}}$? Are the units of $\vec{\mathbf{R}}$ correct?

Although the head to tail method of adding vectors works well, it suffers from two disadvantages. First, some people find using the laws of cosines and sines to be awkward. Second, a triangle only results if you are adding two vectors. If you are adding three or more vectors, the resulting geometric shape is usually not a triangle. In Section 3.4, we explore a new method of adding vectors that will address both of these disadvantages.

WHAT IF? Suppose the trip were taken with the two vectors in reverse order: 35.0 km at 60.0° west of north first and then 20.0 km due north. How would the magnitude and the direction of the resultant vector change?

Answer They would not change. The commutative law for vector addition tells us that the order of vectors in an addition is irrelevant. Graphically, Figure 3.11b shows that the vectors added in the reverse order give us the same resultant vector.

3.4 Components of a Vector and Unit Vectors

The graphical method of adding vectors is not recommended whenever high accuracy is required or in three-dimensional problems. In this section, we describe a method of adding vectors that makes use of the projections of vectors along coordinate axes. These projections are called the **components** of the vector or its **rectangular components.** Any vector can be completely described by its components.

Consider a vector $\vec{\mathbf{A}}$ lying in the xy plane and making an arbitrary angle θ with the positive x axis as shown in Figure 3.12a. This vector can be expressed as the sum of two other *component vectors* $\vec{\mathbf{A}}_x$, which is parallel to the x axis, and $\vec{\mathbf{A}}_y$, which is parallel to the y axis. From Figure 3.12b, we see that the three vectors form a right triangle and that $\vec{\mathbf{A}} = \vec{\mathbf{A}}_x + \vec{\mathbf{A}}_y$. We shall often refer to the "components of a vector $\vec{\mathbf{A}}$," written A_x and A_y (without the boldface notation). The component A_x represents the projection of $\vec{\mathbf{A}}$ along the x axis, and the component A_y represents the projection of $\vec{\mathbf{A}}$ along the y axis. These components can be positive or negative. The component A_x is positive if the component vector $\vec{\mathbf{A}}_x$ points in the positive x direction and is negative if $\vec{\mathbf{A}}_x$ points in the negative x direction. A similar statement is made for the component A_y.

Pitfall Prevention 3.2

***x* and *y* Components**

Equations 3.8 and 3.9 associate the cosine of the angle with the x component and the sine of the angle with the y component. This association is true *only* because we measured the angle θ with respect to the x axis, so do not memorize these equations. If θ is measured with respect to the y axis (as in some problems), these equations will be incorrect. Think about which side of the triangle containing the components is adjacent to the angle and which side is opposite and then assign the cosine and sine accordingly.

From Figure 3.12 and the definition of sine and cosine, we see that $\cos\theta = A_x/A$ and that $\sin\theta = A_y/A$. Hence, the components of $\vec{\mathbf{A}}$ are

◀ **Components of the vector $\vec{\mathbf{A}}$**

$$A_x = A\cos\theta \quad (3.8)$$

$$A_y = A\sin\theta \quad (3.9)$$

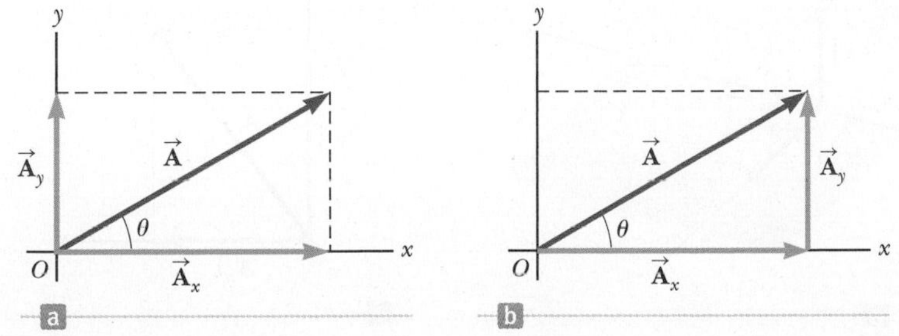

Figure 3.12 (a) A vector $\vec{\mathbf{A}}$ lying in the xy plane can be represented by its component vectors $\vec{\mathbf{A}}_x$ and $\vec{\mathbf{A}}_y$. (b) The y component vector $\vec{\mathbf{A}}_y$ can be moved to the right so that it adds to $\vec{\mathbf{A}}_x$. The vector sum of the component vectors is $\vec{\mathbf{A}}$. These three vectors form a right triangle.

The magnitudes of these components are the lengths of the two sides of a right triangle with a hypotenuse of length A. Therefore, the magnitude and direction of $\vec{\mathbf{A}}$ are related to its components through the expressions

$$A = \sqrt{A_x^{\,2} + A_y^{\,2}} \tag{3.10}$$

$$\theta = \tan^{-1}\left(\frac{A_y}{A_x}\right) \tag{3.11}$$

Notice that the signs of the components A_x and A_y depend on the angle θ. For example, if $\theta = 120°$, A_x is negative and A_y is positive. If $\theta = 225°$, both A_x and A_y are negative. Figure 3.13 summarizes the signs of the components when $\vec{\mathbf{A}}$ lies in the various quadrants.

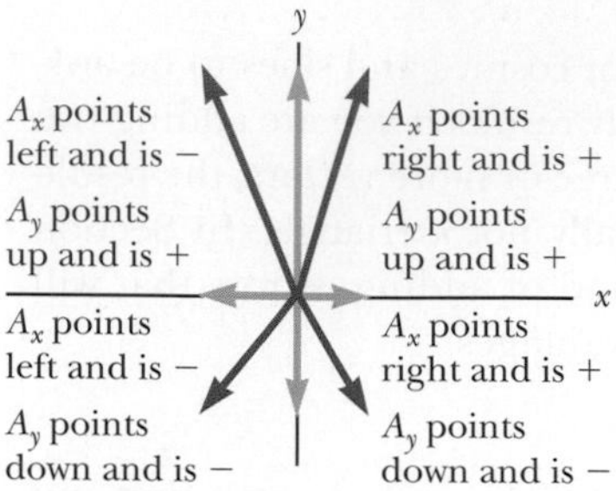

Figure 3.13 The signs of the components of a vector $\vec{\mathbf{A}}$ depend on the quadrant in which the vector is located.

When solving problems, you can specify a vector $\vec{\mathbf{A}}$ either with its components A_x and A_y or with its magnitude and direction A and θ.

Suppose you are working a physics problem that requires resolving a vector into its components. In many applications, it is convenient to express the components in a coordinate system having axes that are not horizontal and vertical but that are still perpendicular to each other. For example, we will consider the motion of objects sliding down inclined planes. For these examples, it is often convenient to orient the x axis parallel to the plane and the y axis perpendicular to the plane.

***Quick Quiz* 3.4** Choose the correct response to make the sentence true: A component of a vector is **(a)** always, **(b)** never, or **(c)** sometimes larger than the magnitude of the vector.

Unit Vectors

Vector quantities often are expressed in terms of unit vectors. A **unit vector** is a dimensionless vector having a magnitude of exactly 1. Unit vectors are used to specify a given direction and have no other physical significance. They are used solely as a bookkeeping convenience in describing a direction in space. We shall use the symbols $\hat{\mathbf{i}}$, $\hat{\mathbf{j}}$, and $\hat{\mathbf{k}}$ to represent unit vectors pointing in the positive x, y, and z directions, respectively. (The "hats," or circumflexes, on the symbols are a standard notation for unit vectors.) The unit vectors $\hat{\mathbf{i}}$, $\hat{\mathbf{j}}$, and $\hat{\mathbf{k}}$ form a set of mutually perpendicular vectors in a right-handed coordinate system as shown in Active Figure 3.14a. The magnitude of each unit vector equals 1; that is, $|\hat{\mathbf{i}}| = |\hat{\mathbf{j}}| = |\hat{\mathbf{k}}| = 1$.

Consider a vector $\vec{\mathbf{A}}$ lying in the xy plane as shown in Active Figure 3.14b. The product of the component A_x and the unit vector $\hat{\mathbf{i}}$ is the component vector $\vec{\mathbf{A}}_x = A_x\hat{\mathbf{i}}$, which lies on the x axis and has magnitude $|A_x|$. Likewise, $\vec{\mathbf{A}}_y = A_y\hat{\mathbf{j}}$ is the component vector of magnitude $|A_y|$ lying on the y axis. Therefore, the unit-vector notation for the vector $\vec{\mathbf{A}}$ is

$$\vec{\mathbf{A}} = A_x\hat{\mathbf{i}} + A_y\hat{\mathbf{j}} \tag{3.12}$$

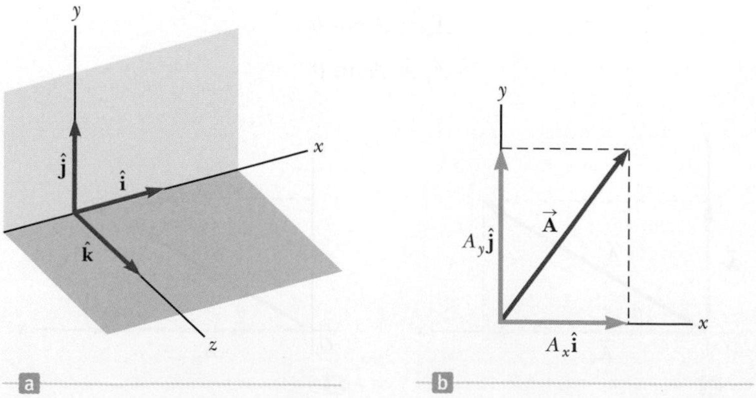

ACTIVE FIGURE 3.14

(a) The unit vectors $\hat{\mathbf{i}}$, $\hat{\mathbf{j}}$, and $\hat{\mathbf{k}}$ are directed along the x, y, and z axes, respectively. (b) Vector $\vec{\mathbf{A}} = A_x\hat{\mathbf{i}} + A_y\hat{\mathbf{j}}$ lying in the xy plane has components A_x and A_y.

For example, consider a point lying in the xy plane and having Cartesian coordinates (x, y) as in Figure 3.15. The point can be specified by the **position vector** $\vec{\mathbf{r}}$, which in unit-vector form is given by

$$\vec{\mathbf{r}} = x\hat{\mathbf{i}} + y\hat{\mathbf{j}} \tag{3.13}$$

This notation tells us that the components of $\vec{\mathbf{r}}$ are the coordinates x and y.

Now let us see how to use components to add vectors when the graphical method is not sufficiently accurate. Suppose we wish to add vector $\vec{\mathbf{B}}$ to vector $\vec{\mathbf{A}}$ in Equation 3.12, where vector $\vec{\mathbf{B}}$ has components B_x and B_y. Because of the bookkeeping convenience of the unit vectors, all we do is add the x and y components separately. The resultant vector $\vec{\mathbf{R}} = \vec{\mathbf{A}} + \vec{\mathbf{B}}$ is

$$\vec{\mathbf{R}} = (A_x\hat{\mathbf{i}} + A_y\hat{\mathbf{j}}) + (B_x\hat{\mathbf{i}} + B_y\hat{\mathbf{j}})$$

or

$$\vec{\mathbf{R}} = (A_x + B_x)\hat{\mathbf{i}} + (A_y + B_y)\hat{\mathbf{j}} \tag{3.14}$$

Because $\vec{\mathbf{R}} = R_x\hat{\mathbf{i}} + R_y\hat{\mathbf{j}}$, we see that the components of the resultant vector are

$$R_x = A_x + B_x$$
$$R_y = A_y + B_y \tag{3.15}$$

Therefore, we see that in the component method of adding vectors, we add all the x components together to find the x component of the resultant vector and use the same process for the y components. We can check this addition by components with a geometric construction as shown in Figure 3.16.

The magnitude of $\vec{\mathbf{R}}$ and the angle it makes with the x axis are obtained from its components using the relationships

$$R = \sqrt{R_x^{\,2} + R_y^{\,2}} = \sqrt{(A_x + B_x)^2 + (A_y + B_y)^2} \tag{3.16}$$

$$\tan\theta = \frac{R_y}{R_x} = \frac{A_y + B_y}{A_x + B_x} \tag{3.17}$$

At times, we need to consider situations involving motion in three component directions. The extension of our methods to three-dimensional vectors is straightforward. If $\vec{\mathbf{A}}$ and $\vec{\mathbf{B}}$ both have x, y, and z components, they can be expressed in the form

$$\vec{\mathbf{A}} = A_x\hat{\mathbf{i}} + A_y\hat{\mathbf{j}} + A_z\hat{\mathbf{k}} \tag{3.18}$$

$$\vec{\mathbf{B}} = B_x\hat{\mathbf{i}} + B_y\hat{\mathbf{j}} + B_z\hat{\mathbf{k}} \tag{3.19}$$

The sum of $\vec{\mathbf{A}}$ and $\vec{\mathbf{B}}$ is

$$\vec{\mathbf{R}} = (A_x + B_x)\hat{\mathbf{i}} + (A_y + B_y)\hat{\mathbf{j}} + (A_z + B_z)\hat{\mathbf{k}} \tag{3.20}$$

Notice that Equation 3.20 differs from Equation 3.14: in Equation 3.20, the resultant vector also has a z component $R_z = A_z + B_z$. If a vector $\vec{\mathbf{R}}$ has x, y, and z components, the magnitude of the vector is $R = \sqrt{R_x^{\,2} + R_y^{\,2} + R_z^{\,2}}$. The angle θ_x that $\vec{\mathbf{R}}$ makes with the x axis is found from the expression $\cos\theta_x = R_x/R$, with similar expressions for the angles with respect to the y and z axes.

The extension of our method to adding more than two vectors is also straightforward. For example, $\vec{\mathbf{A}} + \vec{\mathbf{B}} + \vec{\mathbf{C}} = (A_x + B_x + C_x)\hat{\mathbf{i}} + (A_y + B_y + C_y)\hat{\mathbf{j}} + (A_z + B_z + C_z)\hat{\mathbf{k}}$. We have described adding displacement vectors in this section because these types of vectors are easy to visualize. We can also add other types of vectors, such as velocity, force, and electric field vectors, which we will do in later chapters.

***Quick Quiz* 3.5** For which of the following vectors is the magnitude of the vector equal to one of the components of the vector? **(a)** $\vec{\mathbf{A}} = 2\hat{\mathbf{i}} + 5\hat{\mathbf{j}}$ **(b)** $\vec{\mathbf{B}} = -3\hat{\mathbf{j}}$ **(c)** $\vec{\mathbf{C}} = +5\hat{\mathbf{k}}$

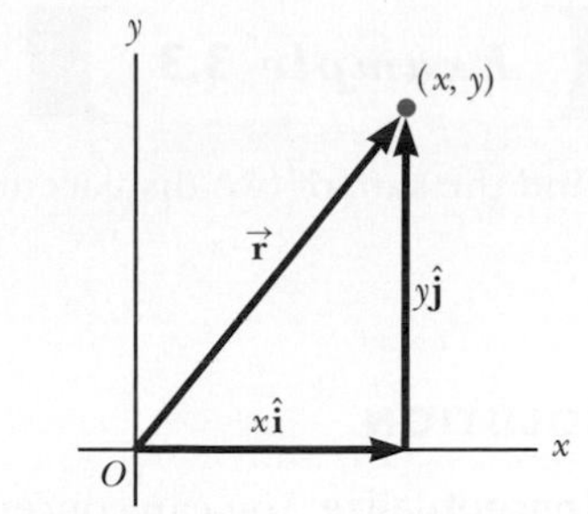

Figure 3.15 The point whose Cartesian coordinates are (x, y) can be represented by the position vector $\vec{\mathbf{r}} = x\hat{\mathbf{i}} + y\hat{\mathbf{j}}$.

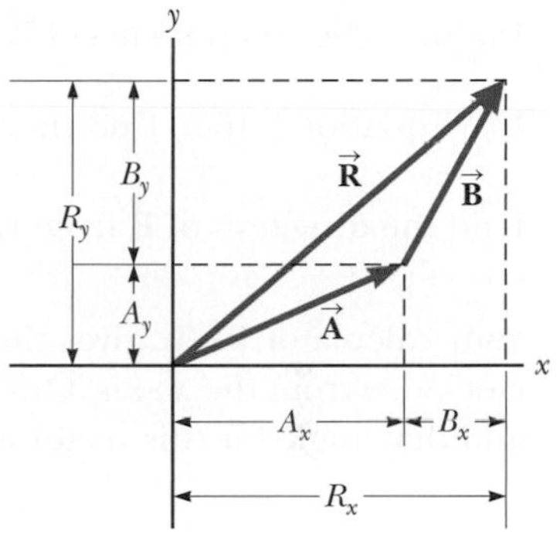

Figure 3.16 This geometric construction for the sum of two vectors shows the relationship between the components of the resultant $\vec{\mathbf{R}}$ and the components of the individual vectors.

Pitfall Prevention 3.3

Tangents on Calculators

Equation 3.17 involves the calculation of an angle by means of a tangent function. Generally, the inverse tangent function on calculators provides an angle between −90° and +90°. As a consequence, if the vector you are studying lies in the second or third quadrant, the angle measured from the positive x axis will be the angle your calculator returns plus 180°.

Example 3.3 The Sum of Two Vectors

Find the sum of two displacement vectors $\vec{\mathbf{A}}$ and $\vec{\mathbf{B}}$ lying in the xy plane and given by

$$\vec{\mathbf{A}} = (2.0\hat{\mathbf{i}} + 2.0\hat{\mathbf{j}})\text{ m} \quad \text{and} \quad \vec{\mathbf{B}} = (2.0\hat{\mathbf{i}} - 4.0\hat{\mathbf{j}})\text{ m}$$

SOLUTION

Conceptualize You can conceptualize the situation by drawing the vectors on graph paper.

Categorize We categorize this example as a simple substitution problem. Comparing this expression for $\vec{\mathbf{A}}$ with the general expression $\vec{\mathbf{A}} = A_x\hat{\mathbf{i}} + A_y\hat{\mathbf{j}} + A_z\hat{\mathbf{k}}$, we see that $A_x = 2.0$ m, $A_y = 2.0$ m, and $A_z = 0$. Likewise, $B_x = 2.0$ m, $B_y = -4.0$ m, and $B_z = 0$. We can use a two-dimensional approach because there are no z components.

Use Equation 3.14 to obtain the resultant vector $\vec{\mathbf{R}}$:

$$\vec{\mathbf{R}} = \vec{\mathbf{A}} + \vec{\mathbf{B}} = (2.0 + 2.0)\hat{\mathbf{i}}\text{ m} + (2.0 - 4.0)\hat{\mathbf{j}}\text{ m}$$

Evaluate the components of $\vec{\mathbf{R}}$:

$$R_x = 4.0\text{ m} \qquad R_y = -2.0\text{ m}$$

Use Equation 3.16 to find the magnitude of $\vec{\mathbf{R}}$:

$$R = \sqrt{R_x^2 + R_y^2} = \sqrt{(4.0\text{ m})^2 + (-2.0\text{ m})^2} = \sqrt{20}\text{ m} = \boxed{4.5\text{ m}}$$

Find the direction of $\vec{\mathbf{R}}$ from Equation 3.17:

$$\tan\theta = \frac{R_y}{R_x} = \frac{-2.0\text{ m}}{4.0\text{ m}} = -0.50$$

Your calculator likely gives the answer $-27°$ for $\theta = \tan^{-1}(-0.50)$. This answer is correct if we interpret it to mean 27° clockwise from the x axis. Our standard form has been to quote the angles measured counterclockwise from the $+x$ axis, and that angle for this vector is $\theta = \boxed{333°}$.

Example 3.4 The Resultant Displacement

A particle undergoes three consecutive displacements: $\Delta\vec{\mathbf{r}}_1 = (15\hat{\mathbf{i}} + 30\hat{\mathbf{j}} + 12\hat{\mathbf{k}})$ cm, $\Delta\vec{\mathbf{r}}_2 = (23\hat{\mathbf{i}} - 14\hat{\mathbf{j}} - 5.0\hat{\mathbf{k}})$ cm, and $\Delta\vec{\mathbf{r}}_3 = (-13\hat{\mathbf{i}} + 15\hat{\mathbf{j}})$ cm. Find unit-vector notation for the resultant displacement and its magnitude.

SOLUTION

Conceptualize Although x is sufficient to locate a point in one dimension, we need a vector $\vec{\mathbf{r}}$ to locate a point in two or three dimensions. The notation $\Delta\vec{\mathbf{r}}$ is a generalization of the one-dimensional displacement Δx in Equation 2.1. Three-dimensional displacements are more difficult to conceptualize than those in two dimensions because the latter can be drawn on paper.

For this problem, let us imagine that you start with your pencil at the origin of a piece of graph paper on which you have drawn x and y axes. Move your pencil 15 cm to the right along the x axis, then 30 cm upward along the y axis, and then 12 cm *perpendicularly toward you away* from the graph paper. This procedure provides the displacement described by $\Delta\vec{\mathbf{r}}_1$. From this point, move your pencil 23 cm to the right parallel to the x axis, then 14 cm parallel to the graph paper in the $-y$ direction, and then 5.0 cm perpendicularly away from you toward the graph paper. You are now at the displacement from the origin described by $\Delta\vec{\mathbf{r}}_1 + \Delta\vec{\mathbf{r}}_2$. From this point, move your pencil 13 cm to the left in the $-x$ direction, and (finally!) 15 cm parallel to the graph paper along the y axis. Your final position is at a displacement $\Delta\vec{\mathbf{r}}_1 + \Delta\vec{\mathbf{r}}_2 + \Delta\vec{\mathbf{r}}_3$ from the origin.

Categorize Despite the difficulty in conceptualizing in three dimensions, we can categorize this problem as a substitution problem because of the careful bookkeeping methods that we have developed for vectors. The mathematical manipulation keeps track of this motion along the three perpendicular axes in an organized, compact way, as we see below.

To find the resultant displacement, add the three vectors:

$$\begin{aligned}\Delta\vec{\mathbf{r}} &= \Delta\vec{\mathbf{r}}_1 + \Delta\vec{\mathbf{r}}_2 + \Delta\vec{\mathbf{r}}_3 \\ &= (15 + 23 - 13)\hat{\mathbf{i}}\text{ cm} + (30 - 14 + 15)\hat{\mathbf{j}}\text{ cm} + (12 - 5.0 + 0)\hat{\mathbf{k}}\text{ cm} \\ &= \boxed{(25\hat{\mathbf{i}} + 31\hat{\mathbf{j}} + 7.0\hat{\mathbf{k}})\text{ cm}}\end{aligned}$$

Find the magnitude of the resultant vector:

$$\begin{aligned}R &= \sqrt{R_x^2 + R_y^2 + R_z^2} \\ &= \sqrt{(25\text{ cm})^2 + (31\text{ cm})^2 + (7.0\text{ cm})^2} = \boxed{40\text{ cm}}\end{aligned}$$

Example 3.5 Taking a Hike

A hiker begins a trip by first walking 25.0 km southeast from her car. She stops and sets up her tent for the night. On the second day, she walks 40.0 km in a direction 60.0° north of east, at which point she discovers a forest ranger's tower.

(A) Determine the components of the hiker's displacement for each day.

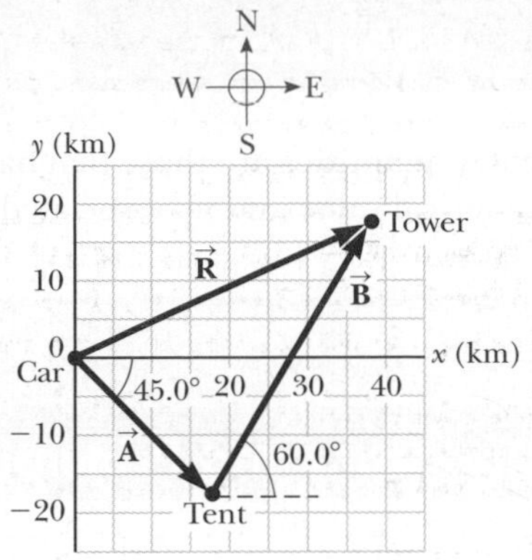

Figure 3.17 (Example 3.5) The total displacement of the hiker is the vector $\vec{R} = \vec{A} + \vec{B}$.

SOLUTION

Conceptualize We conceptualize the problem by drawing a sketch as in Figure 3.17. If we denote the displacement vectors on the first and second days by $\vec{A}$ and $\vec{B}$, respectively, and use the car as the origin of coordinates, we obtain the vectors shown in Figure 3.17.

Categorize Drawing the resultant $\vec{R}$, we can now categorize this problem as one we've solved before: an addition of two vectors. You should now have a hint of the power of categorization in that many new problems are very similar to problems we have already solved if we are careful to conceptualize them. Once we have drawn the displacement vectors and categorized the problem, this problem is no longer about a hiker, a walk, a car, a tent, or a tower. It is a problem about vector addition, one that we have already solved.

Analyze Displacement $\vec{A}$ has a magnitude of 25.0 km and is directed 45.0° below the positive *x* axis.

Find the components of $\vec{A}$ using Equations 3.8 and 3.9:

$$A_x = A\cos(\ \ 45.0°) = (25.0\text{ km})(0.707) = \boxed{17.7\text{ km}}$$

$$A_y = A\sin(-45.0°) = (25.0\text{ km})(-0.707) = \boxed{-17.7\text{ km}}$$

The negative value of A_y indicates that the hiker walks in the negative *y* direction on the first day. The signs of A_x and A_y also are evident from Figure 3.17.

Find the components of $\vec{B}$ using Equations 3.8 and 3.9:

$$B_x = B\cos 60.0° = (40.0\text{ km})(0.500) = \boxed{20.0\text{ km}}$$

$$B_y = B\sin 60.0° = (40.0\text{ km})(0.866) = \boxed{34.6\text{ km}}$$

(B) Determine the components of the hiker's resultant displacement $\vec{R}$ for the trip. Find an expression for $\vec{R}$ in terms of unit vectors.

SOLUTION

Use Equation 3.15 to find the components of the resultant displacement $\vec{R} = \vec{A} + \vec{B}$:

$$R_x = A_x + B_x = 17.7\text{ km} + 20.0\text{ km} = \boxed{37.7\text{ km}}$$

$$R_y = A_y + B_y = -17.7\text{ km} + 34.6\text{ km} = \boxed{17.0\text{ km}}$$

Write the total displacement in unit-vector form:

$$\vec{R} = \boxed{(37.7\hat{\mathbf{i}} + 17.0\hat{\mathbf{j}})\text{ km}}$$

Finalize Looking at the graphical representation in Figure 3.17, we estimate the position of the tower to be about (38 km, 17 km), which is consistent with the components of $\vec{R}$ in our result for the final position of the hiker. Also, both components of $\vec{R}$ are positive, putting the final position in the first quadrant of the coordinate system, which is also consistent with Figure 3.17.

WHAT IF? After reaching the tower, the hiker wishes to return to her car along a single straight line. What are the components of the vector representing this hike? What should the direction of the hike be?

Answer The desired vector $\vec{R}_{car}$ is the negative of vector $\vec{R}$:

$$\vec{R}_{car} = -\vec{R} = (-37.7\hat{\mathbf{i}}\ \ 17.0\hat{\mathbf{j}})\text{ km}$$

The direction is found by calculating the angle that the vector makes with the *x* axis:

$$\tan\theta = \frac{R_{car,y}}{R_{car,x}} = \frac{-17.0\text{ km}}{-37.7\text{ km}} = 0.450$$

which gives an angle of $\theta = 204.2°$, or 24.2° south of west.

Summary

Definitions

Scalar quantities are those that have only a numerical value and no associated direction.

Vector quantities have both magnitude and direction and obey the laws of vector addition. The magnitude of a vector is *always* a positive number.

Concepts and Principles

When two or more vectors are added together, they must all have the same units and they all must be the same type of quantity. We can add two vectors $\vec{\mathbf{A}}$ and $\vec{\mathbf{B}}$ graphically. In this method (Active Fig. 3.6), the resultant vector $\vec{\mathbf{R}} = \vec{\mathbf{A}} + \vec{\mathbf{B}}$ runs from the tail of $\vec{\mathbf{A}}$ to the tip of $\vec{\mathbf{B}}$.

A second method of adding vectors involves **components** of the vectors. The x component A_x of the vector $\vec{\mathbf{A}}$ is equal to the projection of $\vec{\mathbf{A}}$ along the x axis of a coordinate system, where $A_x = A \cos \theta$. The y component A_y of $\vec{\mathbf{A}}$ is the projection of $\vec{\mathbf{A}}$ along the y axis, where $A_y = A \sin \theta$.

If a vector $\vec{\mathbf{A}}$ has an x component A_x and a y component A_y, the vector can be expressed in unit-vector form as $\vec{\mathbf{A}} = A_x\hat{\mathbf{i}} + A_y\hat{\mathbf{j}}$. In this notation, $\hat{\mathbf{i}}$ is a unit vector pointing in the positive x direction and $\hat{\mathbf{j}}$ is a unit vector pointing in the positive y direction. Because $\hat{\mathbf{i}}$ and $\hat{\mathbf{j}}$ are unit vectors, $|\hat{\mathbf{i}}| = |\hat{\mathbf{j}}| = 1$.

We can find the resultant of two or more vectors by resolving all vectors into their x and y components, adding their resultant x and y components, and then using the Pythagorean theorem to find the magnitude of the resultant vector. We can find the angle that the resultant vector makes with respect to the x axis by using a suitable trigonometric function.

Objective Questions

☐ denotes answer available in Student Solutions Manual/Study Guide

1. Yes or no: Is each of the following quantities a vector? (a) force (b) temperature (c) the volume of water in a can (d) the ratings of a TV show (e) the height of a building (f) the velocity of a sports car (g) the age of the Universe

2. A submarine dives from the water surface at an angle of 30° below the horizontal, following a straight path 50 m long. How far is the submarine then below the water surface? (a) 50 m (b) (50 m)/sin 30° (c) (50 m) sin 30° (d) (50 m) cos 30° (e) none of those answers

3. Figure OQ3.3 shows two vectors $\vec{\mathbf{D}}_1$ and $\vec{\mathbf{D}}_2$. Which of the possibilities (a) through (d) is the vector $\vec{\mathbf{D}}_2 - 2\vec{\mathbf{D}}_1$, or (e) is it none of them?

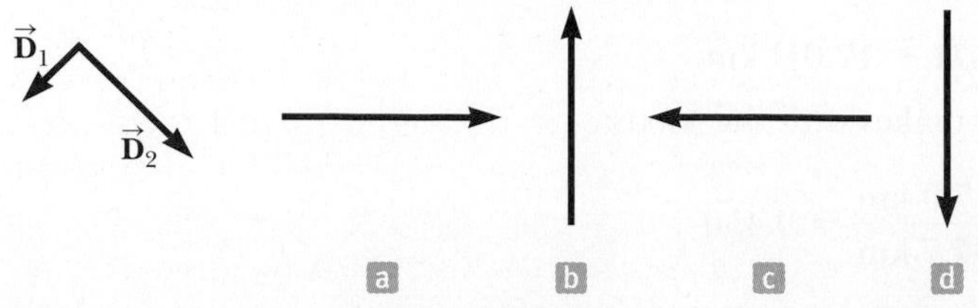

Figure OQ3.3

4. The cutting tool on a lathe is given two displacements, one of magnitude 4 cm and one of magnitude 3 cm, in each one of five situations (a) through (e) diagrammed in Figure OQ3.4. Rank these situations according to the magnitude of the total displacement of the tool, putting the situation with the greatest resultant magnitude first. If the total displacement is the same size in two situations, give those letters equal ranks.

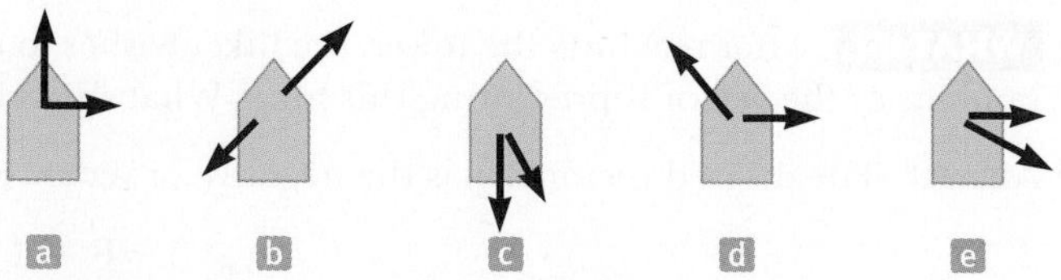

Figure OQ3.4

5. A vector points from the origin into the second quadrant of the xy plane. What can you conclude about its components? (a) Both components are positive. (b) The x component is positive, and the y component is negative. (c) The x component is negative, and the y component is positive. (d) Both components are negative. (e) More than one answer is possible.

6. Let vector $\vec{\mathbf{A}}$ point from the origin into the second quadrant of the xy plane and vector $\vec{\mathbf{B}}$ point from the origin into the fourth quadrant. The vector $\vec{\mathbf{B}} - \vec{\mathbf{A}}$ must be in which quadrant, (a) the first, (b) the second, (c) the third, or (d) the fourth, or (e) is more than one answer possible?

7. What is the magnitude of the vector $(10\hat{\mathbf{i}} - 10\hat{\mathbf{k}})$ m/s? (a) 0 (b) 10 m/s (c) −10 m/s (d) 10 (e) 14.1 m/s

8. What is the y component of the vector $(3\hat{\mathbf{i}} - 8\hat{\mathbf{k}})$ m/s? (a) 3 m/s (b) −8 m/s (c) 0 (d) 8 m/s (e) none of those answers

9. What is the x component of the vector shown in Figure OQ3.9? (a) 3 cm (b) 6 cm (c) −4 cm (d) −6 cm (e) none of those answers

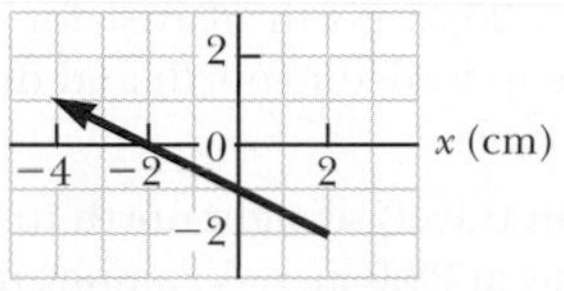

Figure OQ3.9 Objective Questions 9 and 10.

10. What is the y component of the vector shown in Figure OQ3.9? (a) 3 cm (b) 6 cm (c) −4 cm (d) −6 cm (e) none of those answers

11. Vector $\vec{\mathbf{A}}$ lies in the xy plane. Both of its components will be negative if it points from the origin into which quadrant? (a) the first quadrant (b) the second quadrant (c) the third quadrant (d) the fourth quadrant (e) the second or fourth quadrants

12. A vector lying in the xy plane has components of opposite sign. The vector must lie in which quadrant? (a) the first quadrant (b) the second quadrant (c) the third quadrant (d) the fourth quadrant (e) either the second or the fourth quadrant

13. The magnitude of vector $\vec{\mathbf{A}}$ is 8 km, and the magnitude of $\vec{\mathbf{B}}$ is 6 km. Which of the following are possible values for the magnitude of $\vec{\mathbf{A}} + \vec{\mathbf{B}}$? Choose all possible answers. (a) 10 km (b) 8 km (c) 2 km (d) 0 (e) −2 km

Conceptual Questions

☐ denotes answer available in Student Solutions Manual/Study Guide

1. A book is moved once around the perimeter of a tabletop with the dimensions 1.0 m by 2.0 m. The book ends up at its initial position. (a) What is its displacement? (b) What is the distance traveled?

2. If the component of vector $\vec{\mathbf{A}}$ along the direction of vector $\vec{\mathbf{B}}$ is zero, what can you conclude about the two vectors?

3. Is it possible to add a vector quantity to a scalar quantity? Explain.

4. Can the magnitude of a vector have a negative value? Explain.

5. On a certain calculator, the inverse tangent function returns a value between −90° and +90°. In what cases will this value correctly state the direction of a vector in the xy plane, by giving its angle measured counterclockwise from the positive x axis? In what cases will it be incorrect?

Problems

ENHANCED WebAssign The problems found in this chapter may be assigned online in Enhanced WebAssign

1. denotes straightforward problem; 2. denotes intermediate problem; 3. denotes challenging problem

1. full solution available in the Student Solutions Manual/Study Guide

1. denotes problems most often assigned in Enhanced WebAssign; these provide students with targeted feedback and either a Master It tutorial or a Watch It solution video.

QC denotes asking for quantitative and conceptual reasoning

S denotes symbolic reasoning problem

M denotes Master It tutorial available in Enhanced WebAssign

GP denotes guided problem

shaded denotes "paired problems" that develop reasoning with symbols and numerical values

Section 3.1 Coordinate Systems

1. The polar coordinates of a point are $r = 5.50$ m and $\theta = 240°$. What are the Cartesian coordinates of this point?

2. Two points in a plane have polar coordinates (2.50 m, 30.0°) and (3.80 m, 120.0°). Determine (a) the Cartesian coordinates of these points and (b) the distance between them.

3. A fly lands on one wall of a room. The lower-left corner of the wall is selected as the origin of a two-dimensional Cartesian coordinate system. If the fly is located at the point having coordinates (2.00, 1.00) m, (a) how far is it from the origin? (b) What is its location in polar coordinates?

4. The rectangular coordinates of a point are given by (2, y), and its polar coordinates are (r, 30°). Determine (a) the value of y and (b) the value of r.

5. The polar coordinates of a certain point are (r = 4.30 cm, θ = 214°). (a) Find its Cartesian coordinates x and y. Find the polar coordinates of the points with Cartesian coordinates (b) $(-x, y)$, (c) $(-2x, -2y)$, and (d) $(3x, -3y)$.

6. **S** Let the polar coordinates of the point (x, y) be (r, θ). Determine the polar coordinates for the points (a) $(-x, y)$, (b) $(-2x, -2y)$, and (c) $(3x, -3y)$.

Section 3.2 Vector and Scalar Quantities

Section 3.3 Some Properties of Vectors

7. A surveyor measures the distance across a straight river by the following method (Fig. P3.7). Starting directly across from a tree on the opposite bank, she walks d = 100 m along the riverbank to establish a baseline. Then she sights across to the tree. The angle from her baseline to the tree is θ = 35.0°. How wide is the river?

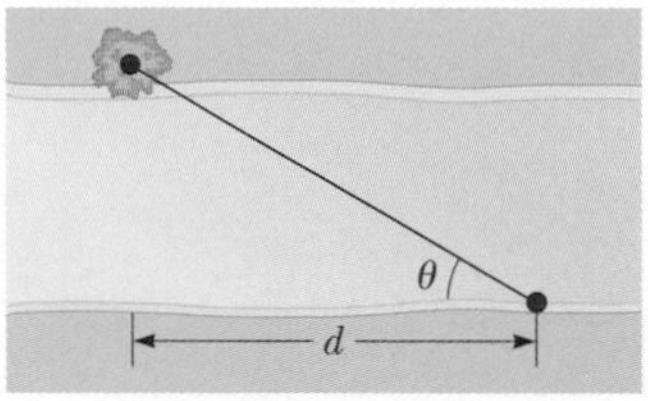

Figure P3.7

8. Vector $\vec{\mathbf{A}}$ has a magnitude of 29 units and points in the positive y direction. When vector $\vec{\mathbf{B}}$ is added to $\vec{\mathbf{A}}$, the resultant vector $\vec{\mathbf{A}} + \vec{\mathbf{B}}$ points in the negative y direction with a magnitude of 14 units. Find the magnitude and direction of $\vec{\mathbf{B}}$.

9. *Why is the following situation impossible?* A skater glides along a circular path. She defines a certain point on the circle as her origin. Later on, she passes through a point at which the distance she has traveled along the path from the origin is smaller than the magnitude of her displacement vector from the origin.

10. A force $\vec{\mathbf{F}}_1$ of magnitude 6.00 units acts on an object at the origin in a direction θ = 30.0° above the positive x axis (Fig. P3.10). A second force $\vec{\mathbf{F}}_2$ of magnitude 5.00 units acts on the object in the direction of the positive y axis. Find graphically the magnitude and direction of the resultant force $\vec{\mathbf{F}}_1 + \vec{\mathbf{F}}_2$.

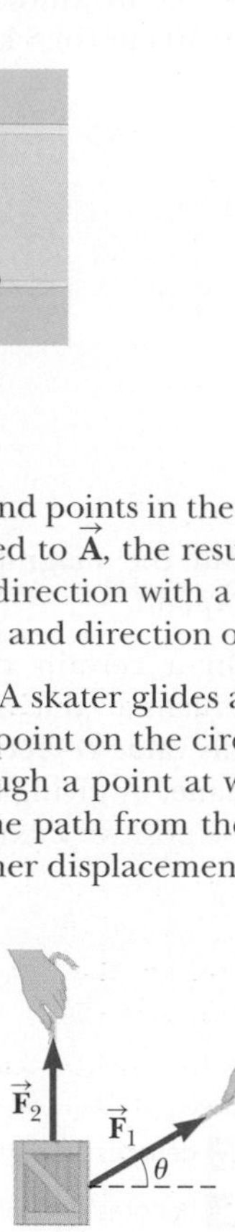

Figure P3.10

11. **M** The displacement vectors $\vec{\mathbf{A}}$ and $\vec{\mathbf{B}}$ shown in Figure P3.11 both have magnitudes of 3.00 m. The direction of vector $\vec{\mathbf{A}}$ is θ = 30.0°. Find graphically (a) $\vec{\mathbf{A}} + \vec{\mathbf{B}}$, (b) $\vec{\mathbf{A}} - \vec{\mathbf{B}}$, (c) $\vec{\mathbf{B}} - \vec{\mathbf{A}}$, and (d) $\vec{\mathbf{A}} - 2\vec{\mathbf{B}}$. (Report all angles counterclockwise from the positive x axis.)

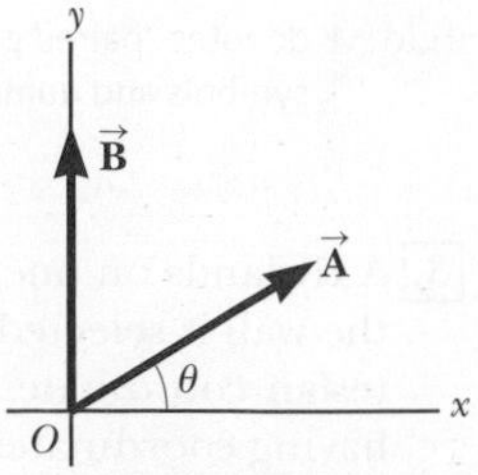

Figure P3.11
Problems 11 and 20.

12. **QC** Three displacements are $\vec{\mathbf{A}}$ = 200 m due south, $\vec{\mathbf{B}}$ = 250 m due west, and $\vec{\mathbf{C}}$ = 150 m at 30.0° east of north. (a) Construct a separate diagram for each of the following possible ways of adding these vectors: $\vec{\mathbf{R}}_1 = \vec{\mathbf{A}} + \vec{\mathbf{B}} + \vec{\mathbf{C}}$; $\vec{\mathbf{R}}_2 = \vec{\mathbf{B}} + \vec{\mathbf{C}} + \vec{\mathbf{A}}$; $\vec{\mathbf{R}}_3 = \vec{\mathbf{C}} + \vec{\mathbf{B}} + \vec{\mathbf{A}}$. (b) Explain what you can conclude from comparing the diagrams.

13. A roller-coaster car moves 200 ft horizontally and then rises 135 ft at an angle of 30.0° above the horizontal. It next travels 135 ft at an angle of 40.0° downward. What is its displacement from its starting point? Use graphical techniques.

Section 3.4 Components of a Vector and Unit Vectors

14. Vector $\vec{\mathbf{A}}$ has a magnitude of 35.0 units and points in the direction 325° counterclockwise from the positive x axis. Calculate the x and y components of this vector.

15. A vector has an x component of −25.0 units and a y component of 40.0 units. Find the magnitude and direction of this vector.

16. A person walks 25.0° north of east for 3.10 km. How far would she have to walk due north and due east to arrive at the same location?

17. **QC** A minivan travels straight north in the right lane of a divided highway at 28.0 m/s. A camper passes the minivan and then changes from the left lane into the right lane. As it does so, the camper's path on the road is a straight displacement at 8.50° east of north. To avoid cutting off the minivan, the north–south distance between the camper's back bumper and the minivan's front bumper should not decrease. (a) Can the camper be driven to satisfy this requirement? (b) Explain your answer.

18. A girl delivering newspapers covers her route by traveling 3.00 blocks west, 4.00 blocks north, and then 6.00 blocks east. (a) What is her resultant displacement? (b) What is the total distance she travels?

19. **M** Obtain expressions in component form for the position vectors having the polar coordinates (a) 12.8 m, 150°; (b) 3.30 cm, 60.0°; and (c) 22.0 in., 215°.

20. Use the component method to add the vectors $\vec{\mathbf{A}}$ and $\vec{\mathbf{B}}$ shown in Figure P3.11. Both vectors have magnitudes of 3.00 m and vector $\vec{\mathbf{A}}$ makes an angle of θ = 30.0° with the x axis. Express the resultant $\vec{\mathbf{A}} + \vec{\mathbf{B}}$ in unit-vector notation.

21. While exploring a cave, a spelunker starts at the entrance and moves the following distances in a horizontal plane. She goes 75.0 m north, 250 m east, 125 m at an angle θ = 30.0° north of east, and 150 m south. Find her resultant displacement from the cave entrance. Figure P3.21 suggests the situation but is not drawn to scale.

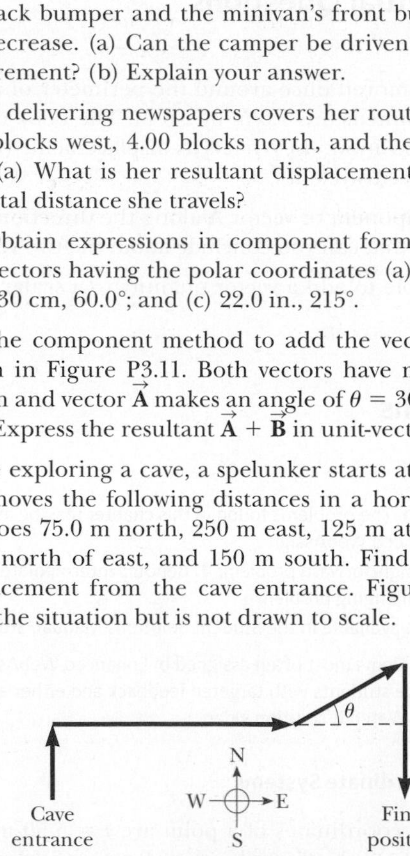

Figure P3.21

22. A map suggests that Atlanta is 730 miles in a direction of 5.00° north of east from Dallas. The same map shows that Chicago is 560 miles in a direction of 21.0° west of north from Atlanta. Figure P3.22 shows the locations of these three cities. Modeling the Earth as flat, use this information to find the displacement from Dallas to Chicago.

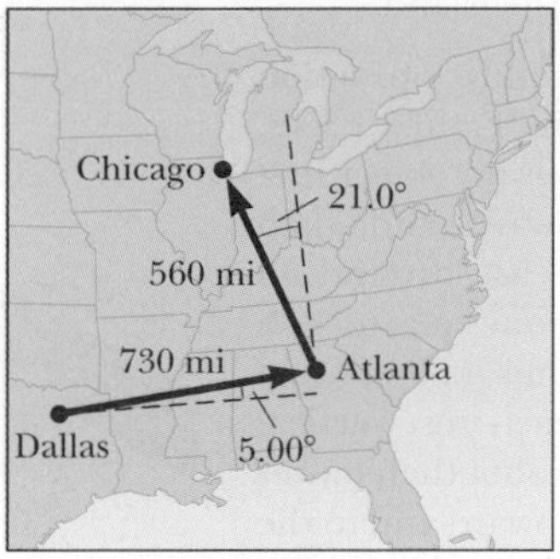

Figure P3.22

23. Consider the two vectors $\vec{\mathbf{A}} = 3\hat{\mathbf{i}} - 2\hat{\mathbf{j}}$ and $\vec{\mathbf{B}} = -\hat{\mathbf{i}} - 4\hat{\mathbf{j}}$. Calculate (a) $\vec{\mathbf{A}} + \vec{\mathbf{B}}$, (b) $\vec{\mathbf{A}} - \vec{\mathbf{B}}$, (c) $|\vec{\mathbf{A}} + \vec{\mathbf{B}}|$, (d) $|\vec{\mathbf{A}} - \vec{\mathbf{B}}|$, and (e) the directions of $\vec{\mathbf{A}} + \vec{\mathbf{B}}$ and $\vec{\mathbf{A}} - \vec{\mathbf{B}}$.

24. Given the vectors $\vec{\mathbf{A}} = 2.00\hat{\mathbf{i}} + 6.00\hat{\mathbf{j}}$ and $\vec{\mathbf{B}} = 3.00\hat{\mathbf{i}} - 2.00\hat{\mathbf{j}}$, (a) draw the vector sum $\vec{\mathbf{C}} = \vec{\mathbf{A}} + \vec{\mathbf{B}}$ and the vector difference $\vec{\mathbf{D}} = \vec{\mathbf{A}} - \vec{\mathbf{B}}$. (b) Calculate $\vec{\mathbf{C}}$ and $\vec{\mathbf{D}}$, in terms of unit vectors. (c) Calculate $\vec{\mathbf{C}}$ and $\vec{\mathbf{D}}$ in terms of polar coordinates, with angles measured with respect to the positive x axis.

25. **M** Your dog is running around the grass in your back yard. He undergoes successive displacements 3.50 m south, 8.20 m northeast, and 15.0 m west. What is the resultant displacement?

26. A snow-covered ski slope makes an angle of 35.0° with the horizontal. When a ski jumper plummets onto the hill, a parcel of splashed snow is thrown up to a maximum displacement of 1.50 m at 16.0° from the vertical in the uphill direction as shown in Figure P3.26. Find the components of its maximum displacement (a) parallel to the surface and (b) perpendicular to the surface.

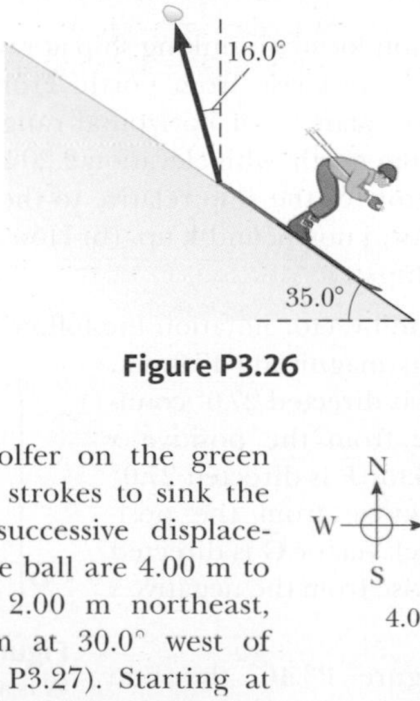

Figure P3.26

27. A novice golfer on the green takes three strokes to sink the ball. The successive displacements of the ball are 4.00 m to the north, 2.00 m northeast, and 1.00 m at 30.0° west of south (Fig. P3.27). Starting at the same initial point, an expert golfer could make the hole in what single displacement?

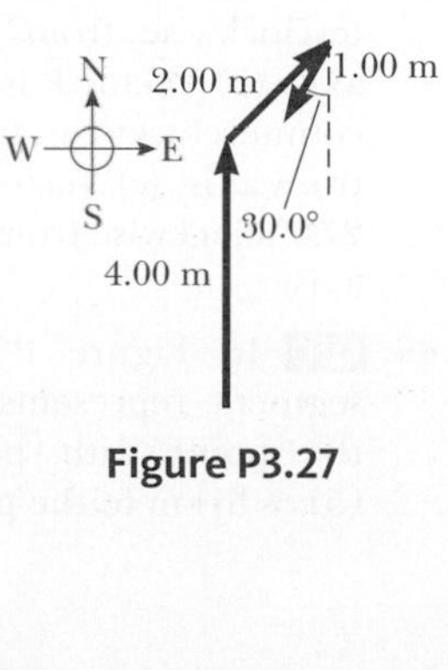

Figure P3.27

28. In a game of American football, a quarterback takes the ball from the line of scrimmage, runs backward a distance of 10.0 yards, and then runs sideways parallel to the line of scrimmage for 15.0 yards. At this point, he throws a forward pass downfield 50.0 yards perpendicular to the line of scrimmage. What is the magnitude of the football's resultant displacement?

29. The helicopter view in Fig. P3.29 shows two people pulling on a stubborn mule. The person on the right pulls with a force $\vec{\mathbf{F}}_1$ of magnitude 120 N and direction of $\theta_1 = 60.0°$. The person on the left pulls with a force $\vec{\mathbf{F}}_2$ of magnitude 80.0 N and direction of $\theta_2 = 75.0°$. Find (a) the single force that is equivalent to the two forces shown and (b) the force that a third person would have to exert on the mule to make the resultant force equal to zero. The forces are measured in units of newtons (symbolized N).

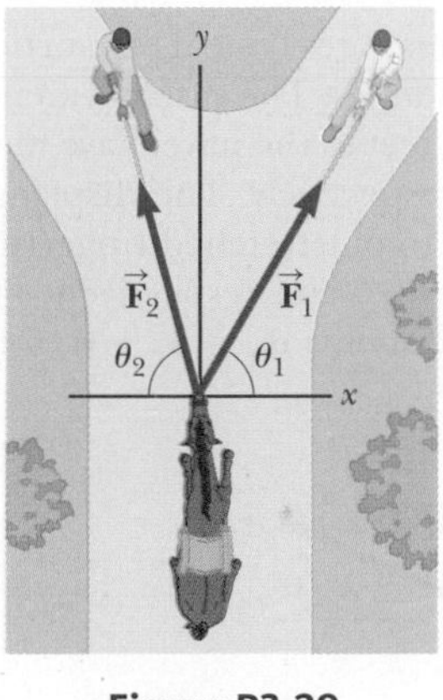

Figure P3.29

30. Vector $\vec{\mathbf{A}}$ has x and y components of −8.70 cm and 15.0 cm, respectively; vector $\vec{\mathbf{B}}$ has x and y components of 13.2 cm and −6.60 cm, respectively. If $\vec{\mathbf{A}} - \vec{\mathbf{B}} + 3\vec{\mathbf{C}} = 0$, what are the components of $\vec{\mathbf{C}}$?

31. Consider the three displacement vectors $\vec{\mathbf{A}} = (3\hat{\mathbf{i}} - 3\hat{\mathbf{j}})$ m, $\vec{\mathbf{B}} = (\hat{\mathbf{i}} - 4\hat{\mathbf{j}})$ m, and $\vec{\mathbf{C}} = (-2\hat{\mathbf{i}} + 5\hat{\mathbf{j}})$ m. Use the component method to determine (a) the magnitude and direction of the vector $\vec{\mathbf{D}} = \vec{\mathbf{A}} + \vec{\mathbf{B}} + \vec{\mathbf{C}}$ and (b) the magnitude and direction of $\vec{\mathbf{E}} = -\vec{\mathbf{A}} - \vec{\mathbf{B}} + \vec{\mathbf{C}}$.

32. Given the displacement vectors $\vec{\mathbf{A}} = (3\hat{\mathbf{i}} - 4\hat{\mathbf{j}} + 4\hat{\mathbf{k}})$ m and $\vec{\mathbf{B}} = (2\hat{\mathbf{i}} + 3\hat{\mathbf{j}} - 7\hat{\mathbf{k}})$ m, find the magnitudes of the following vectors and express each in terms of its rectangular components. (a) $\vec{\mathbf{C}} = \vec{\mathbf{A}} + \vec{\mathbf{B}}$ (b) $\vec{\mathbf{D}} = 2\vec{\mathbf{A}} - \vec{\mathbf{B}}$

33. **M** The vector $\vec{\mathbf{A}}$ has x, y, and z components of 8.00, 12.0, and −4.00 units, respectively. (a) Write a vector expression for $\vec{\mathbf{A}}$ in unit-vector notation. (b) Obtain a unit-vector expression for a vector $\vec{\mathbf{B}}$ one-fourth the length of $\vec{\mathbf{A}}$ pointing in the same direction as $\vec{\mathbf{A}}$. (c) Obtain a unit-vector expression for a vector $\vec{\mathbf{C}}$ three times the length of $\vec{\mathbf{A}}$ pointing in the direction opposite the direction of $\vec{\mathbf{A}}$.

34. Vector $\vec{\mathbf{B}}$ has x, y, and z components of 4.00, 6.00, and 3.00 units, respectively. Calculate (a) the magnitude of $\vec{\mathbf{B}}$ and (b) the angle that $\vec{\mathbf{B}}$ makes with each coordinate axis.

35. **M** Vector $\vec{\mathbf{A}}$ has a negative x component 3.00 units in length and a positive y component 2.00 units in length. (a) Determine an expression for $\vec{\mathbf{A}}$ in unit-vector notation. (b) Determine the magnitude and direction of $\vec{\mathbf{A}}$. (c) What vector $\vec{\mathbf{B}}$ when added to $\vec{\mathbf{A}}$ gives a resultant vector with no

x component and a negative y component 4.00 units in length?

36. Three displacement vectors of a croquet ball are shown in Figure P3.36, where $|\vec{\mathbf{A}}| = 20.0$ units, $|\vec{\mathbf{B}}| = 40.0$ units, and $|\vec{\mathbf{C}}| = 30.0$ units. Find (a) the resultant in unit-vector notation and (b) the magnitude and direction of the resultant displacement.

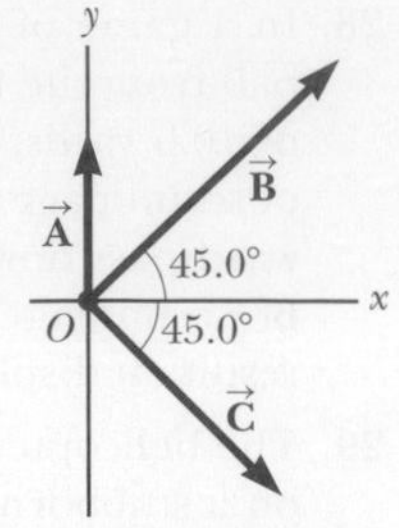

Figure P3.36

37. Q|C (a) Taking $\vec{\mathbf{A}} = (6.00\hat{\mathbf{i}} - 8.00\hat{\mathbf{j}})$ units, $\vec{\mathbf{B}} = (-8.00\hat{\mathbf{i}} + 3.00\hat{\mathbf{j}})$ units, and $\vec{\mathbf{C}} = (26.0\hat{\mathbf{i}} + 19.0\hat{\mathbf{j}})$ units, determine a and b such that $a\vec{\mathbf{A}} + b\vec{\mathbf{B}} + \vec{\mathbf{C}} = 0$. (b) A student has learned that a single equation cannot be solved to determine values for more than one unknown in it. How would you explain to him that both a and b can be determined from the single equation used in part (a)?

38. Figure P3.38 illustrates typical proportions of male (m) and female (f) anatomies. The displacements $\vec{\mathbf{d}}_{1m}$ and $\vec{\mathbf{d}}_{1f}$ from the soles of the feet to the navel have magnitudes of 104 cm and 84.0 cm, respectively. The displacements $\vec{\mathbf{d}}_{2m}$ and $\vec{\mathbf{d}}_{2f}$ from the navel to outstretched fingertips have magnitudes of 100 cm and 86.0 cm, respectively. Find the vector sum of these displacements $\vec{\mathbf{d}}_3 = \vec{\mathbf{d}}_1 + \vec{\mathbf{d}}_2$ for both people.

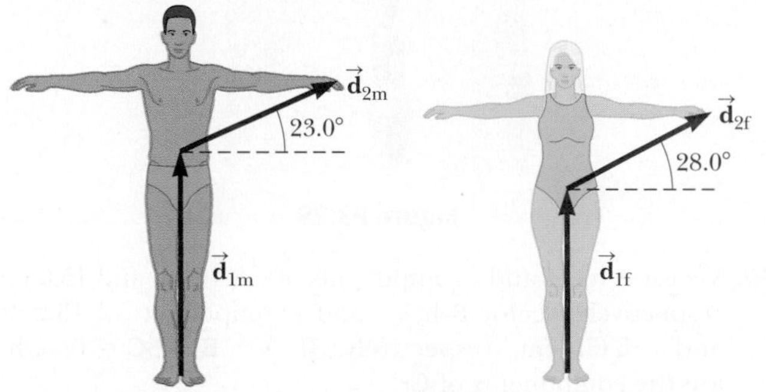

Figure P3.38

39. M A man pushing a mop across a floor causes it to undergo two displacements. The first has a magnitude of 150 cm and makes an angle of 120° with the positive x axis. The resultant displacement has a magnitude of 140 cm and is directed at an angle of 35.0° to the positive x axis. Find the magnitude and direction of the second displacement.

40. *Why is the following situation impossible?* A shopper pushing a cart through a market follows directions to the canned goods and moves through a displacement $8.00\hat{\mathbf{i}}$ m down one aisle. He then makes a 90.0° turn and moves 3.00 m along the y axis. He then makes another 90.0° turn and moves 4.00 m along the x axis. *Every* shopper who follows these directions correctly ends up 5.00 m from the starting point.

41. GP **Review.** As it passes over Grand Bahama Island, the eye of a hurricane is moving in a direction 60.0° north of west with a speed of 41.0 km/h. (a) What is the unit-vector expression for the velocity of the hurricane? It maintains this velocity for 3.00 h, at which time the course of the hurricane suddenly shifts due north, and its speed slows to a constant 25.0 km/h. This new velocity is maintained for 1.50 h. (b) What is the unit-vector expression for the new velocity of the hurricane? (c) What is the unit-vector expression for the displacement of the hurricane during the first 3.00 h? (d) What is the unit-vector expression for the displacement of the hurricane during the latter 1.50 h? (e) How far from Grand Bahama is the eye 4.50 h after it passes over the island?

42. In an assembly operation illustrated in Figure P3.42, a robot moves an object first straight upward and then also to the east, around an arc forming one-quarter of a circle of radius 4.80 cm that lies in an east–west vertical plane. The robot then moves the object upward and to the north, through one-quarter of a circle of radius 3.70 cm that lies in a north–south vertical plane. Find (a) the magnitude of the total displacement of the object and (b) the angle the total displacement makes with the vertical.

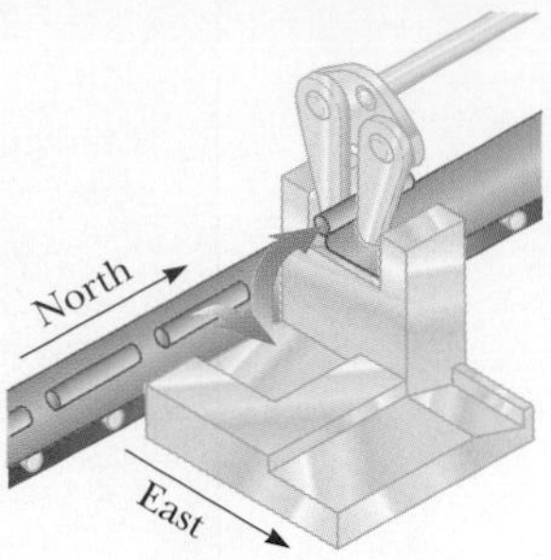

Figure P3.42

43. **Review.** You are standing on the ground at the origin of a coordinate system. An airplane flies over you with constant velocity parallel to the x axis and at a fixed height of 7.60×10^3 m. At time $t = 0$, the airplane is directly above you so that the vector leading from you to it is $\vec{\mathbf{P}}_0 = 7.60 \times 10^3\hat{\mathbf{j}}$ m. At $t = 30.0$ s, the position vector leading from you to the airplane is $\vec{\mathbf{P}}_{30} = (8.04 \times 10^3\hat{\mathbf{i}} + 7.60 \times 10^3\hat{\mathbf{j}})$ m as suggested in Figure P3.43. Determine the magnitude and orientation of the airplane's position vector at $t = 45.0$ s.

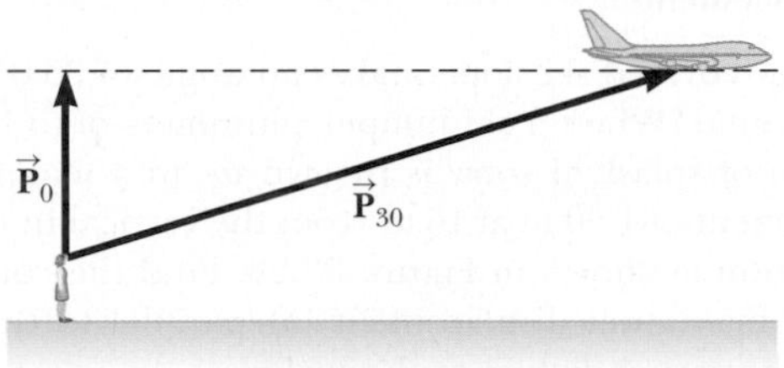

Figure P3.43

44. A radar station locates a sinking ship at range 17.3 km and bearing 136° clockwise from north. From the same station, a rescue plane is at horizontal range 19.6 km, 153° clockwise from north, with elevation 2.20 km. (a) Write the position vector for the ship relative to the plane, letting $\hat{\mathbf{i}}$ represent east, $\hat{\mathbf{j}}$ north, and $\hat{\mathbf{k}}$ up. (b) How far apart are the plane and ship?

45. Express in unit-vector notation the following vectors, each of which has magnitude 17.0 cm. (a) Vector $\vec{\mathbf{E}}$ is directed 27.0° counterclockwise from the positive x axis. (b) Vector $\vec{\mathbf{F}}$ is directed 27.0° counterclockwise from the positive y axis. (c) Vector $\vec{\mathbf{G}}$ is directed 27.0° clockwise from the negative y axis.

46. Q|C In Figure P3.46, the line segment represents a path from the point with position vector $(5\hat{\mathbf{i}} + 3\hat{\mathbf{j}})$ m to the point with loca-

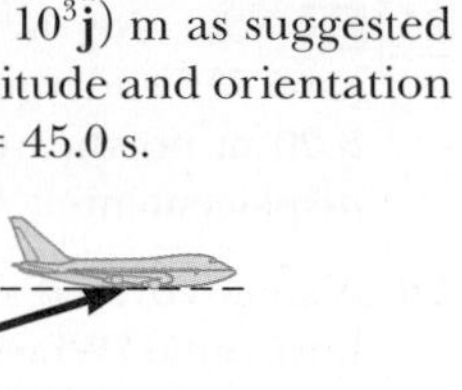

Figure P3.46 Point Ⓐ is a fraction f of the distance from the initial point (5, 3) to the final point (16, 12).

tion $(16\hat{\mathbf{i}} + 12\hat{\mathbf{j}})$ m. Point Ⓐ is along this path, a fraction f of the way to the destination. (a) Find the position vector of point Ⓐ in terms of f. (b) Evaluate the expression from part (a) for $f = 0$. (c) Explain whether the result in part (b) is reasonable. (d) Evaluate the expression for $f = 1$. (e) Explain whether the result in part (d) is reasonable.

Additional Problems

47. A person going for a walk follows the path shown in Figure P3.47. The total trip consists of four straight-line paths. At the end of the walk, what is the person's resultant displacement measured from the starting point?

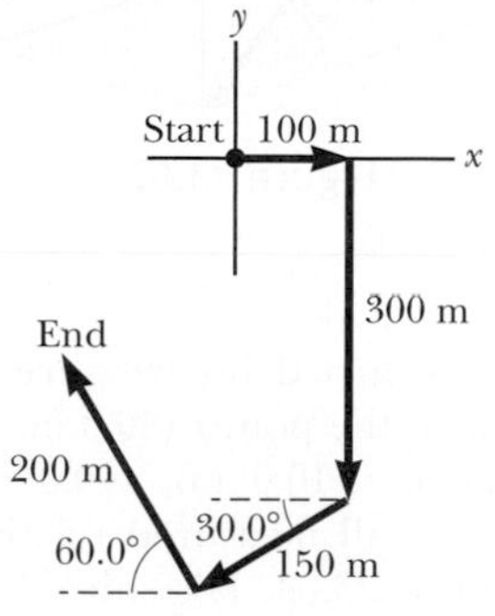

Figure P3.47

48. A jet airliner, moving initially at 300 mi/h to the east, suddenly enters a region where the wind is blowing at 100 mi/h toward the direction 30.0° north of east. What are the new speed and direction of the aircraft relative to the ground?

49. As she picks up her riders, a bus driver traverses four successive displacements represented by the expression

$$(-6.30\text{ b})\hat{\mathbf{i}} - (4.00\text{ b}\cos 40°)\hat{\mathbf{i}} - (4.00\text{ b}\sin 40°)\hat{\mathbf{j}}$$
$$+ (3.00\text{ b}\cos 50°)\hat{\mathbf{i}} - (3.00\text{ b}\sin 50°)\hat{\mathbf{j}} - (5.00\text{ b})\hat{\mathbf{j}}$$

Here b represents one city block, a convenient unit of distance of uniform size; $\hat{\mathbf{i}}$ is east; and $\hat{\mathbf{j}}$ is north. The displacements at 40° and 50° represent travel on roadways in the city that are at these angles to the main east–west and north–south streets. (a) Draw a map of the successive displacements. (b) What total distance did she travel? (c) Compute the magnitude and direction of her total displacement. The logical structure of this problem and of several problems in later chapters was suggested by Alan Van Heuvelen and David Maloney, *American Journal of Physics* **67**(3) 252–256, March 1999.

50. An air-traffic controller observes two aircraft on his radar screen. The first is at altitude 800 m, horizontal distance 19.2 km, and 25.0° south of west. The second aircraft is at altitude 1 100 m, horizontal distance 17.6 km, and 20.0° south of west. What is the distance between the two aircraft? (Place the x axis west, the y axis south, and the z axis vertical.)

51. **Review.** The biggest stuffed animal in the world is a snake 420 m long, constructed by Norwegian children. Suppose the snake is laid out in a park as shown in Figure P3.51, forming two straight sides of a 105° angle, with one side 240 m long. Olaf and Inge run a race they invent. Inge runs directly from the tail of the snake to its head, and Olaf starts from the same place at the same moment but runs along the snake. (a) If both children run steadily at 12.0 km/h, Inge reaches the head of the snake how much earlier than Olaf? (b) If Inge runs the race again at a constant speed of 12.0 km/h, at what constant speed must Olaf run to reach the end of the snake at the same time as Inge?

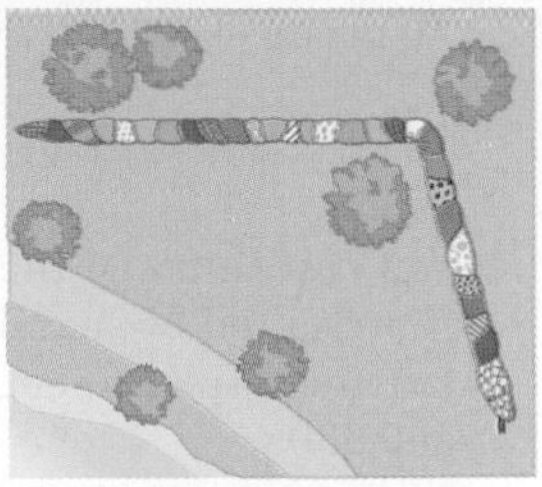

Figure P3.51

52. A ferry transports tourists between three islands. It sails from the first island to the second island, 4.76 km away, in a direction 37.0° north of east. It then sails from the second island to the third island in a direction 69.0° west of north. Finally it returns to the first island, sailing in a direction 28.0° east of south. Calculate the distance between (a) the second and third islands and (b) the first and third islands.

53. A vector is given by $\vec{\mathbf{R}} = 2\hat{\mathbf{i}} + \hat{\mathbf{j}} + 3\hat{\mathbf{k}}$. Find (a) the magnitudes of the x, y, and z components; (b) the magnitude of $\vec{\mathbf{R}}$; and (c) the angles between $\vec{\mathbf{R}}$ and the x, y, and z axes.

54. The rectangle shown in Figure P3.54 has sides parallel to the x and y axes. The position vectors of two corners are $\vec{\mathbf{A}}$ = 10.0 m at 50.0° and $\vec{\mathbf{B}}$ = 12.0 m at 30.0°. (a) Find the perimeter of the rectangle. (b) Find the magnitude and direction of the vector from the origin to the upper-right corner of the rectangle.

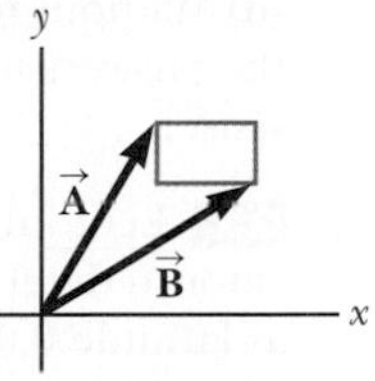

Figure P3.54

55. In Figure P3.55, a spider is resting after starting to spin its web. The gravitational force on the spider makes it exert a downward force of 0.150 N on the junction of the three strands of silk. The junction is supported by different tension forces in the two strands above it so that the resultant force on the junction is zero. The two sloping strands are perpendicular, and we have chosen the x and y directions to be along them. The tension T_x is 0.127 N. Find (a) the tension T_y, (b) the angle the x axis makes with the horizontal, and (c) the angle the y axis makes with the horizontal.

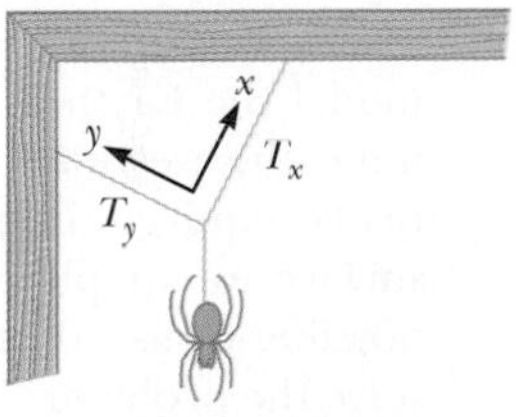

Figure P3.55

56. After a ball rolls off the edge of a horizontal table at time $t = 0$, its velocity as a function of time is given by

$$\vec{\mathbf{v}} = 1.2\hat{\mathbf{i}} - 9.8t\hat{\mathbf{j}}$$

where $\vec{\mathbf{v}}$ is in meters per second and t is in seconds. The ball's displacement away from the edge of the table, during the time interval of 0.380 s for which the ball is in flight, is given by

$$\Delta\vec{\mathbf{r}} = \int_0^{0.380\text{ s}} \vec{\mathbf{v}}\,dt$$

To perform the integral, you can use the calculus theorem

$$\int [A + Bf(x)]dx = \int A\,dx + B\int f(x)\,dx$$

You can think of the units and unit vectors as constants, represented by A and B. Perform the integration to calculate the displacement of the ball from the edge of the table at 0.380 s.

57. Two vectors $\vec{\mathbf{A}}$ and $\vec{\mathbf{B}}$ have precisely equal magnitudes. For the magnitude of $\vec{\mathbf{A}} + \vec{\mathbf{B}}$ to be 100 times larger than the magnitude of $\vec{\mathbf{A}} - \vec{\mathbf{B}}$, what must be the angle between them?

58. S Two vectors $\vec{\mathbf{A}}$ and $\vec{\mathbf{B}}$ have precisely equal magnitudes. For the magnitude of $\vec{\mathbf{A}} + \vec{\mathbf{B}}$ to be larger than the magnitude of $\vec{\mathbf{A}} - \vec{\mathbf{B}}$ by the factor n, what must be the angle between them?

59. Q|C Let $\vec{\mathbf{A}}$ = 60.0 cm at 270° measured from the horizontal. Let $\vec{\mathbf{B}}$ = 80.0 cm at some angle θ. (a) Find the magnitude of $\vec{\mathbf{A}} + \vec{\mathbf{B}}$ as a function of θ. (b) From the answer to part (a), for what value of θ does $|\vec{\mathbf{A}} + \vec{\mathbf{B}}|$ take on its maximum value? What is this maximum value? (c) From the answer to part (a), for what value of θ does $|\vec{\mathbf{A}} + \vec{\mathbf{B}}|$ take on its minimum value? What is this minimum value? (d) Without reference to the answer to part (a), argue that the answers to each of parts (b) and (c) do or do not make sense.

60. Q|C Ecotourists use their global positioning system indicator to determine their location inside a botanical garden as latitude 0.002 43 degree south of the equator, longitude 75.642 38 degrees west. They wish to visit a tree at latitude 0.001 62 degree north, longitude 75.644 26 degrees west. (a) Determine the straight-line distance and the direction in which they can walk to reach the tree as follows. First model the Earth as a sphere of radius 6.37 Mm to determine the westward and northward displacement components required, in meters. Then model the Earth as a flat surface to complete the calculation. (b) Explain why it is possible to use these two geometrical models together to solve the problem.

61. Q|C **Review.** The instantaneous position of an object is specified by its position vector leading from a fixed origin to the location of the object, modeled as a particle. Suppose for a certain object the position vector is a function of time given by $\vec{\mathbf{r}} = 4\hat{\mathbf{i}} + 3\hat{\mathbf{j}} - 2t\hat{\mathbf{k}}$, where $\vec{\mathbf{r}}$ is in meters and t is in seconds. (a) Evaluate $d\vec{\mathbf{r}}/dt$. (b) What physical quantity does $d\vec{\mathbf{r}}/dt$ represent about the object?

62. Vectors $\vec{\mathbf{A}}$ and $\vec{\mathbf{B}}$ have equal magnitudes of 5.00. The sum of $\vec{\mathbf{A}}$ and $\vec{\mathbf{B}}$ is the vector $6.00\hat{\mathbf{j}}$. Determine the angle between $\vec{\mathbf{A}}$ and $\vec{\mathbf{B}}$.

63. S A rectangular parallelepiped has dimensions a, b, and c as shown in Figure P3.63. (a) Obtain a vector expression for the face diagonal vector $\vec{\mathbf{R}}_1$. (b) What is the magnitude of this vector? (c) Notice that $\vec{\mathbf{R}}_1$, $c\hat{\mathbf{k}}$, and $\vec{\mathbf{R}}_2$ make a right triangle. Obtain a vector expression for the body diagonal vector $\vec{\mathbf{R}}_2$.

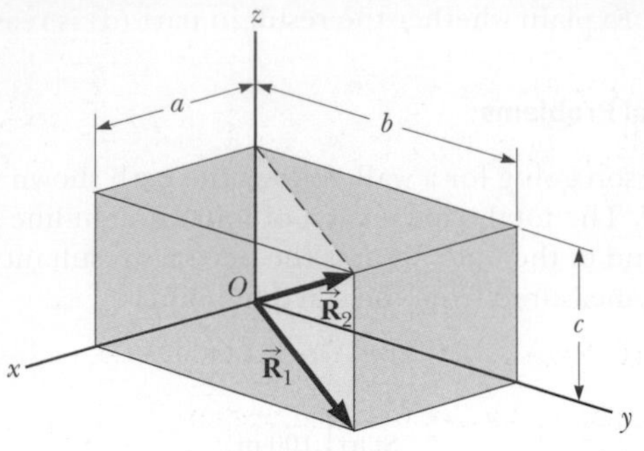

Figure P3.63

Challenge Problem

64. Q|C A pirate has buried his treasure on an island with five trees located at the points (30.0 m, −20.0 m), (60.0 m, 80.0 m), (−10.0 m, −10.0 m), (40.0 m, −30.0 m), and (−70.0 m, 60.0 m), all measured relative to some origin, as shown in Figure P3.64. His ship's log instructs you to start at tree A and move toward tree B, but to cover only one-half the distance between A and B. Then move toward tree C, covering one-third the distance between your current location and C. Next move toward tree D, covering one-fourth the distance between where you are and D. Finally move toward tree E, covering one-fifth the distance between you and E, stop, and dig. (a) Assume you have correctly determined the order in which the pirate labeled the trees as A, B, C, D, and E as shown in the figure. What are the coordinates of the point where his treasure is buried? (b) **What If?** What if you do not really know the way the pirate labeled the trees? What would happen to the answer if you rearranged the order of the trees, for instance, to B (30 m, −20 m), A (60 m, 80 m), E (−10 m, −10 m), C (40 m, −30 m), and D (−70 m, 60 m)? State reasoning to show that the answer does not depend on the order in which the trees are labeled.

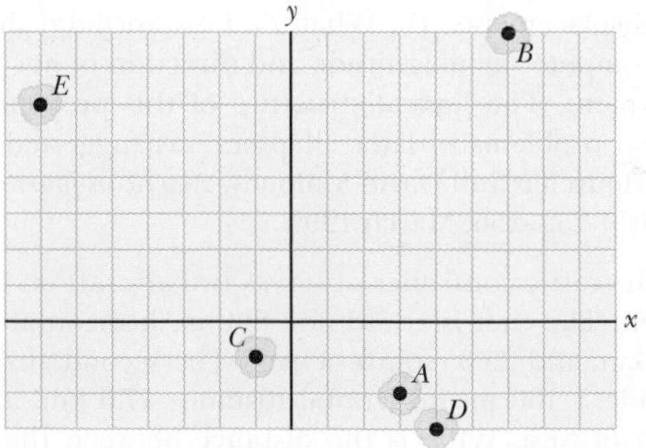

Figure P3.64

chapter **4**

Motion in Two Dimensions

Fireworks erupt from the Sydney Harbour Bridge in New South Wales, Australia. Notice the parabolic paths of embers projected into the air. All projectiles follow a parabolic path in the absence of air resistance. *(Graham Monro/Photolibrary/Jupiter Images)*

In this chapter, we explore the kinematics of a particle moving in two dimensions. Knowing the basics of two-dimensional motion will allow us—in future chapters—to examine a variety of situations, ranging from the motion of satellites in orbit to the motion of electrons in a uniform electric field. We begin by studying in greater detail the vector nature of position, velocity, and acceleration. We then treat projectile motion and uniform circular motion as special cases of motion in two dimensions. We also discuss the concept of relative motion, which shows why observers in different frames of reference may measure different positions and velocities for a given particle.

4.1 The Position, Velocity, and Acceleration Vectors

In Chapter 2, we found that the motion of a particle along a straight line such as the x axis is completely known if its position is known as a function of time. Let us now extend this idea to two-dimensional motion of a particle in the xy plane. We begin by describing the position of the particle by its **position vector** $\vec{\mathbf{r}}$, drawn from the origin of some coordinate system to the location of the particle in the xy plane as in Figure 4.1. At time t_i, the particle is at point Ⓐ, described by position vector $\vec{\mathbf{r}}_i$. At some later time t_f, it is at point Ⓑ, described by position vector $\vec{\mathbf{r}}_f$. The path from Ⓐ to Ⓑ is not necessarily a straight line. As the particle moves from Ⓐ to Ⓑ in the time interval $\Delta t = t_f - t_i$, its position vector changes from $\vec{\mathbf{r}}_i$ to $\vec{\mathbf{r}}_f$. As we learned in Chapter 2, displacement is a vector, and the displacement of the particle is the difference between its final position and its initial position. We now define the **displacement vector** $\Delta\vec{\mathbf{r}}$ for a particle such as the one in Figure 4.1 as being the difference between its final position vector and its initial position vector:

Displacement vector ▶

$$\Delta\vec{\mathbf{r}} \equiv \vec{\mathbf{r}}_f - \vec{\mathbf{r}}_i \tag{4.1}$$

The direction of $\Delta\vec{\mathbf{r}}$ is indicated in Figure 4.1. As we see from the figure, the magnitude of $\Delta\vec{\mathbf{r}}$ is *less* than the distance traveled along the curved path followed by the particle.

As we saw in Chapter 2, it is often useful to quantify motion by looking at the displacement divided by the time interval during which that displacement occurs, which gives the rate of change of position. Two-dimensional (or three-dimensional) kinematics is similar to one-dimensional kinematics, but we must now use full vector notation rather than positive and negative signs to indicate the direction of motion.

We define the **average velocity** $\vec{\mathbf{v}}_{\text{avg}}$ of a particle during the time interval Δt as the displacement of the particle divided by the time interval:

Average velocity ▶

$$\vec{\mathbf{v}}_{\text{avg}} \equiv \frac{\Delta\vec{\mathbf{r}}}{\Delta t} \tag{4.2}$$

Multiplying or dividing a vector quantity by a positive scalar quantity such as Δt changes only the magnitude of the vector, not its direction. Because displacement is a vector quantity and the time interval is a positive scalar quantity, we conclude that the average velocity is a vector quantity directed along $\Delta\vec{\mathbf{r}}$. Compare Equation 4.2 with its one-dimensional counterpart, Equation 2.2.

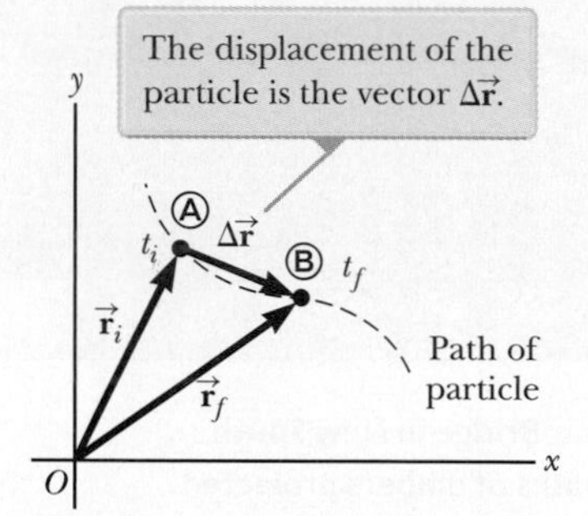

Figure 4.1 A particle moving in the xy plane is located with the position vector $\vec{\mathbf{r}}$ drawn from the origin to the particle. The displacement of the particle as it moves from Ⓐ to Ⓑ in the time interval $\Delta t = t_f - t_i$ is equal to the vector $\Delta\vec{\mathbf{r}} = \vec{\mathbf{r}}_f - \vec{\mathbf{r}}_i$.

The average velocity between points is *independent of the path* taken. That is because average velocity is proportional to displacement, which depends only on the initial and final position vectors and not on the path taken. As with one-dimensional motion, we conclude that if a particle starts its motion at some point and returns to this point via any path, its average velocity is zero for this trip because its displacement is zero. Consider again our basketball players on the court in Figure 2.2 (page 22). We previously considered only their one-dimensional motion back and forth between the baskets. In reality, however, they move over a two-dimensional surface, running back and forth between the baskets as well as left and right across the width of the court. Starting from one basket, a given player may follow a very complicated two-dimensional path. Upon returning to the original basket, however, a player's average velocity is zero because the player's displacement for the whole trip is zero.

Consider again the motion of a particle between two points in the xy plane as shown in Figure 4.2. As the time interval over which we observe the motion becomes smaller and smaller—that is, as Ⓑ is moved to Ⓑ′ and then to Ⓑ″ and so on—the direction of the displacement approaches that of the line tangent to the path at Ⓐ. The **instantaneous velocity** $\vec{\mathbf{v}}$ is defined as the limit of the average velocity $\Delta\vec{\mathbf{r}}/\Delta t$ as Δt approaches zero:

Instantaneous velocity ▶

$$\vec{\mathbf{v}} \equiv \lim_{\Delta t \to 0} \frac{\Delta\vec{\mathbf{r}}}{\Delta t} = \frac{d\vec{\mathbf{r}}}{dt} \tag{4.3}$$

That is, the instantaneous velocity equals the derivative of the position vector with respect to time. The direction of the instantaneous velocity vector at any point in a particle's path is along a line tangent to the path at that point and in the direction

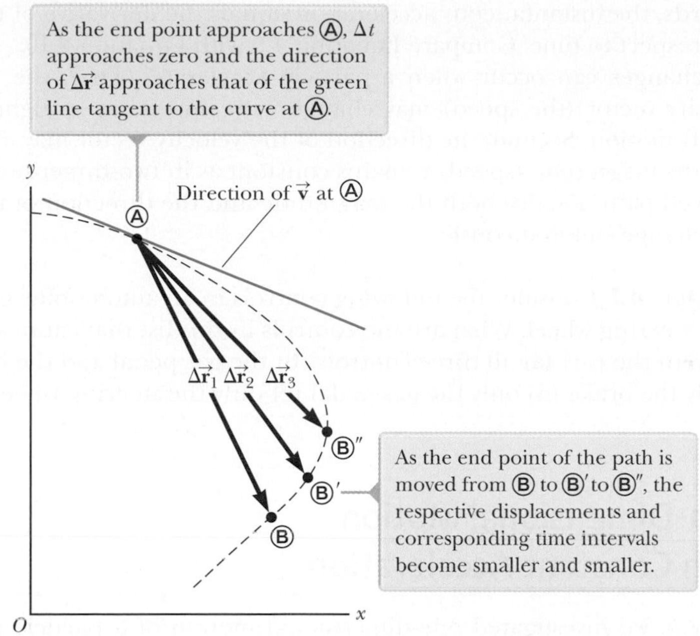

Figure 4.2 As a particle moves between two points, its average velocity is in the direction of the displacement vector $\Delta\vec{\mathbf{r}}$. By definition, the instantaneous velocity at Ⓐ is directed along the line tangent to the curve at Ⓐ.

of motion. Compare Equation 4.3 with the corresponding one-dimensional version, Equation 2.5.

The magnitude of the instantaneous velocity vector $v = |\vec{\mathbf{v}}|$ of a particle is called the *speed* of the particle, which is a scalar quantity.

As a particle moves from one point to another along some path, its instantaneous velocity vector changes from $\vec{\mathbf{v}}_i$ at time t_i to $\vec{\mathbf{v}}_f$ at time t_f. Knowing the velocity at these points allows us to determine the average acceleration of the particle. The **average acceleration** $\vec{\mathbf{a}}_{avg}$ of a particle is defined as the change in its instantaneous velocity vector $\Delta\vec{\mathbf{v}}$ divided by the time interval Δt during which that change occurs:

$$\vec{\mathbf{a}}_{avg} \equiv \frac{\Delta\vec{\mathbf{v}}}{\Delta t} = \frac{\vec{\mathbf{v}}_f - \vec{\mathbf{v}}_i}{t_f - t_i} \qquad (4.4)$$

◀ **Average acceleration**

Because $\vec{\mathbf{a}}_{avg}$ is the ratio of a vector quantity $\Delta\vec{\mathbf{v}}$ and a positive scalar quantity Δt, we conclude that average acceleration is a vector quantity directed along $\Delta\vec{\mathbf{v}}$. As indicated in Figure 4.3, the direction of $\Delta\vec{\mathbf{v}}$ is found by adding the vector $-\vec{\mathbf{v}}_i$ (the negative of $\vec{\mathbf{v}}_i$) to the vector $\vec{\mathbf{v}}_f$ because, by definition, $\Delta\vec{\mathbf{v}} = \vec{\mathbf{v}}_f - \vec{\mathbf{v}}_i$. Compare Equation 4.4 with Equation 2.9.

When the average acceleration of a particle changes during different time intervals, it is useful to define its instantaneous acceleration. The **instantaneous acceleration $\vec{\mathbf{a}}$** is defined as the limiting value of the ratio $\Delta\vec{\mathbf{v}}/\Delta t$ as Δt approaches zero:

$$\vec{\mathbf{a}} \equiv \lim_{\Delta t \to 0} \frac{\Delta\vec{\mathbf{v}}}{\Delta t} = \frac{d\vec{\mathbf{v}}}{dt} \qquad (4.5)$$

◀ **Instantaneous acceleration**

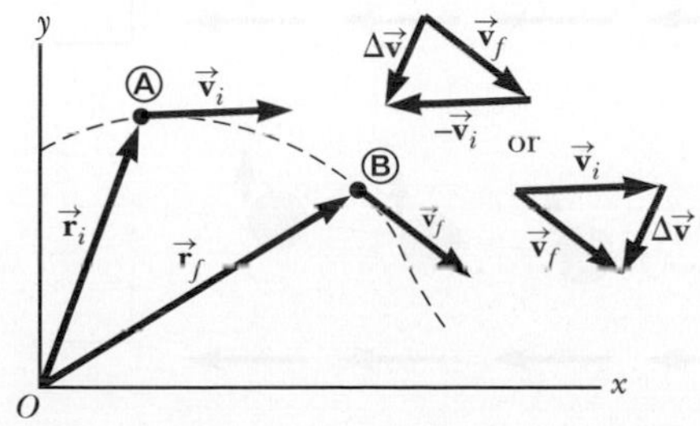

Figure 4.3 A particle moves from position Ⓐ to position Ⓑ. Its velocity vector changes from $\vec{\mathbf{v}}_i$ to $\vec{\mathbf{v}}_f$. The vector diagrams at the upper right show two ways of determining the vector $\Delta\vec{\mathbf{v}}$ from the initial and final velocities.

Pitfall Prevention 4.1

Vector Addition
Although the vector addition discussed in Chapter 3 involves *displacement* vectors, vector addition can be applied to *any* type of vector quantity. Figure 4.3, for example, shows the addition of *velocity* vectors using the graphical approach.

In other words, the instantaneous acceleration equals the derivative of the velocity vector with respect to time. Compare Equation 4.5 with Equation 2.10.

Various changes can occur when a particle accelerates. First, the magnitude of the velocity vector (the speed) may change with time as in straight-line (one-dimensional) motion. Second, the direction of the velocity vector may change with time even if its magnitude (speed) remains constant as in two-dimensional motion along a curved path. Finally, both the magnitude and the direction of the velocity vector may change simultaneously.

***Quick Quiz* 4.1** Consider the following controls in an automobile: gas pedal, brake, steering wheel. What are the controls in this list that cause an acceleration of the car? **(a)** all three controls **(b)** the gas pedal and the brake **(c)** only the brake **(d)** only the gas pedal **(e)** only the steering wheel

4.2 Two-Dimensional Motion with Constant Acceleration

In Section 2.5, we investigated one-dimensional motion of a particle under constant acceleration. Let us now consider two-dimensional motion during which the acceleration of a particle remains constant in both magnitude and direction. As we shall see, this approach is useful for analyzing some common types of motion.

Before embarking on this investigation, we need to emphasize an important point regarding two-dimensional motion. Imagine an air hockey puck moving in a straight line along a perfectly level, friction-free surface of an air hockey table. Figure 4.4a shows a motion diagram from an overhead point of view of this puck. Recall that in Section 2.4 we related the acceleration of an object to a force on the object. Because there are no forces on the puck in the horizontal plane, it moves with constant velocity in the x direction. Now suppose you blow a puff of air on the puck as it passes your position, with the force from your puff of air *exactly* in the y direction. Because the force from this puff of air has no component in the x direction, it causes no acceleration in the x direction. It only causes a momentary acceleration in the y direction, causing the puck to have a constant y component of velocity once the force from the puff of air is removed. After your puff of air on the puck, its velocity component in the x direction is unchanged as shown in Figure 4.4b. The generalization of this simple experiment is that **motion in two dimensions can be modeled as two *independent* motions in each of the two perpendicular directions associated with the x and y axes. That is, any influence in the y direction does not affect the motion in the x direction and vice versa.**

The position vector for a particle moving in the xy plane can be written

$$\vec{\mathbf{r}} = x\hat{\mathbf{i}} + y\hat{\mathbf{j}} \quad (4.6)$$

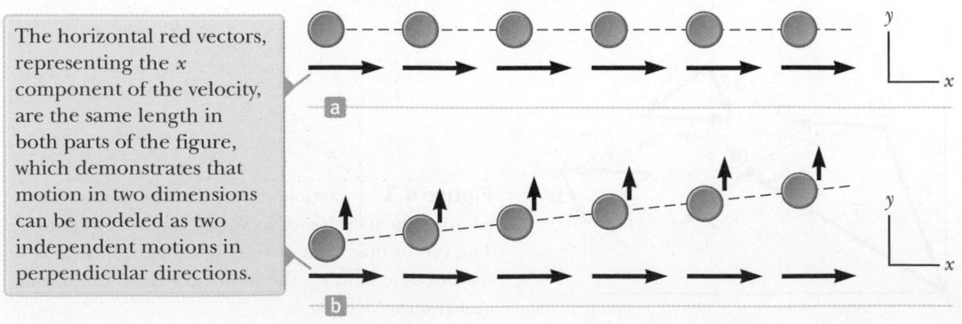

Figure 4.4 (a) A puck moves across a horizontal air hockey table at constant velocity in the x direction. (b) After a puff of air in the y direction is applied to the puck, the puck has gained a y component of velocity, but the x component is unaffected by the force in the perpendicular direction.

where x, y, and $\vec{\mathbf{r}}$ change with time as the particle moves while the unit vectors $\hat{\mathbf{i}}$ and $\hat{\mathbf{j}}$ remain constant. If the position vector is known, the velocity of the particle can be obtained from Equations 4.3 and 4.6, which give

$$\vec{\mathbf{v}} = \frac{d\vec{\mathbf{r}}}{dt} = \frac{dx}{dt}\hat{\mathbf{i}} + \frac{dy}{dt}\hat{\mathbf{j}} = v_x\hat{\mathbf{i}} + v_y\hat{\mathbf{j}} \quad \textbf{(4.7)}$$

Because the acceleration $\vec{\mathbf{a}}$ of the particle is assumed constant in this discussion, its components a_x and a_y also are constants. Therefore, we can model the particle as a particle under constant acceleration independently in each of the two directions and apply the equations of kinematics separately to the x and y components of the velocity vector. Substituting, from Equation 2.13, $v_{xf} = v_{xi} + a_x t$ and $v_{yf} = v_{yi} + a_y t$ into Equation 4.7 to determine the final velocity at any time t, we obtain

$$\vec{\mathbf{v}}_f = (v_{xi} + a_x t)\hat{\mathbf{i}} + (v_{yi} + a_y t)\hat{\mathbf{j}} = (v_{xi}\hat{\mathbf{i}} + v_{yi}\hat{\mathbf{j}}) + (a_x\hat{\mathbf{i}} + a_y\hat{\mathbf{j}})t$$

$$\vec{\mathbf{v}}_f = \vec{\mathbf{v}}_i + \vec{\mathbf{a}}t \quad \textbf{(4.8)}$$

◀ **Velocity vector as a function of time**

This result states that the velocity of a particle at some time t equals the vector sum of its initial velocity $\vec{\mathbf{v}}_i$ at time $t = 0$ and the additional velocity $\vec{\mathbf{a}}t$ acquired at time t as a result of constant acceleration. Equation 4.8 is the vector version of Equation 2.13.

Similarly, from Equation 2.16 we know that the x and y coordinates of a particle moving with constant acceleration are

$$x_f = x_i + v_{xi}t + \tfrac{1}{2}a_x t^2 \qquad y_f = y_i + v_{yi}t + \tfrac{1}{2}a_y t^2$$

Substituting these expressions into Equation 4.6 (and labeling the final position vector $\vec{\mathbf{r}}_f$) gives

$$\vec{\mathbf{r}}_f = (x_i + v_{xi}t + \tfrac{1}{2}a_x t^2)\hat{\mathbf{i}} + (y_i + v_{yi}t + \tfrac{1}{2}a_y t^2)\hat{\mathbf{j}}$$

$$= (x_i\hat{\mathbf{i}} + y_i\hat{\mathbf{j}}) + (v_{xi}\hat{\mathbf{i}} + v_{yi}\hat{\mathbf{j}})t + \tfrac{1}{2}(a_x\hat{\mathbf{i}} + a_y\hat{\mathbf{j}})t^2$$

$$\vec{\mathbf{r}}_f = \vec{\mathbf{r}}_i + \vec{\mathbf{v}}_i t + \tfrac{1}{2}\vec{\mathbf{a}}t^2 \quad \textbf{(4.9)}$$

◀ **Position vector as a function of time**

which is the vector version of Equation 2.16. Equation 4.9 tells us that the position vector $\vec{\mathbf{r}}_f$ of a particle is the vector sum of the original position $\vec{\mathbf{r}}_i$, a displacement $\vec{\mathbf{v}}_i t$ arising from the initial velocity of the particle, and a displacement $\tfrac{1}{2}\vec{\mathbf{a}}t^2$ resulting from the constant acceleration of the particle.

Graphical representations of Equations 4.8 and 4.9 are shown in Active Figure 4.5. The components of the position and velocity vectors are also illustrated in the figure. Notice from Active Figure 4.5a that $\vec{\mathbf{v}}_f$ is generally not along the direction of either $\vec{\mathbf{v}}_i$ or $\vec{\mathbf{a}}$ because the relationship between these quantities is a vector

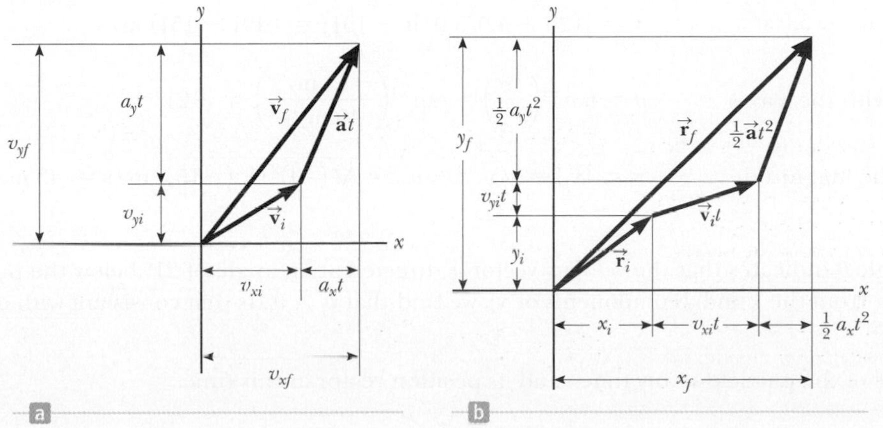

ACTIVE FIGURE 4.5

Vector representations and components of (a) the velocity and (b) the position of a particle moving with a constant acceleration $\vec{\mathbf{a}}$.

expression. For the same reason, from Active Figure 4.5b we see that $\vec{\mathbf{r}}_f$ is generally not along the direction of $\vec{\mathbf{r}}_i$, $\vec{\mathbf{v}}_i$, or $\vec{\mathbf{a}}$. Finally, notice that $\vec{\mathbf{v}}_f$ and $\vec{\mathbf{r}}_f$ are generally not in the same direction.

Example 4.1 Motion in a Plane

A particle moves in the xy plane, starting from the origin at $t = 0$ with an initial velocity having an x component of 20 m/s and a y component of −15 m/s. The particle experiences an acceleration in the x direction, given by $a_x = 4.0$ m/s^2.

(A) Determine the total velocity vector at any time.

SOLUTION

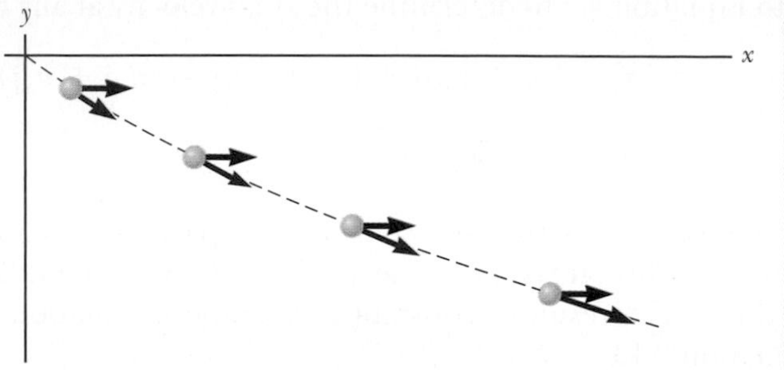

Figure 4.6 (Example 4.1) Motion diagram for the particle.

Conceptualize The components of the initial velocity tell us that the particle starts by moving toward the right and downward. The x component of velocity starts at 20 m/s and increases by 4.0 m/s every second. The y component of velocity never changes from its initial value of −15 m/s. We sketch a motion diagram of the situation in Figure 4.6. Because the particle is accelerating in the $+x$ direction, its velocity component in this direction increases and the path curves as shown in the diagram. Notice that the spacing between successive images increases as time goes on because the speed is increasing. The placement of the acceleration and velocity vectors in Figure 4.6 helps us further conceptualize the situation.

Categorize Because the initial velocity has components in both the x and y directions, we categorize this problem as one involving a particle moving in two dimensions. Because the particle only has an x component of acceleration, we model it as a particle under constant acceleration in the x direction and a particle under constant velocity in the y direction.

Analyze To begin the mathematical analysis, we set $v_{xi} = 20$ m/s, $v_{yi} = -15$ m/s, $a_x = 4.0$ m/s^2, and $a_y = 0$.

Use Equation 4.8 for the velocity vector:

$$\vec{\mathbf{v}}_f = \vec{\mathbf{v}}_i + \vec{\mathbf{a}}t = (v_{xi} + a_x t)\hat{\mathbf{i}} + (v_{yi} + a_y t)\hat{\mathbf{j}}$$

Substitute numerical values with the velocity in meters per second and the time in seconds:

$$\vec{\mathbf{v}}_f = [20 + (4.0)t]\hat{\mathbf{i}} + [-15 + (0)t]\hat{\mathbf{j}}$$

$$(1)\quad \vec{\mathbf{v}}_f = [(20 + 4.0t)\hat{\mathbf{i}} - 15\hat{\mathbf{j}}]$$

Finalize Notice that the x component of velocity increases in time while the y component remains constant; this result is consistent with our prediction.

(B) Calculate the velocity and speed of the particle at $t = 5.0$ s and the angle the velocity vector makes with the x axis.

SOLUTION

Analyze

Evaluate the result from Equation (1) at $t = 5.0$ s:

$$\vec{\mathbf{v}}_f = [(20 + 4.0(5.0))\hat{\mathbf{i}} - 15\hat{\mathbf{j}}] = (40\hat{\mathbf{i}} - 15\hat{\mathbf{j}})\text{ m/s}$$

Determine the angle θ that $\vec{\mathbf{v}}_f$ makes with the x axis at $t = 5.0$ s:

$$\theta = \tan^{-1}\left(\frac{v_{yf}}{v_{xf}}\right) = \tan^{-1}\left(\frac{-15\text{ m/s}}{40\text{ m/s}}\right) = -21°$$

Evaluate the speed of the particle as the magnitude of $\vec{\mathbf{v}}_f$:

$$v_f = |\vec{\mathbf{v}}_f| = \sqrt{v_{xf}^2 + v_{yf}^2} = \sqrt{(40)^2 + (-15)^2}\text{ m/s} = 43\text{ m/s}$$

Finalize The negative sign for the angle θ indicates that the velocity vector is directed at an angle of 21° below the positive x axis. Notice that if we calculate v_i from the x and y components of $\vec{\mathbf{v}}_i$, we find that $v_f > v_i$. Is that consistent with our prediction?

(C) Determine the x and y coordinates of the particle at any time t and its position vector at this time.

4.1 *cont.*

SOLUTION

Analyze

Use the components of Equation 4.9 with $x_i = y_i = 0$ at $t = 0$ with x and y in meters and t in seconds:

$$x_f = v_{xi}t + \tfrac{1}{2}a_x t^2 = \boxed{20t + 2.0t^2}$$

$$y_f = v_{yi}t = \boxed{-15t}$$

Express the position vector of the particle at any time t:

$$\vec{\mathbf{r}}_f = x_f\hat{\mathbf{i}} + y_f\hat{\mathbf{j}} = \boxed{(20t + 2.0t^2)\hat{\mathbf{i}} - 15t\hat{\mathbf{j}}}$$

Finalize Let us now consider a limiting case for very large values of t.

WHAT IF? What if we wait a very long time and then observe the motion of the particle? How would we describe the motion of the particle for large values of the time?

Answer Looking at Figure 4.6, we see the path of the particle curving toward the x axis. There is no reason to assume this tendency will change, which suggests that the path will become more and more parallel to the x axis as time grows large. Mathematically, Equation (1) shows that the y component of the velocity remains constant while the x component grows linearly with t. Therefore, when t is very large, the x component of the velocity will be much larger than the y component, suggesting that the velocity vector becomes more and more parallel to the x axis. Both x_f and y_f continue to grow with time, although x_f grows much faster.

4.3 Projectile Motion

Anyone who has observed a baseball in motion has observed projectile motion. The ball moves in a curved path and returns to the ground. **Projectile motion** of an object is simple to analyze if we make two assumptions: (1) the free-fall acceleration is constant over the range of motion and is directed downward,[1] and (2) the effect of air resistance is negligible.[2] With these assumptions, we find that the path of a projectile, which we call its *trajectory,* is *always* a parabola as shown in Active Figure 4.7 (page 80). **We use these assumptions throughout this chapter.**

The expression for the position vector of the projectile as a function of time follows directly from Equation 4.9, with its acceleration being that due to gravity, $\vec{\mathbf{a}} = \vec{\mathbf{g}}$:

$$\vec{\mathbf{r}}_f = \vec{\mathbf{r}}_i + \vec{\mathbf{v}}_i t + \tfrac{1}{2}\vec{\mathbf{g}}t^2 \quad \textbf{(4.10)}$$

where the initial x and y components of the velocity of the projectile are

$$v_{xi} = v_i \cos\theta_i \qquad v_{yi} = v_i \sin\theta_i \quad \textbf{(4.11)}$$

The expression in Equation 4.10 is plotted in Figure 4.8 (page 80), for a projectile launched from the origin, so that $\vec{\mathbf{r}}_i = 0$. The final position of a particle can be considered to be the superposition of its initial position $\vec{\mathbf{r}}_i$; the term $\vec{\mathbf{v}}_i t$, which is its displacement if no acceleration were present; and the term $\frac{1}{2}\vec{\mathbf{g}}t^2$ that arises from its acceleration due to gravity. In other words, if there were no gravitational acceleration, the particle would continue to move along a straight path in the direction of $\vec{\mathbf{v}}_i$. Therefore, the vertical distance $\frac{1}{2}\vec{\mathbf{g}}t^2$ through which the particle "falls" off the

Pitfall Prevention 4.2

Acceleration at the Highest Point
As discussed in Pitfall Prevention 2.8, many people claim that the acceleration of a projectile at the topmost point of its trajectory is zero. This mistake arises from confusion between zero vertical velocity and zero acceleration. If the projectile were to experience zero acceleration at the highest point, its velocity at that point would not change; rather, the projectile would move horizontally at constant speed from then on! That does not happen, however, because the acceleration is *not* zero anywhere along the trajectory.

Lester Lefkowitz/Taxi/Getty Images

A welder cuts holes through a heavy metal construction beam with a hot torch. The sparks generated in the process follow parabolic paths.

[1]This assumption is reasonable as long as the range of motion is small compared with the radius of the Earth (6.4×10^6 m). In effect, this assumption is equivalent to assuming the Earth is flat over the range of motion considered.

[2]This assumption is often *not* justified, especially at high velocities. In addition, any spin imparted to a projectile, such as that applied when a pitcher throws a curve ball, can give rise to some very interesting effects associated with aerodynamic forces, which will be discussed in Chapter 14.

ACTIVE FIGURE 4.7

The parabolic path of a projectile that leaves the origin with a velocity $\vec{\mathbf{v}}_i$. The velocity vector $\vec{\mathbf{v}}$ changes with time in both magnitude and direction. This change is the result of acceleration $\vec{\mathbf{a}} = \vec{\mathbf{g}}$ in the negative y direction.

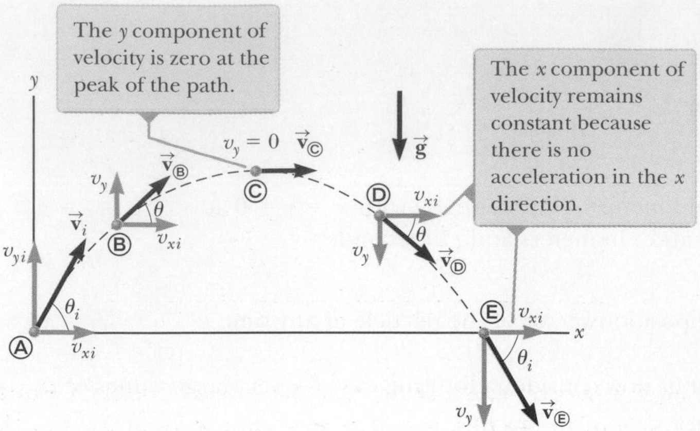

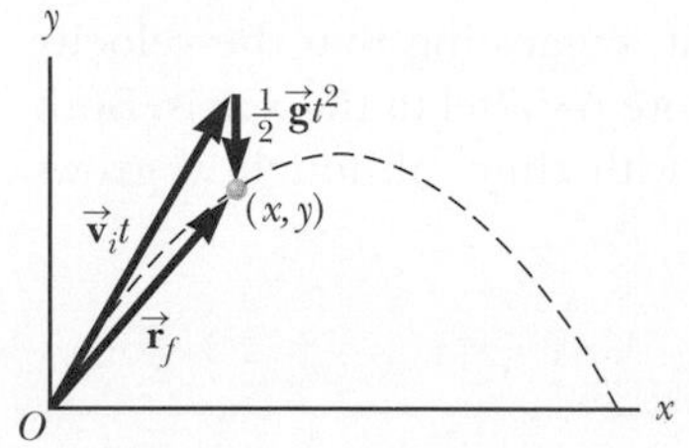

Figure 4.8 The position vector $\vec{\mathbf{r}}_f$ of a projectile launched from the origin whose initial velocity at the origin is $\vec{\mathbf{v}}_i$. The vector $\vec{\mathbf{v}}_i t$ would be the displacement of the projectile if gravity were absent, and the vector $\frac{1}{2}\vec{\mathbf{g}}t^2$ is its vertical displacement from a straight-line path due to its downward gravitational acceleration.

straight-line path is the same distance that an object dropped from rest would fall during the same time interval.

In Section 4.2, we stated that two-dimensional motion with constant acceleration can be analyzed as a combination of two independent motions in the x and y directions, with accelerations a_x and a_y. Projectile motion can also be handled in this way, with zero acceleration in the x direction and a constant acceleration in the y direction, $a_y = -g$. Therefore, when analyzing projectile motion, model it to be the superposition of two motions: (1) motion of a particle under constant velocity in the horizontal direction and (2) motion of a particle under constant acceleration (free fall) in the vertical direction. The horizontal and vertical components of a projectile's motion are completely independent of each other and can be handled separately, with time t as the common variable for both components.

***Quick Quiz* 4.2** **(i)** As a projectile thrown upward moves in its parabolic path (such as in Fig. 4.8), at what point along its path are the velocity and acceleration vectors for the projectile perpendicular to each other? (a) nowhere (b) the highest point (c) the launch point **(ii)** From the same choices, at what point are the velocity and acceleration vectors for the projectile parallel to each other?

Horizontal Range and Maximum Height of a Projectile

Let us assume a projectile is launched from the origin at $t_i = 0$ with a positive v_{yi} component as shown in Figure 4.9 and returns to the same horizontal level. This situation is common in sports, where baseballs, footballs, and golf balls often land at the same level from which they were launched.

Two points in this motion are especially interesting to analyze: the peak point Ⓐ, which has Cartesian coordinates $(R/2, h)$, and the point Ⓑ, which has coordinates $(R, 0)$. The distance R is called the *horizontal range* of the projectile, and the distance h is its *maximum height*. Let us find h and R mathematically in terms of v_i, θ_i, and g.

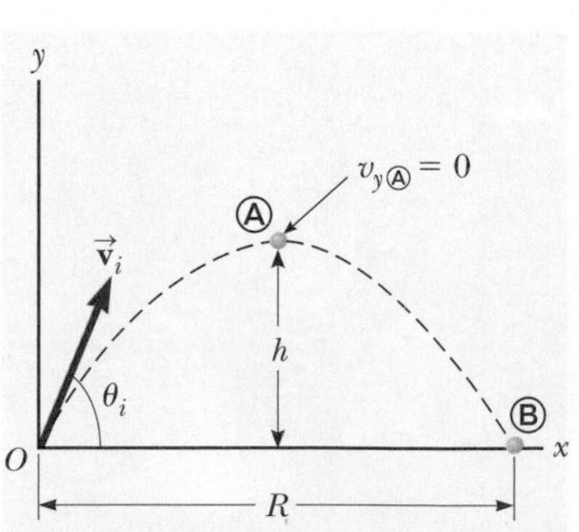

Figure 4.9 A projectile launched over a flat surface from the origin at $t_i = 0$ with an initial velocity $\vec{\mathbf{v}}_i$. The maximum height of the projectile is h, and the horizontal range is R. At Ⓐ, the peak of the trajectory, the particle has coordinates $(R/2, h)$.

We can determine h by noting that at the peak $v_{yⒶ} = 0$. Therefore, we can use the y component of Equation 4.8 to determine the time $t_{Ⓐ}$ at which the projectile reaches the peak:

$$v_{yf} = v_{yi} + a_y t$$

$$0 = v_i \sin\theta_i - g t_{Ⓐ}$$

$$t_{Ⓐ} = \frac{v_i \sin\theta_i}{g}$$

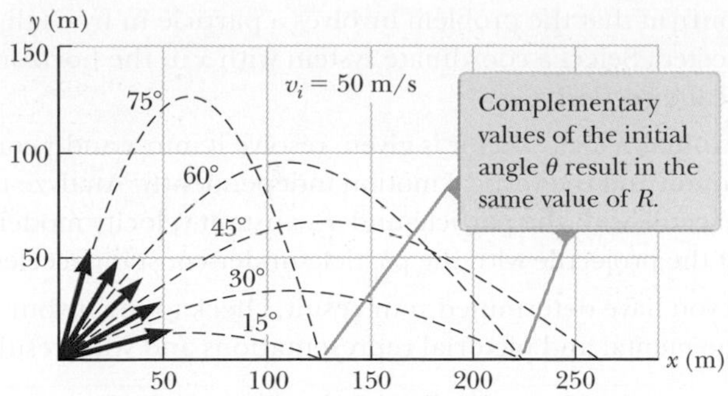

ACTIVE FIGURE 4.10

A projectile launched over a flat surface from the origin with an initial speed of 50 m/s at various angles of projection.

Substituting this expression for $t_{Ⓐ}$ into the y component of Equation 4.9 and replacing $y = y_{Ⓐ}$ with h, we obtain an expression for h in terms of the magnitude and direction of the initial velocity vector:

$$h = (v_i \sin \theta_i)\frac{v_i \sin \theta_i}{g} - \tfrac{1}{2}g\left(\frac{v_i \sin \theta_i}{g}\right)^2$$

$$h = \frac{v_i^2 \sin^2 \theta_i}{2g} \quad \textbf{(4.12)}$$

The range R is the horizontal position of the projectile at a time that is twice the time at which it reaches its peak, that is, at time $t_{Ⓑ} = 2t_{Ⓐ}$. Using the x component of Equation 4.9, noting that $v_{xi} = v_{xⒷ} = v_i \cos \theta_i$, and setting $x_{Ⓑ} = R$ at $t = 2t_{Ⓐ}$, we find that

$$R = v_{xi}t_{Ⓑ} = (v_i \cos \theta_i)2t_{Ⓐ}$$

$$= (v_i \cos \theta_i)\frac{2v_i \sin \theta_i}{g} = \frac{2v_i^2 \sin \theta_i \cos \theta_i}{g}$$

Using the identity $\sin 2\theta = 2 \sin \theta \cos \theta$ (see Appendix B.4), we can write R in the more compact form

$$R = \frac{v_i^2 \sin 2\theta_i}{g} \quad \textbf{(4.13)}$$

The maximum value of R from Equation 4.13 is $R_{max} = v_i^2/g$. This result makes sense because the maximum value of $\sin 2\theta_i$ is 1, which occurs when $2\theta_i = 90°$. Therefore, R is a maximum when $\theta_i = 45°$.

Active Figure 4.10 illustrates various trajectories for a projectile having a given initial speed but launched at different angles. As you can see, the range is a maximum for $\theta_i = 45°$. In addition, for any θ_i other than 45°, a point having Cartesian coordinates $(R, 0)$ can be reached by using either one of two complementary values of θ_i, such as 75° and 15°. Of course, the maximum height and time of flight for one of these values of θ_i are different from the maximum height and time of flight for the complementary value.

Pitfall Prevention 4.3

The Height and Range Equations

Equation 4.13 is useful for calculating R only for a symmetric path as shown in Active Figure 4.10. If the path is not symmetric, *do not use this equation.* The general expressions given by Equations 4.8 and 4.9 are the *more important* results because they give the position and velocity components of *any* particle moving in two dimensions at *any* time t.

***Quick Quiz* 4.3** Rank the launch angles for the five paths in Active Figure 4.10 with respect to time of flight from the shortest time of flight to the longest.

Problem-Solving Strategy

PROJECTILE MOTION

We suggest you use the following approach when solving projectile motion problems.

1. Conceptualize. Think about what is going on physically in the problem. Establish the mental representation by imagining the projectile moving along its trajectory.

2. Categorize. Confirm that the problem involves a particle in free fall and that air resistance is neglected. Select a coordinate system with x in the horizontal direction and y in the vertical direction.

3. Analyze. If the initial velocity vector is given, resolve it into x and y components. Treat the horizontal motion and the vertical motion independently. Analyze the horizontal motion of the projectile with the particle-under-constant-velocity model. Analyze the vertical motion of the projectile with the particle-under-constant-acceleration model.

4. Finalize. Once you have determined your result, check to see if your answers are consistent with the mental and pictorial representations and your results are realistic.

Example 4.2 The Long Jump

A long jumper (Fig. 4.11) leaves the ground at an angle of 20.0° above the horizontal and at a speed of 11.0 m/s.

(A) How far does he jump in the horizontal direction?

Figure 4.11 (Example 4.2) Romain Barras of France competes in the men's decathlon long jump at the 2008 Beijing Olympic Games.

SOLUTION

Conceptualize The arms and legs of a long jumper move in a complicated way, but we will ignore this motion. We conceptualize the motion of the long jumper as equivalent to that of a simple projectile.

Categorize We categorize this example as a projectile motion problem. Because the initial speed and launch angle are given and because the final height is the same as the initial height, we further categorize this problem as satisfying the conditions for which Equations 4.12 and 4.13 can be used. This approach is the most direct way to analyze this problem, although the general methods that have been described will always give the correct answer.

Analyze

Use Equation 4.13 to find the range of the jumper:

$$R = \frac{v_i^2 \sin 2\theta_i}{g} = \frac{(11.0 \text{ m/s})^2 \sin 2(20.0°)}{9.80 \text{ m/s}^2} = \boxed{7.94 \text{ m}}$$

(B) What is the maximum height reached?

SOLUTION

Analyze

Find the maximum height reached by using Equation 4.12:

$$h = \frac{v_i^2 \sin^2\theta_i}{2g} = \frac{(11.0 \text{ m/s})^2(\sin 20.0°)^2}{2(9.80 \text{ m/s}^2)} = \boxed{0.722 \text{ m}}$$

Finalize Find the answers to parts (A) and (B) using the general method. The results should agree. Treating the long jumper as a particle is an oversimplification. Nevertheless, the values obtained are consistent with experience in sports. We can model a complicated system such as a long jumper as a particle and still obtain reasonable results.

Example 4.3 A Bull's-Eye Every Time

In a popular lecture demonstration, a projectile is fired at a target in such a way that the projectile leaves the gun at the same time the target is dropped from rest. Show that if the gun is initially aimed at the stationary target, the projectile hits the falling target as shown in Figure 4.12a.

SOLUTION

Conceptualize We conceptualize the problem by studying Figure 4.12a. Notice that the problem does not ask for numerical values. The expected result must involve an algebraic argument.

4.3 *cont.*

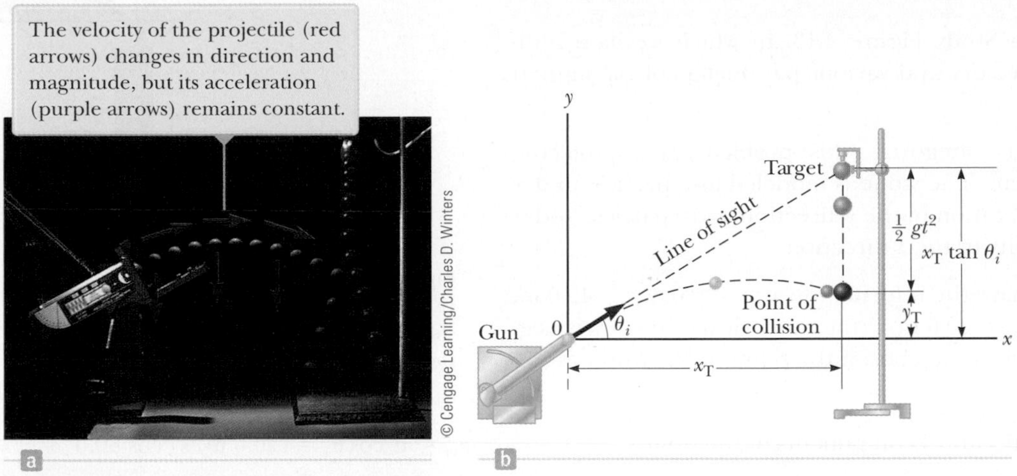

Figure 4.12 (Example 4.3) (a) Multiflash photograph of the projectile–target demonstration. If the gun is aimed directly at the target and is fired at the same instant the target begins to fall, the projectile will hit the target. (b) Schematic diagram of the projectile–target demonstration.

Categorize Because both objects are subject only to gravity, we categorize this problem as one involving two objects in free fall, the target moving in one dimension and the projectile moving in two.

Analyze The target T is modeled as a particle under constant acceleration in one dimension. Figure 4.12b shows that the initial y coordinate y_{iT} of the target is $x_T \tan\theta_i$ and its initial velocity is zero. It falls with acceleration $a_y = -g$. The projectile P is modeled as a particle under constant acceleration in the y direction and a particle under constant velocity in the x direction.

Write an expression for the y coordinate of the target at any moment after release, noting that its initial velocity is zero:

$$(1)\quad y_T = y_{iT} + (0)t - \tfrac{1}{2}gt^2 = x_T \tan\theta_i - \tfrac{1}{2}gt^2$$

Write an expression for the y coordinate of the projectile at any moment:

$$(2)\quad y_P = y_{iP} + v_{yiP}t - \tfrac{1}{2}gt^2 = 0 + (v_{iP}\sin\theta_i)t - \tfrac{1}{2}gt^2 = (v_{iP}\sin\theta_i)t - \tfrac{1}{2}gt^2$$

Write an expression for the x coordinate of the projectile at any moment:

$$x_P = x_{iP} + v_{xiP}t = 0 + (v_{iP}\cos\theta_i)t = (v_{iP}\cos\theta_i)t$$

Solve this expression for time as a function of the horizontal position of the projectile:

$$t = \frac{x_P}{v_{iP}\cos\theta_i}$$

Substitute this expression into Equation (2):

$$(3)\quad y_P = (v_{iP}\sin\theta_i)\left(\frac{x_P}{v_{iP}\cos\theta_i}\right) - \tfrac{1}{2}gt^2 = x_P\tan\theta_i - \tfrac{1}{2}gt^2$$

Compare Equations (1) and (3). We see that when the x coordinates of the projectile and target are the same—that is, when $x_T = x_P$—their y coordinates given by Equations (1) and (3) are the same and a collision results.

Finalize Note that a collision can result only when $v_{iP}\sin\theta_i \geq \sqrt{gd/2}$, where d is the initial elevation of the target above the floor. If $v_{iP}\sin\theta_i$ is less than this value, the projectile strikes the floor before reaching the target.

Example 4.4 That's Quite an Arm!

A stone is thrown from the top of a building upward at an angle of 30.0° to the horizontal with an initial speed of 20.0 m/s as shown in Figure 4.13 (page 84). The height from which the stone is thrown is 45.0 m above the ground.

(A) How long does it take the stone to reach the ground?

continued

4.4 *cont.*

SOLUTION

Conceptualize Study Figure 4.13, in which we have indicated the trajectory and various parameters of the motion of the stone.

Categorize We categorize this problem as a projectile motion problem. The stone is modeled as a particle under constant acceleration in the y direction and a particle under constant velocity in the x direction.

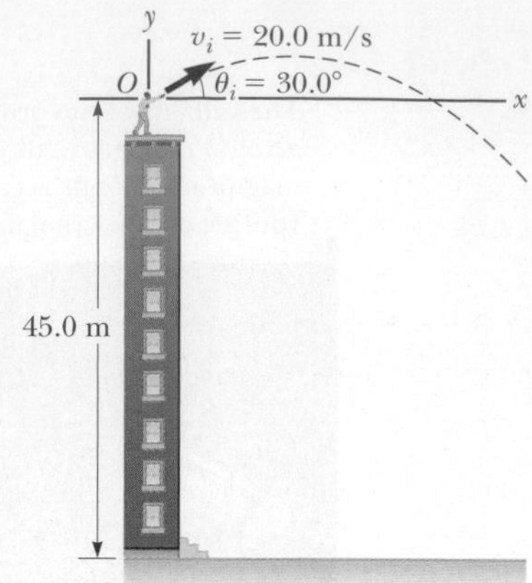

Figure 4.13 (Example 4.4) A stone is thrown from the top of a building.

Analyze We have the information $x_i = y_i = 0$, $y_f = -45.0$ m, $a_y = -g$, and $v_i = 20.0$ m/s (the numerical value of y_f is negative because we have chosen the point of the throw as the origin).

Find the initial x and y components of the stone's velocity:

$$v_{xi} = v_i \cos\theta_i = (20.0 \text{ m/s}) \cos 30.0° = 17.3 \text{ m/s}$$

$$v_{yi} = v_i \sin\theta_i = (20.0 \text{ m/s}) \sin 30.0° = 10.0 \text{ m/s}$$

Express the vertical position of the stone from the vertical component of Equation 4.9:

$$y_f = y_i + v_{yi}t + \tfrac{1}{2}a_y t^2$$

Substitute numerical values:

$$-45.0 \text{ m} = 0 + (10.0 \text{ m/s})t + \tfrac{1}{2}(-9.80 \text{ m/s}^2)t^2$$

Solve the quadratic equation for t:

$$t = \boxed{4.22 \text{ s}}$$

(B) What is the speed of the stone just before it strikes the ground?

SOLUTION

Analyze Use the y component of Equation 4.8 to obtain the y component of the velocity of the stone just before it strikes the ground:

$$v_{yf} = v_{yi} + a_y t$$

Substitute numerical values, using $t = 4.22$ s:

$$v_{yf} = 10.0 \text{ m/s} + (-9.80 \text{ m/s}^2)(4.22 \text{ s}) = -31.3 \text{ m/s}$$

Use this component with the horizontal component $v_{xf} = v_{xi} = 17.3$ m/s to find the speed of the stone at $t = 4.22$ s:

$$v_f = \sqrt{v_{xf}^2 + v_{yf}^2} = \sqrt{(17.3 \text{ m/s})^2 + (-31.3 \text{ m/s})^2} = \boxed{35.8 \text{ m/s}}$$

Finalize Is it reasonable that the y component of the final velocity is negative? Is it reasonable that the final speed is larger than the initial speed of 20.0 m/s?

WHAT IF? What if a horizontal wind is blowing in the same direction as the stone is thrown and it causes the stone to have a horizontal acceleration component $a_x = 0.500$ m/s^2? Which part of this example, (A) or (B), will have a different answer?

Answer Recall that the motions in the x and y directions are independent. Therefore, the horizontal wind cannot affect the vertical motion. The vertical motion determines the time of the projectile in the air, so the answer to part (A) does not change. The wind causes the horizontal velocity component to increase with time, so the final speed will be larger in part (B). Taking $a_x = 0.500$ m/s^2, we find $v_{xf} = 19.4$ m/s and $v_f = 36.9$ m/s.

Example 4.5 The End of the Ski Jump

A ski jumper leaves the ski track moving in the horizontal direction with a speed of 25.0 m/s as shown in Figure 4.14. The landing incline below her falls off with a slope of 35.0°. Where does she land on the incline?

4.5 cont.

SOLUTION

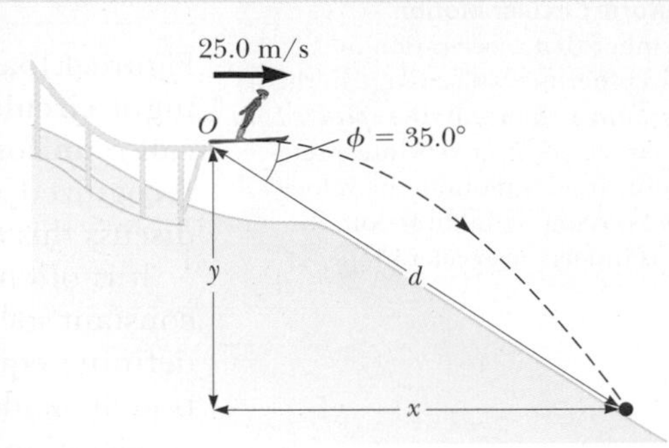

Figure 4.14 (Example 4.5) A ski jumper leaves the track moving in a horizontal direction.

Conceptualize We can conceptualize this problem based on memories of observing winter Olympic ski competitions. We estimate the skier to be airborne for perhaps 4 s and to travel a distance of about 100 m horizontally. We should expect the value of d, the distance traveled along the incline, to be of the same order of magnitude.

Categorize We categorize the problem as one of a particle in projectile motion.

Analyze It is convenient to select the beginning of the jump as the origin. The initial velocity components are $v_{xi} = 25.0$ m/s and $v_{yi} = 0$. From the right triangle in Figure 4.14, we see that the jumper's x and y coordinates at the landing point are given by $x_f = d\cos\phi$ and $y_f = -d\sin\phi$.

Express the coordinates of the jumper as a function of time:

$$(1)\quad x_f = v_{xi}t$$

$$(2)\quad y_f = v_{yi}t + \tfrac{1}{2}a_y t^2 = -\tfrac{1}{2}gt^2$$

Substitute the values of x_f and y_f at the landing point:

$$(3)\quad d\cos\phi = v_{xi}t$$

$$(4)\quad -d\sin\phi = -\tfrac{1}{2}gt^2$$

Solve Equation (3) for t and substitute the result into Equation (4):

$$-d\sin\phi = -\tfrac{1}{2}g\left(\frac{d\cos\phi}{v_{xi}}\right)^2$$

Solve for d:

$$d = \frac{2v_{xi}^2\sin\phi}{g\cos^2\phi} = \frac{2(25.0\text{ m/s})^2\sin 35.0°}{(9.80\text{ m/s}^2)\cos^2 35.0°} = 109\text{ m}$$

Evaluate the x and y coordinates of the point at which the skier lands:

$$x_f = d\cos\phi = (109\text{ m})\cos 35.0° = \boxed{89.3\text{ m}}$$

$$y_f = -d\sin\phi = -(109\text{ m})\sin 35.0° = \boxed{-62.5\text{ m}}$$

Finalize Let us compare these results with our expectations. We expected the horizontal distance to be on the order of 100 m, and our result of 89.3 m is indeed on this order of magnitude. It might be useful to calculate the time interval that the jumper is in the air and compare it with our estimate of about 4 s.

WHAT IF? Suppose everything in this example is the same except the ski jump is curved so that the jumper is projected upward at an angle from the end of the track. Is this design better in terms of maximizing the length of the jump?

Answer If the initial velocity has an upward component, the skier will be in the air longer and should therefore travel farther. Tilting the initial velocity vector upward, however, will reduce the horizontal component of the initial velocity. Therefore, angling the end of the ski track upward at a *large* angle may actually *reduce* the distance. Consider the extreme case: the skier is projected at 90° to the horizontal and simply goes up and comes back down at the end of the ski track! This argument suggests that there must be an optimal angle between 0° and 90° that represents a balance between making the flight time longer and the horizontal velocity component smaller.

Let us find this optimal angle mathematically. We modify Equations (1) through (4) in the following way, assuming the skier is projected at an angle θ with respect to the horizontal over a landing incline sloped with an arbitrary angle ϕ:

$$(1)\text{ and }(3)\quad\rightarrow\quad x_f = (v_i\cos\theta)t = d\cos\phi$$

$$(2)\text{ and }(4)\quad\rightarrow\quad y_f = (v_i\sin\theta)t - \tfrac{1}{2}gt^2 = -d\sin\phi$$

By eliminating the time t between these equations and using differentiation to maximize d in terms of θ, we arrive (after several steps; see Problem 72) at the following equation for the angle θ that gives the maximum value of d:

$$\theta = 45° - \frac{\phi}{2}$$

For the slope angle in Figure 4.14, $\phi = 35.0°$; this equation results in an optimal launch angle of $\theta = 27.5°$. For a slope angle of $\phi = 0°$, which represents a horizontal plane, this equation gives an optimal launch angle of $\theta = 45°$, as we would expect (see Active Figure 4.10).

Pitfall Prevention 4.4

Acceleration of a Particle in Uniform Circular Motion
Remember that acceleration in physics is defined as a change in the *velocity,* not a change in the *speed* (contrary to the everyday interpretation). In circular motion, the velocity vector is changing in direction, so there is indeed an acceleration.

4.4 Analysis Model: Particle in Uniform Circular Motion

Figure 4.15a shows a car moving in a circular path; we describe this motion by calling it **circular motion.** If the car is moving on this path with *constant speed v,* we call it **uniform circular motion.** Because it occurs so often, this type of motion is recognized as an analysis model called the **particle in uniform circular motion.** We discuss this model in this section.

It is often surprising to students to find that even though an object moves at a constant speed in a circular path, *it still has an acceleration.* To see why, consider the defining equation for acceleration, $\vec{\mathbf{a}} = d\vec{\mathbf{v}}/dt$ (Eq. 4.5). Notice that the acceleration depends on the change in the *velocity.* Because velocity is a vector quantity, an acceleration can occur in two ways as mentioned in Section 4.1: by a change in the *magnitude* of the velocity and by a change in the *direction* of the velocity. The latter situation occurs for an object moving with constant speed in a circular path. The constant-magnitude velocity vector is always tangent to the path of the object and perpendicular to the radius of the circular path.

We now show that the acceleration vector in uniform circular motion is always perpendicular to the path and always points toward the center of the circle. If that were not true, there would be a component of the acceleration parallel to the path and therefore parallel to the velocity vector. Such an acceleration component would lead to a change in the speed of the particle along the path. This situation, however, is inconsistent with our setup of the situation: the particle moves with constant speed along the path. Therefore, for *uniform* circular motion, the acceleration vector can only have a component perpendicular to the path, which is toward the center of the circle.

Let us now find the magnitude of the acceleration of the particle. Consider the diagram of the position and velocity vectors in Figure 4.15b. The figure also shows the vector representing the change in position $\Delta\vec{\mathbf{r}}$ for an arbitrary time interval. The particle follows a circular path of radius r, part of which is shown by the dashed curve. The particle is at Ⓐ at time t_i, and its velocity at that time is $\vec{\mathbf{v}}_i$; it is at Ⓑ at some later time t_f, and its velocity at that time is $\vec{\mathbf{v}}_f$. Let us also assume $\vec{\mathbf{v}}_i$ and $\vec{\mathbf{v}}_f$ differ only in direction; their magnitudes are the same (that is, $v_i = v_f = v$ because it is *uniform* circular motion).

In Figure 4.15c, the velocity vectors in Figure 4.15b have been redrawn tail to tail. The vector $\Delta\vec{\mathbf{v}}$ connects the tips of the vectors, representing the vector addition $\vec{\mathbf{v}}_f = \vec{\mathbf{v}}_i + \Delta\vec{\mathbf{v}}$. In both Figures 4.15b and 4.15c, we can identify triangles that help us analyze the motion. The angle $\Delta\theta$ between the two position vectors in Figure 4.15b is the same as the angle between the velocity vectors in Figure 4.15c because the velocity vector $\vec{\mathbf{v}}$ is always perpendicular to the position vector $\vec{\mathbf{r}}$. Therefore, the two triangles are *similar.* (Two triangles are similar if the angle between any two sides is the same for both triangles and if the ratio of the lengths of these sides is the same.) We can now write a relationship between the lengths of the sides for the two triangles in Figures 4.15b and 4.15c:

$$\frac{|\Delta\vec{\mathbf{v}}|}{v} = \frac{|\Delta\vec{\mathbf{r}}|}{r}$$

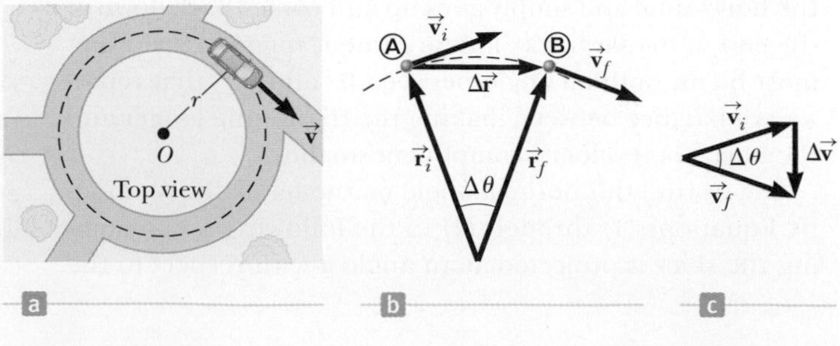

Figure 4.15 (a) A car moving along a circular path at constant speed experiences uniform circular motion. (b) As a particle moves along a portion of a circular path from Ⓐ to Ⓑ, its velocity vector changes from $\vec{\mathbf{v}}_i$ to $\vec{\mathbf{v}}_f$. (c) The construction for determining the direction of the change in velocity $\Delta\vec{\mathbf{v}}$, which is toward the center of the circle for small $\Delta\vec{\mathbf{r}}$.

where $v = v_i = v_f$ and $r = r_i = r_f$. This equation can be solved for $|\Delta\vec{\mathbf{v}}|$, and the expression obtained can be substituted into Equation 4.4, $\vec{\mathbf{a}}_{avg} = \Delta\vec{\mathbf{v}}/\Delta t$, to give the magnitude of the average acceleration over the time interval for the particle to move from Ⓐ to Ⓑ:

$$|\vec{\mathbf{a}}_{avg}| = \frac{|\Delta\vec{\mathbf{v}}|}{|\Delta t|} = \frac{v|\Delta\vec{\mathbf{r}}|}{r\ \Delta t}$$

Now imagine that points Ⓐ and Ⓑ in Figure 4.15b become extremely close together. As Ⓐ and Ⓑ approach each other, Δt approaches zero, $|\Delta\vec{\mathbf{r}}|$ approaches the distance traveled by the particle along the circular path, and the ratio $|\Delta\vec{\mathbf{r}}|/\Delta t$ approaches the speed v. In addition, the average acceleration becomes the instantaneous acceleration at point Ⓐ. Hence, in the limit $\Delta t \rightarrow 0$, the magnitude of the acceleration is

$$a_c = \frac{v^2}{r} \tag{4.14}$$

◀ **Centripetal acceleration**

An acceleration of this nature is called a **centripetal acceleration** (*centripetal* means *center-seeking*). The subscript on the acceleration symbol reminds us that the acceleration is centripetal.

In many situations, it is convenient to describe the motion of a particle moving with constant speed in a circle of radius r in terms of the **period** T, which is defined as the time interval required for one complete revolution of the particle. In the time interval T, the particle moves a distance of $2\pi r$, which is equal to the circumference of the particle's circular path. Therefore, because its speed is equal to the circumference of the circular path divided by the period, or $v = 2\pi r/T$, it follows that

$$T = \frac{2\pi r}{v} \tag{4.15}$$

◀ **Period of circular motion**

Equations 4.14 and 4.15 are to be used when the particle-in-uniform-circular-motion model is identified as appropriate for a given situation.

Pitfall Prevention 4.5

Centripetal Acceleration Is Not Constant

We derived the magnitude of the centripetal acceleration vector and found it to be constant for uniform circular motion, but *the centripetal acceleration vector is not constant*. It always points toward the center of the circle, but it continuously changes direction as the object moves around the circular path.

***Quick Quiz* 4.4** A particle moves in a circular path of radius r with speed v. It then increases its speed to $2v$ while traveling along the same circular path. **(i)** The centripetal acceleration of the particle has changed by what factor? Choose one: (a) 0.25 (b) 0.5 (c) 2 (d) 4 (e) impossible to determine **(ii)** From the same choices, by what factor has the period of the particle changed?

Example 4.6 The Centripetal Acceleration of the Earth

What is the centripetal acceleration of the Earth as it moves in its orbit around the Sun?

SOLUTION

Conceptualize Think about a mental image of the Earth in a circular orbit around the Sun. We will model the Earth as a particle and approximate the Earth's orbit as circular (it's actually slightly elliptical, as we discuss in Chapter 13).

Categorize The Conceptualize step allows us to categorize this problem as one of a particle in uniform circular motion.

Analyze We do not know the orbital speed of the Earth to substitute into Equation 4.14. With the help of Equation 4.15, however, we can recast Equation 4.14 in terms of the period of the Earth's orbit, which we know is one year, and the radius of the Earth's orbit around the Sun, which is 1.496×10^{11} m.

Combine Equations 4.14 and 4.15:

$$a_c = \frac{v^2}{r} = \frac{\left(\frac{2\pi r}{T}\right)^2}{r} = \frac{4\pi^2 r}{T^2}$$

continued

4.6 *cont.*

Substitute numerical values:

$$a_c = \frac{4\pi^2(1.496 \times 10^{11}\ \text{m})}{(1\ \text{yr})^2}\left(\frac{1\ \text{yr}}{3.156 \times 10^7\ \text{s}}\right)^2 = 5.93 \times 10^{-3}\ \text{m/s}^2$$

Finalize This acceleration is much smaller than the free-fall acceleration on the surface of the Earth. An important technique we learned here is replacing the speed v in Equation 4.14 in terms of the period T of the motion. In many problems, it is more likely that T is known rather than v.

4.5 Tangential and Radial Acceleration

Let us consider a more general motion than that presented in Section 4.4. A particle moves to the right along a curved path, and its velocity changes both in direction and in magnitude as described in Active Figure 4.16. In this situation, the velocity vector is always tangent to the path; the acceleration vector $\vec{\mathbf{a}}$, however, is at some angle to the path. At each of three points Ⓐ, Ⓑ, and Ⓒ in Active Figure 4.16, the dashed blue circles represent the curvature of the actual path at each point. The radius of each circle is equal to the path's radius of curvature at each point.

As the particle moves along the curved path in Active Figure 4.16, the direction of the total acceleration vector $\vec{\mathbf{a}}$ changes from point to point. At any instant, this vector can be resolved into two components based on an origin at the center of the dashed circle corresponding to that instant: a radial component a_r along the radius of the circle and a tangential component a_t perpendicular to this radius. The *total* acceleration vector $\vec{\mathbf{a}}$ can be written as the vector sum of the component vectors:

Total acceleration ▶

$$\vec{\mathbf{a}} = \vec{\mathbf{a}}_r + \vec{\mathbf{a}}_t \quad \textbf{(4.16)}$$

The tangential acceleration component causes a change in the speed v of the particle. This component is parallel to the instantaneous velocity, and its magnitude is given by

Tangential acceleration ▶

$$a_t = \left|\frac{dv}{dt}\right| \quad \textbf{(4.17)}$$

The radial acceleration component arises from a change in direction of the velocity vector and is given by

Radial acceleration ▶

$$a_r = -a_c = -\frac{v^2}{r} \quad \textbf{(4.18)}$$

where r is the radius of curvature of the path at the point in question. We recognize the magnitude of the radial component of the acceleration as the centripetal acceleration discussed in Section 4.4. The negative sign in Equation 4.18 indicates that the direction of the centripetal acceleration is toward the center of the circle rep-

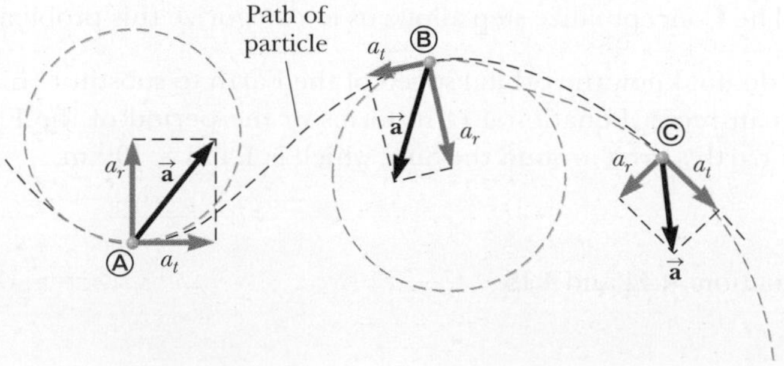

ACTIVE FIGURE 4.16

The motion of a particle along an arbitrary curved path lying in the xy plane. If the velocity vector $\vec{\mathbf{v}}$ (always tangent to the path) changes in direction and magnitude, the components of the acceleration $\vec{\mathbf{a}}$ are a tangential component a_t and a radial component a_r.

resenting the radius of curvature. The direction is opposite that of the radial unit vector $\hat{\mathbf{r}}$, which always points away from the origin at the center of the circle.

Because $\vec{\mathbf{a}}_r$ and $\vec{\mathbf{a}}_t$ are perpendicular component vectors of $\vec{\mathbf{a}}$, it follows that the magnitude of $\vec{\mathbf{a}}$ is $a = \sqrt{a_r^2 + a_t^2}$. At a given speed, a_r is large when the radius of curvature is small (as at points Ⓐ and Ⓑ in Active Fig. 4.16) and small when r is large (as at point Ⓒ). The direction of $\vec{\mathbf{a}}_t$ is either in the same direction as $\vec{\mathbf{v}}$ (if v is increasing) or opposite $\vec{\mathbf{v}}$ (if v is decreasing, as at point Ⓑ).

In uniform circular motion, where v is constant, $a_t = 0$ and the acceleration is always completely radial as described in Section 4.4. In other words, uniform circular motion is a special case of motion along a general curved path. Furthermore, if the direction of $\vec{\mathbf{v}}$ does not change, there is no radial acceleration and the motion is one dimensional (in this case, $a_r = 0$, but a_t may not be zero).

***Quick Quiz* 4.5** A particle moves along a path, and its speed increases with time. **(i)** In which of the following cases are its acceleration and velocity vectors parallel? (a) when the path is circular (b) when the path is straight (c) when the path is a parabola (d) never **(ii)** From the same choices, in which case are its acceleration and velocity vectors perpendicular everywhere along the path?

Example 4.7 Over the Rise

A car leaves a stop sign and exhibits a constant acceleration of 0.300 m/s² parallel to the roadway. The car passes over a rise in the roadway such that the top of the rise is shaped like a circle of radius 500 m. At the moment the car is at the top of the rise, its velocity vector is horizontal and has a magnitude of 6.00 m/s. What are the magnitude and direction of the total acceleration vector for the car at this instant?

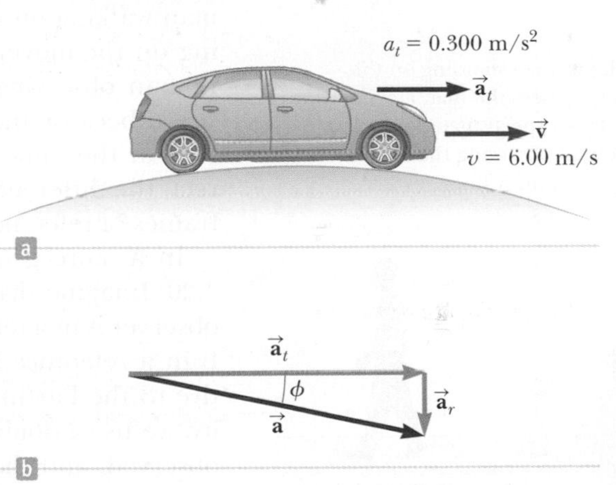

Figure 4.17 (Example 4.7) (a) A car passes over a rise that is shaped like a circle. (b) The total acceleration vector $\vec{\mathbf{a}}$ is the sum of the tangential and radial acceleration vectors $\vec{\mathbf{a}}_t$ and $\vec{\mathbf{a}}_r$.

SOLUTION

Conceptualize Conceptualize the situation using Figure 4.17a and any experiences you have had in driving over rises on a roadway.

Categorize Because the accelerating car is moving along a curved path, we categorize this problem as one involving a particle experiencing both tangential and radial acceleration. We recognize that it is a relatively simple substitution problem.

The radial acceleration is given by Equation 4.18, with v = 6.00 m/s and r = 500 m. The radial acceleration vector is directed straight downward, and the tangential acceleration vector has magnitude 0.300 m/s² and is horizontal.

Evaluate the radial acceleration:

$$a_r = -\frac{v^2}{r} = -\frac{(6.00\ \text{m/s})^2}{500\ \text{m}} = -0.072\,0\ \text{m/s}^2$$

Find the magnitude of $\vec{\mathbf{a}}$:

$$\sqrt{a_r^2 + a_t^2} = \sqrt{(-0.072\,0\ \text{m/s}^2)^2 + (0.300\ \text{m/s}^2)^2}$$
$$= 0.309\ \text{m/s}^2$$

Find the angle ϕ (see Fig. 4.17b) between $\vec{\mathbf{a}}$ and the horizontal:

$$\phi = \tan^{-1}\frac{a_r}{a_t} = \tan^{-1}\left(\frac{-0.072\,0\ \text{m/s}^2}{0.300\ \text{m/s}^2}\right) = -13.5°$$

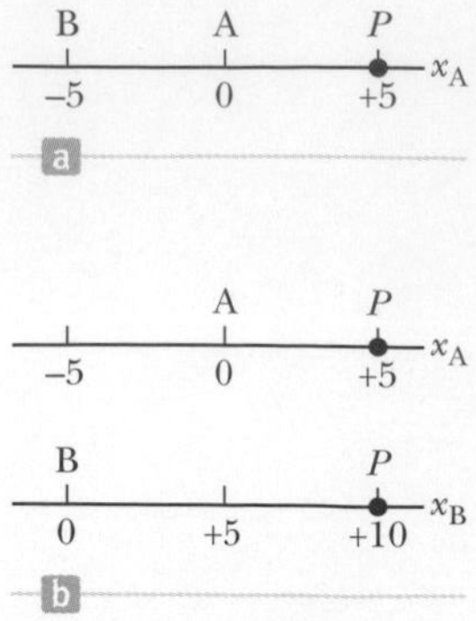

Figure 4.18 Different observers make different measurements. (a) Observer A is located at the origin, and Observer B is at a position of −5. Both observers measure the position of a particle at *P*. (b) If both observers see themselves at the origin of their own coordinate system, they disagree on the value of the position of the particle at *P*.

Figure 4.19 Two observers measure the speed of a man walking on a moving beltway.

4.6 Relative Velocity and Relative Acceleration

In this section, we describe how observations made by different observers in different frames of reference are related to one another. A frame of reference can be described by a Cartesian coordinate system for which an observer is at rest with respect to the origin.

Let us conceptualize a sample situation in which there will be different observations for different observers. Consider the two observers A and B along the number line in Figure 4.18a. Observer A is located at the origin of a one-dimensional x_A axis, while observer B is at the position $x_A = -5$. We denote the position variable as x_A because observer A is at the origin of this axis. Both observers measure the position of point *P*, which is located at $x_A = +5$. Suppose observer B decides that he is located at the origin of an x_B axis as in Figure 4.18b. Notice that the two observers disagree on the value of the position of point *P*. Observer A claims point *P* is located at a position with a value of +5, whereas observer B claims it is located at a position with a value of +10. Both observers are correct, even though they make different measurements. Their measurements differ because they are making the measurement from different frames of reference.

Imagine now that observer B in Figure 4.18b is moving to the right along the x_B axis. Now the two measurements are even more different. Observer A claims point *P* remains at rest at a position with a value of +5, whereas observer B claims the position of *P* continuously changes with time, even passing him and moving behind him! Again, both observers are correct, with the difference in their measurements arising from their different frames of reference.

We explore this phenomenon further by considering two observers watching a man walking on a moving beltway at an airport in Figure 4.19. The woman standing on the moving beltway sees the man moving at a normal walking speed. The woman observing from the stationary floor sees the man moving with a higher speed because the beltway speed combines with his walking speed. Both observers look at the same man and arrive at different values for his speed. Both are correct; the difference in their measurements results from the relative velocity of their frames of reference.

In a more general situation, consider a particle located at point *P* in Figure 4.20. Imagine that the motion of this particle is being described by two observers, observer A in a reference frame S_A fixed relative to the Earth and a second observer B in a reference frame S_B moving to the right relative to S_A (and therefore relative to the Earth) with a constant velocity $\vec{\mathbf{v}}_{BA}$. In this discussion of relative velocity, we use a double-subscript notation; the first subscript represents what is being observed, and the second represents who is doing the observing. Therefore, the notation $\vec{\mathbf{v}}_{BA}$ means the velocity of observer B (and the attached frame S_B) as measured by observer A. With this notation, observer B measures A to be moving to the left with a velocity $\vec{\mathbf{v}}_{AB} = -\vec{\mathbf{v}}_{BA}$. For purposes of this discussion, let us place each observer at her or his respective origin.

We define the time $t = 0$ as the instant at which the origins of the two reference frames coincide in space. Therefore, at time t, the origins of the reference frames will be separated by a distance $v_{BA}t$. We label the position *P* of the particle relative to observer A with the position vector $\vec{\mathbf{r}}_{PA}$ and that relative to observer B with the position vector $\vec{\mathbf{r}}_{PB}$, both at time t. From Figure 4.20, we see that the vectors $\vec{\mathbf{r}}_{PA}$ and $\vec{\mathbf{r}}_{PB}$ are related to each other through the expression

$$\vec{\mathbf{r}}_{PA} = \vec{\mathbf{r}}_{PB} + \vec{\mathbf{v}}_{BA}t \tag{4.19}$$

By differentiating Equation 4.19 with respect to time, noting that $\vec{\mathbf{v}}_{BA}$ is constant, we obtain

$$\frac{d\vec{\mathbf{r}}_{PA}}{dt} = \frac{d\vec{\mathbf{r}}_{PB}}{dt} + \vec{\mathbf{v}}_{BA}$$

Galilean velocity transformation ▶

$$\vec{\mathbf{u}}_{PA} = \vec{\mathbf{u}}_{PB} + \vec{\mathbf{v}}_{BA} \tag{4.20}$$

where $\vec{\mathbf{u}}_{PA}$ is the velocity of the particle at P measured by observer A and $\vec{\mathbf{u}}_{PB}$ is its velocity measured by B. (We use the symbol $\vec{\mathbf{u}}$ for particle velocity rather than $\vec{\mathbf{v}}$, which we have already used for the relative velocity of two reference frames.) Equations 4.19 and 4.20 are known as **Galilean transformation equations.** They relate the position and velocity of a particle as measured by observers in relative motion. Notice the pattern of the subscripts in Equation 4.20. When relative velocities are added, the inner subscripts (B) are the same and the outer ones (P, A) match the subscripts on the velocity on the left of the equation.

Although observers in two frames measure different velocities for the particle, they measure the *same acceleration* when $\vec{\mathbf{v}}_{BA}$ is constant. We can verify that by taking the time derivative of Equation 4.20:

$$\frac{d\vec{\mathbf{u}}_{PA}}{dt} = \frac{d\vec{\mathbf{u}}_{PB}}{dt} + \frac{d\vec{\mathbf{v}}_{BA}}{dt}$$

Because $\vec{\mathbf{v}}_{BA}$ is constant, $d\vec{\mathbf{v}}_{BA}/dt = 0$. Therefore, we conclude that $\vec{\mathbf{a}}_{PA} = \vec{\mathbf{a}}_{PB}$ because $\vec{\mathbf{a}}_{PA} = d\vec{\mathbf{u}}_{PA}/dt$ and $\vec{\mathbf{a}}_{PB} = d\vec{\mathbf{u}}_{PB}/dt$. That is, the acceleration of the particle measured by an observer in one frame of reference is the same as that measured by any other observer moving with constant velocity relative to the first frame.

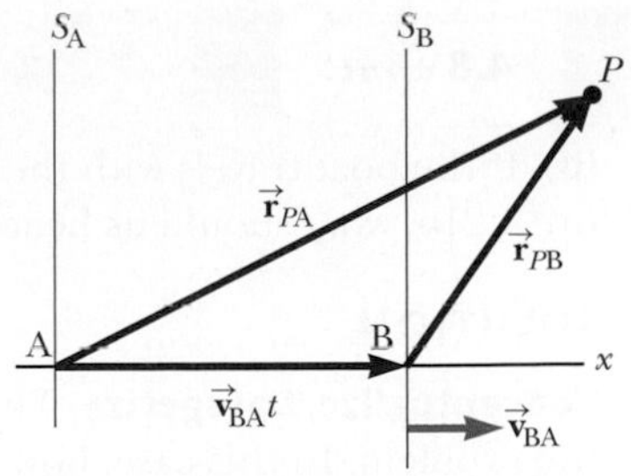

Figure 4.20 A particle located at P is described by two observers, one in the fixed frame of reference S_A and the other in the frame S_B, which moves to the right with a constant velocity $\vec{\mathbf{v}}_{BA}$. The vector $\vec{\mathbf{r}}_{PA}$ is the particle's position vector relative to S_A, and $\vec{\mathbf{r}}_{PB}$ is its position vector relative to S_B.

Example 4.8 A Boat Crossing a River

A boat crossing a wide river moves with a speed of 10.0 km/h relative to the water. The water in the river has a uniform speed of 5.00 km/h due east relative to the Earth.

(A) If the boat heads due north, determine the velocity of the boat relative to an observer standing on either bank.

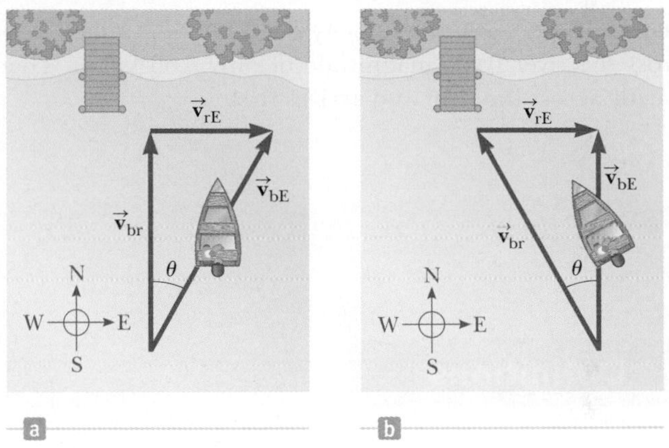

Figure 4.21 (Example 4.8) (a) A boat aims directly across a river and ends up downstream. (b) To move directly across the river, the boat must aim upstream.

SOLUTION

Conceptualize Imagine moving in a boat across a river while the current pushes you down the river. You will not be able to move directly across the river, but will end up downstream as suggested in Figure 4.21a.

Categorize Because of the combined velocities of you relative to the river and the river relative to the Earth, we can categorize this problem as one involving relative velocities.

Analyze We know $\vec{\mathbf{v}}_{br}$, the velocity of the *boat* relative to the *river,* and $\vec{\mathbf{v}}_{rE}$, the velocity of the *river* relative to the *Earth.* What we must find is $\vec{\mathbf{v}}_{bE}$, the velocity of the *boat* relative to the *Earth.* The relationship between these three quantities is $\vec{\mathbf{v}}_{bE} = \vec{\mathbf{v}}_{br} + \vec{\mathbf{v}}_{rE}$. The terms in the equation must be manipulated as vector quantities; the vectors are shown in Figure 4.21a. The quantity $\vec{\mathbf{v}}_{br}$ is due north; $\vec{\mathbf{v}}_{rE}$ is due east; and the vector sum of the two, $\vec{\mathbf{v}}_{bE}$, is at an angle θ as defined in Figure 4.21a.

Find the speed v_{bE} of the boat relative to the Earth using the Pythagorean theorem:

$$v_{bE} = \sqrt{v_{br}^2 + v_{rE}^2} = \sqrt{(10.0 \text{ km/h})^2 + (5.00 \text{ km/h})^2}$$

$$= 11.2 \text{ km/h}$$

Find the direction of $\vec{\mathbf{v}}_{bE}$:

$$\theta = \tan^{-1}\left(\frac{v_{rE}}{v_{br}}\right) = \tan^{-1}\left(\frac{5.00}{10.0}\right) = 26.6°$$

Finalize The boat is moving at a speed of 11.2 km/h in the direction 26.6° east of north relative to the Earth. Notice that the speed of 11.2 km/h is faster than your boat speed of 10.0 km/h. The current velocity adds to yours to give you a higher speed. Notice in Figure 4.21a that your resultant velocity is at an angle to the direction straight across the river, so you will end up downstream, as we predicted.

continued

4.8 cont.

(B) If the boat travels with the same speed of 10.0 km/h relative to the river and is to travel due north as shown in Figure 4.21b, what should its heading be?

SOLUTION

Conceptualize/Categorize This question is an extension of part (A), so we have already conceptualized and categorized the problem. In this case, however, we must aim the boat upstream so as to go straight across the river.

Analyze The analysis now involves the new triangle shown in Figure 4.21b. As in part (A), we know $\vec{\mathbf{v}}_{rE}$ and the magnitude of the vector $\vec{\mathbf{v}}_{br}$, and we want $\vec{\mathbf{v}}_{bE}$ to be directed across the river. Notice the difference between the triangle in Figure 4.21a and the one in Figure 4.21b: the hypotenuse in Figure 4.21b is no longer $\vec{\mathbf{v}}_{bE}$.

Use the Pythagorean theorem to find v_{bE}:

$$v_{bE} = \sqrt{v_{br}^2 - v_{rE}^2} = \sqrt{(10.0 \text{ km/h})^2 - (5.00 \text{ km/h})^2} = 8.66 \text{ km/h}$$

Find the direction in which the boat is heading:

$$\theta = \tan^{-1}\left(\frac{v_{rE}}{v_{bE}}\right) = \tan^{-1}\left(\frac{5.00}{8.66}\right) = 30.0°$$

Finalize The boat must head upstream so as to travel directly northward across the river. For the given situation, the boat must steer a course 30.0° west of north. For faster currents, the boat must be aimed upstream at larger angles.

WHAT IF? Imagine that the two boats in parts (A) and (B) are racing across the river. Which boat arrives at the opposite bank first?

Answer In part (A), the velocity of 10 km/h is aimed directly across the river. In part (B), the velocity that is directed across the river has a magnitude of only 8.66 km/h. Therefore, the boat in part (A) has a larger velocity component directly across the river and arrives first.

Summary

Definitions

The **displacement vector** $\Delta\vec{\mathbf{r}}$ for a particle is the difference between its final position vector and its initial position vector:

$$\Delta\vec{\mathbf{r}} \equiv \vec{\mathbf{r}}_f - \vec{\mathbf{r}}_i \qquad (4.1)$$

The **average velocity** of a particle during the time interval Δt is defined as the displacement of the particle divided by the time interval:

$$\vec{\mathbf{v}}_{avg} \equiv \frac{\Delta\vec{\mathbf{r}}}{\Delta t} \qquad (4.2)$$

The **instantaneous velocity** of a particle is defined as the limit of the average velocity as Δt approaches zero:

$$\vec{\mathbf{v}} \equiv \lim_{\Delta t \to 0} \frac{\Delta\vec{\mathbf{r}}}{\Delta t} = \frac{d\vec{\mathbf{r}}}{dt} \qquad (4.3)$$

The **average acceleration** of a particle is defined as the change in its instantaneous velocity vector divided by the time interval Δt during which that change occurs:

$$\vec{\mathbf{a}}_{avg} \equiv \frac{\Delta\vec{\mathbf{v}}}{\Delta t} = \frac{\vec{\mathbf{v}}_f - \vec{\mathbf{v}}_i}{t_f - t_i} \qquad (4.4)$$

The **instantaneous acceleration** of a particle is defined as the limiting value of the average acceleration as Δt approaches zero:

$$\vec{\mathbf{a}} \equiv \lim_{\Delta t \to 0} \frac{\Delta\vec{\mathbf{v}}}{\Delta t} = \frac{d\vec{\mathbf{v}}}{dt} \qquad (4.5)$$

Projectile motion is one type of two-dimensional motion, exhibited by an object launched into the air near the Earth's surface and experiencing free fall. This common motion can be analyzed by applying the particle-under-constant-velocity model to the motion of the projectile in the x direction and the particle-under-constant-acceleration model ($a_y = -g$) in the y direction.

A particle moving in a circular path with constant speed is exhibiting **uniform circular motion.**

Concepts and Principles

If a particle moves with *constant* acceleration $\vec{\mathbf{a}}$ and has velocity $\vec{\mathbf{v}}_i$ and position $\vec{\mathbf{r}}_i$ at $t = 0$, its velocity and position vectors at some later time t are

$$\vec{\mathbf{v}}_f = \vec{\mathbf{v}}_i + \vec{\mathbf{a}}t \tag{4.8}$$

$$\vec{\mathbf{r}}_f = \vec{\mathbf{r}}_i + \vec{\mathbf{v}}_i t + \tfrac{1}{2}\vec{\mathbf{a}}t^2 \tag{4.9}$$

For two-dimensional motion in the xy plane under constant acceleration, each of these vector expressions is equivalent to two component expressions: one for the motion in the x direction and one for the motion in the y direction.

It is useful to think of projectile motion in terms of a combination of two analysis models: (1) the particle-under-constant-velocity model in the x direction and (2) the particle-under-constant-acceleration model in the vertical direction with a constant downward acceleration of magnitude $g = 9.80 \text{ m/s}^2$.

A particle in uniform circular motion undergoes a radial acceleration $\vec{\mathbf{a}}_r$ because the direction of $\vec{\mathbf{v}}$ changes in time. This acceleration is called **centripetal acceleration,** and its direction is always toward the center of the circle.

If a particle moves along a curved path in such a way that both the magnitude and the direction of $\vec{\mathbf{v}}$ change in time, the particle has an acceleration vector that can be described by two component vectors: (1) a radial component vector $\vec{\mathbf{a}}_r$ that causes the change in direction of $\vec{\mathbf{v}}$ and (2) a tangential component vector $\vec{\mathbf{a}}_t$ that causes the change in magnitude of $\vec{\mathbf{v}}$. The magnitude of $\vec{\mathbf{a}}_r$ is v^2/r, and the magnitude of $\vec{\mathbf{a}}_t$ is $|dv/dt|$.

The velocity $\vec{\mathbf{u}}_{PA}$ of a particle measured in a fixed frame of reference S_A can be related to the velocity $\vec{\mathbf{u}}_{PB}$ of the same particle measured in a moving frame of reference S_B by

$$\vec{\mathbf{u}}_{PA} = \vec{\mathbf{u}}_{PB} + \vec{\mathbf{v}}_{BA} \tag{4.20}$$

where $\vec{\mathbf{v}}_{BA}$ is the velocity of S_B relative to S_A.

Analysis Model for Problem Solving

Particle in Uniform Circular Motion If a particle moves in a circular path of radius r with a constant speed v, the magnitude of its centripetal acceleration is given by

$$a_c = \frac{v^2}{r} \tag{4.14}$$

and the **period** of the particle's motion is given by

$$T = \frac{2\pi r}{v} \tag{4.15}$$

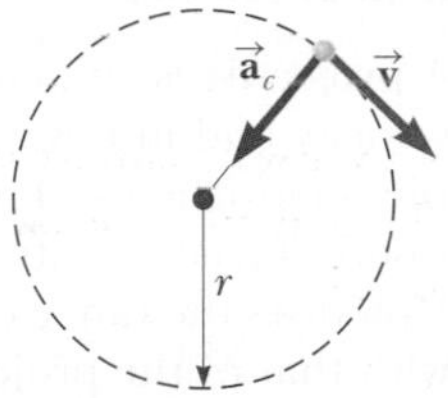

Objective Questions

☐ denotes answer available in Student Solutions Manual/Study Guide

1. Figure OQ4.1 shows a bird's-eye view of a car going around a highway curve. As the car moves from point 1 to point 2, its speed doubles. Which of the vectors (a) through (e) shows the direction of the car's average acceleration between these two points?

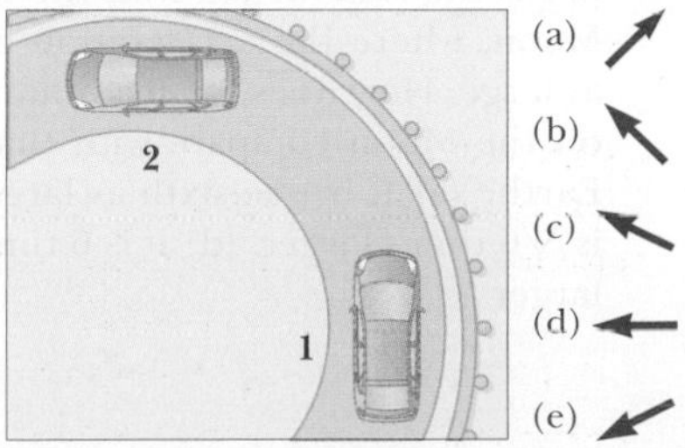

Figure OQ4.1

2. Entering his dorm room, a student tosses his book bag to the right and upward at an angle of 45° with the horizontal (Fig. OQ4.2). Air resistance does not affect the bag. The bag moves through point Ⓐ immediately after it leaves the student's hand, through point Ⓑ at the top of its flight, and through point Ⓒ immediately before it lands on the top bunk bed. **(i)** Rank the following horizontal and vertical velocity components from the largest to the smallest. (a) $v_{\text{Ⓐ}x}$ (b) $v_{\text{Ⓐ}y}$ (c) $v_{\text{Ⓑ}x}$ (d) $v_{\text{Ⓑ}y}$ (e) $v_{\text{Ⓒ}y}$. Note that zero is larger than a negative number. If two quantities are equal, show them as equal in your list. If any quantity is equal to zero, show that fact in your list. **(ii)** Similarly, rank the following acceleration components. (a) $a_{\text{Ⓐ}x}$ (b) $a_{\text{Ⓐ}y}$ (c) $a_{\text{Ⓑ}x}$ (d) $a_{\text{Ⓑ}y}$ (e) $a_{\text{Ⓒ}y}$.

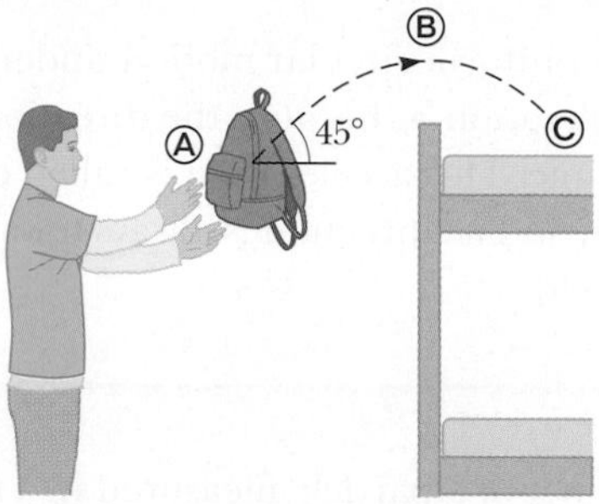

Figure OQ4.2

3. In which of the following situations is the moving object appropriately modeled as a projectile? Choose all correct answers. (a) A shoe is tossed in an arbitrary direction. (b) A jet airplane crosses the sky with its engines thrusting the plane forward. (c) A rocket leaves the launch pad. (d) A rocket moves through the sky, at much less than the speed of sound, after its fuel has been used up. (e) A diver throws a stone under water.

4. An astronaut hits a golf ball on the Moon. Which of the following quantities, if any, remain constant as a ball travels through the vacuum there? (a) speed (b) acceleration (c) horizontal component of velocity (d) vertical component of velocity (e) velocity

5. A projectile is launched on the Earth with a certain initial velocity and moves without air resistance. Another projectile is launched with the same initial velocity on the Moon, where the acceleration due to gravity is one-sixth as large. How does the range of the projectile on the Moon compare with that of the projectile on the Earth? (a) It is one-sixth as large. (b) It is the same. (c) It is $\sqrt{6}$ times larger. (d) It is 6 times larger. (e) It is 36 times larger.

6. A projectile is launched on the Earth with a certain initial velocity and moves without air resistance. Another projectile is launched with the same initial velocity on the Moon, where the acceleration due to gravity is one-sixth as large. How does the maximum altitude of the projectile on the Moon compare with that of the projectile on the Earth? (a) It is one-sixth as large. (b) It is the same. (c) It is $\sqrt{6}$ times larger. (d) It is 6 times larger. (e) It is 36 times larger.

7. Does a car moving around a circular track with constant speed have (a) zero acceleration, (b) an acceleration in the direction of its velocity, (c) an acceleration directed away from the center of its path, (d) an acceleration directed toward the center of its path, or (e) an acceleration with a direction that cannot be determined from the given information?

8. A rubber stopper on the end of a string is swung steadily in a horizontal circle. In one trial, it moves at speed v in a circle of radius r. In a second trial, it moves at a higher speed $3v$ in a circle of radius $3r$. In this second trial, is its acceleration (a) the same as in the first trial, (b) three times larger, (c) one-third as large, (d) nine times larger, or (e) one-ninth as large?

9. A set of keys on the end of a string is swung steadily in a horizontal circle. In one trial, it moves at speed v in a circle of radius r. In a second trial, it moves at a higher speed $4v$ in a circle of radius $4r$. In the second trial, how does the period of its motion compare with its period in the first trial? (a) It is the same as in the first trial. (b) It is 4 times larger. (c) It is one-fourth as large. (d) It is 16 times larger. (e) It is one-sixteenth as large.

10. A certain light truck can go around a curve having a radius of 150 m with a maximum speed of 32.0 m/s. To have the same acceleration, at what maximum speed can it go around a curve having a radius of 75.0 m? (a) 64 m/s (b) 45 m/s (c) 32 m/s (d) 23 m/s (e) 16 m/s

11. A sailor drops a wrench from the top of a sailboat's vertical mast while the boat is moving rapidly and steadily straight forward. Where will the wrench hit the deck? (a) ahead of the base of the mast (b) at the base of the mast (c) behind the base of the mast (d) on the windward side of the base of the mast (e) None of the choices (a) through (d) is true.

12. A girl, moving at 8 m/s on in-line skates, is overtaking a boy moving at 5 m/s as they both skate on a straight path. The boy tosses a ball backward toward the girl, giving it speed 12 m/s relative to him. What is the speed of the ball relative to the girl, who catches it? (a) $(8 + 5 + 12)$ m/s (b) $(8 - 5 - 12)$ m/s (c) $(8 + 5 - 12)$ m/s (d) $(8 - 5 + 12)$ m/s (e) $(-8 + 5 + 12)$ m/s

13. A student throws a heavy red ball horizontally from a balcony of a tall building with an initial speed v_i. At the same time, a second student drops a lighter blue ball from the balcony. Neglecting air resistance, which statement is true? (a) The blue ball reaches the ground first. (b) The balls reach the ground at the same instant. (c) The red ball reaches the ground first. (d) Both balls hit the ground with the same speed. (e) None of statements (a) through (d) is true.

14. A baseball is thrown from the outfield toward the catcher. When the ball reaches its highest point, which statement is true? (a) Its velocity and its acceleration are both zero. (b) Its velocity is not zero, but its acceleration is zero. (c) Its velocity is perpendicular to its acceleration. (d) Its acceleration depends on the angle at which the ball was thrown. (e) None of statements (a) through (d) is true.

Conceptual Questions

□ denotes answer available in Student Solutions Manual/Study Guide

1. If you know the position vectors of a particle at two points along its path and also know the time interval during which it moved from one point to the other, can you determine the particle's instantaneous velocity? Its average velocity? Explain.

2. Construct motion diagrams showing the velocity and acceleration of a projectile at several points along its path, assuming (a) the projectile is launched horizontally and (b) the projectile is launched at an angle θ with the horizontal.

3. A spacecraft drifts through space at a constant velocity. Suddenly, a gas leak in the side of the spacecraft gives it a constant acceleration in a direction perpendicular to the initial velocity. The orientation of the spacecraft does not change, so the acceleration remains perpendicular to the original direction of the velocity. What is the shape of the path followed by the spacecraft in this situation?

4. Describe how a driver can steer a car traveling at constant speed so that (a) the acceleration is zero or (b) the magnitude of the acceleration remains constant.

5. A projectile is launched at some angle to the horizontal with some initial speed v_i, and air resistance is negligible. (a) Is the projectile a freely falling body? (b) What is its acceleration in the vertical direction? (c) What is its acceleration in the horizontal direction?

6. An ice skater is executing a figure eight, consisting of two identically shaped, tangent circular paths. Throughout the first loop she increases her speed uniformly, and during the second loop she moves at a constant speed. Draw a motion diagram showing her velocity and acceleration vectors at several points along the path of motion.

7. Explain whether or not the following particles have an acceleration: (a) a particle moving in a straight line with constant speed and (b) a particle moving around a curve with constant speed.

Problems

ENHANCED WebAssign The problems found in this chapter may be assigned online in Enhanced WebAssign

1. denotes straightforward problem; 2. denotes intermediate problem;
3. denotes challenging problem

□1. full solution available in the Student Solutions Manual/Study Guide

1. denotes problems most often assigned in Enhanced WebAssign; these provide students with targeted feedback and either a Master It tutorial or a Watch It solution video.

QIC denotes asking for quantitative and conceptual reasoning

S denotes symbolic reasoning problem

M denotes Master It tutorial available in Enhanced WebAssign

GP denotes guided problem

shaded denotes "paired problems" that develop reasoning with symbols and numerical values

Section 4.1 The Position, Velocity, and Acceleration Vectors

1. A motorist drives south at 20.0 m/s for 3.00 min, then turns west and travels at 25.0 m/s for 2.00 min, and finally travels northwest at 30.0 m/s for 1.00 min. For this 6.00-min trip, find (a) the total vector displacement, (b) the average speed, and (c) the average velocity. Let the positive x axis point east.

2. When the Sun is directly overhead, a hawk dives toward the ground with a constant velocity of 5.00 m/s at 60.0° below the horizontal. Calculate the speed of its shadow on the level ground.

3. A golf ball is hit off a tee at the edge of a cliff. Its x and y coordinates as functions of time are given by $x = 18.0t$ and $y = 4.00t - 4.90t^2$, where x and y are in meters and t is in seconds. (a) Write a vector expression for the ball's position as a function of time, using the unit vectors $\hat{\mathbf{i}}$ and $\hat{\mathbf{j}}$. tBy taking derivatives, obtain expressions for (b) the velocity vector $\vec{\mathbf{v}}$ as a function of time and (c) the acceleration vector $\vec{\mathbf{a}}$ as a function of time. (d) Next use unit-vector notation to write expressions for the position, the velocity, and the acceleration of the golf ball at $t = 3.00$ s.

4. QIC The coordinates of an object moving in the xy plane vary with time according to the equations $x = -5.00 \sin \omega t$ and $y = 4.00 - 5.00 \cos \omega t$, where ω is a constant, x and y are in meters, and t is in seconds. (a) Determine the components of velocity of the object at $t = 0$. (b) Determine the components of acceleration of the object at $t = 0$. (c) Write expressions for the position vector, the velocity vector, and the acceleration vector of the object at any time $t > 0$. (d) Describe the path of the object in an xy plot.

Section 4.2 Two-Dimensional Motion with Constant Acceleration

5. The vector position of a particle varies in time according to the expression $\vec{\mathbf{r}} = 3.00\hat{\mathbf{i}} - 6.00t^2\hat{\mathbf{j}}$, where $\vec{\mathbf{r}}$ is in meters and t is in seconds. (a) Find an expression for the velocity of the particle as a function of time. (b) Determine the acceleration of the particle as a function of time. (c) Calculate the particle's position and velocity at $t = 1.00$ s.

6. A particle initially located at the origin has an acceleration of $\vec{\mathbf{a}} = 3.00\hat{\mathbf{j}}$ m/s^2 and an initial velocity of $\vec{\mathbf{v}}_i = 5.00\hat{\mathbf{i}}$ m/s. Find (a) the vector position of the particle at any time t, (b) the velocity of the particle at any time t, (c) the coordinates of the particle at $t = 2.00$ s, and (d) the speed of the particle at $t = 2.00$ s.

7. A fish swimming in a horizontal plane has velocity $\vec{\mathbf{v}}_i = (4.00\hat{\mathbf{i}} + 1.00\hat{\mathbf{j}})$ m/s at a point in the ocean where the

position relative to a certain rock is $\vec{\mathbf{r}}_i = (10.0\hat{\mathbf{i}} - 4.00\hat{\mathbf{j}})$ m. After the fish swims with constant acceleration for 20.0 s, its velocity is $\vec{\mathbf{v}} = (20.0\hat{\mathbf{i}} - 5.00\hat{\mathbf{j}})$ m/s. (a) What are the components of the acceleration of the fish? (b) What is the direction of its acceleration with respect to unit vector $\hat{\mathbf{i}}$? (c) If the fish maintains constant acceleration, where is it at $t = 25.0$ s and in what direction is it moving?

8. **Review.** A snowmobile is originally at the point with position vector 29.0 m at 95.0° counterclockwise from the x axis, moving with velocity 4.50 m/s at 40.0°. It moves with constant acceleration 1.90 m/s^2 at 200°. After 5.00 s have elapsed, find (a) its velocity and (b) its position vector.

Section 4.3 Projectile Motion

Note: Ignore air resistance in all problems and take $g = 9.80$ m/s^2 at the Earth's surface.

9. In a local bar, a customer slides an empty beer mug down the counter for a refill. The height of the counter is 1.22 m. The mug slides off the counter and strikes the floor 1.40 m from the base of the counter. (a) With what velocity did the mug leave the counter? (b) What was the direction of the mug's velocity just before it hit the floor?

10. **S** In a local bar, a customer slides an empty beer mug down the counter for a refill. The height of the counter is h. The mug slides off the counter and strikes the floor at distance d from the base of the counter. (a) With what velocity did the mug leave the counter? (b) What was the direction of the mug's velocity just before it hit the floor?

11. A projectile is fired in such a way that its horizontal range is equal to three times its maximumv height. What is the angle of projection?

12. To start an avalanche on a mountain slope, an artillery shell is fired with an initial velocity of 300 m/s at 55.0° above the horizontal. It explodes on the mountainside 42.0 s after firing. What are the x and y coordinates of the shell where it explodes, relative to its firing point?

13. Chinook salmon are able to move through water especially fast by jumping out of the water periodically. This behavior is called *porpoising.* Suppose a salmon swimming in still water jumps out of the water with velocity 6.26 m/s at 45.0° above the horizontal, sails through the air a distance L before returning to the water, and then swims the same distance L underwater in a straight, horizontal line with velocity 3.58 m/s before jumping out again. (a) Determine the average velocity of the fish for the entire process of jumping and swimming underwater. (b) Consider the time interval required to travel the entire distance of $2L$. By what percentage is this time interval reduced by the jumping/swimming process compared with simply swimming underwater at 3.58 m/s?

14. **QC** **S** A rock is thrown upward from level ground in such a way that the maximum height of its flight is equal to its horizontal range R. (a) At what angle θ is the rock thrown? (b) In terms of its original range R, what is the range R_{max} the rock can attain if it is launched at the same speed but at the optimal angle for maximum range? (c) **What If?** Would your answer to part (a) be different if the rock is thrown with the same speed on a different planet? Explain.

15. **S** A firefighter, a distance d from a burning building, directs a stream of water from a fire hose at angle θ_i above the horizontal as shown in Figure P4.15. If the initial speed of the stream is v_i, at what height h does the water strike the building?

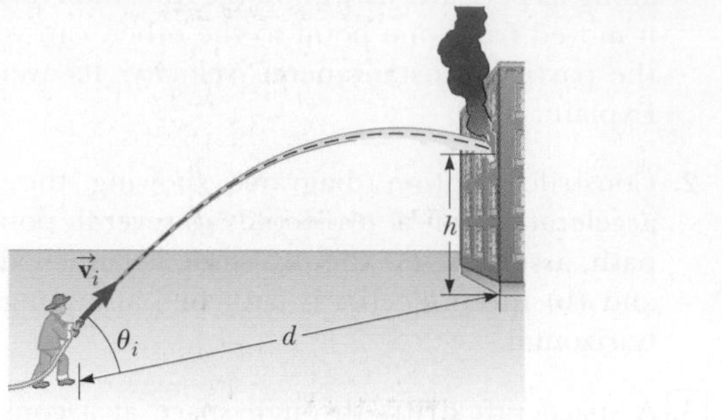

Figure P4.15

16. A ball is tossed from an upper-story window of a building. The ball is given an initial velocity of 8.00 m/s at an angle of 20.0° below the horizontal. It strikes the ground 3.00 s later. (a) How far horizontally from the base of the building does the ball strike the ground? (b) Find the height from which the ball was thrown. (c) How long does it take the ball to reach a point 10.0 m below the level of launching?

17. A placekicker must kick a football from a point 36.0 m (about 40 yards) from the goal. Half the crowd hopes the ball will clear the crossbar, which is 3.05 m high. When kicked, the ball leaves the ground with a speed of 20.0 m/s at an angle of 53.0° to the horizontal. (a) By how much does the ball clear or fall short of clearing the crossbar? (b) Does the ball approach the crossbar while still rising or while falling?

18. A landscape architect is planning an artificial waterfall in a city park. Water flowing at 1.70 m/s will leave the end of a horizontal channel at the top of a vertical wall h = 2.35 m high, and from there it will fall into a pool (Fig. P4.18). (a) Will the space behind the waterfall be wide enough for a pedestrian walkway? (b) To sell her plan to the city council, the architect wants to build a model to standard scale, which is one-twelfth actual size. How fast should the water flow in the channel in the model?

Figure P4.18

19. A playground is on the flat roof of a city school, 6.00 m above the street below (Fig. P4.19). The vertical wall of

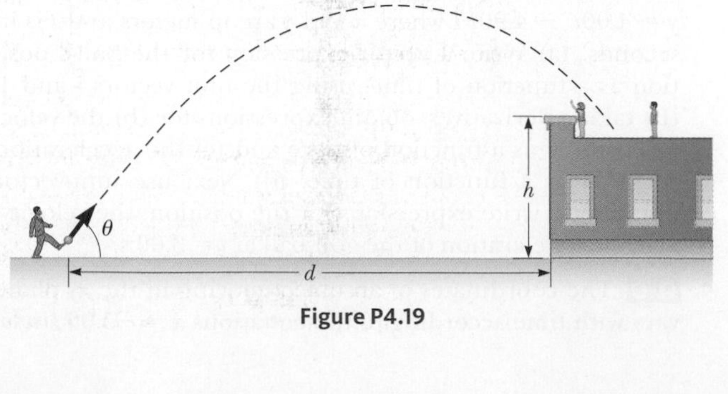

Figure P4.19

the building is $h = 7.00$ m high, forming a 1-m-high railing around the playground. A ball has fallen to the street below, and a passerby returns it by launching it at an angle of $\theta = 53.0°$ above the horizontal at a point $d = 24.0$ m from the base of the building wall. The ball takes 2.20 s to reach a point vertically above the wall. (a) Find the speed at which the ball was launched. (b) Find the vertical distance by which the ball clears the wall. (c) Find the horizontal distance from the wall to the point on the roof where the ball lands.

20. A basketball star covers 2.80 m horizontally in a jump to dunk the ball (Fig. P4.20a). His motion through space can be modeled precisely as that of a particle at his *center of mass*, which we will define in Chapter 9. His center of mass is at elevation 1.02 m when he leaves the floor. It reaches a maximum height of 1.85 m above the floor and is at elevation 0.900 m when he touches down again. Determine (a) his time of flight (his "hang time"), (b) his horizontal and (c) vertical velocity components at the instant of takeoff, and (d) his takeoff angle. (e) For comparison, determine the hang time of a whitetail deer making a jump (Fig. P4.20b) with center-of-mass elevations $y_i = 1.20$ m, $y_{max} = 2.50$ m, and $y_f = 0.700$ m.

Figure P4.20

21. A soccer player kicks a rock horizontally off a 40.0-m-high cliff into a pool of water. If the player hears the sound of the splash 3.00 s later, what was the initial speed given to the rock? Assume the speed of sound in air is 343 m/s.

22. The motion of a human body through space can be modeled as the motion of a particle at the body's center of mass as we will study in Chapter 9. The components of the displacement of an athlete's center of mass from the beginning to the end of a certain jump are described by the equations

$$x_f = 0 + (11.2 \text{ m/s})(\cos 18.5°)t$$

$$0.360 \text{ m} = 0.840 \text{ m} + (11.2 \text{ m/s})(\sin 18.5°)t - \tfrac{1}{2}(9.80 \text{ m/s}^2)t^2$$

where t is in seconds and is the time at which the athlete ends the jump. Identify (a) the athlete's position and (b) his vector velocity at the takeoff point. (c) How far did he jump?

23. GP A student stands at the edge of a cliff and throws a stone horizontally over the edge with a speed of $v_i = 18.0$ m/s. The cliff is $h = 50.0$ m above a body of water as shown in Figure P4.23. (a) What are the coordinates of the initial position of the stone? (b) What are the components of the initial velocity of the stone? (c) What is the appropriate analysis model for the vertical motion of the stone? (d) What is the appropriate analysis model for the horizontal motion of the stone? (e) Write symbolic equations for the x and y components of the velocity of the stone as a function of time. (f) Write symbolic equations for the position of the stone as a function of time. (g) How long after being released does the stone strike the water below the cliff? (h) With what speed and angle of impact does the stone land?

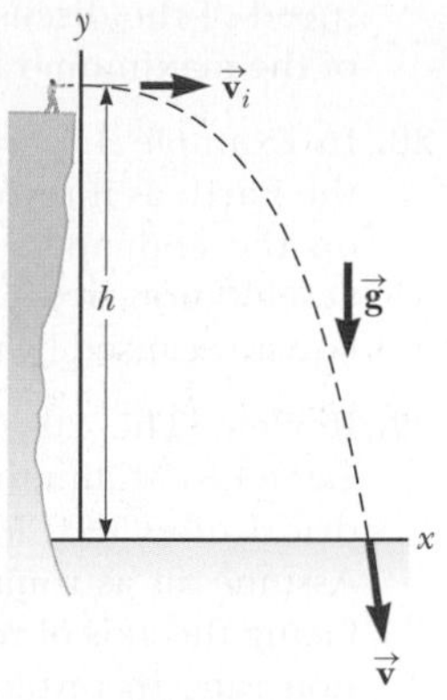

Figure P4.23

24. QIC The record distance in the sport of throwing cowpats is 81.1 m. This record toss was set by Steve Uren of the United States in 1981. Assuming the initial launch angle was 45° and neglecting air resistance, determine (a) the initial speed of the projectile and (b) the total time interval the projectile was in flight. (c) How would the answers change if the range were the same but the launch angle were greater than 45°? Explain.

25. A boy stands on a diving board and tosses a stone into a swimming pool. The stone is thrown from a height of 2.50 m above the water surface with a velocity of 4.00 m/s at an angle of 60.0° above the horizontal. As the stone strikes the water surface, it immediately slows down to exactly half the speed it had when it struck the water and maintains that speed while in the water. After the stone enters the water, it moves in a straight line in the direction of the velocity it had when it struck the water. If the pool is 3.00 m deep, how much time elapses between when the stone is thrown and when it strikes the bottom of the pool?

26. S A projectile is fired from the top of a cliff of height h above the ocean below. The projectile is fired at an angle θ above the horizontal and with an initial speed v_i. (a) Find a symbolic expression in terms of the variables v_i, g, and θ for the time at which the projectile reaches its maximum height. (b) Using the result of part (a), find an expression for the maximum height h_{max} above the ocean attained by the projectile in terms of h, v_i, g, and θ.

Section 4.4 Analysis Model: Particle in Uniform Circular Motion

Note: Problems 6 and 13 in Chapter 6 can also be assigned with this section.

27. The athlete shown in Figure P4.27 rotates a 1.00-kg discus along a circular path of radius 1.06 m. The maximum

Figure P4.27

speed of the discus is 20.0 m/s. Determine the magnitude of the maximum radial acceleration of the discus.

28. In Example 4.6, we found the centripetal acceleration of the Earth as it revolves around the Sun. From information on the endpapers of this book, compute the centripetal acceleration of a point on the surface of the Earth at the equator caused by the rotation of the Earth about its axis.

29. **Review.** The 20-*g* centrifuge at NASA's Ames Research Center in Mountain View, California, is a horizontal, cylindrical tube 58 ft long and is represented in Figure P4.29. Assume an astronaut in training sits in a seat at one end, facing the axis of rotation 29.0 ft away. Determine the rotation rate, in revolutions per second, required to give the astronaut a centripetal acceleration of 20.0*g*.

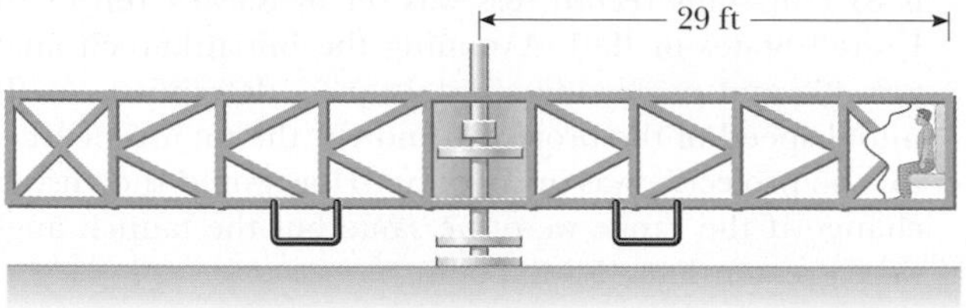

Figure P4.29

30. An athlete swings a ball, connected to the end of a chain, in a horizontal circle. The athlete is able to rotate the ball at the rate of 8.00 rev/s when the length of the chain is 0.600 m. When he increases the length to 0.900 m, he is able to rotate the ball only 6.00 rev/s. (a) Which rate of rotation gives the greater speed for the ball? (b) What is the centripetal acceleration of the ball at 8.00 rev/s? (c) What is the centripetal acceleration at 6.00 rev/s?

Section 4.5 Tangential and Radial Acceleration

31. M A train slows down as it rounds a sharp horizontal turn, going from 90.0 km/h to 50.0 km/h in the 15.0 s it takes to round the bend. The radius of the curve is 150 m. Compute the acceleration at the moment the train speed reaches 50.0 km/h. Assume the train continues to slow down at this time at the same rate.

32. Figure P4.32 represents the total acceleration of a particle moving clockwise in a circle of radius 2.50 m at a certain instant of time. For that instant, find (a) the radial acceleration of the particle, (b) the speed of the particle, and (c) its tangential acceleration.

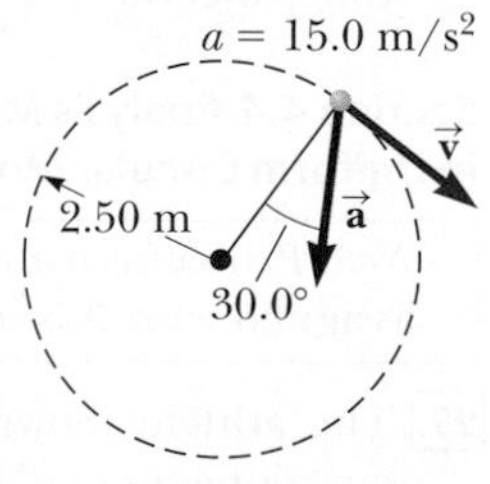

Figure P4.32

33. (a) Can a particle moving with instantaneous speed 3.00 m/s on a path with radius of curvature 2.00 m have an acceleration of magnitude 6.00 m/s^2? (b) Can it have an acceleration of magnitude 4.00 m/s^2? In each case, if the answer is yes, explain how it can happen; if the answer is no, explain why not.

34. A ball swings in a vertical circle at the end of a rope 1.50 m long. When the ball is 36.9° past the lowest point on its way up, its total acceleration is $(-22.5\hat{\mathbf{i}} + 20.2\hat{\mathbf{j}})$ m/s^2. For that instant, (a) sketch a vector diagram showing the components of its acceleration, (b) determine the magnitude of its radial acceleration, and (c) determine the speed and velocity of the ball.

Section 4.6 Relative Velocity and Relative Acceleration

35. A police car traveling at 95.0 km/h is traveling west, chasing a motorist traveling at 80.0 km/h. (a) What is the velocity of the motorist relative to the police car? (b) What is the velocity of the police car relative to the motorist? (c) If they are originally 250 m apart, in what time interval will the police car overtake the motorist?

36. S A moving beltway at an airport has a speed v_1 and a length L. A woman stands on the beltway as it moves from one end to the other, while a man in a hurry to reach his flight walks on the beltway with a speed of v_2 relative to the moving beltway. (a) What time interval is required for the woman to travel the distance L? (b) What time interval is required for the man to travel this distance? (c) A second beltway is located next to the first one. It is identical to the first one but moves in the opposite direction at speed v_1. Just as the man steps onto the beginning of the beltway and begins to walk at speed v_2 relative to his beltway, a child steps on the other end of the adjacent beltway. The child stands at rest relative to this second beltway. How long after stepping on the beltway does the man pass the child?

37. An airplane maintains a speed of 630 km/h relative to the air it is flying through as it makes a trip to a city 750 km away to the north. (a) What time interval is required for the trip if the plane flies through a headwind blowing at 35.0 km/h toward the south? (b) What time interval is required if there is a tailwind with the same speed? (c) What time interval is required if there is a crosswind blowing at 35.0 km/h to the east relative to the ground?

38. A car travels due east with a speed of 50.0 km/h. Raindrops are falling at a constant speed vertically with respect to the Earth. The traces of the rain on the side windows of the car make an angle of 60.0° with the vertical. Find the velocity of the rain with respect to (a) the car and (b) the Earth.

39. Q|C A bolt drops from the ceiling of a moving train car that is accelerating northward at a rate of 2.50 m/s^2. (a) What is the acceleration of the bolt relative to the train car? (b) What is the acceleration of the bolt relative to the Earth? (c) Describe the trajectory of the bolt as seen by an observer inside the train car. (d) Describe the trajectory of the bolt as seen by an observer fixed on the Earth.

40. Q|C M A river has a steady speed of 0.500 m/s. A student swims upstream a distance of 1.00 km and swims back to the starting point. (a) If the student can swim at a speed of 1.20 m/s in still water, how long does the trip take? (b) How much time is required in still water for the same length swim? (c) Intuitively, why does the swim take longer when there is a current?

41. Q|C S A river flows with a steady speed v. A student swims upstream a distance d and then back to the starting point. The student can swim at speed c in still water. (a) In terms of d, v, and c, what time interval is required for the round trip? (b) What time interval would be required if the water were still? (c) Which time interval is larger? Explain whether it is always larger.

42. **Q|C** A farm truck moves due east with a constant velocity of 9.50 m/s on a limitless, horizontal stretch of road. A boy riding on the back of the truck throws a can of soda upward (Fig. P4.42) and catches the projectile at the same location on the truck bed, but 16.0 m farther down the road. (a) In the frame of reference of the truck, at what angle to the vertical does the boy throw the can? (b) What is the initial speed of the can relative to the truck? (c) What is the shape of the can's trajectory as seen by the boy? An observer on the ground watches the boy throw the can and catch it. In this observer's frame of reference, (d) describe the shape of the can's path and (e) determine the initial velocity of the can.

Figure P4.42

43. **M** A science student is riding on a flatcar of a train traveling along a straight, horizontal track at a constant speed of 10.0 m/s. The student throws a ball into the air along a path that he judges to make an initial angle of 60.0° with the horizontal and to be in line with the track. The student's professor, who is standing on the ground nearby, observes the ball to rise vertically. How high does she see the ball rise?

44. A Coast Guard cutter detects an unidentified ship at a distance of 20.0 km in the direction 15.0° east of north. The ship is traveling at 26.0 km/h on a course at 40.0° east of north. The Coast Guard wishes to send a speedboat to intercept and investigate the vessel. If the speedboat travels at 50.0 km/h, in what direction should it head? Express the direction as a compass bearing with respect to due north.

Additional Problems

45. *Why is the following situation impossible?* A normally proportioned adult walks briskly along a straight line in the $+x$ direction, standing straight up and holding his right arm vertical and next to his body so that the arm does not swing. His right hand holds a ball at his side a distance h above the floor. When the ball passes above a point marked as $x = 0$ on the horizontal floor, he opens his fingers to release the ball from rest relative to his hand. The ball strikes the ground for the first time at position $x = 7.00h$.

46. A particle starts from the origin with velocity $5\hat{\mathbf{i}}$ m/s at $t = 0$ and moves in the xy plane with a varying acceleration given by $\vec{\mathbf{a}} = (6\sqrt{t}\,\hat{\mathbf{j}})$, where $\vec{\mathbf{a}}$ is in meters per second squared and t is in seconds. (a) Determine the velocity of the particle as a function of time. (b) Determine the position of the particle as a function of time.

47. *The "Vomit Comet."* In microgravity astronaut training and equipment testing, NASA flies a KC135A aircraft along a parabolic flight path. As shown in Figure P4.47, the aircraft climbs from 24 000 ft to 31 000 ft, where it enters a parabolic path with a velocity of 143 m/s nose high at 45.0° and exits with velocity 143 m/s at 45.0° nose low. During this portion of the flight, the aircraft and objects inside its padded cabin are in free fall; astronauts and equipment float freely as if there were no gravity. What are the aircraft's (a) speed and (b) altitude at the top of the maneuver? (c) What is the time interval spent in microgravity?

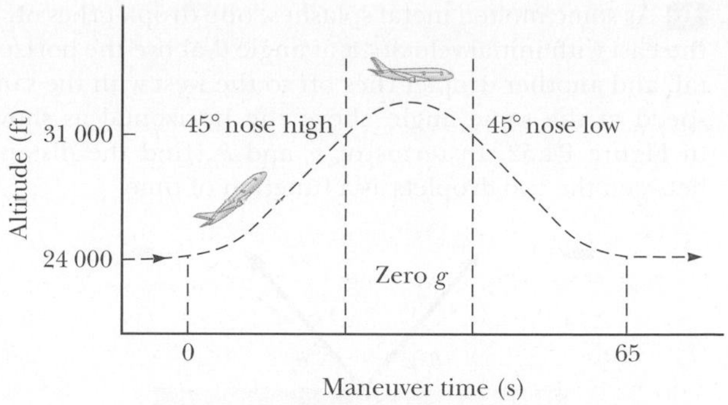

Figure P4.47

48. **S** A boy throws a stone horizontally from the top of a cliff of height h toward the ocean below. The stone strikes the ocean at distance d from the base of the cliff. In terms of h, d, and g, find expressions for (a) the time t at which the stone lands in the ocean, (b) the initial speed of the stone, (c) the speed of the stone immediately before it reaches the ocean, and (d) the direction of the stone's velocity immediately before it reaches the ocean.

49. Lisa in her Lamborghini accelerates at the rate of $(3.00\hat{\mathbf{i}} - 2.00\hat{\mathbf{j}})$ m/s^2, while Jill in her Jaguar accelerates at $(1.00\hat{\mathbf{i}} + 3.00\hat{\mathbf{j}})$ m/s^2. They both start from rest at the origin of an xy coordinate system. After 5.00 s, (a) what is Lisa's speed with respect to Jill, (b) how far apart are they, and (c) what is Lisa's acceleration relative to Jill?

50. A basketball player is standing on the floor 10.0 m from the basket as in Figure P4.50. The height of the basket is 3.05 m, and he shoots the ball at a 40.0° angle with the horizontal from a height of 2.00 m. (a) What is the acceleration of the basketball at the highest point in its trajectory? (b) At what speed must the player throw the basketball so that the ball goes through the hoop without striking the backboard?

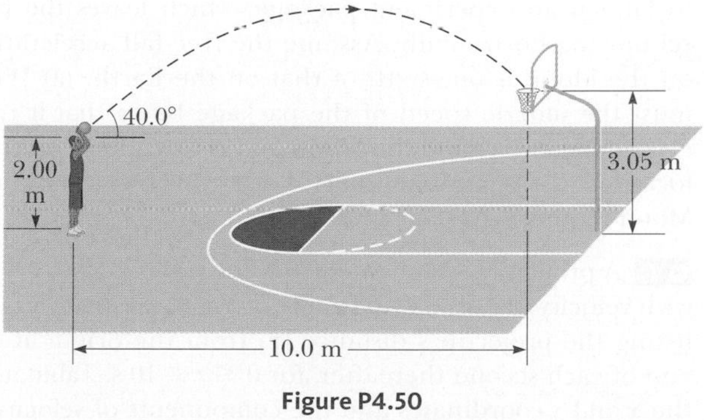

Figure P4.50

51. A flea is at point Ⓐ on a horizontal turntable, 10.0 cm from the center. The turntable is rotating at 33.3 rev/min in the clockwise direction. The flea jumps straight up to a height of 5.00 cm. At takeoff, it gives itself no horizontal velocity relative to the turntable. The flea lands on the turntable at point Ⓑ. Choose the origin of coordinates to be at the center of the turntable and the positive x axis passing through Ⓐ at the moment of takeoff. Then the original position of the flea is $10.0\hat{\mathbf{i}}$ cm. (a) Find the position of point Ⓐ when the flea lands. (b) Find the position of point Ⓑ when the flea lands.

52. **S** As some molten metal splashes, one droplet flies off to the east with initial velocity v_i at angle θ_i above the horizontal, and another droplet flies off to the west with the same speed at the same angle above the horizontal as shown in Figure P4.52. In terms of v_i and θ_i, find the distance between the two droplets as a function of time.

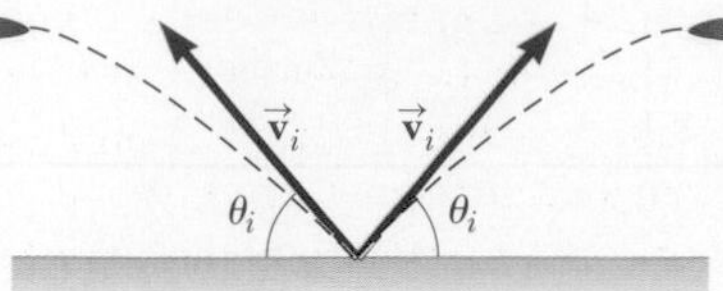

Figure P4.52

53. *Why is the following situation impossible?* Manny Ramírez hits a home run so that the baseball just clears the top row of bleachers, 24.0 m high, located 130 m from home plate. The ball is hit at 41.7 m/s at an angle of 35.0° to the horizontal, and air resistance is negligible.

54. A pendulum with a cord of length r = 1.00 m swings in a vertical plane (Fig. P4.54). When the pendulum is in the two horizontal positions θ = 90.0° and θ = 270°, its speed is 5.00 m/s. Find the magnitude of (a) the radial acceleration and (b) the tangential acceleration for these positions. (c) Draw vector diagrams to determine the direction of the total acceleration for these two positions. (d) Calculate the magnitude and direction of the total acceleration at these two positions.

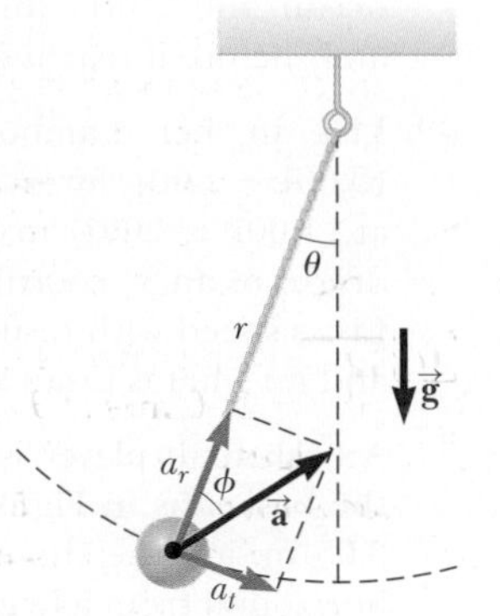

Figure P4.54

55. An astronaut on the surface of the Moon fires a cannon to launch an experiment package, which leaves the barrel moving horizontally. Assume the free-fall acceleration on the Moon is one-sixth of that on the Earth. (a) What must the muzzle speed of the package be so that it travels completely around the Moon and returns to its original location? (b) What time interval does this trip around the Moon require?

56. **QIC** A projectile is launched from the point (x = 0, y = 0), with velocity $(12.0\hat{i} + 49.0\hat{j})$ m/s, at t = 0. (a) Make a table listing the projectile's distance $|\vec{r}|$ from the origin at the end of each second thereafter, for $0 \leq t \leq 10$ s. Tabulating the x and y coordinates and the components of velocity v_x and v_y will also be useful. (b) Notice that the projectile's distance from its starting point increases with time, goes through a maximum, and starts to decrease. Prove that the distance is a maximum when the position vector is perpendicular to the velocity. *Suggestion:* Argue that if $\vec{v}$ is not perpendicular to $\vec{r}$, then $|\vec{r}|$ must be increasing or decreasing. (c) Determine the magnitude of the maximum displacement. (d) Explain your method for solving part (c).

57. **QIC** A spring cannon is located at the edge of a table that is 1.20 m above the floor. A steel ball is launched from the cannon with speed v_i at 35.0° above the horizontal. (a) Find the horizontal position of the ball as a function of v_i at the instant it lands on the floor. We write this function as $x(v_i)$. Evaluate x for (b) v_i = 0.100 m/s and for (c) v_i = 100 m/s. (d) Assume v_i is close to but not equal to zero. Show that one term in the answer to part (a) dominates so that the function $x(v_i)$ reduces to a simpler form. (e) If v_i is very large, what is the approximate form of $x(v_i)$? (f) Describe the overall shape of the graph of the function $x(v_i)$.

58. An outfielder throws a baseball to his catcher in an attempt to throw out a runner at home plate. The ball bounces once before reaching the catcher. Assume the angle at which the bounced ball leaves the ground is the same as the angle at which the outfielder threw it as shown in Figure P4.58, but that the ball's speed after the bounce is one-half of what it was before the bounce. (a) Assume the ball is always thrown with the same initial speed and ignore air resistance. At what angle θ should the fielder throw the ball to make it go the same distance D with one bounce (blue path) as a ball thrown upward at 45.0° with no bounce (green path)? (b) Determine the ratio of the time interval for the one-bounce throw to the flight time for the no-bounce throw.

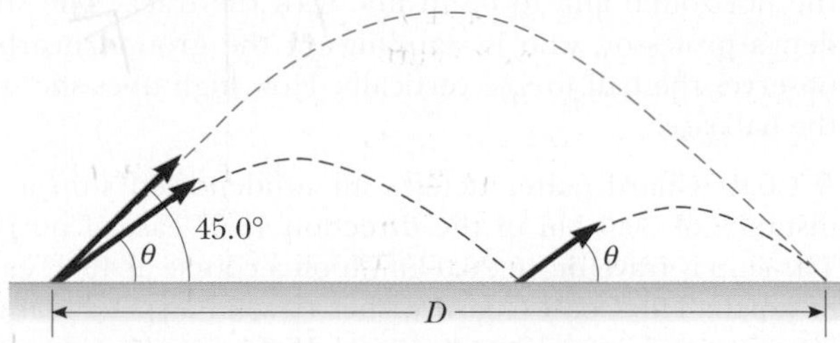

Figure P4.58

59. A World War II bomber flies horizontally over level terrain with a speed of 275 m/s relative to the ground and at an altitude of 3.00 km. The bombardier releases one bomb. (a) How far does the bomb travel horizontally between its release and its impact on the ground? Ignore the effects of air resistance. (b) The pilot maintains the plane's original course, altitude, and speed through a storm of flak. Where is the plane when the bomb hits the ground? (c) The bomb hits the target seen in the telescopic bombsight at the moment of the bomb's release. At what angle from the vertical was the bombsight set?

60. A truck loaded with cannonball watermelons stops suddenly to avoid running over the edge of a washed-out bridge (Fig. P4.60). The quick stop causes a number of melons to fly off the truck. One melon leaves the hood of the truck with an initial speed v_i = 10.0 m/s in the horizontal direction.

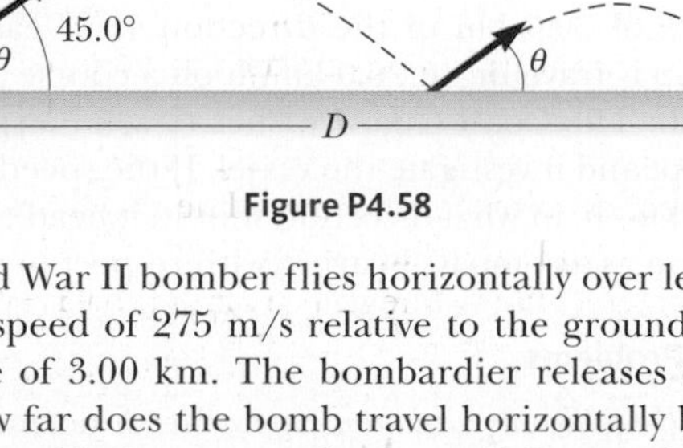

Figure P4.60 The blue dashed curve shows the parabolic shape of the bank.

A cross section of the bank has the shape of the bottom half of a parabola, with its vertex at the initial location of the projected watermelon and with the equation $y^2 = 16x$, where x and y are measured in meters. What are the x and y coordinates of the melon when it splatters on the bank?

61. **M** A hawk is flying horizontally at 10.0 m/s in a straight line, 200 m above the ground. A mouse it has been carrying struggles free from its talons. The hawk continues on its path at the same speed for 2.00 s before attempting to retrieve its prey. To accomplish the retrieval, it dives in a straight line at constant speed and recaptures the mouse 3.00 m above the ground. (a) Assuming no air resistance acts on the mouse, find the diving speed of the hawk. (b) What angle did the hawk make with the horizontal during its descent? (c) For what time interval did the mouse experience free fall?

62. An aging coyote cannot run fast enough to catch a roadrunner. He purchases on eBay a set of jet-powered roller skates, which provide a constant horizontal acceleration of 15.0 m/s^2 (Fig. P4.62). The coyote starts at rest 70.0 m from the edge of a cliff at the instant the roadrunner zips past in the direction of the cliff. (a) Determine the minimum constant speed the roadrunner must have to reach the cliff before the coyote. At the edge of the cliff, the roadrunner escapes by making a sudden turn, while the coyote continues straight ahead. The coyote's skates remain horizontal and continue to operate while he is in flight, so his acceleration while in the air is $(15.0\hat{\mathbf{i}} - 9.80\hat{\mathbf{j}})$ m/s^2. (b) The cliff is 100 m above the flat floor of the desert. Determine how far from the base of the vertical cliff the coyote lands. (c) Determine the components of the coyote's impact velocity.

Figure P4.62

63. **M** A car is parked on a steep incline, making an angle of 37.0° below the horizontal and overlooking the ocean, when its brakes fail and it begins to roll. Starting from rest at $t = 0$, the car rolls down the incline with a constant acceleration of 4.00 m/s^2, traveling 50.0 m to the edge of a vertical cliff. The cliff is 30.0 m above the ocean. Find (a) the speed of the car when it reaches the edge of the cliff, (b) the time interval elapsed when it arrives there, (c) the velocity of the car when it lands in the ocean, (d) the total time interval the car is in motion, and (e) the position of the car when it lands in the ocean, relative to the base of the cliff.

64. **Q|C** Do not hurt yourself; do not strike your hand against anything. Within these limitations, describe what you do to give your hand a large acceleration. Compute an order-of-magnitude estimate of this acceleration, stating the quantities you measure or estimate and their values.

65. A fisherman sets out upstream on a river. His small boat, powered by an outboard motor, travels at a constant speed v in still water. The water flows at a lower constant speed v_w. The fisherman has traveled upstream for 2.00 km when his ice chest falls out of the boat. He notices that the chest is missing only after he has gone upstream for another 15.0 min. At that point, he turns around and heads back downstream, all the time traveling at the same speed relative to the water. He catches up with the floating ice chest just as he returns to his starting point. How fast is the river flowing? Solve this problem in two ways. (a) First, use the Earth as a reference frame. With respect to the Earth, the boat travels upstream at speed $v - v_w$ and downstream at $v + v_w$. (b) A second much simpler and more elegant solution is obtained by using the water as the reference frame. This approach has important applications in many more complicated problems; examples are calculating the motion of rockets and satellites and analyzing the scattering of subatomic particles from massive targets.

Challenge Problems

66. **S** Two swimmers, Chris and Sarah, start together at the same point on the bank of a wide stream that flows with a speed v. Both move at the same speed c (where $c > v$) relative to the water. Chris swims downstream a distance L and then upstream the same distance. Sarah swims so that her motion relative to the Earth is perpendicular to the banks of the stream. She swims the distance L and then back the same distance, with both swimmers returning to the starting point. In terms of L, c, and v, find the time intervals required (a) for Chris's round trip and (b) for Sarah's round trip. (c) Explain which swimmer returns first.

67. **Q|C** A skier leaves the ramp of a ski jump with a velocity of $v = 10.0$ m/s at $\theta = 15.0°$ above the horizontal as shown in Figure P4.67. The slope where she will land is inclined downward at $\phi = 50.0°$, and air resistance is negligible. Find (a) the distance from the end of the ramp to where the jumper lands and (b) her velocity components just before the landing. (c) Explain how you think the results might be affected if air resistance were included.

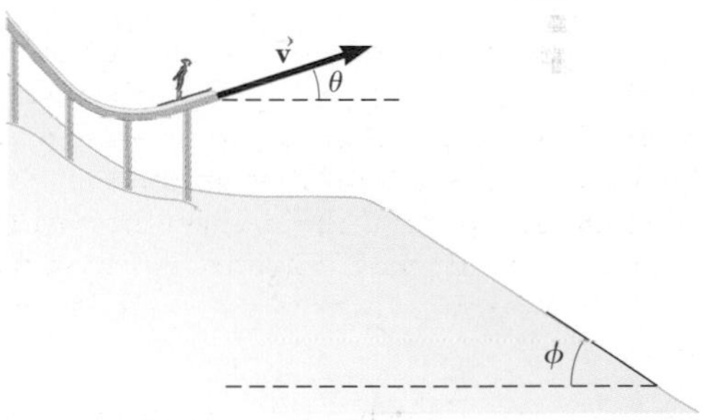

Figure P4.67

68. **S** A person standing at the top of a hemispherical rock of radius R kicks a ball (initially at rest on the top of the rock) to give it horizontal velocity $\vec{\mathbf{v}}_i$ as shown in Figure P4.68.

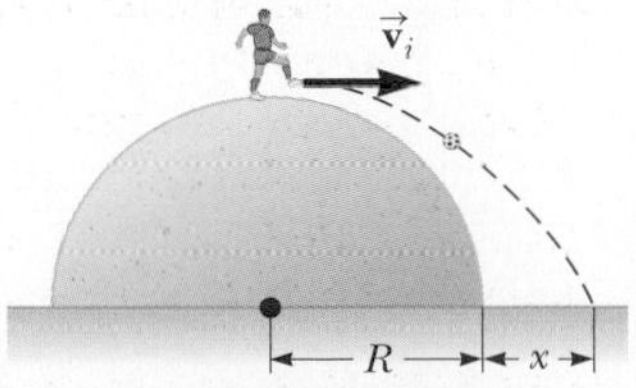

Figure P4.68

(a) What must be its minimum initial speed if the ball is never to hit the rock after it is kicked? (b) With this initial speed, how far from the base of the rock does the ball hit the ground?

69. A dive-bomber has a velocity of 280 m/s at an angle θ below the horizontal. When the altitude of the aircraft is 2.15 km, it releases a bomb, which subsequently hits a target on the ground. The magnitude of the displacement from the point of release of the bomb to the target is 3.25 km. Find the angle θ.

70. S A fireworks rocket explodes at height h, the peak of its vertical trajectory. It throws out burning fragments in all directions, but all at the same speed v. Pellets of solidified metal fall to the ground without air resistance. Find the smallest angle that the final velocity of an impacting fragment makes with the horizontal.

71. An enemy ship is on the east side of a mountain island as shown in Figure P4.71. The enemy ship has maneuvered to within 2 500 m of the 1 800-m-high mountain peak and can shoot projectiles with an initial speed of 250 m/s. If the western shoreline is horizontally 300 m from the peak, what are the distances from the western shore at which a ship can be safe from the bombardment of the enemy ship?

72. S In the What If? section of Example 4.5, it was claimed that the maximum range of a ski jumper occurs for a launch angle θ given by

$$\theta = 45° - \frac{\phi}{2}$$

where ϕ is the angle the hill makes with the horizontal in Figure 4.14. Prove this claim by deriving the equation above.

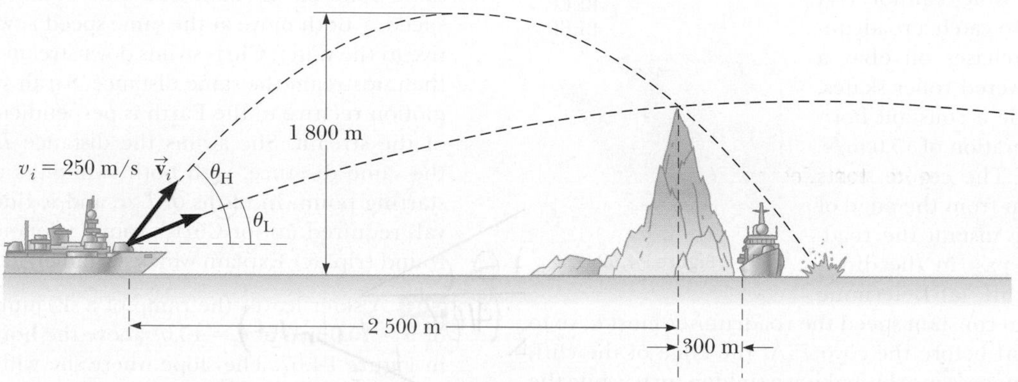

Figure P4.71

chapter 5

The Laws of Motion

A person sculls on a calm waterway. The water exerts forces on the oars to accelerate the boat. *(Tetra Images/Getty Images)*

In Chapters 2 and 4, we *described* the motion of an object in terms of its position, velocity, and acceleration without considering what might *influence* that motion. Now we consider that influence: Why does the motion of an object change? What might cause one object to remain at rest and another object to accelerate? Why is it generally easier to move a small object than a large object? The two main factors we need to consider are the *forces* acting on an object and the *mass* of the object. In this chapter, we begin our study of *dynamics* by discussing the three basic laws of motion, which deal with forces and masses and were formulated more than three centuries ago by Isaac Newton.

5.1 The Concept of Force

Everyone has a basic understanding of the concept of force from everyday experience. When you push your empty dinner plate away, you exert a force on it. Similarly, you exert a force on a ball when you throw or kick it. In these examples, the word *force* refers to an interaction with an object by means of muscular activity and some change in the object's velocity. Forces do not always cause motion, however. For example, when you are sitting, a gravitational force acts on your body and yet

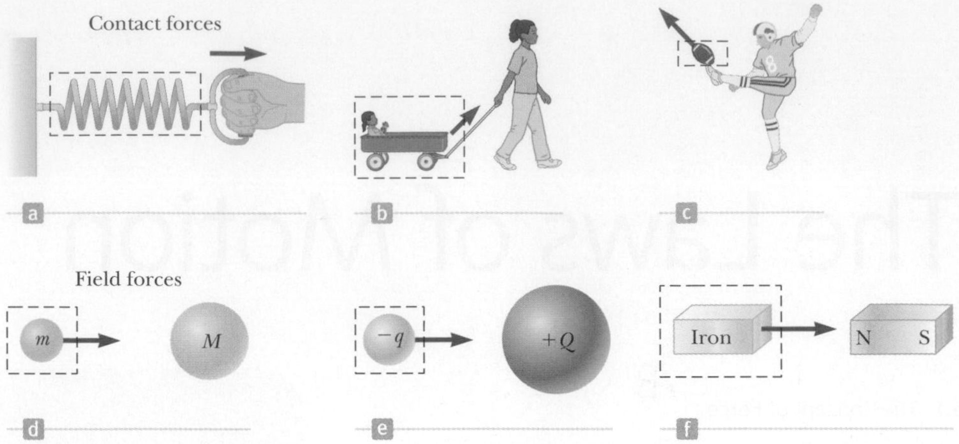

Figure 5.1 Some examples of applied forces. In each case, a force is exerted on the object within the boxed area. Some agent in the environment external to the boxed area exerts a force on the object.

you remain stationary. As a second example, you can push (in other words, exert a force) on a large boulder and not be able to move it.

What force (if any) causes the Moon to orbit the Earth? Newton answered this and related questions by stating that forces are what cause any change in the velocity of an object. The Moon's velocity changes in direction as it moves in a nearly circular orbit around the Earth. This change in velocity is caused by the gravitational force exerted by the Earth on the Moon.

When a coiled spring is pulled, as in Figure 5.1a, the spring stretches. When a stationary cart is pulled, as in Figure 5.1b, the cart moves. When a football is kicked, as in Figure 5.1c, it is both deformed and set in motion. These situations are all examples of a class of forces called *contact forces.* That is, they involve physical contact between two objects. Other examples of contact forces are the force exerted by gas molecules on the walls of a container and the force exerted by your feet on the floor.

Another class of forces, known as *field forces,* does not involve physical contact between two objects. These forces act through empty space. The gravitational force of attraction between two objects with mass, illustrated in Figure 5.1d, is an example of this class of force. The gravitational force keeps objects bound to the Earth and the planets in orbit around the Sun. Another common field force is the electric force that one electric charge exerts on another (Fig. 5.1e), such as the charges of an electron and proton that form a hydrogen atom. A third example of a field force is the force a bar magnet exerts on a piece of iron (Fig. 5.1f).

The distinction between contact forces and field forces is not as sharp as you may have been led to believe by the previous discussion. When examined at the atomic level, all the forces we classify as contact forces turn out to be caused by electric (field) forces of the type illustrated in Figure 5.1e. Nevertheless, in developing models for macroscopic phenomena, it is convenient to use both classifications of forces. The only known *fundamental* forces in nature are all field forces: (1) *gravitational forces* between objects, (2) *electromagnetic forces* between electric charges, (3) *strong forces* between subatomic particles, and (4) *weak forces* that arise in certain radioactive decay processes. In classical physics, we are concerned only with gravitational and electromagnetic forces. We will discuss strong and weak forces in Chapter 46.

Bridgeman-Giraudon/Art Resource, NY

Isaac Newton
English physicist and mathematician (1642–1727)
Isaac Newton was one of the most brilliant scientists in history. Before the age of 30, he formulated the basic concepts and laws of mechanics, discovered the law of universal gravitation, and invented the mathematical methods of calculus. As a consequence of his theories, Newton was able to explain the motions of the planets, the ebb and flow of the tides, and many special features of the motions of the Moon and the Earth. He also interpreted many fundamental observations concerning the nature of light. His contributions to physical theories dominated scientific thought for two centuries and remain important today.

The Vector Nature of Force

It is possible to use the deformation of a spring to measure force. Suppose a vertical force is applied to a spring scale that has a fixed upper end as shown in Figure 5.2a. The spring elongates when the force is applied, and a pointer on the scale reads the extension of the spring. We can calibrate the spring by defining a reference force $\vec{\mathbf{F}}_1$ as the force that produces a pointer reading of 1.00 cm. If we now apply a different downward force $\vec{\mathbf{F}}_2$ whose magnitude is twice that of the reference force

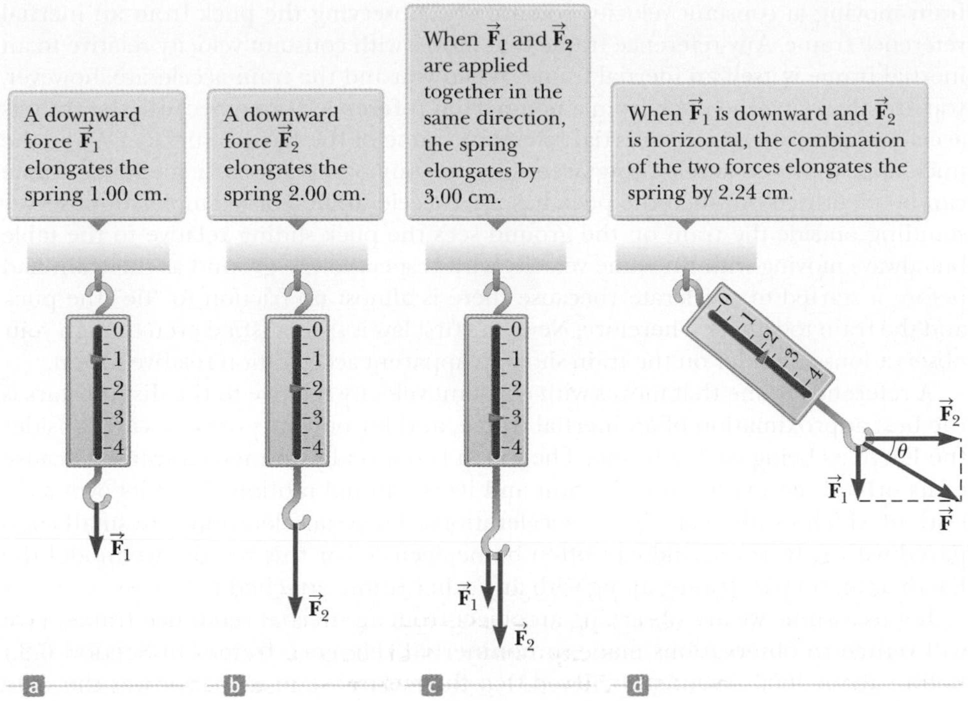

Figure 5.2 The vector nature of a force is tested with a spring scale.

$\vec{F}_1$ as seen in Figure 5.2b, the pointer moves to 2.00 cm. Figure 5.2c shows that the combined effect of the two collinear forces is the sum of the effects of the individual forces.

Now suppose the two forces are applied simultaneously with $\vec{F}_1$ downward and $\vec{F}_2$ horizontal as illustrated in Figure 5.2d. In this case, the pointer reads 2.24 cm. The single force $\vec{F}$ that would produce this same reading is the sum of the two vectors $\vec{F}_1$ and $\vec{F}_2$ as described in Figure 5.2d. That is, $|\vec{F}_1| = \sqrt{F_1^2 + F_2^2} = 2.24$ units, and its direction is $\theta = \tan^{-1}(-0.500) = -26.6°$. Because forces have been experimentally verified to behave as vectors, you *must* use the rules of vector addition to obtain the net force on an object.

5.2 Newton's First Law and Inertial Frames

We begin our study of forces by imagining some physical situations involving a puck on a perfectly level air hockey table (Fig. 5.3). You expect that the puck will remain stationary when it is placed gently at rest on the table. Now imagine your air hockey table is located on a train moving with constant velocity along a perfectly smooth track. If the puck is placed on the table, the puck again remains where it is placed. If the train were to accelerate, however, the puck would start moving along the table opposite the direction of the train's acceleration, just as a set of papers on your dashboard falls onto the floor of your car when you step on the accelerator.

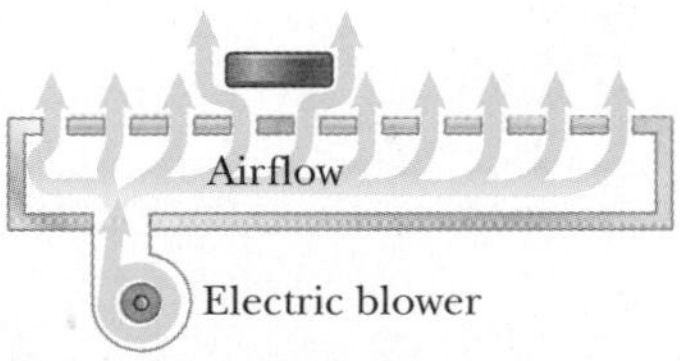

Figure 5.3 On an air hockey table, air blown through holes in the surface allows the puck to move almost without friction. If the table is not accelerating, a puck placed on the table will remain at rest.

As we saw in Section 4.6, a moving object can be observed from any number of reference frames. **Newton's first law of motion,** sometimes called the *law of inertia,* defines a special set of reference frames called *inertial frames.* This law can be stated as follows:

◀ **Newton's first law**

> If an object does not interact with other objects, it is possible to identify a reference frame in which the object has zero acceleration.

◀ **Inertial frame of reference**

Such a reference frame is called an **inertial frame of reference.** When the puck is on the air hockey table located on the ground, you are observing it from an inertial reference frame; there are no horizontal interactions of the puck with any other objects, and you observe it to have zero acceleration in that direction. When you are on the train moving at constant velocity, you are also observing the puck from an inertial reference frame. Any reference frame that moves with constant velocity relative to an inertial frame is itself an inertial frame. When you and the train accelerate, however, you are observing the puck from a **noninertial reference frame** because the train is accelerating relative to the inertial reference frame of the Earth's surface. While the puck appears to be accelerating according to your observations, a reference frame can be identified in which the puck has zero acceleration. For example, an observer standing outside the train on the ground sees the puck sliding relative to the table but always moving with the same velocity with respect to the ground as the train had before it started to accelerate (because there is almost no friction to "tie" the puck and the train together). Therefore, Newton's first law is still satisfied even though your observations as a rider on the train show an apparent acceleration relative to you.

A reference frame that moves with constant velocity relative to the distant stars is the best approximation of an inertial frame, and for our purposes we can consider the Earth as being such a frame. The Earth is not really an inertial frame because of its orbital motion around the Sun and its rotational motion about its own axis, both of which involve centripetal accelerations. These accelerations are small compared with g, however, and can often be neglected. For this reason, we model the Earth as an inertial frame, along with any other frame attached to it.

Pitfall Prevention 5.1

Newton's First Law

Newton's first law does *not* say what happens for an object with *zero net force,* that is, multiple forces that cancel; it says what happens *in the absence of external forces.* This subtle but important difference allows us to define force as that which causes a change in the motion. The description of an object under the effect of forces that balance is covered by Newton's second law.

Let us assume we are observing an object from an inertial reference frame. (We will return to observations made in noninertial reference frames in Section 6.3.) Before about 1600, scientists believed that the natural state of matter was the state of rest. Observations showed that moving objects eventually stopped moving. Galileo was the first to take a different approach to motion and the natural state of matter. He devised thought experiments and concluded that it is not the nature of an object to stop once set in motion: rather, it is its nature to *resist changes in its motion.* In his words, "Any velocity once imparted to a moving body will be rigidly maintained as long as the external causes of retardation are removed." For example, a spacecraft drifting through empty space with its engine turned off will keep moving forever. It would *not* seek a "natural state" of rest.

Given our discussion of observations made from inertial reference frames, we can pose a more practical statement of Newton's first law of motion:

Another statement of Newton's first law ▶

In the absence of external forces and when viewed from an inertial reference frame, an object at rest remains at rest and an object in motion continues in motion with a constant velocity (that is, with a constant speed in a straight line).

In other words, **when no force acts on an object, the acceleration of the object is zero.** From the first law, we conclude that any *isolated object* (one that does not interact with its environment) is either at rest or moving with constant velocity. The tendency of an object to resist any attempt to change its velocity is called **inertia.** Given the statement of the first law above, we can conclude that an object that is accelerating must be experiencing a force. In turn, from the first law, we can define **force** as **that which causes a change in motion of an object.**

Definition of force ▶

***Quick Quiz* 5.1** Which of the following statements is correct? **(a)** It is possible for an object to have motion in the absence of forces on the object. **(b)** It is possible to have forces on an object in the absence of motion of the object. **(c)** Neither statement **(a)** nor statement **(b)** is correct. **(d)** Both statements **(a)** and **(b)** are correct.

5.3 Mass

Imagine playing catch with either a basketball or a bowling ball. Which ball is more likely to keep moving when you try to catch it? Which ball requires more effort to throw it? The bowling ball requires more effort. In the language of physics, we say that the bowling ball is more resistant to changes in its velocity than the basketball. How can we quantify this concept?

◀ **Definition of mass**

Mass is that property of an object that specifies how much resistance an object exhibits to changes in its velocity, and as we learned in Section 1.1, the SI unit of mass is the kilogram. Experiments show that the greater the mass of an object, the less that object accelerates under the action of a given applied force.

To describe mass quantitatively, we conduct experiments in which we compare the accelerations a given force produces on different objects. Suppose a force acting on an object of mass m_1 produces a change in motion of the object that we can quantify with the object's acceleration $\vec{\mathbf{a}}_1$, and the *same force* acting on an object of mass m_2 produces an acceleration $\vec{\mathbf{a}}_2$. The ratio of the two masses is defined as the *inverse* ratio of the magnitudes of the accelerations produced by the force:

$$\frac{m_1}{m_2} \equiv \frac{a_2}{a_1} \tag{5.1}$$

For example, if a given force acting on a 3-kg object produces an acceleration of 4 m/s^2, the same force applied to a 6-kg object produces an acceleration of 2 m/s^2. According to a huge number of similar observations, we conclude that the magnitude of the acceleration of an object is inversely proportional to its mass when acted on by a given force. If one object has a known mass, the mass of the other object can be obtained from acceleration measurements.

Mass is an inherent property of an object and is independent of the object's surroundings and of the method used to measure it. Also, mass is a scalar quantity and thus obeys the rules of ordinary arithmetic. For example, if you combine a 3-kg mass with a 5-kg mass, the total mass is 8 kg. This result can be verified experimentally by comparing the acceleration that a known force gives to several objects separately with the acceleration that the same force gives to the same objects combined as one unit.

◀ **Mass and weight are different quantities**

Mass should not be confused with weight. Mass and weight are two different quantities. The weight of an object is equal to the magnitude of the gravitational force exerted on the object and varies with location (see Section 5.5). For example, a person weighing 180 lb on the Earth weighs only about 30 lb on the Moon. On the other hand, the mass of an object is the same everywhere: an object having a mass of 2 kg on the Earth also has a mass of 2 kg on the Moon.

5.4 Newton's Second Law

Newton's first law explains what happens to an object when no forces act on it: it either remains at rest or moves in a straight line with constant speed. Newton's second law answers the question of what happens to an object when one or more forces act on it.

Imagine performing an experiment in which you push a block of mass m across a frictionless, horizontal surface. When you exert some horizontal force $\vec{\mathbf{F}}$ on the block, it moves with some acceleration $\vec{\mathbf{a}}$. If you apply a force twice as great on the same block, experimental results show that the acceleration of the block doubles; if you increase the applied force to $3\vec{\mathbf{F}}$, the acceleration triples; and so on. From such observations, we conclude that the acceleration of an object is directly proportional to the force acting on it: $\vec{\mathbf{F}} \propto \vec{\mathbf{a}}$. This idea was first introduced in Section 2.4 when we discussed the direction of the acceleration of an object. We also know from the preceding section that the magnitude of the acceleration of an object is inversely proportional to its mass: $|\vec{\mathbf{a}}| \propto 1/m$.

These experimental observations are summarized in **Newton's second law:**

> **Pitfall Prevention 5.2**
>
> **Force Is the Cause of Changes in Motion**
>
> An object can have motion in the absence of forces as described in Newton's first law. Therefore, don't interpret force as the cause of *motion.* Force is the cause of *changes* in motion as measured by acceleration.

When viewed from an inertial reference frame, the acceleration of an object is directly proportional to the net force acting on it and inversely proportional to its mass:

$$\vec{\mathbf{a}} \propto \frac{\sum \vec{\mathbf{F}}}{m}$$

If we choose a proportionality constant of 1, we can relate mass, acceleration, and force through the following mathematical statement of Newton's second law:[1]

Newton's second law ▶

$$\sum \vec{\mathbf{F}} = m\vec{\mathbf{a}} \tag{5.2}$$

In both the textual and mathematical statements of Newton's second law, we have indicated that the acceleration is due to the *net force* $\sum \vec{\mathbf{F}}$ acting on an object. The **net force** on an object is the vector sum of all forces acting on the object. (We sometimes refer to the net force as the *total force,* the *resultant force,* or the *unbalanced force.*) In solving a problem using Newton's second law, it is imperative to determine the correct net force on an object. Many forces may be acting on an object, but there is only one acceleration.

Equation 5.2 is a vector expression and hence is equivalent to three component equations:

Newton's second law: component form ▶

$$\sum F_x = ma_x \qquad \sum F_y = ma_y \qquad \sum F_z = ma_z \tag{5.3}$$

***Quick Quiz* 5.2** An object experiences no acceleration. Which of the following *cannot* be true for the object? **(a)** A single force acts on the object. **(b)** No forces act on the object. **(c)** Forces act on the object, but the forces cancel.

***Quick Quiz* 5.3** You push an object, initially at rest, across a frictionless floor with a constant force for a time interval Δt, resulting in a final speed of v for the object. You then repeat the experiment, but with a force that is twice as large. What time interval is now required to reach the same final speed v? **(a)** $4\,\Delta t$ **(b)** $2\,\Delta t$ **(c)** Δt **(d)** $\Delta t/2$ **(e)** $\Delta t/4$

Pitfall Prevention 5.3

$m\vec{a}$ Is Not a Force

Equation 5.2 does *not* say that the product $m\vec{\mathbf{a}}$ is a force. All forces on an object are added vectorially to generate the net force on the left side of the equation. This net force is then equated to the product of the mass of the object and the acceleration that results from the net force. Do *not* include an "$m\vec{\mathbf{a}}$ force" in your analysis of the forces on an object.

The SI unit of force is the **newton** (N). A force of 1 N is the force that, when acting on an object of mass 1 kg, produces an acceleration of 1 m/s^2. From this definition and Newton's second law, we see that the newton can be expressed in terms of the following fundamental units of mass, length, and time:

Definition of the newton ▶

$$1\text{ N} \equiv 1\text{ kg}\cdot\text{m/s}^2 \tag{5.4}$$

In the U.S. customary system, the unit of force is the **pound** (lb). A force of 1 lb is the force that, when acting on a 1-slug mass,[2] produces an acceleration of 1 ft/s^2:

$$1\text{ lb} \equiv 1\text{ slug}\cdot\text{ft/s}^2 \tag{5.5}$$

A convenient approximation is $1\text{ N} \approx \frac{1}{4}\text{ lb}$.

Example 5.1 An Accelerating Hockey Puck

A hockey puck having a mass of 0.30 kg slides on the frictionless, horizontal surface of an ice rink. Two hockey sticks strike the puck simultaneously, exerting the forces on the puck shown in Figure 5.4. The force $\vec{\mathbf{F}}_1$ has a magnitude of 5.0 N, and the force $\vec{\mathbf{F}}_2$ has a magnitude of 8.0 N. Determine both the magnitude and the direction of the puck's acceleration.

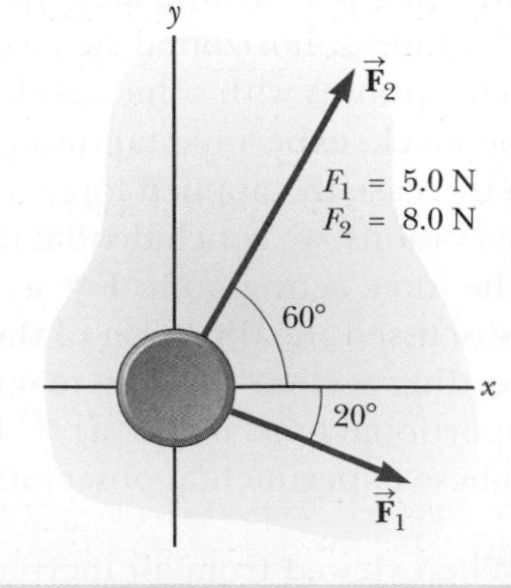

Figure 5.4 (Example 5.1) A hockey puck moving on a frictionless surface is subject to two forces $\vec{\mathbf{F}}_1$ and $\vec{\mathbf{F}}_2$.

SOLUTION

Conceptualize Study Figure 5.4. Using your expertise in vector addition from Chapter 3, predict the approximate direction of the net force vector on the puck. The acceleration of the puck will be in the same direction.

[1]Equation 5.2 is valid only when the speed of the object is much less than the speed of light. We treat the relativistic situation in Chapter 39.

[2]The *slug* is the unit of mass in the U.S. customary system and is that system's counterpart of the SI unit the *kilogram.* Because most of the calculations in our study of classical mechanics are in SI units, the slug is seldom used in this text.

5.1 *cont.*

Categorize Because we can determine a net force and we want an acceleration, this problem is categorized as one that may be solved using Newton's second law.

Analyze Find the component of the net force acting on the puck in the x direction:

$$\sum F_x = F_{1x} + F_{2x} = F_1 \cos(-20°) + F_2 \cos 60°$$
$$= (5.0 \text{ N})(0.940) + (8.0 \text{ N})(0.500) = 8.7 \text{ N}$$

Find the component of the net force acting on the puck in the y direction:

$$\sum F_y = F_{1y} + F_{2y} = F_1 \sin(-20°) + F_2 \sin 60°$$
$$= (5.0 \text{ N})(-0.342) + (8.0 \text{ N})(0.866) = 5.2 \text{ N}$$

Use Newton's second law in component form (Eq. 5.3) to find the x and y components of the puck's acceleration:

$$a_x = \frac{\sum F_x}{m} = \frac{8.7 \text{ N}}{0.30 \text{ kg}} = 29 \text{ m/s}^2$$

$$a_y = \frac{\sum F_y}{m} = \frac{5.2 \text{ N}}{0.30 \text{ kg}} = 17 \text{ m/s}^2$$

Find the magnitude of the acceleration:

$$a = \sqrt{(29 \text{ m/s}^2)^2 + (17 \text{ m/s}^2)^2} = 34 \text{ m/s}^2$$

Find the direction of the acceleration relative to the positive x axis:

$$\theta = \tan^{-1}\left(\frac{a_y}{a_x}\right) = \tan^{-1}\left(\frac{17}{29}\right) = 31°$$

Finalize The vectors in Figure 5.4 can be added graphically to check the reasonableness of our answer. Because the acceleration vector is along the direction of the resultant force, a drawing showing the resultant force vector helps us check the validity of the answer. (Try it!)

WHAT IF? Suppose three hockey sticks strike the puck simultaneously, with two of them exerting the forces shown in Figure 5.4. The result of the three forces is that the hockey puck shows *no* acceleration. What must be the components of the third force?

Answer If there is zero acceleration, the net force acting on the puck must be zero. Therefore, the three forces must cancel. We have found the components of the combination of the first two forces. The components of the third force must be of equal magnitude and opposite sign so that all the components add to zero. Therefore, $F_{3x} = -8.7$ N and $F_{3y} = -5.2$ N.

5.5 The Gravitational Force and Weight

All objects are attracted to the Earth. The attractive force exerted by the Earth on an object is called the **gravitational force** $\vec{\mathbf{F}}_g$. This force is directed toward the center of the Earth,[3] and its magnitude is called the **weight** of the object.

We saw in Section 2.6 that a freely falling object experiences an acceleration $\vec{\mathbf{g}}$ acting toward the center of the Earth. Applying Newton's second law $\sum \vec{\mathbf{F}} = m\vec{\mathbf{a}}$ to a freely falling object of mass m, with $\vec{\mathbf{a}} = \vec{\mathbf{g}}$ and $\sum \vec{\mathbf{F}} = \vec{\mathbf{F}}_g$, gives

$$\vec{\mathbf{F}}_g = m\vec{\mathbf{g}}$$

Therefore, the weight of an object, being defined as the magnitude of $\vec{\mathbf{F}}_g$, is equal to mg:

$$F_g = mg \tag{5.6}$$

Because it depends on g, weight varies with geographic location. Because g decreases with increasing distance from the center of the Earth, objects weigh less at higher altitudes than at sea level. For example, a 1 000-kg pallet of bricks used in the construction of the Empire State Building in New York City weighed 9 800 N at street level, but weighed about 1 N less by the time it was lifted from sidewalk

Pitfall Prevention 5.4

"Weight of an Object"

We are familiar with the everyday phrase, the "weight of an object." Weight, however, is not an inherent property of an object; rather, it is a measure of the gravitational force between the object and the Earth (or other planet). Therefore, weight is a property of a *system* of items: the object and the Earth.

Pitfall Prevention 5.5

Kilogram Is Not a Unit of Weight

You may have seen the "conversion" 1 kg = 2.2 lb. Despite popular statements of weights expressed in kilograms, the kilogram is not a unit of *weight*, it is a unit of *mass*. The conversion statement is not an equality; it is an *equivalence* that is valid only on the Earth's surface.

[3]This statement ignores that the mass distribution of the Earth is not perfectly spherical.

NASA/Eugene Cernan

The life-support unit strapped to the back of astronaut Harrison Schmitt weighed 300 lb on the Earth and had a mass of 136 kg. During his training, a 50-lb mock-up with a mass of 23 kg was used. Although this strategy effectively simulated the reduced weight the unit would have on the Moon, it did not correctly mimic the unchanging mass. It was more difficult to accelerate the 136-kg unit (perhaps by jumping or twisting suddenly) on the Moon than it was to accelerate the 23-kg unit on the Earth.

level to the top of the building. As another example, suppose a student has a mass of 70.0 kg. The student's weight in a location where $g = 9.80\ \text{m/s}^2$ is 686 N (about 150 lb). At the top of a mountain, however, where $g = 9.77\ \text{m/s}^2$, the student's weight is only 684 N. Therefore, if you want to lose weight without going on a diet, climb a mountain or weigh yourself at 30 000 ft during an airplane flight!

Equation 5.6 quantifies the gravitational force on the object, but notice that this equation does not require the object to be moving. Even for a stationary object or for an object on which several forces act, Equation 5.6 can be used to calculate the magnitude of the gravitational force. The result is a subtle shift in the interpretation of m in the equation. The mass m in Equation 5.6 determines the strength of the gravitational attraction between the object and the Earth. This role is completely different from that previously described for mass, that of measuring the resistance to changes in motion in response to an external force. In that role, mass is also called **inertial mass.** We call m in Equation 5.6 the **gravitational mass.** Even though this quantity is different in behavior from inertial mass, it is one of the experimental conclusions in Newtonian dynamics that gravitational mass and inertial mass have the same value.

Although this discussion has focused on the gravitational force on an object due to the Earth, the concept is generally valid on any planet. The value of g will vary from one planet to the next, but the magnitude of the gravitational force will always be given by the value of mg.

Quick Quiz **5.4** Suppose you are talking by interplanetary telephone to a friend who lives on the Moon. He tells you that he has just won a newton of gold in a contest. Excitedly, you tell him that you entered the Earth version of the same contest and also won a newton of gold! Who is richer? **(a)** You are. **(b)** Your friend is. **(c)** You are equally rich.

Conceptual Example 5.2 How Much Do You Weigh in an Elevator?

You have most likely been in an elevator that accelerates upward as it moves toward a higher floor. In this case, you feel heavier. In fact, if you are standing on a bathroom scale at the time, the scale measures a force having a magnitude that is greater than your weight. Therefore, you have tactile and measured evidence that leads you to believe you are heavier in this situation. *Are* you heavier?

SOLUTION

No; your weight is unchanged. Your experiences are due to your being in a noninertial reference frame. To provide the acceleration upward, the floor or scale must exert on your feet an upward force that is greater in magnitude than your weight. It is this greater force you feel, which you interpret as feeling heavier. The scale reads this upward force, not your weight, and so its reading increases.

5.6 Newton's Third Law

If you press against a corner of this textbook with your fingertip, the book pushes back and makes a small dent in your skin. If you push harder, the book does the same and the dent in your skin is a little larger. This simple activity illustrates that forces are *interactions* between two objects: when your finger pushes on the book, the book pushes back on your finger. This important principle is known as **Newton's third law:**

Newton's third law ▶

If two objects interact, the force $\vec{\mathbf{F}}_{12}$ exerted by object 1 on object 2 is equal in magnitude and opposite in direction to the force $\vec{\mathbf{F}}_{21}$ exerted by object 2 on object 1:

$$\vec{\mathbf{F}}_{12} = -\vec{\mathbf{F}}_{21} \tag{5.7}$$

When it is important to designate forces as interactions between two objects, we will use this subscript notation, where $\vec{\mathbf{F}}_{ab}$ means "the force exerted *by* a *on* b." The third law is illustrated in Figure 5.5a. The force that object 1 exerts on object 2 is popularly called the *action force,* and the force of object 2 on object 1 is called the *reaction force.* These italicized terms are not scientific terms; furthermore, either force can be labeled the action or reaction force. We will use these terms for convenience. In all cases, the action and reaction forces act on *different* objects and must be of the same type (gravitational, electrical, etc.). For example, the force acting on a freely falling projectile is the gravitational force exerted by the Earth on the projectile $\vec{\mathbf{F}}_g = \vec{\mathbf{F}}_{Ep}$ (E = Earth, p = projectile), and the magnitude of this force is mg. The reaction to this force is the gravitational force exerted by the projectile on the Earth $\vec{\mathbf{F}}_{pE} = -\vec{\mathbf{F}}_{Ep}$. The reaction force $\vec{\mathbf{F}}_{pE}$ must accelerate the Earth toward the projectile just as the action force $\vec{\mathbf{F}}_{Ep}$ accelerates the projectile toward the Earth. Because the Earth has such a large mass, however, its acceleration due to this reaction force is negligibly small.

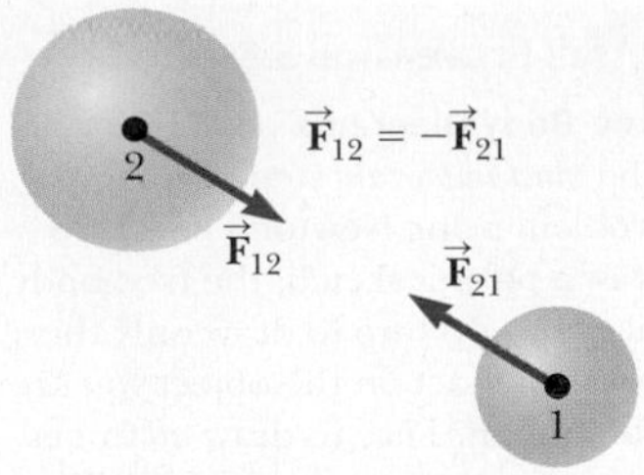

Figure 5.5 Newton's third law. The force $\vec{\mathbf{F}}_{12}$ exerted by object 1 on object 2 is equal in magnitude and opposite in direction to the force $\vec{\mathbf{F}}_{21}$ exerted by object 2 on object 1.

Consider a computer monitor at rest on a table as in Figure 5.6a. The reaction force to the gravitational force $\vec{\mathbf{F}}_g = \vec{\mathbf{F}}_{Em}$ on the monitor is the force $\vec{\mathbf{F}}_{mE} = -\vec{\mathbf{F}}_{Em}$ exerted by the monitor on the Earth. The monitor does not accelerate because it is held up by the table. The table exerts on the monitor an upward force $\vec{\mathbf{n}} = \vec{\mathbf{F}}_{tm}$, called the **normal force.** (*Normal* in this context means *perpendicular.*) This force, which prevents the monitor from falling through the table, can have any value needed, up to the point of breaking the table. Because the monitor has zero acceleration, Newton's second law applied to the monitor gives us $\sum \vec{\mathbf{F}} = \vec{\mathbf{n}} + m\vec{\mathbf{g}} = 0$, so $n\hat{\mathbf{j}} - mg\hat{\mathbf{j}} = 0$, or $n = mg$. The normal force balances the gravitational force on the monitor, so the net force on the monitor is zero. The reaction force to $\vec{\mathbf{n}}$ is the force exerted by the monitor downward on the table, $\vec{\mathbf{F}}_{mt} = -\vec{\mathbf{F}}_{tm} = -\vec{\mathbf{n}}$.

Notice that the forces acting on the monitor are $\vec{\mathbf{F}}_g$ and $\vec{\mathbf{n}}$ as shown in Figure 5.6b. The two forces $\vec{\mathbf{F}}_{mE}$ and $\vec{\mathbf{F}}_{mt}$ are exerted on objects other than the monitor.

Figure 5.6 illustrates an extremely important step in solving problems involving forces. Figure 5.6a shows many of the forces in the situation: those acting on the monitor, one acting on the table, and one acting on the Earth. Figure 5.6b, by contrast, shows only the forces acting on *one object,* the monitor, and is called a **force diagram** or a *diagram showing the forces on the object.* The important pictorial representation in Figure 5.6c is called a **free-body diagram.** In a free-body diagram, the particle model is used by representing the object as a dot and showing the forces that act on the object as being applied to the dot. When analyzing an object subject to forces, we are interested in the net force acting on one object, which we will model as a particle. Therefore, a free-body diagram helps us isolate only those forces on the object and eliminate the other forces from our analysis.

Pitfall Prevention 5.6

n* Does Not Always Equal *mg

In the situation shown in Figure 5.6 and in many others, we find that $n = mg$ (the normal force has the same magnitude as the gravitational force). This result, however, is *not* generally true. If an object is on an incline, if there are applied forces with vertical components, or if there is a vertical acceleration of the system, then $n \neq mg$. *Always* apply Newton's second law to find the relationship between n and mg.

Pitfall Prevention 5.7

Newton's Third Law

Remember that Newton's third-law action and reaction forces act on *different* objects. For example, in Figure 5.6, $\vec{\mathbf{n}} = \vec{\mathbf{F}}_{tm} = -m\vec{\mathbf{g}} = -\vec{\mathbf{F}}_{Em}$. The forces $\vec{\mathbf{n}}$ and $m\vec{\mathbf{g}}$ are equal in magnitude and opposite in direction, but they do not represent an action–reaction pair because both forces act on the *same* object, the monitor.

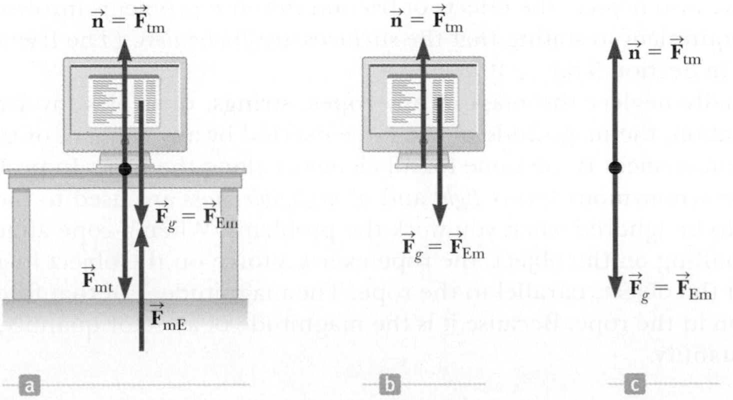

Figure 5.6 (a) When a computer monitor is at rest on a table, the forces acting on the monitor are the normal force $\vec{\mathbf{n}}$ and the gravitational force $\vec{\mathbf{F}}_g$. The reaction to $\vec{\mathbf{n}}$ is the force $\vec{\mathbf{F}}_{mt}$ exerted by the monitor on the table. The reaction to $\vec{\mathbf{F}}_g$ is the force $\vec{\mathbf{F}}_{mE}$ exerted by the monitor on the Earth. (b) A diagram showing the forces on the monitor. (c) A free-body diagram shows the monitor as a black dot with the forces acting on it.

Pitfall Prevention 5.8

Free-Body Diagrams
The *most important* step in solving a problem using Newton's laws is to draw a proper sketch, the free-body diagram. Be sure to draw only those forces that act on the object you are isolating. Be sure to draw *all* forces acting on the object, including any field forces, such as the gravitational force.

Quick Quiz **5.5** **(i)** If a fly collides with the windshield of a fast-moving bus, which experiences an impact force with a larger magnitude? (a) The fly. (b) The bus. (c) The same force is experienced by both. **(ii)** Which experiences the greater acceleration? (a) The fly. (b) The bus. (c) The same acceleration is experienced by both.

Conceptual Example 5.3 You Push Me and I'll Push You

A large man and a small boy stand facing each other on frictionless ice. They put their hands together and push against each other so that they move apart.

(A) Who moves away with the higher speed?

SOLUTION

This situation is similar to what we saw in Quick Quiz 5.5. According to Newton's third law, the force exerted by the man on the boy and the force exerted by the boy on the man are a third-law pair of forces, so they must be equal in magnitude. (A bathroom scale placed between their hands would read the same, regardless of which way it faced.) Therefore, the boy, having the smaller mass, experiences the greater acceleration. Both individuals accelerate for the same amount of time, but the greater acceleration of the boy over this time interval results in his moving away from the interaction with the higher speed.

(B) Who moves farther while their hands are in contact?

SOLUTION

Because the boy has the greater acceleration and therefore the greater average velocity, he moves farther than the man during the time interval during which their hands are in contact.

5.7 Analysis Models Using Newton's Second Law

In this section, we discuss two analysis models for solving problems in which objects are either in equilibrium ($\vec{\mathbf{a}} = 0$) or accelerating along a straight line under the action of constant external forces. Remember that when Newton's laws are applied to an object, we are interested only in external forces that act on the object. If the objects are modeled as particles, we need not worry about rotational motion. For now, we also neglect the effects of friction in those problems involving motion, which is equivalent to stating that the surfaces are *frictionless.* (The friction force is discussed in Section 5.8.)

We usually neglect the mass of any ropes, strings, or cables involved. In this approximation, the magnitude of the force exerted by any element of the rope on the adjacent element is the same for all elements along the rope. In problem statements, the synonymous terms *light* and *of negligible mass* are used to indicate that a mass is to be ignored when you work the problems. When a rope attached to an object is pulling on the object, the rope exerts a force on the object in a direction away from the object, parallel to the rope. The magnitude T of that force is called the **tension** in the rope. Because it is the magnitude of a vector quantity, tension is a scalar quantity.

Analysis Model: The Particle in Equilibrium

If the acceleration of an object modeled as a particle is zero, the object is treated with the **particle in equilibrium** model. In this model, the net force on the object is zero:

$$\sum \vec{\mathbf{F}} = 0 \tag{5.8}$$

Consider a lamp suspended from a light chain fastened to the ceiling as in Figure 5.7a. The force diagram for the lamp (Fig. 5.7b) shows that the forces acting on the lamp are the downward gravitational force $\vec{\mathbf{F}}_g$ and the upward force $\vec{\mathbf{T}}$ exerted by the chain. Because there are no forces in the x direction, $\Sigma F_x = 0$ provides no helpful information. The condition $\Sigma F_y = 0$ gives

$$\sum F_y = T - F_g = 0 \text{ or } T = F_g$$

Again, notice that $\vec{\mathbf{T}}$ and $\vec{\mathbf{F}}_g$ are *not* an action–reaction pair because they act on the same object, the lamp. The reaction force to $\vec{\mathbf{T}}$ is a downward force exerted by the lamp on the chain.

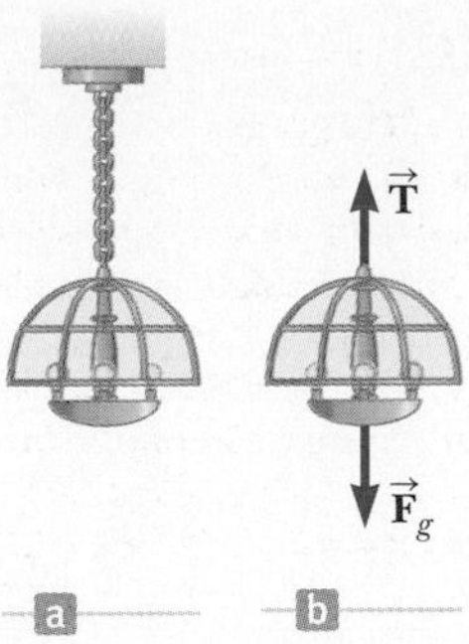

Figure 5.7 (a) A lamp suspended from a ceiling by a chain of negligible mass. (b) The forces acting on the lamp are the gravitational force $\vec{\mathbf{F}}_g$ and the force $\vec{\mathbf{T}}$ exerted by the chain.

Analysis Model: The Particle Under a Net Force

If an object experiences an acceleration, its motion can be analyzed with the **particle under a net force** model. The appropriate equation for this model is Newton's second law, Equation 5.2:

$$\sum \vec{\mathbf{F}} = m\vec{\mathbf{a}} \tag{5.2}$$

Consider a crate being pulled to the right on a frictionless, horizontal floor as in Figure 5.8a. Of course, the floor directly under the boy must have friction; otherwise, his feet would simply slip when he tries to pull on the crate! Suppose you wish to find the acceleration of the crate and the force the floor exerts on it. The forces acting on the crate are illustrated in the free-body diagram in Figure 5.8b. Notice that the horizontal force $\vec{\mathbf{T}}$ being applied to the crate acts through the rope. The magnitude of $\vec{\mathbf{T}}$ is equal to the tension in the rope. In addition to the force $\vec{\mathbf{T}}$, the free-body diagram for the crate includes the gravitational force $\vec{\mathbf{F}}_g$ and the normal force $\vec{\mathbf{n}}$ exerted by the floor on the crate.

We can now apply Newton's second law in component form to the crate. The only force acting in the x direction is $\vec{\mathbf{T}}$. Applying $\Sigma F_x = ma_x$ to the horizontal motion gives

$$\sum F_x = T = ma_x \quad \text{or} \quad a_x = \frac{T}{m}$$

No acceleration occurs in the y direction because the crate moves only horizontally. Therefore, we use the particle in equilibrium model in the y direction. Applying the y component of Equation 5.8 yields

$$\sum F_y = n + (-F_g) = 0 \quad \text{or} \quad n = F_g$$

That is, the normal force has the same magnitude as the gravitational force but acts in the opposite direction.

If $\vec{\mathbf{T}}$ is a constant force, the acceleration $a_x = T/m$ also is constant. Hence, the crate is also modeled as a particle under constant acceleration in the x direction, and the equations of kinematics from Chapter 2 can be used to obtain the crate's position x and velocity v_x as functions of time.

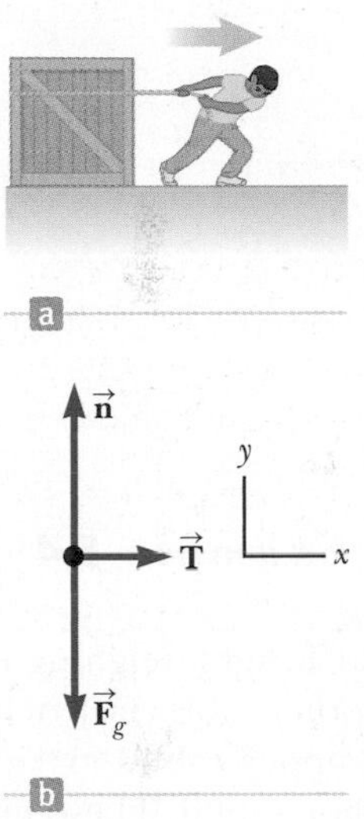

Figure 5.8 (a) A crate being pulled to the right on a frictionless floor. (b) The free-body diagram representing the external forces acting on the crate.

In the situation just described, the magnitude of the normal force $\vec{\mathbf{n}}$ is equal to the magnitude of $\vec{\mathbf{F}}_g$, but that is not always the case, as noted in Pitfall Prevention 5.6. For example, suppose a book is lying on a table and you push down on the book with a force $\vec{\mathbf{F}}$ as in Figure 5.9. Because the book is at rest and therefore not accelerating, $\Sigma F_y = 0$, which gives $n - F_g - F = 0$, or $n = F_g + F = mg + F$. In this situation, the normal force is *greater* than the gravitational force. Other examples in which $n \neq F_g$ are presented later.

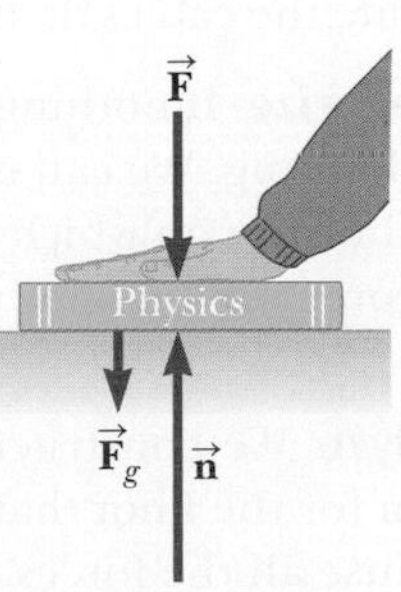

Figure 5.9 When a force $\vec{\mathbf{F}}$ pushes vertically downward on another object, the normal force $\vec{\mathbf{n}}$ on the object is greater than the gravitational force: $n = F_g + F$.

Problem-Solving Strategy

APPLYING NEWTON'S LAWS

The following procedure is recommended when dealing with problems involving Newton's laws:

1. Conceptualize. Draw a simple, neat diagram of the system. The diagram helps establish the mental representation. Establish convenient coordinate axes for each object in the system.

2. Categorize. If an acceleration component for an object is zero, the object is modeled as a particle in equilibrium in this direction and $\Sigma F = 0$. If not, the object is modeled as a particle under a net force in this direction and $\Sigma F = ma$.

3. Analyze. Isolate the object whose motion is being analyzed. Draw a free-body diagram for this object. For systems containing more than one object, draw *separate* free-body diagrams for each object. *Do not* include in the free-body diagram forces exerted by the object on its surroundings.

Find the components of the forces along the coordinate axes. Apply the appropriate model from the Categorize step for each direction. Check your dimensions to make sure that all terms have units of force.

Solve the component equations for the unknowns. Remember that you generally must have as many independent equations as you have unknowns to obtain a complete solution.

4. Finalize. Make sure your results are consistent with the free-body diagram. Also check the predictions of your solutions for extreme values of the variables. By doing so, you can often detect errors in your results.

Example 5.4 A Traffic Light at Rest

A traffic light weighing 122 N hangs from a cable tied to two other cables fastened to a support as in Figure 5.10a. The upper cables make angles of 37.0° and 53.0° with the horizontal. These upper cables are not as strong as the vertical cable and will break if the tension in them exceeds 100 N. Does the traffic light remain hanging in this situation, or will one of the cables break?

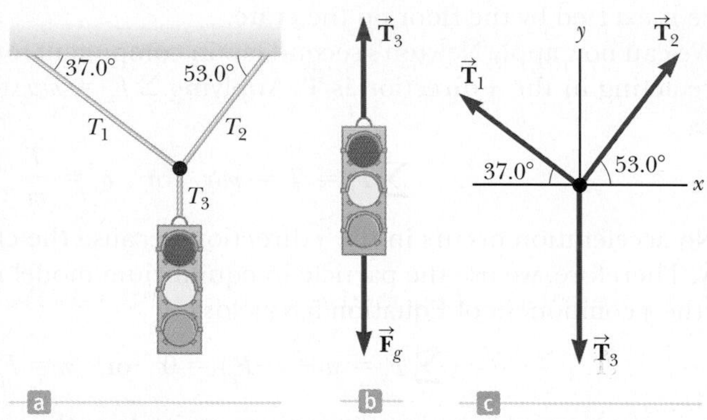

Figure 5.10 (Example 5.4) (a) A traffic light suspended by cables. (b) The forces acting on the traffic light. (c) The free-body diagram for the knot where the three cables are joined.

SOLUTION

Conceptualize Inspect the drawing in Figure 5.10a. Let us assume the cables do not break and nothing is moving.

Categorize If nothing is moving, no part of the system is accelerating. We can now model the light as a particle in equilibrium on which the net force is zero. Similarly, the net force on the knot (Fig. 5.10c) is zero.

Analyze We construct a diagram of the forces acting on the traffic light, shown in Figure 5.10b, and a free-body diagram for the knot that holds the three cables together, shown in Figure 5.10c. This knot is a convenient object to choose because all the forces of interest act along lines passing through the knot.

Apply Equation 5.8 for the traffic light in the y direction:

$$\sum F_y = 0 \rightarrow T_3 - F_g = 0$$

$$T_3 = F_g = 122 \text{ N}$$

5.4 *cont.*

Choose the coordinate axes as shown in Figure 5.10c and resolve the forces acting on the knot into their components:

Force	*x* Component	*y* Component
$\vec{\mathbf{T}}_1$	$-T_1 \cos 37.0°$	$T_1 \sin 37.0°$
$\vec{\mathbf{T}}_2$	$T_2 \cos 53.0°$	$T_2 \sin 53.0°$
$\vec{\mathbf{T}}_3$	0	-122 N

Apply the particle in equilibrium model to the knot:

$$(1)\ \sum F_x = -T_1 \cos 37.0° + T_2 \cos 53.0° = 0$$

$$(2)\ \sum F_y = T_1 \sin 37.0° + T_2 \sin 53.0° + (-122\text{ N}) = 0$$

Equation (1) shows that the horizontal components of $\vec{\mathbf{T}}_1$ and $\vec{\mathbf{T}}_2$ must be equal in magnitude, and Equation (2) shows that the sum of the vertical components of $\vec{\mathbf{T}}_1$ and $\vec{\mathbf{T}}_2$ must balance the downward force $\vec{\mathbf{T}}_3$, which is equal in magnitude to the weight of the light.

Solve Equation (1) for T_2 in terms of T_1:

$$(3)\ T_2 = T_1\left(\frac{\cos 37.0°}{\cos 53.0°}\right) = 1.33T_1$$

Substitute this value for T_2 into Equation (2):

$$T_1 \sin 37.0° + (1.33T_1)(\sin 53.0°) - 122\text{ N} = 0$$

$$T_1 = 73.4\text{ N}$$

$$T_2 = 1.33T_1 = 97.4\text{ N}$$

Both values are less than 100 N (just barely for T_2), so the cables will not break.

Finalize Let us finalize this problem by imagining a change in the system, as in the following What If?

WHAT IF? Suppose the two angles in Figure 5.10a are equal. What would be the relationship between T_1 and T_2?

Answer We can argue from the symmetry of the problem that the two tensions T_1 and T_2 would be equal to each other. Mathematically, if the equal angles are called θ, Equation (3) becomes

$$T_2 = T_1\left(\frac{\cos\theta}{\cos\theta}\right) = T_1$$

which also tells us that the tensions are equal. Without knowing the specific value of θ, we cannot find the values of T_1 and T_2. The tensions will be equal to each other, however, regardless of the value of θ.

Conceptual Example 5.5 Forces Between Cars in a Train

Train cars are connected by *couplers,* which are under tension as the locomotive pulls the train. Imagine you are on a train speeding up with a constant acceleration. As you move through the train from the locomotive to the last car, measuring the tension in each set of couplers, does the tension increase, decrease, or stay the same? When the engineer applies the brakes, the couplers are under compression. How does this compression force vary from the locomotive to the last car? (Assume only the brakes on the wheels of the engine are applied.)

SOLUTION

While the train is speeding up, tension decreases from the front of the train to the back. The coupler between the locomotive and the first car must apply enough force to accelerate the rest of the cars. As you move back along the train, each coupler is accelerating less mass behind it. The last coupler has to accelerate only the last car, and so it is under the least tension.

When the brakes are applied, the force again decreases from front to back. The coupler connecting the locomotive to the first car must apply a large force to slow down the rest of the cars, but the final coupler must apply a force large enough to slow down only the last car.

Example 5.6 The Runaway Car

A car of mass m is on an icy driveway inclined at an angle θ as in Figure 5.11a.

(A) Find the acceleration of the car, assuming the driveway is frictionless.

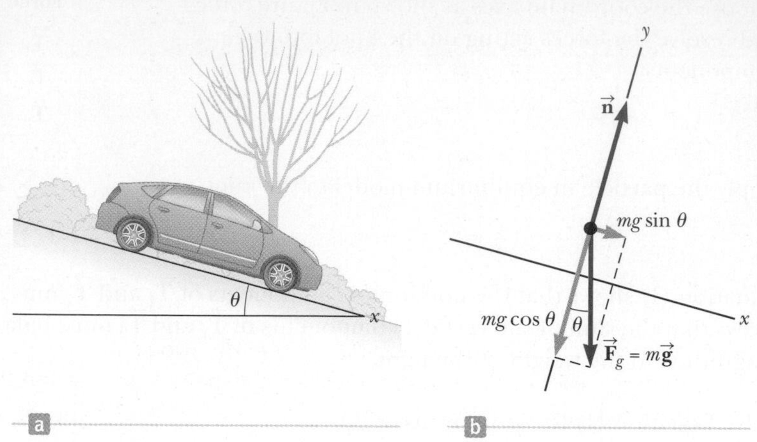

Figure 5.11 (Example 5.6) (a) A car on a frictionless incline. (b) The free-body diagram for the car. The black dot represents the position of the center of mass of the car. We will learn about center of mass in Chapter 9.

SOLUTION

Conceptualize Use Figure 5.11a to conceptualize the situation. From everyday experience, we know that a car on an icy incline will accelerate down the incline. (The same thing happens to a car on a hill with its brakes not set.)

Categorize We categorize the car as a particle under a net force because it accelerates. Furthermore, this example belongs to a very common category of problems in which an object moves under the influence of gravity on an inclined plane.

Analyze Figure 5.11b shows the free-body diagram for the car. The only forces acting on the car are the normal force $\vec{n}$ exerted by the inclined plane, which acts perpendicular to the plane, and the gravitational force $\vec{F}_g = m\vec{g}$, which acts vertically downward. For problems involving inclined planes, it is convenient to choose the coordinate axes with x along the incline and y perpendicular to it as in Figure 5.11b. With these axes, we represent the gravitational force by a component of magnitude $mg \sin\theta$ along the positive x axis and one of magnitude $mg\cos\theta$ along the negative y axis. Our choice of axes results in the car being modeled as a particle under a net force in the x direction and a particle in equilibrium in the y direction.

Apply these models to the car:

$$(1)\quad \sum F_x = mg\sin\theta = ma_x$$

$$(2)\quad \sum F_y = n - mg\cos\theta = 0$$

Solve Equation (1) for a_x:

$$(3)\quad a_x = \boxed{g\sin\theta}$$

Finalize Note that the acceleration component a_x is independent of the mass of the car! It depends only on the angle of inclination and on g.

From Equation (2), we conclude that the component of $\vec{F}_g$ perpendicular to the incline is balanced by the normal force; that is, $n = mg\cos\theta$. This situation is another case in which the normal force is *not* equal in magnitude to the weight of the object.

It is possible, although inconvenient, to solve the problem with "standard" horizontal and vertical axes. You may want to try it, just for practice.

(B) Suppose the car is released from rest at the top of the incline and the distance from the car's front bumper to the bottom of the incline is d. How long does it take the front bumper to reach the bottom of the hill, and what is the car's speed as it arrives there?

SOLUTION

Conceptualize Imagine that the car is sliding down the hill and you use a stopwatch to measure the entire time interval until it reaches the bottom.

Categorize This part of the problem belongs to kinematics rather than to dynamics, and Equation (3) shows that the acceleration a_x is constant. Therefore, you should categorize the car in this part of the problem as a particle under constant acceleration.

5.6 *cont.*

Analyze Defining the initial position of the front bumper as $x_i = 0$ and its final position as $x_f = d$, and recognizing that $v_{xi} = 0$, apply Equation 2.16, $x_f = x_i + v_{xi}t + \frac{1}{2}a_x t^2$:

$$d = \tfrac{1}{2}a_x t^2$$

Solve for t:

$$(4)\ t = \sqrt{\frac{2d}{a_x}} = \sqrt{\frac{2d}{g\sin\theta}}$$

Use Equation 2.17, with $v_{xi} = 0$, to find the final velocity of the car:

$$v_{xf}^{\ 2} = 2a_x d$$

$$(5)\ v_{xf} = \sqrt{2a_x d} = \sqrt{2gd\sin\theta}$$

Finalize We see from Equations (4) and (5) that the time t at which the car reaches the bottom and its final speed v_{xf} are independent of the car's mass, as was its acceleration. Noticc that wc havc combined techniques from Chapter 2 with new techniques from this chapter in this example. As we learn more techniques in later chapters, this process of combining analysis models and information from several parts of the book will occur more often. In these cases, use the General Problem-Solving Strategy to help you identify what analysis models you will need.

WHAT IF? What previously solved problem does this situation become if $\theta = 90°$?

Answer Imagine θ going to 90° in Figure 5.11. The inclined plane becomes vertical, and the car is an object in free fall! Equation (3) becomes

$$a_x = g\sin\theta = g\sin 90° = g$$

which is indeed the free-fall acceleration. (We find $a_x = g$ rather than $a_x = -g$ because we have chosen positive x to be downward in Fig. 5.11.) Notice also that the condition $n = mg\cos\theta$ gives us $n = mg\cos 90° = 0$. That is consistent with the car falling downward *next to* the vertical plane, in which case there is no contact force between the car and the plane.

Example 5.7 One Block Pushes Another

Two blocks of masses m_1 and m_2, with $m_1 > m_2$, are placed in contact with each other on a frictionless, horizontal surface as in Active Figure 5.12a. A constant horizontal force $\vec{\mathbf{F}}$ is applied to m_1 as shown.

(A) Find the magnitude of the acceleration of the system.

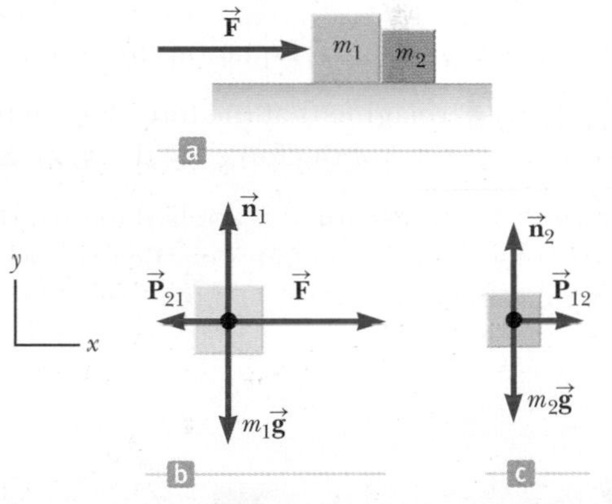

ACTIVE FIGURE 5.12

(Example 5.7) (a) A force is applied to a block of mass m_1, which pushes on a second block of mass m_2. (b) The forces acting on m_1. (c) The forces acting on m_2.

SOLUTION

Conceptualize Conceptualize the situation by using Active Figure 5.12a and realize that both blocks must experience the *same* acceleration because they are in contact with each other and remain in contact throughout thc motion.

Categorize We categorize this problem as one involving a particle under a net force because a force is applied to a system of blocks and we are looking for the acceleration of the system.

Analyze First model the combination of two blocks as a single particle under a net force. Apply Newton's second law to the combination in the x direction to find the acceleration:

$$\sum F_x = F = (m_1 + m_2)a_x$$

$$(1)\ a_x = \frac{F}{m_1 + m_2}$$

continued

5.7 *cont.*

Finalize The acceleration given by Equation (1) is the same as that of a single object of mass $m_1 + m_2$ and subject to the same force.

(B) Determine the magnitude of the contact force between the two blocks.

SOLUTION

Conceptualize The contact force is internal to the system of two blocks. Therefore, we cannot find this force by modeling the whole system (the two blocks) as a single particle.

Categorize Now consider each of the two blocks individually by categorizing each as a particle under a net force.

Analyze We construct a diagram of forces acting on the object for each block as shown in Active Figures 5.12b and 5.12c, where the contact force is denoted by $\vec{\mathbf{P}}$. From Active Figure 5.12c, we see that the only horizontal force acting on m_2 is the contact force $\vec{\mathbf{P}}_{12}$ (the force exerted by m_1 on m_2), which is directed to the right.

Apply Newton's second law to m_2:

$$(2)\quad \sum F_x = P_{12} = m_2 a_x$$

Substitute the value of the acceleration a_x given by Equation (1) into Equation (2):

$$(3)\quad P_{12} = m_2 a_x = \left(\frac{m_2}{m_1 + m_2}\right)F$$

Finalize This result shows that the contact force P_{12} is *less* than the applied force F. The force required to accelerate block 2 alone must be less than the force required to produce the same acceleration for the two-block system.

To finalize further, let us check this expression for P_{12} by considering the forces acting on m_1, shown in Active Figure 5.12b. The horizontal forces acting on m_1 are the applied force $\vec{\mathbf{F}}$ to the right and the contact force $\vec{\mathbf{P}}_{21}$ to the left (the force exerted by m_2 on m_1). From Newton's third law, $\vec{\mathbf{P}}_{21}$ is the reaction force to $\vec{\mathbf{P}}_{12}$, so $P_{21} = P_{12}$.

Apply Newton's second law to m_1:

$$(4)\quad \sum F_x = F - P_{21} = F - P_{12} = m_1 a_x$$

Solve for P_{12} and substitute the value of a_x from Equation (1):

$$P_{12} = F - m_1 a_x = F - m_1\left(\frac{F}{m_1 + m_2}\right) = \left(\frac{m_2}{m_1 + m_2}\right)F$$

This result agrees with Equation (3), as it must.

WHAT IF? Imagine that the force $\vec{\mathbf{F}}$ in Active Figure 5.12 is applied toward the left on the right-hand block of mass m_2. Is the magnitude of the force $\vec{\mathbf{P}}_{12}$ the same as it was when the force was applied toward the right on m_1?

Answer When the force is applied toward the left on m_2, the contact force must accelerate m_1. In the original situation, the contact force accelerates m_2. Because $m_1 > m_2$, more force is required, so the magnitude of $\vec{\mathbf{P}}_{12}$ is greater than in the original situation.

Example 5.8 Weighing a Fish in an Elevator

A person weighs a fish of mass m on a spring scale attached to the ceiling of an elevator as illustrated in Figure 5.13.

(A) Show that if the elevator accelerates either upward or downward, the spring scale gives a reading that is different from the weight of the fish.

SOLUTION

Conceptualize The reading on the scale is related to the extension of the spring in the scale, which is related to the force on the end of the spring as in Figure 5.2. Imagine that the fish is hanging on a string attached to the end of the spring.

5.8 *cont.*

In this case, the magnitude of the force exerted on the spring is equal to the tension T in the string. Therefore, we are looking for T. The force $\vec{T}$ pulls down on the string and pulls up on the fish.

Categorize We can categorize this problem by identifying the fish as a particle under a net force.

Analyze Inspect the diagrams of the forces acting on the fish in Figure 5.13 and notice that the external forces acting on the fish are the downward gravitational force $\vec{F}_g = m\vec{g}$ and the force $\vec{T}$ exerted by the string. If the elevator is either at rest or moving at constant velocity, the fish is a particle in equilibrium, so $\sum F_y = T - F_g = 0$ or $T = F_g = mg$. (Remember that the scalar mg is the weight of the fish.)

Now suppose the elevator is moving with an acceleration $\vec{a}$ relative to an observer standing outside the elevator in an inertial frame. The fish is now a particle under a net force.

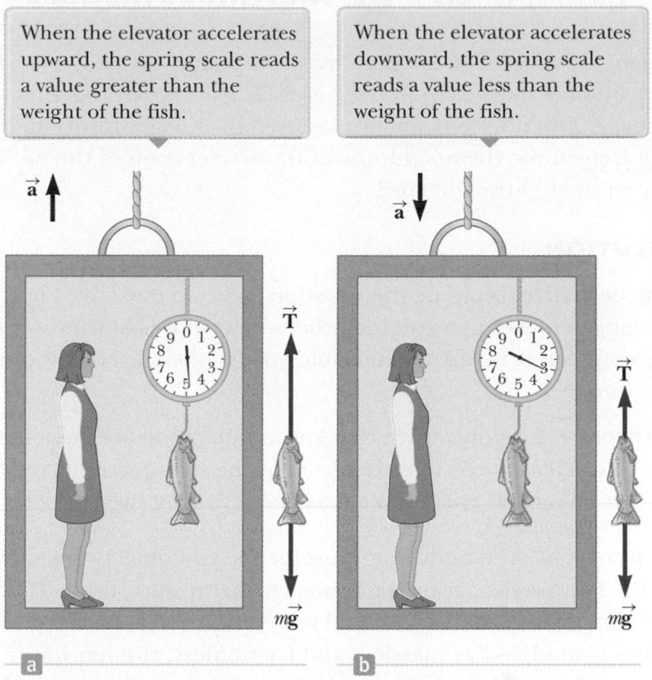

Figure 5.13 (Example 5.8) A fish is weighed on a spring scale in an accelerating elevator car.

Apply Newton's second law to the fish:

$$\sum F_y = T - mg = ma_y$$

Solve for T:

$$(1)\quad T = ma_y + mg = mg\left(\frac{a_y}{g} + 1\right) = F_g\left(\frac{a_y}{g} + 1\right)$$

where we have chosen upward as the positive y direction. We conclude from Equation (1) that the scale reading T is greater than the fish's weight mg if $\vec{a}$ is upward, so a_y is positive (Fig. 5.13a), and that the reading is less than mg if $\vec{a}$ is downward, so a_y is negative (Fig. 5.13b).

(B) Evaluate the scale readings for a 40.0-N fish if the elevator moves with an acceleration $a_y = \pm 2.00 \text{ m/s}^2$.

SOLUTION

Evaluate the scale reading from Equation (1) if $\vec{a}$ is upward:

$$T = (40.0 \text{ N})\left(\frac{2.00 \text{ m/s}^2}{9.80 \text{ m/s}^2} + 1\right) = \boxed{48.2 \text{ N}}$$

Evaluate the scale reading from Equation (1) if $\vec{a}$ is downward:

$$T = (40.0 \text{ N})\left(\frac{-2.00 \text{ m/s}^2}{9.80 \text{ m/s}^2} + 1\right) = \boxed{31.8 \text{ N}}$$

Finalize Take this advice: if you buy a fish in an elevator, make sure the fish is weighed while the elevator is either at rest or accelerating downward! Furthermore, notice that from the information given here, one cannot determine the direction of motion of the elevator.

WHAT IF? Suppose the elevator cable breaks and the elevator and its contents are in free fall. What happens to the reading on the scale?

Answer If the elevator falls freely, its acceleration is $a_y = -g$. We see from Equation (1) that the scale reading T is zero in this case; that is, the fish *appears* to be weightless.

Example 5.9 The Atwood Machine

When two objects of unequal mass are hung vertically over a frictionless pulley of negligible mass as in Active Figure 5.14a, the arrangement is called an *Atwood machine*. The device is sometimes used in the laboratory to determine the value of g. Determine the magnitude of the acceleration of the two objects and the tension in the lightweight cord.

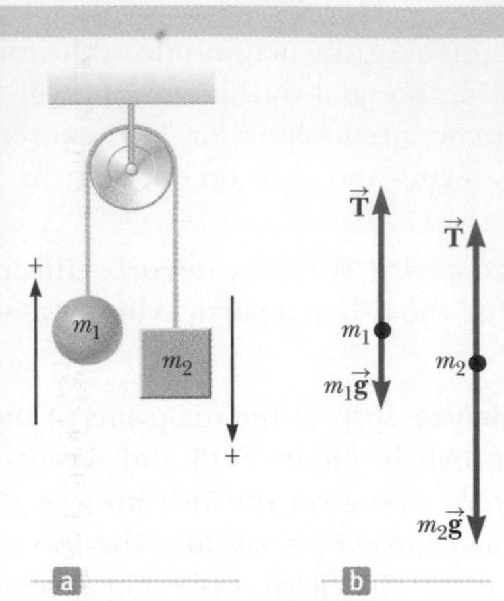

ACTIVE FIGURE 5.14

(Example 5.9) The Atwood machine. (a) Two objects connected by a massless inextensible cord over a frictionless pulley. (b) The free-body diagrams for the two objects.

SOLUTION

Conceptualize Imagine the situation pictured in Active Figure 5.14a in action: as one object moves upward, the other object moves downward. Because the objects are connected by an inextensible string, their accelerations must be of equal magnitude.

Categorize The objects in the Atwood machine are subject to the gravitational force as well as to the forces exerted by the strings connected to them. Therefore, we can categorize this problem as one involving two particles under a net force.

Analyze The free-body diagrams for the two objects are shown in Active Figure 5.14b. Two forces act on each object: the upward force $\vec{\mathbf{T}}$ exerted by the string and the downward gravitational force. In problems such as this one in which the pulley is modeled as massless and frictionless, the tension in the string on both sides of the pulley is the same. If the pulley has mass or is subject to friction, the tensions on either side are not the same and the situation requires techniques we will learn in Chapter 10.

We must be very careful with signs in problems such as this one. In Active Figure 5.14a, notice that if object 1 accelerates upward, object 2 accelerates downward. Therefore, for consistency with signs, if we define the upward direction as positive for object 1, we must define the downward direction as positive for object 2. With this sign convention, both objects accelerate in the same direction as defined by the choice of sign. Furthermore, according to this sign convention, the y component of the net force exerted on object 1 is $T - m_1g$, and the y component of the net force exerted on object 2 is $m_2g - T$.

Apply Newton's second law to object 1:

$$(1)\quad \sum F_y = T - m_1g = m_1a_y$$

Apply Newton's second law to object 2:

$$(2)\quad \sum F_y = m_2g - T = m_2a_y$$

Add Equation (2) to Equation (1), noticing that T cancels:

$$-m_1g + m_2g = m_1a_y + m_2a_y$$

Solve for the acceleration:

$$(3)\quad a_y = \left(\frac{m_2 - m_1}{m_1 + m_2}\right)g$$

Substitute Equation (3) into Equation (1) to find T:

$$(4)\quad T = m_1(g + a_y) = \left(\frac{2m_1m_2}{m_1 + m_2}\right)g$$

Finalize The acceleration given by Equation (3) can be interpreted as the ratio of the magnitude of the unbalanced force on the system $(m_2 - m_1)g$ to the total mass of the system $(m_1 + m_2)$, as expected from Newton's second law. Notice that the sign of the acceleration depends on the relative masses of the two objects.

WHAT IF? Describe the motion of the system if the objects have equal masses, that is, $m_1 = m_2$.

Answer If we have the same mass on both sides, the system is balanced and should not accelerate. Mathematically, we see that if $m_1 = m_2$, Equation (3) gives us $a_y = 0$.

WHAT IF? What if one of the masses is much larger than the other: $m_1 >> m_2$?

Answer In the case in which one mass is infinitely larger than the other, we can ignore the effect of the smaller mass. Therefore, the larger mass should simply fall as if the smaller mass were not there. We see that if $m_1 >> m_2$, Equation (3) gives us $a_y = -g$.

Example 5.10 Acceleration of Two Objects Connected by a Cord

A ball of mass m_1 and a block of mass m_2 are attached by a lightweight cord that passes over a frictionless pulley of negligible mass as in Figure 5.15a. The block lies on a frictionless incline of angle θ. Find the magnitude of the acceleration of the two objects and the tension in the cord.

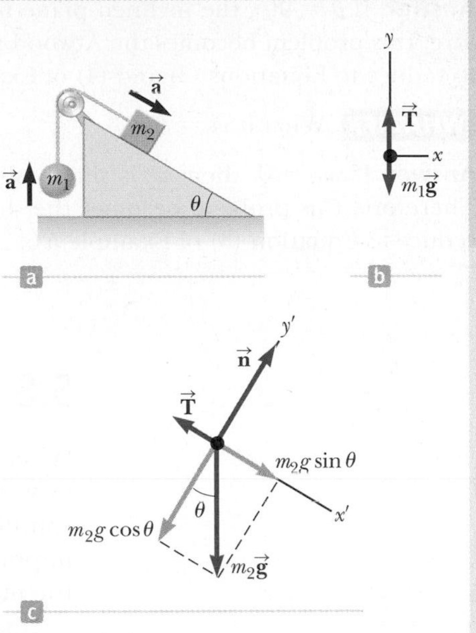

Figure 5.15 (Example 5.10) (a) Two objects connected by a lightweight cord strung over a frictionless pulley. (b) The free-body diagram for the ball. (c) The free-body diagram for the block. (The incline is frictionless.)

SOLUTION

Conceptualize Imagine the objects in Figure 5.15 in motion. If m_2 moves down the incline, then m_1 moves upward. Because the objects are connected by a cord (which we assume does not stretch), their accelerations have the same magnitude.

Categorize We can identify forces on each of the two objects and we are looking for an acceleration, so we categorize the objects as particles under a net force.

Analyze Consider the free-body diagrams shown in Figures 5.15b and 5.15c.

Apply Newton's second law in component form to the ball, choosing the upward direction as positive:

$$(1)\ \sum F_x = 0$$

$$(2)\ \sum F_y = T - m_1 g = m_1 a_y = m_1 a$$

For the ball to accelerate upward, it is necessary that $T > m_1 g$. In Equation (2), we replaced a_y with a because the acceleration has only a y component.

For the block, it is convenient to choose the positive x' axis along the incline as in Figure 5.15c. For consistency with our choice for the ball, we choose the positive direction to be down the incline.

Apply Newton's second law in component form to the block:

$$(3)\ \sum F_{x'} = m_2 g \sin\theta - T = m_2 a_{x'} = m_2 a$$

$$(4)\ \sum F_{y'} = n - m_2 g \cos\theta = 0$$

In Equation (3), we replaced $a_{x'}$ with a because the two objects have accelerations of equal magnitude a.

Solve Equation (2) for T:

$$(5)\ T = m_1(g + a)$$

Substitute this expression for T into Equation (3):

$$m_2 g \sin\theta - m_1(g + a) = m_2 a$$

Solve for a:

$$(6)\ a = \left(\frac{m_2 \sin\theta - m_1}{m_1 + m_2}\right) g$$

Substitute this expression for a into Equation (5) to find T:

$$(7)\ T = \left(\frac{m_1 m_2(\sin\theta + 1)}{m_1 + m_2}\right) g$$

Finalize The block accelerates down the incline only if $m_2 \sin\theta > m_1$. If $m_1 > m_2 \sin\theta$, the acceleration is up the incline for the block and downward for the ball. Also notice that the result for the acceleration, Equation (6), can be interpreted as the magnitude of the net external force acting on the ball–block system divided by the total mass of the system; this result is consistent with Newton's second law.

WHAT IF? What happens in this situation if $\theta = 90°$?

continued

5.10 *cont.*

Answer If $\theta = 90°$, the inclined plane becomes vertical and there is no interaction between its surface and m_2. Therefore, this problem becomes the Atwood machine of Example 5.9. Letting $\theta \rightarrow 90°$ in Equations (6) and (7) causes them to reduce to Equations (3) and (4) of Example 5.9!

WHAT IF? What if $m_1 = 0$?

Answer If $m_1 = 0$, then m_2 is simply sliding down an inclined plane without interacting with m_1 through the string. Therefore, this problem becomes the sliding car problem in Example 5.6. Letting $m_1 \rightarrow 0$ in Equation (6) causes it to reduce to Equation (3) of Example 5.6!

5.8 Forces of Friction

When an object is in motion either on a surface or in a viscous medium such as air or water, there is resistance to the motion because the object interacts with its surroundings. We call such resistance a **force of friction.** Forces of friction are very important in our everyday lives. They allow us to walk or run and are necessary for the motion of wheeled vehicles.

Imagine that you are working in your garden and have filled a trash can with yard clippings. You then try to drag the trash can across the surface of your concrete patio as in Active Figure 5.16a. This surface is *real,* not an idealized, frictionless surface. If we apply an external horizontal force $\vec{\mathbf{F}}$ to the trash can, acting to the right, the trash can remains stationary when $\vec{\mathbf{F}}$ is small. The force on the trash can that counteracts $\vec{\mathbf{F}}$ and keeps it from moving acts toward the left and is called

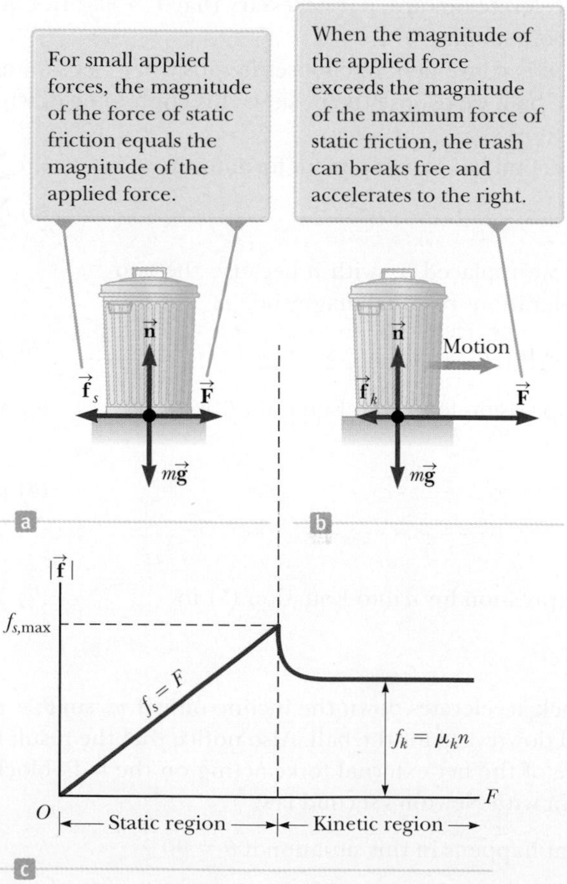

ACTIVE FIGURE 5.16

(a) and (b) When pulling on a trash can, the direction of the force of friction $\vec{\mathbf{f}}$ between the can and a rough surface is opposite the direction of the applied force $\vec{\mathbf{F}}$. (c) A graph of friction force versus applied force. Notice that $f_{s,\max} > f_k$.

the **force of static friction $\vec{\mathbf{f}}_s$**. As long as the trash can is not moving, $f_s = F$. Therefore, if $\vec{\mathbf{F}}$ is increased, $\vec{\mathbf{f}}_s$ also increases. Likewise, if $\vec{\mathbf{F}}$ decreases, $\vec{\mathbf{f}}_s$ also decreases.

◀ Force of static friction

Experiments show that the friction force arises from the nature of the two surfaces: because of their roughness, contact is made only at a few locations where peaks of the material touch. At these locations, the friction force arises in part because one peak physically blocks the motion of a peak from the opposing surface and in part from chemical bonding ("spot welds") of opposing peaks as they come into contact. Although the details of friction are quite complex at the atomic level, this force ultimately involves an electrical interaction between atoms or molecules.

If we increase the magnitude of $\vec{\mathbf{F}}$ as in Active Figure 5.16b, the trash can eventually slips. When the trash can is on the verge of slipping, f_s has its maximum value $f_{s,\max}$ as shown in Active Figure 5.16c. When F exceeds $f_{s,\max}$, the trash can moves and accelerates to the right. We call the friction force for an object in motion the **force of kinetic friction $\vec{\mathbf{f}}_k$**. When the trash can is in motion, the force of kinetic friction on the can is less than $f_{s,\max}$ (Active Fig. 5.16c). The net force $F - f_k$ in the x direction produces an acceleration to the right, according to Newton's second law. If $F = f_k$, the acceleration is zero and the trash can moves to the right with constant speed. If the applied force $\vec{\mathbf{F}}$ is removed from the moving can, the friction force $\vec{\mathbf{f}}_k$ acting to the left provides an acceleration of the trash can in the $-x$ direction and eventually brings it to rest, again consistent with Newton's second law.

◀ Force of kinetic friction

Experimentally, we find that, to a good approximation, both $f_{s,\max}$ and f_k are proportional to the magnitude of the normal force exerted on an object by the surface. The following descriptions of the force of friction are based on experimental observations and serve as the model we shall use for forces of friction in problem solving:

- The magnitude of the force of static friction between any two surfaces in contact can have the values

$$f_s \leq \mu_s n \tag{5.9}$$

  where the dimensionless constant μ_s is called the **coefficient of static friction** and n is the magnitude of the normal force exerted by one surface on the other. The equality in Equation 5.9 holds when the surfaces are on the verge of slipping, that is, when $f_s = f_{s,\max} = \mu_s n$. This situation is called *impending motion*. The inequality holds when the surfaces are not on the verge of slipping.
- The magnitude of the force of kinetic friction acting between two surfaces is

$$f_k = \mu_k n \tag{5.10}$$

  where μ_k is the **coefficient of kinetic friction.** Although the coefficient of kinetic friction can vary with speed, we shall usually neglect any such variations in this text.
- The values of μ_k and μ_s depend on the nature of the surfaces, but μ_k is generally less than μ_s. Typical values range from around 0.03 to 1.0. Table 5.1 (page 124) lists some reported values.
- The direction of the friction force on an object is parallel to the surface with which the object is in contact and opposite to the actual motion (kinetic friction) or the impending motion (static friction) of the object relative to the surface.
- The coefficients of friction are nearly independent of the area of contact between the surfaces. We might expect that placing an object on the side having the most area might increase the friction force. Although this method provides more points in contact, the weight of the object is spread out over a larger area and the individual points are not pressed together as tightly. Because these effects approximately compensate for each other, the friction force is independent of the area.

Pitfall Prevention 5.9

The Equal Sign Is Used in Limited Situations

In Equation 5.9, the equal sign is used *only* in the case in which the surfaces are just about to break free and begin sliding. Do not fall into the common trap of using $f_s = \mu_s n$ in *any* static situation.

Pitfall Prevention 5.10

Friction Equations

Equations 5.9 and 5.10 are *not* vector equations. They are relationships between the *magnitudes* of the vectors representing the friction and normal forces. Because the friction and normal forces are perpendicular to each other, the vectors cannot be related by a multiplicative constant.

Pitfall Prevention 5.11

The Direction of the Friction Force

Sometimes, an incorrect statement about the friction force between an object and a surface is made—"the friction force on an object is opposite to its motion or impending motion"—rather than the correct phrasing, "the friction force on an object is opposite to its motion or impending motion *relative to the surface*."

TABLE 5.1 *Coefficients of Friction*

	μ_s	μ_k
Rubber on concrete	1.0	0.8
Steel on steel	0.74	0.57
Aluminum on steel	0.61	0.47
Glass on glass	0.94	0.4
Copper on steel	0.53	0.36
Wood on wood	0.25–0.5	0.2
Waxed wood on wet snow	0.14	0.1
Waxed wood on dry snow	—	0.04
Metal on metal (lubricated)	0.15	0.06
Teflon on Teflon	0.04	0.04
Ice on ice	0.1	0.03
Synovial joints in humans	0.01	0.003

Note: All values are approximate. In some cases, the coefficient of friction can exceed 1.0.

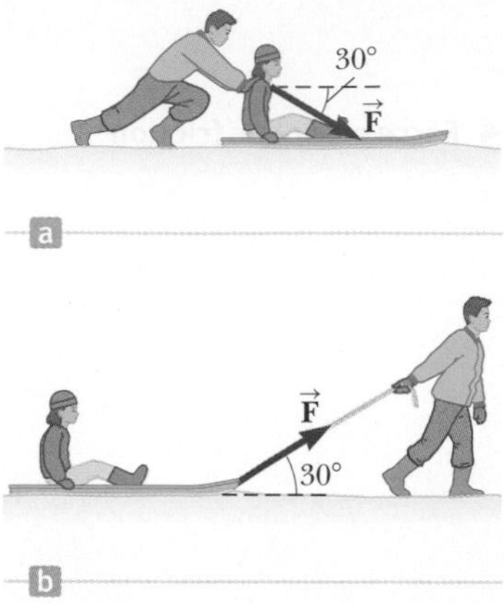

Figure 5.17 (Quick Quiz 5.7) A father slides his daughter on a sled either by (a) pushing down on her shoulders or (b) pulling up on a rope.

***Quick Quiz* 5.6** You press your physics textbook flat against a vertical wall with your hand. What is the direction of the friction force exerted by the wall on the book? **(a)** downward **(b)** upward **(c)** out from the wall **(d)** into the wall

***Quick Quiz* 5.7** You are playing with your daughter in the snow. She sits on a sled and asks you to slide her across a flat, horizontal field. You have a choice of **(a)** pushing her from behind by applying a force downward on her shoulders at 30° below the horizontal (Fig. 5.17a) or **(b)** attaching a rope to the front of the sled and pulling with a force at 30° above the horizontal (Fig. 5.17b). Which would be easier for you and why?

Example 5.11 Experimental Determination of μ_s and μ_k

The following is a simple method of measuring coefficients of friction. Suppose a block is placed on a rough surface inclined relative to the horizontal as shown in Active Figure 5.18. The incline angle is increased until the block starts to move. Show that you can obtain μ_s by measuring the critical angle θ_c at which this slipping just occurs.

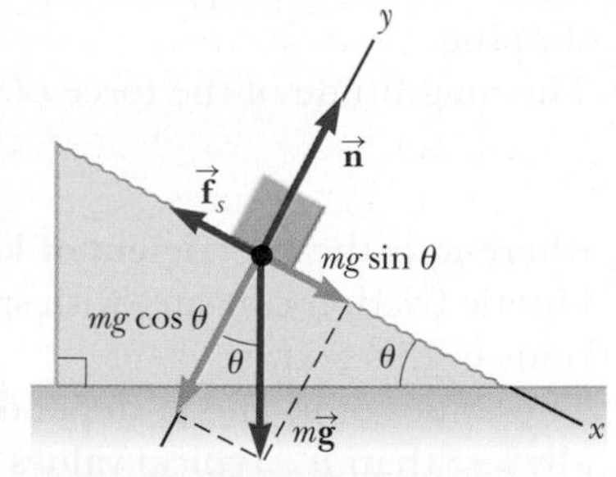

ACTIVE FIGURE 5.18

(Example 5.11) The external forces exerted on a block lying on a rough incline are the gravitational force $m\vec{g}$, the normal force $\vec{n}$, and the force of friction $\vec{f}_s$. For convenience, the gravitational force is resolved into a component $mg \sin \theta$ along the incline and a component $mg \cos \theta$ perpendicular to the incline.

SOLUTION

Conceptualize Consider Active Figure 5.18 and imagine that the block tends to slide down the incline due to the gravitational force. To simulate the situation, place a coin on this book's cover and tilt the book until the coin begins to slide. Notice how this example differs from Example 5.6. When there is no friction on an incline, *any* angle of the incline will cause a stationary object to begin moving. When there is friction, however, there is no movement of the object for angles less than the critical angle.

Categorize The block is subject to various forces. Because we are raising the plane to the angle at which the block is just ready to begin to move but is not moving, we categorize the block as a particle in equilibrium.

Analyze The diagram in Active Figure 5.18 shows the forces on the block: the gravitational force $m\vec{g}$, the normal force $\vec{n}$, and the force of static friction $\vec{f}_s$. We choose x to be parallel to the plane and y perpendicular to it.

5.11 cont.

Apply Equation 5.8 to the block in both the x and y directions:

$$(1)\quad \sum F_x = mg\sin\theta - f_s = 0$$

$$(2)\quad \sum F_y = n - mg\cos\theta = 0$$

Substitute $mg = n/\cos\theta$ from Equation (2) into Equation (1):

$$(3)\quad f_s = mg\sin\theta = \left(\frac{n}{\cos\theta}\right)\sin\theta = n\tan\theta$$

When the incline angle is increased until the block is on the verge of slipping, the force of static friction has reached its maximum value $\mu_s n$. The angle θ in this situation is the critical angle θ_c. Make these substitutions in Equation (3):

$$\mu_s n = n\tan\theta_c$$

$$\mu_s = \tan\theta_c$$

For example, if the block just slips at $\theta_c = 20.0°$, we find that $\mu_s = \tan 20.0° = 0.364$.

Finalize Once the block starts to move at $\theta \geq \theta_c$, it accelerates down the incline and the force of friction is $f_k = \mu_k n$. If θ is reduced to a value less than θ_c, however, it may be possible to find an angle θ'_c such that the block moves down the incline with constant speed as a particle in equilibrium again ($a_x = 0$). In this case, use Equations (1) and (2) with f_s replaced by f_k to find μ_k: $\mu_k = \tan\theta'_c$, where $\theta'_c < \theta_c$.

Example 5.12 The Sliding Hockey Puck

A hockey puck on a frozen pond is given an initial speed of 20.0 m/s. If the puck always remains on the ice and slides 115 m before coming to rest, determine the coefficient of kinetic friction between the puck and ice.

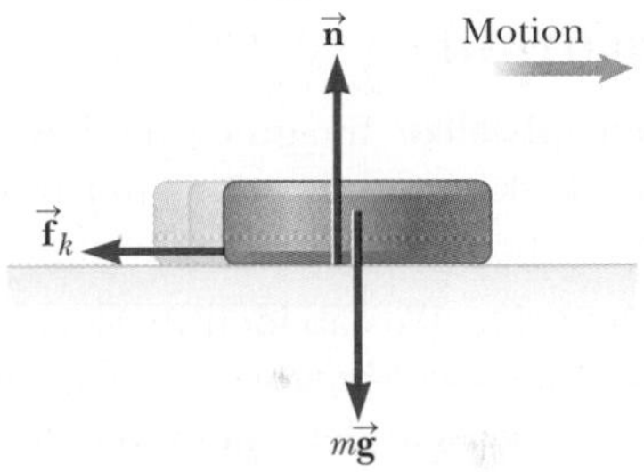

Figure 5.19 (Example 5.12) After the puck is given an initial velocity to the right, the only external forces acting on it are the gravitational force $m\vec{g}$, the normal force $\vec{n}$, and the force of kinetic friction $\vec{f}_k$.

SOLUTION

Conceptualize Imagine that the puck in Figure 5.19 slides to the right and eventually comes to rest due to the force of kinetic friction.

Categorize The forces acting on the puck are identified in Figure 5.19, but the text of the problem provides kinematic variables. Therefore, we categorize the problem in two ways. First, it involves a particle under a net force: kinetic friction causes the puck to accelerate. Furthermore, because we model the force of kinetic friction as independent of speed, the acceleration of the puck is constant. So, we can also categorize this problem as one involving a particle under constant acceleration.

Analyze First, let's find the acceleration algebraically in terms of the coefficient of kinetic friction, using Newton's second law. Once we know the acceleration of the puck and the distance it travels, the equations of kinematics can be used to find the numerical value of the coefficient of kinetic friction. The diagram in Figure 5.19 shows the forces on the puck.

Apply the particle under a net force model in the x direction to the puck:

$$(1)\quad \sum F_x = -f_k = ma_x$$

Apply the particle in equilibrium model in the y direction to the puck:

$$(2)\quad \sum F_y = n - mg = 0$$

Substitute $n = mg$ from Equation (2) and $f_k = \mu_k n$ into Equation (1):

$$-\mu_k n = -\mu_k mg = ma_x$$

$$a_x = -\mu_k g$$

The negative sign means the acceleration is to the left in Figure 5.19. Because the velocity of the puck is to the right, the puck is slowing down. The acceleration is independent of the mass of the puck and is constant because we assume μ_k remains constant.

continued

5.12 *cont.*

Apply the particle under constant acceleration model to the puck, using Equation 2.17, $v_{xf}^2 = v_{xi}^2 + 2a_x(x_f - x_i)$, with $x_i = 0$ and $v_f = 0$:

$$0 = v_{xi}^2 + 2a_x x_f = v_{xi}^2 - 2\mu_k g x_f$$

Solve for the coefficient of kinetic friction:

$$\mu_k = \frac{v_{xi}^2}{2gx_f}$$

Substitute the numerical values:

$$\mu_k = \frac{(20.0\ \text{m/s})^2}{2(9.80\ \text{m/s}^2)(115\ \text{m})} = \boxed{0.177}$$

Finalize Notice that μ_k is dimensionless, as it should be, and that it has a low value, consistent with an object sliding on ice.

Example 5.13 Acceleration of Two Connected Objects When Friction Is Present

A block of mass m_2 on a rough, horizontal surface is connected to a ball of mass m_1 by a lightweight cord over a lightweight, frictionless pulley as shown in Figure 5.20a. A force of magnitude F at an angle θ with the horizontal is applied to the block as shown, and the block slides to the right. The coefficient of kinetic friction between the block and surface is μ_k. Determine the magnitude of the acceleration of the two objects.

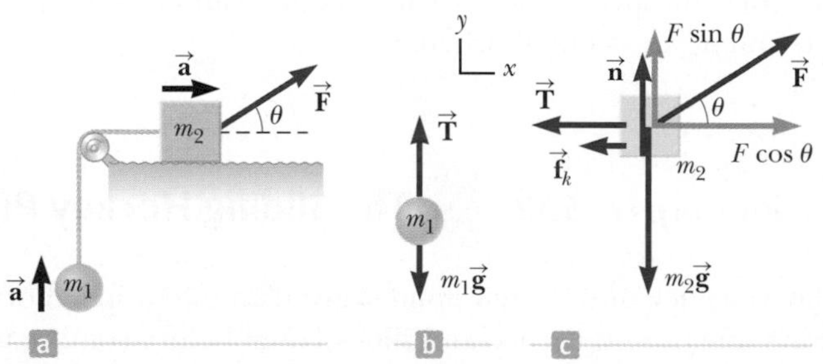

Figure 5.20 (Example 5.13) (a) The external force $\vec{\mathbf{F}}$ applied as shown can cause the block to accelerate to the right. (b, c) Diagrams showing the forces on the two objects, assuming the block accelerates to the right and the ball accelerates upward.

SOLUTION

Conceptualize Imagine what happens as $\vec{\mathbf{F}}$ is applied to the block. Assuming $\vec{\mathbf{F}}$ is not large enough to lift the block, the block slides to the right and the ball rises.

Categorize We can identify forces and we want an acceleration, so we categorize this problem as one involving two particles under a net force, the ball and the block.

Analyze First draw force diagrams for the two objects as shown in Figures 5.20b and 5.20c. Notice that the string exerts a force of magnitude T on both objects. The applied force $\vec{\mathbf{F}}$ has x and y components $F\cos\theta$ and $F\sin\theta$, respectively. Because the two objects are connected, we can equate the magnitudes of the x component of the acceleration of the block and the y component of the acceleration of the ball and call them both a. Let us assume the motion of the block is to the right.

Apply the particle under a net force model to the block in the horizontal direction:

$$(1)\ \sum F_x = F\cos\theta - f_k - T = m_2 a_x = m_2 a$$

Because the block moves only horizontally, apply the particle in equilibrium model to the block in the vertical direction:

$$(2)\ \sum F_y = n + F\sin\theta - m_2 g = 0$$

Apply the particle under a net force model to the ball in the vertical direction:

$$(3)\ \sum F_y = T - m_1 g = m_1 a_y = m_1 a$$

Solve Equation (2) for n:

$$n = m_2 g - F\sin\theta$$

Substitute n into $f_k = \mu_k n$ from Equation 5.10:

$$(4)\ f_k = \mu_k(m_2 g - F\sin\theta)$$

Substitute Equation (4) and the value of T from Equation (3) into Equation (1):

$$F\cos\theta - \mu_k(m_2 g - F\sin\theta) - m_1(a + g) = m_2 a$$

Solve for a:

$$(5)\ a = \boxed{\frac{F(\cos\theta + \mu_k\sin\theta) - (m_1 + \mu_k m_2)g}{m_1 + m_2}}$$

5.13 *cont.*

Finalize The acceleration of the block can be either to the right or to the left depending on the sign of the numerator in Equation (5). If the motion is to the left, we must reverse the sign of f_k in Equation (1) because the force of kinetic friction must oppose the motion of the block relative to the surface. In this case, the value of a is the same as in Equation (5), with the two plus signs in the numerator changed to minus signs.

What does Equation (5) reduce to if the force $\vec{\mathbf{F}}$ is removed and the surface becomes frictionless? Call this expression Equation (6). Does this algebraic expression match your intuition about the physical situation in this case? Now go back to Example 5.10 and let angle θ go to zero in Equation (6) of that example. How does the resulting equation compare with your Equation (6) here in Example 5.13? Should the algebraic expressions compare in this way based on the physical situations?

Summary

Definitions

An **inertial frame of reference** is a frame in which an object that does not interact with other objects experiences zero acceleration. Any frame moving with constant velocity relative to an inertial frame is also an inertial frame.

We define **force** as **that which causes a change in motion of an object.**

Concepts and Principles

Newton's first law states that it is possible to find an inertial frame in which an object that does not interact with other objects experiences zero acceleration, or, equivalently, in the absence of an external force, when viewed from an inertial frame, an object at rest remains at rest and an object in uniform motion in a straight line maintains that motion.

Newton's second law states that the acceleration of an object is directly proportional to the net force acting on it and inversely proportional to its mass.

Newton's third law states that if two objects interact, the force exerted by object 1 on object 2 is equal in magnitude and opposite in direction to the force exerted by object 2 on object 1.

The **gravitational force** exerted on an object is equal to the product of its mass (a scalar quantity) and the free-fall acceleration: $\vec{\mathbf{F}}_g = m\vec{\mathbf{g}}$.

The **weight** of an object is the magnitude of the gravitational force acting on the object.

The maximum **force of static friction** $\vec{\mathbf{f}}_{s,\max}$ between an object and a surface is proportional to the normal force acting on the object. In general, $f_s \le \mu_s n$, where μ_s is the **coefficient of static friction** and n is the magnitude of the normal force.

When an object slides over a surface, the magnitude of the **force of kinetic friction** $\vec{\mathbf{f}}_k$ is given by $f_k = \mu_k n$, where μ_k is the **coefficient of kinetic friction.**

continued

Analysis Models for Problem Solving

Particle Under a Net Force If a particle of mass m experiences a nonzero net force, its acceleration is related to the net force by Newton's second law:

$$\sum \vec{\mathbf{F}} = m\vec{\mathbf{a}} \qquad (5.2)$$

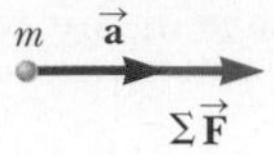

Particle in Equilibrium If a particle maintains a constant velocity (so that $\vec{\mathbf{a}} = 0$), which could include a velocity of zero, the forces on the particle balance and Newton's second law reduces to

$$\sum \vec{\mathbf{F}} = 0 \qquad (5.8)$$

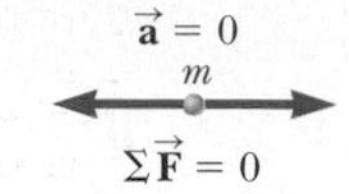

Objective Questions

☐ denotes answer available in Student Solutions Manual/Study Guide

1. An experiment is performed on a puck on a level air hockey table, where friction is negligible. A constant horizontal force is applied to the puck, and the puck's acceleration is measured. Now the same puck is transported far into outer space, where both friction and gravity are negligible. The same constant force is applied to the puck (through a spring scale that stretches the same amount), and the puck's acceleration (relative to the distant stars) is measured. What is the puck's acceleration in outer space? (a) It is somewhat greater than its acceleration on the Earth. (b) It is the same as its acceleration on the Earth. (c) It is less than its acceleration on the Earth. (d) It is infinite because neither friction nor gravity constrains it. (e) It is very large because acceleration is inversely proportional to weight and the puck's weight is very small but not zero.

2. In Figure OQ5.2, a locomotive has broken through the wall of a train station. During the collision, what can be said about the force exerted by the locomotive on the wall? (a) The force exerted by the locomotive on the wall was larger than the force the wall could exert on the locomotive. (b) The force exerted by the locomotive on the wall was the same in magnitude as the force exerted by the wall on the locomotive. (c) The force exerted by the locomotive on the wall was less than the force exerted by the wall on the locomotive. (d) The wall cannot be said to "exert" a force; after all, it broke.

Studio Lévy and Sons

Figure OQ5.2

3. The third graders are on one side of a schoolyard, and the fourth graders are on the other. They are throwing snowballs at each other. Between them, snowballs of various masses are moving with different velocities as shown in Figure OQ5.3. Rank the snowballs (a) through (e) according to the magnitude of the total force exerted on each one. Ignore air resistance. If two snowballs rank together, make that fact clear.

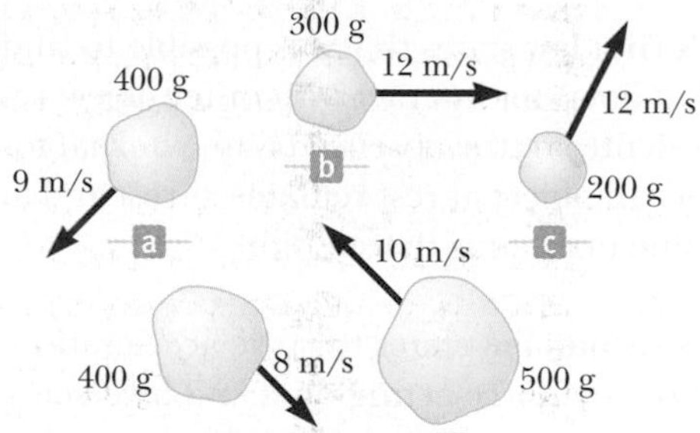

Figure OQ5.3

4. The manager of a department store is pushing horizontally with a force of magnitude 200 N on a box of shirts. The box is sliding across the horizontal floor with a forward acceleration. Nothing else touches the box. What must be true about the magnitude of the force of kinetic friction acting on the box (choose one)? (a) It is greater than 200 N. (b) It is less than 200 N. (c) It is equal to 200 N. (d) None of those statements is necessarily true.

5. The driver of a speeding empty truck slams on the brakes and skids to a stop through a distance d. On a second trial, the truck carries a load that doubles its mass. What

will now be the truck's "skidding distance"? (a) $4d$ (b) $2d$ (c) $\sqrt{2}d$ (d) d (e) $d/2$

6. The driver of a speeding truck slams on the brakes and skids to a stop through a distance d. On another trial, the initial speed of the truck is half as large. What now will be the truck's skidding distance? (a) $2d$ (b) $\sqrt{2}d$ (c) d (d) $d/2$ (e) $d/4$

7. An object of mass m moves with acceleration $\vec{a}$ down a rough incline. Which of the following forces should appear in a free-body diagram of the object? Choose all correct answers. (a) the gravitational force exerted by the planet (b) $m\vec{a}$ in the direction of motion (c) the normal force exerted by the incline (d) the friction force exerted by the incline (e) the force exerted by the object on the incline

8. A large crate of mass m is place on the flatbed of a truck but not tied down. As the truck accelerates forward with acceleration a, the crate remains at rest relative to the truck. What force causes the crate to accelerate? (a) the normal force (b) the gravitational force (c) the friction force (d) the ma force exerted by the crate (e) No force is required.

9. A crate remains stationary after it has been placed on a ramp inclined at an angle with the horizontal. Which of the following statements is or are correct about the magnitude of the friction force that acts on the crate? Choose all that are true. (a) It is larger than the weight of the crate. (b) It is equal to $\mu_s n$. (c) It is greater than the component of the gravitational force acting down the ramp. (d) It is equal to the component of the gravitational force acting down the ramp. (e) It is less than the component of the gravitational force acting down the ramp.

10. An object of mass m is sliding with speed v_i at some instant across a level tabletop, with which its coefficient of kinetic friction is μ. It then moves through a distance d and comes to rest. Which of the following equations for the speed v_i is reasonable? (a) $v_i = \sqrt{-2\mu mgd}$ (b) $v_i = \sqrt{2\mu mgd}$ (c) $v_i = \sqrt{-2\mu gd}$ (d) $v_i = \sqrt{2\mu gd}$ (e) $v_i = \sqrt{2\mu d}$

11. If an object is in equilibrium, which of the following statements is *not* true? (a) The speed of the object remains constant. (b) The acceleration of the object is zero. (c) The net force acting on the object is zero. (d) The object must be at rest. (e) There are at least two forces acting on the object.

12. A truck loaded with sand accelerates along a highway. The driving force on the truck remains constant. What happens to the acceleration of the truck if its trailer leaks sand at a constant rate through a hole in its bottom? (a) It decreases at a steady rate. (b) It increases at a steady rate. (c) It increases and then decreases. (d) It decreases and then increases. (e) It remains constant.

13. Two objects are connected by a string that passes over a frictionless pulley as in Active Figure 5.14a, where $m_1 < m_2$ and a_1 and a_2 are the magnitudes of the respective accelerations. Which mathematical statement is true regarding the magnitude of the acceleration a_2 of the mass m_2? (a) $a_2 < g$ (b) $a_2 > g$ (c) $a_2 = g$ (d) $a_2 < a_1$ (e) $a_2 > a_1$

Conceptual Questions

☐ denotes answer available in Student Solutions Manual/Study Guide

1. A person holds a ball in her hand. (a) Identify all the external forces acting on the ball and the Newton's third-law reaction force to each one. (b) If the ball is dropped, what force is exerted on it while it is falling? Identify the reaction force in this case. (Ignore air resistance.)

2. If a car is traveling due westward with a constant speed of 20 m/s, what is the resultant force acting on it?

3. In the motion picture *It Happened One Night* (Columbia Pictures, 1934), Clark Gable is standing inside a stationary bus in front of Claudette Colbert, who is seated. The bus suddenly starts moving forward and Clark falls into Claudette's lap. Why did this happen?

4. Your hands are wet, and the restroom towel dispenser is empty. What do you do to get drops of water off your hands? How does the motion of the drops exemplify one of Newton's laws? Which one?

5. A passenger sitting in the rear of a bus claims that she was injured when the driver slammed on the brakes, causing a suitcase to come flying toward her from the front of the bus. If you were the judge in this case, what disposition would you make? Why?

6. A spherical rubber balloon inflated with air is held stationary, with its opening, on the west side, pinched shut. (a) Describe the forces exerted by the air inside and outside the balloon on sections of the rubber. (b) After the balloon is released, it takes off toward the east, gaining speed rapidly. Explain this motion in terms of the forces now acting on the rubber. (c) Account for the motion of a skyrocket taking off from its launch pad.

7. If you hold a horizontal metal bar several centimeters above the ground and move it through grass, each leaf of grass bends out of the way. If you increase the speed of the bar, each leaf of grass will bend more quickly. How then does a rotary power lawn mower manage to cut grass? How can it exert enough force on a leaf of grass to shear it off?

8. A child tosses a ball straight up. She says that the ball is moving away from her hand because the ball feels an upward "force of the throw" as well as the gravitational force. (a) Can the "force of the throw" exceed the gravitational force? How would the ball move if it did? (b) Can the "force of the throw" be equal in magnitude to the gravitational force? Explain. (c) What strength can accurately be attributed to the "force of the throw"? Explain. (d) Why does the ball move away from the child's hand?

9. A rubber ball is dropped onto the floor. What force causes the ball to bounce?

10. The mayor of a city reprimands some city employees because they will not remove the obvious sags from the cables that support the city traffic lights. What explanation can the employees give? How do you think the case will be settled in mediation?

11. Balancing carefully, three boys inch out onto a horizontal tree branch above a pond, each planning to dive in separately. The third boy in line notices that the branch is barely strong enough to support them. He decides to jump straight up and land back on the branch to break it, spilling all three into the pond. When he starts to carry out his plan, at what precise moment does the branch break? Explain. *Suggestion*: Pretend to be the third boy and imitate what he does in slow motion. If you are still unsure, stand on a bathroom scale and repeat the suggestion.

12. When you push on a box with a 200-N force instead of a 50-N force, you can feel that you are making a greater effort. When a table exerts a 200-N normal force instead of one of smaller magnitude, is the table really doing anything differently?

13. A weightlifter stands on a bathroom scale. He pumps a barbell up and down. What happens to the reading on the scale as he does so? **What If?** What if he is strong enough to actually *throw* the barbell upward? How does the reading on the scale vary now?

14. Give reasons for the answers to each of the following questions: (a) Can a normal force be horizontal? (b) Can a normal force be directed vertically downward? (c) Consider a tennis ball in contact with a stationary floor and with nothing else. Can the normal force be different in magnitude from the gravitational force exerted on the ball? (d) Can the force exerted by the floor on the ball be different in magnitude from the force the ball exerts on the floor?

15. A car is moving forward slowly and is speeding up. A student claims that "the car exerts a force on itself" or that "the car's engine exerts a force on the car." (a) Argue that this idea cannot be accurate and that friction exerted by the road is the propulsive force on the car. Make your evidence and reasoning as persuasive as possible. (b) Is it static or kinetic friction? *Suggestions*: Consider a road covered with light gravel. Consider a sharp print of the tire tread on an asphalt road, obtained by coating the tread with dust.

16. In Figure CQ5.16, the light, taut, unstretchable cord B joins block 1 and the larger-mass block 2. Cord A exerts a force on block 1 to make it accelerate forward. (a) How does the magnitude of the force exerted by cord A on block 1 compare with the magnitude of the force exerted by cord B on block 2? Is it larger, smaller, or equal? (b) How does the acceleration of block 1 compare with the acceleration (if any) of block 2? (c) Does cord B exert a force on block 1? If so, is it forward or backward? Is it larger, smaller, or equal in magnitude to the force exerted by cord B on block 2?

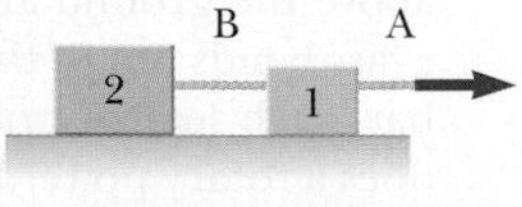

Figure CQ5.16

17. Identify action–reaction pairs in the following situations: (a) a man takes a step (b) a snowball hits a girl in the back (c) a baseball player catches a ball (d) a gust of wind strikes a window

18. Twenty people participate in a tug-of-war. The two teams of ten people are so evenly matched that neither team wins. After the game they notice that a car is stuck in the mud. They attach the tug-of-war rope to the bumper of the car, and all the people pull on the rope. The heavy car has just moved a couple of decimeters when the rope breaks. Why did the rope break in this situation when it did not break when the same twenty people pulled on it in a tug-of-war?

19. An athlete grips a light rope that passes over a low-friction pulley attached to the ceiling of a gym. A sack of sand precisely equal in weight to the athlete is tied to the other end of the rope. Both the sand and the athlete are initially at rest. The athlete climbs the rope, sometimes speeding up and slowing down as he does so. What happens to the sack of sand? Explain.

20. Can an object exert a force on itself? Argue for your answer.

21. Describe two examples in which the force of friction exerted on an object is in the direction of motion of the object.

22. As shown in Figure CQ5.22, student A, a 55-kg girl, sits on one chair with metal runners, at rest on a classroom floor. Student B, an 80-kg boy, sits on an identical chair. Both students keep their feet off the floor. A rope runs from student A's hands around a light pulley and then over her shoulder to the hands of a teacher standing on the floor behind her. The low-friction axle of the pulley is attached to a second rope held by student B. All ropes run parallel to the chair runners. (a) If student A pulls on her end of the rope, will her chair or will B's chair slide on the floor? Explain why. (b) If instead the teacher pulls on his rope end, which chair slides? Why this one? (c) If student B pulls on his rope, which chair slides? Why? (d) Now the teacher ties his end of the rope to student A's chair. Student A pulls on the end of the rope in her hands. Which chair slides and why?

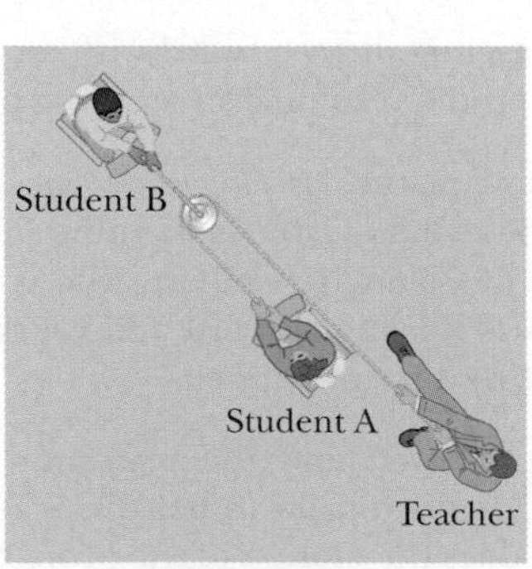

Figure CQ5.22

23. Suppose you are driving a classic car. Why should you avoid slamming on your brakes when you want to stop in the shortest possible distance? (Many modern cars have antilock brakes that avoid this problem.)

Problems

WebAssign The problems found in this chapter may be assigned online in Enhanced WebAssign

1. denotes straightforward problem; **2.** denotes intermediate problem; **3.** denotes challenging problem

1. full solution available in the Student Solutions Manual/Study Guide

1. denotes problems most often assigned in Enhanced WebAssign; these provide students with targeted feedback and either a Master It tutorial or a Watch It solution video.

QC denotes asking for quantitative and conceptual reasoning

S denotes symbolic reasoning problem

M denotes Master It tutorial available in Enhanced WebAssign

GP denotes guided problem

shaded denotes "paired problems" that develop reasoning with symbols and numerical values

Sections 5.1 through 5.6

1. A 3.00-kg object undergoes an acceleration given by $\vec{a} = (2.00\hat{i} + 5.00\hat{j})$ m/s^2. Find (a) the resultant force acting on the object and (b) the magnitude of the resultant force.

2. The average speed of a nitrogen molecule in air is about 6.70×10^2 m/s, and its mass is 4.68×10^{-26} kg. (a) If it takes 3.00×10^{-13} s for a nitrogen molecule to hit a wall and rebound with the same speed but moving in the opposite direction, what is the average acceleration of the molecule during this time interval? (b) What average force does the molecule exert on the wall?

3. A toy rocket engine is securely fastened to a large puck that can glide with negligible friction over a horizontal surface, taken as the *xy* plane. The 4.00-kg puck has a velocity of $3.00\hat{i}$ m/s at one instant. Eight seconds later, its velocity is $(8\hat{i} + 10\hat{j})$ m/s. Assuming the rocket engine exerts a constant horizontal force, find (a) the components of the force and (b) its magnitude.

4. A certain orthodontist uses a wire brace to align a patient's crooked tooth as in Figure P5.4. The tension in the wire is adjusted to have a magnitude of 18.0 N. Find the magnitude of the net force exerted by the wire on the crooked tooth.

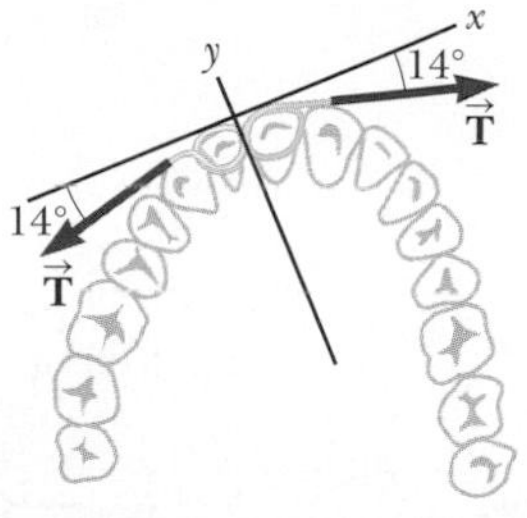

Figure P5.4

5. Review. The gravitational force exerted on a baseball is 2.21 N down. A pitcher throws the ball horizontally with velocity 18.0 m/s by uniformly accelerating it along a straight horizontal line for a time interval of 170 ms. The ball starts from rest. (a) Through what distance does it move before its release? (b) What are the magnitude and direction of the force the pitcher exerts on the ball?

6. **S** **Review.** The gravitational force exerted on a baseball is $-F_g\hat{j}$. A pitcher throws the ball with velocity $v\hat{i}$ by uniformly accelerating it along a straight horizontal line for a time interval of $\Delta t = t - 0 = t$. (a) Starting from rest, through what distance does the ball move before its release? (b) What force does the pitcher exert on the ball?

7. **M** **Review.** An electron of mass 9.11×10^{-31} kg has an initial speed of 3.00×10^5 m/s. It travels in a straight line, and its speed increases to 7.00×10^5 m/s in a distance of 5.00 cm. Assuming its acceleration is constant, (a) determine the magnitude of the force exerted on the electron and (b) compare this force with the weight of the electron, which we ignored.

8. Besides the gravitational force, a 2.80-kg object is subjected to one other constant force. The object starts from rest and in 1.20 s experiences a displacement of $(4.20\hat{i} - 3.30\hat{j})$ m, where the direction of $\hat{j}$ is the upward vertical direction. Determine the other force.

9. One or more external forces, large enough to be easily measured, are exerted on each object enclosed in a dashed box shown in Figure 5.1. Identify the reaction to each of these forces.

10. A brick of mass *M* has been placed on a rubber cushion of mass *m*. Together they are sliding to the right at constant velocity on an ice-covered parking lot. (a) Draw a free-body diagram of the brick and identify each force acting on it. (b) Draw a free-body diagram of the cushion and identify each force acting on it. (c) Identify all of the action–reaction pairs of forces in the brick–cushion–planet system.

11. **S** An object of mass *m* is dropped at $t = 0$ from the roof of a building of height *h*. While the object is falling, a wind blowing parallel to the face of the building exerts a constant horizontal force *F* on the object. (a) At what time *t* does the object strike the ground? Express *t* in terms of *g* and *h*. (b) Find an expression in terms of *m* and *F* for the acceleration a_x of the object in the horizontal direction (taken as the positive *x* direction). (c) How far is the object displaced horizontally before hitting the ground? Answer in terms of *m*, *g*, *F*, and *h*. (d) Find the magnitude of the object's acceleration while it is falling, using the variables *F*, *m*, and *g*.

12. A force $\vec{F}$ applied to an object of mass m_1 produces an acceleration of 3.00 m/s^2. The same force applied to a second object of mass m_2 produces an acceleration of 1.00 m/s^2. (a) What is the value of the ratio m_1/m_2? (b) If m_1 and m_2 are combined into one object, find its acceleration under the action of the force $\vec{F}$.

13. Two forces $\vec{\mathbf{F}}_1$ and $\vec{\mathbf{F}}_2$ act on a 5.00-kg object. Taking F_1 = 20.0 N and F_2 = 15.0 N, find the accelerations of the object for the configurations of forces shown in parts (a) and (b) of Figure P5.13.

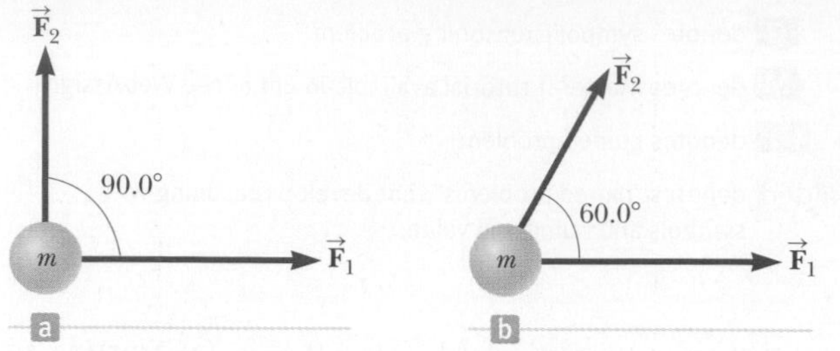

Figure P5.13

14. You stand on the seat of a chair and then hop off. (a) During the time interval you are in flight down to the floor, the Earth moves toward you with an acceleration of what order of magnitude? In your solution, explain your logic. Model the Earth as a perfectly solid object. (b) The Earth moves toward you through a distance of what order of magnitude?

15. A 15.0-lb block rests on the floor. (a) What force does the floor exert on the block? (b) A rope is tied to the block and is run vertically over a pulley. The other end is attached to a free-hanging 10.0-lb object. What now is the force exerted by the floor on the 15.0-lb block? (c) If the 10.0-lb object in part (b) is replaced with a 20.0-lb object, what is the force exerted by the floor on the 15.0-lb block?

16. Review. Three forces acting on an object are given by $\vec{\mathbf{F}}_1 = (-2.00\hat{\mathbf{i}} + 2.00\hat{\mathbf{j}})$ N, $\vec{\mathbf{F}}_2 = (5.00\hat{\mathbf{i}} - 3.00\hat{\mathbf{j}})$ N, and $\vec{\mathbf{F}}_3 = (-45.0\hat{\mathbf{i}})$ N. The object experiences an acceleration of magnitude 3.75 m/s^2. (a) What is the direction of the acceleration? (b) What is the mass of the object? (c) If the object is initially at rest, what is its speed after 10.0 s? (d) What are the velocity components of the object after 10.0 s?

Section 5.7 Some Applications of Newton's Laws

17. Review. Figure P5.17 shows a worker poling a boat—a very efficient mode of transportation—across a shallow lake. He pushes parallel to the length of the light pole, exerting a force of magnitude 240 N on the bottom of the lake. Assume the pole lies in the vertical plane containing the keel of the boat. At one moment, the pole makes an angle of 35.0° with the vertical and the water exerts a horizontal drag force of 47.5 N on the boat, opposite to its forward velocity of magnitude 0.857 m/s. The mass of the boat including its cargo and the worker is 370 kg. (a) The water exerts a buoyant force vertically upward on the boat. Find the magnitude of this force. (b) Model the forces as constant over a short interval of time to find the velocity of the boat 0.450 s after the moment described.

AP Images

Figure P5.17

18. An iron bolt of mass 65.0 g hangs from a string 35.7 cm long. The top end of the string is fixed. Without touching it, a magnet attracts the bolt so that it remains stationary, but is displaced horizontally 28.0 cm to the right from the previously vertical line of the string. The magnet is located to the right of the bolt and on the same vertical level as the bolt in the final configuration. (a) Draw a free-body diagram of the bolt. (b) Find the tension in the string. (c) Find the magnetic force on the bolt.

19. QC Figure P5.19 shows the horizontal forces acting on a sailboat moving north at constant velocity, seen from a point straight above its mast. At the particular speed of the sailboat, the water exerts a 220-N drag force on its hull and θ = 40.0°. For each of the situations (a) and (b) described below, write two component equations representing Newton's second law. Then solve the equations for P (the force exerted by the wind on the sail) and for n (the force exerted by the water on the keel). (a) Choose the x direction as east and the y direction as north. (b) Now choose the x direction as θ = 40.0° north of east and the y direction as θ = 40.0° west of north. (c) Compare your solutions to parts (a) and (b). Do the results agree? Is one method significantly easier?

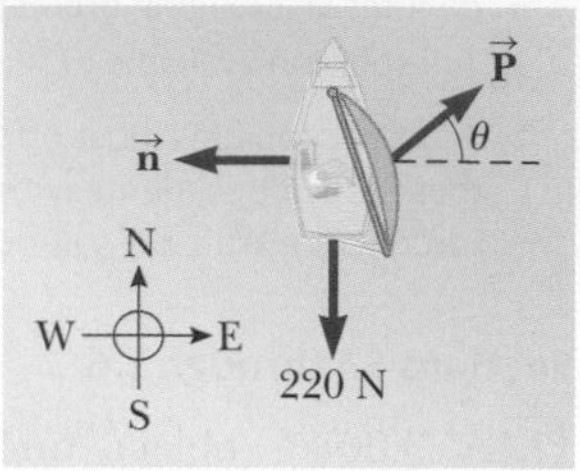

Figure P5.19

20. The systems shown in Figure P5.20 are in equilibrium. If the spring scales are calibrated in newtons, what do they read? Ignore the masses of the pulleys and strings and assume the pulleys and the incline in Figure P5.20d are frictionless.

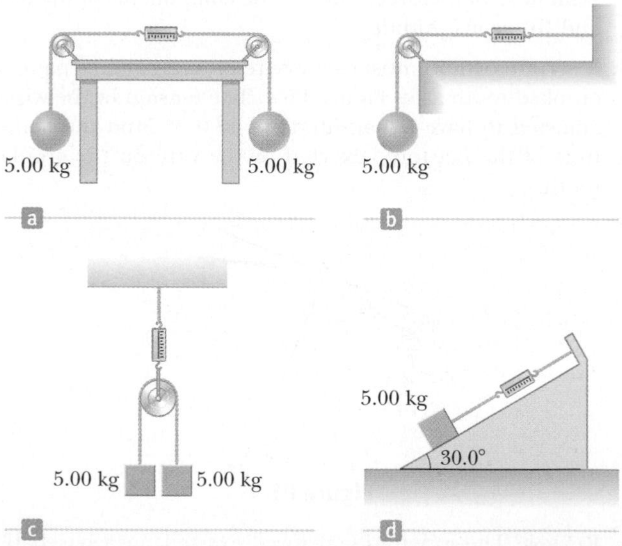

Figure P5.20

21. A block slides down a frictionless plane having an inclination of θ = 15.0°. The block starts from rest at the top, and the length of the incline is 2.00 m. (a) Draw a free-body diagram of the block. Find (b) the acceleration of the block and (c) its speed when it reaches the bottom of the incline.

22. A 3.00-kg object is moving in a plane, with its x and y coordinates given by $x = 5t^2 - 1$ and $y = 3t^3 + 2$, where x and y are in meters and t is in seconds. Find the magnitude of the net force acting on this object at t = 2.00 s.

23. The distance between two telephone poles is 50.0 m. When a 1.00-kg bird lands on the telephone wire midway between the poles, the wire sags 0.200 m. (a) Draw a free-body diagram of the bird. (b) How much tension does the bird produce in the wire? Ignore the weight of the wire.

24. A bag of cement weighing 325 N hangs in equilibrium from three wires as suggested in Figure P5.24. Two of the wires make angles $\theta_1 = 60.0°$ and $\theta_2 = 40.0°$ with the horizontal. Assuming the system is in equilibrium, find the tensions T_1, T_2, and T_3 in the wires.

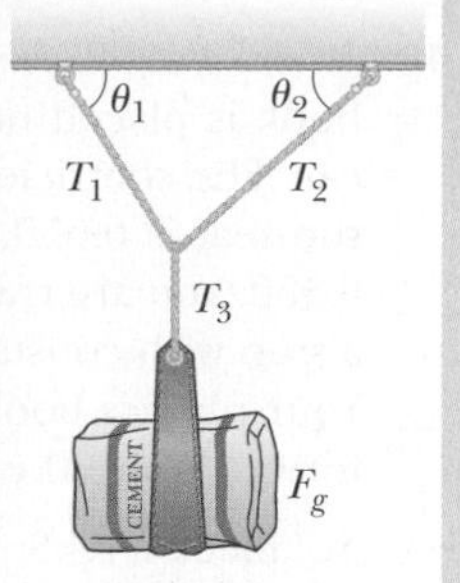

Figure P5.24
Problems 24 and 25.

25. **S** A bag of cement whose weight is F_g hangs in equilibrium from three wires as shown in Figure P5.24. Two of the wires make angles θ_1 and θ_2 with the horizontal. Assuming the system is in equilibrium, show that the tension in the left-hand wire is

$$T_1 = \frac{F_g \cos \theta_2}{\sin(\theta_1 + \theta_2)}$$

26. A setup similar to the one shown in Figure P5.26 is often used in hospitals to support and apply a horizontal traction force to an injured leg. (a) Determine the force of tension in the rope supporting the leg. (b) What is the traction force exerted to the right on the leg?

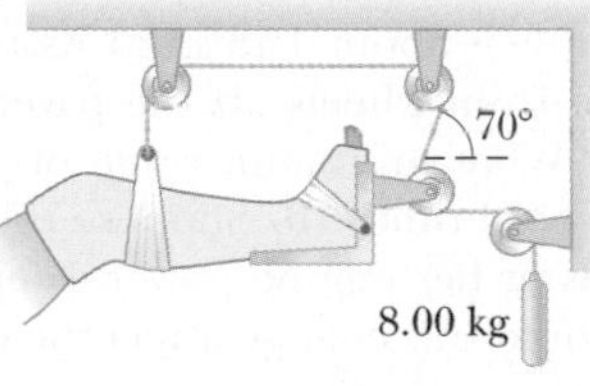

Figure P5.26

27. An object of mass $m = 1.00$ kg is observed to have an acceleration $\vec{\mathbf{a}}$ with a magnitude of 10.0 m/s^2 in a direction 60.0° east of north. Figure P5.27 shows a view of the object from above. The force $\vec{\mathbf{F}}_2$ acting on the object has a magnitude of 5.00 N and is directed north. Determine the magnitude and direction of the one other horizontal force $\vec{\mathbf{F}}_1$ acting on the object.

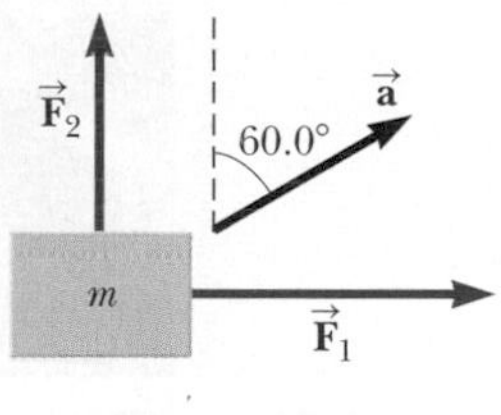

Figure P5.27

28. An object of mass $m_1 = 5.00$ kg placed on a frictionless, horizontal table is connected to a string that passes over a pulley and then is fastened to a hanging object of mass $m_2 = 9.00$ kg as shown in Figure P5.28. (a) Draw free-body diagrams of both objects. Find (b) the magnitude of the acceleration of the objects and (c) the tension in the string.

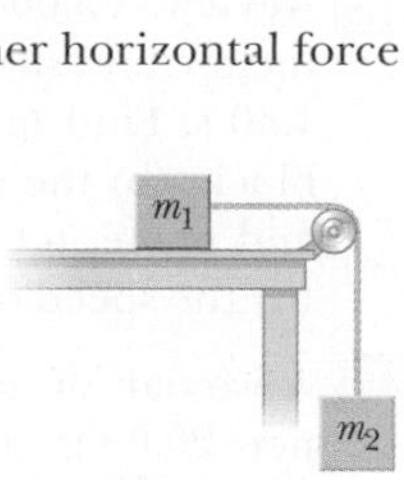

Figure P5.28
Problems 28 and 45.

29. Figure P5.29 shows the speed of a person's body as he does a chin-up. Assume the motion is vertical and the mass of the person's body is 64.0 kg. Determine the force exerted by the chin-up bar on his body at (a) $t = 0$, (b) $t = 0.5$ s, (c) $t = 1.1$ s, and (d) $t = 1.6$ s.

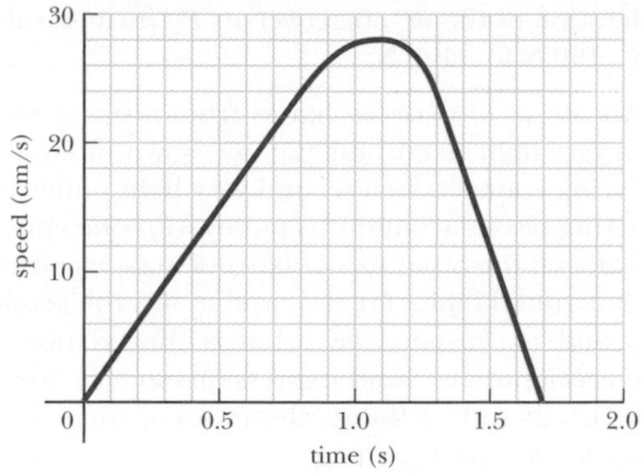

Figure P5.29

30. Two objects are connected by a light string that passes over a frictionless pulley as shown in Figure P5.30. Assume the incline is frictionless and take $m_1 = 2.00$ kg, $m_2 = 6.00$ kg, and $\theta = 55.0°$. (a) Draw free-body diagrams of both objects. Find (b) the magnitude of the acceleration of the objects, (c) the tension in the string, and (d) the speed of each object 2.00 s after it is released from rest.

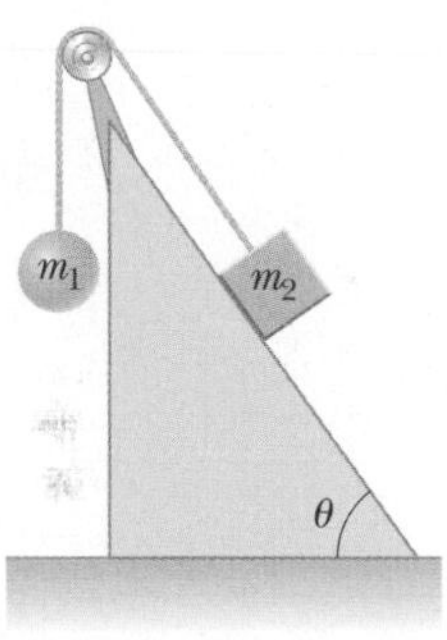

Figure P5.30

31. Two blocks, each of mass $m = 3.50$ kg, are hung from the ceiling of an elevator as in Figure P5.31. (a) If the elevator moves with an upward acceleration $\vec{\mathbf{a}}$ of magnitude 1.60 m/s^2, find the tensions T_1 and T_2 in the upper and lower strings. (b) If the strings can withstand a maximum tension of 85.0 N, what maximum acceleration can the elevator have before a string breaks?

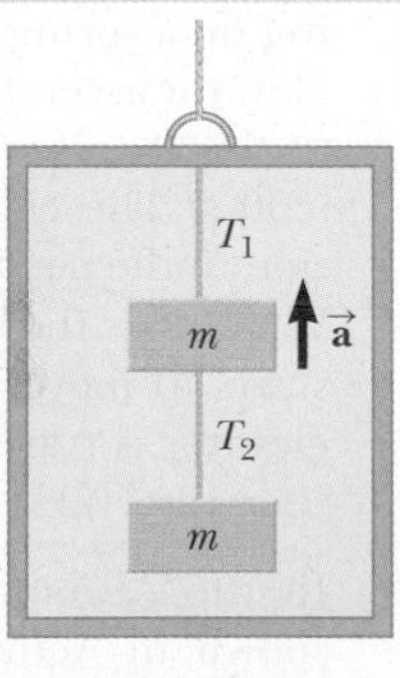

Figure P5.31
Problems 31 and 32.

32. **S** Two blocks, each of mass m, are hung from the ceiling of an elevator as in Figure P5.31. The elevator has an upward acceleration a. The strings have negligible mass. (a) Find the tensions T_1 and T_2 in the upper and lower strings in terms of m, a, and g. (b) Compare the two tensions and determine which string would break first if a is made sufficiently large. (c) What are the tensions if the cable supporting the elevator breaks?

33. **M** In the system shown in Figure P5.33, a horizontal force $\vec{\mathbf{F}}_x$ acts on an object of mass $m_2 = 8.00$ kg. The horizontal

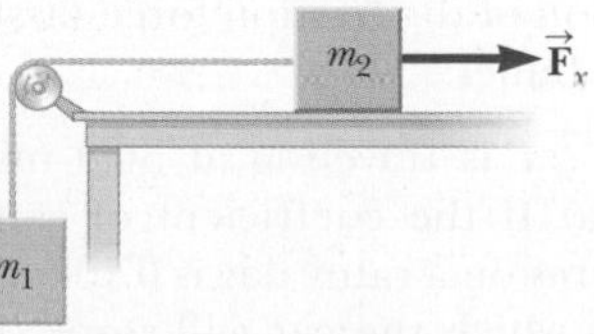

Figure P5.33

surface is frictionless. Consider the acceleration of the sliding object as a function of F_x. (a) For what values of F_x does the object of mass $m_1 = 2.00$ kg accelerate upward? (b) For what values of F_x is the tension in the cord zero? (c) Plot the acceleration of the m_2 object versus F_x. Include values of F_x from -100 N to $+100$ N.

34. **S** An object of mass m_1 hangs from a string that passes over a very light fixed pulley P_1 as shown in Figure P5.34. The string connects to a second very light pulley P_2. A second string passes around this pulley with one end attached to a wall and the other to an object of mass m_2 on a frictionless, horizontal table. (a) If a_1 and a_2 are the accelerations of m_1 and m_2, respectively, what is the relation between these accelerations? Find expressions for (b) the tensions in the strings and (c) the accelerations a_1 and a_2 in terms of the masses m_1 and m_2, and g.

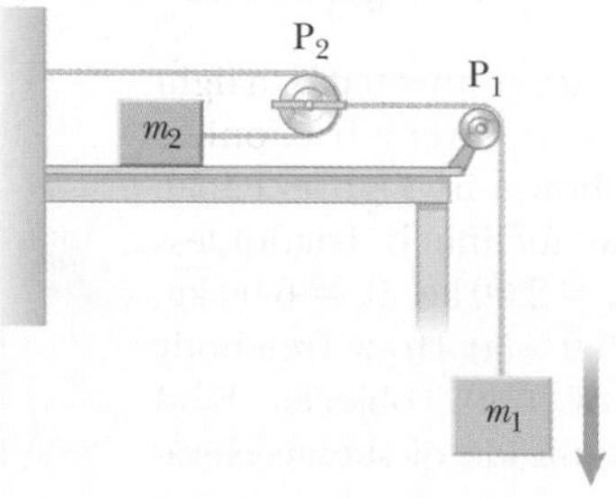

Figure P5.34

35. **M** In Example 5.8, we investigated the apparent weight of a fish in an elevator. Now consider a 72.0-kg man standing on a spring scale in an elevator. Starting from rest, the elevator ascends, attaining its maximum speed of 1.20 m/s in 0.800 s. It travels with this constant speed for the next 5.00 s. The elevator then undergoes a uniform acceleration in the negative y direction for 1.50 s and comes to rest. What does the spring scale register (a) before the elevator starts to move, (b) during the first 0.800 s, (c) while the elevator is traveling at constant speed, and (d) during the time interval it is slowing down?

36. In the Atwood machine discussed in Example 5.9 and shown in Active Figure 5.14a, $m_1 = 2.00$ kg and $m_2 = 7.00$ kg. The masses of the pulley and string are negligible by comparison. The pulley turns without friction, and the string does not stretch. The lighter object is released with a sharp push that sets it into motion at $v_i = 2.40$ m/s downward. (a) How far will m_1 descend below its initial level? (b) Find the velocity of m_1 after 1.80 s.

Section 5.8 Forces of Friction

37. **Review.** A rifle bullet with a mass of 12.0 g traveling toward the right at 260 m/s strikes a large bag of sand and penetrates it to a depth of 23.0 cm. Determine the magnitude and direction of the friction force (assumed constant) that acts on the bullet.

38. **Review.** A car is traveling at 50.0 mi/h on a horizontal highway. (a) If the coefficient of static friction between road and tires on a rainy day is 0.100, what is the minimum distance in which the car will stop? (b) What is the stopping distance when the surface is dry and $\mu_s = 0.600$?

39. A 25.0-kg block is initially at rest on a horizontal surface. A horizontal force of 75.0 N is required to set the block in motion, after which a horizontal force of 60.0 N is required to keep the block moving with constant speed. Find (a) the coefficient of static friction and (b) the coefficient of kinetic friction between the block and the surface.

40. *Why is the following situation impossible?* Your 3.80-kg physics book is placed next to you on the horizontal seat of your car. The coefficient of static friction between the book and the seat is 0.650, and the coefficient of kinetic friction is 0.550. You are traveling forward at 72.0 km/h and brake to a stop with constant acceleration over a distance of 30.0 m. Your physics book remains on the seat rather than sliding forward onto the floor.

41. To meet a U.S. Postal Service requirement, employees' footwear must have a coefficient of static friction of 0.5 or more on a specified tile surface. A typical athletic shoe has a coefficient of static friction of 0.800. In an emergency, what is the minimum time interval in which a person starting from rest can move 3.00 m on the tile surface if she is wearing (a) footwear meeting the Postal Service minimum and (b) a typical athletic shoe?

42. **Q|C** Before 1960, people believed that the maximum attainable coefficient of static friction for an automobile tire on a roadway was $\mu_s = 1$. Around 1962, three companies independently developed racing tires with coefficients of 1.6. This problem shows that tires have improved further since then. The shortest time interval in which a piston-engine car initially at rest has covered a distance of one-quarter mile is about 4.43 s. (a) Assume the car's rear wheels lift the front wheels off the pavement as shown in Figure P5.42. What minimum value of μ_s is necessary to achieve the record time? (b) Suppose the driver were able to increase his or her engine power, keeping other things equal. How would this change affect the elapsed time?

Jamie Squire/Allsport/Getty Images

Figure P5.42

43. **Review.** A 3.00-kg block starts from rest at the top of a 30.0° incline and slides a distance of 2.00 m down the incline in 1.50 s. Find (a) the magnitude of the acceleration of the block, (b) the coefficient of kinetic friction between block and plane, (c) the friction force acting on the block, and (d) the speed of the block after it has slid 2.00 m.

44. A woman at an airport is towing her 20.0-kg suitcase at constant speed by pulling on a strap at an angle θ above the horizontal (Fig. P5.44). She pulls on the strap with a 35.0-N force, and the friction force on the suitcase is 20.0 N. (a) Draw a free-body diagram of the suitcase. (b) What angle does

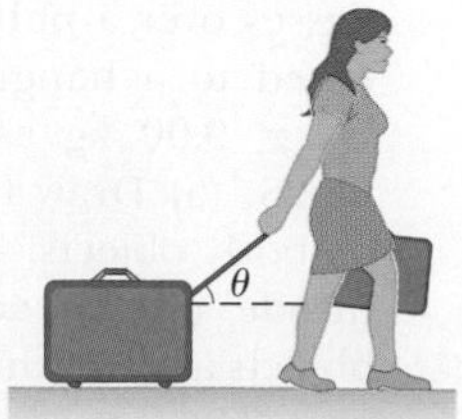

Figure P5.44

the strap make with the horizontal? (c) What is the magnitude of the normal force that the ground exerts on the suitcase?

45. A 9.00-kg hanging object is connected by a light, inextensible cord over a light, frictionless pulley to a 5.00-kg block that is sliding on a flat table (Fig. P5.28). Taking the coefficient of kinetic friction as 0.200, find the tension in the string.

46. **Q|C** Three objects are connected on a table as shown in Figure P5.46. The coefficient of kinetic friction between the block of mass m_2 and the table is 0.350. The objects have masses of $m_1 = 4.00$ kg, $m_2 = 1.00$ kg, and $m_3 = 2.00$ kg, and the pulleys are frictionless. (a) Draw a free-body diagram of each object. (b) Determine the acceleration of each object, including its direction. (c) Determine the tensions in the two cords. **What If?** (d) If the tabletop were smooth, would the tensions increase, decrease, or remain the same? Explain.

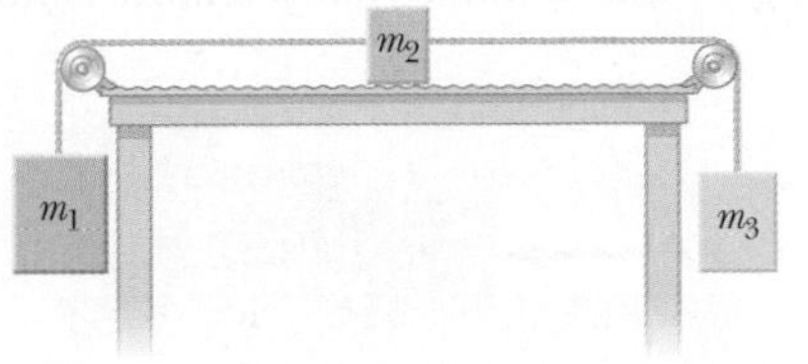

Figure P5.46

47. Two blocks connected by a rope of negligible mass are being dragged by a horizontal force (Fig. P5.47). Suppose $F = 68.0$ N, $m_1 = 12.0$ kg, $m_2 = 18.0$ kg, and the coefficient of kinetic friction between each block and the surface is 0.100. (a) Draw a free-body diagram for each block. Determine (b) the acceleration of the system and (c) the tension T in the rope.

Figure P5.47

48. **Q|C** A block of mass 3.00 kg is pushed up against a wall by a force $\vec{P}$ that makes an angle of $\theta = 50.0°$ with the horizontal as shown in Figure P5.48. The coefficient of static friction between the block and the wall is 0.250. (a) Determine the possible values for the magnitude of $\vec{P}$ that allow the block to remain stationary. (b) Describe what happens if $|\vec{P}|$ has a larger value and what happens if it is smaller. (c) Repeat parts (a) and (b), assuming the force makes an angle of $\theta = 13.0°$ with the horizontal.

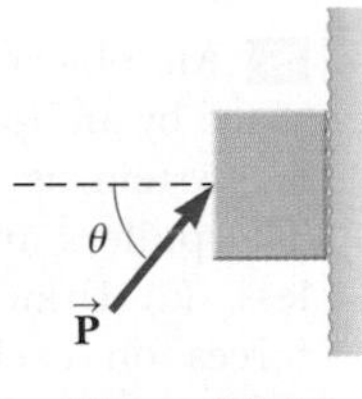

Figure P5.48

49. **Review.** One side of the roof of a house slopes up at 37.0°. A roofer kicks a round, flat rock that has been thrown onto the roof by a neighborhood child. The rock slides straight up the incline with an initial speed of 15.0 m/s. The coefficient of kinetic friction between the rock and the roof is 0.400. The rock slides 10.0 m up the roof to its peak. It crosses the ridge and goes into free fall, following a parabolic trajectory above the far side of the roof, with negligible air resistance. Determine the maximum height the rock reaches above the point where it was kicked.

50. **Review.** A Chinook salmon can swim underwater at 3.58 m/s, and it can also jump vertically upward, leaving the water with a speed of 6.26 m/s. A record salmon has length 1.50 m and mass 61.0 kg. Consider the fish swimming straight upward in the water below the surface of a lake. The gravitational force exerted on it is very nearly canceled out by a buoyant force exerted by the water as we will study in Chapter 14. The fish experiences an upward force P exerted by the water on its threshing tail fin and a downward fluid friction force that we model as acting on its front end. Assume the fluid friction force disappears as soon as the fish's head breaks the water surface and assume the force on its tail is constant. Model the gravitational force as suddenly switching full on when half the length of the fish is out of the water. Find the value of P.

51. **Review.** A magician pulls a tablecloth from under a 200-g mug located 30.0 cm from the edge of the cloth. The cloth exerts a friction force of 0.100 N on the mug, and the cloth is pulled with a constant acceleration of 3.00 m/s². How far does the mug move relative to the horizontal tabletop before the cloth is completely out from under it? Note that the cloth must move more than 30 cm relative to the tabletop during the process.

Additional Problems

52. A black aluminum glider floats on a film of air above a level aluminum air track. Aluminum feels essentially no force in a magnetic field, and air resistance is negligible. A strong magnet is attached to the top of the glider, forming a total mass of 240 g. A piece of scrap iron attached to one end stop on the track attracts the magnet with a force of 0.823 N when the iron and the magnet are separated by 2.50 cm. (a) Find the acceleration of the glider at this instant. (b) The scrap iron is now attached to another green glider, forming total mass 120 g. Find the acceleration of each glider when the gliders are simultaneously released at 2.50-cm separation.

53. **S** **Review.** A hockey puck struck by a hockey stick is given an initial speed v_i in the positive x direction. The coefficient of kinetic friction between the ice and the puck is μ_k. (a) Obtain an expression for the acceleration of the puck as it slides across the ice. (b) Use the result of part (a) to obtain an expression for the distance d the puck slides. The answer should be in terms of the variables v_i, μ_k, and g only.

54. *Why is the following situation impossible?* A book sits on an inclined plane on the surface of the Earth. The angle of the plane with the horizontal is 60.0°. The coefficient of kinetic friction between the book and the plane is 0.300. At time $t = 0$, the book is released from rest. The book then slides through a distance of 1.00 m, measured along the plane, in a time interval of 0.483 s.

55. **S** **GP** Two blocks of masses m_1 and m_2 are placed on a table in contact with each other as discussed in Example 5.7 and shown in Active Figure 5.12a. The coefficient of kinetic friction between the block of mass m_1 and the table is μ_1, and that between the block of mass m_2 and the table is μ_2. A horizontal force of magnitude F is applied to the block of mass m_1. We wish to find P, the magnitude of the contact force between the blocks. (a) Draw diagrams showing the forces for each block. (b) What is the net force on the system of two

blocks? (c) What is the net force acting on m_1? (d) What is the net force acting on m_2? (e) Write Newton's second law in the x direction for each block. (f) Solve the two equations in two unknowns for the acceleration a of the blocks in terms of the masses, the applied force F, the coefficients of friction, and g. (g) Find the magnitude P of the contact force between the blocks in terms of the same quantities.

56. **Q|C** **S** A rope with mass m_r is attached to a block with mass m_b as in Figure P5.56. The block rests on a frictionless, horizontal surface. The rope does not stretch. The free end of the rope is pulled to the right with a horizontal force $\vec{F}$. (a) Draw force diagrams for the rope and the block, noting that the tension in the rope is not uniform. (b) Find the acceleration of the system in terms of m_b, m_r, and F. (c) Find the magnitude of the force the rope exerts on the block. (d) What happens to the force on the block as the rope's mass approaches zero? What can you state about the tension in a *light* cord joining a pair of moving objects?

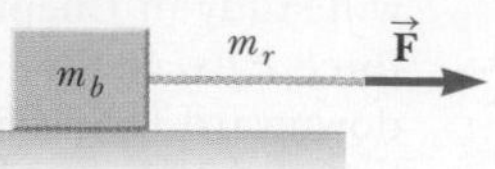

Figure P5.56

57. An inventive child named Nick wants to reach an apple in a tree without climbing the tree. Sitting in a chair connected to a rope that passes over a frictionless pulley (Fig. P5.57), Nick pulls on the loose end of the rope with such a force that the spring scale reads 250 N. Nick's true weight is 320 N, and the chair weighs 160 N. Nick's feet are not touching the ground. (a) Draw one pair of diagrams showing the forces for Nick and the chair considered as separate systems and another diagram for Nick and the chair considered as one system. (b) Show that the acceleration of the system is *upward* and find its magnitude. (c) Find the force Nick exerts on the chair.

Figure P5.57 Problems 57 and 58.

58. **Q|C** In the situation described in Problem 57 and Figure P5.57, the masses of the rope, spring balance, and pulley are negligible. Nick's feet are not touching the ground. (a) Assume Nick is momentarily at rest when he stops pulling down on the rope and passes the end of the rope to another child, of weight 440 N, who is standing on the ground next to him. The rope does not break. Describe the ensuing motion. (b) Instead, assume Nick is momentarily at rest when he ties the end of the rope to a strong hook projecting from the tree trunk. Explain why this action can make the rope break.

59. **Q|C** In Example 5.7, we pushed on two blocks on a table. Suppose three blocks are in contact with one another on a frictionless, horizontal surface as shown in Figure P5.59. A horizontal force $\vec{F}$ is applied to m_1. Take $m_1 = 2.00$ kg, $m_2 = 3.00$ kg, $m_3 = 4.00$ kg, and $F = 18.0$ N. (a) Draw a separate free-body diagram for each block. (b) Determine the acceleration of the blocks. (c) Find the *resultant* force on each block. (d) Find the magnitudes of the contact forces between the blocks. (e) You are working on a construction project. A coworker is nailing up plasterboard on one side of a light partition, and you are on the opposite side, providing "backing" by leaning against the wall with your back pushing on it. Every hammer blow makes your back sting. The supervisor helps you put a heavy block of wood between the wall and your back. Using the situation analyzed in parts (a) through (d) as a model, explain how this change works to make your job more comfortable.

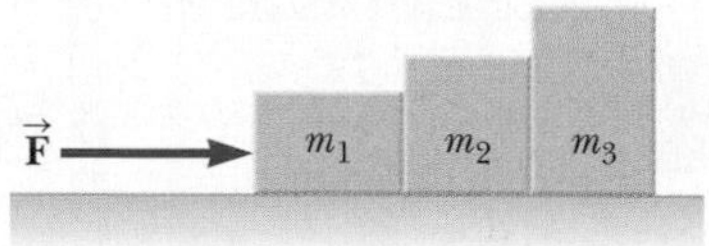

Figure P5.59

60. **Q|C** On a single, light, vertical cable that does not stretch, a crane is lifting a 1 207-kg Ferrari and, below it, a 1 461-kg BMW Z8. The Ferrari is moving upward with speed 3.50 m/s and acceleration 1.25 m/s^2. (a) How do the velocity and acceleration of the BMW compare with those of the Ferrari? (b) Find the tension in the cable between the BMW and the Ferrari. (c) Find the tension in the cable above the Ferrari.

61. **S** An object of mass M is held in place by an applied force $\vec{F}$ and a pulley system as shown in Figure P5.61. The pulleys are massless and frictionless. (a) Draw diagrams showing the forces on each pulley. Find (b) the tension in each section of rope, T_1, T_2, T_3, T_4, and T_5 and (c) the magnitude of $\vec{F}$.

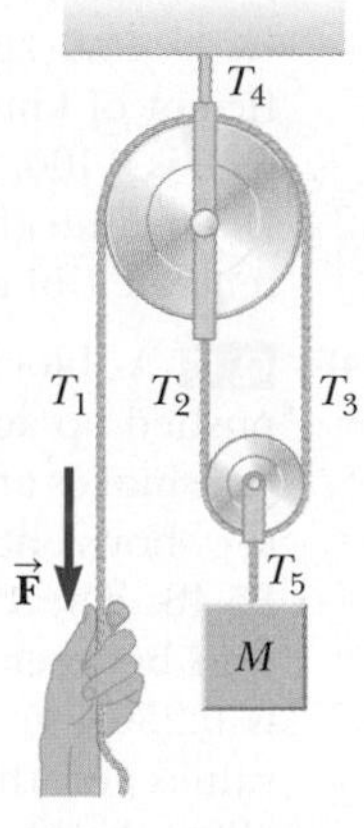

Figure P5.61

62. An aluminum block of mass $m_1 = 2.00$ kg and a copper block of mass $m_2 = 6.00$ kg are connected by a light string over a frictionless pulley. They sit on a steel surface as shown in Figure P5.62, where $\theta = 30.0°$. (a) When they are released from rest, will they start to move? If they do, determine (b) their acceleration and (c) the tension in the string. If they do not move, determine (d) the sum of the magnitudes of the forces of friction acting on the blocks.

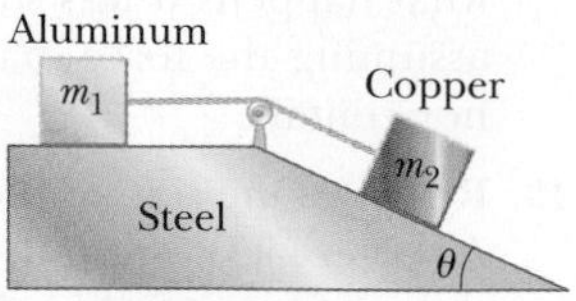

Figure P5.62

63. S A crate of weight F_g is pushed by a force $\vec{\mathbf{P}}$ on a horizontal floor as shown in Figure P5.63. The coefficient of static friction is μ_s, and $\vec{\mathbf{P}}$ is directed at angle θ below the horizontal. (a) Show that the minimum value of P that will move the crate is given by

$$P = \frac{\mu_s F_g \sec\theta}{1 - \mu_s \tan\theta}$$

(b) Find the condition on θ in terms of μ_s for which motion of the crate is impossible for any value of P.

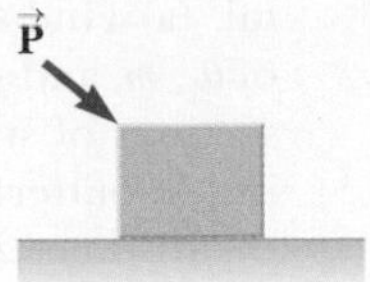

Figure P5.63

64. Q|C A student is asked to measure the acceleration of a glider on a frictionless, inclined plane, using an air track, a stopwatch, and a meterstick. The top of the track is measured to be 1.774 cm higher than the bottom of the track, and the length of the track is d = 127.1 cm. The cart is released from rest at the top of the incline, taken as x = 0, and its position x along the incline is measured as a function of time. For x values of 10.0 cm, 20.0 cm, 35.0 cm, 50.0 cm, 75.0 cm, and 100 cm, the measured times at which these positions are reached (averaged over five runs) are 1.02 s, 1.53 s, 2.01 s, 2.64 s, 3.30 s, and 3.75 s, respectively. (a) Construct a graph of x versus t^2, with a best-fit straight line to describe the data. (b) Determine the acceleration of the cart from the slope of this graph. (c) Explain how your answer to part (b) compares with the theoretical value you calculate using $a = g\sin\theta$ as derived in Example 5.6.

65. Q|C A flat cushion of mass m is released from rest at the corner of the roof of a building, at height h. A wind blowing along the side of the building exerts a constant horizontal force of magnitude F on the cushion as it drops as shown in Figure P5.65. The air exerts no vertical force. (a) Show that the path of the cushion is a straight line. (b) Does the cushion fall with constant velocity? Explain. (c) If m = 1.20 kg, h = 8.00 m, and F = 2.40 N, how far from the building will the cushion hit the level ground? **What If?** (d) If the cushion is thrown downward with a nonzero speed at the top of the building, what will be the shape of its trajectory? Explain.

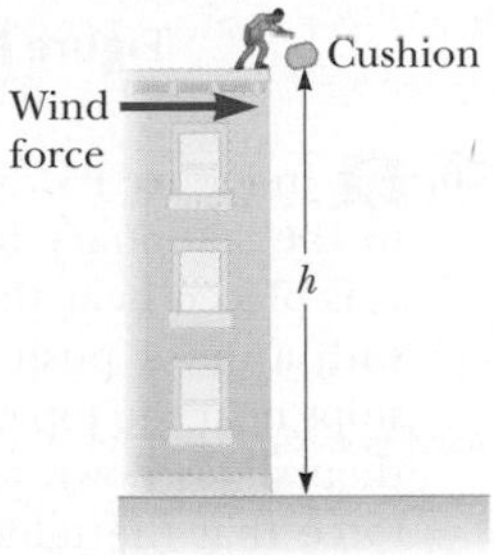

Figure P5.65

66. Q|C In Figure P5.66, the pulleys and the cord are light, all surfaces are frictionless, and the cord does not stretch. (a) How does the acceleration of block 1 compare with the acceleration of block 2? Explain your reasoning. (b) The mass of block 2 is 1.30 kg. Find its acceleration as it depends on the mass m_1 of block 1. (c) **What If?** What does the result of part (b) predict if m_1 is very much less than 1.30 kg? (d) What does the result of part (b) predict if m_1 approaches infinity? (e) In this last case, what is the tension in the cord? (f) Could you anticipate the answers to parts (c), (d), and (e) without first doing part (b)? Explain.

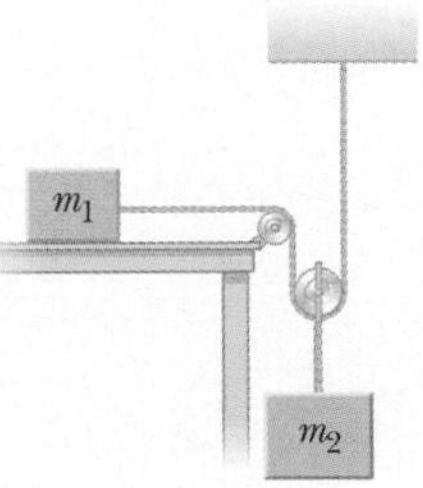

Figure P5.66

67. S What horizontal force must be applied to a large block of mass M shown in Figure P5.67 so that the tan blocks remain stationary relative to M? Assume all surfaces and the pulley are frictionless. Notice that the force exerted by the string accelerates m_2.

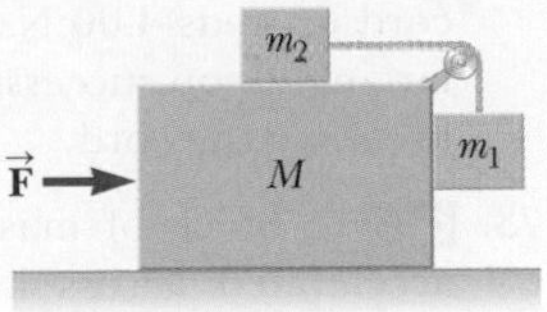

Figure P5.67

68. Q|C An 8.40-kg object slides down a fixed, frictionless, inclined plane. Use a computer to determine and tabulate (a) the normal force exerted on the object and (b) its acceleration for a series of incline angles (measured from the horizontal) ranging from 0° to 90° in 5° increments. (c) Plot a graph of the normal force and the acceleration as functions of the incline angle. (d) In the limiting cases of 0° and 90°, are your results consistent with the known behavior?

69. M A car accelerates down a hill (Fig. P5.69), going from rest to 30.0 m/s in 6.00 s. A toy inside the car hangs by a string from the car's ceiling. The ball in the figure represents the toy, of mass 0.100 kg. The acceleration is such that the string remains perpendicular to the ceiling. Determine (a) the angle θ and (b) the tension in the string.

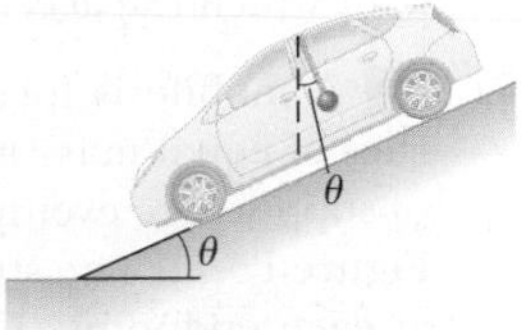

Figure P5.69

Challenge Problems

70. A time-dependent force, $\vec{\mathbf{F}} = (8.00\hat{\mathbf{i}} - 4.00t\hat{\mathbf{j}})$, where $\vec{\mathbf{F}}$ is in newtons and t is in seconds, is exerted on a 2.00-kg object initially at rest. (a) At what time will the object be moving with a speed of 15.0 m/s? (b) How far is the object from its initial position when its speed is 15.0 m/s? (c) Through what total displacement has the object traveled at this moment?

71. The board sandwiched between two other boards in Figure P5.71 weighs 95.5 N. If the coefficient of static friction between the boards is 0.663, what must be the magnitude of the compression forces (assumed horizontal) acting on both sides of the center board to keep it from slipping?

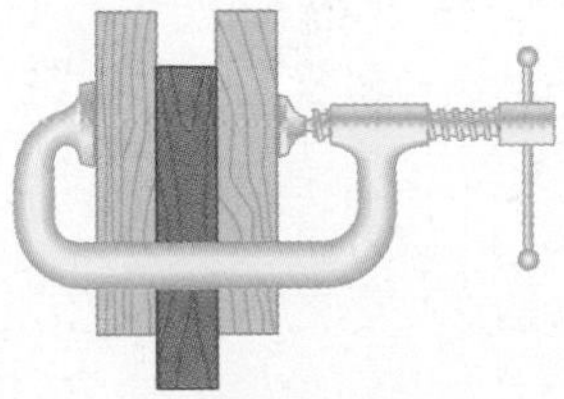

Figure P5.71

72. *Why is the following situation impossible?* A 1.30-kg toaster is not plugged in. The coefficient of static friction between the toaster and a horizontal countertop is 0.350. To make the toaster start moving, you carelessly pull on its electric cord. Unfortunately, the cord has become frayed from your previous similar actions and will break if the tension in the cord exceeds 4.00 N. By pulling on the cord at a particular angle, you successfully start the toaster moving without breaking the cord.

73. **Q|C** A block of mass 2.20 kg is accelerated across a rough surface by a light cord passing over a small pulley as shown in Figure P5.73. The tension T in the cord is maintained at 10.0 N, and the pulley is 0.100 m above the top of the block. The coefficient of kinetic friction is 0.400. (a) Determine the acceleration of the block when $x = 0.400$ m. (b) Describe the general behavior of the acceleration as the block slides from a location where x is large to $x = 0$. (c) Find the maximum value of the acceleration and the position x for which it occurs. (d) Find the value of x for which the acceleration is zero.

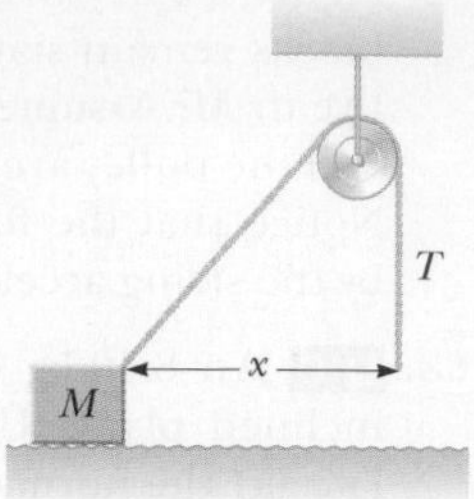

Figure P5.73

74. **S** A mobile is formed by supporting four metal butterflies of equal mass m from a string of length L. The points of support are evenly spaced a distance ℓ apart as shown in Figure P5.74. The string forms an angle θ_1 with the ceiling at each endpoint. The center section of string is horizontal. (a) Find the tension in each section of string in terms of θ_1, m, and g. (b) In terms of θ_1, find the angle θ_2 that the sections of string between the outside butterflies and the inside butterflies form with the horizontal. (c) Show that the distance D between the endpoints of the string is

$$D = \frac{L}{5}\left\{2\cos\theta_1 + 2\cos\left[\tan^{-1}\left(\tfrac{1}{2}\tan\theta_1\right)\right] + 1\right\}$$

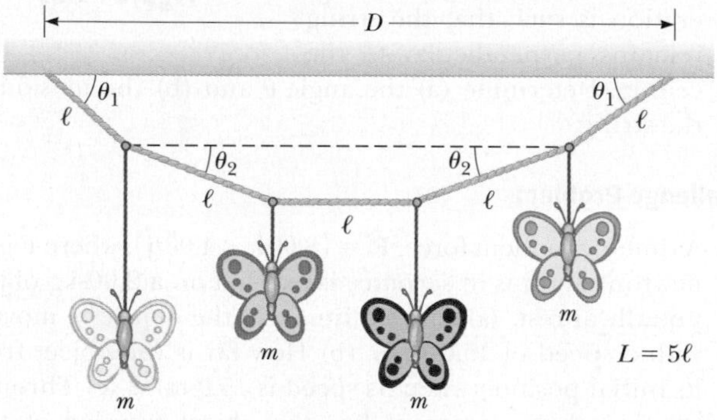

Figure P5.74

75. **Review.** A block of mass $m = 2.00$ kg is released from rest at $h = 0.500$ m above the surface of a table, at the top of a $\theta = 30.0°$ incline as shown in Figure P5.75. The frictionless incline is fixed on a table of height $H = 2.00$ m. (a) Determine the acceleration of the block as it slides down the incline. (b) What is the velocity of the block as it leaves the incline? (c) How far from the table will the block hit the floor? (d) What time interval elapses between when the block is released and when it hits the floor? (e) Does the mass of the block affect any of the above calculations?

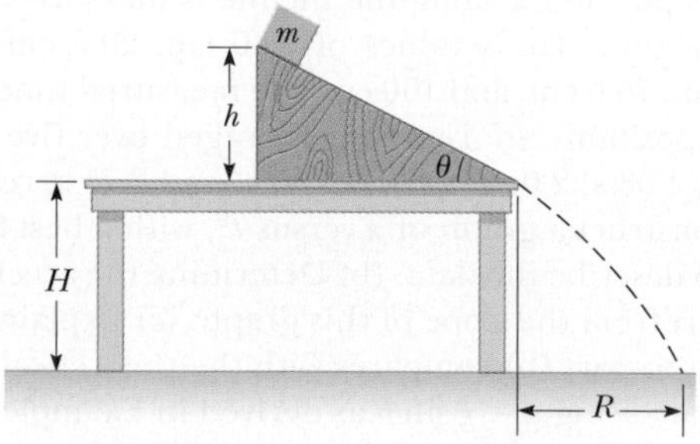

Figure P5.75 Problems 75 and 76.

76. **S** In Figure P5.75, the incline has mass M and is fastened to the stationary horizontal tabletop. The block of mass m is placed near the bottom of the incline and is released with a quick push that sets it sliding upward. The block stops near the top of the incline as shown in the figure and then slides down again, always without friction. Find the force that the tabletop exerts on the incline throughout this motion in terms of m, M, g, and θ.

chapter 6

Circular Motion and Other Applications of Newton's Laws

Kyle Busch, driver of the #18 Snickers Toyota, leads Jeff Gordon, driver of the #24 Dupont Chevrolet, during the NASCAR Sprint Cup Series Kobalt Tools 500 at the Atlanta Motor Speedway on March 9, 2008 in Hampton, Georgia. The cars travel on a banked roadway to help them undergo circular motion on the turns. *(Chris Graythen/Getty Images for NASCAR)*

In the preceding chapter, we introduced Newton's laws of motion and incorporated them into two analysis models involving linear motion. Now we discuss motion that is slightly more complicated. For example, we shall apply Newton's laws to objects traveling in circular paths. We shall also discuss motion observed from an accelerating frame of reference and motion of an object through a viscous medium. For the most part, this chapter consists of a series of examples selected to illustrate the application of Newton's laws to a variety of new circumstances.

6.1 Extending the Particle in Uniform Circular Motion Model

In Section 4.4, we discussed the analysis model of a particle in uniform circular motion, in which a particle moves with constant speed v in a circular path having a radius r. The particle experiences an acceleration that has a magnitude

$$a_c = \frac{v^2}{r}$$

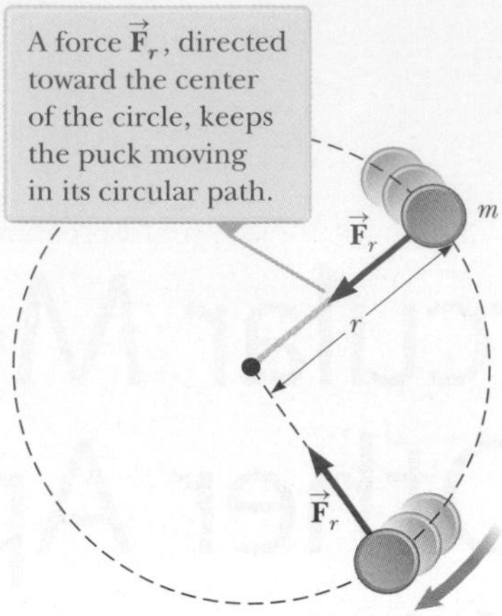

Figure 6.1 An overhead view of a puck moving in a circular path in a horizontal plane.

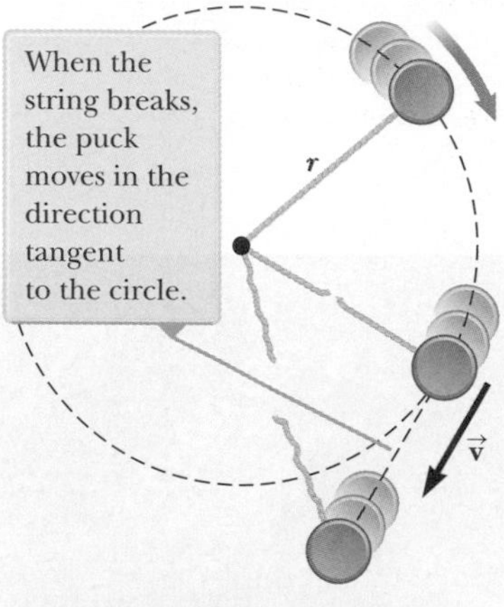

ACTIVE FIGURE 6.2

The string holding the puck in its circular path breaks.

The acceleration is called *centripetal acceleration* because $\vec{a}_c$ is directed toward the center of the circle. Furthermore, $\vec{a}_c$ is *always* perpendicular to $\vec{v}$. (If there were a component of acceleration parallel to $\vec{v}$, the particle's speed would be changing.)

Let us now extend the particle in uniform circular motion model from Section 4.4 by incorporating the concept of force. Consider a puck of mass m that is tied to a string of length r and moves at constant speed in a horizontal, circular path as illustrated in Figure 6.1. Its weight is supported by a frictionless table, and the string is anchored to a peg at the center of the circular path of the puck. Why does the puck move in a circle? According to Newton's first law, the puck would move in a straight line if there were no force on it; the string, however, prevents motion along a straight line by exerting on the puck a radial force $\vec{F}_r$ that makes it follow the circular path. This force is directed along the string toward the center of the circle as shown in Figure 6.1.

If Newton's second law is applied along the radial direction, the net force causing the centripetal acceleration can be related to the acceleration as follows:

Force causing centripetal acceleration ▶

$$\sum F = ma_c = m\frac{v^2}{r} \qquad \textbf{(6.1)}$$

A force causing a centripetal acceleration acts toward the center of the circular path and causes a change in the direction of the velocity vector. If that force should vanish, the object would no longer move in its circular path; instead, it would move along a straight-line path tangent to the circle. This idea is illustrated in Active Figure 6.2 for the puck moving in a circular path at the end of a string in a horizontal plane. If the string breaks at some instant, the puck moves along the straight-line path that is tangent to the circle at the position of the puck at this instant.

Pitfall Prevention 6.1

Direction of Travel When the String Is Cut

Study Active Figure 6.2 very carefully. Many students (wrongly) think that the puck will move *radially* away from the center of the circle when the string is cut. The velocity of the puck is *tangent* to the circle. By Newton's first law, the puck continues to move in the same direction in which it is moving just as the force from the string disappears.

***Quick Quiz* 6.1** You are riding on a Ferris wheel that is rotating with constant speed. The car in which you are riding always maintains its correct upward orientation; it does not invert. **(i)** What is the direction of the normal force on you from the seat when you are at the top of the wheel? (a) upward (b) downward (c) impossible to determine **(ii)** From the same choices, what is the direction of the net force on you when you are at the top of the wheel?

Example 6.1 The Conical Pendulum

A small ball of mass m is suspended from a string of length L. The ball revolves with constant speed v in a horizontal circle of radius r as shown in Figure 6.3. (Because the string sweeps out the surface of a cone, the system is known as a *conical pendulum*.) Find an expression for v.

6.1 *cont.*

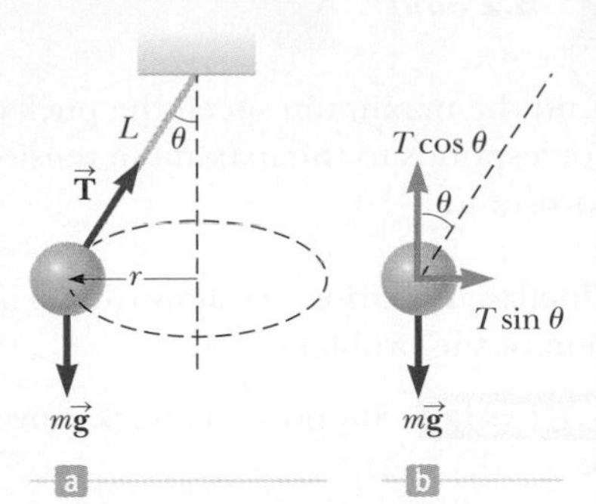

Figure 6.3 (Example 6.1) (a) A conical pendulum. The path of the ball is a horizontal circle. (b) The forces acting on the ball.

SOLUTION

Conceptualize Imagine the motion of the ball in Figure 6.3a and convince yourself that the string sweeps out a cone and that the ball moves in a horizontal circle.

Categorize The ball in Figure 6.3 does not accelerate vertically. Therefore, we model it as a particle in equilibrium in the vertical direction. It experiences a centripetal acceleration in the horizontal direction, so it is modeled as a particle in uniform circular motion in this direction.

Analyze Let θ represent the angle between the string and the vertical. In the diagram of forces acting on the ball in Figure 6.3b, the force $\vec{\mathbf{T}}$ exerted by the string on the ball is resolved into a vertical component $T\cos\theta$ and a horizontal component $T\sin\theta$ acting toward the center of the circular path.

Apply the particle in equilibrium model in the vertical direction:

$$\sum F_y = T\cos\theta - mg = 0$$

$$(1)\quad T\cos\theta = mg$$

Use Equation 6.1 from the particle in uniform circular motion model in the horizontal direction:

$$(2)\quad \sum F_x = T\sin\theta = ma_c = \frac{mv^2}{r}$$

Divide Equation (2) by Equation (1) and use $\sin\theta/\cos\theta = \tan\theta$:

$$\tan\theta = \frac{v^2}{rg}$$

Solve for v:

$$v = \sqrt{rg\tan\theta}$$

Incorporate $r = L\sin\theta$ from the geometry in Figure 6.3a:

$$v = \boxed{\sqrt{Lg\sin\theta\tan\theta}}$$

Finalize Notice that the speed is independent of the mass of the ball. Consider what happens when θ goes to 90° so that the string is horizontal. Because the tangent of 90° is infinite, the speed v is infinite, which tells us the string cannot possibly be horizontal. If it were, there would be no vertical component of the force $\vec{\mathbf{T}}$ to balance the gravitational force on the ball. That is why we mentioned in regard to Figure 6.1 that the puck's weight in the figure is supported by a frictionless table.

Example 6.2 How Fast Can It Spin?

A puck of mass 0.500 kg is attached to the end of a cord 1.50 m long. The puck moves in a horizontal circle as shown in Figure 6.1. If the cord can withstand a maximum tension of 50.0 N, what is the maximum speed at which the puck can move before the cord breaks? Assume the string remains horizontal during the motion.

SOLUTION

Conceptualize It makes sense that the stronger the cord, the faster the puck can move before the cord breaks. Also, we expect a more massive puck to break the cord at a lower speed. (Imagine whirling a bowling ball on the cord!)

Categorize Because the puck moves in a circular path, we model it as a particle in uniform circular motion.

Analyze Incorporate the tension and the centripetal acceleration into Newton's second law as described by Equation 6.1:

$$T = m\frac{v^2}{r}$$

Solve for v:

$$(1)\quad v = \sqrt{\frac{Tr}{m}}$$

continued

6.2 *cont.*

Find the maximum speed the puck can have, which corresponds to the maximum tension the string can withstand:

$$v_{max} = \sqrt{\frac{T_{max}r}{m}} = \sqrt{\frac{(50.0\text{ N})(1.50\text{ m})}{0.500\text{ kg}}} = 12.2\text{ m/s}$$

Finalize Equation (1) shows that v increases with T and decreases with larger m, as we expected from our conceptualization of the problem.

WHAT IF? Suppose the puck moves in a circle of larger radius at the same speed v. Is the cord more likely or less likely to break?

Answer The larger radius means that the change in the direction of the velocity vector will be smaller in a given time interval. Therefore, the acceleration is smaller and the required tension in the string is smaller. As a result, the string is less likely to break when the puck travels in a circle of larger radius.

Example 6.3 What Is the Maximum Speed of the Car?

A 1 500-kg car moving on a flat, horizontal road negotiates a curve as shown in Figure 6.4a. If the radius of the curve is 35.0 m and the coefficient of static friction between the tires and dry pavement is 0.523, find the maximum speed the car can have and still make the turn successfully.

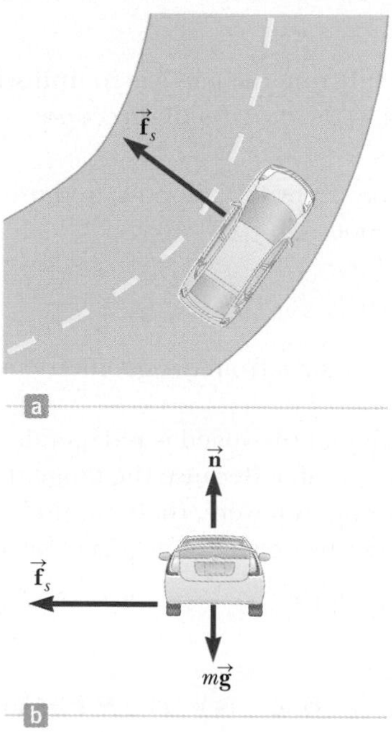

Figure 6.4 (Example 6.3) (a) The force of static friction directed toward the center of the curve keeps the car moving in a circular path. (b) The forces acting on the car.

SOLUTION

Conceptualize Imagine that the curved roadway is part of a large circle so that the car is moving in a circular path.

Categorize Based on the conceptualize step of the problem, we model the car as a particle in uniform circular motion in the horizontal direction. The car is not accelerating vertically, so it is modeled as a particle in equilibrium in the vertical direction.

Analyze The force that enables the car to remain in its circular path is the force of static friction. (It is *static* because no slipping occurs at the point of contact between road and tires. If this force of static friction were zero—for example, if the car were on an icy road—the car would continue in a straight line and slide off the curved road.) The maximum speed v_{max} the car can have around the curve is the speed at which it is on the verge of skidding outward. At this point, the friction force has its maximum value $f_{s,max} = \mu_s n$.

Apply Equation 6.1 in the radial direction for the maximum speed condition:

$$(1)\quad f_{s,max} = \mu_s n = m\frac{v_{max}^2}{r}$$

Apply the particle in equilibrium model to the car in the vertical direction:

$$\sum F_y = 0 \rightarrow n - mg = 0 \rightarrow n = mg$$

Solve Equation (1) for the maximum speed and substitute for n:

$$(2)\quad v_{max} = \sqrt{\frac{\mu_s nr}{m}} = \sqrt{\frac{\mu_s mgr}{m}} = \sqrt{\mu_s gr}$$

Substitute numerical values:

$$v_{max} = \sqrt{(0.523)(9.80\text{ m/s}^2)(35.0\text{ m})} = 13.4\text{ m/s}$$

6.3 *cont.*

Finalize This speed is equivalent to 30.0 mi/h. Therefore, if the speed limit on this roadway is higher than 30 mi/h this roadway could benefit greatly from some banking, as in the next example! Notice that the maximum speed does not depend on the mass of the car, which is why curved highways do not need multiple speed limits to cover the various masses of vehicles using the road.

WHAT IF? Suppose a car travels this curve on a wet day and begins to skid on the curve when its speed reaches only 8.00 m/s. What can we say about the coefficient of static friction in this case?

Answer The coefficient of static friction between the tires and a wet road should be smaller than that between the tires and a dry road. This expectation is consistent with experience with driving because a skid is more likely on a wet road than a dry road.

To check our suspicion, we can solve Equation (2) for the coefficient of static friction:

$$\mu_s = \frac{v_{\text{max}}^2}{gr}$$

Substituting the numerical values gives

$$\mu_s = \frac{v_{\text{max}}^2}{gr} = \frac{(8.00\text{ m/s})^2}{(9.80\text{ m/s}^2)(35.0\text{ m})} = 0.187$$

which is indeed smaller than the coefficient of 0.523 for the dry road.

Example 6.4 The Banked Roadway

A civil engineer wishes to redesign the curved roadway in Example 6.3 in such a way that a car will not have to rely on friction to round the curve without skidding. In other words, a car moving at the designated speed can negotiate the curve even when the road is covered with ice. Such a ramp is usually *banked,* which means that the roadway is tilted toward the inside of the curve as seen in the opening photograph for this chapter. Suppose the designated speed for the ramp is to be 13.4 m/s (30.0 mi/h) and the radius of the curve is 35.0 m. At what angle should the curve be banked?

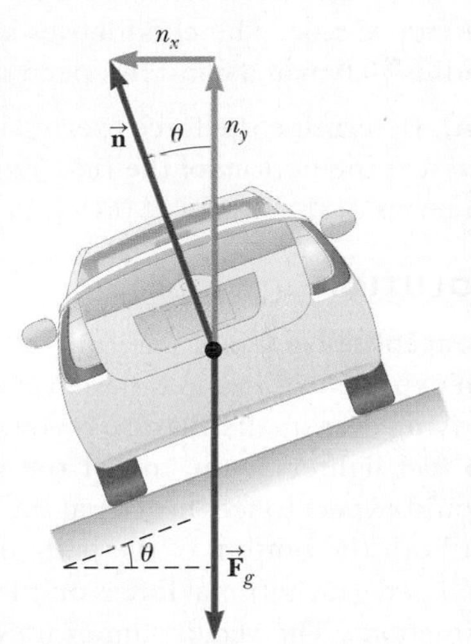

Figure 6.5 (Example 6.4) A car moves into the page and is rounding a curve on a road banked at an angle θ to the horizontal. When friction is neglected, the force that causes the centripetal acceleration and keeps the car moving in its circular path is the horizontal component of the normal force.

SOLUTION

Conceptualize The difference between this example and Example 6.3 is that the car is no longer moving on a flat roadway. Figure 6.5 shows the banked roadway, with the center of the circular path of the car far to the left of the figure. Notice that the horizontal component of the normal force participates in causing the car's centripetal acceleration.

Categorize As in Example 6.3, the car is modeled as a particle in equilibrium in the vertical direction and a particle in uniform circular motion in the horizontal direction.

Analyze On a level (unbanked) road, the force that causes the centripetal acceleration is the force of static friction between car and road as we saw in the preceding example. If the road is banked at an angle θ as in Figure 6.5, however, the normal force $\vec{\mathbf{n}}$ has a horizontal component toward the center of the curve. Because the ramp is to be designed so that the force of static friction is zero, only the component $n_x = n \sin\theta$ causes the centripetal acceleration.

Write Newton's second law for the car in the radial direction, which is the x direction:

$$(1)\quad \sum F_r = n \sin\theta = \frac{mv^2}{r}$$

Apply the particle in equilibrium model to the car in the vertical direction:

$$\sum F_y = n\cos\theta - mg = 0$$

$$(2)\quad n\cos\theta = mg$$

continued

6.4 *cont.*

Divide Equation (1) by Equation (2): $$(3)\ \tan\theta = \frac{v^2}{rg}$$

Solve for the angle θ: $$\theta = \tan^{-1}\left[\frac{(13.4\ \text{m/s})^2}{(35.0\ \text{m})(9.80\ \text{m/s}^2)}\right] = 27.6°$$

Finalize Equation (3) shows that the banking angle is independent of the mass of the vehicle negotiating the curve. If a car rounds the curve at a speed less than 13.4 m/s, friction is needed to keep it from sliding down the bank (to the left in Fig. 6.5). A driver attempting to negotiate the curve at a speed greater than 13.4 m/s has to depend on friction to keep from sliding up the bank (to the right in Fig. 6.5).

WHAT IF? Imagine that this same roadway were built on Mars in the future to connect different colony centers. Could it be traveled at the same speed?

Answer The reduced gravitational force on Mars would mean that the car is not pressed as tightly to the roadway. The reduced normal force results in a smaller component of the normal force toward the center of the circle. This smaller component would not be sufficient to provide the centripetal acceleration associated with the original speed. The centripetal acceleration must be reduced, which can be done by reducing the speed v.

Mathematically, notice that Equation (3) shows that the speed v is proportional to the square root of g for a roadway of fixed radius r banked at a fixed angle θ. Therefore, if g is smaller, as it is on Mars, the speed v with which the roadway can be safely traveled is also smaller.

Example 6.5 Riding the Ferris Wheel

A child of mass m rides on a Ferris wheel as shown in Figure 6.6a. The child moves in a vertical circle of radius 10.0 m at a constant speed of 3.00 m/s.

(A) Determine the force exerted by the seat on the child at the bottom of the ride. Express your answer in terms of the weight of the child, mg.

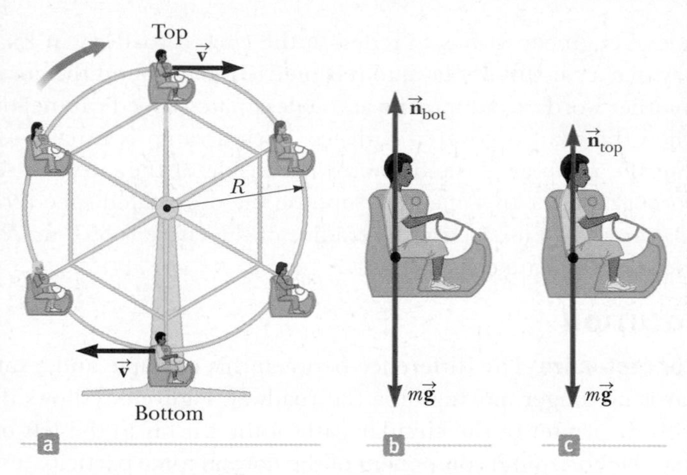

Figure 6.6 (Example 6.5) (a) A child rides on a Ferris wheel. (b) The forces acting on the child at the bottom of the path. (c) The forces acting on the child at the top of the path.

SOLUTION

Conceptualize Look carefully at Figure 6.6a. Based on experiences you may have had on a Ferris wheel or driving over small hills on a roadway, you would expect to feel lighter at the top of the path. Similarly, you would expect to feel heavier at the bottom of the path. At both the bottom of the path and the top, the normal and gravitational forces on the child act in *opposite* directions. The vector sum of these two forces gives a force of constant magnitude that keeps the child moving in a circular path at a constant speed. To yield net force vectors with the same magnitude, the normal force at the bottom must be greater than that at the top.

Categorize Because the speed of the child is constant, we can categorize this problem as one involving a particle (the child) in uniform circular motion, complicated by the gravitational force acting at all times on the child.

Analyze We draw a diagram of forces acting on the child at the bottom of the ride as shown in Figure 6.6b. The only forces acting on him are the downward gravitational force $\vec{\mathbf{F}}_g = m\vec{\mathbf{g}}$ and the upward force $\vec{\mathbf{n}}_{bot}$ exerted by the seat. The net upward force on the child that provides his centripetal acceleration has a magnitude $n_{bot} - mg$.

Apply Newton's second law to the child in the radial direction: $$\sum F = n_{bot} - mg = m\frac{v^2}{r}$$

Solve for the force exerted by the seat on the child: $$n_{bot} = mg + m\frac{v^2}{r} = mg\left(1 + \frac{v^2}{rg}\right)$$

6.5 cont.

Substitute the values given for the speed and radius:

$$n_{bot} = mg\left[1 + \frac{(3.00\text{ m/s})^2}{(10.0\text{ m})(9.80\text{ m/s}^2)}\right]$$

$$= 1.09\,mg$$

Hence, the magnitude of the force $\vec{\mathbf{n}}_{bot}$ exerted by the seat on the child is *greater* than the weight of the child by a factor of 1.09. So, the child experiences an apparent weight that is greater than his true weight by a factor of 1.09.

(B) Determine the force exerted by the seat on the child at the top of the ride.

SOLUTION

Analyze The diagram of forces acting on the child at the top of the ride is shown in Figure 6.6c. The net downward force that provides the centripetal acceleration has a magnitude $mg - n_{top}$.

Apply Newton's second law to the child at this position:

$$\sum F = mg - n_{top} = m\frac{v^2}{r}$$

Solve for the force exerted by the seat on the child:

$$n_{top} = mg - m\frac{v^2}{r} = mg\left(1 - \frac{v^2}{rg}\right)$$

Substitute numerical values:

$$n_{top} = mg\left[1 - \frac{(3.00\text{ m/s})^2}{(10.0\text{ m})(9.80\text{ m/s}^2)}\right]$$

$$= 0.908\,mg$$

In this case, the magnitude of the force exerted by the seat on the child is *less* than his true weight by a factor of 0.908, and the child feels lighter.

Finalize The variations in the normal force are consistent with our prediction in the Conceptualize step of the problem.

WHAT IF? Suppose a defect in the Ferris wheel mechanism causes the speed of the child to increase to 10.0 m/s. What does the child experience at the top of the ride in this case?

Answer If the calculation above is performed with v = 10.0 m/s, the magnitude of the normal force at the top of the ride is negative, which is impossible. We interpret it to mean that the required centripetal acceleration of the child is larger than that due to gravity. As a result, the child will lose contact with the seat and will only stay in his circular path if there is a safety bar that provides a downward force on him to keep him in his seat. At the bottom of the ride, the normal force is 2.02 mg, which would be uncomfortable.

6.2 Nonuniform Circular Motion

In Chapter 4, we found that if a particle moves with varying speed in a circular path, there is, in addition to the radial component of acceleration, a tangential component having magnitude $|dv/dt|$. Therefore, the force acting on the particle must also have a tangential and a radial component. Because the total acceleration is $\vec{\mathbf{a}} = \vec{\mathbf{a}}_r + \vec{\mathbf{a}}_t$, the total force exerted on the particle is $\sum\vec{\mathbf{F}} = \sum\vec{\mathbf{F}}_r + \sum\vec{\mathbf{F}}_t$ as shown in Active Figure 6.7. (We express the radial and tangential forces as net forces with the summation notation because each force could consist of multiple forces that combine.) The vector $\sum\vec{\mathbf{F}}_r$ is directed toward the center of the circle and is responsible for the centripetal acceleration. The vector $\sum\vec{\mathbf{F}}_t$ tangent to the circle is responsible for the tangential acceleration, which represents a change in the particle's speed with time.

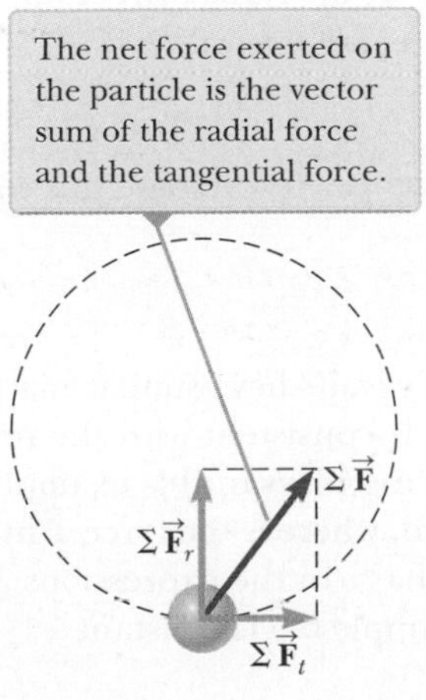

ACTIVE FIGURE 6.7

When the net force acting on a particle moving in a circular path has a tangential component ΣF_t, the particle's speed changes.

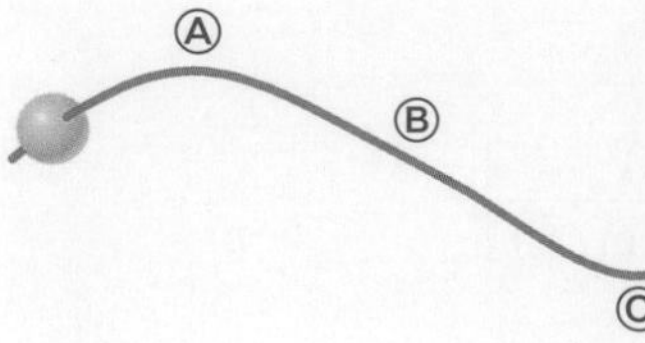

Figure 6.8 (Quick Quiz 6.2) A bead slides along a curved wire.

***Quick Quiz* 6.2** A bead slides freely along a curved wire lying on a horizontal surface at constant speed as shown in Figure 6.8. **(a)** Draw the vectors representing the force exerted by the wire on the bead at points Ⓐ, Ⓑ, and Ⓒ. **(b)** Suppose the bead in Figure 6.8 speeds up with constant tangential acceleration as it moves toward the right. Draw the vectors representing the force on the bead at points Ⓐ, Ⓑ, and Ⓒ.

Example 6.6 Keep Your Eye on the Ball

A small sphere of mass m is attached to the end of a cord of length R and set into motion in a *vertical* circle about a fixed point O as illustrated in Figure 6.9. Determine the tangential acceleration of the sphere and the tension in the cord at any instant when the speed of the sphere is v and the cord makes an angle θ with the vertical.

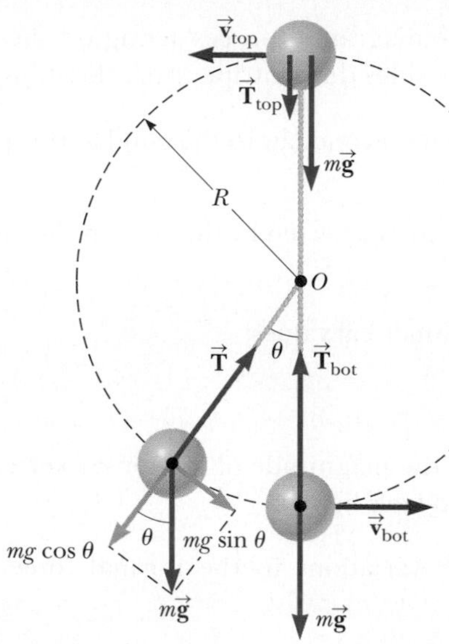

Figure 6.9 (Example 6.6) The forces acting on a sphere of mass m connected to a cord of length R and rotating in a vertical circle centered at O. Forces acting on the sphere are shown when the sphere is at the top and bottom of the circle and at an arbitrary location.

SOLUTION

Conceptualize Compare the motion of the sphere in Figure 6.9 with that of the child in Figure 6.6a associated with Example 6.5. Both objects travel in a circular path. Unlike the child in Example 6.5, however, the speed of the sphere is *not* uniform in this example because, at most points along the path, a tangential component of acceleration arises from the gravitational force exerted on the sphere.

Categorize We model the sphere as a particle under a net force and moving in a circular path, but it is not a particle in *uniform* circular motion. We need to use the techniques discussed in this section on nonuniform circular motion.

Analyze From the force diagram in Figure 6.9, we see that the only forces acting on the sphere are the gravitational force $\vec{\mathbf{F}}_g = m\vec{\mathbf{g}}$ exerted by the Earth and the force $\vec{\mathbf{T}}$ exerted by the cord. We resolve $\vec{\mathbf{F}}_g$ into a tangential component $mg\sin\theta$ and a radial component $mg\cos\theta$.

Apply Newton's second law to the sphere in the tangential direction:

$$\sum F_t = mg\sin\theta = ma_t$$

$$a_t = \boxed{g\sin\theta}$$

Apply Newton's second law to the forces acting on the sphere in the radial direction, noting that both $\vec{\mathbf{T}}$ and $\vec{\mathbf{a}}_r$ are directed toward O:

$$\sum F_r = T - mg\cos\theta = \frac{mv^2}{R}$$

$$T = \boxed{mg\left(\frac{v^2}{Rg} + \cos\theta\right)}$$

Finalize Let us evaluate this result at the top and bottom of the circular path (Fig. 6.9):

$$T_{top} = mg\left(\frac{v_{top}^2}{Rg} - 1\right) \qquad T_{bot} = mg\left(\frac{v_{bot}^2}{Rg} + 1\right)$$

These results have similar mathematical forms as those for the normal forces n_{top} and n_{bot} on the child in Example 6.5, which is consistent with the normal force on the child playing a similar physical role in Example 6.5 as the tension in the string plays in this example. Keep in mind, however, that the normal force $\vec{\mathbf{n}}$ on the child in Example 6.5 is always upward, whereas the force $\vec{\mathbf{T}}$ in this example changes direction because it must always point inward along the string. Also note that v in the expressions above varies for different positions of the sphere, as indicated by the subscripts, whereas v in Example 6.5 is constant.

6.6 *cont.*

WHAT IF? What if the ball is set in motion with a slower speed?

(A) What speed would the ball have as it passes over the top of the circle if the tension in the cord goes to zero instantaneously at this point?

Answer Let us set the tension equal to zero in the expression for T_{top}:

$$0 = mg\left(\frac{v_{top}^2}{Rg} - 1\right) \rightarrow v_{top} = \sqrt{gR}$$

(B) What if the ball is set in motion such that the speed at the top is less than this value? What happens?

Answer In this case, the ball never reaches the top of the circle. At some point on the way up, the tension in the string goes to zero and the ball becomes a projectile. It follows a segment of a parabolic path over the top of its motion, rejoining the circular path on the other side when the tension becomes nonzero again.

6.3 Motion in Accelerated Frames

Newton's laws of motion, which we introduced in Chapter 5, describe observations that are made in an inertial frame of reference. In this section, we analyze how Newton's laws are applied by an observer in a noninertial frame of reference, that is, one that is accelerating. For example, recall the discussion of the air hockey table on a train in Section 5.2. The train moving at constant velocity represents an inertial frame. An observer on the train sees the puck at rest remain at rest, and Newton's first law appears to be obeyed. The accelerating train is not an inertial frame. According to you as the observer on this train, there appears to be no force on the puck, yet it accelerates from rest toward the back of the train, appearing to violate Newton's first law. This property is a general property of observations made in noninertial frames: there appear to be unexplained accelerations of objects that are not "fastened" to the frame. Newton's first law is not violated, of course. It only appears to be violated because of observations made from a noninertial frame. In general, the direction of the unexplained acceleration is opposite the direction of the acceleration of the noninertial frame.

On the accelerating train, as you watch the puck accelerating toward the back of the train, you might conclude based on your belief in Newton's second law that a force has acted on the puck to cause it to accelerate. We call an apparent force such as this one a **fictitious force** because it is due to observations made in an accelerated reference frame. A fictitious force appears to act on an object in the same way as a real force. Real forces are always interactions between two objects, however, and you cannot identify a second object for a fictitious force. (What second object is interacting with the puck to cause it to accelerate?) In general, simple fictitious forces appear to act in the direction *opposite* that of the acceleration of the noninertial frame. For example, the train accelerates forward and there appears to be a fictitious force causing the puck to slide toward the back of the train.

The train example describes a fictitious force due to a change in the train's speed. Another fictitious force is due to the change in the *direction* of the velocity vector. To understand the motion of a system that is noninertial because of a change in direction, consider a car traveling along a highway at a high speed and approaching a curved exit ramp on the left as shown in Figure 6.10a (page 148). As the car takes the sharp left turn on the ramp, a person sitting in the passenger seat leans or slides to the right and hits the door. At that point the force exerted by the door on the passenger keeps her from being ejected from the car. What causes her to move toward the door? A popular but incorrect explanation is that a force acting toward the right in Figure 6.10b pushes the passenger outward from the

Pitfall Prevention 6.2

Centrifugal Force

The commonly heard phrase "centrifugal force" is described as a force pulling *outward* on an object moving in a circular path. If you are feeling a "centrifugal force" on a rotating carnival ride, what is the other object with which you are interacting? You cannot identify another object because it is a fictitious force that occurs when you are in a noninertial reference frame.

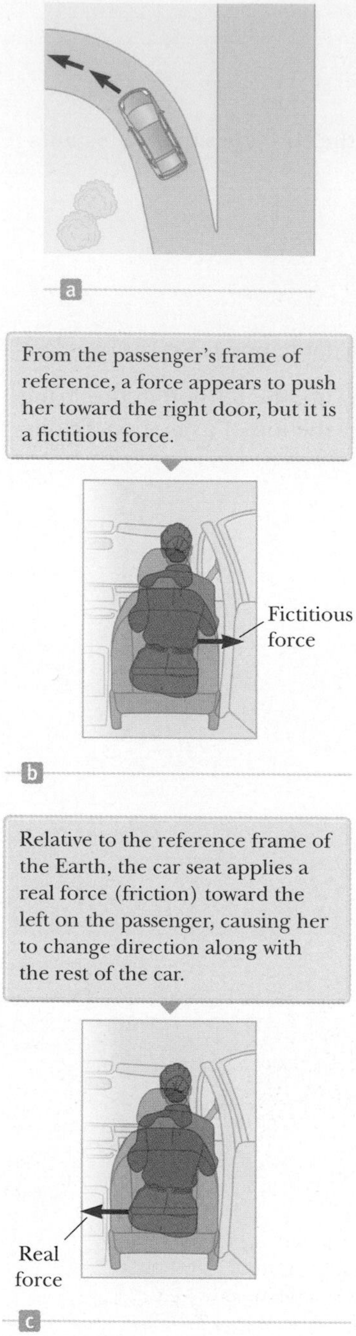

Figure 6.10 (a) A car approaching a curved exit ramp. What causes a passenger in the front seat to move toward the right-hand door? (b) Passenger's frame of reference. (c) Reference frame of the Earth.

center of the circular path. Although often called the "centrifugal force," it is a fictitious force. The car represents a noninertial reference frame that has a centripetal acceleration toward the center of its circular path. As a result, the passenger feels an apparent force which is outward from the center of the circular path, or to the right in Figure 6.10b, in the direction opposite that of the acceleration.

Let us address this phenomenon in terms of Newton's laws. Before the car enters the ramp, the passenger is moving in a straight-line path. As the car enters the ramp and travels a curved path, the passenger tends to move along the original straight-line path, which is in accordance with Newton's first law: the natural tendency of an object is to continue moving in a straight line. If a sufficiently large force (toward the center of curvature) acts on the passenger as in Figure 6.10c, however, she moves in a curved path along with the car. This force is the force of friction between her and the car seat. If this friction force is not large enough, the seat follows a curved path while the passenger tends to continue in the straight-line path of the car before the car began the turn. Therefore, from the point of view of an observer in the car, the passenger leans or slides to the right relative to the seat. Eventually, she encounters the door, which provides a force large enough to enable her to follow the same curved path as the car.

Another interesting fictitious force is the "Coriolis force." It is an apparent force caused by changing the radial position of an object in a rotating coordinate system.

For example, suppose you and a friend are on opposite sides of a rotating circular platform and you decide to throw a baseball to your friend. Active Figure 6.11a represents what an observer would see if the ball is viewed while the observer is hovering at rest above the rotating platform. According to this observer, who is in an inertial frame, the ball follows a straight line as it must according to Newton's first law. At $t = 0$ you throw the ball toward your friend, but by the time t_f when the ball has crossed the platform, your friend has moved to a new position and can't catch the ball. Now, however, consider the situation from your friend's viewpoint. Your friend is in a noninertial reference frame because he is undergoing a centripetal acceleration relative to the inertial frame of the Earth's surface. He starts off seeing the baseball coming toward him, but as it crosses the platform, it veers to one side as shown in Active Figure 6.11b. Therefore, your friend on the rotating platform states that the ball does not obey Newton's first law and claims that a sideways force is causing the ball to follow a curved path. This fictitious force is called the Coriolis force.

Fictitious forces may not be real forces, but they can have real effects. An object on your dashboard *really* slides off if you press the accelerator of your car. As you ride on a merry-go-round, you feel pushed toward the outside as if due to the fictitious "centrifugal force." You are likely to fall over and injure yourself due to the Coriolis force if you walk along a radial line while a merry-go-round rotates. (One of the authors did so and suffered a separation of the ligaments from his ribs when he fell over.) The Coriolis force due to the rotation of the Earth is responsible for rotations of hurricanes and for large-scale ocean currents.

***Quick Quiz* 6.3** Consider the passenger in the car making a left turn in Figure 6.10. Which of the following is correct about forces in the horizontal direction if she is making contact with the right-hand door? **(a)** The passenger is in equilibrium between real forces acting to the right and real forces acting to the left. **(b)** The passenger is subject only to real forces acting to the right. **(c)** The passenger is subject only to real forces acting to the left. **(d)** None of those statements is true.

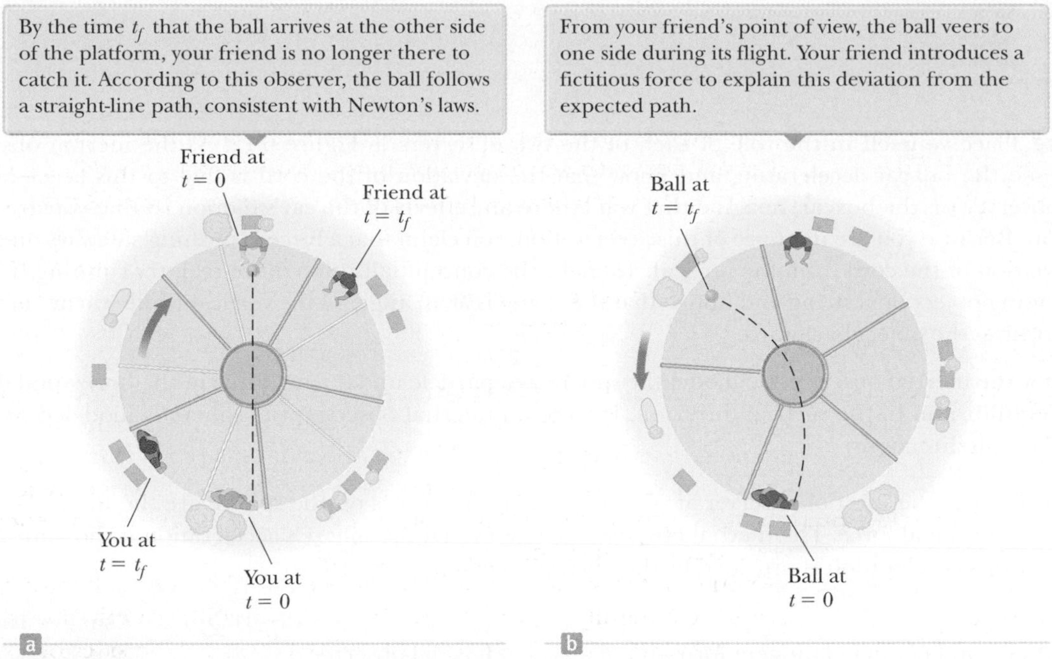

ACTIVE FIGURE 6.11

You and your friend stand at the edge of a rotating circular platform. You throw the ball at $t = 0$ in the direction of your friend. (a) Overhead view observed by someone in an inertial reference frame attached to the Earth. The ground appears stationary, and the platform rotates clockwise. (b) Overhead view observed by someone in an inertial reference frame attached to the platform. The platform appears stationary, and the ground rotates counterclockwise.

Example 6.7 Fictitious Forces in Linear Motion

A small sphere of mass m hangs by a cord from the ceiling of a boxcar that is accelerating to the right as shown in Figure 6.12. Both the inertial observer on the ground in Figure 6.12a and the noninertial observer on the train in Figure 6.12b agree that the cord makes an angle θ with respect to the vertical. The noninertial observer claims that a force, which we know to be fictitious, causes the observed deviation of the cord from the vertical. How is the magnitude of this force related to the boxcar's acceleration measured by the inertial observer in Figure 6.12a?

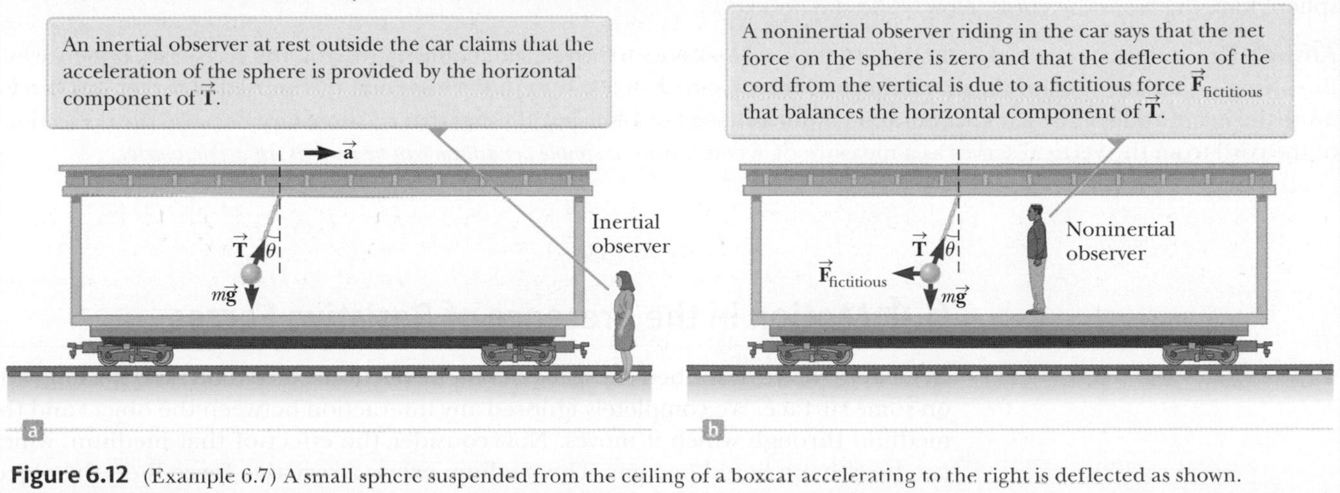

Figure 6.12 (Example 6.7) A small sphere suspended from the ceiling of a boxcar accelerating to the right is deflected as shown.

continued

6.7 cont.

SOLUTION

Conceptualize Place yourself in the role of each of the two observers in Figure 6.12. As the inertial observer on the ground, you see the boxcar accelerating and know that the deviation of the cord is due to this acceleration. As the noninertial observer on the boxcar, imagine that you ignore any effects of the car's motion so that you are not aware of its acceleration. Because you are unaware of this acceleration, you claim that a force is pushing sideways on the sphere to cause the deviation of the cord from the vertical. To make the conceptualization more real, try running from rest while holding a hanging object on a string and notice that the string is at an angle to the vertical while you are accelerating, as if a force is pushing the object backward.

Categorize For the inertial observer, we model the sphere as a particle under a net force in the horizontal direction and a particle in equilibrium in the vertical direction. For the noninertial observer, the sphere is modeled as a particle in equilibrium in both directions.

Analyze According to the inertial observer at rest (Fig. 6.12a), the forces on the sphere are the force $\vec{\mathbf{T}}$ exerted by the cord and the gravitational force. The inertial observer concludes that the sphere's acceleration is the same as that of the boxcar and that this acceleration is provided by the horizontal component of $\vec{\mathbf{T}}$.

Apply Newton's second law in component form to the sphere according to the inertial observer:

Inertial observer
$$\begin{cases} (1)\ \sum F_x = T\sin\theta = ma \\ (2)\ \sum F_y = T\cos\theta - mg = 0 \end{cases}$$

According to the noninertial observer riding in the car (Fig. 6.12b), the cord also makes an angle θ with the vertical; to that observer, however, the sphere is at rest and so its acceleration is zero. Therefore, the noninertial observer introduces a force (which we know to be fictitious) in the horizontal direction to balance the horizontal component of $\vec{\mathbf{T}}$ and claims that the net force on the sphere is zero.

Apply Newton's second law in component form to the sphere according to the noninertial observer:

Noninertial observer
$$\begin{cases} \sum F'_x = T\sin\theta - F_{\text{fictitious}} = 0 \\ \sum F'_y = T\cos\theta - mg = 0 \end{cases}$$

These expressions are equivalent to Equations (1) and (2) if $F_{\text{fictitious}} = ma$, where a is the acceleration according to the inertial observer.

Finalize If we make this substitution in the equation for $\Sigma F'_x$ above, we obtain the same mathematical results as the inertial observer. The physical interpretation of the cord's deflection, however, differs in the two frames of reference.

WHAT IF? Suppose the inertial observer wants to measure the acceleration of the train by means of the pendulum (the sphere hanging from the cord). How could she do so?

Answer Our intuition tells us that the angle θ the cord makes with the vertical should increase as the acceleration increases. By solving Equations (1) and (2) simultaneously for a, we find that $a = g\tan\theta$. Therefore, the inertial observer can determine the magnitude of the car's acceleration by measuring the angle θ and using that relationship. Because the deflection of the cord from the vertical serves as a measure of acceleration, *a simple pendulum can be used as an accelerometer.*

6.4 Motion in the Presence of Resistive Forces

In Chapter 5, we described the force of kinetic friction exerted on an object moving on some surface. We completely ignored any interaction between the object and the medium through which it moves. Now consider the effect of that medium, which can be either a liquid or a gas. The medium exerts a **resistive force $\vec{\mathbf{R}}$** on the object moving through it. Some examples are the air resistance associated with moving vehicles (sometimes called *air drag*) and the viscous forces that act on objects moving through a liquid. The magnitude of $\vec{\mathbf{R}}$ depends on factors such as the speed of the object, and the direction of $\vec{\mathbf{R}}$ is always opposite the direction of the object's motion relative to the medium. This direction may or may not be in the direction opposite the object's velocity according to the observer. For example, if a marble is dropped into a bottle of shampoo, the marble moves downward and the resistive force is upward, resisting the falling of the marble. In contrast, imagine the moment at which there is no wind and you are looking at a flag hanging limply on a flagpole. When a breeze begins to blow toward the right, the flag moves toward the right. In this case, the drag force on the flag from the moving air is to the right and the motion of the flag in response is also to the right, the *same* direction as the drag force. Because the air moves toward the right with respect to the flag, the flag moves to the left relative to the air. Therefore, the direction of the drag force is indeed opposite to the direction of the motion of the flag with respect to the air!

The magnitude of the resistive force can depend on speed in a complex way, and here we consider only two simplified models. In the first model, we assume the resistive force is proportional to the speed of the moving object; this model is valid for objects falling slowly through a liquid and for very small objects, such as dust particles, moving through air. In the second model, we assume a resistive force that is proportional to the square of the speed of the moving object; large objects, such as skydivers moving through air in free fall, experience such a force.

Model 1: Resistive Force Proportional to Object Velocity

If we model the resistive force acting on an object moving through a liquid or gas as proportional to the object's velocity, the resistive force can be expressed as

$$\vec{\mathbf{R}} = -b\vec{\mathbf{v}} \tag{6.2}$$

where b is a constant whose value depends on the properties of the medium and on the shape and dimensions of the object and $\vec{\mathbf{v}}$ is the velocity of the object relative to the medium. The negative sign indicates that $\vec{\mathbf{R}}$ is in the opposite direction to $\vec{\mathbf{v}}$.

Consider a small sphere of mass m released from rest in a liquid as in Active Figure 6.13a. Assuming the only forces acting on the sphere are the resistive force $\vec{\mathbf{R}} = -b\vec{\mathbf{v}}$ and the gravitational force $\vec{\mathbf{F}}_g$, let us describe its motion.[1] Applying Newton's second law to the vertical motion, choosing the downward direction to be positive, and noting that $\Sigma F_y = mg - bv$, we obtain

$$mg - bv = ma \tag{6.3}$$

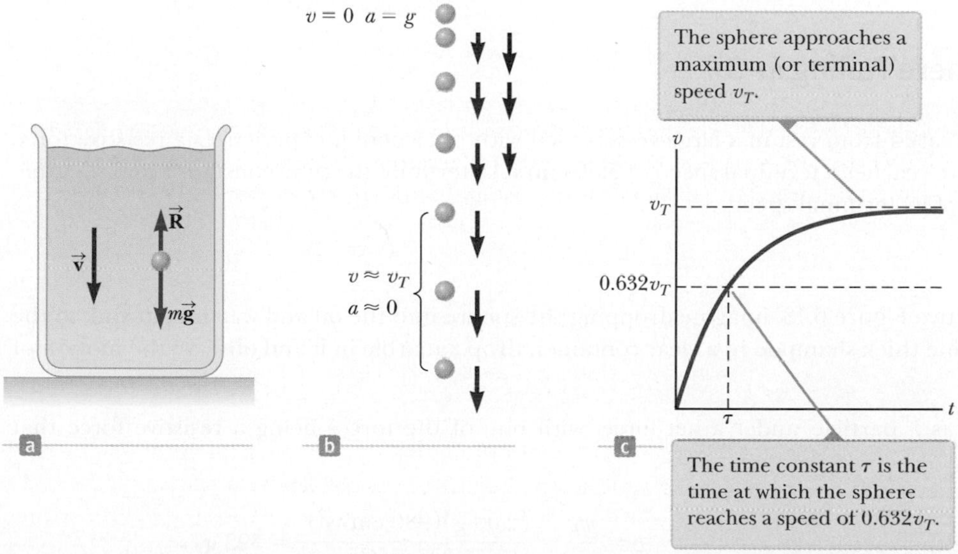

ACTIVE FIGURE 6.13

(a) A small sphere falling through a liquid. (b) A motion diagram of the sphere as it falls. Velocity vectors (red) and acceleration vectors (violet) are shown for each image after the first one. (c) A speed–time graph for the sphere.

[1]A *buoyant force* is also acting on the submerged object. This force is constant, and its magnitude is equal to the weight of the displaced liquid. This force changes the apparent weight of the sphere by a constant factor, so we will ignore the force here. We will discuss buoyant forces in Chapter 14.

where the acceleration of the sphere is downward. Noting that the acceleration a is equal to dv/dt gives

$$\frac{dv}{dt} = g - \frac{b}{m}v \tag{6.4}$$

This equation is called a *differential equation,* and the methods of solving it may not be familiar to you as yet. Notice, however, that initially when $v = 0$, the magnitude of the resistive force is also zero and the acceleration of the sphere is simply g. As t increases, the magnitude of the resistive force increases and the acceleration decreases. The acceleration approaches zero when the magnitude of the resistive force approaches the sphere's weight. In this situation, the speed of the sphere approaches its **terminal speed** v_T.

Terminal speed ▶

The terminal speed is obtained from Equation 6.4 by setting $dv/dt = 0$, which gives

$$mg - bv_T = 0 \quad \text{or} \quad v_T = \frac{mg}{b}$$

Because you may not be familiar with differential equations yet, we won't show the details of the solution that gives the expression for v for all times t. If $v = 0$ at $t = 0$, this expression is

$$v = \frac{mg}{b}(1 - e^{-bt/m}) = v_T(1 - e^{-t/\tau}) \tag{6.5}$$

This function is plotted in Active Figure 6.13c. The symbol e represents the base of the natural logarithm and is also called *Euler's number:* $e = 2.718\ 28$. The **time constant** $\tau = m/b$ (Greek letter tau) is the time at which the sphere released from rest at $t = 0$ reaches 63.2% of its terminal speed; when $t = \tau$, Equation 6.5 yields $v = 0.632v_T$. (The number 0.632 is $1 - e^{-1}$.)

We can check that Equation 6.5 is a solution to Equation 6.4 by direct differentiation:

$$\frac{dv}{dt} = \frac{d}{dt}\left[\frac{mg}{b}(1 - e^{-bt/m})\right] = \frac{mg}{b}\left(0 + \frac{b}{m}e^{-bt/m}\right) = ge^{-bt/m}$$

(See Appendix Table B.4 for the derivative of e raised to some power.) Substituting into Equation 6.4 both this expression for dv/dt and the expression for v given by Equation 6.5 shows that our solution satisfies the differential equation.

Example 6.8 Sphere Falling in Oil

A small sphere of mass 2.00 g is released from rest in a large vessel filled with oil, where it experiences a resistive force proportional to its speed. The sphere reaches a terminal speed of 5.00 cm/s. Determine the time constant τ and the time at which the sphere reaches 90.0% of its terminal speed.

SOLUTION

Conceptualize With the help of Active Figure 6.13, imagine dropping the sphere into the oil and watching it sink to the bottom of the vessel. If you have some thick shampoo in a clear container, drop a marble in it and observe the motion of the marble.

Categorize We model the sphere as a particle under a net force, with one of the forces being a resistive force that depends on the speed of the sphere.

Analyze From $v_T = mg/b$, evaluate the coefficient b:

$$b = \frac{mg}{v_T} = \frac{(2.00\text{ g})(980\text{ cm/s}^2)}{5.00\text{ cm/s}} = 392\text{ g/s}$$

Evaluate the time constant τ:

$$\tau = \frac{m}{b} = \frac{2.00\text{ g}}{392\text{ g/s}} = \boxed{5.10 \times 10^{-3}\text{ s}}$$

6.8 *cont.*

Find the time t at which the sphere reaches a speed of $0.900v_T$ by setting $v = 0.900v_T$ in Equation 6.5 and solving for t:

$$0.900v_T = v_T(1 - e^{-t/\tau})$$
$$1 - e^{-t/\tau} = 0.900$$
$$e^{-t/\tau} = 0.100$$
$$-\frac{t}{\tau} = \ln(0.100) = -2.30$$
$$t = 2.30\tau = 2.30(5.10 \times 10^{-3}\text{ s}) = 11.7 \times 10^{-3}\text{ s}$$
$$= \boxed{11.7\text{ ms}}$$

Finalize The sphere reaches 90.0% of its terminal speed in a very short time interval. You should have also seen this behavior if you performed the activity with the marble and the shampoo. Because of the short time interval required to reach terminal velocity, you may not have noticed the time interval at all. The marble may have appeared to immediately begin moving through the shampoo at a constant velocity.

Model 2: Resistive Force Proportional to Object Speed Squared

For objects moving at high speeds through air, such as airplanes, skydivers, cars, and baseballs, the resistive force is reasonably well modeled as proportional to the square of the speed. In these situations, the magnitude of the resistive force can be expressed as

$$R = \tfrac{1}{2}D\rho Av^2 \quad \textbf{(6.6)}$$

where D is a dimensionless empirical quantity called the *drag coefficient,* ρ is the density of air, and A is the cross-sectional area of the moving object measured in a plane perpendicular to its velocity. The drag coefficient has a value of about 0.5 for spherical objects but can have a value as great as 2 for irregularly shaped objects.

Let us analyze the motion of a falling object subject to an upward air resistive force of magnitude $R = \frac{1}{2}D\rho Av^2$. Suppose an object of mass m is released from rest. As Figure 6.14 shows, the object experiences two external forces:[2] the downward gravitational force $\vec{\mathbf{F}}_g = m\vec{\mathbf{g}}$ and the upward resistive force $\vec{\mathbf{R}}$. Hence, the magnitude of the net force is

$$\sum F = mg - \tfrac{1}{2}D\rho Av^2 \quad \textbf{(6.7)}$$

where we have taken downward to be the positive vertical direction. Using the force in Equation 6.7 in Newton's second law, we find that the object has a downward acceleration of magnitude

$$a = g - \left(\frac{D\rho A}{2m}\right)v^2 \quad \textbf{(6.8)}$$

We can calculate the terminal speed v_T by noticing that when the gravitational force is balanced by the resistive force, the net force on the object is zero and therefore its acceleration is zero. Setting $a = 0$ in Equation 6.8 gives

$$g - \left(\frac{D\rho A}{2m}\right)v_T^{\,2} = 0$$

so

$$v_T = \sqrt{\frac{2mg}{D\rho A}} \quad \textbf{(6.9)}$$

Table 6.1 (page 154) lists the terminal speeds for several objects falling through air.

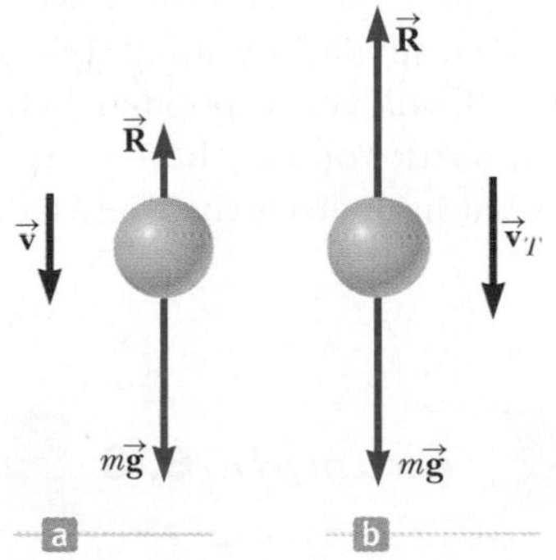

Figure 6.14 (a) An object falling through air experiences a resistive force $\vec{\mathbf{R}}$ and a gravitational force $\vec{\mathbf{F}}_g = m\vec{\mathbf{g}}$. (b) The object reaches terminal speed when the net force acting on it is zero, that is, when $\vec{\mathbf{R}} = -\vec{\mathbf{F}}_g$ or $R = mg$.

[2]As with Model 1, there is also an upward buoyant force that we neglect.

TABLE 6.1 *Terminal Speed for Various Objects Falling Through Air*

Object	Mass (kg)	Cross-Sectional Area (m^2)	v_T (m/s)
Skydiver	75	0.70	60
Baseball (radius 3.7 cm)	0.145	4.2×10^{-3}	43
Golf ball (radius 2.1 cm)	0.046	1.4×10^{-3}	44
Hailstone (radius 0.50 cm)	4.8×10^{-4}	7.9×10^{-5}	14
Raindrop (radius 0.20 cm)	3.4×10^{-5}	1.3×10^{-5}	9.0

***Quick Quiz* 6.4** A baseball and a basketball, having the same mass, are dropped through air from rest such that their bottoms are initially at the same height above the ground, on the order of 1 m or more. Which one strikes the ground first? **(a)** The baseball strikes the ground first. **(b)** The basketball strikes the ground first. **(c)** Both strike the ground at the same time.

Conceptual Example 6.9 The Skysurfer

Consider a skysurfer (Fig. 6.15) who jumps from a plane with her feet attached firmly to her surfboard, does some tricks, and then opens her parachute. Describe the forces acting on her during these maneuvers.

Jump Run Productions/The Image Bank/Getty Images

Figure 6.15 (Conceptual Example 6.9) A skysurfer.

SOLUTION

When the surfer first steps out of the plane, she has no vertical velocity. The downward gravitational force causes her to accelerate toward the ground. As her downward speed increases, so does the upward resistive force exerted by the air on her body and the board. This upward force reduces their acceleration, and so their speed increases more slowly. Eventually, they are going so fast that the upward resistive force matches the downward gravitational force. Now the net force is zero and they no longer accelerate, but instead reach their terminal speed. At some point after reaching terminal speed, she opens her parachute, resulting in a drastic increase in the upward resistive force. The net force (and thus the acceleration) is now upward, in the direction opposite the direction of the velocity. The downward velocity therefore decreases rapidly, and the resistive force on the parachute also decreases. Eventually, the upward resistive force and the downward gravitational force balance each other and a much smaller terminal speed is reached, permitting a safe landing.

(Contrary to popular belief, the velocity vector of a skydiver never points upward. You may have seen a video in which a skydiver appears to "rocket" upward once the parachute opens. In fact, what happens is that the skydiver slows down but the person holding the camera continues falling at high speed.)

Example 6.10 Falling Coffee Filters

The dependence of resistive force on the square of the speed is a model. Let's test the model for a specific situation. Imagine an experiment in which we drop a series of bowl-shaped, pleated coffee filters and measure their terminal speeds. Table 6.2 presents typical terminal speed data from a real experiment using these coffee filters as they fall through the air. The time constant τ is small, so a dropped filter quickly reaches terminal speed. Each filter has a mass of 1.64 g. When the filters are nested together, they combine in such a way that the front-facing surface area does not increase. Determine the relationship between the resistive force exerted by the air and the speed of the falling filters.

6.10 *cont.*

SOLUTION

Conceptualize Imagine dropping the coffee filters through the air. (If you have some coffee filters, try dropping them.) Because of the relatively small mass of the coffee filter, you probably won't notice the time interval during which there is an acceleration. The filters will appear to fall at constant velocity immediately upon leaving your hand.

Categorize Because a filter moves at constant velocity, we model it as a particle in equilibrium.

Analyze At terminal speed, the upward resistive force on the filter balances the downward gravitational force so that $R = mg$.

Evaluate the magnitude of the resistive force: $$R = mg = (1.64\text{ g})\left(\frac{1\text{ kg}}{1\,000\text{ g}}\right)(9.80\text{ m/s}^2) = 0.016\,1\text{ N}$$

Likewise, two filters nested together experience 0.032 2 N of resistive force, and so forth. These values of resistive force are shown in the far right column of Table 6.2. A graph of the resistive force on the filters as a function of terminal speed is shown in Figure 6.16a. A straight line is not a good fit, indicating that the resistive force is *not* proportional to the speed. The behavior is more clearly seen in Figure 6.16b, in which the resistive force is plotted as a function of the square of the terminal speed. This graph indicates that the resistive force is proportional to the *square* of the speed as suggested by Equation 6.6.

Finalize Here is a good opportunity for you to take some actual data at home on real coffee filters and see if you can reproduce the results shown in Figure 6.16. If you have shampoo and a marble as mentioned in Example 6.8, take data on that system too and see if the resistive force is appropriately modeled as being proportional to the speed.

TABLE 6.2 *Terminal Speed and Resistive Force for Nested Coffee Filters*

Number of Filters	v_T (m/s)[a]	R (N)
1	1.01	0.016 1
2	1.40	0.032 2
3	1.63	0.048 3
4	2.00	0.064 4
5	2.25	0.080 5
6	2.40	0.096 6
7	2.57	0.112 7
8	2.80	0.128 8
9	3.05	0.144 9
10	3.22	0.161 0

[a]All values of v_T are approximate.

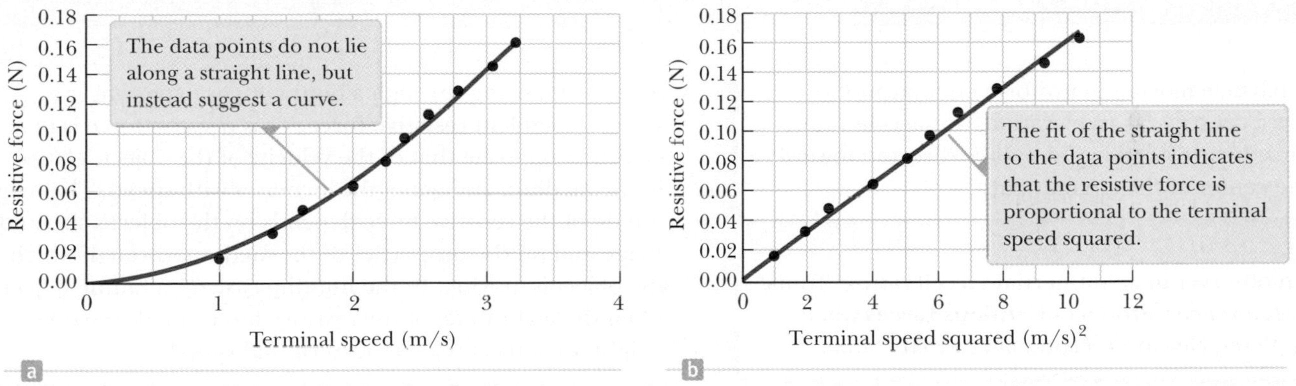

Figure 6.16 (Example 6.10) (a) Relationship between the resistive force acting on falling coffee filters and their terminal speed. (b) Graph relating the resistive force to the square of the terminal speed.

Example 6.11 Resistive Force Exerted on a Baseball

A pitcher hurls a 0.145-kg baseball past a batter at 40.2 m/s (= 90 mi/h). Find the resistive force acting on the ball at this speed.

SOLUTION

Conceptualize This example is different from the previous ones in that the object is now moving horizontally through the air instead of moving vertically under the influence of gravity and the resistive force. The resistive force causes the

continued

6.11 *cont.*

ball to slow down, and gravity causes its trajectory to curve downward. We simplify the situation by assuming the velocity vector is exactly horizontal at the instant it is traveling at 40.2 m/s.

Categorize In general, the ball is a particle under a net force. Because we are considering only one instant of time, however, we are not concerned about acceleration, so the problem involves only finding the value of one of the forces.

Analyze To determine the drag coefficient D, imagine that we drop the baseball and allow it to reach terminal speed. Solve Equation 6.9 for D and substitute the appropriate values for m, v_T, and A from Table 6.1, taking the density of air as 1.20 kg/m^3:

$$D = \frac{2mg}{v_T^{\,2}\rho A} = \frac{2(0.145\text{ kg})(9.80\text{ m/s}^2)}{(43\text{ m/s})^2(1.20\text{ kg/m}^3)(4.2\times10^{-3}\text{ m}^2)}$$

$$= 0.305$$

Use this value for D in Equation 6.6 to find the magnitude of the resistive force:

$$R = \tfrac{1}{2}D\rho A v^2$$

$$= \tfrac{1}{2}(0.305)(1.20\text{ kg/m}^3)(4.2\times10^{-3}\text{ m}^2)(40.2\text{ m/s})^2$$

$$= \boxed{1.2\text{ N}}$$

Finalize The magnitude of the resistive force is similar in magnitude to the weight of the baseball, which is about 1.4 N. Therefore, air resistance plays a major role in the motion of the ball, as evidenced by the variety of curve balls, floaters, sinkers, and the like thrown by baseball pitchers.

Summary

Concepts and Principles

A particle moving in uniform circular motion has a centripetal acceleration; this acceleration must be provided by a net force directed toward the center of the circular path.

An observer in a noninertial (accelerating) frame of reference introduces **fictitious forces** when applying Newton's second law in that frame.

An object moving through a liquid or gas experiences a speed-dependent **resistive force.** This resistive force is in a direction opposite that of the velocity of the object relative to the medium and generally increases with speed. The magnitude of the resistive force depends on the object's size and shape and on the properties of the medium through which the object is moving. In the limiting case for a falling object, when the magnitude of the resistive force equals the object's weight, the object reaches its **terminal speed.**

Analysis Model for Problem-Solving

Particle in Uniform Circular Motion (Extension) With our new knowledge of forces, we can extend the model of a particle in uniform circular motion, first introduced in Chapter 4. Newton's second law applied to a particle moving in uniform circular motion states that the net force causing the particle to undergo a centripetal acceleration (Eq. 4.15) is related to the acceleration according to

$$\sum F = ma_c = m\frac{v^2}{r} \qquad \textbf{(6.1)}$$

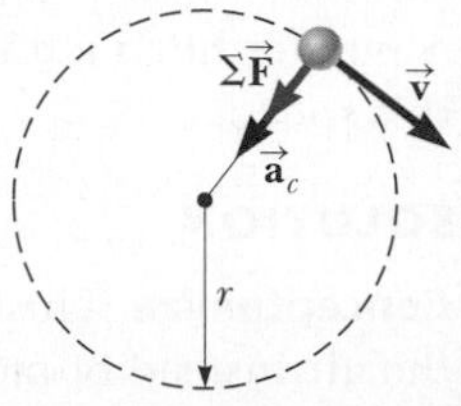

Objective Questions

☐ denotes answer available in Student Solutions Manual/Study Guide

1. A door in a hospital has a pneumatic closer that pulls the door shut such that the doorknob moves with constant speed over most of its path. In this part of its motion, (a) does the doorknob experience a centripetal acceleration? (b) Does it experience a tangential acceleration?

2. An office door is given a sharp push and swings open against a pneumatic device that slows the door down and then reverses its motion. At the moment the door is open the widest, (a) does the doorknob have a centripetal acceleration? (b) Does it have a tangential acceleration?

[3.] A child is practicing for a BMX race. His speed remains constant as he goes counterclockwise around a level track with two straight sections and two nearly semicircular sections as shown in the aerial view of Figure OQ6.3. (a) Rank the magnitudes of his acceleration at the points *A*, *B*, *C*, *D*, and *E* from largest to smallest. If his acceleration is the same size at two points, display that fact in your ranking. If his acceleration is zero, display that fact. (b) What are the directions of his velocity at points *A*, *B*, and *C*? For each point, choose one: north, south, east, west, or nonexistent. (c) What are the directions of his acceleration at points *A*, *B*, and *C*?

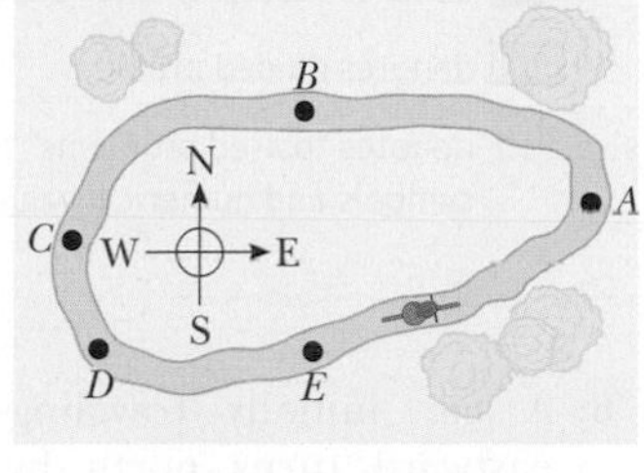

Figure OQ6.3

4. A pendulum consists of a small object called a bob hanging from a light cord of fixed length, with the top end of the cord fixed, as represented in Figure OQ6.4. The bob moves without friction, swinging equally high on both sides. It moves from its turning point *A* through point *B* and reaches its maximum speed at point *C*. (a) Of these points, is there a point where the bob has nonzero radial acceleration and zero tangential acceleration? If so, which point? What is the direction of its total acceleration at this point? (b) Of these points, is there a point where the bob has nonzero tangential acceleration and zero radial acceleration? If so, which point? What is the direction of its total acceleration at this point? (c) Is there a point where the bob has no acceleration? If so, which point? (d) Is there a point where the bob has both nonzero tangential and radial acceleration? If so, which point? What is the direction of its total acceleration at this point?

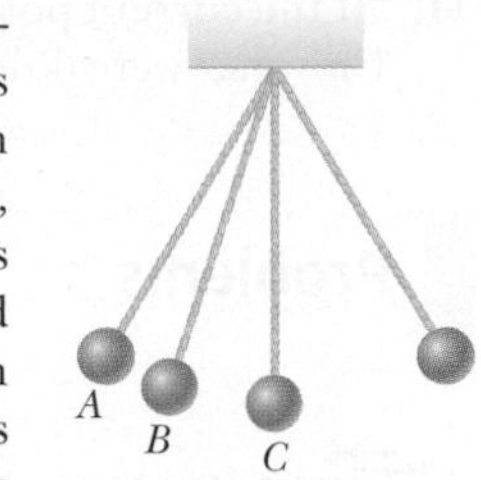

Figure OQ6.4

5. Before takeoff on an airplane, an inquisitive student on the plane dangles an iPod by its earphone wire. It hangs straight down as the plane is at rest waiting to take off. The plane then gains speed rapidly as it moves down the runway. **(i)** Relative to the student's hand, does the iPod (a) shift toward the front of the plane, (b) continue to hang straight down, or (c) shift toward the back of the plane? **(ii)** The speed of the plane increases at a constant rate over a time interval of several seconds. During this interval, does the angle the earphone wire makes with the vertical (a) increase, (b) stay constant, or (c) decrease?

6. Consider a skydiver who has stepped from a helicopter and is falling through air. Before she reaches terminal speed and long before she opens her parachute, does her speed (a) increase, (b) decrease, or (c) stay constant?

7. As a raindrop falls through the atmosphere, its speed initially changes as it falls toward the Earth. Before the raindrop reaches its terminal speed, does the magnitude of its acceleration (a) increase, (b) decrease, (c) stay constant at zero, (d) stay constant at 9.80 m/s^2, or (e) stay constant at some other value?

Conceptual Questions

☐ denotes answer available in Student Solutions Manual/Study Guide

1. The observer in the accelerating elevator of Example 5.8 would claim that the "weight" of the fish is *T*, the scale reading, but this answer is obviously wrong. Why does this observation differ from that of a person outside the elevator, at rest with respect to the Earth?

2. Describe the path of a moving body in the event that (a) its acceleration is constant in magnitude at all times and perpendicular to the velocity, and (b) its acceleration is constant in magnitude at all times and parallel to the velocity.

3. An object executes circular motion with constant speed whenever a net force of constant magnitude acts perpendicular to the velocity. What happens to the speed if the force is not perpendicular to the velocity?

[4.] A falling skydiver reaches terminal speed with her parachute closed. After the parachute is opened, what parameters change to decrease this terminal speed?

5. What forces cause (a) an automobile, (b) a propeller-driven airplane, and (c) a rowboat to move?

6. Consider a small raindrop and a large raindrop falling through the atmosphere. (a) Compare their terminal speeds. (b) What are their accelerations when they reach terminal speed?

[7.] It has been suggested that rotating cylinders about 20 km in length and 8 km in diameter be placed in space and used as colonies. The purpose of the rotation is to simulate gravity for the inhabitants. Explain this concept for producing an effective imitation of gravity.

8. If someone told you that astronauts are weightless in orbit because they are beyond the pull of gravity, would you accept the statement? Explain.

[9.] Why does a pilot tend to black out when pulling out of a steep dive?

10. A pail of water can be whirled in a vertical path such that no water is spilled. Why does the water stay in the pail, even when the pail is above your head?

11. "If the current position and velocity of every particle in the Universe were known, together with the laws describing the forces that particles exert on one another, the whole future of the Universe could be calculated. The future is determinate and preordained. Free will is an illusion." Do you agree with this thesis? Argue for or against it.

Problems

ENHANCED WebAssign The problems found in this chapter may be assigned online in Enhanced WebAssign

1. denotes straightforward problem; **2.** denotes intermediate problem;
3. denotes challenging problem

1. full solution available in the Student Solutions Manual/Study Guide

1. denotes problems most often assigned in Enhanced WebAssign; these provide students with targeted feedback and either a Master It tutorial or a Watch It solution video.

QC denotes asking for quantitative and conceptual reasoning

S denotes symbolic reasoning problem

M denotes Master It tutorial available in Enhanced WebAssign

GP denotes guided problem

shaded denotes "paired problems" that develop reasoning with symbols and numerical values

Section 6.1 Extending the Particle in Uniform Circular Motion Model

1. **M** A light string can support a stationary hanging load of 25.0 kg before breaking. An object of mass m = 3.00 kg attached to the string rotates on a frictionless, horizontal table in a circle of radius r = 0.800 m, and the other end of the string is held fixed as in Figure P6.1. What range of speeds can the object have before the string breaks?

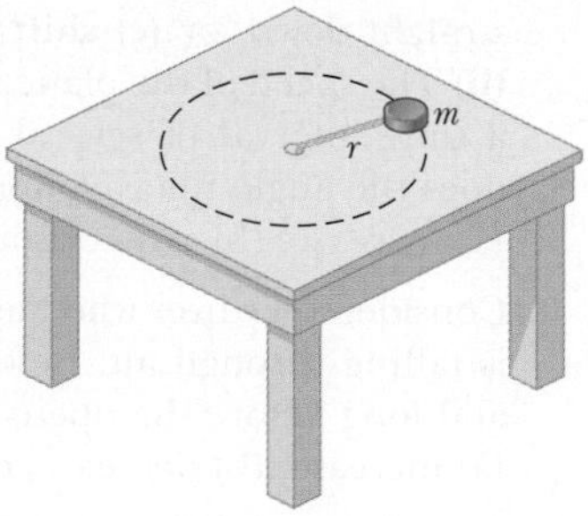

Figure P6.1

2. A curve in a road forms part of a horizontal circle. As a car goes around it at constant speed 14.0 m/s, the total horizontal force on the driver has magnitude 130 N. What is the total horizontal force on the driver if the speed on the same curve is 18.0 m/s instead?

3. In the Bohr model of the hydrogen atom, an electron moves in a circular path around a proton. The speed of the electron is approximately 2.20×10^6 m/s. Find (a) the force acting on the electron as it revolves in a circular orbit of radius 0.530×10^{-10} m and (b) the centripetal acceleration of the electron.

4. Whenever two *Apollo* astronauts were on the surface of the Moon, a third astronaut orbited the Moon. Assume the orbit to be circular and 100 km above the surface of the Moon, where the acceleration due to gravity is 1.52 m/s^2. The radius of the Moon is 1.70×10^6 m. Determine (a) the astronaut's orbital speed and (b) the period of the orbit.

5. In a cyclotron (one type of particle accelerator), a deuteron (of mass 2.00 u) reaches a final speed of 10.0% of the speed of light while moving in a circular path of radius 0.480 m. What magnitude of magnetic force is required to maintain the deuteron in a circular path?

6. A car initially traveling eastward turns north by traveling in a circular path at uniform speed as shown in Figure P6.6. The length of the arc ABC is 235 m, and the car completes the turn in 36.0 s. (a) What is the acceleration when the car is at B located at an angle of 35.0°? Express your answer in terms of the unit vectors $\hat{\mathbf{i}}$ and $\hat{\mathbf{j}}$. Determine (b) the car's average speed and (c) its average acceleration during the 36.0-s interval.

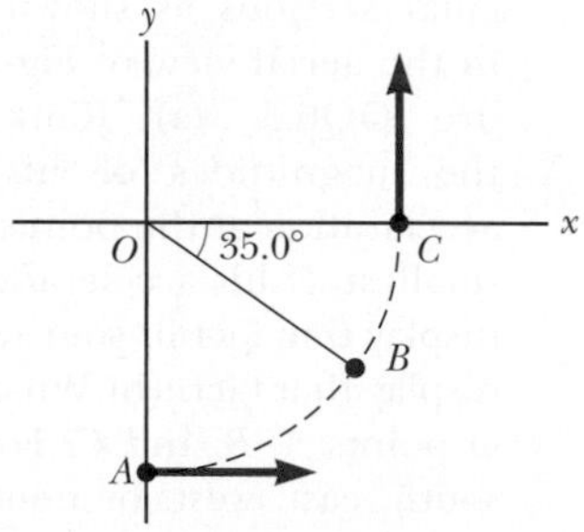

Figure P6.6

7. A space station, in the form of a wheel 120 m in diameter, rotates to provide an "artificial gravity" of 3.00 m/s^2 for persons who walk around on the inner wall of the outer rim. Find the rate of the wheel's rotation in revolutions per minute that will produce this effect.

8. Consider a conical pendulum (Fig. P6.8) with a bob of mass m = 80.0 kg on a string of length L = 10.0 m that makes an angle of θ = 5.00° with the vertical. Determine (a) the horizontal and vertical components of the force exerted by the string on the pendulum and (b) the radial acceleration of the bob.

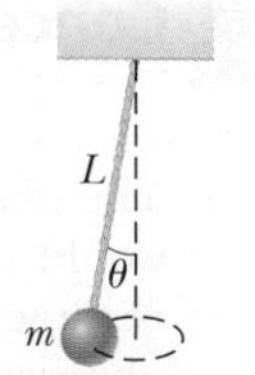

Figure P6.8

9. A crate of eggs is located in the middle of the flatbed of a pickup truck as the truck negotiates a curve in the flat road. The curve may be regarded as an arc of a circle of radius 35.0 m. If the coefficient of static friction between crate and truck is 0.600, how fast can the truck be moving without the crate sliding?

10. *Why is the following situation impossible?* The object of mass m = 4.00 kg in Figure P6.10 is attached to a ver-

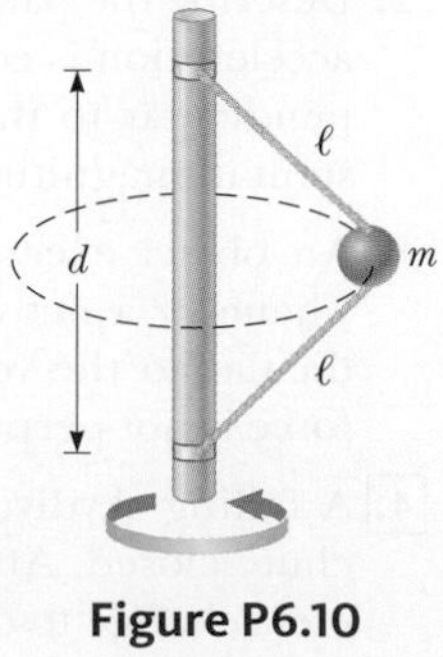

Figure P6.10

tical rod by two strings of length $\ell = 2.00$ m. The strings are attached to the rod at points a distance $d = 3.00$ m apart. The object rotates in a horizontal circle at a constant speed of $v = 3.00$ m/s, and the strings remain taut. The rod rotates along with the object so that the strings do not wrap onto the rod. **What If?** Could this situation be possible on another planet?

11. A coin placed 30.0 cm from the center of a rotating, horizontal turntable slips when its speed is 50.0 cm/s. (a) What force causes the centripetal acceleration when the coin is stationary relative to the turntable? (b) What is the coefficient of static friction between coin and turntable?

Section 6.2 Nonuniform Circular Motion

12. **Q|C** A pail of water is rotated in a vertical circle of radius 1.00 m. (a) What two external forces act on the water in the pail? (b) Which of the two forces is most important in causing the water to move in a circle? (c) What is the pail's minimum speed at the top of the circle if no water is to spill out? (d) Assume the pail with the speed found in part (c) were to suddenly disappear at the top of the circle. Describe the subsequent motion of the water. Would it differ from the motion of a projectile?

13. A hawk flies in a horizontal arc of radius 12.0 m at constant speed 4.00 m/s. (a) Find its centripetal acceleration. (b) It continues to fly along the same horizontal arc, but increases its speed at the rate of 1.20 m/s^2. Find the acceleration (magnitude and direction) in this situation at the moment the hawk's speed is 4.00 m/s.

14. A 40.0-kg child swings in a swing supported by two chains, each 3.00 m long. The tension in each chain at the lowest point is 350 N. Find (a) the child's speed at the lowest point and (b) the force exerted by the seat on the child at the lowest point. (Ignore the mass of the seat.)

15. **S** A child of mass m swings in a swing supported by two chains, each of length R. If the tension in each chain at the lowest point is T, find (a) the child's speed at the lowest point and (b) the force exerted by the seat on the child at the lowest point. (Ignore the mass of the seat.)

16. A roller-coaster car (Fig. P6.16) has a mass of 500 kg when fully loaded with passengers. The path of the coaster from its initial point shown in the figure to point Ⓑ involves only up-and-down motion (as seen by the riders), with no motion to the left or right. (a) If the vehicle has a speed of 20.0 m/s at point Ⓐ, what is the force exerted by the track on the car at this point? (b) What is the maximum speed the vehicle can have at point Ⓑ and still remain on the track? Assume the roller-coaster tracks at points Ⓐ and Ⓑ are parts of vertical circles of radius $r_1 = 10.0$ m and $r_2 = 15.0$ m, respectively.

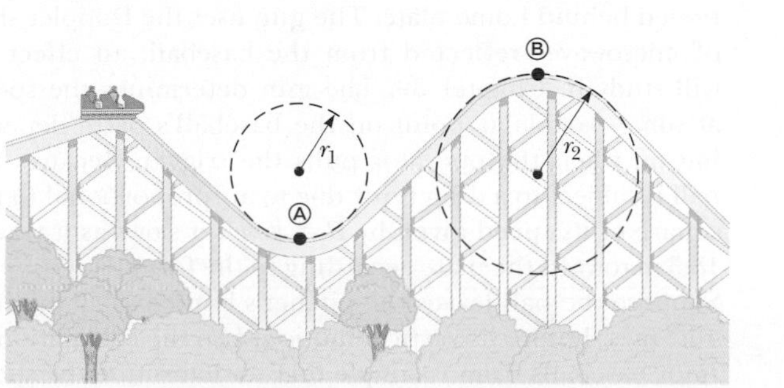

Figure P6.16 Problems 16 and 38.

17. An adventurous archeologist ($m = 85.0$ kg) tries to cross a river by swinging from a vine. The vine is 10.0 m long, and his speed at the bottom of the swing is 8.00 m/s. The archeologist doesn't know that the vine has a breaking strength of 1 000 N. Does he make it across the river without falling in?

18. **Q|C** One end of a cord is fixed and a small 0.500-kg object is attached to the other end, where it swings in a section of a vertical circle of radius 2.00 m as shown in Figure P6.18. When $\theta = 20.0°$, the speed of the object is 8.00 m/s. At this instant, find (a) the tension in the string, (b) the tangential and radial components of acceleration, and (c) the total acceleration. (d) Is your answer changed if the object is swinging down toward its lowest point instead of swinging up? (e) Explain your answer to part (d).

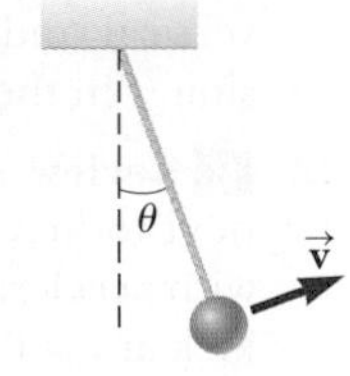

Figure P6.18

19. **Q|C** A roller coaster at the Six Flags Great America amusement park in Gurnee, Illinois, incorporates some clever design technology and some basic physics. Each vertical loop, instead of being circular, is shaped like a teardrop (Fig. P6.19). The cars ride on the inside of the loop at the top, and the speeds are fast enough to ensure the cars remain on the track. The biggest loop is 40.0 m high. Suppose the speed at the top of the loop is 13.0 m/s and the corresponding centripetal acceleration of the riders is $2g$. (a) What is the radius of the arc of the teardrop at the top? (b) If the total mass of a car plus the riders is M, what force does the rail exert on the car at the top? (c) Suppose the roller coaster had a circular loop of radius 20.0 m. If the cars have the same speed, 13.0 m/s at the top, what is the centripetal acceleration of the riders at the top? (d) Comment on the normal force at the top in the situation described in part (c) and on the advantages of having teardrop-shaped loops.

© Frank Cezus

Figure P6.19

Section 6.3 Motion in Accelerated Frames

20. An object of mass $m = 5.00$ kg, attached to a spring scale, rests on a frictionless, horizontal surface as shown in Figure P6.20. The spring scale, attached to the front end of a boxcar, reads zero when the car is at rest. (a) Determine the acceleration of the car if the spring scale has a constant reading of 18.0 N when the car is in motion. (b) What constant reading will the spring

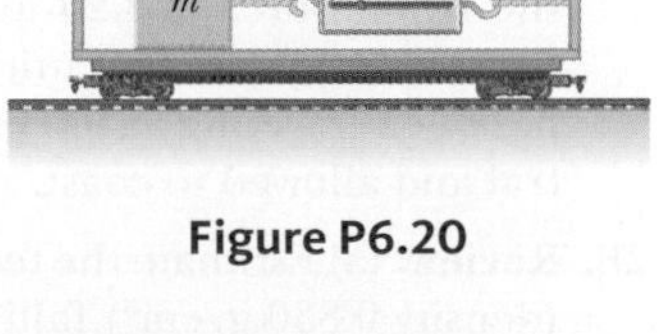

Figure P6.20

scale show if the car moves with constant velocity? Describe the forces on the object as observed (c) by someone in the car and (d) by someone at rest outside the car.

21. An object of mass $m = 0.500$ kg is suspended from the ceiling of an accelerating truck as shown in Figure P6.21. Taking $a = 3.00$ m/s^2, find (a) the angle θ that the string makes with the vertical and (b) the tension T in the string.

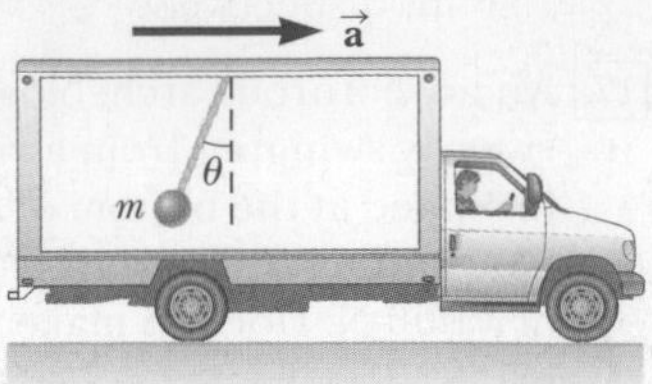

Figure P6.21

22. **S** **Review.** A student, along with her backpack on the floor next to her, are in an elevator that is accelerating upward with acceleration a. The student gives her backpack a quick kick at $t = 0$, imparting to it speed v and causing it to slide across the elevator floor. At time t, the backpack hits the opposite wall a distance L away from the student. Find the coefficient of kinetic friction μ_k between the backpack and the elevator floor.

23. A person stands on a scale in an elevator. As the elevator starts, the scale has a constant reading of 591 N. As the elevator later stops, the scale reading is 391 N. Assuming the magnitude of the acceleration is the same during starting and stopping, determine (a) the weight of the person, (b) the person's mass, and (c) the acceleration of the elevator.

24. A child lying on her back experiences 55.0 N tension in the muscles on both sides of her neck when she raises her head to look past her toes. Later, sliding feet first down a water slide at terminal speed 5.70 m/s and riding high on the outside wall of a horizontal curve of radius 2.40 m, she raises her head again to look forward past her toes. Find the tension in the muscles on both sides of her neck while she is sliding.

25. A small container of water is placed on a turntable inside a microwave oven, at a radius of 12.0 cm from the center. The turntable rotates steadily, turning one revolution in each 7.25 s. What angle does the water surface make with the horizontal?

Section 6.4 Motion in the Presence of Resistive Forces

26. A skydiver of mass 80.0 kg jumps from a slow-moving aircraft and reaches a terminal speed of 50.0 m/s. (a) What is her acceleration when her speed is 30.0 m/s? What is the drag force on the skydiver when her speed is (b) 50.0 m/s and (c) 30.0 m/s?

27. The mass of a sports car is 1 200 kg. The shape of the body is such that the aerodynamic drag coefficient is 0.250 and the frontal area is 2.20 m^2. Ignoring all other sources of friction, calculate the initial acceleration the car has if it has been traveling at 100 km/h and is now shifted into neutral and allowed to coast.

28. **Review.** (a) Estimate the terminal speed of a wooden sphere (density 0.830 g/cm^3) falling through air, taking its radius as 8.00 cm and its drag coefficient as 0.500. (b) From what height would a freely falling object reach this speed in the absence of air resistance?

29. Calculate the force required to pull a copper ball of radius 2.00 cm upward through a fluid at the constant speed 9.00 cm/s. Take the drag force to be proportional to the speed, with proportionality constant 0.950 kg/s. Ignore the buoyant force.

30. A small piece of Styrofoam packing material is dropped from a height of 2.00 m above the ground. Until it reaches terminal speed, the magnitude of its acceleration is given by $a = g - Bv$. After falling 0.500 m, the Styrofoam effectively reaches terminal speed and then takes 5.00 s more to reach the ground. (a) What is the value of the constant B? (b) What is the acceleration at $t = 0$? (c) What is the acceleration when the speed is 0.150 m/s?

31. **M** A small, spherical bead of mass 3.00 g is released from rest at $t = 0$ from a point under the surface of a viscous liquid. The terminal speed is observed to be $v_T = 2.00$ cm/s. Find (a) the value of the constant b that appears in Equation 6.2, (b) the time t at which the bead reaches $0.632v_T$, and (c) the value of the resistive force when the bead reaches terminal speed.

32. **Review.** A window washer pulls a rubber squeegee down a very tall vertical window. The squeegee has mass 160 g and is mounted on the end of a light rod. The coefficient of kinetic friction between the squeegee and the dry glass is 0.900. The window washer presses it against the window with a force having a horizontal component of 4.00 N. (a) If she pulls the squeegee down the window at constant velocity, what vertical force component must she exert? (b) The window washer increases the downward force component by 25.0%, while all other forces remain the same. Find the squeegee's acceleration in this situation. (c) The squeegee is moved into a wet portion of the window, where its motion is resisted by a fluid drag force R proportional to its velocity according to $R = -20.0v$, where R is in newtons and v is in meters per second. Find the terminal velocity that the squeegee approaches, assuming the window washer exerts the same force described in part (b).

33. **S** Assume the resistive force acting on a speed skater is proportional to the square of the skater's speed v and is given by $f = -kmv^2$, where k is a constant and m is the skater's mass. The skater crosses the finish line of a straight-line race with speed v_i and then slows down by coasting on his skates. Show that the skater's speed at any time t after crossing the finish line is $v(t) = v_i/(1 + ktv_i)$.

34. At major league baseball games, it is commonplace to flash on the scoreboard a speed for each pitch. This speed is determined with a radar gun aimed by an operator positioned behind home plate. The gun uses the Doppler shift of microwaves reflected from the baseball, an effect we will study in Chapter 39. The gun determines the speed at some particular point on the baseball's path, depending on when the operator pulls the trigger. Because the ball is subject to a drag force due to air proportional to the square of its speed given by $R = kmv^2$, it slows as it travels 18.3 m toward the plate according to the formula $v = v_ie^{-kx}$. Suppose the ball leaves the pitcher's hand at 90.0 mi/h = 40.2 m/s. Ignore its vertical motion. Use the calculation of R for baseballs from Example 6.11 to determine the speed of the pitch when the ball crosses the plate.

35. A motorboat cuts its engine when its speed is 10.0 m/s and then coasts to rest. The equation describing the motion of the motorboat during this period is $v = v_i e^{-ct}$, where v is the speed at time t, v_i is the initial speed at $t = 0$, and c is a constant. At $t = 20.0$ s, the speed is 5.00 m/s. (a) Find the constant c. (b) What is the speed at $t = 40.0$ s? (c) Differentiate the expression for $v(t)$ and thus show that the acceleration of the boat is proportional to the speed at any time.

36. You can feel a force of air drag on your hand if you stretch your arm out of the open window of a speeding car. *Note:* Do not endanger yourself. What is the order of magnitude of this force? In your solution, state the quantities you measure or estimate and their values.

Additional Problems

37. A car travels clockwise at constant speed around a circular section of a horizontal road as shown in the aerial view of Figure P6.37. Find the directions of its velocity and acceleration at (a) position Ⓐ and (b) position Ⓑ.

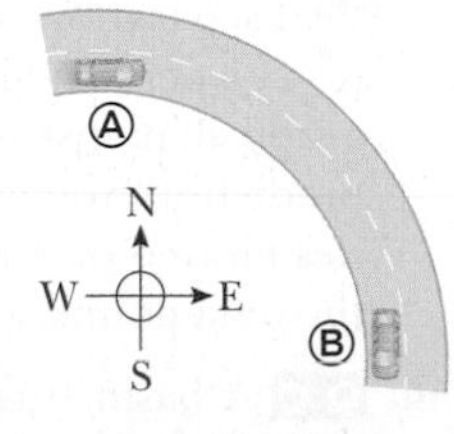

Figure P6.37

38. The mass of a roller-coaster car, including its passengers, is 500 kg. Its speed at the bottom of the track in Figure P6.16 is 19 m/s. The radius of this section of the track is $r_1 = 25$ m. Find the force that a seat in the roller-coaster car exerts on a 50-kg passenger at the lowest point.

39. A string under a tension of 50.0 N is used to whirl a rock in a horizontal circle of radius 2.50 m at a speed of 20.4 m/s on a frictionless surface as shown in Figure P6.39. As the string is pulled in, the speed of the rock increases. When the string is 1.00 m long and the speed of the rock is 51.0 m/s, the string breaks. What is the breaking strength, in newtons, of the string?

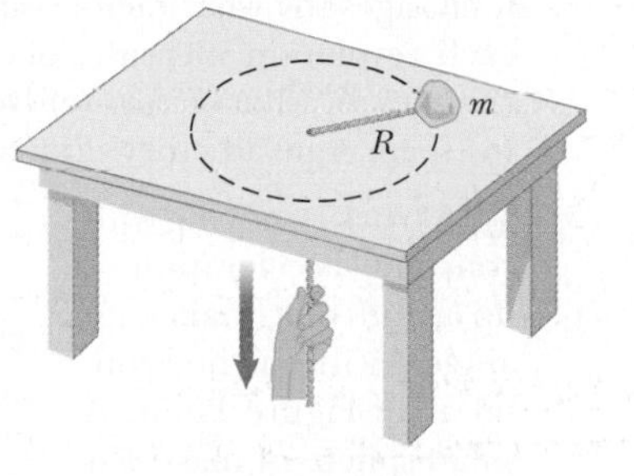

Figure P6.39

40. Disturbed by speeding cars outside his workplace, Nobel laureate Arthur Holly Compton designed a speed bump (called the "Holly hump") and had it installed. Suppose a 1 800-kg car passes over a hump in a roadway that follows the arc of a circle of radius 20.4 m as shown in Figure P6.40. (a) If the car travels at 30.0 km/h, what force does the road exert on the car as the car passes the highest point of the hump? (b) **What If?** What is the maximum speed the car can have without losing contact with the road as it passes this highest point?

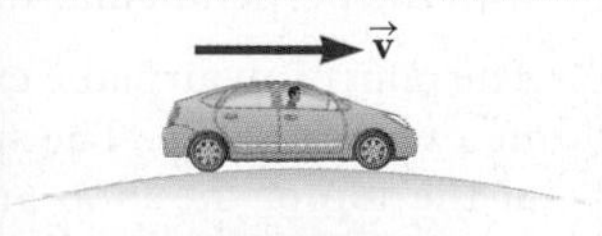

Figure P6.40 Problems 40 and 41.

41. **S** A car of mass m passes over a hump in a road that follows the arc of a circle of radius R as shown in Figure P6.40. (a) If the car travels at a speed v, what force does the road exert on the car as the car passes the highest point of the hump? (b) **What If?** What is the maximum speed the car can have without losing contact with the road as it passes this highest point?

42. **S** A child's toy consists of a small wedge that has an acute angle θ (Fig. P6.42). The sloping side of the wedge is frictionless, and an object of mass m on it remains at constant height if the wedge is spun at a certain constant speed. The wedge is spun by rotating, as an axis, a vertical rod that is firmly attached to the wedge at the bottom end. Show that, when the object sits at rest at a point at distance L up along the wedge, the speed of the object must be $v = (gL \sin\theta)^{1/2}$.

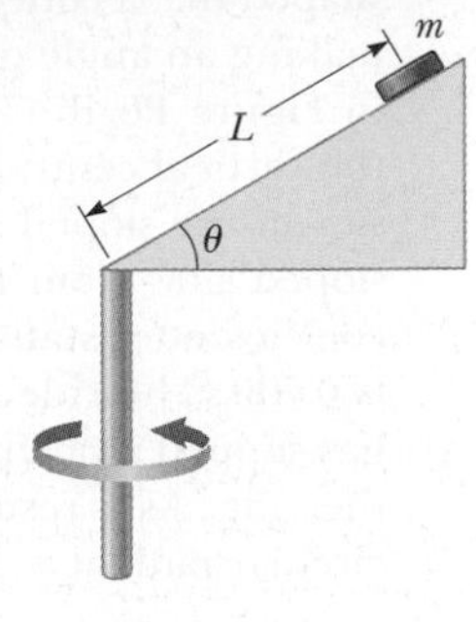

Figure P6.42

43. **S** A seaplane of total mass m lands on a lake with initial speed $v_i\hat{\mathbf{i}}$. The only horizontal force on it is a resistive force on its pontoons from the water. The resistive force is proportional to the velocity of the seaplane: $\vec{\mathbf{R}} = -b\vec{\mathbf{v}}$. Newton's second law applied to the plane is $-bv\hat{\mathbf{i}} = m(dv/dt)\hat{\mathbf{i}}$. From the fundamental theorem of calculus, this differential equation implies that the speed changes according to

$$\int_{v_i}^{v} \frac{dv}{v} = -\frac{b}{m}\int_0^t dt$$

(a) Carry out the integration to determine the speed of the seaplane as a function of time. (b) Sketch a graph of the speed as a function of time. (c) Does the seaplane come to a complete stop after a finite interval of time? (d) Does the seaplane travel a finite distance in stopping?

44. An object of mass $m_1 = 4.00$ kg is tied to an object of mass $m_2 = 3.00$ kg with String 1 of length $\ell = 0.500$ m. The combination is swung in a vertical circular path on a second string, String 2, of length $\ell = 0.500$ m. During the motion, the two strings are collinear at all times as shown in Figure P6.44. At the top of its motion, m_2 is traveling at $v = 4.00$ m/s. (a) What is the tension in String 1 at this instant? (b) What is the tension in String 2 at this instant? (c) Which string will break first if the combination is rotated faster and faster?

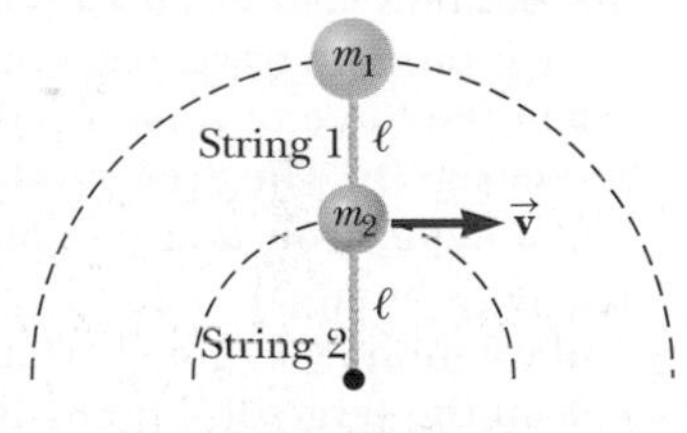

Figure P6.44

45. A ball of mass $m = 0.275$ kg swings in a vertical circular path on a string $L = 0.850$ m long as in Figure P6.45. (a) What are the forces acting on the ball at any point on the path? (b) Draw force diagrams for the ball when it is at the bottom of the circle and when it is at the top. (c) If its speed is 5.20 m/s at the top of the circle, what is the tension in the string there? (d) If the string breaks when its tension exceeds 22.5 N, what is the maximum speed the ball can have at the bottom before that happens?

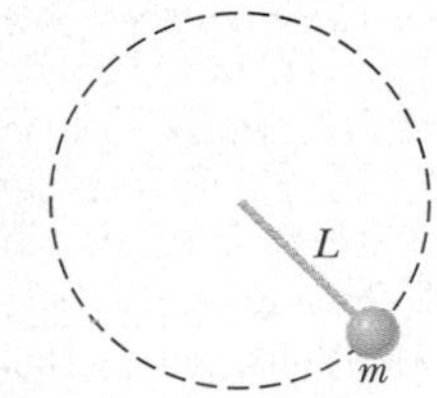

Figure P6.45

46. *Why is the following situation impossible?* A mischievous child goes to an amusement park with his family. On one ride, after a severe scolding from his mother, he slips out of his seat and climbs to the top of the ride's structure, which is shaped like a cone with its axis vertical and its sloped sides making an angle of $\theta = 20.0°$ with the horizontal as shown in Figure P6.46. This part of the structure rotates about the vertical central axis when the ride operates. The child sits on the sloped surface at a point $d = 5.32$ m down the sloped side from the center of the cone and pouts. The coefficient of static friction between the boy and the cone is 0.700. The ride operator does not notice that the child has slipped away from his seat and so continues to operate the ride. As a result, the sitting, pouting boy rotates in a circular path at a speed of 3.75 m/s.

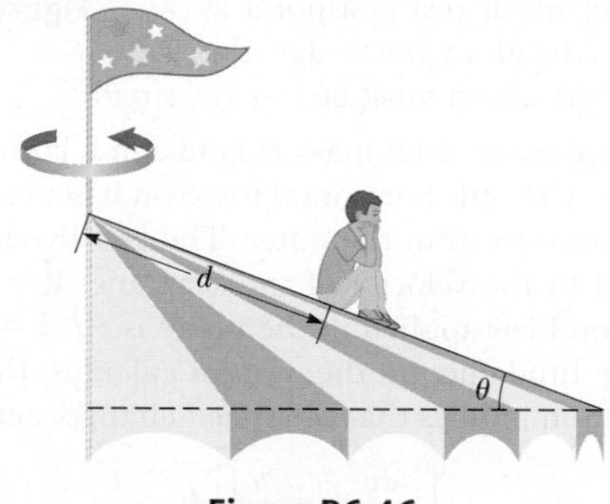

Figure P6.46

47. (a) A luggage carousel at an airport has the form of a section of a large cone, steadily rotating about its vertical axis. Its metallic surface slopes downward toward the outside, making an angle of 20.0° with the horizontal. A piece of luggage having mass 30.0 kg is placed on the carousel at a position 7.46 m measured horizontally from the axis of rotation. The travel bag goes around once in 38.0 s. Calculate the force of static friction exerted by the carousel on the bag. (b) The drive motor is shifted to turn the carousel at a higher constant rate of rotation, and the piece of luggage is bumped to another position, 7.94 m from the axis of rotation. Now going around once in every 34.0 s, the bag is on the verge of slipping down the sloped surface. Calculate the coefficient of static friction between the bag and the carousel.

48. In a home laundry dryer, a cylindrical tub containing wet clothes is rotated steadily about a horizontal axis as shown in Figure P6.48. So that the clothes will dry uniformly, they are made to tumble. The rate of rotation of the smooth-walled tub is chosen so that a small piece of cloth will lose contact with the tub when the cloth is at an angle of $\theta = 68.0°$ above the horizontal. If the radius of the tub is $r = 0.330$ m, what rate of revolution is needed?

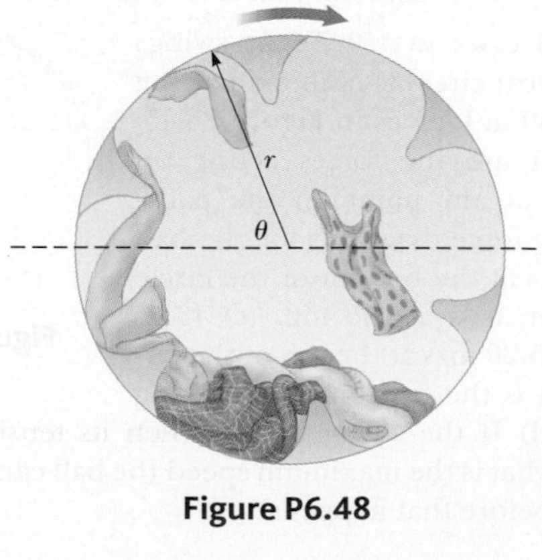

Figure P6.48

49. Interpret the graph in Figure 6.16(b), which describes the results for falling coffee filters discussed in Example 6.10. Proceed as follows. (a) Find the slope of the straight line, including its units. (b) From Equation 6.6, $R = \frac{1}{2}D\rho Av^2$, identify the theoretical slope of a graph of resistive force versus squared speed. (c) Set the experimental and theoretical slopes equal to each other and proceed to calculate the drag coefficient of the filters. Model the cross-sectional area of the filters as that of a circle of radius 10.5 cm and take the density of air to be 1.20 kg/m^3. (d) Arbitrarily choose the eighth data point on the graph and find its vertical separation from the line of best fit. Express this scatter as a percentage. (e) In a short paragraph, state what the graph demonstrates and compare it with the theoretical prediction. You will need to make reference to the quantities plotted on the axes, to the shape of the graph line, to the data points, and to the results of parts (c) and (d).

50. **Q|C** A basin surrounding a drain has the shape of a circular cone opening upward, having everywhere an angle of 35.0° with the horizontal. A 25.0-g ice cube is set sliding around the cone without friction in a horizontal circle of radius R. (a) Find the speed the ice cube must have as a function of R. (b) Is any piece of data unnecessary for the solution? Suppose R is made two times larger. (c) Will the required speed increase, decrease, or stay constant? If it changes, by what factor? (d) Will the time required for each revolution increase, decrease, or stay constant? If it changes, by what factor? (e) Do the answers to parts (c) and (d) seem contradictory? Explain.

51. **S** A truck is moving with constant acceleration a up a hill that makes an angle ϕ with the horizontal as in Figure P6.51. A small sphere of mass m is suspended from the ceiling of the truck by a light cord. If the pendulum makes a constant angle θ with the perpendicular to the ceiling, what is a?

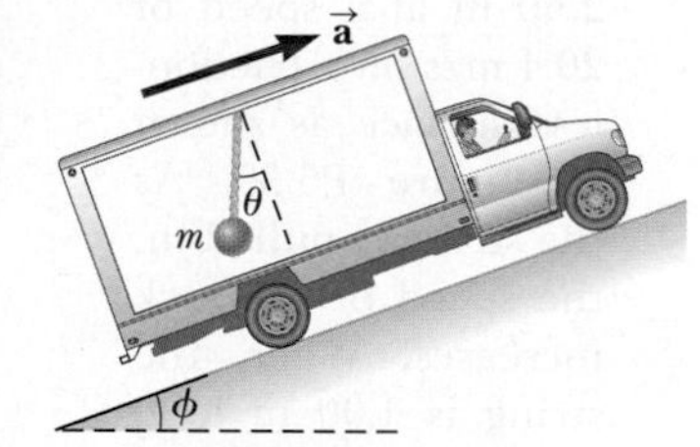

Figure P6.51

52. The pilot of an airplane executes a loop-the-loop maneuver in a vertical circle. The speed of the airplane is 300 mi/h at the top of the loop and 450 mi/h at the bottom, and the radius of the circle is 1 200 ft. (a) What is the pilot's apparent weight at the lowest point if his true weight is 160 lb? (b) What is his apparent weight at the highest point? (c) **What If?** Describe how the pilot could experience weightlessness if both the radius and the speed can be varied. *Note:* His apparent weight is equal to the magnitude of the force exerted by the seat on his body.

53. **M** Because the Earth rotates about its axis, a point on the equator experiences a centripetal acceleration of 0.033 7 m/s^2, whereas a point at the poles experiences no centripetal acceleration. If a person at the equator has a mass of 75.0 kg, calculate (a) the gravitational force (true weight) on the person and (b) the normal force (apparent

weight) on the person. (c) Which force is greater? Assume the Earth is a uniform sphere and take $g = 9.800\ \text{m/s}^2$.

54. **Q|C** **S** A puck of mass m_1 is tied to a string and allowed to revolve in a circle of radius R on a frictionless, horizontal table. The other end of the string passes through a small hole in the center of the table, and an object of mass m_2 is tied to it (Fig. P6.54). The suspended object remains in equilibrium while the puck on the tabletop revolves. Find symbolic expressions for (a) the tension in the string, (b) the radial force acting on the puck, and (c) the speed of the puck. (d) Qualitatively describe what will happen in the motion of the puck if the value of m_2 is increased by placing a small additional load on the puck. (e) Qualitatively describe what will happen in the motion of the puck if the value of m_2 is instead decreased by removing a part of the hanging load.

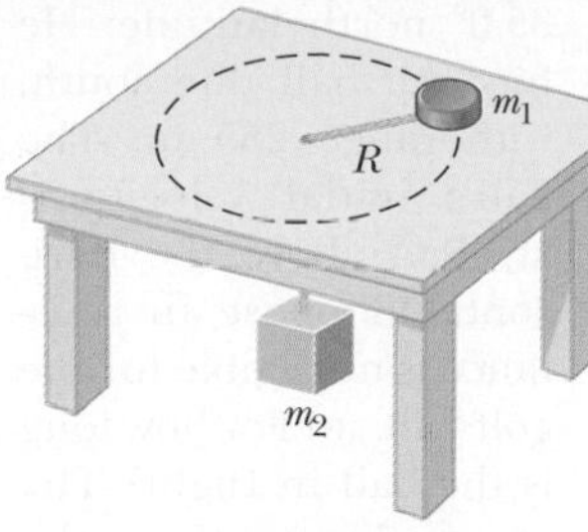

Figure P6.54

55. **Q|C** **Review.** While learning to drive, you are in a 1 200-kg car moving at 20.0 m/s across a large, vacant, level parking lot. Suddenly you realize you are heading straight toward the brick sidewall of a large supermarket and are in danger of running into it. The pavement can exert a maximum horizontal force of 7 000 N on the car. (a) Explain why you should expect the force to have a well-defined maximum value. (b) Suppose you apply the brakes and do not turn the steering wheel. Find the minimum distance you must be from the wall to avoid a collision. (c) If you do not brake but instead maintain constant speed and turn the steering wheel, what is the minimum distance you must be from the wall to avoid a collision? (d) Of the two methods in parts (b) and (c), which is better for avoiding a collision? Or should you use both the brakes and the steering wheel, or neither? Explain. (e) Does the conclusion in part (d) depend on the numerical values given in this problem, or is it true in general? Explain.

56. In Example 6.5, we investigated the forces a child experiences on a Ferris wheel. Assume the data in that example applies to this problem. What force (magnitude and direction) does the seat exert on a 40.0-kg child when the child is halfway between top and bottom?

57. Figure P6.57 shows a photo of a swing ride at an amusement park. The structure consists of a horizontal, rotating, circular platform of diameter D from which seats of mass m are suspended at the end of massless chains of length d. When the system rotates at constant speed, the chains swing outward and make an angle θ with the vertical. Consider such a ride with the following parameters: $D = 8.00$ m, $d = 2.50$ m, $m = 10.0$ kg, and $\theta = 28.0°$. (a) What is the speed of each seat? (b) Draw a diagram of forces acting on a 40.0-kg child riding in a seat and (c) find the tension in the chain.

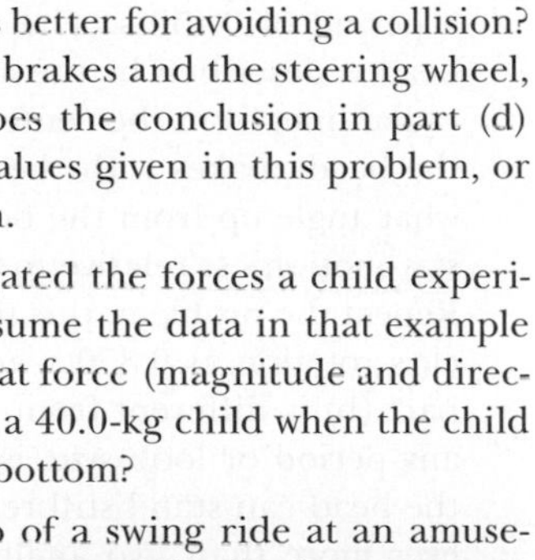

Stuart Gregory/Photographer's Choice/Getty Images

Figure P6.57

58. **GP** **S** **Review.** A piece of putty is initially located at point A on the rim of a grinding wheel rotating at constant angular speed about a horizontal axis. The putty is dislodged from point A when the diameter through A is horizontal. It then rises vertically and returns to A at the instant the wheel completes one revolution. From this information, we wish to find the speed v of the putty when it leaves the wheel and the force holding it to the wheel. (a) What analysis model is appropriate for the motion of the putty as it rises and falls? (b) Use this model to find a symbolic expression for the time interval between when the putty leaves point A and when it arrives back at A, in terms of v and g. (c) What is the appropriate analysis model to describe point A on the wheel? (d) Find the period of the motion of point A in terms of the tangential speed v and the radius R of the wheel. (e) Set the time interval from part (b) equal to the period from part (d) and solve for the speed v of the putty as it leaves the wheel. (f) If the mass of the putty is m, what is the magnitude of the force that held it to the wheel before it was released?

59. **Q|C** **S** An amusement park ride consists of a large vertical cylinder that spins about its axis fast enough that any person inside is held up against the wall when the floor drops away (Fig. P6.59). The coefficient of static friction between person and wall is μ_s, and the radius of the cylinder is R. (a) Show that the maximum period of revolution necessary to keep the person from falling is $T = (4\pi^2 R\mu_s/g)^{1/2}$. (b) If the rate of revolution of the cylinder is made to be somewhat larger, what happens to the magnitude of each one of the forces acting on the person? What happens in the motion of the person? (c) If the rate of revolution of the cylinder is instead made to be somewhat smaller, what happens to the magnitude of each one of the forces acting on the person? How does the motion of the person change?

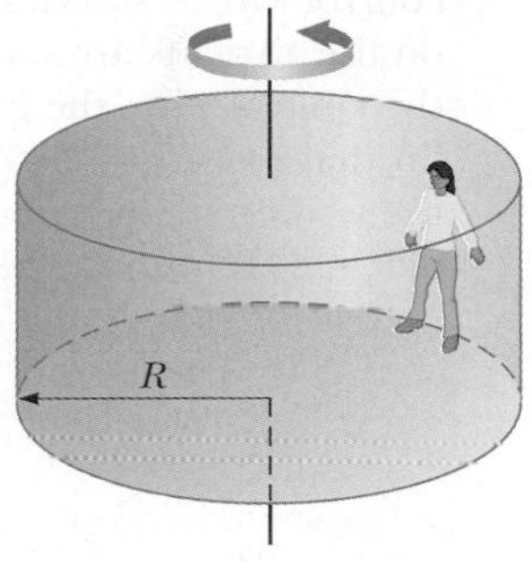

Figure P6.59

60. Members of a skydiving club were given the following data to use in planning their jumps. In the table, d is the distance fallen from rest by a skydiver in a "free-fall stable spread position" versus the time of fall t. (a) Convert the distances in feet into meters. (b) Graph d (in meters) versus t. (c) Determine the value of the terminal speed v_T by finding the slope of the straight portion of the curve. Use a least-squares fit to determine this slope.

t (s)	d (ft)	t (s)	d (ft)	t (s)	d (ft)
0	0	7	652	14	1 831
1	16	8	808	15	2 005
2	62	9	971	16	2 179
3	138	10	1 138	17	2 353
4	242	11	1 309	18	2 527
5	366	12	1 483	19	2 701
6	504	13	1 657	20	2 875

61. **S** A car rounds a banked curve as discussed in Example 6.4 and shown in Figure 6.5. The radius of curvature of the road is R, the banking angle is θ, and the coefficient of static friction is μ_s. (a) Determine the range of speeds the car can have without slipping up or down the road. (b) Find the minimum value for μ_s such that the minimum speed is zero.

62. **Q|C** **S** Galileo thought about whether acceleration should be defined as the rate of change of velocity over time or as the rate of change in velocity over distance. He chose the former, so let's use the name "vroomosity" for the rate of change of velocity over distance. For motion of a particle on a straight line with constant acceleration, the equation $v = v_i + at$ gives its velocity v as a function of time. Similarly, for a particle's linear motion with constant vroomosity k, the equation $v = v_i + kx$ gives the velocity as a function of the position x if the particle's speed is v_i at $x = 0$. (a) Find the law describing the total force acting on this object of mass m. (b) Describe an example of such a motion or explain why it is unrealistic. Consider (c) the possibility of k positive and (d) the possibility of k negative.

63. **M** A model airplane of mass 0.750 kg flies with a speed of 35.0 m/s in a horizontal circle at the end of a 60.0-m-long control wire as shown in Figure P6.63a. The forces exerted on the airplane are shown in Figure P6.63b: the tension in the control wire, the gravitational force, and aerodynamic lift that acts at $\theta = 20.0°$ inward from the vertical. Compute the tension in the wire, assuming it makes a constant angle of $\theta = 20.0°$ with the horizontal.

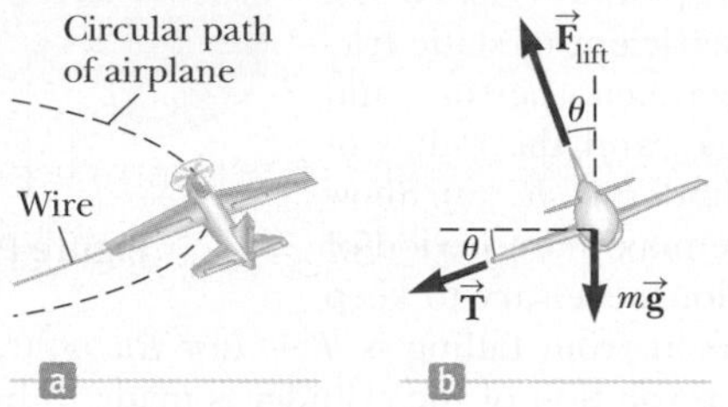

Figure P6.63

Challenge Problems

64. Because of the Earth's rotation, a plumb bob does not hang exactly along a line directed to the center of the Earth. How much does the plumb bob deviate from a radial line at 35.0° north latitude? Assume the Earth is spherical.

65. A 9.00-kg object starting from rest falls through a viscous medium and experiences a resistive force given by Equation 6.2. The object reaches one half its terminal speed in 5.54 s. (a) Determine the terminal speed. (b) At what time is the speed of the object three-fourths the terminal speed? (c) How far has the object traveled in the first 5.54 s of motion?

66. **S** For $t < 0$, an object of mass m experiences no force and moves in the positive x direction with a constant speed v_i. Beginning at $t = 0$, when the object passes position $x = 0$, it experiences a net resistive force proportional to the square of its speed: $\vec{\mathbf{F}}_{net} = -mkv^2\hat{\mathbf{i}}$, where k is a constant. The speed of the object after $t = 0$ is given by $v = v_i/(1 + kv_it)$. (a) Find the position x of the object as a function of time. (b) Find the object's velocity as a function of position.

67. A golfer tees off from a location precisely at $\phi_i = 35.0°$ north latitude. He hits the ball due south, with range 285 m. The ball's initial velocity is at 48.0° above the horizontal. Suppose air resistance is negligible for the golf ball. (a) For how long is the ball in flight? The cup is due south of the golfer's location, and the golfer would have a hole-in-one if the Earth were not rotating. The Earth's rotation makes the tee move in a circle of radius $R_E \cos \phi_i = (6.37 \times 10^6 \text{ m}) \cos 35.0°$ as shown in Figure P6.67. The tee completes one revolution each day. (b) Find the eastward speed of the tee relative to the stars. The hole is also moving east, but it is 285 m farther south and thus at a slightly lower latitude ϕ_f. Because the hole moves in a slightly larger circle, its speed must be greater than that of the tee. (c) By how much does the hole's speed exceed that of the tee? During the time interval the ball is in flight, it moves upward and downward as well as southward with the projectile motion you studied in Chapter 4, but it also moves eastward with the speed you found in part (b). The hole moves to the east at a faster speed, however, pulling ahead of the ball with the relative speed you found in part (c). (d) How far to the west of the hole does the ball land?

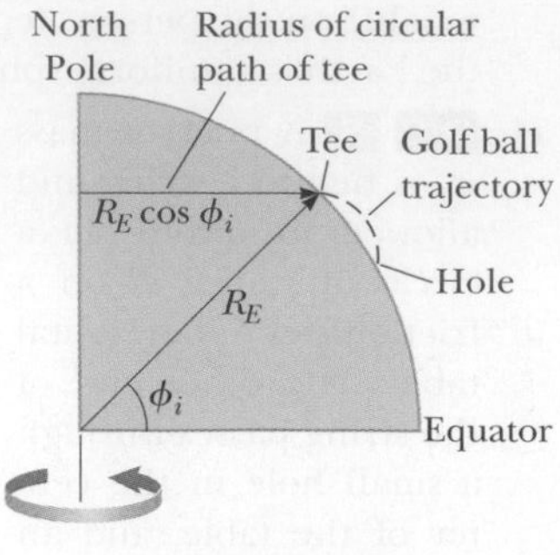

Figure P6.67

68. **Q|C** A single bead can slide with negligible friction on a stiff wire that has been bent into a circular loop of radius 15.0 cm as shown in Figure P6.68. The circle is always in a vertical plane and rotates steadily about its vertical diameter with a period of 0.450 s. The position of the bead is described by the angle θ that the radial line, from the center of the loop to the bead, makes with the vertical. (a) At what angle up from the bottom of the circle can the bead stay motionless relative to the turning circle? (b) **What If?** Repeat the problem, this time taking the period of the circle's rotation as 0.850 s. (c) Describe how the solution to part (b) is different from the solution to part (a). (d) For any period or loop size, is there always an angle at which the bead can stand still relative to the loop? (e) Are there ever more than two angles? Arnold Arons suggested the idea for this problem.

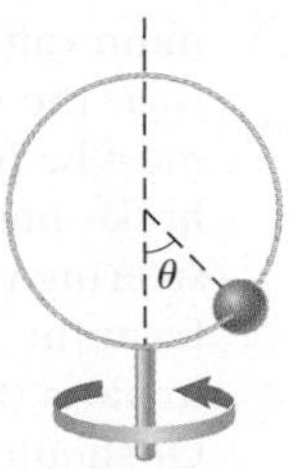

Figure P6.68

69. The expression $F = arv + br^2v^2$ gives the magnitude of the resistive force (in newtons) exerted on a sphere of radius r (in meters) by a stream of air moving at speed v (in meters per second), where a and b are constants with appropriate SI units. Their numerical values are $a = 3.10 \times 10^{-4}$ and $b = 0.870$. Using this expression, find the terminal speed for water droplets falling under their own weight in air, taking the following values for the drop radii: (a) 10.0 μm, (b) 100 μm, (c) 1.00 mm. For parts (a) and (c), you can obtain accurate answers without solving a quadratic equation by considering which of the two contributions to the air resistance is dominant and ignoring the lesser contribution.

chapter 7

Energy of a System

On a wind farm at the mouth of the River Mersey in Liverpool, England, the moving air does work on the blades of the windmills, causing the blades and the rotor of an electrical generator to rotate. Energy is transferred out of the system of the windmill by means of electricity. *(Christopher Furlong/Getty Images)*

The definitions of quantities such as position, velocity, acceleration, and force and associated principles such as Newton's second law have allowed us to solve a variety of problems. Some problems that could theoretically be solved with Newton's laws, however, are very difficult in practice, but they can be made much simpler with a different approach. Here and in the following chapters, we will investigate this new approach, which will include definitions of quantities that may not be familiar to you. Other quantities may sound familiar, but they may have more specific meanings in physics than in everyday life. We begin this discussion by exploring the notion of *energy.*

The concept of energy is one of the most important topics in science and engineering. In everyday life, we think of energy in terms of fuel for transportation and heating, electricity for lights and appliances, and foods for consumption. These ideas, however, do not truly define energy. They merely tell us that fuels are needed to do a job and that those fuels provide us with something we call energy.

Energy is present in the Universe in various forms. *Every* physical process that occurs in the Universe involves energy and energy transfers or transformations. Unfortunately, despite its extreme importance, energy cannot be easily defined. The variables in previous chapters were relatively concrete; we have everyday experience with velocities and forces, for example. Although we have *experiences* with energy, such as running out of gasoline or losing our electrical service following a violent storm, the *notion* of energy is more abstract.

The concept of energy can be applied to mechanical systems without resorting to Newton's laws. Furthermore, the energy approach allows us to understand thermal and electrical phenomena in later chapters of the book.

Our analysis models presented in earlier chapters were based on the motion of a *particle* or an object that could be modeled as a particle. We begin our new approach by focusing our attention on a *system* and analysis models based on the model of a system. These analysis models will be formally introduced in Chapter 8. In this chapter, we introduce systems and three ways to store energy in a system.

7.1 Systems and Environments

In the system model, we focus our attention on a small portion of the Universe—the **system**—and ignore details of the rest of the Universe outside of the system. A critical skill in applying the system model to problems is *identifying the system.*

> **Pitfall Prevention 7.1**
> **Identify the System**
> The most important *first* step to take in solving a problem using the energy approach is to identify the appropriate system of interest.

A valid system

- may be a single object or particle
- may be a collection of objects or particles
- may be a region of space (such as the interior of an automobile engine combustion cylinder)
- may vary with time in size and shape (such as a rubber ball, which deforms upon striking a wall)

Identifying the need for a system approach to solving a problem (as opposed to a particle approach) is part of the Categorize step in the General Problem-Solving Strategy outlined in Chapter 2. Identifying the particular system is a second part of this step.

No matter what the particular system is in a given problem, we identify a **system boundary,** an imaginary surface (not necessarily coinciding with a physical surface) that divides the Universe into the system and the **environment** surrounding the system.

As an example, imagine a force applied to an object in empty space. We can define the object as the system and its surface as the system boundary. The force applied to it is an influence on the system from the environment that acts across the system boundary. We will see how to analyze this situation from a system approach in a subsequent section of this chapter.

Another example was seen in Example 5.10, where the system can be defined as the combination of the ball, the block, and the cord. The influence from the environment includes the gravitational forces on the ball and the block, the normal and friction forces on the block, and the force exerted by the pulley on the cord. The forces exerted by the cord on the ball and the block are internal to the system and therefore are not included as an influence from the environment.

There are a number of mechanisms by which a system can be influenced by its environment. The first one we shall investigate is *work.*

Figure 7.1 An eraser being pushed along a chalkboard tray by a force acting at different angles with respect to the horizontal direction.

7.2 Work Done by a Constant Force

Almost all the terms we have used thus far—velocity, acceleration, force, and so on—convey a similar meaning in physics as they do in everyday life. Now, however, we encounter a term whose meaning in physics is distinctly different from its everyday meaning: work.

To understand what work as an influence on a system means to the physicist, consider the situation illustrated in Figure 7.1. A force $\vec{\mathbf{F}}$ is applied to a chalkboard eraser, which we identify as the system, and the eraser slides along the tray. If we want to know how effective the force is in moving the eraser, we must consider not only the magnitude of the force but also its direction. Notice that the finger in Figure 7.1 applies forces in three different directions on the eraser. Assuming the magnitude of the applied force is the same in all three photographs, the push applied in Figure 7.1b does more to move the eraser than the push in Figure 7.1a. On the other hand, Figure 7.1c shows a situation in which the applied force does not move the eraser at all, regardless of how hard it is pushed (unless, of course, we apply a force so great that we break the chalkboard tray!). These results suggest that when analyzing forces to determine the influence they have on the system, we must consider the vector nature of forces. We must also consider the magnitude of the force. Moving a force with a magnitude of $|\vec{\mathbf{F}}| = 2$ N through a displacement represents a greater influence on the system than moving a force of magnitude 1 N through the same displacement. The magnitude of the displacement is also important. Moving the eraser 3 m along the tray represents a greater influence than moving it 2 cm if the same force is used in both cases.

Pitfall Prevention 7.2

Work Is Done by . . . on . . .
Not only must you identify the system, you must also identify what agent in the environment is doing work on the system. When discussing work, always use the phrase, "the work done by . . . on . . . ." After "by," insert the part of the environment that is interacting directly with the system. After "on," insert the system. For example, "the work done by the hammer on the nail" identifies the nail as the system, and the force from the hammer represents the influence from the environment.

Let us examine the situation in Figure 7.2, where the object (the system) undergoes a displacement along a straight line while acted on by a constant force of magnitude F that makes an angle θ with the direction of the displacement.

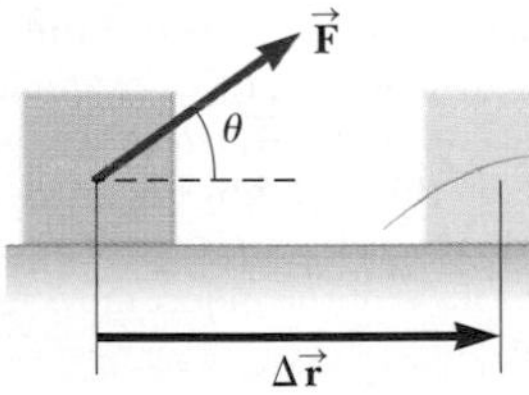

Figure 7.2 An object undergoes a displacement $\Delta\vec{\mathbf{r}}$ under the action of a constant force $\vec{\mathbf{F}}$.

The **work** W done on a system by an agent exerting a constant force on the system is the product of the magnitude F of the force, the magnitude Δr of the displacement of the point of application of the force, and $\cos\theta$, where θ is the angle between the force and displacement vectors:

$$W \equiv F\,\Delta r \cos\theta \tag{7.1}$$

◀ **Work done by a constant force**

Notice in Equation 7.1 that work is a scalar, even though it is defined in terms of two vectors, a force $\vec{\mathbf{F}}$ and a displacement $\Delta\vec{\mathbf{r}}$. In Section 7.3, we explore how to combine two vectors to generate a scalar quantity.

Notice also that the displacement in Equation 7.1 is that of *the point of application of the force*. If the force is applied to a particle or a rigid object that can be modeled as a particle, this displacement is the same as that of the particle. For a deformable system, however, these displacements are not the same. For example, imagine pressing in on the sides of a balloon with both hands. The center of the balloon moves

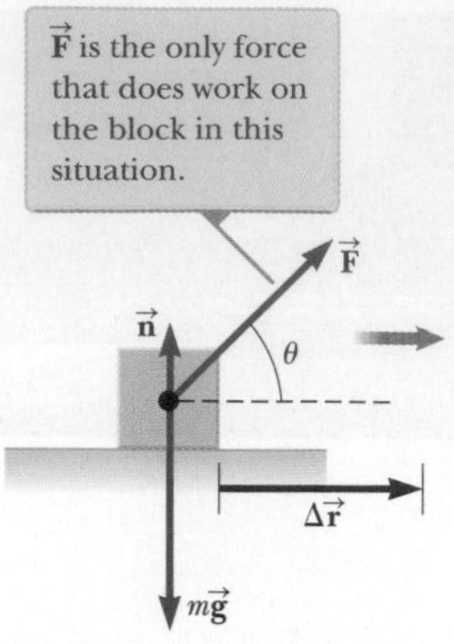

Figure 7.3 An object is displaced on a frictionless, horizontal surface. The normal force $\vec{n}$ and the gravitational force $m\vec{g}$ do no work on the object.

through zero displacement. The points of application of the forces from your hands on the sides of the balloon, however, do indeed move through a displacement as the balloon is compressed, and that is the displacement to be used in Equation 7.1. We will see other examples of deformable systems, such as springs and samples of gas contained in a vessel.

As an example of the distinction between the definition of work and our everyday understanding of the word, consider holding a heavy chair at arm's length for 3 min. At the end of this time interval, your tired arms may lead you to think you have done a considerable amount of work on the chair. According to our definition, however, you have done no work on it whatsoever. You exert a force to support the chair, but you do not move it. A force does no work on an object if the force does not move through a displacement. If $\Delta r = 0$, Equation 7.1 gives $W = 0$, which is the situation depicted in Figure 7.1c.

Also notice from Equation 7.1 that the work done by a force on a moving object is zero when the force applied is perpendicular to the displacement of its point of application. That is, if $\theta = 90°$, then $W = 0$ because $\cos 90° = 0$. For example, in Figure 7.3, the work done by the normal force on the object and the work done by the gravitational force on the object are both zero because both forces are perpendicular to the displacement and have zero components along an axis in the direction of $\Delta\vec{r}$.

The sign of the work also depends on the direction of $\vec{F}$ relative to $\Delta\vec{r}$. The work done by the applied force on a system is positive when the projection of $\vec{F}$ onto $\Delta\vec{r}$ is in the same direction as the displacement. For example, when an object is lifted, the work done by the applied force on the object is positive because the direction of that force is upward, in the same direction as the displacement of its point of application. When the projection of $\vec{F}$ onto $\Delta\vec{r}$ is in the direction opposite the displacement, W is negative. For example, as an object is lifted, the work done by the gravitational force on the object is negative. The factor $\cos\theta$ in the definition of W (Eq. 7.1) automatically takes care of the sign.

If an applied force $\vec{F}$ is in the same direction as the displacement $\Delta\vec{r}$, then $\theta = 0$ and $\cos 0 = 1$. In this case, Equation 7.1 gives

$$W = F\Delta r$$

The units of work are those of force multiplied by those of length. Therefore, the SI unit of work is the **newton · meter** ($N \cdot m = kg \cdot m^2/s^2$). This combination of units is used so frequently that it has been given a name of its own, the **joule** (J).

An important consideration for a system approach to problems is that **work is an energy transfer.** If W is the work done on a system and W is positive, energy is transferred *to* the system; if W is negative, energy is transferred *from* the system. Therefore, if a system interacts with its environment, this interaction can be described as a transfer of energy across the system boundary. The result is a change in the energy stored in the system. We will learn about the first type of energy storage in Section 7.5, after we investigate more aspects of work.

Pitfall Prevention 7.3

Cause of the Displacement
We can calculate the work done by a force on an object, but that force is *not* necessarily the cause of the object's displacement. For example, if you lift an object, (negative) work is done on the object by the gravitational force, although gravity is not the cause of the object moving upward!

***Quick Quiz* 7.1** The gravitational force exerted by the Sun on the Earth holds the Earth in an orbit around the Sun. Let us assume that the orbit is perfectly circular. The work done by this gravitational force during a short time interval in which the Earth moves through a displacement in its orbital path is **(a)** zero **(b)** positive **(c)** negative **(d)** impossible to determine

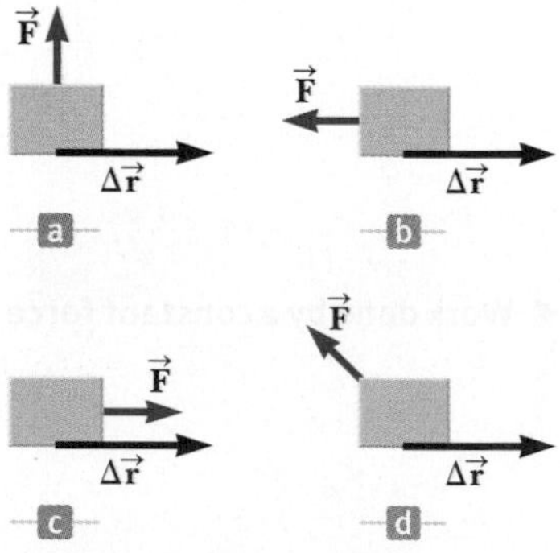

Figure 7.4 (Quick Quiz 7.2) A block is pulled by a force in four different directions. In each case, the displacement of the block is to the right and of the same magnitude.

***Quick Quiz* 7.2** Figure 7.4 shows four situations in which a force is applied to an object. In all four cases, the force has the same magnitude, and the displacement of the object is to the right and of the same magnitude. Rank the situations in order of the work done by the force on the object, from most positive to most negative.

Example 7.1 Mr. Clean

A man cleaning a floor pulls a vacuum cleaner with a force of magnitude $F = 50.0$ N at an angle of 30.0° with the horizontal (Fig. 7.5). Calculate the work done by the force on the vacuum cleaner as the vacuum cleaner is displaced 3.00 m to the right.

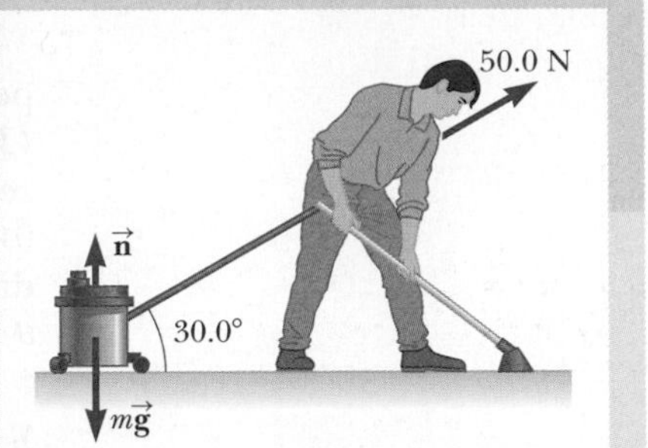

Figure 7.5 (Example 7.1) A vacuum cleaner being pulled at an angle of 30.0° from the horizontal.

SOLUTION

Conceptualize Figure 7.5 helps conceptualize the situation. Think about an experience in your life in which you pulled an object across the floor with a rope or cord.

Categorize We are asked for the work done on an object by a force and are given the force on the object, the displacement of the object, and the angle between the two vectors, so we categorize this example as a substitution problem. We identify the vacuum cleaner as the system.

Use the definition of work (Eq. 7.1):

$$W = F\Delta r\cos\theta = (50.0\text{ N})(3.00\text{ m})(\cos 30.0°)$$
$$= \boxed{130\text{ J}}$$

Notice in this situation that the normal force $\vec{n}$ and the gravitational $\vec{F}_g = m\vec{g}$ do no work on the vacuum cleaner because these forces are perpendicular to its displacement. Furthermore, there was no mention of whether there was friction between the vacuum cleaner and the floor. The presence or absence of friction is not important when calculating the work done by the applied force. In addition, this work does not depend on whether the vacuum moved at constant velocity or if it accelerated.

7.3 The Scalar Product of Two Vectors

Because of the way the force and displacement vectors are combined in Equation 7.1, it is helpful to use a convenient mathematical tool called the **scalar product** of two vectors. We write this scalar product of vectors $\vec{A}$ and $\vec{B}$ as $\vec{A}\cdot\vec{B}$. (Because of the dot symbol, the scalar product is often called the **dot product.**)

The scalar product of any two vectors $\vec{A}$ and $\vec{B}$ is defined as a scalar quantity equal to the product of the magnitudes of the two vectors and the cosine of the angle θ between them:

$$\vec{A}\cdot\vec{B} \equiv AB\cos\theta \tag{7.2}$$

◀ **Scalar product of any two vectors $\vec{A}$ and $\vec{B}$**

As is the case with any multiplication, $\vec{A}$ and $\vec{B}$ need not have the same units.

By comparing this definition with Equation 7.1, we can express Equation 7.1 as a scalar product:

$$W = F\Delta r\cos\theta = \vec{F}\cdot\Delta\vec{r} \tag{7.3}$$

In other words, $\vec{F}\cdot\Delta\vec{r}$ is a shorthand notation for $F\Delta r\cos\theta$.

Before continuing with our discussion of work, let us investigate some properties of the dot product. Figure 7.6 shows two vectors $\vec{A}$ and $\vec{B}$ and the angle θ between them used in the definition of the dot product. In Figure 7.6, $B\cos\theta$ is the projection of $\vec{B}$ onto $\vec{A}$. Therefore, Equation 7.2 means that $\vec{A}\cdot\vec{B}$ is the product of the magnitude of $\vec{A}$ and the projection of $\vec{B}$ onto $\vec{A}$.[1]

From the right-hand side of Equation 7.2, we also see that the scalar product is **commutative.**[2] That is,

$$\vec{A}\cdot\vec{B} = \vec{B}\cdot\vec{A}$$

Pitfall Prevention 7.4

Work Is a Scalar

Although Equation 7.3 defines the work in terms of two vectors, *work is a scalar;* there is no direction associated with it. *All* types of energy and energy transfer are scalars. This fact is a major advantage of the energy approach because we don't need vector calculations!

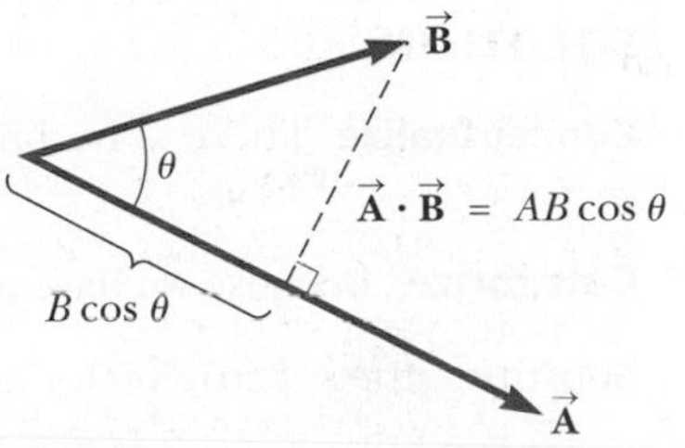

Figure 7.6 The scalar product $\vec{A}\cdot\vec{B}$ equals the magnitude of $\vec{A}$ multiplied by $B\cos\theta$, which is the projection of $\vec{B}$ onto $\vec{A}$.

[1]This statement is equivalent to stating that $\vec{A}\cdot\vec{B}$ equals the product of the magnitude of $\vec{B}$ and the projection of $\vec{A}$ onto $\vec{B}$.

[2]In Chapter 11, you will see another way of combining vectors that proves useful in physics and is not commutative.

Finally, the scalar product obeys the **distributive law of multiplication,** so

$$\vec{\mathbf{A}} \cdot (\vec{\mathbf{B}} + \vec{\mathbf{C}}) = \vec{\mathbf{A}} \cdot \vec{\mathbf{B}} + \vec{\mathbf{A}} \cdot \vec{\mathbf{C}}$$

The scalar product is simple to evaluate from Equation 7.2 when $\vec{\mathbf{A}}$ is either perpendicular or parallel to $\vec{\mathbf{B}}$. If $\vec{\mathbf{A}}$ is perpendicular to $\vec{\mathbf{B}}$ ($\theta = 90°$), then $\vec{\mathbf{A}} \cdot \vec{\mathbf{B}} = 0$. (The equality $\vec{\mathbf{A}} \cdot \vec{\mathbf{B}} = 0$ also holds in the more trivial case in which either $\vec{\mathbf{A}}$ or $\vec{\mathbf{B}}$ is zero.) If vector $\vec{\mathbf{A}}$ is parallel to vector $\vec{\mathbf{B}}$ and the two point in the same direction ($\theta = 0$), then $\vec{\mathbf{A}} \cdot \vec{\mathbf{B}} = AB$. If vector $\vec{\mathbf{A}}$ is parallel to vector $\vec{\mathbf{B}}$ but the two point in opposite directions ($\theta = 180°$), then $\vec{\mathbf{A}} \cdot \vec{\mathbf{B}} = -AB$. The scalar product is negative when $90° < \theta \leq 180°$.

The unit vectors **i**, **j**, and **k**, which were defined in Chapter 3, lie in the positive *x*, *y*, and *z* directions, respectively, of a right-handed coordinate system. Therefore, it follows from the definition of $\vec{\mathbf{A}} \cdot \vec{\mathbf{B}}$ that the scalar products of these unit vectors are

Scalar products of unit vectors ▶

$$\hat{\mathbf{i}} \cdot \hat{\mathbf{i}} = \hat{\mathbf{j}} \cdot \hat{\mathbf{j}} = \hat{\mathbf{k}} \cdot \hat{\mathbf{k}} = 1 \qquad \textbf{(7.4)}$$

$$\hat{\mathbf{i}} \cdot \hat{\mathbf{j}} = \hat{\mathbf{i}} \cdot \hat{\mathbf{k}} = \hat{\mathbf{j}} \cdot \hat{\mathbf{k}} = 0 \qquad \textbf{(7.5)}$$

Equations 3.18 and 3.19 state that two vectors $\vec{\mathbf{A}}$ and $\vec{\mathbf{B}}$ can be expressed in unit-vector form as

$$\vec{\mathbf{A}} = A_x\hat{\mathbf{i}} + A_y\hat{\mathbf{j}} + A_z\hat{\mathbf{k}}$$

$$\vec{\mathbf{B}} = B_x\hat{\mathbf{i}} + B_y\hat{\mathbf{j}} + B_z\hat{\mathbf{k}}$$

Using these expressions for the vectors and the information given in Equations 7.4 and 7.5 shows that the scalar product of $\vec{\mathbf{A}}$ and $\vec{\mathbf{B}}$ reduces to

$$\vec{\mathbf{A}} \cdot \vec{\mathbf{B}} = A_xB_x + A_yB_y + A_zB_z \qquad \textbf{(7.6)}$$

(Details of the derivation are left for you in Problem 7 at the end of the chapter.) In the special case in which $\vec{\mathbf{A}} = \vec{\mathbf{B}}$, we see that

$$\vec{\mathbf{A}} \cdot \vec{\mathbf{A}} = A_x^2 + A_y^2 + A_z^2 = A^2$$

***Quick Quiz* 7.3** Which of the following statements is true about the relationship between the dot product of two vectors and the product of the magnitudes of the vectors? **(a)** $\vec{\mathbf{A}} \cdot \vec{\mathbf{B}}$ is larger than AB. **(b)** $\vec{\mathbf{A}} \cdot \vec{\mathbf{B}}$ is smaller than AB. **(c)** $\vec{\mathbf{A}} \cdot \vec{\mathbf{B}}$ could be larger or smaller than AB, depending on the angle between the vectors. **(d)** $\vec{\mathbf{A}} \cdot \vec{\mathbf{B}}$ could be equal to AB.

Example 7.2 The Scalar Product

The vectors $\vec{\mathbf{A}}$ and $\vec{\mathbf{B}}$ are given by $\vec{\mathbf{A}} = 2\hat{\mathbf{i}} + 3\hat{\mathbf{j}}$ and $\vec{\mathbf{B}} = -\hat{\mathbf{i}} + 2\hat{\mathbf{j}}$.

(A) Determine the scalar product $\vec{\mathbf{A}} \cdot \vec{\mathbf{B}}$.

SOLUTION

Conceptualize There is no physical system to imagine here. Rather, it is purely a mathematical exercise involving two vectors.

Categorize Because we have a definition for the scalar product, we categorize this example as a substitution problem.

Substitute the specific vector expressions for $\vec{\mathbf{A}}$ and $\vec{\mathbf{B}}$:

$$\vec{\mathbf{A}} \cdot \vec{\mathbf{B}} = (2\hat{\mathbf{i}} + 3\hat{\mathbf{j}}) \cdot (-\hat{\mathbf{i}} + 2\hat{\mathbf{j}})$$

$$= -2\hat{\mathbf{i}} \cdot \hat{\mathbf{i}} + 2\hat{\mathbf{i}} \cdot 2\hat{\mathbf{j}} - 3\hat{\mathbf{j}} \cdot \hat{\mathbf{i}} + 3\hat{\mathbf{j}} \cdot 2\hat{\mathbf{j}}$$

$$= -2(1) + 4(0) - 3(0) + 6(1) = -2 + 6 = \boxed{4}$$

The same result is obtained when we use Equation 7.6 directly, where $A_x = 2$, $A_y = 3$, $B_x = -1$, and $B_y = 2$.

7.2 *cont.*

(B) Find the angle θ between $\vec{\mathbf{A}}$ and $\vec{\mathbf{B}}$.

SOLUTION

Evaluate the magnitudes of $\vec{\mathbf{A}}$ and $\vec{\mathbf{B}}$ using the Pythagorean theorem:

$$A = \sqrt{A_x^2 + A_y^2} = \sqrt{(2)^2 + (3)^2} = \sqrt{13}$$

$$B = \sqrt{B_x^2 + B_y^2} = \sqrt{(-1)^2 + (2)^2} = \sqrt{5}$$

Use Equation 7.2 and the result from part (A) to find the angle:

$$\cos\theta = \frac{\vec{\mathbf{A}}\cdot\vec{\mathbf{B}}}{AB} = \frac{4}{\sqrt{13}\sqrt{5}} = \frac{4}{\sqrt{65}}$$

$$\theta = \cos^{-1}\frac{4}{\sqrt{65}} = \boxed{60.3^\circ}$$

Example 7.3 Work Done by a Constant Force

A particle moving in the xy plane undergoes a displacement given by $\Delta\vec{\mathbf{r}} = (2.0\hat{\mathbf{i}} + 3.0\hat{\mathbf{j}})$ m as a constant force $\vec{\mathbf{F}} = (5.0\hat{\mathbf{i}} + 2.0\hat{\mathbf{j}})$ N acts on the particle. Calculate the work done by $\vec{\mathbf{F}}$ on the particle.

SOLUTION

Conceptualize Although this example is a little more physical than the previous one in that it identifies a force and a displacement, it is similar in terms of its mathematical structure.

Categorize Because we are given force and displacement vectors and asked to find the work done by this force on the particle, we categorize this example as a substitution problem.

Substitute the expressions for $\vec{\mathbf{F}}$ and $\Delta\vec{\mathbf{r}}$ into Equation 7.3 and use Equations 7.4 and 7.5:

$$W = \vec{\mathbf{F}}\cdot\Delta\vec{\mathbf{r}} = [(5.0\hat{\mathbf{i}} + 2.0\hat{\mathbf{j}})\text{ N}]\cdot[(2.0\hat{\mathbf{i}} + 3.0\hat{\mathbf{j}})\text{ m}]$$

$$= (5.0\hat{\mathbf{i}}\cdot 2.0\hat{\mathbf{i}} + 5.0\hat{\mathbf{i}}\cdot 3.0\hat{\mathbf{j}} + 2.0\hat{\mathbf{j}}\cdot 2.0\hat{\mathbf{i}} + 2.0\hat{\mathbf{j}}\cdot 3.0\hat{\mathbf{j}})\text{ N}\cdot\text{m}$$

$$= [10 + 0 + 0 + 6]\text{ N}\cdot\text{m} = \boxed{16\text{ J}}$$

7.4 Work Done by a Varying Force

Consider a particle being displaced along the x axis under the action of a force that varies with position. The particle is displaced in the direction of increasing x from $x = x_i$ to $x = x_f$. In such a situation, we cannot use $W = F\,\Delta r\cos\theta$ to calculate the work done by the force because this relationship applies only when $\vec{\mathbf{F}}$ is constant in magnitude and direction. If, however, we imagine that the particle undergoes a very small displacement Δx, shown in Figure 7.7a (page 172), the x component F_x of the force is approximately constant over this small interval; for this small displacement, we can approximate the work done on the particle by the force as

$$W \approx F_x\,\Delta x$$

which is the area of the shaded rectangle in Figure 7.7a. If we imagine the F_x versus x curve divided into a large number of such intervals, the total work done for the displacement from x_i to x_f is approximately equal to the sum of a large number of such terms:

$$W \approx \sum_{x_i}^{x_f} F_x\,\Delta x$$

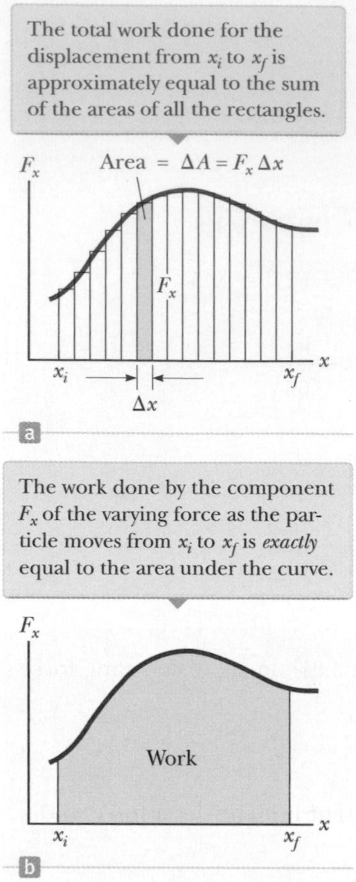

Figure 7.7 (a) The work done on a particle by the force component F_x for the small displacement Δx is $F_x \Delta x$, which equals the area of the shaded rectangle. (b) The width Δx of each rectangle is shrunk to zero.

If the size of the small displacements is allowed to approach zero, the number of terms in the sum increases without limit but the value of the sum approaches a definite value equal to the area bounded by the F_x curve and the x axis:

$$\lim_{\Delta x \to 0} \sum_{x_i}^{x_f} F_x \, \Delta x = \int_{x_i}^{x_f} F_x \, dx$$

Therefore, we can express the work done by F_x on the particle as it moves from x_i to x_f as

$$W = \int_{x_i}^{x_f} F_x \, dx \qquad \textbf{(7.7)}$$

This equation reduces to Equation 7.1 when the component $F_x = F \cos \theta$ remains constant.

If more than one force acts on a system *and the system can be modeled as a particle,* the total work done on the system is just the work done by the net force. If we express the net force in the x direction as $\sum F_x$, the total work, or *net work,* done as the particle moves from x_i to x_f is

$$\sum W = W_{\text{ext}} = \int_{x_i}^{x_f} \left(\sum F_x\right) dx \quad \text{(particle)}$$

For the general case of a net force $\sum \vec{\mathbf{F}}$ whose magnitude and direction may vary, we use the scalar product,

$$\sum W = W_{\text{ext}} = \int \left(\sum \vec{\mathbf{F}}\right) \cdot d\vec{\mathbf{r}} \quad \text{(particle)} \qquad \textbf{(7.8)}$$

where the integral is calculated over the path that the particle takes through space. The subscript "ext" on work reminds us that the net work is done by an *external* agent on the system. We will use this notation in this chapter as a reminder and to differentiate this work from an *internal* work to be described shortly.

If the system cannot be modeled as a particle (for example, if the system is deformable), we cannot use Equation 7.8 because different forces on the system may move through different displacements. In this case, we must evaluate the work done by each force separately and then add the works algebraically to find the net work done on the system:

$$\sum W = W_{\text{ext}} = \sum_{\text{forces}} \left(\int \vec{\mathbf{F}} \cdot d\vec{\mathbf{r}} \right) \quad \text{(deformable system)}$$

Example 7.4 Calculating Total Work Done from a Graph

A force acting on a particle varies with x as shown in Figure 7.8. Calculate the work done by the force on the particle as it moves from $x = 0$ to $x = 6.0$ m.

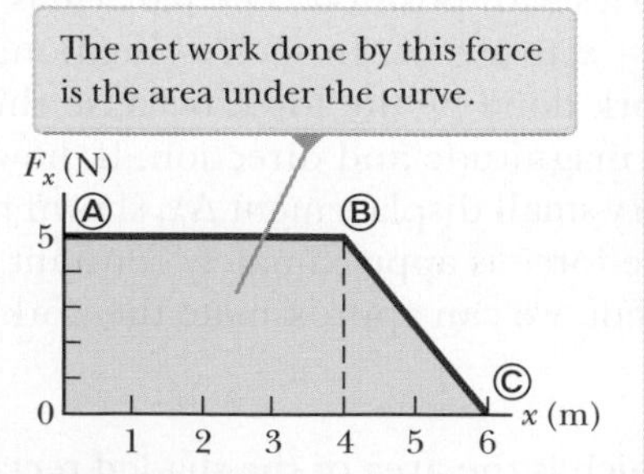

Figure 7.8 (Example 7.4) The force acting on a particle is constant for the first 4.0 m of motion and then decreases linearly with x from $x_{Ⓑ} = 4.0$ m to $x_{Ⓒ} = 6.0$ m.

SOLUTION

Conceptualize Imagine a particle subject to the force in Figure 7.8. Notice that the force remains constant as the particle moves through the first 4.0 m and then decreases linearly to zero at 6.0 m. In terms of earlier discussions of motion, the particle could be modeled as a particle under constant acceleration for the first 4.0 m because the force is constant. Between 4.0 m and 6.0 m, however, the motion does not fit into one of our earlier analysis models because the acceleration of the particle is changing. If the particle starts from rest, its speed increases throughout the motion and the particle is always moving in the positive x direction. These details about its speed and direction are not necessary for the calculation of the work done, however.

7.4 *cont.*

Categorize Because the force varies during the entire motion of the particle, we must use the techniques for work done by varying forces. In this case, the graphical representation in Figure 7.8 can be used to evaluate the work done.

Analyze The work done by the force is equal to the area under the curve from $x_{\text{Ⓐ}} = 0$ to $x_{\text{Ⓒ}} = 6.0$ m. This area is equal to the area of the rectangular section from Ⓐ to Ⓑ plus the area of the triangular section from Ⓑ to Ⓒ.

Evaluate the area of the rectangle: $W_{\text{Ⓐ to Ⓑ}} = (5.0 \text{ N})(4.0 \text{ m}) = 20 \text{ J}$

Evaluate the area of the triangle: $W_{\text{Ⓑ to Ⓒ}} = \frac{1}{2}(5.0 \text{ N})(2.0 \text{ m}) = 5.0 \text{ J}$

Find the total work done by the force on the particle: $W_{\text{Ⓐ to Ⓒ}} = W_{\text{Ⓐ to Ⓑ}} + W_{\text{Ⓑ to Ⓒ}} = 20 \text{ J} + 5.0 \text{ J} = \boxed{25 \text{ J}}$

Finalize Because the graph of the force consists of straight lines, we can use rules for finding the areas of simple geometric shapes to evaluate the total work done in this example. In a case in which the force does not vary linearly, such rules cannot be used and the force function must be integrated as in Equation 7.7 or 7.8.

Work Done by a Spring

A model of a common physical system on which the force varies with position is shown in Active Figure 7.9. The system is a block on a frictionless, horizontal surface and connected to a spring. For many springs, if the spring is either stretched or compressed a small distance from its unstretched (equilibrium) configuration, it exerts on the block a force that can be mathematically modeled as

$$F_s = -kx \qquad (7.9)$$

◀ **Spring force**

where x is the position of the block relative to its equilibrium ($x = 0$) position and k is a positive constant called the **force constant** or the **spring constant** of the spring. In other words, the force required to stretch or compress a spring is proportional to the amount of stretch or compression x. This force law for springs is known as

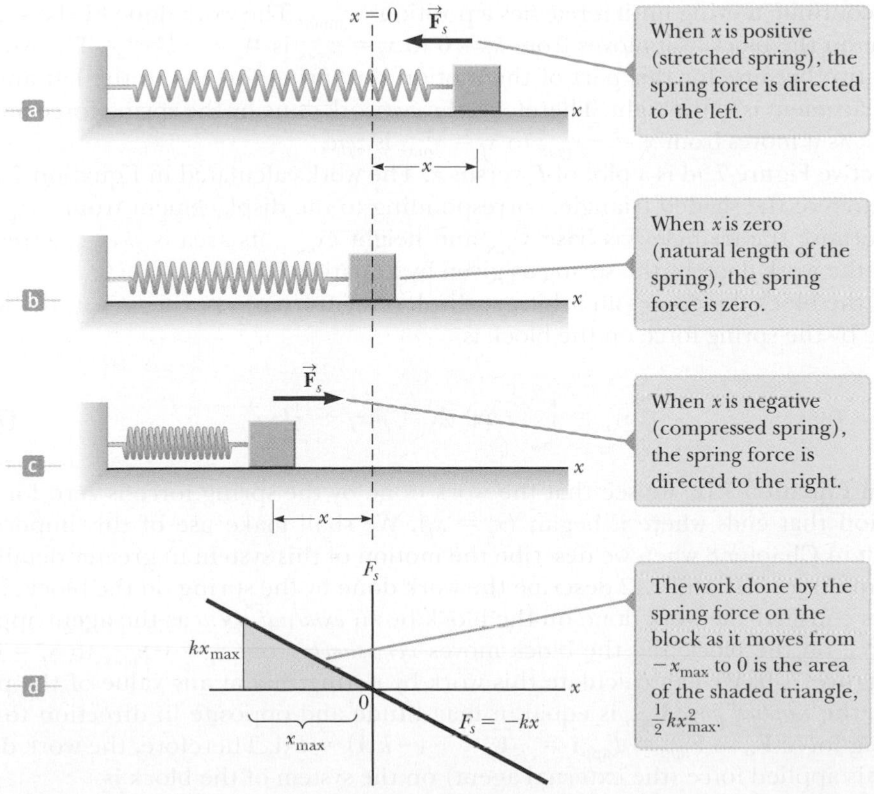

ACTIVE FIGURE 7.9

The force exerted by a spring on a block varies with the block's position x relative to the equilibrium position $x = 0$. (a) x is positive. (b) x is zero. (c) x is negative. (d) Graph of F_s versus x for the block–spring system.

Hooke's law. The value of k is a measure of the *stiffness* of the spring. Stiff springs have large k values, and soft springs have small k values. As can be seen from Equation 7.9, the units of k are N/m.

The vector form of Equation 7.9 is

$$\vec{\mathbf{F}}_s = F_s\hat{\mathbf{i}} = -kx\hat{\mathbf{i}} \quad \textbf{(7.10)}$$

where we have chosen the x axis to lie along the direction the spring extends or compresses.

The negative sign in Equations 7.9 and 7.10 signifies that the force exerted by the spring is always directed *opposite* the displacement from equilibrium. When $x > 0$ as in Active Figure 7.9a so that the block is to the right of the equilibrium position, the spring force is directed to the left, in the negative x direction. When $x < 0$ as in Active Figure 7.9c, the block is to the left of equilibrium and the spring force is directed to the right, in the positive x direction. When $x = 0$ as in Active Figure 7.9b, the spring is unstretched and $F_s = 0$. Because the spring force always acts toward the equilibrium position ($x = 0$), it is sometimes called a *restoring force.*

If the spring is compressed until the block is at the point $-x_{max}$ and is then released, the block moves from $-x_{max}$ through zero to $+x_{max}$. It then reverses direction, returns to $-x_{max}$, and continues oscillating back and forth. We will study these oscillations in more detail in Chapter 15. For now, let's investigate the work done by the spring on the block over small portions of one oscillation.

Suppose the block has been pushed to the left to a position $-x_{max}$ and is then released. We identify the block as our system and calculate the work W_s done by the spring force on the block as the block moves from $x_i = -x_{max}$ to $x_f = 0$. Applying Equation 7.8 and assuming the block may be modeled as a particle, we obtain

$$W_s = \int \vec{\mathbf{F}}_s \cdot d\vec{\mathbf{r}} = \int_{x_i}^{x_f} (-kx\hat{\mathbf{i}}) \cdot (dx\hat{\mathbf{i}}) = \int_{-x_{max}}^{0} (-kx)\, dx = \tfrac{1}{2}kx_{max}^2 \quad \textbf{(7.11)}$$

where we have used the integral $\int x^n\, dx = x^{n+1}/(n+1)$ with $n = 1$. The work done by the spring force is positive because the force is in the same direction as its displacement (both are to the right). Because the block arrives at $x = 0$ with some speed, it will continue moving until it reaches a position $+x_{max}$. The work done by the spring force on the block as it moves from $x_i = 0$ to $x_f = x_{max}$ is $W_s = -\frac{1}{2}kx_{max}^2$. The work is negative because for this part of the motion the spring force is to the left and its displacement is to the right. Therefore, the *net* work done by the spring force on the block as it moves from $x_i = -x_{max}$ to $x_f = x_{max}$ is *zero.*

Active Figure 7.9d is a plot of F_s versus x. The work calculated in Equation 7.11 is the area of the shaded triangle, corresponding to the displacement from $-x_{max}$ to 0. Because the triangle has base x_{max} and height kx_{max}, its area is $\frac{1}{2}kx_{max}^2$, agreeing with the work done by the spring as given by Equation 7.11.

If the block undergoes an arbitrary displacement from $x = x_i$ to $x = x_f$, the work done by the spring force on the block is

Work done by a spring ▶

$$W_s = \int_{x_i}^{x_f} (-kx)\, dx = \tfrac{1}{2}kx_i^2 - \tfrac{1}{2}kx_f^2 \quad \textbf{(7.12)}$$

From Equation 7.12, we see that the work done by the spring force is zero for any motion that ends where it began ($x_i = x_f$). We shall make use of this important result in Chapter 8 when we describe the motion of this system in greater detail.

Equations 7.11 and 7.12 describe the work done by the spring on the block. Now let us consider the work done on the block by an *external agent* as the agent applies a force on the block and the block moves *very slowly* from $x_i = -x_{max}$ to $x_f = 0$ as in Figure 7.10. We can calculate this work by noting that at any value of the position, the *applied force* $\vec{\mathbf{F}}_{app}$ is equal in magnitude and opposite in direction to the spring force $\vec{\mathbf{F}}_s$, so $\vec{\mathbf{F}}_{app} = F_{app}\hat{\mathbf{i}} = -\vec{\mathbf{F}}_s = -(-kx\hat{\mathbf{i}}) = kx\hat{\mathbf{i}}$. Therefore, the work done by this applied force (the external agent) on the system of the block is

$$W_{app} = \int \vec{\mathbf{F}}_{app} \cdot d\vec{\mathbf{r}} = \int_{x_i}^{x_f} (kx\hat{\mathbf{i}}) \cdot (dx\hat{\mathbf{i}}) = \int_{-x_{max}}^{0} kx\,dx = -\tfrac{1}{2}kx_{max}^2$$

This work is equal to the negative of the work done by the spring force for this displacement (Eq. 7.11). The work is negative because the external agent must push inward on the spring to prevent it from expanding, and this direction is opposite the direction of the displacement of the point of application of the force as the block moves from $-x_{max}$ to 0.

For an arbitrary displacement of the block, the work done on the system by the external agent is

$$W_{app} = \int_{x_i}^{x_f} kx\,dx = \tfrac{1}{2}kx_f^2 - \tfrac{1}{2}kx_i^2 \quad (7.13)$$

Notice that this equation is the negative of Equation 7.12.

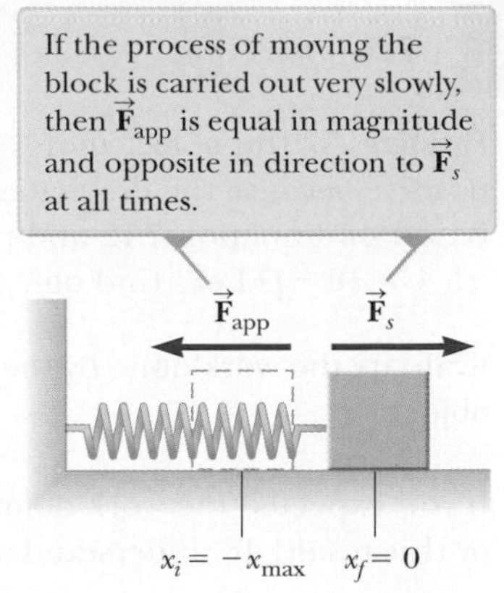

Figure 7.10 A block moves from $x_i = -x_{max}$ to $x_f = 0$ on a frictionless surface as a force $\vec{\mathbf{F}}_{app}$ is applied to the block.

***Quick Quiz* 7.4** A dart is inserted into a spring-loaded dart gun by pushing the spring in by a distance x. For the next loading, the spring is compressed a distance $2x$. How much work is required to load the second dart compared with that required to load the first? **(a)** four times as much **(b)** two times as much **(c)** the same **(d)** half as much **(e)** one-fourth as much

Example 7.5 Measuring *k* for a Spring

A common technique used to measure the force constant of a spring is demonstrated by the setup in Figure 7.11. The spring is hung vertically (Fig. 7.11a), and an object of mass m is attached to its lower end. Under the action of the "load" mg, the spring stretches a distance d from its equilibrium position (Fig. 7.11b).

(A) If a spring is stretched 2.0 cm by a suspended object having a mass of 0.55 kg, what is the force constant of the spring?

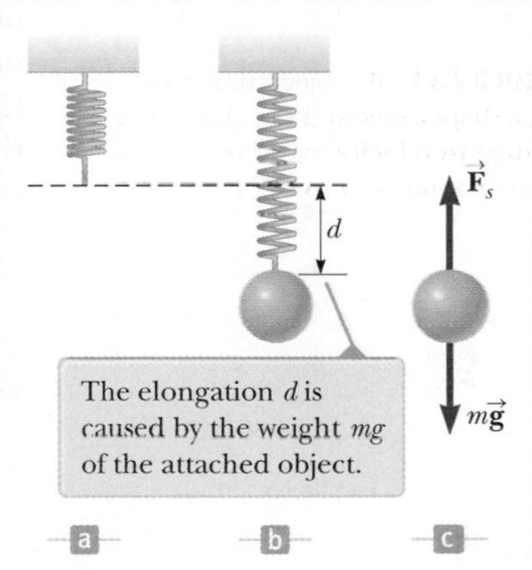

Figure 7.11 (Example 7.5) Determining the force constant k of a spring.

SOLUTION

Conceptualize Consider Figure 7.11b, which shows what happens to the spring when the object is attached to it. Simulate this situation by hanging an object on a rubber band.

Categorize The object in Figure 7.11b is not accelerating, so it is modeled as a particle in equilibrium.

Analyze Because the object is in equilibrium, the net force on it is zero and the upward spring force balances the downward gravitational force $m\vec{\mathbf{g}}$ (Fig. 7.11c).

Apply the particle in equilibrium model to the object: $\vec{\mathbf{F}}_s + m\vec{\mathbf{g}} = 0 \rightarrow F_s - mg = 0 \rightarrow F_s = mg$

Apply Hooke's law to give $F_s = kd$ and solve for k: $k = \dfrac{mg}{d} = \dfrac{(0.55\text{ kg})(9.80\text{ m/s}^2)}{2.0 \times 10^{-2}\text{ m}} = 2.7 \times 10^2\text{ N/m}$

(B) How much work is done by the spring on the object as it stretches through this distance?

SOLUTION

Use Equation 7.12 to find the work done by the spring on the object:

$$W_s = 0 - \tfrac{1}{2}kd^2 = -\tfrac{1}{2}(2.7 \times 10^2\text{ N/m})(2.0 \times 10^{-2}\text{ m})^2$$
$$= -5.4 \times 10^{-2}\text{ J}$$

continued

7.5 cont.

Finalize As the object moves through the 2.0-cm distance, the gravitational force also does work on it. This work is positive because the gravitational force is downward and so is the displacement of the point of application of this force. Based on Equation 7.12 and the discussion afterward, would we expect the work done by the gravitational force to be $+5.4 \times 10^{-2}$ J? Let's find out.

Evaluate the work done by the gravitational force on the object:

$$W = \vec{\mathbf{F}} \cdot \Delta\vec{\mathbf{r}} = (mg)(d)\cos 0 = mgd$$

$$= (0.55 \text{ kg})(9.80 \text{ m/s}^2)(2.0 \times 10^{-2} \text{ m}) = 1.1 \times 10^{-1} \text{ J}$$

If you expected the work done by gravity simply to be that done by the spring with a positive sign, you may be surprised by this result! To understand why that is not the case, we need to explore further, as we do in the next section.

7.5 Kinetic Energy and the Work–Kinetic Energy Theorem

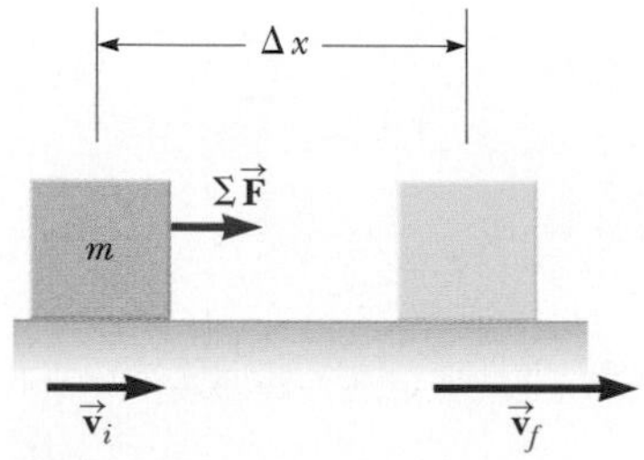

Figure 7.12 An object undergoing a displacement $\Delta\vec{\mathbf{r}} = \Delta x\hat{\mathbf{i}}$ and a change in velocity under the action of a constant net force $\sum \vec{\mathbf{F}}$.

We have investigated work and identified it as a mechanism for transferring energy into a system. We have stated that work is an influence on a system from the environment, but we have not yet discussed the *result* of this influence on the system. One possible result of doing work on a system is that the system changes its speed. In this section, we investigate this situation and introduce our first type of energy that a system can possess, called *kinetic energy.*

Consider a system consisting of a single object. Figure 7.12 shows a block of mass m moving through a displacement directed to the right under the action of a net force $\sum \vec{\mathbf{F}}$, also directed to the right. We know from Newton's second law that the block moves with an acceleration $\vec{\mathbf{a}}$. If the block (and therefore the force) moves through a displacement $\Delta\vec{\mathbf{r}} = \Delta x\hat{\mathbf{i}} = (x_f - x_i)\hat{\mathbf{i}}$, the net work done on the block by the external net force $\sum \vec{\mathbf{F}}$ is

$$W_{\text{ext}} = \int_{x_i}^{x_f} \sum F\, dx \quad \textbf{(7.14)}$$

Using Newton's second law, we substitute for the magnitude of the net force $\sum F = ma$ and then perform the following chain-rule manipulations on the integrand:

$$W_{\text{ext}} = \int_{x_i}^{x_f} ma\, dx = \int_{x_i}^{x_f} m\frac{dv}{dt}\, dx = \int_{x_i}^{x_f} m\frac{dv}{dx}\frac{dx}{dt}\, dx = \int_{v_i}^{v_f} mv\, dv$$

$$W_{\text{ext}} = \tfrac{1}{2}mv_f^{\,2} - \tfrac{1}{2}mv_i^{\,2} \quad \textbf{(7.15)}$$

where v_i is the speed of the block at $x = x_i$ and v_f is its speed at x_f.

Equation 7.15 was generated for the specific situation of one-dimensional motion, but it is a general result. It tells us that the work done by the net force on a particle of mass m is equal to the difference between the initial and final values of a quantity $\frac{1}{2}mv^2$. This quantity is so important that it has been given a special name, **kinetic energy:**

Kinetic energy ▶

$$K \equiv \tfrac{1}{2}mv^2 \quad \textbf{(7.16)}$$

Kinetic energy represents the energy associated with the motion of the particle. Kinetic energy is a scalar quantity and has the same units as work. For example, a 2.0-kg object moving with a speed of 4.0 m/s has a kinetic energy of 16 J. Table 7.1 lists the kinetic energies for various objects.

Equation 7.15 states that the work done on a particle by a net force $\sum \vec{\mathbf{F}}$ acting on it equals the change in kinetic energy of the particle. It is often convenient to write Equation 7.15 in the form

$$W_{\text{ext}} = K_f - K_i = \Delta K \quad \textbf{(7.17)}$$

TABLE 7.1 *Kinetic Energies for Various Objects*

Object	Mass (kg)	Speed (m/s)	Kinetic Energy (J)
Earth orbiting the Sun	5.97×10^{24}	2.98×10^{4}	2.65×10^{33}
Moon orbiting the Earth	7.35×10^{22}	1.02×10^{3}	3.82×10^{28}
Rocket moving at escape speed[a]	500	1.12×10^{4}	3.14×10^{10}
Automobile at 65 mi/h	2 000	29	8.4×10^{5}
Running athlete	70	10	3 500
Stone dropped from 10 m	1.0	14	98
Golf ball at terminal speed	0.046	44	45
Raindrop at terminal speed	3.5×10^{-5}	9.0	1.4×10^{-3}
Oxygen molecule in air	5.3×10^{-26}	500	6.6×10^{-21}

[a]Escape speed is the minimum speed an object must reach near the Earth's surface to move infinitely far away from the Earth.

Another way to write it is $K_f = K_i + W_{ext}$, which tells us that the final kinetic energy of an object is equal to its initial kinetic energy plus the change in energy due to the net work done on it.

We have generated Equation 7.17 by imagining doing work on a particle. We could also do work on a deformable system, in which parts of the system move with respect to one another. In this case, we also find that Equation 7.17 is valid as long as the net work is found by adding up the works done by each force and adding, as discussed earlier with regard to Equation 7.8.

Equation 7.17 is an important result known as the **work–kinetic energy theorem:**

◀ **Work–kinetic energy theorem**

When work is done on a system and the only change in the system is in its speed, the net work done on the system equals the change in kinetic energy of the system.

The work–kinetic energy theorem indicates that the speed of a system *increases* if the net work done on it is *positive* because the final kinetic energy is greater than the initial kinetic energy. The speed *decreases* if the net work is *negative* because the final kinetic energy is less than the initial kinetic energy.

Because we have so far only investigated translational motion through space, we arrived at the work–kinetic energy theorem by analyzing situations involving translational motion. Another type of motion is *rotational motion,* in which an object spins about an axis. We will study this type of motion in Chapter 10. The work–kinetic energy theorem is also valid for systems that undergo a change in the rotational speed due to work done on the system. The windmill in the photograph at the beginning of this chapter is an example of work causing rotational motion.

The work–kinetic energy theorem will clarify a result seen earlier in this chapter that may have seemed odd. In Section 7.4, we arrived at a result of zero net work done when we let a spring push a block from $x_i = -x_{max}$ to $x_f = x_{max}$. Notice that because the speed of the block is continually changing, it may seem complicated to analyze this process. The quantity ΔK in the work–kinetic energy theorem, however, only refers to the initial and final points for the speeds; it does not depend on details of the path followed between these points. Therefore, because the speed is zero at both the initial and final points of the motion, the net work done on the block is zero. We will often see this concept of path independence in similar approaches to problems.

Let us also return to the mystery in the Finalize step at the end of Example 7.5. Why was the work done by gravity not just the value of the work done by the spring with a positive sign? Notice that the work done by gravity is larger than the magnitude of the work done by the spring. Therefore, the total work done by all forces on the object is positive. Imagine now how to create the situation in which the *only* forces on the object are the spring force and the gravitational force. You must support the object at the highest point and then remove your hand and let the object

Pitfall Prevention 7.5

Conditions for the Work–Kinetic Energy Theorem

The work–kinetic energy theorem is important but limited in its application; it is not a general principle. In many situations, other changes in the system occur besides its speed, and there are other interactions with the environment besides work. A more general principle involving energy is *conservation of energy* in Section 8.1.

Pitfall Prevention 7.6

The Work–Kinetic Energy Theorem: Speed, Not Velocity

The work–kinetic energy theorem relates work to a change in the *speed* of a system, not a change in its velocity. For example, if an object is in uniform circular motion, its speed is constant. Even though its velocity is changing, no work is done on the object by the force causing the circular motion.

fall. If you do so, you know that when the object reaches a position 2.0 cm below your hand, it will be *moving*, which is consistent with Equation 7.17. Positive net work is done on the object, and the result is that it has a kinetic energy as it passes through the 2.0-cm point.

The only way to prevent the object from having a kinetic energy after moving through 2.0 cm is to slowly lower it with your hand. Then, however, there is a third force doing work on the object, the normal force from your hand. If this work is calculated and added to that done by the spring force and the gravitational force, the net work done on the object is zero, which is consistent because it is not moving at the 2.0-cm point.

Earlier, we indicated that work can be considered as a mechanism for transferring energy into a system. Equation 7.17 is a mathematical statement of this concept. When work W_{net} is done on a system, the result is a transfer of energy across the boundary of the system. The result on the system, in the case of Equation 7.17, is a change ΔK in kinetic energy. In the next section, we investigate another type of energy that can be stored in a system as a result of doing work on the system.

***Quick Quiz* 7.5** A dart is inserted into a spring-loaded dart gun by pushing the spring in by a distance x. For the next loading, the spring is compressed a distance $2x$. How much faster does the second dart leave the gun compared with the first? **(a)** four times as fast **(b)** two times as fast **(c)** the same **(d)** half as fast **(e)** one-fourth as fast

Example 7.6 A Block Pulled on a Frictionless Surface

A 6.0-kg block initially at rest is pulled to the right along a frictionless, horizontal surface by a constant horizontal force of 12 N. Find the block's speed after it has moved 3.0 m.

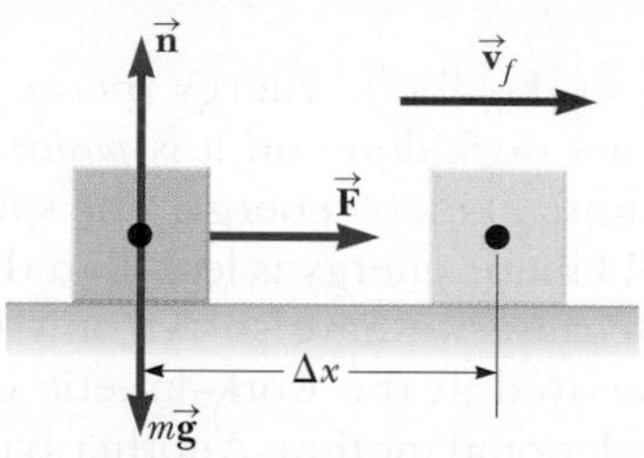

Figure 7.13 (Example 7.6) A block pulled to the right on a frictionless surface by a constant horizontal force.

SOLUTION

Conceptualize Figure 7.13 illustrates this situation. Imagine pulling a toy car across a table with a horizontal rubber band attached to the front of the car. The force is maintained constant by ensuring that the stretched rubber band always has the same length.

Categorize We could apply the equations of kinematics to determine the answer, but let us practice the energy approach. The block is the system, and three external forces act on the system. The normal force balances the gravitational force on the block, and neither of these vertically acting forces does work on the block because their points of application are horizontally displaced.

Analyze The net external force acting on the block is the horizontal 12-N force.

Use the work–kinetic energy theorem for the block, noting that its initial kinetic energy is zero:

$$W_{\text{ext}} = K_f - K_i = \tfrac{1}{2}mv_f^2 - 0 = \tfrac{1}{2}mv_f^2$$

Solve for v_f and use Equation 7.1 for the work done on the block by $\vec{F}$:

$$v_f = \sqrt{\frac{2W_{\text{ext}}}{m}} = \sqrt{\frac{2F\,\Delta x}{m}}$$

Substitute numerical values:

$$v_f = \sqrt{\frac{2(12\text{ N})(3.0\text{ m})}{6.0\text{ kg}}} = 3.5\text{ m/s}$$

Finalize It would be useful for you to solve this problem again by modeling the block as a particle under a net force to find its acceleration and then as a particle under constant acceleration to find its final velocity.

7.6 cont.

WHAT IF? Suppose the magnitude of the force in this example is doubled to $F' = 2F$. The 6.0-kg block accelerates to 3.5 m/s due to this applied force while moving through a displacement $\Delta x'$. How does the displacement $\Delta x'$ compare with the original displacement Δx?

Answer If we pull harder, the block should accelerate to a given speed in a shorter distance, so we expect that $\Delta x' < \Delta x$. In both cases, the block experiences the same change in kinetic energy ΔK. Mathematically, from the work–kinetic energy theorem, we find that

$$W_{\text{ext}} = F'\Delta x' = \Delta K = F\Delta x$$

$$\Delta x' = \frac{F}{F'}\Delta x = \frac{F}{2F}\Delta x = \tfrac{1}{2}\Delta x$$

and the distance is shorter as suggested by our conceptual argument.

Conceptual Example 7.7 Does the Ramp Lessen the Work Required?

A man wishes to load a refrigerator onto a truck using a ramp at angle θ as shown in Figure 7.14. He claims that less work would be required to load the truck if the length L of the ramp were increased. Is his claim valid?

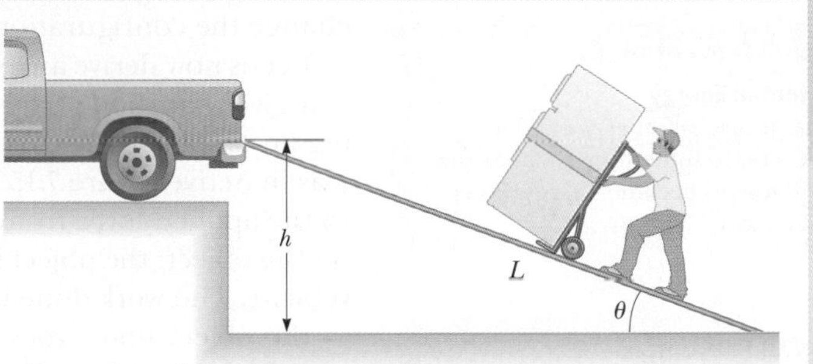

Figure 7.14 (Conceptual Example 7.7) A refrigerator attached to a frictionless, wheeled hand truck is moved up a ramp at constant speed.

SOLUTION

No. Suppose the refrigerator is wheeled on a hand truck up the ramp at constant speed. In this case, for the system of the refrigerator and the hand truck, $\Delta K = 0$. The normal force exerted by the ramp on the system is directed at 90° to the displacement of its point of application and so does no work on the system. Because $\Delta K = 0$, the work–kinetic energy theorem gives

$$W_{\text{ext}} = W_{\text{by man}} + W_{\text{by gravity}} = 0$$

The work done by the gravitational force equals the product of the weight mg of the system, the distance L through which the refrigerator is displaced, and $\cos(\theta + 90°)$. Therefore,

$$W_{\text{by man}} = -W_{\text{by gravity}} = -(mg)(L)[\cos(\theta + 90°)]$$
$$= mgL\sin\theta = mgh$$

where $h = L\sin\theta$ is the height of the ramp. Therefore, the man must do the same amount of work mgh on the system *regardless* of the length of the ramp. The work depends only on the height of the ramp. Although less force is required with a longer ramp, the point of application of that force moves through a greater displacement.

7.6 Potential Energy of a System

So far in this chapter, we have defined a system in general, but have focused our attention primarily on single particles or objects under the influence of external forces. Let us now consider systems of two or more particles or objects interacting via a force that is *internal* to the system. The kinetic energy of such a system is the algebraic sum of the kinetic energies of all members of the system. There may be systems, however, in which one object is so massive that it can be modeled as stationary and its kinetic energy can be neglected. For example, if we consider a ball–Earth system as the ball falls to the Earth, the kinetic energy of the system can be considered as just the kinetic energy of the ball. The Earth moves so slowly in this

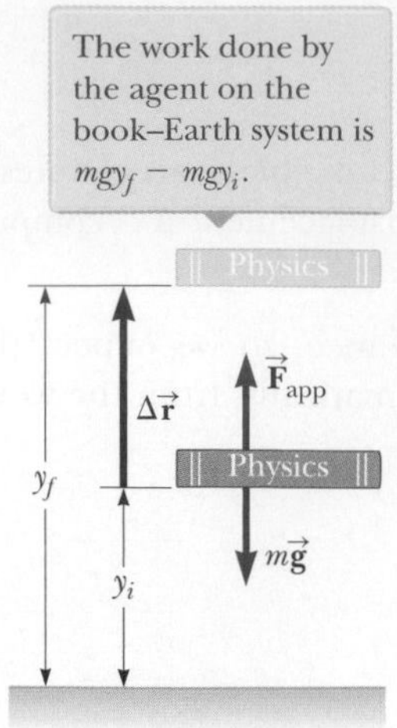

ACTIVE FIGURE 7.15

An external agent lifts a book slowly from a height y_i to a height y_f.

Pitfall Prevention 7.7

Potential Energy

The phrase *potential energy* does not refer to something that has the potential to become energy. Potential energy *is* energy.

Pitfall Prevention 7.8

Potential Energy Belongs to a System

Potential energy is always associated with a *system* of two or more interacting objects. When a small object moves near the surface of the Earth under the influence of gravity, we may sometimes refer to the potential energy "associated with the object" rather than the more proper "associated with the system" because the Earth does not move significantly. We will not, however, refer to the potential energy "of the object" because this wording ignores the role of the Earth.

process that we can ignore its kinetic energy. On the other hand, the kinetic energy of a system of two electrons must include the kinetic energies of both particles.

Let us imagine a system consisting of a book and the Earth, interacting via the gravitational force. We do some work on the system by lifting the book slowly from rest through a vertical displacement $\Delta\vec{r} = (y_f - y_i)\hat{j}$ as in Active Figure 7.15. According to our discussion of work as an energy transfer, this work done on the system must appear as an increase in energy of the system. The book is at rest before we perform the work and is at rest after we perform the work. Therefore, there is no change in the kinetic energy of the system.

Because the energy change of the system is not in the form of kinetic energy, it must appear as some other form of energy storage. After lifting the book, we could release it and let it fall back to the position y_i. Notice that the book (and therefore, the system) now has kinetic energy and that its source is in the work that was done in lifting the book. While the book was at the highest point, the system had the *potential* to possess kinetic energy, but it did not do so until the book was allowed to fall. Therefore, we call the energy storage mechanism before the book is released **potential energy.** We will find that the potential energy of a system can only be associated with specific types of forces acting between members of a system. The amount of potential energy in the system is determined by the *configuration* of the system. Moving members of the system to different positions or rotating them may change the configuration of the system and therefore its potential energy.

Let us now derive an expression for the potential energy associated with an object at a given location above the surface of the Earth. Consider an external agent lifting an object of mass m from an initial height y_i above the ground to a final height y_f as in Active Figure 7.15. We assume the lifting is done slowly, with no acceleration, so the applied force from the agent is equal in magnitude to the gravitational force on the object: the object is modeled as a particle in equilibrium moving at constant velocity. The work done by the external agent on the system (object and the Earth) as the object undergoes this upward displacement is given by the product of the upward applied force $\vec{F}_{app}$ and the upward displacement of this force, $\Delta\vec{r} = \Delta y\hat{j}$:

$$W_{ext} = (\vec{F}_{app}) \cdot \Delta\vec{r} = (mg\hat{j}) \cdot [(y_f - y_i)\hat{j}] = mgy_f - mgy_i \quad \textbf{(7.18)}$$

where this result is the net work done on the system because the applied force is the only force on the system from the environment. (Remember that the gravitational force is *internal* to the system.) Notice the similarity between Equation 7.18 and Equation 7.15. In each equation, the work done on a system equals a difference between the final and initial values of a quantity. In Equation 7.15, the work represents a transfer of energy into the system and the increase in energy of the system is kinetic in form. In Equation 7.18, the work represents a transfer of energy into the system and the system energy appears in a different form, which we have called potential energy.

Therefore, we can identify the quantity mgy as the **gravitational potential energy** U_g:

Gravitational potential energy ▶

$$U_g \equiv mgy \quad \textbf{(7.19)}$$

The units of gravitational potential energy are joules, the same as the units of work and kinetic energy. Potential energy, like work and kinetic energy, is a scalar quantity. Notice that Equation 7.19 is valid only for objects near the surface of the Earth, where g is approximately constant.[3]

Using our definition of gravitational potential energy, Equation 7.18 can now be rewritten as

$$W_{ext} = \Delta U_g \quad \textbf{(7.20)}$$

which mathematically describes that the net external work done on the system in this situation appears as a change in the gravitational potential energy of the system.

[3]The assumption that g is constant is valid as long as the vertical displacement of the object is small compared with the Earth's radius.

Gravitational potential energy depends only on the vertical height of the object above the surface of the Earth. The same amount of work must be done on an object–Earth system whether the object is lifted vertically from the Earth or is pushed starting from the same point up a frictionless incline, ending up at the same height. We verified this statement for a specific situation of rolling a refrigerator up a ramp in Conceptual Example 7.7. This statement can be shown to be true in general by calculating the work done on an object by an agent moving the object through a displacement having both vertical and horizontal components:

$$W_{ext} = (\vec{\mathbf{F}}_{app}) \cdot \Delta\vec{\mathbf{r}} = (mg\hat{\mathbf{j}}) \cdot [(x_f - x_i)\hat{\mathbf{i}} + (y_f - y_i)\hat{\mathbf{j}}] = mgy_f - mgy_i$$

where there is no term involving x in the final result because $\hat{\mathbf{j}} \cdot \hat{\mathbf{i}} = 0$.

In solving problems, you must choose a reference configuration for which the gravitational potential energy of the system is set equal to some reference value, which is normally zero. The choice of reference configuration is completely arbitrary because the important quantity is the *difference* in potential energy, and this difference is independent of the choice of reference configuration.

It is often convenient to choose as the reference configuration for zero gravitational potential energy the configuration in which an object is at the surface of the Earth, but this choice is not essential. Often, the statement of the problem suggests a convenient configuration to use.

***Quick Quiz* 7.6** Choose the correct answer. The gravitational potential energy of a system **(a)** is always positive **(b)** is always negative **(c)** can be negative or positive

Example 7.8 The Proud Athlete and the Sore Toe

A trophy being shown off by a careless athlete slips from the athlete's hands and drops on his toe. Choosing floor level as the $y = 0$ point of your coordinate system, estimate the change in gravitational potential energy of the trophy–Earth system as the trophy falls. Repeat the calculation, using the top of the athlete's head as the origin of coordinates.

SOLUTION

Conceptualize The trophy changes its vertical position with respect to the surface of the Earth. Associated with this change in position is a change in the gravitational potential energy of the trophy–Earth system.

Categorize We evaluate a change in gravitational potential energy defined in this section, so we categorize this example as a substitution problem. Because there are no numbers, it is also an estimation problem.

The problem statement tells us that the reference configuration of the trophy–Earth system corresponding to zero potential energy is when the bottom of the trophy is at the floor. To find the change in potential energy for the system, we need to estimate a few values. Let's say the trophy has a mass of approximately 2 kg, and the top of a person's toe is about 0.03 m above the floor. Also, let's assume the trophy falls from a height of 0.5 m.

Calculate the gravitational potential energy of the trophy–Earth system just before the trophy is released:

$$U_i = mgy_i = (2\text{ kg})(9.80\text{ m/s}^2)(0.5\text{ m}) = 9.80\text{ J}$$

Calculate the gravitational potential energy of the trophy–Earth system when the trophy reaches the athlete's toe:

$$U_f = mgy_f = (2\text{ kg})(9.80\text{ m/s}^2)(0.03\text{ m}) = 0.588\text{ J}$$

Evaluate the change in gravitational potential energy of the trophy–Earth system:

$$\Delta U_g = 0.588\text{ J} - 9.80\text{ J} = -9.21\text{ J}$$

We should probably keep only one digit because of the roughness of our estimates; therefore, we estimate that the change in gravitational potential energy is -9 J. The system had about 10 J of gravitational potential energy before the trophy began its fall and approximately 1 J of potential energy as the trophy reaches the top of the toe.

continued

7.8 *cont.*

The second case presented indicates that the reference configuration of the system for zero potential energy is chosen to be when the trophy is at the athlete's head (even though the trophy is never at this position in its motion). We estimate this position to be 1.50 m above the floor).

Calculate the gravitational potential energy of the trophy–Earth system just before the trophy is released from its position 1 m below the athlete's head:

$$U_i = mgy_i = (2\text{ kg})(9.80\text{ m/s}^2)(-1\text{ m}) = -19.6\text{ J}$$

Calculate the gravitational potential energy of the trophy–Earth system when the trophy reaches the athlete's toe located 1.47 m below the athlete's head:

$$U_f = mgy_f = (2\text{ kg})(9.80\text{ m/s}^2)(-1.47\text{ m}) = -28.8\text{ J}$$

Evaluate the change in gravitational potential energy of the trophy–Earth system:

$$\Delta U_g = -28.8\text{ J} - (-19.6\text{ J}) = -9.2\text{ J} \approx -9\text{ J}$$

This value is the same as before, as it must be.

Elastic Potential Energy

Because members of a system can interact with one another by means of different types of forces, it is possible that there are different types of potential energy in a system. We are familiar with gravitational potential energy of a system in which members interact via the gravitational force. Let us explore a second type of potential energy that a system can possess.

Consider a system consisting of a block and a spring as shown in Active Figure 7.16. In Section 7.4, we identified *only* the block as the system. Now we include both the block and the spring in the system and recognize that the spring force is the interaction between the two members of the system. The force that the spring exerts on the block is given by $F_s = -kx$ (Eq. 7.9). The work done by an external applied force F_{app} on a system consisting of a block connected to the spring is given by Equation 7.13:

$$W_{\text{app}} = \tfrac{1}{2}kx_f^2 - \tfrac{1}{2}kx_i^2 \quad \textbf{(7.21)}$$

In this situation, the initial and final x coordinates of the block are measured from its equilibrium position, $x = 0$. Again (as in the gravitational case) we see that the work done on the system is equal to the difference between the initial and final values of an expression related to the system's configuration. The **elastic potential energy** function associated with the block–spring system is defined by

Elastic potential energy ▶

$$U_s \equiv \tfrac{1}{2}kx^2 \quad \textbf{(7.22)}$$

The elastic potential energy of the system can be thought of as the energy stored in the deformed spring (one that is either compressed or stretched from its equilibrium position). The elastic potential energy stored in a spring is zero whenever the spring is undeformed ($x = 0$). Energy is stored in the spring only when the spring is either stretched or compressed. Because the elastic potential energy is proportional to x^2, we see that U_s is always positive in a deformed spring. Everyday examples of the storage of elastic potential energy can be found in old-style clocks or watches that operate from a wound-up spring and small wind-up toys for children.

Consider Active Figure 7.16, which shows a spring on a frictionless, horizontal surface. When a block is pushed against the spring by an external agent, the elastic potential energy and the total energy of the system increase as indicated in Figure 7.16b. When the spring is compressed a distance x_{max} (Active Fig. 7.16c), the elastic potential energy stored in the spring is $\frac{1}{2}kx^2_{\text{max}}$. When the block is released from rest, the spring exerts a force on the block and pushes the block to the right. The elastic potential energy of the system decreases, whereas the kinetic energy increases and

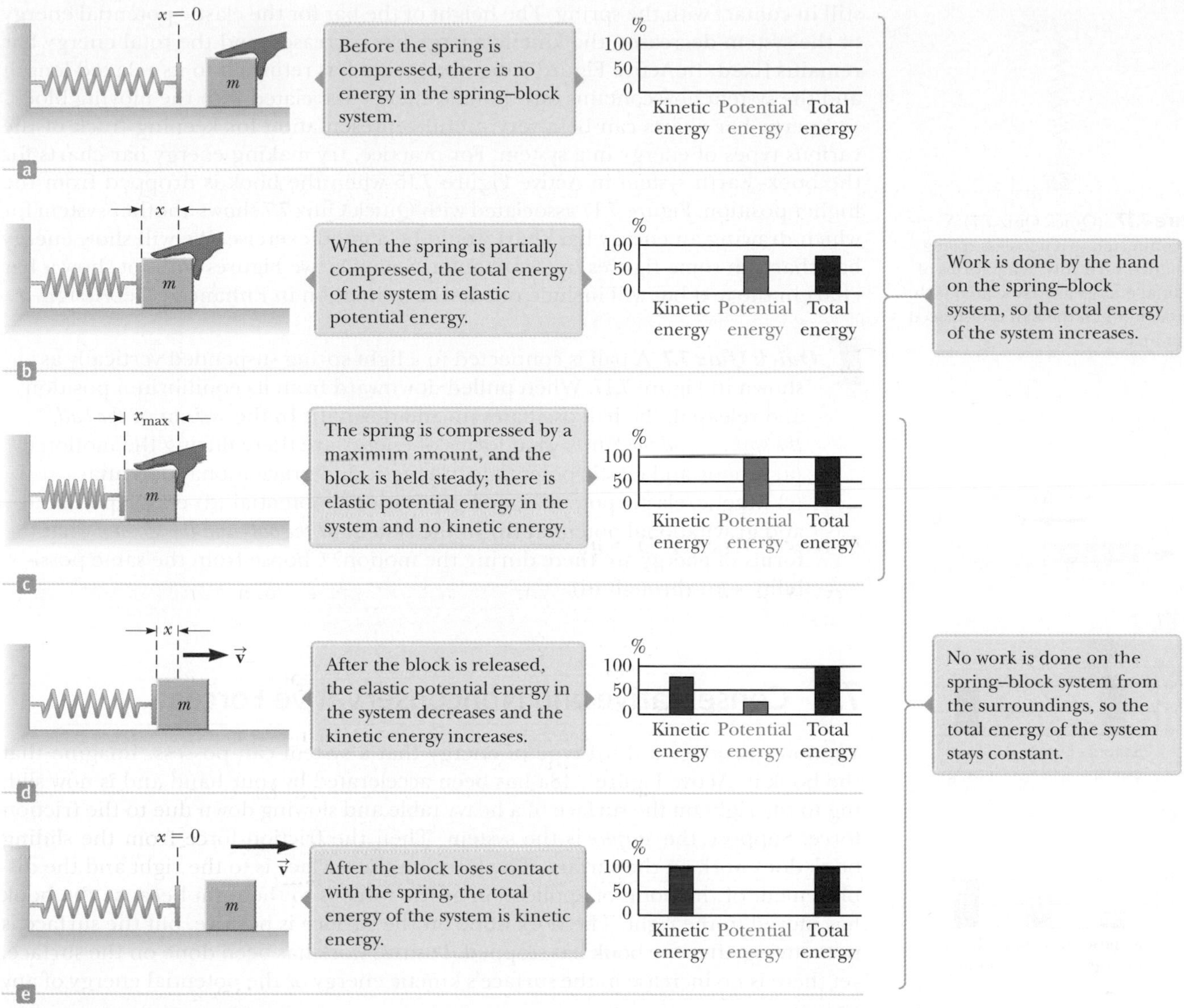

ACTIVE FIGURE 7.16

A spring on a frictionless, horizontal surface is compressed a distance x_{max} when a block of mass m is pushed against it. The block is then released and the spring pushes it to the right, where the block eventually loses contact with the spring. Parts (a) through (e) show various instants in the process. Energy bar charts on the right of each part of the figure help keep track of the energy in the system.

the total energy remains fixed (Fig. 7.16d). When the spring returns to its original length, the stored elastic potential energy is completely transformed into kinetic energy of the block (Active Fig. 7.16e).

Energy Bar Charts

Active Figure 7.16 shows an important graphical representation of information related to energy of systems called an **energy bar chart.** The vertical axis represents the amount of energy of a given type in the system. The horizontal axis shows the types of energy in the system. The bar chart in Active Figure 7.16a shows that the system contains zero energy because the spring is relaxed and the block is not moving. Between Active Figure 7.16a and Active Figure 7.16c, the hand does work on the system, compressing the spring and storing elastic potential energy in the system. In Active Figure 7.16d, the block has been released and is moving to the right while

Figure 7.17 (Quick Quiz 7.7) A ball connected to a massless spring suspended vertically. What forms of potential energy are associated with the system when the ball is displaced downward?

still in contact with the spring. The height of the bar for the elastic potential energy of the system decreases, the kinetic energy bar increases, and the total energy bar remains fixed. In Active Figure 7.16e, the spring has returned to its relaxed length and the system now contains only kinetic energy associated with the moving block.

Energy bar charts can be a very useful representation for keeping track of the various types of energy in a system. For practice, try making energy bar charts for the book–Earth system in Active Figure 7.15 when the book is dropped from the higher position. Figure 7.17 associated with Quick Quiz 7.7 shows another system for which drawing an energy bar chart would be a good exercise. We will show energy bar charts in some figures in this chapter. Some Active Figures will not show a bar chart in the text but will include one in the animation in Enhanced WebAssign.

***Quick Quiz* 7.7** A ball is connected to a light spring suspended vertically as shown in Figure 7.17. When pulled downward from its equilibrium position and released, the ball oscillates up and down. **(i)** In the system of *the ball, the spring, and the Earth,* what forms of energy are there during the motion? (a) kinetic and elastic potential (b) kinetic and gravitational potential (c) kinetic, elastic potential, and gravitational potential (d) elastic potential and gravitational potential **(ii)** In the system of *the ball and the spring,* what forms of energy are there during the motion? Choose from the same possibilities (a) through (d).

7.7 Conservative and Nonconservative Forces

We now introduce a third type of energy that a system can possess. Imagine that the book in Active Figure 7.18a has been accelerated by your hand and is now sliding to the right on the surface of a heavy table and slowing down due to the friction force. Suppose the *surface* is the system. Then the friction force from the sliding book does work on the surface. The force on the surface is to the right and the displacement of the point of application of the force is to the right because the book has moved to the right. The work done on the surface is positive, but the surface is not moving after the book has stopped. Positive work has been done on the surface, yet there is no increase in the surface's kinetic energy or the potential energy of any system.

From your everyday experience with sliding over surfaces with friction, you can probably guess that the surface will be *warmer* after the book slides over it. (Rub your hands together briskly to find out!) The work that was done on the surface has gone into warming the surface rather than increasing its speed or changing the configuration of a system. We call the energy associated with the temperature of a system its **internal energy,** symbolized E_{int}. (We will define internal energy more generally in Chapter 20.) In this case, the work done on the surface does indeed represent energy transferred into the system, but it appears in the system as internal energy rather than kinetic or potential energy.

Consider the book and the surface in Active Figure 7.18a together as a system. Initially, the system has kinetic energy because the book is moving. While the book is sliding, the internal energy of the system increases: the book and the surface are warmer than before. When the book stops, the kinetic energy has been completely transformed to internal energy. We can consider the work done by friction within the system—that is, between the book and the surface—as a *transformation mechanism* for energy. This work transforms the kinetic energy of the system into internal energy. Similarly, when a book falls straight down with no air resistance, the work done by the gravitational force within the book–Earth system transforms gravitational potential energy of the system to kinetic energy.

Active Figures 7.18b through 7.18d show energy bar charts for the situation in Active Figure 7.18a. In Active Figure 7.18b, the bar chart shows that the system con-

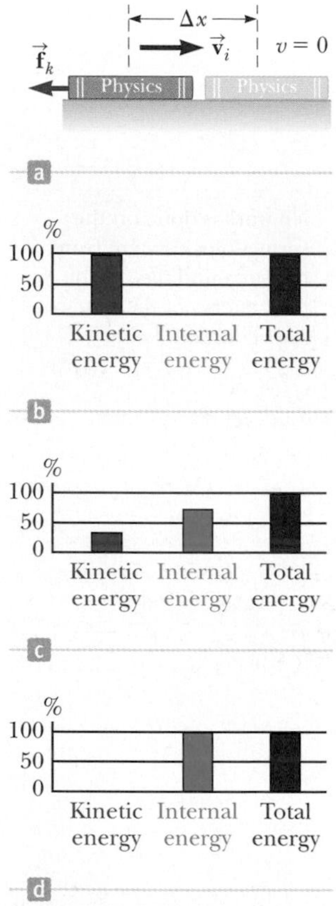

ACTIVE FIGURE 7.18

(a) A book sliding to the right on a horizontal surface slows down in the presence of a force of kinetic friction acting to the left. (b) An energy bar chart showing the energy in the system of the book and the surface at the initial instant of time. The energy of the system is all kinetic energy. (c) While the book is sliding, the kinetic energy of the system decreases as it is transformed to internal energy. (d) After the book has stopped, the energy of the system is all internal energy.

tains kinetic energy at the instant the book is released by your hand. We define the reference amount of internal energy in the system as zero at this instant. Active Figure 7.18c shows the kinetic energy transforming to internal energy as the book slows down due to the friction force. In Active Figure 7.18d, after the book has stopped sliding, the kinetic energy is zero and the system now contains only internal energy. Notice that the total energy bar in red has not changed during the process. The amount of internal energy in the system after the book has stopped is equal to the amount of kinetic energy in the system at the initial instant. This equality is described by an important principle called *conservation of energy.* We will explore this principle in Chapter 8.

Now consider in more detail an object moving downward near the surface of the Earth. The work done by the gravitational force on the object does not depend on whether it falls vertically or slides down a sloping incline with friction. All that matters is the change in the object's elevation. The energy transformation to internal energy due to friction on that incline, however, depends very much on the distance the object slides. The longer the incline, the more potential energy is transformed to internal energy. In other words, the path makes no difference when we consider the work done by the gravitational force, but it does make a difference when we consider the energy transformation due to friction forces. We can use this varying dependence on path to classify forces as either conservative or nonconservative. Of the two forces just mentioned, the gravitational force is conservative and the friction force is nonconservative.

Conservative Forces

Conservative forces have these two equivalent properties:

◀ **Properties of conservative forces**

1. The work done by a conservative force on a particle moving between any two points is independent of the path taken by the particle.
2. The work done by a conservative force on a particle moving through any closed path is zero. (A closed path is one for which the beginning point and the endpoint are identical.)

The gravitational force is one example of a conservative force; the force that an ideal spring exerts on any object attached to the spring is another. The work done by the gravitational force on an object moving between any two points near the Earth's surface is $W_g = -mg\mathbf{\hat{j}} \cdot [(y_f - y_i)\mathbf{\hat{j}}] = mgy_i - mgy_f$. From this equation, notice that W_g depends only on the initial and final y coordinates of the object and hence is independent of the path. Furthermore, W_g is zero when the object moves over any closed path (where $y_i = y_f$).

For the case of the object–spring system, the work W_s done by the spring force is given by $W_s = \frac{1}{2}kx_i^2 - \frac{1}{2}kx_f^2$ (Eq. 7.12). We see that the spring force is conservative because W_s depends only on the initial and final x coordinates of the object and is zero for any closed path.

We can associate a potential energy for a system with a force acting between members of the system, but we can do so only if the force is conservative. In general, the work W_{int} done by a conservative force on an object that is a member of a system as the system changes from one configuration to another is equal to the initial value of the potential energy of the system minus the final value:

$$W_{\text{int}} = U_i - U_f = -\Delta U \quad \textbf{(7.23)}$$

The subscript "int" in Equation 7.23 reminds us that the work we are discussing is done by one member of the system on another member and is therefore *internal* to the system. It is different from the work W_{ext} done *on* the system as a whole by an external agent. As an example, compare Equation 7.23 with the specific equation for the work done by the spring force (Eq. 7.12) as the extension of the spring changes.

Pitfall Prevention 7.9

Similar Equation Warning

Compare Equation 7.23 with Equation 7.20. These equations are similar except for the negative sign, which is a common source of confusion. Equation 7.20 tells us that positive work done *by an outside agent* on a system causes an increase in the potential energy of the system (with no change in the kinetic or internal energy). Equation 7.23 states that work done *on a component of a system by a conservative force internal to the system* causes a decrease in the potential energy of the system.

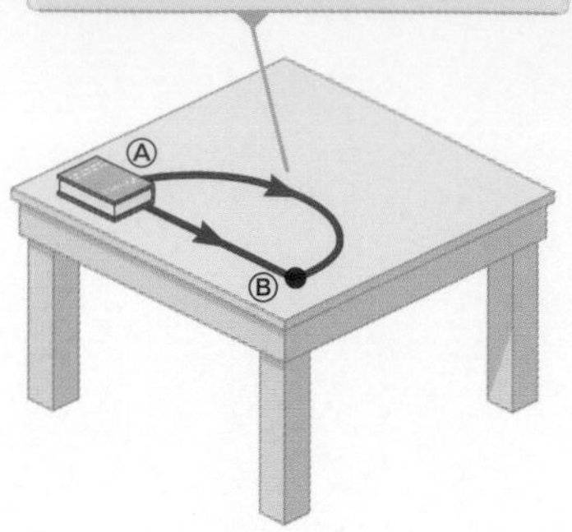

Figure 7.19 The work done against the force of kinetic friction depends on the path taken as the book is moved from Ⓐ to Ⓑ.

Nonconservative Forces

A force is **nonconservative** if it does not satisfy properties 1 and 2 for conservative forces. We define the sum of the kinetic and potential energies of a system as the **mechanical energy** of the system:

$$E_{\text{mech}} \equiv K + U \tag{7.24}$$

where K includes the kinetic energy of all moving members of the system and U includes all types of potential energy in the system. For a book falling under the action of the gravitational force, the mechanical energy of the book–Earth system remains fixed; gravitational potential energy transforms to kinetic energy, and the total energy of the system remains constant. Nonconservative forces acting within a system, however, cause a *change* in the mechanical energy of the system. For example, for a book sent sliding on a horizontal surface that is not frictionless, the mechanical energy of the book–surface system is transformed to internal energy as we discussed earlier. Only part of the book's kinetic energy is transformed to internal energy in the book. The rest appears as internal energy in the surface. (When you trip and slide across a gymnasium floor, not only does the skin on your knees warm up, so does the floor!) Because the force of kinetic friction transforms the mechanical energy of a system into internal energy, it is a nonconservative force.

As an example of the path dependence of the work for a nonconservative force, consider Figure 7.19. Suppose you displace a book between two points on a table. If the book is displaced in a straight line along the blue path between points Ⓐ and Ⓑ in Figure 7.19, you do a certain amount of work against the kinetic friction force to keep the book moving at a constant speed. Now, imagine that you push the book along the brown semicircular path in Figure 7.19. You perform more work against friction along this curved path than along the straight path because the curved path is longer. The work done on the book depends on the path, so the friction force *cannot* be conservative.

7.8 Relationship Between Conservative Forces and Potential Energy

In the preceding section, we found that the work done on a member of a system by a conservative force between the members of the system does not depend on the path taken by the moving member. The work depends only on the initial and final coordinates. For such a system, we can define a **potential energy function** U such that the work done within the system by the conservative force equals the decrease in the potential energy of the system. Let us imagine a system of particles in which a conservative force $\vec{\mathbf{F}}$ acts between the particles. Imagine also that the configuration of the system changes due to the motion of one particle along the x axis. The work done by the force $\vec{\mathbf{F}}$ as the particle moves along the x axis is[4]

$$W_{\text{int}} = \int_{x_i}^{x_f} F_x\,dx = -\Delta U \tag{7.25}$$

where F_x is the component of $\vec{\mathbf{F}}$ in the direction of the displacement. That is, the work done by a conservative force acting between members of a system equals the negative of the change in the potential energy of the system associated with that force when the system's configuration changes. We can also express Equation 7.25 as

$$\Delta U = U_f - U_i = -\int_{x_i}^{x_f} F_x\,dx \tag{7.26}$$

[4]For a general displacement, the work done in two or three dimensions also equals $-\Delta U$, where $U = U(x, y, z)$. We write this equation formally as $W_{\text{int}} = \int_i^f \vec{\mathbf{F}} \cdot d\vec{\mathbf{r}} = U_i - U_f$.

Therefore, ΔU is negative when F_x and dx are in the same direction, as when an object is lowered in a gravitational field or when a spring pushes an object toward equilibrium.

It is often convenient to establish some particular location x_i of one member of a system as representing a reference configuration and measure all potential energy differences with respect to it. We can then define the potential energy function as

$$U_f(x) = -\int_{x_i}^{x_f} F_x\,dx + U_i \tag{7.27}$$

The value of U_i is often taken to be zero for the reference configuration. It does not matter what value we assign to U_i because any nonzero value merely shifts $U_f(x)$ by a constant amount and only the *change* in potential energy is physically meaningful.

If the point of application of the force undergoes an infinitesimal displacement dx, we can express the infinitesimal change in the potential energy of the system dU as

$$dU = -F_x\,dx$$

Therefore, the conservative force is related to the potential energy function through the relationship[5]

$$F_x = -\frac{dU}{dx} \tag{7.28}$$

◀ Relation of force between members of a system to the potential energy of the system

That is, the x component of a conservative force acting on a member within a system equals the negative derivative of the potential energy of the system with respect to x.

We can easily check Equation 7.28 for the two examples already discussed. In the case of the deformed spring, $U_s = \frac{1}{2}kx^2$; therefore,

$$F_s = -\frac{dU_s}{dx} = -\frac{d}{dx}\left(\tfrac{1}{2}kx^2\right) = -kx$$

which corresponds to the restoring force in the spring (Hooke's law). Because the gravitational potential energy function is $U_g = mgy$, it follows from Equation 7.28 that $F_g = -mg$ when we differentiate U_g with respect to y instead of x.

We now see that U is an important function because a conservative force can be derived from it. Furthermore, Equation 7.28 should clarify that adding a constant to the potential energy is unimportant because the derivative of a constant is zero.

***Quick Quiz* 7.8** What does the slope of a graph of $U(x)$ versus x represent? **(a)** the magnitude of the force on the object **(b)** the negative of the magnitude of the force on the object **(c)** the x component of the force on the object **(d)** the negative of the x component of the force on the object

7.9 Energy Diagrams and Equilibrium of a System

The motion of a system can often be understood qualitatively through a graph of its potential energy versus the position of a member of the system. Consider the potential energy function for a block–spring system, given by $U_s = \frac{1}{2}kx^2$. This function is plotted versus x in Active Figure 7.20a, where x is the position of the block. The force F_s exerted by the spring on the block is related to U_s through Equation 7.28:

$$F_s = -\frac{dU_s}{dx} = -kx$$

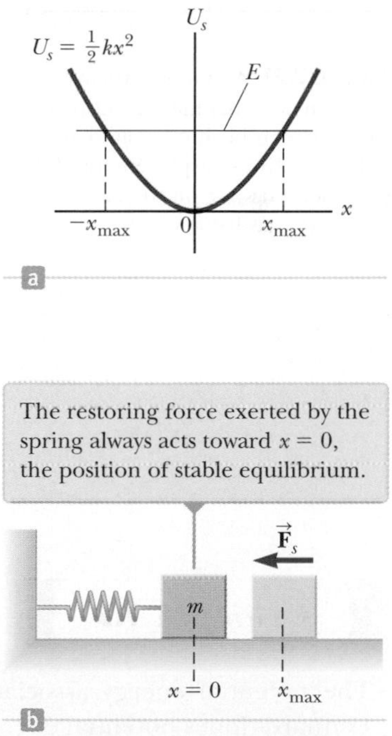

ACTIVE FIGURE 7.20

(a) Potential energy as a function of x for the frictionless block–spring system shown in (b). For a given energy E of the system, the block oscillates between the turning points, which have the coordinates $x = \pm x_{max}$.

[5]In three dimensions, the expression is

$$\vec{\mathbf{F}} = -\frac{\partial U}{\partial x}\hat{\mathbf{i}} - \frac{\partial U}{\partial y}\hat{\mathbf{j}} - \frac{\partial U}{\partial z}\hat{\mathbf{k}}$$

where $(\partial U/\partial x)$ and so forth are partial derivatives. In the language of vector calculus, $\vec{\mathbf{F}}$ equals the negative of the *gradient* of the scalar quantity $U(x, y, z)$.

Pitfall Prevention 7.10

Energy Diagrams
A common mistake is to think that potential energy on the graph in an energy diagram represents the height of some object. For example, that is not the case in Active Figure 7.20, where the block is only moving horizontally.

As we saw in Quick Quiz 7.8, the x component of the force is equal to the negative of the slope of the U-versus-x curve. When the block is placed at rest at the equilibrium position of the spring ($x = 0$), where $F_s = 0$, it will remain there unless some external force F_{ext} acts on it. If this external force stretches the spring from equilibrium, x is positive and the slope dU/dx is positive; therefore, the force F_s exerted by the spring is negative and the block accelerates back toward $x = 0$ when released. If the external force compresses the spring, x is negative and the slope is negative; therefore, F_s is positive and again the mass accelerates toward $x = 0$ upon release.

From this analysis, we conclude that the $x = 0$ position for a block–spring system is one of **stable equilibrium.** That is, any movement away from this position results in a force directed back toward $x = 0$. In general, configurations of a system in stable equilibrium correspond to those for which $U(x)$ for the system is a minimum.

If the block in Active Figure 7.20 is moved to an initial position x_{max} and then released from rest, its total energy initially is the potential energy $\frac{1}{2}kx_{max}^2$ stored in the spring. As the block starts to move, the system acquires kinetic energy and loses potential energy. The block oscillates (moves back and forth) between the two points $x = -x_{max}$ and $x = +x_{max}$, called the *turning points.* In fact, because no energy is transformed to internal energy due to friction, the block oscillates between $-x_{max}$ and $+x_{max}$ forever. (We will discuss these oscillations further in Chapter 15.)

Another simple mechanical system with a configuration of stable equilibrium is a ball rolling about in the bottom of a bowl. Anytime the ball is displaced from its lowest position, it tends to return to that position when released.

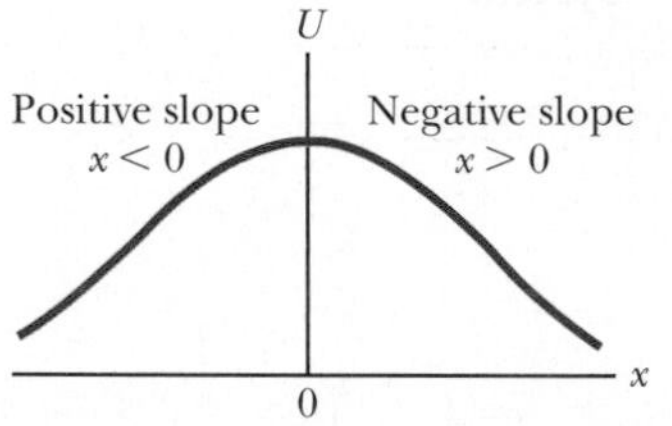

Figure 7.21 A plot of U versus x for a particle that has a position of unstable equilibrium located at $x = 0$. For any finite displacement of the particle, the force on the particle is directed away from $x = 0$.

Now consider a particle moving along the x axis under the influence of a conservative force F_x, where the U-versus-x curve is as shown in Figure 7.21. Once again, $F_x = 0$ at $x = 0$, and so the particle is in equilibrium at this point. This position, however, is one of **unstable equilibrium** for the following reason. Suppose the particle is displaced to the right ($x > 0$). Because the slope is negative for $x > 0$, $F_x = -dU/dx$ is positive and the particle accelerates away from $x = 0$. If instead the particle is at $x = 0$ and is displaced to the left ($x < 0$), the force is negative because the slope is positive for $x < 0$ and the particle again accelerates away from the equilibrium position. The position $x = 0$ in this situation is one of unstable equilibrium because for any displacement from this point, the force pushes the particle farther away from equilibrium and toward a position of lower potential energy. A pencil balanced on its point is in a position of unstable equilibrium. If the pencil is displaced slightly from its absolutely vertical position and is then released, it will surely fall over. In general, configurations of a system in unstable equilibrium correspond to those for which $U(x)$ for the system is a maximum.

Finally, a configuration called **neutral equilibrium** arises when U is constant over some region. Small displacements of an object from a position in this region produce neither restoring nor disrupting forces. A ball lying on a flat, horizontal surface is an example of an object in neutral equilibrium.

Example 7.9 Force and Energy on an Atomic Scale

The potential energy associated with the force between two neutral atoms in a molecule can be modeled by the Lennard–Jones potential energy function:

$$U(x) = 4\epsilon\left[\left(\frac{\sigma}{x}\right)^{12} - \left(\frac{\sigma}{x}\right)^{6}\right]$$

where x is the separation of the atoms. The function $U(x)$ contains two parameters σ and ϵ that are determined from experiments. Sample values for the interaction between two atoms in a molecule are $\sigma = 0.263$ nm and $\epsilon = 1.51 \times 10^{-22}$ J. Using a spreadsheet or similar tool, graph this function and find the most likely distance between the two atoms.

7.9 *cont.*

SOLUTION

Conceptualize We identify the two atoms in the molecule as a system. Based on our understanding that stable molecules exist, we expect to find stable equilibrium when the two atoms are separated by some equilibrium distance.

Categorize Because a potential energy function exists, we categorize the force between the atoms as conservative. For a conservative force, Equation 7.28 describes the relationship between the force and the potential energy function.

Analyze Stable equilibrium exists for a separation distance at which the potential energy of the system of two atoms (the molecule) is a minimum.

Take the derivative of the function $U(x)$:

$$\frac{dU(x)}{dx} = 4\epsilon \frac{d}{dx}\left[\left(\frac{\sigma}{x}\right)^{12} - \left(\frac{\sigma}{x}\right)^{6}\right] = 4\epsilon\left[\frac{-12\sigma^{12}}{x^{13}} + \frac{6\sigma^{6}}{x^{7}}\right]$$

Minimize the function $U(x)$ by setting its derivative equal to zero:

$$4\epsilon\left[\frac{-12\sigma^{12}}{x_{\text{eq}}^{13}} + \frac{6\sigma^{6}}{x_{\text{eq}}^{7}}\right] = 0 \;\rightarrow\; x_{\text{eq}} = (2)^{1/6}\sigma$$

Evaluate x_{eq}, the equilibrium separation of the two atoms in the molecule:

$$x_{\text{eq}} = (2)^{1/6}(0.263 \text{ nm}) = \boxed{2.95 \times 10^{-10} \text{ m}}$$

We graph the Lennard–Jones function on both sides of this critical value to create our energy diagram as shown in Figure 7.22.

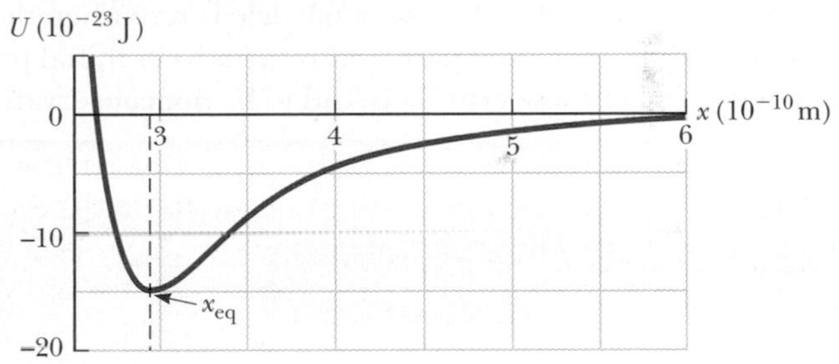

Figure 7.22 (Example 7.9) Potential energy curve associated with a molecule. The distance x is the separation between the two atoms making up the molecule.

Finalize Notice that $U(x)$ is extremely large when the atoms are very close together, is a minimum when the atoms are at their critical separation, and then increases again as the atoms move apart. When $U(x)$ is a minimum, the atoms are in stable equilibrium, indicating that the most likely separation between them occurs at this point.

Summary

Definitions

A **system** is most often a single particle, a collection of particles, or a region of space, and may vary in size and shape. A **system boundary** separates the system from the **environment.**

The **work** W done on a system by an agent exerting a constant force $\vec{\mathbf{F}}$ on the system is the product of the magnitude Δr of the displacement of the point of application of the force and the component $F\cos\theta$ of the force along the direction of the displacement $\Delta\vec{\mathbf{r}}$:

$$W \equiv F\,\Delta r\cos\theta \qquad \textbf{(7.1)}$$

continued

If a varying force does work on a particle as the particle moves along the x axis from x_i to x_f, the work done by the force on the particle is given by

$$W = \int_{x_i}^{x_f} F_x\,dx \quad (7.7)$$

where F_x is the component of force in the x direction.

The **scalar product** (dot product) of two vectors $\vec{\mathbf{A}}$ and $\vec{\mathbf{B}}$ is defined by the relationship

$$\vec{\mathbf{A}} \cdot \vec{\mathbf{B}} \equiv AB\cos\theta \quad (7.2)$$

where the result is a scalar quantity and θ is the angle between the two vectors. The scalar product obeys the commutative and distributive laws.

The **kinetic energy** of a particle of mass m moving with a speed v is

$$K \equiv \tfrac{1}{2}mv^2 \quad (7.16)$$

If a particle of mass m is at a distance y above the Earth's surface, the **gravitational potential energy** of the particle–Earth system is

$$U_g \equiv mgy \quad (7.19)$$

The **elastic potential energy** stored in a spring of force constant k is

$$U_s \equiv \tfrac{1}{2}kx^2 \quad (7.22)$$

A force is **conservative** if the work it does on a particle that is a member of the system as the particle moves between two points is independent of the path the particle takes between the two points. Furthermore, a force is conservative if the work it does on a particle is zero when the particle moves through an arbitrary closed path and returns to its initial position. A force that does not meet these criteria is said to be **nonconservative.**

The **total mechanical energy of a system** is defined as the sum of the kinetic energy and the potential energy:

$$E_{\text{mech}} \equiv K + U \quad (7.24)$$

Concepts and Principles

The **work–kinetic energy theorem** states that if work is done on a system by external forces and the only change in the system is in its speed,

$$W_{\text{ext}} = K_f - K_i = \Delta K = \tfrac{1}{2}mv_f^{\,2} - \tfrac{1}{2}mv_i^{\,2} \quad (7.15, 7.17)$$

A **potential energy function** U can be associated only with a conservative force. If a conservative force $\vec{\mathbf{F}}$ acts between members of a system while one member moves along the x axis from x_i to x_f, the change in the potential energy of the system equals the negative of the work done by that force:

$$U_f - U_i = -\int_{x_i}^{x_f} F_x\,dx \quad (7.26)$$

Systems can be in three types of equilibrium configurations when the net force on a member of the system is zero. Configurations of **stable equilibrium** correspond to those for which $U(x)$ is a minimum.

Configurations of **unstable equilibrium** correspond to those for which $U(x)$ is a maximum.

Neutral equilibrium arises when U is constant as a member of the system moves over some region.

Objective Questions

☐ denotes answer available in Student Solutions Manual/Study Guide

1. A worker pushes a wheelbarrow with a horizontal force of 50 N on level ground over a distance of 5.0 m. If a friction force of 43 N acts on the wheelbarrow in a direction opposite that of the worker, what work is done on the wheelbarrow by the worker? (a) 250 J (b) 215 J (c) 35 J (d) 10 J (e) None of those answers is correct.

2. Is the work required to be done by an external force on an object on a frictionless, horizontal surface to accelerate it from a speed v to a speed $2v$ (a) equal to the work required to accelerate the object from $v = 0$ to v, (b) twice the work required to accelerate the object from $v = 0$ to v, (c) three times the work required to accelerate the object from $v = 0$

to v, (d) four times the work required to accelerate the object from 0 to v, or (e) not known without knowledge of the acceleration?

3. Alex and John are loading identical cabinets onto a truck. Alex lifts his cabinet straight up from the ground to the bed of the truck, whereas John slides his cabinet up a rough ramp to the truck. Which statement is correct about the work done on the cabinet–Earth system? (a) Alex and John do the same amount of work. (b) Alex does more work than John. (c) John does more work than Alex. (d) None of those statements is necessarily true because the force of friction is unknown. (e) None of those statements is necessarily true because the angle of the incline is unknown.

4. Mark and David are loading identical cement blocks onto David's pickup truck. Mark lifts his block straight up from the ground to the truck, whereas David slides his block up a ramp containing frictionless rollers. Which statement is true about the work done on the block–Earth system? (a) Mark does more work than David. (b) Mark and David do the same amount of work. (c) David does more work than Mark. (d) None of those statements is necessarily true because the angle of the incline is unknown. (e) None of those statements is necessarily true because the mass of one block is not given.

5. Bullet 2 has twice the mass of bullet 1. Both are fired so that they have the same speed. If the kinetic energy of bullet 1 is K, is the kinetic energy of bullet 2 (a) $0.25K$, (b) $0.5K$, (c) $0.71K$, (d) K, or (e) $2K$?

6. If the net work done by external forces on a particle is zero, which of the following statements about the particle must be true? (a) Its velocity is zero. (b) Its velocity is decreased. (c) Its velocity is unchanged. (d) Its speed is unchanged. (e) More information is needed.

7. A block of mass m is dropped from the fourth floor of an office building and hits the sidewalk below at speed v. From what floor should the mass be dropped to double that impact speed? (a) the sixth floor (b) the eighth floor (c) the tenth floor (d) the twelfth floor (e) the sixteenth floor

8. As a simple pendulum swings back and forth, the forces acting on the suspended object are (a) the gravitational force, (b) the tension in the supporting cord, and (c) air resistance. **(i)** Which of these forces, if any, does no work on the pendulum at any time? **(ii)** Which of these forces does negative work on the pendulum at all times during its motion?

9. Let $\hat{\mathbf{N}}$ represent the direction horizontally north, $\widehat{\mathbf{NE}}$ represent northeast (halfway between north and east), and so on. Each direction specification can be thought of as a unit vector. Rank from the largest to the smallest the following dot products. Note that zero is larger than a negative number. If two quantities are equal, display that fact in your ranking. (a) $\hat{\mathbf{N}} \cdot \hat{\mathbf{N}}$ (b) $\hat{\mathbf{N}} \cdot \widehat{\mathbf{NE}}$ (c) $\hat{\mathbf{N}} \cdot \hat{\mathbf{S}}$ (d) $\hat{\mathbf{N}} \cdot \hat{\mathbf{E}}$ (e) $\widehat{\mathbf{SE}} \cdot \hat{\mathbf{S}}$

10. Figure OQ7.10 shows a light extended spring exerting a force F_s to the left on a block. **(i)** Does the block exert a force on the spring? Choose every correct answer. (a) No, it doesn't. (b) Yes, it does, to the left. (c) Yes, it does, to the right. (d) Yes, it does, and its magnitude is larger than F_s. (e) Yes, it does, and its magnitude is equal to F_s. **(ii)** Does the spring exert a force on the wall? Choose your answers from the same list (a) through (e).

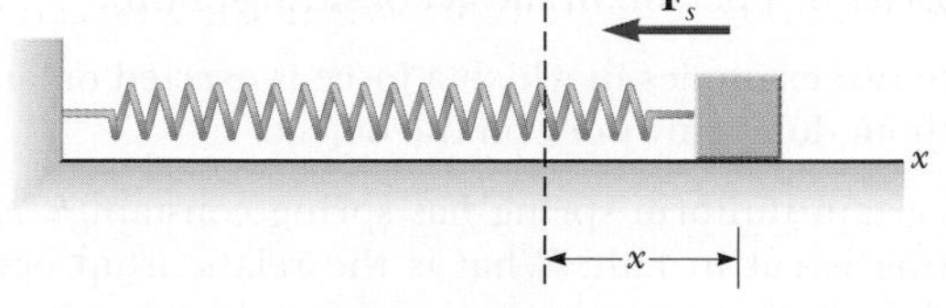

Figure OQ7.10

11. If the speed of a particle is doubled, what happens to its kinetic energy? (a) It becomes four times larger. (b) It becomes two times larger. (c) It becomes $\sqrt{2}$ times larger. (d) It is unchanged. (e) It becomes half as large.

12. A cart is set rolling across a level table, at the same speed on every trial. If it runs into a patch of sand, the cart exerts on the sand an average horizontal force of 6 N and travels a distance of 6 cm through the sand as it comes to a stop. If instead the cart runs into a patch of gravel on which the cart exerts an average horizontal force of 9 N, how far into the gravel will the cart roll before stopping? (a) 9 cm (b) 6 cm (c) 4 cm (d) 3 cm (e) none of those answers

13. A cart is set rolling across a level table, at the same speed on every trial. If it runs into a patch of sand, the cart exerts on the sand an average horizontal force of 6 N and travels a distance of 6 cm through the sand as it comes to a stop. If instead the cart runs into a patch of flour, it rolls an average of 18 cm before stopping. What is the average magnitude of the horizontal force the cart exerts on the flour? (a) 2 N (b) 3 N (c) 6 N (d) 18 N (e) none of those answers

14. A certain spring that obeys Hooke's law is stretched by an external agent. The work done in stretching the spring by 10 cm is 4 J. How much additional work is required to stretch the spring an additional 10 cm? (a) 2 J (b) 4 J (c) 8 J (d) 12 J (e) 16 J

15. **(i)** Rank the gravitational accelerations you would measure for the following falling objects: (a) a 2-kg object 5 cm above the floor, (b) a 2-kg object 120 cm above the floor, (c) a 3-kg object 120 cm above the floor, and (d) a 3-kg object 80 cm above the floor. List the one with the largest magnitude of acceleration first. If any are equal, show their equality in your list. **(ii)** Rank the gravitational forces on the same four objects, listing the one with the largest magnitude first. **(iii)** Rank the gravitational potential energies (of the object–Earth system) for the same four objects, largest first, taking $y = 0$ at the floor.

16. An ice cube has been given a push and slides without friction on a level table. Which is correct? (a) It is in stable equilibrium. (b) It is in unstable equilibrium. (c) It is in neutral equilibrium. (d) It is not in equilibrium.

Conceptual Questions

☐ denotes answer available in Student Solutions Manual/Study Guide

1. Discuss whether any work is being done by each of the following agents and, if so, whether the work is positive or negative. (a) a chicken scratching the ground (b) a person studying (c) a crane lifting a bucket of concrete (d) the gravitational force on the bucket in part (c) (e) the leg muscles of a person in the act of sitting down

2. Cite two examples in which a force is exerted on an object without doing any work on the object.

3. A certain uniform spring has spring constant k. Now the spring is cut in half. What is the relationship between k and the spring constant k' of each resulting smaller spring? Explain your reasoning.

4. (a) For what values of the angle θ between two vectors is their scalar product positive? (b) For what values of θ is their scalar product negative?

[5.] Can kinetic energy be negative? Explain.

6. Discuss the work done by a pitcher throwing a baseball. What is the approximate distance through which the force acts as the ball is thrown?

7. Can a normal force do work? If not, why not? If so, give an example.

8. If only one external force acts on a particle, does it necessarily change the particle's (a) kinetic energy? (b) Its velocity?

9. Does the kinetic energy of an object depend on the frame of reference in which its motion is measured? Provide an example to prove this point.

10. You are reshelving books in a library. You lift a book from the floor to the top shelf. The kinetic energy of the book on the floor was zero and the kinetic energy of the book on the top shelf is zero, so no change occurs in the kinetic energy, yet you did some work in lifting the book. Is the work–kinetic energy theorem violated? Explain.

11. A student has the idea that the total work done on an object is equal to its final kinetic energy. Is this idea true always, sometimes, or never? If it is sometimes true, under what circumstances? If it is always or never true, explain why.

12. What shape would the graph of U versus x have if a particle were in a region of neutral equilibrium?

13. Preparing to clean them, you pop all the removable keys off a computer keyboard. Each key has the shape of a tiny box with one side open. By accident, you spill the keys onto the floor. Explain why many more keys land letter-side down than land open-side down.

14. Object 1 pushes on object 2 as the objects move together, like a bulldozer pushing a stone. Assume object 1 does 15.0 J of work on object 2. Does object 2 do work on object 1? Explain your answer. If possible, determine how much work and explain your reasoning.

Problems

ENHANCED WebAssign The problems found in this chapter may be assigned online in Enhanced WebAssign

1. denotes straightforward problem; **2.** denotes intermediate problem;
3. denotes challenging problem

[1.] full solution available in the Student Solutions Manual/Study Guide

1. denotes problems most often assigned in Enhanced WebAssign; these provide students with targeted feedback and either a Master It tutorial or a Watch It solution video.

Q|C denotes asking for quantitative and conceptual reasoning

S denotes symbolic reasoning problem

M denotes Master It tutorial available in Enhanced WebAssign

GP denotes guided problem

shaded denotes "paired problems" that develop reasoning with symbols and numerical values

Section 7.2 Work Done by a Constant Force

[1.] A block of mass $m = 2.50$ kg is pushed a distance $d = 2.20$ m along a frictionless, horizontal table by a constant applied force of magnitude $F = 16.0$ N directed at an angle $\theta = 25.0°$ below the horizontal as shown in Figure P7.1. Determine the work done on the block by (a) the applied force, (b) the normal force exerted by the table, (c) the gravitational force, and (d) the net force on the block.

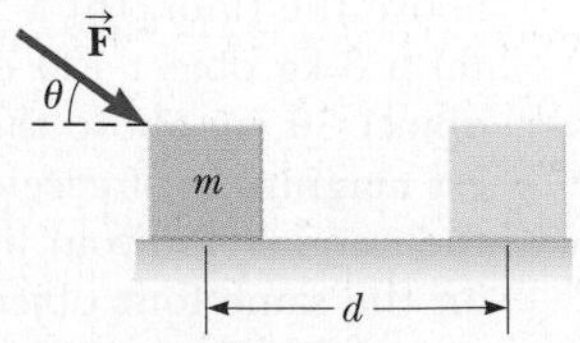

Figure P7.1

2. A raindrop of mass 3.35×10^{-5} kg falls vertically at constant speed under the influence of gravity and air resistance. Model the drop as a particle. As it falls 100 m, what is the work done on the raindrop (a) by the gravitational force and (b) by air resistance?

3. In 1990, Walter Arfeuille of Belgium lifted a 281.5-kg object through a distance of 17.1 cm using only his teeth. (a) How much work was done on the object by Arfeuille in this lift, assuming the object was lifted at constant speed? (b) What total force was exerted on Arfeuille's teeth during the lift?

4. The record number of boat lifts, including the boat and its ten crew members, was achieved by Sami Heinonen and Juha Räsänen of Sweden in 2000. They lifted a total mass of 653.2 kg approximately 4 in. off the ground a total of 24 times. Estimate the total work done by the two men on the boat in this record lift, ignoring the negative work done by the men when they lowered the boat back to the ground.

5. QIC A shopper in a supermarket pushes a cart with a force of 35 N directed at an angle of 25° below the horizontal. The force is just sufficient to balance various friction forces, so the cart moves at constant speed. (a) Find the work done by the shopper on the cart as she moves down a 50.0-m-long aisle. (b) What is the net work done on the cart by all forces? Why? (c) The shopper goes down the next aisle, pushing horizontally and maintaining the same speed as before. If the friction force doesn't change, would the shopper's applied force be larger, smaller, or the same? (d) What about the work done on the cart by the shopper?

6. M Spiderman, whose mass is 80.0 kg, is dangling on the free end of a 12.0-m-long rope, the other end of which is fixed to a tree limb above. By repeatedly bending at the waist, he is able to get the rope in motion, eventually getting it to swing enough that he can reach a ledge when the rope makes a 60.0° angle with the vertical. How much work was done by the gravitational force on Spiderman in this maneuver?

Section 7.3 The Scalar Product of Two Vectors

7. S For any two vectors $\vec{A}$ and $\vec{B}$, show that $\vec{A} \cdot \vec{B} = A_xB_x + A_yB_y + A_zB_z$. *Suggestions:* Write $\vec{A}$ and $\vec{B}$ in unit-vector form and use Equations 7.4 and 7.5.

8. Vector $\vec{A}$ has a magnitude of 5.00 units, and vector $\vec{B}$ has a magnitude of 9.00 units. The two vectors make an angle of 50.0° with each other. Find $\vec{A} \cdot \vec{B}$.

Note: In Problems 9 through 12, calculate numerical answers to three significant figures as usual.

9. M A force $\vec{F} = (6\hat{i} - 2\hat{j})$ N acts on a particle that undergoes a displacement $\Delta\vec{r} = (3\hat{i} + \hat{j})$ m. Find (a) the work done by the force on the particle and (b) the angle between $\vec{F}$ and $\Delta\vec{r}$.

10. Find the scalar product of the vectors in Figure P7.10.

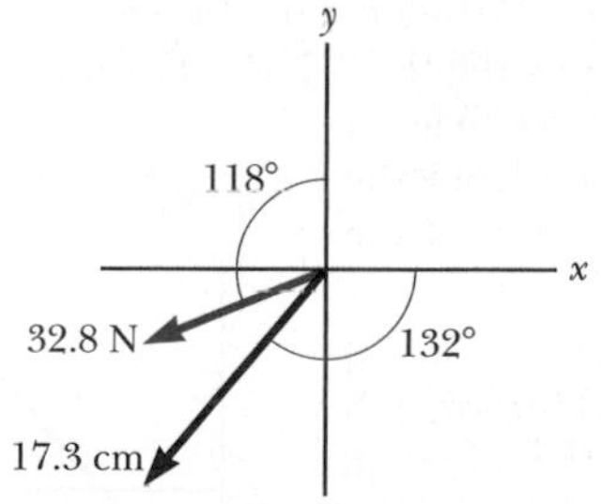

Figure P7.10

11. For $\vec{A} = 3\hat{i} + \hat{j} - \hat{k}$, $\vec{B} = -\hat{i} + 2\hat{j} + 5\hat{k}$, and $\vec{C} = 2\hat{j} - 3\hat{k}$, find $\vec{C} \cdot (\vec{A} - \vec{B})$.

12. Using the definition of the scalar product, find the angles between (a) $\vec{A} = 3\hat{i} - 2\hat{j}$ and $\vec{B} = 4\hat{i} - 4\hat{j}$, (b) $\vec{A} = -2\hat{i} + 4\hat{j}$ and $\vec{B} = 3\hat{i} - 4\hat{j} + 2\hat{k}$, and (c) $\vec{A} = \hat{i} - 2\hat{j} + 2\hat{k}$ and $\vec{B} = 3\hat{j} + 4\hat{k}$.

13. Let $\vec{B} = 5.00$ m at 60.0°. Let the vector $\vec{C}$ have the same magnitude as $\vec{A}$ and a direction angle greater than that of $\vec{A}$ by 25.0°. Let $\vec{A} \cdot \vec{B} = 30.0\ \text{m}^2$ and $\vec{B} \cdot \vec{C} = 35.0\ \text{m}^2$. Find the magnitude and direction of $\vec{A}$.

Section 7.4 Work Done by a Varying Force

14. The force acting on a particle varies as shown in Figure P7.14. Find the work done by the force on the particle as it moves (a) from $x = 0$ to $x = 8.00$ m, (b) from $x = 8.00$ m to $x = 10.0$ m, and (c) from $x = 0$ to $x = 10.0$ m.

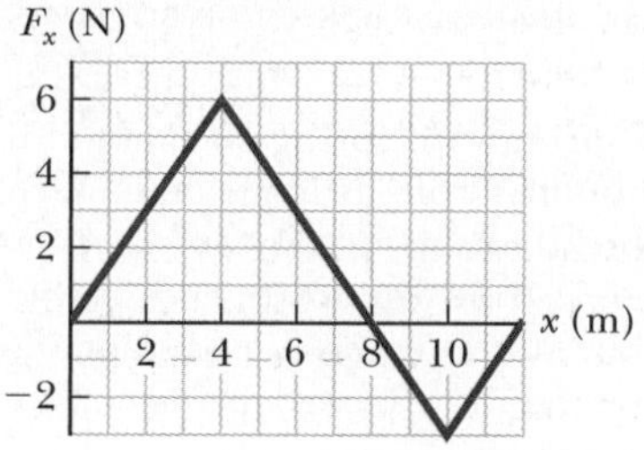

Figure P7.14

15. A particle is subject to a force F_x that varies with position as shown in Figure P7.15. Find the work done by the force on the particle as it moves (a) from $x = 0$ to $x = 5.00$ m, (b) from $x = 5.00$ m to $x = 10.0$ m, and (c) from $x = 10.0$ m to $x = 15.0$ m. (d) What is the total work done by the force over the distance $x = 0$ to $x = 15.0$ m?

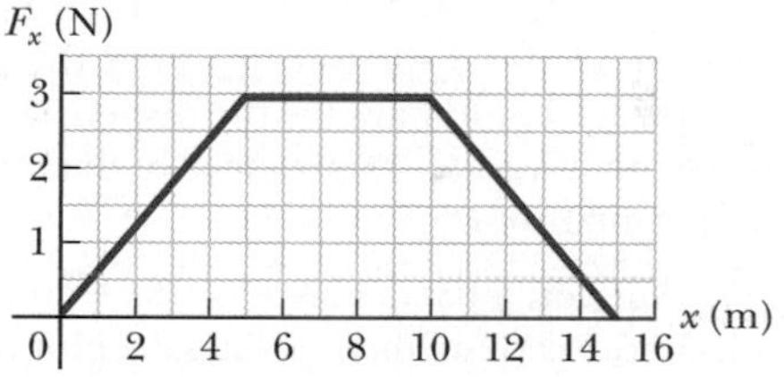

Figure P7.15 Problems 15 and 34.

16. An archer pulls her bowstring back 0.400 m by exerting a force that increases uniformly from zero to 230 N. (a) What is the equivalent spring constant of the bow? (b) How much work does the archer do on the string in drawing the bow?

17. When a 4.00-kg object is hung vertically on a certain light spring that obeys Hooke's law, the spring stretches 2.50 cm. If the 4.00-kg object is removed, (a) how far will the spring stretch if a 1.50-kg block is hung on it? (b) How much work must an external agent do to stretch the same spring 4.00 cm from its unstretched position?

18. Hooke's law describes a certain light spring of unstretched length 35.0 cm. When one end is attached to the top of a doorframe and a 7.50-kg object is hung from the other end, the length of the spring is 41.5 cm. (a) Find its spring constant. (b) The load and the spring are taken down. Two people pull in opposite directions on the ends of the spring, each with a force of 190 N. Find the length of the spring in this situation.

19. In a control system, an accelerometer consists of a 4.70-g object sliding on a calibrated horizontal rail. A low-mass spring attaches the object to a flange at one end of the rail. Grease on the rail makes static friction negligible, but rapidly damps out vibrations of the sliding object. When subject to a steady acceleration of 0.800*g*, the object should be at a location 0.500 cm away from its equilibrium position. Find the force constant of the spring required for the calibration to be correct.

20. A light spring with spring constant 1 200 N/m is hung from an elevated support. From its lower end hangs a second light spring, which has spring constant 1 800 N/m. An object of mass 1.50 kg is hung at rest from the lower end of the second spring. (a) Find the total extension distance of the pair of springs. (b) Find the effective spring constant of the pair of springs as a system. We describe these springs as *in series.*

21. **S** A light spring with spring constant k_1 is hung from an elevated support. From its lower end a second light spring is hung, which has spring constant k_2. An object of mass m is hung at rest from the lower end of the second spring. (a) Find the total extension distance of the pair of springs. (b) Find the effective spring constant of the pair of springs as a system.

22. **S** Express the units of the force constant of a spring in SI fundamental units.

23. A cafeteria tray dispenser supports a stack of trays on a shelf that hangs from four identical spiral springs under tension, one near each corner of the shelf. Each tray is rectangular, 45.3 cm by 35.6 cm, 0.450 cm thick, and with mass 580 g. (a) Demonstrate that the top tray in the stack can always be at the same height above the floor, however many trays are in the dispenser. (b) Find the spring constant each spring should have for the dispenser to function in this convenient way. (c) Is any piece of data unnecessary for this determination?

24. The force acting on a particle is $F_x = (8x - 16)$, where F is in newtons and x is in meters. (a) Make a plot of this force versus x from $x = 0$ to $x = 3.00$ m. (b) From your graph, find the net work done by this force on the particle as it moves from $x = 0$ to $x = 3.00$ m.

25. A force $\vec{\mathbf{F}} = (4x\hat{\mathbf{i}} + 3y\hat{\mathbf{j}})$, where $\vec{\mathbf{F}}$ is in newtons and x and y are in meters, acts on an object as the object moves in the x direction from the origin to $x = 5.00$ m. Find the work $W = \int \vec{\mathbf{F}} \cdot d\vec{\mathbf{r}}$ done by the force on the object.

26. A light spring with force constant 3.85 N/m is compressed by 8.00 cm as it is held between a 0.250-kg block on the left and a 0.500-kg block on the right, both resting on a horizontal surface. The spring exerts a force on each block, tending to push the blocks apart. The blocks are simultaneously released from rest. Find the acceleration with which each block starts to move, given that the coefficient of kinetic friction between each block and the surface is (a) 0, (b) 0.100, and (c) 0.462.

27. A 6 000-kg freight car rolls along rails with negligible friction. The car is brought to rest by a combination of two coiled springs as illustrated in Figure P7.27. Both springs are described by Hooke's law and have spring constants $k_1 = 1\,600$ N/m and $k_2 = 3\,400$ N/m. After the first spring compresses a distance of 30.0 cm, the second spring acts with the first to increase the force as additional compression occurs as shown in the graph. The car comes to rest 50.0 cm after first contacting the two-spring system. Find the car's initial speed.

28. A 100-g bullet is fired from a rifle having a barrel 0.600 m long. Choose the origin to be at the location where the bullet begins to move. Then the force (in newtons) exerted by the expanding gas on the bullet is $15\,000 + 10\,000x - 25\,000x^2$, where x is in meters. (a) Determine the work done by the gas on the bullet as the bullet travels the length of the barrel. (b) **What If?** If the barrel is 1.00 m long, how much work is done, and (c) how does this value compare with the work calculated in part (a)?

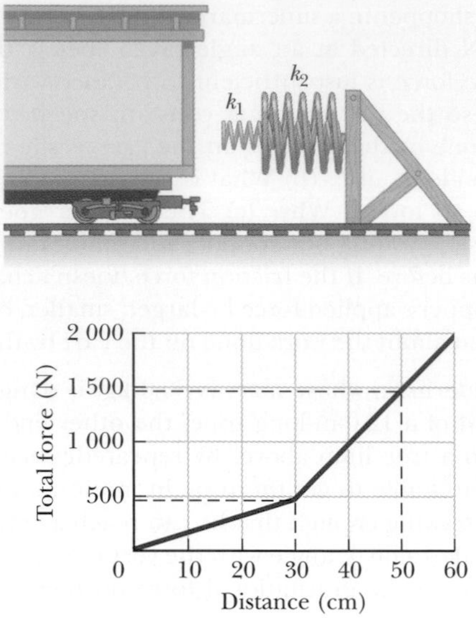

Figure P7.27

29. **S** A small particle of mass m is pulled to the top of a frictionless half-cylinder (of radius R) by a light cord that passes over the top of the cylinder as illustrated in Figure P7.29. (a) Assuming the particle moves at a constant speed, show that $F = mg \cos \theta$. *Note:* If the particle moves at constant speed, the component of its acceleration tangent to the cylinder must be zero at all times. (b) By directly integrating $W = \int \vec{\mathbf{F}} \cdot d\vec{\mathbf{r}}$, find the work done in moving the particle at constant speed from the bottom to the top of the half-cylinder.

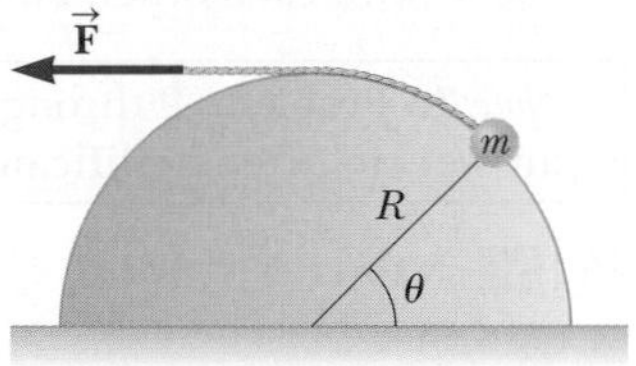

Figure P7.29

30. **Review.** The graph in Figure P7.30 specifies a functional relationship between the two variables u and v. (a) Find $\int_a^b u\,dv$. (b) Find $\int_b^a u\,dv$. (c) Find $\int_a^b v\,du$.

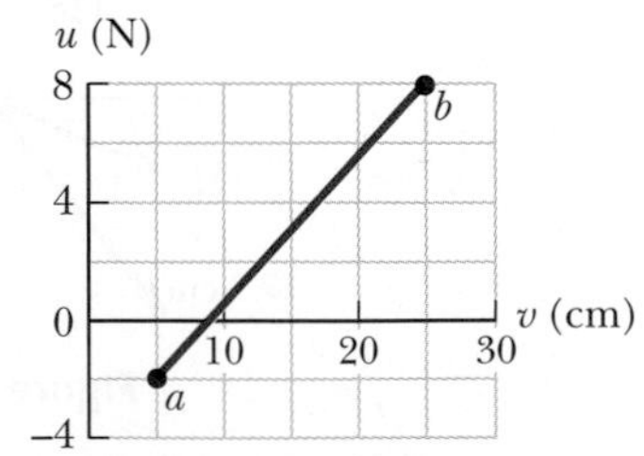

Figure P7.30

Section 7.5 Kinetic Energy and the Work–Kinetic Energy Theorem

31. A 0.600-kg particle has a speed of 2.00 m/s at point Ⓐ and kinetic energy of 7.50 J at point Ⓑ. What is (a) its kinetic energy at Ⓐ, (b) its speed at Ⓑ, and (c) the net work done on the particle by external forces as it moves from Ⓐ to Ⓑ?

32. **Q|C** A worker pushing a 35.0-kg wooden crate at a constant speed for 12.0 m along a wood floor does 350 J of work by applying a constant horizontal force of magnitude F on the crate. (a) Determine the value of F. (b) If the worker now applies a force greater than F, describe the subsequent motion of the crate. (c) Describe what would happen to the crate if the applied force is less than F.

33. A 3.00-kg object has a velocity $(6.00\hat{\mathbf{i}} - 1.00\hat{\mathbf{j}})$ m/s. (a) What is its kinetic energy at this moment? (b) What is the net work done on the object if its velocity changes to $(8.00\hat{\mathbf{i}} + 4.00\hat{\mathbf{j}})$ m/s. (*Note:* From the definition of the dot product, $v^2 = \vec{\mathbf{v}} \cdot \vec{\mathbf{v}}$.)

34. A 4.00-kg particle is subject to a net force that varies with position as shown in Figure P7.15. The particle starts from rest at $x = 0$. What is its speed at (a) $x = 5.00$ m, (b) $x = 10.0$ m, and (c) $x = 15.0$ m?

35. **M** A 2 100-kg pile driver is used to drive a steel I-beam into the ground. The pile driver falls 5.00 m before coming into contact with the top of the beam, and it drives the beam 12.0 cm farther into the ground before coming to rest. Using energy considerations, calculate the average force the beam exerts on the pile driver while the pile driver is brought to rest.

36. **Review.** In an electron microscope, there is an electron gun that contains two charged metallic plates 2.80 cm apart. An electric force accelerates each electron in the beam from rest to 9.60% of the speed of light over this distance. (a) Determine the kinetic energy of the electron as it leaves the electron gun. Electrons carry this energy to a phosphorescent viewing screen where the microscope's image is formed, making it glow. For an electron passing between the plates in the electron gun, determine (b) the magnitude of the constant electric force acting on the electron, (c) the acceleration of the electron, and (d) the time interval the electron spends between the plates.

37. **Review.** A 5.75-kg object passes through the origin at time $t = 0$ such that its x component of velocity is 5.00 m/s and its y component of velocity is -3.00 m/s. (a) What is the kinetic energy of the object at this time? (b) At a later time $t = 2.00$ s, the particle is located at $x = 8.50$ m and $y = 5.00$ m. What constant force acted on the object during this time interval? (c) What is the speed of the particle at $t = 2.00$ s?

38. **Review.** A 7.80-g bullet moving at 575 m/s strikes the hand of a superhero, causing the hand to move 5.50 cm in the direction of the bullet's velocity before stopping. (a) Use work and energy considerations to find the average force that stops the bullet. (b) Assuming the force is constant, determine how much time elapses between the moment the bullet strikes the hand and the moment it stops moving.

39. **GP** **Q|C** **Review.** You can think of the work–kinetic energy theorem as a second theory of motion, parallel to Newton's laws in describing how outside influences affect the motion of an object. In this problem, solve parts (a), (b), and (c) separately from parts (d) and (e) so you can compare the predictions of the two theories. A 15.0-g bullet is accelerated from rest to a speed of 780 m/s in a rifle barrel of length 72.0 cm. (a) Find the kinetic energy of the bullet as it leaves the barrel. (b) Use the work–kinetic energy theorem to find the net work that is done on the bullet. (c) Use your result to part (b) to find the magnitude of the average net force that acted on the bullet while it was in the barrel. (d) Now model the bullet as a particle under constant acceleration. Find the constant acceleration of a bullet that starts from rest and gains a speed of 780 m/s over a distance of 72.0 cm. (e) Modeling the bullet as a particle under a net force, find the net force that acted on it during its acceleration. (f) What conclusion can you draw from comparing your results of parts (c) and (e)?

Section 7.6 Potential Energy of a System

40. A 1 000-kg roller coaster car is initially at the top of a rise, at point Ⓐ. It then moves 135 ft, at an angle of 40.0° below the horizontal, to a lower point Ⓑ. (a) Choose the car at point Ⓑ to be the zero configuration for gravitational potential energy of the roller coaster–Earth system. Find the potential energy of the system when the car is at points Ⓐ and Ⓑ, and the change in potential energy as the car moves between these points. (b) Repeat part (a), setting the zero configuration with the car at point Ⓐ.

41. A 0.20-kg stone is held 1.3 m above the top edge of a water well and then dropped into it. The well has a depth of 5.0 m. Relative to the configuration with the stone at the top edge of the well, what is the gravitational potential energy of the stone–Earth system (a) before the stone is released and (b) when it reaches the bottom of the well? (c) What is the change in gravitational potential energy of the system from release to reaching the bottom of the well?

42. A 400-N child is in a swing that is attached to a pair of ropes 2.00 m long. Find the gravitational potential energy of the child–Earth system relative to the child's lowest position when (a) the ropes are horizontal, (b) the ropes make a 30.0° angle with the vertical, and (c) the child is at the bottom of the circular arc.

Section 7.7 Conservative and Nonconservative Forces

43. **Q|C** A 4.00-kg particle moves from the origin to position Ⓒ, having coordinates $x = 5.00$ m and $y = 5.00$ m (Fig. P7.43). One force on the particle is the gravitational force acting in the negative y direction. Using Equation 7.3, calculate the work done by the gravitational force on the particle as it goes from O to Ⓒ along (a) the purple path, (b) the red path, and (c) the blue path. (d) Your results should all be identical. Why?

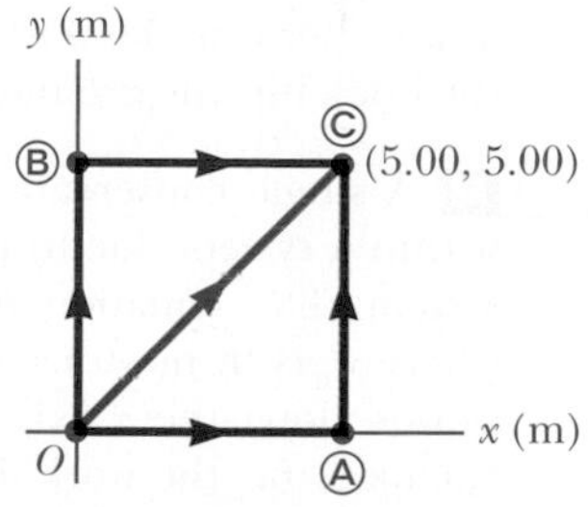

Figure P7.43
Problems 43 through 46.

44. (a) Suppose a constant force acts on an object. The force does not vary with time or with the position or the velocity of the object. Start with the general definition for work done by a force

$$W = \int_i^f \vec{\mathbf{F}} \cdot d\vec{\mathbf{r}}$$

and show that the force is conservative. (b) As a special case, suppose the force $\vec{\mathbf{F}} = (3\hat{\mathbf{i}} + 4\hat{\mathbf{j}})$ N acts on a particle that moves from O to © in Figure P7.43. Calculate the work done by $\vec{F}$ on the particle as it moves along each one of the three paths shown in the figure and show that the work done along the three paths is identical.

45. **Q|C** A force acting on a particle moving in the xy plane is given by $\vec{F} = (2y\hat{i} + x^2\hat{j})$, where $\vec{F}$ is in newtons and x and y are in meters. The particle moves from the origin to a final position having coordinates $x = 5.00$ m and $y = 5.00$ m as shown in Figure P7.43. Calculate the work done by $\vec{F}$ on the particle as it moves along (a) the purple path, (b) the red path, and (c) the blue path. (d) Is $\vec{F}$ conservative or nonconservative? (e) Explain your answer to part (d).

46. **Q|C** An object moves in the xy plane in Figure P7.43 and experiences a friction force with constant magnitude 3.00 N, always acting in the direction opposite the object's velocity. Calculate the work that you must do to slide the object at constant speed against the friction force as the object moves along (a) the purple path O to Ⓐ followed by a return purple path to O, (b) the purple path O to © followed by a return blue path to O, and (c) the blue path O to © followed by a return blue path to O. (d) Each of your three answers should be nonzero. What is the significance of this observation?

Section 7.8 Relationship Between Conservative Forces and Potential Energy

47. **S** The potential energy of a system of two particles separated by a distance r is given by $U(r) = A/r$, where A is a constant. Find the radial force $\vec{F}_r$ that each particle exerts on the other.

48. *Why is the following situation impossible?* A librarian lifts a book from the ground to a high shelf, doing 20.0 J of work in the lifting process. As he turns his back, the book falls off the shelf back to the ground. The gravitational force from the Earth on the book does 20.0 J of work on the book while it falls. Because the work done was 20.0 J + 20.0 J = 40.0 J, the book hits the ground with 40.0 J of kinetic energy.

49. **M** A single conservative force acts on a 5.00-kg particle within a system due to its interaction with the rest of the system. The equation $F_x = 2x + 4$ describes the force, where F_x is in newtons and x is in meters. As the particle moves along the x axis from $x = 1.00$ m to $x = 5.00$ m, calculate (a) the work done by this force on the particle, (b) the change in the potential energy of the system, and (c) the kinetic energy the particle has at $x = 5.00$ m if its speed is 3.00 m/s at $x = 1.00$ m.

50. A single conservative force acting on a particle within a system varies as $\vec{F} = (-Ax + Bx^2)\hat{i}$, where A and B are constants, $\vec{F}$ is in newtons, and x is in meters. (a) Calculate the potential energy function $U(x)$ associated with this force for the system, taking $U = 0$ at $x = 0$. Find (b) the change in potential energy and (c) the change in kinetic energy of the system as the particle moves from $x = 2.00$ m to $x = 3.00$ m.

51. A potential energy function for a system in which a two-dimensional force acts is of the form $U = 3x^3y - 7x$. Find the force that acts at the point (x, y).

Section 7.9 Energy Diagrams and Equilibrium of a System

52. For the potential energy curve shown in Figure P7.52, (a) determine whether the force F_x is positive, negative, or zero at the five points indicated. (b) Indicate points of stable, unstable, and neutral equilibrium. (c) Sketch the curve for F_x versus x from $x = 0$ to $x = 9.5$ m.

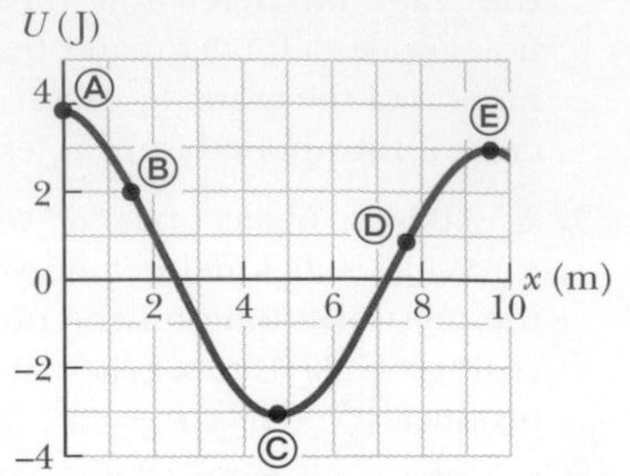

Figure P7.52

53. A right circular cone can theoretically be balanced on a horizontal surface in three different ways. Sketch these three equilibrium configurations and identify them as positions of stable, unstable, or neutral equilibrium.

Additional Problems

54. **S** When an object is displaced by an amount x from stable equilibrium, a restoring force acts on it, tending to return the object to its equilibrium position. The magnitude of the restoring force can be a complicated function of x. In such cases, we can generally imagine the force function $F(x)$ to be expressed as a power series in x as $F(x) = -(k_1x + k_2x^2 + k_3x^3 + \ldots)$. The first term here is just Hooke's law, which describes the force exerted by a simple spring for small displacements. For small excursions from equilibrium, we generally ignore the higher-order terms, but in some cases it may be desirable to keep the second term as well. If we model the restoring force as $F = -(k_1x + k_2x^2)$, how much work is done on an object in displacing it from $x = 0$ to $x = x_{max}$ by an applied force $-F$?

55. **Review.** A baseball outfielder throws a 0.150-kg baseball at a speed of 40.0 m/s and an initial angle of 30.0° to the horizontal. What is the kinetic energy of the baseball at the highest point of its trajectory?

56. The spring constant of an automotive suspension spring increases with increasing load due to a spring coil that is widest at the bottom, smoothly tapering to a smaller diameter near the top. The result is a softer ride on normal road surfaces from the wider coils, but the car does not bottom out on bumps because when the lower coils collapse, the stiffer coils near the top absorb the load. For such springs, the force exerted by the spring can be empirically found to be given by $F = ax^b$. For a tapered spiral spring that compresses 12.9 cm with a 1 000-N load and 31.5 cm with a 5 000-N load, (a) evaluate the constants a and b in the empirical equation for F and (b) find the work needed to compress the spring 25.0 cm.

57. **Q|C** Two identical steel balls, each of diameter 25.4 mm and moving in opposite directions at 5 m/s, run into each other head-on and bounce apart. Prior to the collision, one of the balls is squeezed in a vise while precise measurements are made of the resulting amount of compression. The results show that Hooke's law is a fair model of the ball's elastic behavior. For one datum, a force of 16 kN exerted by each jaw of the vise results in a 0.2-mm reduction in the diameter. The diameter returns to its original value when the force is removed. (a) Modeling the ball as

a spring, find its spring constant. (b) Does the interaction of the balls during the collision last only for an instant or for a nonzero time interval? State your evidence. (c) Compute an estimate for the kinetic energy of each of the balls before they collide. (d) Compute an estimate for the maximum amount of compression each ball undergoes when the balls collide. (e) Compute an order-of-magnitude estimate for the time interval for which the balls are in contact. (In Chapter 15, you will learn to calculate the contact time interval precisely.)

58. The potential energy function for a system of particles is given by $U(x) = -x^3 + 2x^2 + 3x$, where x is the position of one particle in the system. (a) Determine the force F_x on the particle as a function of x. (b) For what values of x is the force equal to zero? (c) Plot $U(x)$ versus x and F_x versus x and indicate points of stable and unstable equilibrium.

59. **Q|C** (a) Take $U = 5$ for a system with a particle at position $x = 0$ and calculate the potential energy of the system as a function of the particle position x. The force on the particle is given by $(8e^{-2x})\hat{\mathbf{i}}$. (b) Explain whether the force is conservative or nonconservative and how you can tell.

60. *Why is the following situation impossible?* In a new casino, a supersized pinball machine is introduced. Casino advertising boasts that a professional basketball player can lie on top of the machine and his head and feet will not hang off the edge! The ball launcher in the machine sends metal balls up one side of the machine and then into play. The spring in the launcher (Fig. P7.60) has a force constant of 1.20 N/cm. The surface on which the ball moves is inclined $\theta = 10.0°$ with respect to the horizontal. The spring is initially compressed its maximum distance $d = 5.00$ cm. A ball of mass 100 g is projected into play by releasing the plunger. Casino visitors find the play of the giant machine quite exciting.

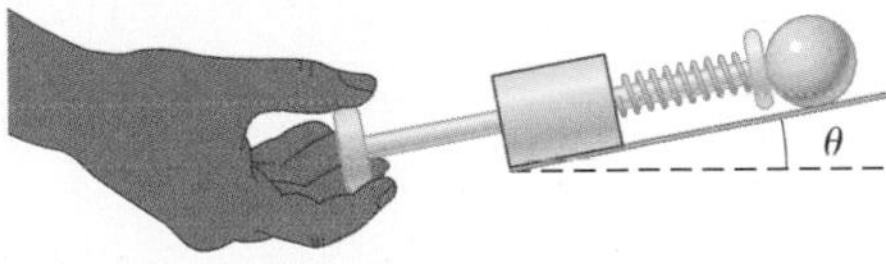

Figure P7.60

61. **Q|C** **Review.** Two constant forces act on an object of mass $m = 5.00$ kg moving in the xy plane as shown in Figure P7.61. Force $\vec{\mathbf{F}}_1$ is 25.0 N at 35.0°, and force $\vec{\mathbf{F}}_2$ is 42.0 N at 150°. At time $t = 0$, the object is at the origin and has velocity $(4.00\hat{\mathbf{i}} + 2.50\hat{\mathbf{j}})$ m/s. (a) Express the two forces in unit-vector notation. Use unit-vector notation for your other answers. (b) Find the total force exerted on the object. (c) Find the object's acceleration. Now, considering the instant $t = 3.00$ s, find (d) the object's velocity, (e) its position, (f) its kinetic energy from $\frac{1}{2}mv_f^2$, and (g) its kinetic energy from $\frac{1}{2}mv_i^2 + \sum \vec{\mathbf{F}} \cdot \Delta\vec{\mathbf{r}}$. (h) What conclusion can you draw by comparing the answers to parts (f) and (g)?

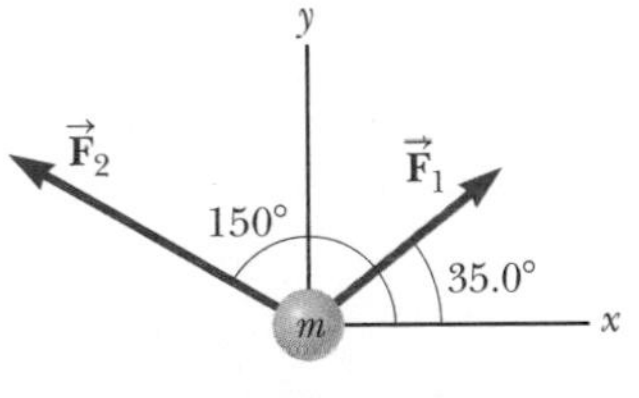

Figure P7.61

62. A particle moves along the x axis from $x = 12.8$ m to $x = 23.7$ m under the influence of a force

$$F = \frac{375}{x^3 + 3.75x}$$

where F is in newtons and x is in meters. Using numerical integration, determine the work done by this force on the particle during this displacement. Your result should be accurate to within 2%.

63. An inclined plane of angle $\theta = 20.0°$ has a spring of force constant $k = 500$ N/m fastened securely at the bottom so that the spring is parallel to the surface as shown in Figure P7.63. A block of mass $m = 2.50$ kg is placed on the plane at a distance $d = 0.300$ m from the spring. From this position, the block is projected downward toward the spring with speed $v = 0.750$ m/s. By what distance is the spring compressed when the block momentarily comes to rest?

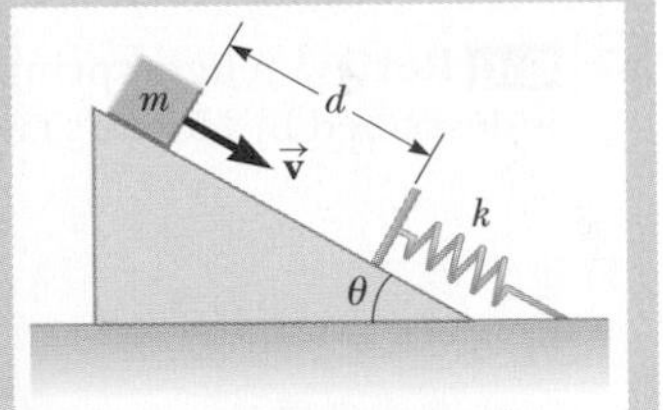

Figure P7.63
Problems 63 and 64.

64. **S** An inclined plane of angle θ has a spring of force constant k fastened securely at the bottom so that the spring is parallel to the surface. A block of mass m is placed on the plane at a distance d from the spring. From this position, the block is projected downward toward the spring with speed v as shown in Figure P7.63. By what distance is the spring compressed when the block momentarily comes to rest?

65. **Q|C** When different loads hang on a spring, the spring stretches to different lengths as shown in the following table. (a) Make a graph of the applied force versus the extension of the spring. (b) By least-squares fitting, determine the straight line that best fits the data. (c) To complete part (b), do you want to use all the data points, or should you ignore some of them? Explain. (d) From the slope of the best-fit line, find the spring constant k. (e) If the spring is extended to 105 mm, what force does it exert on the suspended object?

F **(N)**	2.0	4.0	6.0	8.0	10	12	14	16	18	20	22
L **(mm)**	15	32	49	64	79	98	112	126	149	175	190

Challenge Problems

66. A particle of mass $m = 1.18$ kg is attached between two identical springs on a frictionless, horizontal tabletop. Both springs have spring constant k and are initially unstressed, and the particle is at $x = 0$. (a) The particle is pulled a distance x along a direction perpendicular to the initial configuration of the springs as shown in Figure P7.66. Show that the force exerted by the springs on the particle is

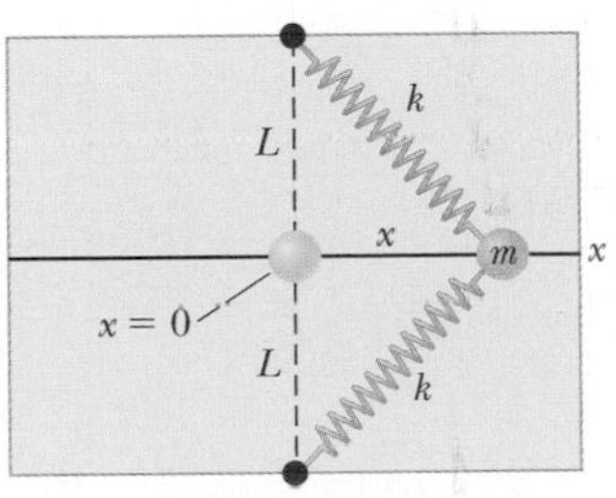

Figure P7.66

$$\vec{\mathbf{F}} = -2kx\left(1 - \frac{L}{\sqrt{x^2 + L^2}}\right)\hat{\mathbf{i}}$$

(b) Show that the potential energy of the system is

$$U(x) = kx^2 + 2kL(L - \sqrt{x^2 + L^2})$$

(c) Make a plot of $U(x)$ versus x and identify all equilibrium points. Assume $L = 1.20$ m and $k = 40.0$ N/m. (d) If the particle is pulled 0.500 m to the right and then released, what is its speed when it reaches $x = 0$?

67. **Q|C Review.** A light spring has unstressed length 15.5 cm. It is described by Hooke's law with spring constant 4.30 N/m. One end of the horizontal spring is held on a fixed vertical axle, and the other end is attached to a puck of mass m that can move without friction over a horizontal surface. The puck is set into motion in a circle with a period of 1.30 s. (a) Find the extension of the spring x as it depends on m. Evaluate x for (b) $m = 0.070\,0$ kg, (c) $m = 0.140$ kg, (d) $m = 0.180$ kg, and (e) $m = 0.190$ kg. (f) Describe the pattern of variation of x as it depends on m.

chapter 8

Conservation of Energy

Three youngsters enjoy the transformation of potential energy to kinetic energy on a waterslide. We can analyze processes such as these with the techniques developed in this chapter. *(Jade Lee/Asia Images/Getty Images)*

In Chapter 7, we introduced three methods for storing energy in a system: kinetic energy, associated with movement of members of the system; potential energy, determined by the configuration of the system; and internal energy, which is related to the temperature of the system.

We now consider analyzing physical situations using the energy approach for two types of systems: *nonisolated* and *isolated* systems. For nonisolated systems, we shall investigate ways that energy can cross the boundary of the system, resulting in a change in the system's total energy. This analysis leads to a critically important principle called *conservation of energy*. The conservation of energy principle extends well beyond physics and can be applied to biological organisms, technological systems, and engineering situations.

In isolated systems, energy does not cross the boundary of the system. For these systems, the total energy of the system is constant. If no nonconservative forces act within the system, we can use *conservation of mechanical energy* to solve a variety of problems.

Situations involving the transformation of mechanical energy to internal energy due to nonconservative forces require special handling. We investigate the procedures for these types of problems.

Finally, we recognize that energy can cross the boundary of a system at different rates. We describe the rate of energy transfer with the quantity *power.*

8.1 Analysis Model: Nonisolated System (Energy)

As we have seen, an object, modeled as a particle, can be acted on by various forces, resulting in a change in its kinetic energy. If we choose the object as the system, this very simple situation is the first example of a *nonisolated system,* for which energy crosses the boundary of the system during some time interval due to an interaction with the environment. This scenario is common in physics problems. If a system does not interact with its environment, it is an isolated system, which we will study in Section 8.2.

The work–kinetic energy theorem from Chapter 7 is our first example of an energy equation appropriate for a nonisolated system. In the case of that theorem, the interaction of the system with its environment is the work done by the external force, and the quantity in the system that changes is the kinetic energy.

So far, we have seen only one way to transfer energy into a system: work. We mention below a few other ways to transfer energy into or out of a system. The details of these processes will be studied in other sections of the book. We illustrate mechanisms to transfer energy in Figure 8.1 and summarize them as follows.

Work, as we have learned in Chapter 7, is a method of transferring energy to a system by applying a force to the system such that the point of application of the force undergoes a displacement (Fig. 8.1a).

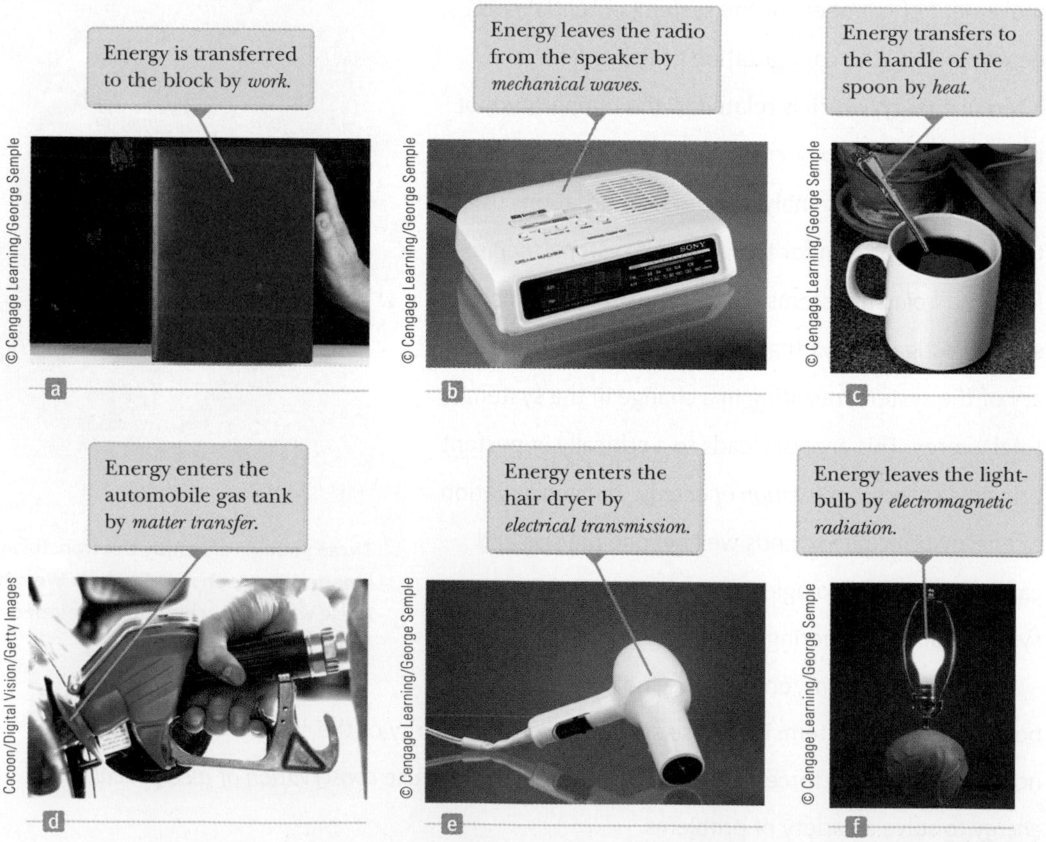

Figure 8.1 Energy transfer mechanisms. In each case, the system into which or from which energy is transferred is indicated.

Mechanical waves (Chapters 16–18) are a means of transferring energy by allowing a disturbance to propagate through air or another medium. It is the method by which energy (which you detect as sound) leaves the system of your clock radio through the loudspeaker and enters your ears to stimulate the hearing process (Fig. 8.1b). Other examples of mechanical waves are seismic waves and ocean waves.

Heat (Chapter 20) is a mechanism of energy transfer that is driven by a temperature difference between a system and its environment. For example, imagine dividing a metal spoon into two parts: the handle, which we identify as the system, and the portion submerged in a cup of coffee, which is part of the environment (Fig. 8.1c). The handle of the spoon becomes hot because fast-moving electrons and atoms in the submerged portion bump into slower ones in the nearby part of the handle. These particles move faster because of the collisions and bump into the next group of slow particles. Therefore, the internal energy of the spoon handle rises from energy transfer due to this collision process.

Pitfall Prevention 8.1

Heat Is Not a Form of Energy

The word *heat* is one of the most misused words in our popular language. Heat is a method of *transferring* energy, *not* a form of storing energy. Therefore, phrases such as "heat content," "the heat of the summer," and "the heat escaped" all represent uses of this word that are inconsistent with our physics definition. See Chapter 20.

Matter transfer (Chapter 20) involves situations in which matter physically crosses the boundary of a system, carrying energy with it. Examples include filling your automobile tank with gasoline (Fig. 8.1d) and carrying energy to the rooms of your home by circulating warm air from the furnace, a process called *convection*.

Electrical transmission (Chapters 27 and 28) involves energy transfer into or out of a system by means of electric currents. It is how energy transfers into your hair dryer (Fig. 8.1e), stereo system, or any other electrical device.

Electromagnetic radiation (Chapter 34) refers to electromagnetic waves such as light, microwaves, and radio waves (Fig. 8.1f) crossing the boundary of a system. Examples of this method of transfer include cooking a baked potato in your microwave oven and light energy traveling from the Sun to the Earth through space.[1]

A central feature of the energy approach is the notion that we can neither create nor destroy energy, that energy is always *conserved.* This feature has been tested in countless experiments, and no experiment has ever shown this statement to be incorrect. Therefore, **if the total amount of energy in a system changes, it can *only* be because energy has crossed the boundary of the system by a transfer mechanism such as one of the methods listed above.**

Energy is one of several quantities in physics that are conserved. We will see other conserved quantities in subsequent chapters. There are many physical quantities that do not obey a conservation principle. For example, there is no conservation of force principle or conservation of velocity principle. Similarly, in areas other than physical quantities, such as in everyday life, some quantities are conserved and some are not. For example, the money in the system of your bank account is a conserved quantity. The only way the account balance changes is if money crosses the boundary of the system by deposits or withdrawals. On the other hand, the number of people in the system of a country is not conserved. Although people indeed cross the boundary of the system, which changes the total population, the population can also change by people dying and by giving birth to new babies. Even if no people cross the system boundary, the births and deaths will change the number of people in the system. There is no equivalent in the concept of energy to dying or giving birth. The general statement of the principle of **conservation of energy** can be described mathematically with the **conservation of energy equation** as follows:

$$\Delta E_{\text{system}} = \sum T \tag{8.1}$$

◀ **Conservation of energy**

where E_{system} is the total energy of the system, including all methods of energy storage (kinetic, potential, and internal), and T (for *transfer*) is the amount of energy transferred across the system boundary by some mechanism. Two of our transfer mechanisms have well-established symbolic notations. For work, $T_{\text{work}} = W$ as discussed in Chapter 7, and for heat, $T_{\text{heat}} = Q$ as defined in Chapter 20. (Now that we

[1]Electromagnetic radiation and work done by field forces are the only energy transfer mechanisms that do not require molecules of the environment to be available at the system boundary. Therefore, systems surrounded by a vacuum (such as planets) can only exchange energy with the environment by means of these two possibilities.

are familiar with work, we can simplify the appearance of equations by letting the simple symbol W represent the external work W_{ext} on a system. For internal work, we will always use W_{int} to differentiate it from W.) The other four members of our list do not have established symbols, so we will call them T_{MW} (mechanical waves), T_{MT} (matter transfer), T_{ET} (electrical transmission), and T_{ER} (electromagnetic radiation).

The full expansion of Equation 8.1 is

$$\Delta K + \Delta U + \Delta E_{int} = W + Q + T_{MW} + T_{MT} + T_{ET} + T_{ER} \quad (8.2)$$

which is the primary mathematical representation of the energy version of the analysis model of the **nonisolated system.** (We will see other versions of the nonisolated system model, involving linear momentum and angular momentum, in later chapters.) In most cases, Equation 8.2 reduces to a much simpler one because some of the terms are zero. If, for a given system, all terms on the right side of the conservation of energy equation are zero, the system is an *isolated system,* which we study in the next section.

The conservation of energy equation is no more complicated in theory than the process of balancing your checking account statement. If your account is the system, the change in the account balance for a given month is the sum of all the transfers: deposits, withdrawals, fees, interest, and checks written. You may find it useful to think of energy as the *currency of nature!*

Suppose a force is applied to a nonisolated system and the point of application of the force moves through a displacement. Then suppose the only effect on the system is to change its speed. In this case, the only transfer mechanism is work (so that the right side of Eq. 8.2 reduces to just W) and the only kind of energy in the system that changes is the kinetic energy (so that ΔE_{system} reduces to just ΔK). Equation 8.2 then becomes

$$\Delta K = W$$

which is the work–kinetic energy theorem. This theorem is a special case of the more general principle of conservation of energy. We shall see several more special cases in future chapters.

***Quick Quiz* 8.1** By what transfer mechanisms does energy enter and leave **(a)** your television set? **(b)** Your gasoline-powered lawn mower? **(c)** Your hand-cranked pencil sharpener?

***Quick Quiz* 8.2** Consider a block sliding over a horizontal surface with friction. Ignore any sound the sliding might make. **(i)** If the system is the *block,* this system is (a) isolated (b) nonisolated (c) impossible to determine **(ii)** If the system is the *surface,* describe the system from the same set of choices. **(iii)** If the system is the *block and the surface,* describe the system from the same set of choices.

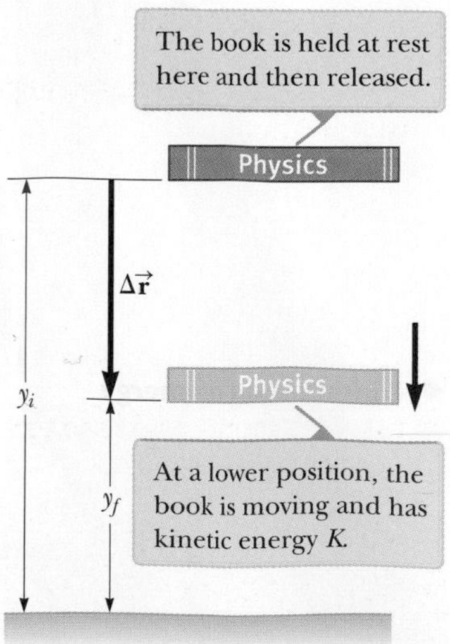

Figure 8.2 A book is released from rest and falls due to work done by the gravitational force on the book.

8.2 Analysis Model: Isolated System (Energy)

In this section, we study another very common scenario in physics problems: a system is chosen such that no energy crosses the system boundary by any method. We begin by considering a gravitational situation. Think about the book–Earth system in Active Figure 7.15 in the preceding chapter. After we have lifted the book, there is gravitational potential energy stored in the system, which can be calculated from the work done by the external agent on the system, using $W = \Delta U_g$.

Let us now shift our focus to the work done *on the book alone* by the gravitational force (Fig. 8.2) as the book falls back to its original height. As the book falls from y_i to y_f, the work done by the gravitational force on the book is

$$W_{on\ book} = (m\vec{g}) \cdot \Delta\vec{r} = (-mg\hat{j}) \cdot [(y_f - y_i)\hat{j}] = mgy_i - mgy_f \quad (8.3)$$

From the work–kinetic energy theorem of Chapter 7, the work done on the book is equal to the change in the kinetic energy of the book:

$$W_{\text{on book}} = \Delta K_{\text{book}}$$

We can equate these two expressions for the work done on the book:

$$\Delta K_{\text{book}} = mgy_i - mgy_f \tag{8.4}$$

Let us now relate each side of this equation to the *system* of the book and the Earth. For the right-hand side,

$$mgy_i - mgy_f = -(mgy_f - mgy_i) = -\Delta U_g$$

where $U_g = mgy$ is the gravitational potential energy of the system. For the left-hand side of Equation 8.4, because the book is the only part of the system that is moving, we see that $\Delta K_{\text{book}} = \Delta K$, where K is the kinetic energy of the system. Therefore, with each side of Equation 8.4 replaced with its system equivalent, the equation becomes

$$\Delta K = -\Delta U_g \tag{8.5}$$

This equation can be manipulated to provide a very important general result for solving problems. First, we move the change in potential energy to the left side of the equation:

$$\Delta K + \Delta U_g = 0$$

The left side represents a sum of changes of the energy stored in the system. The right-hand side is zero because there are no transfers of energy across the boundary of the system; the book–Earth system is *isolated* from the environment. We developed this equation for a gravitational system, but it can be shown to be valid for a system with any type of potential energy. Therefore, for an isolated system,

$$\Delta K + \Delta U = 0 \tag{8.6}$$

We defined in Chapter 7 the sum of the kinetic and potential energies of a system as its mechanical energy:

$$E_{\text{mech}} \equiv K + U \tag{8.7}$$

◀ **Mechanical energy of a system**

where U represents the total of *all* types of potential energy. Because the system under consideration is isolated, Equations 8.6 and 8.7 tell us that the mechanical energy of the system is conserved:

$$\Delta E_{\text{mech}} = 0 \tag{8.8}$$

◀ **The mechanical energy of an isolated system with no nonconservative forces acting is conserved.**

Equation 8.8 is a statement of **conservation of mechanical energy** for an isolated system with no nonconservative forces acting. The mechanical energy in such a system is conserved: the sum of the kinetic and potential energies remains constant.

If there are nonconservative forces acting within the system, mechanical energy is transformed to internal energy as discussed in Section 7.7. If nonconservative forces act in an isolated system, the total energy of the system is conserved although the mechanical energy is not. In that case, we can express the conservation of energy of the system as

$$\Delta E_{\text{system}} = 0 \tag{8.9}$$

◀ **The total energy of an isolated system is conserved.**

where E_{system} includes all kinetic, potential, and internal energies. This equation is the most general statement of the energy version of the **isolated system** model. It is equivalent to Equation 8.2 with all terms on the right-hand side equal to zero.

Let us now write the changes in energy in Equation 8.6 explicitly:

$$(K_f - K_i) + (U_f - U_i) = 0$$

$$K_f + U_f = K_i + U_i \tag{8.10}$$

Pitfall Prevention 8.2

Conditions on Equation 8.10
Equation 8.10 is only true for a system in which conservative forces act. We will see how to handle nonconservative forces in Sections 8.3 and 8.4.

For the gravitational situation of the falling book, Equation 8.10 can be written as

$$\tfrac{1}{2}mv_f^{\,2} + mgy_f = \tfrac{1}{2}mv_i^{\,2} + mgy_i$$

As the book falls to the Earth, the book–Earth system loses potential energy and gains kinetic energy such that the total of the two types of energy always remains constant.

***Quick Quiz* 8.3** A rock of mass m is dropped to the ground from a height h. A second rock, with mass $2m$, is dropped from the same height. When the second rock strikes the ground, what is its kinetic energy? **(a)** twice that of the first rock **(b)** four times that of the first rock **(c)** the same as that of the first rock **(d)** half as much as that of the first rock **(e)** impossible to determine

***Quick Quiz* 8.4** Three identical balls are thrown from the top of a building, all with the same initial speed. As shown in Active Figure 8.3, the first is thrown horizontally, the second at some angle above the horizontal, and the third at some angle below the horizontal. Neglecting air resistance, rank the speeds of the balls at the instant each hits the ground.

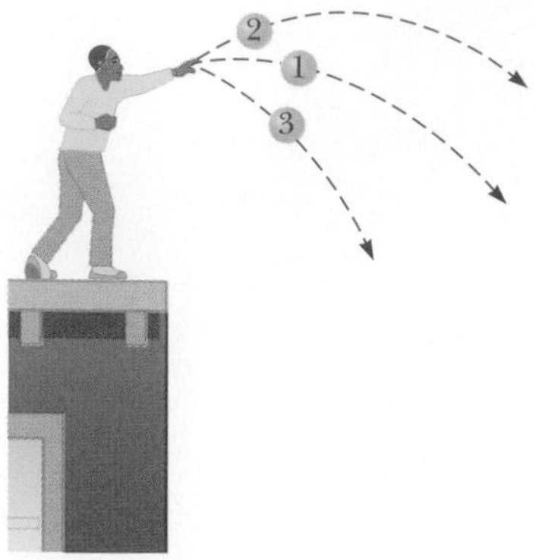

ACTIVE FIGURE 8.3

(Quick Quiz 8.4) Three identical balls are thrown with the same initial speed from the top of a building.

Problem-Solving Strategy

ISOLATED SYSTEMS WITH NO NONCONSERVATIVE FORCES: CONSERVATION OF MECHANICAL ENERGY

Many problems in physics can be solved using the principle of conservation of energy for an isolated system. The following procedure should be used when you apply this principle:

1. Conceptualize. Study the physical situation carefully and form a mental representation of what is happening. As you become more proficient working energy problems, you will begin to be comfortable imagining the types of energy that are changing in the system.

2. Categorize. Define your system, which may consist of more than one object and may or may not include springs or other possibilities for storing potential energy. Determine if any energy transfers occur across the boundary of your system. If so, use the nonisolated system model, $\Delta E_{\text{system}} = \Sigma\, T$, from Section 8.1. If not, use the isolated system model, $\Delta E_{\text{system}} = 0$.

Determine whether any nonconservative forces are present within the system. If so, use the techniques of Sections 8.3 and 8.4. If not, use the principle of conservation of mechanical energy as outlined below.

3. Analyze. Choose configurations to represent the initial and final conditions of the system. For each object that changes elevation, select a reference position for the object that defines the zero configuration of gravitational potential energy for the system. For an object on a spring, the zero configuration for elastic potential energy is when the object is at its equilibrium position. If there is more than one conservative force, write an expression for the potential energy associated with each force.

Write the total initial mechanical energy E_i of the system for some configuration as the sum of the kinetic and potential energies associated with the configuration. Then write a similar expression for the total mechanical energy E_f of the system for the final configuration that is of interest. Because mechanical energy is *conserved,* equate the two total energies and solve for the quantity that is unknown.

4. Finalize. Make sure your results are consistent with your mental representation. Also make sure the values of your results are reasonable and consistent with connections to everyday experience.

Example 8.1 Ball in Free Fall

A ball of mass m is dropped from a height h above the ground as shown in Active Figure 8.4.

(A) Neglecting air resistance, determine the speed of the ball when it is at a height y above the ground.

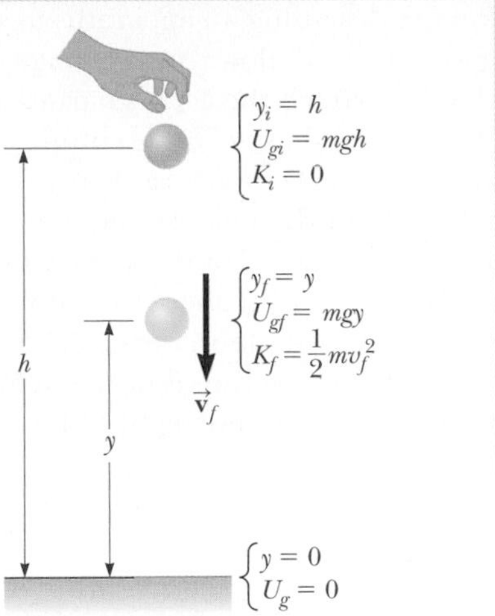

ACTIVE FIGURE 8.4

(Example 8.1) A ball is dropped from a height h above the ground. Initially, the total energy of the ball–Earth system is gravitational potential energy, equal to mgh relative to the ground. At the position y, the total energy is the sum of the kinetic and potential energies.

SOLUTION

Conceptualize Active Figure 8.4 and our everyday experience with falling objects allow us to conceptualize the situation. Although we can readily solve this problem with the techniques of Chapter 2, let us practice an energy approach.

Categorize We identify the system as the ball and the Earth. Because there is neither air resistance nor any other interaction between the system and the environment, the system is isolated and we use the isolated system model. The only force between members of the system is the gravitational force, which is conservative.

Analyze Because the system is isolated and there are no nonconservative forces acting within the system, we apply the principle of conservation of mechanical energy to the ball–Earth system. At the instant the ball is released, its kinetic energy is $K_i = 0$ and the gravitational potential energy of the system is $U_{gi} = mgh$. When the ball is at a position y above the ground, its kinetic energy is $K_f = \frac{1}{2}mv_f^2$ and the potential energy relative to the ground is $U_{gf} = mgy$.

Apply Equation 8.10: $K_f + U_{gf} = K_i + U_{gi}$

$$\tfrac{1}{2}mv_f^2 + mgy = 0 + mgh$$

Solve for v_f: $v_f^2 = 2g(h - y) \rightarrow v_f = \sqrt{2g(h - y)}$

The speed is always positive. If you had been asked to find the ball's velocity, you would use the negative value of the square root as the y component to indicate the downward motion.

(B) Determine the speed of the ball at y if at the instant of release it already has an initial upward speed v_i at the initial altitude h.

SOLUTION

Analyze In this case, the initial energy includes kinetic energy equal to $\frac{1}{2}mv_i^2$.

Apply Equation 8.10: $\frac{1}{2}mv_f^2 + mgy = \frac{1}{2}mv_i^2 + mgh$

Solve for v_f: $v_f^2 = v_i^2 + 2g(h - y) \rightarrow v_f = \sqrt{v_i^2 + 2g(h - y)}$

Finalize This result for the final speed is consistent with the expression $v_{yf}^2 = v_{yi}^2 - 2g(y_f - y_i)$ from the particle under constant acceleration model for a falling object, where $y_i = h$. Furthermore, this result is valid even if the initial velocity is at an angle to the horizontal (Quick Quiz 8.4) for two reasons: (1) the kinetic energy, a scalar, depends only on the magnitude of the velocity; and (2) the change in the gravitational potential energy of the system depends only on the change in position of the ball in the vertical direction.

WHAT IF? What if the initial velocity $\vec{\mathbf{v}}_i$ in part (B) were downward? How would that affect the speed of the ball at position y?

Answer You might claim that throwing the ball downward would result in it having a higher speed at y than if you threw it upward. Conservation of mechanical energy, however, depends on kinetic and potential energies, which are scalars. Therefore, the direction of the initial velocity vector has no bearing on the final speed.

Example 8.2 A Grand Entrance

You are designing an apparatus to support an actor of mass 65 kg who is to "fly" down to the stage during the performance of a play. You attach the actor's harness to a 130-kg sandbag by means of a lightweight steel cable running smoothly over two frictionless pulleys as in Figure 8.5a. You need 3.0 m of cable between the harness and the nearest pulley so that the pulley can be hidden behind a curtain. For the apparatus to work successfully, the sandbag must never lift above the floor as the actor swings from above the stage to the floor. Let us call the initial angle that the actor's cable makes with the vertical θ. What is the maximum value θ can have before the sandbag lifts off the floor?

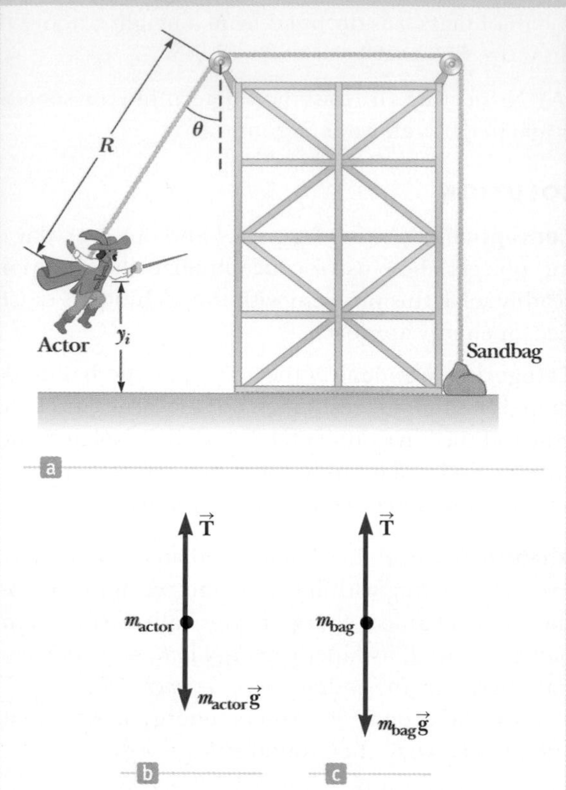

Figure 8.5 (Example 8.2) (a) An actor uses some clever staging to make his entrance. (b) The free-body diagram for the actor at the bottom of the circular path. (c) The free-body diagram for the sandbag if the normal force from the floor goes to zero.

SOLUTION

Conceptualize We must use several concepts to solve this problem. Imagine what happens as the actor approaches the bottom of the swing. At the bottom, the cable is vertical and must support his weight as well as provide centripetal acceleration of his body in the upward direction. At this point in his swing, the tension in the cable is the highest and the sandbag is most likely to lift off the floor.

Categorize Looking first at the swinging of the actor from the initial point to the lowest point, we model the actor and the Earth as an isolated system. We ignore air resistance, so there are no nonconservative forces acting. You might initially be tempted to model the system as nonisolated because of the interaction of the system with the cable, which is in the environment. The force applied to the actor by the cable, however, is always perpendicular to each element of the displacement of the actor and hence does no work. Therefore, in terms of energy transfers across the boundary, the system is isolated.

Analyze We first find the actor's speed as he arrives at the floor as a function of the initial angle θ and the radius R of the circular path through which he swings.

From the isolated system model, apply conservation of mechanical energy to the actor–Earth system:

$$K_f + U_f = K_i + U_i$$

Let y_i be the initial height of the actor above the floor and v_f be his speed at the instant before he lands. (Notice that $K_i = 0$ because the actor starts from rest and that $U_f = 0$ because we define the configuration of the actor at the floor as having a gravitational potential energy of zero.)

$$(1)\quad \tfrac{1}{2}m_{\text{actor}}\,v_f^2 + 0 = 0 + m_{\text{actor}}\,g y_i$$

From the geometry in Figure 8.5a, notice that $y_f = 0$, so $y_i = R - R\cos\theta = R(1 - \cos\theta)$. Use this relationship in Equation (1) and solve for v_f^2:

$$(2)\quad v_f^2 = 2gR(1 - \cos\theta)$$

Categorize Next, focus on the instant the actor is at the lowest point. Because the tension in the cable is transferred as a force applied to the sandbag, we model the actor at this instant as a particle under a net force. Because the actor moves along a circular arc, he experiences at the bottom of the swing a centripetal acceleration of v_f^2/r directed upward.

8.2 *cont.*

Analyze Apply Newton's second law from the particle under a net force model to the actor at the bottom of his path, using the free-body diagram in Figure 8.5b as a guide:

$$\sum F_y = T - m_{\text{actor}}g = m_{\text{actor}}\frac{v_f^2}{R}$$

$$(3)\quad T = m_{\text{actor}}g + m_{\text{actor}}\frac{v_f^2}{R}$$

Categorize Finally, notice that the sandbag lifts off the floor when the upward force exerted on it by the cable exceeds the gravitational force acting on it; the normal force is zero when that happens. We do *not,* however, want the sandbag to lift off the floor. The sandbag must remain at rest, so we model it as a particle in equilibrium.

Analyze A force T of the magnitude given by Equation (3) is transmitted by the cable to the sandbag. If the sandbag remains at rest but is just ready to be lifted off the floor if any more force were applied by the cable, the normal force on it becomes zero and the particle in equilibrium model tells us that $T = m_{\text{bag}}g$ as in Figure 8.5c.

Substitute this condition and Equation (2) into Equation (3):

$$m_{\text{bag}}g = m_{\text{actor}}g + m_{\text{actor}}\frac{2gR(1 - \cos\theta)}{R}$$

Solve for $\cos\theta$ and substitute the given parameters:

$$\cos\theta = \frac{3m_{\text{actor}} - m_{\text{bag}}}{2m_{\text{actor}}} = \frac{3(65\text{ kg}) - 130\text{ kg}}{2(65\text{ kg})} = 0.50$$

$$\theta = \boxed{60°}$$

Finalize Here we had to combine several analysis models from different areas of our study. Furthermore, notice that the length R of the cable from the actor's harness to the leftmost pulley did not appear in the final algebraic equation. Therefore, the final answer is independent of R.

Example 8.3 The Spring-Loaded Popgun

The launching mechanism of a popgun consists of a trigger-released spring (Active Fig. 8.6a on page 208). The spring is compressed to a position $y_Ⓐ$, and the trigger is fired. The projectile of mass m rises to a position $y_Ⓒ$ above the position at which it leaves the spring, indicated in Active Figure 8.6b as position $y_Ⓑ = 0$. Consider a firing of the gun for which $m = 35.0$ g, $y_Ⓐ = -0.120$ m, and $y_Ⓒ = 20.0$ m.

(A) Neglecting all resistive forces, determine the spring constant.

SOLUTION

Conceptualize Imagine the process illustrated in parts (a) and (b) of Active Figure 8.6. The projectile starts from rest, speeds up as the spring pushes upward on it, leaves the spring, and then slows down as the gravitational force pulls downward on it.

Categorize We identify the system as the projectile, the spring, and the Earth. We ignore air resistance on the projectile and friction in the gun, so we model the system as isolated with no nonconservative forces acting.

Analyze Because the projectile starts from rest, its initial kinetic energy is zero. We choose the zero configuration for the gravitational potential energy of the system to be when the projectile leaves the spring. For this configuration, the elastic potential energy is also zero.

After the gun is fired, the projectile rises to a maximum height $y_Ⓒ$. The final kinetic energy of the projectile is zero.

From the isolated system model, write a conservation of mechanical energy equation for the system between points Ⓐ and Ⓒ:

$$K_Ⓒ + U_{gⒸ} + U_{sⒸ} = K_Ⓐ + U_{gⒶ} + U_{sⒶ}$$

Substitute for each energy:

$$0 + mgy_Ⓒ + 0 = 0 + mgy_Ⓐ + \tfrac{1}{2}kx^2$$

Solve for k:

$$k = \frac{2mg(y_Ⓒ - y_Ⓐ)}{x^2}$$

continued

8.3 cont.

ACTIVE FIGURE 8.6

(Example 8.3) A spring-loaded popgun (a) before firing and (b) when the spring extends to its relaxed length. (c) An energy bar chart for the popgun-projectile-Earth system before the popgun is loaded. The energy in the system is zero. (d) The popgun is loaded by means of an external agent doing work on the system to push the spring downward. Therefore the system is nonisolated during this process. After the popgun is loaded, elastic potential energy is stored in the spring and the gravitational potential energy of the system is lower because the projectile is below point Ⓑ. (e) as the projectile passes through point Ⓑ, all of the energy of the isolated system is kinetic. (f) When the projectile reaches point Ⓒ, all of the energy of the isolated system is gravitational potential.

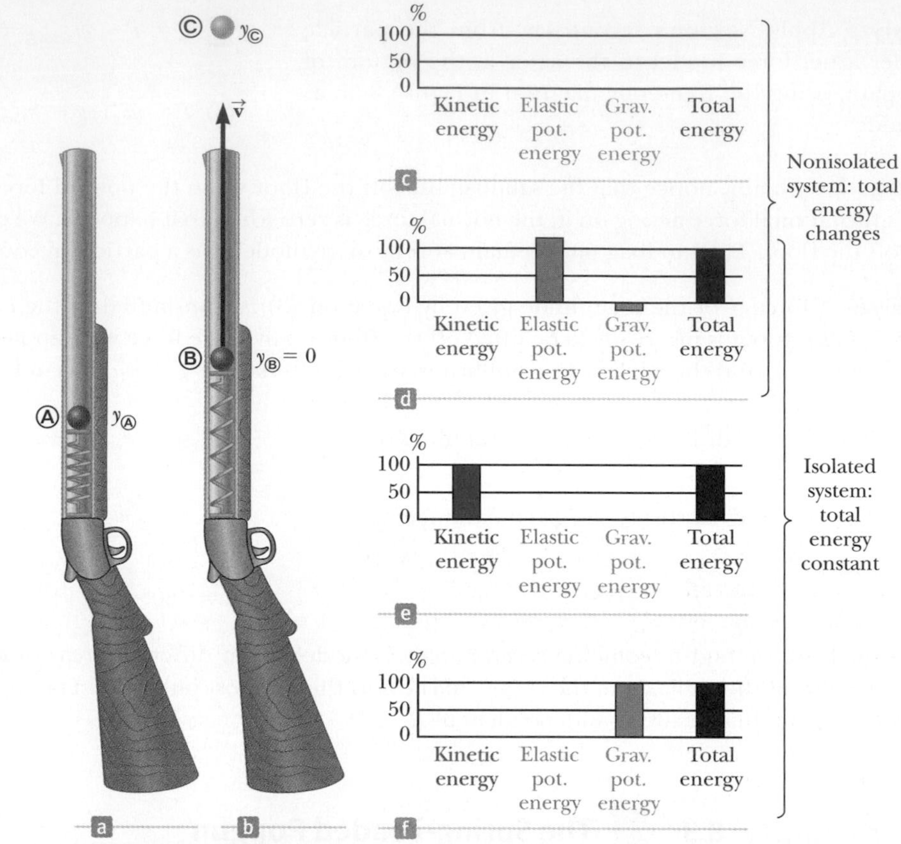

Substitute numerical values:

$$k = \frac{2(0.035\,0\text{ kg})(9.80\text{ m/s}^2)[20.0\text{ m} - (-0.120\text{ m})]}{(0.120\text{ m})^2} = \boxed{958\text{ N/m}}$$

(B) Find the speed of the projectile as it moves through the equilibrium position Ⓑ of the spring as shown in Active Figure 8.6b.

SOLUTION

Analyze The energy of the system as the projectile moves through the equilibrium position of the spring includes only the kinetic energy of the projectile $\frac{1}{2}mv_Ⓑ^2$. Both types of potential energy are equal to zero for this configuration of the system.

Write a conservation of mechanical energy equation for the system between points Ⓐ and Ⓑ:

$$K_Ⓑ + U_{gⒷ} + U_{sⒷ} = K_Ⓐ + U_{gⒶ} + U_{sⒶ}$$

Substitute for each energy:

$$\tfrac{1}{2}mv_Ⓑ^2 + 0 + 0 = 0 + mgy_Ⓐ + \tfrac{1}{2}kx^2$$

Solve for $v_Ⓑ$:

$$v_Ⓑ = \sqrt{\frac{kx^2}{m} + 2gy_Ⓐ}$$

Substitute numerical values:

$$v_Ⓑ = \sqrt{\frac{(958\text{ N/m})(0.120\text{ m})^2}{(0.035\,0\text{ kg})} + 2(9.80\text{ m/s}^2)(-0.120\text{ m})} = \boxed{19.8\text{ m/s}}$$

Finalize This example is the first one we have seen in which we must include two different types of potential energy. Notice in part (A) that we never needed to consider anything about the speed of the ball between points Ⓐ and Ⓒ, which is part of the power of the energy approach: changes in kinetic and potential energy depend only on the initial and final values, not on what happens between the configurations corresponding to these values.

8.3 Situations Involving Kinetic Friction

Consider again the book in Active Figure 7.18a sliding to the right on the surface of a heavy table and slowing down due to the friction force. Work is done by the friction force because there is a force and a displacement. Keep in mind, however, that our equations for work involve the displacement *of the point of application of the force.* A simple model of the friction force between the book and the surface is shown in Figure 8.7a. We have represented the entire friction force between the book and surface as being due to two identical teeth that have been spot-welded together.[2] One tooth projects upward from the surface, the other downward from the book, and they are welded at the points where they touch. The friction force acts at the junction of the two teeth. Imagine that the book slides a small distance d to the right as in Figure 8.7b. Because the teeth are modeled as identical, the junction of the teeth moves to the right by a distance $d/2$. Therefore, the displacement of the point of application of the friction force is $d/2$, but the displacement of the book is d!

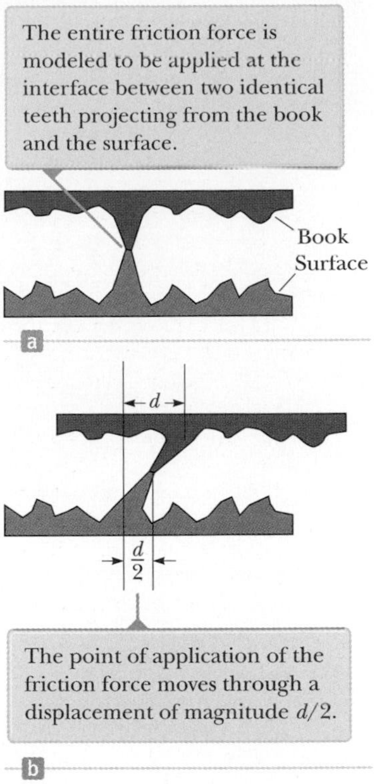

Figure 8.7 (a) A simplified model of friction between a book and a surface. (b) The book is moved to the right by a distance d.

In reality, the friction force is spread out over the entire contact area of an object sliding on a surface, so the force is not localized at a point. In addition, because the magnitudes of the friction forces at various points are constantly changing as individual spot welds occur, the surface and the book deform locally, and so on, the displacement of the point of application of the friction force is not at all the same as the displacement of the book. In fact, the displacement of the point of application of the friction force is not calculable and so neither is the work done by the friction force.

The work–kinetic energy theorem is valid for a particle or an object that can be modeled as a particle. When a friction force acts, however, we cannot calculate the work done by friction. For such situations, Newton's second law is still valid for the system even though the work–kinetic energy theorem is not. The case of a nondeformable object like our book sliding on the surface[3] can be handled in a relatively straightforward way.

Starting from a situation in which forces, including friction, are applied to the book, we can follow a similar procedure to that done in developing Equation 7.17. Let us start by writing Equation 7.8 for all forces other than friction:

$$\sum W_{\text{other forces}} = \int \left(\sum \vec{\mathbf{F}}_{\text{other forces}}\right) \cdot d\vec{\mathbf{r}} \tag{8.11}$$

The $d\vec{\mathbf{r}}$ in this equation is the displacement of the object because for forces other than friction, under the assumption that these forces do not deform the object, this displacement is the same as the displacement of the point of application of the forces. To each side of Equation 8.11 let us add the integral of the scalar product of the force of kinetic friction and $d\vec{\mathbf{r}}$. In doing so, we are not defining this quantity as work! We are simply saying that it is a quantity that can be calculated mathematically and will turn out to be useful to us in what follows.

$$\sum W_{\text{other forces}} + \int \vec{\mathbf{f}}_k \cdot d\vec{\mathbf{r}} = \int \left(\sum \vec{\mathbf{F}}_{\text{other forces}}\right) \cdot d\vec{\mathbf{r}} + \int \vec{\mathbf{f}}_k \cdot d\vec{\mathbf{r}}$$

$$= \int \left(\sum \vec{\mathbf{F}}_{\text{other forces}} + \vec{\mathbf{f}}_k\right) \cdot d\vec{\mathbf{r}}$$

The integrand on the right side of this equation is the net force $\sum \vec{\mathbf{F}}$, so

$$\sum W_{\text{other forces}} + \int \vec{\mathbf{f}}_k \cdot d\vec{\mathbf{r}} = \int \sum \vec{\mathbf{F}} \cdot d\vec{\mathbf{r}}$$

Incorporating Newton's second law $\sum \vec{\mathbf{F}} = m\vec{\mathbf{a}}$ gives

$$\sum W_{\text{other forces}} + \int \vec{\mathbf{f}}_k \cdot d\vec{\mathbf{r}} = \int m\vec{\mathbf{a}} \cdot d\vec{\mathbf{r}} = \int m\frac{d\vec{\mathbf{v}}}{dt} \cdot d\vec{\mathbf{r}} = \int_{t_i}^{t_f} m\frac{d\vec{\mathbf{v}}}{dt} \cdot \vec{\mathbf{v}}\, dt \tag{8.12}$$

[2]Figure 8.7 and its discussion are inspired by a classic article on friction: B. A. Sherwood and W. H. Bernard, "Work and heat transfer in the presence of sliding friction," *American Journal of Physics,* **52:**1001, 1984.

[3]The overall shape of the book remains the same, which is why we say it is nondeformable. On a microscopic level, however, there is deformation of the book's face as it slides over the surface.

where we have used Equation 4.3 to rewrite $d\vec{\mathbf{r}}$ as $\vec{\mathbf{v}}\,dt$. The scalar product obeys the product rule for differentiation (See Eq. B.30 in Appendix B.6), so the derivative of the scalar product of $\vec{\mathbf{v}}$ with itself can be written

$$\frac{d}{dt}(\vec{\mathbf{v}}\cdot\vec{\mathbf{v}}) = \frac{d\vec{\mathbf{v}}}{dt}\cdot\vec{\mathbf{v}} + \vec{\mathbf{v}}\cdot\frac{d\vec{\mathbf{v}}}{dt} = 2\frac{d\vec{\mathbf{v}}}{dt}\cdot\vec{\mathbf{v}}$$

where we have used the commutative property of the scalar product to justify the final expression in this equation. Consequently,

$$\frac{d\vec{\mathbf{v}}}{dt}\cdot\vec{\mathbf{v}} = \tfrac{1}{2}\frac{d}{dt}(\vec{\mathbf{v}}\cdot\vec{\mathbf{v}}) = \tfrac{1}{2}\frac{dv^2}{dt}$$

Substituting this result into Equation 8.12 gives

$$\sum W_{\text{other forces}} + \int \vec{\mathbf{f}}_k\cdot d\vec{\mathbf{r}} = \int_{t_i}^{t_f} m\left(\tfrac{1}{2}\frac{dv^2}{dt}\right)dt = \tfrac{1}{2}m\int_{v_i}^{v_f} d(v^2) = \tfrac{1}{2}mv_f^{\,2} - \tfrac{1}{2}mv_i^{\,2} = \Delta K$$

Looking at the left side of this equation, notice that in the inertial frame of the surface, $\vec{\mathbf{f}}_k$ and $d\vec{\mathbf{r}}$ will be in opposite directions for every increment $d\vec{\mathbf{r}}$ of the path followed by the object. Therefore, $\vec{\mathbf{f}}_k\cdot d\vec{\mathbf{r}} = -f_k\,dr$. The previous expression now becomes

$$\sum W_{\text{other forces}} - \int f_k\,dr = \Delta K$$

In our model for friction, the magnitude of the kinetic friction force is constant, so f_k can be brought out of the integral. The remaining integral $\int dr$ is simply the sum of increments of length along the path, which is the total path length d. Therefore,

$$\sum W_{\text{other forces}} - f_k d = \Delta K \qquad \textbf{(8.13)}$$

or

$$K_f = K_i - f_k d + \sum W_{\text{other forces}} \qquad \textbf{(8.14)}$$

Equation 8.13 can be used when a friction force acts on an object. The change in kinetic energy is equal to the work done by all forces other than friction minus a term $f_k d$ associated with the friction force.

Considering the sliding book situation again, let's identify the larger system of the book *and* the surface as the book slows down under the influence of a friction force alone. There is no work done across the boundary of this system because the system does not interact with the environment. There are no other types of energy transfer occurring across the boundary of the system, assuming we ignore the inevitable sound the sliding book makes! In this case, Equation 8.2 becomes

$$\Delta E_{\text{system}} = \Delta K + \Delta E_{\text{int}} = 0$$

The change in kinetic energy of this book–surface system is the same as the change in kinetic energy of the book alone because the book is the only part of the system that is moving. Therefore, incorporating Equation 8.13 gives

$$-f_k d + \Delta E_{\text{int}} = 0$$

Change in internal energy due to a constant friction force within the system ▶

$$\Delta E_{\text{int}} = f_k d \qquad \textbf{(8.15)}$$

The increase in internal energy of the system is therefore equal to the product of the friction force and the path length through which the block moves. In summary, a friction force transforms kinetic energy in a system to internal energy, and the increase in internal energy of the system is equal to its decrease in kinetic energy. Equation 8.13, with the help of Equation 8.15, can be written as

$$\sum W_{\text{other forces}} = W = \Delta K + \Delta E_{\text{int}}$$

which is a reduced form of Equation 8.2 and represents the nonisolated system model for a system within which a nonconservative force acts.

Quick Quiz **8.5** You are traveling along a freeway at 65 mi/h. Your car has kinetic energy. You suddenly skid to a stop because of congestion in traffic. Where is the kinetic energy your car once had? **(a)** It is all in internal energy in the road. **(b)** It is all in internal energy in the tires. **(c)** Some of it has transformed to internal energy and some of it transferred away by mechanical waves. **(d)** It is all transferred away from your car by various mechanisms.

Example 8.4 A Block Pulled on a Rough Surface

A 6.0-kg block initially at rest is pulled to the right along a horizontal surface by a constant horizontal force of 12 N.

(A) Find the speed of the block after it has moved 3.0 m if the surfaces in contact have a coefficient of kinetic friction of 0.15.

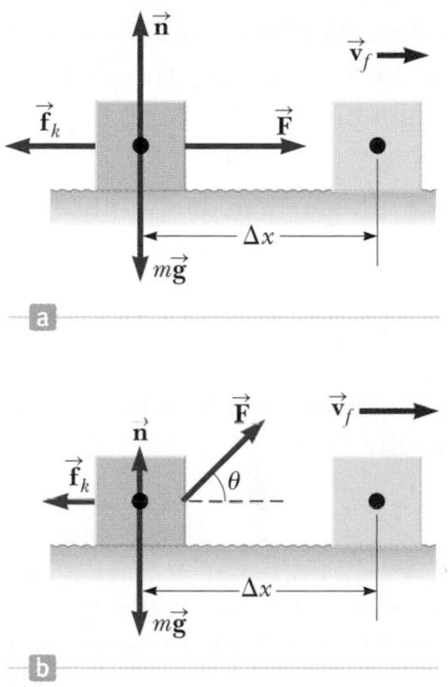

ACTIVE FIGURE 8.8

(Example 8.4) (a) A block pulled to the right on a rough surface by a constant horizontal force. (b) The applied force is at an angle θ to the horizontal.

SOLUTION

Conceptualize This example is Example 7.6 (page 178), modified so that the surface is no longer frictionless. The rough surface applies a friction force on the block opposite to the applied force. As a result, we expect the speed to be lower than that found in Example 7.6.

Categorize The block is pulled by a force and the surface is rough, so we model the block–surface system as nonisolated with a nonconservative force acting.

Analyze Active Figure 8.8a illustrates this situation. Neither the normal force nor the gravitational force does work on the system because their points of application are displaced horizontally.

Find the work done on the system by the applied force just as in Example 7.6:

$$\sum W_{\text{other forces}} = W_F = F\Delta x$$

Apply the particle in equilibrium model to the block in the vertical direction:

$$\sum F_y = 0 \rightarrow n - mg = 0 \rightarrow n = mg$$

Find the magnitude of the friction force:

$$f_k = \mu_k n = \mu_k mg = (0.15)(6.0 \text{ kg})(9.80 \text{ m/s}^2) = 8.82 \text{ N}$$

Find the final speed of the block from Equation 8.14:

$$\tfrac{1}{2}mv_f^{\,2} = \tfrac{1}{2}mv_i^{\,2} - f_k d + W_F$$

$$v_f = \sqrt{v_i^{\,2} + \frac{2}{m}(-f_k d + F\Delta x)}$$

Substitute numerical values:

$$v_f = \sqrt{0 + \frac{2}{6.0 \text{ kg}}[-(8.82 \text{ N})(3.0 \text{ m}) + (12 \text{ N})(3.0 \text{ m})]} = \boxed{1.8 \text{ m/s}}$$

Finalize As expected, this value is less than the 3.5 m/s found in the case of the block sliding on a frictionless surface (see Example 7.6). The difference in kinetic energies between the block in Example 7.6 and the block in this example is equal to the increase in internal energy of the block–surface system in this example.

(B) Suppose the force $\vec{F}$ is applied at an angle θ as shown in Active Figure 8.8b. At what angle should the force be applied to achieve the largest possible speed after the block has moved 3.0 m to the right?

continued

8.4 *cont.*

SOLUTION

Conceptualize You might guess that $\theta = 0$ would give the largest speed because the force would have the largest component possible in the direction parallel to the surface. Think about $\vec{\mathbf{F}}$ applied at an arbitrary nonzero angle, however. Although the horizontal component of the force would be reduced, the vertical component of the force would reduce the normal force, in turn reducing the force of friction, which suggests that the speed could be maximized by pulling at an angle other than $\theta = 0$.

Categorize As in part (A), we model the block–surface system as nonisolated with a nonconservative force acting.

Analyze Find the work done by the applied force, noting that $\Delta x = d$ because the path followed by the block is a straight line:

$$\sum W_{\text{other forces}} = W_F = F\Delta x \cos\theta = Fd\cos\theta$$

Apply the particle in equilibrium model to the block in the vertical direction:

$$\sum F_y = n + F\sin\theta - mg = 0$$

Solve for n:

$$n = mg - F\sin\theta$$

Use Equation 8.14 to find the final kinetic energy for this situation:

$$K_f = K_i - f_k d + W_F$$
$$= 0 - \mu_k nd + Fd\cos\theta = -\mu_k(mg - F\sin\theta)d + Fd\cos\theta$$

Maximizing the speed is equivalent to maximizing the final kinetic energy. Consequently, differentiate K_f with respect to θ and set the result equal to zero:

$$\frac{dK_f}{d\theta} = -\mu_k(0 - F\cos\theta)d - Fd\sin\theta = 0$$
$$\mu_k\cos\theta - \sin\theta = 0$$
$$\tan\theta = \mu_k$$

Evaluate θ for $\mu_k = 0.15$:

$$\theta = \tan^{-1}(\mu_k) = \tan^{-1}(0.15) = \boxed{8.5^\circ}$$

Finalize Notice that the angle at which the speed of the block is a maximum is indeed not $\theta = 0$. When the angle exceeds 8.5°, the horizontal component of the applied force is too small to be compensated by the reduced friction force and the speed of the block begins to decrease from its maximum value.

Conceptual Example 8.5 Useful Physics for Safer Driving

A car traveling at an initial speed v slides a distance d to a halt after its brakes lock. If the car's initial speed is instead $2v$ at the moment the brakes lock, estimate the distance it slides.

SOLUTION

Let us assume the force of kinetic friction between the car and the road surface is constant and the same for both speeds. According to Equation 8.14, the friction force multiplied by the distance d is equal to the initial kinetic energy of the car (because $K_f = 0$ and there is no work done by other forces). If the speed is doubled, as it is in this example, the kinetic energy is quadrupled. For a given friction force, the distance traveled is four times as great when the initial speed is doubled, and so the estimated distance the car slides is $4d$.

Example 8.6 A Block–Spring System

A block of mass 1.6 kg is attached to a horizontal spring that has a force constant of 1 000 N/m as shown in Figure 8.9. The spring is compressed 2.0 cm and is then released from rest.

(A) Calculate the speed of the block as it passes through the equilibrium position $x = 0$ if the surface is frictionless.

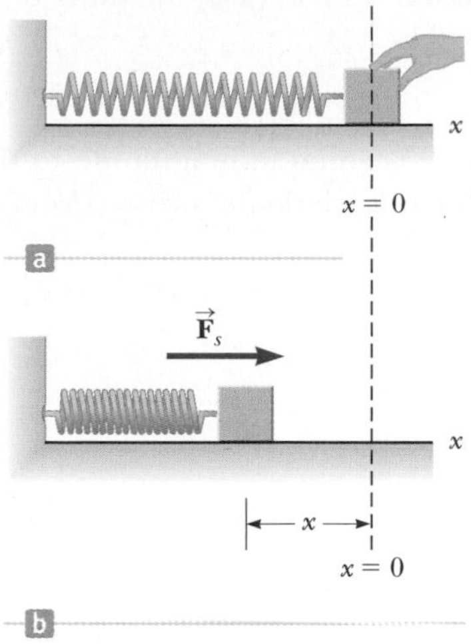

Figure 8.9 (Example 8.6) (a) A block attached to a spring is pushed inward from an initial position $x = 0$ by an external agent. (b) At position x, the block is released from rest and the spring pushes it to the right.

SOLUTION

Conceptualize This situation has been discussed before, and it is easy to visualize the block being pushed to the right by the spring and moving off with some speed.

Categorize We identify the system as the block and model the block as a nonisolated system.

Analyze In this situation, the block starts with $v_i = 0$ at $x_i = -2.0$ cm, and we want to find v_f at $x_f = 0$.

Use Equation 7.11 to find the work done by the spring on the system with $x_{max} = x_i$:

$$\sum W_{\text{other forces}} = W_s = \tfrac{1}{2}kx_{max}^2$$

Work is done on the block, and its speed changes. The conservation of energy equation, Equation 8.2, reduces to the work–kinetic energy theorem. Use that theorem to find the speed at $x = 0$:

$$W_s = \tfrac{1}{2}mv_f^2 - \tfrac{1}{2}mv_i^2$$

$$v_f = \sqrt{v_i^2 + \frac{2}{m}W_s}$$

Substitute numerical values:

$$v_f = \sqrt{0 + \frac{2}{1.6\text{ kg}}\left[\tfrac{1}{2}(1\,000\text{ N/m})(0.020\text{ m})^2\right]} = \boxed{0.50\text{ m/s}}$$

Finalize Although this problem could have been solved in Chapter 7, it is presented here to provide contrast with the following part (B), which requires the techniques of this chapter.

(B) Calculate the speed of the block as it passes through the equilibrium position if a constant friction force of 4.0 N retards its motion from the moment it is released.

SOLUTION

Conceptualize The correct answer must be less than that found in part (A) because the friction force retards the motion.

Categorize We identify the system as the block and the surface. The system is nonisolated because of the work done by the spring, and there is a nonconservative force acting: the friction between the block and the surface.

Analyze Write Equation 8.14:

$$(1)\quad K_f = K_i - f_k d + W_s$$

Substitute numerical values:

$$K_f = 0 - (4.0\text{ N})(0.020\text{ m}) + \tfrac{1}{2}(1\,000\text{ N/m})(0.020\text{ m})^2 = 0.12\text{ J}$$

Write the definition of kinetic energy:

$$K_f = \tfrac{1}{2}mv_f^2$$

Solve for v_f and substitute numerical values:

$$v_f = \sqrt{\frac{2K_f}{m}} = \sqrt{\frac{2(0.12\text{ J})}{1.6\text{ kg}}} = \boxed{0.39\text{ m/s}}$$

Finalize As expected, this value is less than the 0.50 m/s found in part (A).

WHAT IF? What if the friction force were increased to 10.0 N? What is the block's speed at $x = 0$?

continued

8.6 *cont.*

Answer In this case, the value of $f_k d$ as the block moves to $x = 0$ is

$$f_k d = (10.0 \text{ N})(0.020 \text{ m}) = 0.20 \text{ J}$$

which is equal in magnitude to the kinetic energy at $x = 0$ for the frictionless case. (Verify it!). Therefore, all the kinetic energy has been transformed to internal energy by friction when the block arrives at $x = 0$, and its speed at this point is $v = 0$.

In this situation as well as that in part (B), the speed of the block reaches a maximum at some position other than $x = 0$. Problem 57 asks you to locate these positions.

8.4 Changes in Mechanical Energy for Nonconservative Forces

Consider the book sliding across the surface in the preceding section. As the book moves through a distance d, the only force that does work on it is the force of kinetic friction. This force causes a change $-f_k d$ in the kinetic energy of the book as described by Equation 8.13.

Now, however, suppose the book is part of a system that also exhibits a change in potential energy. In this case, $-f_k d$ is the amount by which the mechanical energy of the system changes because of the force of kinetic friction. For example, if the book moves on an incline that is not frictionless, there is a change in both the kinetic energy and the gravitational potential energy of the book–Earth system. Consequently,

$$\Delta E_{\text{mech}} = \Delta K + \Delta U_g = -f_k d$$

In general, if a friction force acts within an isolated system,

Change in mechanical energy of a system due to friction within the system ▶

$$\Delta E_{\text{mech}} = \Delta K + \Delta U = -f_k d \tag{8.16}$$

where ΔU is the change in all forms of potential energy. Notice that Equation 8.16 reduces to Equation 8.10 if the friction force is zero.

If the system in which nonconservative forces act is nonisolated and the external influence on the system is by means of work, the generalization of Equation 8.13 is

$$\Delta E_{\text{mech}} = -f_k d + \sum W_{\text{other forces}} \tag{8.17}$$

Equation 8.17, with the help of Equations 8.7 and 8.15, can be written as

$$\sum W_{\text{other forces}} = W = \Delta K + \Delta U + \Delta E_{\text{int}}$$

This reduced form of Equation 8.2 represents the nonisolated system model for a system that possesses potential energy and within which a nonconservative force acts. In practice during problem solving, you do not need to use equations like Equation 8.15 or Equation 8.17. You can simply use Equation 8.2 and keep only those terms in the equation that correspond to the physical situation. See Example 8.9 for a sample of this approach.

Problem-Solving Strategy

SYSTEMS WITH NONCONSERVATIVE FORCES

The following procedure should be used when you face a problem involving a system in which nonconservative forces act:

1. Conceptualize. Study the physical situation carefully and form a mental representation of what is happening.

2. Categorize. Define your system, which may consist of more than one object. The system could include springs or other possibilities for storage of potential energy. Determine whether any nonconservative forces are present. If not, use the principle of conservation of mechanical energy as outlined in Section 8.2. If so, use the procedure discussed below.

Determine if any work is done across the boundary of your system by forces other than friction. If so, use Equation 8.17 to analyze the problem. If not, use Equation 8.16.

3. Analyze. Choose configurations to represent the initial and final conditions of the system. For each object that changes elevation, select a reference position for the object that defines the zero configuration of gravitational potential energy for the system. For an object on a spring, the zero configuration for elastic potential energy is when the object is at its equilibrium position. If there is more than one conservative force, write an expression for the potential energy associated with each force.

Use either Equation 8.16 or Equation 8.17 to establish a mathematical representation of the problem. Solve for the unknown.

4. Finalize. Make sure your results are consistent with your mental representation. Also make sure the values of your results are reasonable and consistent with connections to everyday experience.

Example 8.7 Crate Sliding Down a Ramp

A 3.00-kg crate slides down a ramp. The ramp is 1.00 m in length and inclined at an angle of 30.0° as shown in Figure 8.10. The crate starts from rest at the top, experiences a constant friction force of magnitude 5.00 N, and continues to move a short distance on the horizontal floor after it leaves the ramp.

(A) Use energy methods to determine the speed of the crate at the bottom of the ramp.

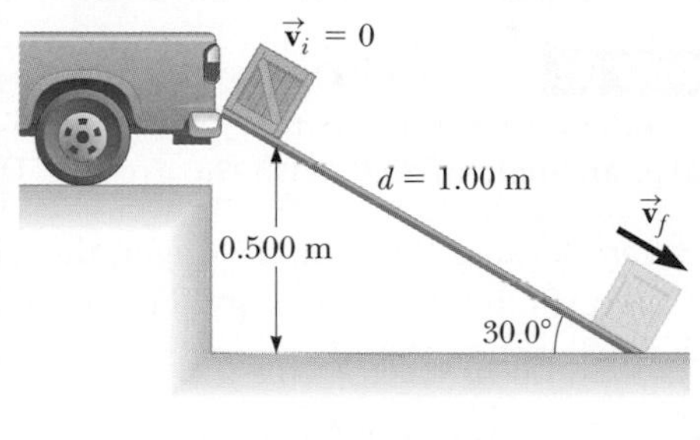

Figure 8.10 (Example 8.7) A crate slides down a ramp under the influence of gravity. The potential energy of the system decreases, whereas the kinetic energy increases.

SOLUTION

Conceptualize Imagine the crate sliding down the ramp in Figure 8.10. The larger the friction force, the more slowly the crate will slide.

Categorize We identify the crate, the surface, and the Earth as the system. The system is categorized as isolated with a nonconservative force acting.

Analyze Because $v_i = 0$, the initial kinetic energy of the system when the crate is at the top of the ramp is zero. If the y coordinate is measured from the bottom of the ramp (the final position of the crate, for which we choose the gravitational potential energy of the system to be zero) with the upward direction being positive, then $y_i = 0.500$ m.

Write the expression for the total mechanical energy of the system when the crate is at the top:

$$E_i = K_i + U_i = 0 + U_i = mgy_i$$

Write an expression for the final mechanical energy:

$$E_f = K_f + U_f = \tfrac{1}{2}mv_f^2 + 0$$

Apply Equation 8.16:

$$\Delta E_{\text{mech}} = E_f - E_i = \tfrac{1}{2}mv_f^2 - mgy_i = -f_k d$$

Solve for v_f:

$$(1)\quad v_f = \sqrt{\frac{2}{m}(mgy_i - f_k d)}$$

Substitute numerical values:

$$v_f = \sqrt{\frac{2}{3.00\text{ kg}}[(3.00\text{ kg})(9.80\text{ m/s}^2)(0.500\text{ m}) - (5.00\text{ N})(1.00\text{ m})]} = \boxed{2.54\text{ m/s}}$$

continued

8.7 *cont.*

(B) How far does the crate slide on the horizontal floor if it continues to experience a friction force of magnitude 5.00 N?

SOLUTION

Analyze This part of the problem is handled in exactly the same way as part (A), but in this case we can consider the mechanical energy of the system to consist only of kinetic energy because the potential energy of the system remains fixed.

Write an expression for the mechanical energy of the system when the crate leaves the bottom of the ramp:

$$E_i = K_i = \tfrac{1}{2}mv_i^2$$

Apply Equation 8.16 with $E_f = 0$:

$$E_f - E_i = 0 - \tfrac{1}{2}mv^2 = -f_k d \rightarrow \tfrac{1}{2}mv^2 = f_k d$$

Solve for the distance d and substitute numerical values:

$$d = \frac{mv^2}{2f_k} = \frac{(3.00\text{ kg})(2.54\text{ m/s})^2}{2(5.00\text{ N})} = \boxed{1.94\text{ m}}$$

Finalize For comparison, you may want to calculate the speed of the crate at the bottom of the ramp in the case in which the ramp is frictionless. Also notice that the increase in internal energy of the system as the crate slides down the ramp is $f_k d = (5.00\text{ N})(1.00\text{ m}) = 5.00\text{ J}$. This energy is shared between the crate and the surface, each of which is a bit warmer than before.

Also notice that the distance d the object slides on the horizontal surface is infinite if the surface is frictionless. Is that consistent with your conceptualization of the situation?

WHAT IF? A cautious worker decides that the speed of the crate when it arrives at the bottom of the ramp may be so large that its contents may be damaged. Therefore, he replaces the ramp with a longer one such that the new ramp makes an angle of 25.0° with the ground. Does this new ramp reduce the speed of the crate as it reaches the ground?

Answer Because the ramp is longer, the friction force acts over a longer distance and transforms more of the mechanical energy into internal energy. The result is a reduction in the kinetic energy of the crate, and we expect a lower speed as it reaches the ground.

Find the length d of the new ramp:

$$\sin 25.0° = \frac{0.500\text{ m}}{d} \rightarrow d = \frac{0.500\text{ m}}{\sin 25.0°} = 1.18\text{ m}$$

Find v_f from Equation (1) in part (A):

$$v_f = \sqrt{\frac{2}{3.00\text{ kg}}[(3.00\text{ kg})(9.80\text{ m/s}^2)(0.500\text{ m}) - (5.00\text{ N})(1.18\text{ m})]} = 2.42\text{ m/s}$$

The final speed is indeed lower than in the higher-angle case.

Example 8.8 Block–Spring Collision

A block having a mass of 0.80 kg is given an initial velocity $v_{Ⓐ} = 1.2$ m/s to the right and collides with a spring whose mass is negligible and whose force constant is $k = 50$ N/m as shown in Figure 8.11.

(A) Assuming the surface to be frictionless, calculate the maximum compression of the spring after the collision.

SOLUTION

Conceptualize The various parts of Figure 8.11 help us imagine what the block will do in this situation. All motion takes place in a horizontal plane, so we do not need to consider changes in gravitational potential energy.

Categorize We identify the system to be the block and the spring. The block–spring system is isolated with no nonconservative forces acting.

8.8 *cont.*

Analyze Before the collision, when the block is at Ⓐ, it has kinetic energy and the spring is uncompressed, so the elastic potential energy stored in the system is zero. Therefore, the total mechanical energy of the system before the collision is just $\frac{1}{2}mv_{Ⓐ}^2$. After the collision, when the block is at Ⓒ, the spring is fully compressed; now the block is at rest and so has zero kinetic energy. The elastic potential energy stored in the system, however, has its maximum value $\frac{1}{2}kx^2 = \frac{1}{2}kx_{\max}^2$, where the origin of coordinates $x = 0$ is chosen to be the equilibrium position of the spring and $x_{\max}$ is the maximum compression of the spring, which in this case happens to be $x_{Ⓒ}$. The total mechanical energy of the system is conserved because no nonconservative forces act on objects within the isolated system.

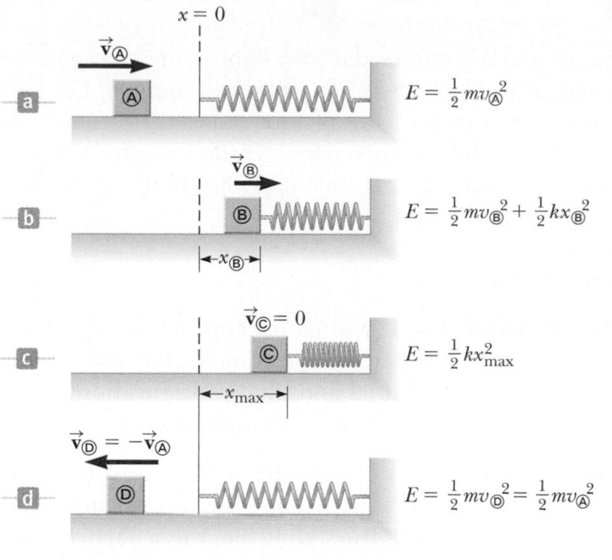

Figure 8.11 (Example 8.8) A block sliding on a frictionless, horizontal surface collides with a light spring. (a) Initially, the mechanical energy is all kinetic energy. (b) The mechanical energy is the sum of the kinetic energy of the block and the elastic potential energy in the spring. (c) The energy is entirely potential energy. (d) The energy is transformed back to the kinetic energy of the block. The total energy of the system remains constant throughout the motion.

Write a conservation of mechanical energy equation:

$$K_{Ⓒ} + U_{sⒸ} = K_{Ⓐ} + U_{sⒶ}$$

$$0 + \tfrac{1}{2}kx_{\max}^2 = \tfrac{1}{2}mv_{Ⓐ}^2 + 0$$

Solve for $x_{\max}$ and evaluate:

$$x_{\max} = \sqrt{\frac{m}{k}}\,v_{Ⓐ} = \sqrt{\frac{0.80\text{ kg}}{50\text{ N/m}}}(1.2\text{ m/s}) = \boxed{0.15\text{ m}}$$

(B) Suppose a constant force of kinetic friction acts between the block and the surface, with $\mu_k = 0.50$. If the speed of the block at the moment it collides with the spring is $v_{Ⓐ} = 1.2$ m/s, what is the maximum compression $x_{Ⓒ}$ in the spring?

SOLUTION

Conceptualize Because of the friction force, we expect the compression of the spring to be smaller than in part (A) because some of the block's kinetic energy is transformed to internal energy in the block and the surface.

Categorize We identify the system as the block, the surface, and the spring. This system is isolated but now involves a nonconservative force.

Analyze In this case, the mechanical energy $E_{\text{mech}} = K + U_s$ of the system is *not* conserved because a friction force acts on the block. From the particle in equilibrium model in the vertical direction, we see that $n = mg$.

Evaluate the magnitude of the friction force:

$$f_k = \mu_k n = \mu_k mg$$

Write the change in the mechanical energy of the system due to friction as the block is displaced from $x = 0$ to $x_{Ⓒ}$:

$$\Delta E_{\text{mech}} = -f_k x_{Ⓒ}$$

Substitute the initial and final energies:

$$\Delta E_{\text{mech}} = E_f - E_i = (0 + \tfrac{1}{2}kx_{Ⓒ}^2) - (\tfrac{1}{2}mv_{Ⓐ}^2 + 0) = -f_k x_{Ⓒ}$$

$$\tfrac{1}{2}kx_{Ⓒ}^2 - \tfrac{1}{2}mv_{Ⓐ}^2 = -\mu_k mgx_{Ⓒ}$$

Substitute numerical values:

$$\tfrac{1}{2}(50)x_{Ⓒ}^2 - \tfrac{1}{2}(0.80)(1.2)^2 = -(0.50)(0.80\text{ kg})(9.80\text{ m/s}^2)x_{Ⓒ}$$

$$25x_{Ⓒ}^2 + 3.9x_{Ⓒ} - 0.58 = 0$$

Solving the quadratic equation for $x_{Ⓒ}$ gives $x_{Ⓒ} = 0.093$ m and $x_{Ⓒ} = -0.25$ m. The physically meaningful root is $x_{Ⓒ} = \boxed{0.093\text{ m.}}$

Finalize The negative root does not apply to this situation because the block must be to the right of the origin (positive value of x) when it comes to rest. Notice that the value of 0.093 m is less than the distance obtained in the frictionless case of part (A) as we expected.

Example 8.9 Connected Blocks in Motion

Two blocks are connected by a light string that passes over a frictionless pulley as shown in Figure 8.12. The block of mass m_1 lies on a horizontal surface and is connected to a spring of force constant k. The system is released from rest when the spring is unstretched. If the hanging block of mass m_2 falls a distance h before coming to rest, calculate the coefficient of kinetic friction between the block of mass m_1 and the surface.

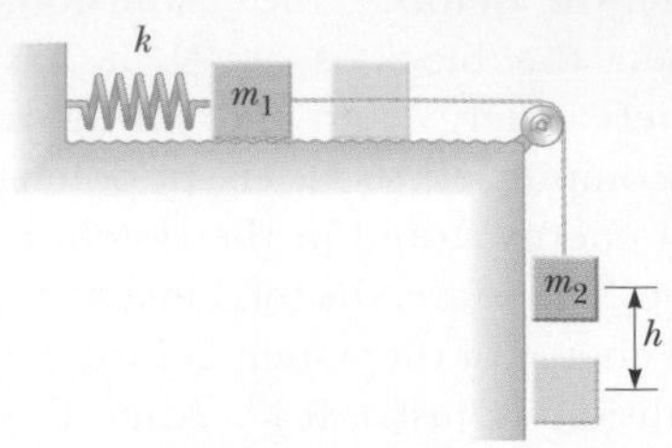

Figure 8.12 (Example 8.9) As the hanging block moves from its highest elevation to its lowest, the system loses gravitational potential energy but gains elastic potential energy in the spring. Some mechanical energy is transformed to internal energy because of friction between the sliding block and the surface.

SOLUTION

Conceptualize The key word *rest* appears twice in the problem statement. This word suggests that the configurations of the system associated with rest are good candidates for the initial and final configurations because the kinetic energy of the system is zero for these configurations.

Categorize In this situation, the system consists of the two blocks, the spring, the surface, and the Earth. The system is isolated with a nonconservative force acting. We also model the sliding block as a particle in equilibrium in the vertical direction, leading to $n = m_1 g$.

Analyze We need to consider two forms of potential energy for the system, gravitational and elastic: $\Delta U_g = U_{gf} - U_{gi}$ is the change in the system's gravitational potential energy, and $\Delta U_s = U_{sf} - U_{si}$ is the change in the system's elastic potential energy. The change in the gravitational potential energy of the system is associated with only the falling block because the vertical coordinate of the horizontally sliding block does not change. The initial and final kinetic energies of the system are zero, so $\Delta K = 0$.

For this example, let us start from Equation 8.2 to show how this approach would work in practice. Because the system is isolated, the entire right side of Equation 8.2 is zero. Based on the physical situation described in the problem, we see that there could be changes of kinetic energy, potential energy, and internal energy in the system. Write the corresponding reduction of equation 8.2:

$$\Delta K + \Delta U + \Delta E_{\text{int}} = 0$$

Incorporate into this equation that $\Delta K = 0$ and that there are two types of potential energy:

$$(1)\ \Delta U_g + \Delta U_s + \Delta E_{\text{int}} = 0$$

Use Equation 8.15 to find the change in internal energy in the system due to friction between the horizontally sliding block and the surface, noticing that as the hanging block falls a distance h, the horizontally moving block moves the same distance h to the right:

$$(2)\ \Delta E_{\text{int}} = f_k h = (\mu_k n)h = \mu_k m_1 g h$$

Evaluate the change in gravitational potential energy of the system, choosing the configuration with the hanging block at the lowest position to represent zero potential energy:

$$(3)\ \Delta U_g = U_{gf} - U_{gi} = 0 - m_2 g h$$

Evaluate the change in the elastic potential energy of the system:

$$(4)\ \Delta U_s = U_{sf} - U_{si} = \tfrac{1}{2}kh^2 - 0$$

Substitute Equations (2), (3), and (4) into Equation (1):

$$-m_2 g h + \tfrac{1}{2}kh^2 + \mu_k m_1 g h = 0$$

Solve for μ_k:

$$\mu_k = \frac{m_2 g - \frac{1}{2}kh}{m_1 g}$$

Finalize This setup represents a method of measuring the coefficient of kinetic friction between an object and some surface. Notice that we do not need to remember which energy equation goes with which type of problem with this approach. You can always begin with Equation 8.2 and then tailor it to the physical situation. This process may include deleting terms, such as the kinetic energy term and all terms on the right-hand side in this example. It can also include expanding terms, such as rewriting ΔU due to two types of potential energy in this example.

Conceptual Example 8.10 Interpreting the Energy Bars

The energy bar charts in Figure 8.13 show three instants in the motion of the system in Figure 8.12 and described in Example 8.9. For each bar chart, identify the configuration of the system that corresponds to the chart.

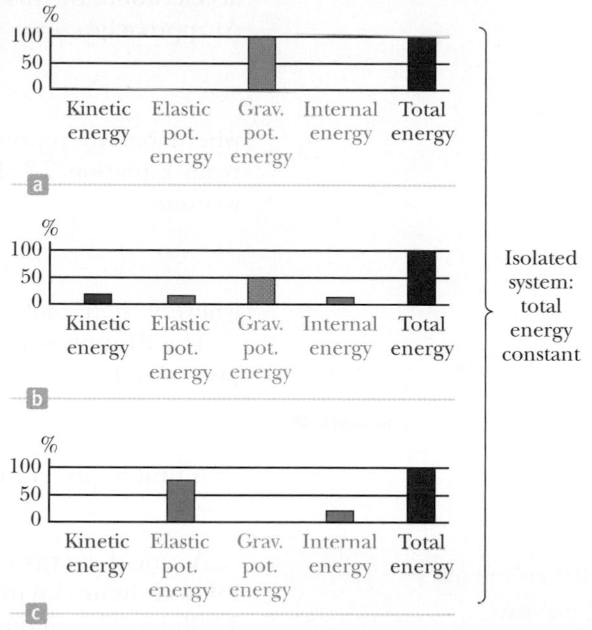

Figure 8.13 (Conceptual Example 8.10) Three energy bar charts are shown for the system in Figure 8.12.

SOLUTION

In Figure 8.13a, there is no kinetic energy in the system. Therefore, nothing in the system is moving. The bar chart shows that the system contains only gravitational potential energy and no internal energy yet, which corresponds to the configuration with the darker blocks in Figure 8.12 and represents the instant just after the system is released.

In Figure 8.13b, the system contains four types of energy. The height of the gravitational potential energy bar is at 50%, which tells us that the hanging block has moved halfway between its position corresponding to Figure 8.13a and the position defined as $y = 0$. Therefore, in this configuration, the hanging block is between the dark and light images of the hanging block in Figure 8.12. The system has gained kinetic energy because the blocks are moving, elastic potential energy because the spring is stretching, and internal energy because of friction between the block of mass m_1 and the surface.

In Figure 8.13c, the height of the gravitational potential energy bar is zero, telling us that the hanging block is at $y = 0$. In addition, the height of the kinetic energy bar is zero, indicating that the blocks have stopped moving momentarily. Therefore, the configuration of the system is that shown by the light images of the blocks in Figure 8.12. The height of the elastic potential energy bar is high because the spring is stretched its maximum amount. The height of the internal energy bar is higher than in Figure 8.13b because the block of mass m_1 has continued to slide over the surface.

8.5 Power

Consider Conceptual Example 7.7 again, which involved rolling a refrigerator up a ramp into a truck. Suppose the man is not convinced the work is the same regardless of the ramp's length and sets up a long ramp with a gentle rise. Although he does the same amount of work as someone using a shorter ramp, he takes longer to do the work because he has to move the refrigerator over a greater distance. Although the work done on both ramps is the same, there is *something* different about the tasks: the *time interval* during which the work is done.

The time rate of energy transfer is called the **instantaneous power** P and is defined as

$$P \equiv \frac{dE}{dt} \quad (8.18)$$

◀ **Definition of power**

We will focus on work as the energy transfer method in this discussion, but keep in mind that the notion of power is valid for *any* means of energy transfer discussed in Section 8.1. If an external force is applied to an object (which we model as a particle) and if the work done by this force on the object in the time interval Δt is W, the **average power** during this interval is

$$P_{avg} = \frac{W}{\Delta t}$$

Therefore, in Conceptual Example 7.7, although the same work is done in rolling the refrigerator up both ramps, less power is required for the longer ramp.

In a manner similar to the way we approached the definition of velocity and acceleration, the instantaneous power is the limiting value of the average power as Δt approaches zero:

$$P = \lim_{\Delta t \to 0} \frac{W}{\Delta t} = \frac{dW}{dt}$$

where we have represented the infinitesimal value of the work done by dW. We find from Equation 7.3 that $dW = \vec{\mathbf{F}} \cdot d\vec{\mathbf{r}}$. Therefore, the instantaneous power can be written

$$P = \frac{dW}{dt} = \vec{\mathbf{F}} \cdot \frac{d\vec{\mathbf{r}}}{dt} = \vec{\mathbf{F}} \cdot \vec{\mathbf{v}} \qquad \textbf{(8.19)}$$

where $\vec{\mathbf{v}} = d\vec{\mathbf{r}}/dt$.

The SI unit of power is joules per second (J/s), also called the **watt** (W) after James Watt:

The watt ▶

$$1 \text{ W} = 1 \text{ J/s} = 1 \text{ kg} \cdot \text{m}^2/\text{s}^3$$

A unit of power in the U.S. customary system is the **horsepower** (hp):

$$1 \text{ hp} = 746 \text{ W}$$

A unit of energy (or work) can now be defined in terms of the unit of power. One **kilowatt-hour** (kWh) is the energy transferred in 1 h at the constant rate of 1 kW = 1 000 J/s. The amount of energy represented by 1 kWh is

$$1 \text{ kWh} = (10^3 \text{ W})(3\,600 \text{ s}) = 3.60 \times 10^6 \text{ J}$$

A kilowatt-hour is a unit of energy, not power. When you pay your electric bill, you are buying energy, and the amount of energy transferred by electrical transmission into a home during the period represented by the electric bill is usually expressed in kilowatt-hours. For example, your bill may state that you used 900 kWh of energy during a month and that you are being charged at the rate of 10¢ per kilowatt-hour. Your obligation is then $90 for this amount of energy. As another example, suppose an electric bulb is rated at 100 W. In 1.00 h of operation, it would have energy transferred to it by electrical transmission in the amount of (0.100 kW)(1.00 h) = 0.100 kWh = 3.60×10^5 J.

Pitfall Prevention 8.3

W, *W*, and watts

Do not confuse the symbol W for the watt with the italic symbol *W* for work. Also, remember that the watt already represents a rate of energy transfer, so "watts per second" does not make sense. The watt is *the same as* a joule per second.

Example 8.11 Power Delivered by an Elevator Motor

An elevator car (Fig. 8.14a) has a mass of 1 600 kg and is carrying passengers having a combined mass of 200 kg. A constant friction force of 4 000 N retards its motion.

(A) How much power must a motor deliver to lift the elevator car and its passengers at a constant speed of 3.00 m/s?

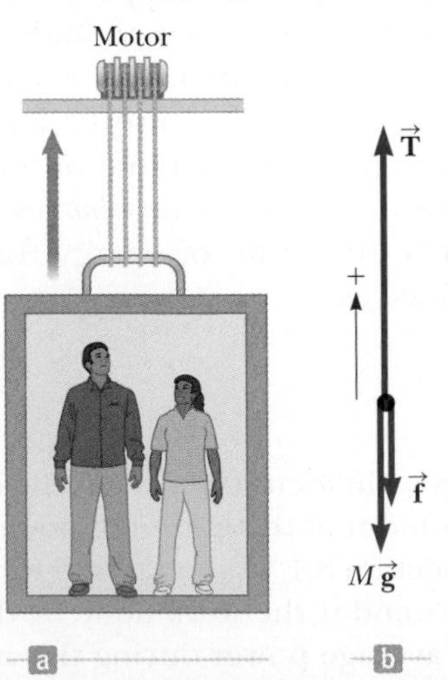

Figure 8.14 (Example 8.11) (a) The motor exerts an upward force $\vec{\mathbf{T}}$ on the elevator car. The magnitude of this force is the tension T in the cable connecting the car and motor. The downward forces acting on the car are a friction force $\vec{\mathbf{f}}$ and the gravitational force $\vec{\mathbf{F}}_g = M\vec{\mathbf{g}}$. (b) The free-body diagram for the elevator car.

SOLUTION

Conceptualize The motor must supply the force of magnitude T that pulls the elevator car upward.

Categorize The friction force increases the power necessary to lift the elevator. The problem states that the speed of the elevator is constant, which tells us that $a = 0$. We model the elevator as a particle in equilibrium.

Analyze The free-body diagram in Figure 8.14b specifies the upward direction as positive. The *total* mass M of the system (car plus passengers) is equal to 1 800 kg.

8.11 *cont.*

Using the particle in equilibrium model, apply Newton's second law to the car: $\sum F_y = T - f - Mg = 0$

Solve for T: $T = f + Mg$

Use Equation 8.19 and that $\vec{\mathbf{T}}$ is in the same direction as $\vec{\mathbf{v}}$ to find the power: $P = \vec{\mathbf{T}} \cdot \vec{\mathbf{v}} = Tv = (f + Mg)v$

Substitute numerical values: $P = [(4\,000 \text{ N}) + (1\,800 \text{ kg})(9.80 \text{ m/s}^2)](3.00 \text{ m/s}) = \boxed{6.49 \times 10^4 \text{ W}}$

(B) What power must the motor deliver at the instant the speed of the elevator is v if the motor is designed to provide the elevator car with an upward acceleration of 1.00 m/s^2?

SOLUTION

Conceptualize In this case, the motor must supply the force of magnitude T that pulls the elevator car upward with an increasing speed. We expect that more power will be required to do that than in part (A) because the motor must now perform the additional task of accelerating the car.

Categorize In this case, we model the elevator car as a particle under a net force because it is accelerating.

Analyze Using the particle under a net force model, apply Newton's second law to the car: $\sum F_y = T - f - Mg = Ma$

Solve for T: $T = M(a + g) + f$

Use Equation 8.19 to obtain the required power: $P = Tv = [M(a + g) + f]v$

Substitute numerical values: $P = [(1\,800 \text{ kg})(1.00 \text{ m/s}^2 + 9.80 \text{ m/s}^2) + 4\,000 \text{ N}]v$

$= \boxed{(2.34 \times 10^4)v}$

where v is the instantaneous speed of the car in meters per second and P is in watts.

Finalize To compare with part (A), let $v = 3.00$ m/s, giving a power of

$$P = (2.34 \times 10^4 \text{ N})(3.00 \text{ m/s}) = 7.02 \times 10^4 \text{ W}$$

which is larger than the power found in part (A), as expected.

Summary

Definitions

A **nonisolated system** is one for which energy crosses the boundary of the system. An **isolated system** is one for which no energy crosses the boundary of the system.

The **instantaneous power** P is defined as the time rate of energy transfer:

$$P \equiv \frac{dE}{dt} \quad \textbf{(8.18)}$$

continued

Concepts and Principles

For a nonisolated system, we can equate the change in the total energy stored in the system to the sum of all the transfers of energy across the system boundary, which is a statement of **conservation of energy.** For an isolated system, the total energy is constant.

If a system is isolated and if no nonconservative forces are acting on objects inside the system, the total mechanical energy of the system is constant:

$$K_f + U_f = K_i + U_i \quad \textbf{(8.10)}$$

If nonconservative forces (such as friction) act between objects inside a system, mechanical energy is not conserved. In these situations, the difference between the total final mechanical energy and the total initial mechanical energy of the system equals the energy transformed to internal energy by the nonconservative forces.

If a friction force acts within an isolated system, the mechanical energy of the system is reduced and the appropriate equation to be applied is

$$\Delta E_{\text{mech}} = \Delta K + \Delta U = -f_k d \quad \textbf{(8.16)}$$

If a friction force acts within a nonisolated system, the appropriate equation to be applied is

$$\Delta E_{\text{mech}} = -f_k d + \sum W_{\text{other forces}} \quad \textbf{(8.17)}$$

Analysis Models for Problem Solving

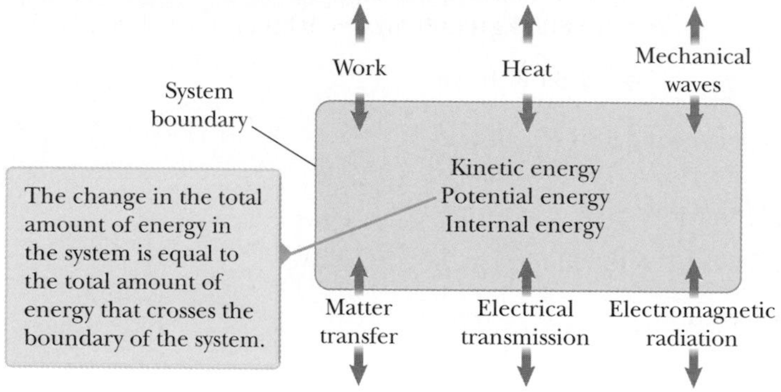

Nonisolated System (Energy). The most general statement describing the behavior of a nonisolated system is the **conservation of energy equation:**

$$\Delta E_{\text{system}} = \Sigma\, T \quad \textbf{(8.1)}$$

Including the types of energy storage and energy transfer that we have discussed gives

$$\Delta K + \Delta U + \Delta E_{\text{int}} = W + Q + T_{\text{MW}} + T_{\text{MT}} + T_{\text{ET}} + T_{\text{ER}} \quad \textbf{(8.2)}$$

For a specific problem, this equation is generally reduced to a smaller number of terms by eliminating the terms that are not appropriate to the situation.

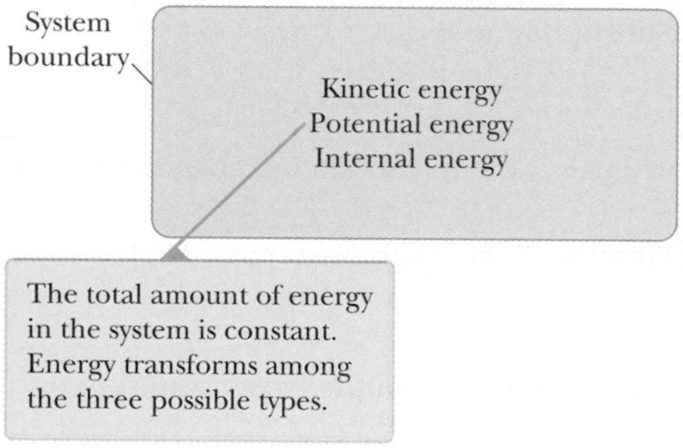

Isolated System (Energy). The total energy of an isolated system is conserved, so

$$\Delta E_{\text{system}} = 0 \quad \textbf{(8.9)}$$

If no nonconservative forces act within the isolated system, the mechanical energy of the system is conserved, so

$$\Delta E_{\text{mech}} = 0 \quad \textbf{(8.8)}$$

Objective Questions

☐ denotes answer available in Student Solutions Manual/Study Guide

1. What average power is generated by a 70.0-kg mountain climber who climbs a summit of height 325 m in 95.0 min? (a) 39.1 W (b) 54.6 W (c) 25.5 W (d) 67.0 W (e) 88.4 W

2. An athlete jumping vertically on a trampoline leaves the surface with a velocity of 8.5 m/s upward. What maximum height does she reach? (a) 13 m (b) 2.3 m (c) 3.7 m (d) 0.27 m (e) The answer can't be determined because the mass of the athlete isn't given.

3. A pile driver drives posts into the ground by repeatedly dropping a heavy object on them. Assume the object is dropped from the same height each time. By what factor does the energy of the pile driver–Earth system change when the mass of the object being dropped is doubled? (a) $\frac{1}{2}$ (b) 1; the energy is the same (c) 2 (d) 4

4. Two children stand on a platform at the top of a curving slide next to a backyard swimming pool. At the same moment the smaller child hops off to jump straight down into the pool, the bigger child releases herself at the top of the frictionless slide. **(i)** Upon reaching the water, the kinetic energy of the smaller child compared with that of the larger child is (a) greater (b) less (c) equal. **(ii)** Upon reaching the water, the speed of the smaller child compared with that of the larger child is (a) greater (b) less (c) equal. **(iii)** During their motions from the platform to the water, the average acceleration of the smaller child compared with that of the larger child is (a) greater (b) less (c) equal.

5. Answer yes or no to each of the following questions. (a) Can an object–Earth system have kinetic energy and not gravitational potential energy? (b) Can it have gravitational potential energy and not kinetic energy? (c) Can it have both types of energy at the same moment? (d) Can it have neither?

6. A ball of clay falls freely to the hard floor. It does not bounce noticeably, and it very quickly comes to rest. What, then, has happened to the energy the ball had while it was falling? (a) It has been used up in producing the downward motion. (b) It has been transformed back into potential energy. (c) It has been transferred into the ball by heat. (d) It is in the ball and floor (and walls) as energy of invisible molecular motion. (e) Most of it went into sound.

7. You hold a slingshot at arm's length, pull the light elastic band back to your chin, and release it to launch a pebble horizontally with speed 200 cm/s. With the same procedure, you fire a bean with speed 600 cm/s. What is the ratio of the mass of the bean to the mass of the pebble? (a) $\frac{1}{9}$ (b) $\frac{1}{3}$ (c) 1 (d) 3 (e) 9

8. In a laboratory model of cars skidding to a stop, data are measured for four trials using two blocks. The blocks have identical masses but different coefficients of kinetic friction with a table: μ_k = 0.2 and 0.8. Each block is launched with speed v_i = 1 m/s and slides across the level table as the block comes to rest. This process represents the first two trials. For the next two trials, the procedure is repeated but the blocks are launched with speed v_i = 2 m/s. Rank the four trials (a) through (d) according to the stopping distance from largest to smallest. If the stopping distance is the same in two cases, give them equal rank. (a) v_i = 1 m/s, μ_k = 0.2 (b) v_i = 1 m/s, μ_k = 0.8 (c) v_i = 2 m/s, μ_k = 0.2 (d) v_i = 2 m/s, μ_k = 0.8

9. At the bottom of an air track tilted at angle θ, a glider of mass m is given a push to make it coast a distance d up the slope as it slows down and stops. Then the glider comes back down the track to its starting point. Now the experiment is repeated with the same original speed but with a second identical glider set on top of the first. The airflow from the track is strong enough to support the stacked pair of gliders so that the combination moves over the track with negligible friction. Static friction holds the second glider stationary relative to the first glider throughout the motion. The coefficient of static friction between the two gliders is μ_s. What is the change in mechanical energy of the two-glider–Earth system in the up- and down-slope motion after the pair of gliders is released? Choose one. (a) $-2\mu_s mg$ (b) $-2mgd \cos \theta$ (c) $-2\mu_s mgd \cos \theta$ (d) 0 (e) $+2\mu_s mgd \cos \theta$

Conceptual Questions

☐ denotes answer available in Student Solutions Manual/Study Guide

1. Does everything have energy? Give the reasoning for your answer.

2. In Chapter 7, the work–kinetic energy theorem, $W_{net} = \Delta K$, was introduced. This equation states that work done on a system appears as a change in kinetic energy. It is a special-case equation, valid if there are no changes in any other type of energy such as potential or internal. Give two or three examples in which work is done on a system but the change in energy of the system is not a change in kinetic energy.

3. One person drops a ball from the top of a building while another person at the bottom observes its motion. Will these two people agree (a) on the value of the gravitational potential energy of the ball–Earth system? (b) On the change in potential energy? (c) On the kinetic energy of the ball at some point in its motion?

4. You ride a bicycle. In what sense is your bicycle solar-powered?

5. A bowling ball is suspended from the ceiling of a lecture hall by a strong cord. The ball is drawn away from its equilibrium position and released from rest at the tip of the demonstrator's nose as shown in Figure CQ8.5. The demonstrator remains stationary.

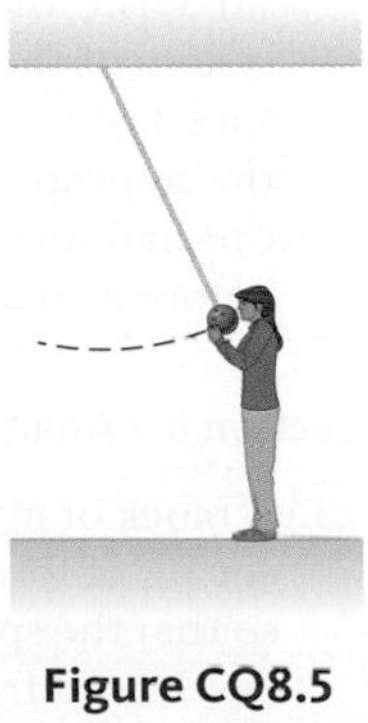

Figure CQ8.5

(a) Explain why the ball does not strike her on its return swing. (b) Would this demonstrator be safe if the ball were given a push from its starting position at her nose?

6. Can a force of static friction do work? If not, why not? If so, give an example.

7. A block is connected to a spring that is suspended from the ceiling. Assuming air resistance is ignored, describe the energy transformations that occur within the system consisting of the block, the Earth, and the spring when the block is set into vertical motion.

8. Consider the energy transfers and transformations listed below in parts (a) through (e). For each part, **(i)** describe human-made devices designed to produce each of the energy transfers or transformations and, **(ii)** whenever possible, describe a natural process in which the energy transfer or transformation occurs. Give details to defend your choices, such as identifying the system and identifying other output energy if the device or natural process has limited efficiency. (a) Chemical potential energy transforms into internal energy. (b) Energy transferred by electrical transmission becomes gravitational potential energy. (c) Elastic potential energy transfers out of a system by heat. (d) Energy transferred by mechanical waves does work on a system. (e) Energy carried by electromagnetic waves becomes kinetic energy in a system.

9. In the general conservation of energy equation, state which terms predominate in describing each of the following devices and processes. For a process going on continuously, you may consider what happens in a 10-s time interval. State which terms in the equation represent original and final forms of energy, which would be inputs, and which outputs. (a) a slingshot firing a pebble (b) a fire burning (c) a portable radio operating (d) a car braking to a stop (e) the surface of the Sun shining visibly (f) a person jumping up onto a chair

10. A car salesperson claims that a 300-hp engine is a necessary option in a compact car, in place of the conventional 130-hp engine. Suppose you intend to drive the car within speed limits ($\leq$ 65 mi/h) on flat terrain. How would you counter this sales pitch?

Problems

ENHANCED WebAssign The problems found in this chapter may be assigned online in Enhanced WebAssign

1. denotes straightforward problem; **2.** denotes intermediate problem;
3. denotes challenging problem

[1.] full solution available in the Student Solutions Manual/Study Guide

1. denotes problems most often assigned in Enhanced WebAssign; these provide students with targeted feedback and either a Master It tutorial or a Watch It solution video.

QIC denotes asking for quantitative and conceptual reasoning

S denotes symbolic reasoning problem

M denotes Master It tutorial available in Enhanced WebAssign

GP denotes guided problem

shaded denotes "paired problems" that develop reasoning with symbols and numerical values

Section 8.1 Analysis Model: Nonisolated System (Energy)

1. **S** For each of the following systems and time intervals, write the appropriate expanded version of Equation 8.2, the conservation of energy equation. (a) the heating coils in your toaster during the first five seconds after you turn the toaster on (b) your automobile from just before you fill it with gasoline until you pull away from the gas station at speed v (c) your body while you sit quietly and eat a peanut butter and jelly sandwich for lunch (d) your home during five minutes of a sunny afternoon while the temperature in the home remains fixed

2. **S** A ball of mass m falls from a height h to the floor. (a) Write the appropriate version of Equation 8.2 for the system of the ball and the Earth and use it to calculate the speed of the ball just before it strikes the Earth. (b) Write the appropriate version of Equation 8.2 for the system of the ball and use it to calculate the speed of the ball just before it strikes the Earth.

Section 8.2 Analysis Model: Isolated System (Energy)

[3.] A block of mass 0.250 kg is placed on top of a light, vertical spring of force constant 5 000 N/m and pushed downward so that the spring is compressed by 0.100 m. After the block is released from rest, it travels upward and then leaves the spring. To what maximum height above the point of release does it rise?

4. At 11:00 a.m. on September 7, 2001, more than one million British schoolchildren jumped up and down for one minute to simulate an earthquake. (a) Find the energy stored in the children's bodies that was converted into internal energy in the ground and their bodies and propagated into the ground by seismic waves during the experiment. Assume 1 050 000 children of average mass 36.0 kg jumped 12 times each, raising their centers of mass by 25.0 cm each time and briefly resting between one jump and the next. (b) Of the energy that propagated into the ground, most produced high-frequency "microtremor" vibrations that were rapidly damped and did not travel far. Assume 0.01% of the total energy was carried away by long-range seismic waves. The magnitude of an earthquake on the Richter scale is given by

$$M = \frac{\log E - 4.8}{1.5}$$

where E is the seismic wave energy in joules. According to this model, what was the magnitude of the demonstration quake?

[5.] **Review.** A bead slides without friction around a loop-the-loop (Fig. P8.5). The bead is released from rest at a height

$h = 3.50R$. (a) What is its speed at point Ⓐ? (b) How large is the normal force on the bead at point Ⓐ if its mass is 5.00 g?

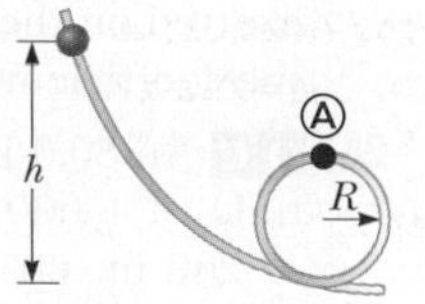

Figure P8.5

6. A block of mass $m = 5.00$ kg is released from point Ⓐ and slides on the frictionless track shown in Figure P8.6. Determine (a) the block's speed at points Ⓑ and Ⓒ and (b) the net work done by the gravitational force on the block as it moves from point Ⓐ to point Ⓒ.

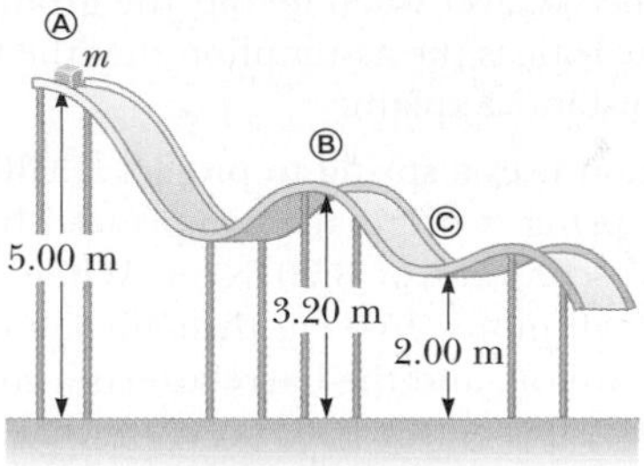

Figure P8.6

7. Two objects are connected by a light string passing over a light, frictionless pulley as shown in Figure P8.7. The object of mass $m_1 = 5.00$ kg is released from rest at a height $h = 4.00$ m above the table. Using the isolated system model, (a) determine the speed of the object of mass $m_2 = 3.00$ kg just as the 5.00-kg object hits the table and (b) find the maximum height above the table to which the 3.00-kg object rises.

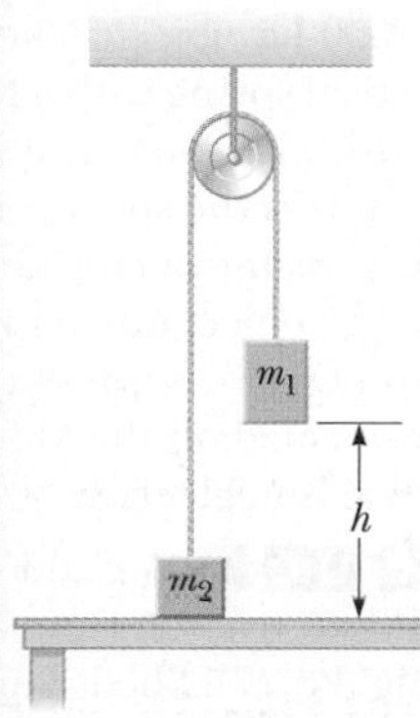

Figure P8.7
Problems 7 and 8.

8. **S** Two objects are connected by a light string passing over a light, frictionless pulley as shown in Figure P8.7. The object of mass m_1 is released from rest at height h above the table. Using the isolated system model, (a) determine the speed of m_2 just as m_1 hits the table and (b) find the maximum height above the table to which m_2 rises.

9. A light, rigid rod is 77.0 cm long. Its top end is pivoted on a frictionless, horizontal axle. The rod hangs straight down at rest with a small, massive ball attached to its bottom end. You strike the ball, suddenly giving it a horizontal velocity so that it swings around in a full circle. What minimum speed at the bottom is required to make the ball go over the top of the circle?

10. A 20.0-kg cannonball is fired from a cannon with muzzle speed of 1 000 m/s at an angle of 37.0° with the horizontal. A second ball is fired at an angle of 90.0°. Use the isolated system model to find (a) the maximum height reached by each ball and (b) the total mechanical energy of the ball–Earth system at the maximum height for each ball. Let $y = 0$ at the cannon.

11. **S** **Review.** The system shown in Figure P8.11 consists of a light, inextensible cord, light, frictionless pulleys, and blocks of equal mass. Notice that block B is attached to one of the pulleys. The system is initially held at rest so that the blocks are at the same height above the ground. The blocks are then released. Find the speed of block A at the moment the vertical separation of the blocks is h.

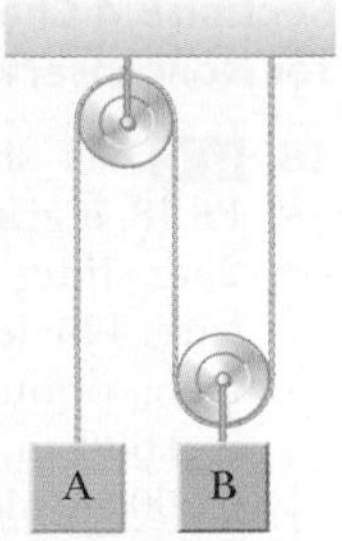

Figure P8.11

Section 8.3 Situations Involving Kinetic Friction

12. A sled of mass m is given a kick on a frozen pond. The kick imparts to the sled an initial speed of 2.00 m/s. The coefficient of kinetic friction between sled and ice is 0.100. Use energy considerations to find the distance the sled moves before it stops.

13. **S** A sled of mass m is given a kick on a frozen pond. The kick imparts to the sled an initial speed of v. The coefficient of kinetic friction between sled and ice is μ_k. Use energy considerations to find the distance the sled moves before it stops.

14. A 40.0-kg box initially at rest is pushed 5.00 m along a rough, horizontal floor with a constant applied horizontal force of 130 N. The coefficient of friction between box and floor is 0.300. Find (a) the work done by the applied force, (b) the increase in internal energy in the box–floor system as a result of friction, (c) the work done by the normal force, (d) the work done by the gravitational force, (e) the change in kinetic energy of the box, and (f) the final speed of the box.

15. A block of mass $m = 2.00$ kg is attached to a spring of force constant $k = 500$ N/m as shown in Figure P8.15. The block is pulled to a position $x_i = 5.00$ cm to the right of equilibrium and released from rest. Find the speed the block has as it passes through equilibrium if (a) the horizontal surface is frictionless and (b) the coefficient of friction between block and surface is $\mu_k = 0.350$.

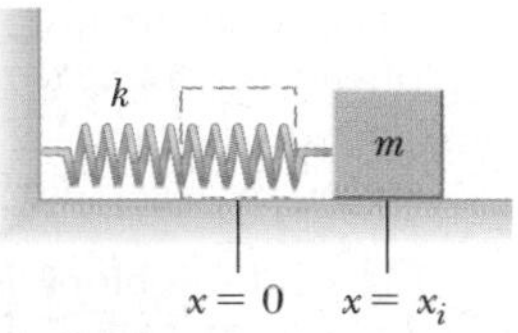

Figure P8.15

16. A crate of mass 10.0 kg is pulled up a rough incline with an initial speed of 1.50 m/s. The pulling force is 100 N parallel to the incline, which makes an angle of 20.0° with the horizontal. The coefficient of kinetic friction is 0.400, and the crate is pulled 5.00 m. (a) How much work is done by the gravitational force on the crate? (b) Determine the increase in internal energy of the crate–incline system owing to friction. (c) How much work is done by the 100-N force on the crate? (d) What is the change in kinetic energy of the crate? (e) What is the speed of the crate after being pulled 5.00 m?

17. A smooth circular hoop with a radius of 0.500 m is placed flat on the floor. A 0.400-kg particle slides around the inside edge of the hoop. The particle is given an initial speed of 8.00 m/s. After one revolution, its speed has dropped to 6.00 m/s because of friction with the floor. (a) Find the energy transformed from mechanical to internal in the particle–hoop–floor system as a result of friction in one revolution. (b) What is the total number of revolutions the particle makes before stopping? Assume the friction force remains constant during the entire motion.

Section 8.4 Changes in Mechanical Energy for Nonconservative Forces

18. **Q|C** As shown in Figure P8.18, a green bead of mass 25 g slides along a straight wire. The length of the wire from point Ⓐ to point Ⓑ is 0.600 m, and point Ⓐ is 0.200 m higher than point Ⓑ. A constant friction force of magnitude 0.025 0 N acts on the bead. (a) If the bead is released from rest at point Ⓐ, what is its speed at point Ⓑ? (b) A red bead of mass 25 g slides along a curved wire, subject to a friction force with the same constant magnitude as that on the green bead. If the green and red beads are released simultaneously from rest at point Ⓐ, which bead reaches point Ⓑ with a higher speed? Explain.

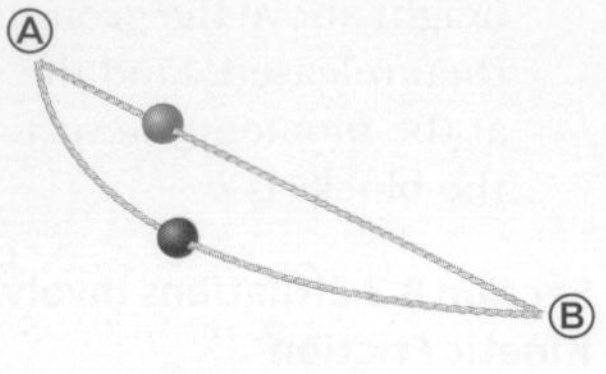

Figure P8.18

19. A boy in a wheelchair (total mass 47.0 kg) has speed 1.40 m/s at the crest of a slope 2.60 m high and 12.4 m long. At the bottom of the slope his speed is 6.20 m/s. Assume air resistance and rolling resistance can be modeled as a constant friction force of 41.0 N. Find the work he did in pushing forward on his wheels during the downhill ride.

20. **Q|C** At time t_i, the kinetic energy of a particle is 30.0 J and the potential energy of the system to which it belongs is 10.0 J. At some later time t_f, the kinetic energy of the particle is 18.0 J. (a) If only conservative forces act on the particle, what are the potential energy and the total energy of the system at time t_f? (b) If the potential energy of the system at time t_f is 5.00 J, are any nonconservative forces acting on the particle? (c) Explain your answer to part (b).

21. **M** A 200-g block is pressed against a spring of force constant 1.40 kN/m until the block compresses the spring 10.0 cm. The spring rests at the bottom of a ramp inclined at 60.0° to the horizontal. Using energy considerations, determine how far up the incline the block moves from its initial position before it stops (a) if the ramp exerts no friction force on the block and (b) if the coefficient of kinetic friction is 0.400.

22. The coefficient of friction between the block of mass $m_1 =$ 3.00 kg and the surface in Figure P8.22 is $\mu_k = 0.400$. The system starts from rest. What is the speed of the ball of mass m_2 = 5.00 kg when it has fallen a distance $h = 1.50$ m?

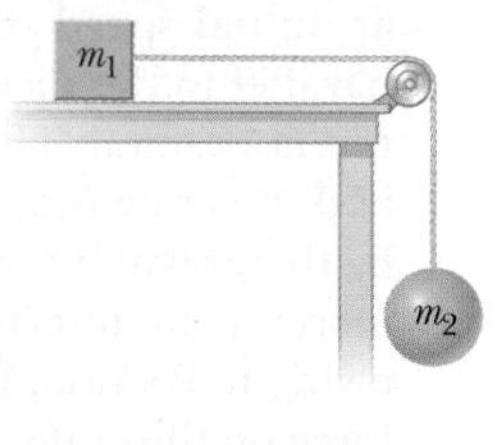

Figure P8.22

23. A 5.00-kg block is set into motion up an inclined plane with an initial speed of $v_i = 8.00$ m/s (Fig. P8.23). The block comes to rest after traveling $d = 3.00$ m along the plane, which is inclined at an angle of $\theta = 30.0°$ to the horizontal. For this motion, determine (a) the change in the block's kinetic energy, (b) the change in the potential energy of the block–Earth system, and (c) the friction force exerted on the block (assumed to be constant). (d) What is the coefficient of kinetic friction?

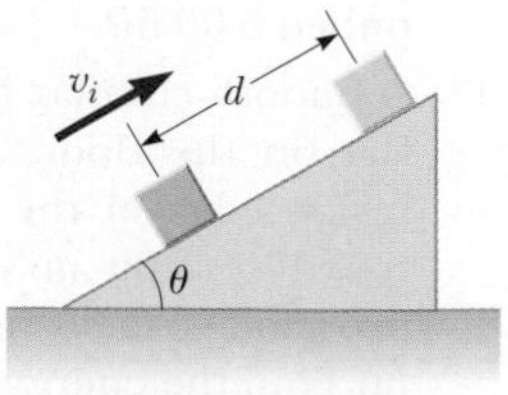

Figure P8.23

24. **Q|C** An 80.0-kg skydiver jumps out of a balloon at an altitude of 1 000 m and opens his parachute at an altitude of 200 m. (a) Assuming the total retarding force on the skydiver is constant at 50.0 N with the parachute closed and constant at 3 600 N with the parachute open, find the speed of the skydiver when he lands on the ground. (b) Do you think the skydiver will be injured? Explain. (c) At what height should the parachute be opened so that the final speed of the skydiver when he hits the ground is 5.00 m/s? (d) How realistic is the assumption that the total retarding force is constant? Explain.

25. A toy cannon uses a spring to project a 5.30-g soft rubber ball. The spring is originally compressed by 5.00 cm and has a force constant of 8.00 N/m. When the cannon is fired, the ball moves 15.0 cm through the horizontal barrel of the cannon, and the barrel exerts a constant friction force of 0.032 0 N on the ball. (a) With what speed does the projectile leave the barrel of the cannon? (b) At what point does the ball have maximum speed? (c) What is this maximum speed?

26. A 1.50-kg object is held 1.20 m above a relaxed massless, vertical spring with a force constant of 320 N/m. The object is dropped onto the spring. (a) How far does the object compress the spring? (b) **What If?** Repeat part (a), but this time assume a constant air-resistance force of 0.700 N acts on the object during its motion. (c) **What If?** How far does the object compress the spring if the same experiment is performed on the Moon, where $g = 1.63$ m/s^2 and air resistance is neglected?

27. **GP** **Q|C** **S** A child of mass m starts from rest and slides without friction from a height h along a slide next to a pool (Fig. P8.27). She is launched from a height $h/5$ into the air over the pool. We wish to find the maximum height she reaches above the water in her projectile motion. (a) Is the child–Earth system isolated or nonisolated? Why? (b) Is there a nonconservative force acting within the system? (c) Define the configuration of the system when the child is at the water level as having zero gravitational potential energy. Express the total energy of the system when the child is at the top of the waterslide. (d) Express the total energy of the system when the child is at the launching point. (e) Express the total energy of the system when the child is at the highest point in her projectile motion. (f) From parts (c) and (d), determine her initial speed v_i at the launch point in terms of g and h. (g) From parts (d), (e), and (f), determine her maximum airborne height y_{max} in terms of h and the launch angle θ. (h) Would your answers be the same if the waterslide were not frictionless? Explain.

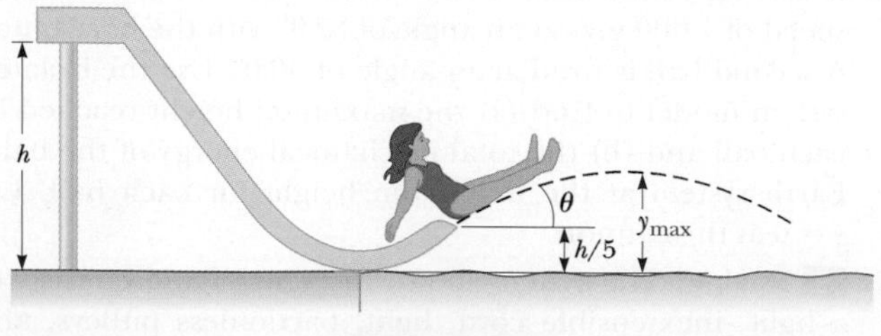

Figure P8.27

Section 8.5 Power

28. A certain rain cloud at an altitude of 1.75 km contains 3.20×10^7 kg of water vapor. How long would it take a 2.70-kW pump to raise the same amount of water from the Earth's surface to the cloud's position?

29. An 820-N Marine in basic training climbs a 12.0-m vertical rope at a constant speed in 8.00 s. What is his power output?

30. **Q|C** The electric motor of a model train accelerates the train from rest to 0.620 m/s in 21.0 ms. The total mass of the train is 875 g. (a) Find the minimum power delivered to the train by electrical transmission from the metal rails during the acceleration. (b) Why is it the minimum power?

31. Make an order-of-magnitude estimate of the power a car engine contributes to speeding the car up to highway speed. In your solution, state the physical quantities you take as data and the values you measure or estimate for them. The mass of a vehicle is often given in the owner's manual.

32. Sewage at a certain pumping station is raised vertically by 5.49 m at the rate of 1 890 000 liters each day. The sewage, of density 1 050 kg/m^3, enters and leaves the pump at atmospheric pressure and through pipes of equal diameter. (a) Find the output mechanical power of the lift station. (b) Assume an electric motor continuously operating with average power 5.90 kW runs the pump. Find its efficiency.

33. An energy-efficient lightbulb, taking in 28.0 W of power, can produce the same level of brightness as a conventional lightbulb operating at power 100 W. The lifetime of the energy-efficient bulb is 10 000 h and its purchase price is \$4.50, whereas the conventional bulb has a lifetime of 750 h and costs \$0.42. Determine the total savings obtained by using one energy-efficient bulb over its lifetime as opposed to using conventional bulbs over the same time interval. Assume an energy cost of \$0.200 per kilowatt-hour.

34. An electric scooter has a battery capable of supplying 120 Wh of energy. If friction forces and other losses account for 60.0% of the energy usage, what altitude change can a rider achieve when driving in hilly terrain if the rider and scooter have a combined weight of 890 N?

35. When an automobile moves with constant speed down a highway, most of the power developed by the engine is used to compensate for the energy transformations due to friction forces exerted on the car by the air and the road. If the power developed by an engine is 175 hp, estimate the total friction force acting on the car when it is moving at a speed of 29 m/s. One horsepower equals 746 W.

36. **S** An older-model car accelerates from 0 to speed v in a time interval of Δt. A newer, more powerful sports car accelerates from 0 to $2v$ in the same time period. Assuming the energy coming from the engine appears only as kinetic energy of the cars, compare the power of the two cars.

37. A 3.50-kN piano is lifted by three workers at constant speed to an apartment 25.0 m above the street using a pulley system fastened to the roof of the building. Each worker is able to deliver 165 W of power, and the pulley system is 75.0% efficient (so that 25.0% of the mechanical energy is transformed to other forms due to friction in the pulley). Neglecting the mass of the pulley, find the time required to lift the piano from the street to the apartment.

38. A 650-kg elevator starts from rest. It moves upward for 3.00 s with constant acceleration until it reaches its cruising speed of 1.75 m/s. (a) What is the average power of the elevator motor during this time interval? (b) How does this power compare with the motor power when the elevator moves at its cruising speed?

39. A loaded ore car has a mass of 950 kg and rolls on rails with negligible friction. It starts from rest and is pulled up a mine shaft by a cable connected to a winch. The shaft is inclined at 30.0° above the horizontal. The car accelerates uniformly to a speed of 2.20 m/s in 12.0 s and then continues at constant speed. (a) What power must the winch motor provide when the car is moving at constant speed? (b) What maximum power must the winch motor provide? (c) What total energy has transferred out of the motor by work by the time the car moves off the end of the track, which is of length 1 250 m?

40. Energy is conventionally measured in Calories as well as in joules. One Calorie in nutrition is one kilocalorie, defined as 1 kcal = 4 186 J. Metabolizing 1 g of fat can release 9.00 kcal. A student decides to try to lose weight by exercising. He plans to run up and down the stairs in a football stadium as fast as he can and as many times as necessary. To evaluate the program, suppose he runs up a flight of 80 steps, each 0.150 m high, in 65.0 s. For simplicity, ignore the energy he uses in coming down (which is small). Assume a typical efficiency for human muscles is 20.0%. This statement means that when your body converts 100 J from metabolizing fat, 20 J goes into doing mechanical work (here, climbing stairs). The remainder goes into extra internal energy. Assume the student's mass is 75.0 kg. (a) How many times must the student run the flight of stairs to lose 1.00 kg of fat? (b) What is his average power output, in watts and in horsepower, as he runs up the stairs? (c) Is this activity in itself a practical way to lose weight?

Additional Problems

41. A small block of mass m = 200 g is released from rest at point Ⓐ along the horizontal diameter on the inside of a frictionless, hemispherical bowl of radius R = 30.0 cm (Fig. P8.41). Calculate (a) the gravitational potential energy of the block–Earth system when the block is at point Ⓐ relative to point Ⓑ, (b) the kinetic energy of the block at point Ⓑ, (c) its speed at point Ⓑ, and (d) its kinetic energy and the potential energy when the block is at point Ⓒ.

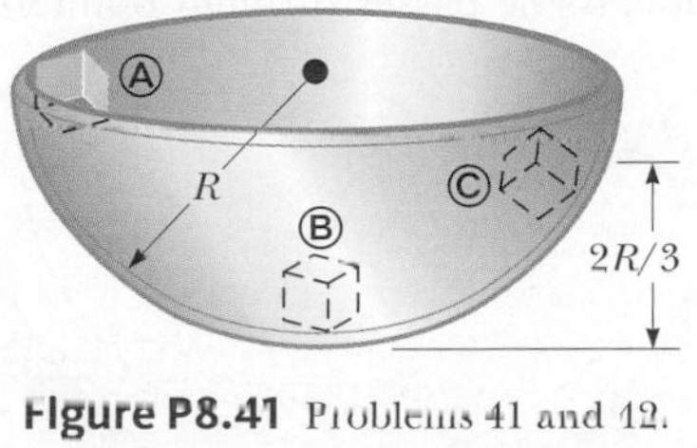

Figure P8.41 Problems 41 and 42.

42. **Q|C** **What If?** The block of mass m = 200 g described in Problem 41 (Fig. P8.41) is released from rest at point Ⓐ, and the surface of the bowl is rough. The block's speed at point Ⓑ is 1.50 m/s. (a) What is its kinetic energy at point Ⓑ? (b) How much mechanical energy is transformed into internal energy as the block moves from point Ⓐ to point Ⓑ? (c) Is it possible to determine the coefficient of friction from these results in any simple manner? (d) Explain your answer to part (c).

43. **S** **Review.** A boy starts at rest and slides down a frictionless slide as in Figure P8.43. The bottom of the track is a height h above the ground. The boy then leaves the track horizontally, striking the ground at a distance d as shown. Using energy methods, determine the initial height H of the boy above the ground in terms of h and d.

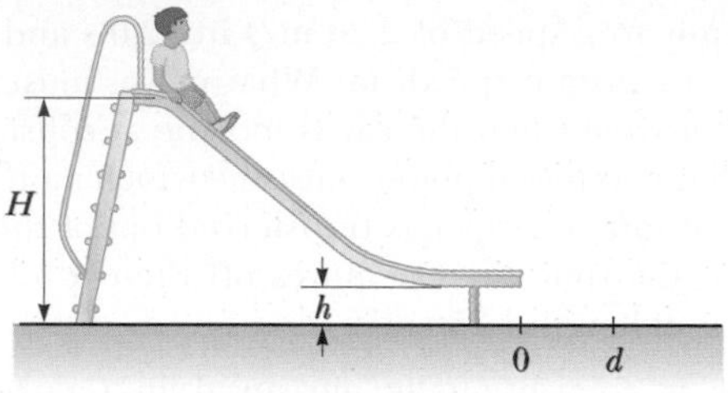

Figure P8.43

44. *Why is the following situation impossible?* A softball pitcher has a strange technique: she begins with her hand at rest at the highest point she can reach and then quickly rotates her arm backward so that the ball moves through a half-circle path. She releases the ball when her hand reaches the bottom of the path. The pitcher maintains a component of force on the 0.180-kg ball of constant magnitude 12.0 N in the direction of motion around the complete path. As the ball arrives at the bottom of the path, it leaves her hand with a speed of 25.0 m/s.

45. A skateboarder with his board can be modeled as a particle of mass 76.0 kg, located at his center of mass (which we will study in Chapter 9). As shown in Figure P8.45, the skateboarder starts from rest in a crouching position at one lip of a half-pipe (point Ⓐ). The half-pipe is one half of a cylinder of radius 6.80 m with its axis horizontal. On his descent, the skateboarder moves without friction so that his center of mass moves through one quarter of a circle of radius 6.30 m. (a) Find his speed at the bottom of the half-pipe (point Ⓑ). (b) Immediately after passing point Ⓑ, he stands up and raises his arms, lifting his center of mass from 0.500 m to 0.950 m above the concrete (point Ⓒ). Next, the skateboarder glides upward with his center of mass moving in a quarter circle of radius 5.85 m. His body is horizontal when he passes point Ⓓ, the far lip of the half-pipe. As he passes through point Ⓓ, the speed of the skateboarder is 5.14 m/s. How much chemical potential energy in the body of the skateboarder was converted to mechanical energy in the skateboarder–Earth system when he stood up at point Ⓑ? (c) How high above point Ⓓ does he rise? *Caution:* Do not try this stunt yourself without the required skill and protective equipment.

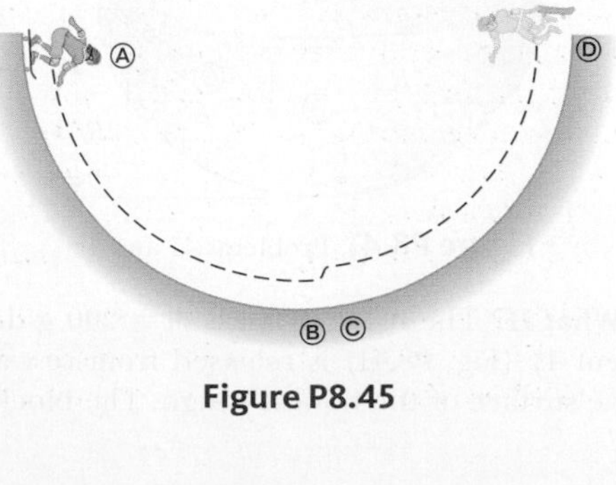

Figure P8.45

46. **Q|C** **Review.** As shown in Figure P8.46, a light string that does not stretch changes from horizontal to vertical as it passes over the edge of a table. The string connects m_1, a 3.50-kg block originally at rest on the horizontal table at a height h = 1.20 m above the floor, to m_2, a hanging 1.90-kg block originally a distance d = 0.900 m above the floor. Neither the surface of the table nor its edge exerts a force of kinetic friction. The blocks start to move from rest. The sliding block m_1 is projected horizontally after reaching the edge of the table. The hanging block m_2 stops without bouncing when it strikes the floor. Consider the two blocks plus the Earth as the system. (a) Find the speed at which m_1 leaves the edge of the table. (b) Find the impact speed of m_1 on the floor. (c) What is the shortest length of the string so that it does not go taut while m_1 is in flight? (d) Is the energy of the system when it is released from rest equal to the energy of the system just before m_1 strikes the ground? (e) Why or why not?

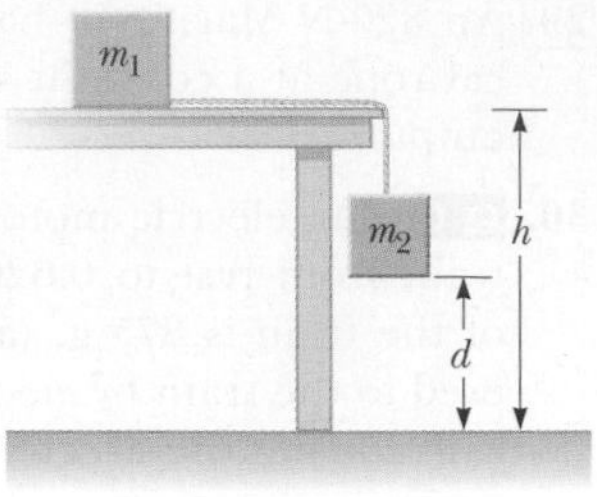

Figure P8.46

47. **M** A 4.00-kg particle moves along the x axis. Its position varies with time according to $x = t + 2.0t^3$, where x is in meters and t is in seconds. Find (a) the kinetic energy of the particle at any time t, (b) the acceleration of the particle and the force acting on it at time t, (c) the power being delivered to the particle at time t, and (d) the work done on the particle in the interval t = 0 to t = 2.00 s.

48. **Q|C** Heedless of danger, a child leaps onto a pile of old mattresses to use them as a trampoline. His motion between two particular points is described by the energy conservation equation

$$\tfrac{1}{2}(46.0\text{ kg})(2.40\text{ m/s})^2 + (46.0\text{ kg})(9.80\text{ m/s}^2)(2.80\text{ m} + x) = \tfrac{1}{2}(1.94 \times 10^4\text{ N/m})x^2$$

(a) Solve the equation for x. (b) Compose the statement of a problem, including data, for which this equation gives the solution. (c) Add the two values of x obtained in part (a) and divide by 2. (d) What is the significance of the resulting value in part (c)?

49. Jonathan is riding a bicycle and encounters a hill of height 7.30 m. At the base of the hill, he is traveling at 6.00 m/s. When he reaches the top of the hill, he is traveling at 1.00 m/s. Jonathan and his bicycle together have a mass of 85.0 kg. Ignore friction in the bicycle mechanism and between the bicycle tires and the road. (a) What is the total external work done on the system of Jonathan and the bicycle between the time he starts up the hill and the time he reaches the top? (b) What is the change in potential energy stored in Jonathan's body during this process? (c) How much work does Jonathan do on the bicycle pedals within the Jonathan–bicycle–Earth system during this process?

50. S Jonathan is riding a bicycle and encounters a hill of height h. At the base of the hill, he is traveling at a speed v_i. When he reaches the top of the hill, he is traveling at a speed v_f. Jonathan and his bicycle together have a mass m. Ignore friction in the bicycle mechanism and between the bicycle tires and the road. (a) What is the total external work done on the system of Jonathan and the bicycle between the time he starts up the hill and the time he reaches the top? (b) What is the change in potential energy stored in Jonathan's body during this process? (c) How much work does Jonathan do on the bicycle pedals within the Jonathan–bicycle–Earth system during this process?

51. Q|C As the driver steps on the gas pedal, a car of mass 1 160 kg accelerates from rest. During the first few seconds of motion, the car's acceleration increases with time according to the expression

$$a = 1.16t - 0.210t^2 + 0.240t^3$$

where t is in seconds and a is in m/s^2. (a) What is the change in kinetic energy of the car during the interval from $t = 0$ to $t = 2.50$ s? (b) What is the minimum average power output of the engine over this time interval? (c) Why is the value in part (b) described as the *minimum* value?

52. **Review.** *Why is the following situation impossible?* A new high-speed roller coaster is claimed to be so safe that the passengers do not need to wear seat belts or any other restraining device. The coaster is designed with a vertical circular section over which the coaster travels on the inside of the circle so that the passengers are upside down for a short time interval. The radius of the circular section is 12.0 m, and the coaster enters the bottom of the circular section at a speed of 22.0 m/s. Assume the coaster moves without friction on the track and model the coaster as a particle.

53. Q|C A horizontal spring attached to a wall has a force constant of $k = 850$ N/m. A block of mass $m = 1.00$ kg is attached to the spring and rests on a frictionless, horizontal surface as in Figure P8.53. (a) The block is pulled to a position $x_i = 6.00$ cm from equilibrium and released. Find the elastic potential energy stored in the spring when the block is 6.00 cm from equilibrium and when the block passes through equilibrium. (b) Find the speed of the block as it passes through the equilibrium point. (c) What is the speed of the block when it is at a position $x_i/2 = 3.00$ cm? (d) Why isn't the answer to part (c) half the answer to part (b)?

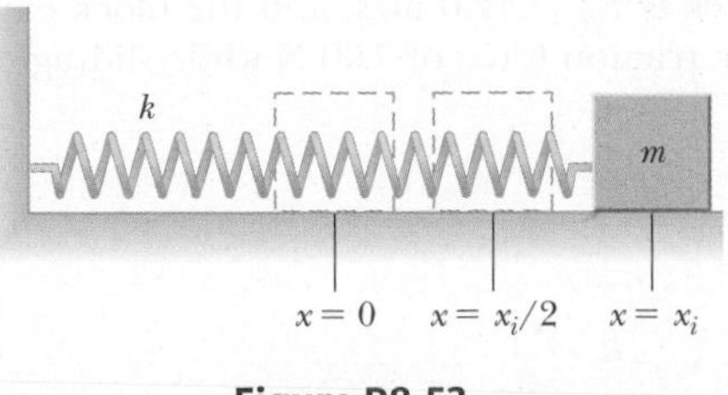

Figure P8.53

54. Q|C S As it plows a parking lot, a snowplow pushes an ever-growing pile of snow in front of it. Suppose a car moving through the air is similarly modeled as a cylinder of area A pushing a growing disk of air in front of it. The originally stationary air is set into motion at the constant speed v of the cylinder as shown in Figure P8.54. In a time interval Δt, a new disk of air of mass Δm must be moved a distance $v\,\Delta t$ and hence must be given a kinetic energy $\frac{1}{2}(\Delta m)v^2$. Using this model, show that the car's power loss owing to air resistance is $\frac{1}{2}\rho A v^3$ and that the resistive force acting on the car is $\frac{1}{2}\rho A v^2$, where ρ is the density of air. Compare this result with the empirical expression $\frac{1}{2}D\rho A v^2$ for the resistive force.

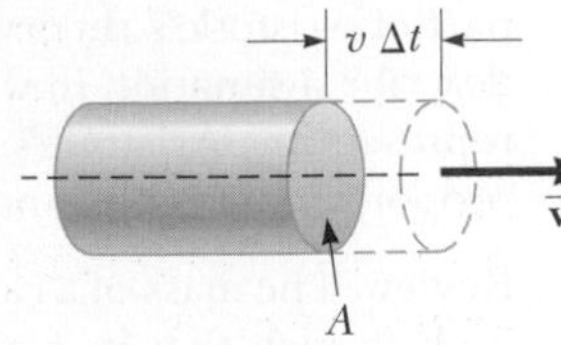

Figure P8.54

55. A wind turbine on a wind farm turns in response to a force of high-speed air resistance, $R = \frac{1}{2}D\rho A v^2$. The power available is $P = Rv = \frac{1}{2}D\rho \pi r^2 v^3$, where v is the wind speed and we have assumed a circular face for the wind turbine of radius r. Take the drag coefficient as $D = 1.00$ and the density of air from the front endpaper. For a wind turbine having $r = 1.50$ m, calculate the power available with (a) $v = 8.00$ m/s and (b) $v = 24.0$ m/s. The power delivered to the generator is limited by the efficiency of the system, about 25%. For comparison, a large American home uses about 2 kW of electric power.

56. Consider the popgun in Example 8.3. Suppose the projectile mass, compression distance, and spring constant remain the same as given or calculated in the example. Suppose, however, there is a friction force of magnitude 2.00 N acting on the projectile as it rubs against the interior of the barrel. The vertical length from point Ⓐ to the end of the barrel is 0.600 m. (a) After the spring is compressed and the popgun fired, to what height does the projectile rise above point Ⓑ? (b) Draw four energy bar charts for this situation, analogous to those in Figures 8.6c–d.

57. Consider the block–spring–surface system in part (B) of Example 8.6. (a) Using an energy approach, find the position x of the block at which its speed is a maximum. (b) In the **What If?** section of this example, we explored the effects of an increased friction force of 10.0 N. At what position of the block does its maximum speed occur in this situation?

58. Q|C S More than 2 300 years ago, the Greek teacher Aristotle wrote the first book called *Physics*. Put into more precise terminology, this passage is from the end of its Section Eta:

> Let P be the power of an agent causing motion; w, the load moved; d, the distance covered; and Δt, the time interval required. Then (1) a power equal to P will in an interval of time equal to Δt move $w/2$ a distance $2d$; or (2) it will move $w/2$ the given distance d in the time interval $\Delta t/2$. Also, if (3) the given power P moves the given load w a distance $d/2$ in time interval $\Delta t/2$, then (4) $P/2$ will move $w/2$ the given distance d in the given time interval Δt.

(a) Show that Aristotle's proportions are included in the equation $P\Delta t = bwd$, where b is a proportionality constant. (b) Show that our theory of motion includes this part of Aristotle's theory as one special case. In particular, describe a situation in which it is true, derive the equation representing Aristotle's proportions, and determine the proportionality constant.

59. **Review.** The mass of a car is 1 500 kg. The shape of the car's body is such that its aerodynamic drag coefficient is $D = 0.330$ and its frontal area is 2.50 m^2. Assuming the drag force is proportional to v^2 and ignoring other sources of friction, calculate the power required to maintain a speed of 100 km/h as the car climbs a long hill sloping at 3.20°.

60. **Review.** As a prank, someone has balanced a pumpkin at the highest point of a grain silo. The silo is topped with a hemispherical cap that is frictionless when wet. The line from the center of curvature of the cap to the pumpkin makes an angle $\theta_i = 0°$ with the vertical. While you happen to be standing nearby in the middle of a rainy night, a breath of wind makes the pumpkin start sliding downward from rest. It loses contact with the cap when the line from the center of the hemisphere to the pumpkin makes a certain angle with the vertical. What is this angle?

61. A child's pogo stick (Fig. P8.61) stores energy in a spring with a force constant of 2.50×10^4 N/m. At position Ⓐ ($x_{Ⓐ} = -0.100$ m), the spring compression is a maximum and the child is momentarily at rest. At position Ⓑ ($x_{Ⓑ} = 0$), the spring is relaxed and the child is moving upward. At position Ⓒ, the child is again momentarily at rest at the top of the jump. The combined mass of child and pogo stick is 25.0 kg. Although the boy must lean forward to remain balanced, the angle is small, so let's assume the pogo stick is vertical. Also assume the boy does not bend his legs during the motion. (a) Calculate the total energy of the child–stick–Earth system, taking both gravitational and elastic potential energies as zero for $x = 0$. (b) Determine $x_{Ⓒ}$. (c) Calculate the speed of the child at $x = 0$. (d) Determine the value of x for which the kinetic energy of the system is a maximum. (e) Calculate the child's maximum upward speed.

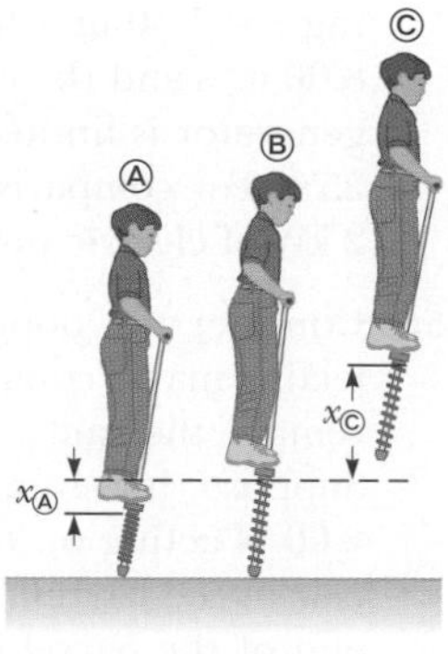

Figure P8.61

62. A 1.00-kg object slides to the right on a surface having a coefficient of kinetic friction 0.250 (Fig. P8.62a). The object has a speed of $v_i = 3.00$ m/s when it makes contact with a light spring (Fig. P8.62b) that has a force constant of 50.0 N/m. The object comes to rest after the spring has been compressed a distance d (Fig. P8.62c). The object is then forced toward the left by the spring (Fig. P8.62d) and continues to move in that direction beyond the spring's unstretched position. Finally, the object comes to rest a distance D to the left of the unstretched spring (Fig. P8.62e). Find (a) the distance of compression d, (b) the speed v at the unstretched position when the object is moving to the left (Fig. P8.62d), and (c) the distance D where the object comes to rest.

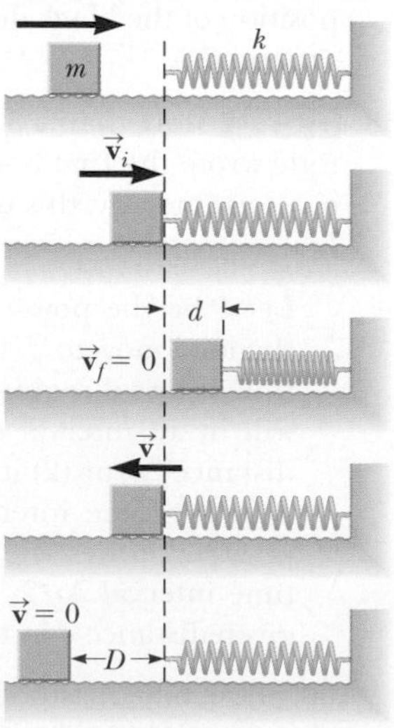

Figure P8.62

63. A 10.0-kg block is released from rest at point Ⓐ in Figure P8.63. The track is frictionless except for the portion between points Ⓑ and Ⓒ, which has a length of 6.00 m. The block travels down the track, hits a spring of force constant 2 250 N/m, and compresses the spring 0.300 m from its equilibrium position before coming to rest momentarily. Determine the coefficient of kinetic friction between the block and the rough surface between points Ⓑ and Ⓒ.

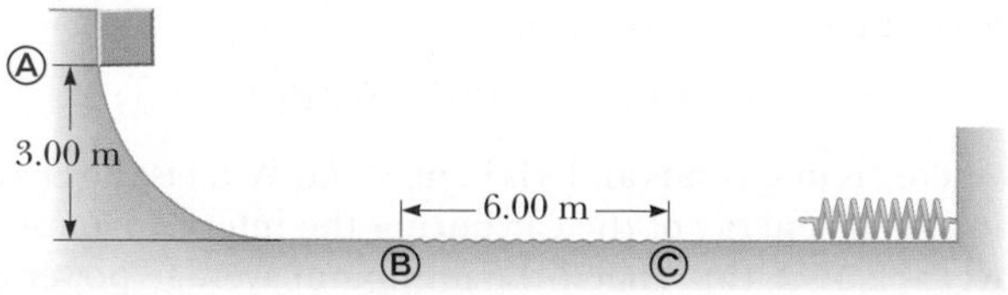

Figure P8.63

64. **M** A block of mass $m_1 = 20.0$ kg is connected to a block of mass $m_2 = 30.0$ kg by a massless string that passes over a light, frictionless pulley. The 30.0-kg block is connected to a spring that has negligible mass and a force constant of $k = 250$ N/m as shown in Figure P8.64. The spring is unstretched when the system is as shown in the figure, and the incline is frictionless. The 20.0-kg block is pulled a distance $h = 20.0$ cm down the incline of angle $\theta = 40.0°$ and released from rest. Find the speed of each block when the spring is again unstretched.

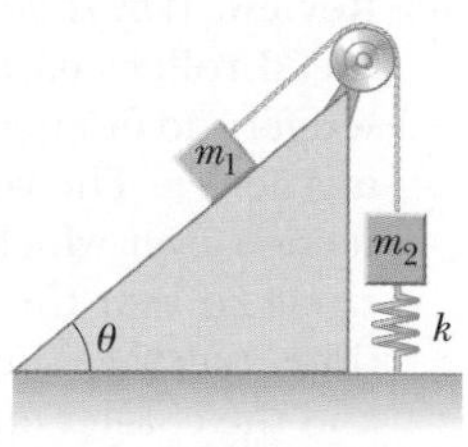

Figure P8.64

65. **Q|C** A block of mass 0.500 kg is pushed against a horizontal spring of negligible mass until the spring is compressed a distance x (Fig. P8.65). The force constant of the spring is 450 N/m. When it is released, the block travels along a frictionless, horizontal surface to point Ⓐ, the bottom of a vertical circular track of radius $R = 1.00$ m, and continues to move up the track. The block's speed at the bottom of the track is $v_{Ⓐ} = 12.0$ m/s, and the block experiences an average friction force of 7.00 N while sliding up the track.

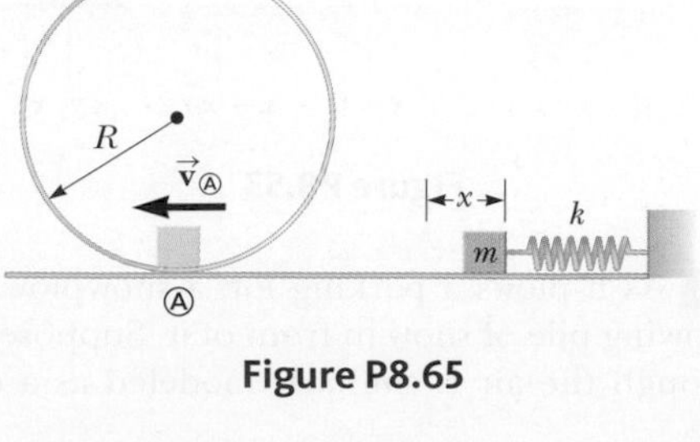

Figure P8.65

(a) What is x? (b) If the block were to reach the top of the track, what would be its speed at that point? (c) Does the block actually reach the top of the track, or does it fall off before reaching the top?

66. **S** A block of mass M rests on a table. It is fastened to the lower end of a light, vertical spring. The upper end of the spring is fastened to a block of mass m. The upper block is pushed down by an additional force $3mg$, so the spring compression is $4mg/k$. In this configuration, the upper block is released from rest. The spring lifts the lower block off the table. In terms of m, what is the greatest possible value for M?

67. **S** A pendulum, comprising a light string of length L and a small sphere, swings in the vertical plane. The string hits a peg located a distance d below the point of suspension (Fig. P8.67). (a) Show that if the sphere is released from a height below that of the peg, it will return to this height after the string strikes the peg. (b) Show that if the pendulum is released from rest at the horizontal position ($\theta = 90°$) and is to swing in a complete circle centered on the peg, the minimum value of d must be $3L/5$.

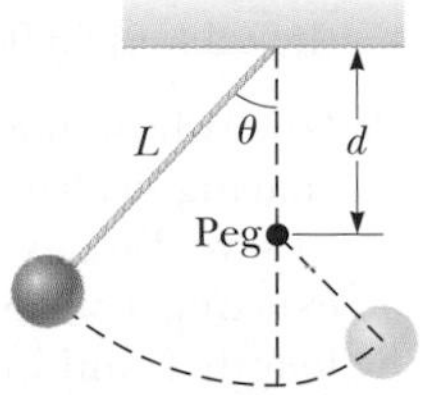

Figure P8.67

68. **Review.** *Why is the following situation impossible?* An athlete tests her hand strength by having an assistant hang weights from her belt as she hangs onto a horizontal bar with her hands. When the weights hanging on her belt have increased to 80% of her body weight, her hands can no longer support her and she drops to the floor. Frustrated at not meeting her hand-strength goal, she decides to swing on a trapeze. The trapeze consists of a bar suspended by two parallel ropes, each of length ℓ, allowing performers to swing in a vertical circular arc (Fig. P8.68). The athlete holds the bar and steps off an elevated platform, starting from rest with the ropes at an angle $\theta_i = 60.0°$ with respect to the vertical. As she swings several times back and forth in a circular arc, she forgets her frustration related to the hand-strength test. Assume the size of the performer's body is small compared to the length ℓ and air resistance is negligible.

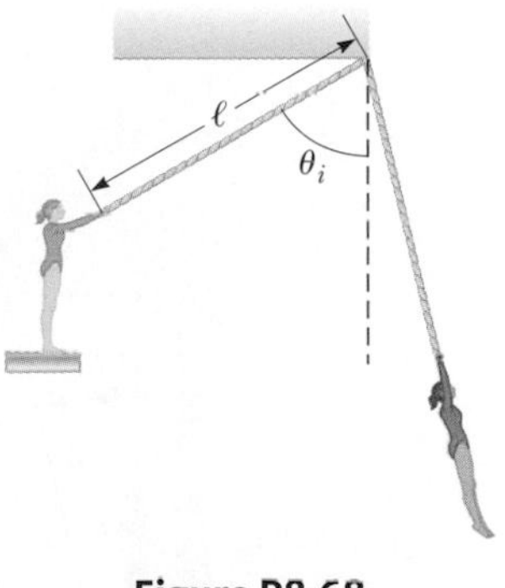

Figure P8.68

69. **S** A ball whirls around in a *vertical* circle at the end of a string. The other end of the string is fixed at the center of the circle. Assuming the total energy of the ball–Earth system remains constant, show that the tension in the string at the bottom is greater than the tension at the top by six times the ball's weight.

70. **QC** An airplane of mass 1.50×10^4 kg is in level flight, initially moving at 60.0 m/s. The resistive force exerted by air on the airplane has a magnitude of 4.0×10^4 N. By Newton's third law, if the engines exert a force on the exhaust gases to expel them out of the back of the engine, the exhaust gases exert a force on the engines in the direction of the airplane's travel. This force is called thrust, and the value of the thrust in this situation is 7.50×10^4 N. (a) Is the work done by the exhaust gases on the airplane during some time interval equal to the change in the airplane's kinetic energy? Explain. (b) Find the speed of the airplane after it has traveled 5.0×10^2 m.

71. Consider the block–spring collision discussed in Example 8.8. (a) For the situation in part (B), in which the surface exerts a friction force on the block, show that the block never arrives back at $x = 0$. (b) What is the maximum value of the coefficient of friction that would allow the block to return to $x = 0$?

72. **S** A roller-coaster car shown in Figure P8.72 is released from rest from a height h and then moves freely with negligible friction. The roller-coaster track includes a circular loop of radius R in a vertical plane. (a) First suppose the car barely makes it around the loop; at the top of the loop, the riders are upside down and feel weightless. Find the required height h of the release point above the bottom of the loop in terms of R. (b) Now assume the release point is at or above the minimum required height. Show that the normal force on the car at the bottom of the loop exceeds the normal force at the top of the loop by six times the car's weight. The normal force on each rider follows the same rule. Such a large normal force is dangerous and very uncomfortable for the riders. Roller coasters are therefore not built with circular loops in vertical planes. Figure P6.19 (page 159) shows an actual design.

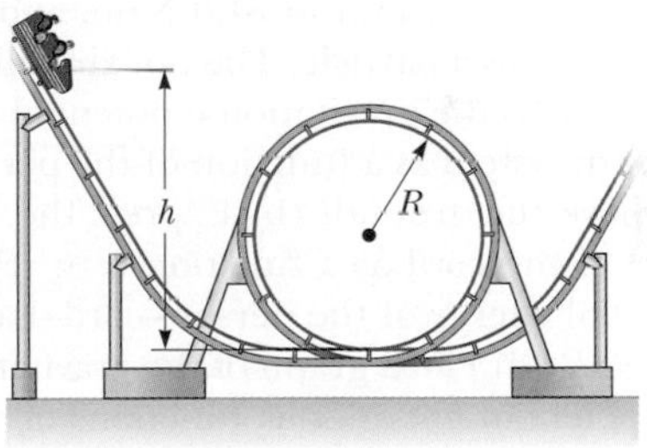

Figure P8.72

73. **Review.** In 1887 in Bridgeport, Connecticut, C. J. Belknap built the water slide shown in Figure P8.73 (page 232). A rider on a small sled, of total mass 80.0 kg, pushed off to start at the top of the slide (point Ⓐ) with a speed of 2.50 m/s. The chute was 9.76 m high at the top and 54.3 m long. Along its length, 725 small wheels made friction negligible. Upon leaving the chute horizontally at its bottom end (point Ⓒ), the rider skimmed across the water of Long Island Sound for as much as 50 m, "skipping along like a flat pebble," before at last coming to rest and swimming ashore, pulling his sled after him. (a) Find the speed of the sled and rider at point Ⓒ. (b) Model the force of water friction as a constant retarding force acting on a particle. Find the magnitude of the friction force the water exerts on the sled. (c) Find the magnitude of the force the chute exerts on the sled at point Ⓑ. (d) At point Ⓒ, the chute is horizon-

tal but curving in the vertical plane. Assume its radius of curvature is 20.0 m. Find the force the chute exerts on the sled at point Ⓒ.

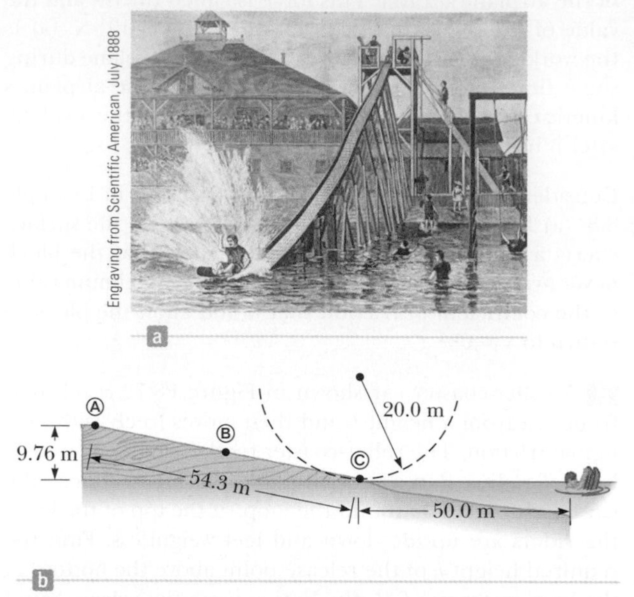

Figure P8.73

Challenge Problems

74. Q|C Starting from rest, a 64.0-kg person bungee jumps from a tethered hot-air balloon 65.0 m above the ground. The bungee cord has negligible mass and unstretched length 25.8 m. One end is tied to the basket of the balloon and the other end to a harness around the person's body. The cord is modeled as a spring that obeys Hooke's law with a spring constant of 81.0 N/m, and the person's body is modeled as a particle. The hot-air balloon does not move. (a) Express the gravitational potential energy of the person–Earth system as a function of the person's variable height y above the ground. (b) Express the elastic potential energy of the cord as a function of y. (c) Express the total potential energy of the person–cord–Earth system as a function of y. (d) Plot a graph of the gravitational, elastic, and total potential energies as functions of y. (e) Assume air resistance is negligible. Determine the minimum height of the person above the ground during his plunge. (f) Does the potential energy graph show any equilibrium position or positions? If so, at what elevations? Are they stable or unstable? (g) Determine the jumper's maximum speed.

75. S **Review.** A uniform board of length L is sliding along a smooth, frictionless, horizontal plane as shown in Figure P8.75a. The board then slides across the boundary with a rough horizontal surface. The coefficient of kinetic friction between the board and the second surface is μ_k. (a) Find the acceleration of the board at the moment its front end has traveled a distance x beyond the boundary. (b) The board stops at the moment its back end reaches the boundary as shown in Figure P8.75b. Find the initial speed v of the board.

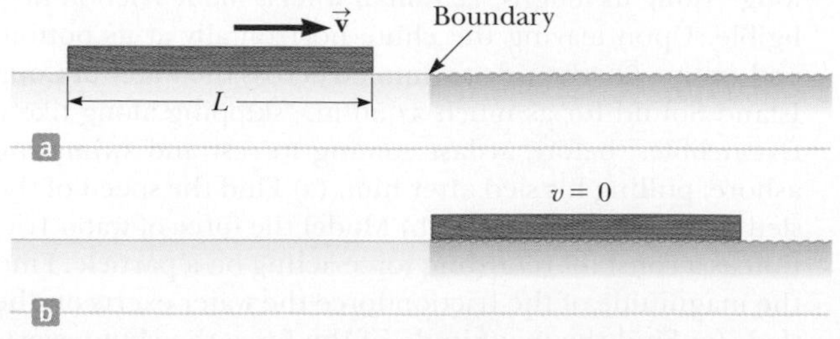

Figure P8.75

76. A uniform chain of length 8.00 m initially lies stretched out on a horizontal table. (a) Assuming the coefficient of static friction between chain and table is 0.600, show that the chain will begin to slide off the table if at least 3.00 m of it hangs over the edge of the table. (b) Determine the speed of the chain as its last link leaves the table, given that the coefficient of kinetic friction between the chain and the table is 0.400.

77. Jane, whose mass is 50.0 kg, needs to swing across a river (having width D) filled with person-eating crocodiles to save Tarzan from danger. She must swing into a wind exerting constant horizontal force $\vec{\mathbf{F}}$, on a vine having length L and initially making an angle θ with the vertical (Fig. P8.77). Take $D = 50.0$ m, $F = 110$ N, $L = 40.0$ m, and $\theta = 50.0°$. (a) With what minimum speed must Jane begin her swing to just make it to the other side? (b) Once the rescue is complete, Tarzan and Jane must swing back across the river. With what minimum speed must they begin their swing? Assume Tarzan has a mass of 80.0 kg.

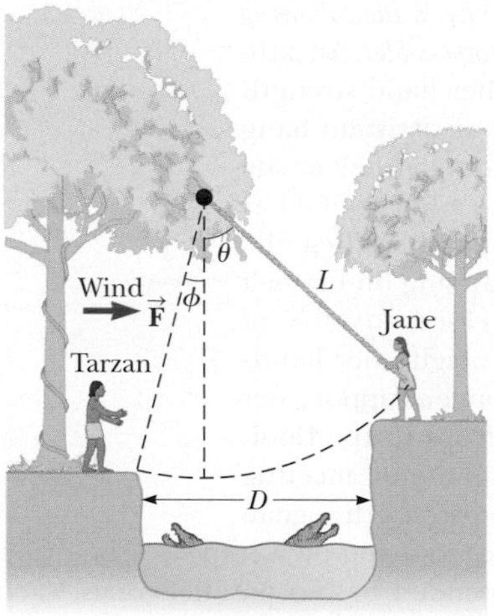

Figure P8.77

78. A ball of mass $m = 300$ g is connected by a strong string of length $L = 80.0$ cm to a pivot and held in place with the string vertical. A wind exerts constant force F to the right on the ball as shown in Figure P8.78. The ball is released

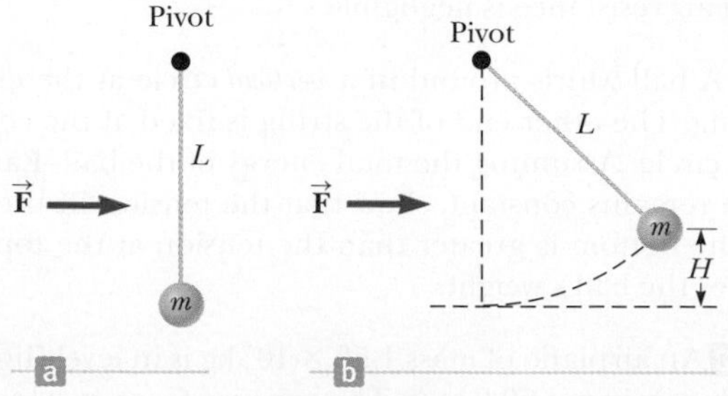

Figure P8.78

from rest. The wind makes it swing up to attain maximum height H above its starting point before it swings down again. (a) Find H as a function of F. Evaluate H for (b) $F = 1.00$ N and (c) $F = 10.0$ N. How does H behave (d) as F approaches zero and (e) as F approaches infinity? (f) Now consider the equilibrium height of the ball with the wind blowing. Determine it as a function of F. Evaluate the equilibrium height for (g) $F = 10$ N and (h) F going to infinity.

79. A daredevil plans to bungee jump from a balloon 65.0 m above the ground. He will use a uniform elastic cord, tied to a harness around his body, to stop his fall at a point 10.0 m above the ground. Model his body as a particle and the cord as having negligible mass and obeying Hooke's law. In a preliminary test he finds that when hanging at rest from a 5.00-m length of the cord, his body weight stretches it by 1.50 m. He will drop from rest at the point where the top end of a longer section of the cord is attached to the stationary balloon. (a) What length of cord should he use? (b) What maximum acceleration will he experience?

80. Q|C **What If?** Consider the roller coaster described in Problem 52. Because of some friction between the coaster and the track, the coaster enters the circular section at a speed of 15.0 m/s rather than the 22.0 m/s in Problem 52. Is this situation *more* or *less* dangerous for the passengers than that in Problem 52? Assume the circular section is still frictionless.

chapter 9

Linear Momentum and Collisions

On a pool table, a cue ball collides with a collection of other balls. Even though the ensuing process is complicated, it is possible to analyze it with principles involving momentum, which we study in this chapter. *(David Leah/Stone/Getty Images)*

Consider what happens when a cue ball on a pool table strikes the rest of the balls, as in the photograph at the right. The white cue ball comes in from the top of the photograph and strikes the colored balls. The colored balls, initially at rest, leave the site with various velocities. Because the average force exerted on a particular ball during the collision is large (resulting in a large acceleration), the ball can achieve a large velocity very rapidly over the very short time interval that the ball is in contact with another ball.

Although the force and acceleration of a given ball are related by Newton's second law, they vary in time, making for a complicated situation! One of the main objectives of this chapter is to enable you to understand and analyze such events in a simple way. First, we introduce the concept of *momentum,* which is useful for describing objects in motion. The momentum of an object is related to both its mass and its velocity. The concept of momentum leads us to a second conservation law, that of conservation of momentum. In turn, we identify new momentum versions of analysis models for isolated and nonisolated system. These models are especially useful for treating problems that involve collisions between objects and for analyzing rocket propulsion. This chapter also introduces the concept of the center of mass of a system of particles. We find that the motion of a system of particles can be described by the motion of one representative particle located at the center of mass.

9.1 Linear Momentum

In Chapter 8, we studied situations that are difficult to analyze with Newton's laws. We were able to solve problems involving these situations by identifying a system and applying a conservation principle, conservation of energy. Let us consider another situation and see if we can solve it with the models we have developed so far:

> A 60-kg archer stands at rest on frictionless ice and fires a 0.50-kg arrow horizontally at 50 m/s. With what velocity does the archer move across the ice after firing the arrow?

From Newton's third law, we know that the force that the bow exerts on the arrow is matched by a force in the opposite direction on the bow (and the archer). This force causes the archer to slide backward on the ice with the speed requested in the problem. We cannot determine this speed using motion models such as the particle under constant acceleration because we don't have any information about the acceleration of the archer. We cannot use force models such as the particle under a net force because we don't know anything about forces in this situation. Energy models are of no help because we know nothing about the work done in pulling the bowstring back or the elastic potential energy in the system related to the taut bowstring.

Despite our inability to solve the archer problem using models learned so far, this problem is very simple to solve if we introduce a new quantity that describes motion, *linear momentum.* To generate this new quantity, consider an isolated system of two particles (Fig. 9.1) with masses m_1 and m_2 moving with velocities $\vec{\mathbf{v}}_1$ and $\vec{\mathbf{v}}_2$ at an instant of time. Because the system is isolated, the only force on one particle is that from the other particle. If a force from particle 1 (for example, a gravitational force) acts on particle 2, there must be a second force—equal in magnitude but opposite in direction—that particle 2 exerts on particle 1. That is, the forces on the particles form a Newton's third law action–reaction pair, and $\vec{\mathbf{F}}_{12} = -\vec{\mathbf{F}}_{21}$. We can express this condition as

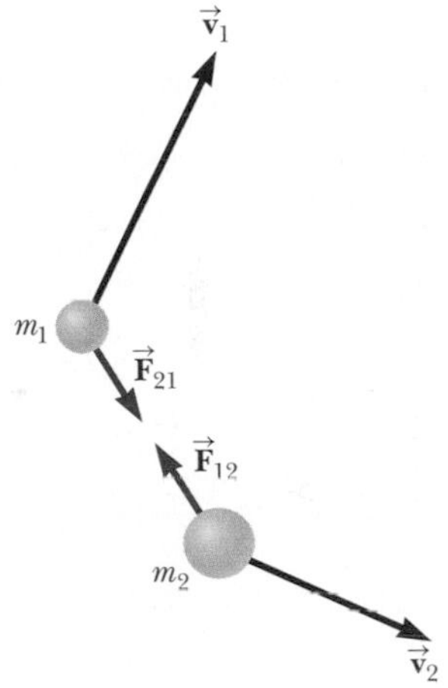

Figure 9.1 Two particles interact with each other. According to Newton's third law, we must have $\vec{\mathbf{F}}_{12} = -\vec{\mathbf{F}}_{21}$.

$$\vec{\mathbf{F}}_{21} + \vec{\mathbf{F}}_{12} = 0$$

Let us further analyze this situation by incorporating Newton's second law. At the instant shown in Figure 9.1, the interacting particles in the system have accelerations corresponding to the forces on them. Therefore, replacing the force on each particle with $m\vec{\mathbf{a}}$ for the particle gives

$$m_1\vec{\mathbf{a}}_1 + m_2\vec{\mathbf{a}}_2 = 0$$

Now we replace each acceleration with its definition from Equation 4.5:

$$m_1\frac{d\vec{\mathbf{v}}_1}{dt} + m_2\frac{d\vec{\mathbf{v}}_2}{dt} = 0$$

If the masses m_1 and m_2 are constant, we can bring them inside the derivative operation, which gives

$$\frac{d(m_1\vec{\mathbf{v}}_1)}{dt} + \frac{d(m_2\vec{\mathbf{v}}_2)}{dt} = 0$$

$$\frac{d}{dt}(m_1\vec{\mathbf{v}}_1 + m_2\vec{\mathbf{v}}_2) = 0 \tag{9.1}$$

Notice that the derivative of the sum $m_1\vec{\mathbf{v}}_1 + m_2\vec{\mathbf{v}}_2$ with respect to time is zero. Consequently, this sum must be constant. We learn from this discussion that the quantity $m\vec{\mathbf{v}}$ for a particle is important in that the sum of these quantities for an isolated system of particles is conserved. We call this quantity *linear momentum:*

Definition of linear momentum of a particle ▶

The **linear momentum** of a particle or an object that can be modeled as a particle of mass m moving with a velocity $\vec{\mathbf{v}}$ is defined to be the product of the mass and velocity of the particle:

$$\vec{\mathbf{p}} \equiv m\vec{\mathbf{v}} \tag{9.2}$$

Linear momentum is a vector quantity because it equals the product of a scalar quantity m and a vector quantity $\vec{\mathbf{v}}$. Its direction is along $\vec{\mathbf{v}}$, it has dimensions ML/T, and its SI unit is kg · m/s.

If a particle is moving in an arbitrary direction, $\vec{\mathbf{p}}$ has three components, and Equation 9.2 is equivalent to the component equations

$$p_x = mv_x \qquad p_y = mv_y \qquad p_z = mv_z$$

As you can see from its definition, the concept of momentum[1] provides a quantitative distinction between heavy and light particles moving at the same velocity. For example, the momentum of a bowling ball is much greater than that of a tennis ball moving at the same speed. Newton called the product $m\vec{\mathbf{v}}$ *quantity of motion;* this term is perhaps a more graphic description than our present-day word *momentum,* which comes from the Latin word for movement.

We have seen another quantity, kinetic energy, that is a combination of mass and speed. It would be a legitimate question to ask why we need another quantity, momentum, based on mass and velocity. There are clear differences between kinetic energy and momentum. First, kinetic energy is a scalar, whereas momentum is a vector. Consider a system of two equal-mass particles heading toward each other along a line with equal speeds. There is kinetic energy associated with this system because members of the system are moving. Because of the vector nature of momentum, however, the momentum of this system is zero. A second major difference is that kinetic energy can transform to other types of energy, such as potential energy or internal energy. There is only one type of linear momentum, so we see no such transformations when using a momentum approach to a problem. These differences are sufficient to make models based on momentum separate from those based on energy, providing an independent tool to use in solving problems.

Using Newton's second law of motion, we can relate the linear momentum of a particle to the resultant force acting on the particle. We start with Newton's second law and substitute the definition of acceleration:

$$\sum \vec{\mathbf{F}} = m\vec{\mathbf{a}} = m\frac{d\vec{\mathbf{v}}}{dt}$$

In Newton's second law, the mass m is assumed to be constant. Therefore, we can bring m inside the derivative operation to give us

Newton's second law for a particle ▶

$$\sum \vec{\mathbf{F}} = \frac{d(m\vec{\mathbf{v}})}{dt} = \frac{d\vec{\mathbf{p}}}{dt} \tag{9.3}$$

This equation shows that **the time rate of change of the linear momentum of a particle is equal to the net force acting on the particle.**

This alternative form of Newton's second law is the form in which Newton presented the law, and it is actually more general than the form introduced in Chapter 5. In addition to situations in which the velocity vector varies with time, we can use

[1] In this chapter, the terms *momentum* and *linear momentum* have the same meaning. Later, in Chapter 11, we shall use the term *angular momentum* for a different quantity when dealing with rotational motion.

Equation 9.3 to study phenomena in which the mass changes. For example, the mass of a rocket changes as fuel is burned and ejected from the rocket. We cannot use $\sum \vec{\mathbf{F}} = m\vec{\mathbf{a}}$ to analyze rocket propulsion; we must use a momentum approach, as we will show in Section 9.9.

***Quick Quiz* 9.1** Two objects have equal kinetic energies. How do the magnitudes of their momenta compare? **(a)** $p_1 < p_2$ **(b)** $p_1 = p_2$ **(c)** $p_1 > p_2$ **(d)** not enough information to tell

***Quick Quiz* 9.2** Your physical education teacher throws a baseball to you at a certain speed and you catch it. The teacher is next going to throw you a medicine ball whose mass is ten times the mass of the baseball. You are given the following choices: You can have the medicine ball thrown with **(a)** the same speed as the baseball, **(b)** the same momentum, or **(c)** the same kinetic energy. Rank these choices from easiest to hardest to catch.

9.2 Analysis Model: Isolated System (Momentum)

Using the definition of momentum, Equation 9.1 can be written

$$\frac{d}{dt}(\vec{\mathbf{p}}_1 + \vec{\mathbf{p}}_2) = 0$$

Because the time derivative of the total momentum $\vec{\mathbf{p}}_{\text{tot}} = \vec{\mathbf{p}}_1 + \vec{\mathbf{p}}_2$ is *zero,* we conclude that the *total* momentum of the isolated system of the two particles in Figure 9.1 must remain constant:

$$\vec{\mathbf{p}}_{\text{tot}} = \text{constant} \quad \textbf{(9.4)}$$

or, equivalently,

$$\vec{\mathbf{p}}_{1i} + \vec{\mathbf{p}}_{2i} = \vec{\mathbf{p}}_{1f} + \vec{\mathbf{p}}_{2f} \quad \textbf{(9.5)}$$

where $\vec{\mathbf{p}}_{1i}$ and $\vec{\mathbf{p}}_{2i}$ are the initial values and $\vec{\mathbf{p}}_{1f}$ and $\vec{\mathbf{p}}_{2f}$ are the final values of the momenta for the two particles for the time interval during which the particles interact. Equation 9.5 in component form demonstrates that the total momenta in the *x*, *y*, and *z* directions are all independently conserved:

$$p_{1ix} + p_{2ix} = p_{1fx} + p_{2fx} \qquad p_{1iy} + p_{2iy} = p_{1fy} + p_{2fy} \qquad p_{1iz} + p_{2iz} = p_{1fz} + p_{2fz} \quad \textbf{(9.6)}$$

Equation 9.5 is the mathematical statement of a new analysis model, the **isolated system (momentum).** It can be extended to any number of particles in an isolated system as we show in Section 9.7. We studied the energy version of the isolated system model in Chapter 8 and now we have a momentum version. In general, Equation 9.5 can be stated in words as follows:

> Whenever two or more particles in an isolated system interact, the total momentum of the system remains constant.

◀ **The momentum version of the isolated system model**

This statement tells us that the total momentum of an isolated system at all times equals its initial momentum.

Notice that we have made no statement concerning the type of forces acting on the particles of the system. Furthermore, we have not specified whether the forces are conservative or nonconservative. We have also not indicated whether or not the forces are constant. The only requirement is that the forces must be *internal* to the system. This single requirement should give you a hint about the power of this new model.

Pitfall Prevention 9.1

Momentum of an Isolated *System* Is Conserved

Although the momentum of an isolated *system* is conserved, the momentum of one particle within an isolated system is not necessarily conserved because other particles in the system may be interacting with it. Avoid applying conservation of momentum to a single particle.

Example 9.1 The Archer

Let us consider the situation proposed at the beginning of Section 9.1. A 60-kg archer stands at rest on frictionless ice and fires a 0.50-kg arrow horizontally at 50 m/s (Fig. 9.2). With what velocity does the archer move across the ice after firing the arrow?

Figure 9.2 (Example 9.1) An archer fires an arrow horizontally to the right. Because he is standing on frictionless ice, he will begin to slide to the left across the ice.

SOLUTION

Conceptualize You may have conceptualized this problem already when it was introduced at the beginning of Section 9.1. Imagine the arrow being fired one way and the archer recoiling in the opposite direction.

Categorize As discussed in Section 9.1, we cannot solve this problem with models based on motion, force, or energy. Nonetheless, we *can* solve this problem very easily with an approach involving momentum.

Let us take the system to consist of the archer (including the bow) and the arrow. The system is not isolated because the gravitational force and the normal force from the ice act on the system. These forces, however, are vertical and perpendicular to the motion of the system. Therefore, there are no external forces in the horizontal direction, and we can apply the isolated system (momentum) model in terms of momentum components in this direction.

Analyze The total horizontal momentum of the system before the arrow is fired is zero because nothing in the system is moving. Therefore, the total horizontal momentum of the system after the arrow is fired must also be zero. We choose the direction of firing of the arrow as the positive x direction. Identifying the archer as particle 1 and the arrow as particle 2, we have $m_1 = 60$ kg, $m_2 = 0.50$ kg, and $\vec{\mathbf{v}}_{2f} = 50\hat{\mathbf{i}}$ m/s.

Using the isolated system (momentum) model, set the final momentum of the system equal to the initial value of zero:

$$m_1\vec{\mathbf{v}}_{1f} + m_2\vec{\mathbf{v}}_{2f} = 0$$

Solve this equation for $\vec{\mathbf{v}}_{1f}$ and substitute numerical values:

$$\vec{\mathbf{v}}_{1f} = -\frac{m_2}{m_1}\vec{\mathbf{v}}_{2f} = -\left(\frac{0.50 \text{ kg}}{60 \text{ kg}}\right)(50\hat{\mathbf{i}} \text{ m/s}) = \boxed{-0.42\hat{\mathbf{i}} \text{ m/s}}$$

Finalize The negative sign for $\vec{\mathbf{v}}_{1f}$ indicates that the archer is moving to the left in Figure 9.2 after the arrow is fired, in the direction opposite the direction of motion of the arrow, in accordance with Newton's third law. Because the archer is much more massive than the arrow, his acceleration and consequent velocity are much smaller than the acceleration and velocity of the arrow. Notice that this problem sounds very simple, but we could not solve it with models based on motion, force, or energy. Our new momentum model, however, shows us that it not only *sounds* simple, it *is* simple!

WHAT IF? What if the arrow were fired in a direction that makes an angle θ with the horizontal? How will that change the recoil velocity of the archer?

Answer The recoil velocity should decrease in magnitude because only a component of the velocity of the arrow is in the x direction. Conservation of momentum in the x direction gives

$$m_1v_{1f} + m_2v_{2f}\cos\theta = 0$$

leading to

$$v_{1f} = -\frac{m_2}{m_1}v_{2f}\cos\theta$$

For $\theta = 0$, $\cos\theta = 1$ and the final velocity of the archer reduces to the value when the arrow is fired horizontally. For nonzero values of θ, the cosine function is less than 1 and the recoil velocity is less than the value calculated for $\theta = 0$. If $\theta = 90°$, then $\cos\theta = 0$ and $v_{1f} = 0$, so there is no recoil velocity. In this case, the archer is simply pushed downward harder against the ice as the arrow is fired.

Example 9.2 Can We Really Ignore the Kinetic Energy of the Earth?

In Section 7.6, we claimed that we can ignore the kinetic energy of the Earth when considering the energy of a system consisting of the Earth and a dropped ball. Verify this claim.

SOLUTION

Conceptualize Imagine dropping a ball at the surface of the Earth. From your point of view, the ball falls while the Earth remains stationary. By Newton's third law, however, the Earth experiences an upward force and therefore an upward acceleration while the ball falls. In the calculation below, we will show that this motion is extremely small and can be ignored.

Categorize We identify the system as the ball and the Earth. We assume there are no forces on the system from outer space, so the system is isolated. Let's use the momentum version of the isolated system model.

Analyze We begin by setting up a ratio of the kinetic energy of the Earth to that of the ball. We identify v_E and v_b as the speeds of the Earth and the ball, respectively, after the ball has fallen through some distance.

Use the definition of kinetic energy to set up this ratio:

$$(1)\quad \frac{K_E}{K_b} = \frac{\frac{1}{2}m_E v_E^{\,2}}{\frac{1}{2}m_b v_b^{\,2}} = \left(\frac{m_E}{m_b}\right)\left(\frac{v_E}{v_b}\right)^2$$

Apply the isolated system (momentum) model: the initial momentum of the system is zero, so set the final momentum equal to zero:

$$p_i = p_f \quad \rightarrow \quad 0 = m_b v_b + m_E v_E$$

Solve the equation for the ratio of speeds:

$$\frac{v_E}{v_b} = -\frac{m_b}{m_E}$$

Substitute this expression for v_E/v_b in Equation (1):

$$\frac{K_E}{K_b} = \left(\frac{m_E}{m_b}\right)\left(-\frac{m_b}{m_E}\right)^2 = \frac{m_b}{m_E}$$

Substitute order-of-magnitude numbers for the masses:

$$\frac{K_E}{K_b} = \frac{m_b}{m_E} \sim \frac{1\text{ kg}}{10^{25}\text{ kg}} \sim 10^{-25}$$

Finalize The kinetic energy of the Earth is a very small fraction of the kinetic energy of the ball, so we are justified in ignoring it in the kinetic energy of the system.

9.3 Analysis Model: Nonisolated System (Momentum)

According to Equation 9.3, the momentum of a particle changes if a net force acts on the particle. The same can be said about a net force applied to a system as we will show explicitly in Section 9.7: the momentum of a system changes if a net force from the environment acts on the system. This may sound similar to our discussion of energy in Chapter 8: the energy of a system changes if energy crosses the boundary of the system to or from the environment. In this section, we consider a *nonisolated system.* For energy considerations, a system is nonisolated if energy transfers across the boundary of the system by any of the means listed in Section 8.1. For momentum considerations, a system is nonisolated if a net force acts on the system for a time interval. In this case, we can imagine momentum being transferred to the system from the environment by means of the net force. Knowing the change in momentum caused by a force is useful in solving some types of problems. To build a better understanding of this important concept, let us assume a net force $\sum \vec{\mathbf{F}}$ acts on a particle and this force may vary with time. According to Newton's second law, $\sum \vec{\mathbf{F}} = d\vec{\mathbf{p}}/dt$, or

$$d\vec{\mathbf{p}} = \sum \vec{\mathbf{F}}\, dt \qquad (9.7)$$

We can integrate[2] this expression to find the change in the momentum of a particle when the force acts over some time interval. If the momentum of the particle changes from $\vec{\mathbf{p}}_i$ at time t_i to $\vec{\mathbf{p}}_f$ at time t_f, integrating Equation 9.7 gives

$$\Delta\vec{\mathbf{p}} = \vec{\mathbf{p}}_f - \vec{\mathbf{p}}_i = \int_{t_i}^{t_f} \sum \vec{\mathbf{F}}\, dt \qquad \textbf{(9.8)}$$

To evaluate the integral, we need to know how the net force varies with time. The quantity on the right side of this equation is a vector called the **impulse** of the net force $\sum \vec{\mathbf{F}}$ acting on a particle over the time interval $\Delta t = t_f - t_i$:

Impulse of a force ▶

$$\vec{\mathbf{I}} \equiv \int_{t_i}^{t_f} \sum \vec{\mathbf{F}}\, dt \qquad \textbf{(9.9)}$$

From its definition, we see that impulse $\vec{\mathbf{I}}$ is a vector quantity having a magnitude equal to the area under the force–time curve as described in Figure 9.3a. It is assumed the force varies in time in the general manner shown in the figure and is nonzero in the time interval $\Delta t = t_f - t_i$. The direction of the impulse vector is the same as the direction of the change in momentum. Impulse has the dimensions of momentum, that is, ML/T. Impulse is *not* a property of a particle; rather, it is a measure of the degree to which an external force changes the particle's momentum.

Combining Equations 9.8 and 9.9 gives us an important statement known as the **impulse–momentum theorem:**

Impulse–momentum theorem for a particle ▶

The change in the momentum of a particle is equal to the impulse of the net force acting on the particle:

$$\Delta\vec{\mathbf{p}} = \vec{\mathbf{I}} \qquad \textbf{(9.10)}$$

This statement is equivalent to Newton's second law. When we say that an impulse is given to a particle, we mean that momentum is transferred from an external agent to that particle. Equation 9.10 is identical in form to the conservation of energy equation, Equation 8.1, and its full expansion, Equation 8.2. Equation 9.10 is the most general statement of the principle of **conservation of momentum** and is called the **conservation of momentum equation.** In the case of a momentum approach, isolated systems tend to appear in problems more often than nonisolated systems, so, in practice, the conservation of momentum equation is often identified as the special case of Equation 9.5.

The left side of Equation 9.10 represents the change in the momentum of the system, which in this case is a single particle. The right side is a measure of how much momentum crosses the boundary of the system due to the net force being applied to the system. Equation 9.10 is the mathematical statement of a new analysis model, the **nonisolated system (momentum)** model. Although this equation is similar in form to Equation 8.1, there are several differences in its application to problems. First, Equation 9.10 is a vector equation, whereas Equation 8.1 is a scalar equation. Therefore, directions are important for Equation 9.10. Second, there is only one type of momentum and therefore only one way to store momentum in a system. In contrast, as we see from Equation 8.2, there are three ways to store energy in a system: kinetic, potential, and internal. Third, there is only one way to transfer momentum into a system: by the application of a force on the system over a time interval. Equation 8.2 shows six ways we have identified as transferring energy into a system. Therefore, there is no expansion of Equation 9.10 analogous to Equation 8.2.

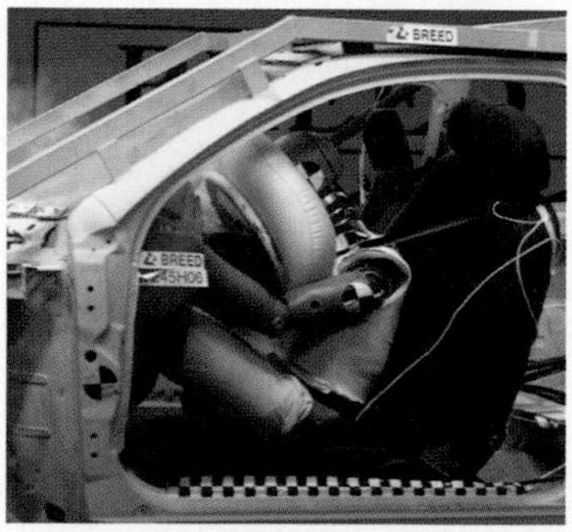

© David Woods/Terra/Corbis

Air bags in automobiles have saved countless lives in accidents. The air bag increases the time interval during which the passenger is brought to rest, thereby decreasing the force on (and resultant injury to) the passenger.

Because the net force imparting an impulse to a particle can generally vary in time, it is convenient to define a time-averaged net force:

$$\left(\sum \vec{\mathbf{F}}\right)_{\text{avg}} \equiv \frac{1}{\Delta t}\int_{t_i}^{t_f} \sum \vec{\mathbf{F}}\, dt \qquad \textbf{(9.11)}$$

[2]Here we are integrating force with respect to time. Compare this strategy with our efforts in Chapter 7, where we integrated force with respect to position to find the work done by the force.

where $\Delta t = t_f - t_i$. (This equation is an application of the mean value theorem of calculus.) Therefore, we can express Equation 9.9 as

$$\vec{\mathbf{I}} = \left(\sum \vec{\mathbf{F}}\right)_{\text{avg}} \Delta t \tag{9.12}$$

This time-averaged force, shown in Figure 9.3b, can be interpreted as the constant force that would give to the particle in the time interval Δt the same impulse that the time-varying force gives over this same interval.

In principle, if $\sum \vec{\mathbf{F}}$ is known as a function of time, the impulse can be calculated from Equation 9.9. The calculation becomes especially simple if the force acting on the particle is constant. In this case, $\left(\sum \vec{\mathbf{F}}\right)_{\text{avg}} = \sum \vec{\mathbf{F}}$, where $\sum \vec{\mathbf{F}}$ is the constant net force, and Equation 9.12 becomes

$$\vec{\mathbf{I}} = \sum \vec{\mathbf{F}}\, \Delta t \tag{9.13}$$

In many physical situations, we shall use what is called the **impulse approximation,** in which we assume one of the forces exerted on a particle acts for a short time but is much greater than any other force present. In this case, the net force $\sum \vec{\mathbf{F}}$ in Equation 9.9 is replaced with a single force $\vec{\mathbf{F}}$ to find the impulse on the particle. This approximation is especially useful in treating collisions in which the duration of the collision is very short. When this approximation is made, the single force is referred to as an *impulsive force.* For example, when a baseball is struck with a bat, the time of the collision is about 0.01 s and the average force that the bat exerts on the ball in this time is typically several thousand newtons. Because this contact force is much greater than the magnitude of the gravitational force, the impulse approximation justifies our ignoring the gravitational forces exerted on the ball and bat. When we use this approximation, it is important to remember that $\vec{\mathbf{p}}_i$ and $\vec{\mathbf{p}}_f$ represent the momenta *immediately* before and after the collision, respectively. Therefore, in any situation in which it is proper to use the impulse approximation, the particle moves very little during the collision.

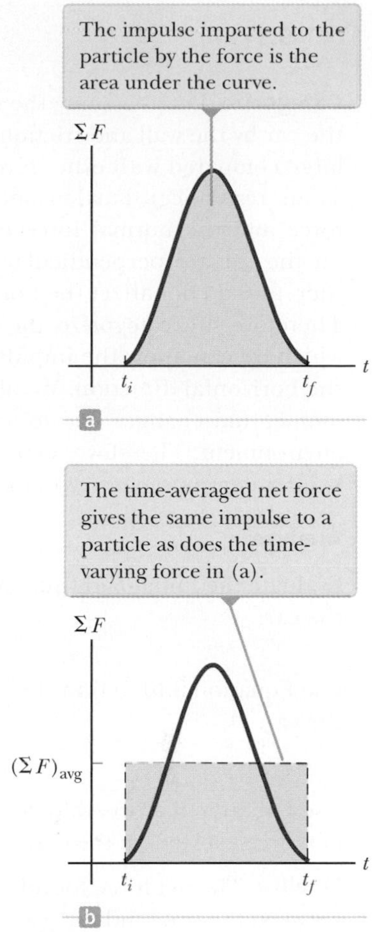

Figure 9.3 (a) A net force acting on a particle may vary in time. (b) The value of the constant force $(\Sigma F)_{\text{avg}}$ (horizontal dashed line) is chosen so that the area $(\Sigma F)_{\text{avg}}\,\Delta t$ of the rectangle is the same as the area under the curve in (a).

***Quick Quiz* 9.3** Two objects are at rest on a frictionless surface. Object 1 has a greater mass than object 2. **(i)** When a constant force is applied to object 1, it accelerates through a distance d in a straight line. The force is removed from object 1 and is applied to object 2. At the moment when object 2 has accelerated through the same distance d, which statements are true? (a) $p_1 < p_2$ (b) $p_1 = p_2$ (c) $p_1 > p_2$ (d) $K_1 < K_2$ (e) $K_1 = K_2$ (f) $K_1 > K_2$ **(ii)** When a force is applied to object 1, it accelerates for a time interval Δt. The force is removed from object 1 and is applied to object 2. From the same list of choices, which statements are true after object 2 has accelerated for the same time interval Δt?

***Quick Quiz* 9.4** Rank an automobile dashboard, seat belt, and air bag in terms of (a) the impulse and (b) the average force each delivers to a front-seat passenger during a collision, from greatest to least.

Example 9.3 How Good Are the Bumpers?

In a particular crash test, a car of mass 1 500 kg collides with a wall as shown in Figure 9.4 (page 242). The initial and final velocities of the car are $\vec{\mathbf{v}}_i = -15.0\hat{\mathbf{i}}$ m/s and $\vec{\mathbf{v}}_f = 2.60\hat{\mathbf{i}}$ m/s, respectively. If the collision lasts 0.150 s, find the impulse caused by the collision and the average net force exerted on the car.

SOLUTION

Conceptualize The collision time is short, so we can imagine the car being brought to rest very rapidly and then moving back in the opposite direction with a reduced speed.

continued

9.3 cont.

Categorize Let us assume the net force exerted on the car by the wall and friction from the ground is large compared with other forces on the car (such as air resistance). Furthermore, the gravitational force and the normal force exerted by the road on the car are perpendicular to the motion and therefore do not affect the horizontal momentum. Therefore, we categorize the problem as one in which we can apply the impulse approximation in the horizontal direction. We also see that the car's momentum changes due to an impulse from the environment. Therefore, we can apply the nonisolated system (momentum) model.

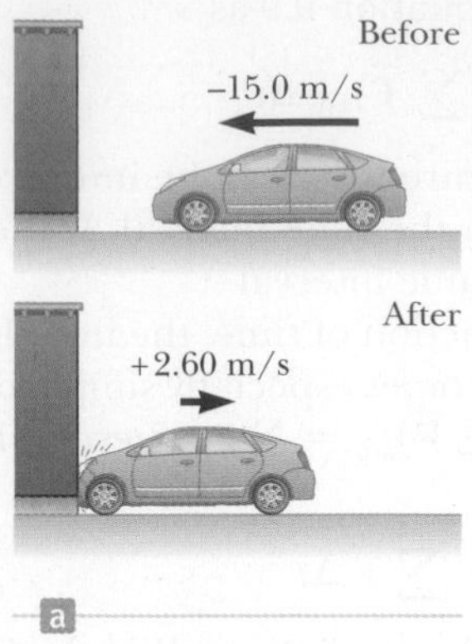

Figure 9.4 (Example 9.3) (a) This car's momentum changes as a result of its collision with the wall. (b) In a crash test, much of the car's initial kinetic energy is transformed into energy associated with the damage to the car.

Analyze

Evaluate the initial and final momenta of the car:

$$\vec{\mathbf{p}}_i = m\vec{\mathbf{v}}_i = (1\,500 \text{ kg})(-15.0\hat{\mathbf{i}} \text{ m/s}) = -2.25 \times 10^4\hat{\mathbf{i}} \text{ kg}\cdot\text{m/s}$$

$$\vec{\mathbf{p}}_f = m\vec{\mathbf{v}}_f = (1\,500 \text{ kg})(2.60\hat{\mathbf{i}} \text{ m/s}) = 0.39 \times 10^4\hat{\mathbf{i}} \text{ kg}\cdot\text{m/s}$$

Use Equation 9.10 to find the impulse on the car:

$$\vec{\mathbf{I}} = \Delta\vec{\mathbf{p}} = \vec{\mathbf{p}}_f - \vec{\mathbf{p}}_i = 0.39 \times 10^4\hat{\mathbf{i}} \text{ kg}\cdot\text{m/s} - (-2.25 \times 10^4\hat{\mathbf{i}} \text{ kg}\cdot\text{m/s})$$

$$= 2.64 \times 10^4\hat{\mathbf{i}} \text{ kg}\cdot\text{m/s}$$

Use Equation 9.12 to evaluate the average net force exerted on the car:

$$\left(\sum \vec{\mathbf{F}}\right)_{\text{avg}} = \frac{\vec{\mathbf{I}}}{\Delta t} = \frac{2.64 \times 10^4\hat{\mathbf{i}} \text{ kg}\cdot\text{m/s}}{0.150 \text{ s}} = 1.76 \times 10^5\hat{\mathbf{i}} \text{ N}$$

Finalize The net force found above is a combination of the normal force on the car from the wall and any friction force between the tires and the ground as the front of the car crumples. If the brakes are not operating while the crash occurs and the crumpling metal does not interfere with the free rotation of the tires, this friction force could be relatively small due to the freely rotating wheels. Notice that the signs of the velocities in this example indicate the reversal of directions. What would the mathematics be describing if both the initial and final velocities had the same sign?

WHAT IF? What if the car did not rebound from the wall? Suppose the final velocity of the car is zero and the time interval of the collision remains at 0.150 s. Would that represent a larger or a smaller net force on the car?

Answer In the original situation in which the car rebounds, the net force on the car does two things during the time interval: (1) it stops the car, and (2) it causes the car to move away from the wall at 2.60 m/s after the collision. If the car does not rebound, the net force is only doing the first of these steps—stopping the car—which requires a *smaller* force.

Mathematically, in the case of the car that does not rebound, the impulse is

$$\vec{\mathbf{I}} = \Delta\vec{\mathbf{p}} = \vec{\mathbf{p}}_f - \vec{\mathbf{p}}_i = 0 - (-2.25 \times 10^4\hat{\mathbf{i}} \text{ kg}\cdot\text{m/s}) = 2.25 \times 10^4\hat{\mathbf{i}} \text{ kg}\cdot\text{m/s}$$

The average net force exerted on the car is

$$\left(\sum \vec{\mathbf{F}}\right)_{\text{avg}} = \frac{\vec{\mathbf{I}}}{\Delta t} = \frac{2.25 \times 10^4\hat{\mathbf{i}} \text{ kg}\cdot\text{m/s}}{0.150 \text{ s}} = 1.50 \times 10^5\hat{\mathbf{i}} \text{ N}$$

which is indeed smaller than the previously calculated value, as was argued conceptually.

9.4 Collisions in One Dimension

In this section, we use the isolated system (momentum) model to describe what happens when two particles collide. The term **collision** represents an event during which two particles come close to each other and interact by means of forces. The interaction forces are assumed to be much greater than any external forces present, so we can use the impulse approximation.

A collision may involve physical contact between two macroscopic objects as described in Active Figure 9.5a, but the notion of what is meant by a collision must be generalized because "physical contact" on a submicroscopic scale is ill-defined and hence meaningless. To understand this concept, consider a collision on an atomic scale (Active Fig. 9.5b) such as the collision of a proton with an alpha particle (the nucleus of a helium atom). Because the particles are both positively charged, they repel each other due to the strong electrostatic force between them at close separations and never come into "physical contact."

When two particles of masses m_1 and m_2 collide as shown in Active Figure 9.5, the impulsive forces may vary in time in complicated ways, such as that shown in Figure 9.3. Regardless of the complexity of the time behavior of the impulsive force, however, this force is internal to the system of two particles. Therefore, the two particles form an isolated system and the momentum of the system must be conserved in *any* collision.

In contrast, the total kinetic energy of the system of particles may or may not be conserved, depending on the type of collision. In fact, collisions are categorized as being either *elastic* or *inelastic* depending on whether or not kinetic energy is conserved.

An **elastic collision** between two objects is one in which the total kinetic energy (as well as total momentum) of the system is the same before and after the collision. Collisions between certain objects in the macroscopic world, such as billiard balls, are only *approximately* elastic because some deformation and loss of kinetic energy take place. For example, you can hear a billiard ball collision, so you know that some of the energy is being transferred away from the system by sound. An elastic collision must be perfectly silent! *Truly* elastic collisions occur between atomic and subatomic particles. These collisions are described by the isolated system model for both energy and momentum. Furthermore, there must be no transformation of kinetic energy into other types of energy within the system.

An **inelastic collision** is one in which the total kinetic energy of the system is not the same before and after the collision (even though the momentum of the system is conserved). Inelastic collisions are of two types. When the objects stick together after they collide, as happens when a meteorite collides with the Earth, the collision is called **perfectly inelastic.** When the colliding objects do not stick together but some kinetic energy is transformed or transferred away, as in the case of a rubber ball colliding with a hard surface, the collision is called **inelastic** (with no modifying adverb). When the rubber ball collides with the hard surface, some of the ball's kinetic energy is transformed when the ball is deformed while it is in contact with the surface. Inelastic collisions are described by the momentum version of the isolated system model. The system could be isolated for energy, with kinetic energy transformed to potential or internal energy. If the system is nonisolated, there could be energy leaving the system by some means. In this latter case, there could also be some transformation of energy within the system. In either of these cases, the kinetic energy of the system changes.

In the remainder of this section, we investigate the mathematical details for collisions in one dimension and consider the two extreme cases, perfectly inelastic and elastic collisions.

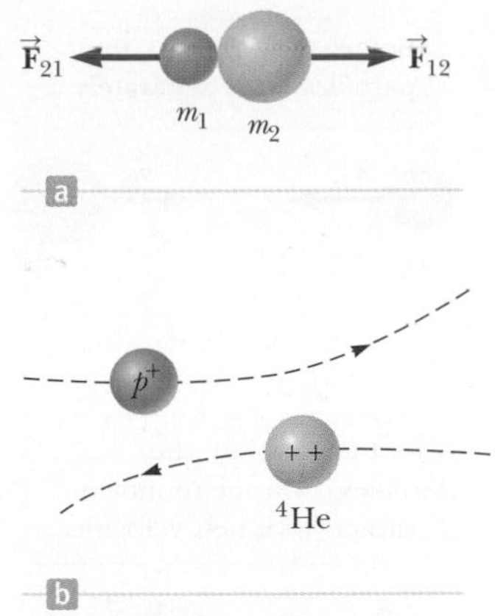

ACTIVE FIGURE 9.5

(a) The collision between two objects as the result of direct contact. (b) The "collision" between two charged particles.

Pitfall Prevention 9.2

Inelastic Collisions

Generally, inelastic collisions are hard to analyze without additional information. Lack of this information appears in the mathematical representation as having more unknowns than equations.

Perfectly Inelastic Collisions

Consider two particles of masses m_1 and m_2 moving with initial velocities $\vec{\mathbf{v}}_{1i}$ and $\vec{\mathbf{v}}_{2i}$ along the same straight line as shown in Active Figure 9.6. The two particles collide head-on, stick together, and then move with some common velocity $\vec{\mathbf{v}}_f$ after the collision. Because the momentum of an isolated system is conserved in *any* collision, we can say that the total momentum before the collision equals the total momentum of the composite system after the collision:

$$m_1\vec{\mathbf{v}}_{1i} + m_2\vec{\mathbf{v}}_{2i} = (m_1 + m_2)\vec{\mathbf{v}}_f \qquad \textbf{(9.14)}$$

Solving for the final velocity gives

$$\vec{\mathbf{v}}_f = \frac{m_1\vec{\mathbf{v}}_{1i} + m_2\vec{\mathbf{v}}_{2i}}{m_1 + m_2} \qquad \textbf{(9.15)}$$

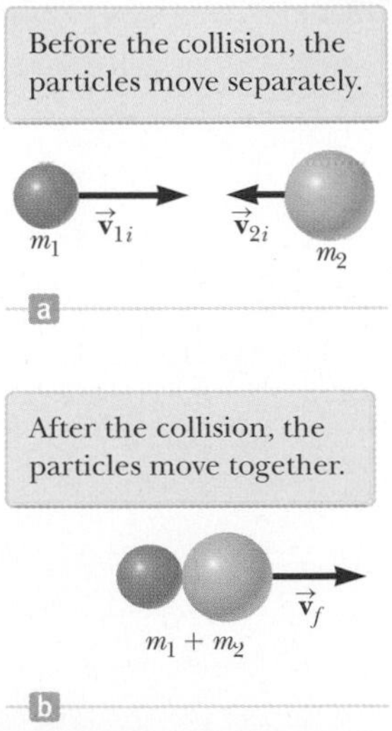

ACTIVE FIGURE 9.6

Schematic representation of a perfectly inelastic head-on collision between two particles.

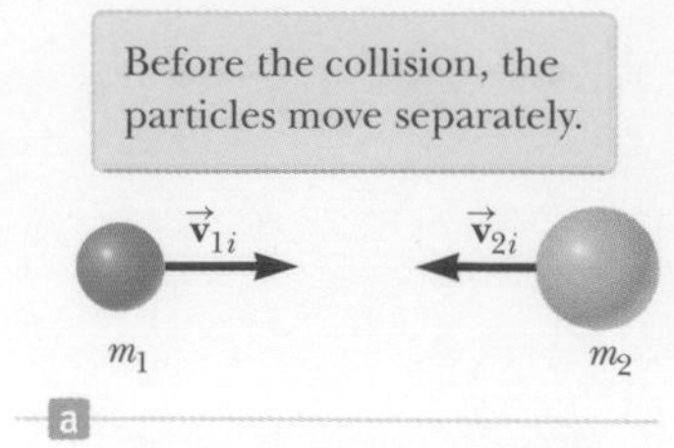

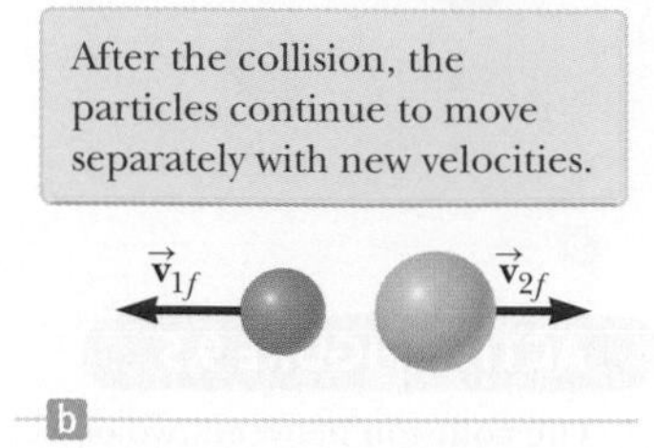

ACTIVE FIGURE 9.7

Schematic representation of an elastic head-on collision between two particles.

Elastic Collisions

Consider two particles of masses m_1 and m_2 moving with initial velocities $\vec{\mathbf{v}}_{1i}$ and $\vec{\mathbf{v}}_{2i}$ along the same straight line as shown in Active Figure 9.7. The two particles collide head-on and then leave the collision site with different velocities, $\vec{\mathbf{v}}_{1f}$ and $\vec{\mathbf{v}}_{2f}$. In an elastic collision, both the momentum and kinetic energy of the system are conserved. Therefore, considering velocities along the horizontal direction in Active Figure 9.7, we have

$$m_1v_{1i} + m_2v_{2i} = m_1v_{1f} + m_2v_{2f} \tag{9.16}$$

$$\tfrac{1}{2}m_1v_{1i}^2 + \tfrac{1}{2}m_2v_{2i}^2 = \tfrac{1}{2}m_1v_{1f}^2 + \tfrac{1}{2}m_2v_{2f}^2 \tag{9.17}$$

Because all velocities in Active Figure 9.7 are either to the left or the right, they can be represented by the corresponding speeds along with algebraic signs indicating direction. We shall indicate v as positive if a particle moves to the right and negative if it moves to the left.

In a typical problem involving elastic collisions, there are two unknown quantities, and Equations 9.16 and 9.17 can be solved simultaneously to find them. An alternative approach, however—one that involves a little mathematical manipulation of Equation 9.17—often simplifies this process. To see how, let us cancel the factor $\frac{1}{2}$ in Equation 9.17 and rewrite it as

$$m_1(v_{1i}^2 - v_{1f}^2) = m_2(v_{2f}^2 - v_{2i}^2)$$

Factoring both sides of this equation gives

$$m_1(v_{1i} - v_{1f})(v_{1i} + v_{1f}) = m_2(v_{2f} - v_{2i})(v_{2f} + v_{2i}) \tag{9.18}$$

Next, let us separate the terms containing m_1 and m_2 in Equation 9.16 to obtain

$$m_1(v_{1i} - v_{1f}) = m_2(v_{2f} - v_{2i}) \tag{9.19}$$

To obtain our final result, we divide Equation 9.18 by Equation 9.19 and obtain

$$v_{1i} + v_{1f} = v_{2f} + v_{2i}$$

$$v_{1i} - v_{2i} = -(v_{1f} - v_{2f}) \tag{9.20}$$

This equation, in combination with Equation 9.16, can be used to solve problems dealing with elastic collisions. This pair of equations (Eqs. 9.16 and 9.20) is easier to handle than the pair of Equations 9.16 and 9.17 because there are no quadratic terms like there are in Equation 9.17. According to Equation 9.20, the *relative* velocity of the two particles before the collision, $v_{1i} - v_{2i}$, equals the negative of their relative velocity after the collision, $-(v_{1f} - v_{2f})$.

Suppose the masses and initial velocities of both particles are known. Equations 9.16 and 9.20 can be solved for the final velocities in terms of the initial velocities because there are two equations and two unknowns:

$$v_{1f} = \left(\frac{m_1 - m_2}{m_1 + m_2}\right)v_{1i} + \left(\frac{2m_2}{m_1 + m_2}\right)v_{2i} \tag{9.21}$$

$$v_{2f} = \left(\frac{2m_1}{m_1 + m_2}\right)v_{1i} + \left(\frac{m_2 - m_1}{m_1 + m_2}\right)v_{2i} \tag{9.22}$$

It is important to use the appropriate signs for v_{1i} and v_{2i} in Equations 9.21 and 9.22.

Let us consider some special cases. If $m_1 = m_2$, Equations 9.21 and 9.22 show that $v_{1f} = v_{2i}$ and $v_{2f} = v_{1i}$, which means that the particles exchange velocities if they have equal masses. That is approximately what one observes in head-on billiard ball collisions: the cue ball stops and the struck ball moves away from the collision with the same velocity the cue ball had.

Pitfall Prevention 9.3

Not a General Equation

Equation 9.20 can only be used in a very *specific* situation, a one-dimensional, elastic collision between two objects. The *general* concept is conservation of momentum (and conservation of kinetic energy if the collision is elastic) for an isolated system.

If particle 2 is initially at rest, then $v_{2i} = 0$, and Equations 9.21 and 9.22 become

$$v_{1f} = \left(\frac{m_1 - m_2}{m_1 + m_2}\right)v_{1i} \quad (9.23)$$

$$v_{2f} = \left(\frac{2m_1}{m_1 + m_2}\right)v_{1i} \quad (9.24)$$

◀ **Elastic collision: particle 2 initially at rest**

If m_1 is much greater than m_2 and $v_{2i} = 0$, we see from Equations 9.23 and 9.24 that $v_{1f} \approx v_{1i}$ and $v_{2f} \approx 2v_{1i}$. That is, when a very heavy particle collides head-on with a very light one that is initially at rest, the heavy particle continues its motion unaltered after the collision and the light particle rebounds with a speed equal to about twice the initial speed of the heavy particle. An example of such a collision is that of a moving heavy atom, such as uranium, striking a light atom, such as hydrogen.

If m_2 is much greater than m_1 and particle 2 is initially at rest, then $v_{1f} \approx -v_{1i}$ and $v_{2f} \approx 0$. That is, when a very light particle collides head-on with a very heavy particle that is initially at rest, the light particle has its velocity reversed and the heavy one remains approximately at rest.

***Quick Quiz* 9.5** In a perfectly inelastic one-dimensional collision between two moving objects, what condition alone is necessary so that the final kinetic energy of the system is zero after the collision? **(a)** The objects must have initial momenta with the same magnitude but opposite directions. **(b)** The objects must have the same mass. **(c)** The objects must have the same initial velocity. **(d)** The objects must have the same initial speed, with velocity vectors in opposite directions.

***Quick Quiz* 9.6** A table-tennis ball is thrown at a stationary bowling ball. The table-tennis ball makes a one-dimensional elastic collision and bounces back along the same line. Compared with the bowling ball after the collision, does the table-tennis ball have **(a)** a larger magnitude of momentum and more kinetic energy, **(b)** a smaller magnitude of momentum and more kinetic energy, **(c)** a larger magnitude of momentum and less kinetic energy, **(d)** a smaller magnitude of momentum and less kinetic energy, or **(e)** the same magnitude of momentum and the same kinetic energy?

Problem-Solving Strategy

ONE-DIMENSIONAL COLLISIONS

You should use the following approach when solving collision problems in one dimension:

1. Conceptualize. Imagine the collision occurring in your mind. Draw simple diagrams of the particles before and after the collision and include appropriate velocity vectors. At first, you may have to guess at the directions of the final velocity vectors.

2. Categorize. Is the system of particles isolated? If so, categorize the collision as elastic, inelastic, or perfectly inelastic.

3. Analyze. Set up the appropriate mathematical representation for the problem. If the collision is perfectly inelastic, use Equation 9.15. If the collision is elastic, use Equations 9.16 and 9.20. If the collision is inelastic, use Equation 9.16. To find the final velocities in this case, you will need some additional information.

4. Finalize. Once you have determined your result, check to see if your answers are consistent with the mental and pictorial representations and that your results are realistic.

Example 9.4 The Executive Stress Reliever

An ingenious device that illustrates conservation of momentum and kinetic energy is shown in Figure 9.8. It consists of five identical hard balls supported by strings of equal lengths. When ball 1 is pulled out and released, after the almost-elastic collision between it and ball 2, ball 1 stops and ball 5 moves out as shown in Figure 9.8b. If balls 1 and 2 are pulled out and released, they stop after the collision and balls 4 and 5 swing out, and so forth. Is it ever possible that when ball 1 is released, it stops after the collision and balls 4 and 5 will swing out on the opposite side and travel with half the speed of ball 1 as in Figure 9.8c?

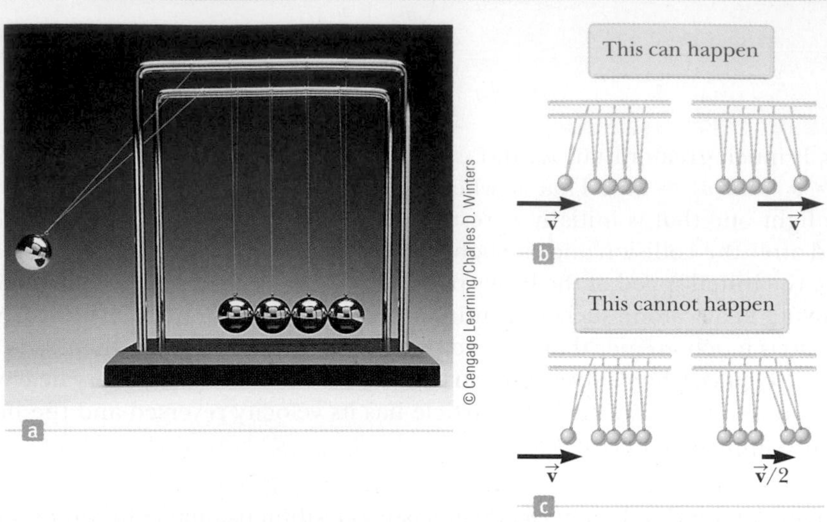

Figure 9.8 (Example 9.4) (a) An executive stress reliever. (b) If one ball swings down, we see one ball swing out at the other end. (c) Is it possible for one ball to swing down and two balls to leave the other end with half the speed of the first ball? In (b) and (c), the velocity vectors shown represent those of the balls immediately before and immediately after the collision.

SOLUTION

Conceptualize With the help of Figure 9.8c, imagine one ball coming in from the left and two balls exiting the collision on the right. That is the phenomenon we want to test to see if it could ever happen.

Categorize Because of the very short time interval between the arrival of the ball from the left and the departure of the ball(s) from the right, we can use the impulse approximation to ignore the gravitational forces on the balls and categorize the system of five balls as isolated in terms of both momentum and energy. Because the balls are hard, we can categorize the collisions between them as elastic for purposes of calculation.

Analyze The momentum of the system before the collision is mv, where m is the mass of ball 1 and v is its speed immediately before the collision. After the collision, we imagine that ball 1 stops and balls 4 and 5 swing out, each moving with speed $v/2$. The total momentum of the system after the collision would be $m(v/2) + m(v/2) = mv$. Therefore, the momentum of the system is conserved.

The kinetic energy of the system immediately before the collision is $K_i = \frac{1}{2}mv^2$ and that after the collision is $K_f = \frac{1}{2}m(v/2)^2 + \frac{1}{2}m(v/2)^2 = \frac{1}{4}mv^2$. That shows that the kinetic energy of the system is *not* conserved, which is inconsistent with our assumption that the collisions are elastic.

Finalize Our analysis shows that it is *not* possible for balls 4 and 5 to swing out when only ball 1 is released. The only way to conserve both momentum and kinetic energy of the system is for one ball to move out when one ball is released, two balls to move out when two are released, and so on.

WHAT IF? Consider what would happen if balls 4 and 5 are glued together. Now what happens when ball 1 is pulled out and released?

Answer In this situation, balls 4 and 5 *must* move together as a single object after the collision. We have argued that both momentum and energy of the system cannot be conserved in this case. We assumed, however, ball 1 stopped after striking ball 2. What if we do not make this assumption? Consider the conservation equations with the assumption that ball 1 moves after the collision. For conservation of momentum,

$$p_i = p_f$$

$$mv_{1i} = mv_{1f} + 2mv_{4,5}$$

where $v_{4,5}$ refers to the final speed of the ball 4–ball 5 combination. Conservation of kinetic energy gives us

$$K_i = K_f$$

$$\tfrac{1}{2}mv_{1i}^{\ 2} = \tfrac{1}{2}mv_{1f}^{\ 2} + \tfrac{1}{2}(2m)v_{4,5}^{\ 2}$$

9.4 *cont.*

Combining these equations gives

$$v_{4,5} = \tfrac{2}{3}v_{1i} \qquad v_{1f} = -\tfrac{1}{3}v_{1i}$$

Therefore, balls 4 and 5 move together as one object after the collision while ball 1 bounces back from the collision with one third of its original speed.

Example 9.5 Carry Collision Insurance!

An 1 800-kg car stopped at a traffic light is struck from the rear by a 900-kg car. The two cars become entangled, moving along the same path as that of the originally moving car. If the smaller car were moving at 20.0 m/s before the collision, what is the velocity of the entangled cars after the collision?

SOLUTION

Conceptualize This kind of collision is easily visualized, and one can predict that after the collision both cars will be moving in the same direction as that of the initially moving car. Because the initially moving car has only half the mass of the stationary car, we expect the final velocity of the cars to be relatively small.

Categorize We identify the system of two cars as isolated in terms of momentum in the horizontal direction and apply the impulse approximation during the short time interval of the collision. The phrase "become entangled" tells us to categorize the collision as perfectly inelastic.

Analyze The magnitude of the total momentum of the system before the collision is equal to that of the smaller car because the larger car is initially at rest.

Set the initial momentum of the system equal to the final momentum of the system:

$$p_i = p_f \;\rightarrow\; m_1 v_i = (m_1 + m_2)v_f$$

Solve for v_f and substitute numerical values:

$$v_f = \frac{m_1 v_i}{m_1 + m_2} = \frac{(900\text{ kg})(20.0\text{ m/s})}{900\text{ kg} + 1\,800\text{ kg}} = \boxed{6.67\text{ m/s}}$$

Finalize Because the final velocity is positive, the direction of the final velocity of the combination is the same as the velocity of the initially moving car as predicted. The speed of the combination is also much lower than the initial speed of the moving car.

WHAT IF? Suppose we reverse the masses of the cars. What if a stationary 900-kg car is struck by a moving 1 800-kg car? Is the final speed the same as before?

Answer Intuitively, we can guess that the final speed of the combination is higher than 6.67 m/s if the initially moving car is the more massive car. Mathematically, that should be the case because the system has a larger momentum if the initially moving car is the more massive one. Solving for the new final velocity, we find

$$v_f = \frac{m_1 v_i}{m_1 + m_2} = \frac{(1\,800\text{ kg})(20.0\text{ m/s})}{1\,800\text{ kg} + 900\text{ kg}} = 13.3\text{ m/s}$$

which is two times greater than the previous final velocity.

Example 9.6 The Ballistic Pendulum

The ballistic pendulum (Fig. 9.9, page 248) is an apparatus used to measure the speed of a fast-moving projectile such as a bullet. A projectile of mass m_1 is fired into a large block of wood of mass m_2 suspended from some light wires. The projectile embeds in the block, and the entire system swings through a height h. How can we determine the speed of the projectile from a measurement of h?

continued

9.6 cont.

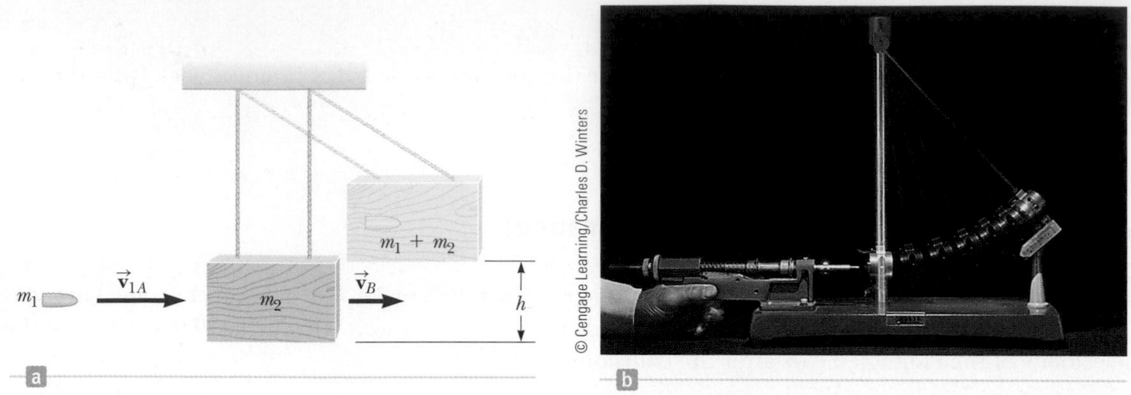

Figure 9.9 (Example 9.6) (a) Diagram of a ballistic pendulum. Notice that $\vec{\mathbf{v}}_{1A}$ is the velocity of the projectile immediately before the collision and $\vec{\mathbf{v}}_B$ is the velocity of the projectile–block system immediately after the perfectly inelastic collision. (b) Multiflash photograph of a ballistic pendulum used in the laboratory.

SOLUTION

Conceptualize Figure 9.9a helps conceptualize the situation. Run the animation in your mind: the projectile enters the pendulum, which swings up to some height at which it momentarily comes to rest.

Categorize The projectile and the block form an isolated system in terms of momentum if we identify configuration *A* as immediately before the collision and configuration *B* as immediately after the collision. Because the projectile imbeds in the block, we can categorize the collision between them as perfectly inelastic.

Analyze To analyze the collision, we use Equation 9.15, which gives the speed of the system immediately after the collision when we assume the impulse approximation.

Noting that $v_{2A} = 0$, solve Equation 9.15 for v_B:

$$(1)\quad v_B = \frac{m_1 v_{1A}}{m_1 + m_2}$$

Categorize For the process during which the projectile–block combination swings upward to height h (ending at a configuration we'll call C), we focus on a *different* system, that of the projectile, the block, and the Earth. We categorize this part of the problem as one involving an isolated system for energy with no nonconservative forces acting.

Analyze Write an expression for the total kinetic energy of the system immediately after the collision:

$$(2)\quad K_B = \tfrac{1}{2}(m_1 + m_2){v_B}^2$$

Substitute the value of v_B from Equation (1) into Equation (2):

$$K_B = \frac{{m_1}^2{v_{1A}}^2}{2(m_1 + m_2)}$$

This kinetic energy of the system immediately after the collision is *less* than the initial kinetic energy of the projectile as is expected in an inelastic collision.

We define the gravitational potential energy of the system for configuration B to be zero. Therefore, $U_B = 0$, whereas $U_C = (m_1 + m_2)gh$.

Apply the conservation of mechanical energy principle to the system:

$$K_B + U_B = K_C + U_C$$

$$\frac{{m_1}^2{v_{1A}}^2}{2(m_1 + m_2)} + 0 = 0 + (m_1 + m_2)gh$$

Solve for v_{1A}:

$$v_{1A} = \left(\frac{m_1 + m_2}{m_1}\right)\sqrt{2gh}$$

9.6 *cont.*

Finalize We had to solve this problem in two steps. Each step involved a different system and a different analysis model: isolated system (momentum) for the first step and isolated system (energy) for the second. Because the collision was assumed to be perfectly inelastic, some mechanical energy was transformed to internal energy during the collision. Therefore, it would have been *incorrect* to apply the isolated system (energy) model to the entire process by equating the initial kinetic energy of the incoming projectile with the final gravitational potential energy of the projectile–block–Earth combination.

Example 9.7 A Two-Body Collision with a Spring

A block of mass $m_1 = 1.60$ kg initially moving to the right with a speed of 4.00 m/s on a frictionless, horizontal track collides with a light spring attached to a second block of mass $m_2 = 2.10$ kg initially moving to the left with a speed of 2.50 m/s as shown in Figure 9.10a. The spring constant is 600 N/m.

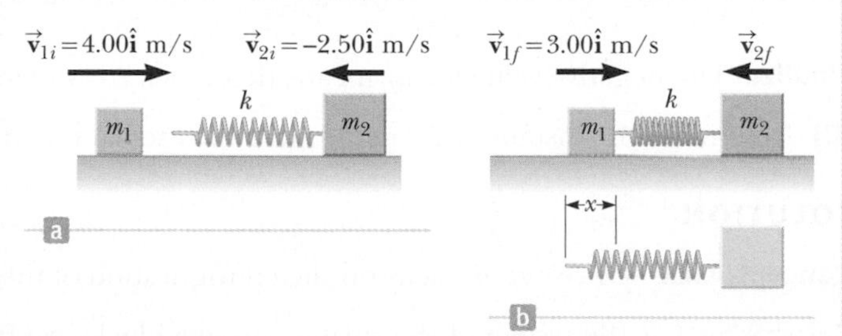

Figure 9.10 (Example 9.7) A moving block approaches a second moving block that is attached to a spring.

(A) Find the velocities of the two blocks after the collision.

SOLUTION

Conceptualize With the help of Figure 9.10a, run an animation of the collision in your mind. Figure 9.10b shows an instant during the collision when the spring is compressed. Eventually, block 1 and the spring will again separate, so the system will look like Figure 9.10a again but with different velocity vectors for the two blocks.

Categorize Because the spring force is conservative, kinetic energy in the system of two blocks and the spring is not transformed to internal energy during the compression of the spring. Ignoring any sound made when the block hits the spring, we can categorize the collision as being elastic and the system as being isolated for both energy and momentum.

Analyze Because momentum of the system is conserved, apply Equation 9.16:

$$(1)\quad m_1v_{1i} + m_2v_{2i} = m_1v_{1f} + m_2v_{2f}$$

Because the collision is elastic, apply Equation 9.20:

$$(2)\quad v_{1i} - v_{2i} = -(v_{1f} - v_{2f})$$

Multiply Equation (2) by m_1:

$$(3)\quad m_1v_{1i} - m_1v_{2i} = -m_1v_{1f} + m_1v_{2f}$$

Add Equations (1) and (3):

$$2m_1v_{1i} + (m_2 - m_1)v_{2i} = (m_1 + m_2)v_{2f}$$

Solve for v_{2f}:

$$v_{2f} = \frac{2m_1v_{1i} + (m_2 - m_1)v_{2i}}{m_1 + m_2}$$

Substitute numerical values:

$$v_{2f} = \frac{2(1.60\text{ kg})(4.00\text{ m/s}) + (2.10\text{ kg} - 1.60\text{ kg})(-2.50\text{ m/s})}{1.60\text{ kg} + 2.10\text{ kg}} = \boxed{3.12\text{ m/s}}$$

Solve Equation (2) for v_{1f} and substitute numerical values:

$$v_{1f} = v_{2f} - v_{1i} + v_{2i} = 3.12\text{ m/s} - 4.00\text{ m/s} + (-2.50\text{ m/s}) = \boxed{-3.38\text{ m/s}}$$

(B) Determine the velocity of block 2 during the collision, at the instant block 1 is moving to the right with a velocity of +3.00 m/s as in Figure 9.10b.

SOLUTION

Conceptualize Focus your attention now on Figure 9.10b, which represents the final configuration of the system for the time interval of interest.

continued

9.7 cont.

Categorize Because the momentum and mechanical energy of the system of two blocks and the spring are conserved *throughout* the collision, the collision can be categorized as elastic for *any* final instant of time. Let us now choose the final instant to be when block 1 is moving with a velocity of +3.00 m/s.

Analyze Apply Equation 9.16:

$$m_1v_{1i} + m_2v_{2i} = m_1v_{1f} + m_2v_{2f}$$

Solve for v_{2f}:

$$v_{2f} = \frac{m_1v_{1i} + m_2v_{2i} - m_1v_{1f}}{m_2}$$

Substitute numerical values:

$$v_{2f} = \frac{(1.60\text{ kg})(4.00\text{ m/s}) + (2.10\text{ kg})(-2.50\text{ m/s}) - (1.60\text{ kg})(3.00\text{ m/s})}{2.10\text{ kg}}$$

$$= -1.74\text{ m/s}$$

Finalize The negative value for v_{2f} means that block 2 is still moving to the left at the instant we are considering.

(C) Determine the distance the spring is compressed at that instant.

SOLUTION

Conceptualize Once again, focus on the configuration of the system shown in Figure 9.10b.

Categorize For the system of the spring and two blocks, no friction or other nonconservative forces act within the system. Therefore, we categorize the system as isolated in terms of energy with no nonconservative forces acting. The system also remains isolated in terms of momentum.

Analyze We choose the initial configuration of the system to be that existing immediately before block 1 strikes the spring and the final configuration to be that when block 1 is moving to the right at 3.00 m/s.

Write a conservation of mechanical energy equation for the system:

$$K_i + U_i = K_f + U_f$$

Evaluate the energies, recognizing that two objects in the system have kinetic energy and that the potential energy is elastic:

$$\tfrac{1}{2}m_1v_{1i}^2 + \tfrac{1}{2}m_2v_{2i}^2 + 0 = \tfrac{1}{2}m_1v_{1f}^2 + \tfrac{1}{2}m_2v_{2f}^2 + \tfrac{1}{2}kx^2$$

Substitute the known values and the result of part (B):

$$\tfrac{1}{2}(1.60\text{ kg})(4.00\text{ m/s})^2 + \tfrac{1}{2}(2.10\text{ kg})(2.50\text{ m/s})^2 + 0$$

$$= \tfrac{1}{2}(1.60\text{ kg})(3.00\text{ m/s})^2 + \tfrac{1}{2}(2.10\text{ kg})(1.74\text{ m/s})^2 + \tfrac{1}{2}(600\text{ N/m})x^2$$

Solve for x:

$$x = 0.173\text{ m}$$

Finalize This answer is not the maximum compression of the spring because the two blocks are still moving toward each other at the instant shown in Figure 9.10b. Can you determine the maximum compression of the spring?

9.5 Collisions in Two Dimensions

In Section 9.2, we showed that the momentum of a system of two particles is conserved when the system is isolated. For any collision of two particles, this result implies that the momentum in each of the directions x, y, and z is conserved. An important subset of collisions takes place in a plane. The game of billiards is a familiar example involving multiple collisions of objects moving on a two-dimensional surface. For such two-dimensional collisions, we obtain two component equations for conservation of momentum:

$$m_1v_{1ix} + m_2v_{2ix} = m_1v_{1fx} + m_2v_{2fx}$$

$$m_1v_{1iy} + m_2v_{2iy} = m_1v_{1fy} + m_2v_{2fy}$$

where the three subscripts on the velocity components in these equations represent, respectively, the identification of the object (1, 2), initial and final values (i, f), and the velocity component (x, y).

Let us consider a specific two-dimensional problem in which particle 1 of mass m_1 collides with particle 2 of mass m_2 initially at rest as in Active Figure 9.11. After the collision (Active Fig. 9.11b), particle 1 moves at an angle θ with respect to the horizontal and particle 2 moves at an angle ϕ with respect to the horizontal. This event is called a *glancing* collision. Applying the law of conservation of momentum in component form and noting that the initial y component of the momentum of the two-particle system is zero gives

$$m_1v_{1i} = m_1v_{1f}\cos\theta + m_2v_{2f}\cos\phi \quad \textbf{(9.25)}$$

$$0 = m_1v_{1f}\sin\theta - m_2v_{2f}\sin\phi \quad \textbf{(9.26)}$$

where the minus sign in Equation 9.26 is included because after the collision particle 2 has a y component of velocity that is downward. (The symbols v in these particular equations are speeds, not velocity components. The direction of the component vector is indicated explicitly with plus or minus signs.) We now have two independent equations. As long as no more than two of the seven quantities in Equations 9.25 and 9.26 are unknown, we can solve the problem.

If the collision is elastic, we can also use Equation 9.17 (conservation of kinetic energy) with $v_{2i} = 0$:

$$\tfrac{1}{2}m_1{v_{1i}}^2 = \tfrac{1}{2}m_1{v_{1f}}^2 + \tfrac{1}{2}m_2{v_{2f}}^2 \quad \textbf{(9.27)}$$

Knowing the initial speed of particle 1 and both masses, we are left with four unknowns (v_{1f}, v_{2f}, θ, and ϕ). Because we have only three equations, one of the four remaining quantities must be given to determine the motion after the elastic collision from conservation principles alone.

If the collision is inelastic, kinetic energy is *not* conserved and Equation 9.27 does *not* apply.

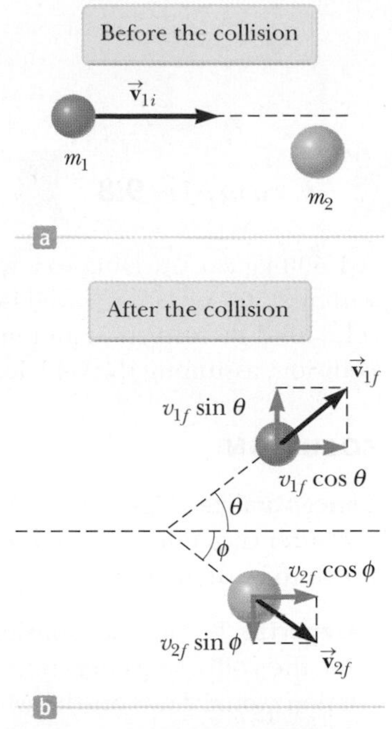

ACTIVE FIGURE 9.11

An elastic, glancing collision between two particles.

Pitfall Prevention 9.4

Don't Use Equation 9.20

Equation 9.20, relating the initial and final relative velocities of two colliding objects, is only valid for one-dimensional elastic collisions. Do not use this equation when analyzing two-dimensional collisions.

Problem-Solving Strategy

TWO-DIMENSIONAL COLLISIONS

The following procedure is recommended when dealing with problems involving collisions between two particles in two dimensions.

1. Conceptualize. Imagine the collisions occurring and predict the approximate directions in which the particles will move after the collision. Set up a coordinate system and define your velocities in terms of that system. It is convenient to have the x axis coincide with one of the initial velocities. Sketch the coordinate system, draw and label all velocity vectors, and include all the given information.

2. Categorize. Is the system of particles truly isolated? If so, categorize the collision as elastic, inelastic, or perfectly inelastic.

3. Analyze. Write expressions for the x and y components of the momentum of each object before and after the collision. Remember to include the appropriate signs for the components of the velocity vectors and pay careful attention to signs throughout the calculation.

Write expressions for the *total* momentum in the x direction *before* and *after* the collision and equate the two. Repeat this procedure for the total momentum in the y direction.

Proceed to solve the momentum equations for the unknown quantities. If the collision is inelastic, kinetic energy is *not* conserved and additional information is probably required. If the collision is perfectly inelastic, the final velocities of the two objects are equal.

If the collision is elastic, kinetic energy is conserved and you can equate the total kinetic energy of the system before the collision to the total kinetic energy after the collision, providing an additional relationship between the velocity magnitudes.

4. Finalize. Once you have determined your result, check to see if your answers are consistent with the mental and pictorial representations and that your results are realistic.

Example 9.8 Collision at an Intersection

A 1 500-kg car traveling east with a speed of 25.0 m/s collides at an intersection with a 2 500-kg truck traveling north at a speed of 20.0 m/s as shown in Figure 9.12. Find the direction and magnitude of the velocity of the wreckage after the collision, assuming the vehicles stick together after the collision.

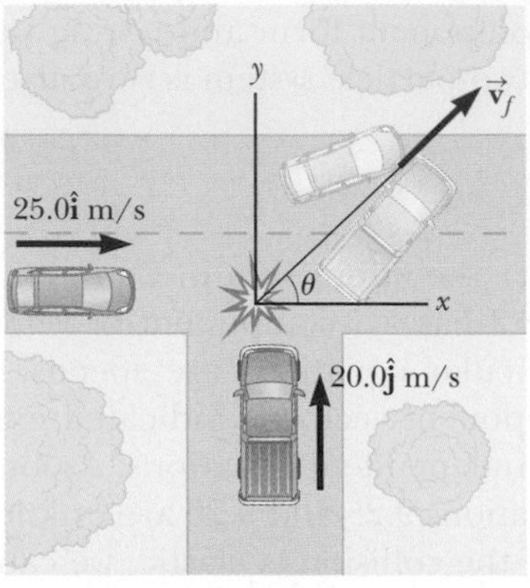

Figure 9.12 (Example 9.8) An eastbound car colliding with a northbound truck.

SOLUTION

Conceptualize Figure 9.12 should help you conceptualize the situation before and after the collision. Let us choose east to be along the positive x direction and north to be along the positive y direction.

Categorize Because we consider moments immediately before and immediately after the collision as defining our time interval, we ignore the small effect that friction would have on the wheels of the vehicles and model the system of two vehicles as isolated in terms of momentum. We also ignore the vehicles' sizes and model them as particles. The collision is perfectly inelastic because the car and the truck stick together after the collision.

Analyze Before the collision, the only object having momentum in the x direction is the car. Therefore, the magnitude of the total initial momentum of the system (car plus truck) in the x direction is that of only the car. Similarly, the total initial momentum of the system in the y direction is that of the truck. After the collision, let us assume the wreckage moves at an angle θ with respect to the x axis with speed v_f.

Equate the initial and final momenta of the system in the x direction:

$$\sum p_{xi} = \sum p_{xf} \rightarrow \quad (1)\ m_1 v_{1i} = (m_1 + m_2) v_f \cos\theta$$

Equate the initial and final momenta of the system in the y direction:

$$\sum p_{yi} = \sum p_{yf} \rightarrow \quad (2)\ m_2 v_{2i} = (m_1 + m_2) v_f \sin\theta$$

Divide Equation (2) by Equation (1):

$$\frac{m_2 v_{2i}}{m_1 v_{1i}} = \frac{\sin\theta}{\cos\theta} = \tan\theta$$

Solve for θ and substitute numerical values:

$$\theta = \tan^{-1}\left(\frac{m_2 v_{2i}}{m_1 v_{1i}}\right) = \tan^{-1}\left[\frac{(2\,500\text{ kg})(20.0\text{ m/s})}{(1\,500\text{ kg})(25.0\text{ m/s})}\right] = \boxed{53.1^\circ}$$

Use Equation (2) to find the value of v_f and substitute numerical values:

$$v_f = \frac{m_2 v_{2i}}{(m_1 + m_2)\sin\theta} = \frac{(2\,500\text{ kg})(20.0\text{ m/s})}{(1\,500\text{ kg} + 2\,500\text{ kg})\sin 53.1^\circ} = \boxed{15.6\text{ m/s}}$$

Finalize Notice that the angle θ is qualitatively in agreement with Figure 9.12. Also notice that the final speed of the combination is less than the initial speeds of the two cars. This result is consistent with the kinetic energy of the system being reduced in an inelastic collision. It might help if you draw the momentum vectors of each vehicle before the collision and the two vehicles together after the collision.

Example 9.9 Proton–Proton Collision

A proton collides elastically with another proton that is initially at rest. The incoming proton has an initial speed of 3.50×10^5 m/s and makes a glancing collision with the second proton as in Active Figure 9.11. (At close separations, the protons exert a repulsive electrostatic force on each other.) After the collision, one proton moves off at an angle of 37.0° to the original direction of motion and the second deflects at an angle of ϕ to the same axis. Find the final speeds of the two protons and the angle ϕ.

9.9 *cont.*

SOLUTION

Conceptualize This collision is like that shown in Active Figure 9.11, which will help you conceptualize the behavior of the system. We define the *x* axis to be along the direction of the velocity vector of the initially moving proton.

Categorize The pair of protons form an isolated system. Both momentum and kinetic energy of the system are conserved in this glancing elastic collision.

Analyze Using the isolated system model for both momentum and energy for a two-dimensional elastic collision, set up the mathematical representation with Equations 9.25 through 9.27:

$$(1)\quad v_{1f}\cos\theta + v_{2f}\cos\phi = v_{1i}$$

$$(2)\quad v_{1f}\sin\theta - v_{2f}\sin\phi = 0$$

$$(3)\quad v_{1f}^{\,2} + v_{2f}^{\,2} = v_{1i}^{\,2}$$

Rearrange Equations (1) and (2):

$$v_{2f}\cos\phi = v_{1i} - v_{1f}\cos\theta$$

$$v_{2f}\sin\phi = v_{1f}\sin\theta$$

Square these two equations and add them:

$$v_{2f}^{\,2}\cos^2\phi + v_{2f}^{\,2}\sin^2\phi = v_{1i}^{\,2} - 2v_{1i}v_{1f}\cos\theta + v_{1f}^{\,2}\cos^2\theta + v_{1f}^{\,2}\sin^2\theta$$

Incorporate that the sum of the squares of sine and cosine for *any* angle is equal to 1:

$$(4)\quad v_{2f}^{\,2} = v_{1i}^{\,2} - 2v_{1i}v_{1f}\cos\theta + v_{1f}^{\,2}$$

Substitute Equation (4) into Equation (3):

$$v_{1f}^{\,2} + (v_{1i}^{\,2} - 2v_{1i}v_{1f}\cos\theta + v_{1f}^{\,2}) = v_{1i}^{\,2}$$

$$(5)\quad v_{1f}^{\,2} - v_{1i}v_{1f}\cos\theta = 0$$

One possible solution of Equation (5) is $v_{1f} = 0$, which corresponds to a head-on, one-dimensional collision in which the first proton stops and the second continues with the same speed in the same direction. That is not the solution we want.

Divide both sides of Equation (5) by v_{1f} and solve for the remaining factor of v_{1f}:

$$v_{1f} = v_{1i}\cos\theta = (3.50\times10^5\ \text{m/s})\cos 37.0^\circ = \boxed{2.80\times10^5\ \text{m/s}}$$

Use Equation (3) to find v_{2f}:

$$v_{2f} = \sqrt{v_{1i}^{\,2} - v_{1f}^{\,2}} = \sqrt{(3.50\times10^5\ \text{m/s})^2 - (2.80\times10^5\ \text{m/s})^2}$$

$$= \boxed{2.11\times10^5\ \text{m/s}}$$

Use Equation (2) to find ϕ:

$$(2)\quad \phi = \sin^{-1}\left(\frac{v_{1f}\sin\theta}{v_{2f}}\right) = \sin^{-1}\left[\frac{(2.80\times10^5\ \text{m/s})\sin 37.0^\circ}{(2.11\times10^5\ \text{m/s})}\right]$$

$$= \boxed{53.0^\circ}$$

Finalize It is interesting that $\theta + \phi = 90^\circ$. This result is *not* accidental. Whenever two objects of equal mass collide elastically in a glancing collision and one of them is initially at rest, their final velocities are perpendicular to each other.

9.6 The Center of Mass

In this section, we describe the overall motion of a system in terms of a special point called the **center of mass** of the system. The system can be either a group of particles, such as a collection of atoms in a container, or an extended object, such as a gymnast leaping through the air. We shall see that the translational motion of the center of mass of the system is the same as if all the mass of the system were concentrated at that point. That is, the system moves as if the net external force were applied to a single particle located at the center of mass. This behavior is independent of other motion, such as rotation or vibration of the system or deformation of the system (for instance, when a gymnast folds her body). This model, the *particle model,* was introduced in Chapter 2.

Consider a system consisting of a pair of particles that have different masses and are connected by a light, rigid rod (Active Fig. 9.13 on page 254). The position of the center of mass of a system can be described as being the *average position* of the

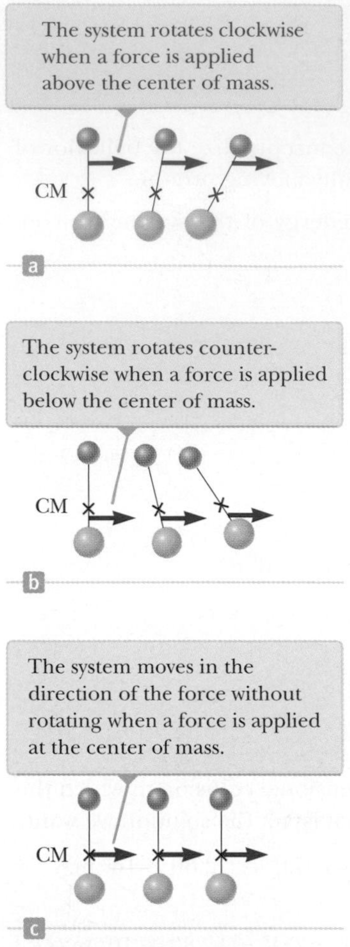

ACTIVE FIGURE 9.13

A force is applied to a system of two particles of unequal mass connected by a light, rigid rod.

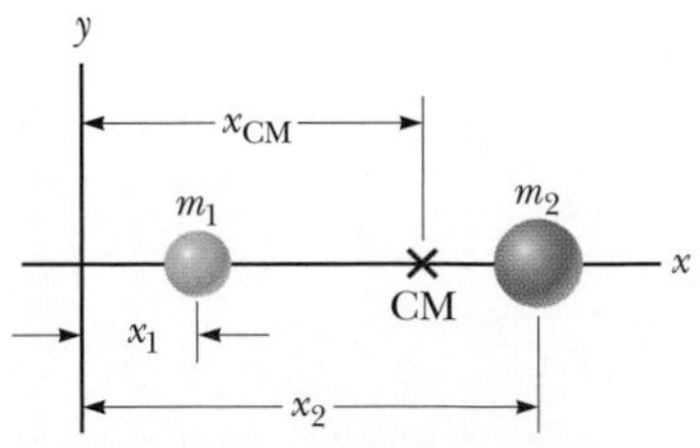

ACTIVE FIGURE 9.14

The center of mass of two particles of unequal mass on the x axis is located at x_{CM}, a point between the particles, closer to the one having the larger mass.

system's mass. The center of mass of the system is located somewhere on the line joining the two particles and is closer to the particle having the larger mass. If a single force is applied at a point on the rod above the center of mass, the system rotates clockwise (see Active Fig. 9.13a). If the force is applied at a point on the rod below the center of mass, the system rotates counterclockwise (see Active Fig. 9.13b). If the force is applied at the center of mass, the system moves in the direction of the force without rotating (see Active Fig. 9.13c). The center of mass of an object can be located with this procedure.

The center of mass of the pair of particles described in Active Figure 9.14 is located on the x axis and lies somewhere between the particles. Its x coordinate is given by

$$x_{CM} \equiv \frac{m_1 x_1 + m_2 x_2}{m_1 + m_2} \qquad \textbf{(9.28)}$$

For example, if $x_1 = 0$, $x_2 = d$, and $m_2 = 2m_1$, we find that $x_{CM} = \frac{2}{3}d$. That is, the center of mass lies closer to the more massive particle. If the two masses are equal, the center of mass lies midway between the particles.

We can extend this concept to a system of many particles with masses m_i in three dimensions. The x coordinate of the center of mass of n particles is defined to be

$$x_{CM} \equiv \frac{m_1 x_1 + m_2 x_2 + m_3 x_3 + \cdots + m_n x_n}{m_1 + m_2 + m_3 + \cdots + m_n} = \frac{\sum_i m_i x_i}{\sum_i m_i} = \frac{\sum_i m_i x_i}{M} = \frac{1}{M}\sum_i m_i x_i \qquad \textbf{(9.29)}$$

where x_i is the x coordinate of the ith particle and the total mass is $M \equiv \Sigma_i\, m_i$ where the sum runs over all n particles. The y and z coordinates of the center of mass are similarly defined by the equations

$$y_{CM} \equiv \frac{1}{M}\sum_i m_i y_i \quad \text{and} \quad z_{CM} \equiv \frac{1}{M}\sum_i m_i z_i \qquad \textbf{(9.30)}$$

The center of mass can be located in three dimensions by its position vector $\vec{\mathbf{r}}_{CM}$. The components of this vector are x_{CM}, y_{CM}, and z_{CM}, defined in Equations 9.29 and 9.30. Therefore,

$$\vec{\mathbf{r}}_{CM} = x_{CM}\hat{\mathbf{i}} + y_{CM}\hat{\mathbf{j}} + z_{CM}\hat{\mathbf{k}} = \frac{1}{M}\sum_i m_i x_i\hat{\mathbf{i}} + \frac{1}{M}\sum_i m_i y_i\hat{\mathbf{j}} + \frac{1}{M}\sum_i m_i z_i\hat{\mathbf{k}}$$

$$\vec{\mathbf{r}}_{CM} \equiv \frac{1}{M}\sum_i m_i \vec{\mathbf{r}}_i \qquad \textbf{(9.31)}$$

where $\vec{\mathbf{r}}_i$ is the position vector of the ith particle, defined by

$$\vec{\mathbf{r}}_i \equiv x_i\hat{\mathbf{i}} + y_i\hat{\mathbf{j}} + z_i\hat{\mathbf{k}}$$

Although locating the center of mass for an extended object is somewhat more cumbersome than locating the center of mass of a system of particles, the basic ideas we have discussed still apply. Think of an extended object as a system containing a large number of small mass elements such as the cube in Figure 9.15. Because the separation between elements is very small, the object can be considered to have a continuous mass distribution. By dividing the object into elements of mass Δm_i with coordinates x_i, y_i, z_i, we see that the x coordinate of the center of mass is approximately

$$x_{CM} \approx \frac{1}{M}\sum_i x_i\,\Delta m_i$$

with similar expressions for y_{CM} and z_{CM}. If we let the number of elements n approach infinity, the size of each element approaches zero and x_{CM} is given precisely. In this limit, we replace the sum by an integral and Δm_i by the differential element dm:

$$x_{CM} = \lim_{\Delta m_i \to 0} \frac{1}{M}\sum_i x_i\,\Delta m_i = \frac{1}{M}\int x\,dm \qquad \textbf{(9.32)}$$

Likewise, for y_{CM} and z_{CM} we obtain

$$y_{CM} = \frac{1}{M}\int y\,dm \quad \text{and} \quad z_{CM} = \frac{1}{M}\int z\,dm \tag{9.33}$$

We can express the vector position of the center of mass of an extended object in the form

$$\vec{\mathbf{r}}_{CM} = \frac{1}{M}\int \vec{\mathbf{r}}\,dm \tag{9.34}$$

which is equivalent to the three expressions given by Equations 9.32 and 9.33.

The center of mass of any symmetric object of uniform density lies on an axis of symmetry and on any plane of symmetry. For example, the center of mass of a uniform rod lies in the rod, midway between its ends. The center of mass of a sphere or a cube lies at its geometric center.

Because an extended object is a continuous distribution of mass, each small mass element is acted upon by the gravitational force. The net effect of all these forces is equivalent to the effect of a single force $M\vec{\mathbf{g}}$ acting through a special point, called the **center of gravity.** If $\vec{\mathbf{g}}$ is constant over the mass distribution, the center of gravity coincides with the center of mass. If an extended object is pivoted at its center of gravity, it balances in any orientation.

The center of gravity of an irregularly shaped object such as a wrench can be determined by suspending the object first from one point and then from another. In Figure 9.16, a wrench is hung from point *A* and a vertical line *AB* (which can be established with a plumb bob) is drawn when the wrench has stopped swinging. The wrench is then hung from point *C*, and a second vertical line *CD* is drawn. The center of gravity is halfway through the thickness of the wrench, under the intersection of these two lines. In general, if the wrench is hung freely from any point, the vertical line through this point must pass through the center of gravity.

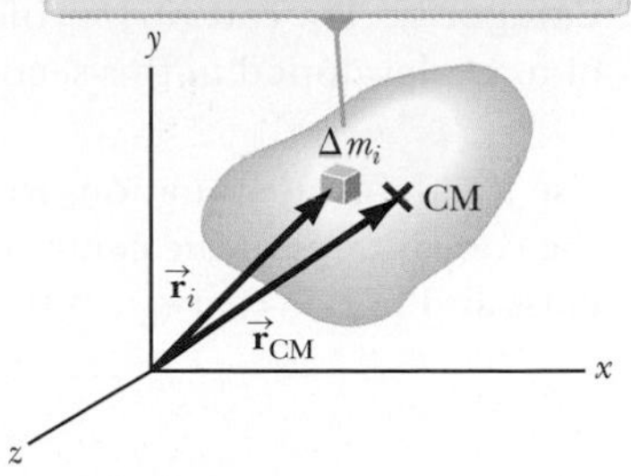

Figure 9.15 The center of mass is located at the vector position $\vec{\mathbf{r}}_{CM}$, which has coordinates x_{CM}, y_{CM}, and z_{CM}.

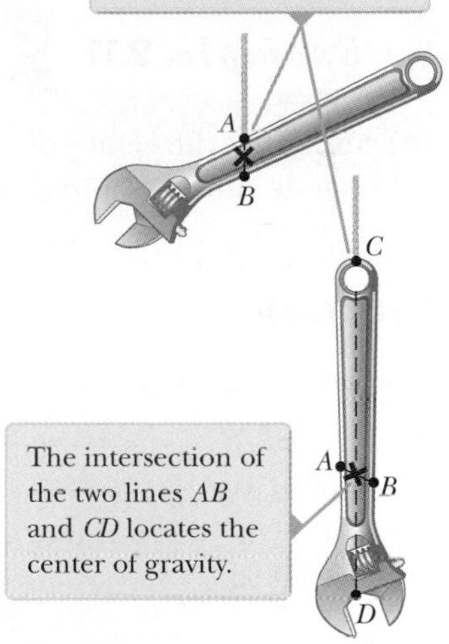

Figure 9.16 An experimental technique for determining the center of gravity of a wrench.

***Quick Quiz* 9.7** A baseball bat of uniform density is cut at the location of its center of mass as shown in Figure 9.17. Which piece has the smaller mass? **(a)** the piece on the right **(b)** the piece on the left **(c)** both pieces have the same mass **(d)** impossible to determine

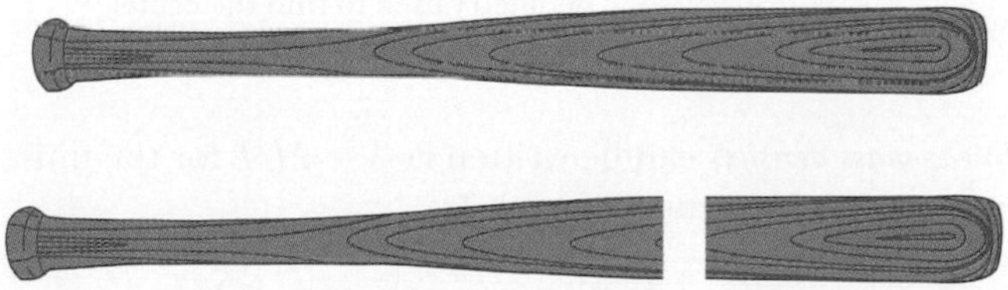

Figure 9.17 (Quick Quiz 9.7) A baseball bat cut at the location of its center of mass.

Example 9.10 The Center of Mass of Three Particles

A system consists of three particles located as shown in Figure 9.18. Find the center of mass of the system. The masses of the particles are $m_1 = m_2 = 1.0$ kg and $m_3 = 2.0$ kg.

SOLUTION

Conceptualize Figure 9.18 shows the three masses. Your intuition should tell you that the center of mass is located somewhere in the region between the blue particle and the pair of tan particles as shown in the figure.

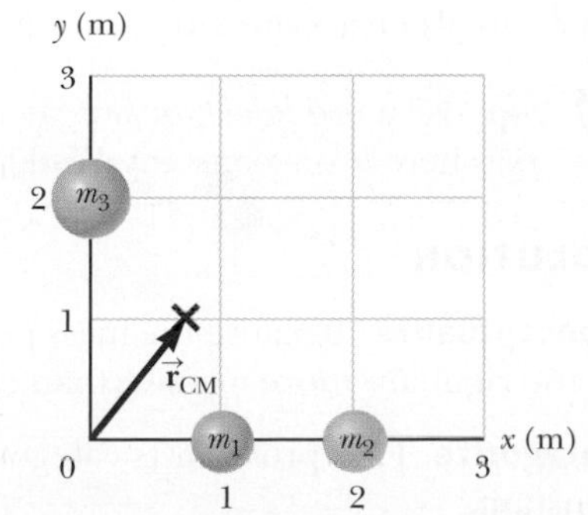

Figure 9.18 (Example 9.10) Two particles are located on the *x* axis, and a single particle is located on the *y* axis as shown. The vector indicates the location of the system's center of mass.

continued

9.10 *cont.*

Categorize We categorize this example as a substitution problem because we will be using the equations for the center of mass developed in this section.

Use the defining equations for the coordinates of the center of mass and notice that $z_{CM} = 0$:

$$x_{CM} = \frac{1}{M}\sum_i m_i x_i = \frac{m_1x_1 + m_2x_2 + m_3x_3}{m_1 + m_2 + m_3}$$

$$= \frac{(1.0 \text{ kg})(1.0 \text{ m}) + (1.0 \text{ kg})(2.0 \text{ m}) + (2.0 \text{ kg})(0)}{1.0 \text{ kg} + 1.0 \text{ kg} + 2.0 \text{ kg}} = \frac{3.0 \text{ kg}\cdot\text{m}}{4.0 \text{ kg}} = 0.75 \text{ m}$$

$$y_{CM} = \frac{1}{M}\sum_i m_i y_i = \frac{m_1y_1 + m_2y_2 + m_3y_3}{m_1 + m_2 + m_3}$$

$$= \frac{(1.0 \text{ kg})(0) + (1.0 \text{ kg})(0) + (2.0 \text{ kg})(2.0 \text{ m})}{4.0 \text{ kg}} = \frac{4.0 \text{ kg}\cdot\text{m}}{4.0 \text{ kg}} = 1.0 \text{ m}$$

Write the position vector of the center of mass:

$$\vec{\mathbf{r}}_{CM} \equiv x_{CM}\hat{\mathbf{i}} + y_{CM}\hat{\mathbf{j}} = \boxed{(0.75\hat{\mathbf{i}} + 1.0\hat{\mathbf{j}}) \text{ m}}$$

Example 9.11 The Center of Mass of a Rod

(A) Show that the center of mass of a rod of mass M and length L lies midway between its ends, assuming the rod has a uniform mass per unit length.

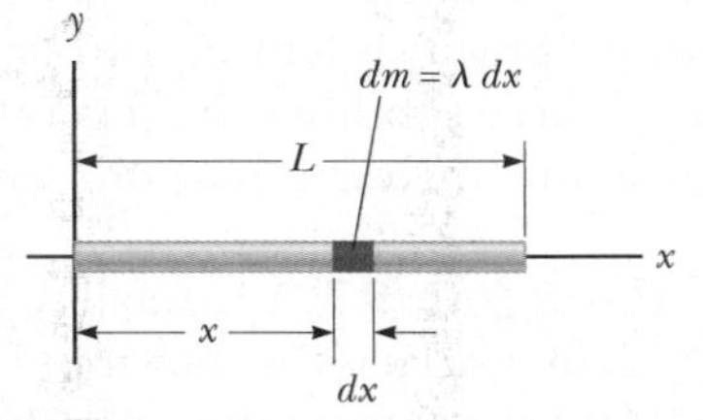

Figure 9.19 (Example 9.11) The geometry used to find the center of mass of a uniform rod.

SOLUTION

Conceptualize The rod is shown aligned along the x axis in Figure 9.19, so $y_{CM} = z_{CM} = 0$.

Categorize We categorize this example as an analysis problem because we need to divide the rod into small mass elements to perform the integration in Equation 9.32.

Analyze The mass per unit length (this quantity is called the *linear mass density*) can be written as $\lambda = M/L$ for the uniform rod. If the rod is divided into elements of length dx, the mass of each element is $dm = \lambda\, dx$.

Use Equation 9.32 to find an expression for x_{CM}:

$$x_{CM} = \frac{1}{M}\int x\, dm = \frac{1}{M}\int_0^L x\lambda\, dx = \frac{\lambda}{M}\frac{x^2}{2}\bigg|_0^L = \frac{\lambda L^2}{2M}$$

Substitute $\lambda = M/L$:

$$x_{CM} = \frac{L^2}{2M}\left(\frac{M}{L}\right) = \boxed{\tfrac{1}{2}L}$$

One can also use symmetry arguments to obtain the same result.

(B) Suppose a rod is *nonuniform* such that its mass per unit length varies linearly with x according to the expression $\lambda = \alpha x$, where α is a constant. Find the x coordinate of the center of mass as a fraction of L.

SOLUTION

Conceptualize Because the mass per unit length is not constant in this case but is proportional to x, elements of the rod to the right are more massive than elements near the left end of the rod.

Categorize This problem is categorized similarly to part (A), with the added twist that the linear mass density is not constant.

Analyze In this case, we replace dm in Equation 9.32 by $\lambda\, dx$, where $\lambda = \alpha x$.

9.11 *cont.*

Use Equation 9.32 to find an expression for x_{CM}:

$$x_{CM} = \frac{1}{M}\int x\,dm = \frac{1}{M}\int_0^L x\lambda\,dx = \frac{1}{M}\int_0^L x\alpha x\,dx$$

$$= \frac{\alpha}{M}\int_0^L x^2\,dx = \frac{\alpha L^3}{3M}$$

Find the total mass of the rod:

$$M = \int dm = \int_0^L \lambda\,dx = \int_0^L \alpha x\,dx = \frac{\alpha L^2}{2}$$

Substitute M into the expression for x_{CM}:

$$x_{CM} = \frac{\alpha L^3}{3\alpha L^2/2} = \boxed{\tfrac{2}{3}L}$$

Finalize Notice that the center of mass in part (B) is farther to the right than that in part (A). That result is reasonable because the elements of the rod become more massive as one moves to the right along the rod in part (B).

Example 9.12 The Center of Mass of a Right Triangle

You have been asked to hang a metal sign from a single vertical string. The sign has the triangular shape shown in Figure 9.20a. The bottom of the sign is to be parallel to the ground. At what distance from the left end of the sign should you attach the support string?

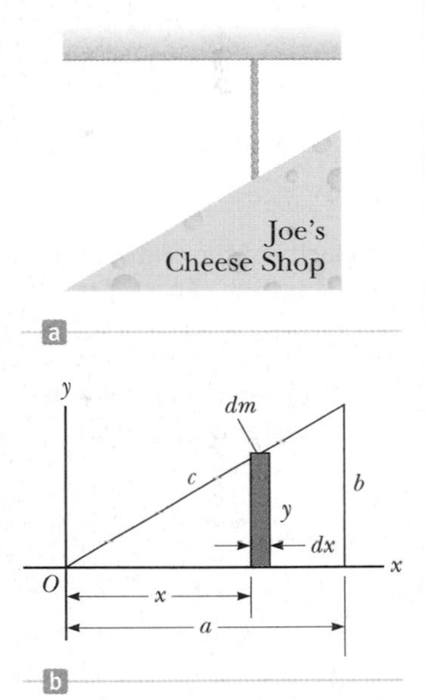

Figure 9.20 (Example 9.12) (a) A triangular sign to be hung from a single string. (b) Geometric construction for locating the center of mass.

SOLUTION

Conceptualize Figure 9.20a shows the sign hanging from the string. The string must be attached at a point directly above the center of gravity of the sign, which is the same as its center of mass because it is in a uniform gravitational field.

Categorize As in the case of Example 9.11, we categorize this example as an analysis problem because it is necessary to identify infinitesimal mass elements of the sign to perform the integration in Equation 9.32.

Analyze We assume the triangular sign has a uniform density and total mass M. Because the sign is a continuous distribution of mass, we must use the integral expression in Equation 9.32 to find the x coordinate of the center of mass.

We divide the triangle into narrow strips of width dx and height y as shown in Figure 9.20b, where y is the height of the hypotenuse of the triangle above the x axis for a given value of x. The mass of each strip is the product of the volume of the strip and the density ρ of the material from which the sign is made: $dm = \rho yt\,dx$, where t is the thickness of the metal sign. The density of the material is the total mass of the sign divided by its total volume (area of the triangle times thickness).

Evaluate dm:

$$dm = \rho yt\,dx = \left(\frac{M}{\frac{1}{2}abt}\right)yt\,dx = \frac{2My}{ab}\,dx$$

Use Equation 9.32 to find the x coordinate of the center of mass:

$$(1)\quad x_{CM} = \frac{1}{M}\int x\,dm = \frac{1}{M}\int_0^a x\frac{2My}{ab}\,dx = \frac{2}{ab}\int_0^a xy\,dx$$

To proceed further and evaluate the integral, we must express y in terms of x. The line representing the hypotenuse of the triangle in Figure 9.20b has a slope of b/a and passes through the origin, so the equation of this line is $y = (b/a)x$.

continued

9.12 *cont.*

Substitute for y in Equation (1):

$$x_{CM} = \frac{2}{ab}\int_0^a x\left(\frac{b}{a}x\right)dx = \frac{2}{a^2}\int_0^a x^2\,dx = \frac{2}{a^2}\left[\frac{x^3}{3}\right]_0^a$$

$$= \tfrac{2}{3}a$$

Therefore, the string must be attached to the sign at a distance two-thirds of the length of the bottom edge from the left end.

Finalize This answer is identical to that in part (B) of Example 9.11. For the triangular sign, the linear increase in height y with position x means that elements in the sign increase in mass linearly along the x axis, just like the linear increase in mass density in Example 9.11. We could also find the y coordinate of the center of mass of the sign, but that is not needed to determine where the string should be attached. You might try cutting a right triangle out of cardboard and hanging it from a string so that the long base is horizontal. Does the string need to be attached at $\frac{2}{3}a$?

9.7 Systems of Many Particles

Consider a system of two or more particles for which we have identified the center of mass. We can begin to understand the physical significance and utility of the center of mass concept by taking the time derivative of the position vector for the center of mass given by Equation 9.31. From Section 4.1, we know that the time derivative of a position vector is by definition the velocity vector. Assuming M remains constant for a system of particles—that is, no particles enter or leave the system—we obtain the following expression for the **velocity of the center of mass** of the system:

Velocity of the center of mass of a system of particles ▶

$$\vec{\mathbf{v}}_{CM} = \frac{d\vec{\mathbf{r}}_{CM}}{dt} = \frac{1}{M}\sum_i m_i \frac{d\vec{\mathbf{r}}_i}{dt} = \frac{1}{M}\sum_i m_i\vec{\mathbf{v}}_i \qquad \textbf{(9.35)}$$

where $\vec{\mathbf{v}}_i$ is the velocity of the ith particle. Rearranging Equation 9.35 gives

Total momentum of a system of particles ▶

$$M\vec{\mathbf{v}}_{CM} = \sum_i m_i\vec{\mathbf{v}}_i = \sum_i \vec{\mathbf{p}}_i = \vec{\mathbf{p}}_{tot} \qquad \textbf{(9.36)}$$

Therefore, the total linear momentum of the system equals the total mass multiplied by the velocity of the center of mass. In other words, the total linear momentum of the system is equal to that of a single particle of mass M moving with a velocity $\vec{\mathbf{v}}_{CM}$.

Differentiating Equation 9.35 with respect to time, we obtain the **acceleration of the center of mass** of the system:

Acceleration of the center of mass of a system of particles ▶

$$\vec{\mathbf{a}}_{CM} = \frac{d\vec{\mathbf{v}}_{CM}}{dt} = \frac{1}{M}\sum_i m_i \frac{d\vec{\mathbf{v}}_i}{dt} = \frac{1}{M}\sum_i m_i\vec{\mathbf{a}}_i \qquad \textbf{(9.37)}$$

Rearranging this expression and using Newton's second law gives

$$M\vec{\mathbf{a}}_{CM} = \sum_i m_i\vec{\mathbf{a}}_i = \sum_i \vec{\mathbf{F}}_i \qquad \textbf{(9.38)}$$

where $\vec{\mathbf{F}}_i$ is the net force on particle i.

The forces on any particle in the system may include both external forces (from outside the system) and internal forces (from within the system). By Newton's third law, however, the internal force exerted by particle 1 on particle 2, for example, is equal in magnitude and opposite in direction to the internal force exerted by particle 2 on particle 1. Therefore, when we sum over all internal force vectors in Equation 9.38, they cancel in pairs and we find that the net force on the system is caused *only* by external forces. We can then write Equation 9.38 in the form

Newton's second law for a system of particles ▶

$$\sum \vec{\mathbf{F}}_{ext} = M\vec{\mathbf{a}}_{CM} \qquad \textbf{(9.39)}$$

That is, the net external force on a system of particles equals the total mass of the system multiplied by the acceleration of the center of mass. Comparing Equation 9.39 with Newton's second law for a single particle, we see that the particle model we have used in several chapters can be described in terms of the center of mass:

The center of mass of a system of particles having combined mass M moves like an equivalent particle of mass M would move under the influence of the net external force on the system.

Let us integrate Equation 9.39 over a finite time interval:

$$\int \sum \vec{\mathbf{F}}_{\text{ext}}\, dt = \int M\vec{\mathbf{a}}_{\text{CM}}\, dt = \int M\frac{d\vec{\mathbf{v}}_{\text{CM}}}{dt}\, dt = M\int d\vec{\mathbf{v}}_{\text{CM}} = M\,\Delta\vec{\mathbf{v}}_{\text{CM}}$$

Notice that this equation can be written as

$$\Delta\vec{\mathbf{p}}_{\text{tot}} = \vec{\mathbf{I}} \qquad (9.40)$$

◀ **Impulse–momentum theorem for a system of particles**

where $\vec{\mathbf{I}}$ is the impulse imparted to the system by external forces and $\vec{\mathbf{p}}_{\text{tot}}$ is the momentum of the system. Equation 9.40 is the generalization of the impulse–momentum theorem for a particle (Eq. 9.10) to a system of many particles. It is also the mathematical representation of the nonisolated system (momentum) model for a system of many particles.

Finally, if the net external force on a system is zero, it follows from Equation 9.39 that

$$M\vec{\mathbf{a}}_{\text{CM}} = M\frac{d\vec{\mathbf{v}}_{\text{CM}}}{dt} = 0$$

so

$$M\vec{\mathbf{v}}_{\text{CM}} = \vec{\mathbf{p}}_{\text{tot}} = \text{constant} \quad \left(\text{when } \sum\vec{\mathbf{F}}_{\text{ext}} = 0\right) \qquad (9.41)$$

That is, the total linear momentum of a system of particles is conserved if no net external force is acting on the system. It follows that for an isolated system of particles, both the total momentum and the velocity of the center of mass are constant in time. This statement is a generalization of the isolated system (momentum) model for a many-particle system.

Suppose the center of mass of an isolated system consisting of two or more members is at rest. The center of mass of the system remains at rest if there is no net force on the system. For example, consider a system of a swimmer standing on a raft, with the system initially at rest. When the swimmer dives horizontally off the raft, the raft moves in the direction opposite that of the swimmer and the center of mass of the system remains at rest (if we neglect friction between raft and water). Furthermore, the linear momentum of the diver is equal in magnitude to that of the raft, but opposite in direction.

***Quick Quiz* 9.8** A cruise ship is moving at constant speed through the water. The vacationers on the ship are eager to arrive at their next destination. They decide to try to speed up the cruise ship by gathering at the bow (the front) and running together toward the stern (the back) of the ship. **(i)** While they are running toward the stern, is the speed of the ship (a) higher than it was before, (b) unchanged, (c) lower than it was before, or (d) impossible to determine? **(ii)** The vacationers stop running when they reach the stern of the ship. After they have all stopped running, is the speed of the ship (a) higher than it was before they started running, (b) unchanged from what it was before they started running, (c) lower than it was before they started running, or (d) impossible to determine?

Conceptual Example 9.13 Exploding Projectile

A projectile fired into the air suddenly explodes into several fragments (Fig. 9.21).

(A) What can be said about the motion of the center of mass of the system made up of all the fragments after the explosion?

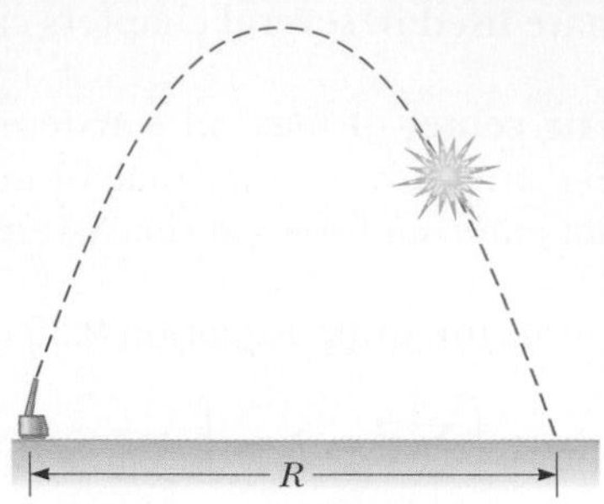

FIGURE 9.21 (Conceptual Example 9.13) When a projectile explodes into several fragments, the center of mass of the system made up of all the fragments follows the same parabolic path the projectile would have taken had there been no explosion.

SOLUTION

Neglecting air resistance, the only external force on the projectile is the gravitational force. Therefore, if the projectile did not explode, it would continue to move along the parabolic path indicated by the dashed line in Figure 9.21. Because the forces caused by the explosion are internal, they do not affect the motion of the center of mass of the system (the fragments). Therefore, after the explosion, the center of mass of the fragments follows the same parabolic path the projectile would have followed if no explosion had occurred.

(B) If the projectile did not explode, it would land at a distance R from its launch point. Suppose the projectile explodes and splits into two pieces of equal mass. One piece lands at a distance $2R$ from the launch point. Where does the other piece land?

SOLUTION

As discussed in part (A), the center of mass of the two-piece system lands at a distance R from the launch point. One of the pieces lands at a farther distance R from the landing point (or a distance $2R$ from the launch point), to the right in Figure 9.21. Because the two pieces have the same mass, the other piece must land a distance R to the left of the landing point in Figure 9.21, which places this piece right back at the launch point!

Example 9.14 The Exploding Rocket

A rocket is fired vertically upward. At the instant it reaches an altitude of 1 000 m and a speed of $v_i = 300$ m/s, it explodes into three fragments having equal mass. One fragment moves upward with a speed of $v_1 = 450$ m/s following the explosion. The second fragment has a speed of $v_2 = 240$ m/s and is moving east right after the explosion. What is the velocity of the third fragment immediately after the explosion?

SOLUTION

Conceptualize Picture the explosion in your mind, with one piece going upward and a second piece moving horizontally toward the east. Do you have an intuitive feeling about the direction in which the third piece moves?

Categorize This example is a two-dimensional problem because we have two fragments moving in perpendicular directions after the explosion as well as a third fragment moving in an unknown direction in the plane defined by the velocity vectors of the other two fragments. We assume the time interval of the explosion is very short, so we use the impulse approximation in which we ignore the gravitational force and air resistance. Because the forces of the explosion are internal to the system (the rocket), the system is modeled as isolated in terms of momentum. Therefore, the total momentum $\vec{\mathbf{p}}_i$ of the rocket immediately before the explosion must equal the total momentum $\vec{\mathbf{p}}_f$ of the fragments immediately after the explosion.

Analyze Because the three fragments have equal mass, the mass of each fragment is $M/3$, where M is the total mass of the rocket. We will let $\vec{\mathbf{v}}_3$ represent the unknown velocity of the third fragment.

Using the isolated system (momentum) model, equate the initial and final momenta of the system and express the momenta in terms of masses and velocities:

$$\vec{\mathbf{p}}_i = \vec{\mathbf{p}}_f \rightarrow M\vec{\mathbf{v}}_i = \frac{M}{3}\vec{\mathbf{v}}_1 + \frac{M}{3}\vec{\mathbf{v}}_2 + \frac{M}{3}\vec{\mathbf{v}}_3$$

Solve for $\vec{\mathbf{v}}_3$:

$$\vec{\mathbf{v}}_3 = 3\vec{\mathbf{v}}_i - \vec{\mathbf{v}}_1 - \vec{\mathbf{v}}_2$$

9.14 cont.

Substitute the numerical values: $\vec{\mathbf{v}}_3 = 3(300\hat{\mathbf{j}}\ \text{m/s}) - (450\hat{\mathbf{j}}\ \text{m/s}) - (240\hat{\mathbf{i}}\ \text{m/s}) = \boxed{(-240\hat{\mathbf{i}} + 450\hat{\mathbf{j}})\ \text{m/s}}$

Finalize Notice that this event is the reverse of a perfectly inelastic collision. There is one object before the collision and three objects afterward. Imagine running a movie of the event backward: the three objects would come together and become a single object. In a perfectly inelastic collision, the kinetic energy of the system decreases. If you were to calculate the kinetic energy before and after the event in this example, you would find that the kinetic energy of the system increases. (Try it!) This increase in kinetic energy comes from the potential energy stored in whatever fuel exploded to cause the breakup of the rocket.

9.8 Deformable Systems

So far in our discussion of mechanics, we have analyzed the motion of particles or nondeformable systems that can be modeled as particles. The discussion in Section 9.7 can be applied to an analysis of the motion of deformable systems. For example, suppose you stand on a skateboard and push off a wall, setting yourself in motion away from the wall. How would we describe this event?

The force from the wall on your hands moves through no displacement; the force is always located at the interface between the wall and your hands. Therefore, the force does no work on the system, which is you and your skateboard. Pushing off the wall, however, does indeed result in a change in the kinetic energy of the system. If you try to use the work–kinetic energy theorem, $W = \Delta K$, to describe this event, you will notice that the left side of the equation is zero but the right side is not zero. The work–kinetic energy theorem is not valid for this event and is often not valid for systems that are deformable. Your body has deformed during this event: your arms were bent before the event, and they straightened out while you pushed off the wall.

To analyze the motion of deformable systems, we appeal to Equation 8.2, the conservation of energy equation, and Equation 9.40, the impulse–momentum theorem. For the example of you pushing off the wall on your skateboard, identifying the system as you and the skateboard, Equation 8.2 gives

$$\Delta E_{\text{system}} = \sum T \quad \rightarrow \quad \Delta K + \Delta U = 0$$

where ΔK is the change in kinetic energy due to the increased speed of the system and ΔU is the decrease in potential energy stored in the body from previous meals. This equation tells us that the system transformed potential energy into kinetic energy by virtue of the muscular exertion necessary to push off the wall. Notice that the system is isolated in terms of energy but nonisolated in terms of momentum.

Applying Equation 9.40 to the system in this situation gives us

$$\Delta\vec{\mathbf{p}}_{\text{tot}} = \vec{\mathbf{I}} \quad \rightarrow \quad m\,\Delta\vec{\mathbf{v}} = \int \vec{\mathbf{F}}_{\text{wall}}\,dt$$

where $\vec{\mathbf{F}}_{\text{wall}}$ is the force exerted by the wall on your hands, m is the mass of you and the skateboard, and $\Delta\vec{\mathbf{v}}$ is the change in the velocity of the system during the event. To evaluate the right side of this equation, we would need to know how the force from the wall varies in time. In general, this process might be complicated. In the case of constant forces, or well-behaved forces, however, the integral on the right side of the equation can be evaluated.

Example 9.15 Pushing on a Spring[3]

As shown in Figure 9.22a, two blocks are at rest on a frictionless, level table. Both blocks have the same mass m, and they are connected by a spring of negligible mass. The separation distance of the blocks when the spring is relaxed is L. During a time interval Δt, a constant force of magnitude F is applied horizontally to the left block, moving it through a distance x_1 as shown in Figure 9.22b. During this time interval, the right block moves through a distance x_2. At the end of this time interval, the force F is removed.

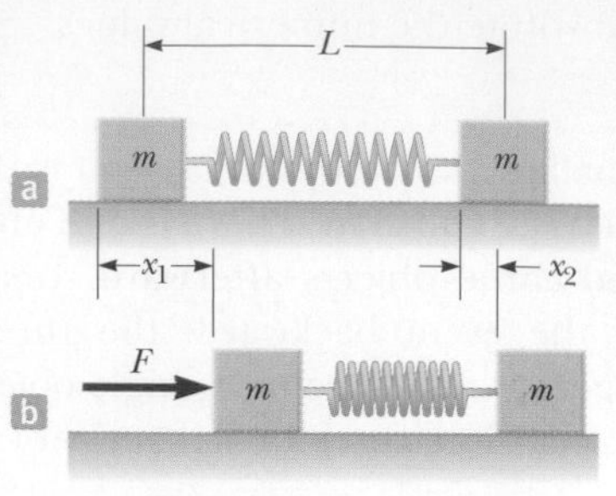

Figure 9.22 (Example 9.15) (a) Two blocks of equal mass are connected by a spring. (b) The left block is pushed with a constant force of magnitude F and moves a distance x_1 during some time interval. During this same time interval, the right block moves through a distance x_2.

(A) Find the resulting speed $\vec{\mathbf{v}}_{CM}$ of the center of mass of the system.

SOLUTION

Conceptualize Imagine what happens as you push on the left block. It begins to move to the right in Figure 9.22, and the spring begins to compress. As a result, the spring pushes to the right on the right block, which begins to move to the right. At any given time, the blocks are generally moving with different velocities. As the center of mass of the system moves to the right after the force is removed, the two blocks oscillate back and forth with respect to the center of mass.

Categorize We apply three analysis models in this problem: the deformable system of two blocks and a spring is modeled as a nonisolated system in terms of energy because work is being done on it by the applied force. It is also modeled as a nonisolated system in terms of momentum because of the force acting on the system during a time interval. Because the applied force on the system is constant, the acceleration of its center of mass is constant and the center of mass is modeled as a particle under constant acceleration.

Analyze Using the nonisolated system (momentum) model, we apply the impulse–momentum theorem to the system of two blocks, recognizing that the force F is constant during the time interval Δt while the force is applied.

Write Equation 9.40 for the system:

$$(1)\quad F\Delta t = (2m)(v_{CM} - 0) = 2mv_{CM}$$

During the time interval Δt, the center of mass of the system moves a distance $\frac{1}{2}(x_1 + x_2)$. Use this fact to express the time interval in terms of $v_{CM,avg}$:

$$\Delta t = \frac{\frac{1}{2}(x_1 + x_2)}{v_{CM,avg}}$$

Because the center of mass is modeled as a particle under constant acceleration, the average velocity of the center of mass is the average of the initial velocity, which is zero, and the final velocity v_{CM}:

$$\Delta t = \frac{\frac{1}{2}(x_1 + x_2)}{\frac{1}{2}(0 + v_{CM})} = \frac{(x_1 + x_2)}{v_{CM}}$$

Substitute this expression into Equation (1):

$$F\frac{(x_1 + x_2)}{v_{CM}} = 2mv_{CM}$$

Solve for v_{CM}:

$$v_{CM} = \sqrt{F\frac{(x_1 + x_2)}{2m}}$$

(B) Find the total energy of the system associated with vibration relative to its center of mass after the force F is removed.

SOLUTION

Analyze The vibrational energy is all the energy of the system other than the kinetic energy associated with translational motion of the center of mass. To find the vibrational energy, we apply the conservation of energy equation. The kinetic energy of the system can be expressed as $K = K_{CM} + K_{vib}$, where K_{vib} is the kinetic energy of the blocks relative to

[3]Example 9.15 was inspired in part by C. E. Mungan, "A primer on work–energy relationships for introductory physics," *The Physics Teacher* **43**:10, 2005.

9.15 *cont.*

the center of mass due to their vibration. The potential energy of the system is U_{vib}, which is the potential energy stored in the spring when the separation of the blocks is some value other than L.

From the nonisolated system (energy) model, express Equation 8.2 for this system:

$$(2)\quad \Delta K_{CM} + \Delta K_{vib} + \Delta U_{vib} = W$$

Express Equation (2) in an alternate form, noting that $K_{vib} + U_{vib} = E_{vib}$:

$$\Delta K_{CM} + \Delta E_{vib} = W$$

The initial values of the kinetic energy of the center of mass and the vibrational energy of the system are zero. Use this fact and substitute for the work done on the system by the force F:

$$K_{CM} + E_{vib} = W = Fx_1$$

Solve for the vibrational energy and use the result from part (A):

$$E_{vib} = Fx_1 - K_{CM} = Fx_1 - \tfrac{1}{2}(2m)v_{CM}{}^2 = F\frac{(x_1 - x_2)}{2}$$

Finalize Neither of the two answers in this example depends on the spring length, the spring constant, or the time interval. Notice also that the magnitude x_1 of the displacement of the point of application of the applied force is different from the magnitude $\frac{1}{2}(x_1 + x_2)$ of the displacement of the center of mass of the system. This difference reminds us that the displacement in the definition of work (Eq. 7.1) is that of the point of application of the force.

9.9 Rocket Propulsion

When ordinary vehicles such as cars are propelled, the driving force for the motion is friction. In the case of the car, the driving force is the force exerted by the road on the car. We can model the car as a nonisolated system in terms of momentum. An impulse is applied to the car from the roadway, and the result is a change in the momentum of the car as described by Equation 9.40.

A rocket moving in space, however, has no road to push against. The rocket is an isolated system in terms of momentum. Therefore, the source of the propulsion of a rocket must be something other than an external force. The operation of a rocket depends on the law of conservation of linear momentum as applied to an isolated system, where the system is the rocket plus its ejected fuel.

Rocket propulsion can be understood by first considering our archer standing on frictionless ice in Example 9.1. Imagine that the archer fires several arrows horizontally. For each arrow fired, the archer receives a compensating momentum in the opposite direction. As more arrows are fired, the archer moves faster and faster across the ice. In addition to this analysis in terms of momentum, we can also understand this phenomenon in terms of Newton's second and third laws. Every time the bow pushes an arrow forward, the arrow pushes the bow (and the archer) backward, and these forces result in an acceleration of the archer.

In a similar manner, as a rocket moves in free space, its linear momentum changes when some of its mass is ejected in the form of exhaust gases. Because the gases are given momentum when they are ejected out of the engine, the rocket receives a compensating momentum in the opposite direction. Therefore, the rocket is accelerated as a result of the "push," or thrust, from the exhaust gases. In free space, the center of mass of the system (rocket plus expelled gases) moves uniformly, independent of the propulsion process.[4]

Courtesy of NASA

The force from a nitrogen-propelled hand-controlled device allows an astronaut to move about freely in space without restrictive tethers, using the thrust force from the expelled nitrogen.

[4]The rocket and the archer represent cases of the reverse of a perfectly inelastic collision: momentum is conserved, but the kinetic energy of the rocket–exhaust gas system increases (at the expense of chemical potential energy in the fuel), as does the kinetic energy of the archer–arrow system (at the expense of potential energy from the archer's previous meals).

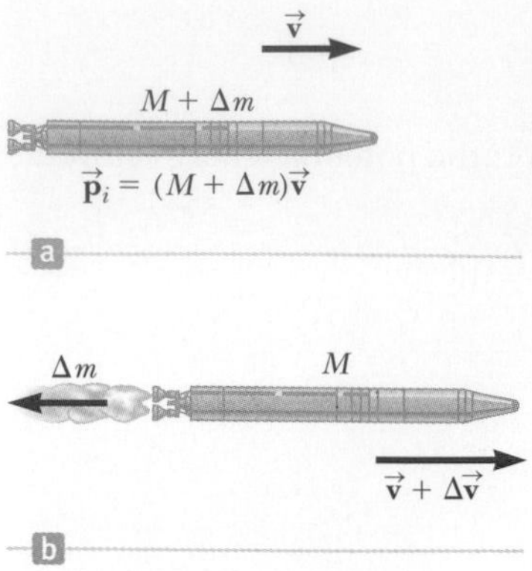

Figure 9.23 Rocket propulsion. (a) The initial mass of the rocket plus all its fuel is $M + \Delta m$ at a time t, and its speed is v. (b) At a time $t + \Delta t$, the rocket's mass has been reduced to M and an amount of fuel Δm has been ejected. The rocket's speed increases by an amount Δv.

Suppose at some time t the magnitude of the momentum of a rocket plus its fuel is $(M + \Delta m)v$, where v is the speed of the rocket relative to the Earth (Fig. 9.23a). Over a short time interval Δt, the rocket ejects fuel of mass Δm. At the end of this interval, the rocket's mass is M and its speed is $v + \Delta v$, where Δv is the change in speed of the rocket (Fig. 9.23b). If the fuel is ejected with a speed v_e relative to the rocket (the subscript e stands for *exhaust,* and v_e is usually called the *exhaust speed*), the velocity of the fuel relative to the Earth is $v - v_e$. Because the system of the rocket and the ejected fuel is isolated, we can equate the total initial momentum of the system to the total final momentum and obtain

$$(M + \Delta m)v = M(v + \Delta v) + \Delta m(v - v_e)$$

Simplifying this expression gives

$$M\,\Delta v = v_e\,\Delta m$$

If we now take the limit as Δt goes to zero, we let $\Delta v \to dv$ and $\Delta m \to dm$. Furthermore, the increase in the exhaust mass dm corresponds to an equal decrease in the rocket mass, so $dm = -dM$. Notice that dM is negative because it represents a decrease in mass, so $-dM$ is a positive number. Using this fact gives

$$M\,dv = v_e\,dm = -v_e\,dM \qquad \textbf{(9.42)}$$

Now divide the equation by M and integrate, taking the initial mass of the rocket plus fuel to be M_i and the final mass of the rocket plus its remaining fuel to be M_f. The result is

$$\int_{v_i}^{v_f} dv = -v_e \int_{M_i}^{M_f} \frac{dM}{M}$$

Expression for rocket propulsion ▶

$$v_f - v_i = v_e \ln\left(\frac{M_i}{M_f}\right) \qquad \textbf{(9.43)}$$

which is the basic expression for rocket propulsion. First, Equation 9.43 tells us that the increase in rocket speed is proportional to the exhaust speed v_e of the ejected gases. Therefore, the exhaust speed should be very high. Second, the increase in rocket speed is proportional to the natural logarithm of the ratio M_i/M_f. Therefore, this ratio should be as large as possible; that is, the mass of the rocket without its fuel should be as small as possible and the rocket should carry as much fuel as possible.

The **thrust** on the rocket is the force exerted on it by the ejected exhaust gases. We obtain the following expression for the thrust from Newton's second law and Equation 9.42:

$$\text{Thrust} = M\frac{dv}{dt} = \left|v_e\frac{dM}{dt}\right| \qquad \textbf{(9.44)}$$

This expression shows that the thrust increases as the exhaust speed increases and as the rate of change of mass (called the *burn rate*) increases.

Example 9.16 Fighting a Fire

Two firefighters must apply a total force of 600 N to steady a hose that is discharging water at the rate of 3 600 L/min. Estimate the speed of the water as it exits the nozzle.

SOLUTION

Conceptualize As the water leaves the hose, it acts in a way similar to the gases being ejected from a rocket engine. As a result, a force (thrust) acts on the firefighters in a direction opposite the direction of motion of the water. In this case, we want the end of the hose to be modeled as a particle in equilibrium rather than to accelerate as in the case of the rocket. Consequently, the firefighters must apply a force of magnitude equal to the thrust in the opposite direction to keep the end of the hose stationary.

9.16 *cont.*

Categorize This example is a substitution problem in which we use given values in an equation derived in this section. The water exits at 3 600 L/min, which is 60 L/s. Knowing that 1 L of water has a mass of 1 kg, we estimate that about 60 kg of water leaves the nozzle each second.

Use Equation 9.44 for the thrust: $\text{Thrust} = \left| v_e \dfrac{dM}{dt} \right|$

Substitute the known values: $600 \text{ N} = |v_e(60 \text{ kg/s})|$

Solve for the exhaust speed: $v_e = \boxed{10 \text{ m/s}}$

Example 9.17 A Rocket in Space

A rocket moving in space, far from all other objects, has a speed of 3.0×10^3 m/s relative to the Earth. Its engines are turned on, and fuel is ejected in a direction opposite the rocket's motion at a speed of 5.0×10^3 m/s relative to the rocket.

(A) What is the speed of the rocket relative to the earth once the rocket's mass is reduced to half its mass before ignition?

SOLUTION

Conceptualize Figure 9.23 shows the situation in this problem. From the discussion in this section and scenes from science fiction movies, we can easily imagine the rocket accelerating to a higher speed as the engine operates.

Categorize This problem is a substitution problem in which we use given values in the equations derived in this section.

Solve Equation 9.43 for the final velocity and substitute the known values:

$$v_f = v_i + v_e \ln\left(\frac{M_i}{M_f}\right)$$

$$= 3.0 \times 10^3 \text{ m/s} + (5.0 \times 10^3 \text{ m/s})\ln\left(\frac{M_i}{0.50M_i}\right)$$

$$= \boxed{6.5 \times 10^3 \text{ m/s}}$$

(B) What is the thrust on the rocket if it burns fuel at the rate of 50 kg/s?

SOLUTION

Use Equation 9.44 and the result from part (A), noting that $dM/dt = 50$ kg/s:

$$\text{Thrust} = \left| v_e \frac{dM}{dt} \right| = (5.0 \times 10^3 \text{ m/s})(50 \text{ kg/s}) = \boxed{2.5 \times 10^5 \text{ N}}$$

Summary

Definitions

The **linear momentum** $\vec{\mathbf{p}}$ of a particle of mass m moving with a velocity $\vec{\mathbf{v}}$ is

$$\vec{\mathbf{p}} \equiv m\vec{\mathbf{v}} \qquad (9.2)$$

The **impulse** imparted to a particle by a net force $\sum \vec{\mathbf{F}}$ is equal to the time integral of the force:

$$\vec{\mathbf{I}} \equiv \int_{t_i}^{t_f} \sum \vec{\mathbf{F}}\, dt \qquad (9.9)$$

continued

An **inelastic collision** is one for which the total kinetic energy of the system of colliding particles is not conserved. A **perfectly inelastic collision** is one in which the colliding particles stick together after the collision. An **elastic collision** is one in which the kinetic energy of the system is conserved.

The position vector of the **center of mass** of a system of particles is defined as

$$\vec{\mathbf{r}}_{CM} \equiv \frac{1}{M}\sum_i m_i \vec{\mathbf{r}}_i \qquad (9.31)$$

where $M = \sum_i m_i$ is the total mass of the system and $\vec{\mathbf{r}}_i$ is the position vector of the ith particle.

Concepts and Principles

The position vector of the center of mass of an extended object can be obtained from the integral expression

$$\vec{\mathbf{r}}_{CM} = \frac{1}{M}\int \vec{\mathbf{r}}\, dm \qquad (9.34)$$

The velocity of the center of mass for a system of particles is

$$\vec{\mathbf{v}}_{CM} = \frac{1}{M}\sum_i m_i \vec{\mathbf{v}}_i \qquad (9.35)$$

The total momentum of a system of particles equals the total mass multiplied by the velocity of the center of mass.

Newton's second law applied to a system of particles is

$$\sum \vec{\mathbf{F}}_{ext} = M\vec{\mathbf{a}}_{CM} \qquad (9.39)$$

where $\vec{\mathbf{a}}_{CM}$ is the acceleration of the center of mass and the sum is over all external forces. The center of mass moves like an imaginary particle of mass M under the influence of the resultant external force on the system.

Analysis Models for Problem Solving

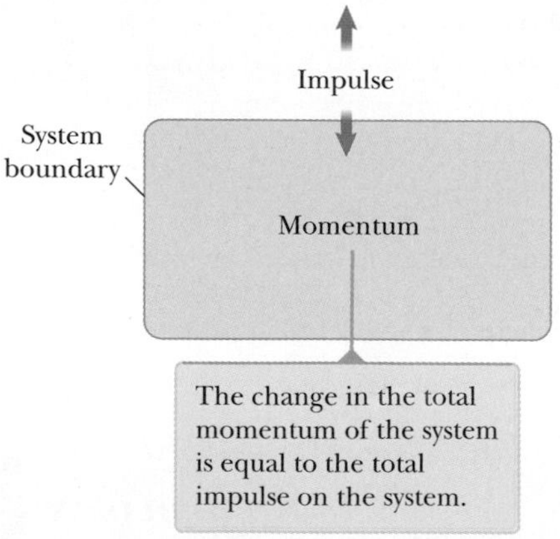

Nonisolated System (Momentum). If a system interacts with its environment in the sense that there is an external force on the system, the behavior of the system is described by the **impulse–momentum theorem:**

$$\Delta\vec{\mathbf{p}}_{tot} = \vec{\mathbf{I}} \qquad (9.40)$$

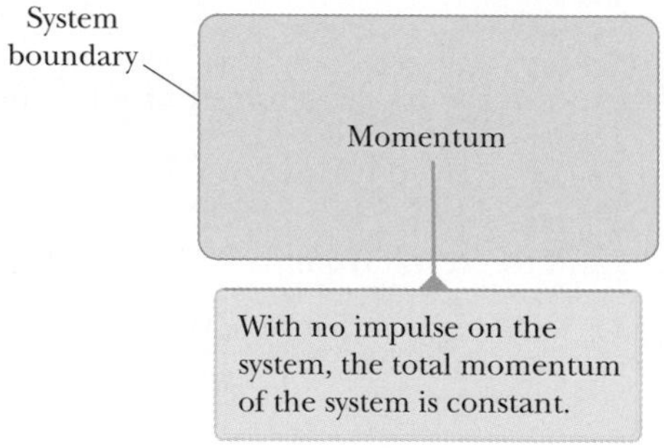

Isolated System (Momentum). The principle of **conservation of linear momentum** indicates that the total momentum of an isolated system (no external forces) is conserved regardless of the nature of the forces between the members of the system:

$$M\vec{\mathbf{v}}_{CM} = \vec{\mathbf{p}}_{tot} = \text{constant} \quad \left(\text{when } \sum \vec{\mathbf{F}}_{ext} = 0\right) \qquad (9.41)$$

In the case of a two-particle system, this principle can be expressed as

$$\vec{\mathbf{p}}_{1i} + \vec{\mathbf{p}}_{2i} = \vec{\mathbf{p}}_{1f} + \vec{\mathbf{p}}_{2f} \qquad (9.5)$$

The system may be isolated in terms of momentum but nonisolated in terms of energy, as in the case of inelastic collisions.

Objective Questions

☐ denotes answer available in Student Solutions Manual/Study Guide

1. A car of mass m traveling at speed v crashes into the rear of a truck of mass $2m$ that is at rest and in neutral at an intersection. If the collision is perfectly inelastic, what is the speed of the combined car and truck after the collision? (a) v (b) $v/2$ (c) $v/3$ (d) $2v$ (e) None of those answers is correct.

2. A head-on, elastic collision occurs between two billiard balls of equal mass. If a red ball is traveling to the right with speed v and a blue ball is traveling to the left with speed $3v$ before the collision, what statement is true concerning their velocities subsequent to the collision? Neglect any effects of spin. (a) The red ball travels to the left with speed v, while the blue ball travels to the right with speed $3v$. (b) The red ball travels to the left with speed v, while the blue ball continues to move to the left with a speed $2v$. (c) The red ball travels to the left with speed $3v$, while the blue ball travels to the right with speed v. (d) Their final velocities cannot be determined because momentum is not conserved in the collision. (e) The velocities cannot be determined without knowing the mass of each ball.

3. A 3-kg object moving to the right on a frictionless, horizontal surface with a speed of 2 m/s collides head-on and sticks to a 2-kg object that is initially moving to the left with a speed of 4 m/s. After the collision, which statement is true? (a) The kinetic energy of the system is 20 J. (b) The momentum of the system is 14 kg · m/s. (c) The kinetic energy of the system is greater than 5 J but less than 20 J. (d) The momentum of the system is −2 kg · m/s. (e) The momentum of the system is less than the momentum of the system before the collision.

4. A 2-kg object moving to the right with a speed of 4 m/s makes a head-on, elastic collision with a 1-kg object that is initially at rest. The velocity of the 1-kg object after the collision is (a) greater than 4 m/s, (b) less than 4 m/s, (c) equal to 4 m/s, (d) zero, or (e) impossible to say based on the information provided.

5. A 5-kg cart moving to the right with a speed of 6 m/s collides with a concrete wall and rebounds with a speed of 2 m/s. What is the change in momentum of the cart? (a) 0 (b) 40 kg · m/s (c) −40 kg · m/s (d) −30 kg · m/s (e) −10 kg · m/s

6. A 57.0-g tennis ball is traveling straight at a player at 21.0 m/s. The player volleys the ball straight back at 25.0 m/s. If the ball remains in contact with the racket for 0.060 0 s, what average force acts on the ball? (a) 22.6 N (b) 32.5 N (c) 43.7 N (d) 72.1 N (e) 102 N

7. The momentum of an object is increased by a factor of 4 in magnitude. By what factor is its kinetic energy changed? (a) 16 (b) 8 (c) 4 (d) 2 (e) 1

8. The kinetic energy of an object is increased by a factor of 4. By what factor is the magnitude of its momentum changed? (a) 16 (b) 8 (c) 4 (d) 2 (e) 1

9. If two particles have equal momenta, are their kinetic energies equal? (a) yes, always (b) no, never (c) no, except when their speeds are the same (d) yes, as long as they move along parallel lines

10. If two particles have equal kinetic energies, are their momenta equal? (a) yes, always (b) no, never (c) yes, as long as their masses are equal (d) yes, if both their masses and directions of motion are the same (e) yes, as long as they move along parallel lines

11. A 10.0-g bullet is fired into a 200-g block of wood at rest on a horizontal surface. After impact, the block slides 8.00 m before coming to rest. If the coefficient of friction between the block and the surface is 0.400, what is the speed of the bullet before impact? (a) 106 m/s (b) 166 m/s (c) 226 m/s (d) 286 m/s (e) none of those answers is correct

12. Two particles of different mass start from rest. The same net force acts on both of them as they move over equal distances. How do their final kinetic energies compare? (a) The particle of larger mass has more kinetic energy. (b) The particle of smaller mass has more kinetic energy. (c) The particles have equal kinetic energies. (d) Either particle might have more kinetic energy.

13. Two particles of different mass start from rest. The same net force acts on both of them as they move over equal distances. How do the magnitudes of their final momenta compare? (a) The particle of larger mass has more momentum. (b) The particle of smaller mass has more momentum. (c) The particles have equal momenta. (d) Either particle might have more momentum.

14. A basketball is tossed up into the air, falls freely, and bounces from the wooden floor. From the moment after the player releases it until the ball reaches the top of its bounce, what is the smallest system for which momentum is conserved? (a) the ball (b) the ball plus player (c) the ball plus floor (d) the ball plus the Earth (e) momentum is not conserved for any system

15. A massive tractor is rolling down a country road. In a perfectly inelastic collision, a small sports car runs into the machine from behind. **(i)** Which vehicle experiences a change in momentum of larger magnitude? (a) The car does. (b) The tractor does. (c) Their momentum changes are the same size. (d) It could be either vehicle. **(ii)** Which vehicle experiences a larger change in kinetic energy? (a) The car does. (b) The tractor does. (c) Their kinetic energy changes are the same size. (d) It could be either vehicle.

16. A ball is suspended by a string that is tied to a fixed point above a wooden block standing on end. The ball is pulled back as shown in Figure OQ9.16 and released. In trial A, the ball rebounds elastically from the block. In trial B, two-sided tape causes the ball to stick to the block. In which case is the ball more likely to knock the block over? (a) It is more likely in trial A. (b) It is more likely in trial B. (c) It makes no difference. (d) It could be either case, depending on other factors.

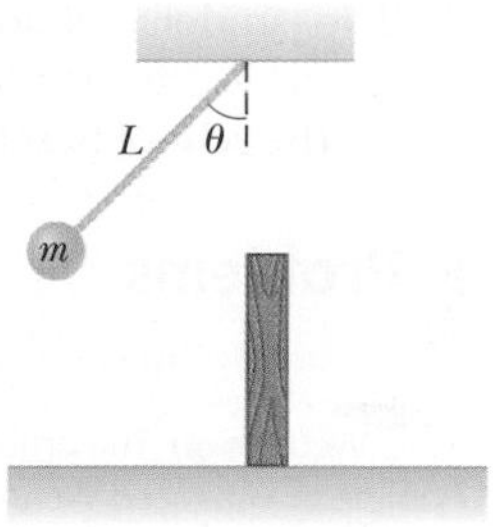

Figure OQ9.16

17. You are standing on a saucer-shaped sled at rest in the middle of a frictionless ice rink. Your lab partner throws

you a heavy Frisbee. You take different actions in successive experimental trials. Rank the following situations according to your final speed from largest to smallest. If your final speed is the same in two cases, give them equal rank. (a) You catch the Frisbee and hold onto it. (b) You catch the Frisbee and throw it back to your partner. (c) You bobble the catch, just touching the Frisbee so that it continues in its original direction more slowly. (d) You catch the Frisbee and throw it so that it moves vertically upward above your head. (e) You catch the Frisbee and set it down so that it remains at rest on the ice.

18. A boxcar at a rail yard is set into motion at the top of a hump. The car rolls down quietly and without friction onto a straight, level track where it couples with a flatcar of smaller mass, originally at rest, so that the two cars then roll together without friction. Consider the two cars as a system from the moment of release of the boxcar until both are rolling together. Answer the following questions yes or no. (a) Is mechanical energy of the system conserved? (b) Is momentum of the system conserved? Next, consider only the process of the boxcar gaining speed as it rolls down the hump. For the boxcar and the Earth as a system, (c) is mechanical energy conserved? (d) Is momentum conserved? Finally, consider the two cars as a system as the boxcar is slowing down in the coupling process. (e) Is mechanical energy of this system conserved? (f) Is momentum of this system conserved?

Conceptual Questions

☐ denotes answer available in Student Solutions Manual/Study Guide

1. Does a larger net force exerted on an object always produce a larger change in the momentum of the object compared with a smaller net force? Explain.
2. Does a larger net force always produce a larger change in kinetic energy than a smaller net force? Explain.
3. A bomb, initially at rest, explodes into several pieces. (a) Is linear momentum of the system (the bomb before the explosion, the pieces after the explosion) conserved? Explain. (b) Is kinetic energy of the system conserved? Explain.
4. While in motion, a pitched baseball carries kinetic energy and momentum. (a) Can we say that it carries a force that it can exert on any object it strikes? (b) Can the baseball deliver more kinetic energy to the bat and batter than the ball carries initially? (c) Can the baseball deliver to the bat and batter more momentum than the ball carries initially? Explain each of your answers.
5. You are standing perfectly still and then take a step forward. Before the step, your momentum was zero, but afterward you have some momentum. Is the principle of conservation of momentum violated in this case? Explain your answer.
6. A sharpshooter fires a rifle while standing with the butt of the gun against her shoulder. If the forward momentum of a bullet is the same as the backward momentum of the gun, why isn't it as dangerous to be hit by the gun as by the bullet?
7. Two students hold a large bed sheet vertically between them. A third student, who happens to be the star pitcher on the school baseball team, throws a raw egg at the center of the sheet. Explain why the egg does not break when it hits the sheet, regardless of its initial speed.
8. A juggler juggles three balls in a continuous cycle. Any one ball is in contact with one of his hands for one fifth of the time. (a) Describe the motion of the center of mass of the three balls. (b) What average force does the juggler exert on one ball while he is touching it?
9. (a) Does the center of mass of a rocket in free space accelerate? Explain. (b) Can the speed of a rocket exceed the exhaust speed of the fuel? Explain.
10. On the subject of the following positions, state your own view and argue to support it. (a) The best theory of motion is that force causes acceleration. (b) The true measure of a force's effectiveness is the work it does, and the best theory of motion is that work done on an object changes its energy. (c) The true measure of a force's effect is impulse, and the best theory of motion is that impulse imparted to an object changes its momentum.
11. An airbag in an automobile inflates when a collision occurs, which protects the passenger from serious injury (see the photo on page 240). Why does the airbag soften the blow? Discuss the physics involved in this dramatic photograph.
12. In golf, novice players are often advised to be sure to "follow through" with their swing. Why does this advice make the ball travel a longer distance? If a shot is taken near the green, very little follow-through is required. Why?
13. An open box slides across a frictionless, icy surface of a frozen lake. What happens to the speed of the box as water from a rain shower falls vertically downward into the box? Explain.

Problems

ENHANCED WebAssign The problems found in this chapter may be assigned online in Enhanced WebAssign

1. denotes straightforward problem; **2.** denotes intermediate problem; **3.** denotes challenging problem

1. (boxed) full solution available in the Student Solutions Manual/Study Guide

1. (shaded) denotes problems most often assigned in Enhanced WebAssign; these provide students with targeted feedback and either a Master It tutorial or a Watch It solution video.

QC denotes asking for quantitative and conceptual reasoning

S denotes symbolic reasoning problem

M denotes Master It tutorial available in Enhanced WebAssign

GP denotes guided problem

shaded denotes "paired problems" that develop reasoning with symbols and numerical values

Section 9.1 Linear Momentum

1. **S** A particle of mass m moves with momentum of magnitude p. (a) Show that the kinetic energy of the particle is $K = p^2/2m$. (b) Express the magnitude of the particle's momentum in terms of its kinetic energy and mass.

2. An object has a kinetic energy of 275 J and a momentum of magnitude 25.0 kg · m/s. Find the speed and mass of the object.

3. At one instant, a 17.5-kg sled is moving over a horizontal surface of snow at 3.50 m/s. After 8.75 s has elapsed, the sled stops. Use a momentum approach to find the average friction force acting on the sled while it was moving.

4. A baseball approaches home plate at a speed of 45.0 m/s, moving horizontally just before being hit by a bat. The batter hits a pop-up such that after hitting the bat, the baseball is moving at 55.0 m/s straight up. The ball has a mass of 145 g and is in contact with the bat for 2.00 ms. What is the average vector force the ball exerts on the bat during their interaction?

Section 9.2 Analysis Model: Isolated System (Momentum)

5. **QC** A 65.0-kg boy and his 40.0-kg sister, both wearing roller blades, face each other at rest. The girl pushes the boy hard, sending him backward with velocity 2.90 m/s toward the west. Ignore friction. (a) Describe the subsequent motion of the girl. (b) How much potential energy in the girl's body is converted into mechanical energy of the boy–girl system? (c) Is the momentum of the boy–girl system conserved in the pushing-apart process? If so, explain how that is possible considering (d) there are large forces acting and (e) there is no motion beforehand and plenty of motion afterward.

6. **M** A 45.0-kg girl is standing on a 150-kg plank. Both are originally at rest on a frozen lake that constitutes a frictionless, flat surface. The girl begins to walk along the plank at a constant velocity of $1.50\hat{\mathbf{i}}$ m/s relative to the plank. (a) What is the velocity of the plank relative to the ice surface? (b) What is the girl's velocity relative to the ice surface?

7. **S** A girl of mass m_g is standing on a plank of mass m_p. Both are originally at rest on a frozen lake that constitutes a frictionless, flat surface. The girl begins to walk along the plank at a constant velocity v_{gp} to the right relative to the plank. (The subscript gp denotes the girl relative to plank.) (a) What is the velocity v_{pi} of the plank relative to the surface of the ice? (b) What is the girl's velocity v_{gi} relative to the ice surface?

8. When you jump straight up as high as you can, what is the order of magnitude of the maximum recoil speed that you give to the Earth? Model the Earth as a perfectly solid object. In your solution, state the physical quantities you take as data and the values you measure or estimate for them.

9. **QC** Two blocks of masses m and $3m$ are placed on a frictionless, horizontal surface. A light spring is attached to the more massive block, and the blocks are pushed together with the spring between them (Fig. P9.9). A cord initially holding the blocks together is burned; after that happens, the block of mass $3m$ moves to the right with a speed of 2.00 m/s. (a) What is the velocity of the block of mass m? (b) Find the system's original elastic potential energy, taking $m = 0.350$ kg. (c) Is the original energy in the spring or in the cord? (d) Explain your answer to part (c). (e) Is the momentum of the system conserved in the bursting-apart process? Explain how that is possible considering (f) there are large forces acting and (g) there is no motion beforehand and plenty of motion afterward?

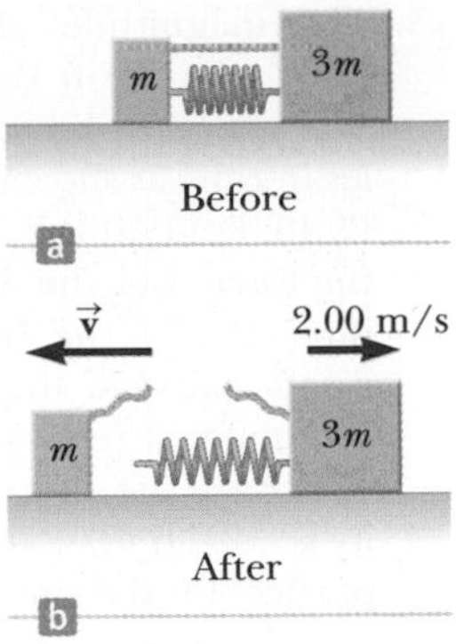

Figure P9.9

Section 9.3 Analysis Model: Nonisolated System (Momentum)

10. **QC** A man claims that he can hold onto a 12.0-kg child in a head-on collision as long as he has his seat belt on. Consider this man in a collision in which he is in one of two identical cars each traveling toward the other at 60.0 mi/h relative to the ground. The car in which he rides is brought to rest in 0.10 s. (a) Find the magnitude of the average force needed to hold onto the child. (b) Based on your result to part (a), is the man's claim valid? (c) What does the answer to this problem say about laws requiring the use of proper safety devices such as seat belts and special toddler seats?

11. An estimated force–time curve for a baseball struck by a bat is shown in Figure P9.11. From this curve, determine (a) the magnitude of the impulse delivered to the ball and (b) the average force exerted on the ball.

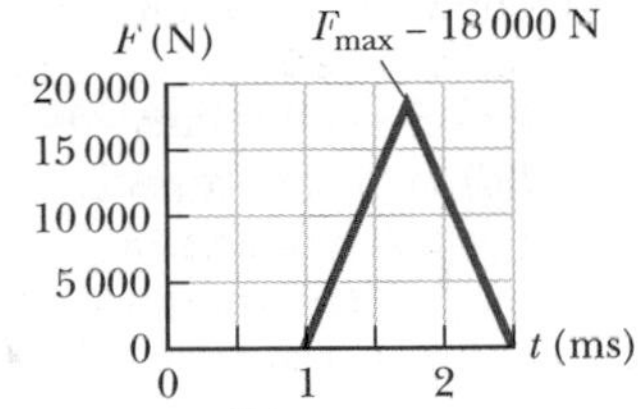

Figure P9.11

12. **Review.** After a 0.300-kg rubber ball is dropped from a height of 1.75 m, it bounces off a concrete floor and rebounds to a height of 1.50 m. (a) Determine the magnitude and direction of the impulse delivered to the ball by the floor. (b) Estimate the time the ball is in contact with the floor and use this estimate to calculate the average force the floor exerts on the ball.

13. **QC** **S** A glider of mass m is free to slide along a horizontal air track. It is pushed against a launcher at one end of the track. Model the launcher as a light spring of force constant k compressed by a distance x. The glider is released from rest. (a) Show that the glider attains a speed of $v = x(k/m)^{1/2}$. (b) Show that the magnitude of the impulse imparted to the glider is given by the expression $I = x(km)^{1/2}$. (c) Is more work done on a cart with a large or a small mass?

14. A tennis player receives a shot with the ball (0.060 0 kg) traveling horizontally at 50.0 m/s and returns the shot with the ball traveling horizontally at 40.0 m/s in the opposite direction. (a) What is the impulse delivered to the ball by the tennis racquet? (b) What work does the racquet do on the ball?

15. The magnitude of the net force exerted in the x direction on a 2.50-kg particle varies in time as shown in Figure P9.15. Find (a) the impulse of the force over the 5.00-s time interval, (b) the final velocity the particle attains if it is originally at rest, (c) its final velocity if its original velocity is $-2.00\hat{\mathbf{i}}$ m/s, and (d) the average force exerted on the particle for the time interval between 0 and 5.00 s.

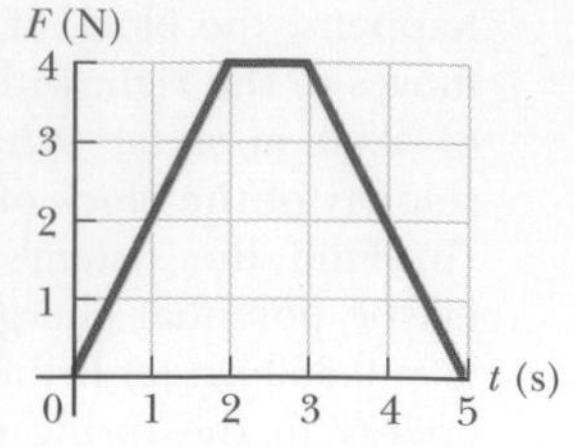

Figure P9.15

16. **Review.** A *force platform* is a tool used to analyze the performance of athletes by measuring the vertical force the athlete exerts on the ground as a function of time. Starting from rest, a 65.0-kg athlete jumps down onto the platform from a height of 0.600 m. While she is in contact with the platform during the time interval $0 < t < 0.800$ s, the force she exerts on it is described by the function

$$F = 9\,200t - 11\,500t^2$$

where F is in newtons and t is in seconds. (a) What impulse did the athlete receive from the platform? (b) With what speed did she reach the platform? (c) With what speed did she leave it? (d) To what height did she jump upon leaving the platform?

17. Water falls without splashing at a rate of 0.250 L/s from a height of 2.60 m into a 0.750-kg bucket on a scale. If the bucket is originally empty, what does the scale read in newtons 3.00 s after water starts to accumulate in it?

Section 9.4 Collisions in One Dimension

18. Q|C A 1 200-kg car traveling initially at $v_{Ci} = 25.0$ m/s in an easterly direction crashes into the back of a 9 000-kg truck moving in the same direction at $v_{Ti} = 20.0$ m/s (Fig. P9.18). The velocity of the car immediately after the collision is $v_{Cf} = 18.0$ m/s to the east. (a) What is the velocity of the truck immediately after the collision? (b) What is the change in mechanical energy of the car–truck system in the collision? (c) Account for this change in mechanical energy.

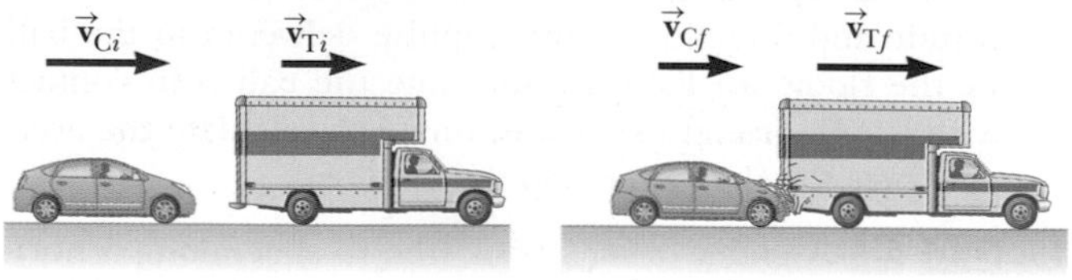

Figure P9.18

19. A 10.0-g bullet is fired into a stationary block of wood having mass $m = 5.00$ kg. The bullet imbeds into the block. The speed of the bullet-plus-wood combination immediately after the collision is 0.600 m/s. What was the original speed of the bullet?

20. S A car of mass m moving at a speed v_1 collides and couples with the back of a truck of mass $2m$ moving initially in the same direction as the car at a lower speed v_2. (a) What is the speed v_f of the two vehicles immediately after the collision? (b) What is the change in kinetic energy of the car–truck system in the collision?

21. M A neutron in a nuclear reactor makes an elastic, head-on collision with the nucleus of a carbon atom initially at rest. (a) What fraction of the neutron's kinetic energy is transferred to the carbon nucleus? (b) The initial kinetic energy of the neutron is 1.60×10^{-13} J. Find its final kinetic energy and the kinetic energy of the carbon nucleus after the collision. (The mass of the carbon nucleus is nearly 12.0 times the mass of the neutron.)

22. Q|C S A tennis ball of mass m_t is held just above a basketball of mass m_b, as shown in Figure P9.22. With their centers vertically aligned, both are released from rest at the same moment so that the bottom of the basketball falls freely through a height h and strikes the floor. Assume an elastic collision with the ground instantaneously reverses the velocity of the basketball while the tennis ball is still moving down because the balls have separated a bit while falling. Next, the two balls meet in an elastic collision. (a) To what height does the tennis ball rebound? (b) How do you account for the height in (a) being larger than h? Does that seem like a violation of conservation of energy?

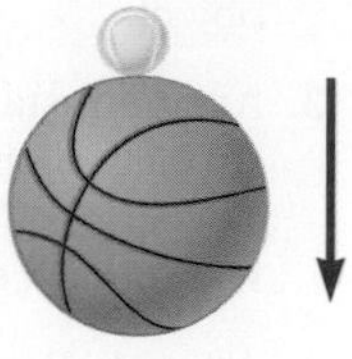
Figure P9.22

23. M A 12.0-g wad of sticky clay is hurled horizontally at a 100-g wooden block initially at rest on a horizontal surface. The clay sticks to the block. After impact, the block slides 7.50 m before coming to rest. If the coefficient of friction between the block and the surface is 0.650, what was the speed of the clay immediately before impact?

24. S A wad of sticky clay of mass m is hurled horizontally at a wooden block of mass M initially at rest on a horizontal surface. The clay sticks to the block. After impact, the block slides a distance d before coming to rest. If the coefficient of friction between the block and the surface is μ, what was the speed of the clay immediately before impact?

25. Q|C (a) Three carts of masses $m_1 = 4.00$ kg, $m_2 = 10.0$ kg, and $m_3 = 3.00$ kg move on a frictionless, horizontal track with speeds of $v_1 = 5.00$ m/s to the right, $v_2 = 3.00$ m/s to the right, and $v_3 = 4.00$ m/s to the left as shown in Figure P9.25. Velcro couplers make the carts stick together after colliding. Find the final velocity of the train of three carts. (b) **What If?** Does your answer in part (a) require that all the carts collide and stick together at the same moment? What if they collide in a different order?

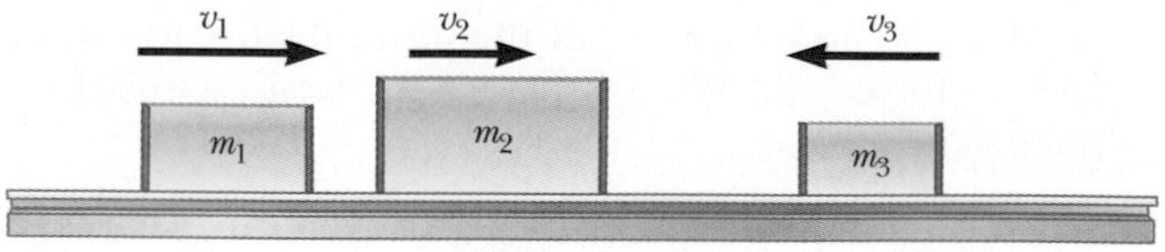

Figure P9.25

26. S As shown in Figure P9.26, a bullet of mass m and speed v passes completely through a pendulum bob of mass M. The bullet emerges with a speed of $v/2$. The pendulum bob is suspended by a stiff rod (*not* a string) of length ℓ and negligible mass. What is

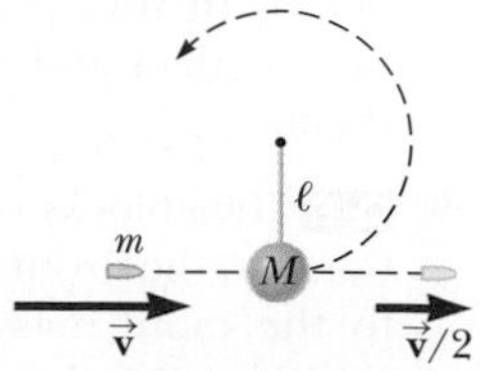

Figure P9.26

the minimum value of v such that the pendulum bob will barely swing through a complete vertical circle?

27. Two blocks are free to slide along the frictionless, wooden track shown in Figure P9.27. The block of mass $m_1 =$ 5.00 kg is released from the position shown, at height $h =$ 5.00 m above the flat part of the track. Protruding from its front end is the north pole of a strong magnet, which repels the north pole of an identical magnet embedded in the back end of the block of mass $m_2 =$ 10.0 kg, initially at rest. The two blocks never touch. Calculate the maximum height to which m_1 rises after the elastic collision.

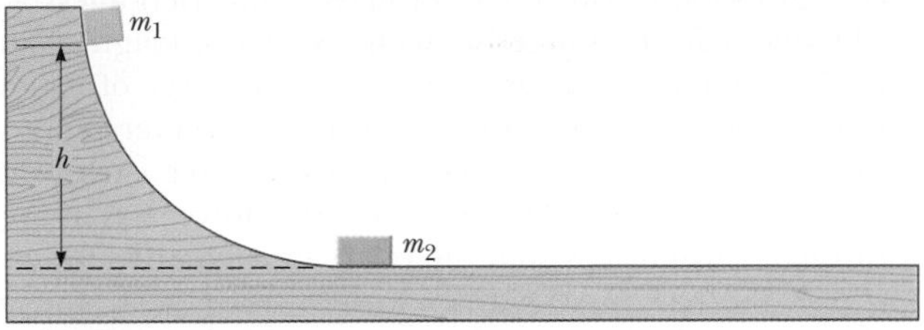

Figure P9.27

Section 9.5 Collisions in Two Dimensions

28. Two automobiles of equal mass approach an intersection. One vehicle is traveling with speed 13.0 m/s toward the east, and the other is traveling north with speed v_{2i}. Neither driver sees the other. The vehicles collide in the intersection and stick together, leaving parallel skid marks at an angle of 55.0° north of east. The speed limit for both roads is 35 mi/h, and the driver of the northward-moving vehicle claims he was within the speed limit when the collision occurred. Is he telling the truth? Explain your reasoning.

29. An object of mass 3.00 kg, moving with an initial velocity of $5.00\hat{\mathbf{i}}$ m/s, collides with and sticks to an object of mass 2.00 kg with an initial velocity of $-3.00\hat{\mathbf{j}}$ m/s. Find the final velocity of the composite object.

30. Two shuffleboard disks of equal mass, one orange and the other yellow, are involved in an elastic, glancing collision. The yellow disk is initially at rest and is struck by the orange disk moving with a speed of 5.00 m/s. After the collision, the orange disk moves along a direction that makes an angle of 37.0° with its initial direction of motion. The velocities of the two disks are perpendicular after the collision. Determine the final speed of each disk.

31. **S** Two shuffleboard disks of equal mass, one orange and the other yellow, are involved in an elastic, glancing collision. The yellow disk is initially at rest and is struck by the orange disk moving with a speed v_i. After the collision, the orange disk moves along a direction that makes an angle θ with its initial direction of motion. The velocities of the two disks are perpendicular after the collision. Determine the final speed of each disk.

32. **Q|C** A 90.0-kg fullback running east with a speed of 5.00 m/s is tackled by a 95.0-kg opponent running north with a speed of 3.00 m/s. (a) Explain why the successful tackle constitutes a perfectly inelastic collision. (b) Calculate the velocity of the players immediately after the tackle. (c) Determine the mechanical energy that disappears as a result of the collision. Account for the missing energy.

33. A billiard ball moving at 5.00 m/s strikes a stationary ball of the same mass. After the collision, the first ball moves at 4.33 m/s at an angle of 30.0° with respect to the original line of motion. Assuming an elastic collision (and ignoring friction and rotational motion), find the struck ball's velocity after the collision.

34. The mass of the blue puck in Figure P9.34 is 20.0% greater than the mass of the green puck. Before colliding, the pucks approach each other with momenta of equal magnitudes and opposite directions, and the green puck has an initial speed of 10.0 m/s. Find the speeds the pucks have after the collision if half the kinetic energy of the system becomes internal energy during the collision.

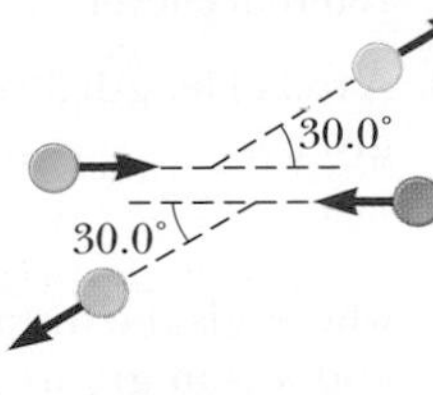

Figure P9.34

35. **M** An unstable atomic nucleus of mass 17.0×10^{-27} kg initially at rest disintegrates into three particles. One of the particles, of mass 5.00×10^{-27} kg, moves in the y direction with a speed of 6.00×10^6 m/s. Another particle, of mass 8.40×10^{-27} kg, moves in the x direction with a speed of 4.00×10^6 m/s. Find (a) the velocity of the third particle and (b) the total kinetic energy increase in the process.

Section 9.6 The Center of Mass

36. The mass of the Earth is 5.97×10^{24} kg, and the mass of the Moon is 7.35×10^{22} kg. The distance of separation, measured between their centers, is 3.84×10^8 m. Locate the center of mass of the Earth–Moon system as measured from the center of the Earth.

37. Four objects are situated along the y axis as follows: a 2.00-kg object is at +3.00 m, a 3.00-kg object is at +2.50 m, a 2.50-kg object is at the origin, and a 4.00-kg object is at −0.500 m. Where is the center of mass of these objects?

38. A uniform piece of sheet metal is shaped as shown in Figure P9.38. Compute the x and y coordinates of the center of mass of the piece.

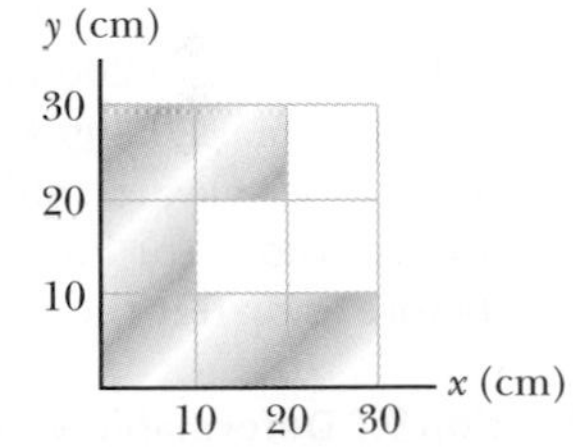

Figure P9.38

39. Explorers in the jungle find an ancient monument in the shape of a large isosceles triangle as shown in Figure P9.39. The monument is made from tens of thousands of small stone blocks of density 3 800 kg/m³. The monument is 15.7 m high and 64.8 m wide at its base and is everywhere 3.60 m thick from front to back. Before the monument was built many years ago, all the stone blocks lay on the ground. How much work did laborers do on the blocks to put them in position while

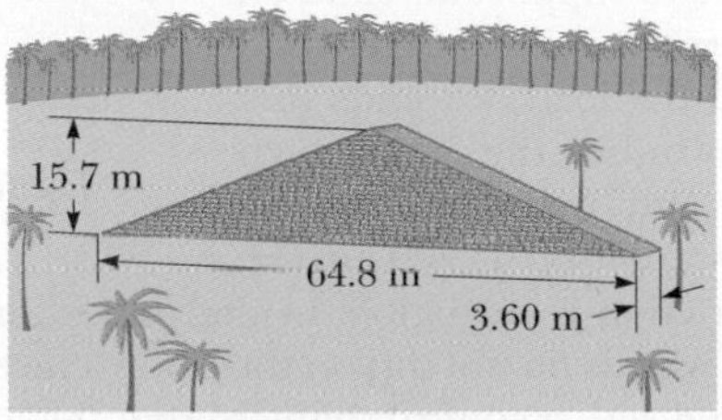

Figure P9.39

building the entire monument? *Note*: The gravitational potential energy of an object–Earth system is given by $U_g = Mgy_{CM}$, where M is the total mass of the object and y_{CM} is the elevation of its center of mass above the chosen reference level.

40. A rod of length 30.0 cm has linear density (mass per length) given by

$$\lambda = 50.0 + 20.0x$$

where x is the distance from one end, measured in meters, and λ is in grams/meter. (a) What is the mass of the rod? (b) How far from the $x = 0$ end is its center of mass?

Section 9.7 Systems of Many Particles

41. A 2.00-kg particle has a velocity $(2.00\hat{\mathbf{i}} - 3.00\hat{\mathbf{j}})$ m/s, and a 3.00-kg particle has a velocity $(1.00\hat{\mathbf{i}} + 6.00\hat{\mathbf{j}})$ m/s. Find (a) the velocity of the center of mass and (b) the total momentum of the system.

42. The vector position of a 3.50-g particle moving in the xy plane varies in time according to $\vec{\mathbf{r}}_1 = (3\hat{\mathbf{i}} + 3\hat{\mathbf{j}})t + 2\hat{\mathbf{j}}t^2$, where t is in seconds and $\vec{\mathbf{r}}$ is in centimeters. At the same time, the vector position of a 5.50-g particle varies as $\vec{\mathbf{r}}_2 = 3\hat{\mathbf{i}} - 2\hat{\mathbf{i}}t^2 - 6\hat{\mathbf{j}}t$. At $t = 2.50$ s, determine (a) the vector position of the center of mass, (b) the linear momentum of the system, (c) the velocity of the center of mass, (d) the acceleration of the center of mass, and (e) the net force exerted on the two-particle system.

43. **M** Romeo (77.0 kg) entertains Juliet (55.0 kg) by playing his guitar from the rear of their boat at rest in still water, 2.70 m away from Juliet, who is in the front of the boat. After the serenade, Juliet carefully moves to the rear of the boat (away from shore) to plant a kiss on Romeo's cheek. How far does the 80.0-kg boat move toward the shore it is facing?

44. A ball of mass 0.200 kg with a velocity of $1.50\hat{\mathbf{i}}$ m/s meets a ball of mass 0.300 kg with a velocity of $-0.400\hat{\mathbf{i}}$ m/s in a head-on, elastic collision. (a) Find their velocities after the collision. (b) Find the velocity of their center of mass before and after the collision.

Section 9.8 Deformable Systems

45. **Q|C** For a technology project, a student has built a vehicle, of total mass 6.00 kg, that moves itself. As shown in Figure P9.45, it runs on four light wheels. A reel is attached to one of the axles, and a cord originally wound on the reel goes up over a pulley attached to the vehicle to support an elevated load. After the vehicle is released from rest, the load descends very slowly, unwinding the cord to turn the axle and make the vehicle move forward (to the left in Fig. P9.45). Friction is negligible in the pulley and axle bearings. The wheels do not slip on the floor. The reel has been constructed with a conical shape so that the load descends at a constant low speed while the vehicle moves horizontally across the floor with constant acceleration, reaching a final velocity of $3.00\hat{\mathbf{i}}$ m/s. (a) Does the floor impart impulse to the vehicle? If so, how much? (b) Does the floor do work on the vehicle? If so, how much? (c) Does it make sense to say that the final momentum of the vehicle came from the floor? If not, where did it come from? (d) Does it make sense to say that the final kinetic energy of the vehicle came from the floor? If not, where did it come from? (e) Can we say that one particular force causes the forward acceleration of the vehicle? What does cause it?

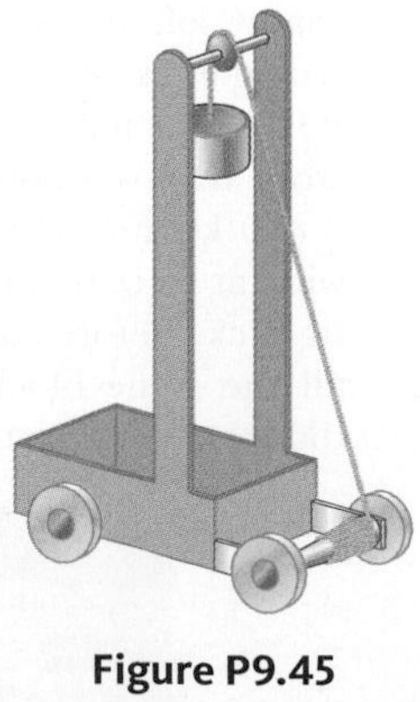

Figure P9.45

46. **S** Figure P9.46a shows an overhead view of the initial configuration of two pucks of mass m on frictionless ice. The pucks are tied together with a string of length ℓ and negligible mass. At time $t = 0$, a constant force of magnitude F begins to pull to the right on the center point of the string. At time t, the moving pucks strike each other and stick together. At this time, the force has moved through a distance d, and the pucks have attained a speed v (Fig. P9.46b). (a) What is v in terms of F, d, ℓ, and m? (b) How much of the energy transferred into the system by work done by the force has been transformed to internal energy?

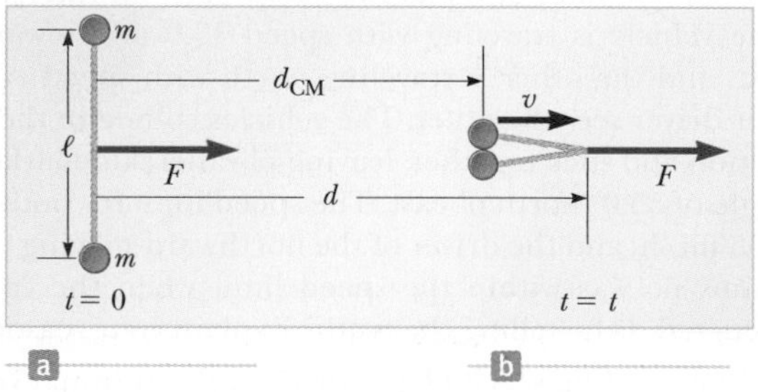

Figure P9.46

47. **Q|C** **S** A particle is suspended from a post on top of a cart by a light string of length L as shown in Figure P9.47a. The cart and particle are initially moving to the right at constant speed v_i, with the string vertical. The cart suddenly comes to rest when it runs into and sticks to a bumper as shown in Figure P9.47b. The suspended particle swings through an angle θ. (a) Show that the original speed of the cart can be computed from $v_i = \sqrt{2gL(1 - \cos\theta)}$. (b) If the bumper is still exerting a horizontal force on the cart when the hanging particle is at its maximum angle *forward* from the vertical, at what moment does the bumper *stop* exerting a horizontal force?

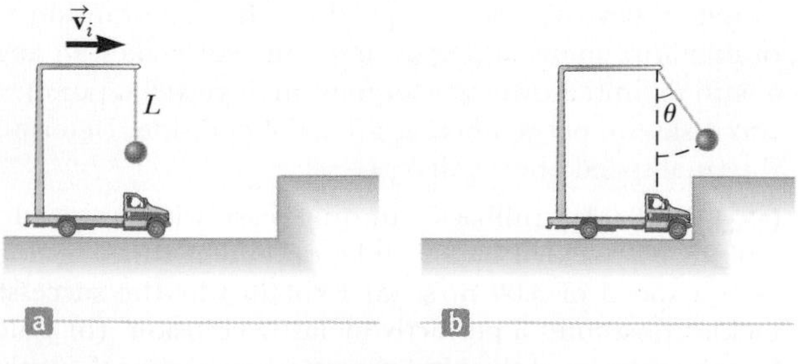

Figure P9.47

48. **Q|C** A 60.0-kg person bends his knees and then jumps straight up. After his feet leave the floor, his motion is unaffected by air resistance and his center of mass rises by a maximum of 15.0 cm. Model the floor as completely solid and motionless. (a) Does the floor impart impulse to the person? (b) Does the floor do work on the person? (c) With what momentum does the person leave the floor? (d) Does it make sense to say that this momentum came from the floor? Explain. (e) With what kinetic energy does the person leave the floor? (f) Does it make sense to say that this energy came from the floor? Explain.

Section 9.9 Rocket Propulsion

49. A model rocket engine has an average thrust of 5.26 N. It has an initial mass of 25.5 g, which includes fuel mass of 12.7 g. The duration of its burn is 1.90 s. (a) What is the average exhaust speed of the engine? (b) This engine is placed in a rocket body of mass 53.5 g. What is the final velocity of the rocket if it were to be fired from rest in outer space by an astronaut on a spacewalk? Assume the fuel burns at a constant rate.

50. **Review.** The first stage of a Saturn V space vehicle consumed fuel and oxidizer at the rate of 1.50×10^4 kg/s with an exhaust speed of 2.60×10^3 m/s. (a) Calculate the thrust produced by this engine. (b) Find the acceleration the vehicle had just as it lifted off the launch pad on the Earth, taking the vehicle's initial mass as 3.00×10^6 kg.

51. **QIC** A rocket for use in deep space is to be capable of boosting a total load (payload plus rocket frame and engine) of 3.00 metric tons to a speed of 10 000 m/s. (a) It has an engine and fuel designed to produce an exhaust speed of 2 000 m/s. How much fuel plus oxidizer is required? (b) If a different fuel and engine design could give an exhaust speed of 5 000 m/s, what amount of fuel and oxidizer would be required for the same task? (c) Noting that the exhaust speed in part (b) is 2.50 times higher than that in part (a), explain why the required fuel mass is not simply smaller by a factor of 2.50.

52. A rocket has total mass $M_i = 360$ kg, including $M_f = 330$ kg of fuel and oxidizer. In interstellar space, it starts from rest at the position $x = 0$, turns on its engine at time $t = 0$, and puts out exhaust with relative speed $v_e = 1\,500$ m/s at the constant rate $k = 2.50$ kg/s. The fuel will last for a burn time of $T_b = M_f/k = 330 \text{ kg}/(2.5 \text{ kg/s}) = 132$ s. (a) Show that during the burn the velocity of the rocket as a function of time is given by

$$v(t) = -v_e \ln\left(1 - \frac{kt}{M_i}\right)$$

(b) Make a graph of the velocity of the rocket as a function of time for times running from 0 to 132 s. (c) Show that the acceleration of the rocket is

$$a(t) = \frac{kv_e}{M_i - kt}$$

(d) Graph the acceleration as a function of time. (e) Show that the position of the rocket is

$$x(t) = v_e\left(\frac{M_i}{k} - t\right)\ln\left(1 - \frac{kt}{M_i}\right) + v_e t$$

(f) Graph the position during the burn as a function of time.

Additional Problems

53. **S** A ball of mass m is thrown straight up into the air with an initial speed v_i. Find the momentum of the ball (a) at its maximum height and (b) halfway to its maximum height.

54. **QIC** **S** An amateur skater of mass M is trapped in the middle of an ice rink and is unable to return to the side where there is no ice. Every motion she makes causes her to slip on the ice and remain in the same spot. She decides to try to return to safety by throwing her gloves of mass m in the direction opposite the safe side. (a) She throws the gloves as hard as she can, and they leave her hand with a horizontal velocity $\vec{\mathbf{v}}_{\text{gloves}}$. Explain whether or not she moves. If she does move, calculate her velocity $\vec{\mathbf{v}}_{\text{girl}}$ relative to the Earth after she throws the gloves. (b) Discuss her motion from the point of view of the forces acting on her.

55. A 3.00-kg steel ball strikes a wall with a speed of 10.0 m/s at an angle of $\theta = 60.0°$ with the surface. It bounces off with the same speed and angle (Fig. P9.55). If the ball is in contact with the wall for 0.200 s, what is the average force exerted by the wall on the ball?

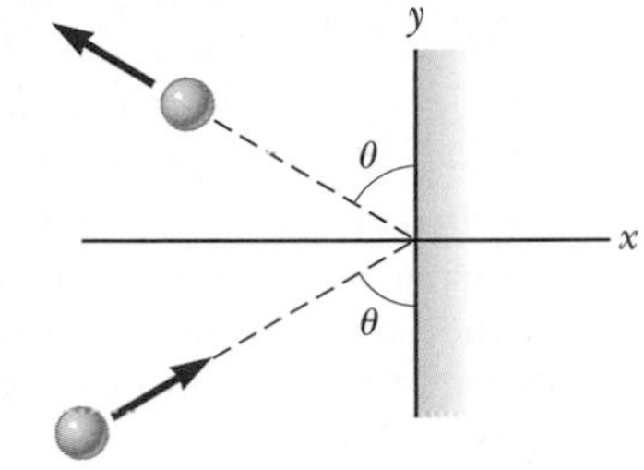

Figure P9.55

56. **QIC** **S** (a) Figure P9.56 shows three points in the operation of the ballistic pendulum discussed in Example 9.6 (and shown in Fig. 9.9b). The projectile approaches the pendulum in Figure P9.56a. Figure P9.56b shows the situation just after the projectile is captured in the pendulum. In Figure P9.56c, the pendulum arm has swung upward and come to rest at a height h above its initial position. Prove that the ratio of the kinetic energy of the projectile–pendulum system immediately after the collision to the kinetic energy immediately before is $m_1/(m_1 + m_2)$. (b) What is the ratio of the momentum of the system immediately after the collision to the momentum immediately before? (c) A student believes that such a large decrease in mechanical energy must be accompanied by at least a small decrease in momentum. How would you convince this student of the truth?

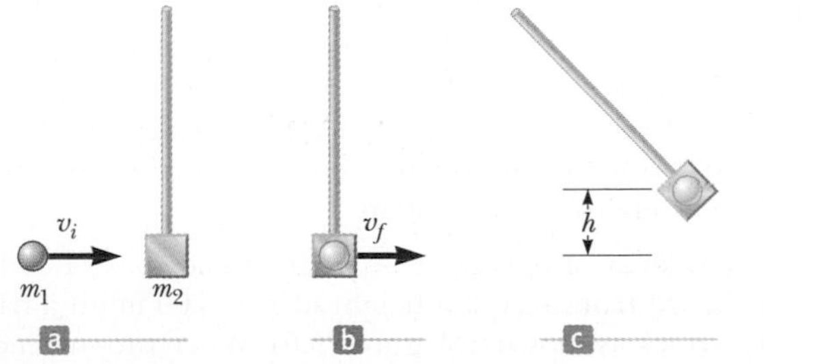

Figure P9.56 Problems 56 and 70. (a) A metal ball moves toward the pendulum. (b) The ball is captured by the pendulum. (c) The ball–pendulum combination swings up through a height h before coming to rest.

57. **QIC** A 1.25-kg wooden block rests on a table over a large hole as in Figure P9.57 on page 274. A 5.00-g bullet with an initial velocity v_i is fired upward into the bottom of the block and remains in the block after the collision. The block and bullet rise to a maximum height of 22.0 cm. (a) Describe how you would find the initial velocity of the bullet using ideas you have learned in this chapter. (b) Calculate the initial velocity of the bullet from the information provided.

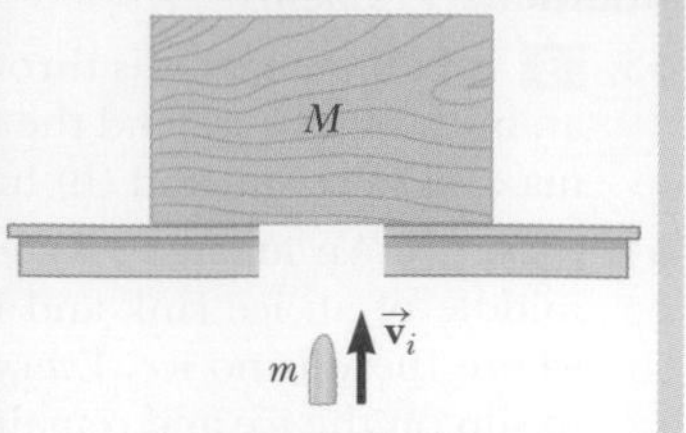

Figure P9.57
Problems 57 and 58.

58. **Q|C** **S** A wooden block of mass M rests on a table over a large hole as in Figure 9.57. A bullet of mass m with an initial velocity of v_i is fired upward into the bottom of the block and remains in the block after the collision. The block and bullet rise to a maximum height of h. (a) Describe how you would find the initial velocity of the bullet using ideas you have learned in this chapter. (b) Find an expression for the initial velocity of the bullet.

59. **S** Two gliders are set in motion on a horizontal air track. A spring of force constant k is attached to the back end of the second glider. As shown in Figure P9.59, the first glider, of mass m_1, moves to the right with speed v_1, and the second glider, of mass m_2, moves more slowly to the right with speed v_2. When m_1 collides with the spring attached to m_2, the spring compresses by a distance x_{max}, and the gliders then move apart again. In terms of v_1, v_2, m_1, m_2, and k, find (a) the speed v at maximum compression, (b) the maximum compression x_{max}, and (c) the velocity of each glider after m_1 has lost contact with the spring.

Figure P9.59

60. Pursued by ferocious wolves, you are in a sleigh with no horses, gliding without friction across an ice-covered lake. You take an action described by the equations

$$(270\text{ kg})(7.50\text{ m/s})\hat{\mathbf{i}} = (15.0\text{ kg})(-v_{1f}\hat{\mathbf{i}}) + (255\text{ kg})(v_{2f}\hat{\mathbf{i}})$$
$$v_{1f} + v_{2f} = 8.00\text{ m/s}$$

(a) Complete the statement of the problem, giving the data and identifying the unknowns. (b) Find the values of v_{1f} and v_{2f}. (c) Find the amount of energy that has been transformed from potential energy stored in your body to kinetic energy of the system.

61. Two blocks of masses $m_1 = 2.00$ kg and $m_2 = 4.00$ kg are released from rest at a height of $h = 5.00$ m on a frictionless track as shown in Figure P9.61. When they meet on the level portion of the track, they undergo a head-on, elastic collision. Determine the maximum heights to which m_1 and m_2 rise on the curved portion of the track after the collision.

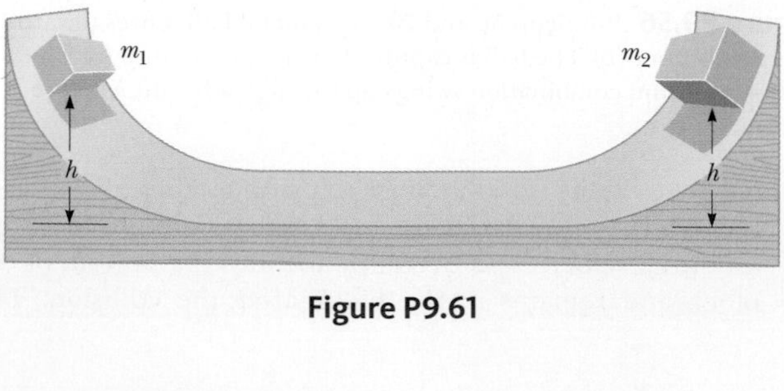

Figure P9.61

62. *Why is the following situation impossible?* An astronaut, together with the equipment he carries, has a mass of 150 kg. He is taking a space walk outside his spacecraft, which is drifting through space with a constant velocity. The astronaut accidentally pushes against the spacecraft and begins moving away at 20.0 m/s, relative to the spacecraft, without a tether. To return, he takes equipment off his space suit and throws it in the direction away from the spacecraft. Because of his bulky space suit, he can throw equipment at a maximum speed of 5.00 m/s relative to himself. After throwing enough equipment, he starts moving back to the spacecraft and is able to grab onto it and climb inside.

63. A 0.400-kg blue bead slides on a frictionless, curved wire, starting from rest at point Ⓐ in Figure P9.63, where $h = 1.50$ m. At point Ⓑ, the blue bead collides elastically with a 0.600-kg green bead at rest. Find the maximum height the green bead rises as it moves up the wire.

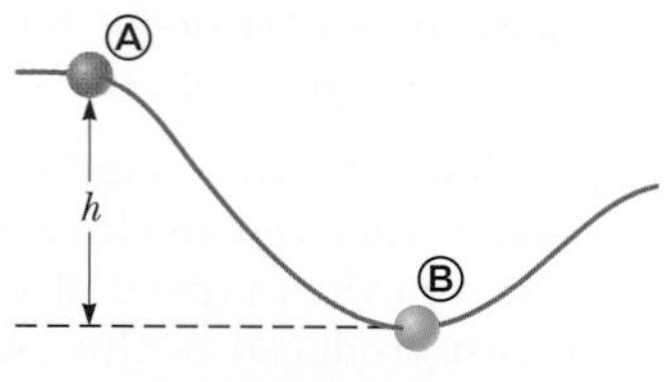

Figure P9.63

64. **Q|C** **Review.** A metal cannonball of mass m rests next to a tree at the very edge of a cliff 36.0 m above the surface of the ocean. In an effort to knock the cannonball off the cliff, some children tie one end of a rope around a stone of mass 80.0 kg and the other end to a tree limb just above the cannonball. They tighten the rope so that the stone just clears the ground and hangs next to the cannonball. The children manage to swing the stone back until it is held at rest 1.80 m above the ground. The children release the stone, which then swings down and makes a head-on, elastic collision with the cannonball, projecting it horizontally off the cliff. The cannonball lands in the ocean a horizontal distance R away from its initial position. (a) Find the horizontal component R of the cannonball's displacement as it depends on m. (b) What is the maximum possible value for R, and (c) to what value of m does it correspond? (d) For the stone–cannonball–Earth system, is mechanical energy conserved throughout the process? Is this principle sufficient to solve the entire problem? Explain. (e) **What if?** Show that R does not depend on the value of the gravitational acceleration. Is this result remarkable? State how one might make sense of it.

65. **S** **Review.** A bullet of mass m is fired into a block of mass M initially at rest at the edge of a frictionless table of height h (Fig. P9.65). The bullet remains in the block, and after

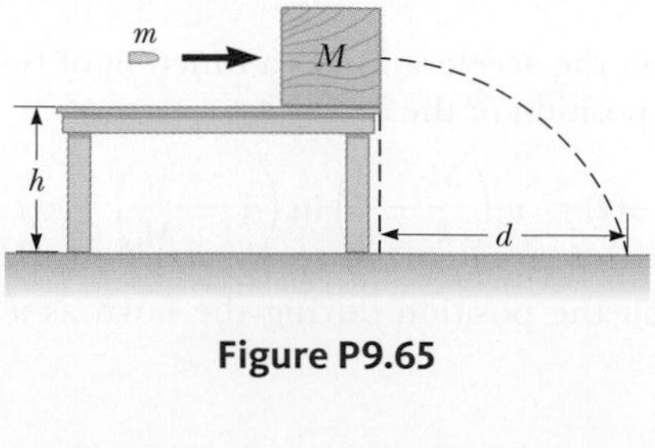

Figure P9.65

impact the block lands a distance d from the bottom of the table. Determine the initial speed of the bullet.

66. A small block of mass $m_1 = 0.500$ kg is released from rest at the top of a frictionless, curve-shaped wedge of mass $m_2 = 3.00$ kg, which sits on a frictionless, horizontal surface as shown in Figure P9.66a. When the block leaves the wedge, its velocity is measured to be 4.00 m/s to the right as shown in Figure P9.66b. (a) What is the velocity of the wedge after the block reaches the horizontal surface? (b) What is the height h of the wedge?

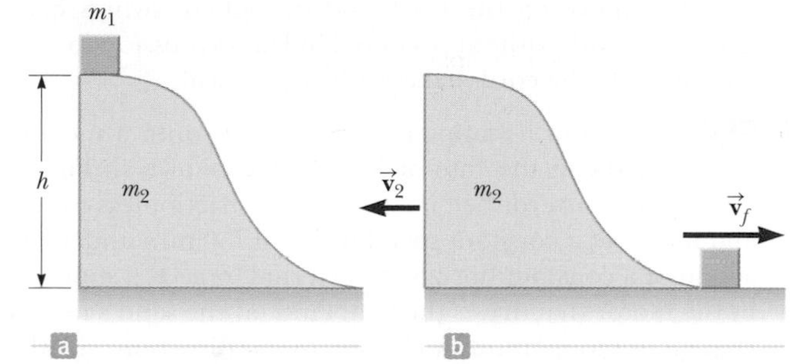

Figure P9.66

67. A 0.500-kg sphere moving with a velocity given by $(2.00\hat{\mathbf{i}} - 3.00\hat{\mathbf{j}} + 1.00\hat{\mathbf{k}})$ m/s strikes another sphere of mass 1.50 kg moving with an initial velocity of $(-1.00\hat{\mathbf{i}} + 2.00\hat{\mathbf{j}} - 3.00\hat{\mathbf{k}})$ m/s. (a) The velocity of the 0.500-kg sphere after the collision is $(-1.00\hat{\mathbf{i}} + 3.00\hat{\mathbf{j}} - 8.00\hat{\mathbf{k}})$ m/s. Find the final velocity of the 1.50-kg sphere and identify the kind of collision (elastic, inelastic, or perfectly inelastic). (b) Now assume the velocity of the 0.500-kg sphere after the collision is $(-0.250\hat{\mathbf{i}} + 0.750\hat{\mathbf{j}} - 2.00\hat{\mathbf{k}})$ m/s. Find the final velocity of the 1.50-kg sphere and identify the kind of collision. (c) **What If?** Take the velocity of the 0.500-kg sphere after the collision as $(-1.00\hat{\mathbf{i}} + 3.00\hat{\mathbf{j}} + a\hat{\mathbf{k}})$ m/s. Find the value of a and the velocity of the 1.50-kg sphere after an elastic collision.

68. A 75.0-kg firefighter slides down a pole while a constant friction force of 300 N retards her motion. A horizontal 20.0-kg platform is supported by a spring at the bottom of the pole to cushion the fall. The firefighter starts from rest 4.00 m above the platform, and the spring constant is 4 000 N/m. Find (a) the firefighter's speed just before she collides with the platform and (b) the maximum distance the spring is compressed. Assume the friction force acts during the entire motion.

69. George of the Jungle, with mass m, swings on a light vine hanging from a stationary tree branch. A second vine of equal length hangs from the same point, and a gorilla of larger mass M swings in the opposite direction on it. Both vines are horizontal when the primates start from rest at the same moment. George and the gorilla meet at the lowest point of their swings. Each is afraid that one vine will break, so they grab each other and hang on. They swing upward together, reaching a point where the vines make an angle of 35.0° with the vertical. Find the value of the ratio m/M.

70. **QC** **Review.** A student performs a ballistic pendulum experiment using an apparatus similar to that discussed in Example 9.6 and shown in Figure P9.56. She obtains the following average data: $h = 8.68$ cm, projectile mass $m_1 = 68.8$ g, and pendulum mass $m_2 = 263$ g. (a) Determine the initial speed v_{1A} of the projectile. (b) The second part of her experiment is to obtain v_{1A} by firing the same projectile horizontally (with the pendulum removed from the path) and measuring its final horizontal position x and distance of fall y (Fig. P9.70). What numerical value does she obtain for v_{1A} based on her measured values of $x = 257$ cm and $y = 85.3$ cm? (c) What factors might account for the difference in this value compared with that obtained in part (a)?

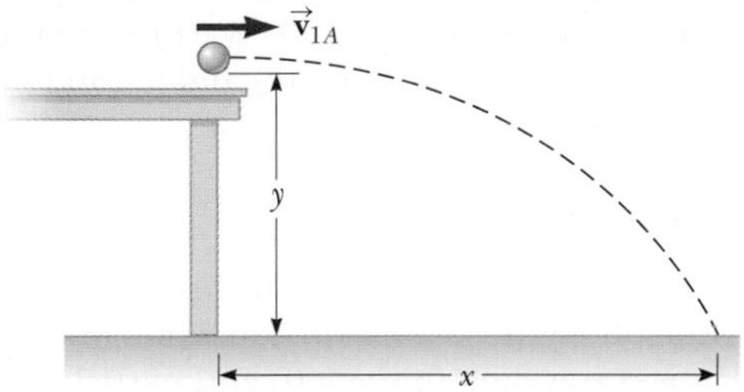

Figure P9.70

71. **Review.** A light spring of force constant 3.85 N/m is compressed by 8.00 cm and held between a 0.250-kg block on the left and a 0.500-kg block on the right. Both blocks are at rest on a horizontal surface. The blocks are released simultaneously so that the spring tends to push them apart. Find the maximum velocity each block attains if the coefficient of kinetic friction between each block and the surface is (a) 0, (b) 0.100, and (c) 0.462. Assume the coefficient of static friction is greater than the coefficient of kinetic friction in every case.

72. Consider as a system the Sun with the Earth in a circular orbit around it. Find the magnitude of the change in the velocity of the Sun relative to the center of mass of the system over a six-month period. Ignore the influence of other celestial objects. You may obtain the necessary astronomical data from the endpapers of the book.

73. **M** A 5.00-g bullet moving with an initial speed of $v_i = 400$ m/s is fired into and passes through a 1.00-kg block as shown in Figure P9.73. The block, initially at rest on a frictionless, horizontal surface, is connected to a spring with force constant 900 N/m. The block moves $d = 5.00$ cm to the right after impact before being brought to rest by the spring. Find (a) the speed at which the bullet emerges from the block and (b) the amount of initial kinetic energy of the bullet that is converted into internal energy in the bullet–block system during the collision.

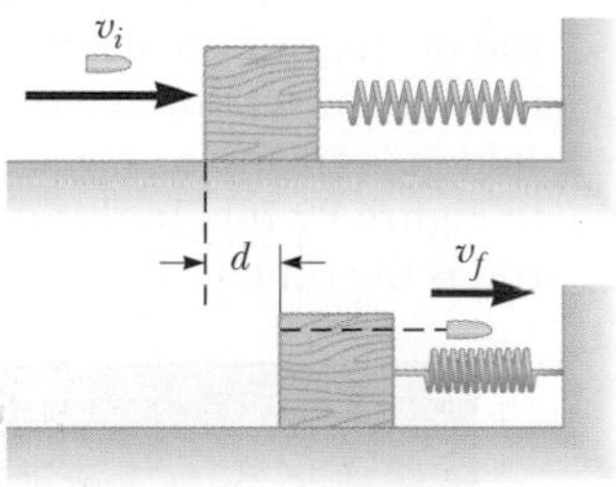

Figure P9.73

74. **QC** **GP** **Review.** There are (one can say) three coequal theories of motion for a single particle: Newton's second law, stating that the total force on the particle causes its acceleration; the work–kinetic energy theorem, stating that the total work on the particle causes its change in kinetic

energy; and the impulse–momentum theorem, stating that the total impulse on the particle causes its change in momentum. In this problem, you compare predictions of the three theories in one particular case. A 3.00-kg object has velocity $7.00\hat{\mathbf{j}}$ m/s. Then, a constant net force $12.0\hat{\mathbf{i}}$ N acts on the object for 5.00 s. (a) Calculate the object's final velocity, using the impulse–momentum theorem. (b) Calculate its acceleration from $\vec{\mathbf{a}} = (\vec{\mathbf{v}}_f - \vec{\mathbf{v}}_i)/\Delta t$. (c) Calculate its acceleration from $\vec{\mathbf{a}} = \sum \vec{\mathbf{F}}/m$. (d) Find the object's vector displacement from $\Delta\mathbf{r} = \vec{\mathbf{v}}_i t + \frac{1}{2}\vec{\mathbf{a}}t^2$. (e) Find the work done on the object from $W = \vec{\mathbf{F}} \cdot \Delta\vec{\mathbf{r}}$. (f) Find the final kinetic energy from $\frac{1}{2}mv_f^2 = \frac{1}{2}m\vec{\mathbf{v}}_f \cdot \vec{\mathbf{v}}_f$. (g) Find the final kinetic energy from $\frac{1}{2}mv_i^2 + W$. (h) State the result of comparing the answers to parts (b) and (c), and the answers to parts (f) and (g).

Challenge Problems

75. Two particles with masses m and $3m$ are moving toward each other along the x axis with the same initial speeds v_i. Particle m is traveling to the left, and particle $3m$ is traveling to the right. They undergo an elastic glancing collision such that particle m is moving in the negative y direction after the collision at a right angle from its initial direction. (a) Find the final speeds of the two particles in terms of v_i. (b) What is the angle θ at which the particle $3m$ is scattered?

76. In the 1968 Olympic games, University of Oregon jumper Dick Fosbury introduced a new technique of high jumping called the "Fosbury flop." It contributed to raising the world record by about 30 cm and is currently used by nearly every world-class jumper. In this technique, the jumper goes over the bar face up while arching her back as much as possible as shown in Figure P9.76a. This action places her center of mass outside her body, below her back. As her body goes over the bar, her center of mass passes below the bar. Because a given energy input implies a certain elevation for her center of mass, the action of arching her back means that her body is higher than if her back were straight. As a model, consider the jumper as a thin uniform rod of length L. When the rod is straight, its center of mass is at its center. Now bend the rod in a circular arc so that it subtends an angle of 90.0° at the center of the arc as shown in Figure P9.76b. In this configuration, how far outside the rod is the center of mass?

Figure P9.76

77. S On a horizontal air track, a glider of mass m carries a Γ-shaped post. The post supports a small dense sphere, also of mass m, hanging just above the top of the glider on a cord of length L. The glider and sphere are initially at rest with the cord vertical. (Figure P9.47a shows a cart and a sphere similarly connected.) A constant horizontal force of magnitude F is applied to the glider, moving it through displacement x_1; then the force is removed. During the time interval when the force is applied, the sphere moves through a displacement with horizontal component x_2. (a) Find the horizontal component of the velocity of the center of mass of the glider–sphere system when the force is removed. (b) After the force is removed, the glider continues to move on the track and the sphere swings back and forth, both without friction. Find an expression for the largest angle the cord makes with the vertical.

78. QC Sand from a stationary hopper falls onto a moving conveyor belt at the rate of 5.00 kg/s as shown in Figure P9.78. The conveyor belt is supported by frictionless rollers and moves at a constant speed of $v = 0.750$ m/s under the action of a constant horizontal external force $\vec{\mathbf{F}}_{ext}$ supplied by the motor that drives the belt. Find (a) the sand's rate of change of momentum in the horizontal direction, (b) the force of friction exerted by the belt on the sand, (c) the external force $\vec{\mathbf{F}}_{ext}$, (d) the work done by $\vec{\mathbf{F}}_{ext}$ in 1 s, and (e) the kinetic energy acquired by the falling sand each second due to the change in its horizontal motion. (f) Why are the answers to parts (d) and (e) different?

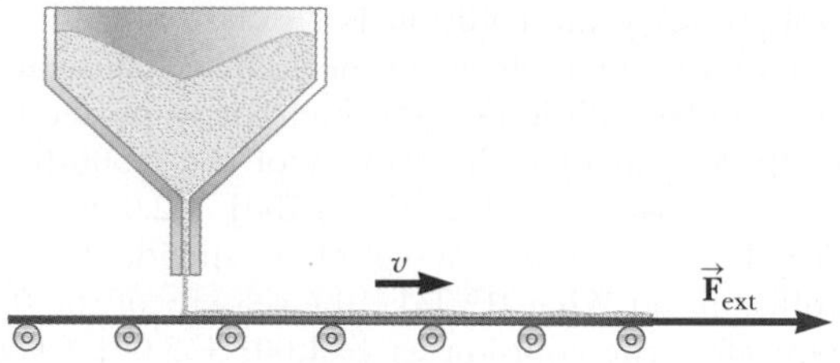

Figure P9.78

79. S **Review.** A chain of length L and total mass M is released from rest with its lower end just touching the top of a table as shown in Figure P9.79a. Find the force exerted by the table on the chain after the chain has fallen through a distance x as shown in Figure P9.79b. (Assume each link comes to rest the instant it reaches the table.)

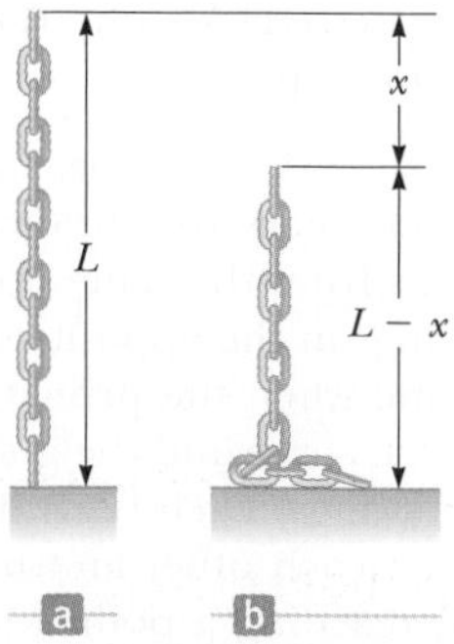

Figure P9.79

chapter

10

Rotation of a Rigid Object About a Fixed Axis

The Malaysian pastime of *gasing* involves the spinning of tops that can have masses up to 5 kg. Professional spinners can spin their tops so that they might rotate for more than an hour before stopping. We will study the rotational motion of objects such as these tops in this chapter. *(Courtesy Tourism Malaysia)*

When an extended object such as a wheel rotates about its axis, the motion cannot be analyzed by modeling the object as a particle because at any given time different parts of the object have different linear velocities and linear accelerations. We can, however, analyze the motion of an extended object by modeling it as a system of many particles, each of which has its own linear velocity and linear acceleration as discussed in Section 9.7.

In dealing with a rotating object, analysis is greatly simplified by assuming the object is rigid. A **rigid object** is one that is nondeformable; that is, the relative locations of all particles of which the object is composed remain constant. All real objects are deformable to some extent; our rigid-object model, however, is useful in many situations in which deformation is negligible. We have developed analysis models based on particles and systems. In this chapter, we introduce another class of analysis models based on the rigid-object model.

10.1 Angular Position, Velocity, and Acceleration

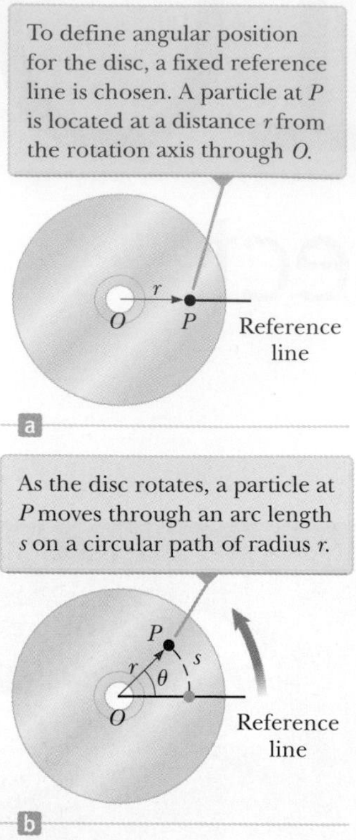

Figure 10.1 A compact disc rotating about a fixed axis through O perpendicular to the plane of the figure.

Figure 10.1 illustrates an overhead view of a rotating compact disc, or CD. The disc rotates about a fixed axis perpendicular to the plane of the figure and passing through the center of the disc at O. A small element of the disc modeled as a particle at P is at a fixed distance r from the origin and rotates about it in a circle of radius r. (In fact, *every* element of the disc undergoes circular motion about O.) It is convenient to represent the position of P with its polar coordinates (r, θ), where r is the distance from the origin to P and θ is measured *counterclockwise* from some reference line fixed in space as shown in Figure 10.1a. In this representation, the angle θ changes in time while r remains constant. As the particle moves along the circle from the reference line, which is at angle $\theta = 0$, it moves through an arc of length s as in Figure 10.1b. The arc length s is related to the angle θ through the relationship

$$s = r\theta \tag{10.1a}$$

$$\theta = \frac{s}{r} \tag{10.1b}$$

Because θ is the ratio of an arc length and the radius of the circle, it is a pure number. Usually, however, we give θ the artificial unit **radian** (rad), where one radian is the angle subtended by an arc length equal to the radius of the arc. Because the circumference of a circle is $2\pi r$, it follows from Equation 10.1b that 360° corresponds to an angle of $(2\pi r/r)$ rad $= 2\pi$ rad. Hence, 1 rad $= 360°/2\pi \approx 57.3°$. To convert an angle in degrees to an angle in radians, we use that π rad $= 180°$, so

$$\theta(\text{rad}) = \frac{\pi}{180°}\theta(\text{deg})$$

For example, 60° equals $\pi/3$ rad and 45° equals $\pi/4$ rad.

Pitfall Prevention 10.1

Remember the Radian

In rotational equations, you *must* use angles expressed in radians. Don't fall into the trap of using angles measured in degrees in rotational equations.

Because the disc in Figure 10.1 is a rigid object, as the particle moves through an angle θ from the reference line, every other particle on the object rotates through the same angle θ. Therefore, we can associate the angle θ with the entire rigid object as well as with an individual particle, which allows us to define the *angular position* of a rigid object in its rotational motion. We choose a reference line on the object, such as a line connecting O and a chosen particle on the object. The **angular position** of the rigid object is the angle θ between this reference line on the object and the fixed reference line in space, which is often chosen as the x axis. Such identification is similar to the way we define the position of an object in translational motion as the distance x between the object and the reference position, which is the origin, $x = 0$. Therefore, the angle θ plays the same role in rotational motion that the position x does in translational motion.

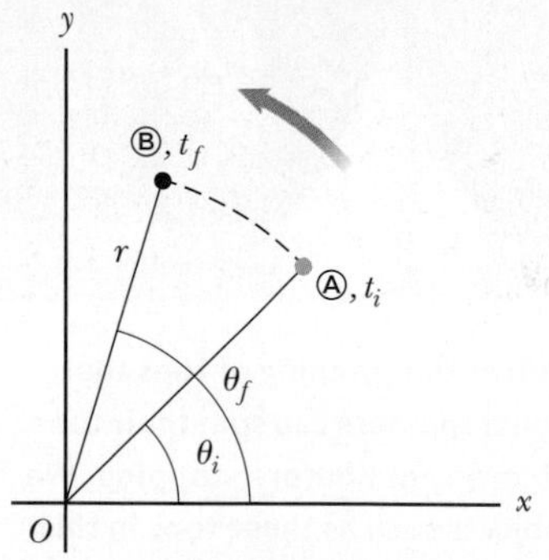

Figure 10.2 A particle on a rotating rigid object moves from Ⓐ to Ⓑ along the arc of a circle. In the time interval $\Delta t = t_f - t_i$, the radial line of length r moves through an angular displacement $\Delta\theta = \theta_f - \theta_i$.

As the particle in question on our rigid object travels from position Ⓐ to position Ⓑ in a time interval Δt as in Figure 10.2, the reference line fixed to the object sweeps out an angle $\Delta\theta = \theta_f - \theta_i$. This quantity $\Delta\theta$ is defined as the **angular displacement** of the rigid object:

$$\Delta\theta \equiv \theta_f - \theta_i$$

The rate at which this angular displacement occurs can vary. If the rigid object spins rapidly, this displacement can occur in a short time interval. If it rotates slowly, this displacement occurs in a longer time interval. These different rotation rates can be quantified by defining the **average angular speed** ω_{avg} (Greek letter omega) as the ratio of the angular displacement of a rigid object to the time interval Δt during which the displacement occurs:

Average angular speed ▶

$$\omega_{avg} \equiv \frac{\theta_f - \theta_i}{t_f - t_i} = \frac{\Delta\theta}{\Delta t} \tag{10.2}$$

In analogy to translational speed, the **instantaneous angular speed** ω is defined as the limit of the average angular speed as Δt approaches zero:

$$\omega \equiv \lim_{\Delta t \to 0} \frac{\Delta \theta}{\Delta t} = \frac{d\theta}{dt} \quad (10.3)$$

◀ **Instantaneous angular speed**

Angular speed has units of radians per second (rad/s), which can be written as s^{-1} because radians are not dimensional. We take ω to be positive when θ is increasing (counterclockwise motion in Fig. 10.2) and negative when θ is decreasing (clockwise motion in Fig. 10.2).

***Quick Quiz* 10.1** A rigid object rotates in a counterclockwise sense around a fixed axis. Each of the following pairs of quantities represents an initial angular position and a final angular position of the rigid object. **(i)** Which of the sets can *only* occur if the rigid object rotates through more than 180°? (a) 3 rad, 6 rad (b) −1 rad, 1 rad (c) 1 rad, 5 rad **(ii)** Suppose the change in angular position for each of these pairs of values occurs in 1 s. Which choice represents the lowest average angular speed?

If the instantaneous angular speed of an object changes from ω_i to ω_f in the time interval Δt, the object has an angular acceleration. The **average angular acceleration** α_{avg} (Greek letter alpha) of a rotating rigid object is defined as the ratio of the change in the angular speed to the time interval Δt during which the change in the angular speed occurs:

$$\alpha_{avg} \equiv \frac{\omega_f - \omega_i}{t_f - t_i} = \frac{\Delta \omega}{\Delta t} \quad (10.4)$$

◀ **Average angular acceleration**

In analogy to translational acceleration, the **instantaneous angular acceleration** is defined as the limit of the average angular acceleration as Δt approaches zero:

$$\alpha = \lim_{\Delta t \to 0} \frac{\Delta \omega}{\Delta t} = \frac{d\omega}{dt} \quad (10.5)$$

◀ **Instantaneous angular acceleration**

Angular acceleration has units of radians per second squared (rad/s^2), or simply s^{-2}. Notice that α is positive when a rigid object rotating counterclockwise is speeding up or when a rigid object rotating clockwise is slowing down during some time interval.

When a rigid object is rotating about a *fixed* axis, every particle on the object rotates through the same angle in a given time interval and has the same angular speed and the same angular acceleration. Therefore, like the angular position θ, the quantities ω and α characterize the rotational motion of the entire rigid object as well as individual particles in the object.

Angular position (θ), angular speed (ω), and angular acceleration (α) are analogous to translational position (x), translational speed (v), and translational acceleration (a). The variables θ, ω, and α differ dimensionally from the variables x, v, and a only by a factor having the unit of length. (See Section 10.3.)

We have not specified any direction for angular speed and angular acceleration. Strictly speaking, ω and α are the magnitudes of the angular velocity and the angular acceleration vectors[1] $\vec{\omega}$ and $\vec{\alpha}$, respectively, and they should always be positive. Because we are considering rotation about a fixed axis, however, we can use nonvector notation and indicate the vectors' directions by assigning a positive or negative sign to ω and α as discussed earlier with regard to Equations 10.3 and 10.5. For rotation about a fixed axis, the only direction that uniquely specifies the rotational motion is the direction along the axis of rotation. Therefore, the directions of $\vec{\omega}$ and $\vec{\alpha}$ are along this axis. If a particle rotates in the xy plane as in Figure 10.2, the

Pitfall Prevention 10.2

Specify Your Axis

In solving rotation problems, you must specify an axis of rotation. This new feature does not exist in our study of translational motion. The choice is arbitrary, but once you make it, you must maintain that choice consistently throughout the problem. In some problems, the physical situation suggests a natural axis, such as the center of an automobile wheel. In other problems, there may not be an obvious choice, and you must exercise judgment.

[1]Although we do not verify it here, the instantaneous angular velocity and instantaneous angular acceleration are vector quantities, but the corresponding average values are not because angular displacements do not add as vector quantities for finite rotations.

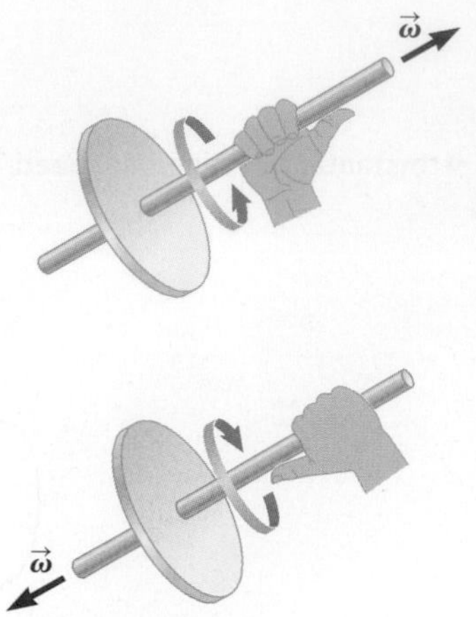

Figure 10.3 The right-hand rule for determining the direction of the angular velocity vector.

direction of $\vec{\omega}$ for the particle is out of the plane of the diagram when the rotation is counterclockwise and into the plane of the diagram when the rotation is clockwise. To illustrate this convention, it is convenient to use the *right-hand rule* demonstrated in Figure 10.3. When the four fingers of the right hand are wrapped in the direction of rotation, the extended right thumb points in the direction of $\vec{\omega}$. The direction of $\vec{\alpha}$ follows from its definition $\vec{\alpha} \equiv d\vec{\omega}/dt$. It is in the same direction as $\vec{\omega}$ if the angular speed is increasing in time, and it is antiparallel to $\vec{\omega}$ if the angular speed is decreasing in time.

10.2 Analysis Model: Rigid Object Under Constant Angular Acceleration

Imagine that a rigid object rotates about a fixed axis and that it has a constant angular acceleration. In this case, we generate a new analysis model for rotational motion called the **rigid object under constant angular acceleration.** This model is the rotational analog to the particle under constant acceleration model. We develop kinematic relationships for this model in this section. Writing Equation 10.5 in the form $d\omega = \alpha\, dt$ and integrating from $t_i = 0$ to $t_f = t$ gives

Rotational kinematic equations ▶

$$\omega_f = \omega_i + \alpha t \quad \text{(for constant } \alpha) \tag{10.6}$$

where ω_i is the angular speed of the rigid object at time $t = 0$. Equation 10.6 allows us to find the angular speed ω_f of the object at any later time t. Substituting Equation 10.6 into Equation 10.3 and integrating once more, we obtain

$$\theta_f = \theta_i + \omega_i t + \tfrac{1}{2}\alpha t^2 \quad \text{(for constant } \alpha) \tag{10.7}$$

where θ_i is the angular position of the rigid object at time $t = 0$. Equation 10.7 allows us to find the angular position θ_f of the object at any later time t. Eliminating t from Equations 10.6 and 10.7 gives

$$\omega_f^2 = \omega_i^2 + 2\alpha(\theta_f - \theta_i) \quad \text{(for constant } \alpha) \tag{10.8}$$

This equation allows us to find the angular speed ω_f of the rigid object for any value of its angular position θ_f. If we eliminate α between Equations 10.6 and 10.7, we obtain

$$\theta_f = \theta_i + \tfrac{1}{2}(\omega_i + \omega_f)t \quad \text{(for constant } \alpha) \tag{10.9}$$

Notice that these kinematic expressions for the rigid object under constant angular acceleration are of the same mathematical form as those for a particle under constant acceleration (Chapter 2). They can be generated from the equations for translational motion by making the substitutions $x \to \theta$, $v \to \omega$, and $a \to \alpha$. Table 10.1 compares the kinematic equations for rotational and translational motion.

Pitfall Prevention 10.3

Just Like Translation?

Equations 10.6 to 10.9 and Table 10.1 might suggest that rotational kinematics is just like translational kinematics. That is almost true, with two key differences. (1) In rotational kinematics, you must specify a rotation axis (per Pitfall Prevention 10.2). (2) In rotational motion, the object keeps returning to its original orientation; therefore, you may be asked for the number of revolutions made by a rigid object. This concept has no analog in translational motion.

***Quick Quiz* 10.2** consider again the pairs of angular positions for the rigid object in Quick Quiz 10.1. If the object starts from rest at the initial angular position, moves counterclockwise with constant angular acceleration, and arrives at the final angular position with the same angular speed in all three cases, for which choice is the angular acceleration the highest?

TABLE 10.1 *Kinematic Equations for Rotational and Translational Motion*

Rigid Body Under Constant Angular Acceleration	Particle Under Constant Acceleration
$\omega_f = \omega_i + \alpha t$	$v_f = v_i + at$
$\theta_f = \theta_i + \omega_i t + \frac{1}{2}\alpha t^2$	$x_f = x_i + v_i t + \frac{1}{2}at^2$
$\omega_f^2 = \omega_i^2 + 2\alpha(\theta_f - \theta_i)$	$v_f^2 = v_i^2 + 2a(x_f - x_i)$
$\theta_f = \theta_i + \frac{1}{2}(\omega_i + \omega_f)t$	$x_f = x_i + \frac{1}{2}(v_i + v_f)t$

Example 10.1 Rotating Wheel

A wheel rotates with a constant angular acceleration of 3.50 rad/s^2.

(A) If the angular speed of the wheel is 2.00 rad/s at $t_i = 0$, through what angular displacement does the wheel rotate in 2.00 s?

SOLUTION

Conceptualize Look again at Figure 10.1. Imagine that the compact disc rotates with its angular speed increasing at a constant rate. You start your stopwatch when the disc is rotating at 2.00 rad/s. This mental image is a model for the motion of the wheel in this example.

Categorize The phrase "with a constant angular acceleration" tells us to use the rigid object under constant angular acceleration model.

Analyze Arrange Equation 10.7 so that it expresses the angular displacement of the object:

$$\Delta\theta = \theta_f - \theta_i = \omega_i t + \tfrac{1}{2}\alpha t^2$$

Substitute the known values to find the angular displacement at $t = 2.00$ s:

$$\Delta\theta = (2.00 \text{ rad/s})(2.00 \text{ s}) + \tfrac{1}{2}(3.50 \text{ rad/s}^2)(2.00 \text{ s})^2$$
$$= \boxed{11.0 \text{ rad}} = (11.0 \text{ rad})(180°/\pi \text{ rad}) = \boxed{630°}$$

(B) Through how many revolutions has the wheel turned during this time interval?

SOLUTION

Multiply the angular displacement found in part (A) by a conversion factor to find the number of revolutions:

$$\Delta\theta = 630°\left(\frac{1 \text{ rev}}{360°}\right) = \boxed{1.75 \text{ rev}}$$

(C) What is the angular speed of the wheel at $t = 2.00$ s?

SOLUTION

Use Equation 10.6 to find the angular speed at $t = 2.00$ s:

$$\omega_f = \omega_i + \alpha t = 2.00 \text{ rad/s} + (3.50 \text{ rad/s}^2)(2.00 \text{ s})$$
$$= \boxed{9.00 \text{ rad/s}}$$

Finalize We could also obtain this result using Equation 10.8 and the results of part (A). (Try it!)

WHAT IF? Suppose a particle moves along a straight line with a constant acceleration of 3.50 m/s^2. If the velocity of the particle is 2.00 m/s at $t_i = 0$, through what displacement does the particle move in 2.00 s? What is the velocity of the particle at $t = 2.00$ s?

Answer Notice that these questions are translational analogs to parts (A) and (C) of the original problem. The mathematical solution follows exactly the same form. For the displacement,

$$\Delta x = x_f - x_i = v_i t + \tfrac{1}{2}at^2$$
$$= (2.00 \text{ m/s})(2.00 \text{ s}) + \tfrac{1}{2}(3.50 \text{ m/s}^2)(2.00 \text{ s})^2 = 11.0 \text{ m}$$

and for the velocity,

$$v_f = v_i + at = 2.00 \text{ m/s} + (3.50 \text{ m/s}^2)(2.00 \text{ s}) = 9.00 \text{ m/s}$$

There is no translational analog to part (B) because translational motion under constant acceleration is not repetitive.

10.3 Angular and Translational Quantities

In this section, we derive some useful relationships between the angular speed and acceleration of a rotating rigid object and the translational speed and acceleration of a point in the object. To do so, we must keep in mind that when a rigid object rotates about a fixed axis as in Active Figure 10.4 (page 282), every particle of the object moves in a circle whose center is on the axis of rotation.

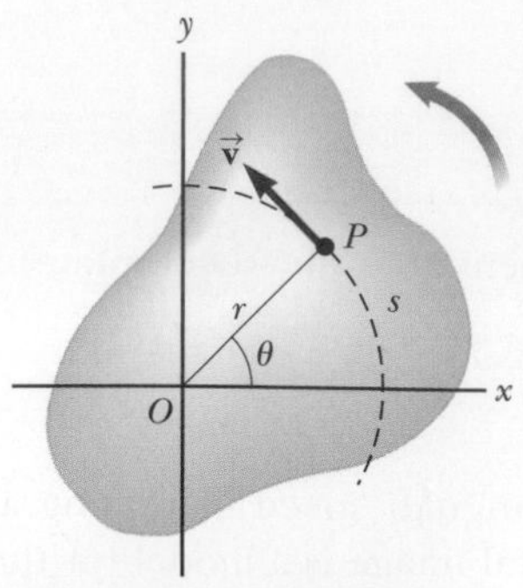

ACTIVE FIGURE 10.4

As a rigid object rotates about the fixed axis (the z axis) through O, the point P has a tangential velocity $\vec{v}$ that is always tangent to the circular path of radius r.

Because point P in Active Figure 10.4 moves in a circle, the translational velocity vector $\vec{v}$ is always tangent to the circular path and hence is called *tangential velocity*. The magnitude of the tangential velocity of the point P is by definition the tangential speed $v = ds/dt$, where s is the distance traveled by this point measured along the circular path. Recalling that $s = r\theta$ (Eq. 10.1a) and noting that r is constant, we obtain

$$v = \frac{ds}{dt} = r\frac{d\theta}{dt}$$

Because $d\theta/dt = \omega$ (see Eq. 10.3), it follows that

$$v = r\omega \tag{10.10}$$

That is, the tangential speed of a point on a rotating rigid object equals the perpendicular distance of that point from the axis of rotation multiplied by the angular speed. Therefore, although every point on the rigid object has the same *angular* speed, not every point has the same *tangential* speed because r is not the same for all points on the object. Equation 10.10 shows that the tangential speed of a point on the rotating object increases as one moves outward from the center of rotation, as we would intuitively expect. For example, the outer end of a swinging golf club moves much faster than the handle.

We can relate the angular acceleration of the rotating rigid object to the tangential acceleration of the point P by taking the time derivative of v:

$$a_t = \frac{dv}{dt} = r\frac{d\omega}{dt}$$

Relation between tangential acceleration and angular acceleration ▶

$$a_t = r\alpha \tag{10.11}$$

That is, the tangential component of the translational acceleration of a point on a rotating rigid object equals the point's perpendicular distance from the axis of rotation multiplied by the angular acceleration.

In Section 4.4, we found that a point moving in a circular path undergoes a radial acceleration a_r directed toward the center of rotation and whose magnitude is that of the centripetal acceleration v^2/r (Fig. 10.5). Because $v = r\omega$ for a point P on a rotating object, we can express the centripetal acceleration at that point in terms of angular speed as

$$a_c = \frac{v^2}{r} = r\omega^2 \tag{10.12}$$

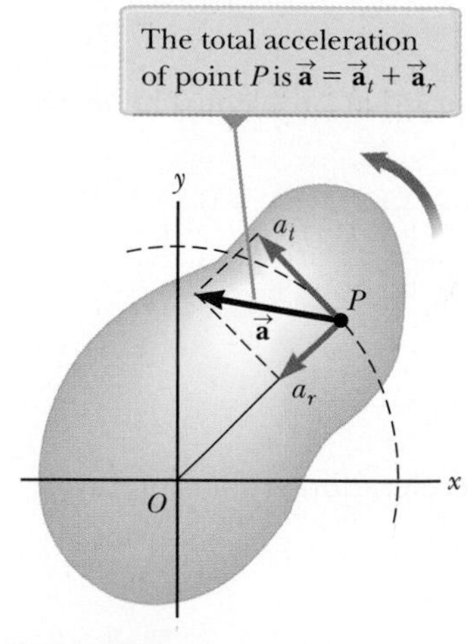

Figure 10.5 As a rigid object rotates about a fixed axis (the z axis) through O, the point P experiences a tangential component of translational acceleration a_t and a radial component of translational acceleration a_r.

The total acceleration vector at the point is $\vec{a} = \vec{a}_t + \vec{a}_r$, where the magnitude of $\vec{a}_r$ is the centripetal acceleration a_c. Because $\vec{a}$ is a vector having a radial and a tangential component, the magnitude of $\vec{a}$ at the point P on the rotating rigid object is

$$a = \sqrt{a_t^2 + a_r^2} = \sqrt{r^2\alpha^2 + r^2\omega^4} = r\sqrt{\alpha^2 + \omega^4} \tag{10.13}$$

Quick Quiz **10.3** Ethan and Joseph are riding on a merry-go-round. Ethan rides on a horse at the outer rim of the circular platform, twice as far from the center of the circular platform as Joseph, who rides on an inner horse. **(i)** When the merry-go-round is rotating at a constant angular speed, what is Ethan's angular speed? (a) twice Joseph's (b) the same as Joseph's (c) half of Joseph's (d) impossible to determine **(ii)** When the merry-go-round is rotating at a constant angular speed, describe Ethan's tangential speed from the same list of choices.

Example 10.2 CD Player

On a compact disc (Fig. 10.6), audio information is stored digitally in a series of pits and flat areas on the surface of the disc. The alternations between pits and flat areas on the surface represent binary ones and zeros to be read by the CD player and converted back to sound waves. The pits and flat areas are detected by a system consisting of a laser and lenses. The length of a string of ones and zeros representing one piece of information is the same everywhere on the disc, whether the information is near the center of the disc or near its outer edge. So that this length of ones and zeros always passes by the laser–lens system in the same time interval, the tangential speed of the disc surface at the location of the lens must be constant. According to Equation 10.10, the angular speed must therefore vary as the laser–lens system moves radially along the disc. In a typical CD player, the constant speed of the surface at the point of the laser–lens system is 1.3 m/s.

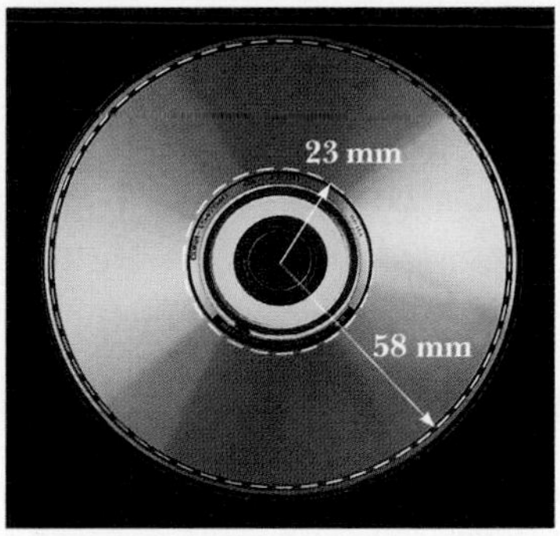

Figure 10.6 (Example 10.2) A compact disc.

(A) Find the angular speed of the disc in revolutions per minute when information is being read from the innermost first track ($r = 23$ mm) and the outermost final track ($r = 58$ mm).

SOLUTION

Conceptualize Figure 10.6 shows a photograph of a compact disc. Trace your finger around the circle marked "23 mm" and mentally estimate the time interval to go around the circle once. Now trace your finger around the circle marked "58 mm," moving your finger across the surface of the page at the same speed as you did when tracing the smaller circle. Notice how much longer in time it takes your finger to go around the larger circle. If your finger represents the laser reading the disc, you can see that the disc rotates once in a longer time interval when the laser reads the information in the outer circle. Therefore, the disc must rotate more slowly when the laser is reading information from this part of the disc.

Categorize This part of the example is categorized as a simple substitution problem. In later parts, we will need to identify analysis models.

Use Equation 10.10 to find the angular speed that gives the required tangential speed at the position of the inner track:

$$\omega_i = \frac{v}{r_i} = \frac{1.3 \text{ m/s}}{2.3 \times 10^{-2} \text{ m}} = 57 \text{ rad/s}$$

$$= (57 \text{ rad/s})\left(\frac{1 \text{ rev}}{2\pi \text{ rad}}\right)\left(\frac{60 \text{ s}}{1 \text{ min}}\right) = \boxed{5.4 \times 10^2 \text{ rev/min}}$$

Do the same for the outer track:

$$\omega_f = \frac{v}{r_f} = \frac{1.3 \text{ m/s}}{5.8 \times 10^{-2} \text{ m}} = 22 \text{ rad/s} = \boxed{2.1 \times 10^2 \text{ rev/min}}$$

The CD player adjusts the angular speed ω of the disc within this range so that information moves past the objective lens at a constant rate.

(B) The maximum playing time of a standard music disc is 74 min and 33 s. How many revolutions does the disc make during that time?

SOLUTION

Categorize From part (A), the angular speed decreases as the disc plays. Let us assume it decreases steadily, with α constant. We can then apply the rigid object under constant angular acceleration model to the disc.

Analyze If $t = 0$ is the instant the disc begins rotating, with angular speed of 57 rad/s, the final value of the time t is (74 min)(60 s/min) + 33 s = 4 473 s. We are looking for the angular displacement $\Delta\theta$ during this time interval.

Use Equation 10.9 to find the angular displacement of the disc at $t = 4\,473$ s:

$$\Delta\theta = \theta_f - \theta_i = \tfrac{1}{2}(\omega_i + \omega_f)t$$

$$= \tfrac{1}{2}(57 \text{ rad/s} + 22 \text{ rad/s})(4\,473 \text{ s}) = 1.8 \times 10^5 \text{ rad}$$

Convert this angular displacement to revolutions:

$$\Delta\theta = (1.8 \times 10^5 \text{ rad})\left(\frac{1 \text{ rev}}{2\pi \text{ rad}}\right) = \boxed{2.8 \times 10^4 \text{ rev}}$$

continued

10.2 *cont.*

(C) What is the angular acceleration of the compact disc over the 4 473-s time interval?

SOLUTION

Categorize We again model the disc as a rigid object under constant angular acceleration. In this case, Equation 10.6 gives the value of the constant angular acceleration. Another approach is to use Equation 10.4 to find the average angular acceleration. In this case, we are not assuming the angular acceleration is constant. The answer is the same from both equations; only the interpretation of the result is different.

Analyze Use Equation 10.6 to find the angular acceleration:

$$\alpha = \frac{\omega_f - \omega_i}{t} = \frac{22\text{ rad/s} - 57\text{ rad/s}}{4\,473\text{ s}} = -7.6 \times 10^{-3}\text{ rad/s}^2$$

Finalize The disc experiences a very gradual decrease in its rotation rate, as expected from the long time interval required for the angular speed to change from the initial value to the final value. In reality, the angular acceleration of the disc is not constant. Problem 82 allows you to explore the actual time behavior of the angular acceleration.

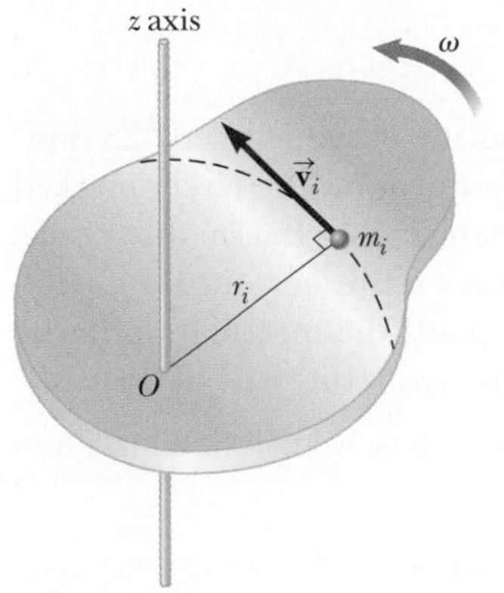

Figure 10.7 A rigid object rotating about the z axis with angular speed ω. The kinetic energy of the particle of mass m_i is $\frac{1}{2}m_iv_i^2$. The total kinetic energy of the object is called its rotational kinetic energy.

10.4 Rotational Kinetic Energy

In Chapter 7, we defined the kinetic energy of an object as the energy associated with its motion through space. An object rotating about a fixed axis remains stationary in space, so there is no kinetic energy associated with translational motion. The individual particles making up the rotating object, however, are moving through space; they follow circular paths. Consequently, there is kinetic energy associated with rotational motion.

Let us consider an object as a system of particles and assume it rotates about a fixed z axis with an angular speed ω. Figure 10.7 shows the rotating object and identifies one particle on the object located at a distance r_i from the rotation axis. If the mass of the ith particle is m_i and its tangential speed is v_i, its kinetic energy is

$$K_i = \tfrac{1}{2}m_iv_i^2$$

To proceed further, recall that although every particle in the rigid object has the same angular speed ω, the individual tangential speeds depend on the distance r_i from the axis of rotation according to Equation 10.10. The *total* kinetic energy of the rotating rigid object is the sum of the kinetic energies of the individual particles:

$$K_R = \sum_i K_i = \sum_i \tfrac{1}{2}m_iv_i^2 = \tfrac{1}{2}\sum_i m_ir_i^2\omega^2$$

We can write this expression in the form

$$K_R = \tfrac{1}{2}\left(\sum_i m_ir_i^2\right)\omega^2 \quad \textbf{(10.14)}$$

where we have factored ω^2 from the sum because it is common to every particle. We simplify this expression by defining the quantity in parentheses as the **moment of inertia** $\boldsymbol{I}$ of the rigid object:

Moment of inertia ▶

$$I \equiv \sum_i m_ir_i^2 \quad \textbf{(10.15)}$$

From the definition of moment of inertia,[2] we see that it has dimensions of ML^2 ($kg \cdot m^2$ in SI units). With this notation, Equation 10.14 becomes

Rotational kinetic energy ▶

$$K_R = \tfrac{1}{2}I\omega^2 \quad \textbf{(10.16)}$$

[2]Civil engineers use moment of inertia to characterize the elastic properties (rigidity) of such structures as loaded beams. Hence, it is often useful even in a nonrotational context.

Although we commonly refer to the quantity $\frac{1}{2}I\omega^2$ as **rotational kinetic energy,** it is not a new form of energy. It is ordinary kinetic energy because it is derived from a sum over individual kinetic energies of the particles contained in the rigid object. The mathematical form of the kinetic energy given by Equation 10.16 is convenient when we are dealing with rotational motion, provided we know how to calculate I.

It is important to recognize the analogy between kinetic energy $\frac{1}{2}mv^2$ associated with translational motion and rotational kinetic energy $\frac{1}{2}I\omega^2$. The quantities I and ω in rotational motion are analogous to m and v in translational motion, respectively. (In fact, I takes the place of m and ω takes the place of v every time we compare a translational motion equation with its rotational counterpart.) Moment of inertia is a measure of the resistance of an object to changes in its rotational motion, just as mass is a measure of the tendency of an object to resist changes in its translational motion. For rotation, this resistance depends not only on the mass of the object, but also on how the mass is distributed around the rotation axis.

Pitfall Prevention 10.4

No Single Moment of Inertia There is one major difference between mass and moment of inertia. Mass is an inherent property of an object. The moment of inertia of an object depends on your choice of rotation axis. Therefore, there is no single value of the moment of inertia for an object. There is a *minimum* value of the moment of inertia, which is that calculated about an axis passing through the center of mass of the object.

Example 10.3 An Unusual Baton

Four tiny spheres are fastened to the ends of two rods of negligible mass lying in the xy plane to form an unusual baton (Fig. 10.8). We shall assume the radii of the spheres are small compared with the dimensions of the rods.

(A) If the system rotates about the y axis (Fig. 10.8a) with an angular speed ω, find the moment of inertia and the rotational kinetic energy of the system about this axis.

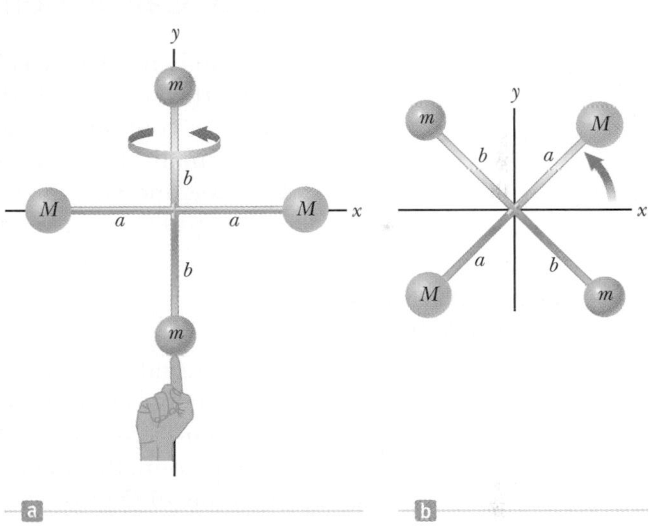

Figure 10.8 (Example 10.3) Four spheres form an unusual baton. (a) The baton is rotated about the y axis. (b) The baton is rotated about the z axis.

SOLUTION

Conceptualize Figure 10.8 is a pictorial representation that helps conceptualize the system of spheres and how it spins.

Categorize This example is a substitution problem because it is a straightforward application of the definitions discussed in this section.

Apply Equation 10.15 to the system:

$$I_y = \sum_i m_i r_i^2 = Ma^2 + Ma^2 = 2Ma^2$$

Evaluate the rotational kinetic energy using Equation 10.16:

$$K_R = \frac{1}{2}I_y\omega^2 = \frac{1}{2}(2Ma^2)\omega^2 = Ma^2\omega^2$$

That the two spheres of mass m do not enter into this result makes sense because they have no motion about the axis of rotation; hence, they have no rotational kinetic energy. By similar logic, we expect the moment of inertia about the x axis to be $I_x = 2mb^2$ with a rotational kinetic energy about that axis of $K_R = mb^2\omega^2$.

(B) Suppose the system rotates in the xy plane about an axis (the z axis) through the center of the baton (Fig. 10.8b). Calculate the moment of inertia and rotational kinetic energy about this axis.

SOLUTION

Apply Equation 10.15 for this new rotation axis:

$$I_z = \sum_i m_i r_i^2 = Ma^2 + Ma^2 + mb^2 + mb^2 = 2Ma^2 + 2mb^2$$

Evaluate the rotational kinetic energy using Equation 10.16:

$$K_R = \frac{1}{2}I_z\omega^2 = \frac{1}{2}(2Ma^2 + 2mb^2)\omega^2 = (Ma^2 + mb^2)\omega^2$$

continued

10.3 *cont.*

Comparing the results for parts (A) and (B), we conclude that the moment of inertia and therefore the rotational kinetic energy associated with a given angular speed depend on the axis of rotation. In part (B), we expect the result to include all four spheres and distances because all four spheres are rotating in the *xy* plane. Based on the work–kinetic energy theorem, the smaller rotational kinetic energy in part (A) than in part (B) indicates it would require less work to set the system into rotation about the *y* axis than about the *z* axis.

WHAT IF? What if the mass M is much larger than m? How do the answers to parts (A) and (B) compare?

Answer If $M >> m$, then m can be neglected and the moment of inertia and the rotational kinetic energy in part (B) become

$$I_z = 2Ma^2 \quad \text{and} \quad K_R = Ma^2\omega^2$$

which are the same as the answers in part (A). If the masses m of the two tan spheres in Figure 10.8 are negligible, these spheres can be removed from the figure and rotations about the *y* and *z* axes are equivalent.

10.5 Calculation of Moments of Inertia

The moment of inertia of a system of discrete particles can be calculated in a straightforward way with Equation 10.15. We can evaluate the moment of inertia of a continuous rigid object by imagining the object to be divided into many small elements, each of which has mass Δm_i. We use the definition $I = \sum_i r_i^2 \, \Delta m_i$ and take the limit of this sum as $\Delta m_i \rightarrow 0$. In this limit, the sum becomes an integral over the volume of the object:

Moment of inertia of a rigid object ▶

$$I = \lim_{\Delta m_i \rightarrow 0} \sum_i r_i^2 \, \Delta m_i = \int r^2 \, dm \qquad \textbf{(10.17)}$$

It is usually easier to calculate moments of inertia in terms of the volume of the elements rather than their mass, and we can easily make that change by using Equation 1.1, $\rho \equiv m/V$, where ρ is the density of the object and V is its volume. From this equation, the mass of a small element is $dm = \rho \, dV$. Substituting this result into Equation 10.17 gives

$$I = \int \rho r^2 \, dV$$

If the object is homogeneous, ρ is constant and the integral can be evaluated for a known geometry. If ρ is not constant, its variation with position must be known to complete the integration.

The density given by $\rho = m/V$ sometimes is referred to as *volumetric mass density* because it represents mass per unit volume. Often we use other ways of expressing density. For instance, when dealing with a sheet of uniform thickness t, we can define a *surface mass density* $\sigma = \rho t$, which represents *mass per unit area*. Finally, when mass is distributed along a rod of uniform cross-sectional area A, we sometimes use *linear mass density* $\lambda = M/L = \rho A$, which is the *mass per unit length*.

Table 10.2 gives the moments of inertia for a number of objects about specific axes. The moments of inertia of rigid objects with simple geometry (high symmetry) are relatively easy to calculate provided the rotation axis coincides with an axis of symmetry, as in the examples below.

***Quick Quiz* 10.4** A section of hollow pipe and a solid cylinder have the same radius, mass, and length. They both rotate about their long central axes with the same angular speed. Which object has the higher rotational kinetic energy? **(a)** The hollow pipe does. **(b)** The solid cylinder does. **(c)** They have the same rotational kinetic energy. **(d)** It is impossible to determine.

TABLE 10.2 *Moments of Inertia of Homogeneous Rigid Objects with Different Geometries*

Hoop or thin cylindrical shell
$I_{CM} = MR^2$

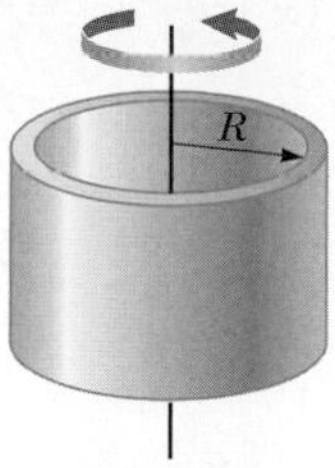

Hollow cylinder
$I_{CM} = \frac{1}{2}M(R_1^2 + R_2^2)$

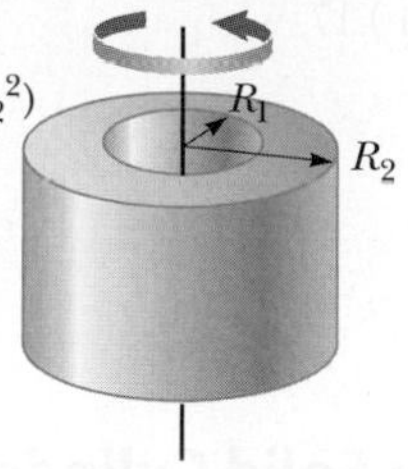

Solid cylinder or disk
$I_{CM} = \frac{1}{2}MR^2$

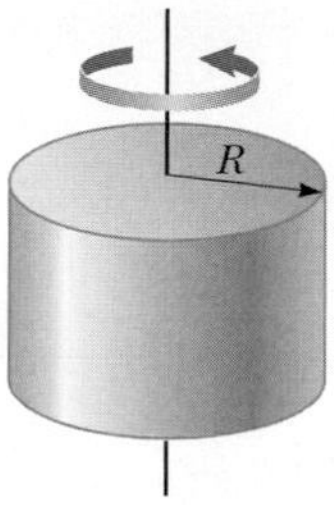

Rectangular plate
$I_{CM} = \frac{1}{12}M(a^2 + b^2)$

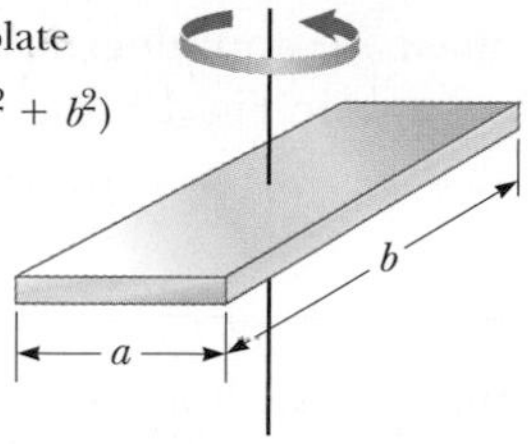

Long, thin rod with rotation axis through center
$I_{CM} = \frac{1}{12}ML^2$

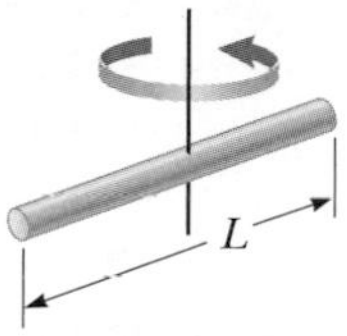

Long, thin rod with rotation axis through end
$I = \frac{1}{3}ML^2$

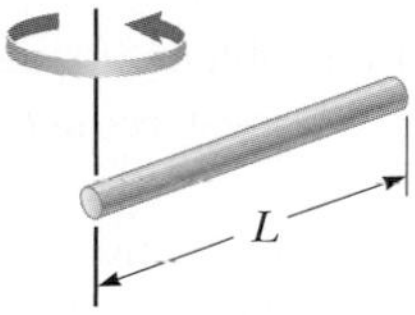

Solid sphere
$I_{CM} = \frac{2}{5}MR^2$

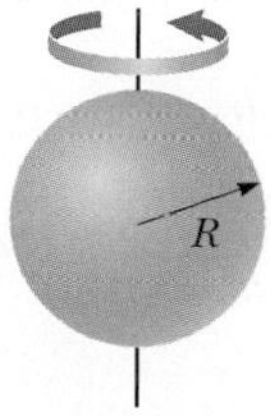

Thin spherical shell
$I_{CM} = \frac{2}{3}MR^2$

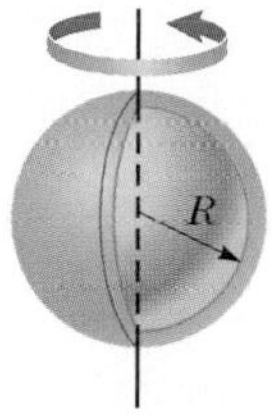

Example 10.4 Uniform Rigid Rod

Calculate the moment of inertia of a uniform rigid rod of length L and mass M (Fig. 10.9) about an axis perpendicular to the rod (the y' axis) and passing through its center of mass.

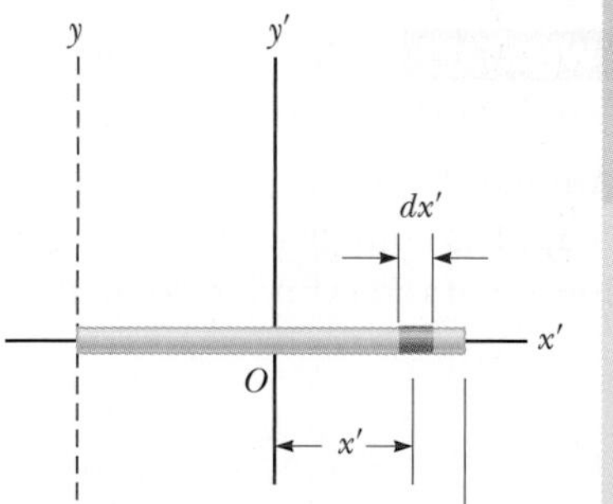

Figure 10.9 (Example 10.4) A uniform rigid rod of length L. The moment of inertia about the y' axis is less than that about the y axis. The latter axis is examined in Example 10.6.

SOLUTION

Conceptualize Imagine twirling the rod in Figure 10.9 with your fingers around its midpoint. If you have a meterstick handy, use it to simulate the spinning of a thin rod and feel the resistance it offers to being spun.

Categorize This example is a substitution problem, using the definition of moment of inertia in Equation 10.17. As with any calculus problem, the solution involves reducing thc integrand to a single variable.

The shaded length element dx' in Figure 10.9 has a mass dm equal to the mass per unit length λ multiplied by dx'.

Express dm in terms of dx':

$$dm = \lambda\, dx' = \frac{M}{L}\, dx'$$

continued

10.4 cont.

Substitute this expression into Equation 10.17, with $r^2 = (x')^2$:

$$I_y = \int r^2\,dm = \int_{-L/2}^{L/2} (x')^2 \frac{M}{L}\,dx' = \frac{M}{L}\int_{-L/2}^{L/2} (x')^2\,dx'$$

$$= \frac{M}{L}\left[\frac{(x')^3}{3}\right]_{-L/2}^{L/2} = \frac{1}{12}ML^2$$

Check this result in Table 10.2.

Example 10.5 Uniform Solid Cylinder

A uniform solid cylinder has a radius R, mass M, and length L. Calculate its moment of inertia about its central axis (the z axis in Fig. 10.10).

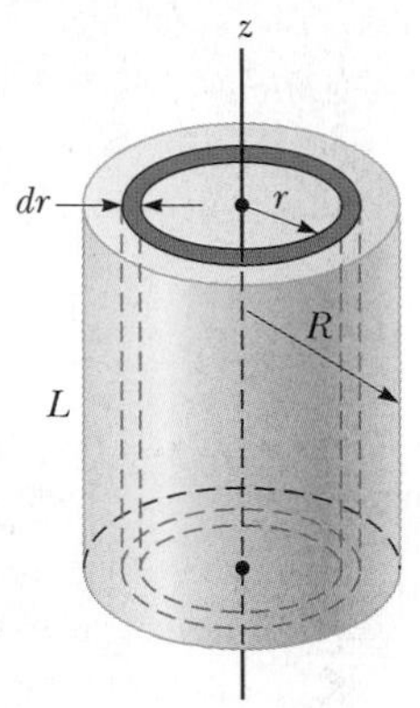

Figure 10.10 (Example 10.5) Calculating I about the z axis for a uniform solid cylinder.

SOLUTION

Conceptualize To simulate this situation, imagine twirling a can of frozen juice around its central axis. Don't twirl a can of vegetable soup; it is not a rigid object! The liquid is able to move relative to the metal can.

Categorize This example is a substitution problem, using the definition of moment of inertia. As with Example 10.4, we must reduce the integrand to a single variable.

It is convenient to divide the cylinder into many cylindrical shells, each having radius r, thickness dr, and length L as shown in Figure 10.10. The density of the cylinder is ρ. The volume dV of each shell is its cross-sectional area multiplied by its length: $dV = L\,dA = L(2\pi r)\,dr$.

Express dm in terms of dr:

$$dm = \rho\,dV = \rho L(2\pi r)\,dr$$

Substitute this expression into Equation 10.17:

$$I_z = \int r^2\,dm = \int r^2[\rho L(2\pi r)\,dr] = 2\pi\rho L\int_0^R r^3\,dr = \frac{1}{2}\pi\rho L R^4$$

Use the total volume $\pi R^2 L$ of the cylinder to express its density:

$$\rho = \frac{M}{V} = \frac{M}{\pi R^2 L}$$

Substitute this value into the expression for I_z:

$$I_z = \frac{1}{2}\pi\left(\frac{M}{\pi R^2 L}\right)LR^4 = \frac{1}{2}MR^2$$

Check this result in Table 10.2.

WHAT IF? What if the length of the cylinder in Figure 10.10 is increased to $2L$, while the mass M and radius R are held fixed? How does that change the moment of inertia of the cylinder?

Answer Notice that the result for the moment of inertia of a cylinder does not depend on L, the length of the cylinder. It applies equally well to a long cylinder and a flat disk having the same mass M and radius R. Therefore, the moment of inertia of the cylinder would not be affected by changing its length.

The calculation of moments of inertia of an object about an arbitrary axis can be cumbersome, even for a highly symmetric object. Fortunately, use of an important theorem, called the **parallel-axis theorem,** often simplifies the calculation.

To generate the parallel-axis theorem, suppose the object in Figure 10.11a rotates about the z axis. The moment of inertia does not depend on how the mass is distributed along the z axis; as we found in Example 10.5, the moment of inertia of a cylinder is independent of its length. Imagine collapsing the three-dimensional object into a planar object as in Figure 10.11b. In this imaginary process, all mass moves parallel to the z axis until it lies in the xy plane. The coordinates of the object's center of mass are now x_{CM}, y_{CM}, and $z_{CM} = 0$. Let the mass element dm have coordinates $(x, y, 0)$ as shown in the view down the z axis in Figure 10.11c. Because

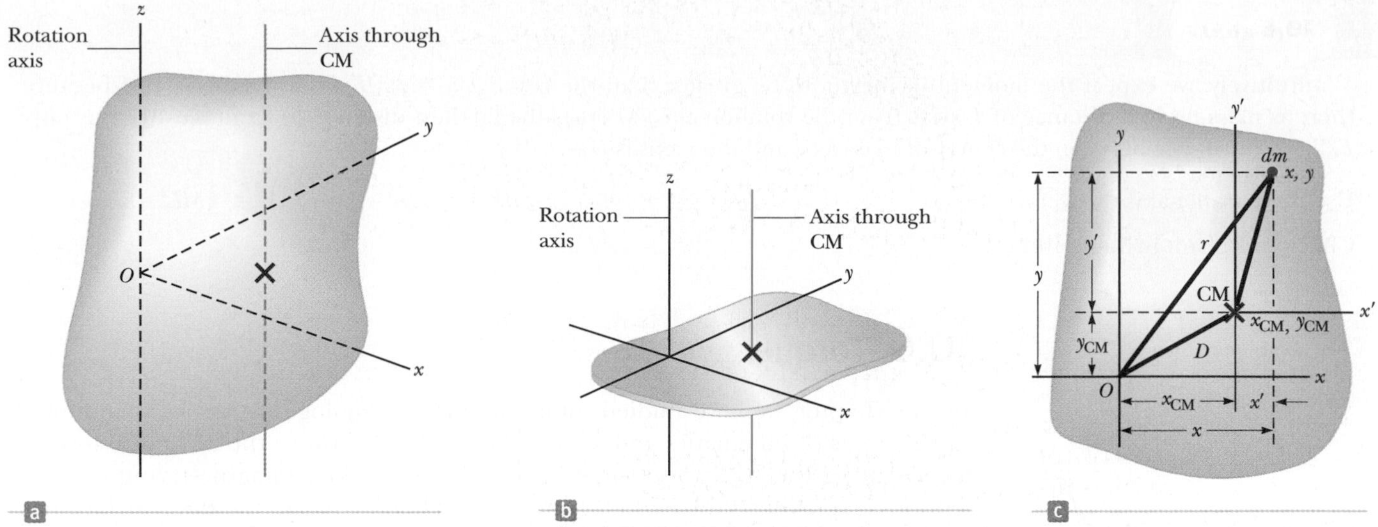

Figure 10.11 (a) An arbitrarily shaped rigid object. The origin of the coordinate system is not at the center of mass of the object. Imagine the object rotating about the *z* axis. (b) All mass elements of the object are collapsed parallel to the *z* axis to form a planar object. (c) An arbitrary mass element *dm* is indicated in blue in this view down the *z* axis. The parallel axis theorem can be used with the geometry shown to determine the moment of inertia of the original object around the *z* axis.

this element is a distance $r = \sqrt{x^2 + y^2}$ from the *z* axis, the moment of inertia of the entire object about the *z* axis is

$$I = \int r^2\, dm = \int (x^2 + y^2)\, dm$$

We can relate the coordinates *x*, *y* of the mass element *dm* to the coordinates of this same element located in a coordinate system having the object's center of mass as its origin. If the coordinates of the center of mass are x_{CM}, y_{CM}, and $z_{CM} = 0$ in the original coordinate system centered on *O*, we see from Figure 10.11c that the relationships between the unprimed and primed coordinates are $x = x' + x_{CM}$, $y = y' + y_{CM}$, and $z = z' = 0$. Therefore,

$$I = \int [(x' + x_{CM})^2 + (y' + y_{CM})^2]\, dm$$

$$= \int [(x')^2 + (y')^2]\, dm + 2x_{CM} \int x'\, dm + 2y_{CM} \int y'\, dm + ({x_{CM}}^2 + {y_{CM}}^2) \int dm$$

The first integral is, by definition, the moment of inertia I_{CM} about an axis that is parallel to the *z* axis and passes through the center of mass. The second two integrals are zero because, by definition of the center of mass, $\int x'\, dm = \int y'\, dm = 0$. The last integral is simply MD^2 because $\int dm = M$ and $D^2 = {x_{CM}}^2 + {y_{CM}}^2$. Therefore, we conclude that

$$I = I_{CM} + MD^2 \qquad (10.18)$$

◀ **Parallel-axis theorem**

Example 10.6 Applying the Parallel-Axis Theorem

Consider once again the uniform rigid rod of mass *M* and length *L* shown in Figure 10.9. Find the moment of inertia of the rod about an axis perpendicular to the rod through one end (the *y* axis in Fig. 10.9).

SOLUTION

Conceptualize Imagine twirling the rod around an endpoint rather than the midpoint. If you have a meterstick handy, try it and notice the degree of difficulty in rotating it around the end compared with rotating it around the center.

Categorize This example is a substitution problem, involving the parallel-axis theorem.

continued

10.6 *cont.*

Intuitively, we expect the moment of inertia to be greater than the result $I_{CM} = \frac{1}{12}ML^2$ from Example 10.4 because there is mass up to a distance of L away from the rotation axis, whereas the farthest distance in Example 10.4 was only $L/2$. The distance between the center-of-mass axis and the y axis is $D = L/2$.

Use the parallel-axis theorem:

$$I = I_{CM} + MD^2 = \frac{1}{12}ML^2 + M\left(\frac{L}{2}\right)^2 = \boxed{\frac{1}{3}ML^2}$$

Check this result in Table 10.2.

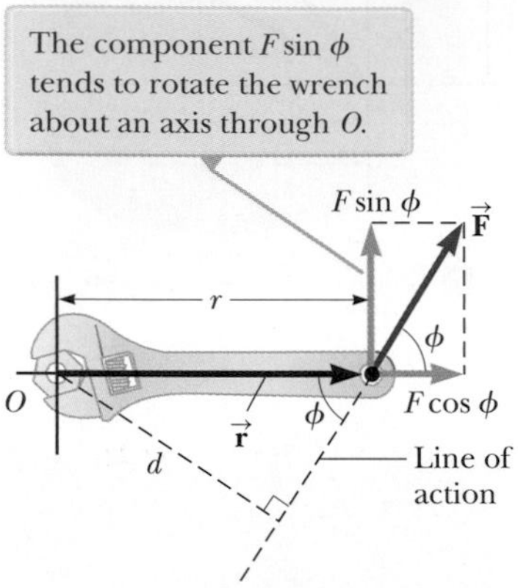

Figure 10.12 The force $\vec{F}$ has a greater rotating tendency about an axis through O as F increases and as the moment arm d increases.

Pitfall Prevention 10.5

Torque Depends on Your Choice of Axis

Like moment of inertia, there is no unique value of the torque on an object. Its value depends on your choice of rotation axis.

10.6 Torque

In this chapter, we have studied rotational motion analogs to translational motion in the areas of kinematics and energy. Let us now consider the analog to force by investigating the cause of changes in rotational motion. Imagine trying to rotate a door by applying a force of magnitude F perpendicular to the door surface near the hinges and then at various distances from the hinges. You will achieve a more rapid rate of rotation for the door by applying the force near the doorknob than by applying it near the hinges.

When a force is exerted on a rigid object pivoted about an axis, the object tends to rotate about that axis. The tendency of a force to rotate an object about some axis is measured by a quantity called **torque** $\vec{\tau}$ (Greek letter tau). Torque is a vector, but we will consider only its magnitude here; we will explore its vector nature in Chapter 11.

Consider the wrench in Figure 10.12 that we wish to rotate around an axis that is perpendicular to the page and passes through the center of the bolt. The applied force $\vec{F}$ acts at an angle ϕ to the horizontal. We define the magnitude of the torque associated with the force $\vec{F}$ around the axis passing through O by the expression

$$\tau \equiv rF \sin \phi = Fd \qquad \textbf{(10.19)}$$

where r is the distance between the rotation axis and the point of application of $\vec{F}$, and d is the perpendicular distance from the rotation axis to the line of action of $\vec{F}$. (The *line of action* of a force is an imaginary line extending out both ends of the vector representing the force. The dashed line extending from the tail of $\vec{F}$ in Fig. 10.12 is part of the line of action of $\vec{F}$.) From the right triangle in Figure 10.12 that has the wrench as its hypotenuse, we see that $d = r \sin \phi$. The quantity d is called the **moment arm** (or *lever arm*) of $\vec{F}$.

Moment arm ▶

In Figure 10.12, the only component of $\vec{F}$ that tends to cause rotation of the wrench around an axis through O is $F \sin \phi$, the component perpendicular to a line drawn from the rotation axis to the point of application of the force. The horizontal component $F \cos \phi$, because its line of action passes through O, has no tendency to produce rotation about an axis passing through O. From the definition of torque, the rotating tendency increases as F increases and as d increases, which explains why it is easier to rotate a door if we push at the doorknob rather than at a point close to the hinges. We also want to apply our push as closely perpendicular to the door as we can so that ϕ is close to 90°. Pushing sideways on the doorknob ($\phi = 0$) will not cause the door to rotate.

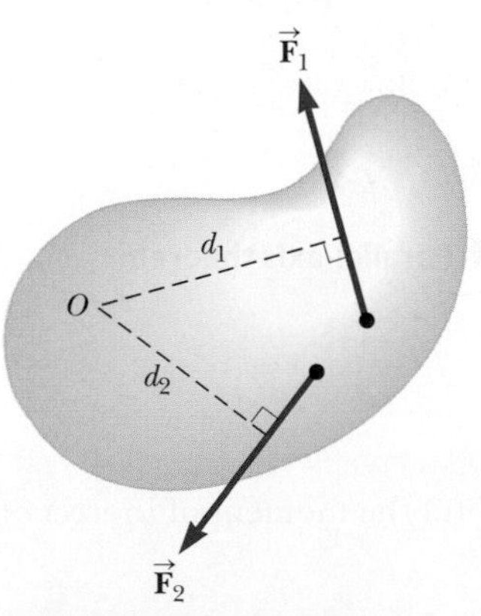

ACTIVE FIGURE 10.13

The force $\vec{F}_1$ tends to rotate the object counterclockwise about an axis through O, and $\vec{F}_2$ tends to rotate it clockwise.

If two or more forces act on a rigid object as in Active Figure 10.13, each tends to produce rotation about the axis through O. In this example, $\vec{F}_2$ tends to rotate the object clockwise and $\vec{F}_1$ tends to rotate it counterclockwise. We use the convention that the sign of the torque resulting from a force is positive if the turning tendency of the force is counterclockwise and negative if the turning tendency is clockwise. For Example, in Active Figure 10.13, the torque resulting from $\vec{F}_1$, which has a moment arm d_1, is positive and equal to $+F_1d_1$; the torque from $\vec{F}_2$ is negative and equal to $-F_2d_2$. Hence, the *net* torque about an axis through O is

$$\sum \tau = \tau_1 + \tau_2 = F_1d_1 - F_2d_2$$

Torque should not be confused with force. Forces can cause a change in translational motion as described by Newton's second law. Forces can also cause a change in rotational motion, but the effectiveness of the forces in causing this change depends on both the magnitudes of the forces and the moment arms of the forces, in the combination we call *torque*. Torque has units of force times length—newton meters in SI units—and should be reported in these units. Do not confuse torque and work, which have the same units but are very different concepts.

***Quick Quiz* 10.5** **(i)** If you are trying to loosen a stubborn screw from a piece of wood with a screwdriver and fail, should you find a screwdriver for which the handle is (a) longer or (b) fatter? **(ii)** If you are trying to loosen a stubborn bolt from a piece of metal with a wrench and fail, should you find a wrench for which the handle is (a) longer or (b) fatter?

Example 10.7 The Net Torque on a Cylinder

A one-piece cylinder is shaped as shown in Figure 10.14, with a core section protruding from the larger drum. The cylinder is free to rotate about the central z axis shown in the drawing. A rope wrapped around the drum, which has radius R_1, exerts a force $\vec{\mathbf{T}}_1$ to the right on the cylinder. A rope wrapped around the core, which has radius R_2, exerts a force $\vec{\mathbf{T}}_2$ downward on the cylinder.

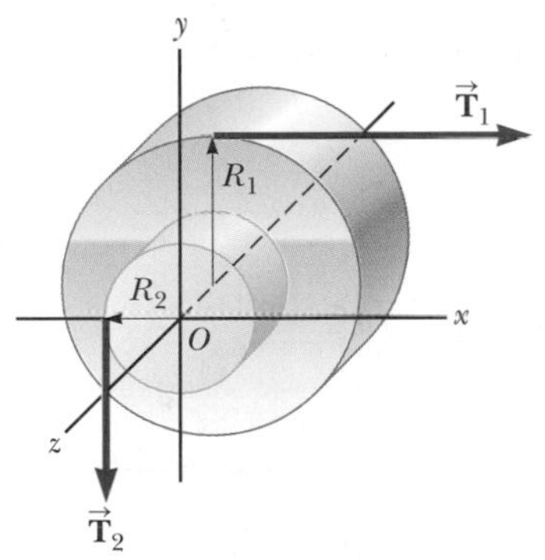

Figure 10.14 (Example 10.7) A solid cylinder pivoted about the z axis through O. The moment arm of $\vec{\mathbf{T}}_1$ is R_1, and the moment arm of $\vec{\mathbf{T}}_2$ is R_2.

(A) What is the net torque acting on the cylinder about the rotation axis (which is the z axis in Fig. 10.14)?

SOLUTION

Conceptualize Imagine that the cylinder in Figure 10.14 is a shaft in a machine. The force $\vec{\mathbf{T}}_1$ could be applied by a drive belt wrapped around the drum. The force $\vec{\mathbf{T}}_2$ could be applied by a friction brake at the surface of the core.

Categorize This example is a substitution problem in which we evaluate the net torque using Equation 10.19.

The torque due to $\vec{\mathbf{T}}_1$ about the rotation axis is $-R_1T_1$. (The sign is negative because the torque tends to produce clockwise rotation.) The torque due to $\vec{\mathbf{T}}_2$ is $+R_2T_2$. (The sign is positive because the torque tends to produce counterclockwise rotation of the cylinder.)

Evaluate the net torque about the rotation axis:
$$\sum \tau = \tau_1 + \tau_2 = \boxed{R_2T_2 - R_1T_1}$$

As a quick check, notice that if the two forces are of equal magnitude, the net torque is negative because $R_1 > R_2$. Starting from rest with both forces of equal magnitude acting on it, the cylinder would rotate clockwise because $\vec{\mathbf{T}}_1$ would be more effective at turning it than would $\vec{\mathbf{T}}_2$.

(B) Suppose $T_1 = 5.0$ N, $R_1 = 1.0$ m, $T_2 = 15$ N, and $R_2 = 0.50$ m. What is the net torque about the rotation axis, and which way does the cylinder rotate starting from rest?

SOLUTION

Substitute the given values:
$$\sum \tau = (0.50 \text{ m})(15 \text{ N}) - (1.0 \text{ m})(5.0 \text{ N}) = \boxed{2.5 \text{ N} \cdot \text{m}}$$

Because this net torque is positive, the cylinder begins to rotate in the counterclockwise direction.

10.7 Analysis Model: Rigid Object Under a Net Torque

In Chapter 5, we learned that a net force on an object causes an acceleration of the object and that the acceleration is proportional to the net force. These facts are the basis of the particle under a net force model whose mathematical representation is Newton's second law. In this section, we show the rotational analog of Newton's second law: the angular acceleration of a rigid object rotating about a fixed axis is proportional to the net torque acting about that axis. Before discussing the more complex case of rigid-object rotation, however, it is instructive first to discuss the case of a particle moving in a circular path about some fixed point under the influence of an external force.

Consider a particle of mass m rotating in a circle of radius r under the influence of a tangential net force $\sum \vec{\mathbf{F}}_t$ and a radial net force $\sum \vec{\mathbf{F}}_r$ as shown in Figure 10.15. The radial net force causes the particle to move in the circular path with a centripetal acceleration. The tangential force provides a tangential acceleration $\vec{\mathbf{a}}_t$, and

$$\sum F_t = ma_t$$

The magnitude of the net torque due to $\sum \vec{\mathbf{F}}_t$ on the particle about an axis perpendicular to the page through the center of the circle is

$$\sum \tau = \sum F_t r = (ma_t)r$$

Because the tangential acceleration is related to the angular acceleration through the relationship $a_t = r\alpha$ (Eq. 10.11), the net torque can be expressed as

$$\sum \tau = (mr\alpha)r = (mr^2)\alpha$$

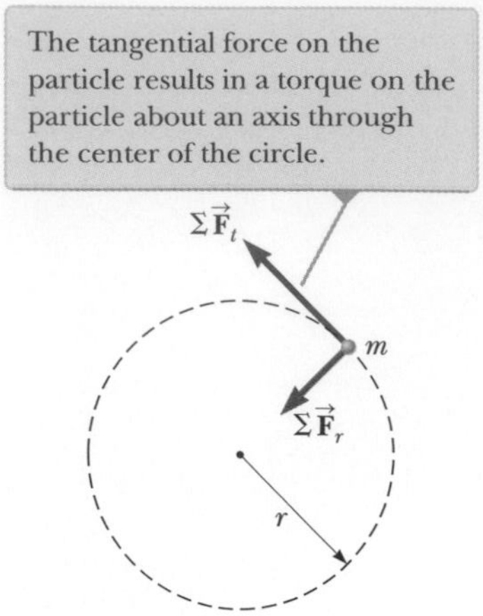

Figure 10.15 A particle rotating in a circle under the influence of a tangential net force $\sum \vec{\mathbf{F}}_t$. A radial net force $\sum \vec{\mathbf{F}}_r$ also must be present to maintain the circular motion.

Recall from Equation 10.15 that mr^2 is the moment of inertia of the particle about the z axis passing through the origin, so

$$\sum \tau = I\alpha \tag{10.20}$$

That is, the net torque acting on the particle is proportional to its angular acceleration, and the proportionality constant is the moment of inertia. Notice that $\sum \tau = I\alpha$ has the same mathematical form as Newton's second law of motion, $\sum F = ma$.

Now let us extend this discussion to a rigid object of arbitrary shape rotating about a fixed axis as in Figure 10.16. The object can be regarded as an infinite number of mass elements dm of infinitesimal size. If we impose a Cartesian coordinate system on the object, each mass element rotates in a circle about the origin and each has a tangential acceleration $\vec{\mathbf{a}}_t$ produced by an external tangential force $d\vec{\mathbf{F}}_t$. For any given element, we know from Newton's second law that

$$dF_t = (dm)a_t$$

The external torque $d\vec{\boldsymbol{\tau}}_{\text{ext}}$ associated with the force $d\vec{\mathbf{F}}_t$ acts about the origin and its magnitude is given by

$$d\tau_{\text{ext}} = r\,dF_t = a_t r\,dm$$

Because $a_t = r\alpha$, the expression for $d\tau_{\text{ext}}$ becomes

$$d\tau_{\text{ext}} = \alpha r^2\,dm$$

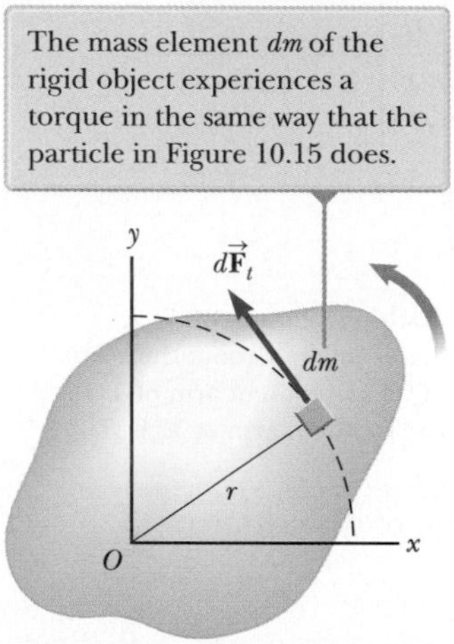

Figure 10.16 A rigid object rotating about an axis through O. Each mass element dm rotates about the axis with the same angular acceleration α.

Although each mass element of the rigid object may have a different translational acceleration $\vec{\mathbf{a}}_t$, they all have the *same* angular acceleration α. With that in mind, we can integrate the above expression to obtain the net external torque $\Sigma\tau_{\text{ext}}$ about an axis through O due to the external forces:

$$\sum \tau_{\text{ext}} = \int \alpha r^2\,dm = \alpha \int r^2\,dm$$

where α can be taken outside the integral because it is common to all mass elements. From Equation 10.17, we know that $\int r^2\,dm$ is the moment of inertia of the object about the rotation axis through O, and so the expression for $\sum \tau_{\text{ext}}$ becomes

Torque on a rigid object is proportional to angular acceleration ▶

$$\sum \tau_{\text{ext}} = I\alpha \tag{10.21}$$

This equation for a rigid object is the same as that found for a particle moving in a circular path (Eq. 10.20). The net torque about the rotation axis is proportional to the angular acceleration of the object, with the proportionality factor being I, a quantity that depends on the axis of rotation and on the size and shape of the object. Equation 10.21 is the mathematical representation of the analysis model of a **rigid object under a net torque,** the rotational analog to the particle under a net force.

Finally, notice that the result $\Sigma\tau_{\text{ext}} = I\alpha$ also applies when the forces acting on the mass elements have radial components as well as tangential components. That is because the line of action of all radial components must pass through the axis of rotation; hence, all radial components produce zero torque about that axis.

***Quick Quiz* 10.6** You turn off your electric drill and find that the time interval for the rotating bit to come to rest due to frictional torque in the drill is Δt. You replace the bit with a larger one that results in a doubling of the moment of inertia of the drill's entire rotating mechanism. When this larger bit is rotated at the same angular speed as the first and the drill is turned off, the frictional torque remains the same as that for the previous situation. What is the time interval for this second bit to come to rest? **(a)** $4\Delta t$ **(b)** $2\Delta t$ **(c)** Δt **(d)** $0.5\Delta t$ **(e)** $0.25\Delta t$ **(f)** impossible to determine

Example 10.8 Rotating Rod

A uniform rod of length L and mass M is attached at one end to a frictionless pivot and is free to rotate about the pivot in the vertical plane as in Figure 10.17. The rod is released from rest in the horizontal position. What are the initial angular acceleration of the rod and the initial translational acceleration of its right end?

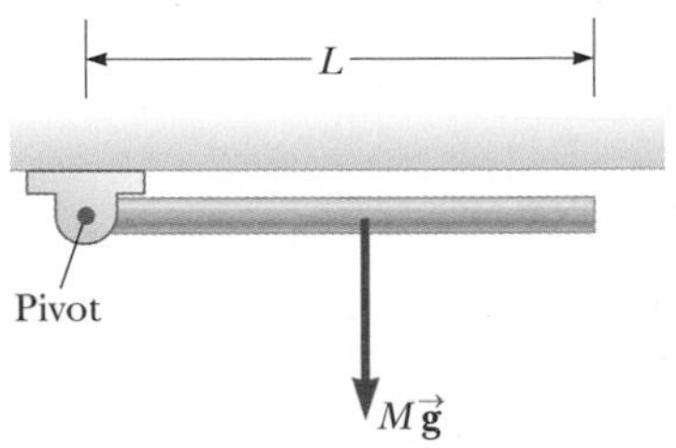

Figure 10.17 (Example 10.8) A rod is free to rotate around a pivot at the left end. The gravitational force on the rod acts at its center of mass.

SOLUTION

Conceptualize Imagine what happens to the rod in Figure 10.17 when it is released. It rotates clockwise around the pivot at the left end.

Categorize The rod is categorized as a rigid object under a net torque. The torque is due only to the gravitational force on the rod if the rotation axis is chosen to pass through the pivot in Figure 10.17. We *cannot* categorize the rod as a rigid object under constant angular acceleration because the torque exerted on the rod and therefore the angular acceleration of the rod vary with its angular position.

Analyze The only force contributing to the torque about an axis through the pivot is the gravitational force $M\vec{g}$ exerted on the rod. (The force exerted by the pivot on the rod has zero torque about the pivot because its moment arm is zero.) To compute the torque on the rod, we assume the gravitational force acts at the center of mass of the rod as shown in Figure 10.17.

Write an expression for the magnitude of the net external torque due to the gravitational force about an axis through the pivot:

$$\sum \tau_{\text{ext}} = Mg\left(\frac{L}{2}\right)$$

Use Equation 10.21 to obtain the angular acceleration of the rod:

$$(1)\ \alpha = \frac{\sum \tau_{\text{ext}}}{I} = \frac{Mg(L/2)}{\frac{1}{3}ML^2} = \boxed{\frac{3g}{2L}}$$

Use Equation 10.11 with $r = L$ to find the initial translational acceleration of the right end of the rod:

$$a_t = L\alpha = \boxed{\tfrac{3}{2}g}$$

Finalize These values are the *initial* values of the angular and translational accelerations. Once the rod begins to rotate, the gravitational force is no longer perpendicular to the rod and the values of the two accelerations decrease, going to zero at the moment the rod passes through the vertical orientation.

WHAT IF? What if we were to place a penny on the end of the rod and then release the rod? Would the penny stay in contact with the rod?

Answer The result for the initial acceleration of a point on the end of the rod shows that $a_t > g$. An unsupported penny falls at acceleration g. So, if we place a penny on the end of the rod and then release the rod, the end of the rod falls faster than the penny does! The penny does not stay in contact with the rod. (Try this with a penny and a meterstick!)

The question now is to find the location on the rod at which we can place a penny that *will* stay in contact as both begin to fall. To find the translational acceleration of an arbitrary point on the rod at a distance $r < L$ from the pivot point, we combine Equation (1) with Equation 10.11:

$$a_t = r\alpha = \frac{3g}{2L}r$$

continued

10.8 *cont.*

For the penny to stay in contact with the rod, the limiting case is that the translational acceleration must be equal to that due to gravity:

$$a_t = g = \frac{3g}{2L} r$$

$$r = \tfrac{2}{3}L$$

Therefore, a penny placed closer to the pivot than two-thirds of the length of the rod stays in contact with the falling rod, but a penny farther out than this point loses contact.

Conceptual Example 10.9 Falling Smokestacks and Tumbling Blocks

When a tall smokestack falls over, it often breaks somewhere along its length before it hits the ground as shown in Figure 10.18. Why?

Figure 10.18 (Conceptual Example 10.9) A falling smokestack breaks at some point along its length.

SOLUTION

As the smokestack rotates around its base, each higher portion of the smokestack falls with a larger tangential acceleration than the portion below it according to Equation 10.11. The angular acceleration increases as the smokestack tips farther. Eventually, higher portions of the smokestack experience an acceleration greater than the acceleration that could result from gravity alone; this situation is similar to that described in Example 10.8. It can happen only if these portions are being pulled downward by a force in addition to the gravitational force. The force that causes that to occur is the shear force from lower portions of the smokestack. Eventually, the shear force that provides this acceleration is greater than the smokestack can withstand, and the smokestack breaks. The same thing happens with a tall tower of children's toy blocks. Borrow some blocks from a child and build such a tower. Push it over and watch it come apart at some point before it strikes the floor.

Example 10.10 Angular Acceleration of a Wheel

A wheel of radius R, mass M, and moment of inertia I is mounted on a frictionless, horizontal axle as in Figure 10.19. A light cord wrapped around the wheel supports an object of mass m. When the wheel is released, the object accelerates downward, the cord unwraps off the wheel, and the wheel rotates with an angular acceleration. Calculate the angular acceleration of the wheel, the translational acceleration of the object, and the tension in the cord.

SOLUTION

Conceptualize Imagine that the object is a bucket in an old-fashioned wishing well. It is tied to a cord that passes around a cylinder equipped with a crank for raising the bucket. After the bucket has been raised, the system is released and the bucket accelerates downward while the cord unwinds off the cylinder.

Categorize We apply two analysis models here. The object is modeled as a particle under a net force. The wheel is modeled as a rigid object under a net torque.

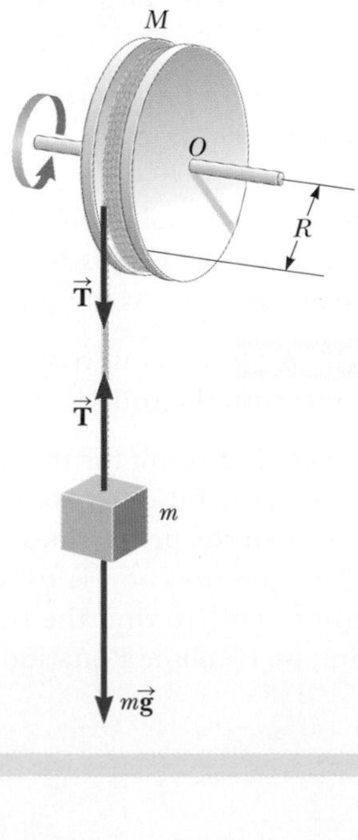

Figure 10.19 (Example 10.10) An object hangs from a cord wrapped around a wheel.

10.10 *cont.*

Analyze The magnitude of the torque acting on the wheel about its axis of rotation is $\tau = TR$, where T is the force exerted by the cord on the rim of the wheel. (The gravitational force exerted by the Earth on the wheel and the normal force exerted by the axle on the wheel both pass through the axis of rotation and therefore produce no torque.)

Write Equation 10.21: $$\sum \tau_{\text{ext}} = I\alpha$$

Solve for α and substitute the net torque: $$(1)\quad \alpha = \frac{\sum \tau_{\text{ext}}}{I} = \frac{TR}{I}$$

Apply Newton's second law to the motion of the object, taking the downward direction to be positive: $$\sum F_y = mg - T = ma$$

Solve for the acceleration a: $$(2)\quad a = \frac{mg - T}{m}$$

Equations (1) and (2) have three unknowns: α, a, and T. Because the object and wheel are connected by a cord that does not slip, the translational acceleration of the suspended object is equal to the tangential acceleration of a point on the wheel's rim. Therefore, the angular acceleration α of the wheel and the translational acceleration of the object are related by $a = R\alpha$.

Use this fact together with Equations (1) and (2): $$(3)\quad a = R\alpha = \frac{TR^2}{I} = \frac{mg - T}{m}$$

Solve for the tension T: $$(4)\quad T = \boxed{\frac{mg}{1 + (mR^2/I)}}$$

Substitute Equation (4) into Equation (2) and solve for a: $$(5)\quad a = \boxed{\frac{g}{1 + (I/mR^2)}}$$

Use $a = R\alpha$ and Equation (5) to solve for α: $$\alpha = \frac{a}{R} = \boxed{\frac{g}{R + (I/mR)}}$$

Finalize We finalize this problem by imagining the behavior of the system in some extreme limits.

WHAT IF? What if the wheel were to become very massive so that I becomes very large? What happens to the acceleration a of the object and the tension T?

Answer If the wheel becomes infinitely massive, we can imagine that the object of mass m will simply hang from the cord without causing the wheel to rotate.

We can show that mathematically by taking the limit $I \to \infty$. Equation (5) then becomes

$$a = \frac{g}{1 + (I/mR^2)} \quad \to \quad 0$$

which agrees with our conceptual conclusion that the object will hang at rest. Also, Equation (4) becomes

$$T = \frac{mg}{1 + (mR^2/I)} \quad \to \quad \frac{mg}{1 + 0} = mg$$

which is consistent because the object simply hangs at rest in equilibrium between the gravitational force and the tension in the string.

10.8 Energy Considerations in Rotational Motion

Up to this point in our discussion of rotational motion in this chapter, we focused primarily on an approach involving force, leading to a description of torque on a rigid object. In Section 10.4, we discussed the rotational kinetic energy of a rigid object. Let us now extend that initial energy discussion and see how an energy approach can be useful in solving rotational problems.

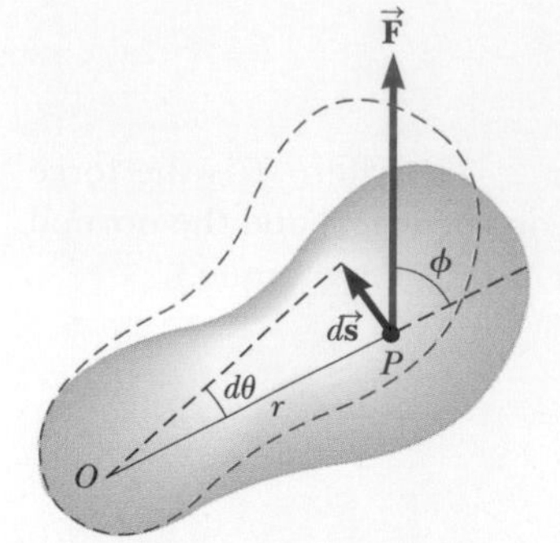

Figure 10.20 A rigid object rotates about an axis through O under the action of an external force $\vec{F}$ applied at P.

We begin by considering the relationship between the torque acting on a rigid object and its resulting rotational motion so as to generate expressions for power and a rotational analog to the work–kinetic energy theorem. Consider the rigid object pivoted at O in Figure 10.20. Suppose a single external force $\vec{F}$ is applied at P, where $\vec{F}$ lies in the plane of the page. The work done on the object by $\vec{F}$ as its point of application rotates through an infinitesimal distance $ds = r\,d\theta$ is

$$dW = \vec{F} \cdot d\vec{s} = (F \sin \phi) r\, d\theta$$

where $F \sin \phi$ is the tangential component of $\vec{F}$, or, in other words, the component of the force along the displacement. Notice that the radial component vector of $\vec{F}$ does no work on the object because it is perpendicular to the displacement of the point of application of $\vec{F}$.

Because the magnitude of the torque due to $\vec{F}$ about an axis through O is defined as $rF \sin \phi$ by Equation 10.19, we can write the work done for the infinitesimal rotation as

$$dW = \tau\, d\theta \tag{10.22}$$

The rate at which work is being done by $\vec{F}$ as the object rotates about the fixed axis through the angle $d\theta$ in a time interval dt is

$$\frac{dW}{dt} = \tau \frac{d\theta}{dt}$$

Because dW/dt is the instantaneous power P (see Section 8.5) delivered by the force and $d\theta/dt = \omega$, this expression reduces to

Power delivered to a rotating rigid object ▶

$$P = \frac{dW}{dt} = \tau\omega \tag{10.23}$$

This equation is analogous to $P = Fv$ in the case of translational motion, and Equation 10.22 is analogous to $dW = F_x\, dx$.

In studying translational motion, we have seen that models based on an energy approach can be extremely useful in describing a system's behavior. From what we learned of translational motion, we expect that when a symmetric object rotates about a fixed axis, the work done by external forces equals the change in the rotational energy of the object.

To prove that fact, let us begin with $\Sigma\tau_{\text{ext}} = I\alpha$. Using the chain rule from calculus, we can express the net torque as

$$\sum \tau_{\text{ext}} = I\alpha = I\frac{d\omega}{dt} = I\frac{d\omega}{d\theta}\frac{d\theta}{dt} = I\frac{d\omega}{d\theta}\,\omega$$

Rearranging this expression and noting that $\Sigma\,\tau_{\text{ext}}\, d\theta = dW$ gives

$$\sum \tau_{\text{ext}}\, d\theta = dW = I\omega\, d\omega$$

Integrating this expression, we obtain for the total work done by the net external force acting on a rotating system

Work–kinetic energy theorem for rotational motion ▶

$$\sum W = \int_{\omega_i}^{\omega_f} I\omega\, d\omega = \tfrac{1}{2}I\omega_f^2 - \tfrac{1}{2}I\omega_i^2 \tag{10.24}$$

where the angular speed changes from ω_i to ω_f. Equation 10.24 is the **work–kinetic energy theorem for rotational motion.** Similar to the work–kinetic energy theorem in translational motion (Section 7.5), this theorem states that the net work done by external forces in rotating a symmetric rigid object about a fixed axis equals the change in the object's rotational energy.

This theorem is a form of the nonisolated system (energy) model discussed in Chapter 8. Work is done on the system of the rigid object, which represents a transfer of energy across the boundary of the system that appears as an increase in the object's rotational kinetic energy.

TABLE 10.3 *Useful Equations in Rotational and Translational Motion*

Rotational Motion About a Fixed Axis	Translational Motion
Angular speed $\omega = d\theta/dt$	Translational speed $v = dx/dt$
Angular acceleration $\alpha = d\omega/dt$	Translational acceleration $a = dv/dt$
Net torque $\Sigma\tau_{ext} = I\alpha$	Net force $\Sigma F = ma$
If $\alpha =$ constant $\begin{cases} \omega_f = \omega_i + \alpha t \\ \theta_f = \theta_i + \omega_i t + \frac{1}{2}\alpha t^2 \\ \omega_f^2 = \omega_i^2 + 2\alpha(\theta_f - \theta_i) \end{cases}$	If $a =$ constant $\begin{cases} v_f = v_i + at \\ x_f = x_i + v_i t + \frac{1}{2}at^2 \\ v_f^2 = v_i^2 + 2a(x_f - x_i) \end{cases}$
Work $W = \int_{\theta_i}^{\theta_f} \tau \, d\theta$	Work $W = \int_{x_i}^{x_f} F_x \, dx$
Rotational kinetic energy $K_R = \frac{1}{2}I\omega^2$	Kinetic energy $K = \frac{1}{2}mv^2$
Power $P = \tau\omega$	Power $P = Fv$
Angular momentum $L = I\omega$	Linear momentum $p = mv$
Net torque $\Sigma\tau = dL/dt$	Net force $\Sigma F = dp/dt$

In general, we can combine this theorem with the translational form of the work–kinetic energy theorem from Chapter 7. Therefore, the net work done by external forces on an object is the change in its *total* kinetic energy, which is the sum of the translational and rotational kinetic energies. For example, when a pitcher throws a baseball, the work done by the pitcher's hands appears as kinetic energy associated with the ball moving through space as well as rotational kinetic energy associated with the spinning of the ball.

In addition to the work–kinetic energy theorem, other energy principles can also be applied to rotational situations. For example, if a system involving rotating objects is isolated and no nonconservative forces act within the system, the isolated system model and the principle of conservation of mechanical energy can be used to analyze the system as in Example 10.11 below.

Finally, in some situations an energy approach does not provide enough information to solve the problem and it must be combined with a momentum approach. Such a case is illustrated in Example 10.14 in Section 10.9.

Table 10.3 lists the various equations we have discussed pertaining to rotational motion together with the analogous expressions for translational motion. Notice the similar mathematical forms of the equations. The last two equations in the left-hand column of Table 10.3, involving angular momentum L, are discussed in chapter 11 and are included here only for the sake of completeness.

Example 10.11 Rotating Rod Revisited

A uniform rod of length L and mass M is free to rotate on a frictionless pin passing through one end (Fig 10.21). The rod is released from rest in the horizontal position.

(A) What is its angular speed when the rod reaches its lowest position?

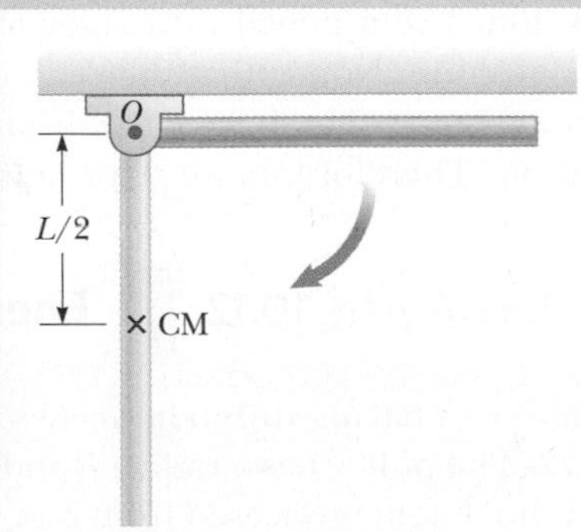

Figure 10.21 (Example 10.11) A uniform rigid rod pivoted at O rotates in a vertical plane under the action of the gravitational force.

SOLUTION

Conceptualize Consider Figure 10.21 and imagine the rod rotating downward through a quarter turn about the pivot at the left end. Also look back at Example 10.8. This physical situation is the same.

Categorize As mentioned in Example 10.8, the angular acceleration of the rod is not constant. Therefore, the kinematic equations for rotation (Section 10.2) cannot be used to solve this example. We categorize the system of the rod and the Earth as an isolated system in terms of energy with no nonconservative forces acting and use the principle of conservation of mechanical energy.

continued

10.11 *cont.*

Analyze We choose the configuration in which the rod is hanging straight down as the reference configuration for gravitational potential energy and assign a value of zero for this configuration. When the rod is in the horizontal position, it has no rotational kinetic energy. The potential energy of the system in this configuration relative to the reference configuration is $MgL/2$ because the center of mass of the rod is at a height $L/2$ higher than its position in the reference configuration. When the rod reaches its lowest position, the energy of the system is entirely rotational energy $\frac{1}{2}I\omega^2$, where I is the moment of inertia of the rod about an axis passing through the pivot.

Using the isolated system (energy) model, write a conservation of mechanical energy equation for the system:

$$K_f + U_f = K_i + U_i$$

Substitute for each of the energies:

$$\tfrac{1}{2}I\omega^2 + 0 = 0 + \tfrac{1}{2}MgL$$

Solve for ω and use $I = \frac{1}{3}ML^2$ (see Table 10.2) for the rod:

$$\omega = \sqrt{\frac{MgL}{I}} = \sqrt{\frac{MgL}{\frac{1}{3}ML^2}} = \sqrt{\frac{3g}{L}}$$

(B) Determine the tangential speed of the center of mass and the tangential speed of the lowest point on the rod when it is in the vertical position.

SOLUTION

Use Equation 10.10 and the result from part (A):

$$v_{CM} = r\omega = \frac{L}{2}\omega = \tfrac{1}{2}\sqrt{3gL}$$

Because r for the lowest point on the rod is twice what it is for the center of mass, the lowest point has a tangential speed twice that of the center of mass:

$$v = 2v_{CM} = \sqrt{3gL}$$

Finalize The initial configuration in this example is the same as that in Example 10.8. In Example 10.8, however, we could only find the initial angular acceleration of the rod. Applying an energy approach in the current example allows us to find additional information, the angular speed of the rod at the lowest point. Convince yourself that you could find the angular speed of the rod at any angular position by knowing the location of the center of mass at this position.

WHAT IF? What if we want to find the angular speed of the rod when the angle it makes with the horizontal is 45.0°? Because this angle is half of 90.0°, for which we solved the problem above, is the angular speed at this configuration half the answer in the calculation above, that is, $\frac{1}{2}\sqrt{3g/L}$?

Answer Imagine the rod in Figure 10.21 at the 45.0° position. Use a pencil or a ruler to represent the rod at this position. Notice that the center of mass has dropped through more than half of the distance $L/2$ in this configuration. Therefore, more than half of the initial gravitational potential energy has been transformed to rotational kinetic energy. So, we should not expect the value of the angular speed to be as simple as proposed above.

Note that the center of mass of the rod drops through a distance of $0.500L$ as the rod reaches the vertical configuration. When the rod is at 45.0° to the horizontal, we can show that the center of mass of the rod drops through a distance of $0.354L$. Continuing the calculation, we find that the angular speed of the rod at this configuration is $0.841\sqrt{3g/L}$, (not $\frac{1}{2}\sqrt{3g/L}$).

Example 10.12 Energy and the Atwood Machine

Two blocks having different masses m_1 and m_2 are connected by a string passing over a pulley as shown in Active Figure 10.22. The pulley has a radius R and moment of inertia I about its axis of rotation. The string does not slip on the pulley, and the system is released from rest. Find the translational speeds of the blocks after block 2 descends through a distance h and find the angular speed of the pulley at this time.

SOLUTION

Conceptualize We have already seen examples involving the Atwood machine, so the motion of the objects in Active Figure 10.22 should be easy to visualize.

10.12 *cont.*

Categorize Because the string does not slip, the pulley rotates about the axle. We can neglect friction in the axle because the axle's radius is small relative to that of the pulley. Hence, the frictional torque is much smaller than the net torque applied by the two blocks provided that their masses are significantly different. Consequently, the system consisting of the two blocks, the pulley, and the Earth is an isolated system in terms of energy with no nonconservative forces acting; therefore, the mechanical energy of the system is conserved.

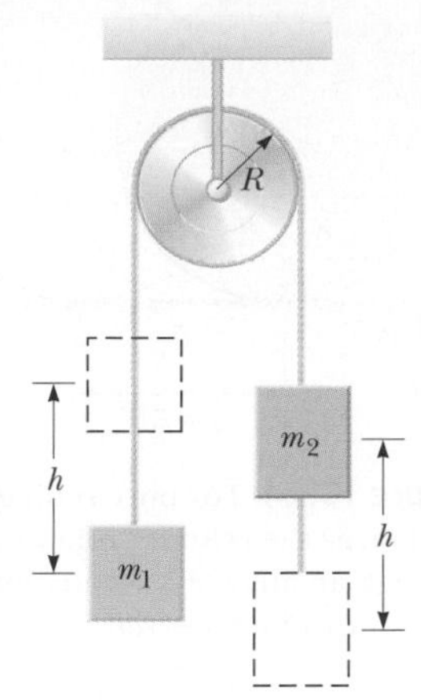

ACTIVE FIGURE 10.22

(Example 10.12) An Atwood machine with a massive pulley.

Analyze We define the zero configuration for gravitational potential energy as that which exists when the system is released. From Active Figure 10.22, we see that the descent of block 2 is associated with a decrease in system potential energy and that the rise of block 1 represents an increase in potential energy.

Using the isolated system (energy) model, write a conservation of mechanical energy equation for the system:

$$K_f + U_f = K_i + U_i$$

Substitute for each of the energies:

$$(\tfrac{1}{2}m_1 v_f^2 + \tfrac{1}{2}m_2 v_f^2 + \tfrac{1}{2}I\omega_f^2) + (m_1 gh - m_2 gh) = 0 + 0$$

Use $v_f = R\omega_f$ to substitute for ω_f:

$$\tfrac{1}{2}m_1 v_f^2 + \tfrac{1}{2}m_2 v_f^2 + \tfrac{1}{2}I\frac{v_f^2}{R^2} = m_2 gh - m_1 gh$$

$$\tfrac{1}{2}\left(m_1 + m_2 + \frac{I}{R^2}\right)v_f^2 = m_2 gh - m_1 gh$$

Solve for v_f:

$$(1)\quad v_f = \left[\frac{2(m_2 - m_1)gh}{m_1 + m_2 + I/R^2}\right]^{1/2}$$

Use $v_f = R\omega_f$ to solve for ω_f:

$$\omega_f = \frac{v_f}{R} = \frac{1}{R}\left[\frac{2(m_2 - m_1)gh}{m_1 + m_2 + I/R^2}\right]^{1/2}$$

Finalize Each block can be modeled as a particle under constant acceleration because it experiences a constant net force. Think about what you would need to do to use Equation (1) to find the acceleration of one of the blocks. Then imagine the pulley becoming massless and determine the acceleration of a block. How does this result compare with the result of Example 5.9?

10.9 Rolling Motion of a Rigid Object

In this section, we treat the motion of a rigid object rolling along a flat surface. In general, such motion is complex. For example, suppose a cylinder is rolling on a straight path such that the axis of rotation remains parallel to its initial orientation in space. As Figure 10.23 shows, a point on the rim of the cylinder moves in a complex path called a *cycloid*. We can simplify matters, however, by focusing on the center of mass rather than on a point on the rim of the rolling object. As shown in Figure 10.23, the center of mass moves in a straight line. If an object such as a

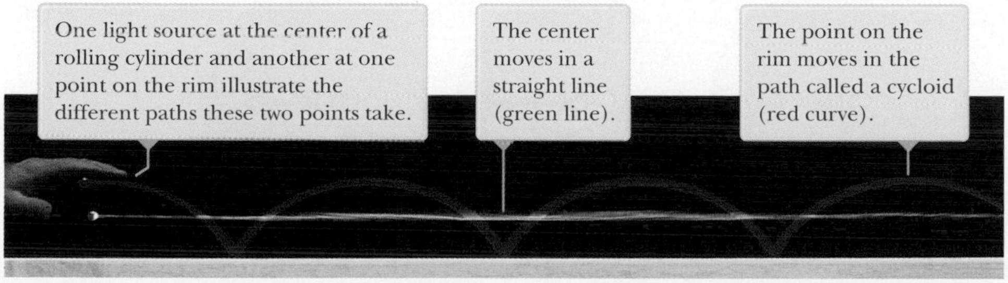

Henry Leap and Jim Lehman

Figure 10.23 Two points on a rolling object take different paths through space.

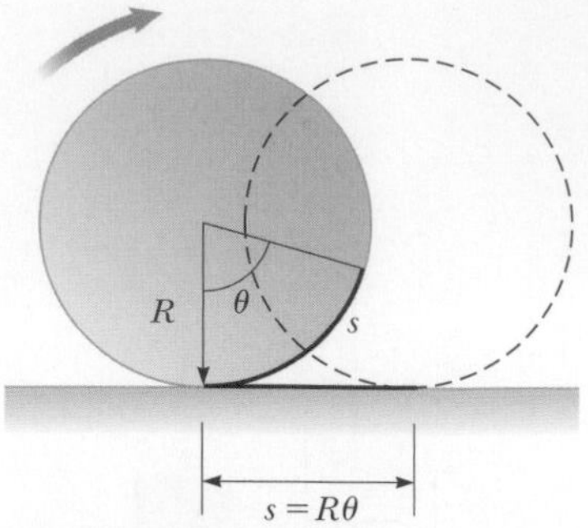

Figure 10.24 For pure rolling motion, as the cylinder rotates through an angle θ its center moves a linear distance $s = R\theta$.

cylinder rolls without slipping on the surface (called *pure rolling motion*), a simple relationship exists between its rotational and translational motions.

Consider a uniform cylinder of radius R rolling without slipping on a horizontal surface (Fig. 10.24). As the cylinder rotates through an angle θ, its center of mass moves a linear distance $s = R\theta$ (see Eq. 10.1a). Therefore, the translational speed of the center of mass for pure rolling motion is given by

$$v_{CM} = \frac{ds}{dt} = R\frac{d\theta}{dt} = R\omega \qquad \textbf{(10.25)}$$

where ω is the angular speed of the cylinder. Equation 10.25 holds whenever a cylinder or sphere rolls without slipping and is the **condition for pure rolling motion.** The magnitude of the linear acceleration of the center of mass for pure rolling motion is

$$a_{CM} = \frac{dv_{CM}}{dt} = R\frac{d\omega}{dt} = R\alpha \qquad \textbf{(10.26)}$$

where α is the angular acceleration of the cylinder.

Pitfall Prevention 10.6

Equation 10.25 Looks Familiar

Equation 10.25 looks very similar to Equation 10.10, so be sure to be clear on the difference. Equation 10.10 gives the *tangential* speed of a point on a *rotating* object located a distance r from a fixed rotation axis if the object is rotating with angular speed ω. Equation 10.25 gives the *translational* speed of the center of mass of a *rolling* object of radius R rotating with angular speed ω.

Imagine that you are moving along with a rolling object at speed v_{CM}, staying in a frame of reference at rest with respect to the center of mass of the object. As you observe the object, you will see the object in pure rotation around its center of mass. Figure 10.25a shows the velocities of points at the top, center, and bottom of the object as observed by you. In addition to these velocities, every point on the object moves in the same direction with speed v_{CM} relative to the surface on which it rolls. Figure 10.25b shows these velocities for a nonrotating object. In the reference frame at rest with respect to the surface, the velocity of a given point on the object is the sum of the velocities shown in Figures 10.25a and 10.25b. Figure 10.25c shows the results of adding these velocities.

Notice that the contact point between the surface and object in Figure 10.25c has a translational speed of zero. At this instant, the rolling object is moving in exactly the same way as if the surface were removed and the object were pivoted at point P and spun about an axis passing through P. We can express the total kinetic energy of this imagined spinning object as

$$K = \tfrac{1}{2}I_P\omega^2 \qquad \textbf{(10.27)}$$

where I_P is the moment of inertia about a rotation axis through P.

Because the motion of the imagined spinning object is the same at this instant as our actual rolling object, Equation 10.27 also gives the kinetic energy of the rolling object. Applying the parallel-axis theorem, we can substitute $I_P = I_{CM} + MR^2$ into Equation 10.27 to obtain

$$K = \tfrac{1}{2}I_{CM}\omega^2 + \tfrac{1}{2}MR^2\omega^2$$

Using $v_{CM} = R\omega$, this equation can be expressed as

Total kinetic energy of a rolling object ▶

$$K = \tfrac{1}{2}I_{CM}\omega^2 + \tfrac{1}{2}Mv_{CM}^2 \qquad \textbf{(10.28)}$$

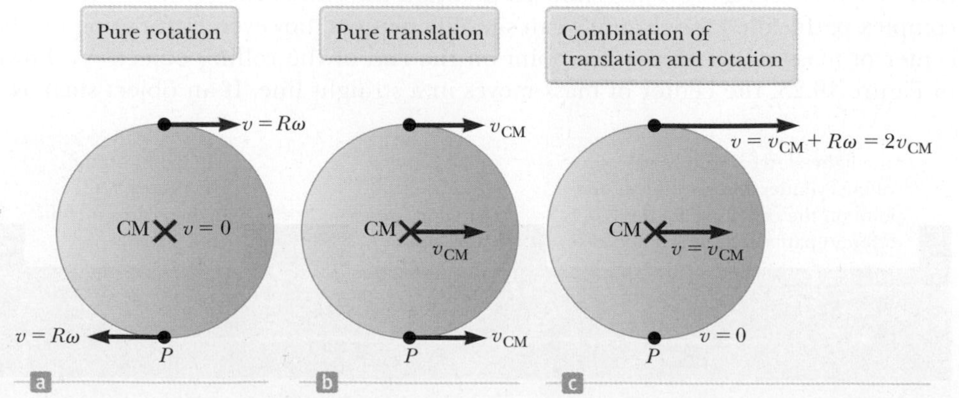

Figure 10.25 The motion of a rolling object can be modeled as a combination of pure translation and pure rotation.

The term $\frac{1}{2}I_{CM}\omega^2$ represents the rotational kinetic energy of the object about its center of mass, and the term $\frac{1}{2}Mv_{CM}{}^2$ represents the kinetic energy the object would have if it were just translating through space without rotating. Therefore, the total kinetic energy of a rolling object is the sum of the rotational kinetic energy about the center of mass and the translational kinetic energy of the center of mass. This statement is consistent with the situation illustrated in Figure 10.25, which shows that the velocity of a point on the object is the sum of the velocity of the center of mass and the tangential velocity around the center of mass.

Energy methods can be used to treat a class of problems concerning the rolling motion of an object on a rough incline. For example, consider Active Figure 10.26, which shows a sphere rolling without slipping after being released from rest at the top of the incline. Accelerated rolling motion is possible only if a friction force is present between the sphere and the incline to produce a net torque about the center of mass. Despite the presence of friction, no loss of mechanical energy occurs because the contact point is at rest relative to the surface at any instant. (On the other hand, if the sphere were to slip, mechanical energy of the sphere–incline–Earth system would decrease due to the nonconservative force of kinetic friction.)

In reality, *rolling friction* causes mechanical energy to transform to internal energy. Rolling friction is due to deformations of the surface and the rolling object. For example, automobile tires flex as they roll on a roadway, representing a transformation of mechanical energy to internal energy. The roadway also deforms a small amount, representing additional rolling friction. In our problem-solving models, we ignore rolling friction unless stated otherwise.

Using $v_{CM} = R\omega$ for pure rolling motion, we can express Equation 10.28 as

$$K = \frac{1}{2}I_{CM}\left(\frac{v_{CM}}{R}\right)^2 + \frac{1}{2}Mv_{CM}{}^2$$

$$K = \frac{1}{2}\left(\frac{I_{CM}}{R^2} + M\right)v_{CM}{}^2 \quad \textbf{(10.29)}$$

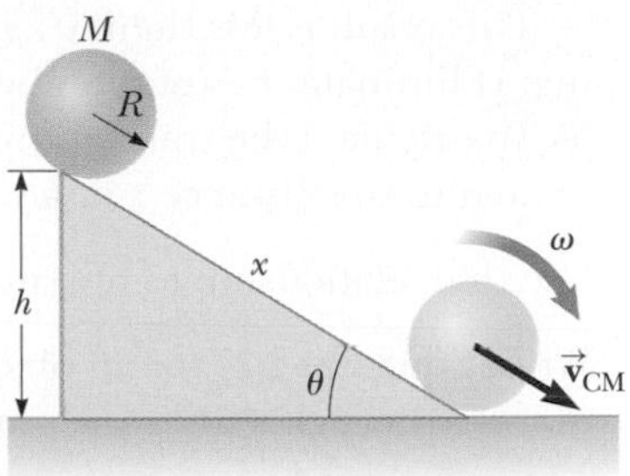

ACTIVE FIGURE 10.26

A sphere rolling down an incline. Mechanical energy of the sphere–Earth system is conserved if no slipping occurs.

For the sphere–Earth system in Active Figure 10.26, we define the zero configuration of gravitational potential energy to be when the sphere is at the bottom of the incline. Therefore, conservation of mechanical energy gives

$$K_f + U_f = K_i + U_i$$

$$\frac{1}{2}\left(\frac{I_{CM}}{R^2} + M\right)v_{CM}{}^2 + 0 = 0 + Mgh$$

$$v_{CM} = \left[\frac{2gh}{1 + (I_{CM}/MR^2)}\right]^{1/2} \quad \textbf{(10.30)}$$

***Quick Quiz* 10.7** A ball rolls without slipping down incline A, starting from rest. At the same time, a box starts from rest and slides down incline B, which is identical to incline A except that it is frictionless. Which arrives at the bottom first? **(a)** The ball arrives first. **(b)** The box arrives first. **(c)** Both arrive at the same time. **(d)** It is impossible to determine.

Example 10.13 Sphere Rolling Down an Incline

For the solid sphere shown in Active Figure 10.26, calculate the translational speed of the center of mass at the bottom of the incline and the magnitude of the translational acceleration of the center of mass.

SOLUTION

Conceptualize Imagine rolling the sphere down the incline. Compare it in your mind to a book sliding down a frictionless incline. You probably have experience with objects rolling down inclines and may be tempted to think that the

continued

10.13 *cont.*

sphere would move down the incline faster than the book. You do *not,* however, have experience with objects sliding down *frictionless* inclines! So, which object will reach the bottom first? (See Quick Quiz 10.7.)

Categorize We model the sphere and the Earth as an isolated system in terms of energy with no nonconservative forces acting. This model is the one that led to Equation 10.30, so we can use that result.

Analyze Evaluate the speed of the center of mass of the sphere from Equation 10.30:

$$(1)\quad v_{\text{CM}} = \left[\frac{2gh}{1 + (\frac{2}{5}MR^2/MR^2)}\right]^{1/2} = (\tfrac{10}{7}gh)^{1/2}$$

This result is less than $\sqrt{2gh}$, which is the speed an object would have if it simply slid down the incline without rotating. (Eliminate the rotation by setting $I_{\text{CM}} = 0$ in Eq. 10.30.)

To calculate the translational acceleration of the center of mass, notice that the vertical displacement of the sphere is related to the distance x it moves along the incline through the relationship $h = x \sin\theta$.

Use this relationship to rewrite Equation (1):

$$v_{\text{CM}}^2 = \tfrac{10}{7}gx\sin\theta$$

Write Equation 2.17 for an object starting from rest and moving through a distance x under constant acceleration:

$$v_{\text{CM}}^2 = 2a_{\text{CM}}x$$

Equate the preceding two expressions to find a_{CM}:

$$a_{\text{CM}} = \tfrac{5}{7}g\sin\theta$$

Finalize Both the speed and the acceleration of the center of mass are *independent* of the mass and the radius of the sphere. That is, all homogeneous solid spheres experience the same speed and acceleration on a given incline. Try to verify this statement experimentally with balls of different sizes, such as a marble and a croquet ball.

If we were to repeat the acceleration calculation for a hollow sphere, a solid cylinder, or a hoop, we would obtain similar results in which only the factor in front of $g\sin\theta$ would differ. The constant factors that appear in the expressions for v_{CM} and a_{CM} depend only on the moment of inertia about the center of mass for the specific object. In all cases, the acceleration of the center of mass is *less* than $g\sin\theta$, the value the acceleration would have if the incline were frictionless and no rolling occurred.

Example 10.14 Pulling on a Spool[3]

A cylindrically symmetric spool of mass m and radius R sits at rest on a horizontal table with friction (Fig. 10.27). With your hand on a massless string wrapped around the axle of radius r, you pull on the spool with a constant horizontal force of magnitude T to the right. As a result, the spool rolls without slipping a distance L along the table with no rolling friction.

(A) Find the final translational speed of the center of mass of the spool.

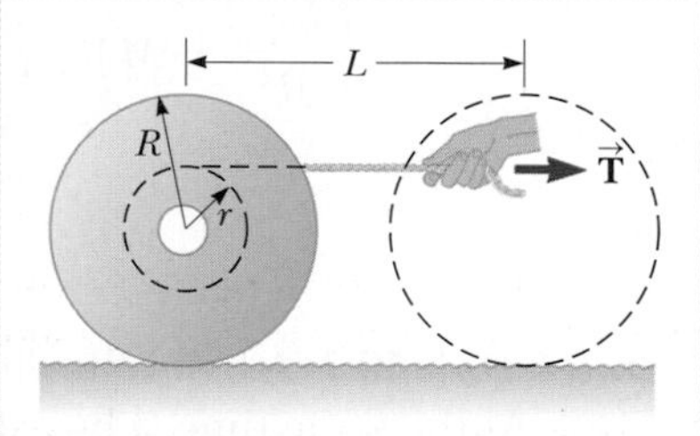

Figure 10.27 (Example 10.14) A spool rests on a horizontal table. A string is wrapped around the axle and is pulled to the right by a hand.

SOLUTION

Conceptualize Use Figure 10.27 to visualize the motion of the spool when you pull the string. For the spool to roll through a distance L, notice that your hand on the string must pull through a distance *different* from L.

Categorize The spool is a rigid object under a net torque, but the net torque includes that due to the friction force at the bottom of the spool, about which we know nothing. Therefore, an approach based on the rigid object under a net torque model will not be successful. Work is done by your hand on the spool and string, which form a nonisolated system in terms of energy. Let's see if an approach based on the nonisolated system (energy) model is fruitful.

Analyze The only type of energy that changes in the system is the kinetic energy of the spool. There is no rolling friction, so there is no change in internal energy. The only way that energy crosses the system's boundary is by the work done

[3]Example 10.14 was inspired in part by C. E. Mungan, "A primer on work–energy relationships for introductory physics," *The Physics Teacher,* **43**:10, 2005.

10.14 *cont.*

by your hand on the string. No work is done by the static force of friction on the bottom of the spool because the point of application of the force moves through no displacement.

Write the appropriate reduction of the conservation of energy equation, Equation 8.2:

$$(1)\quad W = \Delta K = \Delta K_{\text{trans}} + \Delta K_{\text{rot}}$$

where W is the work done on the string by your hand. To find this work, we need to find the displacement of your hand during the process.

We first find the length of string that has unwound off the spool. If the spool rolls through a distance L, the total angle through which it rotates is $\theta = L/R$. The axle also rotates through this angle.

Use Equation 10.1a to find the total arc length through which the axle turns:

$$\ell = r\theta = \frac{r}{R}L$$

This result also gives the length of string pulled off the axle. Your hand will move through this distance *plus* the distance L through which the spool moves. Therefore, the magnitude of the displacement of the point of application of the force applied by your hand is $\ell + L = L(1 + r/R)$.

Evaluate the work done by your hand on the string:

$$(2)\quad W = TL\left(1 + \frac{r}{R}\right)$$

Substitute Equation (2) into Equation (1):

$$TL\left(1 + \frac{r}{R}\right) = \tfrac{1}{2}mv_{\text{CM}}^2 + \tfrac{1}{2}I\omega^2$$

where I is the moment of inertia of the spool about its center of mass and v_{CM} and ω are the final values after the wheel rolls through the distance L.

Apply the nonslip rolling condition $\omega = v_{\text{CM}}/R$:

$$TL\left(1 + \frac{r}{R}\right) = \tfrac{1}{2}mv_{\text{CM}}^2 + \tfrac{1}{2}I\frac{v_{\text{CM}}^2}{R^2}$$

Solve for v_{CM}:

$$(3)\quad v_{\text{CM}} = \sqrt{\frac{2TL(1 + r/R)}{m(1 + I/mR^2)}}$$

(B) Find the value of the friction force f.

SOLUTION

Categorize Because the friction force does no work, we cannot evaluate it from an energy approach. We model the spool as a nonisolated system, but this time in terms of momentum. The string applies a force across the boundary of the system, resulting in an impulse on the system. Because the forces on the spool are constant, we can model the spool's center of mass as a particle under constant acceleration.

Analyze Write the impulse–momentum theorem (Eq. 9.40) for the spool:

$$(4)\quad (T - f)\Delta t = m(v_{\text{CM}} - 0) = mv_{\text{CM}}$$

For a particle under constant acceleration starting from rest, Equation 2.14 tells us that the average velocity of the center of mass is half the final velocity.

Use Equation 2.2 to find the time interval for the center of mass of the spool to move a distance L from rest to a final speed v_{CM}:

$$(5)\quad \Delta t = \frac{L}{v_{\text{CM,avg}}} = \frac{2L}{v_{\text{CM}}}$$

Substitute Equation (5) into Equation (4):

$$(T - f)\frac{2L}{v_{\text{CM}}} = mv_{\text{CM}}$$

Solve for the friction force f:

$$f = T - \frac{mv_{\text{CM}}^2}{2L}$$

continued

10.14 *cont.*

Substitute v_{CM} from Equation (3):

$$f = T - \frac{m}{2L}\left[\frac{2TL(1 + r/R)}{m(1 + I/mR^2)}\right]$$

$$= T - T\frac{(1 + r/R)}{(1 + I/mR^2)} = T\left[\frac{I - mrR}{I + mR^2}\right]$$

Finalize Notice that we could use the impulse–momentum theorem for the translational motion of the spool while ignoring that the spool is rotating! This fact demonstrates the power of our growing list of approaches to solving problems.

Summary

Definitions

The **angular position** of a rigid object is defined as the angle θ between a reference line attached to the object and a reference line fixed in space. The **angular displacement** of a particle moving in a circular path or a rigid object rotating about a fixed axis is $\Delta\theta \equiv \theta_f - \theta_i$.

The **instantaneous angular speed** of a particle moving in a circular path or of a rigid object rotating about a fixed axis is

$$\omega \equiv \frac{d\theta}{dt} \qquad \textbf{(10.3)}$$

The **instantaneous angular acceleration** of a particle moving in a circular path or of a rigid object rotating about a fixed axis is

$$\alpha \equiv \frac{d\omega}{dt} \qquad \textbf{(10.5)}$$

When a rigid object rotates about a fixed axis, every part of the object has the same angular speed and the same angular acceleration.

The **moment of inertia of a system of particles** is defined as

$$I \equiv \sum_i m_i r_i^2 \qquad \textbf{(10.15)}$$

where m_i is the mass of the ith particle and r_i is its distance from the rotation axis.

The magnitude of the **torque** associated with a force $\vec{\mathbf{F}}$ acting on an object at a distance r from the rotation axis is

$$\tau = rF\sin\phi = Fd \qquad \textbf{(10.19)}$$

where ϕ is the angle between the position vector of the point of application of the force and the force vector, and d is the moment arm of the force, which is the perpendicular distance from the rotation axis to the line of action of the force.

Concepts and Principles

When a rigid object rotates about a fixed axis, the angular position, angular speed, and angular acceleration are related to the translational position, translational speed, and translational acceleration through the relationships

$$s = r\theta \qquad \textbf{(10.1a)}$$

$$v = r\omega \qquad \textbf{(10.10)}$$

$$a_t = r\alpha \qquad \textbf{(10.11)}$$

If a rigid object rotates about a fixed axis with angular speed ω, its **rotational kinetic energy** can be written

$$K_R = \tfrac{1}{2}I\omega^2 \qquad \textbf{(10.16)}$$

where I is the moment of inertia of the object about the axis of rotation.

The **moment of inertia of a rigid object** is

$$I = \int r^2\, dm \qquad \textbf{(10.17)}$$

where r is the distance from the mass element dm to the axis of rotation.

The rate at which work is done by an external force in rotating a rigid object about a fixed axis, or the **power** delivered, is

$$P = \tau\omega \qquad \textbf{(10.23)}$$

If work is done on a rigid object and the only result of the work is rotation about a fixed axis, the net work done by external forces in rotating the object equals the change in the rotational kinetic energy of the object:

$$\sum W = \tfrac{1}{2}I\omega_f^2 - \tfrac{1}{2}I\omega_i^2 \qquad \textbf{(10.24)}$$

The **total kinetic energy** of a rigid object rolling on a rough surface without slipping equals the rotational kinetic energy about its center of mass plus the translational kinetic energy of the center of mass:

$$K = \tfrac{1}{2}I_{CM}\omega^2 + \tfrac{1}{2}Mv_{CM}^2 \qquad \textbf{(10.28)}$$

Analysis Models for Problem Solving

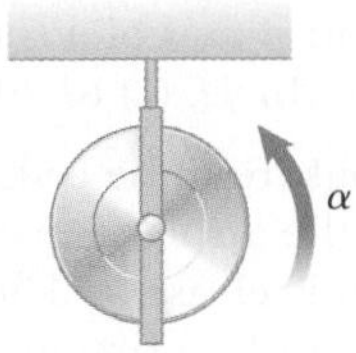

Rigid Object Under Constant Angular Acceleration. If a rigid object rotates about a fixed axis under constant angular acceleration, one can apply equations of kinematics that are analogous to those for translational motion of a particle under constant acceleration:

$$\omega_f = \omega_i + \alpha t \qquad \textbf{(10.6)}$$

$$\theta_f = \theta_i + \omega_i t + \tfrac{1}{2}\alpha t^2 \qquad \textbf{(10.7)}$$

$$\omega_f^2 = \omega_i^2 + 2\alpha(\theta_f - \theta_i) \qquad \textbf{(10.8)}$$

$$\theta_f = \theta_i + \tfrac{1}{2}(\omega_i + \omega_f)t \qquad \textbf{(10.9)}$$

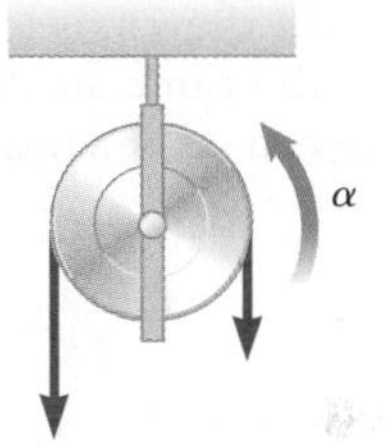

Rigid Object Under a Net Torque. If a rigid object free to rotate about a fixed axis has a net external torque acting on it, the object undergoes an angular acceleration α, where

$$\sum \tau_{ext} = I\alpha \qquad \textbf{(10.21)}$$

This equation is the rotational analog to Newton's second law in the particle under a net force model.

Objective Questions

☐ denotes answer available in Student Solutions Manual/Study Guide

1. A wheel is rotating about a fixed axis with constant angular acceleration 3 rad/s^2. At different moments, its angular speed is −2 rad/s, 0, and +2 rad/s. For a point on the rim of the wheel, consider at these moments the magnitude of the tangential component of acceleration and the magnitude of the radial component of acceleration. Rank the following five items from largest to smallest: (a) $|a_t|$ when $\omega = -2$ rad/s, (b) $|a_r|$ when $\omega = -2$ rad/s, (c) $|a_r|$ when $\omega = 0$, (d) $|a_t|$ when $\omega = 2$ rad/s, and (e) $|a_r|$ when $\omega = 2$ rad/s. If two items are equal, show them as equal in your ranking. If a quantity is equal to zero, show that fact in your ranking.

2. A grindstone increases in angular speed from 4.00 rad/s to 12.00 rad/s in 4.00 s. Through what angle does it turn during that time interval if the angular acceleration is constant? (a) 8.00 rad (b) 12.0 rad (c) 16.0 rad (d) 32.0 rad (e) 64.0 rad

3. A cyclist rides a bicycle with a wheel radius of 0.500 m across campus. A piece of plastic on the front rim makes a clicking sound every time it passes through the fork. If the cyclist counts 320 clicks between her apartment and the cafeteria, how far has she traveled? (a) 0.50 km (b) 0.80 km (c) 1.0 km (d) 1.5 km (e) 1.8 km

4. Consider an object on a rotating disk a distance r from its center, held in place on the disk by static friction. Which of the following statements is *not* true concerning this object? (a) If the angular speed is constant, the object must have constant tangential speed. (b) If the angular speed is

constant, the object is not accelerated. (c) The object has a tangential acceleration only if the disk has an angular acceleration. (d) If the disk has an angular acceleration, the object has both a centripetal acceleration and a tangential acceleration. (e) The object always has a centripetal acceleration except when the angular speed is zero.

5. Suppose a car's standard tires are replaced with tires 1.30 times larger in diameter. **(i)** Will the car's speedometer reading be (a) 1.69 times too high, (b) 1.30 times too high, (c) accurate, (d) 1.30 times too low, (e) 1.69 times too low, or (f) inaccurate by an unpredictable factor? **(ii)** Will the car's fuel economy in miles per gallon or km/L appear to be (a) 1.69 times better, (b) 1.30 times better, (c) essentially the same, (d) 1.30 times worse, or (e) 1.69 times worse?

6. Figure OQ10.6 shows a system of four particles joined by light, rigid rods. Assume $a = b$ and M is larger than m. About which of the coordinate axes does the system have **(i)** the smallest and **(ii)** the largest moment of inertia? (a) the x axis (b) the y axis (c) the z axis. (d) The moment of inertia has the same small value for two axes. (e) The moment of inertia is the same for all three axes.

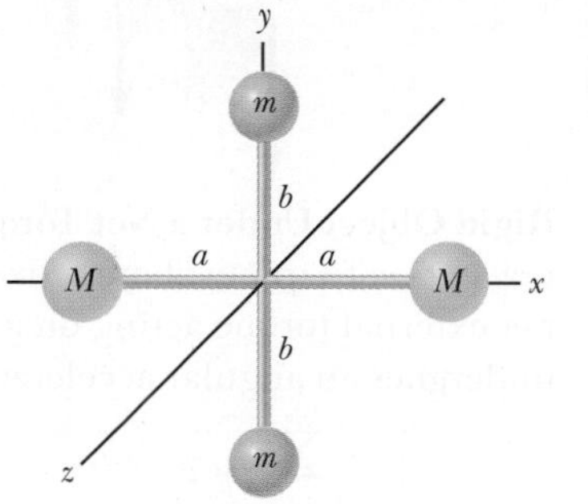

Figure OQ10.6

7. As shown in Figure OQ10.7, a cord is wrapped onto a cylindrical reel mounted on a fixed, frictionless, horizontal axle. When does the reel have a greater magnitude of angular acceleration? (a) When the cord is pulled down with a constant force of 50 N. (b) When an object of weight 50 N is hung from the cord and released. (c) The angular accelerations in parts (a) and (b) are equal. (d) It is impossible to determine.

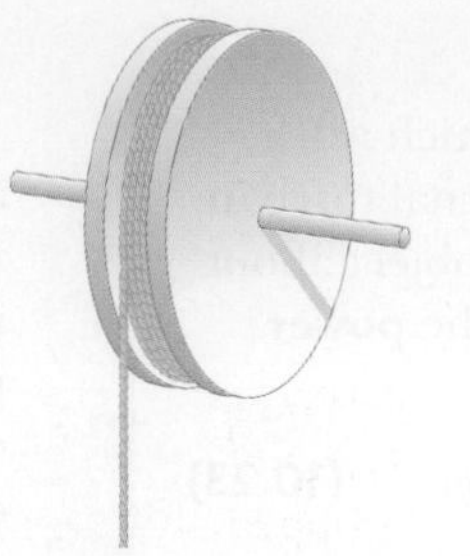

Figure OQ10.7 Objective Question 7 and Conceptual Question 4.

8. A constant net torque is exerted on an object. Which of the following quantities for the object cannot be constant? Choose all that apply. (a) angular position (b) angular velocity (c) angular acceleration (d) moment of inertia (e) kinetic energy

9. A solid aluminum sphere of radius R has moment of inertia I about an axis through its center. Will the moment of inertia about a central axis of a solid aluminum sphere of radius $2R$ be (a) $2I$, (b) $4I$, (c) $8I$, (d) $16I$, or (e) $32I$?

10. A toy airplane hangs from the ceiling at the bottom end of a string. You turn the airplane many times to wind up the string clockwise and release it. The airplane starts to spin counterclockwise, slowly at first and then faster and faster. Take counterclockwise as the positive sense and assume friction is negligible. When the string is entirely unwound, the airplane has its maximum rate of rotation. **(i)** At this moment, is its angular acceleration (a) positive, (b) negative, or (c) zero? **(ii)** The airplane continues to spin, winding the string counterclockwise as it slows down. At the moment it momentarily stops, is its angular acceleration (a) positive, (b) negative, or (c) zero?

11. A basketball rolls across a classroom floor without slipping, with its center of mass moving at a certain speed. A block of ice of the same mass is set sliding across the floor with the same speed along a parallel line. Which object has more **(i)** kinetic energy and **(ii)** momentum? (a) The basketball does. (b) The ice does. (c) The two quantities are equal. **(iii)** The two objects encounter a ramp sloping upward. Which object will travel farther up the ramp? (a) The basketball will. (b) The ice will. (c) They will travel equally far up the ramp.

Conceptual Questions

☐ denotes answer available in Student Solutions Manual/Study Guide

1. (a) What is the angular speed of the second hand of a clock? (b) What is the direction of $\vec{\omega}$ as you view a clock hanging on a vertical wall? (c) What is the magnitude of the angular acceleration vector $\vec{\alpha}$ of the second hand?

2. One blade of a pair of scissors rotates counterclockwise in the xy plane. (a) What is the direction of $\vec{\omega}$ for the blade? (b) What is the direction of $\vec{\alpha}$ if the magnitude of the angular velocity is decreasing in time?

3. Suppose just two external forces act on a stationary, rigid object and the two forces are equal in magnitude and opposite in direction. Under what condition does the object start to rotate?

4. Explain how you might use the apparatus described in Figure OQ10.7 to determine the moment of inertia of the wheel. *Note:* If the wheel does not have a uniform mass density, the moment of inertia is not necessarily equal to $\frac{1}{2}MR^2$.

5. Using the results from Example 10.10, how would you calculate the angular speed of the wheel and the linear speed of the hanging object at $t = 2$ s, assuming the system is released from rest at $t = 0$?

6. Explain why changing the axis of rotation of an object changes its moment of inertia.

7. Suppose you have two eggs, one hard-boiled and the other uncooked. You wish to determine which is the hard-boiled egg without breaking the eggs, which can be done by spinning the two eggs on the floor and comparing the rotational motions. (a) Which egg spins faster? (b) Which egg rotates more uniformly? (c) Which egg begins spinning again after being stopped and then immediately released? Explain your answers to parts (a), (b), and (c).

8. Which of the entries in Table 10.2 applies to finding the moment of inertia (a) of a long, straight sewer pipe rotating about its axis of symmetry? (b) Of an embroidery hoop rotating about an axis through its center and perpendicular to its plane? (c) Of a uniform door turning on its hinges? (d) Of a coin turning about an axis through its center and perpendicular to its faces?

9. Is it possible to change the translational kinetic energy of an object without changing its rotational energy?

10. Must an object be rotating to have a nonzero moment of inertia?

11. If you see an object rotating, is there necessarily a net torque acting on it?

12. If a small sphere of mass M were placed at the end of the rod in Figure 10.21, would the result for ω be greater than, less than, or equal to the value obtained in Example 10.11?

13. Three objects of uniform density—a solid sphere, a solid cylinder, and a hollow cylinder—are placed at the top of an incline (Fig. CQ10.13). They are all released from rest at the same elevation and roll without slipping. (a) Which object reaches the bottom first? (b) Which reaches it last? *Note:* The result is independent of the masses and the radii of the objects. (Try this activity at home!)

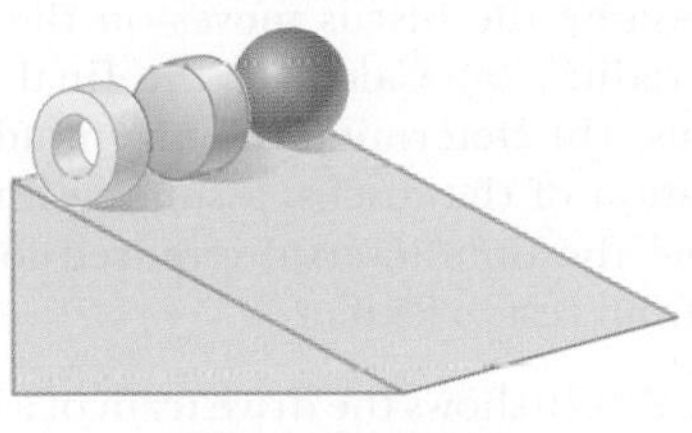

Figure CQ10.13

14. Suppose you set your textbook sliding across a gymnasium floor with a certain initial speed. It quickly stops moving because of a friction force exerted on it by the floor. Next, you start a basketball rolling with the same initial speed. It keeps rolling from one end of the gym to the other. (a) Why does the basketball roll so far? (b) Does friction significantly affect the basketball's motion?

15. Figure CQ10.15 shows a side view of a child's tricycle with rubber tires on a horizontal concrete sidewalk. If a string were attached to the upper pedal on the far side and pulled forward horizontally, the tricycle would start to roll forward. (a) Instead, assume a string is attached to the lower pedal on the near side and pulled forward horizontally as shown by A. Will the tricycle start to roll? If so, which way? Answer the same questions if (b) the string is pulled forward and upward as shown by B, (c) if the string is pulled straight down as shown by C, and (d) if the string is pulled forward and downward as shown by D. (e) **What If?** Suppose the string is instead attached to the rim of the front wheel and pulled upward and backward as shown by E. Which way does the tricycle roll? (f) Explain a pattern of reasoning, based on the figure, that makes it easy to answer questions such as these. What physical quantity must you evaluate?

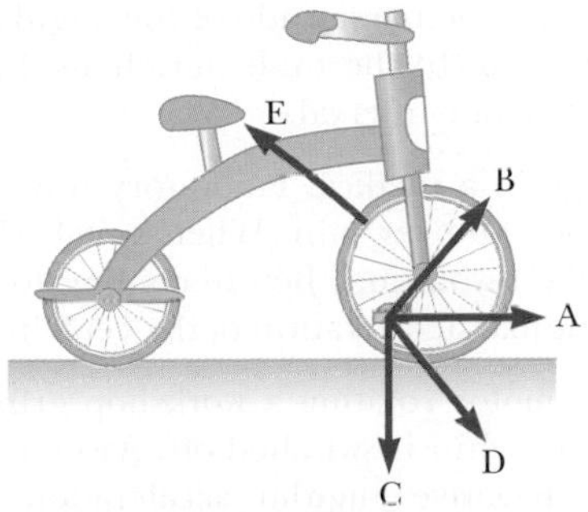

Figure CQ10.15

16. A person balances a meterstick in a horizontal position on the extended index fingers of her right and left hands. She slowly brings the two fingers together. The stick remains balanced, and the two fingers always meet at the 50-cm mark regardless of their original positions. (Try it!) Explain why that occurs.

Problems

ENHANCED WebAssign The problems found in this chapter may be assigned online in Enhanced WebAssign

1. denotes straightforward problem; 2. denotes intermediate problem; 3. denotes challenging problem

1. full solution available in the Student Solutions Manual/Study Guide

1. denotes problems most often assigned in Enhanced WebAssign; these provide students with targeted feedback and either a Master It tutorial or a Watch It solution video.

QC denotes asking for quantitative and conceptual reasoning

S denotes symbolic reasoning problem

M denotes Master It tutorial available in Enhanced WebAssign

GP denotes guided problem

shaded denotes "paired problems" that develop reasoning with symbols and numerical values

Section 10.1 Angular Position, Velocity, and Acceleration

1. QC (a) Find the angular speed of the Earth's rotation about its axis. (b) How does this rotation affect the shape of the Earth?

2. A potter's wheel moves uniformly from rest to an angular speed of 1.00 rev/s in 30.0 s. (a) Find its average angular acceleration in radians per second per second. (b) Would doubling the angular acceleration during the given period have doubled the final angular speed?

3. During a certain time interval, the angular position of a swinging door is described by $\theta = 5.00 + 10.0t + 2.00t^2$, where θ is in radians and t is in seconds. Determine the angular position, angular speed, and angular acceleration of the door (a) at $t = 0$ and (b) at $t = 3.00$ s.

4. A bar on a hinge starts from rest and rotates with an angular acceleration $\alpha = 10 + 6t$, where α is in rad/s^2 and t is in seconds. Determine the angle in radians through which the bar turns in the first 4.00 s.

Section 10.2 Analysis Model: Rigid Object Under Constant Angular Acceleration

5. A wheel starts from rest and rotates with constant angular acceleration to reach an angular speed of 12.0 rad/s in 3.00 s. Find (a) the magnitude of the angular acceleration of the wheel and (b) the angle in radians through which it rotates in this time interval.

6. A centrifuge in a medical laboratory rotates at an angular speed of 3 600 rev/min. When switched off, it rotates through 50.0 revolutions before coming to rest. Find the constant angular acceleration of the centrifuge.

7. An electric motor rotating a workshop grinding wheel at 1.00×10^2 rev/min is switched off. Assume the wheel has a constant negative angular acceleration of magnitude 2.00 rad/s^2. (a) How long does it take the grinding wheel to stop? (b) Through how many radians has the wheel turned during the time interval found in part (a)?

8. QC A machine part rotates at an angular speed of 0.060 rad/s; its speed is then increased to 2.2 rad/s at an angular acceleration of 0.70 rad/s^2. (a) Find the angle through which the part rotates before reaching this final speed. (b) If both the initial and final angular speeds are doubled and the angular acceleration remains the same, by what factor is the angular displacement changed? Why?

9. A rotating wheel requires 3.00 s to rotate through 37.0 revolutions. Its angular speed at the end of the 3.00-s interval is 98.0 rad/s. What is the constant angular acceleration of the wheel?

10. *Why is the following situation impossible?* Starting from rest, a disk rotates around a fixed axis through an angle of 50.0 rad in a time interval of 10.0 s. The angular acceleration of the disk is constant during the entire motion, and its final angular speed is 8.00 rad/s.

11. QC S **Review.** Consider a tall building located on the Earth's equator. As the Earth rotates, a person on the top floor of the building moves faster than someone on the ground with respect to an inertial reference frame because the person on the ground is closer to the Earth's axis. Consequently, if an object is dropped from the top floor to the ground a distance h below, it lands east of the point vertically below where it was dropped. (a) How far to the east will the object land? Express your answer in terms of h, g, and the angular speed ω of the Earth. Ignore air resistance and assume the free-fall acceleration is constant over this range of heights. (b) Evaluate the eastward displacement for $h = 50.0$ m. (c) In your judgment, were we justified in ignoring this aspect of the *Coriolis effect* in our previous study of free fall? (d) Suppose the angular speed of the Earth were to decrease due to tidal friction with constant angular acceleration. Would the eastward displacement of the dropped object increase or decrease compared with that in part (b)?

12. The tub of a washer goes into its spin cycle, starting from rest and gaining angular speed steadily for 8.00 s, at which time it is turning at 5.00 rev/s. At this point, the person doing the laundry opens the lid, and a safety switch turns off the washer. The tub smoothly slows to rest in 12.0 s. Through how many revolutions does the tub turn while it is in motion?

Section 10.3 Angular and Translational Quantities

13. A racing car travels on a circular track of radius 250 m. Assuming the car moves with a constant speed of 45.0 m/s, find (a) its angular speed and (b) the magnitude and direction of its acceleration.

14. Make an order-of-magnitude estimate of the number of revolutions through which a typical automobile tire turns in one year. State the quantities you measure or estimate and their values.

15. A discus thrower (Fig. P4.27, page 97) accelerates a discus from rest to a speed of 25.0 m/s by whirling it through 1.25 rev. Assume the discus moves on the arc of a circle 1.00 m in radius. (a) Calculate the final angular speed of the discus. (b) Determine the magnitude of the angular acceleration of the discus, assuming it to be constant. (c) Calculate the time interval required for the discus to accelerate from rest to 25.0 m/s.

16. QC Figure P10.16 shows the drive train of a bicycle that has wheels 67.3 cm in diameter and pedal cranks 17.5 cm long. The cyclist pedals at a steady cadence of 76.0 rev/min. The chain engages with a front sprocket 15.2 cm in diameter and a rear sprocket 7.00 cm in diameter. Calculate (a) the

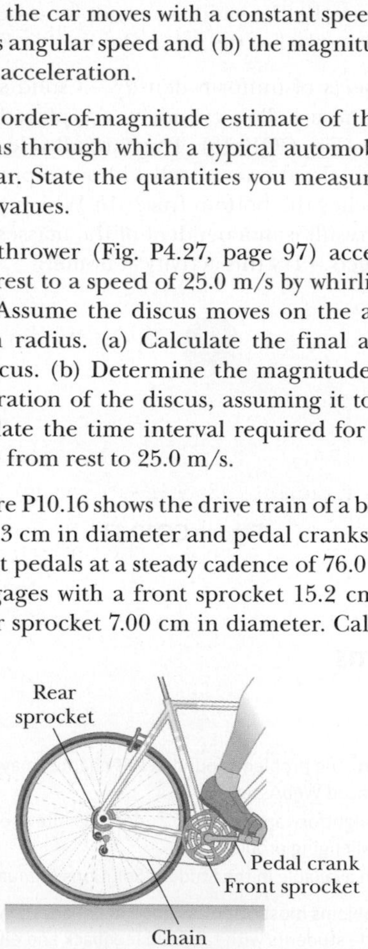

Figure P10.16

speed of a link of the chain relative to the bicycle frame, (b) the angular speed of the bicycle wheels, and (c) the speed of the bicycle relative to the road. (d) What pieces of data, if any, are not necessary for the calculations?

17. **M** A wheel 2.00 m in diameter lies in a vertical plane and rotates about its central axis with a constant angular acceleration of 4.00 rad/s^2. The wheel starts at rest at $t = 0$, and the radius vector of a certain point P on the rim makes an angle of 57.3° with the horizontal at this time. At $t = 2.00$ s, find (a) the angular speed of the wheel and, for point P, (b) the tangential speed, (c) the total acceleration, and (d) the angular position.

18. A car accelerates uniformly from rest and reaches a speed of 22.0 m/s in 9.00 s. Assuming the diameter of a tire is 58.0 cm, (a) find the number of revolutions the tire makes during this motion, assuming that no slipping occurs. (b) What is the final angular speed of a tire in revolutions per second?

19. A disk 8.00 cm in radius rotates at a constant rate of 1 200 rev/min about its central axis. Determine (a) its angular speed in radians per second, (b) the tangential speed at a point 3.00 cm from its center, (c) the radial acceleration of a point on the rim, and (d) the total distance a point on the rim moves in 2.00 s.

20. **Q|C** A straight ladder is leaning against the wall of a house. The ladder has rails 4.90 m long, joined by rungs 0.410 m long. Its bottom end is on solid but sloping ground so that the top of the ladder is 0.690 m to the left of where it should be, and the ladder is unsafe to climb. You want to put a flat rock under one foot of the ladder to compensate for the slope of the ground. (a) What should be the thickness of the rock? (b) Does using ideas from this chapter make it easier to explain the solution to part (a)? Explain your answer.

21. In a manufacturing process, a large, cylindrical roller is used to flatten material fed beneath it. The diameter of the roller is 1.00 m, and, while being driven into rotation around a fixed axis, its angular position is expressed as

$$\theta = 2.50t^2 - 0.600t^3$$

where θ is in radians and t is in seconds. (a) Find the maximum angular speed of the roller. (b) What is the maximum tangential speed of a point on the rim of the roller? (c) At what time t should the driving force be removed from the roller so that the roller does not reverse its direction of rotation? (d) Through how many rotations has the roller turned between $t = 0$ and the time found in part (c)?

22. **Review.** A small object with mass 4.00 kg moves counterclockwise with constant angular speed 1.50 rad/s in a circle of radius 3.00 m centered at the origin. It starts at the point with position vector $3.00\hat{\mathbf{i}}$ m. It then undergoes an angular displacement of 9.00 rad. (a) What is its new position vector? Use unit-vector notation for all vector answers. (b) In what quadrant is the particle located, and what angle does its position vector make with the positive x axis? (c) What is its velocity? (d) In what direction is it moving? (e) What is its acceleration? (f) Make a sketch of its position, velocity, and acceleration vectors. (g) What total force is exerted on the object?

23. A car traveling on a flat (unbanked), circular track accelerates uniformly from rest with a tangential acceleration of 1.70 m/s^2. The car makes it one quarter of the way around the circle before it skids off the track. From these data, determine the coefficient of static friction between the car and the track.

24. **S** A car traveling on a flat (unbanked), circular track accelerates uniformly from rest with a tangential acceleration of a. The car makes it one quarter of the way around the circle before it skids off the track. From these data, determine the coefficient of static friction between the car and the track.

Section 10.4 Rotational Kinetic Energy

25. The four particles in Figure P10.25 are connected by rigid rods of negligible mass. The origin is at the center of the rectangle. The system rotates in the xy plane about the z axis with an angular speed of 6.00 rad/s. Calculate (a) the moment of inertia of the system about the z axis and (b) the rotational kinetic energy of the system.

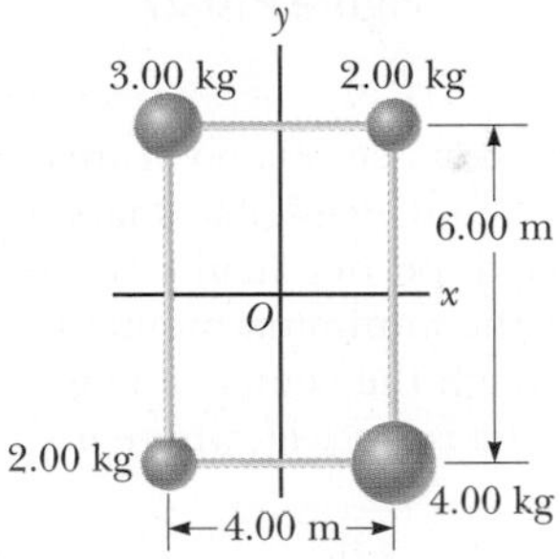

Figure P10.25

26. **Q|C** Rigid rods of negligible mass lying along the y axis connect three particles (Fig. P10.26). The system rotates about the x axis with an angular speed of 2.00 rad/s. Find (a) the moment of inertia about the x axis, (b) the total rotational kinetic energy evaluated from $\frac{1}{2}I\omega^2$, (c) the tangential speed of each particle, and (d) the total kinetic energy evaluated from $\sum \frac{1}{2}m_i v_i^2$. (e) Compare the answers for kinetic energy in parts (a) and (b).

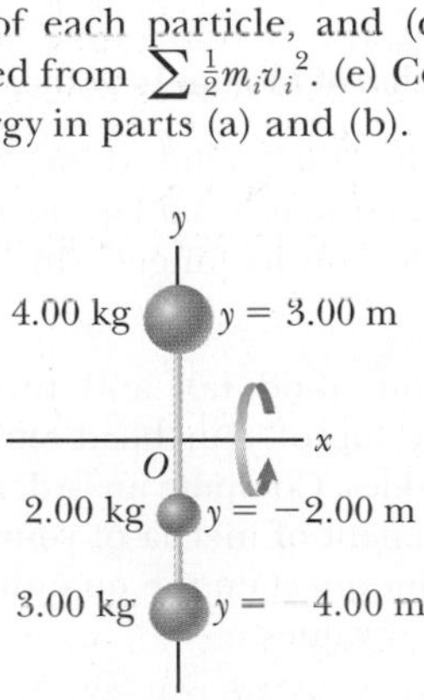

Figure P10.26

27. **Q|C** A *war-wolf* or *trebuchet* is a device used during the Middle Ages to throw rocks at castles and now sometimes used

to fling large vegetables and pianos as a sport. A simple trebuchet is shown in Figure P10.27. Model it as a stiff rod of negligible mass, 3.00 m long, joining particles of mass $m_1 = 0.120$ kg and $m_2 = 60.0$ kg at its ends. It can turn on a frictionless, horizontal axle perpendicular to the rod and 14.0 cm from the large-mass particle. The operator releases the trebuchet from rest in a horizontal orientation. (a) Find the maximum speed that the small-mass object attains. (b) While the small-mass object is gaining speed, does it move with constant acceleration? (c) Does it move with constant tangential acceleration? (d) Does the trebuchet move with constant angular acceleration? (e) Does it have constant momentum? (f) Does the trebuchet–Earth system have constant mechanical energy?

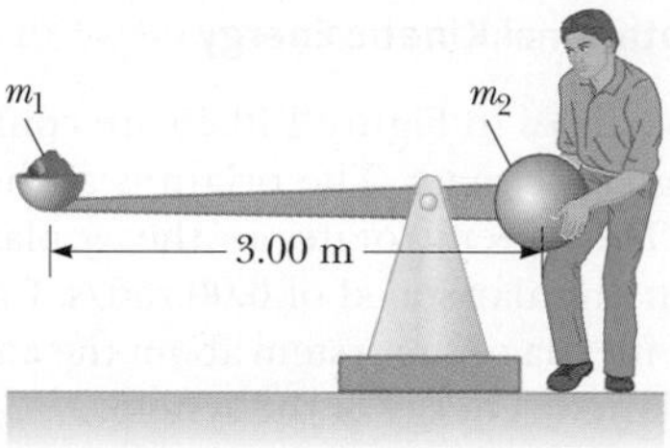

Figure P10.27

28. S Two balls with masses M and m are connected by a rigid rod of length L and negligible mass as shown in Figure P10.28. For an axis perpendicular to the rod, (a) show that the system has the minimum moment of inertia when the axis passes through the center of mass. (b) Show that this moment of inertia is $I = \mu L^2$, where $\mu = mM/(m + M)$.

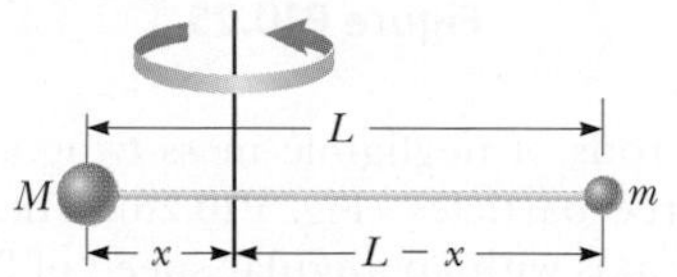

Figure P10.28

Section 10.5 Calculation of Moments of Inertia

29. Q|C A uniform, thin, solid door has height 2.20 m, width 0.870 m, and mass 23.0 kg. (a) Find its moment of inertia for rotation on its hinges. (b) Is any piece of data unnecessary?

30. Imagine that you stand tall and turn about a vertical axis through the top of your head and the point halfway between your ankles. Compute an order-of-magnitude estimate for the moment of inertia of your body for this rotation. In your solution, state the quantities you measure or estimate and their values.

31. Figure P10.31 shows a side view of a car tire before it is mounted on a wheel. Model it as having two sidewalls of uniform thickness 0.635 cm and a tread wall of uniform thickness 2.50 cm and width 20.0 cm. Assume the rubber has uniform density 1.10×10^3 kg/m^3. Find its moment of inertia about an axis perpendicular to the page through its center.

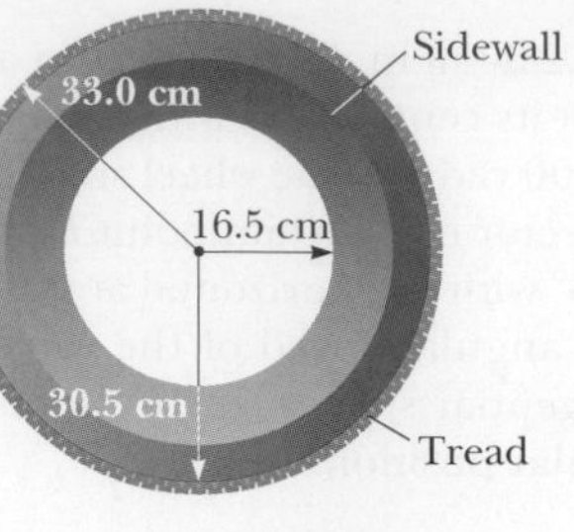

Figure P10.31

32. S Many machines employ cams for various purposes, such as opening and closing valves. In Figure P10.32, the cam is a circular disk of radius R with a hole of diameter R cut through it. As shown in the figure, the hole does not pass through the center of the disk. The cam with the hole cut out has mass M. The cam is mounted on a uniform, solid, cylindrical shaft of diameter R and also of mass M. What is the kinetic energy of the cam–shaft combination when it is rotating with angular speed ω about the shaft's axis?

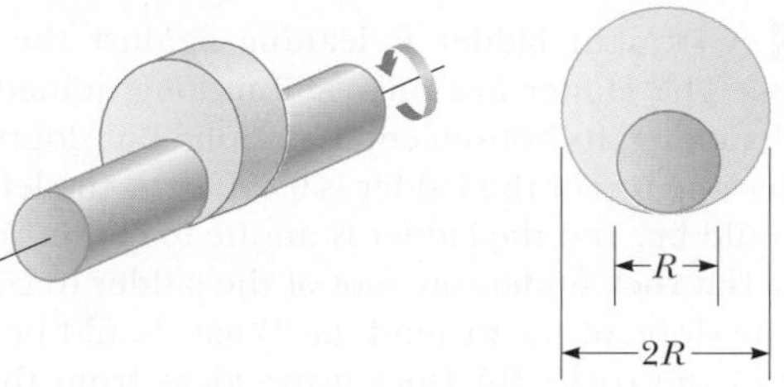

Figure P10.32

33. S Three identical thin rods, each of length L and mass m, are welded perpendicular to one another as shown in Figure P10.33. The assembly is rotated about an axis that passes through the end of one rod and is parallel to another. Determine the moment of inertia of this structure about this axis.

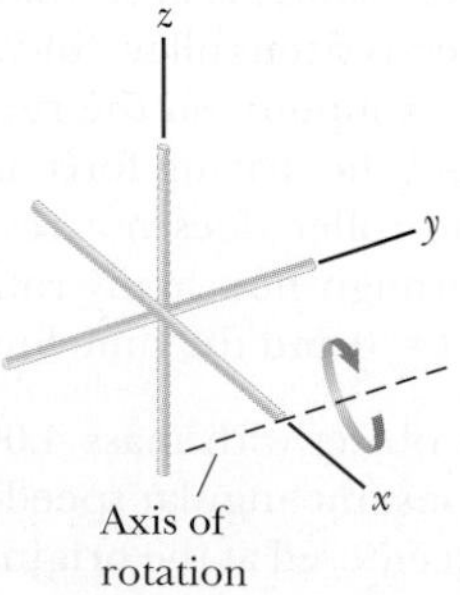

Figure P10.33

34. S Following the procedure used in Example 10.4, prove that the moment of inertia about the y axis of the rigid rod in Figure 10.9 is $\frac{1}{3}ML^2$.

Section 10.6 Torque

35. Find the net torque on the wheel in Figure P10.35 about the axle through O, taking $a = 10.0$ cm and $b = 25.0$ cm.

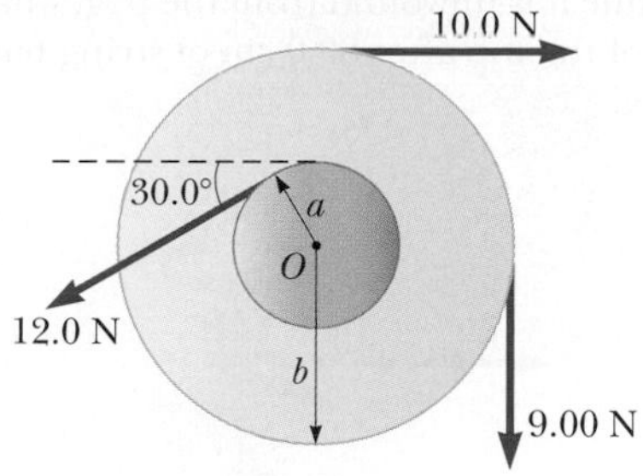

Figure P10.35

36. The fishing pole in Figure P10.36 makes an angle of 20.0° with the horizontal. What is the torque exerted by the fish about an axis perpendicular to the page and passing through the angler's hand if the fish pulls on the fishing line with a force $\vec{\mathbf{F}} = 100$ N at an angle 37.0° below the horizontal? The force is applied at a point 2.00 m from the angler's hands.

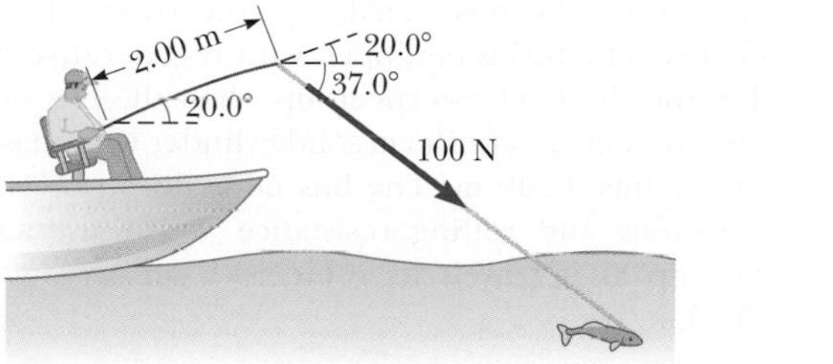

Figure P10.36

Section 10.7 Analysis Model: Rigid Object Under a Net Torque

37. An electric motor turns a flywheel through a drive belt that joins a pulley on the motor and a pulley that is rigidly attached to the flywheel as shown in Figure P10.37. The flywheel is a solid disk with a mass of 80.0 kg and a radius $R = 0.625$ m. It turns on a frictionless axle. Its pulley has much smaller mass and a radius of $r = 0.230$ m. The tension T_u in the upper (taut) segment of the belt is 135 N, and the flywheel has a clockwise angular acceleration of 1.67 rad/s^2. Find the tension in the lower (slack) segment of the belt.

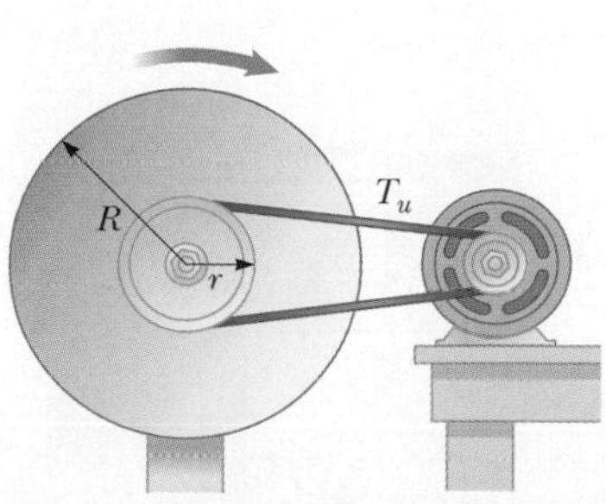

Figure P10.37

38. A grinding wheel is in the form of a uniform solid disk of radius 7.00 cm and mass 2.00 kg. It starts from rest and accelerates uniformly under the action of the constant torque of 0.600 N · m that the motor exerts on the wheel. (a) How long does the wheel take to reach its final operating speed of 1 200 rev/min? (b) Through how many revolutions does it turn while accelerating?

39. **M** A model airplane with mass 0.750 kg is tethered to the ground by a wire so that it flies in a horizontal circle 30.0 m in radius. The airplane engine provides a net thrust of 0.800 N perpendicular to the tethering wire. (a) Find the torque the net thrust produces about the center of the circle. (b) Find the angular acceleration of the airplane. (c) Find the translational acceleration of the airplane tangent to its flight path.

40. **Review.** A block of mass $m_1 = 2.00$ kg and a block of mass $m_2 = 6.00$ kg are connected by a massless string over a pulley in the shape of a solid disk having radius $R = 0.250$ m and mass $M = 10.0$ kg. The fixed, wedge-shaped ramp makes an angle of $\theta = 30.0°$ as shown in Figure P10.40. The coefficient of kinetic friction is 0.360 for both blocks. (a) Draw force diagrams of both blocks and of the pulley. Determine (b) the acceleration of the two blocks and (c) the tensions in the string on both sides of the pulley.

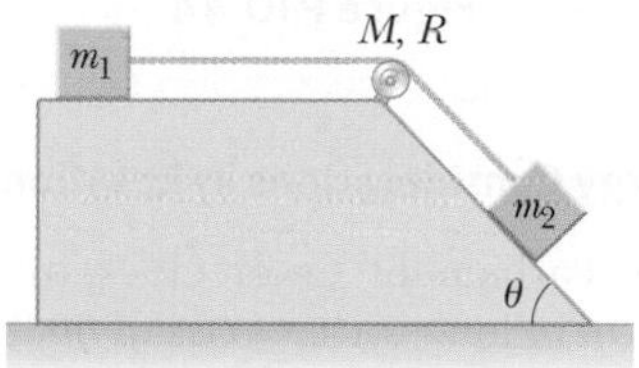

Figure P10.40

41. The combination of an applied force and a friction force produces a constant total torque of 36.0 N · m on a wheel rotating about a fixed axis. The applied force acts for 6.00 s. During this time, the angular speed of the wheel increases from 0 to 10.0 rad/s. The applied force is then removed, and the wheel comes to rest in 60.0 s. Find (a) the moment of inertia of the wheel, (b) the magnitude of the torque due to friction, and (c) the total number of revolutions of the wheel during the entire interval of 66.0 s.

42. **Q|C** A disk having moment of inertia 100 kg · m^2 is free to rotate without friction, starting from rest, about a fixed axis through its center. A tangential force whose magnitude can range from $F = 0$ to $F = 50.0$ N can be applied at any distance ranging from $R = 0$ to $R = 3.00$ m from the axis of rotation. (a) Find a pair of values of F and R that cause the disk to complete 2.00 rev in 10.0 s. (b) Is your answer for part (a) a unique answer? How many answers exist?

43. A potter's wheel—a thick stone disk of radius 0.500 m and mass 100 kg—is freely rotating at 50.0 rev/min. The potter can stop the wheel in 6.00 s by pressing a wet rag against

the rim and exerting a radially inward force of 70.0 N. Find the effective coefficient of kinetic friction between wheel and rag.

44. **Q|C** **Review.** Consider the system shown in Figure P10.44 with $m_1 = 20.0$ kg, $m_2 = 12.5$ kg, $R = 0.200$ m, and the mass of the pulley $M = 5.00$ kg. Object m_2 is resting on the floor, and object m_1 is 4.00 m above the floor when it is released from rest. The pulley axis is frictionless. The cord is light, does not stretch, and does not slip on the pulley. (a) Calculate the time interval required for m_1 to hit the floor. (b) How would your answer change if the pulley were massless?

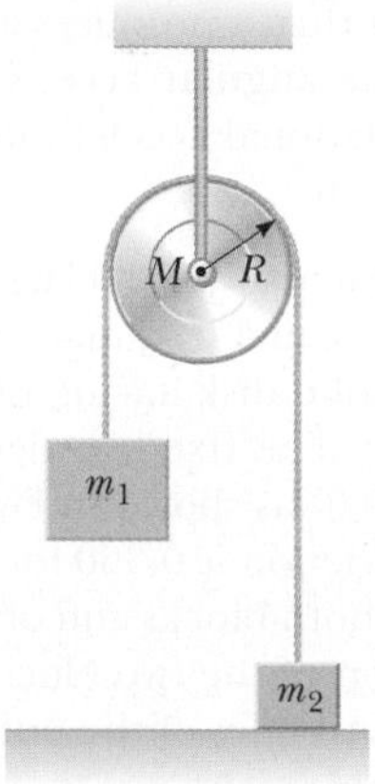

Figure P10.44

Section 10.8 Energy Considerations in Rotational Motion

45. Big Ben, the Parliament tower clock in London, has an hour hand 2.70 m long with a mass of 60.0 kg and a minute hand 4.50 m long with a mass of 100 kg (Fig. P10.45). Calculate the total rotational kinetic energy of the two hands about the axis of rotation. (You may model the hands as long, thin rods rotated about one end. Assume the hour and minute hands are rotating at a constant rate of one revolution per 12 hours and 60 minutes, respectively.)

Travelpix Ltd/Stone/Getty Images

Figure P10.45

46. A horizontal 800-N merry-go-round is a solid disk of radius 1.50 m and is started from rest by a constant horizontal force of 50.0 N applied tangentially to the edge of the disk. Find the kinetic energy of the disk after 3.00 s.

47. The top in Figure P10.47 has a moment of inertia of 4.00×10^{-4} kg · m^2 and is initially at rest. It is free to rotate about the stationary axis AA'. A string, wrapped around a peg along the axis of the top, is pulled in such a manner as to maintain a constant tension of 5.57 N. If the string does not slip while it is unwound from the peg, what is the angular speed of the top after 80.0 cm of string has been pulled off the peg?

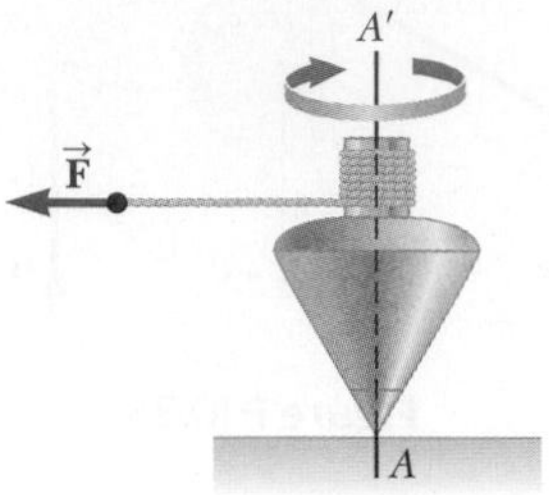

Figure P10.47

48. *Why is the following situation impossible?* In a large city with an air-pollution problem, a bus has no combustion engine. It runs over its citywide route on energy drawn from a large, rapidly rotating flywheel under the floor of the bus. The flywheel is spun up to its maximum rotation rate of 3 000 rev/min by an electric motor at the bus terminal. Every time the bus speeds up, the flywheel slows down slightly. The bus is equipped with regenerative braking so that the flywheel can speed up when the bus slows down. The flywheel is a uniform solid cylinder with mass 1 200 kg and radius 0.500 m. The bus body does work against air resistance and rolling resistance at the average rate of 25.0 hp as it travels its route with an average speed of 35.0 km/h.

49. In Figure P10.49, the hanging object has a mass of $m_1 = 0.420$ kg; the sliding block has a mass of $m_2 = 0.850$ kg; and the pulley is a hollow cylinder with a mass of $M = 0.350$ kg, an inner radius of $R_1 = 0.020\,0$ m, and an outer radius of $R_2 = 0.030\,0$ m. Assume the mass of the spokes is negligible. The coefficient of kinetic friction between the block and the horizontal surface is $\mu_k = 0.250$. The pulley turns without friction on its axle. The light cord does not stretch and does not slip on the pulley. The block has a velocity of $v_i = 0.820$ m/s toward the pulley when it passes a reference point on the table. (a) Use energy methods to predict its speed after it has moved to a second point, 0.700 m away. (b) Find the angular speed of the pulley at the same moment.

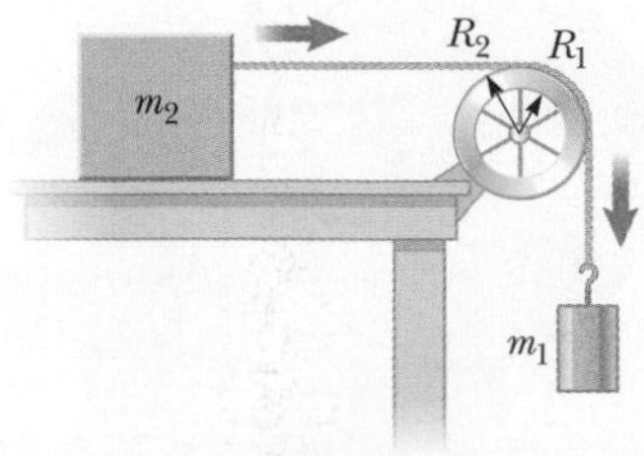

Figure P10.49

50. **Review.** A thin, cylindrical rod $\ell = 24.0$ cm long with mass $m = 1.20$ kg has a ball of diameter $d = 8.00$ cm and mass $M = 2.00$ kg attached to one end. The arrangement is originally vertical and stationary, with the ball at the top as shown in Figure P10.50. The combination is free to pivot about the bottom end of the rod after being given a slight nudge. (a) After the combination rotates through 90 degrees, what is its rotational kinetic energy? (b) What is the angular speed of the rod and ball? (c) What is the linear speed of the center of mass of the ball? (d) How does it compare with the speed had the ball fallen freely through the same distance of 28 cm?

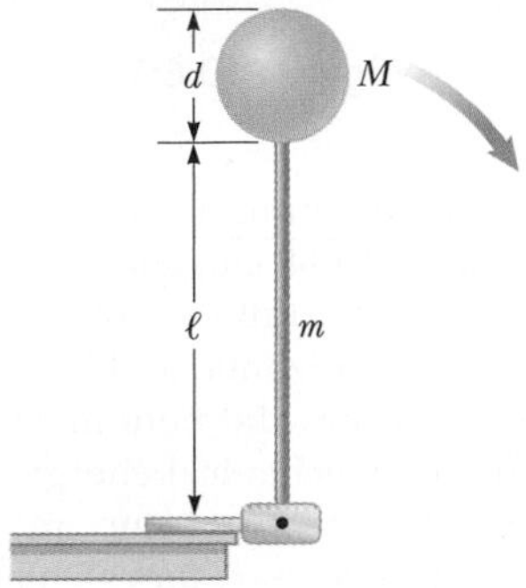

Figure P10.50

51. M **Review.** An object with a mass of $m = 5.10$ kg is attached to the free end of a light string wrapped around a reel of radius $R = 0.250$ m and mass $M = 3.00$ kg. The reel is a solid disk, free to rotate in a vertical plane about the horizontal axis passing through its center as shown in Figure P10.51. The suspended object is released from rest 6.00 m above the floor. Determine (a) the tension in the string, (b) the acceleration of the object, and (c) the speed with which the object hits the floor. (d) Verify your answer to part (c) by using the isolated system (energy) model.

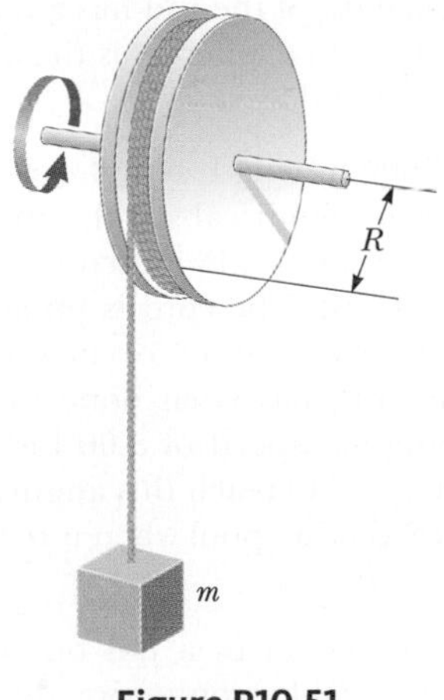

Figure P10.51

52. S This problem describes one experimental method for determining the moment of inertia of an irregularly shaped object such as the payload for a satellite. Figure P10.52 shows a counterweight of mass m suspended by a cord wound around a spool of radius r, forming part of a turntable supporting the object. The turntable can rotate without friction. When the counterweight is released from rest, it descends through a distance h, acquiring a speed v. Show that the moment of inertia I of the rotating apparatus (including the turntable) is $mr^2(2gh/v^2 - 1)$.

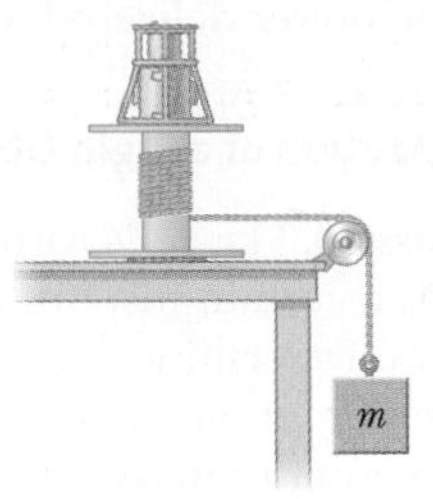

Figure P10.52

53. S A uniform solid disk of radius R and mass M is free to rotate on a frictionless pivot through a point on its rim (Fig. P10.53). If the disk is released from rest in the position shown by the copper-colored circle, (a) what is the speed of its center of mass when the disk reaches the position indicated by the dashed circle? (b) What is the speed of the lowest point on the disk in the dashed position? (c) **What If?** Repeat part (a) using a uniform hoop.

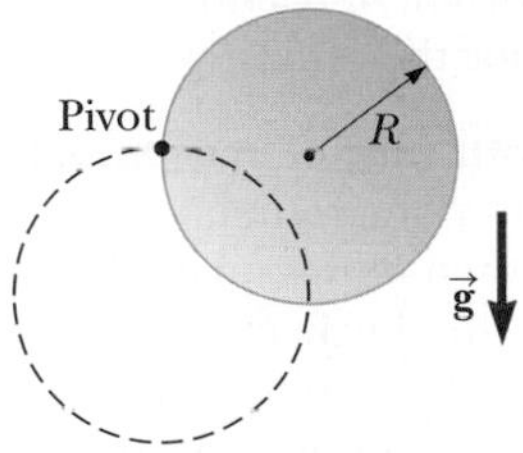

Figure P10.53

54. The head of a grass string trimmer has 100 g of cord wound in a light, cylindrical spool with inside diameter 3.00 cm and outside diameter 18.0 cm as shown in Figure P10.54. The cord has a linear density of 10.0 g/m. A single

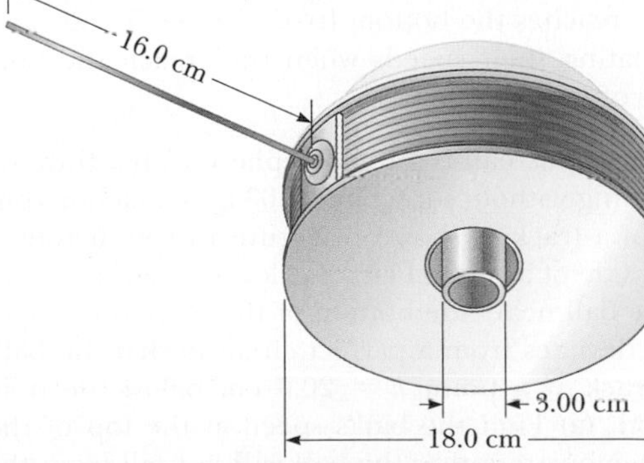

Figure P10.54

strand of the cord extends 16.0 cm from the outer edge of the spool. (a) When switched on, the trimmer speeds up from 0 to 2 500 rev/min in 0.215 s. What average power is delivered to the head by the trimmer motor while it is accelerating? (b) When the trimmer is cutting grass, it spins at 2 000 rev/min and the grass exerts an average tangential force of 7.65 N on the outer end of the cord, which is still at a radial distance of 16.0 cm from the outer edge of the spool. What is the power delivered to the head under load?

Section 10.9 Rolling Motion of a Rigid Object

55. A cylinder of mass 10.0 kg rolls without slipping on a horizontal surface. At a certain instant, its center of mass has a speed of 10.0 m/s. Determine (a) the translational kinetic energy of its center of mass, (b) the rotational kinetic energy about its center of mass, and (c) its total energy.

56. Q|C S A solid sphere is released from height h from the top of an incline making an angle θ with the horizontal. Calculate the speed of the sphere when it reaches the bottom of the incline (a) in the case that it rolls without slipping and (b) in the case that it slides frictionlessly without rolling. (c) Compare the time intervals required to reach the bottom in cases (a) and (b).

57. S (a) Determine the acceleration of the center of mass of a uniform solid disk rolling down an incline making angle θ with the horizontal. (b) Compare the acceleration found in part (a) with that of a uniform hoop. (c) What is the minimum coefficient of friction required to maintain pure rolling motion for the disk?

58. Q|C S A smooth cube of mass m and edge length r slides with speed v on a horizontal surface with negligible friction. The cube then moves up a smooth incline that makes an angle θ with the horizontal. A cylinder of mass m and radius r rolls without slipping with its center of mass moving with speed v and encounters an incline of the same angle of inclination but with sufficient friction that the cylinder continues to roll without slipping. (a) Which object will go the greater distance up the incline? (b) Find the difference between the maximum distances the objects travel up the incline. (c) Explain what accounts for this difference in distances traveled.

59. S A uniform solid disk and a uniform hoop are placed side by side at the top of an incline of height h. (a) If they are released from rest and roll without slipping, which object reaches the bottom first? (b) Verify your answer by calculating their speeds when they reach the bottom in terms of h.

60. Q|C A tennis ball is a hollow sphere with a thin wall. It is set rolling without slipping at 4.03 m/s on a horizontal section of a track as shown in Figure P10.60. It rolls around the inside of a vertical circular loop of radius $r = 45.0$ cm. As the ball nears the bottom of the loop, the shape of the track deviates from a perfect circle so that the ball leaves the track at a point $h = 20.0$ cm below the horizontal section. (a) Find the ball's speed at the top of the loop. (b) Demonstrate that the ball will not fall from the track at the top of the loop. (c) Find the ball's speed as it leaves the track at the bottom. **What If?** (d) Suppose that static friction between ball and track were negligible so that the ball slid instead of rolling. Would its speed then be higher, lower, or the same at the top of the loop? (e) Explain your answer to part (d).

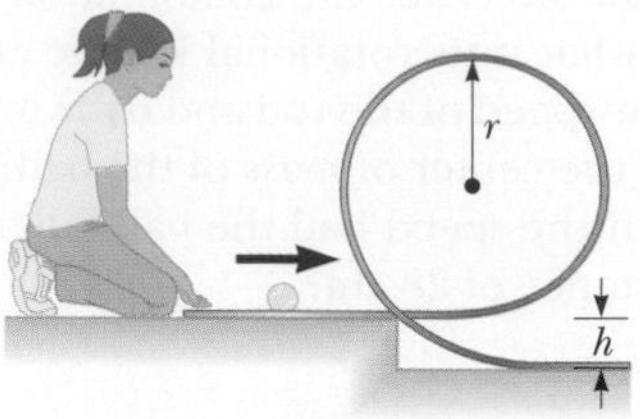

Figure P10.60

61. Q|C A metal can containing condensed mushroom soup has mass 215 g, height 10.8 cm, and diameter 6.38 cm. It is placed at rest on its side at the top of a 3.00-m-long incline that is at 25.0° to the horizontal and is then released to roll straight down. It reaches the bottom of the incline after 1.50 s. (a) Assuming mechanical energy conservation, calculate the moment of inertia of the can. (b) Which pieces of data, if any, are unnecessary for calculating the solution? (c) Why can't the moment of inertia be calculated from $I = \frac{1}{2}mr^2$ for the cylindrical can?

Additional Problems

62. S As shown in Figure 10.18 on page 294, toppling chimneys often break apart in midfall because the mortar between the bricks cannot withstand much shear stress. As the chimney begins to fall, shear forces must act on the topmost sections to accelerate them tangentially so that they can keep up with the rotation of the lower part of the stack. For simplicity, let us model the chimney as a uniform rod of length ℓ pivoted at the lower end. The rod starts at rest in a vertical position (with the frictionless pivot at the bottom) and falls over under the influence of gravity. What fraction of the length of the rod has a tangential acceleration greater than $g \sin \theta$, where θ is the angle the chimney makes with the vertical axis?

63. M **Review.** A 4.00-m length of light nylon cord is wound around a uniform cylindrical spool of radius 0.500 m and mass 1.00 kg. The spool is mounted on a frictionless axle and is initially at rest. The cord is pulled from the spool with a constant acceleration of magnitude 2.50 m/s^2. (a) How much work has been done on the spool when it reaches an angular speed of 8.00 rad/s? (b) How long does it take the spool to reach this angular speed? (c) How much cord is left on the spool when it reaches this angular speed?

64. Q|C An elevator system in a tall building consists of a 800-kg car and a 950-kg counterweight joined by a light cable of constant length that passes over a pulley of mass 280 kg. The pulley, called a sheave, is a solid cylinder of radius 0.700 m turning on a horizontal axle. The cable does not slip on the sheave. A number n of people, each of mass 80.0 kg, are riding in the elevator car, moving

upward at 3.00 m/s and approaching the floor where the car should stop. As an energy-conservation measure, a computer disconnects the elevator motor at just the right moment so that the sheave–car–counterweight system then coasts freely without friction and comes to rest at the floor desired. There it is caught by a simple latch rather than by a massive brake. (a) Determine the distance d the car coasts upward as a function of n. Evaluate the distance for (b) $n = 2$, (c) $n = 12$, and (d) $n = 0$. (e) For what integer values of n does the expression in part (a) apply? (f) Explain your answer to part (e). (g) If an infinite number of people could fit on the elevator, what is the value of d?

65. A shaft is turning at 65.0 rad/s at time $t = 0$. Thereafter, its angular acceleration is given by

$$\alpha = -10.0 - 5.00t$$

where α is in rad/s^2 and t is in seconds. (a) Find the angular speed of the shaft at $t = 3.00$ s. (b) Through what angle does it turn between $t = 0$ and $t = 3.00$ s?

66. **S** A shaft is turning at angular speed ω at time $t = 0$. Thereafter, its angular acceleration is given by

$$\alpha = A + Bt$$

(a) Find the angular speed of the shaft at time t. (b) Through what angle does it turn between $t = 0$ and t?

67. **S** A long, uniform rod of length L and mass M is pivoted about a frictionless, horizontal pin through one end. The rod is nudged from rest in a vertical position as shown in Figure P10.67. At the instant the rod is horizontal, find (a) its angular speed, (b) the magnitude of its angular acceleration, (c) the x and y components of the acceleration of its center of mass, and (d) the components of the reaction force at the pivot.

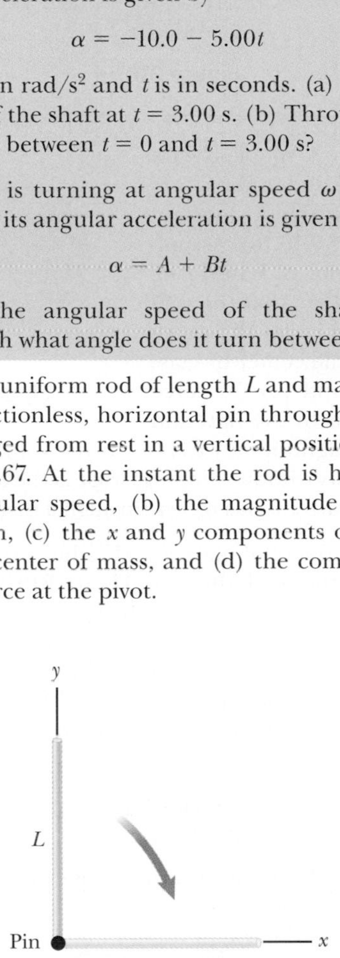

Figure P10.67

68. A bicycle is turned upside down while its owner repairs a flat tire on the rear wheel. A friend spins the front wheel, of radius 0.381 m, and observes that drops of water fly off tangentially in an upward direction when the drops are at the same level as the center of the wheel. She measures the height reached by drops moving vertically (Fig. P10.68). A drop that breaks loose from the tire on one turn rises $h = 54.0$ cm above the tangent point. A drop that breaks loose on the next turn rises 51.0 cm above the tangent point. The height to which the drops rise decreases because the angular speed of the wheel decreases. From this information, determine the magnitude of the average angular acceleration of the wheel.

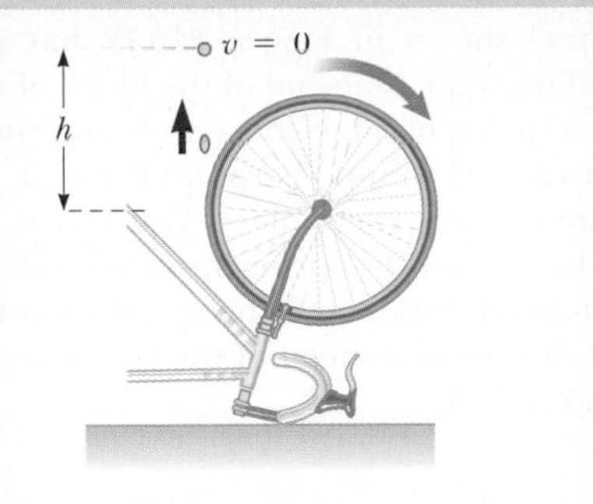

Figure P10.68 Problems 68 and 69.

69. **S** A bicycle is turned upside down while its owner repairs a flat tire on the rear wheel. A friend spins the front wheel, of radius R, and observes that drops of water fly off tangentially in an upward direction when the drops are at the same level as the center of the wheel. She measures the height reached by drops moving vertically (Fig. P10.68). A drop that breaks loose from the tire on one turn rises a distance h_1 above the tangent point. A drop that breaks loose on the next turn rises a distance $h_2 < h_1$ above the tangent point. The height to which the drops rise decreases because the angular speed of the wheel decreases. From this information, determine the magnitude of the average angular acceleration of the wheel.

70. (a) What is the rotational kinetic energy of the Earth about its spin axis? Model the Earth as a uniform sphere and use data from the endpapers of this book. (b) The rotational kinetic energy of the Earth is decreasing steadily because of tidal friction. Assuming the rotational period decreases by 10.0 μs each year, find the change in one day.

71. **GP** **Review.** As shown in Figure P10.71, two blocks are connected by a string of negligible mass passing over a pulley of radius $r = 0.250$ m and moment of inertia I. The block on the frictionless incline is moving with a constant acceleration of magnitude $a = 2.00$ m/s^2. From this information, we wish to find the moment of inertia of the pulley. (a) What analysis model is appropriate for the blocks? (b) What analysis model is appropriate for the pulley? (c) From the analysis model in part (a), find the tension T_1. (d) Similarly, find the tension T_2. (e) From the analysis model in part (b), find a symbolic expression for the moment of inertia of the pulley in terms of the tensions T_1

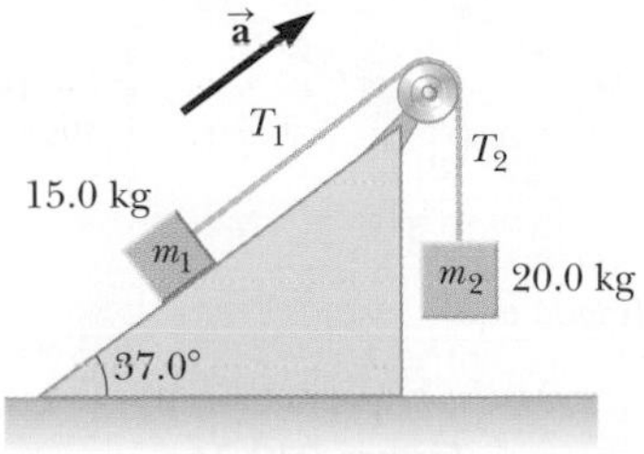

Figure P10.71

and T_2, the pulley radius r, and the acceleration a. (f) Find the numerical value of the moment of inertia of the pulley.

72. **S** The reel shown in Figure P10.72 has radius R and moment of inertia I. One end of the block of mass m is connected to a spring of force constant k, and the other end is fastened to a cord wrapped around the reel. The reel axle and the incline are frictionless. The reel is wound counterclockwise so that the spring stretches a distance d from its unstretched position and the reel is then released from rest. Find the angular speed of the reel when the spring is again unstretched.

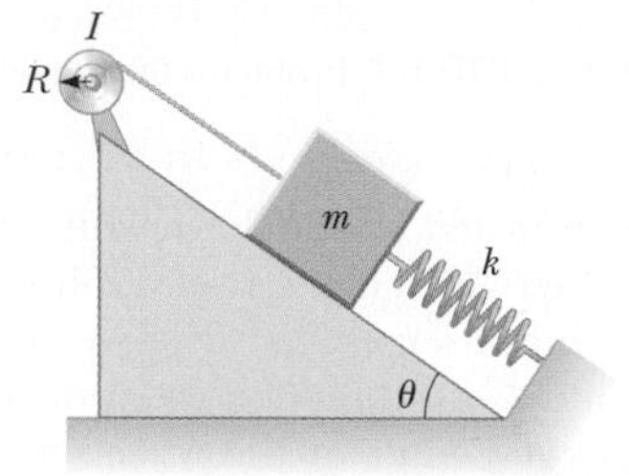

Figure P10.72

73. **S** **Review.** A string is wound around a uniform disk of radius R and mass M. The disk is released from rest with the string vertical and its top end tied to a fixed bar (Fig. P10.73). Show that (a) the tension in the string is one third of the weight of the disk, (b) the magnitude of the acceleration of the center of mass is $2g/3$, and (c) the speed of the center of mass is $(4gh/3)^{1/2}$ after the disk has descended through distance h. (d) Verify your answer to part (c) using the energy approach.

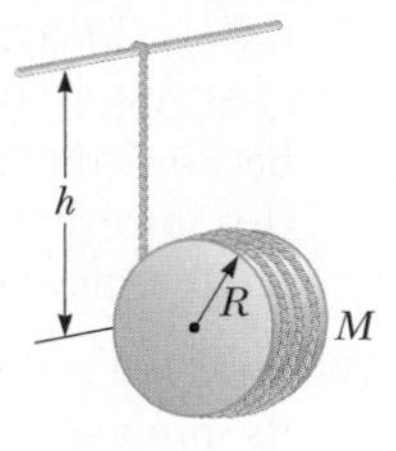

Figure P10.73

74. A common demonstration, illustrated in Figure P10.74, consists of a ball resting at one end of a uniform board of length ℓ that is hinged at the other end and elevated at an angle θ. A light cup is attached to the board at r_c so that it will catch the ball when the support stick is removed suddenly. (a) Show that the ball will lag behind the falling board when θ is less than 35.3°. (b) Assuming the board is 1.00 m long and is supported at this limiting angle, show that the cup must be 18.4 cm from the moving end.

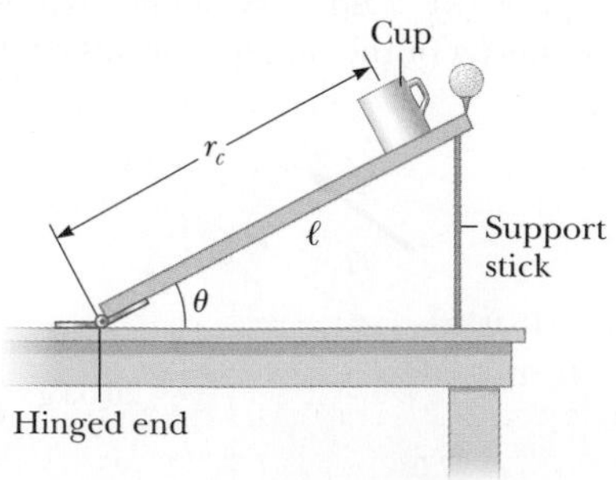

Figure P10.74

75. **S** A uniform solid sphere of radius r is placed on the inside surface of a hemispherical bowl with radius R. The sphere is released from rest at an angle θ to the vertical and rolls without slipping (Fig. P10.75). Determine the angular speed of the sphere when it reaches the bottom of the bowl.

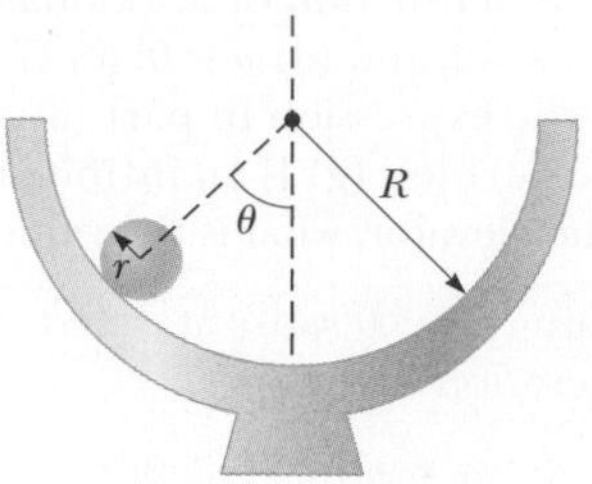

Figure P10.75

76. **S** **Review.** A spool of wire of mass M and radius R is unwound under a constant force $\vec{\mathbf{F}}$ (Fig. P10.76). Assuming the spool is a uniform, solid cylinder that doesn't slip, show that (a) the acceleration of the center of mass is $4\vec{\mathbf{F}}/3M$ and (b) the force of friction is to the *right* and equal in magnitude to $F/3$. (c) If the cylinder starts from rest and rolls without slipping, what is the speed of its center of mass after it has rolled through a distance d?

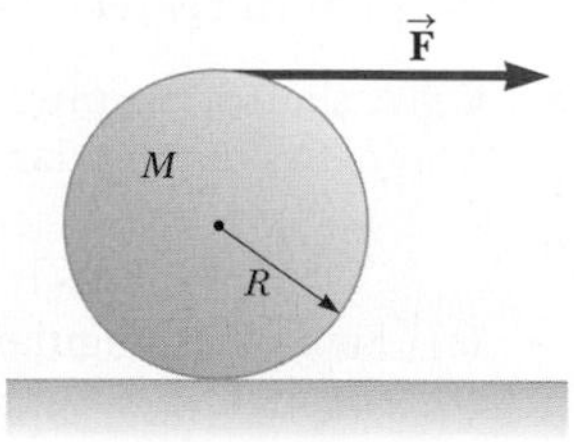

Figure P10.76

77. **S** A solid sphere of mass m and radius r rolls without slipping along the track shown in Figure P10.77. It starts from rest with the lowest point of the sphere at height h above the bottom of the loop of radius R, much larger than r. (a) What is the minimum value of h (in terms of R) such that the sphere completes the loop? (b) What are the force components on the sphere at the point P if $h = 3R$?

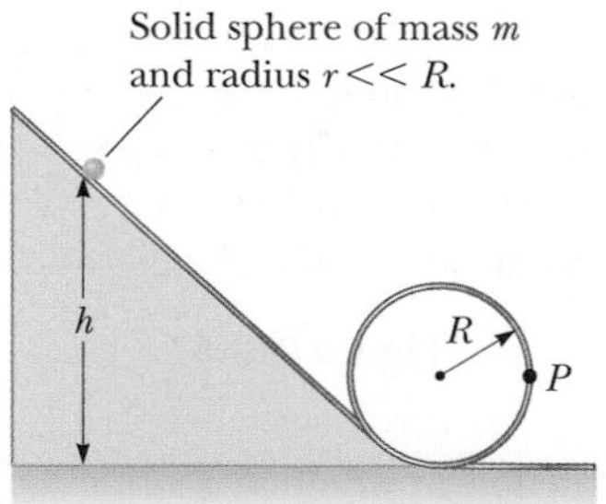

Figure P10.77

78. A thin rod of mass 0.630 kg and length 1.24 m is at rest, hanging vertically from a strong, fixed hinge at its top end. Suddenly, a horizontal impulsive force $14.7\hat{\mathbf{i}}$ N is applied to it. (a) Suppose the force acts at the bottom end of the rod. Find the acceleration of its center of mass and (b) the horizontal force the hinge exerts. (c) Suppose the force acts at

the midpoint of the rod. Find the acceleration of this point and (d) the horizontal hinge reaction force. (e) Where can the impulse be applied so that the hinge will exert no horizontal force? This point is called the *center of percussion.*

79. **S** A thin rod of length h and mass M is held vertically with its lower end resting on a frictionless, horizontal surface. The rod is then released to fall freely. (a) Determine the speed of its center of mass just before it hits the horizontal surface. (b) **What If?** Now suppose the rod has a fixed pivot at its lower end. Determine the speed of the rod's center of mass just before it hits the surface.

80. **Review.** A clown balances a small spherical grape at the top of his bald head, which also has the shape of a sphere. After drawing sufficient applause, the grape starts from rest and rolls down without slipping. It will leave contact with the clown's scalp when the radial line joining it to the center of curvature makes what angle with the vertical?

Challenge Problems

81. **S** A uniform, hollow, cylindrical spool has inside radius $R/2$, outside radius R, and mass M (Fig. P10.81). It is mounted so that it rotates on a fixed, horizontal axle. A counterweight of mass m is connected to the end of a string wound around the spool. The counterweight falls from rest at $t = 0$ to a position y at time t. Show that the torque due to the friction forces between spool and axle is

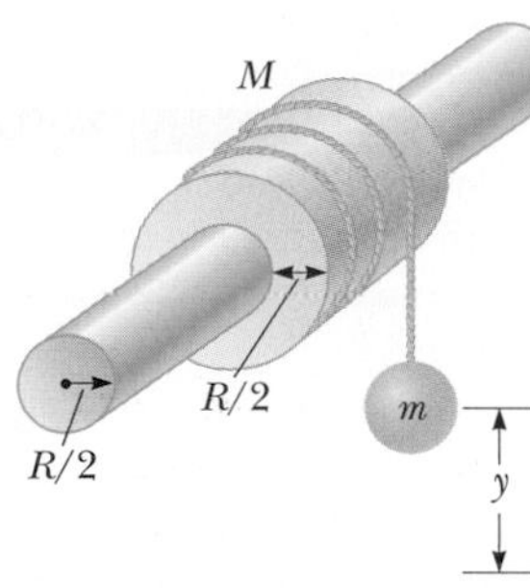

Figure P10.81

$$\tau_f = R\left[m\left(g - \frac{2y}{t^2}\right) - M\frac{5y}{4t^2}\right]$$

82. **S** To find the total angular displacement during the playing time of the compact disc in part (B) of Example 10.2, the disc was modeled as a rigid object under constant angular acceleration. In reality, the angular acceleration of a disc is not constant. In this problem, let us explore the actual time dependence of the angular acceleration. (a) Assume the track on the disc is a spiral such that adjacent loops of the track are separated by a small distance h. Show that the radius r of a given portion of the track is given by

$$r = r_i + \frac{h\theta}{2\pi}$$

where r_i is the radius of the innermost portion of the track and θ is the angle through which the disc turns to arrive at the location of the track of radius r. (b) Show that the rate of change of the angle θ is given by

$$\frac{d\theta}{dt} = \frac{v}{r_i + (h\theta/2\pi)}$$

where v is the constant speed with which the disc surface passes the laser. (c) From the result in part (b), use integration to find an expression for the angle θ as a function of time. (d) From the result in part (c), use differentiation to find the angular acceleration of the disc as a function of time.

83. As a result of friction, the angular speed of a wheel changes with time according to

$$\frac{d\theta}{dt} = \omega_0 e^{-\sigma t}$$

where ω_0 and σ are constants. The angular speed changes from 3.50 rad/s at $t = 0$ to 2.00 rad/s at $t = 9.30$ s. (a) Use this information to determine σ and ω_0. Then determine (b) the magnitude of the angular acceleration at $t = 3.00$ s, (c) the number of revolutions the wheel makes in the first 2.50 s, and (d) the number of revolutions it makes before coming to rest.

84. As a gasoline engine operates, a flywheel turning with the crankshaft stores energy after each fuel explosion, providing the energy required to compress the next charge of fuel and air. For the engine of a certain lawn tractor, suppose a flywheel must be no more than 18.0 cm in diameter. Its thickness, measured along its axis of rotation, must be no larger than 8.00 cm. The flywheel must release energy 60.0 J when its angular speed drops from 800 rev/min to 600 rev/min. Design a sturdy steel (density 7.85×10^3 kg/m^3) flywheel to meet these requirements with the smallest mass you can reasonably attain. Specify the shape and mass of the flywheel.

85. **S** A spool of thread consists of a cylinder of radius R_1 with end caps of radius R_2 as depicted in the end view shown in Figure P10.85. The mass of the spool, including the thread, is m, and its moment of inertia about an axis through its center is I. The spool is placed on a rough, horizontal surface so that it rolls without slipping when a force $\vec{\mathbf{T}}$ acting to the right is applied to the free end of the thread. (a) Show that the magnitude of the friction force exerted by the surface on the spool is given by

$$f = \left(\frac{I + mR_1R_2}{I + mR_2^{\,2}}\right)T$$

(b) Determine the direction of the force of friction.

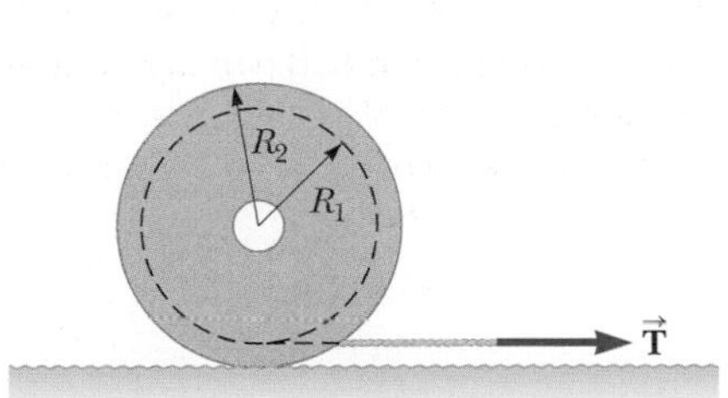

Figure P10.85

86. **S** A cord is wrapped around a pulley that is shaped like a disk of mass m and radius r. The cord's free end is connected to a block of mass M. The block starts from rest and then slides down an incline that makes an angle θ with the horizontal as shown in Figure P10.86. The coefficient of kinetic friction between block and incline is μ. (a) Use

energy methods to show that the block's speed as a function of position d down the incline is

$$v = \sqrt{\frac{4Mgd(\sin\theta - \mu\cos\theta)}{m + 2M}}$$

(b) Find the magnitude of the acceleration of the block in terms of μ, m, M, g, and θ.

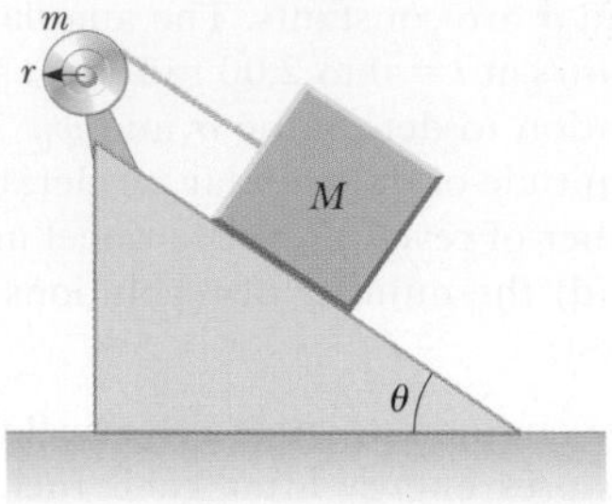

Figure P10.86

87. A merry-go-round is stationary. A dog is running around the merry-go-round on the ground just outside its circumference, moving with a constant angular speed of 0.750 rad/s. The dog does not change his pace when he sees what he has been looking for: a bone resting on the edge of the merry-go-round one-third of a revolution in front of him. At the instant the dog sees the bone ($t = 0$), the merry-go-round begins to move in the direction the dog is running, with a constant angular acceleration of 0.015 0 rad/s^2. (a) At what time will the dog first reach the bone? (b) The confused dog keeps running and passes the bone. How long after the merry-go-round starts to turn do the dog and the bone draw even with each other for the second time?

88. A plank with a mass $M = 6.00$ kg rests on top of two identical, solid, cylindrical rollers that have $R = 5.00$ cm and $m = 2.00$ kg (Fig. P10.88). The plank is pulled by a constant horizontal force $\vec{\mathbf{F}}$ of magnitude 6.00 N applied to the end of the plank and perpendicular to the axes of the cylinders (which are parallel). The cylinders roll without slipping on a flat surface. There is also no slipping between the cylinders and the plank. (a) Find the initial acceleration of the plank at the moment the rollers are equidistant from the ends of the plank. (b) Find the acceleration of the rollers at this moment. (c) What friction forces are acting at this moment?

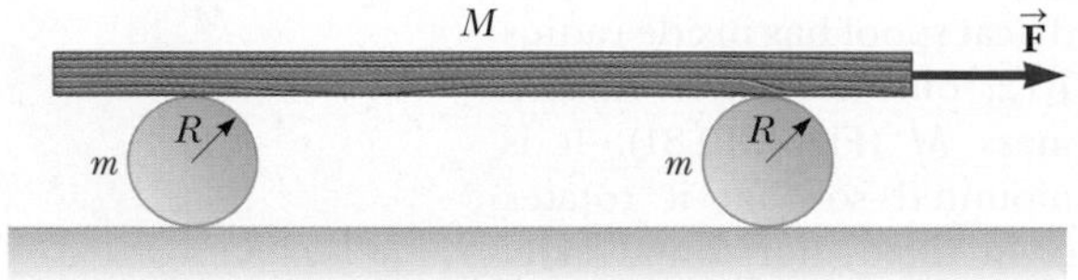

Figure P10.88

chapter 11

Angular Momentum

Two motorcycle racers lean precariously into a turn around a racetrack. The analysis of such a leaning turn is based on principles associated with angular momentum. *(Stuart Westmorland/The Image Bank/Getty Images)*

The central topic of this chapter is angular momentum, a quantity that plays a key role in rotational dynamics. In analogy to the principle of conservation of linear momentum, there is also a principle of conservation of angular momentum. The angular momentum of an isolated system is constant. For angular momentum, an isolated system is one for which no external torques act on the system. If a net external torque acts on a system, it is nonisolated. Like the law of conservation of linear momentum, the law of conservation of angular momentum is a fundamental law of physics, equally valid for relativistic and quantum systems.

11.1 The Vector Product and Torque

An important consideration in defining angular momentum is the process of multiplying two vectors by means of the operation called the *vector product.* We will introduce the vector product by considering the vector nature of torque.

Consider a force $\vec{\mathbf{F}}$ acting on a particle at the vector position $\vec{\mathbf{r}}$ (Active Fig. 11.1 on page 320). As we saw in Section 10.6, the *magnitude* of the torque due to this force about an axis through the origin is $rF\sin\phi$, where ϕ is the angle between $\vec{\mathbf{r}}$ and $\vec{\mathbf{F}}$. The axis about which $\vec{\mathbf{F}}$ tends to produce rotation is perpendicular to the plane formed by $\vec{\mathbf{r}}$ and $\vec{\mathbf{F}}$.

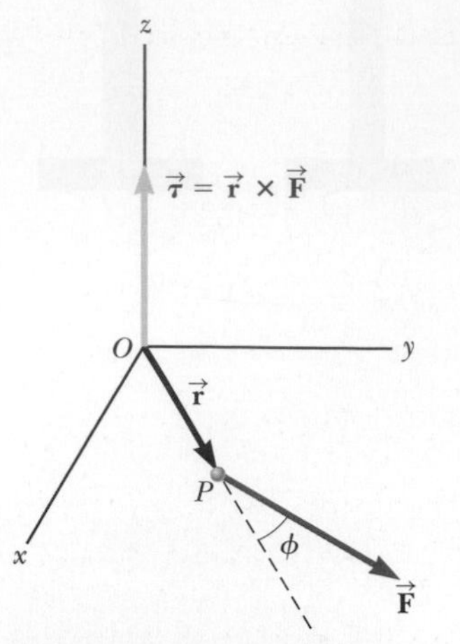

ACTIVE FIGURE 11.1

The torque vector $\vec{\tau}$ lies in a direction perpendicular to the plane formed by the position vector $\vec{r}$ and the applied force vector $\vec{F}$. In the situation shown, $\vec{r}$ and $\vec{F}$ lie in the *xy* plane, so the torque is along the *z* axis.

The torque vector $\vec{\tau}$ is related to the two vectors $\vec{r}$ and $\vec{F}$. We can establish a mathematical relationship between $\vec{\tau}$, $\vec{r}$, and $\vec{F}$ using a mathematical operation called the **vector product:**

$$\vec{\tau} \equiv \vec{r} \times \vec{F} \tag{11.1}$$

We now give a formal definition of the vector product. Given any two vectors $\vec{A}$ and $\vec{B}$, the vector product $\vec{A} \times \vec{B}$ is defined as a third vector $\vec{C}$, which has a magnitude of $AB \sin \theta$, where θ is the angle between $\vec{A}$ and $\vec{B}$. That is, if $\vec{C}$ is given by

$$\vec{C} = \vec{A} \times \vec{B} \tag{11.2}$$

its magnitude is

$$C = AB \sin \theta \tag{11.3}$$

The quantity $AB \sin \theta$ is equal to the area of the parallelogram formed by $\vec{A}$ and $\vec{B}$ as shown in Figure 11.2. The *direction* of $\vec{C}$ is perpendicular to the plane formed by $\vec{A}$ and $\vec{B}$, and the best way to determine this direction is to use the right-hand rule illustrated in Figure 11.2. The four fingers of the right hand are pointed along $\vec{A}$ and then "wrapped" in the direction that would rotate $\vec{A}$ into $\vec{B}$ through the angle θ. The direction of the upright thumb is the direction of $\vec{A} \times \vec{B} = \vec{C}$. Because of the notation, $\vec{A} \times \vec{B}$ is often read "$\vec{A}$ cross $\vec{B}$," so the vector product is also called the **cross product.**

Some properties of the vector product that follow from its definition are as follows:

Properties of the vector product ▶

1. Unlike the scalar product, the vector product is *not* commutative. Instead, the order in which the two vectors are multiplied in a vector product is important:
$$\vec{A} \times \vec{B} = -\vec{B} \times \vec{A} \tag{11.4}$$
Therefore, if you change the order of the vectors in a vector product, you must change the sign. You can easily verify this relationship with the right-hand rule.
2. If $\vec{A}$ is parallel to $\vec{B}$ ($\theta = 0$ or $180°$), then $\vec{A} \times \vec{B} = 0$; therefore, it follows that $\vec{A} \times \vec{A} = 0$.
3. If $\vec{A}$ is perpendicular to $\vec{B}$, then $|\vec{A} \times \vec{B}| = AB$.
4. The vector product obeys the distributive law:
$$\vec{A} \times (\vec{B} + \vec{C}) = \vec{A} \times \vec{B} + \vec{A} \times \vec{C} \tag{11.5}$$
5. The derivative of the vector product with respect to some variable such as t is
$$\frac{d}{dt}(\vec{A} \times \vec{B}) = \frac{d\vec{A}}{dt} \times \vec{B} + \vec{A} \times \frac{d\vec{B}}{dt} \tag{11.6}$$
where it is important to preserve the multiplicative order of the terms on the right side in view of Equation 11.4.

Pitfall Prevention 11.1

The Vector Product Is a Vector
Remember that the result of taking a vector product between two vectors is *a third vector.* Equation 11.3 gives only the magnitude of this vector.

It is left as an exercise (Problem 6) to show from Equations 11.3 and 11.4 and from the definition of unit vectors that the cross products of the unit vectors $\hat{\mathbf{i}}$, $\hat{\mathbf{j}}$, and $\hat{\mathbf{k}}$ obey the following rules:

Cross products of unit vectors ▶

$$\hat{\mathbf{i}} \times \hat{\mathbf{i}} = \hat{\mathbf{j}} \times \hat{\mathbf{j}} = \hat{\mathbf{k}} \times \hat{\mathbf{k}} = 0 \tag{11.7a}$$

$$\hat{\mathbf{i}} \times \hat{\mathbf{j}} = -\hat{\mathbf{j}} \times \hat{\mathbf{i}} = \hat{\mathbf{k}} \tag{11.7b}$$

$$\hat{\mathbf{j}} \times \hat{\mathbf{k}} = -\hat{\mathbf{k}} \times \hat{\mathbf{j}} = \hat{\mathbf{i}} \tag{11.7c}$$

$$\hat{\mathbf{k}} \times \hat{\mathbf{i}} = -\hat{\mathbf{i}} \times \hat{\mathbf{k}} = \hat{\mathbf{j}} \tag{11.7d}$$

Signs are interchangeable in cross products. For example, $\vec{A} \times (-\vec{B}) = -\vec{A} \times \vec{B}$ and $\hat{\mathbf{i}} \times (-\hat{\mathbf{j}}) = -\hat{\mathbf{i}} \times \hat{\mathbf{j}}$.

The cross product of any two vectors $\vec{\mathbf{A}}$ and $\vec{\mathbf{B}}$ can be expressed in the following determinant form:

$$\vec{\mathbf{A}} \times \vec{\mathbf{B}} = \begin{vmatrix} \hat{\mathbf{i}} & \hat{\mathbf{j}} & \hat{\mathbf{k}} \\ A_x & A_y & A_z \\ B_x & B_y & B_z \end{vmatrix} = \begin{vmatrix} A_y & A_z \\ B_y & B_z \end{vmatrix} \hat{\mathbf{i}} + \begin{vmatrix} A_z & A_x \\ B_z & B_x \end{vmatrix} \hat{\mathbf{j}} + \begin{vmatrix} A_x & A_y \\ B_x & B_y \end{vmatrix} \hat{\mathbf{k}}$$

Expanding these determinants gives the result

$$\vec{\mathbf{A}} \times \vec{\mathbf{B}} = (A_yB_z - A_zB_y)\,\hat{\mathbf{i}} + (A_zB_x - A_xB_z)\,\hat{\mathbf{j}} + (A_xB_y - A_yB_x)\,\hat{\mathbf{k}} \quad \textbf{(11.8)}$$

Given the definition of the cross product, we can now assign a direction to the torque vector. If the force lies in the xy plane as in Active Figure 11.1, the torque $\vec{\boldsymbol{\tau}}$ is represented by a vector parallel to the z axis. The force in Active Figure 11.1 creates a torque that tends to rotate the object counterclockwise about the z axis; the direction of $\vec{\boldsymbol{\tau}}$ is toward increasing z, and $\vec{\boldsymbol{\tau}}$ is therefore in the positive z direction. If we reversed the direction of $\vec{\mathbf{F}}$ in Active Figure 11.1, $\vec{\boldsymbol{\tau}}$ would be in the negative z direction.

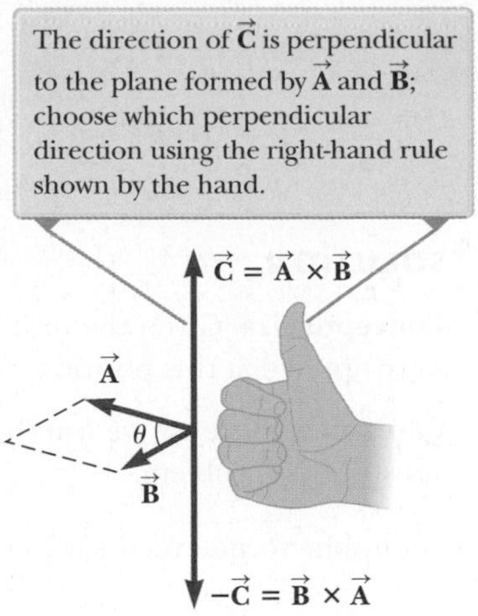

Figure 11.2 The vector product $\vec{\mathbf{A}} \times \vec{\mathbf{B}}$ is a third vector $\vec{\mathbf{C}}$ having a magnitude $AB \sin\theta$ equal to the area of the parallelogram shown.

***Quick Quiz* 11.1** Which of the following statements about the relationship between the magnitude of the cross product of two vectors and the product of the magnitudes of the vectors is true? **(a)** $|\vec{\mathbf{A}} \times \vec{\mathbf{B}}|$ is larger than AB. **(b)** $|\vec{\mathbf{A}} \times \vec{\mathbf{B}}|$ is smaller than AB. **(c)** $|\vec{\mathbf{A}} \times \vec{\mathbf{B}}|$ could be larger or smaller than AB, depending on the angle between the vectors. **(d)** $|\vec{\mathbf{A}} \times \vec{\mathbf{B}}|$ could be equal to AB.

Example 11.1 The Vector Product

Two vectors lying in the xy plane are given by the equations $\vec{\mathbf{A}} = 2\hat{\mathbf{i}} + 3\hat{\mathbf{j}}$ and $\vec{\mathbf{B}} = -\hat{\mathbf{i}} + 2\hat{\mathbf{j}}$. Find $\vec{\mathbf{A}} \times \vec{\mathbf{B}}$ and verify that $\vec{\mathbf{A}} \times \vec{\mathbf{B}} = -\vec{\mathbf{B}} \times \vec{\mathbf{A}}$.

SOLUTION

Conceptualize Given the unit-vector notations of the vectors, think about the directions the vectors point in space. Imagine the parallelogram shown in Figure 11.2 for these vectors.

Categorize Because we use the definition of the cross product discussed in this section, we categorize this example as a substitution problem.

Write the cross product of the two vectors:

$$\vec{\mathbf{A}} \times \vec{\mathbf{B}} = (2\hat{\mathbf{i}} + 3\hat{\mathbf{j}}) \times (-\hat{\mathbf{i}} + 2\hat{\mathbf{j}})$$

Perform the multiplication:

$$\vec{\mathbf{A}} \times \vec{\mathbf{B}} = 2\hat{\mathbf{i}} \times (-\hat{\mathbf{i}}) + 2\hat{\mathbf{i}} \times 2\hat{\mathbf{j}} + 3\hat{\mathbf{j}} \times (-\hat{\mathbf{i}}) + 3\hat{\mathbf{j}} \times 2\hat{\mathbf{j}}$$

Use Equations 11.7a through 11.7d to evaluate the various terms:

$$\vec{\mathbf{A}} \times \vec{\mathbf{B}} = 0 + 4\hat{\mathbf{k}} + 3\hat{\mathbf{k}} + 0 = 7\hat{\mathbf{k}}$$

To verify that $\vec{\mathbf{A}} \times \vec{\mathbf{B}} = -\vec{\mathbf{B}} \times \vec{\mathbf{A}}$, evaluate $\vec{\mathbf{B}} \times \vec{\mathbf{A}}$:

$$\vec{\mathbf{B}} \times \vec{\mathbf{A}} = (-\hat{\mathbf{i}} + 2\hat{\mathbf{j}}) \times (2\hat{\mathbf{i}} + 3\hat{\mathbf{j}})$$

Perform the multiplication:

$$\vec{\mathbf{B}} \times \vec{\mathbf{A}} = (-\hat{\mathbf{i}}) \times 2\hat{\mathbf{i}} + (-\hat{\mathbf{i}}) \times 3\hat{\mathbf{j}} + 2\hat{\mathbf{j}} \times 2\hat{\mathbf{i}} + 2\hat{\mathbf{j}} \times 3\hat{\mathbf{j}}$$

Use Equations 11.7a through 11.7d to evaluate the various terms:

$$\vec{\mathbf{B}} \times \vec{\mathbf{A}} = 0 - 3\hat{\mathbf{k}} - 4\hat{\mathbf{k}} + 0 = -7\hat{\mathbf{k}}$$

Therefore, $\vec{\mathbf{A}} \times \vec{\mathbf{B}} = -\vec{\mathbf{B}} \times \vec{\mathbf{A}}$. As an alternative method for finding $\vec{\mathbf{A}} \times \vec{\mathbf{B}}$, you could use Equation 11.8. Try it!

Example 11.2 The Torque Vector

A force of $\vec{\mathbf{F}} = (2.00\hat{\mathbf{i}} + 3.00\hat{\mathbf{j}})$ N is applied to an object that is pivoted about a fixed axis aligned along the z coordinate axis. The force is applied at a point located at $\vec{\mathbf{r}} = (4.00\hat{\mathbf{i}} + 5.00\hat{\mathbf{j}})$ m. Find the torque $\vec{\boldsymbol{\tau}}$ applied to the object.

SOLUTION

Conceptualize Given the unit-vector notations, think about the directions of the force and position vectors. If this force were applied at this position, in what direction would an object pivoted at the origin turn?

Categorize Because we use the definition of the cross product discussed in this section, we categorize this example as a substitution problem.

Set up the torque vector using Equation 11.1:

$$\vec{\boldsymbol{\tau}} = \vec{\mathbf{r}} \times \vec{\mathbf{F}} = [(4.00\hat{\mathbf{i}} + 5.00\hat{\mathbf{j}})\text{ m}] \times [(2.00\hat{\mathbf{i}} + 3.00\hat{\mathbf{j}})\text{ N}]$$

Perform the multiplication:

$$\vec{\boldsymbol{\tau}} = [(4.00)(2.00)\hat{\mathbf{i}} \times \hat{\mathbf{i}} + (4.00)(3.00)\hat{\mathbf{i}} \times \hat{\mathbf{j}} + (5.00)(2.00)\hat{\mathbf{j}} \times \hat{\mathbf{i}} + (5.00)(3.00)\hat{\mathbf{j}} \times \hat{\mathbf{j}}]\text{ N}\cdot\text{m}$$

Use Equations 11.7a through 11.7d to evaluate the various terms:

$$\vec{\boldsymbol{\tau}} = [0 + 12.0\hat{\mathbf{k}} - 10.0\hat{\mathbf{k}} + 0]\text{ N}\cdot\text{m} = \boxed{2.0\hat{\mathbf{k}}\text{ N}\cdot\text{m}}$$

Notice that both $\vec{\mathbf{r}}$ and $\vec{\mathbf{F}}$ are in the xy plane. As expected, the torque vector is perpendicular to this plane, having only a z component. We have followed the rules for significant figures discussed in Section 1.6, which lead to an answer with two significant figures. We have lost some precision because we ended up subtracting two numbers that are close.

11.2 Analysis Model: Nonisolated System (Angular Momentum)

Imagine a rigid pole sticking up through the ice on a frozen pond (Active Fig. 11.3). A skater glides rapidly toward the pole, aiming a little to the side so that she does not hit it. As she passes the pole, she reaches out to her side and grabs it, an action that causes her to move in a circular path around the pole. Just as the idea of linear momentum helps us analyze translational motion, a rotational analog—*angular momentum*—helps us analyze the motion of this skater and other objects undergoing rotational motion.

In Chapter 9, we developed the mathematical form of linear momentum and then proceeded to show how this new quantity was valuable in problem solving. We will follow a similar procedure for angular momentum.

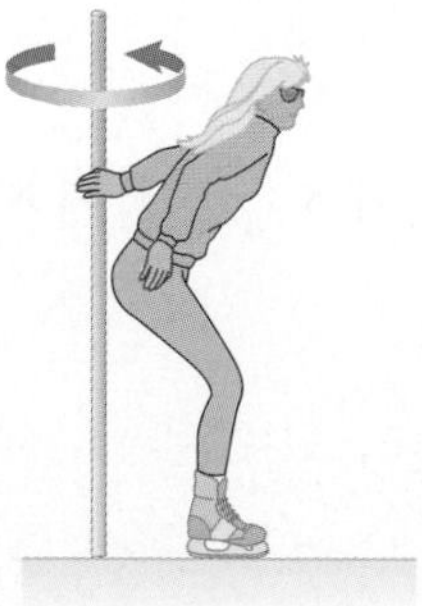

ACTIVE FIGURE 11.3

As the skater passes the pole, she grabs hold of it, which causes her to swing around the pole rapidly in a circular path.

Consider a particle of mass m located at the vector position $\vec{\mathbf{r}}$ and moving with linear momentum $\vec{\mathbf{p}}$ as in Active Figure 11.4. In describing translational motion, we found that the net force on the particle equals the time rate of change of its linear momentum, $\sum \vec{\mathbf{F}} = d\vec{\mathbf{p}}/dt$ (see Eq. 9.3). Let us take the cross product of each side of Equation 9.3 with $\vec{\mathbf{r}}$, which gives the net torque on the particle on the left side of the equation:

$$\vec{\mathbf{r}} \times \sum \vec{\mathbf{F}} = \sum \vec{\boldsymbol{\tau}} = \vec{\mathbf{r}} \times \frac{d\vec{\mathbf{p}}}{dt}$$

Now let's add to the right side the term $(d\vec{\mathbf{r}}/dt) \times \vec{\mathbf{p}}$, which is zero because $d\vec{\mathbf{r}}/dt = \vec{\mathbf{v}}$ and $\vec{\mathbf{v}}$ and $\vec{\mathbf{p}}$ are parallel. Therefore,

$$\sum \vec{\boldsymbol{\tau}} = \vec{\mathbf{r}} \times \frac{d\vec{\mathbf{p}}}{dt} + \frac{d\vec{\mathbf{r}}}{dt} \times \vec{\mathbf{p}}$$

We recognize the right side of this equation as the derivative of $\vec{\mathbf{r}} \times \vec{\mathbf{p}}$ (see Eq. 11.6). Therefore,

$$\sum \vec{\boldsymbol{\tau}} = \frac{d(\vec{\mathbf{r}} \times \vec{\mathbf{p}})}{dt} \quad \textbf{(11.9)}$$

which looks very similar in form to Equation 9.3, $\sum \vec{\mathbf{F}} = d\vec{\mathbf{p}}/dt$. Because torque plays the same role in rotational motion that force plays in translational motion, this result suggests that the combination $\vec{\mathbf{r}} \times \vec{\mathbf{p}}$ should play the same role in rotational motion that $\vec{\mathbf{p}}$ plays in translational motion. We call this combination the *angular momentum* of the particle:

◀ **Angular momentum of a particle**

The instantaneous **angular momentum** $\vec{\mathbf{L}}$ of a particle relative to an axis through the origin O is defined by the cross product of the particle's instantaneous position vector $\vec{\mathbf{r}}$ and its instantaneous linear momentum $\vec{\mathbf{p}}$:

$$\vec{\mathbf{L}} \equiv \vec{\mathbf{r}} \times \vec{\mathbf{p}} \quad \textbf{(11.10)}$$

We can now write Equation 11.9 as

$$\sum \vec{\boldsymbol{\tau}} = \frac{d\vec{\mathbf{L}}}{dt} \quad \textbf{(11.11)}$$

which is the rotational analog of Newton's second law, $\sum \vec{\mathbf{F}} = d\vec{\mathbf{p}}/dt$. Torque causes the angular momentum $\vec{\mathbf{L}}$ to change just as force causes linear momentum $\vec{\mathbf{p}}$ to change.

Notice that Equation 11.11 is valid only if $\sum \vec{\boldsymbol{\tau}}$ and $\vec{\mathbf{L}}$ are measured about the same axis. Furthermore, the expression is valid for any axis fixed in an inertial frame.

The SI unit of angular momentum is $\text{kg} \cdot \text{m}^2/\text{s}$. Notice also that both the magnitude and the direction of $\vec{\mathbf{L}}$ depend on the choice of axis. Following the right-hand rule, we see that the direction of $\vec{\mathbf{L}}$ is perpendicular to the plane formed by $\vec{\mathbf{r}}$ and $\vec{\mathbf{p}}$. In Active Figure 11.4, $\vec{\mathbf{r}}$ and $\vec{\mathbf{p}}$ are in the xy plane, so $\vec{\mathbf{L}}$ points in the z direction. Because $\vec{\mathbf{p}} = m\vec{\mathbf{v}}$, the magnitude of $\vec{\mathbf{L}}$ is

$$L = mvr \sin \phi \quad \textbf{(11.12)}$$

where ϕ is the angle between $\vec{\mathbf{r}}$ and $\vec{\mathbf{p}}$. It follows that L is zero when $\vec{\mathbf{r}}$ is parallel to $\vec{\mathbf{p}}$ ($\phi = 0$ or $180°$). In other words, when the translational velocity of the particle is along a line that passes through the axis, the particle has zero angular momentum with respect to the axis. On the other hand, if $\vec{\mathbf{r}}$ is perpendicular to $\vec{\mathbf{p}}$ ($\phi = 90°$), then $L = mvr$. At that instant, the particle moves exactly as if it were on the rim of a wheel rotating about the axis in a plane defined by $\vec{\mathbf{r}}$ and $\vec{\mathbf{p}}$.

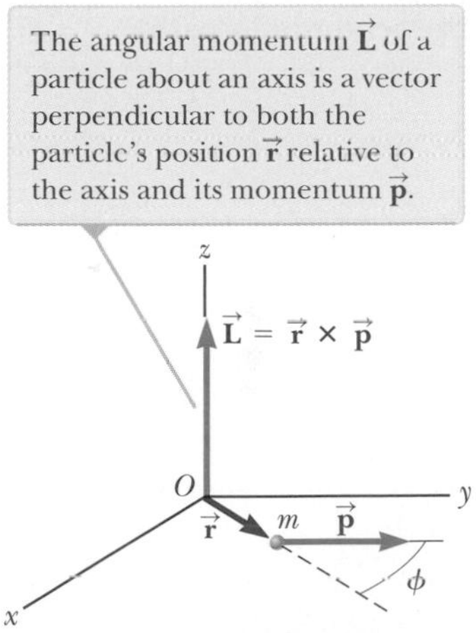

ACTIVE FIGURE 11.4

The angular momentum $\vec{\mathbf{L}}$ of a particle is a vector given by $\vec{\mathbf{L}} = \vec{\mathbf{r}} \times \vec{\mathbf{p}}$.

***Quick Quiz* 11.2** Recall the skater described at the beginning of this section. Let her mass be m. **(i)** What would be her angular momentum relative to the pole at the instant she is a distance d from the pole if she were skating directly toward it at speed v? (a) zero (b) mvd (c) impossible to determine **(ii)** What would be her angular momentum relative to the pole at the instant she is a distance d from the pole if she were skating at speed v along a straight path that is a perpendicular distance a from the pole? (a) zero (b) mvd (c) mva (d) impossible to determine

Pitfall Prevention 11.2

Is Rotation Necessary for Angular Momentum?

We can define angular momentum even if the particle is not moving in a circular path. Even a particle moving in a straight line has angular momentum about any axis displaced from the path of the particle.

Example 11.3 Angular Momentum of a Particle in Circular Motion

A particle moves in the xy plane in a circular path of radius r as shown in Figure 11.5. Find the magnitude and direction of its angular momentum relative to an axis through O when its velocity is $\vec{\mathbf{v}}$.

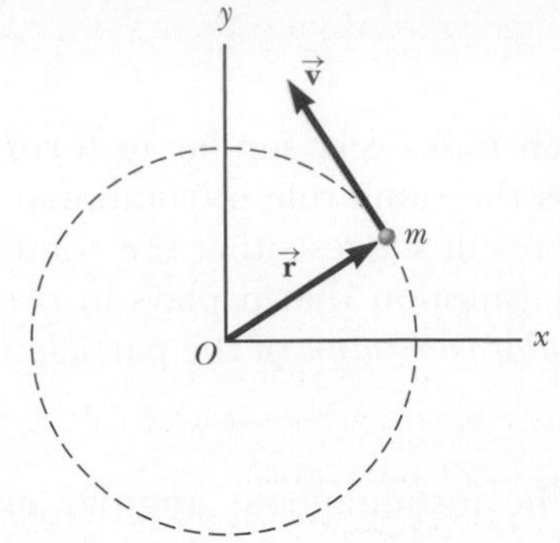

Figure 11.5 (Example 11.3) A particle moving in a circle of radius r has an angular momentum about an axis through O that has magnitude mvr. The vector $\vec{\mathbf{L}} = \vec{\mathbf{r}} \times \vec{\mathbf{p}}$ points *out* of the page.

SOLUTION

Conceptualize The linear momentum of the particle is always changing in direction (but not in magnitude). You might therefore be tempted to conclude that the angular momentum of the particle is always changing. In this situation, however, that is not the case. Let's see why.

Categorize We use the definition of the angular momentum of a particle discussed in this section, so we categorize this example as a substitution problem.

Use Equation 11.12 to evaluate the magnitude of $\vec{\mathbf{L}}$: $L = mvr \sin 90° = \boxed{mvr}$

This value of L is constant because all three factors on the right are constant. The direction of $\vec{\mathbf{L}}$ also is constant, even though the direction of $\vec{\mathbf{p}} = m\vec{\mathbf{v}}$ keeps changing. To verify this statement, apply the right-hand rule to find the direction of $\vec{\mathbf{L}} = \vec{\mathbf{r}} \times \vec{\mathbf{p}} = m\vec{\mathbf{r}} \times \vec{\mathbf{v}}$ in Figure 11.5. Your thumb points out of the page, so that is the direction of $\vec{\mathbf{L}}$. Hence, we can write the vector expression $\vec{\mathbf{L}} = (mvr)\hat{\mathbf{k}}$. If the particle were to move clockwise, $\vec{\mathbf{L}}$ would point downward and into the page and $\vec{\mathbf{L}} = -(mvr)\hat{\mathbf{k}}$. A particle in uniform circular motion has a constant angular momentum about an axis through the center of its path.

Angular Momentum of a System of Particles

Using the techniques of Section 9.7, we can show that Newton's second law for a system of particles is

$$\sum \vec{\mathbf{F}}_{\text{ext}} = \frac{d\vec{\mathbf{p}}_{\text{tot}}}{dt}$$

This equation states that the net external force on a system of particles is equal to the time rate of change of the total linear momentum of the system. Let's see if a similar statement can be made for rotational motion. The total angular momentum of a system of particles about some axis is defined as the vector sum of the angular moment of the individual particles:

$$\vec{\mathbf{L}}_{\text{tot}} = \vec{\mathbf{L}}_1 + \vec{\mathbf{L}}_2 + \cdots + \vec{\mathbf{L}}_n = \sum_i \vec{\mathbf{L}}_i$$

where the vector sum is over all n particles in the system.

Differentiating this equation with respect to time gives

$$\frac{d\vec{\mathbf{L}}_{\text{tot}}}{dt} = \sum_i \frac{d\vec{\mathbf{L}}_i}{dt} = \sum_i \vec{\boldsymbol{\tau}}_i$$

where we have used Equation 11.11 to replace the time rate of change of the angular momentum of each particle with the net torque on the particle.

The torques acting on the particles of the system are those associated with internal forces between particles and those associated with external forces. The net torque associated with all internal forces, however, is zero. Recall that Newton's third law tells us that internal forces between particles of the system are equal in magnitude and opposite in direction. If we assume these forces lie along the line of separation of each pair of particles, the total torque around some axis passing through an origin O due to each action–reaction force pair is zero (that is, the moment arm d from O to the line of action of the forces is equal for both particles, and the forces are in opposite directions). In the summation, therefore, the net internal torque is zero. We conclude that the total angular momentum of a system can vary with time only if a net external torque is acting on the system:

$$\sum \vec{\boldsymbol{\tau}}_{\text{ext}} = \frac{d\vec{\mathbf{L}}_{\text{tot}}}{dt} \quad (11.13)$$

◀ **The net external torque on a system equals the time rate of change of angular momentum of the system**

This equation is indeed the rotational analog of $\sum \vec{\mathbf{F}}_{\text{ext}} = d\vec{\mathbf{p}}_{\text{tot}}/dt$ for a system of particles. Equation 11.13 is the mathematical representation of the **angular momentum version of the nonisolated system model.** If a system is nonisolated in the sense that there is a net torque on it, the torque is equal to the time rate of change of angular momentum.

Although we do not prove it here, this statement is true regardless of the motion of the center of mass. It applies even if the center of mass is accelerating, provided the torque and angular momentum are evaluated relative to an axis through the center of mass.

Equation 11.13 can be rearranged and integrated to give

$$\int \left(\sum \vec{\boldsymbol{\tau}}_{\text{ext}}\right) dt = \Delta \vec{\mathbf{L}}_{\text{tot}}$$

This equation is the rotational analog to Equation 9.40. It represents the *angular impulse–angular momentum theorem.*

Example 11.4 A System of Objects

A sphere of mass m_1 and a block of mass m_2 are connected by a light cord that passes over a pulley as shown in Figure 11.6. The radius of the pulley is R, and the mass of the thin rim is M. The spokes of the pulley have negligible mass. The block slides on a frictionless, horizontal surface. Find an expression for the linear acceleration of the two objects, using the concepts of angular momentum and torque.

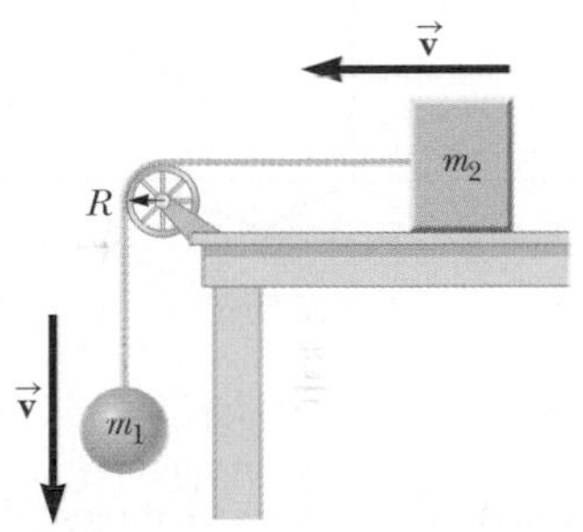

Figure 11.6 (Example 11.4) When the system is released, the sphere moves downward and the block moves to the left.

SOLUTION

Conceptualize When the system is released, the block slides to the left, the sphere drops downward, and the pulley rotates counterclockwise. This situation is similar to problems we have solved earlier except that now we want to use an angular momentum approach.

Categorize We identify the block, pulley, and sphere as a nonisolated system, subject to the external torque due to the gravitational force on the sphere. We shall calculate the angular momentum about an axis that coincides with the axle of the pulley. The angular momentum of the system includes that of two objects moving translationally (the sphere and the block) and one object undergoing pure rotation (the pulley).

Analyze At any instant of time, the sphere and the block have a common speed v, so the angular momentum of the sphere is m_1vR and that of the block is m_2vR. At the same instant, all points on the rim of the pulley also move with speed v, so the angular momentum of the pulley is MvR.

Now let's address the total external torque acting on the system about the pulley axle. Because it has a moment arm of zero, the force exerted by the axle on the pulley does not contribute to the torque. Furthermore, the normal force acting on the block is balanced by the gravitational force $m_2\vec{\mathbf{g}}$, so these forces do not contribute to the torque. The gravitational force $m_1\vec{\mathbf{g}}$ acting on the sphere produces a torque about the axle equal in magnitude to m_1gR, where R is the moment arm of the force about the axle. This result is the total external torque about the pulley axle; that is, $\sum \tau_{\text{ext}} = m_1gR$.

Write an expression for the total angular momentum of the system:

$$(1) \quad L = m_1vR + m_2vR + MvR = (m_1 + m_2 + M)vR$$

continued

11.4 cont.

Substitute this expression and the total external torque into Equation 11.13:

$$\sum \tau_{\text{ext}} = \frac{dL}{dt}$$

$$m_1 gR = \frac{d}{dt}[(m_1 + m_2 + M)vR]$$

$$(2)\quad m_1 gR = (m_1 + m_2 + M)R\frac{dv}{dt}$$

Recognizing that $dv/dt = a$, solve Equation (2) for a:

$$(3)\quad a = \frac{m_1 g}{m_1 + m_2 + M}$$

Finalize When we evaluated the net torque about the axle, we did not include the forces that the cord exerts on the objects because these forces are internal to the system under consideration. Instead, we analyzed the system as a whole. Only *external* torques contribute to the change in the system's angular momentum. Let $M \to 0$ in Equation (3) and call the result Equation A. Now go back to Equation (6) in Example 5.10, let $\theta \to 0$, and call the result Equation B. Do Equations A and B match? Looking at Figures 5.15 and 11.6 in these limits, *should* the two equations match?

11.3 Angular Momentum of a Rotating Rigid Object

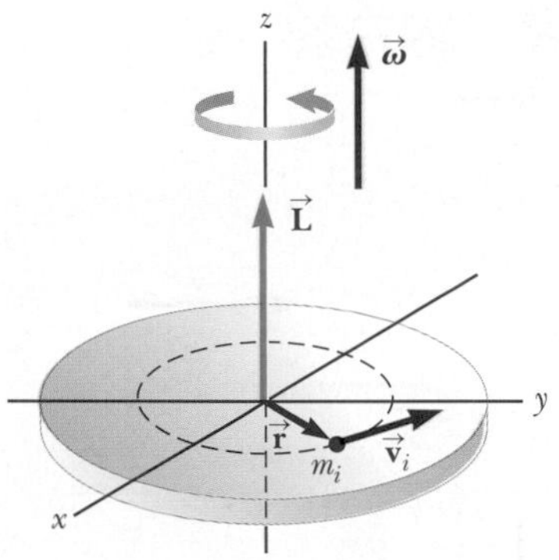

Figure 11.7 When a rigid object rotates about an axis, the angular momentum $\vec{\mathbf{L}}$ is in the same direction as the angular velocity $\vec{\boldsymbol{\omega}}$ according to the expression $\vec{\mathbf{L}} = I\vec{\boldsymbol{\omega}}$.

In Example 11.4, we considered the angular momentum of a deformable system. Let us now restrict our attention to a nondeformable system, a rigid object. Consider a rigid object rotating about a fixed axis that coincides with the z axis of a coordinate system as shown in Figure 11.7. Let's determine the angular momentum of this object. Each *particle* of the object rotates in the xy plane about the z axis with an angular speed ω. The magnitude of the angular momentum of a particle of mass m_i about the z axis is $m_i v_i r_i$. Because $v_i = r_i\omega$ (Eq. 10.10), we can express the magnitude of the angular momentum of this particle as

$$L_i = m_i r_i^2 \omega$$

The vector $\vec{\mathbf{L}}_i$ for this particle is directed along the z axis, as is the vector $\vec{\boldsymbol{\omega}}$.

We can now find the angular momentum (which in this situation has only a z component) of the whole object by taking the sum of L_i over all particles:

$$L_z = \sum_i L_i = \sum_i m_i r_i^2 \omega = \left(\sum_i m_i r_i^2\right)\omega$$

$$L_z = I\omega \qquad \textbf{(11.14)}$$

where we have recognized $\Sigma_i\, m_i r_i^2$ as the moment of inertia I of the object about the z axis (Eq. 10.15).

Now let's differentiate Equation 11.14 with respect to time, noting that I is constant for a rigid object:

$$\frac{dL_z}{dt} = I\frac{d\omega}{dt} = I\alpha \qquad \textbf{(11.15)}$$

where α is the angular acceleration relative to the axis of rotation. Because dL_z/dt is equal to the net external torque (see Eq. 11.13), we can express Equation 11.15 as

Rotational form of Newton's second law ▶

$$\sum \tau_{\text{ext}} = I\alpha \qquad \textbf{(11.16)}$$

That is, the net external torque acting on a rigid object rotating about a fixed axis equals the moment of inertia about the rotation axis multiplied by the object's angular acceleration relative to that axis. This result is the same as Equation 10.21, which was derived using a force approach, but we derived Equation 11.16 using the concept of angular momentum. As we saw in Section 10.7, Equation 11.16 is the mathematical representation of the rigid object under a net torque analysis model. This equation is also valid for a rigid object rotating about a moving axis, provided the moving axis (1) passes through the center of mass and (2) is a symmetry axis.

If a symmetrical object rotates about a fixed axis passing through its center of mass, you can write Equation 11.14 in vector form as $\vec{\mathbf{L}} = I\vec{\boldsymbol{\omega}}$, where $\vec{\mathbf{L}}$ is the total angular momentum of the object measured with respect to the axis of rotation. Furthermore, the expression is valid for any object, regardless of its symmetry, if $\vec{\mathbf{L}}$ stands for the component of angular momentum along the axis of rotation.[1]

Quick Quiz **11.3** A solid sphere and a hollow sphere have the same mass and radius. They are rotating with the same angular speed. Which one has the higher angular momentum? **(a)** the solid sphere **(b)** the hollow sphere **(c)** both have the same angular momentum **(d)** impossible to determine

Example 11.5 Bowling Ball

Estimate the magnitude of the angular momentum of a bowling ball spinning at 10 rev/s as shown in Figure 11.8.

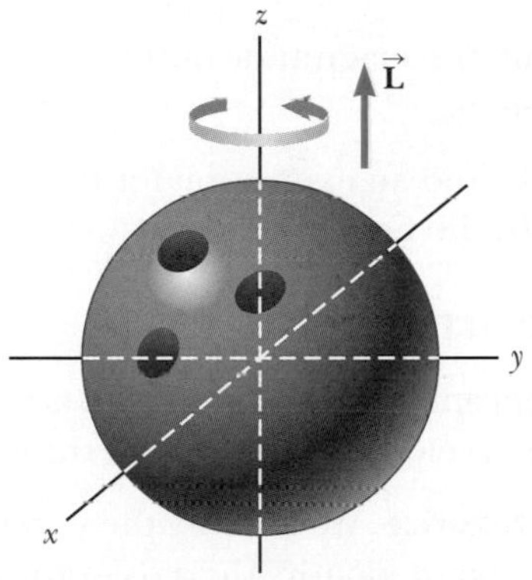

Figure 11.8 (Example 11.5) A bowling ball that rotates about the z axis in the direction shown has an angular momentum $\vec{\mathbf{L}}$ in the positive z direction. If the direction of rotation is reversed, then $\vec{\mathbf{L}}$ points in the negative z direction.

SOLUTION

Conceptualize Imagine spinning a bowling ball on the smooth floor of a bowling alley. Because a bowling ball is relatively heavy, the angular momentum should be relatively large.

Categorize We evaluate the angular momentum using Equation 11.14, so we categorize this example as a substitution problem.

We start by making some estimates of the relevant physical parameters and model the ball as a uniform solid sphere. A typical bowling ball might have a mass of 7.0 kg and a radius of 12 cm.

Evaluate the moment of inertia of the ball about an axis through its center from Table 10.2:

$$I = \frac{2}{5}MR^2 = \frac{2}{5}(7.0\text{ kg})(0.12\text{ m})^2 = 0.040\text{ kg}\cdot\text{m}^2$$

Evaluate the magnitude of the angular momentum from Equation 11.14:

$$L_z = I\omega = (0.040\text{ kg}\cdot\text{m}^2)(10\text{ rev/s})(2\pi\text{ rad/rev}) = 2.53\text{ kg}\cdot\text{m}^2\text{/s}$$

Because of the roughness of our estimates, we should keep only one significant figure, so $L_z = 3\text{ kg}\cdot\text{m}^2\text{/s}$.

Example 11.6 The Seesaw

A father of mass m_f and his daughter of mass m_d sit on opposite ends of a seesaw at equal distances from the pivot at the center (Fig. 11.9, page 328). The seesaw is modeled as a rigid rod of mass M and length ℓ and is pivoted without friction. At a given moment, the combination rotates in a vertical plane with an angular speed ω.

(A) Find an expression for the magnitude of the system's angular momentum.

continued

[1]In general, the expression $\vec{\mathbf{L}} = I\vec{\boldsymbol{\omega}}$ is not always valid. If a rigid object rotates about an *arbitrary* axis, then $\vec{\mathbf{L}}$ and $\vec{\boldsymbol{\omega}}$ may point in different directions. In this case, the moment of inertia cannot be treated as a scalar. Strictly speaking, $\vec{\mathbf{L}} = I\vec{\boldsymbol{\omega}}$ applies only to rigid objects of any shape that rotate about one of three mutually perpendicular axes (called *principal axes*) through the center of mass. This concept is discussed in more advanced texts on mechanics.

11.6 *cont.*

SOLUTION

Conceptualize Identify the z axis through O as the axis of rotation in Figure 11.9. The rotating system has angular momentum about that axis.

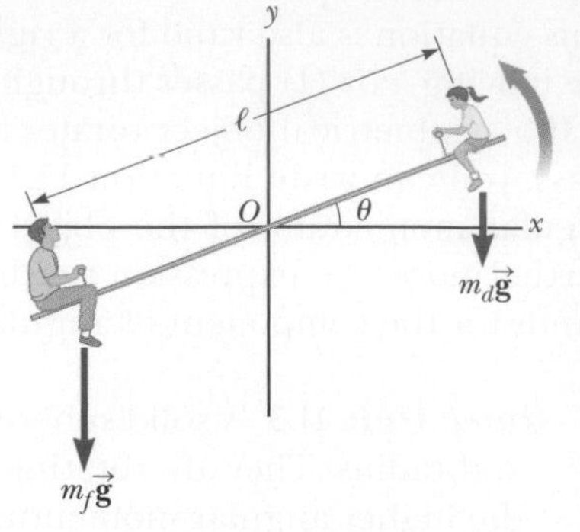

Figure 11.9 (Example 11.6) A father and daughter demonstrate angular momentum on a seesaw.

Categorize Ignore any movement of arms or legs of the father and daughter and model them both as particles. The system is therefore modeled as a rigid object. This first part of the example is categorized as a substitution problem.

The moment of inertia of the system equals the sum of the moments of inertia of the three components: the seesaw and the two individuals. We can refer to Table 10.2 to obtain the expression for the moment of inertia of the rod and use the particle expression $I = mr^2$ for each person.

Find the total moment of inertia of the system about the z axis through O:

$$I = \tfrac{1}{12}M\ell^2 + m_f\left(\frac{\ell}{2}\right)^2 + m_d\left(\frac{\ell}{2}\right)^2 = \frac{\ell^2}{4}\left(\frac{M}{3} + m_f + m_d\right)$$

Find the magnitude of the angular momentum of the system:

$$L = I\omega = \boxed{\frac{\ell^2}{4}\left(\frac{M}{3} + m_f + m_d\right)\omega}$$

(B) Find an expression for the magnitude of the angular acceleration of the system when the seesaw makes an angle θ with the horizontal.

SOLUTION

Conceptualize Generally, fathers are more massive than daughters, so the system is not in equilibrium and has an angular acceleration. We expect the angular acceleration to be positive in Figure 11.9.

Categorize We identify the system as nonisolated because of the external torque associated with the gravitational force. We again identify the axis of rotation as the z axis in Figure 11.9.

Analyze To find the angular acceleration of the system at any angle θ, we first calculate the net torque on the system and then use $\sum \tau_{\text{ext}} = I\alpha$ from the rigid object under a net torque model to obtain an expression for α.

Evaluate the torque due to the gravitational force on the father:

$$\tau_f = m_f g \frac{\ell}{2} \cos\theta \quad (\vec{\tau}_f \text{ out of page})$$

Evaluate the torque due to the gravitational force on the daughter:

$$\tau_d = -m_d g \frac{\ell}{2} \cos\theta \quad (\vec{\tau}_d \text{ into page})$$

Evaluate the net external torque exerted on the system:

$$\sum \tau_{\text{ext}} = \tau_f + \tau_d = \tfrac{1}{2}(m_f - m_d)g\,\ell\cos\theta$$

Use Equation 11.16 and I from part (A) to find α:

$$\alpha = \frac{\sum \tau_{\text{ext}}}{I} = \boxed{\frac{2(m_f - m_d)g\cos\theta}{\ell\left[(M/3) + m_f + m_d\right]}}$$

Finalize For a father more massive than his daughter, the angular acceleration is positive as expected. If the seesaw begins in a horizontal orientation ($\theta = 0$) and is released, the rotation is counterclockwise in Figure 11.9 and the father's end of the seesaw drops, which is consistent with everyday experience.

WHAT IF? Imagine the father moves inward on the seesaw to a distance d from the pivot to try to balance the two sides. What is the angular acceleration of the system in this case when it is released from an arbitrary angle θ?

Answer The angular acceleration of the system should decrease if the system is more balanced.

Find the total moment of inertia about the z axis through O for the modified system:

$$I = \tfrac{1}{12}M\ell^2 + m_f d^2 + m_d\left(\frac{\ell}{2}\right)^2 = \frac{\ell^2}{4}\left(\frac{M}{3} + m_d\right) + m_f d^2$$

11.6 *cont.*

Find the net torque exerted on the system about an axis through O:

$$\sum \tau_{\text{ext}} = \tau_f + \tau_d = m_f g d \cos\theta - \tfrac{1}{2} m_d g \ell \cos\theta$$

Find the new angular acceleration of the system:

$$\alpha = \frac{\sum \tau_{\text{ext}}}{I} = \frac{(m_f d - \frac{1}{2} m_d \ell) g \cos\theta}{(\ell^2/4)\,[(M/3) + m_d] + m_f d^2}$$

The seesaw is balanced when the angular acceleration is zero. In this situation, both father and daughter can push off the ground and rise to the highest possible point.

Find the required position of the father by setting $\alpha = 0$:

$$\alpha = \frac{(m_f d - \frac{1}{2} m_d \ell) g \cos\theta}{(\ell^2/4)[(M/3) + m_d] + m_f d^2} = 0$$

$$m_f d - \tfrac{1}{2} m_d \ell = 0 \quad \rightarrow \quad d = \left(\frac{m_d}{m_f}\right)\frac{\ell}{2}$$

In the rare case that the father and daughter have the same mass, the father is located at the end of the seesaw, $d = \ell/2$.

11.4 Analysis Model: Isolated System (Angular Momentum)

In Chapter 9, we found that the total linear momentum of a system of particles remains constant if the system is isolated, that is, if the net external force acting on the system is zero. We have an analogous conservation law in rotational motion:

> The total angular momentum of a system is constant in both magnitude and direction if the net external torque acting on the system is zero, that is, if the system is isolated.

◀ **Conservation of angular momentum**

This statement is often called[2] the principle of **conservation of angular momentum** and is the basis of the **angular momentum version of the isolated system model.** This principle follows directly from Equation 11.13, which indicates that if

$$\sum \vec{\boldsymbol{\tau}}_{\text{ext}} = \frac{d\vec{\mathbf{L}}_{\text{tot}}}{dt} = 0 \qquad \textbf{(11.17)}$$

then

$$\vec{\mathbf{L}}_{\text{tot}} = \text{constant} \quad \text{or} \quad \vec{\mathbf{L}}_i = \vec{\mathbf{L}}_f \qquad \textbf{(11.18)}$$

For an isolated system consisting of a number of particles, we write this conservation law as $\vec{\mathbf{L}}_{\text{tot}} = \sum \vec{\mathbf{L}}_n = \text{constant}$, where the index n denotes the nth particle in the system.

If an isolated rotating system is deformable so that its mass undergoes redistribution in some way, the system's moment of inertia changes. Because the magnitude of the angular momentum of the system is $L = I\omega$ (Eq. 11.14), conservation of angular momentum requires that the product of I and ω must remain constant. Therefore, a change in I for an isolated system requires a change in ω. In this case, we can express the principle of conservation of angular momentum as

$$I_i \omega_i = I_f \omega_f = \text{constant} \qquad \textbf{(11.19)}$$

[2]The most general conservation of angular momentum equation is Equation 11.13, which describes how the system interacts with its environment.

When his arms and legs are close to his body, the skater's moment of inertia is small and his angular speed is large.

Clive Rose/Getty Images

To slow down for the finish of his spin, the skater moves his arms and legs outward, increasing his moment of inertia.

Al Bello/Getty Images

Figure 11.10 Angular momentum is conserved as Russian gold medalist Evgeni Plushenko performs during the Turin 2006 Winter Olympic Games.

This expression is valid both for rotation about a fixed axis and for rotation about an axis through the center of mass of a moving system as long as that axis remains fixed in direction. We require only that the net external torque be zero.

Many examples demonstrate conservation of angular momentum for a deformable system. You may have observed a figure skater spinning in the finale of a program (Fig. 11.10). The angular speed of the skater is large when his hands and feet are close to the trunk of his body. (Notice the skater's hair!) Ignoring friction between skater and ice, there are no external torques on the skater. The moment of inertia of his body increases as his hands and feet are moved away from his body at the finish of the spin. According to the principle of conservation of angular momentum, his angular speed must decrease. In a similar way, when divers or acrobats wish to make several somersaults, they pull their hands and feet close to their bodies to rotate at a higher rate. In these cases, the external force due to gravity acts through the center of mass and hence exerts no torque about an axis through this point. Therefore, the angular momentum about the center of mass must be conserved; that is, $I_i\omega_i = I_f\omega_f$. For example, when divers wish to double their angular speed, they must reduce their moment of inertia to half its initial value.

In Equation 11.18, we have a third version of the isolated system model. We can now state that the energy, linear momentum, and angular momentum of an isolated system are all constant:

$$E_i = E_f \quad \text{(if there are no energy transfers across the system boundary)}$$

$$\vec{\mathbf{p}}_i = \vec{\mathbf{p}}_f \quad \text{(if the net external force on the system is zero)}$$

$$\vec{\mathbf{L}}_i = \vec{\mathbf{L}}_f \quad \text{(if the net external torque on the system is zero)}$$

A system may be isolated in terms of one of these quantities but not in terms of another. If a system is nonisolated in terms of momentum or angular momentum, it will often be nonisolated also in terms of energy because the system has a net force or torque on it and the net force or torque will do work on the system. We can, however, identify systems that are nonisolated in terms of energy but isolated in terms of momentum. For example, imagine pushing inward on a balloon (the system) between your hands. Work is done in compressing the balloon, so the system is nonisolated in terms of energy, but there is zero net force on the system, so the system is isolated in terms of momentum. A similar statement could be made about twisting the ends of a long, springy piece of metal with both hands. Work is done on the metal (the system), so energy is stored in the nonisolated system as elastic potential energy, but the net torque on the system is zero. Therefore, the system is isolated in terms of angular momentum. Other examples are collisions of macroscopic objects, which represent isolated systems in terms of momentum but nonisolated systems in terms of energy because of the output of energy from the system by mechanical waves (sound).

***Quick Quiz* 11.4** A competitive diver leaves the diving board and falls toward the water with her body straight and rotating slowly. She pulls her arms and legs into a tight tuck position. What happens to her rotational kinetic energy? **(a)** It increases. **(b)** It decreases. **(c)** It stays the same. **(d)** It is impossible to determine.

Example 11.7 Formation of a Neutron Star

A star rotates with a period of 30 days about an axis through its center. The period is the time interval required for a point on the star's equator to make one complete revolution around the axis of rotation. After the star undergoes a supernova explosion, the stellar core, which had a radius of 1.0×10^4 km, collapses into a neutron star of radius 3.0 km. Determine the period of rotation of the neutron star.

11.7 *cont.*

SOLUTION

Conceptualize The change in the neutron star's motion is similar to that of the skater described earlier, but in the reverse direction. As the mass of the star moves closer to the rotation axis, we expect the star to spin faster.

Categorize Let us assume that during the collapse of the stellar core, (1) no external torque acts on it, (2) it remains spherical with the same relative mass distribution, and (3) its mass remains constant. We categorize the star as an isolated system in terms of angular momentum. We do not know the mass distribution of the star, but we have assumed the distribution is symmetric, so the moment of inertia can be expressed as kMR^2, where k is some numerical constant. (From Table 10.2, for example, we see that $k = \frac{2}{5}$ for a solid sphere and $k = \frac{2}{3}$ for a spherical shell.)

Analyze Let's use the symbol T for the period, with T_i being the initial period of the star and T_f being the period of the neutron star. The star's angular speed is given by $\omega = 2\pi/T$.

Write Equation 11.19 for the star:

$$I_i\omega_i = I_f\omega_f$$

Use $\omega = 2\pi/T$ to rewrite this equation in terms of the initial and final periods:

$$I_i\left(\frac{2\pi}{T_i}\right) = I_f\left(\frac{2\pi}{T_f}\right)$$

Substitute the moments of inertia in the preceding equation:

$$kMR_i^2\left(\frac{2\pi}{T_i}\right) = kMR_f^2\left(\frac{2\pi}{T_f}\right)$$

Solve for the final period of the star:

$$T_f = \left(\frac{R_f}{R_i}\right)^2 T_i$$

Substitute numerical values:

$$T_f = \left(\frac{3.0 \text{ km}}{1.0 \times 10^4 \text{ km}}\right)^2(30 \text{ days}) = 2.7 \times 10^{-6} \text{ days} = \boxed{0.23 \text{ s}}$$

Finalize The neutron star does indeed rotate faster after it collapses, as predicted. It moves very fast, in fact, rotating about four times each second.

Example 11.8 The Merry-Go-Round

A horizontal platform in the shape of a circular disk rotates freely in a horizontal plane about a frictionless, vertical axle (Fig. 11.11). The platform has a mass M = 100 kg and a radius R = 2.0 m. A student whose mass is m = 60 kg walks slowly from the rim of the disk toward its center. If the angular speed of the system is 2.0 rad/s when the student is at the rim, what is the angular speed when she reaches a point r = 0.50 m from the center?

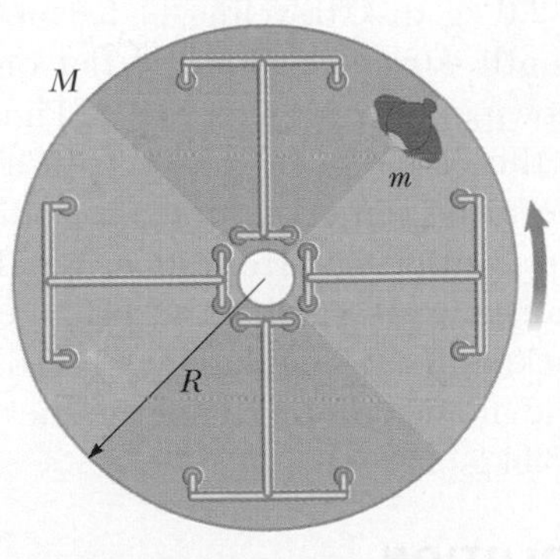

Figure 11.11 (Example 11.8) As the student walks toward the center of the rotating platform, the angular speed of the system increases because the angular momentum of the system remains constant.

SOLUTION

Conceptualize The speed change here is similar to those of the spinning skater and the neutron star in preceding discussions. This problem is different because part of the moment of inertia of the system changes (that of the student) while part remains fixed (that of the platform).

Categorize Because the platform rotates on a frictionless axle, we identify the system of the student and the platform as an isolated system in terms of angular momentum.

Analyze Let us denote the moment of inertia of the platform as I_p and that of the student as I_s. We model the student as a particle.

continued

11.8 *cont.*

Find the initial moment of inertia I_i of the system (student plus platform) about the axis of rotation:

$$I_i = I_{pi} + I_{si} = \tfrac{1}{2}MR^2 + mR^2$$

Find the moment of inertia of the system when the student walks to the position $r < R$:

$$I_f = I_{pf} + I_{sf} = \tfrac{1}{2}MR^2 + mr^2$$

Write Equation 11.19 for the system:

$$I_i\omega_i = I_f\omega_f$$

Substitute the moments of inertia:

$$(\tfrac{1}{2}MR^2 + mR^2)\omega_i = (\tfrac{1}{2}MR^2 + mr^2)\omega_f$$

Solve for the final angular speed:

$$\omega_f = \left(\frac{\tfrac{1}{2}MR^2 + mR^2}{\tfrac{1}{2}MR^2 + mr^2}\right)\omega_i$$

Substitute numerical values:

$$\omega_f = \left[\frac{\tfrac{1}{2}(100\text{ kg})(2.0\text{ m})^2 + (60\text{ kg})(2.0\text{ m})^2}{\tfrac{1}{2}(100\text{ kg})(2.0\text{ m})^2 + (60\text{ kg})(0.50\text{ m})^2}\right](2.0\text{ rad/s}) = 4.1\text{ rad/s}$$

Finalize As expected, the angular speed increases. The fastest that this system could spin would be when the student moves to the center of the platform. Do this calculation to show that this maximum angular speed is 4.4 rad/s. Notice that the activity described in this problem is dangerous as discussed with regard to the Coriolis force in Section 6.3.

WHAT IF? What if you measured the kinetic energy of the system before and after the student walks inward? Are the initial kinetic energy and the final kinetic energy the same?

Answer You may be tempted to say yes because the system is isolated. Remember, however, that energy can be transformed among several forms, so we have to handle an energy question carefully.

Find the initial kinetic energy:

$$K_i = \tfrac{1}{2}I_i\omega_i^2 = \tfrac{1}{2}(440\text{ kg}\cdot\text{m}^2)(2.0\text{ rad/s})^2 = 880\text{ J}$$

Find the final kinetic energy:

$$K_f = \tfrac{1}{2}I_f\omega_f^2 = \tfrac{1}{2}(215\text{ kg}\cdot\text{m}^2)(4.1\text{ rad/s})^2 = 1.80 \times 10^3\text{ J}$$

Therefore, the kinetic energy of the system *increases*. The student must do work to move herself closer to the center of rotation, so this extra kinetic energy comes from chemical potential energy in the student's body. The system is isolated in terms of energy, but a transformation process within the system changes potential energy to kinetic energy.

Example 11.9 Disk and Stick Collision

A 2.0-kg disk traveling at 3.0 m/s strikes a 1.0-kg stick of length 4.0 m that is lying flat on nearly frictionless ice as shown in the overhead view of Figure 11.12a. The disk strikes at the endpoint of the stick, at a distance r = 2.0 m from the stick's center. Assume the collision is elastic and the disk does not deviate from its original line of motion. Find the translational speed of the disk, the translational speed of the stick, and the angular speed of the stick after the collision. The moment of inertia of the stick about its center of mass is 1.33 kg · m².

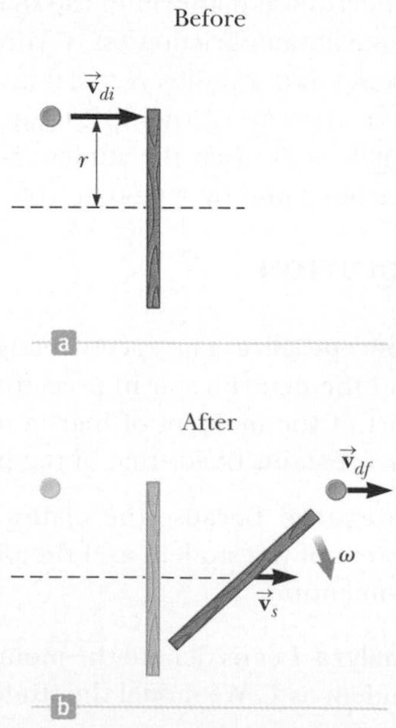

Figure 11.12 (Example 11.9) Overhead view of a disk striking a stick in an elastic collision. (a) Before the collision, the disk moves toward the stick. (b) The collision causes the stick to rotate and move to the right.

SOLUTION

Conceptualize Examine Figure 11.12a and imagine what happens after the disk hits the stick. Figure 11.12b shows what you might expect: the disk continues to move at a slower speed, and the stick is in both translational and rotational motion. We assume the disk does not deviate from its original line of motion because the force exerted by the stick on the disk is parallel to the original path of the disk.

11.9 *cont.*

Categorize Because the ice is frictionless, the disk and stick form an isolated system in terms of momentum and angular momentum. Ignoring the sound made in the collision, we also model the system as isolated in terms of energy. In addition, because the collision is assumed to be elastic, the kinetic energy of the system is constant.

Analyze First notice that we have three unknowns, so we need three equations to solve simultaneously.

Apply the isolated system model for momentum to the system and then rearrange the result:

$$m_d v_{di} = m_d v_{df} + m_s v_s$$

$$(1)\quad m_d(v_{di} - v_{df}) = m_s v_s$$

Apply the isolated system model for angular momentum to the system and rearrange the result. Use an axis passing through the center of the stick as the rotation axis so that the path of the disk is a distance $r = 2.0$ m from the rotation axis:

$$-rm_d v_{di} = -rm_d v_{df} + I\omega$$

$$(2)\quad -rm_d(v_{di} - v_{df}) = I\omega$$

Apply the isolated system model for energy to the system, rearrange the equation, and factor the left side:

$$\tfrac{1}{2}m_d v_{di}^2 = \tfrac{1}{2}m_d v_{df}^2 + \tfrac{1}{2}m_s v_s^2 + \tfrac{1}{2}I\omega^2$$

$$(3)\quad m_d(v_{di} - v_{df})(v_{di} + v_{df}) = m_s v_s^2 + I\omega^2$$

Multiply Equation (1) by r and add to Equation (2):

$$rm_d(v_{di} - v_{df}) = rm_s v_s$$

$$-rm_d(v_{di} - v_{df}) = I\omega$$

$$0 = rm_s v_s + I\omega$$

Solve for ω:

$$(4)\quad \omega = -\frac{rm_s v_s}{I}$$

Divide Equation (3) by Equation (1):

$$\frac{m_d(v_{di} - v_{df})(v_{di} + v_{df})}{m_d(v_{di} - v_{df})} = \frac{m_s v_s^2 + I\omega^2}{m_s v_s}$$

$$(5)\quad v_{di} + v_{df} = v_s + \frac{I\omega^2}{m_s v_s}$$

Substitute Equation (4) into Equation (5):

$$(6)\quad v_{di} + v_{df} = v_s\left(1 + \frac{r^2 m_s}{I}\right)$$

Substitute v_{df} from Equation (1) into Equation (6):

$$v_{di} + \left(v_{di} - \frac{m_s}{m_d}v_s\right) = v_s\left(1 + \frac{r^2 m_s}{I}\right)$$

Solve for v_s and substitute numerical values:

$$v_s = \frac{2v_{di}}{1 + (m_s/m_d) + (r^2 m_s/I)}$$

$$= \frac{2(3.0 \text{ m/s})}{1 + (1.0 \text{ kg}/2.0 \text{ kg}) + [(2.0 \text{ m})^2(1.0 \text{ kg})/1.33 \text{ kg}\cdot\text{m}^2]} = \boxed{1.3 \text{ m/s}}$$

Substitute numerical values into Equation (4):

$$\omega = -\frac{(2.0 \text{ m})(1.0 \text{ kg})(1.3 \text{ m/s})}{1.33 \text{ kg}\cdot\text{m}^2} = \boxed{-2.0 \text{ rad/s}}$$

Solve Equation (1) for v_{df} and substitute numerical values:

$$v_{df} = v_{di} - \frac{m_s}{m_d}v_s = 3.0 \text{ m/s} - \frac{1.0 \text{ kg}}{2.0 \text{ kg}}(1.3 \text{ m/s}) = \boxed{2.3 \text{ m/s}}$$

Finalize These values seem reasonable. The disk is moving more slowly after the collision than it was before the collision, and the stick has a small translational speed. Table 11.1 on page 334 summarizes the initial and final values of variables for the disk and the stick, and it verifies the conservation of linear momentum, angular momentum, and kinetic energy for the isolated system.

continued

11.9 *cont.*

TABLE 11.1 *Comparison of Values in Example 11.9 Before and After the Collision*

	v (m/s)	ω (rad/s)	p (kg · m/s)	L (kg · m²/s)	K_{trans} (J)	K_{rot} (J)
Before						
Disk	3.0	—	6.0	−12	9.0	—
Stick	0	0	0	0	0	0
Total for system	—	—	6.0	−12	9.0	0
After						
Disk	2.3	—	4.7	−9.3	5.4	—
Stick	1.3	−2.0	1.3	−2.7	0.9	2.7
Total for system	—	—	6.0	−12	6.3	2.7

Note: Linear momentum, angular momentum, and total kinetic energy of the system are all conserved.

11.5 The Motion of Gyroscopes and Tops

An unusual and fascinating type of motion you have probably observed is that of a top spinning about its axis of symmetry as shown in Figure 11.13a. If the top spins rapidly, the symmetry axis rotates about the z axis, sweeping out a cone (see Fig. 11.13b). The motion of the symmetry axis about the vertical—known as **precessional motion**—is usually slow relative to the spinning motion of the top.

It is quite natural to wonder why the top does not fall over. Because the center of mass is not directly above the pivot point O, a net torque is acting on the top about an axis passing through O, a torque resulting from the gravitational force $M\vec{g}$. The top would certainly fall over if it were not spinning. Because it is spinning, however, it has an angular momentum $\vec{L}$ directed along its symmetry axis. We shall show that this symmetry axis moves about the z axis (precessional motion occurs) because the torque produces a change in the *direction* of the symmetry axis. This illustration is an excellent example of the importance of the vector nature of angular momentum.

The essential features of precessional motion can be illustrated by considering the simple gyroscope shown in Figure 11.14a. The two forces acting on the gyroscope are shown in Figure 11.14b: the downward gravitational force $M\vec{g}$ and the normal force $\vec{n}$ acting upward at the pivot point O. The normal force produces no torque about an axis passing through the pivot because its moment arm through that point is zero. The gravitational force, however, produces a torque $\vec{\tau} = \vec{r} \times M\vec{g}$ about an axis passing through O, where the direction of $\vec{\tau}$ is perpendicular to the plane formed by $\vec{r}$ and $M\vec{g}$. By necessity, the vector $\vec{\tau}$ lies in a horizontal xy plane perpendicular to the angular momentum vector. The net torque and angular momentum of the gyroscope are related through Equation 11.13:

$$\sum \vec{\tau}_{ext} = \frac{d\vec{L}}{dt}$$

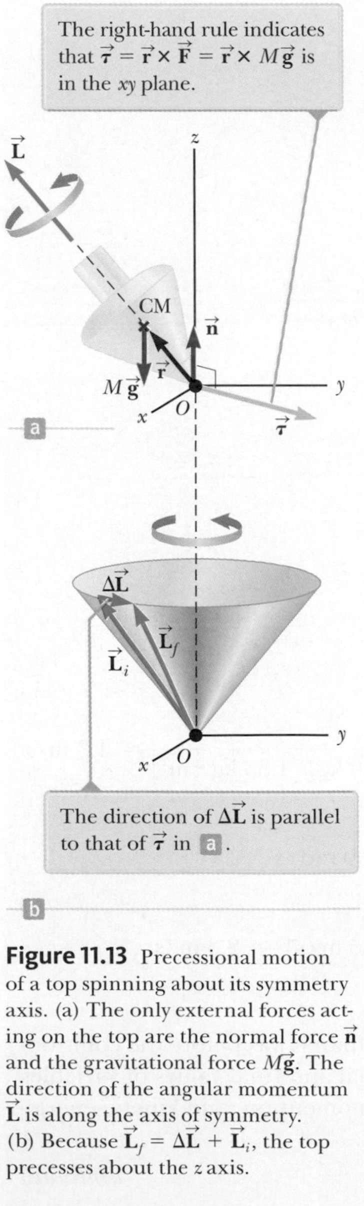

Figure 11.13 Precessional motion of a top spinning about its symmetry axis. (a) The only external forces acting on the top are the normal force $\vec{n}$ and the gravitational force $M\vec{g}$. The direction of the angular momentum $\vec{L}$ is along the axis of symmetry. (b) Because $\vec{L}_f = \Delta\vec{L} + \vec{L}_i$, the top precesses about the z axis.

This expression shows that in the infinitesimal time interval dt, the nonzero torque produces a change in angular momentum $d\vec{L}$, a change that is in the same direction as $\vec{\tau}$. Therefore, like the torque vector, $d\vec{L}$ must also be perpendicular to $\vec{L}$. Figure 11.14c illustrates the resulting precessional motion of the symmetry axis of the gyroscope. In a time interval dt, the change in angular momentum is $d\vec{L} = \vec{L}_f - \vec{L}_i = \vec{\tau}\, dt$. Because $d\vec{L}$ is perpendicular to $\vec{L}$, the magnitude of $\vec{L}$ does not change ($|\vec{L}_i| = |\vec{L}_f|$). Rather, what is changing is the *direction* of $\vec{L}$. Because the change in angular momentum $d\vec{L}$ is in the direction of $\vec{\tau}$, which lies in the xy plane, the gyroscope undergoes precessional motion.

To simplify the description of the system, we assume the total angular momentum of the precessing wheel is the sum of the angular momentum $I\vec{\omega}$ due to the

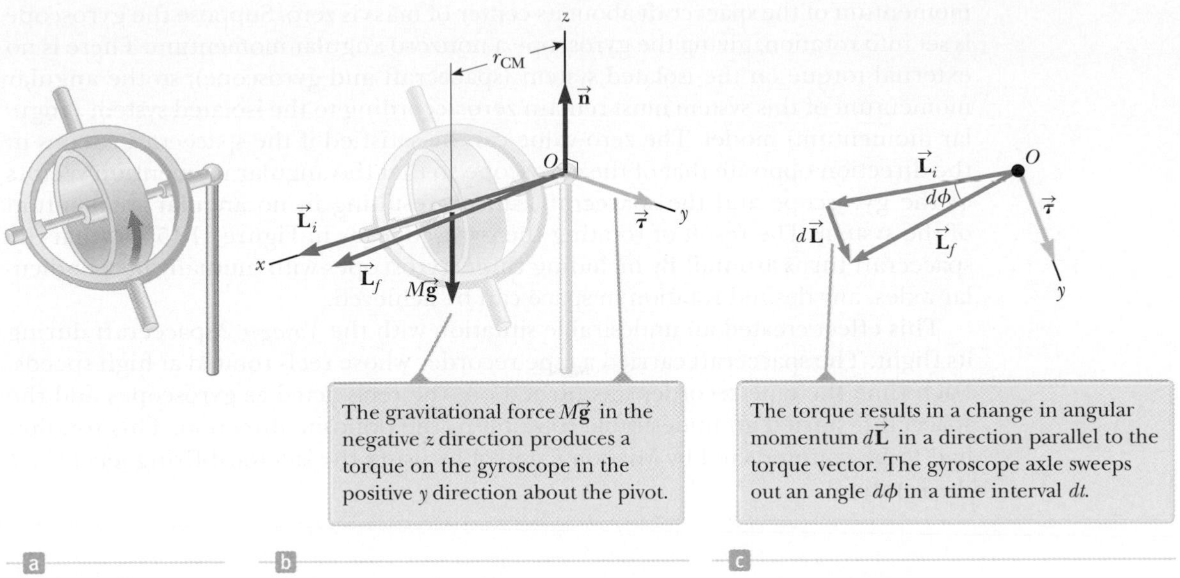

Figure 11.14 (a) A spinning gyroscope is placed on a pivot at the right end. (b) Diagram for the spinning gyroscope showing forces, torque, and angular momentum. (c) Overhead view (looking down the z axis) of the gyroscope's initial and final angular momentum vectors for an infinitesimal time interval dt.

spinning and the angular momentum due to the motion of the center of mass about the pivot. In our treatment, we shall neglect the contribution from the center-of-mass motion and take the total angular momentum to be simply $I\vec{\omega}$. In practice, this approximation is good if $\vec{\omega}$ is made very large.

The vector diagram in Figure 11.14c shows that in the time interval dt, the angular momentum vector rotates through an angle $d\phi$, which is also the angle through which the gyroscope axle rotates. From the vector triangle formed by the vectors $\vec{L}_i$, $\vec{L}_f$, and $d\vec{L}$, we see that

$$d\phi = \frac{dL}{L} = \frac{\sum \tau_{\text{ext}}\, dt}{L} = \frac{(Mgr_{\text{CM}})\, dt}{L}$$

Dividing through by dt and using the relationship $L = I\omega$, we find that the rate at which the axle rotates about the vertical axis is

$$\omega_p = \frac{d\phi}{dt} = \frac{Mgr_{\text{CM}}}{I\omega} \qquad \textbf{(11.20)}$$

The angular speed ω_p is called the **precessional frequency**. This result is valid only when $\omega_p << \omega$. Otherwise, a much more complicated motion is involved. As you can see from Equation 11.20, the condition $\omega_p << \omega$ is met when ω is large, that is, when the wheel spins rapidly. Furthermore, notice that the precessional frequency decreases as ω increases, that is, as the wheel spins faster about its axis of symmetry.

As an example of the usefulness of gyroscopes, suppose you are in a spacecraft in deep space and you need to alter your trajectory. To fire the engines in the correct direction, you need to turn the spacecraft. How, though, do you turn a spacecraft in empty space? One way is to have small rocket engines that fire perpendicularly out the side of the spacecraft, providing a torque around its center of mass. Such a setup is desirable, and many spacecraft have such rockets.

Let us consider another method, however, that is related to angular momentum and does not require the consumption of rocket fuel. Suppose the spacecraft carries a gyroscope that is not rotating as in Figure 11.15a. In this case, the angular

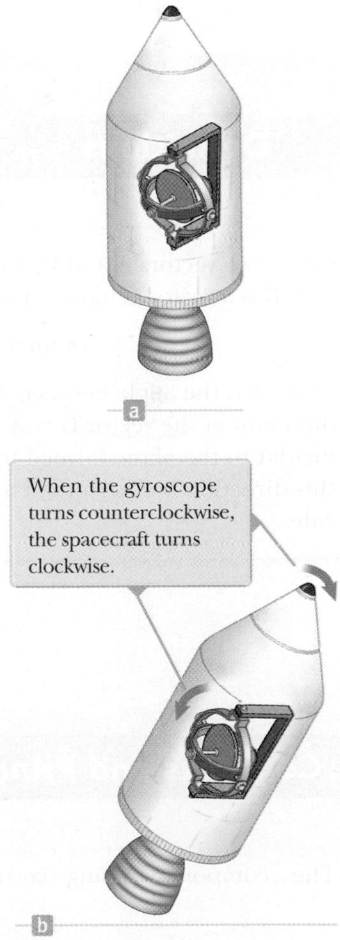

Figure 11.15 (a) A spacecraft carries a gyroscope that is not spinning. (b) The gyroscope is set into rotation.

momentum of the spacecraft about its center of mass is zero. Suppose the gyroscope is set into rotation, giving the gyroscope a nonzero angular momentum. There is no external torque on the isolated system (spacecraft and gyroscope), so the angular momentum of this system must remain zero according to the isolated system (angular momentum) model. The zero value can be satisfied if the spacecraft rotates in the direction opposite that of the gyroscope so that the angular momentum vectors of the gyroscope and the spacecraft cancel, resulting in no angular momentum of the system. The result of rotating the gyroscope, as in Figure 11.15b, is that the spacecraft turns around! By including three gyroscopes with mutually perpendicular axles, any desired rotation in space can be achieved.

This effect created an undesirable situation with the *Voyager 2* spacecraft during its flight. The spacecraft carried a tape recorder whose reels rotated at high speeds. Each time the tape recorder was turned on, the reels acted as gyroscopes and the spacecraft started an undesirable rotation in the opposite direction. This rotation had to be counteracted by Mission Control by using the sideward-firing jets to *stop* the rotation!

Summary

Definitions

Given two vectors $\vec{\mathbf{A}}$ and $\vec{\mathbf{B}}$, the **vector product** $\vec{\mathbf{A}} \times \vec{\mathbf{B}}$ is a vector $\vec{\mathbf{C}}$ having a magnitude

$$C = AB \sin \theta \tag{11.3}$$

where θ is the angle between $\vec{\mathbf{A}}$ and $\vec{\mathbf{B}}$. The direction of the vector $\vec{\mathbf{C}} = \vec{\mathbf{A}} \times \vec{\mathbf{B}}$ is perpendicular to the plane formed by $\vec{\mathbf{A}}$ and $\vec{\mathbf{B}}$, and this direction is determined by the right-hand rule.

The **torque** $\vec{\boldsymbol{\tau}}$ due to a force $\vec{\mathbf{F}}$ about an axis through the origin in an inertial frame is defined to be

$$\vec{\boldsymbol{\tau}} \equiv \vec{\mathbf{r}} \times \vec{\mathbf{F}} \tag{11.1}$$

The **angular momentum** $\vec{\mathbf{L}}$ about an axis through the origin of a particle having linear momentum $\vec{\mathbf{p}} = m\vec{\mathbf{v}}$ is

$$\vec{\mathbf{L}} \equiv \vec{\mathbf{r}} \times \vec{\mathbf{p}} \tag{11.10}$$

where $\vec{\mathbf{r}}$ is the vector position of the particle relative to the origin.

Concepts and Principles

The z component of angular momentum of a rigid object rotating about a fixed z axis is

$$L_z = I\omega \tag{11.14}$$

where I is the moment of inertia of the object about the axis of rotation and ω is its angular speed.

Analysis Models for Problem Solving

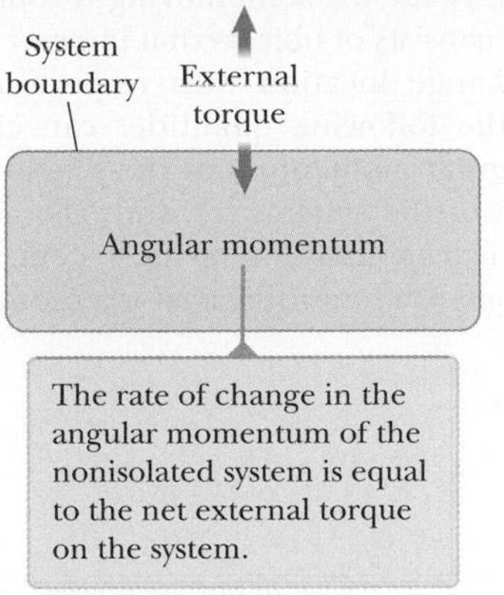

Nonisolated System (Angular Momentum). If a system interacts with its environment in the sense that there is an external torque on the system, the net external torque acting on a system is equal to the time rate of change of its angular momentum:

$$\sum \vec{\boldsymbol{\tau}}_{\text{ext}} = \frac{d\vec{\mathbf{L}}_{\text{tot}}}{dt} \quad \textbf{(11.13)}$$

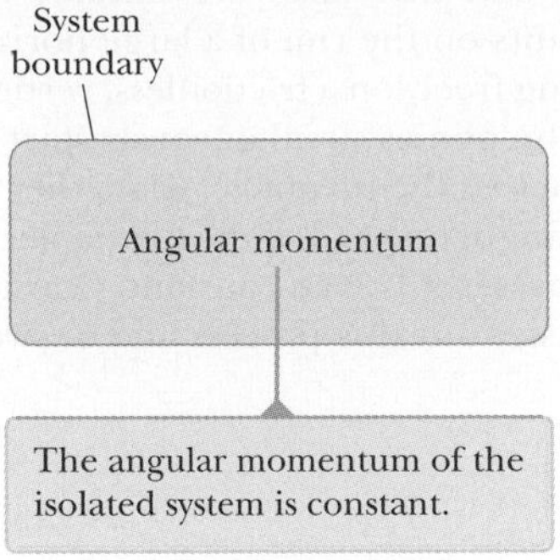

Isolated System (Angular Momentum). If a system experiences no external torque from the environment, the total angular momentum of the system is conserved:

$$\vec{\mathbf{L}}_i = \vec{\mathbf{L}}_f \quad \textbf{(11.18)}$$

Applying this law of conservation of angular momentum to a system whose moment of inertia changes gives

$$I_i\omega_i = I_f\omega_f = \text{constant} \quad \textbf{(11.19)}$$

Objective Questions

☐ denotes answer available in Student Solutions Manual/Study Guide

1. Answer yes or no to the following questions. (a) Is it possible to calculate the torque acting on a rigid object without specifying an axis of rotation? (b) Is the torque independent of the location of the axis of rotation?

2. Vector $\vec{\mathbf{A}}$ is in the negative y direction, and vector $\vec{\mathbf{B}}$ is in the negative x direction. **(i)** What is the direction of $\vec{\mathbf{A}} \times \vec{\mathbf{B}}$? (a) no direction because it is a scalar (b) x (c) $-y$ (d) z (e) $-z$ **(ii)** What is the direction of $\vec{\mathbf{B}} \times \vec{\mathbf{A}}$? Choose from the same possibilities (a) through (e).

3. Let us name three perpendicular directions as right, up, and toward you as you might name them when you are facing a television screen that lies in a vertical plane. Unit vectors for these directions are $\hat{\mathbf{r}}$, $\hat{\mathbf{u}}$, and $\hat{\mathbf{t}}$, respectively. Consider the quantity $(-3\hat{\mathbf{u}} \times 2\hat{\mathbf{t}})$ **(i)** Is the magnitude of this vector (a) 6, (b) 3, (c) 2, or (d) 0? **(ii)** Is the direction of this vector (a) down, (b) toward you, (c) up, (d) away from you, or (e) left?

4. Let the four compass directions north, east, south, and west be represented by unit vectors $\hat{\mathbf{n}}$, $\hat{\mathbf{e}}$, $\hat{\mathbf{s}}$, and $\hat{\mathbf{w}}$, respectively. Vertically up and down are represented as $\hat{\mathbf{u}}$ and $\hat{\mathbf{d}}$. Let us also identify unit vectors that are halfway between these directions such as $\widehat{\mathbf{ne}}$ for northeast. Rank the magnitudes of the following cross products from largest to smallest. If any are equal in magnitude or are equal to zero, show that in your ranking. (a) $\hat{\mathbf{n}} \times \hat{\mathbf{n}}$ (b) $\hat{\mathbf{w}} \times \widehat{\mathbf{ne}}$ (c) $\hat{\mathbf{u}} \times \widehat{\mathbf{ne}}$ (d) $\hat{\mathbf{n}} \times \widehat{\mathbf{nw}}$ (e) $\hat{\mathbf{n}} \times \hat{\mathbf{e}}$

5. An ice skater starts a spin with her arms stretched out to the sides. She balances on the tip of one skate to turn without friction. She then pulls her arms in so that her moment of inertia decreases by a factor of 2. In the process of her doing so, what happens to her kinetic energy? (a) It increases by a factor of 4. (b) It increases by a factor of 2. (c) It remains constant. (d) It decreases by a factor of 2. (e) It decreases by a factor of 4.

6. A pet mouse sleeps near the eastern edge of a stationary, horizontal turntable that is supported by a frictionless, vertical axle through its center. The mouse wakes up and starts to walk north on the turntable. **(i)** As it takes its first steps, what is the direction of the mouse's displacement relative to the stationary ground below? (a) north (b) south (c) no displacement. **(ii)** In this process, the spot on the turntable where the mouse had been snoozing undergoes a displacement in what direction relative to the ground below? (a) north (b) south (c) no displacement. Answer yes or no for the following questions. **(iii)** In this process, is the mechanical energy of the mouse–turntable system constant? **(iv)** Is the momentum of the system constant? **(v)** Is the angular momentum of the system constant?

7. Two ponies of equal mass are initially at diametrically opposite points on the rim of a large horizontal turntable that is turning freely on a frictionless, vertical axle through its center. The ponies simultaneously start walking toward each other across the turntable. **(i)** As they walk, what happens to the angular speed of the turntable? (a) It increases. (b) It decreases. (c) It stays constant. Consider the ponies–turntable system in this process and answer yes or no for the following questions. **(ii)** Is the mechanical energy of the system conserved? **(iii)** Is the momentum of the system conserved? **(iv)** Is the angular momentum of the system conserved?

8. Consider an isolated system moving through empty space. The system consists of objects that interact with each other and can change location with respect to one another. Which of the following quantities can change in time? (a) The angular momentum of the system. (b) The linear momentum of the system. (c) Both the angular momentum and linear momentum of the system. (d) Neither the angular momentum nor linear momentum of the system.

Conceptual Questions

☐ denotes answer available in Student Solutions Manual/Study Guide

1. If the torque acting on a particle about an axis through a certain origin is zero, what can you say about its angular momentum about that axis?

2. A ball is thrown in such a way that it does not spin about its own axis. Does this statement imply that the angular momentum is zero about an arbitrary axis? Explain.

[3.] Why does a long pole help a tightrope walker stay balanced?

[4.] Two children are playing with a roll of paper towels. One child holds the roll between the index fingers of her hands so that it is free to rotate, and the second child pulls at constant speed on the free end of the paper towels. As the child pulls the paper towels, the radius of the roll of remaining towels decreases. (a) How does the torque on the roll change with time? (b) How does the angular speed of the roll change in time? (c) If the child suddenly jerks the end paper towel with a large force, is the towel more likely to break from the others when it is being pulled from a nearly full roll or from a nearly empty roll?

5. Both torque and work are products of force and displacement. How are they different? Do they have the same units?

6. In some motorcycle races, the riders drive over small hills and the motorcycle becomes airborne for a short time interval. If the motorcycle racer keeps the throttle open while leaving the hill and going into the air, the motorcycle tends to nose upward. Why?

7. Stars originate as large bodies of slowly rotating gas. Because of gravity, these clumps of gas slowly decrease in size. What happens to the angular speed of a star as it shrinks? Explain.

8. A scientist arriving at a hotel asks a bellhop to carry a heavy suitcase. When the bellhop rounds a corner, the suitcase suddenly swings away from him for some unknown reason. The alarmed bellhop drops the suitcase and runs away. What might be in the suitcase?

[9.] If global warming continues over the next one hundred years, it is likely that some polar ice will melt and the water will be distributed closer to the equator. (a) How would that change the moment of inertia of the Earth? (b) Would the duration of the day (one revolution) increase or decrease?

10. A cat usually lands on its feet regardless of the position from which it is dropped. A slow-motion film of a cat falling shows that the upper half of its body twists in one direction while the lower half twists in the opposite direction. (See Fig. CQ11.10.) Why does this type of rotation occur?

Figure CQ11.10

11. In Chapters 7 and 8, we made use of energy bar charts to analyze physical situations. Why have we not used bar charts for angular momentum in this chapter?

Problems

WebAssign The problems found in this chapter may be assigned online in Enhanced WebAssign

1. denotes straightforward problem; **2.** denotes intermediate problem; **3.** denotes challenging problem

1. full solution available in the Student Solutions Manual/Study Guide

1. denotes problems most often assigned in Enhanced WebAssign; these provide students with targeted feedback and either a Master It tutorial or a Watch It solution video.

QC denotes asking for quantitative and conceptual reasoning

S denotes symbolic reasoning problem

M denotes Master It tutorial available in Enhanced WebAssign

GP denotes guided problem

shaded denotes "paired problems" that develop reasoning with symbols and numerical values

Section 11.1 The Vector Product and Torque

1. Given $\vec{\mathbf{M}} = 2\hat{\mathbf{i}} - 3\hat{\mathbf{j}} + \hat{\mathbf{k}}$ and $\vec{\mathbf{N}} = 4\hat{\mathbf{i}} + 5\hat{\mathbf{j}} - 2\hat{\mathbf{k}}$, calculate the vector product $\vec{\mathbf{M}} \times \vec{\mathbf{N}}$.

2. The displacement vectors 42.0 cm at 15.0° and 23.0 cm at 65.0° both start from the origin and form two sides of a parallelogram. Both angles are measured counterclockwise from the *x* axis. (a) Find the area of the parallelogram. (b) Find the length of its longer diagonal.

3. Two vectors are given by $\vec{\mathbf{A}} = \hat{\mathbf{i}} + 2\hat{\mathbf{j}}$ and $\vec{\mathbf{B}} = -2\hat{\mathbf{i}} + 3\hat{\mathbf{j}}$. Find (a) $\vec{\mathbf{A}} \times \vec{\mathbf{B}}$ and (b) the angle between $\vec{\mathbf{A}}$ and $\vec{\mathbf{B}}$.

4. Two vectors are given by $\vec{\mathbf{A}} = -3\hat{\mathbf{i}} + 7\hat{\mathbf{j}} - 4\hat{\mathbf{k}}$ and $\vec{\mathbf{B}} = 6\hat{\mathbf{i}} - 10\hat{\mathbf{j}} + 9\hat{\mathbf{k}}$. Evaluate the quantities (a) $\cos^{-1}[\vec{\mathbf{A}} \cdot \vec{\mathbf{B}}/AB]$ and (b) $\sin^{-1}[|\vec{\mathbf{A}} \times \vec{\mathbf{B}}|/AB]$. (c) Which give(s) the angle between the vectors?

5. Calculate the net torque (magnitude and direction) on the beam in Figure P11.5 about (a) an axis through *O* perpendicular to the page and (b) an axis through *C* perpendicular to the page.

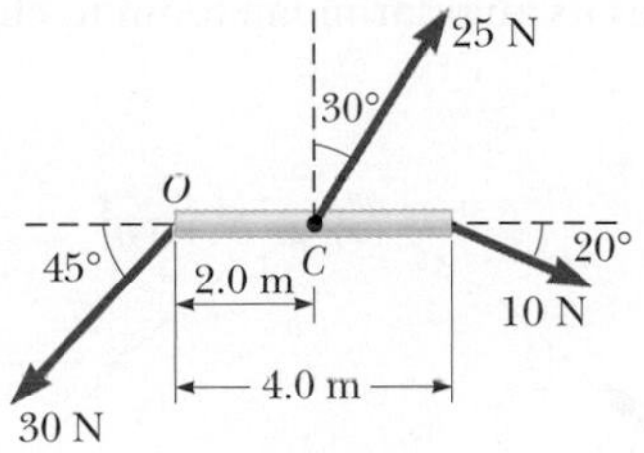

Figure P11.5

6. **S** Use the definition of the vector product and the definitions of the unit vectors $\hat{\mathbf{i}}$, $\hat{\mathbf{j}}$, and $\hat{\mathbf{k}}$ to prove Equations 11.7. You may assume the *x* axis points to the right, the *y* axis up, and the *z* axis horizontally toward you (not away from you). This choice is said to make the coordinate system a *right-handed system.*

7. If $|\vec{\mathbf{A}} \times \vec{\mathbf{B}}| = \vec{\mathbf{A}} \cdot \vec{\mathbf{B}}$, what is the angle between $\vec{\mathbf{A}}$ and $\vec{\mathbf{B}}$?

8. **QC** A student claims that he has found a vector $\vec{\mathbf{A}}$ such that $(2\hat{\mathbf{i}} - 3\hat{\mathbf{j}} + 4\hat{\mathbf{k}}) \times \vec{\mathbf{A}} = (4\hat{\mathbf{i}} + 3\hat{\mathbf{j}} - \hat{\mathbf{k}})$. (a) Do you believe this claim? (b) Explain why or why not.

9. Two forces $\vec{\mathbf{F}}_1$ and $\vec{\mathbf{F}}_2$ act along the two sides of an equilateral triangle as shown in Figure P11.9. Point *O* is the intersection of the altitudes of the triangle. (a) Find a third force $\vec{\mathbf{F}}_3$ to be applied at *B* and along *BC* that will make the total torque zero about the point *O*. (b) **What If?** Will the total torque change if $\vec{\mathbf{F}}_3$ is applied not at *B* but at any other point along *BC*?

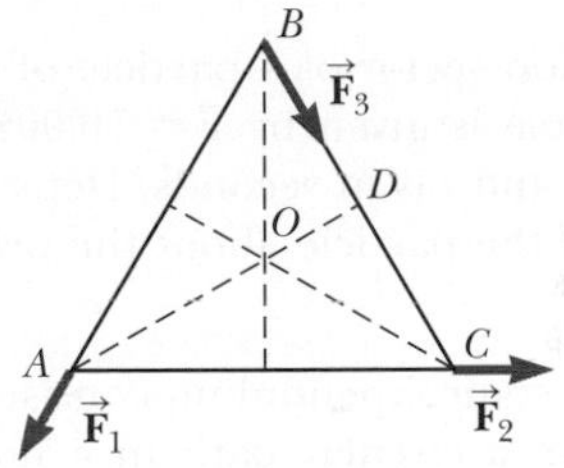

Figure P11.9

10. **QC** A particle is located at the vector position $\vec{\mathbf{r}} = (4.00\hat{\mathbf{i}} + 6.00\hat{\mathbf{j}})$ m, and a force exerted on it is given by $\vec{\mathbf{F}} = (3.00\hat{\mathbf{i}} + 2.00\hat{\mathbf{j}})$ N. (a) What is the torque acting on the particle about the origin? (b) Can there be another point about which the torque caused by this force on this particle will be in the opposite direction and half as large in magnitude? (c) Can there be more than one such point? (d) Can such a point lie on the *y* axis? (e) Can more than one such point lie on the *y* axis? (f) Determine the position vector of one such point.

Section 11.2 Analysis Model: Nonisolated System (Angular Momentum)

11. A light, rigid rod of length $\ell = 1.00$ m joins two particles, with masses $m_1 = 4.00$ kg and $m_2 = 3.00$ kg, at its ends. The combination rotates in the *xy* plane about a pivot through the center of the rod (Fig. P11.11). Determine the angular

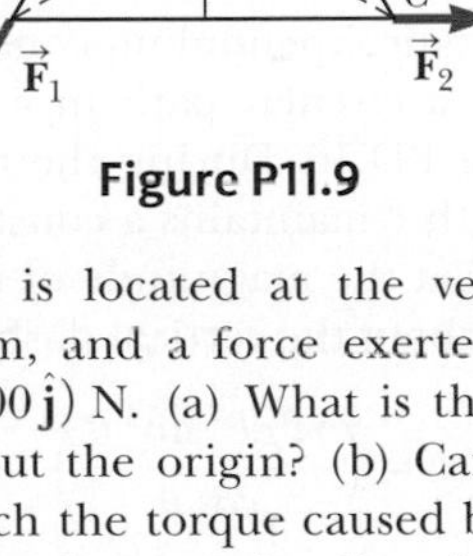

Figure P11.11

momentum of the system about the origin when the speed of each particle is 5.00 m/s.

12. A 1.50-kg particle moves in the xy plane with a velocity of $\vec{\mathbf{v}} = (4.20\hat{\mathbf{i}} - 3.60\hat{\mathbf{j}})$ m/s. Determine the angular momentum of the particle about the origin when its position vector is $\vec{\mathbf{r}} = (1.50\hat{\mathbf{i}} + 2.20\hat{\mathbf{j}})$ m.

13. S A particle of mass m moves in the xy plane with a velocity of $\vec{\mathbf{v}} = v_x\hat{\mathbf{i}} + v_y\hat{\mathbf{j}}$. Determine the angular momentum of the particle about the origin when its position vector is $\vec{\mathbf{r}} = x\hat{\mathbf{i}} + y\hat{\mathbf{j}}$.

14. Heading straight toward the summit of Pike's Peak, an airplane of mass 12 000 kg flies over the plains of Kansas at nearly constant altitude 4.30 km with constant velocity 175 m/s west. (a) What is the airplane's vector angular momentum relative to a wheat farmer on the ground directly below the airplane? (b) Does this value change as the airplane continues its motion along a straight line? (c) **What If?** What is its angular momentum relative to the summit of Pike's Peak?

15. M The position vector of a particle of mass 2.00 kg as a function of time is given by $\vec{\mathbf{r}} = (6.00\hat{\mathbf{i}} + 5.00t\hat{\mathbf{j}})$, where $\vec{\mathbf{r}}$ is in meters and t is in seconds. Determine the angular momentum of the particle about the origin as a function of time.

16. S **Review.** A conical pendulum consists of a bob of mass m in motion in a circular path in a horizontal plane as shown in Figure P11.16. During the motion, the supporting wire of length ℓ maintains a constant angle θ with the vertical. Show that the magnitude of the angular momentum of the bob about the vertical dashed line is

$$L = \left(\frac{m^2 g \ell^3 \sin^4\theta}{\cos\theta}\right)^{1/2}$$

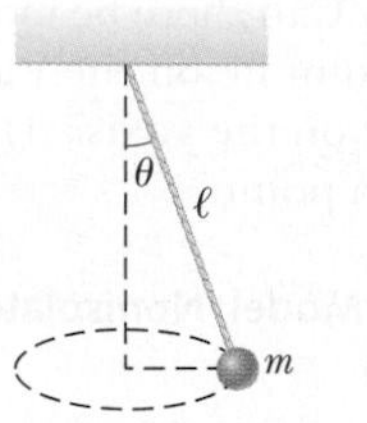

Figure P11.16

17. S A particle of mass m moves in a circle of radius R at a constant speed v as shown in Figure P11.17. The motion begins at point Q at time $t = 0$. Determine the angular momentum of the particle about the axis perpendicular to the page through point P as a function of time.

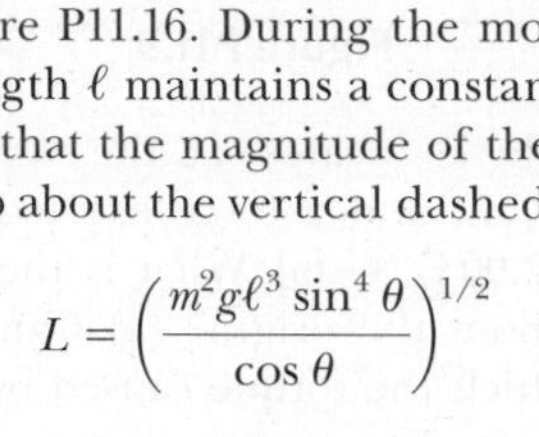

Figure P11.17 Problems 17 and 32.

18. A counterweight of mass $m = 4.00$ kg is attached to a light cord that is wound around a pulley as in Figure P11.18. The pulley is a thin hoop of radius $R = 8.00$ cm and mass $M = 2.00$ kg. The spokes have negligible mass. (a) What is the magnitude of the net torque on the system about the axle of the pulley? (b) When the counterweight has a speed v, the pulley has an angular speed $\omega = v/R$. Determine the magnitude of the total angular momentum of the system about the axle of the pulley. (c) Using your result from part (b) and $\vec{\boldsymbol{\tau}} = d\vec{\mathbf{L}}/dt$, calculate the acceleration of the counterweight.

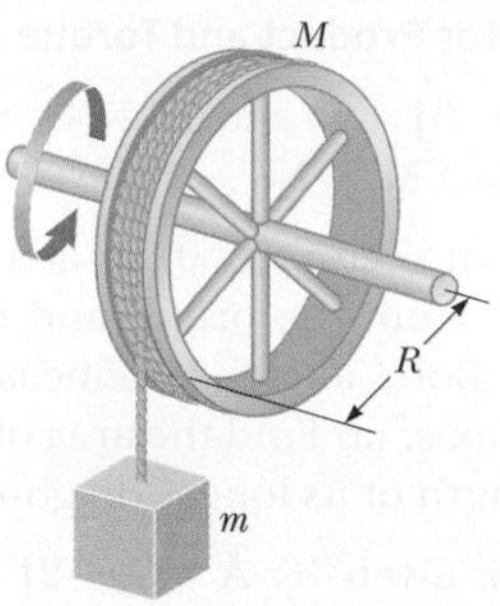

Figure P11.18

19. QC S **Review.** A projectile of mass m is launched with an initial velocity $\vec{\mathbf{v}}_i$ making an angle θ with the horizontal as shown in Figure P11.19. The projectile moves in the gravitational field of the Earth. Find the angular momentum of the projectile about the origin (a) when the projectile is at the origin, (b) when it is at the highest point of its trajectory, and (c) just before it hits the ground. (d) What torque causes its angular momentum to change?

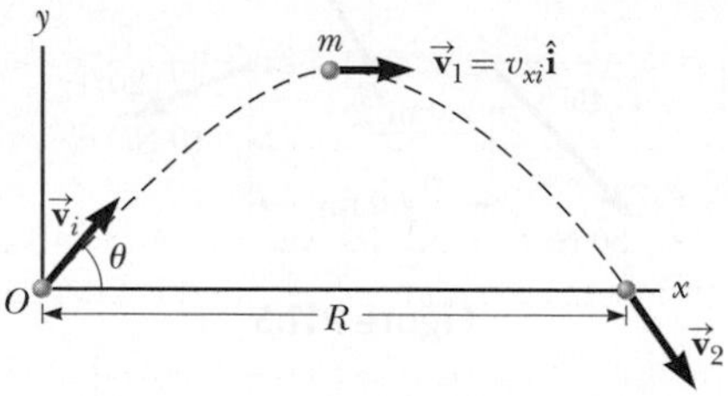

Figure P11.19

20. QC A 5.00-kg particle starts from the origin at time zero. Its velocity as a function of time is given by

$$\vec{\mathbf{v}} = 6t^2\hat{\mathbf{i}} + 2t\hat{\mathbf{j}}$$

where $\vec{\mathbf{v}}$ is in meters per second and t is in seconds. (a) Find its position as a function of time. (b) Describe its motion qualitatively. Find (c) its acceleration as a function of time, (d) the net force exerted on the particle as a function of time, (e) the net torque about the origin exerted on the particle as a function of time, (f) the angular momentum of the particle as a function of time, (g) the kinetic energy of the particle as a function of time, and (h) the power injected into the system of the particle as a function of time.

21. **QC** **S** A ball having mass m is fastened at the end of a flagpole that is connected to the side of a tall building at point P as shown in Figure P11.21. The length of the flagpole is ℓ, and it makes an angle θ with the x axis. The ball becomes loose and starts to fall with acceleration $-g\hat{\mathbf{j}}$. (a) Determine the angular momentum of the ball about point P as a function of time. (b) For what physical reason does the angular momentum change? (c) What is the rate of change of the angular momentum of the ball about point P?

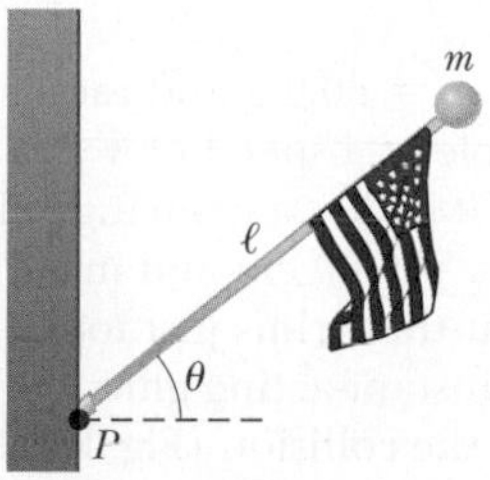

Figure P11.21

Section 11.3 Angular Momentum of a Rotating Rigid Object

22. **S** Show that the kinetic energy of an object rotating about a fixed axis with angular momentum $L = I\omega$ can be written as $K = L^2/2I$.

23. Big Ben (Fig. P10.45, page 312), the Parliament tower clock in London, has hour and minute hands with lengths of 2.70 m and 4.50 m and masses of 60.0 kg and 100 kg, respectively. Calculate the total angular momentum of these hands about the center point. (You may model the hands as long, thin rods rotating about one end. Assume the hour and minute hands are rotating at a constant rate of one revolution per 12 hours and 60 minutes, respectively.)

24. A uniform solid sphere of radius $r = 0.500$ m and mass $m = 15.0$ kg turns counterclockwise about a vertical axis through its center. Find its vector angular momentum about this axis when its angular speed is 3.00 rad/s.

25. A uniform solid disk of mass $m = 3.00$ kg and radius $r = 0.200$ m rotates about a fixed axis perpendicular to its face with angular frequency 6.00 rad/s. Calculate the magnitude of the angular momentum of the disk when the axis of rotation (a) passes through its center of mass and (b) passes through a point midway between the center and the rim.

26. **QC** Model the Earth as a uniform sphere. (a) Calculate the angular momentum of the Earth due to its spinning motion about its axis. (b) Calculate the angular momentum of the Earth due to its orbital motion about the Sun. (c) Explain why the answer in part (b) is larger than that in part (a) even though it takes significantly longer for the Earth to go once around the Sun than to rotate once about its axis.

27. A particle of mass 0.400 kg is attached to the 100-cm mark of a meterstick of mass 0.100 kg. The meterstick rotates on the surface of a frictionless, horizontal table with an angular speed of 4.00 rad/s. Calculate the angular momentum of the system when the stick is pivoted about an axis (a) perpendicular to the table through the 50.0-cm mark and (b) perpendicular to the table through the 0-cm mark.

28. The distance between the centers of the wheels of a motorcycle is 155 cm. The center of mass of the motorcycle, including the rider, is 88.0 cm above the ground and halfway between the wheels. Assume the mass of each wheel is small compared with the body of the motorcycle. The engine drives the rear wheel only. What horizontal acceleration of the motorcycle will make the front wheel rise off the ground?

29. A space station is constructed in the shape of a hollow ring of mass 5.00×10^4 kg. Members of the crew walk on a deck formed by the inner surface of the outer cylindrical wall of the ring, with radius $r = 100$ m. At rest when constructed, the ring is set rotating about its axis so that the people inside experience an effective free-fall acceleration equal to g. (See Fig. P11.29.) The rotation is achieved by firing two small rockets attached tangentially to opposite points on the rim of the ring. (a) What angular momentum does the space station acquire? (b) For what time interval must the rockets be fired if each exerts a thrust of 125 N?

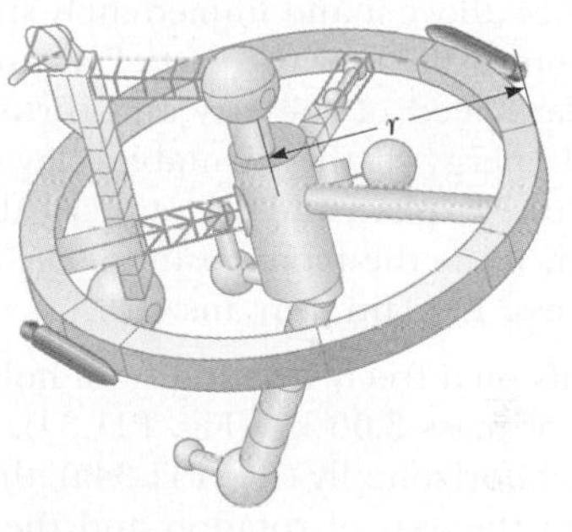

Figure P11.29 Problems 29 and 38.

Section 11.4 Analysis Model: Isolated System (Angular Momentum)

30. **S** A disk with moment of inertia I_1 rotates about a frictionless, vertical axle with angular speed ω_i. A second disk, this one having moment of inertia I_2 and initially not rotating, drops onto the first disk (Fig. P11.30). Because of

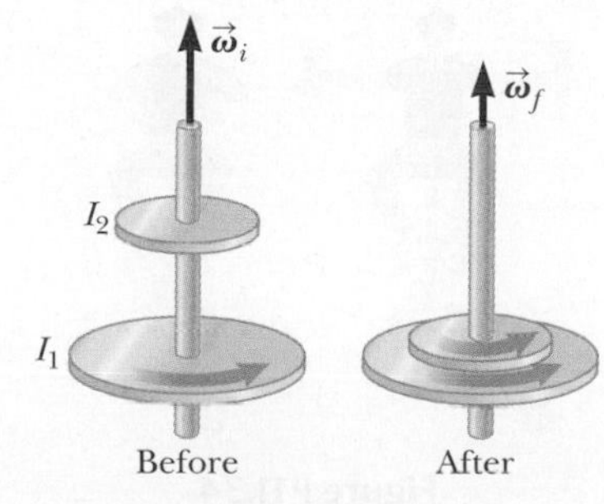

Figure P11.30

friction between the surfaces, the two eventually reach the same angular speed ω_f. (a) Calculate ω_f. (b) Calculate the ratio of the final to the initial rotational energy.

31. A playground merry-go-round of radius $R = 2.00$ m has a moment of inertia $I = 250\ \text{kg} \cdot \text{m}^2$ and is rotating at 10.0 rev/min about a frictionless, vertical axle. Facing the axle, a 25.0-kg child hops onto the merry-go-round and manages to sit down on the edge. What is the new angular speed of the merry-go-round?

32. Q|C Figure P11.17 represents a small, flat puck with mass $m = 2.40$ kg sliding on a frictionless, horizontal surface. It is held in a circular orbit about a fixed axis by a rod with negligible mass and length $R = 1.50$ m, pivoted at one end. Initially, the puck has a speed of $v = 5.00$ m/s. A 1.30-kg ball of putty is dropped vertically onto the puck from a small distance above it and immediately sticks to the puck. (a) What is the new period of rotation? (b) Is the angular momentum of the puck–putty system about the axis of rotation constant in this process? (c) Is the momentum of the system constant in the process of the putty sticking to the puck? (d) Is the mechanical energy of the system constant in the process?

33. Q|C A uniform cylindrical turntable of radius 1.90 m and mass 30.0 kg rotates counterclockwise in a horizontal plane with an initial angular speed of 4π rad/s. The fixed turntable bearing is frictionless. A lump of clay of mass 2.25 kg and negligible size is dropped onto the turntable from a small distance above it and immediately sticks to the turntable at a point 1.80 m to the east of the axis. (a) Find the final angular speed of the clay and turntable. (b) Is the mechanical energy of the turntable–clay system constant in this process? Explain and use numerical results to verify your answer. (c) Is the momentum of the system constant in this process? Explain your answer.

34. A student sits on a freely rotating stool holding two dumbbells, each of mass 3.00 kg (Fig. P11.34). When his arms are extended horizontally (Fig. P11.34a), the dumbbells are 1.00 m from the axis of rotation and the student rotates with an angular speed of 0.750 rad/s. The moment of inertia of the student plus stool is 3.00 kg · m² and is assumed to be constant. The student pulls the dumbbells inward horizontally to a position 0.300 m from the rotation axis (Fig. P11.34b). (a) Find the new angular speed of the student. (b) Find the kinetic energy of the rotating system before and after he pulls the dumbbells inward.

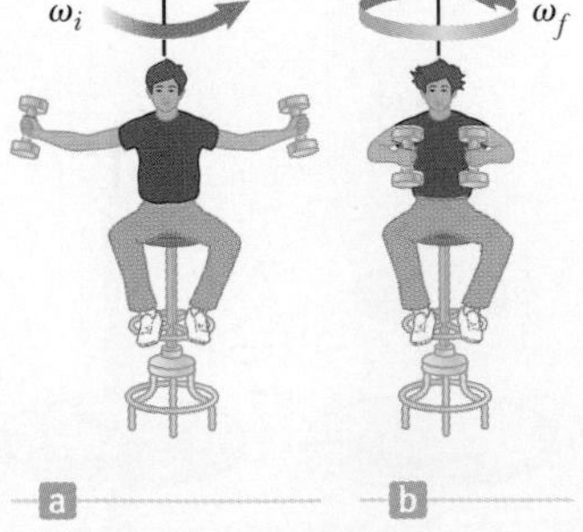

Figure P11.34

35. M Q|C A 60.0-kg woman stands at the western rim of a horizontal turntable having a moment of inertia of 500 kg · m² and a radius of 2.00 m. The turntable is initially at rest and is free to rotate about a frictionless, vertical axle through its center. The woman then starts walking around the rim clockwise (as viewed from above the system) at a constant speed of 1.50 m/s relative to the Earth. Consider the woman–turntable system as motion begins. (a) Is the mechanical energy of the system constant? (b) Is the momentum of the system constant? (c) Is the angular momentum of the system constant? (d) In what direction and with what angular speed does the turntable rotate? (e) How much chemical energy does the woman's body convert into mechanical energy of the woman–turntable system as the woman sets herself and the turntable into motion?

36. A puck of mass $m_1 = 80.0$ g and radius $r_1 = 4.00$ cm glides across an air table at a speed of $\vec{\mathbf{v}} = 1.50$ m/s as shown in Figure P11.36a. It makes a glancing collision with a second puck of radius $r_2 = 6.00$ cm and mass $m_2 = 120$ g (initially at rest) such that their rims just touch. Because their rims are coated with instant-acting glue, the pucks stick together and rotate after the collision (Fig. P11.36b). (a) What is the angular momentum of the system relative to the center of mass? (b) What is the angular speed about the center of mass?

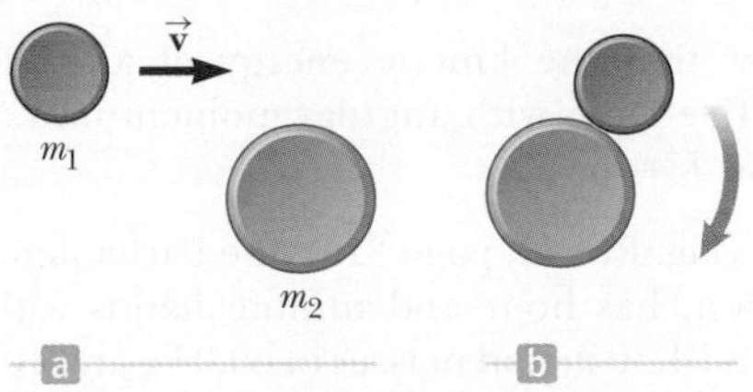

Figure P11.36

37. S A wooden block of mass M resting on a frictionless, horizontal surface is attached to a rigid rod of length ℓ and of negligible mass (Fig. P11.37). The rod is pivoted at the other end. A bullet of mass m traveling parallel to the horizontal surface and perpendicular to the rod with speed v hits the block and becomes embedded in it. (a) What is the angular momentum of the bullet–block system about a vertical axis through the pivot? (b) What fraction of the original kinetic energy of the bullet is converted into internal energy in the system during the collision?

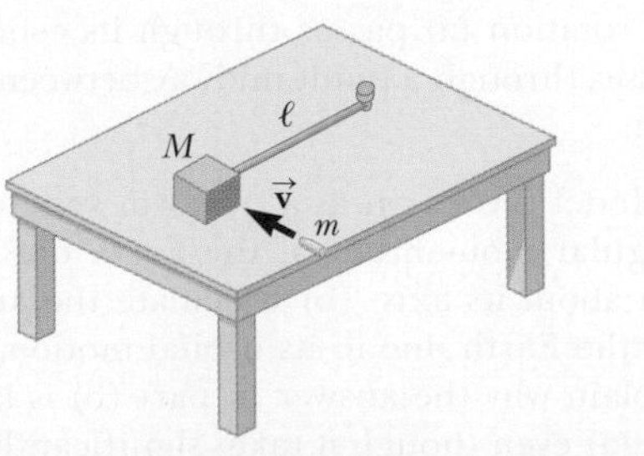

Figure P11.37

38. *Why is the following situation impossible?* A space station shaped like a giant wheel has a radius of $r = 100$ m and a moment of inertia of 5.00×10^8 kg · m^2. A crew of 150 people of average mass 65.0 kg is living on the rim, and the station's rotation causes the crew to experience an apparent free-fall acceleration of g (Fig. P11.29). A research technician is assigned to perform an experiment in which a ball is dropped at the rim of the station every 15 minutes and the time interval for the ball to drop a given distance is measured as a test to make sure the apparent value of g is correctly maintained. One evening, 100 average people move to the center of the station for a union meeting. The research technician, who has already been performing his experiment for an hour before the meeting, is disappointed that he cannot attend the meeting, and his mood sours even further by his boring experiment in which every time interval for the dropped ball is identical for the entire evening.

39. **QC** **S** A wad of sticky clay with mass m and velocity $\vec{\mathbf{v}}_i$ is fired at a solid cylinder of mass M and radius R (Fig. P11.39). The cylinder is initially at rest and is mounted on a fixed horizontal axle that runs through its center of mass. The line of motion of the projectile is perpendicular to the axle and at a distance $d < R$ from the center. (a) Find the angular speed of the system just after the clay strikes and sticks to the surface of the cylinder. (b) Is the mechanical energy of the clay–cylinder system constant in this process? Explain your answer. (c) Is the momentum of the clay–cylinder system constant in this process? Explain your answer.

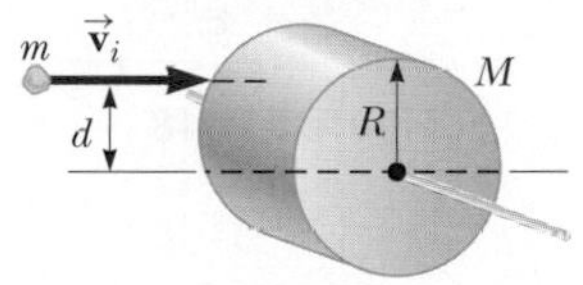

Figure P11.39

40. **Review.** A thin, uniform, rectangular signboard hangs vertically above the door of a shop. The sign is hinged to a stationary horizontal rod along its top edge. The mass of the sign is 2.40 kg, and its vertical dimension is 50.0 cm. The sign is swinging without friction, so it is a tempting target for children armed with snowballs. The maximum angular displacement of the sign is 25.0° on both sides of the vertical. At a moment when the sign is vertical and moving to the left, a snowball of mass 400 g, traveling horizontally with a velocity of 160 cm/s to the right, strikes perpendicularly at the lower edge of the sign and sticks there. (a) Calculate the angular speed of the sign immediately before the impact. (b) Calculate its angular speed immediately after the impact. (c) The spattered sign will swing up through what maximum angle?

41. **QC** A 0.005 00-kg bullet traveling horizontally with speed 1.00×10^3 m/s strikes an 18.0-kg door, imbedding itself 10.0 cm from the side opposite the hinges as shown in Figure P11.41. The 1.00-m wide door is free to swing on its frictionless hinges. (a) Before it hits the door, does the bullet have angular momentum relative to the door's axis of rotation? (b) If so, evaluate this angular momentum. If not, explain why there is no angular momentum. (c) Is the mechanical energy of the bullet–door system constant during this collision? Answer without doing a calculation. (d) At what angular speed does the door swing open immediately after the collision? (e) Calculate the total energy of the bullet–door system and determine whether it is less than or equal to the kinetic energy of the bullet before the collision.

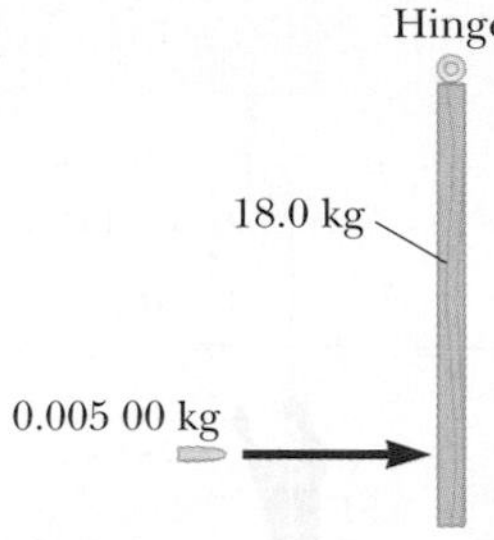

Figure P11.41 An overhead view of a bullet striking a door.

Section 11.5 The Motion of Gyroscopes and Tops

42. A spacecraft is in empty space. It carries on board a gyroscope with a moment of inertia of $I_g = 20.0$ kg · m^2 about the axis of the gyroscope. The moment of inertia of the spacecraft around the same axis is $I_s = 5.00 \times 10^5$ kg · m^2. Neither the spacecraft nor the gyroscope is originally rotating. The gyroscope can be powered up in a negligible period of time to an angular speed of 100 rad/s. If the orientation of the spacecraft is to be changed by 30.0°, for what time interval should the gyroscope be operated?

43. The angular momentum vector of a precessing gyroscope sweeps out a cone as shown in Figure P11.43. The angular speed of the tip of the angular momentum vector, called its precessional frequency, is given by $\omega_p = \tau/L$, where τ is the magnitude of the torque on the gyroscope and L is the magnitude of its angular momentum. In the motion called *precession of the equinoxes*, the Earth's axis of rotation precesses about the perpendicular to its orbital plane with a period of 2.58×10^4 yr. Model the Earth as a uniform sphere and calculate the torque on the Earth that is causing this precession.

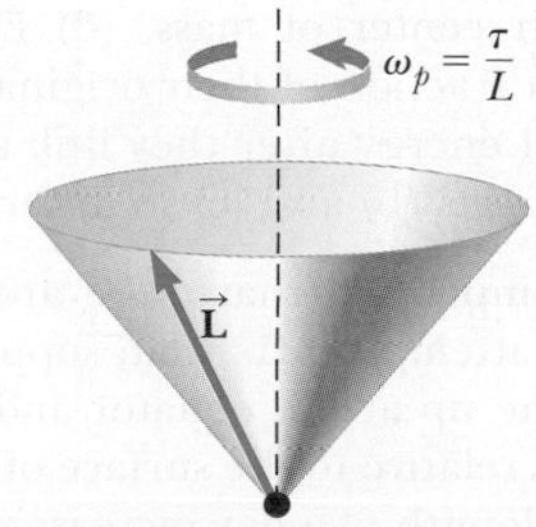

Figure P11.43 A precessing angular momentum vector sweeps out a cone in space.

Additional Problems

44. Q|C S A light rope passes over a light, frictionless pulley. One end is fastened to a bunch of bananas of mass M, and a monkey of mass M clings to the other end (Fig. P11.44). The monkey climbs the rope in an attempt to reach the bananas. (a) Treating the system as consisting of the monkey, bananas, rope, and pulley, find the net torque on the system about the pulley axis. (b) Using the result of part (a), determine the total angular momentum about the pulley axis and describe the motion of the system. (c) Will the monkey reach the bananas?

Figure P11.44

45. Comet Halley moves about the Sun in an elliptical orbit, with its closest approach to the Sun being about 0.590 AU and its greatest distance 35.0 AU (1 AU = the Earth–Sun distance). The angular momentum of the comet about the Sun is constant, and the gravitational force exerted by the Sun has zero moment arm. The comet's speed at closest approach is 54.0 km/s. What is its speed when it is farthest from the Sun?

46. Q|C **Review.** Two boys are sliding toward each other on a frictionless, ice-covered parking lot. Jacob, mass 45.0 kg, is gliding to the right at 8.00 m/s, and Ethan, mass 31.0 kg, is gliding to the left at 11.0 m/s along the same line. When they meet, they grab each other and hang on. (a) What is their velocity immediately thereafter? (b) What fraction of their original kinetic energy is still mechanical energy after their collision? That was so much fun that the boys repeat the collision with the same original velocities, this time moving along parallel lines 1.20 m apart. At closest approach, they lock arms and start rotating about their common center of mass. Model the boys as particles and their arms as a cord that does not stretch. (c) Find the velocity of their center of mass. (d) Find their angular speed. (e) What fraction of their original kinetic energy is still mechanical energy after they link arms? (f) Why are the answers to parts (b) and (e) so different?

47. We have all complained that there aren't enough hours in a day. In an attempt to fix that, suppose all the people in the world line up at the equator and all start running east at 2.50 m/s relative to the surface of the Earth. By how much does the length of a day increase? Assume the world population to be 7.00×10^9 people with an average mass of 55.0 kg each and the Earth to be a solid homogeneous sphere. In addition, depending on the details of your solution, you may need to use the approximation $1/(1 - x) \approx 1 + x$ for small x.

48. Q|C A skateboarder with his board can be modeled as a particle of mass 76.0 kg, located at his center of mass, 0.500 m above the ground. As shown in Figure P11.48, the skateboarder starts from rest in a crouching position at one lip of a half-pipe (point Ⓐ). The half-pipe forms one half of a cylinder of radius 6.80 m with its axis horizontal. On his descent, the skateboarder moves without friction and maintains his crouch so that his center of mass moves through one quarter of a circle. (a) Find his speed at the bottom of the half-pipe (point Ⓑ). (b) Find his angular momentum about the center of curvature at this point. (c) Immediately after passing point Ⓑ, he stands up and raises his arms, lifting his center of gravity to 0.950 m above the concrete (point Ⓒ). Explain why his angular momentum is constant in this maneuver, whereas the kinetic energy of his body is not constant. (d) Find his speed immediately after he stands up. (e) How much chemical energy in the skateboarder's legs was converted into mechanical energy in the skateboarder–Earth system when he stood up?

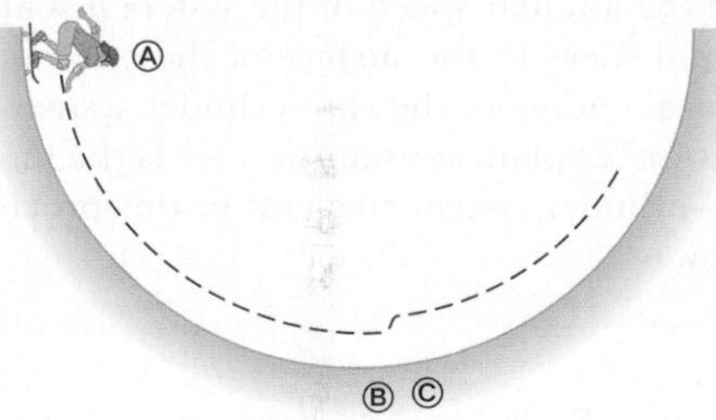

Figure P11.48

49. S A rigid, massless rod has three particles with equal masses attached to it as shown in Figure P11.49. The rod is free to rotate in a vertical plane about a frictionless axle perpendicular to the rod through the point P and is released from rest in the horizontal position at $t = 0$. Assuming m and d are known, find (a) the moment of inertia of the system of three particles about the pivot, (b) the torque acting on the system at $t = 0$, (c) the angular acceleration of the system at $t = 0$, (d) the linear acceleration of the particle labeled 3 at $t = 0$, (e) the maximum kinetic energy of the system, (f) the maximum angular speed reached by the rod, (g) the maximum angular momentum of the system, and (h) the maximum speed reached by the particle labeled 2.

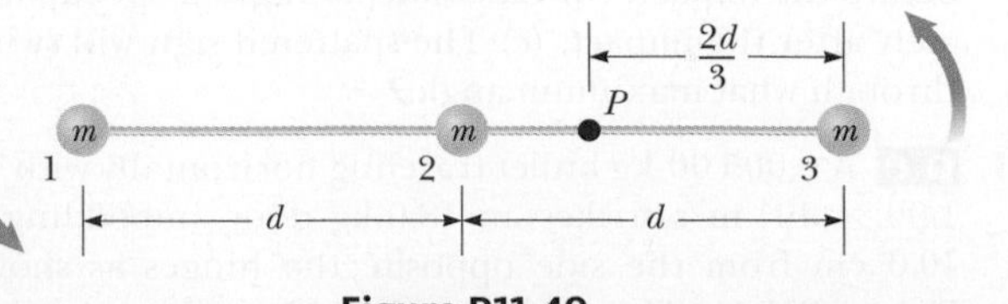

Figure P11.49

50. *Why is the following situation impossible?* A meteoroid strikes the Earth directly on the equator. At the time it lands, it is traveling exactly vertical and downward. Due to the impact, the time for the Earth to rotate once increases by 0.5 s, so the day is 0.5 s longer, undetectable to laypersons. After the impact, people on the Earth ignore the extra half-second each day and life goes on as normal. (Assume the density of the Earth is uniform.)

51. **S** **GP** A projectile of mass m moves to the right with a speed v_i (Fig. P11.51a). The projectile strikes and sticks to the end of a stationary rod of mass M, length d, pivoted about a frictionless axle perpendicular to the page through O (Fig. P11.51b). We wish to find the fractional change of kinetic energy in the system due to the collision. (a) What is the appropriate analysis model to describe the projectile and the rod? (b) What is the angular momentum of the system before the collision about an axis through O? (c) What is the moment of inertia of the system about an axis through O after the projectile sticks to the rod? (d) If the angular speed of the system after the collision is ω, what is the angular momentum of the system after the collision? (e) Find the angular speed ω after the collision in terms of the given quantities. (f) What is the kinetic energy of the system before the collision? (g) What is the kinetic energy of the system after the collision? (h) Determine the fractional change of kinetic energy due to the collision.

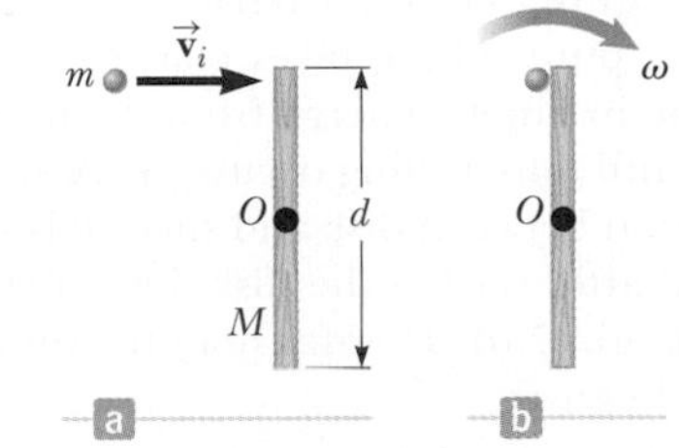

Figure P11.51

52. **M** A puck of mass $m = 50.0$ g is attached to a taut cord passing through a small hole in a frictionless, horizontal surface (Fig. P11.52). The puck is initially orbiting with speed $v_i = 1.50$ m/s in a circle of radius $r_i = 0.300$ m. The cord is then slowly pulled from below, decreasing the radius of the circle to $r = 0.100$ m. (a) What is the puck's speed at the smaller radius? (b) Find the tension in the cord at the smaller radius. (c) How much work is done by the hand in pulling the cord so that the radius of the puck's motion changes from 0.300 m to 0.100 m?

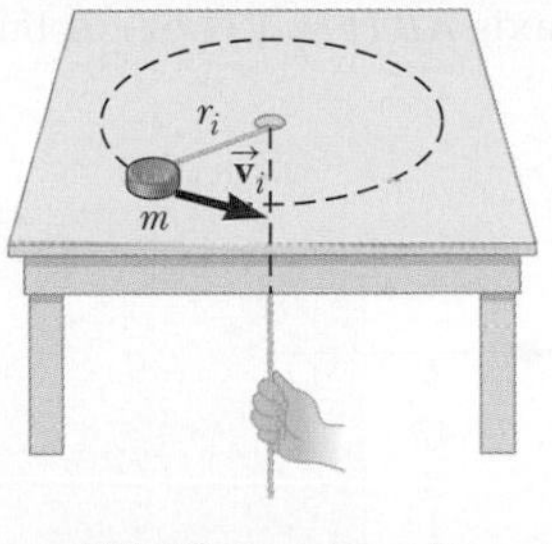

Figure P11.52 Problems 52 and 53.

53. **S** A puck of mass m is attached to a taut cord passing through a small hole in a frictionless, horizontal surface (Fig. P11.52). The puck is initially orbiting with speed v_i in a circle of radius r_i. The cord is then slowly pulled from below, decreasing the radius of the circle to r. (a) What is the puck's speed when the radius is r? (b) Find the tension in the cord as a function of r. (c) How much work is done by the hand in pulling the cord so that the radius of the puck's motion changes from r_i to r?

54. **Q|C** Two children are playing on stools at a restaurant counter. Their feet do not reach the footrests, and the tops of the stools are free to rotate without friction on pedestals fixed to the floor. One of the children catches a tossed ball, in a process described by the equation

$$(0.730 \text{ kg}\cdot\text{m}^2)(2.40\hat{\mathbf{j}} \text{ rad/s}) + (0.120 \text{ kg})(0.350\hat{\mathbf{i}} \text{ m}) \times (4.30\hat{\mathbf{k}} \text{ m/s}) = [0.730 \text{ kg}\cdot\text{m}^2 + (0.120 \text{ kg})(0.350 \text{ m})^2]\vec{\boldsymbol{\omega}}$$

(a) Solve the equation for the unknown $\vec{\boldsymbol{\omega}}$. (b) Complete the statement of the problem to which this equation applies. Your statement must include the given numerical information and specification of the unknown to be determined. (c) Could the equation equally well describe the other child throwing the ball? Explain your answer.

55. Two astronauts (Fig. P11.55), each having a mass of 75.0 kg, are connected by a 10.0-m rope of negligible mass. They are isolated in space, orbiting their center of mass at speeds of 5.00 m/s. Treating the astronauts as particles, calculate (a) the magnitude of the angular momentum of the two-astronaut system and (b) the rotational energy of the system. By pulling on the rope, one astronaut shortens the distance between them to 5.00 m. (c) What is the new angular momentum of the system? (d) What are the astronauts' new speeds? (e) What is the new rotational energy of the system? (f) How much chemical potential energy in the body of the astronaut was converted to mechanical energy in the system when he shortened the rope?

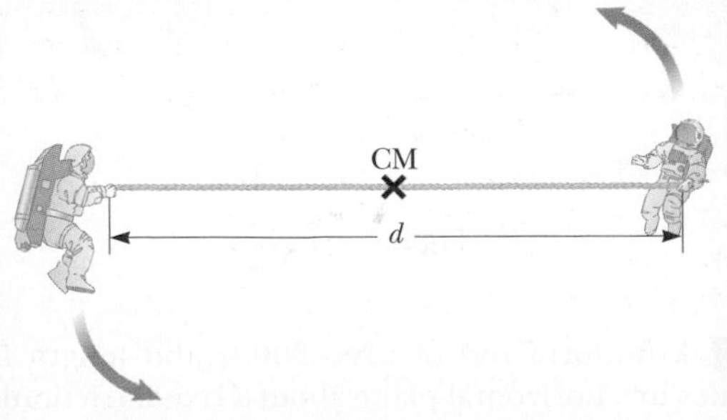

Figure P11.55 Problems 55 and 56.

56. **S** Two astronauts (Fig. P11.55), each having a mass M, are connected by a rope of length d having negligible mass. They are isolated in space, orbiting their center of mass at speeds v. Treating the astronauts as particles, calculate

(a) the magnitude of the angular momentum of the two-astronaut system and (b) the rotational energy of the system. By pulling on the rope, one of the astronauts shortens the distance between them to $d/2$. (c) What is the new angular momentum of the system? (d) What are the astronauts' new speeds? (e) What is the new rotational energy of the system? (f) How much chemical potential energy in the body of the astronaut was converted to mechanical energy in the system when he shortened the rope?

57. Q|C S Native people throughout North and South America used a *bola* to hunt for birds and animals. A bola can consist of three stones, each with mass m, at the ends of three light cords, each with length ℓ. The other ends of the cords are tied together to form a Y. The hunter holds one stone and swings the other two above his head (Figure P11.57a). Both these stones move together in a horizontal circle of radius 2ℓ with speed v_0. At a moment when the horizontal component of their velocity is directed toward the quarry, the hunter releases the stone in his hand. As the bola flies through the air, the cords quickly take a stable arrangement with constant 120-degree angles between them (Fig. P11.57b). In the vertical direction, the bola is in free fall. Gravitational forces exerted by the Earth make the junction of the cords move with the downward acceleration $\vec{g}$. You may ignore the vertical motion as you proceed to describe the horizontal motion of the bola. In terms of m, ℓ, and v_0, calculate (a) the magnitude of the momentum of the bola at the moment of release and, after release, (b) the horizontal speed of the center of mass of the bola and (c) the angular momentum of the bola about its center of mass. (d) Find the angular speed of the bola about its center of mass after it has settled into its Y shape. Calculate the kinetic energy of the bola (e) at the instant of release and (f) in its stable Y shape. (g) Explain how the conservation laws apply to the bola as its configuration changes. Robert Beichner suggested the idea for this problem.

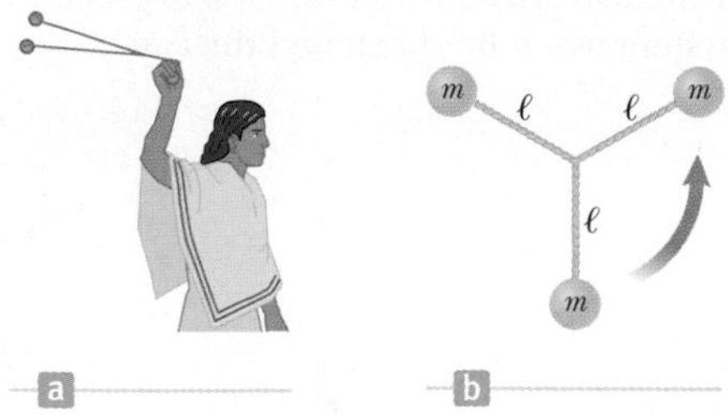

Figure P11.57

58. Q|C A uniform rod of mass 300 g and length 50.0 cm rotates in a horizontal plane about a fixed, frictionless, vertical pin through its center. Two small, dense beads, each of mass m, are mounted on the rod so that they can slide without friction along its length. Initially, the beads are held by catches at positions 10.0 cm on each side of the center and the system is rotating at an angular speed of 36.0 rad/s. The catches are released simultaneously, and the beads slide outward along the rod. (a) Find an expression for the angular speed ω_f of the system at the instant the beads slide off the ends of the rod as it depends on m. (b) What are the maximum and the minimum possible values for ω_f and the values of m to which they correspond?

59. Global warming is a cause for concern because even small changes in the Earth's temperature can have significant consequences. For example, if the Earth's polar ice caps were to melt entirely, the resulting additional water in the oceans would flood many coastal areas. Model the polar ice as having mass 2.30×10^{19} kg and forming two flat disks of radius 6.00×10^{5} m. Assume the water spreads into an unbroken thin, spherical shell after it melts. Calculate the resulting change in the duration of one day both in seconds and as a percentage.

Challenge Problems

60. In Example 11.9, we investigated an elastic collision between a disk and a stick lying on a frictionless surface. Suppose everything is the same as in the example except that the collision is perfectly inelastic so that the disk adheres to the stick at the endpoint at which it strikes. Find (a) the speed of the center of mass of the system and (b) the angular speed of the system after the collision.

61. S A uniform solid disk of radius R is set into rotation with an angular speed ω_i about an axis through its center. While still rotating at this speed, the disk is placed into contact with a horizontal surface and immediately released as shown in Figure P11.61. (a) What is the angular speed of the disk once pure rolling takes place? (b) Find the fractional change in kinetic energy from the moment the disk is set down until pure rolling occurs. (c) Assume the coefficient of friction between disk and surface is μ. What is the time interval after setting the disk down before pure rolling motion begins? (d) How far does the disk travel before pure rolling begins?

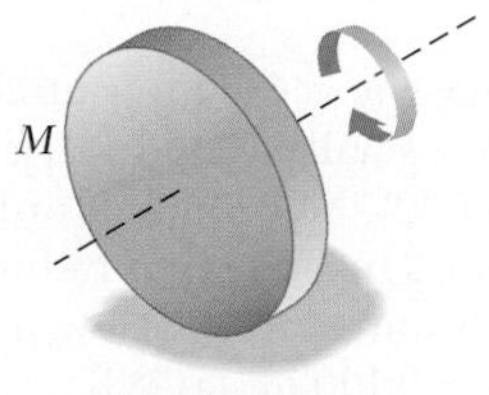

Figure P11.61

62. S A solid cube of wood of side $2a$ and mass M is resting on a horizontal surface. The cube is constrained to rotate about a fixed axis AB (Fig. P11.62). A bullet of mass m and

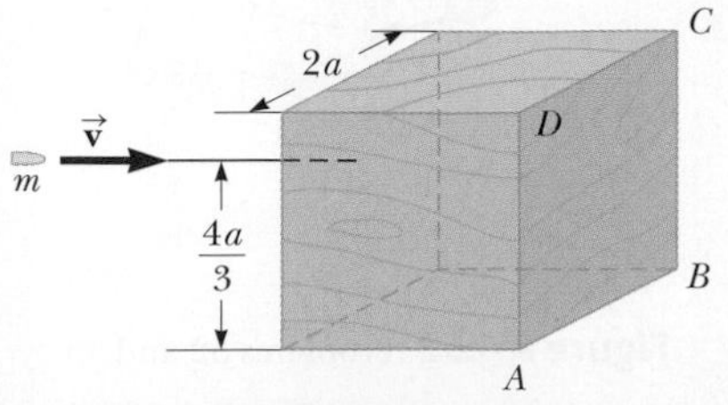

Figure P11.62

speed v is shot at the face opposite *ABCD* at a height of $4a/3$. The bullet becomes embedded in the cube. Find the minimum value of v required to tip the cube so that it falls on face *ABCD*. Assume $m << M$.

63. S A solid cube of side $2a$ and mass M is sliding on a frictionless surface with uniform velocity $\vec{v}$ as shown in Figure P11.63a. It hits a small obstacle at the end of the table that causes the cube to tilt as shown in Figure P11.63b. Find the minimum value of the magnitude of $\vec{v}$ such that the cube tips over and falls off the table. *Note:* The cube undergoes an inelastic collision at the edge.

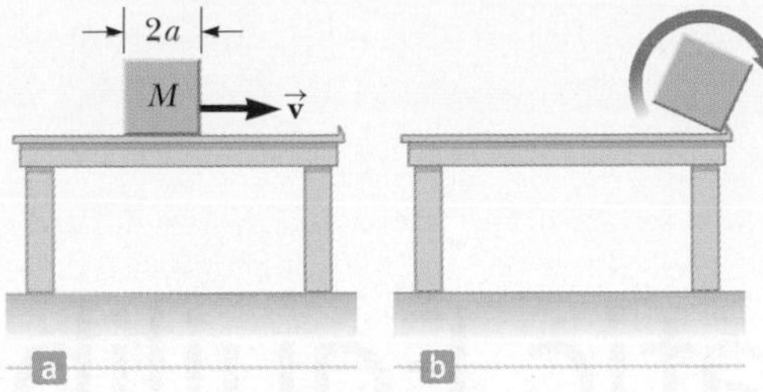

Figure P11.63

chapter

12

Static Equilibrium and Elasticity

Balanced Rock in Arches National Park, Utah, is a 3 000 000-kg boulder that has been in stable equilibrium for several millennia. It had a smaller companion nearby, called "Chip Off the Old Block," that fell during the winter of 1975. Balanced Rock appeared in an early scene of the movie *Indiana Jones and the Last Crusade*. We will study the conditions under which an object is in equilibrium in this chapter. *(John W. Jewett, Jr.)*

In Chapters 10 and 11, we studied the dynamics of rigid objects. Part of this chapter addresses the conditions under which a rigid object is in equilibrium. The term *equilibrium* implies that the object moves with both constant velocity and constant angular velocity relative to an observer in an inertial reference frame. We deal here only with the special case in which both of these velocities are equal to zero. In this case, the object is in what is called *static equilibrium*. Static equilibrium represents a common situation in engineering practice, and the principles it involves are of special interest to civil engineers, architects, and mechanical engineers. If you are an engineering student, you will undoubtedly take an advanced course in statics in the near future.

The last section of this chapter deals with how objects deform under load conditions. An *elastic* object returns to its original shape when the deforming forces are removed. Several elastic constants are defined, each corresponding to a different type of deformation.

12.1 Analysis Model: Rigid Object in Equilibrium

In Chapter 5, we discussed the particle in equilibrium model, in which a particle moves with constant velocity because the net force acting on it is zero. The situation with real (extended) objects is more complex because these objects often cannot be modeled as particles. For an extended object to be in equilibrium, a second condition must be satisfied. This second condition involves the rotational motion of the extended object.

Consider a single force $\vec{\mathbf{F}}$ acting on a rigid object as shown in Figure 12.1. Recall that the torque associated with the force $\vec{\mathbf{F}}$ about an axis through O is given by Equation 11.1:

$$\vec{\boldsymbol{\tau}} = \vec{\mathbf{r}} \times \vec{\mathbf{F}}$$

The magnitude of $\vec{\boldsymbol{\tau}}$ is Fd (see Eq. 10.19), where d is the moment arm shown in Figure 12.1. According to Equation 10.21, the net torque on a rigid object causes it to undergo an angular acceleration.

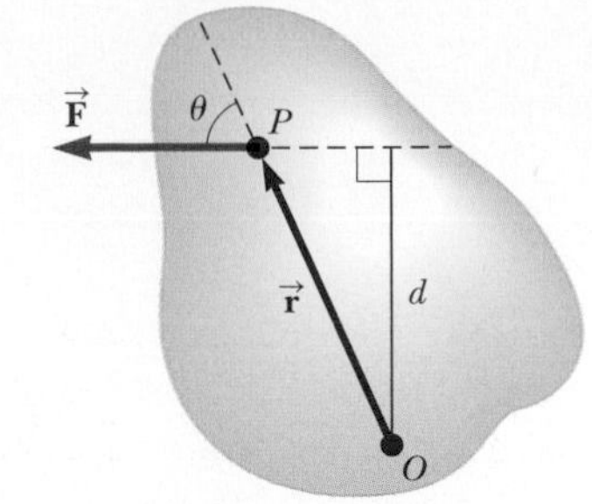

Figure 12.1 A single force $\vec{\mathbf{F}}$ acts on a rigid object at the point P.

In this discussion, we investigate those rotational situations in which the angular acceleration of a rigid object is zero. Such an object is in **rotational equilibrium.** Because $\sum \tau_{ext} = I\alpha$ for rotation about a fixed axis, the necessary condition for rotational equilibrium is that the net torque about any axis must be zero. We now have two necessary conditions for equilibrium of an object:

1. The net external force on the object must equal zero:

$$\sum \vec{\mathbf{F}}_{ext} = 0 \tag{12.1}$$

2. The net external torque on the object about *any* axis must be zero:

$$\sum \vec{\boldsymbol{\tau}}_{ext} = 0 \tag{12.2}$$

Pitfall Prevention 12.1

Zero Torque

Zero net torque does not mean an absence of rotational motion. An object that is rotating at a constant angular speed can be under the influence of a net torque of zero. This possibility is analogous to the translational situation: zero net force does not mean an absence of translational motion.

These conditions describe the **rigid object in equilibrium** analysis model. The first condition is a statement of translational equilibrium; it states that the translational acceleration of the object's center of mass must be zero when viewed from an inertial reference frame. The second condition is a statement of rotational equilibrium; it states that the angular acceleration about any axis must be zero. In the special case of **static equilibrium,** which is the main subject of this chapter, the object in equilibrium is at rest relative to the observer and so has no translational or angular speed (that is, $v_{CM} = 0$ and $\omega = 0$).

***Quick Quiz* 12.1** Consider the object subject to the two forces of equal magnitude in Figure 12.2. Choose the correct statement with regard to this situation. **(a)** The object is in force equilibrium but not torque equilibrium. **(b)** The object is in torque equilibrium but not force equilibrium. **(c)** The object is in both force equilibrium and torque equilibrium. **(d)** The object is in neither force equilibrium nor torque equilibrium.

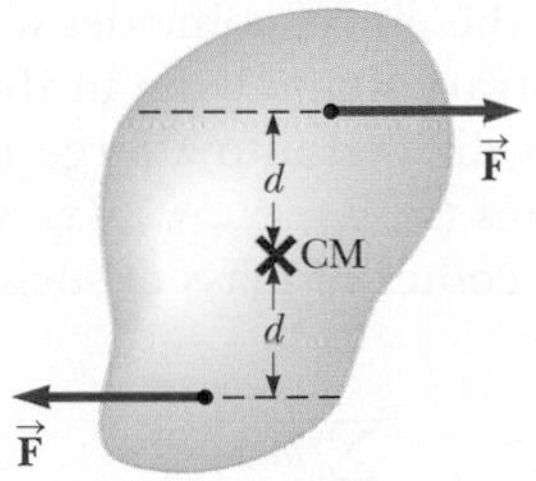

Figure 12.2 (Quick Quiz 12.1) Two forces of equal magnitude are applied at equal distances from the center of mass of a rigid object.

***Quick Quiz* 12.2** Consider the object subject to the three forces in Figure 12.3 (page 350). Choose the correct statement with regard to this situation. **(a)** The object is in force equilibrium but not torque equilibrium. **(b)** The object is in torque equilibrium but not force equilibrium. **(c)** The object is in both force equilibrium and torque equilibrium. **(d)** The object is in neither force equilibrium nor torque equilibrium.

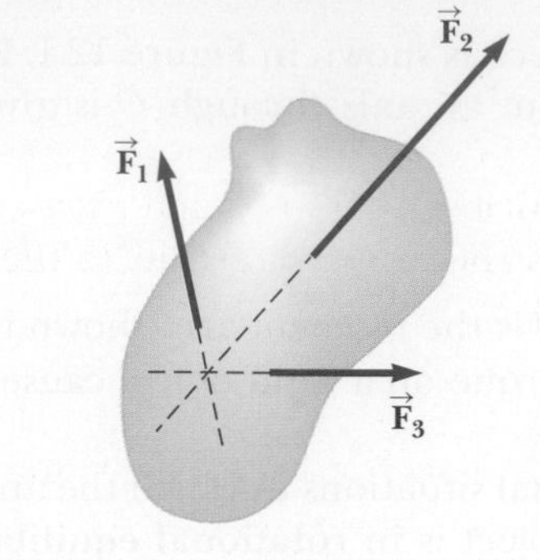

Figure 12.3 (Quick Quiz 12.2) Three forces act on an object. Notice that the lines of action of all three forces pass through a common point.

The two vector expressions given by Equations 12.1 and 12.2 are equivalent, in general, to six scalar equations: three from the first condition for equilibrium and three from the second (corresponding to x, y, and z components). Hence, in a complex system involving several forces acting in various directions, you could be faced with solving a set of equations with many unknowns. Here, we restrict our discussion to situations in which all the forces lie in the xy plane. (Forces whose vector representations are in the same plane are said to be *coplanar.*) With this restriction, we must deal with only three scalar equations. Two come from balancing the forces in the x and y directions. The third comes from the torque equation, namely that the net torque about a perpendicular axis through *any* point in the xy plane must be zero. This perpendicular axis will necessarily be parallel to the z axis, so the two conditions of the rigid object in equilibrium model provide the equations

$$\sum F_x = 0 \quad \sum F_y = 0 \quad \sum \tau_z = 0 \tag{12.3}$$

where the location of the axis of the torque equation is arbitrary.

12.2 More on the Center of Gravity

Whenever we deal with a rigid object, one of the forces we must consider is the gravitational force acting on it, and we must know the point of application of this force. As we learned in Section 9.5, associated with every object is a special point called its center of gravity. The combination of the various gravitational forces acting on all the various mass elements of the object is equivalent to a single gravitational force acting through this point. Therefore, to compute the torque due to the gravitational force on an object of mass M, we need only consider the force $M\vec{g}$ acting at the object's center of gravity.

How do we find this special point? As mentioned in Section 9.5, if we assume $\vec{g}$ is uniform over the object, the center of gravity of the object coincides with its center of mass. To see why, consider an object of arbitrary shape lying in the xy plane as illustrated in Figure 12.4. Suppose the object is divided into a large number of particles of masses $m_1, m_2, m_3, \ldots$ having coordinates (x_1, y_1), (x_2, y_2), (x_3, y_3), ... In Equation 9.29, we defined the x coordinate of the center of mass of such an object to be

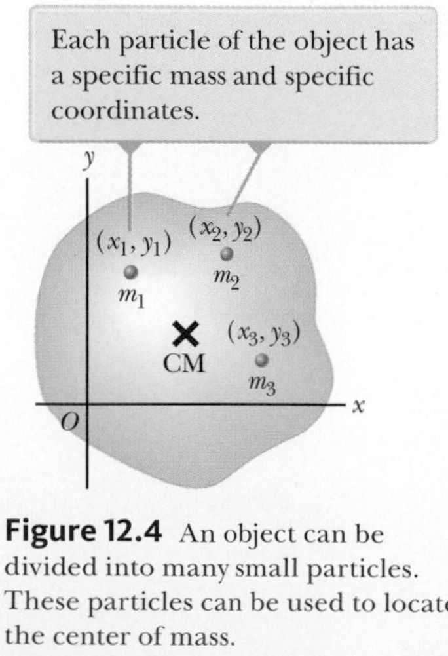

Figure 12.4 An object can be divided into many small particles. These particles can be used to locate the center of mass.

$$x_{\text{CM}} = \frac{m_1 x_1 + m_2 x_2 + m_3 x_3 + \cdots}{m_1 + m_2 + m_3 + \cdots} = \frac{\sum_i m_i x_i}{\sum_i m_i}$$

We use a similar equation to define the y coordinate of the center of mass, replacing each x with its y counterpart.

Let us now examine the situation from another point of view by considering the gravitational force exerted on each particle as shown in Figure 12.5. Each particle contributes a torque about an axis through the origin equal in magnitude to the particle's weight mg multiplied by its moment arm. For example, the magnitude of the torque due to the force $m_1\vec{g}_1$ is $m_1g_1x_1$, where g_1 is the value of the gravitational acceleration at the position of the particle of mass m_1. We wish to locate the center of gravity, the point at which application of the single gravitational force $M\vec{g}_{CG}$ (where $M = m_1 + m_2 + m_3 + \cdots$ is the total mass of the object and $\vec{g}_{CG}$ is the acceleration due to gravity at the location of the center of gravity) has the same effect on rotation as does the combined effect of all the individual gravitational forces $m_i\vec{g}_i$. Equating the torque resulting from $M\vec{g}_{CG}$ acting at the center of gravity to the sum of the torques acting on the individual particles gives

$$(m_1 + m_2 + m_3 + \cdots)g_{CG}\,x_{CG} = m_1g_1x_1 + m_2g_2x_2 + m_3g_3x_3 + \cdots$$

This expression accounts for the possibility that the value of g can in general vary over the object. If we assume uniform g over the object (as is usually the case), the g factors cancel and we obtain

$$x_{CG} = \frac{m_1x_1 + m_2x_2 + m_3x_3 + \cdots}{m_1 + m_2 + m_3 + \cdots} \tag{12.4}$$

Comparing this result with Equation 9.29 shows that the center of gravity is located at the center of mass as long as $\vec{g}$ is uniform over the entire object. Several examples in the next section deal with homogeneous, symmetric objects. The center of gravity for any such object coincides with its geometric center.

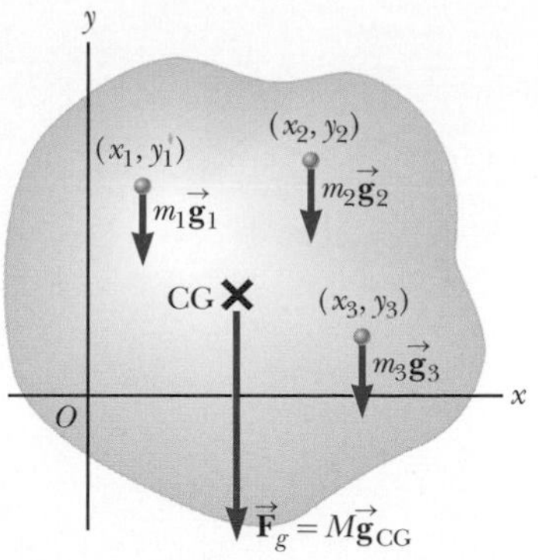

Figure 12.5 By dividing an object into many particles, we can find its center of gravity.

Quick Quiz **12.3** A meterstick is hung from a string tied at the 25-cm mark. A 0.50-kg object is hung from the zero end of the meterstick, and the meterstick is balanced horizontally. What is the mass of the meterstick? **(a)** 0.25 kg **(b)** 0.50 kg **(c)** 0.75 kg **(d)** 1.0 kg **(e)** 2.0 kg **(f)** impossible to determine

12.3 Examples of Rigid Objects in Static Equilibrium

The photograph of the one-bottle wine holder in Figure 12.6 shows one example of a balanced mechanical system that seems to defy gravity. For the system (wine holder plus bottle) to be in equilibrium, the net external force must be zero (see Eq. 12.1) and the net external torque must be zero (see Eq. 12.2). The second condition can be satisfied only when the center of gravity of the system is directly over the support point.

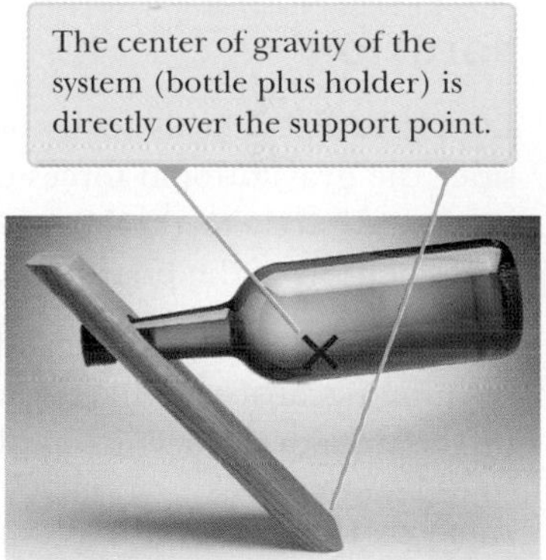

Figure 12.6 This one-bottle wine holder is a surprising display of static equilibrium.

Problem-Solving Strategy

RIGID OBJECT IN EQUILIBRIUM

When analyzing a rigid object in equilibrium under the action of several external forces, use the following procedure.

1. Conceptualize. Think about the object that is in equilibrium and identify all the forces on it. Imagine what effect each force would have on the rotation of the object if it were the only force acting.

2. Categorize. Confirm that the object under consideration is indeed a rigid object in equilibrium. The object must have zero translational acceleration and zero angular acceleration.

3. Analyze. Draw a diagram and label all external forces acting on the object. Try to guess the correct direction for any forces that are not specified. When using the particle under a net force model, the object on which forces act can be represented in a free-body diagram with a dot because it does not matter where on the object the forces are applied. When using the rigid object in equilibrium model, however, we cannot use a dot to represent the object because the location where forces act is important in the calculation. Therefore, in a diagram showing the forces on an object, we must show the actual object or a simplified version of it.

Resolve all forces into rectangular components, choosing a convenient coordinate system. Then apply the first condition for equilibrium, Equation 12.1. Remember to keep track of the signs of the various force components.

Choose a convenient axis for calculating the net torque on the rigid object. Remember that the choice of the axis for the torque equation is arbitrary; therefore, choose an axis that simplifies your calculation as much as possible. Usually, the most convenient axis for calculating torques is one through a point at which several forces act, so their torques around this axis are zero. If you don't know a force or don't need to know a force, it is often beneficial to choose an axis through the point at which this force acts. Apply the second condition for equilibrium, Equation 12.2.

Solve the simultaneous equations for the unknowns in terms of the known quantities.

4. Finalize. Make sure your results are consistent with your diagram. If you selected a direction that leads to a negative sign in your solution for a force, do not be alarmed; it merely means that the direction of the force is the opposite of what you guessed. Add up the vertical and horizontal forces on the object and confirm that each set of components adds to zero. Add up the torques on the object and confirm that the sum equals zero.

Example 12.1 The Seesaw Revisited

A seesaw consisting of a uniform board of mass M and length ℓ supports at rest a father and daughter with masses m_f and m_d, respectively, as shown in Figure 12.7. The support (called the *fulcrum*) is under the center of gravity of the board, the father is a distance d from the center, and the daughter is a distance $\ell/2$ from the center.

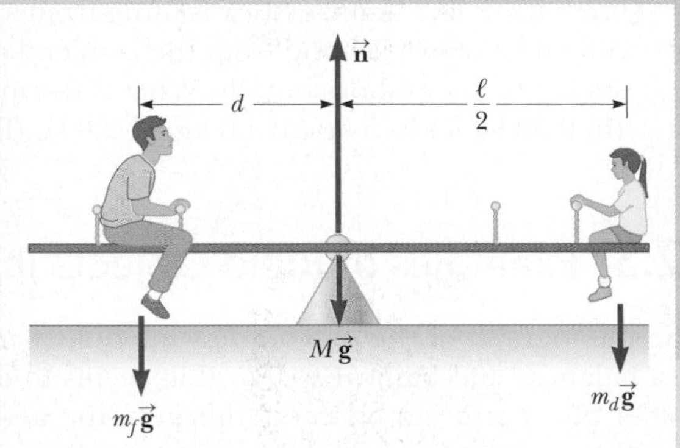

Figure 12.7 (Example 12.1) A balanced system.

(A) Determine the magnitude of the upward force $\vec{\mathbf{n}}$ exerted by the support on the board.

SOLUTION

Conceptualize Let us focus our attention on the board and consider the gravitational forces on the father and daughter as forces applied directly to the board. The daughter would cause a clockwise rotation of the board around the support, whereas the father would cause a counterclockwise rotation.

Categorize Because the text of the problem states that the system is at rest, we model the board as a rigid object in equilibrium. Because we will only need the first condition of equilibrium to solve this part of the problem, however, we model the board as a particle in equilibrium.

Analyze Define upward as the positive y direction and substitute the forces on the board into Equation 12.1:

$$n - m_f g - m_d g - Mg = 0$$

Solve for the magnitude of the force $\vec{\mathbf{n}}$:

$$n = m_f g + m_d g + Mg = (m_f + m_d + M)g$$

(B) Determine where the father should sit to balance the system at rest.

SOLUTION

Categorize This part of the problem requires the introduction of torque to find the position of the father, so we model the board as a rigid object in equilibrium.

Analyze The board's center of gravity is at its geometric center because we are told that the board is uniform. If we choose a rotation axis perpendicular to the page through the center of gravity of the board, the torques produced by $\vec{\mathbf{n}}$ and the gravitational force on the board about this axis are zero.

12.1 *cont.*

Substitute expressions for the torques on the board due to the father and daughter into Equation 12.2:	$(m_f g)(d) - (m_d g)\frac{\ell}{2} = 0$
Solve for d:	$d = \left(\frac{m_d}{m_f}\right)\frac{\ell}{2}$

Finalize This result is the same one we obtained in Example 11.6 by evaluating the angular acceleration of the system and setting the angular acceleration equal to zero.

WHAT IF? Suppose we had chosen another point through which the rotation axis were to pass. For example, suppose the axis is perpendicular to the page and passes through the location of the father. Does that change the results to parts (A) and (B)?

Answer Part (A) is unaffected because the calculation of the net force does not involve a rotation axis. In part (B), we would conceptually expect there to be no change if a different rotation axis is chosen because the second condition of equilibrium claims that the torque is zero about *any* rotation axis.

Let's verify this answer mathematically. Recall that the sign of the torque associated with a force is positive if that force tends to rotate the system counterclockwise, whereas the sign of the torque is negative if the force tends to rotate the system clockwise. Let's choose a rotation axis passing through the location of the father.

Substitute expressions for the torques on the board around this axis into Equation 12.2:	$n(d) - (Mg)(d) - (m_d g)\left(d + \frac{\ell}{2}\right) = 0$
Substitute $n = (m_f + m_d + M)g$ from part (A) and solve for d:	$(m_f + m_d + M)g(d) - (Mg)(d) - (m_d g)\left(d + \frac{\ell}{2}\right) = 0$
	$(m_f g)(d) - (m_d g)\left(\frac{\ell}{2}\right) = 0 \rightarrow d = \left(\frac{m_d}{m_f}\right)\frac{\ell}{2}$

This result is in agreement with the one obtained in part (B).

Example 12.2 Standing on a Horizontal Beam

A uniform horizontal beam with a length of $\ell = 8.00$ m and a weight of $W_b = 200$ N is attached to a wall by a pin connection. Its far end is supported by a cable that makes an angle of $\phi = 53.0°$ with the beam (Fig. 12.8a). A person of weight $W_p = 600$ N stands a distance $d = 2.00$ m from the wall. Find the tension in the cable as well as the magnitude and direction of the force exerted by the wall on the beam.

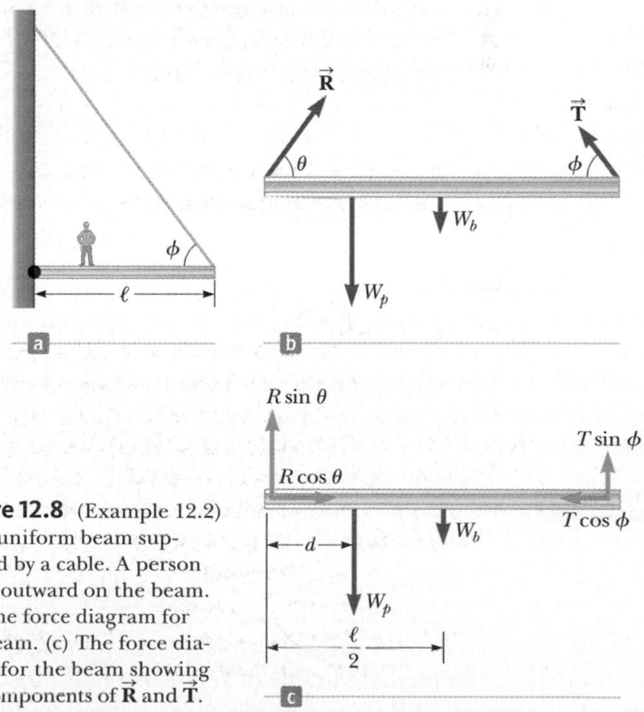

Figure 12.8 (Example 12.2) (a) A uniform beam supported by a cable. A person walks outward on the beam. (b) The force diagram for the beam. (c) The force diagram for the beam showing the components of $\vec{\mathbf{R}}$ and $\vec{\mathbf{T}}$.

SOLUTION

Conceptualize Imagine that the person in Figure 12.8a moves outward on the beam. It seems reasonable that the farther he moves outward, the larger the torque he applies about the pivot and the larger the tension in the cable must be to balance this torque.

Categorize Because the system is at rest, we categorize the beam as a rigid object in equilibrium.

continued

12.2 *cont.*

Analyze We identify all the external forces acting on the beam: the 200-N gravitational force, the force $\vec{\mathbf{T}}$ exerted by the cable, the force $\vec{\mathbf{R}}$ exerted by the wall at the pivot, and the 600-N force that the person exerts on the beam. These forces are all indicated in the force diagram for the beam shown in Figure 12.8b. When we assign directions for forces, it is sometimes helpful to imagine what would happen if a force were suddenly removed. For example, if the wall were to vanish suddenly, the left end of the beam would move to the left as it begins to fall. This scenario tells us that the wall is not only holding the beam up but is also pressing outward against it. Therefore, we draw the vector $\vec{\mathbf{R}}$ in the direction shown in Figure 12.8b. Figure 12.8c shows the horizontal and vertical components of $\vec{\mathbf{T}}$ and $\vec{\mathbf{R}}$.

Substitute expressions for the forces on the beam into Equation 12.1:

$$(1)\ \sum F_x = R\cos\theta - T\cos\phi = 0$$

$$(2)\ \sum F_y = R\sin\theta + T\sin\phi - W_p - W_b = 0$$

where we have chosen rightward and upward as our positive directions. Because R, T, and θ are all unknown, we cannot obtain a solution from these expressions alone. (To solve for the unknowns, the number of simultaneous equations must generally equal the number of unknowns.)

Now let's invoke the condition for rotational equilibrium. A convenient axis to choose for our torque equation is the one that passes through the pin connection. The feature that makes this axis so convenient is that the force $\vec{\mathbf{R}}$ and the horizontal component of $\vec{\mathbf{T}}$ both have a moment arm of zero; hence, these forces produce no torque about this axis.

Substitute expressions for the torques on the beam into Equation 12.2:

$$\sum \tau_z = (T\sin\phi)(\ell) - W_p d - W_b\left(\frac{\ell}{2}\right) = 0$$

This equation contains only T as an unknown because of our choice of rotation axis. Solve for T and substitute numerical values:

$$T = \frac{W_p d + W_b(\ell/2)}{\ell\sin\phi} = \frac{(600\text{ N})(2.00\text{ m}) + (200\text{ N})(4.00\text{ m})}{(8.00\text{ m})\sin 53.0°} = \boxed{313\text{ N}}$$

Rearrange Equations (1) and (2) and then divide:

$$\frac{R\sin\theta}{R\cos\theta} = \tan\theta = \frac{W_p + W_b - T\sin\phi}{T\cos\phi}$$

Solve for θ and substitute numerical values:

$$\theta = \tan^{-1}\left(\frac{W_p + W_b - T\sin\phi}{T\cos\phi}\right)$$

$$= \tan^{-1}\left[\frac{600\text{ N} + 200\text{ N} - (313\text{ N})\sin 53.0°}{(313\text{ N})\cos 53.0°}\right] = \boxed{71.1°}$$

Solve Equation (1) for R and substitute numerical values:

$$R = \frac{T\cos\phi}{\cos\theta} = \frac{(313\text{ N})\cos 53.0°}{\cos 71.1°} = \boxed{581\text{ N}}$$

Finalize The positive value for the angle θ indicates that our estimate of the direction of $\vec{\mathbf{R}}$ was accurate.

Had we selected some other axis for the torque equation, the solution might differ in the details but the answers would be the same. For example, had we chosen an axis through the center of gravity of the beam, the torque equation would involve both T and R. This equation, coupled with Equations (1) and (2), however, could still be solved for the unknowns. Try it!

WHAT IF? What if the person walks farther out on the beam? Does T change? Does R change? Does θ change?

Answer T must increase because the weight of the person exerts a larger torque about the pin connection, which must be countered by a larger torque in the opposite direction due to an increased value of T. If T increases, the vertical component of $\vec{\mathbf{R}}$ decreases to maintain force equilibrium in the vertical direction. Force equilibrium in the horizontal direction, however, requires an increased horizontal component of $\vec{\mathbf{R}}$ to balance the horizontal component of the increased $\vec{\mathbf{T}}$. This fact suggests that θ becomes smaller, but it is hard to predict what happens to R. Problem 66 asks you to explore the behavior of R.

Example 12.3 The Leaning Ladder

A uniform ladder of length ℓ rests against a smooth, vertical wall (Fig. 12.9a). The mass of the ladder is m, and the coefficient of static friction between the ladder and the ground is $\mu_s = 0.40$. Find the minimum angle θ_{min} at which the ladder does not slip.

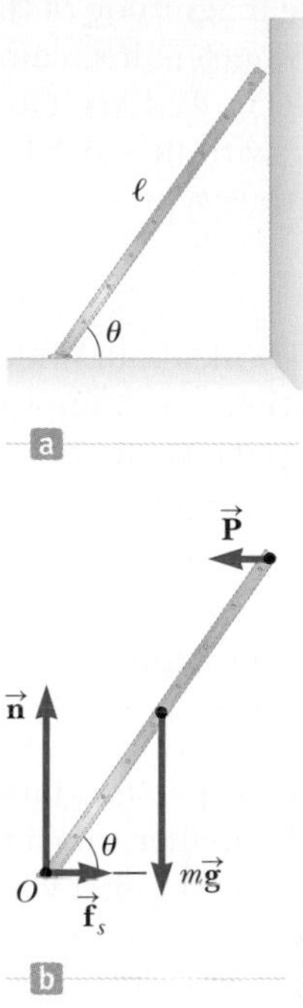

Figure 12.9 (Example 12.3) (a) A uniform ladder at rest, leaning against a smooth wall. The ground is rough. (b) The forces on the ladder.

SOLUTION

Conceptualize Think about any ladders you have climbed. Do you want a large friction force between the bottom of the ladder and the surface or a small one? If the friction force is zero, will the ladder stay up? Simulate a ladder with a ruler leaning against a vertical surface. Does the ruler slip at some angles and stay up at others?

Categorize We do not wish the ladder to slip, so we model it as a rigid object in equilibrium.

Analyze A diagram showing all the external forces acting on the ladder is illustrated in Figure 12.9b. The force exerted by the ground on the ladder is the vector sum of a normal force $\vec{\mathbf{n}}$ and the force of static friction $\vec{\mathbf{f}}_s$. The force $\vec{\mathbf{P}}$ exerted by the wall on the ladder is horizontal because the wall is frictionless.

Apply the first condition for equilibrium to the ladder:

$$(1)\quad \sum F_x = f_s - P = 0$$

$$(2)\quad \sum F_y = n - mg = 0$$

Solve Equation (1) for P:

$$(3)\quad P = f_s$$

Solve Equation (2) for n:

$$(4)\quad n = mg$$

When the ladder is on the verge of slipping, the force of static friction must have its maximum value, which is given by $f_{s,max} = \mu_s n$. Combine this equation with Equations (3) and (4):

$$(5)\quad P = f_{s,max} = \mu_s n = \mu_s mg$$

Apply the second condition for equilibrium to the ladder, taking torques about an axis through O:

$$\sum \tau_O = P\ell \sin\theta_{min} - mg\frac{\ell}{2}\cos\theta_{min} = 0$$

Solve for $\tan\theta_{min}$ and substitute for P from Equation (5):

$$\frac{\sin\theta_{min}}{\cos\theta_{min}} = \tan\theta_{min} = \frac{mg}{2P} = \frac{mg}{2\mu_s mg} = \frac{1}{2\mu_s}$$

Solve for the angle θ_{min}:

$$\theta_{min} = \tan^{-1}\left(\frac{1}{2\mu_s}\right) = \tan^{-1}\left[\frac{1}{2(0.40)}\right] = \boxed{51°}$$

Finalize Notice that the angle depends only on the coefficient of friction, not on the mass or length of the ladder.

Example 12.4 Negotiating a Curb

(A) Estimate the magnitude of the force $\vec{\mathbf{F}}$ a person must apply to a wheelchair's main wheel to roll up over a sidewalk curb (Fig. 12.10a). This main wheel that comes in contact with the curb has a radius r, and the height of the curb is h.

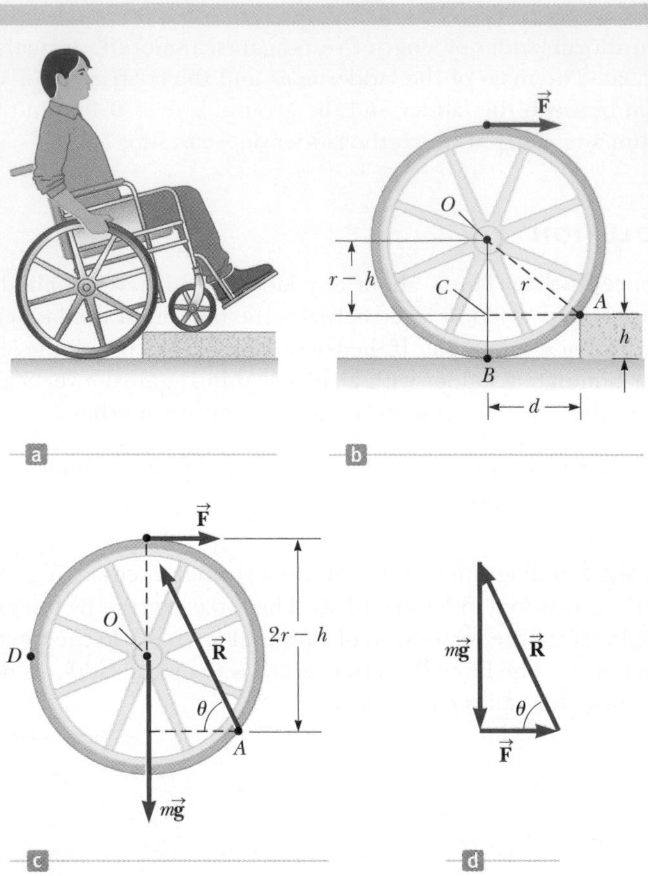

Figure 12.10 (Example 12.4) (a) A person in a wheelchair attempts to roll up over a curb. (b) Details of the wheel and curb. The person applies a force $\vec{\mathbf{F}}$ to the top of the wheel. (c) A force diagram for the wheel when it is just about to be raised. Three forces act on the wheel at this instant: $\vec{\mathbf{F}}$, which is exerted by the hand; $\vec{\mathbf{R}}$, which is exerted by the curb; and the gravitational force $m\vec{\mathbf{g}}$. We estimate that the center of gravity of the system is above the center of the wheel. (d) The vector sum of the three external forces acting on the wheel is zero.

SOLUTION

Conceptualize Think about wheelchair access to buildings. Generally, there are ramps built for individuals in wheelchairs. Steplike structures such as curbs are serious barriers to a wheelchair.

Categorize Imagine that the person exerts enough force so that the bottom of the wheel just loses contact with the lower surface and hovers at rest. We model the wheel in this situation as a rigid object in equilibrium.

Analyze Usually, the person's hands supply the required force to a slightly smaller wheel that is concentric with the main wheel. For simplicity, let's assume the radius of this second wheel is the same as the radius of the main wheel. Let's estimate a combined weight of $mg = 1\,400$ N for the person and the wheelchair and choose a wheel radius of $r = 30$ cm. We also pick a curb height of $h = 10$ cm. Let's also assume the wheelchair and occupant are symmetric and each wheel supports a weight of 700 N. We then proceed to analyze only one of the wheels. Figure 12.10b shows the geometry for a single wheel.

When the wheel is just about to be raised from the street, the normal force exerted by the ground on the wheel at point B goes to zero. Hence, at this time only three forces act on the wheel as shown in the force diagram in Figure 12.10c. The force $\vec{\mathbf{R}}$, which is the force exerted by the curb on the wheel, acts at point A, so if we choose to have our axis of rotation pass through point A, we do not need to include $\vec{\mathbf{R}}$ in our torque equation. The moment arm of $\vec{\mathbf{F}}$ relative to an axis through A is given by $2r - h$ (see Fig. 12.10c).

Use the triangle OAC in Figure 12.10b to find the moment arm d of the gravitational force $m\vec{\mathbf{g}}$ acting on the wheel relative to an axis through point A:

$$(1)\quad d = \sqrt{r^2 - (r-h)^2} = \sqrt{2rh - h^2}$$

Apply the second condition for equilibrium to the wheel, taking torques about an axis through A:

$$(2)\quad \sum \tau_A = mgd - F(2r - h) = 0$$

Substitute for d from Equation (1):

$$mg\sqrt{2rh - h^2} - F(2r - h) = 0$$

Solve for F:

$$F = \frac{mg\sqrt{2rh - h^2}}{2r - h}$$

Substitute the known values:

$$F = \frac{(700\text{ N})\sqrt{2(0.3\text{ m})(0.1\text{ m}) - (0.1\text{ m})^2}}{2(0.3\text{ m}) - 0.1\text{ m}}$$

$$= 3 \times 10^2\text{ N}$$

12.4 *cont.*

(B) Determine the magnitude and direction of $\vec{\mathbf{R}}$.

SOLUTION

Apply the first condition for equilibrium to the wheel:

$$(3)\quad \sum F_x = F - R\cos\theta = 0$$

$$(4)\quad \sum F_y = R\sin\theta - mg = 0$$

Divide Equation (4) by Equation (3):

$$\frac{R\sin\theta}{R\cos\theta} = \tan\theta = \frac{mg}{F}$$

Solve for the angle θ:

$$\theta = \tan^{-1}\left(\frac{mg}{F}\right) = \tan^{-1}\left(\frac{700\text{ N}}{300\text{ N}}\right) = \boxed{70^\circ}$$

Solve Equation (4) for R and substitute numerical values:

$$R = \frac{mg}{\sin\theta} = \frac{700\text{ N}}{\sin 70^\circ} = \boxed{8 \times 10^2\text{ N}}$$

Finalize Notice that we have kept only one digit as significant. (We have written the angle as 70° because $7 \times 10^{1\circ}$ is awkward!) The results indicate that the force that must be applied to each wheel is substantial. You may want to estimate the force required to roll a wheelchair up a typical sidewalk accessibility ramp for comparison.

WHAT IF? Would it be easier to negotiate the curb if the person grabbed the wheel at point D in Figure 12.10c and pulled *upward*?

Answer If the force $\vec{\mathbf{F}}$ in Figure 12.10c is rotated counterclockwise by 90° and applied at D, its moment arm about an axis through A is $d + r$. Let's call the magnitude of this new force F'.

Modify Equation (2) for this situation:

$$\sum \tau_A = mgd - F'(d + r) = 0$$

Solve this equation for F' and substitute for d:

$$F' = \frac{mgd}{d + r} = \frac{mg\sqrt{2rh - h^2}}{\sqrt{2rh - h^2} + r}$$

Take the ratio of this force to the original force that we calculated and express the result in terms of h/r, the ratio of the curb height to the wheel radius:

$$\frac{F'}{F} = \frac{\dfrac{mg\sqrt{2rh - h^2}}{\sqrt{2rh - h^2} + r}}{\dfrac{mg\sqrt{2rh - h^2}}{2r - h}} = \frac{2r - h}{\sqrt{2rh - h^2} + r} = \frac{2 - \left(\dfrac{h}{r}\right)}{\sqrt{2\left(\dfrac{h}{r}\right) - \left(\dfrac{h}{r}\right)^2} + 1}$$

Substitute the ratio $h/r = 0.33$ from the given values:

$$\frac{F'}{F} = \frac{2 - 0.33}{\sqrt{2(0.33) - (0.33)^2} + 1} = 0.96$$

This result tells us that, *for these values,* it is slightly easier to pull upward at D than horizontally at the top of the wheel. For very high curbs, so that h/r is close to 1, the ratio F'/F drops to about 0.5 because point A is located near the right edge of the wheel in Figure 12.10b. The force at D is applied at distance of about $2r$ from A, whereas the force at the top of the wheel has a moment arm of only about r. For high curbs, then, it is best to pull upward at D, although a large value of the force is required. For small curbs, it is best to apply the force at the top of the wheel. The ratio F'/F becomes larger than 1 at about $h/r = 0.3$ because point A is now close to the bottom of the wheel and the force applied at the top of the wheel has a larger moment arm than when applied at D.

Finally, let's comment on the validity of these mathematical results. Consider Figure 12.10d and imagine that the vector $\vec{\mathbf{F}}$ is upward instead of to the right. There is no way the three vectors can add to equal zero as required by the first equilibrium condition. Therefore, our results above may be qualitatively valid, but not exact quantitatively. To cancel the horizontal component of $\vec{\mathbf{R}}$, the force at D must be applied at an angle to the vertical rather than straight upward. This feature makes the calculation more complicated and requires both conditions of equilibrium.

12.4 Elastic Properties of Solids

Except for our discussion about springs in earlier chapters, we have assumed objects remain rigid when external forces act on them. In Section 9.8, we explored deformable systems. In reality, all objects are deformable to some extent. That is, it is possible to change the shape or the size (or both) of an object by applying external forces. As these changes take place, however, internal forces in the object resist the deformation.

We shall discuss the deformation of solids in terms of the concepts of *stress* and *strain*. **Stress** is a quantity that is proportional to the force causing a deformation; more specifically, stress is the external force acting on an object per unit cross-sectional area. The result of a stress is **strain,** which is a measure of the degree of deformation. It is found that, for sufficiently small stresses, stress is proportional to strain; the constant of proportionality depends on the material being deformed and on the nature of the deformation. We call this proportionality constant the **elastic modulus.** The elastic modulus is therefore defined as the ratio of the stress to the resulting strain:

$$\text{Elastic modulus} \equiv \frac{\text{stress}}{\text{strain}} \tag{12.5}$$

The elastic modulus in general relates what is done to a solid object (a force is applied) to how that object responds (it deforms to some extent). It is similar to the spring constant k in Hooke's law (Eq. 7.9) that relates a force applied to a spring and the resultant deformation of the spring, measured by its extension or compression.

We consider three types of deformation and define an elastic modulus for each:

1. **Young's modulus** measures the resistance of a solid to a change in its length.
2. **Shear modulus** measures the resistance to motion of the planes within a solid parallel to each other.
3. **Bulk modulus** measures the resistance of solids or liquids to changes in their volume.

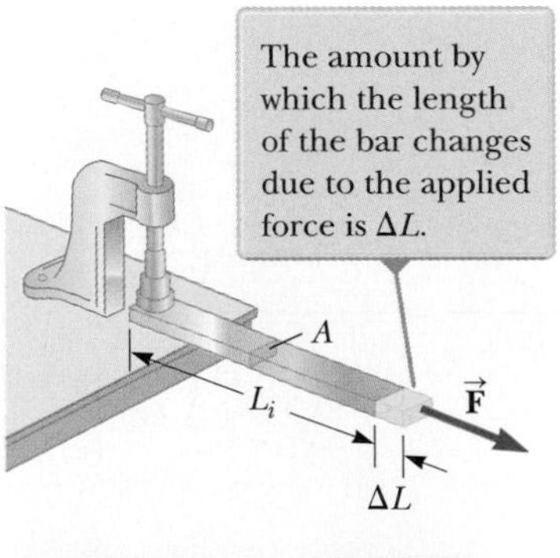

ACTIVE FIGURE 12.11

A force $\vec{\mathbf{F}}$ is applied to the free end of a bar clamped at the other end.

Young's Modulus: Elasticity in Length

Consider a long bar of cross-sectional area A and initial length L_i that is clamped at one end as in Active Figure 12.11. When an external force is applied perpendicular to the cross section, internal forces in the bar resist distortion ("stretching"), but the bar reaches an equilibrium situation in which its final length L_f is greater than L_i and in which the external force is exactly balanced by internal forces. In such a situation, the bar is said to be stressed. We define the **tensile stress** as the ratio of the magnitude of the external force F to the cross-sectional area A. The **tensile strain** in this case is defined as the ratio of the change in length ΔL to the original length L_i. We define **Young's modulus** by a combination of these two ratios:

Young's modulus ▶

$$Y \equiv \frac{\text{tensile stress}}{\text{tensile strain}} = \frac{F/A}{\Delta L/L_i} \tag{12.6}$$

Young's modulus is typically used to characterize a rod or wire stressed under either tension or compression. Because strain is a dimensionless quantity, Y has units of force per unit area. Typical values are given in Table 12.1.

For relatively small stresses, the bar returns to its initial length when the force is removed. The **elastic limit** of a substance is defined as the maximum stress that can be applied to the substance before it becomes permanently deformed and does not return to its initial length. It is possible to exceed the elastic limit of a substance by applying a sufficiently large stress as seen in Figure 12.12. Initially, a stress-versus-

TABLE 12.1 *Typical Values for Elastic Moduli*

Substance	Young's Modulus (N/m^2)	Shear Modulus (N/m^2)	Bulk Modulus (N/m^2)
Tungsten	35×10^{10}	14×10^{10}	20×10^{10}
Steel	20×10^{10}	8.4×10^{10}	6×10^{10}
Copper	11×10^{10}	4.2×10^{10}	14×10^{10}
Brass	9.1×10^{10}	3.5×10^{10}	6.1×10^{10}
Aluminum	7.0×10^{10}	2.5×10^{10}	7.0×10^{10}
Glass	$6.5–7.8 \times 10^{10}$	$2.6–3.2 \times 10^{10}$	$5.0–5.5 \times 10^{10}$
Quartz	5.6×10^{10}	2.6×10^{10}	2.7×10^{10}
Water	—	—	0.21×10^{10}
Mercury	—	—	2.8×10^{10}

strain curve is a straight line. As the stress increases, however, the curve is no longer a straight line. When the stress exceeds the elastic limit, the object is permanently distorted and does not return to its original shape after the stress is removed. As the stress is increased even further, the material ultimately breaks.

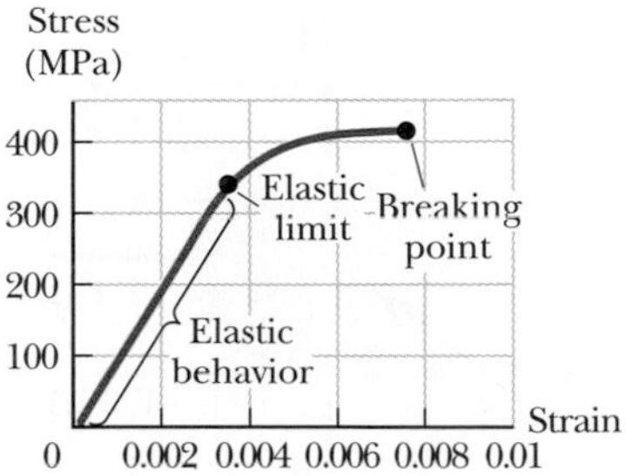

Figure 12.12 Stress-versus-strain curve for an elastic solid.

Shear Modulus: Elasticity of Shape

Another type of deformation occurs when an object is subjected to a force parallel to one of its faces while the opposite face is held fixed by another force (Active Fig. 12.13a). The stress in this case is called a shear stress. If the object is originally a rectangular block, a shear stress results in a shape whose cross section is a parallelogram. A book pushed sideways as shown in Active Figure 12.13b is an example of an object subjected to a shear stress. To a first approximation (for small distortions), no change in volume occurs with this deformation.

We define the **shear stress** as F/A, the ratio of the tangential force to the area A of the face being sheared. The **shear strain** is defined as the ratio $\Delta x/h$, where Δx is the horizontal distance that the sheared face moves and h is the height of the object. In terms of these quantities, the **shear modulus** is

$$S \equiv \frac{\text{shear stress}}{\text{shear strain}} = \frac{F/A}{\Delta x/h} \tag{12.7}$$

◀ **Shear modulus**

Values of the shear modulus for some representative materials are given in Table 12.1. Like Young's modulus, the unit of shear modulus is the ratio of that for force to that for area.

Bulk Modulus: Volume Elasticity

Bulk modulus characterizes the response of an object to changes in a force of uniform magnitude applied perpendicularly over the entire surface of the object as shown in Active Figure 12.14 (page 360). (We assume here the object is made of a single substance.) As we shall see in Chapter 14, such a uniform distribution of

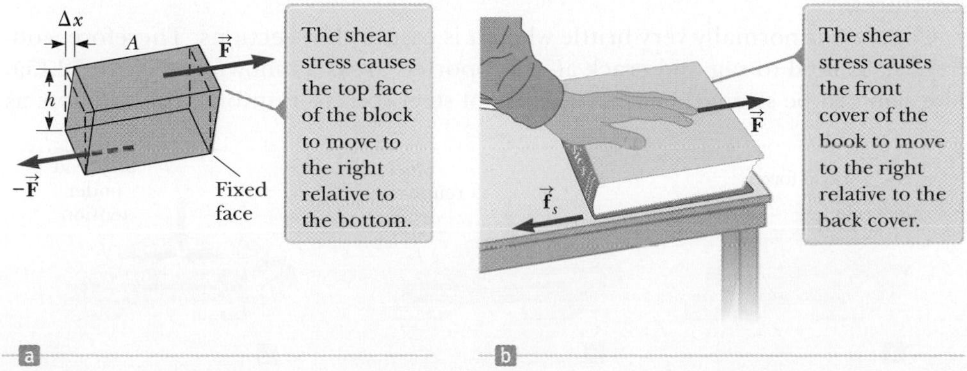

ACTIVE FIGURE 12.13

(a) A shear deformation in which a rectangular block is distorted by two forces of equal magnitude but opposite directions applied to two parallel faces. (b) A book is under shear stress when a hand placed on the cover applies a horizontal force away from the spine.

forces occurs when an object is immersed in a fluid. An object subject to this type of deformation undergoes a change in volume but no change in shape. The **volume stress** is defined as the ratio of the magnitude of the total force F exerted on a surface to the area A of the surface. The quantity $P = F/A$ is called **pressure**, which we shall study in more detail in Chapter 14. If the pressure on an object changes by an amount $\Delta P = \Delta F/A$, the object experiences a volume change ΔV. The **volume strain** is equal to the change in volume ΔV divided by the initial volume V_i. Therefore, from Equation 12.5, we can characterize a volume ("bulk") compression in terms of the **bulk modulus,** which is defined as

Bulk modulus ▶

$$B \equiv \frac{\text{volume stress}}{\text{volume strain}} = -\frac{\Delta F/A}{\Delta V/V_i} = -\frac{\Delta P}{\Delta V/V_i} \tag{12.8}$$

A negative sign is inserted in this defining equation so that B is a positive number. This maneuver is necessary because an increase in pressure (positive ΔP) causes a decrease in volume (negative ΔV) and vice versa.

Table 12.1 lists bulk moduli for some materials. If you look up such values in a different source, you may find the reciprocal of the bulk modulus listed. The reciprocal of the bulk modulus is called the **compressibility** of the material.

Notice from Table 12.1 that both solids and liquids have a bulk modulus. No shear modulus and no Young's modulus are given for liquids, however, because a liquid does not sustain a shearing stress or a tensile stress. If a shearing force or a tensile force is applied to a liquid, the liquid simply flows in response.

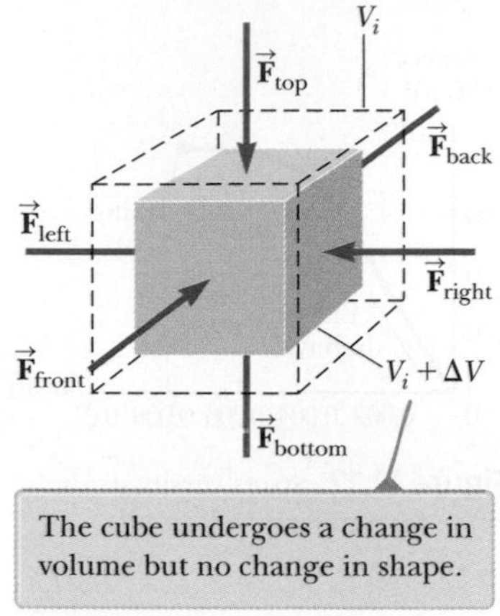

ACTIVE FIGURE 12.14

A cube is under uniform pressure and is therefore compressed on all sides by forces normal to its six faces. The arrowheads of force vectors on the sides of the cube that are not visible are hidden by the cube.

***Quick Quiz* 12.4** For the three parts of this Quick Quiz, choose from the following choices the correct answer for the elastic modulus that describes the relationship between stress and strain for the system of interest, which is in italics: (a) Young's modulus (b) shear modulus (c) bulk modulus (d) none of those choices **(i)** A *block of iron* is sliding across a horizontal floor. The friction force between the block and the floor causes the block to deform. **(ii)** A trapeze artist swings through a circular arc. At the bottom of the swing, the *wires* supporting the trapeze are longer than when the trapeze artist simply hangs from the trapeze due to the increased tension in them. **(iii)** A spacecraft carries a *steel sphere* to a planet on which atmospheric pressure is much higher than on the Earth. The higher pressure causes the radius of the sphere to decrease.

Prestressed Concrete

If the stress on a solid object exceeds a certain value, the object fractures. The maximum stress that can be applied before fracture occurs—called the *tensile strength, compressive strength,* or *shear strength*—depends on the nature of the material and on the type of applied stress. For example, concrete has a tensile strength of about 2×10^6 N/m^2, a compressive strength of 20×10^6 N/m^2, and a shear strength of 2×10^6 N/m^2. If the applied stress exceeds these values, the concrete fractures. It is common practice to use large safety factors to prevent failure in concrete structures.

Concrete is normally very brittle when it is cast in thin sections. Therefore, concrete slabs tend to sag and crack at unsupported areas as shown in Figure 12.15a. The slab can be strengthened by the use of steel rods to reinforce the concrete as

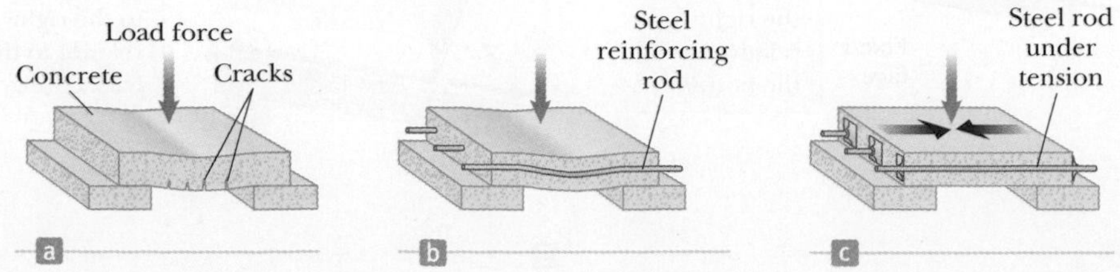

Figure 12.15 (a) A concrete slab with no reinforcement tends to crack under a heavy load. (b) The strength of the concrete is increased by using steel reinforcement rods. (c) The concrete is further strengthened by prestressing it with steel rods under tension.

illustrated in Figure 12.15b. Because concrete is much stronger under compression (squeezing) than under tension (stretching) or shear, vertical columns of concrete can support very heavy loads, whereas horizontal beams of concrete tend to sag and crack. A significant increase in shear strength is achieved, however, if the reinforced concrete is prestressed as shown in Figure 12.15c. As the concrete is being poured, the steel rods are held under tension by external forces. The external forces are released after the concrete cures; the result is a permanent tension in the steel and hence a compressive stress on the concrete. The concrete slab can now support a much heavier load.

Example 12.5 Stage Design

In Example 8.2, we analyzed a cable used to support an actor as he swung onto the stage. Now suppose the tension in the cable is 940 N as the actor reaches the lowest point. What diameter should a 10-m-long steel cable have if we do not want it to stretch more than 0.50 cm under these conditions?

SOLUTION

Conceptualize Look back at Example 8.2 to recall what is happening in this situation. We ignored any stretching of the cable there, but we wish to address this phenomenon in this example.

Categorize We perform a simple calculation involving Equation 12.6, so we categorize this example as a substitution problem.

Solve Equation 12.6 for the cross-sectional area of the cable:

$$A = \frac{FL_i}{Y\Delta L}$$

Substitute the known values:

$$A = \frac{(940\text{ N})(10\text{ m})}{(20 \times 10^{10}\text{ N/m}^2)(0.005\,0\text{ m})} = 9.4 \times 10^{-6}\text{ m}^2$$

Assuming the cross section is circular, find the diameter of the cable from $d = 2r$ and $A = \pi r^2$:

$$d = 2r = 2\sqrt{\frac{A}{\pi}} = 2\sqrt{\frac{9.4 \times 10^{-6}\text{ m}^2}{\pi}} = 3.5 \times 10^{-3}\text{ m} = \boxed{3.5\text{ mm}}$$

To provide a large margin of safety, you would probably use a flexible cable made up of many smaller wires having a total cross-sectional area substantially greater than our calculated value.

Example 12.6 Squeezing a Brass Sphere

A solid brass sphere is initially surrounded by air, and the air pressure exerted on it is 1.0×10^5 N/m^2 (normal atmospheric pressure). The sphere is lowered into the ocean to a depth where the pressure is 2.0×10^7 N/m^2. The volume of the sphere in air is 0.50 m^3. By how much does this volume change once the sphere is submerged?

SOLUTION

Conceptualize Think about movies or television shows you have seen in which divers go to great depths in the water in submersible vessels. These vessels must be very strong to withstand the large pressure under water. This pressure squeezes the vessel and reduces its volume.

Categorize We perform a simple calculation involving Equation 12.8, so we categorize this example as a substitution problem.

Solve Equation 12.8 for the volume change of the sphere:

$$\Delta V = -\frac{V_i \Delta P}{B}$$

continued

12.6 cont.

Substitute numerical values:

$$\Delta V = -\frac{(0.50\ \text{m}^3)(2.0 \times 10^7\ \text{N/m}^2 - 1.0 \times 10^5\ \text{N/m}^2)}{6.1 \times 10^{10}\ \text{N/m}^2}$$

$$= -1.6 \times 10^{-4}\ \text{m}^3$$

The negative sign indicates that the volume of the sphere decreases.

Summary

Definitions

The gravitational force exerted on an object can be considered as acting at a single point called the **center of gravity.** An object's center of gravity coincides with its center of mass if the object is in a uniform gravitational field.

We can describe the elastic properties of a substance using the concepts of stress and strain. **Stress** is a quantity proportional to the force producing a deformation; **strain** is a measure of the degree of deformation. Stress is proportional to strain, and the constant of proportionality is the **elastic modulus:**

$$\text{Elastic modulus} \equiv \frac{\text{stress}}{\text{strain}} \qquad \textbf{(12.5)}$$

Concepts and Principles

Three common types of deformation are represented by (1) the resistance of a solid to elongation under a load, characterized by **Young's modulus** Y; (2) the resistance of a solid to the motion of internal planes sliding past each other, characterized by the **shear modulus** S; and (3) the resistance of a solid or fluid to a volume change, characterized by the **bulk modulus** B.

Analysis Model for Problem Solving

Rigid Object in Equilibrium A rigid object in equilibrium exhibits no translational or angular acceleration. The net external force acting on it is zero, and the net external torque on it is zero about any axis:

$$\sum \vec{\mathbf{F}}_{\text{ext}} = 0 \qquad \textbf{(12.1)}$$

$$\sum \vec{\boldsymbol{\tau}}_{\text{ext}} = 0 \qquad \textbf{(12.2)}$$

The first condition is the condition for translational equilibrium, and the second is the condition for rotational equilibrium.

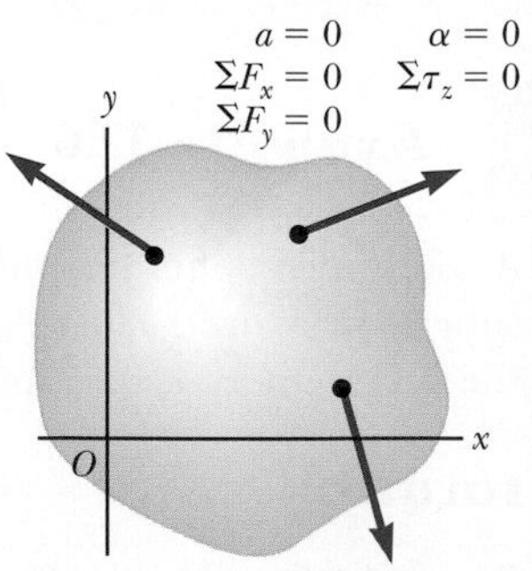

Objective Questions

☐ denotes answer available in Student Solutions Manual/Study Guide

1. Assume a single 300-N force is exerted on a bicycle frame as shown in Figure OQ12.1. Consider the torque produced by this force about axes perpendicular to the plane of the paper and through each of the points A through E, where E is the center of mass of the frame. Rank the torques τ_A, τ_B, τ_C, τ_D, and τ_E from largest to smallest, noting that zero

is greater than a negative quantity. If two torques are equal, note their equality in your ranking.

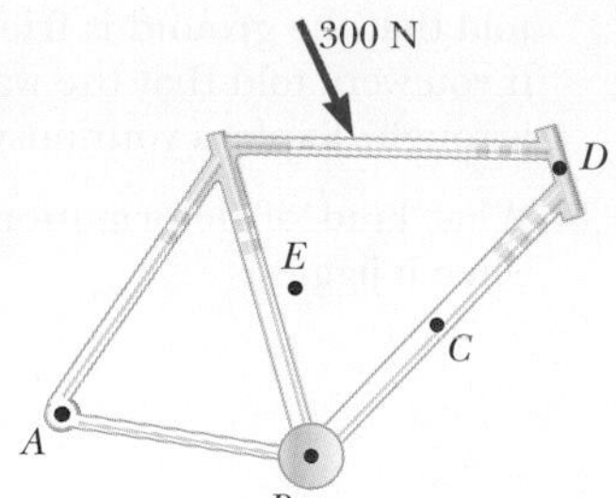

Figure OQ12.1

2. Two forces are acting on an object. Which of the following statements is correct? (a) The object is in equilibrium if the forces are equal in magnitude and opposite in direction. (b) The object is in equilibrium if the net torque on the object is zero. (c) The object is in equilibrium if the forces act at the same point on the object. (d) The object is in equilibrium if the net force and the net torque on the object are both zero. (e) The object cannot be in equilibrium because more than one force acts on it.

3. Consider the object in Figure OQ12.3. A single force is exerted on the object. The line of action of the force does not pass through the object's center of mass. The acceleration of the object's center of mass due to this force (a) is the same as if the force were applied at the center of mass, (b) is larger than the acceleration would be if the force were applied at the center of mass, (c) is smaller than the acceleration would be if the force were applied at the center of mass, or (d) is zero because the force causes only angular acceleration about the center of mass.

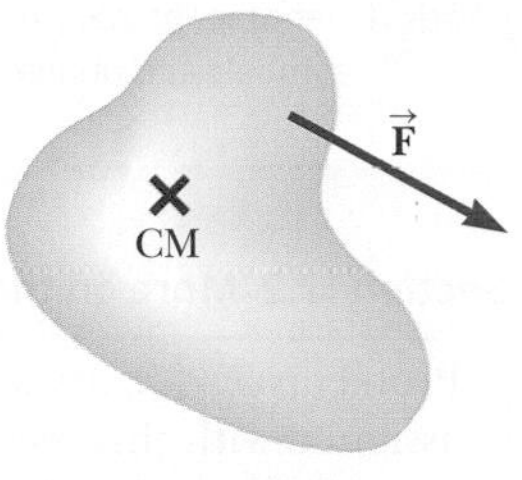

Figure OQ12.3

4. A rod 7.0 m long is pivoted at a point 2.0 m from the left end. A downward force of 50 N acts at the left end, and a downward force of 200 N acts at the right end. At what distance to the right of the pivot can a third force of 300 N acting upward be placed to produce rotational equilibrium? *Note:* Neglect the weight of the rod. (a) 1.0 m (b) 2.0 m (c) 3.0 m (d) 4.0 m (e) 3.5 m

5. In the cabin of a ship, a soda can rests in a saucer-shaped indentation in a built-in counter. The can tilts as the ship slowly rolls. In which case is the can most stable against tipping over? (a) It is most stable when it is full. (b) It is most stable when it is half full. (c) It is most stable when it is empty. (d) It is most stable in two of these cases. (e) It is equally stable in all cases.

6. A 20.0-kg horizontal plank 4.00 m long rests on two supports, one at the left end and a second 1.00 m from the right end. What is the magnitude of the force exerted on the plank by the support near the right end? (a) 32.0 N (b) 45.2 N (c) 112 N (d) 131 N (e) 98.2 N

7. The acceleration due to gravity becomes weaker by about three parts in ten million for each meter of increased elevation above the Earth's surface. Suppose a skyscraper is 100 stories tall, with the same floor plan for each story and with uniform average density. Compare the location of the building's center of mass and the location of its center of gravity. Choose one: (a) Its center of mass is higher by a distance of several meters. (b) Its center of mass is higher by a distance of several millimeters. (c) Its center of mass and its center of gravity are in the same location. (d) Its center of gravity is higher by a distance of several millimeters. (e) Its center of gravity is higher by a distance of several meters.

8. The center of gravity of an ax is on the centerline of the handle, close to the head. Assume you saw across the handle through the center of gravity and weigh the two parts. What will you discover? (a) The handle side is heavier than the head side. (b) The head side is heavier than the handle side. (c) The two parts are equally heavy. (d) Their comparative weights cannot be predicted.

9. A certain wire, 3 m long, stretches by 1.2 mm when under tension 200 N. **(i)** Does an equally thick wire 6 m long, made of the same material and under the same tension, stretch by (a) 4.8 mm, (b) 2.4 mm, (c) 1.2 mm, (d) 0.6 mm, or (e) 0.3 mm? **(ii)** A wire with twice the diameter, 3 m long, made of the same material and under the same tension, stretches by what amount? Choose from the same possibilities (a) through (e).

10. In analyzing the equilibrium of a flat, rigid object, you are about to choose an axis about which you will calculate torques. Which of the following describes the choice you should make? (a) The axis should pass through the object's center of mass. (b) The axis should pass through one end of the object. (c) The axis should be either the *x* axis or the *y* axis. (d) The axis should pass through any point within the object. (e) Any axis within or outside the object can be chosen.

Conceptual Questions

☐ denotes answer available in Student Solutions Manual/Study Guide

1. Can an object be in equilibrium if it is in motion? Explain.

2. Stand with your back against a wall. Why can't you put your heels firmly against the wall and then bend forward without falling?

3. (a) Give an example in which the net force acting on an object is zero and yet the net torque is nonzero. (b) Give an example in which the net torque acting on an object is zero and yet the net force is nonzero.

4. The center of gravity of an object may be located outside the object. Give two examples for which that is the case.

5. An arbitrarily shaped piece of plywood can be suspended from a string attached to the ceiling. Explain how you could use a plumb bob to find its center of gravity.

6. A girl has a large, docile dog she wishes to weigh on a small bathroom scale. She reasons that she can determine her dog's weight with the following method. First she puts the

dog's two front feet on the scale and records the scale reading. Then she places only the dog's two back feet on the scale and records the reading. She thinks that the sum of the readings will be the dog's weight. Is she correct? Explain your answer.

7. A ladder stands on the ground, leaning against a wall. Would you feel safer climbing up the ladder if you were told that the ground is frictionless but the wall is rough or if you were told that the wall is frictionless but the ground is rough? Explain your answer.

8. What kind of deformation does a cube of Jell-O exhibit when it jiggles?

Problems

ENHANCED WebAssign The problems found in this chapter may be assigned online in Enhanced WebAssign

1. denotes straightforward problem; **2.** denotes intermediate problem; **3.** denotes challenging problem

1. full solution available in the Student Solutions Manual/Study Guide

1. denotes problems most often assigned in Enhanced WebAssign; these provide students with targeted feedback and either a Master It tutorial or a Watch It solution video.

QC denotes asking for quantitative and conceptual reasoning

S denotes symbolic reasoning problem

M denotes Master It tutorial available in Enhanced WebAssign

GP denotes guided problem

shaded denotes "paired problems" that develop reasoning with symbols and numerical values

Section 12.1 Analysis Model: Rigid Object in Equilibrium

1. **S** What are the necessary conditions for equilibrium of the object shown in Figure P12.1? Calculate torques about an axis through point O.

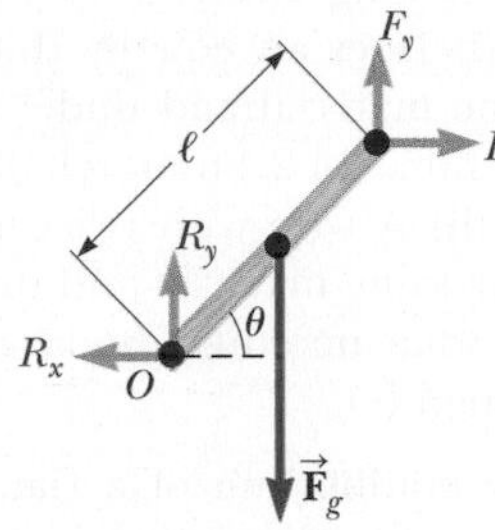

Figure P12.1

2. *Why is the following situation impossible?* A uniform beam of mass $m_b = 3.00$ kg and length $\ell = 1.00$ m supports blocks with masses $m_1 = 5.00$ kg and $m_2 = 15.0$ kg at two positions as shown in Figure P12.2. The beam rests on two triangular blocks, with point P a distance $d = 0.300$ m to the right of the center of gravity of the beam. The position of the object of mass m_2 is adjusted along the length of the beam until the normal force on the beam at O is zero.

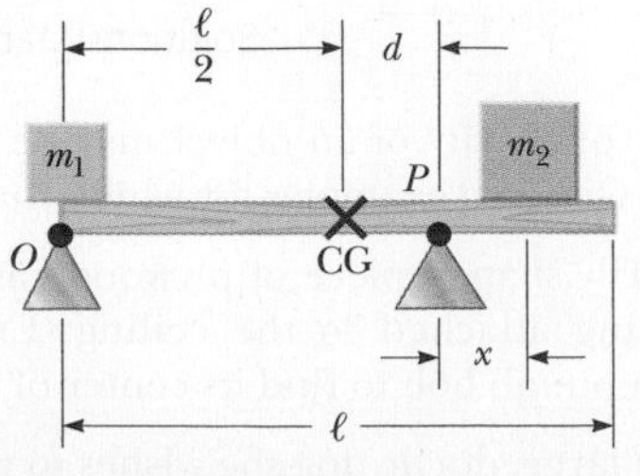

Figure P12.2

Section 12.2 More on the Center of Gravity

Problems 37, 38, 40, and 76 in Chapter 9 can also be assigned with this section.

3. A carpenter's square has the shape of an L as shown in Figure P12.3. Locate its center of gravity.

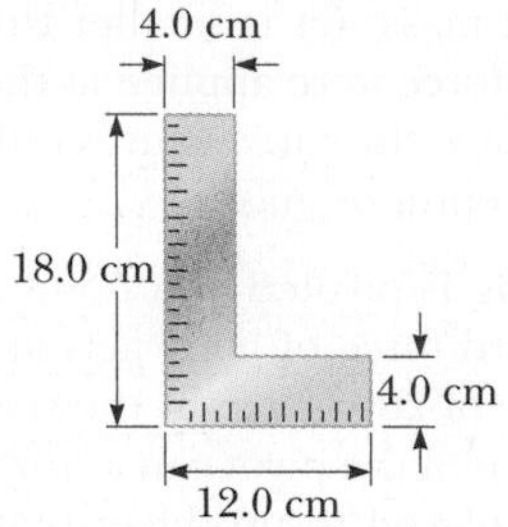

Figure P12.3

4. **M** Consider the following distribution of objects: a 5.00-kg object with its center of gravity at (0, 0) m, a 3.00-kg object at (0, 4.00) m, and a 4.00-kg object at (3.00, 0) m. Where should a fourth object of mass 8.00 kg be placed so that the center of gravity of the four-object arrangement will be at (0, 0)?

5. Pat builds a track for his model car out of solid wood as shown in Figure P12.5. The track is 5.00 cm wide, 1.00 m

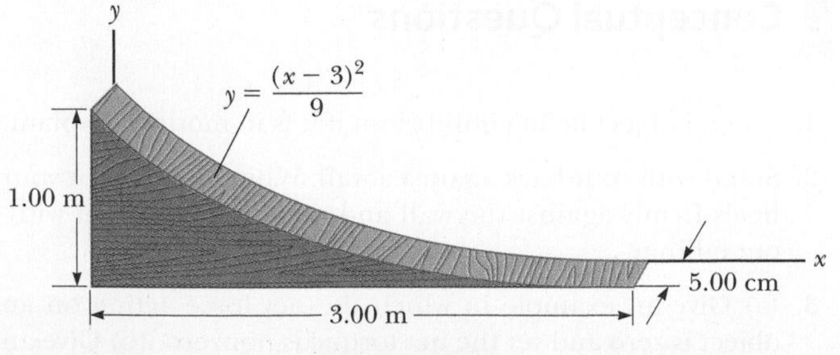

Figure P12.5

high, and 3.00 m long. The runway is cut so that it forms a parabola with the equation $y = (x - 3)^2/9$. Locate the horizontal coordinate of the center of gravity of this track.

6. **S** A circular pizza of radius R has a circular piece of radius $R/2$ removed from one side as shown in Figure P12.6. The center of gravity has moved from C to C' along the x axis. Show that the distance from C to C' is $R/6$. Assume the thickness and density of the pizza are uniform throughout.

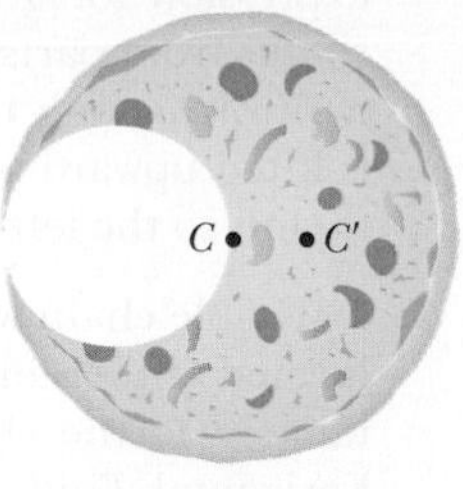

Figure P12.6

7. Figure P12.7 shows three uniform objects: a rod with $m_1 = 6.00$ kg, a right triangle with $m_2 = 3.00$ kg, and a square with $m_3 = 5.00$ kg. Their coordinates in meters are given. Determine the center of gravity for the three-object system.

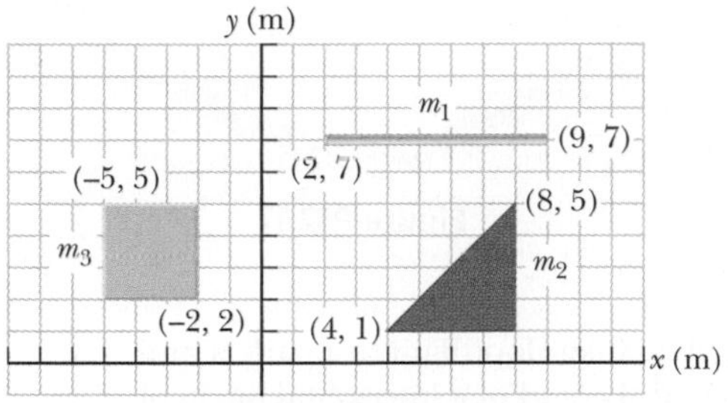

Figure P12.7

Section 12.3 Examples of Rigid Objects in Static Equilibrium

Problems 10, 18, 19, 20, 23, 24, 25, 44, 48, 61, 63, 71, 72, and 74 in Chapter 5 can also be assigned with this section.

8. A mobile is constructed of light rods, light strings, and beach souvenirs as shown in Figure P12.8. If $m_4 = 12.0$ g, find values for (a) m_1, (b) m_2, and (c) m_3.

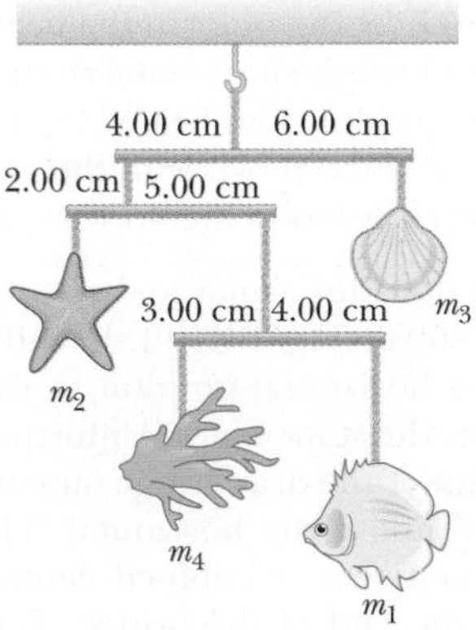

Figure P12.8

9. Find the mass m of the counterweight needed to balance a truck with mass $M = 1\,500$ kg on an incline of $\theta = 45°$ (Fig. P12.9). Assume both pulleys are frictionless and massless.

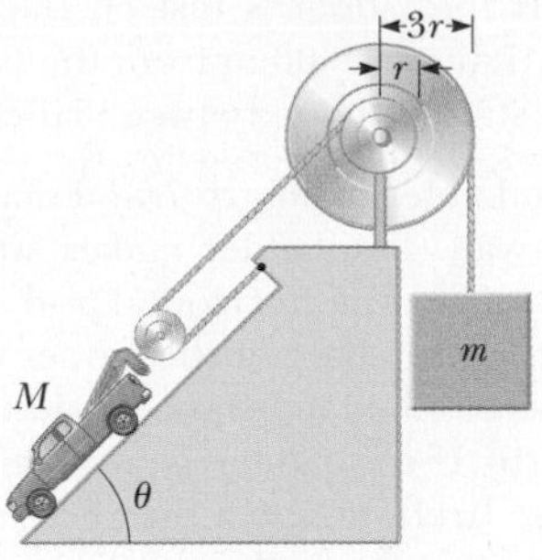

Figure P12.9

10. A 1 500-kg automobile has a wheel base (the distance between the axles) of 3.00 m. The automobile's center of mass is on the centerline at a point 1.20 m behind the front axle. Find the force exerted by the ground on each wheel.

11. A uniform beam of length 7.60 m and weight 4.50×10^2 N is carried by two workers, Sam and Joe, as shown in Figure P12.11. Determine the force that each person exerts on the beam.

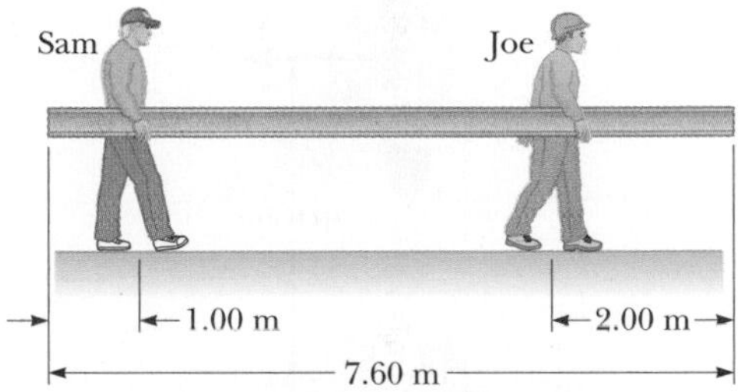

Figure P12.11

12. A vaulter holds a 29.4-N pole in equilibrium by exerting an upward force $\vec{\mathbf{U}}$ with her leading hand and a downward force $\vec{\mathbf{D}}$ with her trailing hand as shown in Figure P12.12. Point C is the center of gravity of the pole. What are the magnitudes of (a) $\vec{\mathbf{U}}$ and (b) $\vec{\mathbf{D}}$?

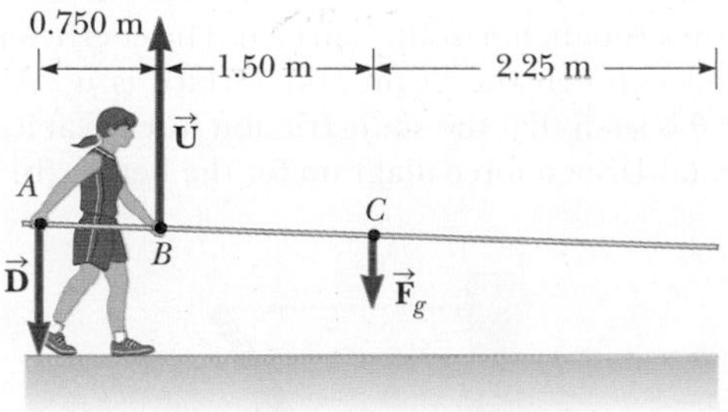

Figure P12.12

13. A 15.0-m uniform ladder weighing 500 N rests against a frictionless wall. The ladder makes a 60.0° angle with the

horizontal. (a) Find the horizontal and vertical forces the ground exerts on the base of the ladder when an 800-N firefighter has climbed 4.00 m along the ladder from the bottom. (b) If the ladder is just on the verge of slipping when the firefighter is 9.00 m from the bottom, what is the coefficient of static friction between ladder and ground?

14. **S** A uniform ladder of length L and mass m_1 rests against a frictionless wall. The ladder makes an angle θ with the horizontal. (a) Find the horizontal and vertical forces the ground exerts on the base of the ladder when a firefighter of mass m_2 has climbed a distance x along the ladder from the bottom. (b) If the ladder is just on the verge of slipping when the firefighter is a distance d along the ladder from the bottom, what is the coefficient of static friction between ladder and ground?

15. Figure P12.15 shows a claw hammer being used to pull a nail out of a horizontal board. The mass of the hammer is 1.00 kg. A force of 150 N is exerted horizontally as shown, and the nail does not yet move relative to the board. Find (a) the force exerted by the hammer claws on the nail and (b) the force exerted by the surface on the point of contact with the hammer head. Assume the force the hammer exerts on the nail is parallel to the nail.

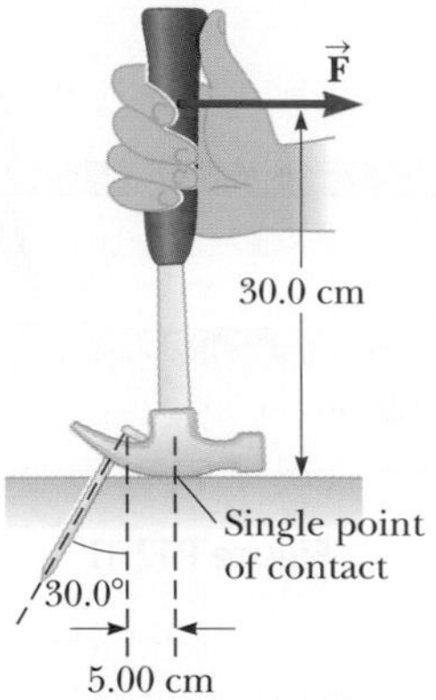

Figure P12.15

16. **Q|C** **S** A uniform beam of length L and mass m shown in Figure P12.16 is inclined at an angle θ to the horizontal. Its upper end is connected to a wall by a rope, and its lower end rests on a rough, horizontal surface. The coefficient of static friction between the beam and surface is μ_s. Assume the angle θ is such that the static friction force is at its *maximum* value. (a) Draw a force diagram for the beam. (b) Using the condition of rotational equilibrium, find an expression for the tension T in the rope in terms of m, g, and θ. (c) Using the condition of translational equilibrium, find a second expression for T in terms of μ_s, m, and g. (d) Using the results from parts (a) through (c), obtain an expression for μ_s involving only the angle θ. (e) What happens if the ladder is lifted upward and its base is placed back on the ground slightly to the left of its position in Figure P12.16? Explain.

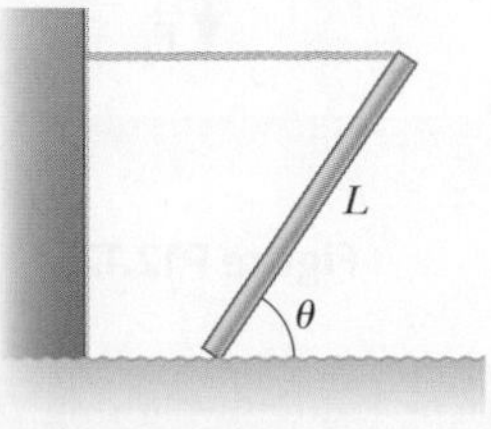

Figure P12.16

17. A flexible chain weighing 40.0 N hangs between two hooks located at the same height (Fig. P12.17). At each hook, the tangent to the chain makes an angle $\theta = 42.0°$ with the horizontal. Find (a) the magnitude of the force each hook exerts on the chain and (b) the tension in the chain at its midpoint. *Suggestion:* For part (b), make a force diagram for half of the chain.

Figure P12.17

18. **Q|C** A 20.0-kg floodlight in a park is supported at the end of a horizontal beam of negligible mass that is hinged to a pole as shown in Figure P12.18. A cable at an angle of $\theta = 30.0°$ with the beam helps support the light. (a) Draw a force diagram for the beam. By computing torques about an axis at the hinge at the left-hand end of the beam, find (b) the tension in the cable, (c) the horizontal component of the force exerted by the pole on the beam, and (d) the vertical component of this force. Now solve the same problem from the force diagram from part (a) by computing torques around the junction between the cable and the beam at the right-hand end of the beam. Find (e) the vertical component of the force exerted by the pole on the beam, (f) the tension in the cable, and (g) the horizontal component of the force exerted by the pole on the beam. (h) Compare the solution to parts (b) through (d) with the solution to parts (e) through (g). Is either solution more accurate?

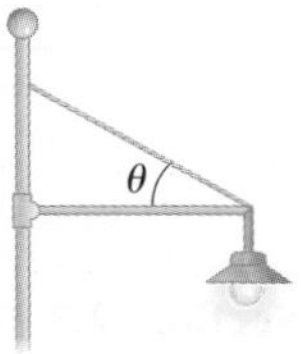

Figure P12.18

19. Sir Lost-a-Lot dons his armor and sets out from the castle on his trusty steed (Fig. P12.19). Usually, the drawbridge is lowered to a horizontal position so that the end of the bridge rests on the stone ledge. Unfortunately, Lost-a-Lot's squire didn't lower the drawbridge far enough and stopped it at $\theta = 20.0°$ above the horizontal. The knight and his horse stop when their combined center of mass is $d = 1.00$ m from the end of the bridge. The uniform bridge is $\ell = 8.00$ m long and has mass 2 000 kg. The lift cable is attached to the bridge 5.00 m from the hinge at the castle end and to a point on the castle wall $h = 12.0$ m above the bridge. Lost-a-Lot's mass combined with his armor and steed is 1 000 kg. Determine (a) the tension in the cable

and (b) the horizontal and (c) the vertical force components acting on the bridge at the hinge.

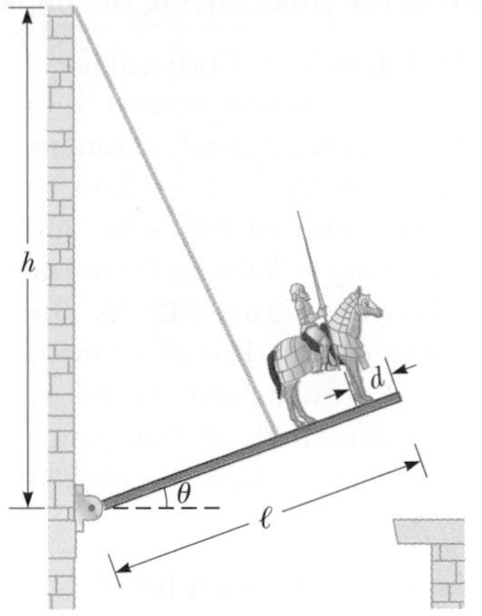

Figure P12.19 Problems 19 and 20.

20. **Review.** While Lost-a-Lot ponders his next move in the situation described in Problem 19 and illustrated in Figure P12.19, the enemy attacks! An incoming projectile breaks off the stone ledge so that the end of the drawbridge can be lowered past the wall where it usually rests. In addition, a fragment of the projectile bounces up and cuts the drawbridge cable! The hinge between the castle wall and the bridge is frictionless, and the bridge swings down freely until it is vertical and smacks into the vertical castle wall below the castle entrance. (a) How long does Lost-a-Lot stay in contact with the bridge while it swings downward? (b) Find the angular acceleration of the bridge just as it starts to move. (c) Find the angular speed of the bridge when it strikes the wall below the hinge. Find the force exerted by the hinge on the bridge (d) immediately after the cable breaks and (e) immediately before it strikes the castle wall.

21. John is pushing his daughter Rachel in a wheelbarrow when it is stopped by a brick 8.00 cm high (Fig. P12.21). The handles make an angle of $\theta = 15.0°$ with the ground. Due to the weight of Rachel and the wheelbarrow, a downward force of 400 N is exerted at the center of the wheel, which has a radius of 20.0 cm. (a) What force must John apply along the handles to just start the wheel over the brick? (b) What is the force (magnitude and direction) that the brick exerts on the wheel just as the wheel begins to lift over the brick? In both parts, assume the brick remains fixed and does not slide along the ground. Also assume the force applied by John is directed exactly toward the center of the wheel.

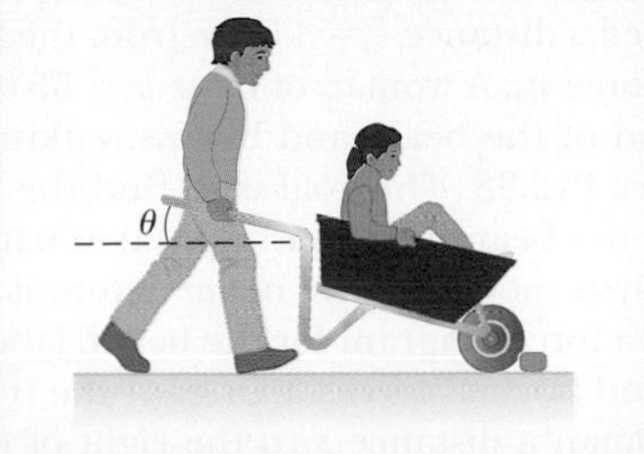

Figure P12.21 Problems 21 and 22.

22. **S** John is pushing his daughter Rachel in a wheelbarrow when it is stopped by a brick of height h (Fig. P12.21). The handles make an angle of θ with the ground. Due to the weight of Rachel and the wheelbarrow, a downward force mg is exerted at the center of the wheel, which has a radius R. (a) What force F must John apply along the handles to just start the wheel over the brick? (b) What are the components of the force that the brick exerts on the wheel just as the wheel begins to lift over the brick? In both parts, assume the brick remains fixed and does not slide along the ground. Also assume the force applied by John is directed exactly toward the center of the wheel.

23. One end of a uniform 4.00-m-long rod of weight F_g is supported by a cable at an angle of $\theta = 37°$ with the rod. The other end rests against the wall, where it is held by friction as shown in Figure P12.23. The coefficient of static friction between the wall and the rod is $\mu_s = 0.500$. Determine the minimum distance x from point A at which an additional object, also with the same weight F_g, can be hung without causing the rod to slip at point A.

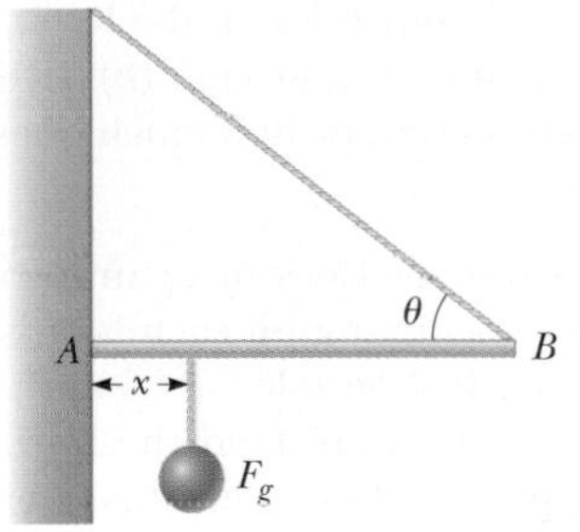

Figure P12.23

24. **Q|C** A 10.0-kg monkey climbs a uniform ladder with weight 1.20×10^2 N and length $L = 3.00$ m as shown in Figure P12.24. The ladder rests against the wall and makes an angle of $\theta = 60.0°$ with the ground. The upper and lower ends of the ladder rest on frictionless surfaces. The lower end is connected to the wall by a horizontal rope that is frayed and can support a maximum tension of only 80.0 N. (a) Draw a force diagram for the ladder. (b) Find the normal force exerted on the bottom of the ladder. (c) Find the tension in the rope when the monkey is two-thirds of the way up the ladder. (d) Find the maximum distance d that the monkey can climb up the ladder before the rope breaks. (e) If the horizontal surface were rough and the rope were removed, how would your analysis of the problem change? What other information would you need to answer parts (c) and (d)?

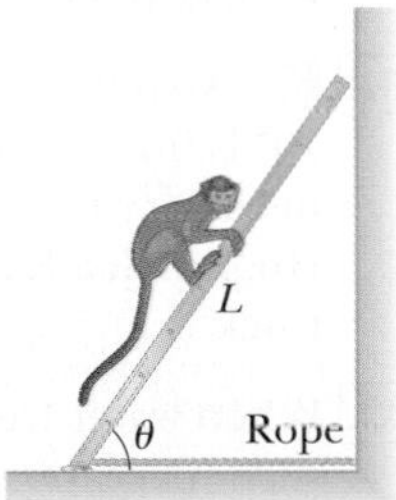

Figure P12.24

25. A uniform plank of length 2.00 m and mass 30.0 kg is supported by three ropes as indicated by the blue vectors in Figure P12.25. Find the tension in each rope when a 700-N person is $d = 0.500$ m from the left end.

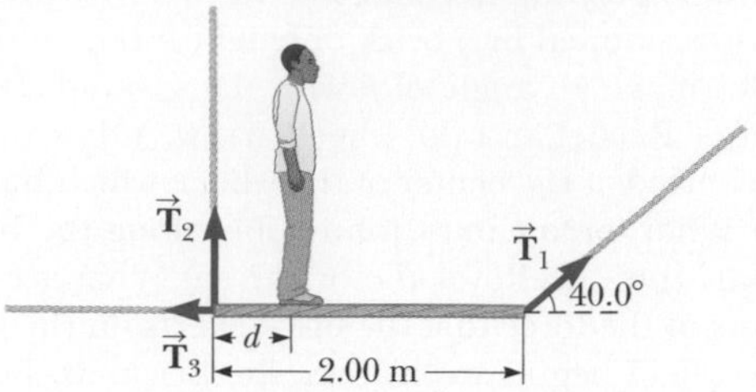

Figure P12.25

Section 12.4 Elastic Properties of Solids

26. Evaluate Young's modulus for the material whose stress–strain curve is shown in Figure 12.12.

27. A 200-kg load is hung on a wire of length 4.00 m, cross-sectional area 0.200×10^{-4} m^2, and Young's modulus 8.00×10^{10} N/m^2. What is its increase in length?

28. Assume Young's modulus for bone is 1.50×10^{10} N/m^2. The bone breaks if stress greater than 1.50×10^8 N/m^2 is imposed on it. (a) What is the maximum force that can be exerted on the femur bone in the leg if it has a minimum effective diameter of 2.50 cm? (b) If this much force is applied compressively, by how much does the 25.0-cm-long bone shorten?

29. A child slides across a floor in a pair of rubber-soled shoes. The friction force acting on each foot is 20.0 N. The footprint area of each shoe sole is 14.0 cm^2, and the thickness of each sole is 5.00 mm. Find the horizontal distance by which the upper and lower surfaces of each sole are offset. The shear modulus of the rubber is 3.00 MN/m^2.

30. A steel wire of diameter 1 mm can support a tension of 0.2 kN. A steel cable to support a tension of 20 kN should have diameter of what order of magnitude?

31. M Assume if the shear stress in steel exceeds about 4.00×10^8 N/m^2 the steel ruptures. Determine the shearing force necessary to (a) shear a steel bolt 1.00 cm in diameter and (b) punch a 1.00-cm-diameter hole in a steel plate 0.500 cm thick.

32. When water freezes, it expands by about 9.00%. What pressure increase would occur inside your automobile engine block if the water in it froze? (The bulk modulus of ice is 2.00×10^9 N/m^2.)

33. Q|C The deepest point in the ocean is in the Mariana Trench, about 11 km deep, in the Pacific. The pressure at this depth is huge, about 1.13×10^8 N/m^2. (a) Calculate the change in volume of 1.00 m^3 of seawater carried from the surface to this deepest point. (b) The density of seawater at the surface is 1.03×10^3 kg/m^3. Find its density at the bottom. (c) Explain whether or when it is a good approximation to think of water as incompressible.

34. **Review.** A 30.0-kg hammer, moving with speed 20.0 m/s, strikes a steel spike 2.30 cm in diameter. The hammer rebounds with speed 10.0 m/s after 0.110 s. What is the average strain in the spike during the impact?

35. **Review.** A 2.00-m-long cylindrical steel wire with a cross-sectional diameter of 4.00 mm is placed over a light, frictionless pulley. An object of mass $m_1 = 5.00$ kg is hung from one end of the wire and an object of mass $m_2 = 3.00$ kg from the other end as shown in Figure P12.35. The objects are released and allowed to move freely. Compared with its length before the objects were attached, by how much has the wire stretched while the objects are in motion?

Figure P12.35

36. A walkway suspended across a hotel lobby is supported at numerous points along its edges by a vertical cable above each point and a vertical column underneath. The steel cable is 1.27 cm in diameter and is 5.75 m long before loading. The aluminum column is a hollow cylinder with an inside diameter of 16.14 cm, an outside diameter of 16.24 cm, and an unloaded length of 3.25 m. When the walkway exerts a load force of 8 500 N on one of the support points, how much does the point move down?

Additional Problems

37. A bridge of length 50.0 m and mass 8.00×10^4 kg is supported on a smooth pier at each end as shown in Figure P12.37. A truck of mass 3.00×10^4 kg is located 15.0 m from one end. What are the forces on the bridge at the points of support?

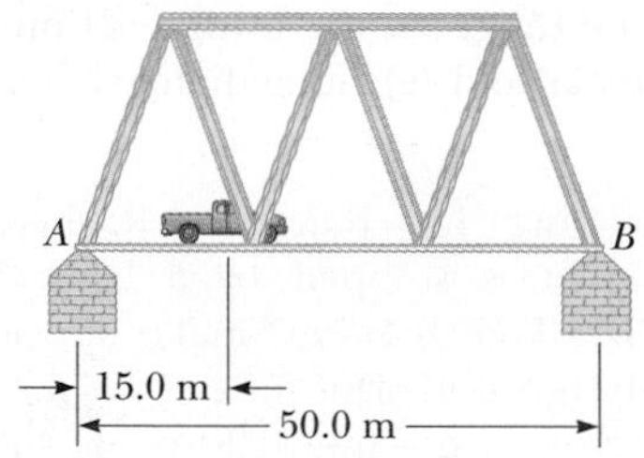

Figure P12.37

38. GP A uniform beam resting on two pivots has a length $L = 6.00$ m and mass $M = 90.0$ kg. The pivot under the left end exerts a normal force n_1 on the beam, and the second pivot located a distance $\ell = 4.00$ m from the left end exerts a normal force n_2. A woman of mass $m = 55.0$ kg steps onto the left end of the beam and begins walking to the right as in Figure P12.38. The goal is to find the woman's position when the beam begins to tip. (a) What is the appropriate analysis model for the beam before it begins to tip? (b) Sketch a force diagram for the beam, labeling the gravitational and normal forces acting on the beam and placing the woman a distance x to the right of the first pivot, which is the origin. (c) Where is the woman when the nor-

mal force n_1 is the greatest? (d) What is n_1 when the beam is about to tip? (e) Use Equation 12.1 to find the value of n_2 when the beam is about to tip. (f) Using the result of part (d) and Equation 12.2, with torques computed around the second pivot, find the woman's position x when the beam is about to tip. (g) Check the answer to part (e) by computing torques around the first pivot point.

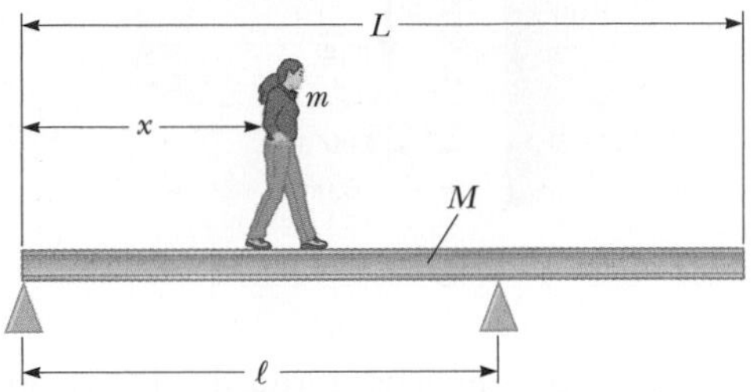

Figure P12.38

39. In exercise physiology studies, it is sometimes important to determine the location of a person's center of mass. This determination can be done with the arrangement shown in Figure P12.39. A light plank rests on two scales, which read F_{g1} = 380 N and F_{g2} = 320 N. A distance of 1.65 m separates the scales. How far from the woman's feet is her center of mass?

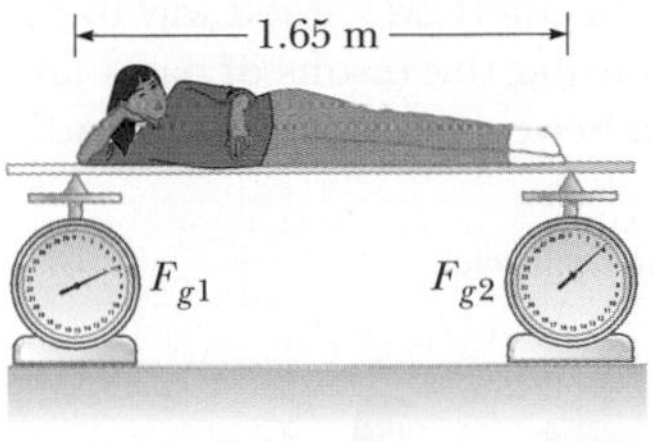

Figure P12.39

40. The lintel of prestressed reinforced concrete in Figure P12.40 is 1.50 m long. The concrete encloses one steel reinforcing rod with cross-sectional area 1.50 cm^2. The rod joins two strong end plates. The cross-sectional area of the concrete perpendicular to the rod is 50.0 cm^2. Young's modulus for the concrete is 30.0×10^9 N/m^2. After the concrete cures and the original tension T_1 in the rod is released, the concrete is to be under compressive stress 8.00×10^6 N/m^2. (a) By what distance will the rod compress the concrete when the original tension in the rod is released? (b) What is the new tension T_2 in the rod? (c) The rod will then be how much longer than its unstressed length? (d) When the concrete was poured, the rod should have been stretched by what extension distance from its unstressed length? (e) Find the required original tension T_1 in the rod.

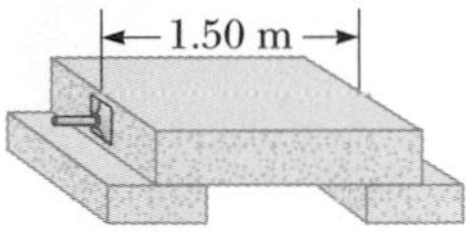

Figure P12.40

41. The arm in Figure P12.41 weighs 41.5 N. The gravitational force on the arm acts through point A. Determine the magnitudes of the tension force $\vec{F}_t$ in the deltoid muscle and the force $\vec{F}_s$ exerted by the shoulder on the humerus (upper-arm bone) to hold the arm in the position shown.

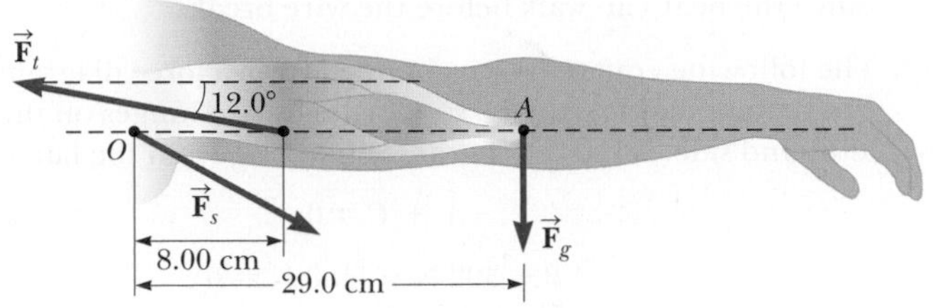

Figure P12.41

42. When a person stands on tiptoe on one foot (a strenuous position), the position of the foot is as shown in Figure P12.42a. The total gravitational force $\vec{F}_g$ on the body is supported by the normal force $\vec{n}$ exerted by the floor on the toes of one foot. A mechanical model of the situation is shown in Figure P12.42b, where $\vec{T}$ is the force exerted on the foot by the Achilles tendon and $\vec{R}$ is the force exerted on the foot by the tibia. Find the values of T, R, and θ when F_g = 700 N.

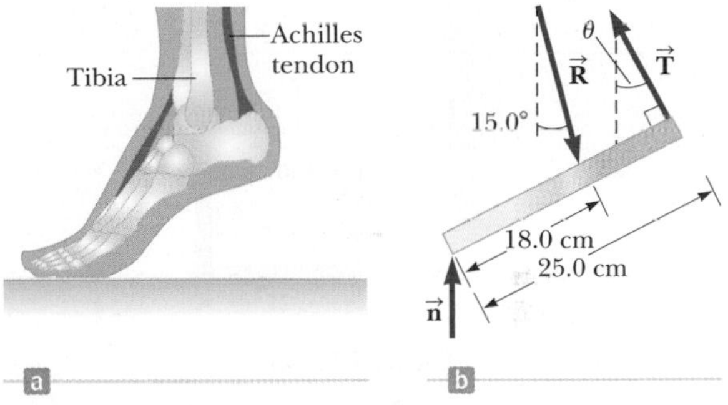

Figure P12.42

43. A hungry bear weighing 700 N walks out on a beam in an attempt to retrieve a basket of goodies hanging at the end of the beam (Fig. P12.43). The beam is uniform, weighs

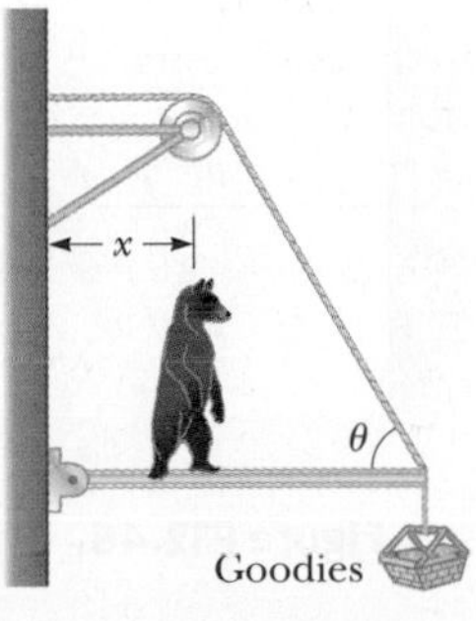

Figure P12.43

200 N, and is 6.00 m long, and it is supported by a wire at an angle of $\theta = 60.0°$. The basket weighs 80.0 N. (a) Draw a force diagram for the beam. (b) When the bear is at $x = 1.00$ m, find the tension in the wire supporting the beam and the components of the force exerted by the wall on the left end of the beam. (c) **What If?** If the wire can withstand a maximum tension of 900 N, what is the maximum distance the bear can walk before the wire breaks?

44. The following equations are obtained from a force diagram of a rectangular farm gate, supported by two hinges on the left-hand side. A bucket of grain is hanging from the latch.

$$-A + C = 0$$

$$+B - 392 \text{ N} - 50.0 \text{ N} = 0$$

$$A(0) + B(0) + C(1.80 \text{ m}) - 392 \text{ N}(1.50 \text{ m}) - 50.0 \text{ N}(3.00 \text{ m}) = 0$$

(a) Draw the force diagram and complete the statement of the problem, specifying the unknowns. (b) Determine the values of the unknowns and state the physical meaning of each.

45. **S** A uniform sign of weight F_g and width $2L$ hangs from a light, horizontal beam hinged at the wall and supported by a cable (Fig. P12.45). Determine (a) the tension in the cable and (b) the components of the reaction force exerted by the wall on the beam in terms of F_g, d, L, and θ.

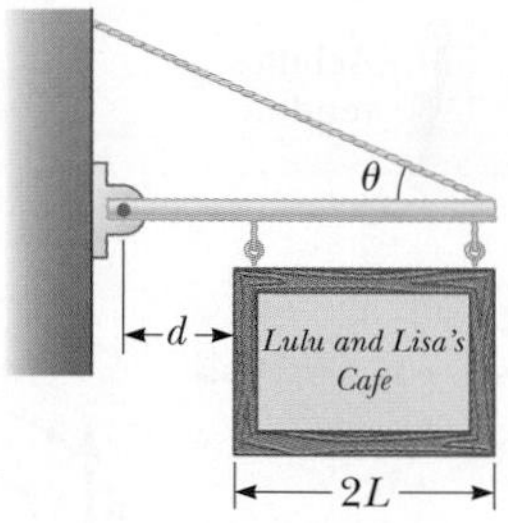

Figure P12.45

46. A 1 200-N uniform boom at $\phi = 65°$ to the vertical is supported by a cable at an angle $\theta = 25.0°$ to the horizontal as shown in Figure P12.46. The boom is pivoted at the bottom, and an object of weight $m = 2\ 000$ N hangs from its top. Find (a) the tension in the support cable and (b) the components of the reaction force exerted by the floor on the boom.

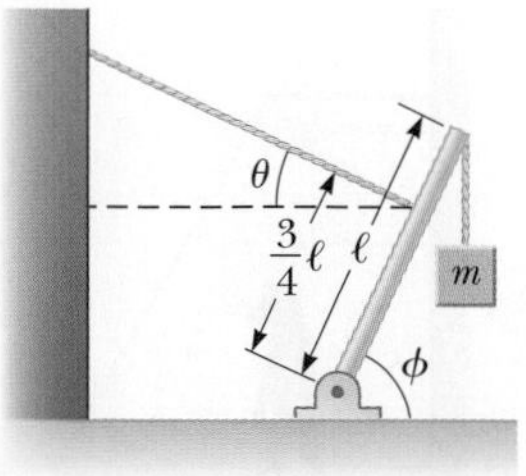

Figure P12.46

47. A crane of mass $m_1 = 3\ 000$ kg supports a load of mass $m_2 = 10\ 000$ kg as shown in Figure P12.47. The crane is pivoted with a frictionless pin at A and rests against a smooth support at B. Find the reaction forces at (a) point A and (b) point B.

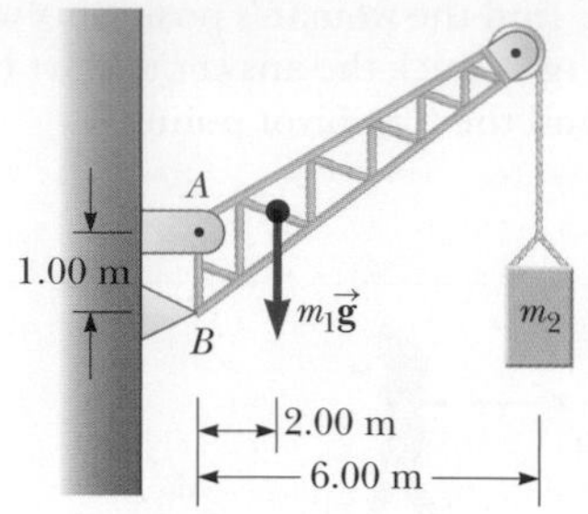

Figure P12.47

48. **Q|C** Assume a person bends forward to lift a load "with his back" as shown in Figure P12.48a. The spine pivots mainly at the fifth lumbar vertebra, with the principal supporting force provided by the erector spinalis muscle in the back. To see the magnitude of the forces involved, consider the model shown in Figure P12.48b for a person bending forward to lift a 200-N object. The spine and upper body are represented as a uniform horizontal rod of weight 350 N, pivoted at the base of the spine. The erector spinalis muscle, attached at a point two-thirds of the way up the spine, maintains the position of the back. The angle between the spine and this muscle is $\theta = 12.0°$. Find (a) the tension T in the back muscle and (b) the compressional force in the spine. (c) Is this method a good way to lift a load? Explain your answer, using the results of parts (a) and (b). (d) Can you suggest a better method to lift a load?

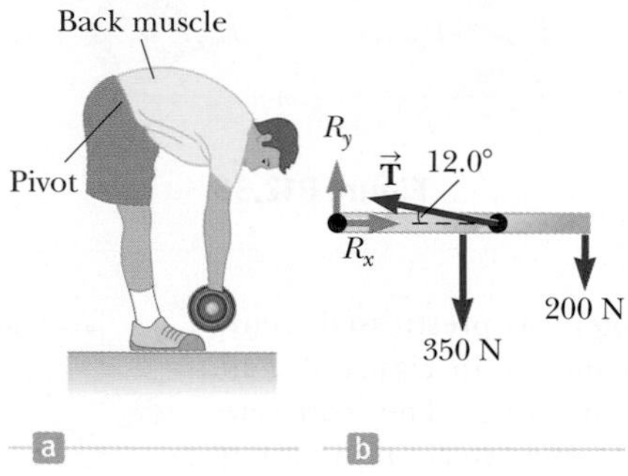

Figure P12.48

49. A 10 000-N shark is supported by a rope attached to a 4.00-m rod that can pivot at the base. (a) Calculate the tension in the cable between the rod and the wall, assuming the cable is holding the system in the position shown in Figure P12.49. Find (b) the horizontal force and (c) the vertical force exerted on the base of the rod. Ignore the weight of the rod.

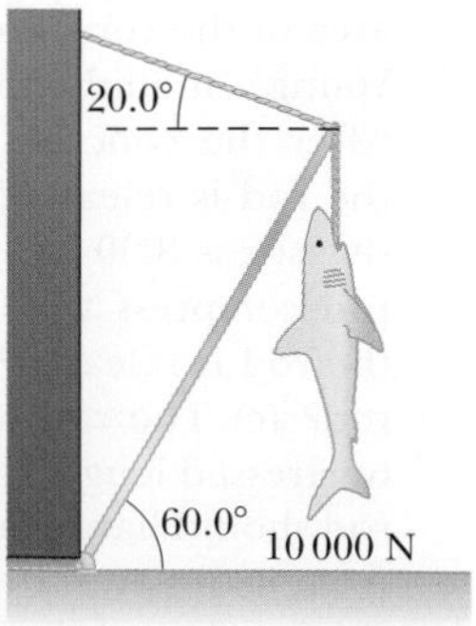

Figure P12.49

50. *Why is the following situation impossible?* A worker in a factory pulls a cabinet across the floor using a rope as shown in Figure P12.50a. The rope make an angle $\theta = 37.0°$ with the floor and is tied $h_1 = 10.0$ cm from the bottom of the cabinet. The uniform rectangular cabinet has height $\ell = 100$ cm and width $w = 60.0$ cm, and it weighs 400 N. The cabinet slides with constant speed when a force $F = 300$ N is applied through the rope. The worker tires of walking backward. He fastens the rope to a point on the cabinet $h_2 = 65.0$ cm off the floor and lays the rope over his shoulder so that he can walk forward and pull as shown in Figure P12.50b. In this way, the rope again makes an angle of $\theta = 37.0°$ with the horizontal and again has a tension of 300 N. Using this technique, the worker is able to slide the cabinet over a long distance on the floor without tiring.

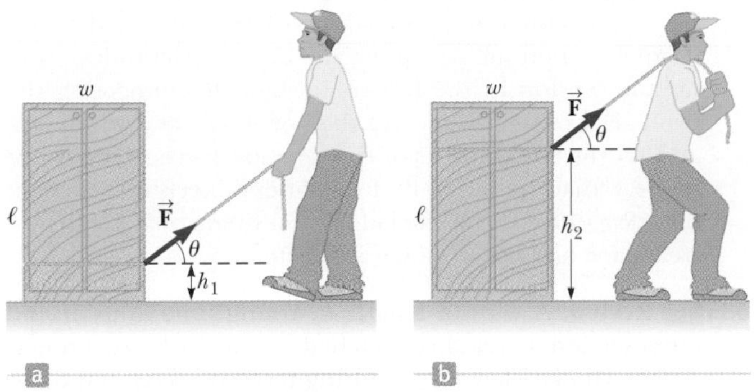

Figure P12.50 Problems 50 and 62.

51. **S** A uniform beam of mass m is inclined at an angle θ to the horizontal. Its upper end (point P) produces a 90° bend in a very rough rope tied to a wall, and its lower end rests on a rough floor (Fig. P12.51). Let μ_s represent the coefficient of static friction between beam and floor. Assume μ_s is less than the cotangent of θ. (a) Find an expression for the maximum mass M that can be suspended from the top before the beam slips. Determine (b) the magnitude of the reaction force at the floor and (c) the magnitude of the force exerted by the beam on the rope at P in terms of m, M, and μ_s.

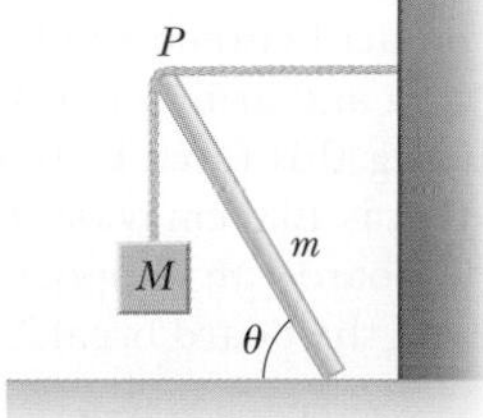

Figure P12.51

52. The large quadriceps muscle in the upper leg terminates at its lower end in a tendon attached to the upper end of the tibia (Fig. P12.52a). The forces on the lower leg when the leg is extended are modeled as in Figure P12.52b, where $\vec{\mathbf{T}}$ is the force in the tendon, $\vec{\mathbf{F}}_{g,\text{leg}}$ is the gravitational force acting on the lower leg, and $\vec{\mathbf{F}}_{g,\text{foot}}$ is the gravitational force acting on the foot. Find T when the tendon is at an angle of $\phi = 25.0°$ with the tibia, assuming $F_{g,\text{leg}} = 30.0$ N, $F_{g,\text{foot}} = 12.5$ N, and the leg is extended at an angle $\theta = 40.0°$ with respect to the vertical. Also assume the center of gravity of the tibia is at its geometric center and the tendon attaches to the lower leg at a position one-fifth of the way down the leg.

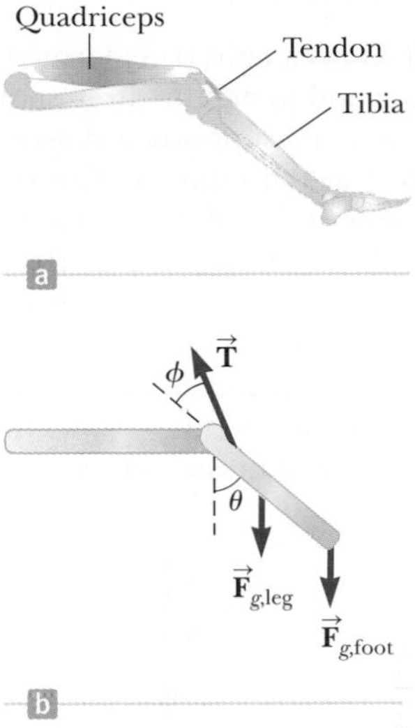

Figure P12.52

53. **Q|C** When a gymnast performing on the rings executes the *iron cross*, he maintains the position at rest shown in Figure P12.53a. In this maneuver, the gymnast's feet (not shown) are off the floor. The primary muscles involved in supporting this position are the latissimus dorsi ("lats") and the pectoralis major ("pecs"). One of the rings exerts an upward force $\vec{\mathbf{F}}_h$ on a hand as shown in Figure P12.53b.

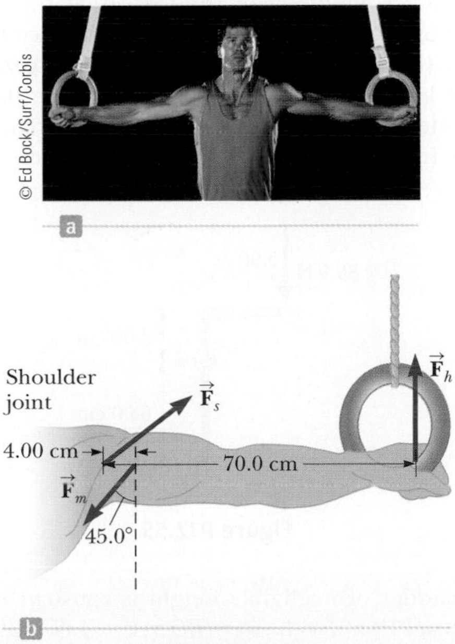

Figure P12.53

The force $\vec{F}_s$ is exerted by the shoulder joint on the arm. The latissimus dorsi and pectoralis major muscles exert a total force $\vec{F}_m$ on the arm. (a) Using the information in the figure, find the magnitude of the force $\vec{F}_m$ for an athlete of weight 750 N. (b) Suppose an athlete in training cannot perform the iron cross but can hold a position similar to the figure in which the arms make a 45° angle with the horizontal rather than being horizontal. Why is this position easier for the athlete?

54. Figure P12.54 shows a light truss formed from three struts lying in a plane and joined by three smooth hinge pins at their ends. The truss supports a downward force of $\vec{F}$ = 1 000 N applied at the point *B*. The truss has negligible weight. The piers at *A* and *C* are smooth. (a) Given θ_1 = 30.0° and θ_2 = 45.0°, find n_A and n_C. (b) One can show that the force any strut exerts on a pin must be directed along the length of the strut as a force of tension or compression. Use that fact to identify the directions of the forces that the struts exert on the pins joining them. Find the force of tension or of compression in each of the three bars.

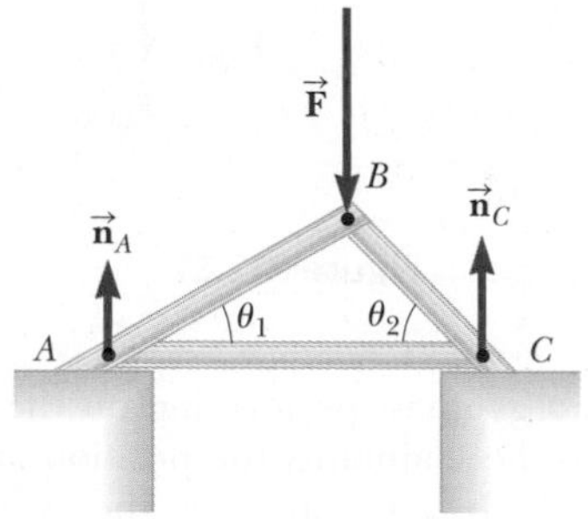

Figure P12.54

55. One side of a plant shelf is supported by a bracket mounted on a vertical wall by a single screw as shown in Figure P12.55. Ignore the weight of the bracket. (a) Find the horizontal component of the force that the screw exerts on the bracket when an 80.0 N vertical force is applied as shown. (b) As your grandfather waters his geraniums, the 80.0-N load force is increasing at the rate 0.150 N/s. At what rate is the force exerted by the screw changing? *Suggestion:* Imagine that the bracket is slightly loose.

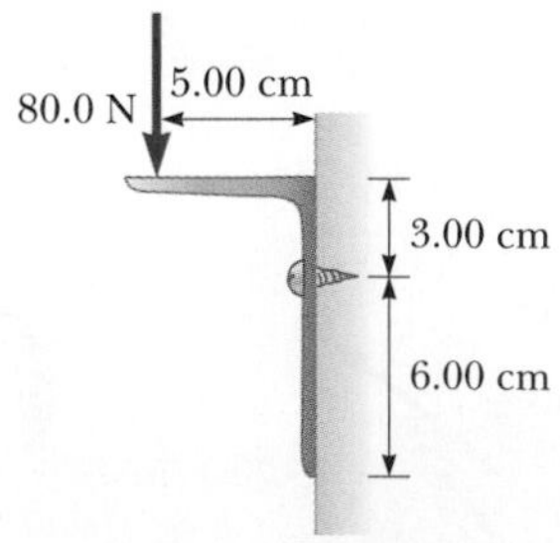

Figure P12.55

56. A stepladder of negligible weight is constructed as shown in Figure P12.56, with $AC = BC = \ell$ = 4.00 m. A painter of mass m = 70.0 kg stands on the ladder d = 3.00 m from the bottom. Assuming the floor is frictionless, find (a) the tension in the horizontal bar *DE* connecting the two halves of the ladder, (b) the normal forces at *A* and *B*, and (c) the components of the reaction force at the single hinge *C* that the left half of the ladder exerts on the right half. *Suggestion:* Treat the ladder as a single object, but also treat each half of the ladder separately.

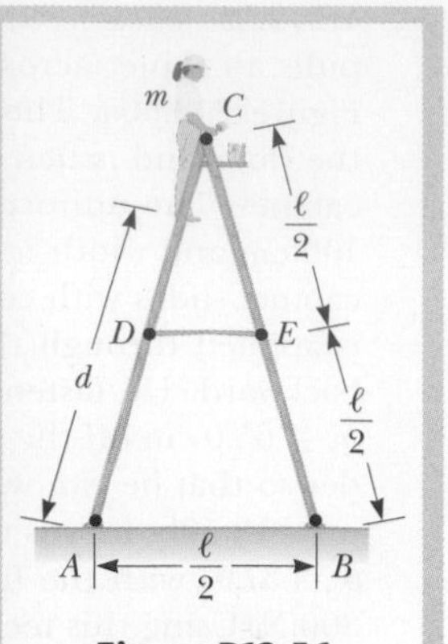

Figure P12.56
Problems 56 and 57.

57. S A stepladder of negligible weight is constructed as shown in Figure P12.56, with $AC = BC = \ell$. A painter of mass m stands on the ladder a distance d from the bottom. Assuming the floor is frictionless, find (a) the tension in the horizontal bar *DE* connecting the two halves of the ladder, (b) the normal forces at *A* and *B*, and (c) the components of the reaction force at the single hinge *C* that the left half of the ladder exerts on the right half. *Suggestion:* Treat the ladder as a single object, but also treat each half of the ladder separately.

58. S **Review.** A wire of length L, Young's modulus Y, and cross-sectional area A is stretched elastically by an amount ΔL. By Hooke's law, the restoring force is $-k\,\Delta L$. (a) Show that $k = YA/L$. (b) Show that the work done in stretching the wire by an amount ΔL is $W = \frac{1}{2}YA(\Delta L)^2/L$.

59. Two racquetballs, each having a mass of 170 g, are placed in a glass jar as shown in Figure P12.59. Their centers lie on a straight line that makes a 45° angle with the horizontal. (a) Assume the walls are frictionless and determine P_1, P_2, and P_3. (b) Determine the magnitude of the force exerted by the left ball on the right ball.

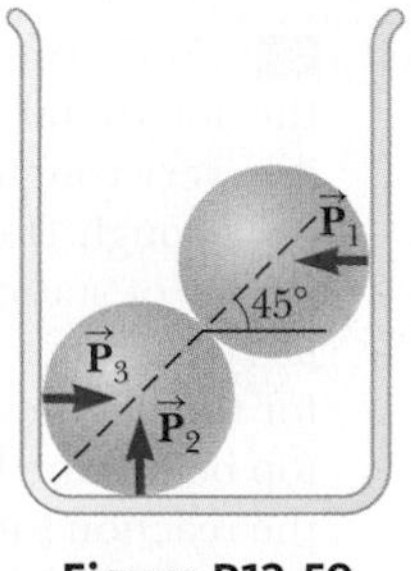

Figure P12.59

60. (a) Estimate the force with which a karate master strikes a board, assuming the hand's speed at the moment of impact is 10.0 m/s and decreases to 1.00 m/s during a 0.002 00-s time interval of contact between the hand and the board. The mass of his hand and arm is 1.00 kg. (b) Estimate the shear stress, assuming this force is exerted on a 1.00-cm-thick pine board that is 10.0 cm wide. (c) If the maximum shear stress a pine board can support before breaking is 3.60×10^6 N/m^2, will the board break?

61. **Review.** An aluminum wire is 0.850 m long and has a circular cross section of diameter 0.780 mm. Fixed at the top end, the wire supports a 1.20-kg object that swings in a horizontal circle. Determine the angular velocity of the object required to produce a strain of 1.00×10^{-3}.

62. Consider the rectangular cabinet of Problem 50 shown in Figure P12.50, but with a force $\vec{F}$ applied horizontally at the upper edge. (a) What is the minimum force required to start to tip the cabinet? (b) What is the minimum coef-

ficient of static friction required for the cabinet not to slide with the application of a force of this magnitude? (c) Find the magnitude and direction of the minimum force required to tip the cabinet if the point of application can be chosen *anywhere* on the cabinet.

63. A steel cable 3.00 cm^2 in cross-sectional area has a mass of 2.40 kg per meter of length. If 500 m of the cable is hung over a vertical cliff, how much does the cable stretch under its own weight? Take $Y_{\text{steel}} = 2.00 \times 10^{11}\ \text{N/m}^2$.

Challenge Problems

64. A uniform rod of weight F_g and length L is supported at its ends by a frictionless trough as shown in Figure P12.64. (a) Show that the center of gravity of the rod must be vertically over point O when the rod is in equilibrium. (b) Determine the equilibrium value of the angle θ. (c) Is the equilibrium of the rod stable or unstable?

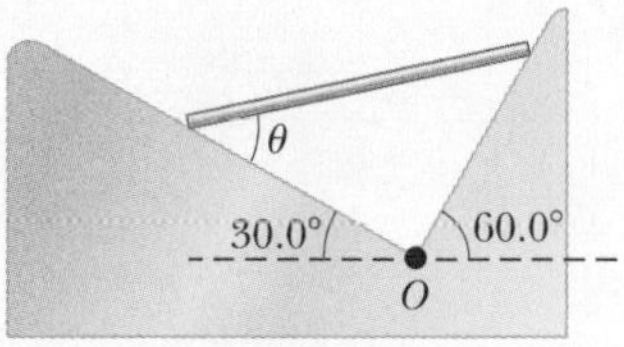

Figure P12.64

65. A uniform pole is propped between the floor and the ceiling of a room. The height of the room is 7.80 ft, and the coefficient of static friction between the pole and the ceiling is 0.576. The coefficient of static friction between the pole and the floor is greater than that between the pole and the ceiling. What is the length of the longest pole that can be propped between the floor and the ceiling?

66. In the What If? section of Example 12.2, let d represent the distance in meters between the person and the hinge at the left end of the beam. (a) Show that the cable tension is given by $T = 93.9d + 125$, with T in newtons. (b) Show that the direction angle θ of the hinge force is described by

$$\tan\theta = \left(\frac{32}{3d+4} - 1\right)\tan 53.0^\circ$$

(c) Show that the magnitude of the hinge force is given by

$$R = \sqrt{8.82 \times 10^3 d^2 - 9.65 \times 10^4 d + 4.96 \times 10^5}$$

(d) Describe how the changes in T, θ, and R as d increases differ from one another.

67. **QC** Figure P12.67 shows a vertical force applied tangentially to a uniform cylinder of weight F_g. The coefficient of static friction between the cylinder and all surfaces is 0.500. The force $\vec{\mathbf{P}}$ is increased in magnitude until the cylinder begins to rotate. In terms of F_g, find the maximum force magnitude P that can be applied without causing the cylinder to rotate. *Suggestion:* Show that both friction forces will be at their maximum values when the cylinder is on the verge of slipping.

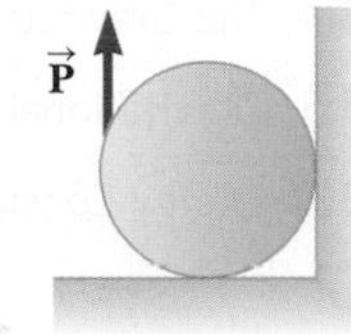

Figure P12.67

chapter 13

Universal Gravitation

The Whirlpool Galaxy (left) was discovered in 1774 by Charles Messier and is listed in his astronomical catalog as M51. It is known for its sharply defined spiral arms. The smaller companion galaxy (right) is at the same distance from Earth as the Whirlpool Galaxy. The visual appearance of the pair indicates a gravitational interaction between them. *(NASA, ESA, S. Beckwith (STScI), and The Hubble Heritage Team (STScI/AURA))*

Before 1687, a large amount of data had been collected on the motions of the Moon and the planets, but a clear understanding of the forces related to these motions was not available. In that year, Isaac Newton provided the key that unlocked the secrets of the heavens. He knew, from his first law, that a net force had to be acting on the Moon because without such a force the Moon would move in a straight-line path rather than in its almost circular orbit. Newton reasoned that this force was the gravitational attraction exerted by the Earth on the Moon. He realized that the forces involved in the Earth–Moon attraction and in the Sun–planet attraction were not something special to those systems, but rather were particular cases of a general and universal attraction between objects. In other words, Newton saw that the same force of attraction that causes the Moon to follow its path around the Earth also causes an apple to fall from a tree. It was the first time that "earthly" and "heavenly" motions were unified.

In this chapter, we study the law of universal gravitation. We emphasize a description of planetary motion because astronomical data provide an important test of this law's validity. We then show that the laws of planetary motion developed by Johannes Kepler follow from the law of universal gravitation and the principle of conservation of angular momentum. We conclude by deriving a general expression for the gravitational potential energy of a system and examining the energetics of planetary and satellite motion.

13.1 Newton's Law of Universal Gravitation

You may have heard the legend that, while napping under a tree, Newton was struck on the head by a falling apple. This alleged accident supposedly prompted him to imagine that perhaps all objects in the Universe were attracted to each other in the same way the apple was attracted to the Earth. Newton analyzed astronomical data on the motion of the Moon around the Earth. From that analysis, he made the bold assertion that the force law governing the motion of planets was the *same* as the force law that attracted a falling apple to the Earth.

In 1687, Newton published his work on the law of gravity in his treatise *Mathematical Principles of Natural Philosophy.* **Newton's law of universal gravitation** states that

◀ **The law of universal gravitation**

> every particle in the Universe attracts every other particle with a force that is directly proportional to the product of their masses and inversely proportional to the square of the distance between them.

If the particles have masses m_1 and m_2 and are separated by a distance r, the magnitude of this gravitational force is

$$F_g = G\frac{m_1 m_2}{r^2} \quad \text{(13.1)}$$

where G is a constant, called the *universal gravitational constant.* Its value in SI units is

$$G = 6.674 \times 10^{-11}\ \text{N}\cdot\text{m}^2/\text{kg}^2 \quad \text{(13.2)}$$

Henry Cavendish (1731–1810) measured the universal gravitational constant in an important 1798 experiment. Cavendish's apparatus consists of two small spheres, each of mass m, fixed to the ends of a light, horizontal rod suspended by a fine fiber or thin metal wire as illustrated in Figure 13.1. When two large spheres, each of mass M, are placed near the smaller ones, the attractive force between smaller and larger spheres causes the rod to rotate and twist the wire suspension to a new equilibrium orientation. The angle of rotation is measured by the deflection of a light beam reflected from a mirror attached to the vertical suspension.

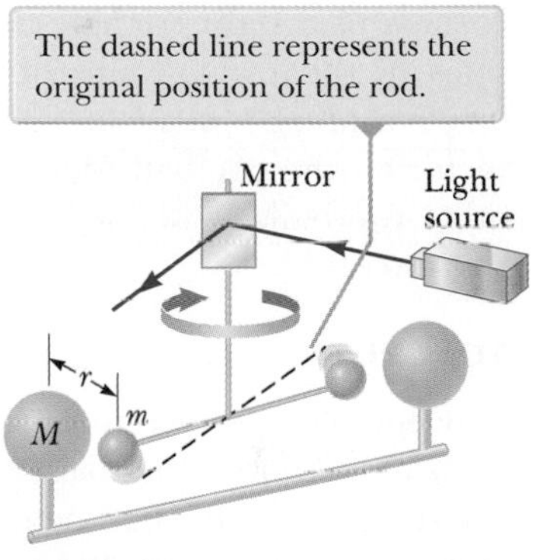

Figure 13.1 Cavendish apparatus for measuring G.

The form of the force law given by Equation 13.1 is often referred to as an **inverse-square law** because the magnitude of the force varies as the inverse square of the separation of the particles.[1] We shall see other examples of this type of force law in subsequent chapters. We can express this force in vector form by defining a unit vector $\hat{\mathbf{r}}_{12}$ (Active Fig. 13.2). Because this unit vector is directed from particle 1 toward particle 2, the force exerted by particle 1 on particle 2 is

$$\vec{\mathbf{F}}_{12} = -G\frac{m_1 m_2}{r^2}\hat{\mathbf{r}}_{12} \quad \text{(13.3)}$$

where the negative sign indicates that particle 2 is attracted to particle 1; hence, the force on particle 2 must be directed toward particle 1. By Newton's third law, the force exerted by particle 2 on particle 1, designated $\vec{\mathbf{F}}_{21}$, is equal in magnitude to $\vec{\mathbf{F}}_{12}$ and in the opposite direction. That is, these forces form an action–reaction pair, and $\vec{\mathbf{F}}_{21} = -\vec{\mathbf{F}}_{12}$.

Two features of Equation 13.3 deserve mention. First, the gravitational force is a field force that always exists between two particles, regardless of the medium that separates them. Because the force varies as the inverse square of the distance between the particles, it decreases rapidly with increasing separation.

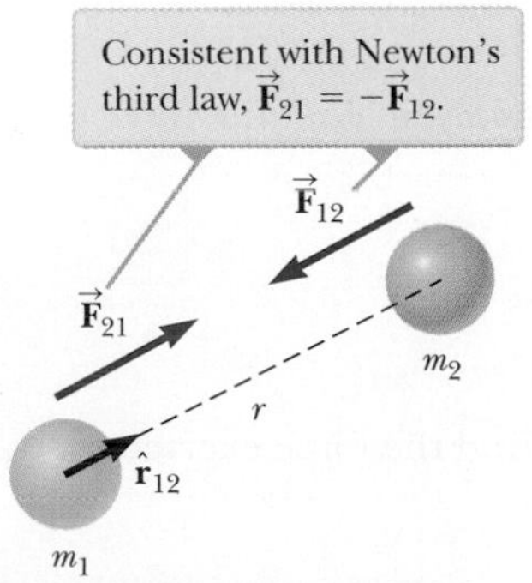

ACTIVE FIGURE 13.2

The gravitational force between two particles is attractive. The unit vector $\hat{\mathbf{r}}_{12}$ is directed from particle 1 toward particle 2.

[1]An *inverse* proportionality between two quantities x and y is one in which $y = k/x$, where k is a constant. A *direct* proportion between x and y exists when $y = kx$.

Pitfall Prevention 13.1

Be Clear on *g* and *G*

The symbol g represents the magnitude of the free-fall acceleration near a planet. At the surface of the Earth, g has an average value of 9.80 m/s^2. On the other hand, G is a universal constant that has the same value everywhere in the Universe.

Equation 13.3 can also be used to show that the gravitational force exerted by a finite-size, spherically symmetric mass distribution on a particle outside the distribution is the same as if the entire mass of the distribution were concentrated at the center. For example, the magnitude of the force exerted by the Earth on a particle of mass m near the Earth's surface is

$$F_g = G\frac{M_E m}{R_E^{\,2}} \tag{13.4}$$

where M_E is the Earth's mass and R_E its radius. This force is directed toward the center of the Earth.

***Quick Quiz* 13.1** A planet has two moons of equal mass. Moon 1 is in a circular orbit of radius r. Moon 2 is in a circular orbit of radius $2r$. What is the magnitude of the gravitational force exerted by the planet on Moon 2? **(a)** four times as large as that on Moon 1 **(b)** twice as large as that on Moon 1 **(c)** equal to that on Moon 1 **(d)** half as large as that on Moon 1 **(e)** one-fourth as large as that on Moon 1

Example 13.1 Billiards, Anyone?

Three 0.300-kg billiard balls are placed on a table at the corners of a right triangle as shown in Figure 13.3. The sides of the triangle are of lengths a = 0.400 m, b = 0.300 m, and c = 0.500 m. Calculate the gravitational force vector on the cue ball (designated m_1) resulting from the other two balls as well as the magnitude and direction of this force.

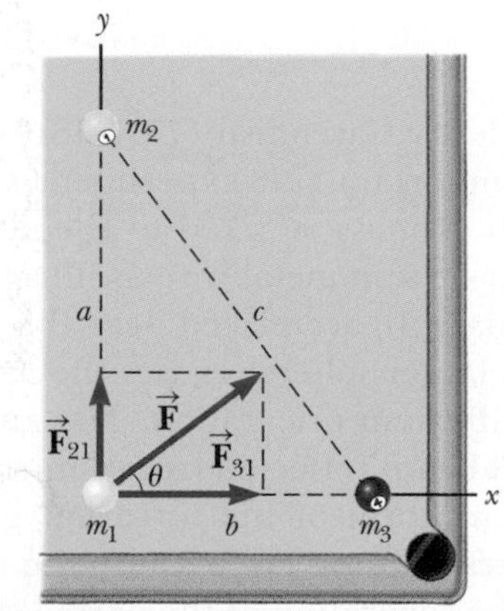

Figure 13.3 (Example 13.1) The resultant gravitational force acting on the cue ball is the vector sum $\vec{F}_{21} + \vec{F}_{31}$.

SOLUTION

Conceptualize Notice in Figure 13.3 that the cue ball is attracted to both other balls by the gravitational force. We can see graphically that the net force should point upward and toward the right. We locate our coordinate axes as shown in Figure 13.3, placing our origin at the position of the cue ball.

Categorize This problem involves evaluating the gravitational forces on the cue ball using Equation 13.3. Once these forces are evaluated, it becomes a vector addition problem to find the net force.

Analyze Find the force exerted by m_2 on the cue ball:

$$\vec{F}_{21} = G\frac{m_2 m_1}{a^2}\hat{j}$$

$$= (6.67 \times 10^{-11}\ \text{N}\cdot\text{m}^2/\text{kg}^2)\frac{(0.300\ \text{kg})(0.300\ \text{kg})}{(0.400\ \text{m})^2}\hat{j}$$

$$= 3.75 \times 10^{-11}\hat{j}\ \text{N}$$

Find the force exerted by m_3 on the cue ball:

$$\vec{F}_{31} = G\frac{m_3 m_1}{b^2}\hat{i}$$

$$= (6.67 \times 10^{-11}\ \text{N}\cdot\text{m}^2/\text{kg}^2)\frac{(0.300\ \text{kg})(0.300\ \text{kg})}{(0.300\ \text{m})^2}\hat{i}$$

$$= 6.67 \times 10^{-11}\hat{i}\ \text{N}$$

Find the net gravitational force on the cue ball by adding these force vectors:

$$\vec{F} = \vec{F}_{31} + \vec{F}_{21} = (6.67\hat{i} + 3.75\hat{j}) \times 10^{-11}\ \text{N}$$

13.1 cont.

Find the magnitude of this force: $F = \sqrt{F_{31}^2 + F_{21}^2} = \sqrt{(6.67)^2 + (3.75)^2} \times 10^{-11}$ N

$= 7.66 \times 10^{-11}$ N

Find the tangent of the angle θ for the net force vector: $\tan\theta = \frac{F_y}{F_x} = \frac{F_{21}}{F_{31}} = \frac{3.75 \times 10^{-11}\text{ N}}{6.67 \times 10^{-11}\text{ N}} = 0.562$

Evaluate the angle θ: $\theta = \tan^{-1}(0.562) = 29.4°$

Finalize The result for F shows that the gravitational forces between everyday objects have extremely small magnitudes.

13.2 Free-Fall Acceleration and the Gravitational Force

We have called the magnitude of the gravitational force on an object near the Earth's surface the *weight* of the object, where the weight is given by Equation 5.6. Equation 13.4 is another expression for this force. Therefore, we can set Equations 5.6 and 13.4 equal to each other to obtain

$$mg = G\frac{M_E m}{R_E^{\,2}}$$

$$g = G\frac{M_E}{R_E^{\,2}} \qquad \textbf{(13.5)}$$

Equation 13.5 relates the free-fall acceleration g to physical parameters of the Earth—its mass and radius—and explains the origin of the value of 9.80 m/s² that we have used in earlier chapters. Now consider an object of mass m located a distance h above the Earth's surface or a distance r from the Earth's center, where $r = R_E + h$. The magnitude of the gravitational force acting on this object is

$$F_g = G\frac{M_E m}{r^2} = G\frac{M_E m}{(R_E + h)^2}$$

The magnitude of the gravitational force acting on the object at this position is also $F_g = mg$, where g is the value of the free-fall acceleration at the altitude h. Substituting this expression for F_g into the last equation shows that g is given by

$$g = \frac{GM_E}{r^2} = \frac{GM_E}{(R_E + h)^2} \qquad \textbf{(13.6)}$$

◀ **Variation of *g* with altitude**

Therefore, it follows that *g decreases* with *increasing altitude.* Values of g at various altitudes are listed in Table 13.1. Because an object's weight is mg, we see that as $r \rightarrow \infty$, the weight of the object approaches zero.

TABLE 13.1

Free-Fall Acceleration g at Various Altitudes Above the Earth's Surface

Altitude h (km)	g (m/s²)
1 000	7.33
2 000	5.68
3 000	4.53
4 000	3.70
5 000	3.08
6 000	2.60
7 000	2.23
8 000	1.93
9 000	1.69
10 000	1.49
50 000	0.13
∞	0

***Quick Quiz* 13.2** Superman stands on top of a very tall mountain and throws a baseball horizontally with a speed such that the baseball goes into a circular orbit around the Earth. While the baseball is in orbit, what is the magnitude of the acceleration of the ball? **(a)** It depends on how fast the baseball is thrown. **(b)** It is zero because the ball does not fall to the ground. **(c)** It is slightly less than 9.80 m/s². **(d)** It is equal to 9.80 m/s².

Example 13.2 Variation of g with Altitude h

The International Space Station operates at an altitude of 350 km. Plans for the final construction show that material of weight 4.22×10^6 N, measured at the Earth's surface, will have been lifted off the surface by various spacecraft. What is the weight of the space station when in orbit?

SOLUTION

Conceptualize The mass of the space station is fixed; it is independent of its location. Based on the discussion in this section, we realize that the value of g will be reduced at the height of the space station's orbit. Therefore, its weight will be smaller than that at the surface of the Earth.

Categorize This example is a relatively simple substitution problem.

Find the mass of the space station from its weight at the surface of the Earth:

$$m = \frac{F_g}{g} = \frac{4.22 \times 10^6 \text{ N}}{9.80 \text{ m/s}^2} = 4.31 \times 10^5 \text{ kg}$$

Use Equation 13.6 with $h = 350$ km to find g at the orbital location:

$$g = \frac{GM_E}{(R_E + h)^2}$$

$$= \frac{(6.67 \times 10^{-11} \text{ N} \cdot \text{m}^2/\text{kg}^2)(5.97 \times 10^{24} \text{ kg})}{(6.37 \times 10^6 \text{ m} + 0.350 \times 10^6 \text{ m})^2} = 8.82 \text{ m/s}^2$$

Use this value of g to find the space station's weight in orbit:

$$mg = (4.31 \times 10^5 \text{ kg})(8.82 \text{ m/s}^2) = \boxed{3.80 \times 10^6 \text{ N}}$$

Example 13.3 The Density of the Earth

Using the known radius of the Earth and that $g = 9.80$ m/s² at the Earth's surface, find the average density of the Earth.

SOLUTION

Conceptualize Assume the Earth is a perfect sphere. The density of material in the Earth varies, but let's adopt a simplified model in which we assume the density to be uniform throughout the Earth. The resulting density is the average density of the Earth.

Categorize This example is a relatively simple substitution problem.

Solve Equation 13.5 for the mass of the Earth:

$$M_E = \frac{gR_E^2}{G}$$

Substitute this mass into the definition of density (Eq. 1.1):

$$\rho_E = \frac{M_E}{V_E} = \frac{gR_E^2/G}{\frac{4}{3}\pi R_E^3} = \frac{3}{4}\frac{g}{\pi G R_E}$$

$$= \frac{3}{4}\frac{9.80 \text{ m/s}^2}{\pi(6.67 \times 10^{-11} \text{ N} \cdot \text{m}^2/\text{kg}^2)(6.37 \times 10^6 \text{ m})} = \boxed{5.51 \times 10^3 \text{ kg/m}^3}$$

WHAT IF? What if you were told that a typical density of granite at the Earth's surface is 2.75×10^3 kg/m³. What would you conclude about the density of the material in the Earth's interior?

Answer Because this value is about half the density we calculated as an average for the entire Earth, we would conclude that the inner core of the Earth has a density much higher than the average value. It is most amazing that the Cavendish experiment—which determines G and can be done on a tabletop—combined with simple free-fall measurements of g provides information about the core of the Earth!

13.3 Kepler's Laws and the Motion of Planets

Humans have observed the movements of the planets, stars, and other celestial objects for thousands of years. In early history, these observations led scientists to regard the Earth as the center of the Universe. This *geocentric model* was elaborated and formalized by the Greek astronomer Claudius Ptolemy (c. 100–c. 170) in the second century and was accepted for the next 1400 years. In 1543, Polish astronomer Nicolaus Copernicus (1473–1543) suggested that the Earth and the other planets revolved in circular orbits around the Sun (the *heliocentric model*).

Danish astronomer Tycho Brahe (1546–1601) wanted to determine how the heavens were constructed and pursued a project to determine the positions of both stars and planets. Those observations of the planets and 777 stars visible to the naked eye were carried out with only a large sextant and a compass. (The telescope had not yet been invented.)

German astronomer Johannes Kepler was Brahe's assistant for a short while before Brahe's death, whereupon he acquired his mentor's astronomical data and spent 16 years trying to deduce a mathematical model for the motion of the planets. Such data are difficult to sort out because the moving planets are observed from a moving Earth. After many laborious calculations, Kepler found that Brahe's data on the revolution of Mars around the Sun led to a successful model.

Kepler's complete analysis of planetary motion is summarized in three statements known as **Kepler's laws:**

Erich Lessing/Art Resource, NY

Johannes Kepler
German astronomer (1571–1630)
Kepler is best known for developing the laws of planetary motion based on the careful observations of Tycho Brahe.

◀ **Kepler's laws**

1. All planets move in elliptical orbits with the Sun at one focus.
2. The radius vector drawn from the Sun to a planet sweeps out equal areas in equal time intervals.
3. The square of the orbital period of any planet is proportional to the cube of the semimajor axis of the elliptical orbit.

Kepler's First Law

We are familiar with circular orbits of objects around gravitational force centers from our discussions in this chapter. Kepler's first law indicates that the circular orbit is a very special case and elliptical orbits are the general situation. This notion was difficult for scientists of the time to accept because they believed that perfect circular orbits of the planets reflected the perfection of heaven.

Active Figure 13.4 shows the geometry of an ellipse, which serves as our model for the elliptical orbit of a planet. An ellipse is mathematically defined by choosing two points F_1 and F_2, each of which is a called a **focus,** and then drawing a curve through points for which the sum of the distances r_1 and r_2 from F_1 and F_2, respectively, is a constant. The longest distance through the center between points on the ellipse (and passing through each focus) is called the **major axis,** and this distance is $2a$. In Active Figure 13.4, the major axis is drawn along the x direction. The distance a is called the **semimajor axis.** Similarly, the shortest distance through the center between points on the ellipse is called the **minor axis** of length $2b$, where the distance b is the **semiminor axis.** Either focus of the ellipse is located at a distance c from the center of the ellipse, where $a^2 = b^2 + c^2$. In the elliptical orbit of a planet around the Sun, the Sun is at one focus of the ellipse. There is nothing at the other focus.

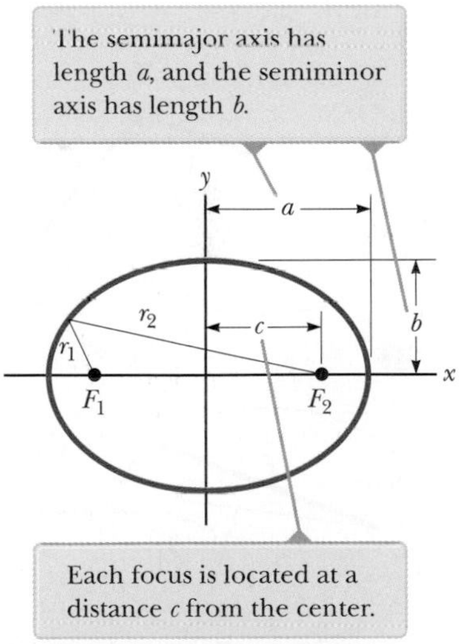

ACTIVE FIGURE 13.4
Plot of an ellipse.

The **eccentricity** of an ellipse is defined as $e = c/a$, and it describes the general shape of the ellipse. For a circle, $c = 0$, and the eccentricity is therefore zero. The smaller b is compared with a, the shorter the ellipse is along the y direction compared with its extent in the x direction in Active Figure 13.4. As b decreases, c increases and the eccentricity e increases. Therefore, higher values of eccentricity correspond to longer and thinner ellipses. The range of values of the eccentricity for an ellipse is $0 < e < 1$.

Pitfall Prevention 13.2
Where Is the Sun?
The Sun is located at one focus of the elliptical orbit of a planet. It is *not* located at the center of the ellipse.

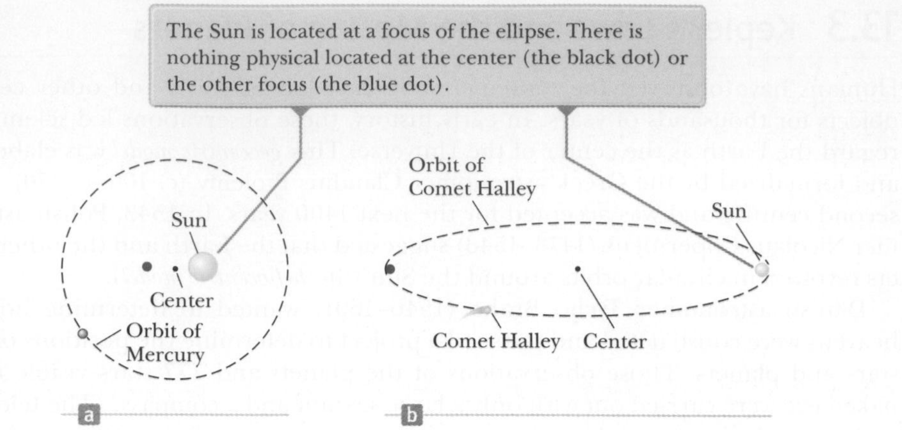

Figure 13.5 (a) The shape of the orbit of Mercury, which has the highest eccentricity ($e = 0.21$) among the eight planets in the solar system. (b) The shape of the orbit of Comet Halley. The shape of the orbit is correct; the comet and the Sun are shown larger than in reality for clarity.

Eccentricities for planetary orbits vary widely in the solar system. The eccentricity of the Earth's orbit is 0.017, which makes it nearly circular. On the other hand, the eccentricity of Mercury's orbit is 0.21, the highest of the eight planets. Figure 13.5a shows an ellipse with an eccentricity equal to that of Mercury's orbit. Notice that even this highest-eccentricity orbit is difficult to distinguish from a circle, which is one reason Kepler's first law is an admirable accomplishment. The eccentricity of the orbit of Comet Halley is 0.97, describing an orbit whose major axis is much longer than its minor axis, as shown in Figure 13.5b. As a result, Comet Halley spends much of its 76-year period far from the Sun and invisible from the Earth. It is only visible to the naked eye during a small part of its orbit when it is near the Sun.

Now imagine a planet in an elliptical orbit such as that shown in Active Figure 13.4, with the Sun at focus F_2. When the planet is at the far left in the diagram, the distance between the planet and the Sun is $a + c$. At this point, called the *aphelion*, the planet is at its maximum distance from the Sun. (For an object in orbit around the Earth, this point is called the *apogee*.) Conversely, when the planet is at the right end of the ellipse, the distance between the planet and the Sun is $a - c$. At this point, called the *perihelion* (for an Earth orbit, the *perigee*), the planet is at its minimum distance from the Sun.

Kepler's first law is a direct result of the inverse-square nature of the gravitational force. We have already discussed circular and elliptical orbits, the allowed shapes of orbits for objects that are *bound* to the gravitational force center. These objects include planets, asteroids, and comets that move repeatedly around the Sun as well as moons orbiting a planet. There are also *unbound* objects, such as a meteoroid from deep space that might pass by the Sun once and then never return. The gravitational force between the Sun and these objects also varies as the inverse square of the separation distance, and the allowed paths for these objects include parabolas ($e = 1$) and hyperbolas ($e > 1$).

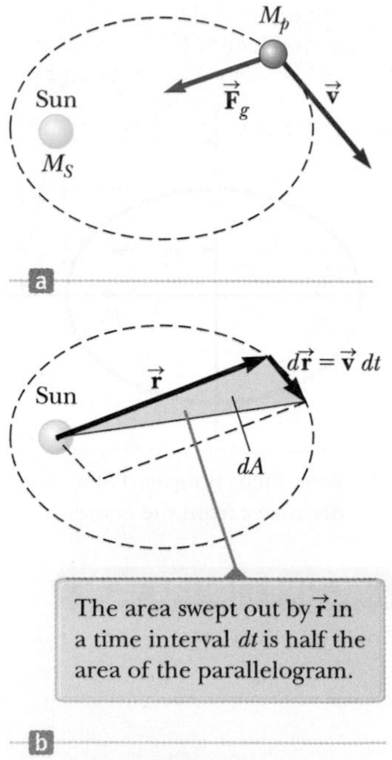

ACTIVE FIGURE 13.6

(a) The gravitational force acting on a planet is directed toward the Sun. (b) During a time interval dt, a parallelogram is formed by the vectors $\vec{r}$ and $d\vec{r} = \vec{v}\,dt$.

Kepler's Second Law

Kepler's second law can be shown to be a consequence of angular momentum conservation for an isolated system as follows. Consider a planet of mass M_p moving about the Sun in an elliptical orbit (Active Fig. 13.6a). Let us consider the planet as a system. We model the Sun to be so much more massive than the planet that the Sun does not move. The gravitational force exerted by the Sun on the planet is a central force, always along the radius vector, directed toward the Sun (Active Fig. 13.6a). The torque on the planet due to this central force is clearly zero because $\vec{F}_g$ is parallel to $\vec{r}$.

Recall that the net external torque on a system equals the time rate of change of angular momentum of the system; that is, $\sum \vec{\tau}_{ext} = d\vec{L}/dt$ (Eq. 11.13). Therefore, because the external torque on the planet is zero, it is modeled as an isolated system for angular momentum, and the angular momentum $\vec{L}$ of the planet is a constant of the motion:

$$\vec{\mathbf{L}} = \vec{\mathbf{r}} \times \vec{\mathbf{p}} = M_p \vec{\mathbf{r}} \times \vec{\mathbf{v}} = \text{constant}$$

We can relate this result to the following geometric consideration. In a time interval dt, the radius vector $\vec{\mathbf{r}}$ in Active Figure 13.6b sweeps out the area dA, which equals half the area $|\vec{\mathbf{r}} \times d\vec{\mathbf{r}}|$ of the parallelogram formed by the vectors $\vec{\mathbf{r}}$ and $d\vec{\mathbf{r}}$. Because the displacement of the planet in the time interval dt is given by $d\vec{\mathbf{r}} = \vec{\mathbf{v}}\,dt$,

$$dA = \tfrac{1}{2}|\vec{\mathbf{r}} \times d\vec{\mathbf{r}}| = \tfrac{1}{2}|\vec{\mathbf{r}} \times \vec{\mathbf{v}}\,dt| = \frac{L}{2M_p}\,dt$$

$$\frac{dA}{dt} = \frac{L}{2M_p} \qquad (13.7)$$

where L and M_p are both constants. This result shows that that the radius vector from the Sun to any planet sweeps out equal areas in equal time intervals as stated in Kepler's second law.

This conclusion is a result of the gravitational force being a central force, which in turn implies that angular momentum of the planet is constant. Therefore, the law applies to *any* situation that involves a central force, whether inverse square or not.

Kepler's Third Law

Kepler's third law can be predicted from the inverse-square law for circular orbits. Consider a planet of mass M_p that is assumed to be moving about the Sun (mass M_S) in a circular orbit as in Figure 13.7. Because the gravitational force provides the centripetal acceleration of the planet as it moves in a circle, we model the planet as a particle under a net force and as a particle in uniform circular motion and incorporate Newton's law of universal gravitation,

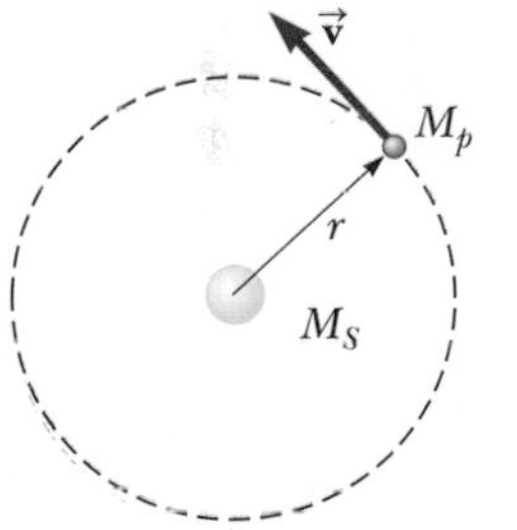

Figure 13.7 A planet of mass M_p moving in a circular orbit around the Sun. The orbits of all planets except Mercury are nearly circular.

$$F_g = M_p a \quad \rightarrow \quad \frac{GM_S M_p}{r^2} = M_p\left(\frac{v^2}{r}\right)$$

The orbital speed of the planet is $2\pi r/T$, where T is the period; therefore, the preceding expression becomes

$$\frac{GM_S}{r^2} = \frac{(2\pi r/T)^2}{r}$$

$$T^2 = \left(\frac{4\pi^2}{GM_S}\right) r^3 = K_S r^3$$

where K_S is a constant given by

$$K_S = \frac{4\pi^2}{GM_S} = 2.97 \times 10^{-19}\ \text{s}^2/\text{m}^3$$

This equation is also valid for elliptical orbits if we replace r with the length a of the semimajor axis (Active Fig. 13.4):

$$T^2 = \left(\frac{4\pi^2}{GM_S}\right) a^3 = K_S a^3 \qquad (13.8)$$

◀ **Kepler's third law**

Equation 13.8 is Kepler's third law, which was stated in words at the beginning of this section. Because the semimajor axis of a circular orbit is its radius, this equation is valid for both circular and elliptical orbits. Notice that the constant of proportionality K_S is independent of the mass of the planet.[2] Equation 13.8 is therefore valid for *any* planet. If we were to consider the orbit of a satellite such as the Moon about the Earth, the constant would have a different value, with the Sun's mass replaced by the Earth's mass; that is, $K_E = 4\pi^2/GM_E$.

[2]Equation 13.8 is indeed a proportion because the ratio of the two quantities T^2 and a^3 is a constant. The variables in a proportion are not required to be limited to the first power only.

TABLE 13.2 *Useful Planetary Data*

Body	Mass (kg)	Mean Radius (m)	Period of Revolution (s)	Mean Distance from the Sun (m)	$\frac{T^2}{r^3}$ (s^2/m^3)
Mercury	3.30×10^{23}	2.44×10^{6}	7.60×10^{6}	5.79×10^{10}	2.98×10^{-19}
Venus	4.87×10^{24}	6.05×10^{6}	1.94×10^{7}	1.08×10^{11}	2.99×10^{-19}
Earth	5.97×10^{24}	6.37×10^{6}	3.156×10^{7}	1.496×10^{11}	2.97×10^{-19}
Mars	6.42×10^{23}	3.39×10^{6}	5.94×10^{7}	2.28×10^{11}	2.98×10^{-19}
Jupiter	1.90×10^{27}	6.99×10^{7}	3.74×10^{8}	7.78×10^{11}	2.97×10^{-19}
Saturn	5.68×10^{26}	5.82×10^{7}	9.29×10^{8}	1.43×10^{12}	2.95×10^{-19}
Uranus	8.68×10^{25}	2.54×10^{7}	2.65×10^{9}	2.87×10^{12}	2.97×10^{-19}
Neptune	1.02×10^{26}	2.46×10^{7}	5.18×10^{9}	4.50×10^{12}	2.94×10^{-19}
Pluto[a]	1.25×10^{22}	1.20×10^{6}	7.82×10^{9}	5.91×10^{12}	2.96×10^{-19}
Moon	7.35×10^{22}	1.74×10^{6}	—	—	—
Sun	1.989×10^{30}	6.96×10^{8}	—	—	—

[a]In August 2006, the International Astronomical Union adopted a definition of a planet that separates Pluto from the other eight planets. Pluto is now defined as a "dwarf planet" like the asteroid Ceres.

Table 13.2 is a collection of useful data for planets and other objects in the solar system. The far-right column verifies that the ratio T^2/r^3 is constant for all objects orbiting the Sun. The small variations in the values in this column are the result of uncertainties in the data measured for the periods and semimajor axes of the objects.

Recent astronomical work has revealed the existence of a large number of solar system objects beyond the orbit of Neptune. In general, these objects lie in the *Kuiper belt*, a region that extends from about 30 AU (the orbital radius of Neptune) to 50 AU. (An AU is an *astronomical unit*, equal to the radius of the Earth's orbit.) Current estimates identify at least 70 000 objects in this region with diameters larger than 100 km. The first Kuiper belt object (KBO) is Pluto, discovered in 1930 and formerly classified as a planet. Starting in 1992, many more have been detected. Several have diameters in the 1 000-km range, such as Varuna (discovered in 2000), Ixion (2001), Quaoar (2002), Sedna (2003), Haumea (2004), Orcus (2004), and Makemake (2005). One KBO, Eris, discovered in 2005, is believed to be significantly larger than Pluto. Other KBOs do not yet have names, but are currently indicated by their year of discovery and a code, such as 2006 QH181 and 2007 UK126.

A subset of about 1 400 KBOs are called "Plutinos" because, like Pluto, they exhibit a resonance phenomenon, orbiting the Sun two times in the same time interval as Neptune revolves three times. The contemporary application of Kepler's laws and such exotic proposals as planetary angular momentum exchange and migrating planets suggest the excitement of this active area of current research.

***Quick Quiz* 13.3** An asteroid is in a highly eccentric elliptical orbit around the Sun. The period of the asteroid's orbit is 90 days. Which of the following statements is true about the possibility of a collision between this asteroid and the Earth? **(a)** There is no possible danger of a collision. **(b)** There is a possibility of a collision. **(c)** There is not enough information to determine whether there is danger of a collision.

Example 13.4 The Mass of the Sun

Calculate the mass of the Sun, noting that the period of the Earth's orbit around the Sun is 3.156×10^7 s and its distance from the Sun is 1.496×10^{11} m.

SOLUTION

Conceptualize Based on Kepler's third law, we realize that the mass of the Sun is related to the orbital size and period of a planet.

13.4 *cont.*

Categorize This example is a relatively simple substitution problem.

Solve Equation 13.8 for the mass of the Sun:

$$M_S = \frac{4\pi^2 r^3}{GT^2}$$

Substitute the known values:

$$M_S = \frac{4\pi^2(1.496 \times 10^{11}\ \text{m})^3}{(6.67 \times 10^{-11}\ \text{N} \cdot \text{m}^2/\text{kg}^2)(3.156 \times 10^7\ \text{s})^2} = 1.99 \times 10^{30}\ \text{kg}$$

In Example 13.3, an understanding of gravitational forces enabled us to find out something about the density of the Earth's core, and now we have used this understanding to determine the mass of the Sun!

Example 13.5 A Geosynchronous Satellite

Consider a satellite of mass m moving in a circular orbit around the Earth at a constant speed v and at an altitude h above the Earth's surface as illustrated in Figure 13.8.

(A) Determine the speed of satellite in terms of G, h, R_E (the radius of the Earth), and M_E (the mass of the Earth).

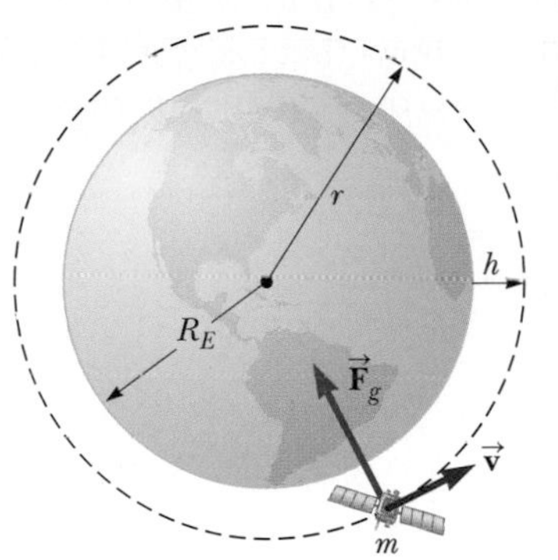

Figure 13.8 (Example 13.5) A satellite of mass m moving around the Earth in a circular orbit of radius r with constant speed v. The only force acting on the satellite is the gravitational force $\vec{\mathbf{F}}_g$. (Not drawn to scale.)

SOLUTION

Conceptualize Imagine the satellite moving around the Earth in a circular orbit under the influence of the gravitational force. This motion is similar to that of the space shuttle, the Hubble Space Telescope, and other objects in orbit around the Earth.

Categorize The satellite must have a centripetal acceleration. Therefore, we categorize the satellite as a particle under a net force and a particle in uniform circular motion.

Analyze The only external force acting on the satellite is the gravitational force, which acts toward the center of the Earth and keeps the satellite in its circular orbit.

Apply the particle under a net force and particle in uniform circular motion models to the satellite:

$$F_g = ma \rightarrow G\frac{M_E m}{r^2} = m\left(\frac{v^2}{r}\right)$$

Solve for v, noting that the distance r from the center of the Earth to the satellite is $r = R_E + h$:

$$(1)\quad v = \sqrt{\frac{GM_E}{r}} = \sqrt{\frac{GM_E}{R_E + h}}$$

(B) If the satellite is to be *geosynchronous* (that is, appearing to remain over a fixed position on the Earth), how fast is it moving through space?

SOLUTION

To appear to remain over a fixed position on the Earth, the period of the satellite must be 24 h = 86 400 s and the satellite must be in orbit directly over the equator.

Solve Kepler's third law (Equation 13.8, with $a = r$ and $M_S \rightarrow M_E$) for r:

$$r = \left(\frac{GM_E T^2}{4\pi^2}\right)^{1/3}$$

Substitute numerical values:

$$r = \left[\frac{(6.67 \times 10^{-11}\ \text{N} \cdot \text{m}^2/\text{kg}^2)(5.97 \times 10^{24}\ \text{kg})(86\ 400\ \text{s})^2}{4\pi^2}\right]^{1/3}$$

$$= 4.22 \times 10^7\ \text{m}$$

continued

13.5 *cont.*

Use Equation (1) to find the speed of the satellite:

$$v = \sqrt{\frac{(6.67 \times 10^{-11}\ \text{N} \cdot \text{m}^2/\text{kg}^2)(5.97 \times 10^{24}\ \text{kg})}{4.22 \times 10^7\ \text{m}}}$$

$$= 3.07 \times 10^3\ \text{m/s}$$

Finalize The value of r calculated here translates to a height of the satellite above the surface of the Earth of almost 36 000 km. Therefore, geosynchronous satellites have the advantage of allowing an earthbound antenna to be aimed in a fixed direction, but there is a disadvantage in that the signals between the Earth and the satellite must travel a long distance. It is difficult to use geosynchronous satellites for optical observation of the Earth's surface because of their high altitude.

WHAT IF? What if the satellite motion in part (A) were taking place at height h above the surface of another planet more massive than the Earth but of the same radius? Would the satellite be moving at a higher speed or a lower speed than it does around the Earth?

Answer If the planet exerts a larger gravitational force on the satellite due to its larger mass, the satellite must move with a higher speed to avoid moving toward the surface. This conclusion is consistent with the predictions of Equation (1), which shows that because the speed v is proportional to the square root of the mass of the planet, the speed increases as the mass of the planet increases.

13.4 The Gravitational Field

When Newton published his theory of universal gravitation, it was considered a success because it satisfactorily explained the motion of the planets. It represented strong evidence that the same laws that describe phenomena on the Earth can be used on large objects like planets and throughout the Universe. Since 1687, Newton's theory has been used to account for the motions of comets, the deflection of a Cavendish balance, the orbits of binary stars, and the rotation of galaxies. Nevertheless, both Newton's contemporaries and his successors found it difficult to accept the concept of a force that acts at a distance. They asked how it was possible for two objects such as the Sun and the Earth to interact when they were not in contact with each other. Newton himself could not answer that question.

An approach to describing interactions between objects that are not in contact came well after Newton's death. This approach enables us to look at the gravitational interaction in a different way, using the concept of a **gravitational field** that exists at every point in space. When a particle of mass m is placed at a point where the gravitational field is $\vec{\mathbf{g}}$, the particle experiences a force $\vec{\mathbf{F}}_g = m\vec{\mathbf{g}}$. In other words, we imagine that the field exerts a force on the particle rather than consider a direct interaction between two particles. The gravitational field $\vec{\mathbf{g}}$ is defined as

Gravitational field ▶

$$\vec{\mathbf{g}} \equiv \frac{\vec{\mathbf{F}}_g}{m} \qquad (13.9)$$

That is, the gravitational field at a point in space equals the gravitational force experienced by a *test particle* placed at that point divided by the mass of the test particle. We call the object creating the field the *source particle.* (Although the Earth is not a particle, it is possible to show that we can model the Earth as a particle for the purpose of finding the gravitational field that it creates.) Notice that the presence of the test particle is not necessary for the field to exist: the source particle creates the gravitational field. We can detect the presence of the field and measure its strength by placing a test particle in the field and noting the force exerted on it. In essence, we are describing the "effect" that any object (in this case, the Earth) has on the empty space around itself in terms of the force that *would* be present *if* a second object were somewhere in that space.[3]

[3]We shall return to this idea of mass affecting the space around it when we discuss Einstein's theory of gravitation in Chapter 39.

As an example of how the field concept works, consider an object of mass m near the Earth's surface. Because the gravitational force acting on the object has a magnitude $GM_E m/r^2$ (see Eq. 13.4), the field $\vec{\mathbf{g}}$ at a distance r from the center of the Earth is

$$\vec{\mathbf{g}} = \frac{\vec{\mathbf{F}}_g}{m} = -\frac{GM_E}{r^2}\hat{\mathbf{r}} \quad (13.10)$$

where the negative sign indicates that the field points toward the center of the Earth as illustrated in Figure 13.9a and $\hat{\mathbf{r}}$ is a unit vector pointing radially outward from the Earth. The field vectors at different points surrounding the Earth vary in both direction and magnitude. In a small region near the Earth's surface, the downward field $\vec{\mathbf{g}}$ is approximately constant and uniform as indicated in Figure 13.9b. Equation 13.10 is valid at all points *outside* the Earth's surface, assuming the Earth is spherical. At the Earth's surface, where $r = R_E$, $\vec{\mathbf{g}}$ has a magnitude of 9.80 N/kg. (The unit N/kg is the same as m/s^2.)

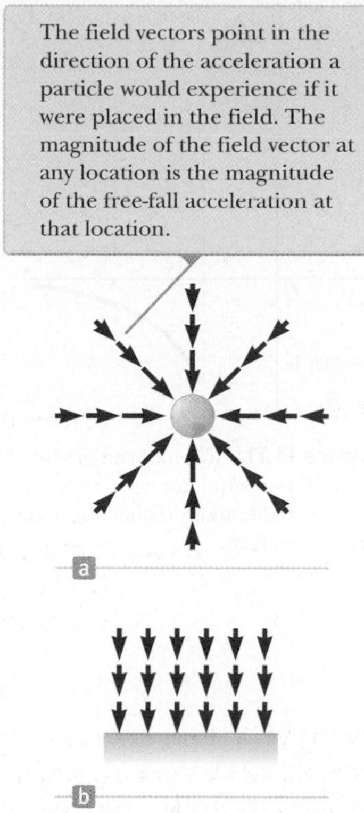

Figure 13.9 (a) The gravitational field vectors in the vicinity of a uniform spherical mass such as the Earth vary in both direction and magnitude. (b) The gravitational field vectors in a small region near the Earth's surface are uniform in both direction and magnitude.

13.5 Gravitational Potential Energy

In Chapter 8, we introduced the concept of gravitational potential energy, which is the energy associated with the configuration of a system of objects interacting via the gravitational force. We emphasized that the gravitational potential energy function $U = mgy$ for a particle–Earth system is valid only when the particle is near the Earth's surface, where the gravitational force is constant. Because the gravitational force between two particles varies as $1/r^2$, we expect that a more general potential energy function—one that is valid without the restriction of having to be near the Earth's surface—will be different from $U = mgy$.

Recall from Equation 7.26 that the change in the potential energy of a system associated with a given displacement of a member of the system is defined as the negative of the internal work done by the force on that member during the displacement:

$$\Delta U = U_f - U_i = -\int_{r_i}^{r_f} F(r)\,dr \quad (13.11)$$

We can use this result to evaluate the general gravitational potential energy function. Consider a particle of mass m moving between two points Ⓐ and Ⓑ above the Earth's surface (Fig. 13.10). The particle is subject to the gravitational force given by Equation 13.1. We can express this force as

$$F(r) = -\frac{GM_E m}{r^2}$$

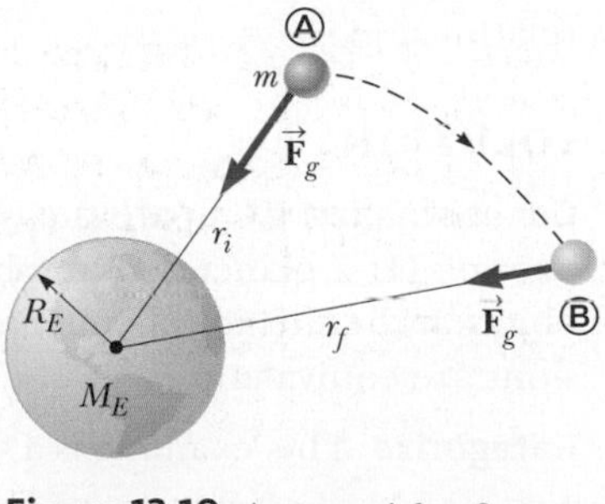

Figure 13.10 As a particle of mass m moves from Ⓐ to Ⓑ above the Earth's surface, the gravitational potential energy of the particle–Earth system changes according to Equation 13.12.

where the negative sign indicates that the force is attractive. Substituting this expression for $F(r)$ into Equation 13.11, we can compute the change in the gravitational potential energy function for the particle–Earth system:

$$U_f - U_i = GM_E m \int_{r_i}^{r_f} \frac{dr}{r^2} = GM_E m \left[-\frac{1}{r}\right]_{r_i}^{r_f}$$

$$U_f - U_i = -GM_E m\left(\frac{1}{r_f} - \frac{1}{r_i}\right) \quad (13.12)$$

As always, the choice of a reference configuration for the potential energy is completely arbitrary. It is customary to choose the reference configuration for zero potential energy to be the same as that for which the force is zero. Taking $U_i = 0$ at $r_i = \infty$, we obtain the important result

$$U(r) = -\frac{GM_E m}{r} \quad (13.13)$$

◀ **Gravitational potential energy of the Earth–particle system**

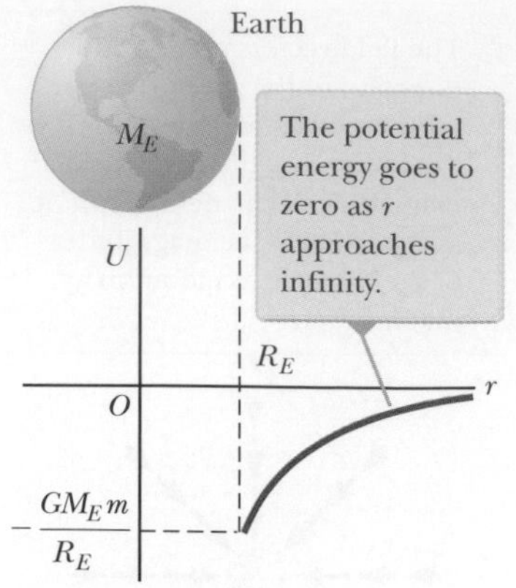

Figure 13.11 Graph of the gravitational potential energy U versus r for the system of an object above the Earth's surface.

This expression applies when the particle is separated from the center of the Earth by a distance r, provided that $r \geq R_E$. The result is not valid for particles inside the Earth, where $r < R_E$. Because of our choice of U_i, the function U is always negative (Fig. 13.11).

Although Equation 13.13 was derived for the particle–Earth system, it can be applied to any two particles. That is, the gravitational potential energy associated with any pair of particles of masses m_1 and m_2 separated by a distance r is

$$U = -\frac{Gm_1m_2}{r} \tag{13.14}$$

This expression shows that the gravitational potential energy for any pair of particles varies as $1/r$, whereas the force between them varies as $1/r^2$. Furthermore, the potential energy is negative because the force is attractive and we have chosen the potential energy as zero when the particle separation is infinite. Because the force between the particles is attractive, an external agent must do positive work to increase the separation between the particles. The work done by the external agent produces an increase in the potential energy as the two particles are separated. That is, U becomes less negative as r increases.

When two particles are at rest and separated by a distance r, an external agent has to supply an energy at least equal to $+Gm_1m_2/r$ to separate the particles to an infinite distance. It is therefore convenient to think of the absolute value of the potential energy as the *binding energy* of the system. If the external agent supplies an energy greater than the binding energy, the excess energy of the system is in the form of kinetic energy of the particles when the particles are at an infinite separation.

We can extend this concept to three or more particles. In this case, the total potential energy of the system is the sum over all pairs of particles. Each pair contributes a term of the form given by Equation 13.14. For example, if the system contains three particles as in Figure 13.12,

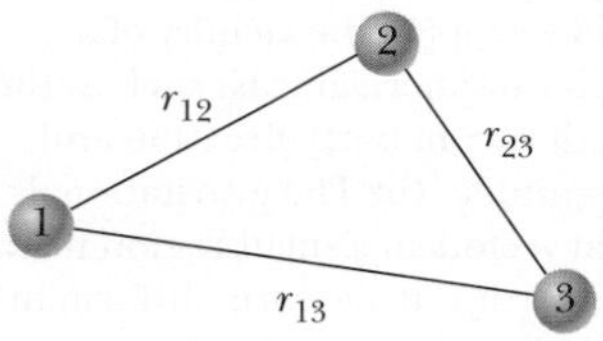

Figure 13.12 Three interacting particles.

$$U_{\text{total}} = U_{12} + U_{13} + U_{23} = -G\left(\frac{m_1m_2}{r_{12}} + \frac{m_1m_3}{r_{13}} + \frac{m_2m_3}{r_{23}}\right) \tag{13.15}$$

The absolute value of U_{total} represents the work needed to separate the particles by an infinite distance.

Example 13.6 The Change in Potential Energy

A particle of mass m is displaced through a small vertical distance Δy near the Earth's surface. Show that in this situation the general expression for the change in gravitational potential energy given by Equation 13.12 reduces to the familiar relationship $\Delta U = mg\,\Delta y$.

SOLUTION

Conceptualize Compare the two different situations for which we have developed expressions for gravitational potential energy: (1) a planet and an object that are far apart for which the energy expression is Equation 13.12 and (2) a small object at the surface of a planet for which the energy expression is Equation 7.19. We wish to show that these two expressions are equivalent.

Categorize This example is a substitution problem.

Combine the fractions in Equation 13.12:

$$(1)\quad \Delta U = -GM_Em\left(\frac{1}{r_f} - \frac{1}{r_i}\right) = GM_Em\left(\frac{r_f - r_i}{r_ir_f}\right)$$

Evaluate $r_f - r_i$ and r_ir_f if both the initial and final positions of the particle are close to the Earth's surface:

$$r_f - r_i = \Delta y \qquad r_ir_f \approx R_E^{\,2}$$

Substitute these expressions into Equation (1):

$$\Delta U \approx \frac{GM_Em}{R_E^{\,2}}\,\Delta y = mg\,\Delta y$$

where $g = GM_E/R_E^{\,2}$ (Eq. 13.5).

13.6 *cont.*

WHAT IF? Suppose you are performing upper-atmosphere studies and are asked by your supervisor to find the height in the Earth's atmosphere at which the "surface equation" $\Delta U = mg\,\Delta y$ gives a 1.0% error in the change in the potential energy. What is this height?

Answer Because the surface equation assumes a constant value for g, it will give a ΔU value that is larger than the value given by the general equation, Equation 13.12.

Set up a ratio reflecting a 1.0% error:

$$\frac{\Delta U_{\text{surface}}}{\Delta U_{\text{general}}} = 1.010$$

Substitute the expressions for each of these changes ΔU:

$$\frac{mg\,\Delta y}{GM_E m(\Delta y/r_i r_f)} = \frac{g r_i r_f}{GM_E} = 1.010$$

Substitute for r_i, r_f, and g from Equation 13.5:

$$\frac{(GM_E/R_E^2)R_E(R_E + \Delta y)}{GM_E} = \frac{R_E + \Delta y}{R_E} = 1 + \frac{\Delta y}{R_E} = 1.010$$

Solve for Δy:

$$\Delta y = 0.010 R_E = 0.010(6.37 \times 10^6 \text{ m}) = 6.37 \times 10^4 \text{ m} = 63.7 \text{ km}$$

13.6 Energy Considerations in Planetary and Satellite Motion

Consider an object of mass m moving with a speed v in the vicinity of a massive object of mass M, where $M >> m$. The system might be a planet moving around the Sun, a satellite in orbit around the Earth, or a comet making a one-time flyby of the Sun. If we assume the object of mass M is at rest in an inertial reference frame, the total mechanical energy E of the two-object system when the objects are separated by a distance r is the sum of the kinetic energy of the object of mass m and the potential energy of the system, given by Equation 13.14:

$$E = K + U$$

$$E = \tfrac{1}{2}mv^2 - \frac{GMm}{r} \qquad (13.16)$$

Equation 13.16 shows that E may be positive, negative, or zero, depending on the value of v. For a bound system such as the Earth–Sun system, however, E is necessarily *less than zero* because we have chosen the convention that $U \to 0$ as $r \to \infty$.

We can easily establish that $E < 0$ for the system consisting of an object of mass m moving in a circular orbit about an object of mass $M >> m$ (Fig. 13.13). Newton's second law applied to the object of mass m gives

$$F_g = ma \quad \rightarrow \quad \frac{GMm}{r^2} = \frac{mv^2}{r}$$

Multiplying both sides by r and dividing by 2 gives

$$\tfrac{1}{2}mv^2 = \frac{GMm}{2r} \qquad (13.17)$$

Substituting this equation into Equation 13.16, we obtain

$$E = \frac{GMm}{2r} - \frac{GMm}{r}$$

$$E = -\frac{GMm}{2r} \quad \text{(circular orbits)} \qquad (13.18)$$

◀ **Total energy for circular orbits**

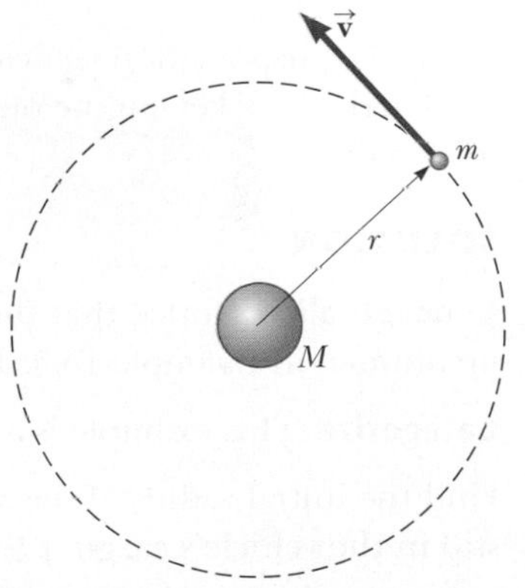

Figure 13.13 An object of mass m moving in a circular orbit about a much larger object of mass M.

This result shows that the total mechanical energy is negative in the case of circular orbits. Notice that the kinetic energy is positive and equal to half the absolute value of the potential energy. The absolute value of E is also equal to the binding energy of the system because this amount of energy must be provided to the system to move the two objects infinitely far apart.

The total mechanical energy is also negative in the case of elliptical orbits. The expression for E for elliptical orbits is the same as Equation 13.18 with r replaced by the semimajor axis length a:

Total energy for elliptical orbits ▶

$$E = -\frac{GMm}{2a} \quad \text{(elliptical orbits)} \tag{13.19}$$

Furthermore, the total energy is constant if we assume the system is isolated. Therefore, as the object of mass m moves from Ⓐ to Ⓑ in Figure 13.10, the total energy remains constant and Equation 13.16 gives

$$E = \tfrac{1}{2}mv_i^2 - \frac{GMm}{r_i} = \tfrac{1}{2}mv_f^2 - \frac{GMm}{r_f} \tag{13.20}$$

Combining this statement of energy conservation with our earlier discussion of conservation of angular momentum, we see that both the total energy and the total angular momentum of a gravitationally bound, two-object system are constants of the motion.

***Quick Quiz* 13.4** A comet moves in an elliptical orbit around the Sun. Which point in its orbit (perihelion or aphelion) represents the highest value of **(a)** the speed of the comet, **(b)** the potential energy of the comet–Sun system, **(c)** the kinetic energy of the comet, and **(d)** the total energy of the comet–Sun system?

Example 13.7 Changing the Orbit of a Satellite

A space transportation vehicle releases a 470-kg communications satellite while in an orbit 280 km above the surface of the Earth. A rocket engine on the satellite boosts it into a geosynchronous orbit. How much energy does the engine have to provide?

SOLUTION

Conceptualize Notice that the height of 280 km is much lower than that for a geosynchronous satellite, 36 000 km, as mentioned in Example 13.5. Therefore, energy must be expended to raise the satellite to this much higher position.

Categorize This example is a substitution problem.

Find the initial radius of the satellite's orbit when it is still in the vehicle's cargo bay:

$$r_i = R_E + 280 \text{ km} = 6.65 \times 10^6 \text{ m}$$

Use Equation 13.18 to find the difference in energies for the satellite–Earth system with the satellite at the initial and final radii:

$$\Delta E = E_f - E_i = -\frac{GM_Em}{2r_f} - \left(-\frac{GM_Em}{2r_i}\right) = -\frac{GM_Em}{2}\left(\frac{1}{r_f} - \frac{1}{r_i}\right)$$

Substitute numerical values, using $r_f = 4.22 \times 10^7$ m from Example 13.5:

$$\Delta E = -\frac{(6.67 \times 10^{-11} \text{ N}\cdot\text{m}^2/\text{kg}^2)(5.97 \times 10^{24} \text{ kg})(470 \text{ kg})}{2}$$
$$\times \left(\frac{1}{4.22 \times 10^7 \text{ m}} - \frac{1}{6.65 \times 10^6 \text{ m}}\right)$$
$$= \boxed{1.19 \times 10^{10} \text{ J}}$$

which is the energy equivalent of 89 gal of gasoline. NASA engineers must account for the changing mass of the spacecraft as it ejects burned fuel, something we have not done here. Would you expect the calculation that includes the effect of this changing mass to yield a greater or a lesser amount of energy required from the engine?

Escape Speed

Suppose an object of mass m is projected vertically upward from the Earth's surface with an initial speed v_i as illustrated in Figure 13.14. We can use energy considerations to find the minimum value of the initial speed needed to allow the object to move infinitely far away from the Earth. Equation 13.16 gives the total energy of the system for any configuration. As the object is projected upward from the surface of the Earth, $v = v_i$ and $r = r_i = R_E$. When the object reaches its maximum altitude, $v = v_f = 0$ and $r = r_f = r_{max}$. Because the total energy of the isolated object–Earth system is constant, substituting these conditions into Equation 13.20 gives

$$\frac{1}{2}mv_i^2 - \frac{GM_E m}{R_E} = -\frac{GM_E m}{r_{max}}$$

Solving for v_i^2 gives

$$v_i^2 = 2GM_E\left(\frac{1}{R_E} - \frac{1}{r_{max}}\right) \quad \textbf{(13.21)}$$

For a given maximum altitude $h = r_{max} - R_E$, we can use this equation to find the required initial speed.

We are now in a position to calculate **escape speed,** which is the minimum speed the object must have at the Earth's surface to approach an infinite separation distance from the Earth. Traveling at this minimum speed, the object continues to move farther and farther away from the Earth as its speed asymptotically approaches zero. Letting $r_{max} \rightarrow \infty$ in Equation 13.21 and taking $v_i = v_{esc}$ gives

$$v_{esc} = \sqrt{\frac{2GM_E}{R_E}} \quad \textbf{(13.22)}$$

This expression for v_{esc} is independent of the mass of the object. In other words, a spacecraft has the same escape speed as a molecule. Furthermore, the result is independent of the direction of the velocity and ignores air resistance.

If the object is given an initial speed equal to v_{esc}, the total energy of the system is equal to zero. Notice that when $r \rightarrow \infty$, the object's kinetic energy and the potential energy of the system are both zero. If v_i is greater than v_{esc}, the total energy of the system is greater than zero and the object has some residual kinetic energy as $r \rightarrow \infty$.

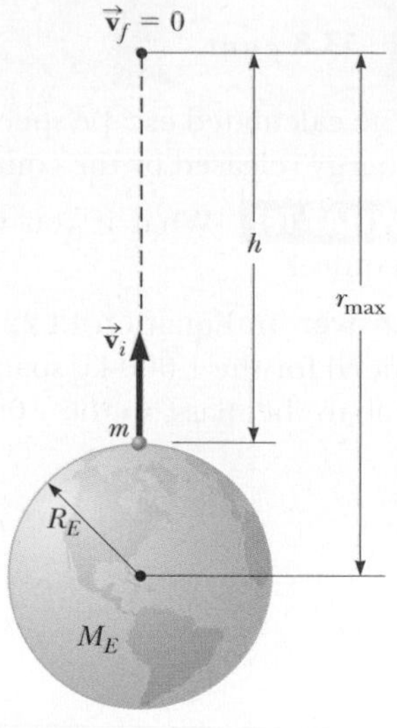

Figure 13.14 An object of mass m projected upward from the Earth's surface with an initial speed v_i reaches a maximum altitude h.

Pitfall Prevention 13.3

You Can't Really Escape
Although Equation 13.22 provides the "escape speed" from the Earth, *complete* escape from the Earth's gravitational influence is impossible because the gravitational force is of infinite range. No matter how far away you are, you will always feel some gravitational force due to the Earth.

Example 13.8 Escape Speed of a Rocket

Calculate the escape speed from the Earth for a 5 000-kg spacecraft and determine the kinetic energy it must have at the Earth's surface to move infinitely far away from the Earth.

SOLUTION

Conceptualize Imagine projecting the spacecraft from the Earth's surface so that it moves farther and farther away, traveling more and more slowly, with its speed approaching zero. Its speed will never reach zero, however, so the object will never turn around and come back.

Categorize This example is a substitution problem.

Use Equation 13.22 to find the escape speed:

$$v_{esc} = \sqrt{\frac{2GM_E}{R_E}} = \sqrt{\frac{2(6.67 \times 10^{-11}\ \text{N}\cdot\text{m}^2/\text{kg}^2)(5.97 \times 10^{24}\ \text{kg})}{6.37 \times 10^6\ \text{m}}}$$
$$= 1.12 \times 10^4\ \text{m/s}$$

Evaluate the kinetic energy of the spacecraft from Equation 7.16:

$$K = \frac{1}{2}mv_{esc}^2 = \frac{1}{2}(5.00 \times 10^3\ \text{kg})(1.12 \times 10^4\ \text{m/s})^2$$
$$= 3.13 \times 10^{11}\ \text{J}$$

continued

13.8 *cont.*

The calculated escape speed corresponds to about 25 000 mi/h. The kinetic energy of the spacecraft is equivalent to the energy released by the combustion of about 2 300 gal of gasoline.

WHAT IF? What if you want to launch a 1 000-kg spacecraft at the escape speed? How much energy would that require?

Answer In Equation 13.22, the mass of the object moving with the escape speed does not appear. Therefore, the escape speed for the 1 000-kg spacecraft is the same as that for the 5 000-kg spacecraft. The only change in the kinetic energy is due to the mass, so the 1 000-kg spacecraft requires one-fifth of the energy of the 5 000-kg spacecraft:

$$K = \frac{1}{5}(3.13 \times 10^{11}\text{ J}) = 6.25 \times 10^{10}\text{ J}$$

Equations 13.21 and 13.22 can be applied to objects projected from any planet. That is, in general, the escape speed from the surface of any planet of mass M and radius R is

$$v_{\text{esc}} = \sqrt{\frac{2GM}{R}} \quad \textbf{(13.23)}$$

TABLE 13.3

Escape Speeds from the Surfaces of the Planets, Moon, and Sun

Planet	v_{esc} (km/s)
Mercury	4.3
Venus	10.3
Earth	11.2
Mars	5.0
Jupiter	60
Saturn	36
Uranus	22
Neptune	24
Moon	2.3
Sun	618

Escape speeds for the planets, the Moon, and the Sun are provided in Table 13.3. The values vary from 2.3 km/s for the Moon to about 618 km/s for the Sun. These results, together with some ideas from the kinetic theory of gases (see Chapter 21), explain why some planets have atmospheres and others do not. As we shall see later, at a given temperature the average kinetic energy of a gas molecule depends only on the mass of the molecule. Lighter molecules, such as hydrogen and helium, have a higher average speed than heavier molecules at the same temperature. When the average speed of the lighter molecules is not much less than the escape speed of a planet, a significant fraction of them have a chance to escape.

This mechanism also explains why the Earth does not retain hydrogen molecules and helium atoms in its atmosphere but does retain heavier molecules, such as oxygen and nitrogen. On the other hand, the very large escape speed for Jupiter enables that planet to retain hydrogen, the primary constituent of its atmosphere.

Black Holes

In Example 11.7, we briefly described a rare event called a supernova, the catastrophic explosion of a very massive star. The material that remains in the central core of such an object continues to collapse, and the core's ultimate fate depends on its mass. If the core has a mass less than 1.4 times the mass of our Sun, it gradually cools down and ends its life as a white dwarf star. If the core's mass is greater than this value, however, it may collapse further due to gravitational forces. What remains is a neutron star, discussed in Example 11.7, in which the mass of a star is compressed to a radius of about 10 km. (On the Earth, a teaspoon of this material would weigh about 5 billion tons!)

An even more unusual star death may occur when the core has a mass greater than about three solar masses. The collapse may continue until the star becomes a very small object in space, commonly referred to as a **black hole.** In effect, black holes are remains of stars that have collapsed under their own gravitational force. If an object such as a spacecraft comes close to a black hole, the object experiences an extremely strong gravitational force and is trapped forever.

The escape speed for a black hole is very high because of the concentration of the star's mass into a sphere of very small radius (see Eq. 13.23). If the escape speed exceeds the speed of light c, radiation from the object (such as visible light) cannot escape and the object appears to be black (hence the origin of the terminology "black hole"). The critical radius R_S at which the escape speed is c is called the **Schwarzschild radius** (Fig. 13.15). The imaginary surface of a sphere of this radius surrounding the black hole is called the **event horizon,** which is the limit of how close you can approach the black hole and hope to escape.

There is evidence that supermassive black holes exist at the centers of galaxies, with masses very much larger than the Sun. (There is strong evidence of a supermassive black hole of mass 2–3 million solar masses at the center of our galaxy.)

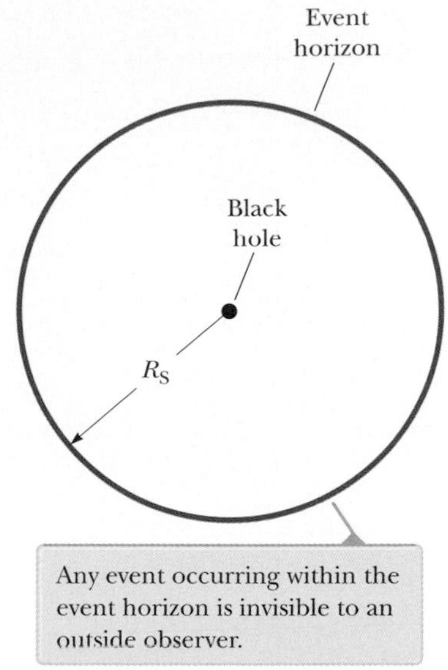

Figure 13.15 A black hole. The distance R_S equals the Schwarzschild radius.

Dark Matter

Equation (1) in Example 13.5 shows that the speed of an object in orbit around the Earth decreases as the object is moved farther away from the Earth:

$$v = \sqrt{\frac{GM_E}{r}} \quad (13.24)$$

Using data in Table 13.2 to find the speeds of planets in their orbits around the Sun, we find the same behavior for the planets. Figure 13.16 shows this behavior for the eight planets of our solar system. The theoretical prediction of the planet speed as a function of distance from the Sun is shown by the red-brown curve, using Equation 13.24 with the mass of the Earth replaced by the mass of the Sun. Data for the individual planets lie right on this curve. This behavior results from the vast majority of the mass of the solar system being concentrated in a small space, i.e., the Sun.

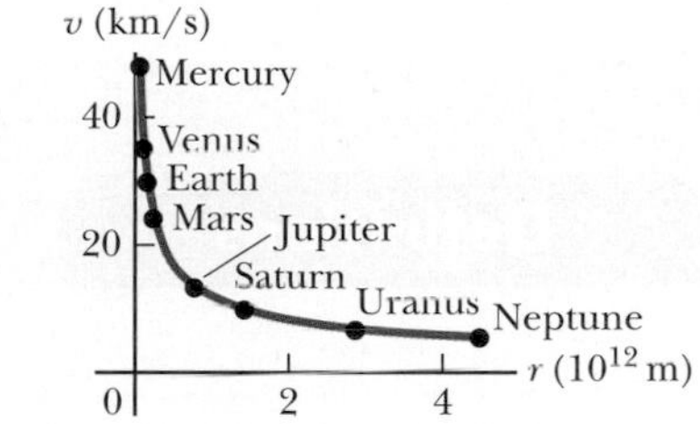

Figure 13.16 The orbital speed v as a function of distance r from the Sun for the eight planets of the solar system. The theoretical curve is in red-brown, and the data points for the planets are in black.

Extending this concept further, we might expect the same behavior in a galaxy. Much of the visible galactic mass, including that of a supermassive black hole, is near the central core of a galaxy. The opening photograph for this chapter shows the central core of the Whirlpool galaxy as a very bright area surrounded by the "arms" of the galaxy, which contain material in orbit around the central core. Based on this distribution of matter in the galaxy, the speed of an object in the outer part of the galaxy would be smaller than that for objects closer to the center, just like for the planets of the solar system.

That is *not* what is observed, however. Figure 13.17 shows the results of measurements of the speeds of objects in the Andromeda galaxy as a function of distance from the galaxy's center.[4] The red-brown curve shows the expected speeds for these objects if they were traveling in circular orbits around the mass concentrated in the central core. The data for the individual objects in the galaxy shown by the black dots are all well above the theoretical curve. These data, as well as an extensive amount of data taken over the past half century, show that for objects outside the central core of the galaxy, the curve of speed versus distance from the center of the galaxy is approximately flat rather than decreasing at larger distances. Therefore, these objects (including our own Solar System in the Milky Way) are rotating faster than can be accounted for by gravity due to the visible galaxy! This surprising result means that there must be additional mass in a more extended distribution, causing these objects to orbit so fast, and has led scientists to propose the existence of **dark matter.** This matter is proposed to exist in a large halo around each galaxy (with a radius up to 10 times as large as the visible galaxy's radius). Because it is not luminous (i.e., does not emit electromagnetic radiation) it must be either very cold or electrically neutral. Therefore, we cannot "see" dark matter, except through its gravitational effects..

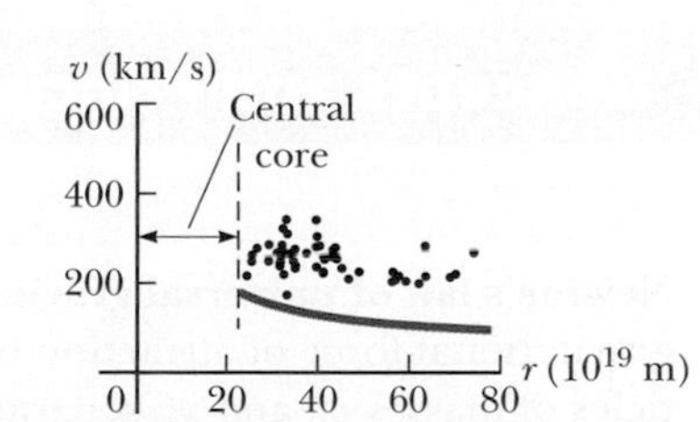

Figure 13.17 The orbital speed v of a galaxy object as a function of distance r from the center of the central core of the Andromeda galaxy. The theoretical curve is in red-brown, and the data points for the galaxy objects are in black. No data are provided on the left because the behavior inside the central core of the galaxy is more complicated.

The proposed existence of dark matter is also implied by earlier observations made on larger gravitationally bound structures known as galaxy clusters.[5] These

[4] V. C. Rubin and W. K. Ford, "Rotation of the Andromeda Nebula from a Spectroscopic Survey of Emission Regions," *Astrophysical Journal* **159:** 379–403 (1970).

[5] F. Zwicky, "On the Masses of Nebulae and of Clusters of Nebulae," *Astrophysical Journal* **86:** 217–246 (1937).

observations show that the orbital speeds of galaxies in a cluster are, on average, too large to be explained by the luminous matter in the cluster alone. The speeds of the individual galaxies are so high, they suggest that there is 50 times as much dark matter in galaxy clusters as in the galaxies themselves!

Why doesn't dark matter affect the orbital speeds of planets like it does those of a galaxy? It seems that a solar system is too small a structure to contain enough dark matter to affect the behavior of orbital speeds. A galaxy or galaxy cluster, on the other hand, contains huge amounts of dark matter, resulting in the surprising behavior.

What, though, *is* dark matter? At this time, no one knows. One theory claims that dark matter is based on a particle called a weakly interacting massive particle, or WIMP. If this theory is correct, calculations show that about 200 WIMPs pass through a human body at any given time. The new Large Hadron Collider in Europe (see Chapter 46) is the first particle accelerator with enough energy to possibly generate and detect the existence of WIMPs, which has generated much current interest in dark matter. Keeping an eye on this research in the future should be exciting.

Summary

Definitions

The **gravitational field** at a point in space is defined as the gravitational force experienced by any test particle located at that point divided by the mass of the test particle:

$$\vec{\mathbf{g}} \equiv \frac{\vec{\mathbf{F}}_g}{m} \qquad \textbf{(13.9)}$$

Concepts and Principles

Newton's law of universal gravitation states that the gravitational force of attraction between any two particles of masses m_1 and m_2 separated by a distance r has the magnitude

$$F_g = G\frac{m_1 m_2}{r^2} \qquad \textbf{(13.1)}$$

where $G = 6.674 \times 10^{-11}\ \text{N} \cdot \text{m}^2/\text{kg}^2$ is the **universal gravitational constant.** This equation enables us to calculate the force of attraction between masses under many circumstances.

An object at a distance h above the Earth's surface experiences a gravitational force of magnitude mg, where g is the free-fall acceleration at that elevation:

$$g = \frac{GM_E}{r^2} = \frac{GM_E}{(R_E + h)^2} \qquad \textbf{(13.6)}$$

In this expression, M_E is the mass of the Earth and R_E is its radius. Therefore, the weight of an object decreases as the object moves away from the Earth's surface.

Kepler's laws of planetary motion state:

1. All planets move in elliptical orbits with the Sun at one focus.
2. The radius vector drawn from the Sun to a planet sweeps out equal areas in equal time intervals.
3. The square of the orbital period of any planet is proportional to the cube of the semimajor axis of the elliptical orbit.

Kepler's third law can be expressed as

$$T^2 = \left(\frac{4\pi^2}{GM_S}\right)a^3 \quad (13.8)$$

where M_S is the mass of the Sun and a is the semimajor axis. For a circular orbit, a can be replaced in Equation 13.8 by the radius r. Most planets have nearly circular orbits around the Sun.

The **gravitational potential energy** associated with a system of two particles separated by a distance r is

$$U = -\frac{Gm_1m_2}{r} \quad (13.14)$$

where U is taken to be zero as $r \to \infty$.

If an isolated system consists of an object of mass m moving with a speed v in the vicinity of a massive object of mass M, the total energy E of the system is the sum of the kinetic and potential energies:

$$E = \tfrac{1}{2}mv^2 - \frac{GMm}{r} \quad (13.16)$$

The total energy of the system is a constant of the motion. If the object moves in an elliptical orbit of semimajor axis a around the massive object and $M >> m$, the total energy of the system is

$$E = -\frac{GMm}{2a} \quad (13.19)$$

For a circular orbit, this same equation applies with $a = r$.

The **escape speed** for an object projected from the surface of a planet of mass M and radius R is

$$v_{esc} = \sqrt{\frac{2GM}{R}} \quad (13.23)$$

Objective Questions

☐ denotes answer available in Student Solutions Manual/Study Guide

1. Rank the magnitudes of the following gravitational forces from largest to smallest. If two forces are equal, show their equality in your list. (a) the force exerted by a 2-kg object on a 3-kg object 1 m away (b) the force exerted by a 2-kg object on a 9-kg object 1 m away (c) the force exerted by a 2-kg object on a 9-kg object 2 m away (d) the force exerted by a 9-kg object on a 2-kg object 2 m away (e) the force exerted by a 4-kg object on another 4-kg object 2 m away

2. The gravitational force exerted on an astronaut on the Earth's surface is 650 N directed downward. When she is in the space station in orbit around the Earth, is the gravitational force on her (a) larger, (b) exactly the same, (c) smaller, (d) nearly but not exactly zero, or (e) exactly zero?

3. Imagine that nitrogen and other atmospheric gases were more soluble in water so that the atmosphere of the Earth is entirely absorbed by the oceans. Atmospheric pressure would then be zero, and outer space would start at the planet's surface. Would the Earth then have a gravitational field? (a) Yes, and at the surface it would be larger in magnitude than 9.8 N/kg. (b) Yes, and it would be essentially the same as the current value. (c) Yes, and it would be somewhat less than 9.8 N/kg. (d) Yes, and it would be much less than 9.8 N/kg. (e) No, it would not.

4. Suppose the gravitational acceleration at the surface of a certain moon A of Jupiter is 2 m/s^2. Moon B has twice the mass and twice the radius of moon A. What is the gravitational acceleration at its surface? Neglect the gravitational acceleration due to Jupiter. (a) 8 m/s^2 (b) 4 m/s^2 (c) 2 m/s^2 (d) 1 m/s^2 (e) 0.5 m/s^2

5. A satellite moves in a circular orbit at a constant speed around the Earth. Which of the following statements is true? (a) No force acts on the satellite. (b) The satellite moves at constant speed and hence doesn't accelerate. (c) The satellite has an acceleration directed away from the Earth. (d) The satellite has an acceleration directed toward the Earth. (e) Work is done on the satellite by the gravitational force.

6. An object of mass m is located on the surface of a spherical planet of mass M and radius R. The escape speed from the planet does not depend on which of the following? (a) M (b) m (c) the density of the planet (d) R (e) the acceleration due to gravity on that planet

7. A satellite originally moves in a circular orbit of radius R around the Earth. Suppose it is moved into a circular orbit

of radius $4R$. **(i)** What does the force exerted on the satellite then become? (a) eight times larger (b) four times larger (c) one-half as large (d) one-eighth as large (e) one-sixteenth as large **(ii)** What happens to the satellite's speed? Choose from the same possibilities (a) through (e). **(iii)** What happens to its period? Choose from the same possibilities (a) through (e).

8. The vernal equinox and the autumnal equinox are associated with two points 180° apart in the Earth's orbit. That is, the Earth is on precisely opposite sides of the Sun when it passes through these two points. From the vernal equinox, 185.4 days elapse before the autumnal equinox. Only 179.8 days elapse from the autumnal equinox until the next vernal equinox. Why is the interval from the March (vernal) to the September (autumnal) equinox (which contains the summer solstice) longer than the interval from the September to the March equinox rather than being equal to that interval? Choose one of the following reasons. (a) They are really the same, but the Earth spins faster during the "summer" interval, so the days are shorter. (b) Over the "summer" interval, the Earth moves slower because it is farther from the Sun. (c) Over the March-to-September interval, the Earth moves slower because it is closer to the Sun. (d) The Earth has less kinetic energy when it is warmer. (e) The Earth has less orbital angular momentum when it is warmer.

9. A system consists of five particles. How many terms appear in the expression for the total gravitational potential energy of the system? (a) 4 (b) 5 (c) 10 (d) 20 (e) 25

10. Rank the following quantities of energy from largest to the smallest. State if any are equal. (a) the absolute value of the average potential energy of the Sun–Earth system (b) the average kinetic energy of the Earth in its orbital motion relative to the Sun (c) the absolute value of the total energy of the Sun–Earth system

11. Halley's comet has a period of approximately 76 years, and it moves in an elliptical orbit in which its distance from the Sun at closest approach is a small fraction of its maximum distance. Estimate the comet's maximum distance from the Sun in astronomical units (AUs) (the distance from the Earth to the Sun). (a) 6 AU (b) 12 AU (c) 20 AU (d) 28 AU (e) 35 AU

Conceptual Questions

☐ denotes answer available in Student Solutions Manual/Study Guide

1. A satellite in low-Earth orbit is not truly traveling through a vacuum. Rather, it moves through very thin air. Does the resulting air friction cause the satellite to slow down?

[2.] Explain why it takes more fuel for a spacecraft to travel from the Earth to the Moon than for the return trip. Estimate the difference.

[3.] Why don't we put a geosynchronous weather satellite in orbit around the 45th parallel? Wouldn't such a satellite be more useful in the United States than one in orbit around the equator?

4. (a) Explain why the force exerted on a particle by a uniform sphere must be directed toward the center of the sphere. (b) Would this statement be true if the mass distribution of the sphere were not spherically symmetric? Explain.

[5.] (a) At what position in its elliptical orbit is the speed of a planet a maximum? (b) At what position is the speed a minimum?

6. You are given the mass and radius of planet X. How would you calculate the free-fall acceleration on this planet's surface?

7. (a) If a hole could be dug to the center of the Earth, would the force on an object of mass m still obey Equation 13.1 there? (b) What do you think the force on m would be at the center of the Earth?

[8.] In his 1798 experiment, Cavendish was said to have "weighed the Earth." Explain this statement.

9. Each *Voyager* spacecraft was accelerated toward escape speed from the Sun by the gravitational force exerted by Jupiter on the spacecraft. (a) Is the gravitational force a conservative or a nonconservative force? (b) Does the interaction of the spacecraft with Jupiter meet the definition of an elastic collision? (c) How could the spacecraft be moving faster after the collision?

Problems

ENHANCED **WebAssign** The problems found in this chapter may be assigned online in Enhanced WebAssign

1. denotes straightforward problem; **2.** denotes intermediate problem; **3.** denotes challenging problem

[1.] full solution available in the Student Solutions Manual/Study Guide

1. denotes problems most often assigned in Enhanced WebAssign; these provide students with targeted feedback and either a Master It tutorial or a Watch It solution video.

Q|C denotes asking for quantitative and conceptual reasoning

S denotes symbolic reasoning problem

M denotes Master It tutorial available in Enhanced WebAssign

GP denotes guided problem

shaded denotes "paired problems" that develop reasoning with symbols and numerical values

Section 13.1 Newton's Law of Universal Gravitation

Problem 10 in Chapter 1 can also be assigned with this section.

1. Two ocean liners, each with a mass of 40 000 metric tons, are moving on parallel courses 100 m apart. What is the magnitude of the acceleration of one of the liners toward the other due to their mutual gravitational attraction? Model the ships as particles.

2. Determine the order of magnitude of the gravitational force that you exert on another person 2 m away. In your solution, state the quantities you measure or estimate and their values.

3. A 200-kg object and a 500-kg object are separated by 4.00 m. (a) Find the net gravitational force exerted by these objects on a 50.0-kg object placed midway between them. (b) At what position (other than an infinitely remote one) can the 50.0-kg object be placed so as to experience a net force of zero from the other two objects?

4. **Q|C** During a solar eclipse, the Moon, the Earth, and the Sun all lie on the same line, with the Moon between the Earth and the Sun. (a) What force is exerted by the Sun on the Moon? (b) What force is exerted by the Earth on the Moon? (c) What force is exerted by the Sun on the Earth? (d) Compare the answers to parts (a) and (b). Why doesn't the Sun capture the Moon away from the Earth?

5. **M** In introductory physics laboratories, a typical Cavendish balance for measuring the gravitational constant G uses lead spheres with masses of 1.50 kg and 15.0 g whose centers are separated by about 4.50 cm. Calculate the gravitational force between these spheres, treating each as a particle located at the sphere's center.

6. Three uniform spheres of masses $m_1 = 2.00$ kg, $m_2 = 4.00$ kg, and $m_3 = 6.00$ kg are placed at the corners of a right triangle as shown in Figure P13.6. Calculate the resultant gravitational force on the object of mass m_2, assuming the spheres are isolated from the rest of the Universe.

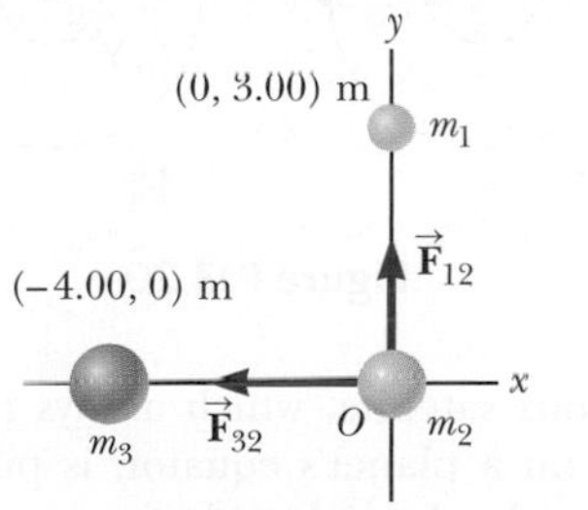

Figure P13.6

7. Two objects attract each other with a gravitational force of magnitude 1.00×10^{-8} N when separated by 20.0 cm. If the total mass of the two objects is 5.00 kg, what is the mass of each?

8. *Why is the following situation impossible?* The centers of two homogeneous spheres are 1.00 m apart. The spheres are each made of the same element from the periodic table. The gravitational force between the spheres is 1.00 N.

9. **Review.** A student proposes to study the gravitational force by suspending two 100.0-kg spherical objects at the lower ends of cables from the ceiling of a tall cathedral and measuring the deflection of the cables from the vertical. The 45.00-m-long cables are attached to the ceiling 1.000 m apart. The first object is suspended, and its position is carefully measured. The second object is suspended, and the two objects attract each other gravitationally. By what distance has the first object moved horizontally from its initial position due to the gravitational attraction to the other object? *Suggestion:* Keep in mind that this distance will be very small and make appropriate approximations.

Section 13.2 Free-Fall Acceleration and the Gravitational Force

10. When a falling meteoroid is at a distance above the Earth's surface of 3.00 times the Earth's radius, what is its acceleration due to the Earth's gravitation?

11. **Review.** Miranda, a satellite of Uranus, is shown in Figure P13.11a. It can be modeled as a sphere of radius 242 km and mass 6.68×10^{19} kg. (a) Find the free-fall acceleration on its surface. (b) A cliff on Miranda is 5.00 km high. It appears on the limb at the 11 o'clock position in Figure P13.11a and is magnified in Figure P13.11b. If a devotee of extreme sports runs horizontally off the top of the cliff at 8.50 m/s, for what time interval is he in flight? (c) How far from the base of the vertical cliff does he strike the icy surface of Miranda? (d) What will be his vector impact velocity?

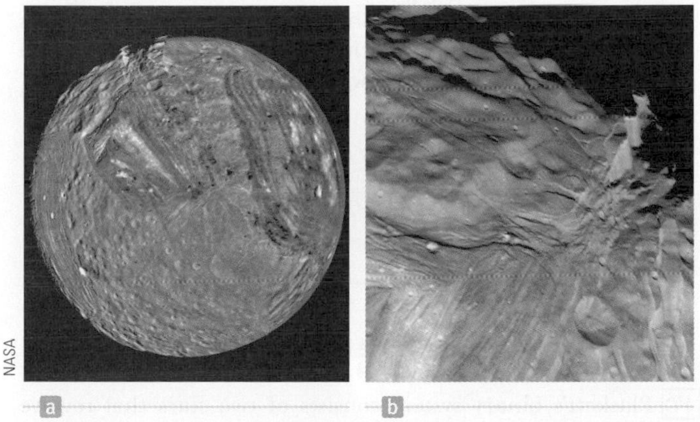

Figure P13.11

12. The free-fall acceleration on the surface of the Moon is about one-sixth that on the surface of the Earth. The radius of the Moon is about $0.250R_E$ (R_E = Earth's radius = 6.37×10^6 m). Find the ratio of their average densities, ρ_{Moon}/ρ_{Earth}.

Section 13.3 Kepler's Laws and the Motion of Planets

13. An artificial satellite circles the Earth in a circular orbit at a location where the acceleration due to gravity is 9.00 m/s^2. Determine the orbital period of the satellite.

14. Io, a satellite of Jupiter, has an orbital period of 1.77 days and an orbital radius of 4.22×10^5 km. From these data, determine the mass of Jupiter.

15. QIC A minimum-energy transfer orbit to an outer planet consists of putting a spacecraft on an elliptical trajectory with the departure planet corresponding to the perihelion of the ellipse, or the closest point to the Sun, and the arrival planet at the aphelion, or the farthest point from the Sun. (a) Use Kepler's third law to calculate how long it would take to go from Earth to Mars on such an orbit as shown in Figure P13.15. (b) Can such an orbit be undertaken at any time? Explain.

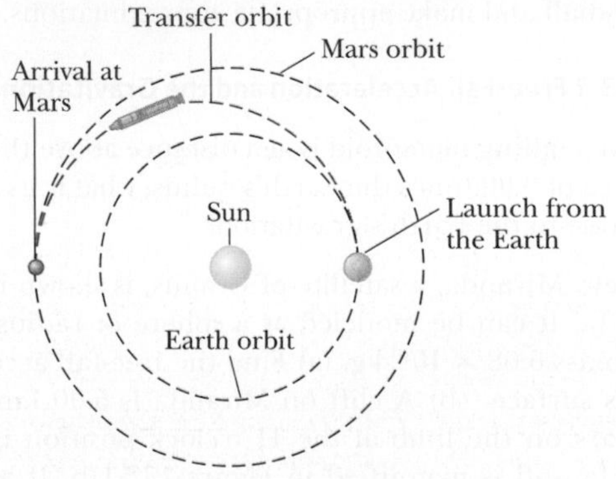

Figure P13.15

16. QIC S A particle of mass m moves along a straight line with constant velocity $\vec{v}_0$ in the x direction, a distance b from the x axis (Fig. P13.16). (a) Does the particle possess any angular momentum about the origin? (b) Explain why the amount of its angular momentum should change or should stay constant. (c) Show that Kepler's second law is satisfied by showing that the two shaded triangles in the figure have the same area when $t_{Ⓓ} - t_{Ⓒ} = t_{Ⓑ} - t_{Ⓐ}$.

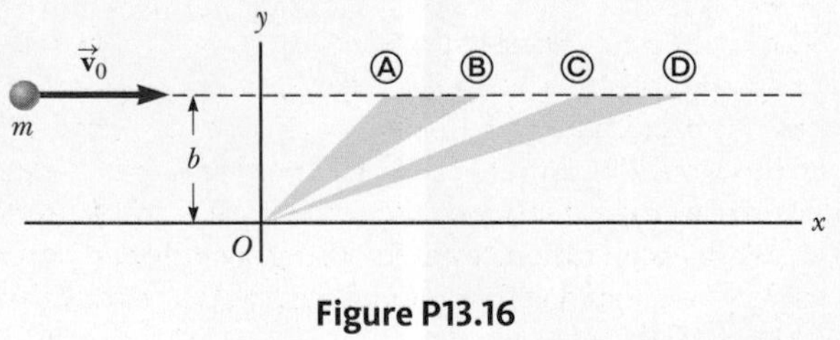

Figure P13.16

17. Plaskett's binary system consists of two stars that revolve in a circular orbit about a center of mass midway between them. This statement implies that the masses of the two stars are equal (Fig. P13.17). Assume the orbital speed of each star is $|\vec{v}| = 220$ km/s and the orbital period of each is 14.4 days. Find the mass M of each star. (For comparison, the mass of our Sun is 1.99×10^{30} kg.)

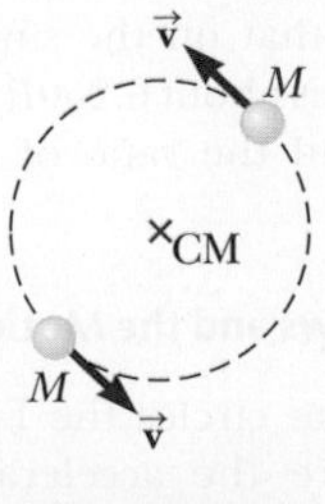

Figure P13.17

18. Comet Halley (Fig. P13.18) approaches the Sun to within 0.570 AU, and its orbital period is 75.6 yr. (AU is the symbol for astronomical unit, where 1 AU = 1.50×10^{11} m is the mean Earth–Sun distance.) How far from the Sun will Halley's comet travel before it starts its return journey?

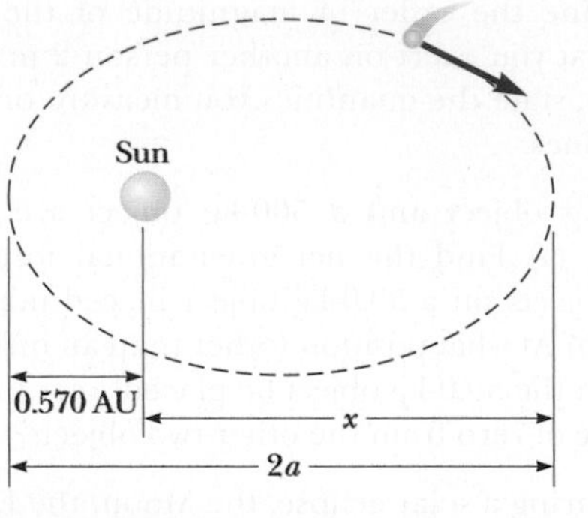

Figure P13.18 (Orbit is not drawn to scale.)

19. Use Kepler's third law to determine how many days it takes a spacecraft to travel in an elliptical orbit from a point 6 670 km from the Earth's center to the Moon, 385 000 km from the Earth's center.

20. Two planets X and Y travel counterclockwise in circular orbits about a star as shown in Figure P13.20. The radii of their orbits are in the ratio 3:1. At one moment, they are aligned as shown in Figure P13.20a, making a straight line with the star. During the next five years, the angular displacement of planet X is 90.0° as shown in Figure P13.20b. What is the angular displacement of planet Y at this moment?

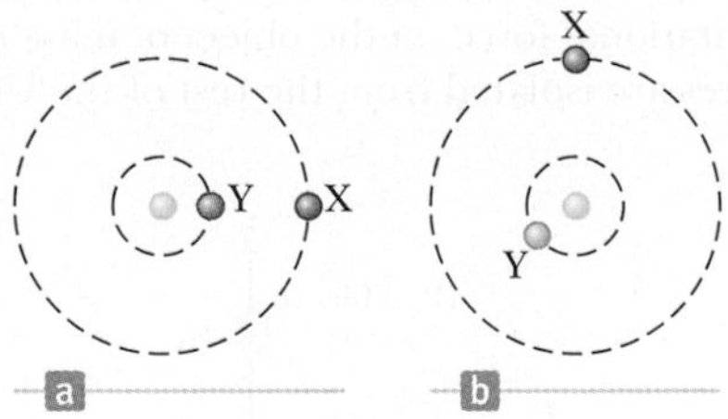

Figure P13.20

21. A synchronous satellite, which always remains above the same point on a planet's equator, is put in orbit around Jupiter to study that planet's famous red spot. Jupiter rotates once every 9.84 h. Use the data of Table 13.2 to find the altitude of the satellite above the surface of the planet.

22. Neutron stars are extremely dense objects formed from the remnants of supernova explosions. Many rotate very rapidly. Suppose the mass of a certain spherical neutron star is twice the mass of the Sun and its radius is 10.0 km. Determine the greatest possible angular speed it can have so that the matter at the surface of the star on its equator is just held in orbit by the gravitational force.

23. Q|C Suppose the Sun's gravity were switched off. The planets would leave their orbits and fly away in straight lines as described by Newton's first law. (a) Would Mercury ever be farther from the Sun than Pluto? (b) If so, find how long it would take Mercury to achieve this passage. If not, give a convincing argument that Pluto is always farther from the Sun than is Mercury.

24. Q|C (a) Given that the period of the Moon's orbit about the Earth is 27.32 days and the nearly constant distance between the center of the Earth and the center of the Moon is 3.84×10^8 m, use Equation 13.8 to calculate the mass of the Earth. (b) Why is the value you calculate a bit too large?

Section 13.4 The Gravitational Field

25. S Three objects of equal mass are located at three corners of a square of edge length ℓ as shown in Figure P13.25. Find the magnitude and direction of the gravitational field at the fourth corner due to these objects.

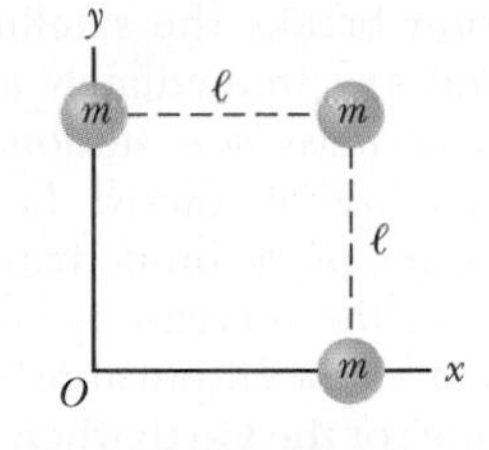

Figure P13.25

26. Q|C S (a) Compute the vector gravitational field at a point P on the perpendicular bisector of the line joining two objects of equal mass separated by a distance $2a$ as shown in Figure P13.26. (b) Explain physically why the field should approach zero as $r \to 0$. (c) Prove mathematically that the answer to part (a) behaves in this way. (d) Explain physically why the magnitude of the field should approach $2GM/r^2$ as $r \to \infty$. (e) Prove mathematically that the answer to part (a) behaves correctly in this limit.

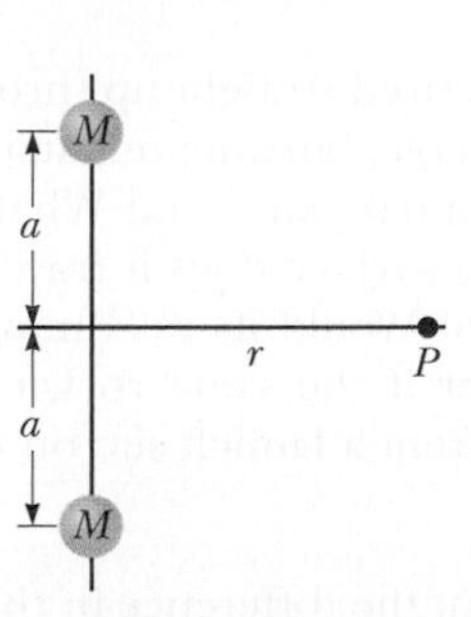

Figure P13.26

27. A spacecraft in the shape of a long cylinder has a length of 100 m, and its mass with occupants is 1 000 kg. It has strayed too close to a black hole having a mass 100 times that of the Sun (Fig. P13.27). The nose of the spacecraft points toward the black hole, and the distance between the nose and the center of the black hole is 10.0 km. (a) Determine the total force on the spacecraft. (b) What is the difference in the gravitational fields acting on the occupants in the nose of the ship and on those in the rear of the ship, farthest from the black hole? (This difference in accelerations grows rapidly as the ship approaches the black hole. It puts the body of the ship under extreme tension and eventually tears it apart.)

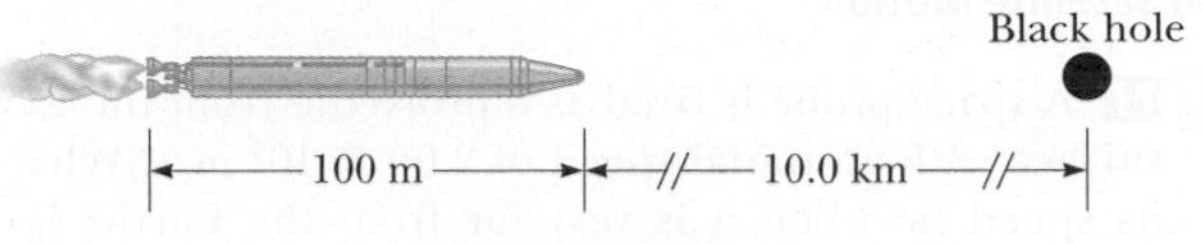

Figure P13.27

Section 13.5 Gravitational Potential Energy

In Problems 28 through 42, assume $U = 0$ at $r = \infty$.

28. A satellite in Earth orbit has a mass of 100 kg and is at an altitude of 2.00×10^6 m. (a) What is the potential energy of the satellite–Earth system? (b) What is the magnitude of the gravitational force exerted by the Earth on the satellite? (c) **What If?** What force, if any, does the satellite exert on the Earth?

29. How much work is done by the Moon's gravitational field on a 1 000-kg meteor as it comes in from outer space and impacts on the Moon's surface?

30. Q|C A system consists of three particles, each of mass 5.00 g, located at the corners of an equilateral triangle with sides of 30.0 cm. (a) Calculate the potential energy of the system. (b) Assume the particles are released simultaneously. Describe the subsequent motion of each. Will any collisions take place? Explain.

31. After the Sun exhausts its nuclear fuel, its ultimate fate will be to collapse to a *white dwarf* state. In this state, it would have approximately the same mass as it has now, but its radius would be equal to the radius of the Earth. Calculate (a) the average density of the white dwarf, (b) the surface free-fall acceleration, and (c) the gravitational potential energy associated with a 1.00-kg object at the surface of the white dwarf.

32. An object is released from rest at an altitude h above the surface of the Earth. (a) Show that its speed at a distance r from the Earth's center, where $R_E \le r \le R_E + h$, is

$$v = \sqrt{2GM_E\left(\frac{1}{r} - \frac{1}{R_E + h}\right)}$$

(b) Assume the release altitude is 500 km. Perform the integral

$$\Delta t = \int_i^f dt = -\int_i^f \frac{dr}{v}$$

to find the time of fall as the object moves from the release point to the Earth's surface. The negative sign appears because the object is moving opposite to the radial direction, so its speed is $v = -dr/dt$. Perform the integral numerically.

33. At the Earth's surface, a projectile is launched straight up at a speed of 10.0 km/s. To what height will it rise? Ignore air resistance and the rotation of the Earth.

Section 13.6 Energy Considerations in Planetary and Satellite Motion

34. M A space probe is fired as a projectile from the Earth's surface with an initial speed of 2.00×10^4 m/s. What will its speed be when it is very far from the Earth? Ignore atmospheric friction and the rotation of the Earth.

35. Q|C A 1 000-kg satellite orbits the Earth at a constant altitude of 100 km. (a) How much energy must be added to the system to move the satellite into a circular orbit with altitude 200 km? What are the changes in the system's (b) kinetic energy and (c) potential energy?

36. S A "treetop satellite" moves in a circular orbit just above the surface of a planet, assumed to offer no air resistance. Show that its orbital speed v and the escape speed from the planet are related by the expression $v_{esc} = \sqrt{2}v$.

37. A comet of mass 1.20×10^{10} kg moves in an elliptical orbit around the Sun. Its distance from the Sun ranges between 0.500 AU and 50.0 AU. (a) What is the eccentricity of its orbit? (b) What is its period? (c) At aphelion, what is the potential energy of the comet–Sun system? *Note:* 1 AU = one astronomical unit = the average distance from the Sun to the Earth = 1.496×10^{11} m.

38. (a) What is the minimum speed, relative to the Sun, necessary for a spacecraft to escape the solar system if it starts at the Earth's orbit? (b) *Voyager 1* achieved a maximum speed of 125 000 km/h on its way to photograph Jupiter. Beyond what distance from the Sun is this speed sufficient to escape the solar system?

39. A satellite of mass 200 kg is placed into Earth orbit at a height of 200 km above the surface. (a) Assuming a circular orbit, how long does the satellite take to complete one orbit? (b) What is the satellite's speed? (c) Starting from the satellite on the Earth's surface, what is the minimum energy input necessary to place this satellite in orbit? Ignore air resistance but include the effect of the planet's daily rotation.

40. S A satellite of mass m, originally on the surface of the Earth, is placed into Earth orbit at an altitude h. (a) Assuming a circular orbit, how long does the satellite take to complete one orbit? (b) What is the satellite's speed? (c) What is the minimum energy input necessary to place this satellite in orbit? Ignore air resistance but include the effect of the planet's daily rotation. Represent the mass and radius of the Earth as M_E and R_E, respectively.

41. Q|C Ganymede is the largest of Jupiter's moons. Consider a rocket on the surface of Ganymede, at the point farthest from the planet (Fig. P13.41). Model the rocket as a particle. (a) Does the presence of Ganymede make Jupiter exert a larger, smaller, or same size force on the rocket compared with the force it would exert if Ganymede were not interposed? (b) Determine the escape speed for the rocket from the planet–satellite system. The radius of Ganymede is 2.64×10^6 m, and its mass is 1.495×10^{23} kg. The distance between Jupiter and Ganymede is 1.071×10^9 m, and the mass of Jupiter is 1.90×10^{27} kg. Ignore the motion of Jupiter and Ganymede as they revolve about their center of mass.

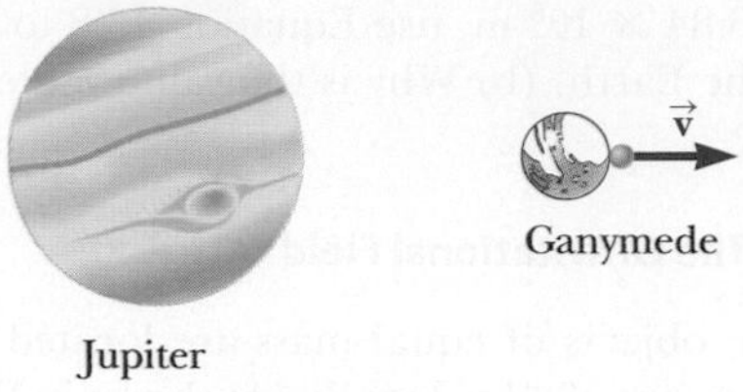

Figure P13.41

42. S A satellite moves around the Earth in a circular orbit of radius r. (a) What is the speed v_i of the satellite? (b) Suddenly, an explosion breaks the satellite into two pieces, with masses m and $4m$. Immediately after the explosion, the smaller piece of mass m is stationary with respect to the Earth and falls directly toward the Earth. What is the speed v of the larger piece immediately after the explosion? (c) Because of the increase in its speed, this larger piece now moves in a new elliptical orbit. Find its distance away from the center of the Earth when it reaches the other end of the ellipse.

Additional Problems

43. **Review.** A cylindrical habitat in space 6.00 km in diameter and 30.0 km long has been proposed (by G. K. O'Neill, 1974). Such a habitat would have cities, land, and lakes on the inside surface and air and clouds in the center. They would all be held in place by rotation of the cylinder about its long axis. How fast would the cylinder have to rotate to imitate the Earth's gravitational field at the walls of the cylinder?

44. Q|C A rocket is fired straight up through the atmosphere from the South Pole, burning out at an altitude of 250 km when traveling at 6.00 km/s. (a) What maximum distance from the Earth's surface does it travel before falling back to the Earth? (b) Would its maximum distance from the surface be larger if the same rocket were fired with the same fuel load from a launch site on the equator? Why or why not?

45. Let Δg_M represent the difference in the gravitational fields produced by the Moon at the points on the Earth's surface nearest to and farthest from the Moon. Find the fraction $\Delta g_M/g$, where g is the Earth's gravitational field. (This difference is responsible for the occurrence of the *lunar tides* on the Earth.)

46. *Why is the following situation impossible?* A spacecraft is launched into a circular orbit around the Earth and circles the Earth once an hour.

47. (a) A space vehicle is launched vertically upward from the Earth's surface with an initial speed of 8.76 km/s, which is less than the escape speed of 11.2 km/s. What maximum height does it attain? (b) A meteoroid falls toward the Earth. It is essentially at rest with respect to the Earth when it is at a height of 2.51×10^7 m above the Earth's surface. With what speed does the meteorite (a meteoroid that survives to impact the Earth's surface) strike the Earth?

48. **S** (a) A space vehicle is launched vertically upward from the Earth's surface with an initial speed of v_i that is comparable to but less than the escape speed v_{esc}. What maximum height does it attain? (b) A meteoroid falls toward the Earth. It is essentially at rest with respect to the Earth when it is at a height h above the Earth's surface. With what speed does the meteorite (a meteoroid that survives to impact the Earth's surface) strike the Earth? (c) **What If?** Assume a baseball is tossed up with an initial speed that is very small compared to the escape speed. Show that the result from part (a) is consistent with Equation 4.12.

49. **Q|C** Assume you are agile enough to run across a horizontal surface at 8.50 m/s, independently of the value of the gravitational field. What would be (a) the radius and (b) the mass of an airless spherical asteroid of uniform density 1.10×10^3 kg/m^3 on which you could launch yourself into orbit by running? (c) What would be your period? (d) Would your running significantly affect the rotation of the asteroid? Explain.

50. **Q|C** A sleeping area for a long space voyage consists of two cabins each connected by a cable to a central hub as shown in Figure P13.50. The cabins are set spinning around the hub axis, which is connected to the rest of the spacecraft to generate artificial gravity in the cabins. A space traveler lies in a bed parallel to the outer wall as shown in Figure P13.50. (a) With $r = 10.0$ m, what would the angular speed of the 60.0-kg traveler need to be if he is to experience half his normal Earth weight? (b) If the astronaut stands up perpendicular to the bed, without holding on to anything with his hands, will his head be moving at a faster, a slower, or the same tangential speed as his feet? Why? (c) Why is the action in part (b) dangerous?

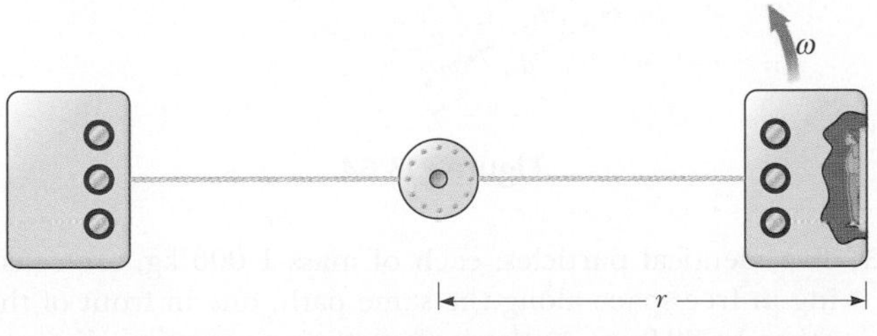

Figure P13.50

51. **M** Two hypothetical planets of masses m_1 and m_2 and radii r_1 and r_2, respectively, are nearly at rest when they are an infinite distance apart. Because of their gravitational attraction, they head toward each other on a collision course. (a) When their center-to-center separation is d, find expressions for the speed of each planet and for their relative speed. (b) Find the kinetic energy of each planet just before they collide, taking $m_1 = 2.00 \times 10^{24}$ kg, $m_2 = 8.00 \times 10^{24}$ kg, $r_1 = 3.00 \times 10^6$ m, and $r_2 = 5.00 \times 10^6$ m. *Note:* Both the energy and momentum of the isolated two-planet system are constant.

52. **S** **GP** Two spheres having masses M and $2M$ and radii R and $3R$, respectively, are simultaneously released from rest when the distance between their centers is $12R$. Assume the two spheres interact only with each other and we wish to find the speeds with which they collide. (a) What *two* isolated system models are appropriate for this system? (b) Write an equation from one of the models and solve it for $\vec{\mathbf{v}}_1$, the velocity of the sphere of mass M at any time after release in terms of $\vec{\mathbf{v}}_2$, the velocity of $2M$. (c) Write an equation from the other model and solve it for speed v_1 in terms of speed v_2 when the spheres collide. (d) Combine the two equations to find the two speeds v_1 and v_2 when the spheres collide.

53. A ring of matter is a familiar structure in planetary and stellar astronomy. Examples include Saturn's rings and a ring nebula. Consider a uniform ring of mass 2.36×10^{20} kg and radius 1.00×10^8 m. An object of mass 1 000 kg is placed at a point A on the axis of the ring, 2.00×10^8 m from the center of the ring (Fig. P13.53). When the object is released, the attraction of the ring makes the object move along the axis toward the center of the ring (point B). (a) Calculate the gravitational potential energy of the object–ring system when the object is at A. (b) Calculate the gravitational potential energy of the system when the object is at B. (c) Calculate the speed of the object as it passes through B.

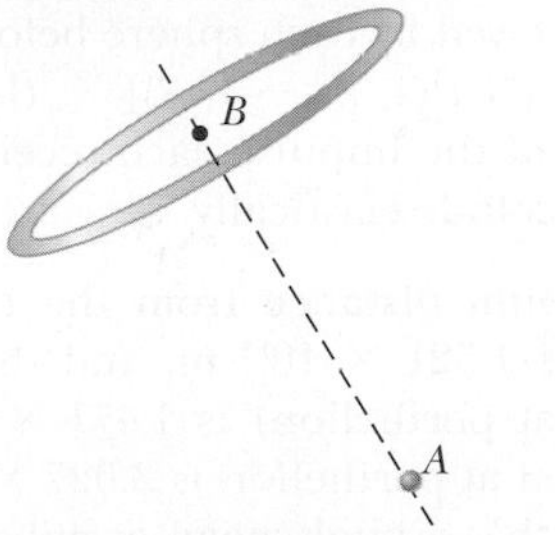

Figure P13.53

54. (a) Show that the rate of change of the free-fall acceleration with vertical position near the Earth's surface is

$$\frac{dg}{dr} = -\frac{2GM_E}{R_E^{\ 3}}$$

This rate of change with position is called a *gradient.* (b) Assuming h is small in comparison to the radius of the

Earth, show that the difference in free-fall acceleration between two points separated by vertical distance h is

$$|\Delta g| = \frac{2GM_E h}{R_E^{\ 3}}$$

(c) Evaluate this difference for $h = 6.00$ m, a typical height for a two-story building.

55. Review. As an astronaut, you observe a small planet to be spherical. After landing on the planet, you set off, walking always straight ahead, and find yourself returning to your spacecraft from the opposite side after completing a lap of 25.0 km. You hold a hammer and a falcon feather at a height of 1.40 m, release them, and observe that they fall together to the surface in 29.2 s. Determine the mass of the planet.

56. **S** A certain quaternary star system consists of three stars, each of mass m, moving in the same circular orbit of radius r about a central star of mass M. The stars orbit in the same sense and are positioned one-third of a revolution apart from one another. Show that the period of each of the three stars is given by

$$T = 2\pi\sqrt{\frac{r^3}{G(M + m/\sqrt{3})}}$$

57. Studies of the relationship of the Sun to our galaxy—the Milky Way—have revealed that the Sun is located near the outer edge of the galactic disc, about 30 000 ly (1 ly = 9.46×10^{15} m) from the center. The Sun has an orbital speed of approximately 250 km/s around the galactic center. (a) What is the period of the Sun's galactic motion? (b) What is the order of magnitude of the mass of the Milky Way galaxy? (c) Suppose the galaxy is made mostly of stars of which the Sun is typical. What is the order of magnitude of the number of stars in the Milky Way?

58. **S** **Review.** Two identical hard spheres, each of mass m and radius r, are released from rest in otherwise empty space with their centers separated by the distance R. They are allowed to collide under the influence of their gravitational attraction. (a) Show that the magnitude of the impulse received by each sphere before they make contact is given by $[Gm^3(1/2r - 1/R)]^{1/2}$. (b) **What If?** Find the magnitude of the impulse each receives during their contact if they collide elastically.

59. The maximum distance from the Earth to the Sun (at aphelion) is 1.521×10^{11} m, and the distance of closest approach (at perihelion) is 1.471×10^{11} m. The Earth's orbital speed at perihelion is 3.027×10^4 m/s. Determine (a) the Earth's orbital speed at aphelion and the kinetic and potential energies of the Earth–Sun system (b) at perihelion, and (c) at aphelion. (d) Is the total energy of the system constant? Explain. Ignore the effect of the Moon and other planets.

60. **Q|C** Many people assume air resistance acting on a moving object will always make the object slow down. It can, however, actually be responsible for making the object speed up. Consider a 100-kg Earth satellite in a circular orbit at an altitude of 200 km. A small force of air resistance makes the satellite drop into a circular orbit with an altitude of 100 km. (a) Calculate the satellite's initial speed. (b) Calculate its final speed in this process. (c) Calculate the initial energy of the satellite–Earth system. (d) Calculate the final energy of the system. (e) Show that the system has lost mechanical energy and find the amount of the loss due to friction. (f) What force makes the satellite's speed increase? *Hint:* You will find a free-body diagram useful in explaining your answer.

61. X-ray pulses from Cygnus X-1, the first black hole to be identified and a celestial x-ray source, have been recorded during high-altitude rocket flights. The signals can be interpreted as originating when a blob of ionized matter orbits a black hole with a period of 5.0 ms. If the blob is in a circular orbit about a black hole whose mass is $20M_{\text{Sun}}$, what is the orbit radius?

62. **S** Show that the minimum period for a satellite in orbit around a spherical planet of uniform density ρ is

$$T_{\min} = \sqrt{\frac{3\pi}{G\rho}}$$

independent of the planet's radius.

63. **S** Astronomers detect a distant meteoroid moving along a straight line that, if extended, would pass at a distance $3R_E$ from the center of the Earth, where R_E is the Earth's radius. What minimum speed must the meteoroid have if it is *not* to collide with the Earth?

64. **S** Two stars of masses M and m, separated by a distance d, revolve in circular orbits about their center of mass (Fig. P13.64). Show that each star has a period given by

$$T^2 = \frac{4\pi^2 d^3}{G(M+m)}$$

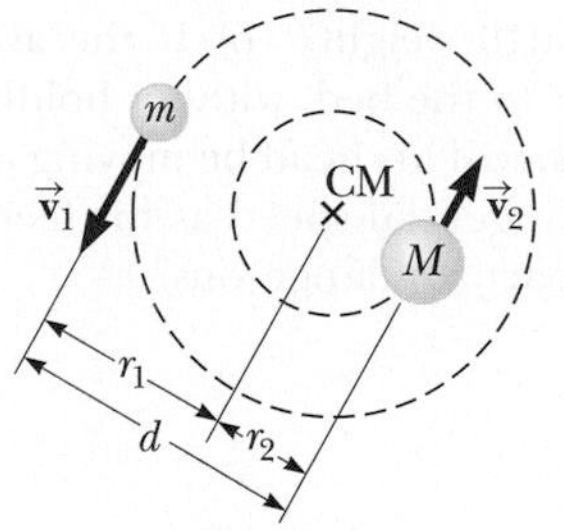

Figure P13.64

65. Two identical particles, each of mass 1 000 kg, are coasting in free space along the same path, one in front of the other by 20.0 m. At the instant their separation distance has this value, each particle has precisely the same velocity of $800\,\hat{\mathbf{i}}$ m/s. What are their precise velocities when they are 2.00 m apart?

66. Consider an object of mass m, not necessarily small compared with the mass of the Earth, released at a distance of 1.20×10^7 m from the center of the Earth. Assume the Earth and the object behave as a pair of particles, isolated from the rest of the Universe. (a) Find the magnitude of the acceleration a_{rel} with which each starts to move relative to the other as a function of m. Evaluate the acceleration (b) for $m = 5.00$ kg, (c) for $m = 2\,000$ kg, and (d) for $m = 2.00 \times 10^{24}$ kg. (e) Describe the pattern of variation of a_{rel} with m.

67. As thermonuclear fusion proceeds in its core, the Sun loses mass at a rate of 3.64×10^9 kg/s. During the 5 000-yr period of recorded history, by how much has the length of the year changed due to the loss of mass from the Sun? *Suggestions:* Assume the Earth's orbit is circular. No external torque acts on the Earth–Sun system, so the angular momentum of the Earth is constant.

Challenge Problems

68. The Solar and Heliospheric Observatory (SOHO) spacecraft has a special orbit, located between the Earth and the Sun along the line joining them, and it is always close enough to the Earth to transmit data easily. Both objects exert gravitational forces on the observatory. It moves around the Sun in a near-circular orbit that is smaller than the Earth's circular orbit. Its period, however, is not less than 1 yr but just equal to 1 yr. Show that its distance from the Earth must be 1.48×10^9 m. In 1772, Joseph Louis Lagrange determined theoretically the special location allowing this orbit. *Suggestions:* Use data that are precise to four digits. The mass of the Earth is 5.974×10^{24} kg. You will not be able to easily solve the equation you generate; instead, use a computer to verify that 1.48×10^9 m is the correct value.

69. The oldest artificial satellite still in orbit is *Vanguard I*, launched March 3, 1958. Its mass is 1.60 kg. Neglecting atmospheric drag, the satellite would still be in its initial orbit, with a minimum distance from the center of the Earth of 7.02 Mm and a speed at this perigee point of 8.23 km/s. For this orbit, find (a) the total energy of the satellite–Earth system and (b) the magnitude of the angular momentum of the satellite. (c) At apogee, find the satellite's speed and its distance from the center of the Earth. (d) Find the semimajor axis of its orbit. (e) Determine its period.

chapter 14

Fluid Mechanics

Fish congregate around a reef in Hawaii searching for food. How do fish such as the yellow butterflyfish in the front control their movements up and down in the water? We'll find out in this chapter. *(Ed Robinson/Pacific Stock/Photolibrary)*

Matter is normally classified as being in one of three states: solid, liquid, or gas. From everyday experience we know that a solid has a definite volume and shape, a liquid has a definite volume but no definite shape, and an unconfined gas has neither a definite volume nor a definite shape. These descriptions help us picture the states of matter, but they are somewhat artificial. For example, asphalt and plastics are normally considered solids, but over long time intervals they tend to flow like liquids. Likewise, most substances can be a solid, a liquid, or a gas (or a combination of any of these three), depending on the temperature and pressure. In general, the time interval required for a particular substance to change its shape in response to an external force determines whether we treat the substance as a solid, a liquid, or a gas.

A **fluid** is a collection of molecules that are randomly arranged and held together by weak cohesive forces and by forces exerted by the walls of a container. Both liquids and gases are fluids.

In our treatment of the mechanics of fluids, we'll be applying principles we have already discussed. First, we consider the mechanics of a fluid at rest, that is, *fluid statics,* and then study fluids in motion, that is, *fluid dynamics.*

14.1 Pressure

Fluids do not sustain shearing stresses or tensile stresses; therefore, the only stress that can be exerted on an object submerged in a static fluid is one that tends to compress the object from all sides. In other words, the force exerted by a static fluid on an object is always perpendicular to the surfaces of the object as shown in Figure 14.1. We discussed this situation in Section 12.4.

The pressure in a fluid can be measured with the device pictured in Figure 14.2. The device consists of an evacuated cylinder that encloses a light piston connected to a spring. As the device is submerged in a fluid, the fluid presses on the top of the piston and compresses the spring until the inward force exerted by the fluid is balanced by the outward force exerted by the spring. The fluid pressure can be measured directly if the spring is calibrated in advance. If F is the magnitude of the force exerted on the piston and A is the surface area of the piston, the **pressure** P of the fluid at the level to which the device has been submerged is defined as the ratio of the force to the area:

$$P \equiv \frac{F}{A} \tag{14.1}$$

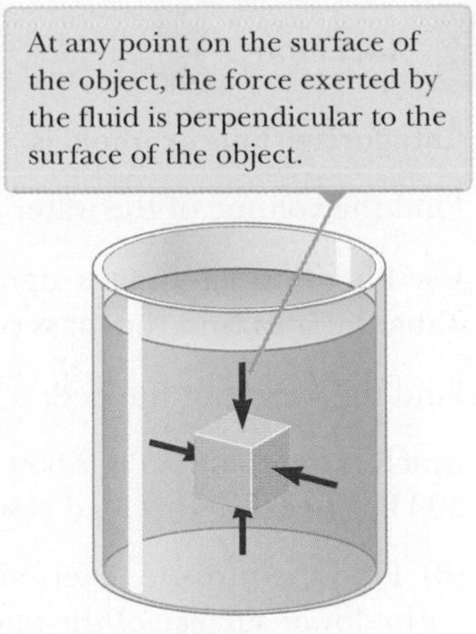

Figure 14.1 The forces exerted by a fluid on the surfaces of a submerged object. (The forces on the front and back sides of the object are not shown.)

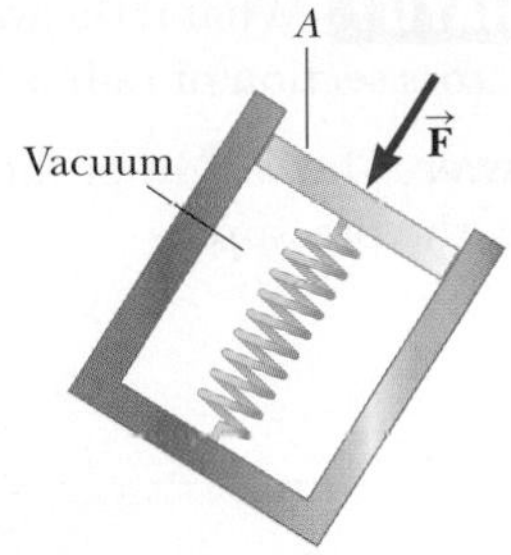

Figure 14.2 A simple device for measuring the pressure exerted by a fluid.

Pressure is a scalar quantity because it is proportional to the magnitude of the force on the piston.

If the pressure varies over an area, the infinitesimal force dF on an infinitesimal surface element of area dA is

$$dF = P\,dA \tag{14.2}$$

where P is the pressure at the location of the area dA. To calculate the total force exerted on a surface of a container, we must integrate Equation 14.2 over the surface.

The units of pressure are newtons per square meter (N/m^2) in the SI system. Another name for the SI unit of pressure is the **pascal** (Pa):

$$1\text{ Pa} \equiv 1\text{ N/m}^2 \tag{14.3}$$

For a tactile demonstration of the definition of pressure, hold a tack between your thumb and forefinger, with the point of the tack on your thumb and the head of the tack on your forefinger. Now *gently* press your thumb and forefinger together. Your thumb will begin to feel pain immediately while your forefinger will not. The tack is exerting the same force on both your thumb and forefinger, but the pressure on your thumb is much larger because of the small area over which the force is applied.

Pitfall Prevention 14.1

Force and Pressure

Equations 14.1 and 14.2 make a clear distinction between force and pressure. Another important distinction is that *force is a vector* and *pressure is a scalar.* There is no direction associated with pressure, but the direction of the force associated with the pressure is perpendicular to the surface on which the pressure acts.

***Quick Quiz* 14.1** Suppose you are standing directly behind someone who steps back and accidentally stomps on your foot with the heel of one shoe. Would you be better off if that person were **(a)** a large, male professional basketball player wearing sneakers or **(b)** a petite woman wearing spike-heeled shoes?

Example 14.1 The Water Bed

The mattress of a water bed is 2.00 m long by 2.00 m wide and 30.0 cm deep.

(A) Find the weight of the water in the mattress.

SOLUTION

Conceptualize Think about carrying a jug of water and how heavy it is. Now imagine a sample of water the size of a water bed. We expect the weight to be relatively large.

continued

14.1 *cont.*

Categorize This example is a substitution problem.

Find the volume of the water filling the mattress: $V = (2.00 \text{ m})(2.00 \text{ m})(0.300 \text{ m}) = 1.20 \text{ m}^3$

Use Equation 1.1 and the density of fresh water (see Table 14.1) to find the mass of the water bed: $M = \rho V = (1\,000 \text{ kg/m}^3)(1.20 \text{ m}^3) = 1.20 \times 10^3 \text{ kg}$

Find the weight of the bed: $Mg = (1.20 \times 10^3 \text{ kg})(9.80 \text{ m/s}^2) = \boxed{1.18 \times 10^4 \text{ N}}$

which is approximately 2 650 lb. (A regular bed, including mattress, box spring, and metal frame, weighs approximately 300 lb.) Because this load is so great, it is best to place a water bed in the basement or on a sturdy, well- supported floor.

(B) Find the pressure exerted by the water bed on the floor when the bed rests in its normal position. Assume the entire lower surface of the bed makes contact with the floor.

SOLUTION

When the water bed is in its normal position, the area in contact with the floor is 4.00 m². Use Equation 14.1 to find the pressure:

$$P = \frac{1.18 \times 10^4 \text{ N}}{4.00 \text{ m}^2} = \boxed{2.94 \times 10^3 \text{ Pa}}$$

WHAT IF? What if the water bed is replaced by a 300-lb regular bed that is supported by four legs? Each leg has a circular cross section of radius 2.00 cm. What pressure does this bed exert on the floor?

Answer The weight of the regular bed is distributed over four circular cross sections at the bottom of the legs. Therefore, the pressure is

$$P = \frac{F}{A} = \frac{mg}{4(\pi r^2)} = \frac{300 \text{ lb}}{4\pi(0.020\,0 \text{ m})^2}\left(\frac{1 \text{ N}}{0.225 \text{ lb}}\right)$$

$$= 2.65 \times 10^5 \text{ Pa}$$

This result is almost 100 times larger than the pressure due to the water bed! The weight of the regular bed, even though it is much less than the weight of the water bed, is applied over the very small area of the four legs. The high pressure on the floor at the feet of a regular bed could cause dents in wood floors or permanently crush carpet pile.

14.2 Variation of Pressure with Depth

As divers well know, water pressure increases with depth. Likewise, atmospheric pressure decreases with increasing altitude; for this reason, aircraft flying at high altitudes must have pressurized cabins for the comfort of the passengers.

TABLE 14.1 *Densities of Some Common Substances at Standard Temperature (0°C) and Pressure (Atmospheric)*

Substance	ρ (kg/m³)	Substance	ρ (kg/m³)
Air	1.29	Iron	7.86×10^3
Air (at 20°C and atmospheric pressure)	1.20	Lead	11.3×10^3
		Mercury	13.6×10^3
Aluminum	2.70×10^3	Nitrogen gas	1.25
Benzene	0.879×10^3	Oak	0.710×10^3
Brass	8.4×10^3	Osmium	22.6×10^3
Copper	8.92×10^3	Oxygen gas	1.43
Ethyl alcohol	0.806×10^3	Pine	0.373×10^3
Fresh water	1.00×10^3	Platinum	21.4×10^3
Glycerin	1.26×10^3	Seawater	1.03×10^3
Gold	19.3×10^3	Silver	10.5×10^3
Helium gas	1.79×10^{-1}	Tin	7.30×10^3
Hydrogen gas	8.99×10^{-2}	Uranium	19.1×10^3
Ice	0.917×10^3		

We now show how the pressure in a liquid increases with depth. As Equation 1.1 describes, the *density* of a substance is defined as its mass per unit volume; Table 14.1 lists the densities of various substances. These values vary slightly with temperature because the volume of a substance is dependent on temperature (as shown in Chapter 19). Under standard conditions (at 0°C and at atmospheric pressure), the densities of gases are about $\frac{1}{1\,000}$ the densities of solids and liquids. This difference in densities implies that the average molecular spacing in a gas under these conditions is about ten times greater than that in a solid or liquid.

Now consider a liquid of density ρ at rest as shown in Figure 14.3. We assume ρ is uniform throughout the liquid, which means the liquid is incompressible. Let us select a parcel of the liquid contained within an imaginary block of cross-sectional area A extending from depth d to depth $d + h$. The liquid external to our parcel exerts forces at all points on the surface of the parcel, perpendicular to the surface. The pressure exerted by the liquid on the bottom face of the parcel is P, and the pressure on the top face is P_0. Therefore, the upward force exerted by the outside fluid on the bottom of the parcel has a magnitude PA, and the downward force exerted on the top has a magnitude P_0A. The mass of liquid in the parcel is $M = \rho V = \rho Ah$; therefore, the weight of the liquid in the parcel is $Mg = \rho Ahg$. Because the parcel is in equilibrium, the net force acting on it must be zero. Choosing upward to be the positive y direction, we see that

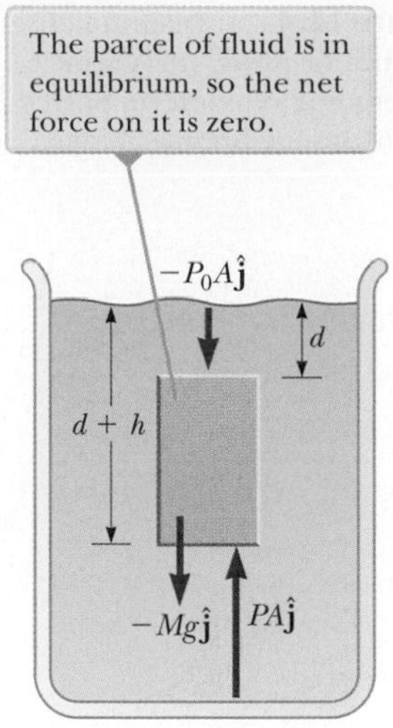

Figure 14.3 A parcel of fluid in a larger volume of fluid is singled out.

$$\sum \vec{\mathbf{F}} = PA\hat{\mathbf{j}} - P_0A\hat{\mathbf{j}} - Mg\hat{\mathbf{j}} = 0$$

or

$$PA - P_0A - \rho Ahg = 0$$

$$P = P_0 + \rho gh \qquad \textbf{(14.4)}$$

◀ **Variation of pressure with depth**

That is, the pressure P at a depth h below a point in the liquid at which the pressure is P_0 is greater by an amount ρgh. If the liquid is open to the atmosphere and P_0 is the pressure at the surface of the liquid, then P_0 is **atmospheric pressure.** In our calculations and working of end-of-chapter problems, we usually take atmospheric pressure to be

$$P_0 = 1.00 \text{ atm} = 1.013 \times 10^5 \text{ Pa}$$

Equation 14.4 implies that the pressure is the same at all points having the same depth, independent of the shape of the container.

Because the pressure in a fluid depends on depth and on the value of P_0, any increase in pressure at the surface must be transmitted to every other point in the fluid. This concept was first recognized by French scientist Blaise Pascal (1623–1662) and is called **Pascal's law: a change in the pressure applied to a fluid is transmitted undiminished to every point of the fluid and to the walls of the container.**

◀ **Pascal's law**

An important application of Pascal's law is the hydraulic press illustrated in Figure 14.4a (page 406). A force of magnitude F_1 is applied to a small piston of surface area A_1. The pressure is transmitted through an incompressible liquid to a larger piston of surface area A_2. Because the pressure must be the same on both sides, $P = F_1/A_1 = F_2/A_2$. Therefore, the force F_2 is greater than the force F_1 by a factor of A_2/A_1. By designing a hydraulic press with appropriate areas A_1 and A_2, a large output force can be applied by means of a small input force. Hydraulic brakes, car lifts, hydraulic jacks, and forklifts all make use of this principle (Fig. 14.4b).

Because liquid is neither added to nor removed from the system, the volume of liquid pushed down on the left in Figure 14.4a as the piston moves downward through a displacement Δx_1 equals the volume of liquid pushed up on the right as the right piston moves upward through a displacement Δx_2. That is, $A_1 \Delta x_1 = A_2 \Delta x_2$; therefore, $A_2/A_1 = \Delta x_1/\Delta x_2$. We have already shown that $A_2/A_1 = F_2/F_1$. Therefore, $F_2/F_1 = \Delta x_1/\Delta x_2$, so $F_1 \Delta x_1 = F_2 \Delta x_2$. Each side of this equation is the work done by the force on its respective piston. Therefore, the work done by $\vec{\mathbf{F}}_1$ on the input piston equals the work done by $\vec{\mathbf{F}}_2$ on the output piston, as it must to conserve energy.

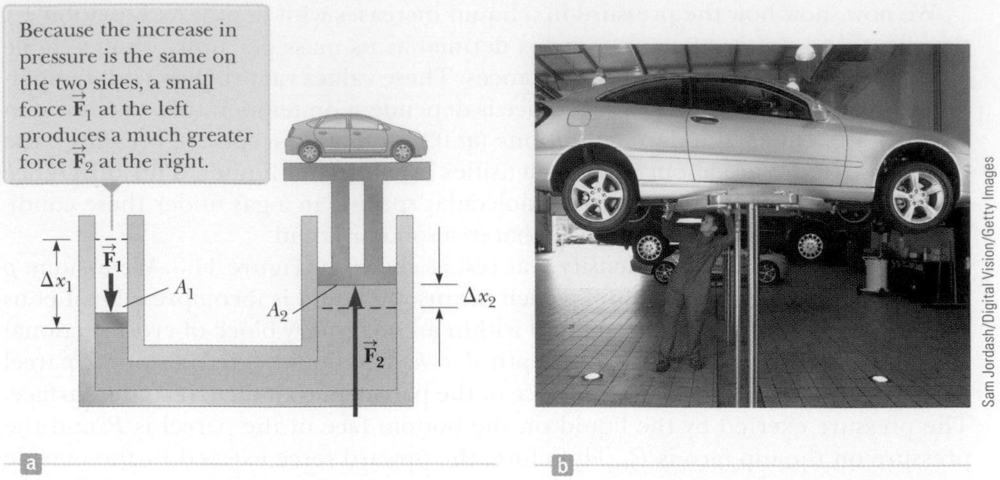

Figure 14.4 (a) Diagram of a hydraulic press. (b) A vehicle undergoing repair is supported by a hydraulic lift in a garage.

***Quick Quiz* 14.2** The pressure at the bottom of a filled glass of water (ρ = 1 000 kg/m^3) is P. The water is poured out, and the glass is filled with ethyl alcohol (ρ = 806 kg/m^3). What is the pressure at the bottom of the glass? **(a)** smaller than P **(b)** equal to P **(c)** larger than P **(d)** indeterminate

Example 14.2 The Car Lift

In a car lift used in a service station, compressed air exerts a force on a small piston that has a circular cross section and a radius of 5.00 cm. This pressure is transmitted by a liquid to a piston that has a radius of 15.0 cm. What force must the compressed air exert to lift a car weighing 13 300 N? What air pressure produces this force?

SOLUTION

Conceptualize Review the material just discussed about Pascal's law to understand the operation of a car lift.

Categorize This example is a substitution problem.

Solve $F_1/A_1 = F_2/A_2$ for F_1:

$$F_1 = \left(\frac{A_1}{A_2}\right)F_2 = \frac{\pi(5.00 \times 10^{-2}\text{ m})^2}{\pi(15.0 \times 10^{-2}\text{ m})^2}(1.33 \times 10^4\text{ N})$$

$$= 1.48 \times 10^3\text{ N}$$

Use Equation 14.1 to find the air pressure that produces this force:

$$P = \frac{F_1}{A_1} = \frac{1.48 \times 10^3\text{ N}}{\pi(5.00 \times 10^{-2}\text{ m})^2}$$

$$= 1.88 \times 10^5\text{ Pa}$$

This pressure is approximately twice atmospheric pressure.

Example 14.3 A Pain in Your Ear

Estimate the force exerted on your eardrum due to the water when you are swimming at the bottom of a pool that is 5.0 m deep.

SOLUTION

Conceptualize As you descend in the water, the pressure increases. You may have noticed this increased pressure in your ears while diving in a swimming pool, a lake, or the ocean. We can find the pressure difference exerted on the eardrum from the depth given in the problem; then, after estimating the ear drum's surface area, we can determine the net force the water exerts on it.

14.3 *cont.*

Categorize This example is a substitution problem.

The air inside the middle ear is normally at atmospheric pressure P_0. Therefore, to find the net force on the eardrum, we must consider the difference between the total pressure at the bottom of the pool and atmospheric pressure. Let's estimate the surface area of the eardrum to be approximately $1\text{ cm}^2 = 1 \times 10^{-4}\text{ m}^2$.

Use Equation 14.4 to find this pressure difference:

$$P_{\text{bot}} - P_0 = \rho g h$$
$$= (1.00 \times 10^3\text{ kg/m}^3)(9.80\text{ m/s}^2)(5.0\text{ m}) = 4.9 \times 10^4\text{ Pa}$$

Use Equation 14.1 to find the net force on the ear:

$$F = (P_{\text{bot}} - P_0)A = (4.9 \times 10^4\text{ Pa})(1 \times 10^{-4}\text{ m}^2) \approx 5\text{ N}$$

Because a force of this magnitude on the eardrum is extremely uncomfortable, swimmers often "pop their ears" while under water, an action that pushes air from the lungs into the middle ear. Using this technique equalizes the pressure on the two sides of the eardrum and relieves the discomfort.

Example 14.4 The Force on a Dam

Water is filled to a height H behind a dam of width w (Fig. 14.5). Determine the resultant force exerted by the water on the dam.

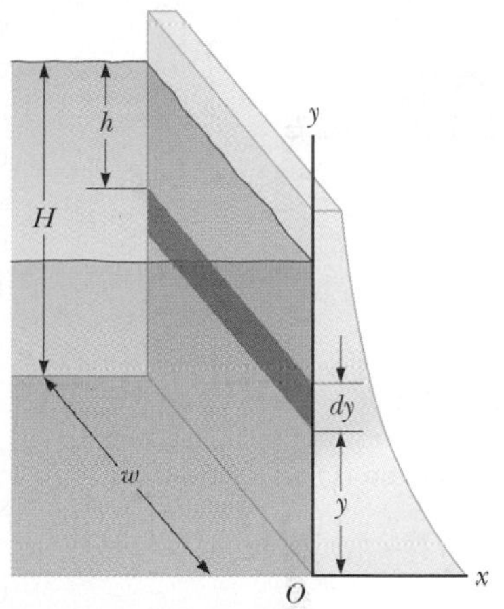

Figure 14.5 (Example 14.4) Water exerts a force on a dam.

SOLUTION

Conceptualize Because pressure varies with depth, we cannot calculate the force simply by multiplying the area by the pressure. As the pressure in the water increases with depth, the force on the adjacent portion of the dam also increases.

Categorize Because of the variation of pressure with depth, we must use integration to solve this example, so we categorize it as an analysis problem.

Analyze Let's imagine a vertical y axis, with $y = 0$ at the bottom of the dam. We divide the face of the dam into narrow horizontal strips at a distance y above the bottom, such as the red strip in Figure 14.5. The pressure on each such strip is due only to the water; atmospheric pressure acts on both sides of the dam.

Use Equation 14.4 to calculate the pressure due to the water at the depth h:

$$P = \rho g h = \rho g(H - y)$$

Use Equation 14.2 to find the force exerted on the shaded strip of area $dA = w\,dy$:

$$dF = P\,dA = \rho g(H - y)w\,dy$$

Integrate to find the total force on the dam:

$$F = \int P\,dA = \int_0^H \rho g(H - y)w\,dy = \tfrac{1}{2}\rho g w H^2$$

Finalize Notice that the thickness of the dam shown in Figure 14.5 increases with depth. This design accounts for the greater force the water exerts on the dam at greater depths.

WHAT IF? What if you were asked to find this force without using calculus? How could you determine its value?

Answer We know from Equation 14.4 that pressure varies linearly with depth. Therefore, the average pressure due to the water over the face of the dam is the average of the pressure at the top and the pressure at the bottom:

$$P_{\text{avg}} = \frac{P_{\text{top}} + P_{\text{bottom}}}{2} = \frac{0 + \rho g H}{2} = \tfrac{1}{2}\rho g H$$

continued

14.4 *cont.*

The total force on the dam is equal to the product of the average pressure and the area of the face of the dam:

$$F = P_{avg}A = (\tfrac{1}{2}\rho g H)(Hw) = \tfrac{1}{2}\rho g w H^2$$

which is the same result we obtained using calculus.

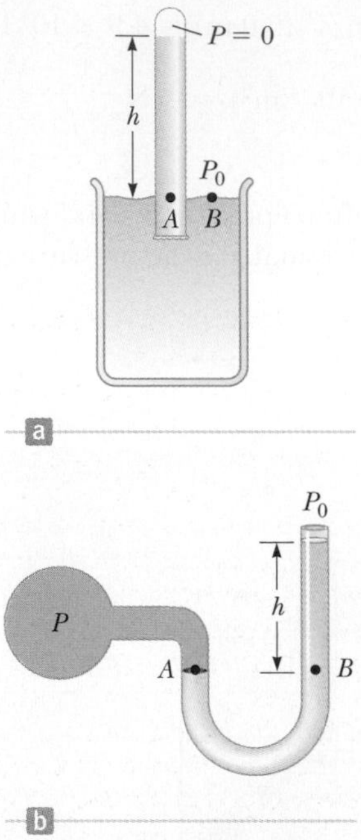

Figure 14.6 Two devices for measuring pressure: (a) a mercury barometer and (b) an open-tube manometer.

14.3 Pressure Measurements

During the weather report on a television news program, the *barometric pressure* is often provided. This reading is the current pressure of the atmosphere, which varies over a small range from the standard value provided earlier. How is this pressure measured?

One instrument used to measure atmospheric pressure is the common barometer, invented by Evangelista Torricelli (1608–1647). A long tube closed at one end is filled with mercury and then inverted into a dish of mercury (Fig. 14.6a). The closed end of the tube is nearly a vacuum, so the pressure at the top of the mercury column can be taken as zero. In Figure 14.6a, the pressure at point *A*, due to the column of mercury, must equal the pressure at point *B*, due to the atmosphere. If that were not the case, there would be a net force that would move mercury from one point to the other until equilibrium is established. Therefore, $P_0 = \rho_{Hg}gh$, where ρ_{Hg} is the density of the mercury and h is the height of the mercury column. As atmospheric pressure varies, the height of the mercury column varies, so the height can be calibrated to measure atmospheric pressure. Let us determine the height of a mercury column for one atmosphere of pressure, $P_0 = 1$ atm $= 1.013 \times 10^5$ Pa:

$$P_0 = \rho_{Hg}gh \rightarrow h = \frac{P_0}{\rho_{Hg}g} = \frac{1.013 \times 10^5 \text{ Pa}}{(13.6 \times 10^3 \text{ kg/m}^3)(9.80 \text{ m/s}^2)} = 0.760 \text{ m}$$

Based on such a calculation, one atmosphere of pressure is defined to be the pressure equivalent of a column of mercury that is exactly 0.760 0 m in height at 0°C.

A device for measuring the pressure of a gas contained in a vessel is the open-tube manometer illustrated in Figure 14.6b. One end of a U-shaped tube containing a liquid is open to the atmosphere, and the other end is connected to a container of gas at pressure *P*. In an equilibrium situation, the pressures at points *A* and *B* must be the same (otherwise, the curved portion of the liquid would experience a net force and would accelerate), and the pressure at *A* is the unknown pressure of the gas. Therefore, equating the unknown pressure *P* to the pressure at point *B*, we see that $P = P_0 + \rho gh$. Again, we can calibrate the height *h* to the pressure *P*.

The difference in the pressures in each part of Figure 14.6 (that is, $P - P_0$) is equal to ρgh. The pressure *P* is called the **absolute pressure,** and the difference $P - P_0$ is called the **gauge pressure.** For example, the pressure you measure in your bicycle tire is gauge pressure.

***Quick Quiz* 14.3** Several common barometers are built, with a variety of fluids. For which of the following fluids will the column of fluid in the barometer be the highest? **(a)** mercury **(b)** water **(c)** ethyl alcohol **(d)** benzene

14.4 Buoyant Forces and Archimedes's Principle

Have you ever tried to push a beach ball down under water (Fig. 14.7a)? It is extremely difficult to do because of the large upward force exerted by the water on the ball. The upward force exerted by a fluid on any immersed object is called a **buoyant force.** We can determine the magnitude of a buoyant force by applying

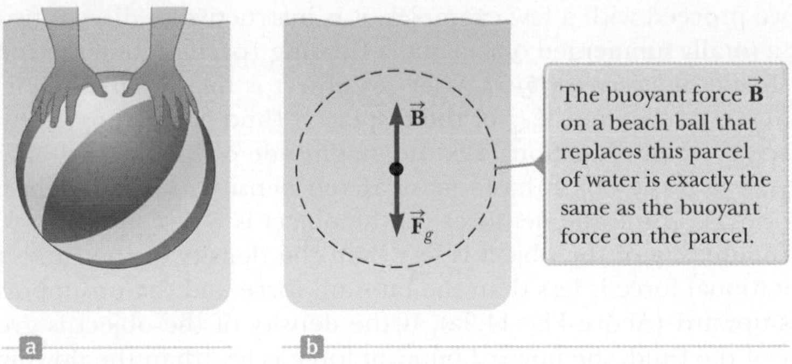

Figure 14.7 (a) A swimmer pushes a beach ball under water. (b) The forces on a beach ball–sized parcel of water.

some logic. Imagine a beach ball–sized parcel of water beneath the water surface as in Figure 14.7b. Because this parcel is in equilibrium, there must be an upward force that balances the downward gravitational force on the parcel. This upward force is the buoyant force, and its magnitude is equal to the weight of the water in the parcel. The buoyant force is the resultant force on the parcel due to all forces applied by the fluid surrounding the parcel.

Now imagine replacing the beach ball–sized parcel of water with a beach ball of the same size. The net force applied by the fluid surrounding the beach ball is the same, regardless of whether it is applied to a beach ball or to a parcel of water. Consequently, **the magnitude of the buoyant force on an object always equals the weight of the fluid displaced by the object.** This statement is known as **Archimedes's principle.**

With the beach ball under water, the buoyant force, equal to the weight of a beach ball–sized parcel of water, is much larger than the weight of the beach ball. Therefore, there is a large net upward force, which explains why it is so hard to hold the beach ball under the water. Note that Archimedes's principle does not refer to the makeup of the object experiencing the buoyant force. The object's composition is not a factor in the buoyant force because the buoyant force is exerted by the surrounding fluid.

To better understand the origin of the buoyant force, consider a cube of solid material immersed in a liquid as in Figure 14.8. According to Equation 14.4, the pressure P_{bot} at the bottom of the cube is greater than the pressure P_{top} at the top by an amount $\rho_{\text{fluid}}gh$, where h is the height of the cube and ρ_{fluid} is the density of the fluid. The pressure at the bottom of the cube causes an *upward* force equal to $P_{\text{bot}}A$, where A is the area of the bottom face. The pressure at the top of the cube causes a *downward* force equal to $P_{\text{top}}A$. The resultant of these two forces is the buoyant force $\vec{\mathbf{B}}$ with magnitude

$$B = (P_{\text{bot}} - P_{\text{top}})A = (\rho_{\text{fluid}}gh)A$$

$$B = \rho_{\text{fluid}}gV_{\text{disp}} \qquad (14.5)$$

where $V_{\text{disp}} = Ah$ is the volume of the fluid displaced by the cube. Because the product $\rho_{\text{fluid}}V_{\text{disp}}$ is equal to the mass of fluid displaced by the object,

$$B = Mg$$

where Mg is the weight of the fluid displaced by the cube. This result is consistent with our initial statement about Archimedes's principle above, based on the discussion of the beach ball.

Under normal conditions, the weight of a fish in the opening photograph for this chapter is slightly greater than the buoyant force on the fish. Hence, the fish would sink if it did not have some mechanism for adjusting the buoyant force. The fish accomplishes that by internally regulating the size of its air-filled swim bladder to increase its volume and the magnitude of the buoyant force acting on it, according to Equation 14.5. In this manner, fish are able to swim to various depths.

Archimedes
Greek Mathematician, Physicist, and Engineer (c. 287–212 BC)
Archimedes was perhaps the greatest scientist of antiquity. He was the first to compute accurately the ratio of a circle's circumference to its diameter, and he also showed how to calculate the volume and surface area of spheres, cylinders, and other geometric shapes. He is well known for discovering the nature of the buoyant force and was also a gifted inventor. One of his practical inventions, still in use today, is Archimedes's screw, an inclined, rotating, coiled tube used originally to lift water from the holds of ships. He also invented the catapult and devised systems of levers, pulleys, and weights for raising heavy loads. Such inventions were successfully used to defend his native city, Syracuse, during a two-year siege by Romans.

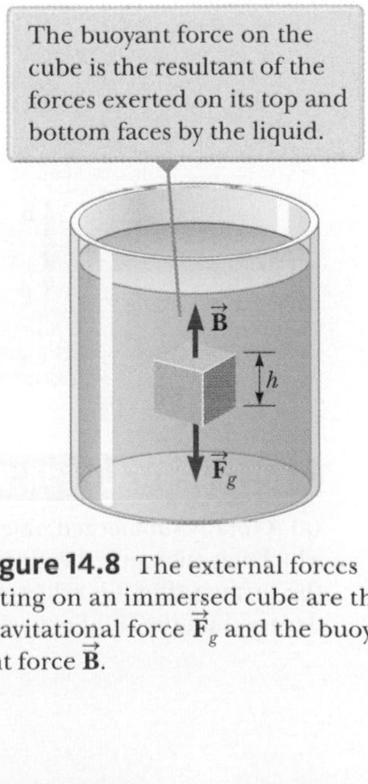

Figure 14.8 The external forces acting on an immersed cube are the gravitational force $\vec{\mathbf{F}}_g$ and the buoyant force $\vec{\mathbf{B}}$.

> **Pitfall Prevention 14.2**
>
> **Buoyant Force Is Exerted by the Fluid** Remember that **the buoyant force is exerted by the fluid.** It is not determined by properties of the object except for the amount of fluid displaced by the object. Therefore, if several objects of different densities but the same volume are immersed in a fluid, they will all experience the same buoyant force. Whether they sink or float is determined by the relationship between the buoyant force and the gravitational force.

Before we proceed with a few examples, it is instructive to discuss two common situations: a totally submerged object and a floating (partly submerged) object.

Case 1: Totally Submerged Object When an object is totally submerged in a fluid of density ρ_{fluid}, the volume V_{disp} of the displaced fluid is equal to the volume V_{obj} of the object; so, from Equation 14.5, the magnitude of the upward buoyant force is $B = \rho_{\text{fluid}} g V_{\text{obj}}$. If the object has a mass M and density ρ_{obj}, its weight is equal to $F_g = Mg = \rho_{\text{obj}} g V_{\text{obj}}$, and the net force on the object is $B - F_g = (\rho_{\text{fluid}} - \rho_{\text{obj}}) g V_{\text{obj}}$. Hence, if the density of the object is less than the density of the fluid, the downward gravitational force is less than the buoyant force and the unsupported object accelerates upward (Active Fig. 14.9a). If the density of the object is greater than the density of the fluid, the upward buoyant force is less than the downward gravitational force and the unsupported object sinks (Active Fig. 14.9b). If the density of the submerged object equals the density of the fluid, the net force on the object is zero and the object remains in equilibrium. Therefore, the direction of motion of an object submerged in a fluid is determined *only* by the densities of the object and the fluid.

Case 2: Floating Object Now consider an object of volume V_{obj} and density $\rho_{\text{obj}} < \rho_{\text{fluid}}$ in static equilibrium floating on the surface of a fluid, that is, an object that is only *partially* submerged (Active Fig. 14.10). In this case, the upward buoyant force is balanced by the downward gravitational force acting on the object. If V_{disp} is the volume of the fluid displaced by the object (this volume is the same as the volume of that part of the object beneath the surface of the fluid), the buoyant force has a magnitude $B = \rho_{\text{fluid}} g V_{\text{disp}}$. Because the weight of the object is $F_g = Mg = \rho_{\text{obj}} g V_{\text{obj}}$ and because $F_g = B$, we see that $\rho_{\text{fluid}} g V_{\text{disp}} = \rho_{\text{obj}} g V_{\text{obj}}$, or

$$\frac{V_{\text{disp}}}{V_{\text{obj}}} = \frac{\rho_{\text{obj}}}{\rho_{\text{fluid}}} \tag{14.6}$$

This equation shows that the fraction of the volume of a floating object that is below the fluid surface is equal to the ratio of the density of the object to that of the fluid.

Quick Quiz **14.4** You are shipwrecked and floating in the middle of the ocean on a raft. Your cargo on the raft includes a treasure chest full of gold that you found before your ship sank, and the raft is just barely afloat. To keep you floating as high as possible in the water, should you **(a)** leave the treasure chest on top of the raft, **(b)** secure the treasure chest to the underside of the raft, or **(c)** hang the treasure chest in the water with a rope attached to the raft? (Assume throwing the treasure chest overboard is not an option you wish to consider.)

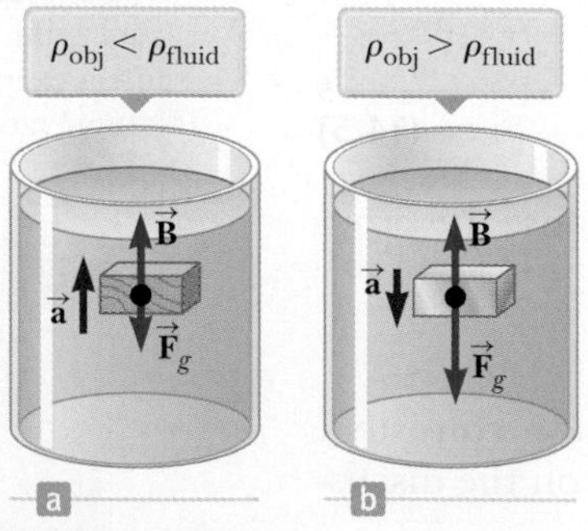

ACTIVE FIGURE 14.9

(a) A totally submerged object that is less dense than the fluid in which it is submerged experiences a net upward force and rises to the surface after it is released. (b) A totally submerged object that is denser than the fluid experiences a net downward force and sinks.

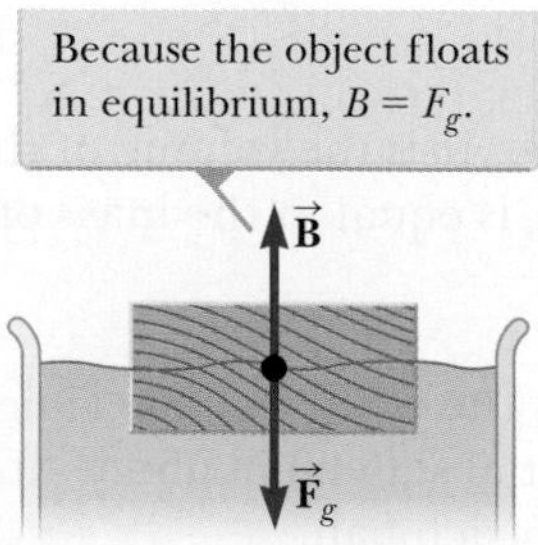

ACTIVE FIGURE 14.10

An object floating on the surface of a fluid experiences two forces, the gravitational force $\vec{F}_g$ and the buoyant force $\vec{B}$.

Example 14.5 Eureka!

Archimedes supposedly was asked to determine whether a crown made for the king consisted of pure gold. According to legend, he solved this problem by weighing the crown first in air and then in water as shown in Figure 14.11. Suppose the scale read 7.84 N when the crown was in air and 6.84 N when it was in water. What should Archimedes have told the king?

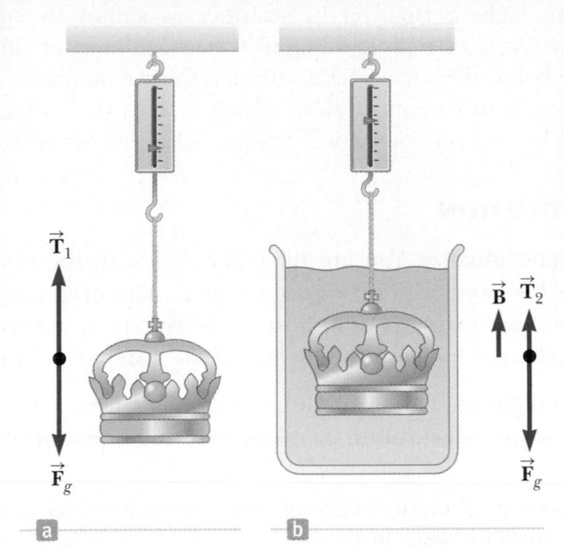

Figure 14.11 (Example 14.5) (a) When the crown is suspended in air, the scale reads its true weight because $T_1 = F_g$ (the buoyancy of air is negligible). (b) When the crown is immersed in water, the buoyant force $\vec{\mathbf{B}}$ changes the scale reading to a lower value $T_2 = F_g - B$.

SOLUTION

Conceptualize Figure 14.11 helps us imagine what is happening in this example. Because of the buoyant force, the scale reading is smaller in Figure 14.11b than in Figure 14.11a.

Categorize This problem is an example of Case 1 discussed earlier because the crown is completely submerged. The scale reading is a measure of one of the forces on the crown, and the crown is stationary. Therefore, we can categorize the crown as a particle in equilibrium.

Analyze When the crown is suspended in air, the scale reads the true weight $T_1 = F_g$ (neglecting the small buoyant force due to the surrounding air). When the crown is immersed in water, the buoyant force $\vec{\mathbf{B}}$ reduces the scale reading to an *apparent* weight of $T_2 = F_g - B$.

Apply the particle in equilibrium model to the crown in water:

$$\sum F = B + T_2 - F_g = 0$$

Solve for B and substitute the known values:

$$B = F_g - T_2 = 7.84\text{ N} - 6.84\text{ N} = 1.00\text{ N}$$

Because this buoyant force is equal in magnitude to the weight of the displaced water, $B = \rho_w g V_{\text{disp}}$, where V_{disp} is the volume of the displaced water and ρ_w is its density. Also, the volume of the crown V_c is equal to the volume of the displaced water because the crown is completely submerged, so $B = \rho_w g V_c$.

Find the density of the crown from Equation 1.1:

$$\rho_c = \frac{m_c}{V_c} = \frac{m_c g}{V_c g} = \frac{m_c g}{(B/\rho_w)} = \frac{m_c g \rho_w}{B}$$

Substitute numerical values:

$$\rho_c = \frac{(7.84\text{ N})(1\,000\text{ kg/m}^3)}{1.00\text{ N}} = 7.84 \times 10^3\text{ kg/m}^3$$

Finalize From Table 14.1, we see that the density of gold is 19.3×10^3 kg/m^3. Therefore, Archimedes should have reported that the king had been cheated. Either the crown was hollow, or it was not made of pure gold.

WHAT IF? Suppose the crown has the same weight but is indeed pure gold and not hollow. What would the scale reading be when the crown is immersed in water?

Answer Find the buoyant force on the crown:

$$B = \rho_w g V_w = \rho_w g V_c = \rho_w g\left(\frac{m_c}{\rho_c}\right) = \rho_w\left(\frac{m_c g}{\rho_c}\right)$$

Substitute numerical values:

$$B = (1.00 \times 10^3\text{ kg/m}^3)\frac{7.84\text{ N}}{19.3 \times 10^3\text{ kg/m}^3} = 0.406\text{ N}$$

Find the tension in the string hanging from the scale:

$$T_2 = F_g - B = 7.84\text{ N} - 0.406\text{ N} = 7.43\text{ N}$$

Example 14.6 A Titanic Surprise

An iceberg floating in seawater as shown in Figure 14.12a is extremely dangerous because most of the ice is below the surface. This hidden ice can damage a ship that is still a considerable distance from the visible ice. What fraction of the iceberg lies below the water level?

© Mark Karrass/Corbis Yellow/Corbis

Figure 14.12 (Example 14.6) (a) Much of the volume of this iceberg is beneath the water. (b) A ship can be damaged even when it is not near the visible ice.

SOLUTION

Conceptualize You are likely familiar with the phrase, "That's only the tip of the iceberg." The origin of this popular saying is that most of the volume of a floating iceberg is beneath the surface of the water (Fig. 14.12b).

Categorize This example corresponds to Case 2. It is also a simple substitution problem involving Equation 14.6.

Evaluate Equation 14.6 using the densities of ice and seawater (Table 14.1):

$$f = \frac{V_{\text{disp}}}{V_{\text{ice}}} = \frac{\rho_{\text{ice}}}{\rho_{\text{seawater}}} = \frac{917 \text{ kg/m}^3}{1\,030 \text{ kg/m}^3} = \boxed{0.890 \text{ or } 89.0\%}$$

Therefore, the visible fraction of ice above the water's surface is about 11%. It is the unseen 89% below the water that represents the danger to a passing ship.

Andy Sacks/Stone/Getty Images

Figure 14.13 Laminar flow around an automobile in a test wind tunnel.

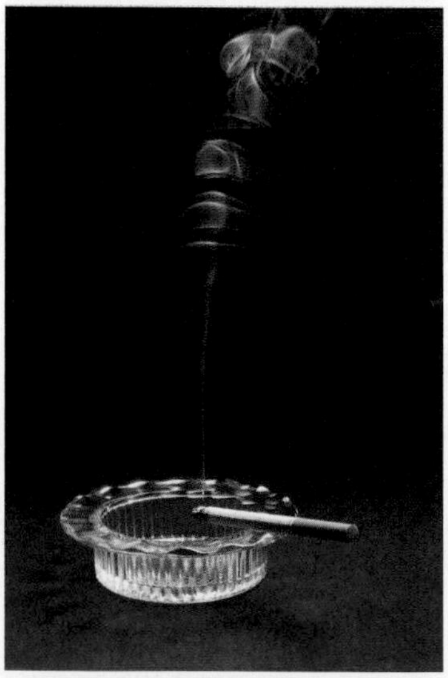

© Cengage Learning/Charles D. Winters

Figure 14.14 Hot gases from a cigarette made visible by smoke particles. The smoke first moves in laminar flow at the bottom and then in turbulent flow above.

14.5 Fluid Dynamics

Thus far, our study of fluids has been restricted to fluids at rest. We now turn our attention to fluids in motion. When fluid is in motion, its flow can be characterized as being one of two main types. The flow is said to be **steady,** or **laminar,** if each particle of the fluid follows a smooth path such that the paths of different particles never cross each other as shown in Figure 14.13. In steady flow, every fluid particle arriving at a given point has the same velocity.

Above a certain critical speed, fluid flow becomes **turbulent.** Turbulent flow is irregular flow characterized by small whirlpool-like regions as shown in Figure 14.14.

The term **viscosity** is commonly used in the description of fluid flow to characterize the degree of internal friction in the fluid. This internal friction, or *viscous force,* is associated with the resistance that two adjacent layers of fluid have to moving relative to each other. Viscosity causes part of the fluid's kinetic energy to be converted to internal energy. This mechanism is similar to the one by which an object sliding on a rough, horizontal surface loses kinetic energy.

Because the motion of real fluids is very complex and not fully understood, we make some simplifying assumptions in our approach. In our model of **ideal fluid flow,** we make the following four assumptions:

1. **The fluid is nonviscous.** In a nonviscous fluid, internal friction is neglected. An object moving through the fluid experiences no viscous force.
2. **The flow is steady.** In steady (laminar) flow, all particles passing through a point have the same velocity.
3. **The fluid is incompressible.** The density of an incompressible fluid is constant.
4. **The flow is irrotational.** In irrotational flow, the fluid has no angular momentum about any point. If a small paddle wheel placed anywhere in the fluid does not rotate about the wheel's center of mass, the flow is irrotational.

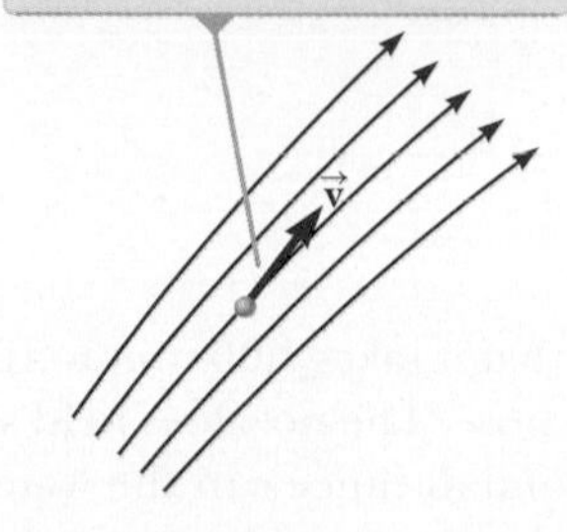

Figure 14.15 A particle in laminar flow follows a streamline.

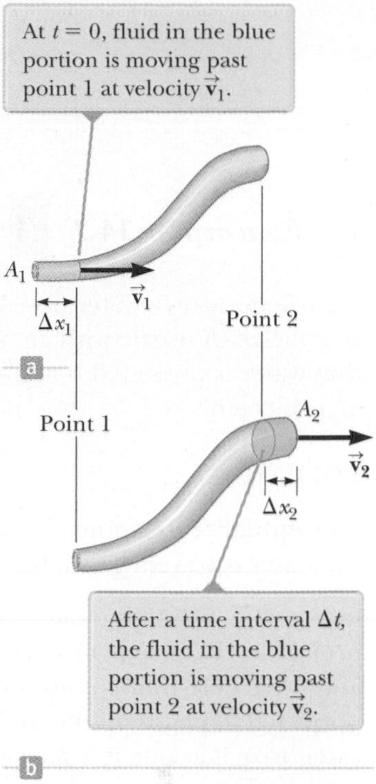

Figure 14.16 A fluid moving with steady flow through a pipe of varying cross-sectional area. (a) At $t = 0$, the small blue-colored portion of the fluid at the left is moving through area A_1. (b) After a time interval Δt, the blue-colored portion is that fluid that has moved through area A_2.

The path taken by a fluid particle under steady flow is called a **streamline.** The velocity of the particle is always tangent to the streamline as shown in Figure 14.15. A set of streamlines like the ones shown in Figure 14.15 form a *tube of flow.* Fluid particles cannot flow into or out of the sides of this tube; if they could, the streamlines would cross one another.

Consider ideal fluid flow through a pipe of nonuniform size as illustrated in Figure 14.16. Let's focus our attention on a segment of fluid in the pipe. Figure 14.16a shows the segment at time $t = 0$ consisting of the gray portion between point 1 and point 2 and the short blue portion to the left of point 1. At this time, the fluid in the short blue portion is flowing through a cross section of area A_1 at speed v_1. During the time interval Δt, the small length Δx_1 of fluid in the blue portion moves past point 1. During the same time, fluid moves past point 2 at the other end of the pipe. Figure 14.16b shows the situation at the end of the time interval Δt. The blue portion at the right end represents the fluid that has moved past point 2 through an area A_2 at a speed v_2.

The mass of fluid contained in the blue portion in Figure 14.16a is given by $m_1 = \rho A_1 \, \Delta x_1 = \rho A_1 v_1 \, \Delta t$, where ρ is the (unchanging) density of the ideal fluid. Similarly, the fluid in the blue portion in Figure 14.16b has a mass $m_2 = \rho A_2 \, \Delta x_2 = \rho A_2 v_2 \, \Delta t$. Because the fluid is incompressible and the flow is steady, however, the mass of fluid that passes point 1 in a time interval Δt must equal the mass that passes point 2 in the same time interval. That is, $m_1 = m_2$ or $\rho A_1 v_1 \, \Delta t = \rho A_2 v_2 \, \Delta t$, which means that

$$A_1 v_1 = A_2 v_2 = \text{constant} \tag{14.7}$$

◀ **Equation of Continuity**

This expression is called the **equation of continuity for fluids.** It states that the product of the area and the fluid speed at all points along a pipe is constant for an incompressible fluid. Equation 14.7 shows that the speed is high where the tube is constricted (small A) and low where the tube is wide (large A). The product Av, which has the dimensions of volume per unit time, is called either the *volume flux* or the *flow rate.* The condition Av = constant is equivalent to the statement that the volume of fluid that enters one end of a tube in a given time interval equals the volume leaving the other end of the tube in the same time interval if no leaks are present.

You demonstrate the equation of continuity each time you water your garden with your thumb over the end of a garden hose as in Figure 14.17. By partially blocking

Figure 14.17 The speed of water spraying from the end of a garden hose increases as the size of the opening is decreased with the thumb.

the opening with your thumb, you reduce the cross-sectional area through which the water passes. As a result, the speed of the water increases as it exits the hose, and the water can be sprayed over a long distance.

Example 14.7 Watering a Garden

A gardener uses a water hose 2.50 cm in diameter to fill a 30.0-L bucket. The gardener notes that it takes 1.00 min to fill the bucket. A nozzle with an opening of cross-sectional area 0.500 cm^2 is then attached to the hose. The nozzle is held so that water is projected horizontally from a point 1.00 m above the ground. Over what horizontal distance can the water be projected?

SOLUTION

Conceptualize Imagine any past experience you have with projecting water from a horizontal hose or a pipe. The faster the water is traveling as it leaves the hose, the farther it will land on the ground from the end of the hose.

Categorize Once the water leaves the hose, it is in free fall. Therefore, we categorize a given element of the water as a projectile. The element is modeled as a particle under constant acceleration (due to gravity) in the vertical direction and a particle under constant velocity in the horizontal direction. The horizontal distance over which the element is projected depends on the speed with which it is projected. This example involves a change in area for the pipe, so we also categorize it as one in which we use the equation of continuity for fluids.

Analyze We first find the speed of the water in the hose from the bucket-filling information.

Find the cross-sectional area of the hose:

$$A = \pi r^2 = \pi \frac{d^2}{4} = \pi\left[\frac{(2.50 \text{ cm})^2}{4}\right] = 4.91 \text{ cm}^2$$

Evaluate the volume flow rate:

$$Av_1 = 30.0 \text{ L/min} = \frac{30.0 \times 10^3 \text{ cm}^3}{60.0 \text{ s}} = 500 \text{ cm}^3/\text{s}$$

Solve for the speed of the water in the hose:

$$v_1 = \frac{500 \text{ cm}^3/\text{s}}{A} = \frac{500 \text{ cm}^3/\text{s}}{4.91 \text{ cm}^2} = 102 \text{ cm/s} = 1.02 \text{ m/s}$$

We have labeled this speed v_1 because we identify point 1 within the hose. We identify point 2 in the air just outside the nozzle. We must find the speed $v_2 = v_{xi}$ with which the water exits the nozzle. The subscript i anticipates that it will be the *initial* velocity component of the water projected from the hose, and the subscript x indicates that the initial velocity vector of the projected water is horizontal.

Solve the continuity equation for fluids for v_2:

$$v_2 = v_{xi} = \frac{A_1}{A_2} v_1$$

Substitute numerical values:

$$v_{xi} = \frac{4.91 \text{ cm}^2}{0.500 \text{ cm}^2}(1.02 \text{ m/s}) = 10.0 \text{ m/s}$$

We now shift our thinking away from fluids and to projectile motion. In the vertical direction, an element of the water starts from rest and falls through a vertical distance of 1.00 m.

Write Equation 2.16 for the vertical position of an element of water, modeled as a particle under constant acceleration:

$$y_f = y_i + v_{yi}t - \tfrac{1}{2}gt^2$$

Substitute numerical values:

$$-1.00 \text{ m} = 0 + 0 - \tfrac{1}{2}(9.80 \text{ m/s}^2)t^2$$

Solve for the time at which the element of water lands on the ground:

$$t = \sqrt{\frac{2(1.00 \text{ m})}{9.80 \text{ m/s}^2}} = 0.452 \text{ s}$$

14.7 *cont.*

Use Equation 2.7 to find the horizontal position of the element at this time, modeled as a particle under constant velocity:

$$x_f = x_i + v_{xi}t = 0 + (10.0 \text{ m/s})(0.452 \text{ s}) = \boxed{4.52 \text{ m}}$$

Finalize The time interval for the element of water to fall to the ground is unchanged if the projection speed is changed because the projection is horizontal. Increasing the projection speed results in the water hitting the ground farther from the end of the hose, but requires the same time interval to strike the ground.

14.6 Bernoulli's Equation

You have probably experienced driving on a highway and having a large truck pass you at high speed. In this situation, you may have had the frightening feeling that your car was being pulled in toward the truck as it passed. We will investigate the origin of this effect in this section.

As a fluid moves through a region where its speed or elevation above the Earth's surface changes, the pressure in the fluid varies with these changes. The relationship between fluid speed, pressure, and elevation was first derived in 1738 by Swiss physicist Daniel Bernoulli. Consider the flow of a segment of an ideal fluid through a nonuniform pipe in a time interval Δt as illustrated in Figure 14.18. This figure is very similar to Figure 14.16, which we used to develop the continuity equation. We have added two features: the forces on the outer ends of the blue portions of fluid and the heights of these portions above the reference position $y = 0$.

The force exerted by the fluid to the left of the blue portion in Figure 14.18a has a magnitude P_1A_1. The work done by this force on the segment in a time interval Δt is $W_1 = F_1\,\Delta x_1 = P_1A_1\,\Delta x_1 = P_1V$, where V is the volume of the blue portion of fluid passing point 1 in Figure 14.18a. In a similar manner, the work done by the fluid to the right of the segment in the same time interval Δt is $W_2 = -P_2A_2\,\Delta x_2 = -P_2V$, where V is the volume of the blue portion of fluid passing point 2 in Figure 14.18b. (The volumes of the blue portions of fluid in Figures 14.18a and 14.18b are equal because the fluid is incompressible.) This work is negative because the force on the segment of fluid is to the left and the displacement of the point of application

Daniel Bernoulli
Swiss physicist (1700–1782)
Bernoulli made important discoveries in fluid dynamics.

Bernoulli's most famous work, *Hydrodynamica*, was published in 1738; it is both a theoretical and a practical study of equilibrium, pressure, and speed in fluids. He showed that as the speed of a fluid increases, its pressure decreases. Referred to as "Bernoulli's principle," Bernoulli's work is used to produce a partial vacuum in chemical laboratories by connecting a vessel to a tube through which water is running rapidly.

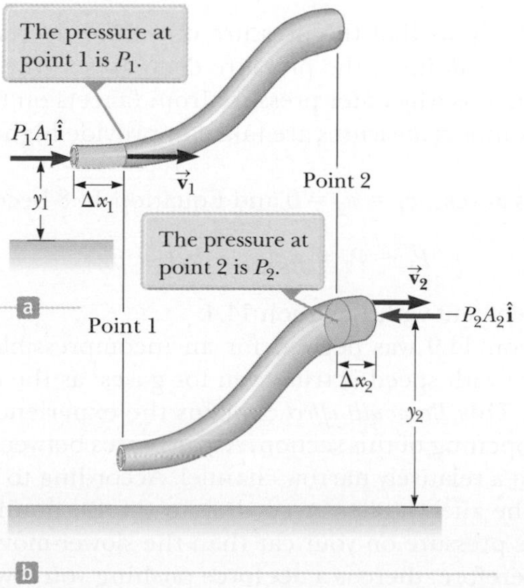

Figure 14.18 A fluid in laminar flow through a pipe. (a) A segment of the fluid at time $t = 0$. A small portion of the blue-colored fluid is at height y_1 above a reference position. (b) After a time interval Δt, the entire segment has moved to the right. The blue-colored portion of the fluid is at height y_2.

of the force is to the right. Therefore, the net work done on the segment by these forces in the time interval Δt is

$$W = (P_1 - P_2)V$$

Part of this work goes into changing the kinetic energy of the segment of fluid, and part goes into changing the gravitational potential energy of the segment–Earth system. Because we are assuming streamline flow, the kinetic energy K_{gray} of the gray portion of the segment is the same in both parts of Figure 14.18. Therefore, the change in the kinetic energy of the segment of fluid is

$$\Delta K = \left(\tfrac{1}{2}mv_2^2 + K_{\text{gray}}\right) - \left(\tfrac{1}{2}mv_1^2 + K_{\text{gray}}\right) = \tfrac{1}{2}mv_2^2 - \tfrac{1}{2}mv_1^2$$

where m is the mass of the blue portions of fluid in both parts of Figure 14.18. (Because the volumes of both portions are the same, they also have the same mass.)

Considering the gravitational potential energy of the segment–Earth system, once again there is no change during the time interval for the gravitational potential energy U_{gray} associated with the gray portion of the fluid. Consequently, the change in gravitational potential energy of the system is

$$\Delta U = \left(mgy_2 + U_{\text{gray}}\right) - \left(mgy_1 + U_{\text{gray}}\right) = mgy_2 - mgy_1$$

From Equation 8.2, the total work done on the system by the fluid outside the segment is equal to the change in mechanical energy of the system: $W = \Delta K + \Delta U$. Substituting for each of these terms gives

$$(P_1 - P_2)V = \tfrac{1}{2}mv_2^2 - \tfrac{1}{2}mv_1^2 + mgy_2 - mgy_1$$

If we divide each term by the portion volume V and recall that $\rho = m/V$, this expression reduces to

$$P_1 - P_2 = \tfrac{1}{2}\rho v_2^2 - \tfrac{1}{2}\rho v_1^2 + \rho g y_2 - \rho g y_1$$

Rearranging terms gives

$$P_1 + \tfrac{1}{2}\rho v_1^2 + \rho g y_1 = P_2 + \tfrac{1}{2}\rho v_2^2 + \rho g y_2 \qquad \textbf{(14.8)}$$

which is **Bernoulli's equation** as applied to an ideal fluid. This equation is often expressed as

Bernoulli's equation ▶

$$P + \tfrac{1}{2}\rho v^2 + \rho g y = \text{constant} \qquad \textbf{(14.9)}$$

Bernoulli's equation shows that the pressure of a fluid decreases as the speed of the fluid increases. In addition, the pressure decreases as the elevation increases. This latter point explains why water pressure from faucets on the upper floors of a tall building is weak unless measures are taken to provide higher pressure for these upper floors.

When the fluid is at rest, $v_1 = v_2 = 0$ and Equation 14.8 becomes

$$P_1 - P_2 = \rho g(y_2 - y_1) = \rho g h$$

This result is in agreement with Equation 14.4.

Although Equation 14.9 was derived for an incompressible fluid, the general behavior of pressure with speed is true even for gases: as the speed increases, the pressure decreases. This *Bernoulli effect* explains the experience with the truck on the highway at the opening of this section. As air passes between you and the truck, it must pass through a relatively narrow channel. According to the continuity equation, the speed of the air is higher. According to the Bernoulli effect, this higher-speed air exerts less pressure on your car than the slower-moving air on the other side of your car. Therefore, there is a net force pushing you toward the truck!

***Quick Quiz* 14.5** You observe two helium balloons floating next to each other at the ends of strings secured to a table. The facing surfaces of the balloons are separated by 1–2 cm. You blow through the small space between the balloons. What happens to the balloons? **(a)** They move toward each other. **(b)** They move away from each other. **(c)** They are unaffected.

Example 14.8 The Venturi Tube

The horizontal constricted pipe illustrated in Figure 14.19, known as a *Venturi tube,* can be used to measure the flow speed of an incompressible fluid. Determine the flow speed at point 2 of Figure 14.19a if the pressure difference $P_1 - P_2$ is known.

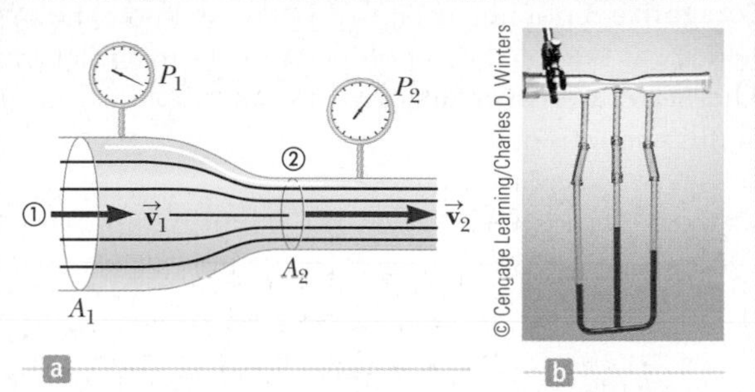

Figure 14.19 (Example 14.8) (a) Pressure P_1 is greater than pressure P_2 because $v_1 < v_2$. This device can be used to measure the speed of fluid flow. (b) A Venturi tube, located at the top of the photograph. The higher level of fluid in the middle column shows that the pressure at the top of the column, which is in the constricted region of the Venturi tube, is lower.

SOLUTION

Conceptualize Bernoulli's equation shows how the pressure of a fluid decreases as its speed increases. Therefore, we should be able to calibrate a device to give us the fluid speed if we can measure pressure.

Categorize Because the problem states that the fluid is incompressible, we can categorize it as one in which we can use the equation of continuity for fluids and Bernoulli's equation.

Analyze Apply Equation 14.8 to points 1 and 2, noting that $y_1 = y_2$ because the pipe is horizontal:

$$(1)\quad P_1 + \tfrac{1}{2}\rho v_1^2 = P_2 + \tfrac{1}{2}\rho v_2^2$$

Solve the equation of continuity for v_1:

$$v_1 = \frac{A_2}{A_1} v_2$$

Substitute this expression into Equation (1):

$$P_1 + \tfrac{1}{2}\rho\left(\frac{A_2}{A_1}\right)^2 v_2^2 = P_2 + \tfrac{1}{2}\rho v_2^2$$

Solve for v_2:

$$v_2 = A_1\sqrt{\frac{2(P_1 - P_2)}{\rho(A_1^2 - A_2^2)}}$$

Finalize From the design of the tube (areas A_1 and A_2) and measurements of the pressure difference $P_1 - P_2$, we can calculate the speed of the fluid with this equation. To see the relationship between fluid speed and pressure difference, place two empty soda cans on their sides about 2 cm apart on a table. Gently blow a stream of air horizontally between the cans and watch them roll together slowly due to a modest pressure difference between the stagnant air on their outside edges and the moving air between them. Now blow more strongly and watch the increased pressure difference move the cans together more rapidly.

Example 14.9 Torricelli's Law

An enclosed tank containing a liquid of density ρ has a hole in its side at a distance y_1 from the tank's bottom (Fig. 14.20 on page 418). The hole is open to the atmosphere, and its diameter is much smaller than the diameter of the tank. The air above the liquid is maintained at a pressure P. Determine the speed of the liquid as it leaves the hole when the liquid's level is a distance h above the hole.

continued

14.9 *cont.*

SOLUTION

Conceptualize Imagine that the tank is a fire extinguisher. When the hole is opened, liquid leaves the hole with a certain speed. If the pressure P at the top of the liquid is increased, the liquid leaves with a higher speed. If the pressure P falls too low, the liquid leaves with a low speed and the extinguisher must be replaced.

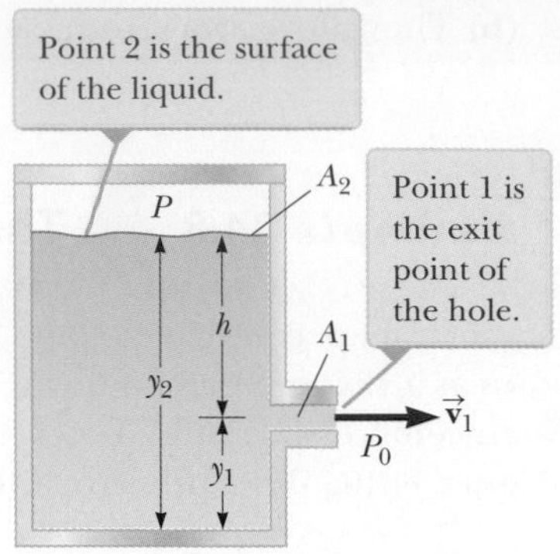

Figure 14.20 (Example 14.9) A liquid leaves a hole in a tank at speed v_1.

Categorize Looking at Figure 14.20, we know the pressure at two points and the velocity at one of those points. We wish to find the velocity at the second point. Therefore, we can categorize this example as one in which we can apply Bernoulli's equation.

Analyze Because $A_2 >> A_1$, the liquid is approximately at rest at the top of the tank, where the pressure is P. At the hole, P_1 is equal to atmospheric pressure P_0.

Apply Bernoulli's equation between points 1 and 2:

$$P_0 + \tfrac{1}{2}\rho v_1^2 + \rho g y_1 = P + \rho g y_2$$

Solve for v_1, noting that $y_2 - y_1 = h$:

$$v_1 = \sqrt{\frac{2(P - P_0)}{\rho} + 2gh}$$

Finalize When P is much greater than P_0 (so that the term $2gh$ can be neglected), the exit speed of the water is mainly a function of P. If the tank is open to the atmosphere, then $P = P_0$ and $v_1 = \sqrt{2gh}$. In other words, for an open tank, the speed of the liquid leaving a hole a distance h below the surface is equal to that acquired by an object falling freely through a vertical distance h. This phenomenon is known as **Torricelli's law.**

WHAT IF? What if the position of the hole in Figure 14.20 could be adjusted vertically? If the tank is open to the atmosphere and sitting on a table, what position of the hole would cause the water to land on the table at the farthest distance from the tank?

Answer Model a parcel of water exiting the hole as a projectile. Find the time at which the parcel strikes the table from a hole at an arbitrary position y_1:

$$y_f = y_i + v_{yi}t - \tfrac{1}{2}gt^2$$

$$0 = y_1 + 0 - \tfrac{1}{2}gt^2$$

$$t = \sqrt{\frac{2y_1}{g}}$$

Find the horizontal position of the parcel at the time it strikes the table:

$$x_f = x_i + v_{xi}t = 0 + \sqrt{2g(y_2 - y_1)}\sqrt{\frac{2y_1}{g}}$$

$$= 2\sqrt{(y_2 y_1 - y_1^2)}$$

Maximize the horizontal position by taking the derivative of x_f with respect to y_1 (because y_1, the height of the hole, is the variable that can be adjusted) and setting it equal to zero:

$$\frac{dx_f}{dy_1} = \tfrac{1}{2}(2)(y_2 y_1 - y_1^2)^{-1/2}(y_2 - 2y_1) = 0$$

Solve for y_1:

$$y_1 = \tfrac{1}{2}y_2$$

Therefore, to maximize the horizontal distance, the hole should be halfway between the bottom of the tank and the upper surface of the water. Below this location, the water is projected at a higher speed but falls for a short time interval, reducing the horizontal range. Above this point, the water is in the air for a longer time interval but is projected with a smaller horizontal speed.

14.7 Other Applications of Fluid Dynamics

Consider the streamlines that flow around an airplane wing as shown in Figure 14.21. Let's assume the airstream approaches the wing horizontally from the right with a velocity $\vec{\mathbf{v}}_1$. The tilt of the wing causes the airstream to be deflected downward with a velocity $\vec{\mathbf{v}}_2$. Because the airstream is deflected by the wing, the wing must exert a force on the airstream. According to Newton's third law, the airstream exerts a force $\vec{\mathbf{F}}$ on the wing that is equal in magnitude and opposite in direction. This force has a vertical component called **lift** (or aerodynamic lift) and a horizontal component called **drag.** The lift depends on several factors, such as the speed of the airplane, the area of the wing, the wing's curvature, and the angle between the wing and the horizontal. The curvature of the wing surfaces causes the pressure above the wing to be lower than that below the wing due to the Bernoulli effect. This pressure difference assists with the lift on the wing. As the angle between the wing and the horizontal increases, turbulent flow can set in above the wing to reduce the lift.

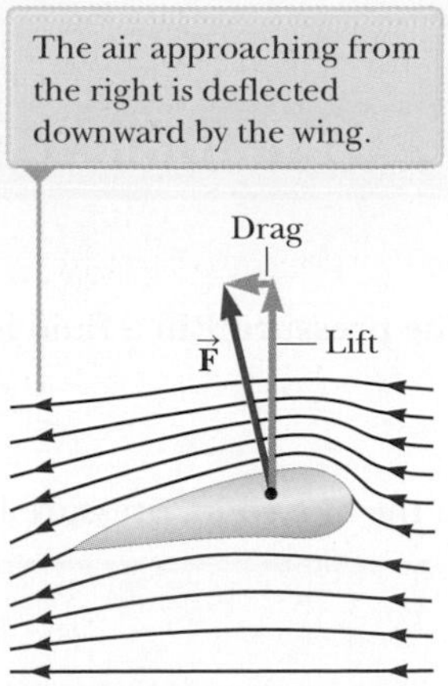

Figure 14.21 Streamline flow around a moving airplane wing. By Newton's third law, the air deflected by the wing results in an upward force on the wing from the air: *lift*. Because of air resistance, there is also a force opposite the velocity of the wing: *drag*.

In general, an object moving through a fluid experiences lift as the result of any effect that causes the fluid to change its direction as it flows past the object. Some factors that influence lift are the shape of the object, its orientation with respect to the fluid flow, any spinning motion it might have, and the texture of its surface. For example, a golf ball struck with a club is given a rapid backspin due to the slant of the club. The dimples on the ball increase the friction force between the ball and the air so that air adheres to the ball's surface. Figure 14.22 shows air adhering to the ball and being deflected downward as a result. Because the ball pushes the air down, the air must push up on the ball. Without the dimples, the friction force is lower and the golf ball does not travel as far. It may seem counterintuitive to increase the range by increasing the friction force, but the lift gained by spinning the ball more than compensates for the loss of range due to the effect of friction on the translational motion of the ball. For the same reason, a baseball's cover helps the spinning ball "grab" the air rushing by and helps deflect it when a "curve ball" is thrown.

A number of devices operate by means of the pressure differentials that result from differences in a fluid's speed. For example, a stream of air passing over one end of an open tube, the other end of which is immersed in a liquid, reduces the pressure above the tube as illustrated in Figure 14.23. This reduction in pressure causes the liquid to rise into the airstream. The liquid is then dispersed into a fine spray of droplets. You might recognize that this *atomizer* is used in perfume bottles and paint sprayers.

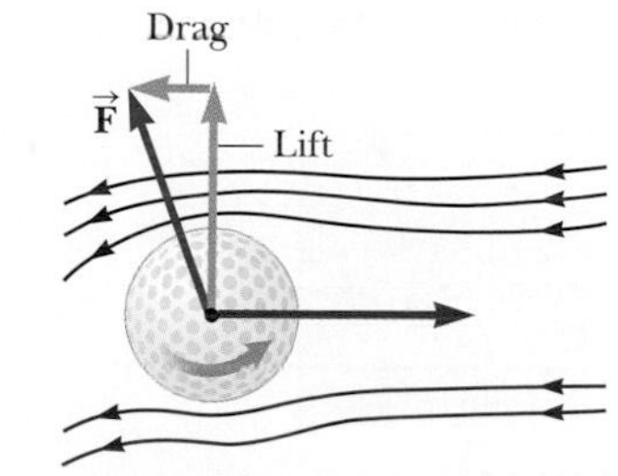

Figure 14.22 Because of the deflection of air, a spinning golf ball experiences a lifting force that allows it to travel much farther than it would if it were not spinning.

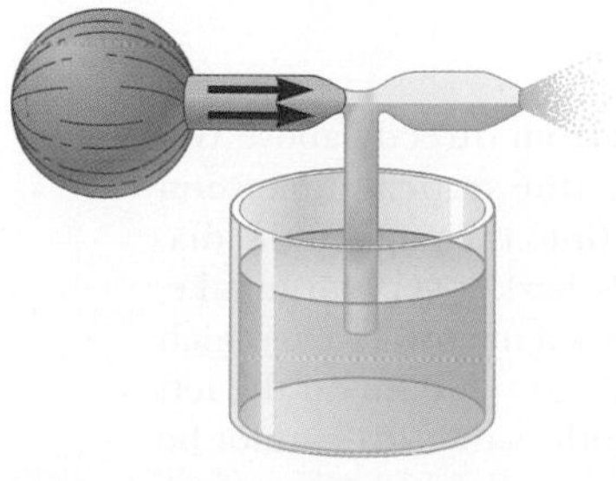

Figure 14.23 A stream of air passing over a tube dipped into a liquid causes the liquid to rise in the tube.

Summary

Definitions

The **pressure** P in a fluid is the force per unit area exerted by the fluid on a surface:

$$P \equiv \frac{F}{A} \quad \textbf{(14.1)}$$

In the SI system, pressure has units of newtons per square meter (N/m^2), and 1 N/m^2 = 1 **pascal** (Pa).

Concepts and Principles

The pressure in a fluid at rest varies with depth h in the fluid according to the expression

$$P = P_0 + \rho g h \quad \textbf{(14.4)}$$

where P_0 is the pressure at $h = 0$ and ρ is the density of the fluid, assumed uniform.

Pascal's law states that when pressure is applied to an enclosed fluid, the pressure is transmitted undiminished to every point in the fluid and to every point on the walls of the container.

When an object is partially or fully submerged in a fluid, the fluid exerts on the object an upward force called the **buoyant force.** According to **Archimedes's principle,** the magnitude of the buoyant force is equal to the weight of the fluid displaced by the object:

$$B = \rho_{\text{fluid}} g V_{\text{disp}} \quad \textbf{(14.5)}$$

The flow rate (volume flux) through a pipe that varies in cross-sectional area is constant; that is equivalent to stating that the product of the cross-sectional area A and the speed v at any point is a constant. This result is expressed in the **equation of continuity for fluids:**

$$A_1 v_1 = A_2 v_2 = \text{constant} \quad \textbf{(14.7)}$$

The sum of the pressure, kinetic energy per unit volume, and gravitational potential energy per unit volume has the same value at all points along a streamline for an ideal fluid. This result is summarized in **Bernoulli's equation:**

$$P + \tfrac{1}{2}\rho v^2 + \rho g y = \text{constant} \quad \textbf{(14.9)}$$

Objective Questions

☐ denotes answer available in Student Solutions Manual/Study Guide

1. Figure OQ14.1 shows aerial views from directly above two dams. Both dams are equally wide (the vertical dimension in the diagram) and equally high (into the page in the diagram). The dam on the left holds back a very large lake, and the dam on the right holds back a narrow river. Which dam has to be built more strongly? (a) the dam on the left (b) the dam on the right (c) both the same (d) cannot be predicted

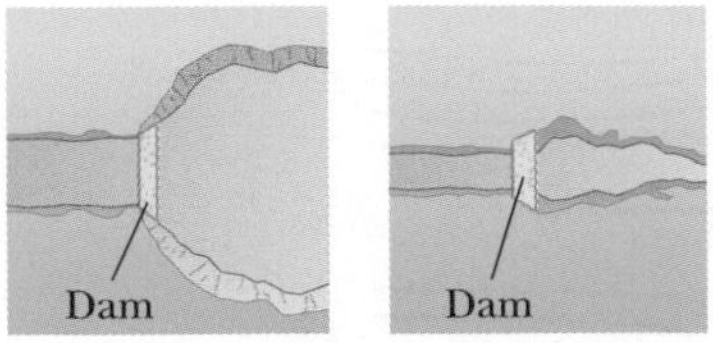

Figure OQ14.1

2. An apple is held completely submerged just below the surface of water in a container. The apple is then moved to a deeper point in the water. Compared with the force needed to hold the apple just below the surface, what is the force needed to hold it at the deeper point? (a) larger (b) the same (c) smaller (d) impossible to determine

3. A wooden block floats in water, and a steel object is attached to the bottom of the block by a string as in Figure OQ14.3. If the block remains floating, which of the following statements are valid? (Choose all correct statements.) (a) The buoyant force on the steel object is equal to its weight. (b) The buoyant force on the block is equal to its weight. (c) The tension in the string is equal to the weight of the steel object. (d) The tension in the string is less than the weight of the steel object. (e) The buoyant force on the block is equal to the volume of water it displaces.

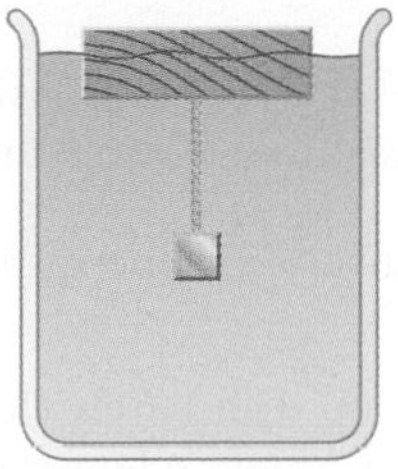

Figure OQ14.3

4. A beach ball filled with air is pushed about 1 m below the surface of a swimming pool and released from rest. Which of the following statements are valid, assuming the size of the ball remains the same? (Choose all correct statements.) (a) As the ball rises in the pool, the buoyant force on it increases. (b) When the ball is released, the buoyant force exceeds the gravitational force, and the ball accelerates upward. (c) The buoyant force on the ball decreases as the ball approaches the surface of the pool. (d) The buoyant force on the ball equals its weight and remains constant as the ball rises. (e) The buoyant force on the ball while it is submerged is approximately equal to the weight of a volume of water that could fill the ball.

5. A boat develops a leak and, after its passengers are rescued, eventually sinks to the bottom of a lake. When the boat is at the bottom, what is the force of the lake bottom on the boat? (a) greater than the weight of the boat (b) equal to the weight of the boat (c) less than the weight of the boat (d) equal to the weight of the displaced water (e) equal to the buoyant force on the boat

6. A solid iron sphere and a solid lead sphere of the same size are each suspended by strings and are submerged in a tank of water. (Note that the density of lead is greater than that of iron.) Which of the following statements are valid? (Choose all correct statements.) (a) The buoyant force on each is the same. (b) The buoyant force on the lead sphere is greater than the buoyant force on the iron sphere because lead has the greater density. (c) The tension in the string supporting the lead sphere is greater than the tension in the string supporting the iron sphere. (d) The buoyant force on the iron sphere is greater than the buoyant force on the lead sphere because lead displaces more water. (e) None of those statements is true.

7. Three vessels of different shapes are filled to the same level with water as in Figure OQ14.7. The area of the base is the same for all three vessels. Which of the following statements are valid? (Choose all correct statements.) (a) The pressure at the top surface of vessel A is greatest because it has the largest surface area. (b) The pressure at the bottom of vessel A is greatest because it contains the most water. (c) The pressure at the bottom of each vessel is the same. (d) The force on the bottom of each vessel is not the same. (e) At a given depth below the surface of each vessel, the pressure on the side of vessel A is greatest because of its slope.

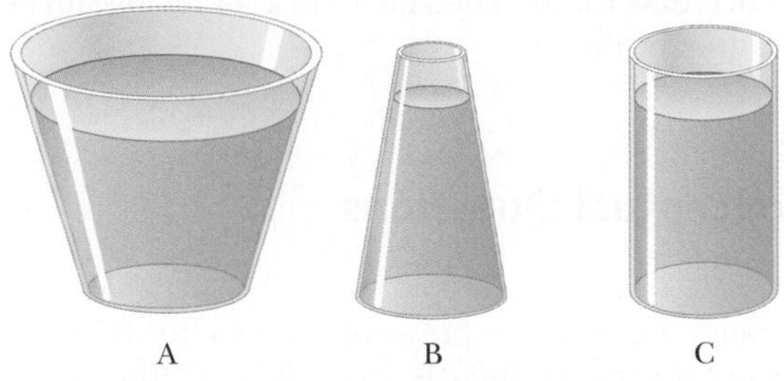

Figure OQ14.7

8. An ideal fluid flows through a horizontal pipe whose diameter varies along its length. Measurements would indicate that the sum of the kinetic energy per unit volume and pressure at different sections of the pipe would (a) decrease as the pipe diameter increases, (b) increase as the pipe diameter increases, (c) increase as the pipe diameter decreases, (d) decrease as the pipe diameter decreases, or (e) remain the same as the pipe diameter changes.

9. A beach ball is made of thin plastic. It has been inflated with air, but the plastic is not stretched. By swimming with fins on, you manage to take the ball from the surface of a pool to the bottom. Once the ball is completely submerged, what happens to the buoyant force exerted on the beach ball as you take it deeper? (a) It increases. (b) It remains constant. (c) It decreases. (d) It is impossible to determine.

10. A small piece of steel is tied to a block of wood. When the wood is placed in a tub of water with the steel on top, half of the block is submerged. Now the block is inverted so that the steel is under water. **(i)** Does the amount of the block submerged (a) increase, (b) decrease, or (c) remain the same? **(ii)** What happens to the water level in the tub when the block is inverted? (a) It rises. (b) It falls. (c) It remains the same.

11. Rank the buoyant forces exerted on the following five objects of equal volume from the largest to the smallest. Assume the objects have been dropped into a swimming pool and allowed to come to mechanical equilibrium. If any buoyant forces are equal, state that in your ranking. (a) a block of solid oak (b) an aluminum block (c) a beach ball made of thin plastic and inflated with air (d) an iron block (e) a thin-walled, sealed bottle of water.

12. A person in a boat floating in a small pond throws an anchor overboard. What happens to the level of the pond? (a) It rises. (b) It falls. (c) It remains the same.

13. A piece of unpainted porous wood barely floats in an open container partly filled with water. The container is then sealed and pressurized above atmospheric pressure. What happens to the wood? (a) It rises in the water. (b) It sinks lower in the water. (c) It remains at the same level.

14. A water supply maintains a constant rate of flow for water in a hose. You want to change the opening of the nozzle so that water leaving the nozzle will reach a height that is four times the current maximum height the water reaches with the nozzle vertical. To do so, should you (a) decrease the area of the opening by a factor of 16, (b) decrease the area by a factor of 8, (c) decrease the area by a factor of 4, (d) decrease the area by a factor of 2, or (e) give up because it cannot be done?

15. A glass of water contains floating ice cubes. When the ice melts, does the water level in the glass (a) go up, (b) go down, or (c) remain the same?

16. One of the predicted problems due to global warming is that ice in the polar ice caps will melt and raise sea levels everywhere in the world. Is that more of a worry for ice (a) at the north pole, where most of the ice floats on water; (b) at the south pole, where most of the ice sits on land; (c) both at the north and south pole equally; or (d) at neither pole?

Conceptual Questions

☐ denotes answer available in Student Solutions Manual/Study Guide

1. Because atmospheric pressure is about 10^5 N/m^2 and the area of a person's chest is about 0.13 m^2, the force of the atmosphere on one's chest is around 13 000 N. In view of this enormous force, why don't our bodies collapse?

2. Two thin-walled drinking glasses having equal base areas but different shapes, with very different cross-sectional areas above the base, are filled to the same level with water. According to the expression $P = P_0 + \rho g h$, the pressure is the same at the bottom of both glasses. In view of this equality, why does one weigh more than the other?

3. When an object is immersed in a liquid at rest, why is the net force on the object in the horizontal direction equal to zero?

4. A fish rests on the bottom of a bucket of water while the bucket is being weighed on a scale. When the fish begins to swim around, does the scale reading change? Explain your answer.

5. You are a passenger on a spacecraft. For your survival and comfort, the interior contains air just like that at the surface of the Earth. The craft is coasting through a very empty region of space. That is, a nearly perfect vacuum exists just outside the wall. Suddenly, a meteoroid pokes a hole, about the size of a large coin, right through the wall next to your seat. (a) What happens? (b) Is there anything you can or should do about it?

6. Does a ship float higher in the water of an inland lake or in the ocean? Why?

7. A water tower is a common sight in many communities. Figure CQ14.7 shows a collection of colorful water towers in Kuwait City, Kuwait. Notice that the large weight of the water results in the center of mass of the system being high above the ground. Why is it desirable for a water tower to have this highly unstable shape rather than being shaped as a tall cylinder?

8. An empty metal soap dish barely floats in water. A bar of Ivory soap floats in water. When the soap is stuck in the soap dish, the combination sinks. Explain why.

9. (a) Is the buoyant force a conservative force? (b) Is a potential energy associated with the buoyant force? (c) Explain your answers to parts (a) and (b).

10. If you release a ball while inside a freely falling elevator, the ball remains in front of you rather than falling to the floor because the ball, the elevator, and you all experience the same downward gravitational acceleration. What happens if you repeat this experiment with a helium-filled balloon?

11. How would you determine the density of an irregularly shaped rock?

12. Place two cans of soft drinks, one regular and one diet, in a container of water. You will find that the diet drink floats while the regular one sinks. Use Archimedes's principle to devise an explanation.

13. The water supply for a city is often provided from reservoirs built on high ground. Water flows from the reservoir, through pipes, and into your home when you turn the tap on your faucet. Why does water flow more rapidly out of a faucet on the first floor of a building than in an apartment on a higher floor?

14. If the airstream from a hair dryer is directed over a table-tennis ball, the ball can be levitated. Explain.

15. When ski jumpers are airborne (Fig. CQ14.15), they bend their bodies forward and keep their hands at their sides. Why?

© iStockPhoto/Klaas Lingbeek-van Kranen

Figure CQ14.7

© iStockPhoto/technotr

Figure CQ14.15

16. Why do airplane pilots prefer to take off with the airplane facing into the wind?

17. Prairie dogs ventilate their burrows by building a mound around one entrance, which is open to a stream of air when wind blows from any direction. A second entrance at ground level is open to almost stagnant air. How does this construction create an airflow through the burrow?

18. In Figure CQ14.18, an airstream moves from right to left through a tube that is constricted at the middle. Three table-tennis balls are levitated in equilibrium above the vertical columns through which the air escapes. (a) Why is the ball at the right higher than the one in the middle? (b) Why is the ball at the left lower than the ball at the right even though the horizontal tube has the same dimensions at these two points?

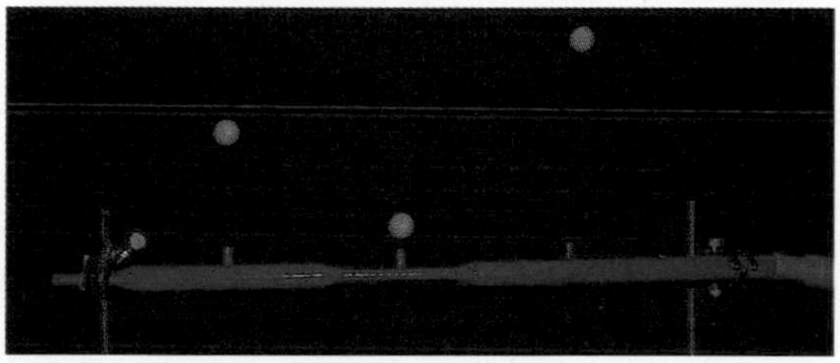
Henry Leap and Jim Lehman

Figure CQ14.18

19. A typical silo on a farm has many metal bands wrapped around its perimeter for support as shown in Figure CQ14.19. Why is the spacing between successive bands smaller for the lower portions of the silo on the left, and why are double bands used at lower portions of the silo on the right?

Henry Leap and Jim Lehman

Figure CQ14.19

Problems

ENHANCED WebAssign The problems found in this chapter may be assigned online in Enhanced WebAssign

1. denotes straightforward problem; **2.** denotes intermediate problem; **3.** denotes challenging problem

[1.] full solution available in the Student Solutions Manual/Study Guide

1. denotes problems most often assigned in Enhanced WebAssign; these provide students with targeted feedback and either a Master It tutorial or a Watch It solution video.

QIC denotes asking for quantitative and conceptual reasoning

S denotes symbolic reasoning problem

M denotes Master It tutorial available in Enhanced WebAssign

GP denotes guided problem

shaded denotes "paired problems" that develop reasoning with symbols and numerical values

In all problems, assume the density of air is the 20°C value from Table 14.1, 1.20 kg/m^3, unless noted otherwise.

Section 14.1 Pressure

[1.] M Calculate the mass of a solid gold rectangular bar that has dimensions of 4.50 cm × 11.0 cm × 26.0 cm.

2. QIC The nucleus of an atom can be modeled as several protons and neutrons closely packed together. Each particle has a mass of 1.67×10^{-27} kg and radius on the order of 10^{-15} m. (a) Use this model and the data provided to estimate the density of the nucleus of an atom. (b) Compare your result with the density of a material such as iron. What do your result and comparison suggest concerning the structure of matter?

[3.] QIC A 50.0-kg woman wearing high-heeled shoes is invited into a home in which the kitchen has vinyl floor covering. The heel on each shoe is circular and has a radius of 0.500 cm. (a) If the woman balances on one heel, what pressure does she exert on the floor? (b) Should the homeowner be concerned? Explain your answer.

4. Estimate the total mass of the Earth's atmosphere. (The radius of the Earth is 6.37×10^6 m, and atmospheric pressure at the surface is 1.013×10^5 Pa.)

5. A large man sits on a four-legged chair with his feet off the floor. The combined mass of the man and chair is 95.0 kg. If the chair legs are circular and have a radius of 0.500 cm at the bottom, what pressure does each leg exert on the floor?

Section 14.2 Variation of Pressure with Depth

6. A swimming pool has dimensions 30.0 m × 10.0 m and a flat bottom. When the pool is filled to a depth of 2.00 m with fresh water, what is the force exerted by the water on (a) the bottom? (b) On each end? (c) On each side?

7. **M** The spring of the pressure gauge shown in Figure P14.7 has a force constant of 1 250 N/m, and the piston has a diameter of 1.20 cm. As the gauge is lowered into water in a lake, what change in depth causes the piston to move in by 0.750 cm?

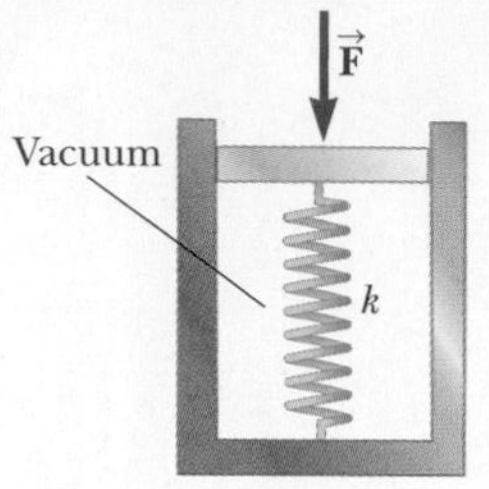

Figure P14.7

8. The small piston of a hydraulic lift (Fig. P14.8) has a cross-sectional area of 3.00 cm^2, and its large piston has a cross-sectional area of 200 cm^2. What downward force of magnitude F_1 must be applied to the small piston for the lift to raise a load whose weight is F_g = 15.0 kN?

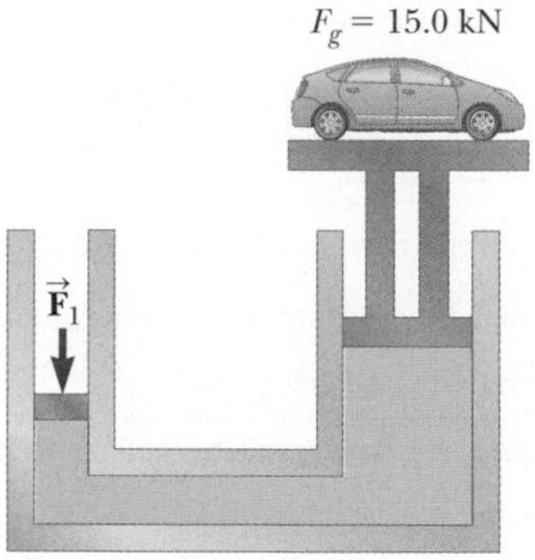

Figure P14.8

9. **M** What must be the contact area between a suction cup (completely evacuated) and a ceiling if the cup is to support the weight of an 80.0-kg student?

10. (a) A very powerful vacuum cleaner has a hose 2.86 cm in diameter. With the end of the hose placed perpendicularly on the flat face of a brick, what is the weight of the heaviest brick that the cleaner can lift? (b) **What If?** An octopus uses one sucker of diameter 2.86 cm on each of the two shells of a clam in an attempt to pull the shells apart. Find the greatest force the octopus can exert on a clamshell in salt water 32.3 m deep.

11. (a) Calculate the absolute pressure at the bottom of a freshwater lake at a point whose depth is 27.5 m. Assume the density of the water is 1.00 × 10^3 kg/m^3 and that the air above is at a pressure of 101.3 kPa. (b) What force is exerted by the water on the window of an underwater vehicle at this depth if the window is circular and has a diameter of 35.0 cm?

12. *Why is the following situation impossible?* Figure P14.12 shows Superman attempting to drink cold water through a straw of length ℓ = 12.0 m. The walls of the tubular straw are very strong and do not collapse. With his great strength, he achieves maximum possible suction and enjoys drinking the cold water.

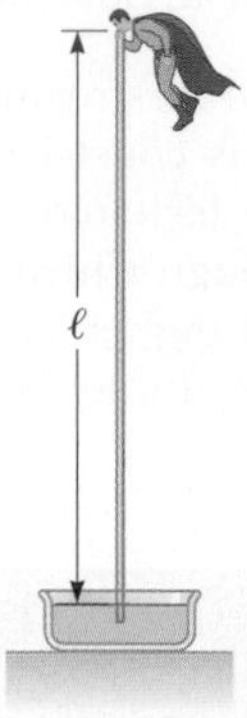

Figure P14.12

13. **Review.** The tank in Figure P14.13 is filled with water of depth d = 2.00 m. At the bottom of one sidewall is a rectangular hatch of height h = 1.00 m and width w = 2.00 m that is hinged at the top of the hatch. (a) Determine the magnitude of the force the water exerts on the hatch. (b) Find the magnitude of the torque exerted by the water about the hinges.

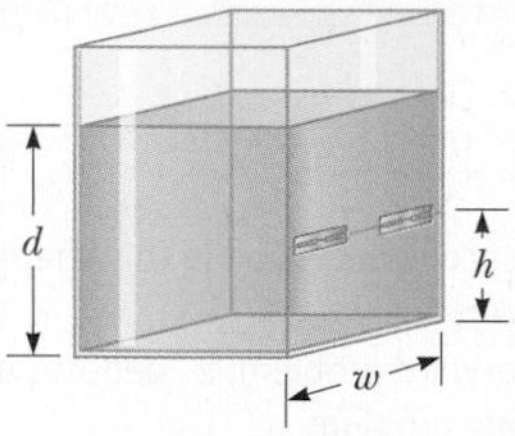

Figure P14.13
Problems 13 and 14.

14. **Review.** **S** The tank in Figure P14.13 is filled with water of depth d. At the bottom of one sidewall is a rectangular hatch of height h and width w that is hinged at the top of the hatch. (a) Determine the magnitude of the force the water exerts on the hatch. (b) Find the magnitude of the torque exerted by the water about the hinges.

15. **Review.** A solid sphere of brass (bulk modulus of 14.0 × 10^{10} N/m^2) with a diameter of 3.00 m is thrown into the ocean. By how much does the diameter of the sphere decrease as it sinks to a depth of 1.0 km?

Section 14.3 Pressure Measurements

16. **Q|C** Blaise Pascal duplicated Torricelli's barometer using a red Bordeaux wine, of density 984 kg/m^3, as the working liquid (Fig. P14.16). (a) What was the height h of the wine column for normal atmospheric pressure? (b) Would you expect the vacuum above the column to be as good as for mercury?

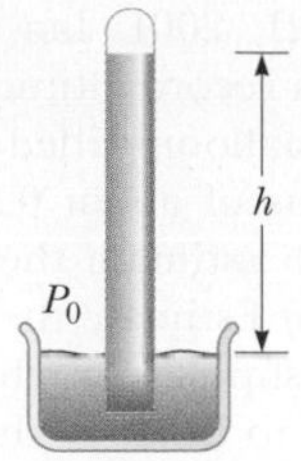

Figure P14.16

17. Normal atmospheric pressure is 1.013×10^5 Pa. The approach of a storm causes the height of a mercury barometer to drop by 20.0 mm from the normal height. What is the atmospheric pressure?

18. **Q|C** The human brain and spinal cord are immersed in the cerebrospinal fluid. The fluid is normally continuous between the cranial and spinal cavities and exerts a pressure of 100 to 200 mm of H_2O above the prevailing atmospheric pressure. In medical work, pressures are often measured in units of millimeters of H_2O because body fluids, including the cerebrospinal fluid, typically have the same density as water. The pressure of the cerebrospinal fluid can be measured by means of a *spinal tap* as illustrated in Figure P14.18. A hollow tube is inserted into the spinal column, and the height to which the fluid rises is observed. If the fluid rises to a height of 160 mm, we write its gauge pressure as 160 mm H_2O. (a) Express this pressure in pascals, in atmospheres, and in millimeters of mercury. (b) Some conditions that block or inhibit the flow of cerebrospinal fluid can be investigated by means of *Queckenstedt's test.* In this procedure, the veins in the patient's neck are compressed to make the blood pressure rise in the brain, which in turn should be transmitted to the cerebrospinal fluid. Explain how the level of fluid in the spinal tap can be used as a diagnostic tool for the condition of the patient's spine.

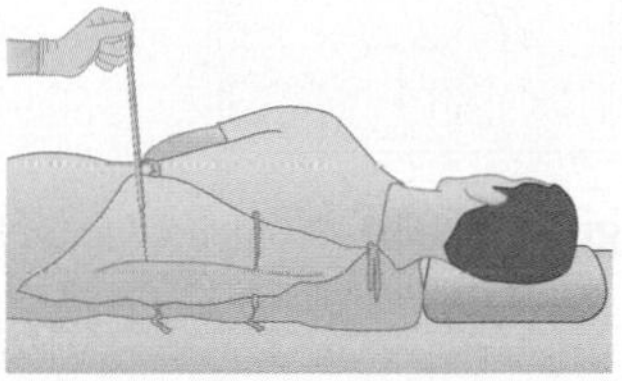

Figure P14.18

19. A backyard swimming pool with a circular base of diameter 6.00 m is filled to depth 1.50 m. (a) Find the absolute pressure at the bottom of the pool. (b) Two persons with combined mass 150 kg enter the pool and float quietly there. No water overflows. Find the pressure increase at the bottom of the pool after they enter the pool and float.

20. **S** A tank with a flat bottom of area A and vertical sides is filled to a depth h with water. The pressure is P_0 at the top surface. (a) What is the absolute pressure at the bottom of the tank? (b) Suppose an object of mass M and density less than the density of water is placed into the tank and floats. No water overflows. What is the resulting increase in pressure at the bottom of the tank?

21. Mercury is poured into a U-tube as shown in Figure P14.21a. The left arm of the tube has cross-sectional area A_1 of 10.0 cm², and the right arm has a cross-sectional area A_2 of 5.00 cm². One hundred grams of water are then poured into the right arm as shown in Figure P14.21b. (a) Determine the length of the water column in the right arm of the U-tube. (b) Given that the density of mercury is 13.6 g/cm³, what distance h does the mercury rise in the left arm?

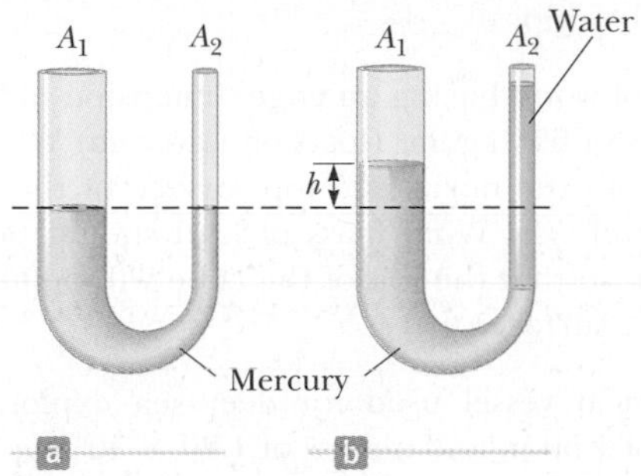

Figure P14.21

Section 14.4 Buoyant Forces and Archimedes's Principle

22. A light balloon is filled with 400 m³ of helium at atmospheric pressure. (a) At 0°C, the balloon can lift a payload of what mass? (b) **What If?** In Table 14.1, observe that the density of hydrogen is nearly half the density of helium. What load can the balloon lift if filled with hydrogen?

23. A table-tennis ball has a diameter of 3.80 cm and average density of 0.084 0 g/cm³. What force is required to hold it completely submerged under water?

24. The gravitational force exerted on a solid object is 5.00 N. When the object is suspended from a spring scale and submerged in water, the scale reads 3.50 N (Fig. P14.24). Find the density of the object.

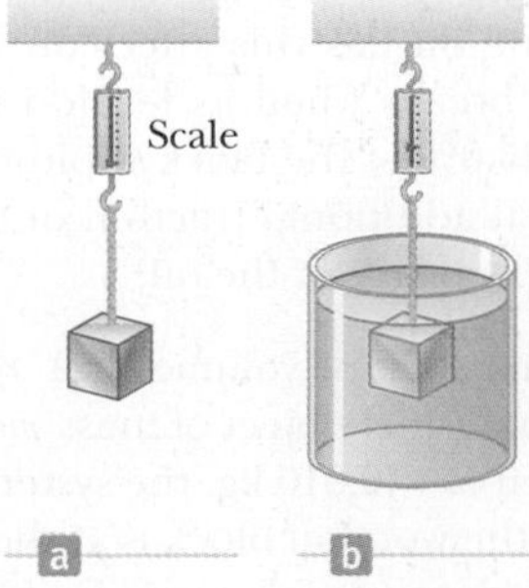

Figure P14.24 Problems 24 and 25.

25. A 10.0-kg block of metal measuring 12.0 cm by 10.0 cm by 10.0 cm is suspended from a scale and immersed in water as shown in Figure P14.24b. The 12.0-cm dimension is vertical, and the top of the block is 5.00 cm below the surface of the water. (a) What are the magnitudes of the forces acting on the top and on the bottom of the block due to the surrounding water? (b) What is the reading of the spring scale? (c) Show that the buoyant force equals the difference between the forces at the top and bottom of the block.

26. The United States possesses the ten largest warships in the world, aircraft carriers of the *Nimitz* class. Suppose one of the ships bobs up to float 11.0 cm higher in the ocean water when 50 fighters take off from it in a time interval of 25 min, at a location where the free-fall acceleration is 9.78 m/s^2. The planes have an average laden mass of 29 000 kg. Find the horizontal area enclosed by the waterline of the ship.

27. A cube of wood having an edge dimension of 20.0 cm and a density of 650 kg/m^3 floats on water. (a) What is the distance from the horizontal top surface of the cube to the water level? (b) What mass of lead should be placed on the cube so that the top of the cube will be just level with the water surface?

28. A spherical vessel used for deep-sea exploration has a radius of 1.50 m and a mass of 1.20×10^4 kg. To dive, the vessel takes on mass in the form of seawater. Determine the mass the vessel must take on if it is to descend at a constant speed of 1.20 m/s, when the resistive force on it is 1 100 N in the upward direction. The density of seawater is equal to 1.03×10^3 kg/m^3.

29. **M** A plastic sphere floats in water with 50.0% of its volume submerged. This same sphere floats in glycerin with 40.0% of its volume submerged. Determine the densities of (a) the glycerin and (b) the sphere.

30. **Q|C** The weight of a rectangular block of low-density material is 15.0 N. With a thin string, the center of the horizontal bottom face of the block is tied to the bottom of a beaker partly filled with water. When 25.0% of the block's volume is submerged, the tension in the string is 10.0 N. (a) Find the buoyant force on the block. (b) Oil of density 800 kg/m^3 is now steadily added to the beaker, forming a layer above the water and surrounding the block. The oil exerts forces on each of the four sidewalls of the block that the oil touches. What are the directions of these forces? (c) What happens to the string tension as the oil is added? Explain how the oil has this effect on the string tension. (d) The string breaks when its tension reaches 60.0 N. At this moment, 25.0% of the block's volume is still below the water line. What additional fraction of the block's volume is below the top surface of the oil?

31. **Q|C** A wooden block of volume 5.24×10^{-4} m^3 floats in water, and a small steel object of mass m is placed on top of the block. When $m = 0.310$ kg, the system is in equilibrium and the top of the wooden block is at the level of the water. (a) What is the density of the wood? (b) What happens to the block when the steel object is replaced by an object whose mass is less than 0.310 kg? (c) What happens to the block when the steel object is replaced by an object whose mass is greater than 0.310 kg?

32. **Q|C** On October 21, 2001, Ian Ashpole of the United Kingdom achieved a record altitude of 3.35 km (11 000 ft) powered by 600 toy balloons filled with helium. Each filled balloon had a radius of about 0.50 m and an estimated mass of 0.30 kg. (a) Estimate the total buoyant force on the 600 balloons. (b) Estimate the net upward force on all 600 balloons. (c) Ashpole parachuted to the Earth after the balloons began to burst at the high altitude and the buoyant force decreased. Why did the balloons burst?

33. **Q|C** A large weather balloon whose mass is 226 kg is filled with helium gas until its volume is 325 m^3. Assume the density of air is 1.20 kg/m^3 and the density of helium is 0.179 kg/m^3. (a) Calculate the buoyant force acting on the balloon. (b) Find the net force on the balloon and determine whether the balloon will rise or fall after it is released. (c) What additional mass can the balloon support in equilibrium?

34. **S** A *hydrometer* is an instrument used to determine liquid density. A simple one is sketched in Figure P14.34. The bulb of a syringe is squeezed and released to let the atmosphere lift a sample of the liquid of interest into a tube containing a calibrated rod of known density. The rod, of length L and average density ρ_0, floats partially immersed in the liquid of density ρ. A length h of the rod protrudes above the surface of the liquid. Show that the density of the liquid is given by

$$\rho = \frac{\rho_0 L}{L - h}$$

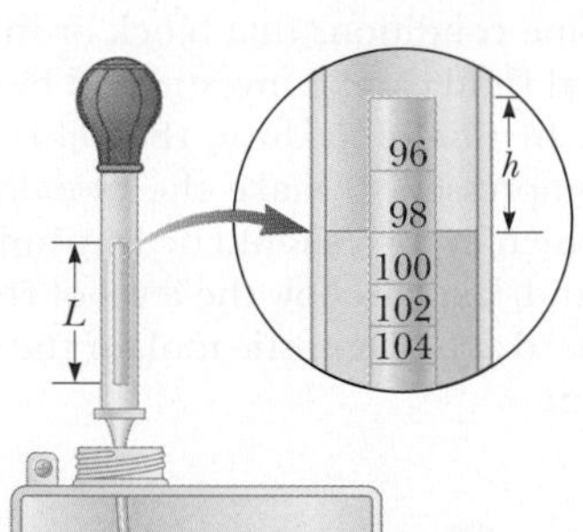

Figure P14.34 Problems 34 and 35.

35. **Q|C** Refer to Problem 34 and Figure P14.34. A hydrometer is to be constructed with a cylindrical floating rod. Nine fiduciary marks are to be placed along the rod to indicate densities of 0.98 g/cm^3, 1.00 g/cm^3, 1.02 g/cm^3, 1.04 g/cm^3, . . . , 1.14 g/cm^3. The row of marks is to start 0.200 cm from the top end of the rod and end 1.80 cm from the top end. (a) What is the required length of the rod? (b) What must be its average density? (c) Should the marks be equally spaced? Explain your answer.

36. **M** How many cubic meters of helium are required to lift a balloon with a 400-kg payload to a height of 8 000 m? Take $\rho_{He} = 0.179$ kg/m^3. Assume the balloon maintains a constant volume and the density of air decreases with the altitude z according to the expression $\rho_{air} = \rho_0 e^{-z/8\,000}$, where z is in meters and $\rho_0 = 1.20\ \text{kg/m}^3$ is the density of air at sea level.

Section 14.5 Fluid Dynamics

Section 14.6 Bernoulli's Equation

37. A large storage tank, open at the top and filled with water, develops a small hole in its side at a point 16.0 m below the water level. The rate of flow from the leak is found to be $2.50 \times 10^{-3}\ \text{m}^3/\text{min}$. Determine (a) the speed at which the water leaves the hole and (b) the diameter of the hole.

38. Water flowing through a garden hose of diameter 2.74 cm fills a 25-L bucket in 1.50 min. (a) What is the speed of the water leaving the end of the hose? (b) A nozzle is now attached to the end of the hose. If the nozzle diameter is one-third the diameter of the hose, what is the speed of the water leaving the nozzle?

39. Water moves through a constricted pipe in steady, ideal flow. At the lower point shown in Figure P14.39, the pressure is $P_1 = 1.75 \times 10^4$ Pa and the pipe diameter is 6.00 cm. At another point $y = 0.250$ m higher, the pressure is $P_2 = 1.20 \times 10^4$ Pa and the pipe diameter is 3.00 cm. Find the speed of flow (a) in the lower section and (b) in the upper section. (c) Find the volume flow rate through the pipe.

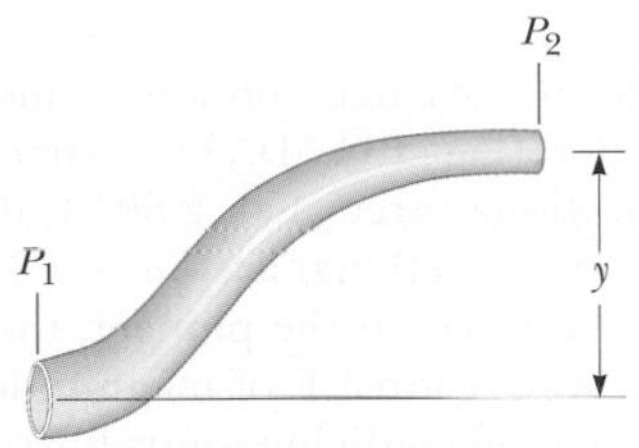

Figure P14.39

40. A village maintains a large tank with an open top, containing water for emergencies. The water can drain from the tank through a hose of diameter 6.60 cm. The hose ends with a nozzle of diameter 2.20 cm. A rubber stopper is inserted into the nozzle. The water level in the tank is kept 7.50 m above the nozzle. (a) Calculate the friction force exerted on the stopper by the nozzle. (b) The stopper is removed. What mass of water flows from the nozzle in 2.00 h? (c) Calculate the gauge pressure of the flowing water in the hose just behind the nozzle.

41. Figure P14.41 shows a stream of water in steady flow from a kitchen faucet. At the faucet, the diameter of the stream is 0.960 cm. The stream fills a 125-cm^3 container in 16.3 s. Find the diameter of the stream 13.0 cm below the opening of the faucet.

Figure P14.41

42. Water falls over a dam of height h with a mass flow rate of R, in units of kilograms per second. (a) Show that the power available from the water is

$$P = Rgh$$

where g is the free-fall acceleration. (b) Each hydroelectric unit at the Grand Coulee Dam takes in water at a rate of 8.50×10^5 kg/s from a height of 87.0 m. The power developed by the falling water is converted to electric power with an efficiency of 85.0%. How much electric power does each hydroelectric unit produce?

43. A legendary Dutch boy saved Holland by plugging a hole of diameter 1.20 cm in a dike with his finger. If the hole was 2.00 m below the surface of the North Sea (density 1 030 kg/m^3), (a) what was the force on his finger? (b) If he pulled his finger out of the hole, during what time interval would the released water fill 1 acre of land to a depth of 1 ft? Assume the hole remained constant in size.

44. **Q|C** In ideal flow, a liquid of density 850 kg/m^3 moves from a horizontal tube of radius 1.00 cm into a second horizontal tube of radius 0.500 cm at the same elevation as the first tube. The pressure differs by ΔP between the liquid in one tube and the liquid in the second tube. (a) Find the volume flow rate as a function of ΔP. Evaluate the volume flow rate for (b) $\Delta P = 6.00$ kPa and (c) $\Delta P = 12.0$ kPa.

45. Water is pumped up from the Colorado River to supply Grand Canyon Village, located on the rim of the canyon. The river is at an elevation of 564 m, and the village is at an elevation of 2 096 m. Imagine that the water is pumped through a single long pipe 15.0 cm in diameter, driven by a single pump at the bottom end. (a) What is the minimum pressure at which the water must be pumped if it is to arrive at the village? (b) If 4 500 m^3 of water are pumped per day, what is the speed of the water in the pipe? *Note:* Assume the free-fall acceleration and the density of air are constant over this range of elevations. The pressures you calculate are too high for an ordinary pipe. The water is actually lifted in stages by several pumps through shorter pipes.

46. **Q|C** **Review.** Old Faithful Geyser in Yellowstone National Park erupts at approximately one-hour intervals, and the height of the water column reaches 40.0 m. (Fig. P14.46, page 428). (a) Model the rising stream as a series of separate droplets. Analyze the free-fall motion of one of the droplets to determine the speed at which the water leaves the ground. (b) **What If?** Model the rising stream as an ideal fluid in streamline flow. Use Bernoulli's equation to determine the speed of the water as it leaves ground level. (c) How does the answer from part (a) compare with the answer from part (b)? (d) What is the pressure (above atmospheric) in the heated underground chamber if its depth is 175 m? Assume the chamber is large compared with the geyser's vent.

Figure P14.46

47. The Venturi tube discussed in Example 14.8 and shown in Figure P14.47 may be used as a fluid flowmeter. Suppose the device is used at a service station to measure the flow rate of gasoline ($\rho = 7.00 \times 10^2$ kg/m^3) through a hose having an outlet radius of 1.20 cm. If the difference in pressure is measured to be $P_1 - P_2 = 1.20$ kPa and the radius of the inlet tube to the meter is 2.40 cm, find (a) the speed of the gasoline as it leaves the hose and (b) the fluid flow rate in cubic meters per second.

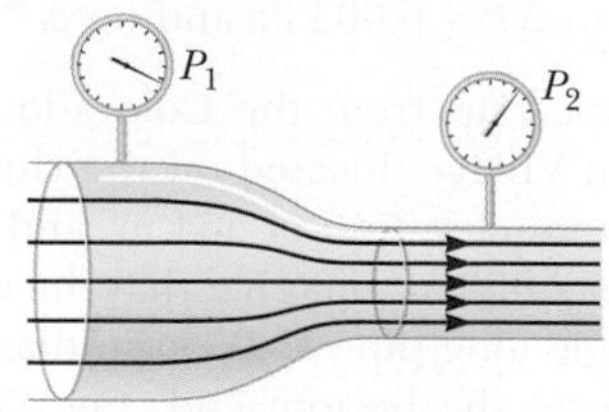

Figure P14.47

Section 14.7 Other Applications of Fluid Dynamics

48. Q|C An airplane has a mass of 1.60×10^4 kg, and each wing has an area of 40.0 m^2. During level flight, the pressure on the lower wing surface is 7.00×10^4 Pa. (a) Suppose the lift on the airplane were due to a pressure difference alone. Determine the pressure on the upper wing surface. (b) More realistically, a significant part of the lift is due to deflection of air downward by the wing. Does the inclusion of this force mean that the pressure in part (a) is higher or lower? Explain.

49. An airplane is cruising at altitude 10 km. The pressure outside the craft is 0.287 atm; within the passenger compartment, the pressure is 1.00 atm and the temperature is 20°C. A small leak occurs in one of the window seals in the passenger compartment. Model the air as an ideal fluid to estimate the speed of the airstream flowing through the leak.

50. A siphon is used to drain water from a tank as illustrated in Figure P14.50. Assume steady flow without friction. (a) If $h = 1.00$ m, find the speed of outflow at the end of the siphon. (b) **What If?** What is the limitation on the height of the top of the siphon above the end of the siphon? *Note:* For the flow of the liquid to be continuous, its pressure must not drop below its vapor pressure. Assume the water is at 20.0°C, at which the vapor pressure is 2.3 kPa.

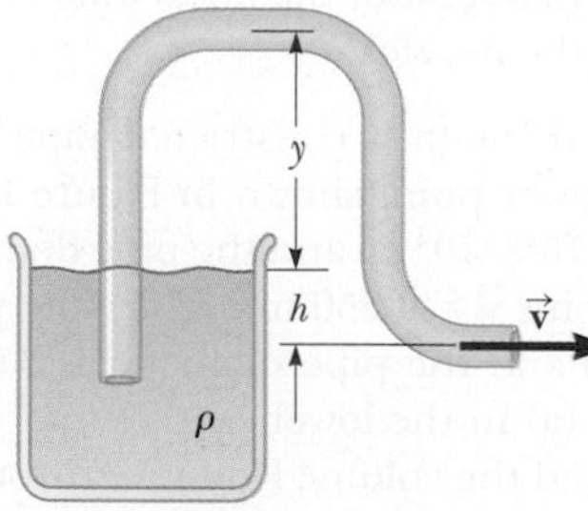

Figure P14.50

51. M A hypodermic syringe contains a medicine with the density of water (Fig. P14.51). The barrel of the syringe has a cross-sectional area $A = 2.50 \times 10^{-5}$ m^2, and the needle has a cross-sectional area $a = 1.00 \times 10^{-8}$ m^2. In the absence of a force on the plunger, the pressure everywhere is 1.00 atm. A force $\vec{F}$ of magnitude 2.00 N acts on the plunger, making medicine squirt horizontally from the needle. Determine the speed of the medicine as it leaves the needle's tip.

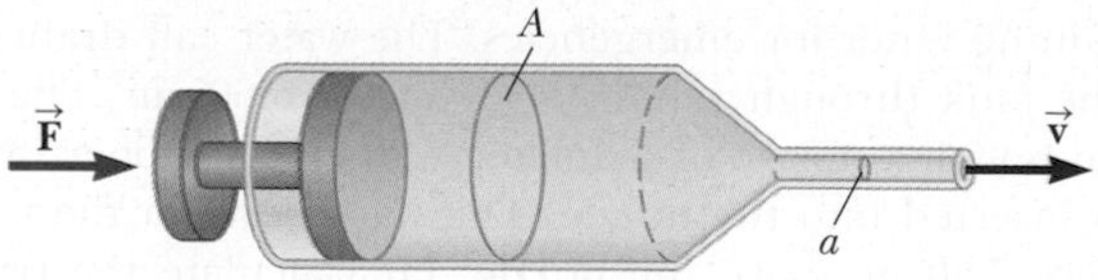

Figure P14.51

52. The Bernoulli effect can have important consequences for the design of buildings. For example, wind can blow around a skyscraper at remarkably high speed, creating low pressure. The higher atmospheric pressure in the still air inside the buildings can cause windows to pop out. As originally constructed, the John Hancock Building in Boston popped windowpanes that fell many stories to the sidewalk below. (a) Suppose a horizontal wind blows with a speed of 11.2 m/s outside a large pane of plate glass with dimensions 4.00 m × 1.50 m. Assume the density of the air to be constant at 1.20 kg/m^3. The air inside the building is at atmospheric pressure. What is the total force exerted by air on the windowpane? (b) **What If?** If a second skyscraper is built nearby, the airspeed can be especially high where wind passes through the narrow separation between the buildings. Solve part (a) again with a wind speed of 22.4 m/s, twice as high.

Additional Problems

53. (a) Calculate the absolute pressure at an ocean depth of 1 000 m. Assume the density of seawater is 1 030 kg/m^3 and the air above exerts a pressure of 101.3 kPa. (b) At this depth, what is the buoyant force on a spherical submarine having a diameter of 5.00 m?

54. **S** In about 1657, Otto von Guericke, inventor of the air pump, evacuated a sphere made of two brass hemispheres (Fig. P14.54). Two teams of eight horses each could pull the hemispheres apart only on some trials and then "with greatest difficulty," with the resulting sound likened to a cannon firing. Find the force F required to pull the thin-walled evacuated hemispheres apart in terms of R, the radius of the hemispheres; P, the pressure inside the hemispheres; and atmospheric pressure P_0.

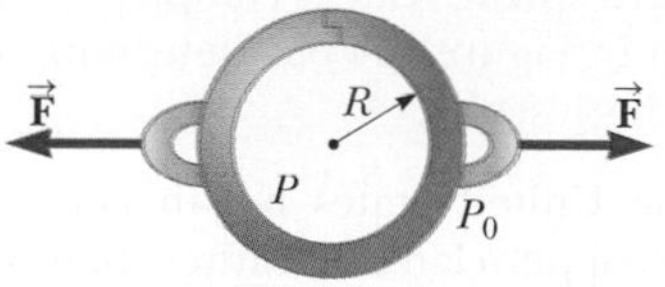

Figure P14.54

55. A spherical aluminum ball of mass 1.26 kg contains an empty spherical cavity that is concentric with the ball. The ball barely floats in water. Calculate (a) the outer radius of the ball and (b) the radius of the cavity.

56. **GP** A helium-filled balloon (whose envelope has a mass of $m_b = 0.250$ kg) is tied to a uniform string of length $\ell = 2.00$ m and mass $m = 0.050\,0$ kg. The balloon is spherical with a radius of $r = 0.400$ m. When released in air of temperature 20°C and density $\rho_{air} = 1.20$ kg/m^3, it lifts a length h of string and then remains stationary as shown in Figure P14.56. We wish to find the length of string lifted by the balloon. (a) When the balloon remains stationary, what is the appropriate analysis model to describe it? (b) Write a force equation for the balloon from this model in terms of the buoyant force B, the weight F_b of the balloon, the weight F_{He} of the helium, and the weight F_s of the segment of string of length h. (c) Make an appropriate substitution for each of these forces and solve symbolically for the mass m_s of the segment of string of length h in terms of m_b, r, ρ_{air}, and the density of helium ρ_{He}. (d) Find the numerical value of the mass m_s. (e) Find the length h numerically.

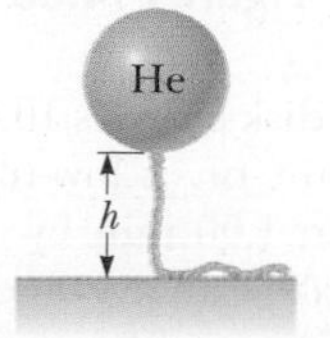

Figure P14.56

57. **Review.** Figure P14.57 shows a valve separating a reservoir from a water tank. If this valve is opened, what is the maximum height above point B attained by the water stream coming out of the right side of the tank? Assume $h = 10.0$ m, $L = 2.00$ m, and $\theta = 30.0°$, and assume the cross-sectional area at A is very large compared with that at B.

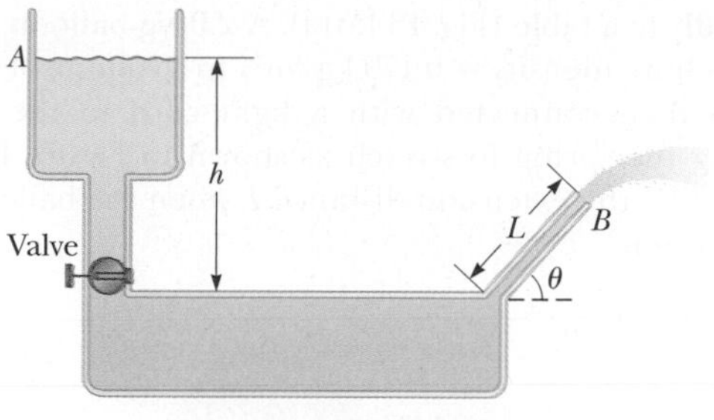

Figure P14.57

58. **S** The true weight of an object can be measured in a vacuum, where buoyant forces are absent. A measurement in air, however, is disturbed by buoyant forces. An object of volume V is weighed in air on an equal-arm balance with the use of counterweights of density ρ. Representing the density of air as ρ_{air} and the balance reading as F'_g, show that the true weight F_g is

$$F_g = F'_g + \left(V - \frac{F'_g}{\rho g}\right)\rho_{air} g$$

59. Water is forced out of a fire extinguisher by air pressure as shown in Figure P14.59. How much gauge air pressure in the tank is required for the water jet to have a speed of 30.0 m/s when the water level is 0.500 m below the nozzle?

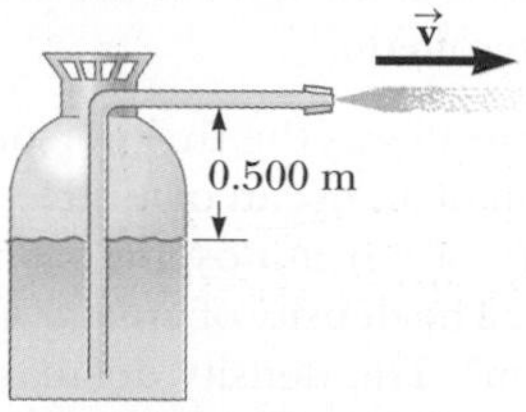

Figure P14.59

60. **QC** **Review.** Assume a certain liquid, with density 1 230 kg/m^3, exerts no friction force on spherical objects. A ball of mass 2.10 kg and radius 9.00 cm is dropped from rest into a deep tank of this liquid from a height of 3.30 m above the surface. (a) Find the speed at which the ball enters the liquid. (b) Evaluate the magnitudes of the two forces that are exerted on the ball as it moves through the liquid. (c) Explain why the ball moves down only a limited distance into the liquid and calculate this distance. (d) With what speed will the ball pop up out of the liquid? (e) How does the time interval Δt_{down}, during which the ball moves from the surface down to its lowest point, compare with the time interval Δt_{up} for the return trip between the same two points? (f) **What If?** Now modify the model to suppose the liquid exerts a small friction force on the ball, opposite in direction to its motion. In this case, how do the time intervals Δt_{down} and Δt_{up} compare? Explain your answer with a conceptual argument rather than a numerical calculation.

61. **Review.** A light spring of constant $k = 90.0$ N/m is attached vertically to a table (Fig. P14.61a). A 2.00-g balloon is filled with helium (density $= 0.179$ kg/m^3) to a volume of 5.00 m^3 and is then connected with a light cord to the spring, causing the spring to stretch as shown in Figure P14.61b. Determine the extension distance L when the balloon is in equilibrium.

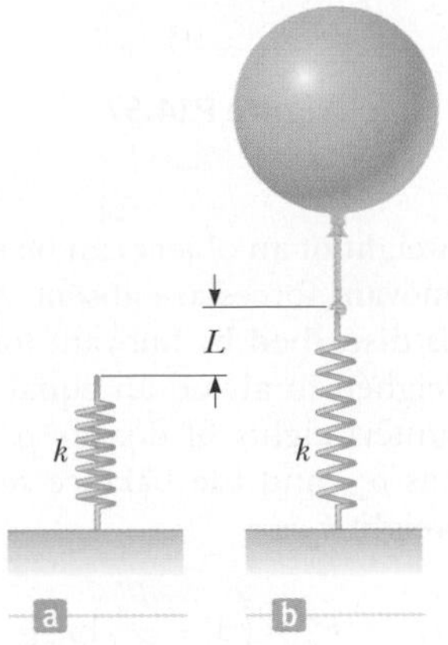

Figure P14.61

62. A 42.0-kg boy uses a solid block of Styrofoam as a raft while fishing on a pond. The Styrofoam has an area of 1.00 m^2 and is 0.050 0 m thick. While sitting on the surface of the raft, the boy finds that the raft just supports him so that the top of the raft is at the level of the pond. Determine the density of the Styrofoam.

63. Evangelista Torricelli was the first person to realize that we live at the bottom of an ocean of air. He correctly surmised that the pressure of our atmosphere is attributable to the weight of the air. The density of air at 0°C at the Earth's surface is 1.29 kg/m^3. The density decreases with increasing altitude (as the atmosphere thins). On the other hand, if we assume the density is constant at 1.29 kg/m^3 up to some altitude h and is zero above that altitude, then h would represent the depth of the ocean of air. (a) Use this model to determine the value of h that gives a pressure of 1.00 atm at the surface of the Earth. (b) Would the peak of Mount Everest rise above the surface of such an atmosphere?

64. S **Review.** With reference to the dam studied in Example 14.4 and shown in Figure 14.5, (a) show that the total torque exerted by the water behind the dam about a horizontal axis through O is $\frac{1}{6}\rho g w H^3$. (b) Show that the effective line of action of the total force exerted by the water is at a distance $\frac{1}{3}H$ above O.

65. A 1.00-kg beaker containing 2.00 kg of oil (density = 916.0 kg/m^3) rests on a scale. A 2.00-kg block of iron suspended from a spring scale is completely submerged in the oil as shown in Figure P14.65. Determine the equilibrium readings of both scales.

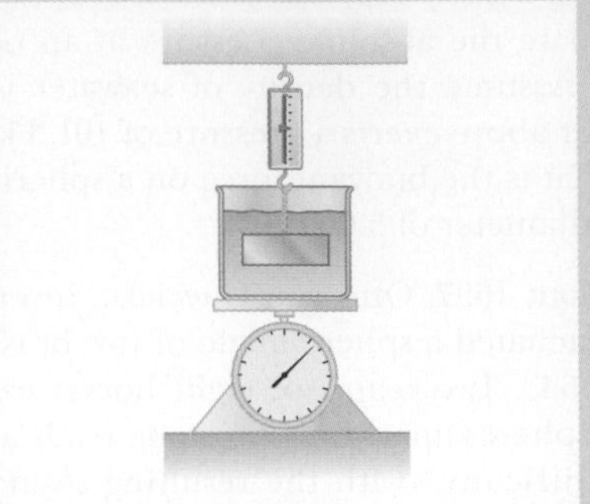

Figure P14.65 Problems 65 and 66.

66. S A beaker of mass m_b containing oil of mass m_o and density ρ_o rests on a scale. A block of iron of mass m_{Fe} suspended from a spring scale is completely submerged in the oil as shown in Figure P14.65. Determine the equilibrium readings of both scales.

67. In 1983, the United States began coining the one-cent piece out of copper-clad zinc rather than pure copper. The mass of the old copper penny is 3.083 g and that of the new cent is 2.517 g. The density of copper is 8.920 g/cm^3 and that of zinc is 7.133 g/cm^3. The new and old coins have the same volume. Calculate the percent of zinc (by volume) in the new cent.

68. **Review.** Figure P14.68 shows the essential parts of a hydraulic brake system. The area of the piston in the master cylinder is 1.8 cm^2 and that of the piston in the brake cylinder is 6.4 cm^2. The coefficient of friction between shoe and wheel drum is 0.50. If the wheel has a radius of 34 cm, determine the frictional torque about the axle when a force of 44 N is exerted on the brake pedal.

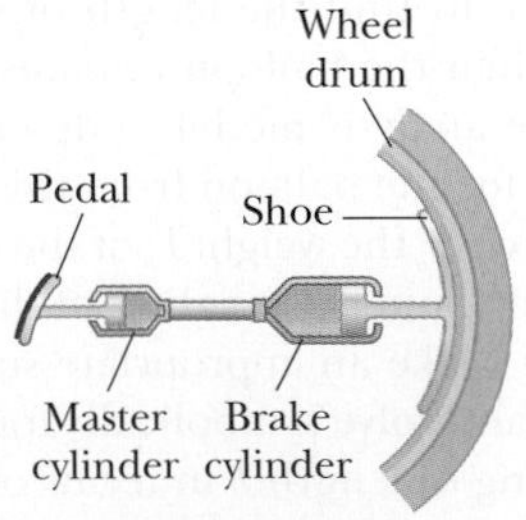

Figure P14.68

69. **Review.** A uniform disk of mass 10.0 kg and radius 0.250 m spins at 300 rev/min on a low-friction axle. It must be brought to a stop in 1.00 min by a brake pad that makes contact with the disk at an average distance 0.220 m from the axis. The coefficient of friction between pad and disk is 0.500. A piston in a cylinder of diameter 5.00 cm presses the brake pad against the disk. Find the pressure required for the brake fluid in the cylinder.

70. **Q|C** **Review.** In a water pistol, a piston drives water through a large tube of area A_1 into a smaller tube of area A_2 as shown in Figure P14.70. The radius of the large tube is 1.00 cm and that of the small tube is 1.00 mm. The smaller tube is 3.00 cm above the larger tube. (a) If the pistol is fired horizontally at a height of 1.50 m, determine the time interval required for the water to travel from the nozzle to the ground. Neglect air resistance and assume atmospheric pressure is 1.00 atm. (b) If the desired range of the stream is 8.00 m, with what speed v_2 must the stream leave the nozzle? (c) At what speed v_1 must the plunger be moved to achieve the desired range? (d) What is the pressure at the nozzle? (e) Find the pressure needed in the larger tube. (f) Calculate the force that must be exerted on the trigger to achieve the desired range. (The force that must be exerted is due to pressure over and above atmospheric pressure.)

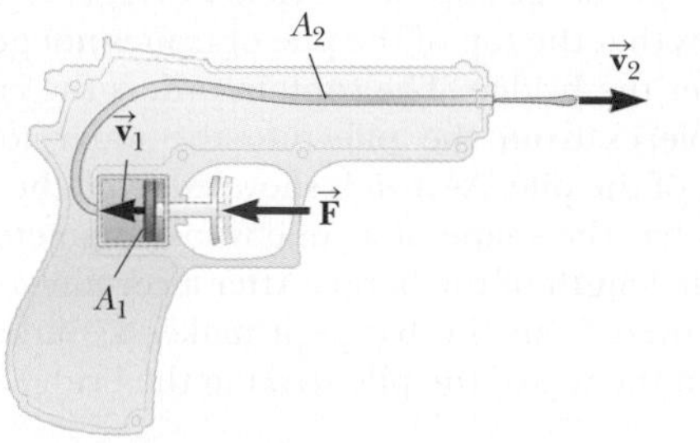

Figure P14.70

71. An incompressible, nonviscous fluid is initially at rest in the vertical portion of the pipe shown in Figure P14.71a, where $L = 2.00$ m. When the valve is opened, the fluid flows into the horizontal section of the pipe. What is the fluid's speed when all the fluid is in the horizontal section as shown in Figure P14.71b? Assume the cross-sectional area of the entire pipe is constant.

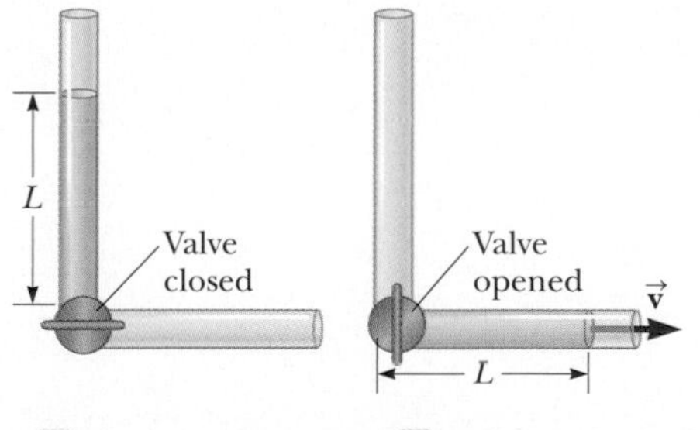

Figure P14.71

72. The water supply of a building is fed through a main pipe 6.00 cm in diameter. A 2.00-cm-diameter faucet tap, located 2.00 m above the main pipe, is observed to fill a 25.0-L container in 30.0 s. (a) What is the speed at which the water leaves the faucet? (b) What is the gauge pressure in the 6-cm main pipe? Assume the faucet is the only "leak" in the building.

73. A U-tube open at both ends is partially filled with water (Fig. P14.73a). Oil having a density 750 kg/m^3 is then poured into the right arm and forms a column $L = 5.00$ cm high (Fig. P14.73b). (a) Determine the difference h in the heights of the two liquid surfaces. (b) The right arm is then shielded from any air motion while air is blown across the top of the left arm until the surfaces of the two liquids are at the same height (Fig. P14.73c). Determine the speed of the air being blown across the left arm. Take the density of air as constant at 1.20 kg/m^3.

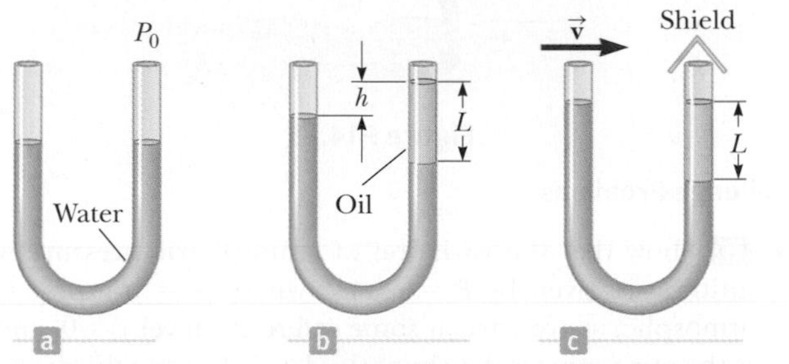

Figure P14.73

74. **S** A woman is draining her fish tank by siphoning the water into an outdoor drain as shown in Figure P14.74. The rectangular tank has footprint area A and depth h. The drain is located a distance d below the surface of the water in the tank, where $d >> h$. The cross-sectional area of the siphon tube is A'. Model the water as flowing without friction. Show that the time interval required to empty the tank is given by

$$\Delta t = \frac{Ah}{A'\sqrt{2gd}}$$

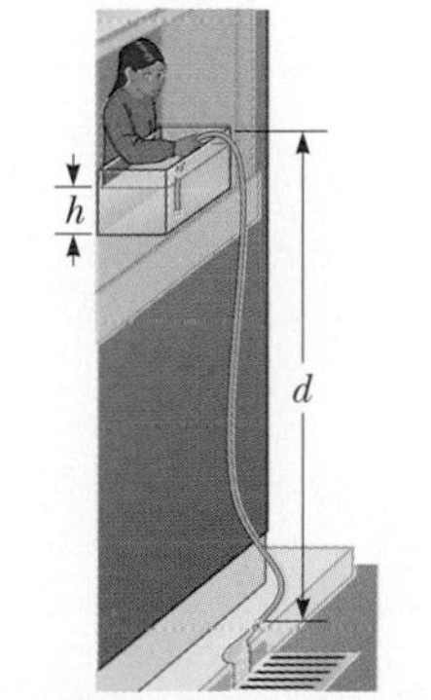

Figure P14.74

75. **S** The hull of an experimental boat is to be lifted above the water by a hydrofoil mounted below its keel as shown in Figure P14.75 on page 432. The hydrofoil has a shape like that of an airplane wing. Its area projected onto a horizontal surface is A. When the boat is towed at sufficiently high speed, water of density ρ moves in streamline flow so that its average speed at the top of the hydrofoil is n times larger than its speed v_b below the hydrofoil. (a) Ignoring

the buoyant force, show that the upward lift force exerted by the water on the hydrofoil has a magnitude

$$F \approx \tfrac{1}{2}(n^2 - 1)\rho v_b{}^2 A$$

(b) The boat has mass M. Show that the liftoff speed is given by

$$v \approx \sqrt{\frac{2Mg}{(n^2 - 1)A\rho}}$$

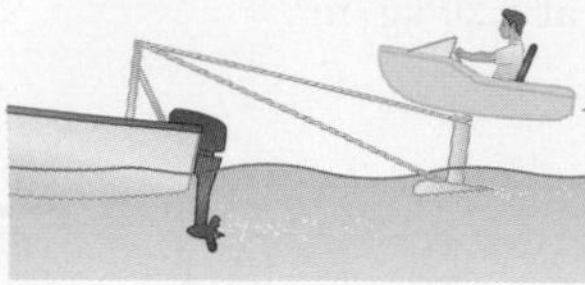

Figure P14.75

Challenge Problems

76. S Show that the variation of atmospheric pressure with altitude is given by $P = P_0 e^{-\alpha y}$, where $\alpha = \rho_0 g / \rho P_0$, P_0 is atmospheric pressure at some reference level $y = 0$, and ρ_0 is the atmospheric density at this level. Assume the decrease in atmospheric pressure over an infinitesimal change in altitude (so that the density is approximately uniform over the infinitesimal change) can be expressed from Equation 14.4 as $dP = -\rho g\, dy$. Also assume the density of air is proportional to the pressure, which, as we will see in Chapter 20, is equivalent to assuming the temperature of the air is the same at all altitudes.

77. An ice cube whose edges measure 20.0 mm is floating in a glass of ice-cold water, and one of the ice cube's faces is parallel to the water's surface. (a) How far below the water surface is the bottom face of the block? (b) Ice-cold ethyl alcohol is gently poured onto the water surface to form a layer 5.00 mm thick above the water. The alcohol does not mix with the water. When the ice cube again attains hydrostatic equilibrium, what is the distance from the top of the water to the bottom face of the block? (c) Additional cold ethyl alcohol is poured onto the water's surface until the top surface of the alcohol coincides with the top surface of the ice cube (in hydrostatic equilibrium). How thick is the required layer of ethyl alcohol?

78. *Why is the following situation impossible?* A barge is carrying a load of small pieces of iron along a river. The iron pile is in the shape of a cone for which the radius r of the base of the cone is equal to the central height h of the cone. The barge is square in shape, with vertical sides of length $2r$, so that the pile of iron comes just up to the edges of the barge. The barge approaches a low bridge, and the captain realizes that the top of the pile of iron is not going to make it under the bridge. The captain orders the crew to shovel iron pieces from the pile into the water to reduce the height of the pile. As iron is shoveled from the pile, the pile always has the shape of a cone whose diameter is equal to the side length of the barge. After a certain volume of iron is removed from the barge, it makes it under the bridge without the top of the pile striking the bridge.

Oscillations and Mechanical Waves

part 2

Falling drops of water cause a water surface to oscillate. These oscillations are associated with circular waves moving away from the point at which the drops fall. In Part 2 of the text, we will explore the principles related to oscillations and waves. *(Marga Buschbell Steeger/Photographer's Choice/Getty Images)*

We begin this new part of the text by studying a special type of motion called *periodic* motion, the repeating motion of an object in which it continues to return to a given position after a fixed time interval. The repetitive movements of such an object are called *oscillations.* We will focus our attention on a special case of periodic motion called *simple harmonic motion.* All periodic motions can be modeled as combinations of simple harmonic motions.

Simple harmonic motion also forms the basis for our understanding of *mechanical waves.* Sound waves, seismic waves, waves on stretched strings, and water waves are all produced by some source of oscillation. As a sound wave travels through the air, elements of the air oscillate back and forth; as a water wave travels across a pond, elements of the water oscillate up and down and backward and forward. The motion of the elements of the medium bears a strong resemblance to the periodic motion of an oscillating pendulum or an object attached to a spring.

To explain many other phenomena in nature, we must understand the concepts of oscillations and waves. For instance, although skyscrapers and bridges appear to be rigid, they actually oscillate, something the architects and engineers who design and build them must take into account. To understand how radio and television work, we must understand the origin and nature of electromagnetic waves and how they propagate through space. Finally, much of what scientists have learned about atomic structure has come from information carried by waves. Therefore, we must first study oscillations and waves if we are to understand the concepts and theories of atomic physics. ■

chapter 15

Oscillatory Motion

To reduce swaying in tall buildings because of the wind, tuned dampers are placed near the top of the building. These mechanisms include an object of large mass that oscillates under computer control at the same frequency as the building, reducing the swaying. The 730-ton suspended sphere in the photograph above is part of the tuned damper system of the Taipei Financial Center, at one time the world's tallest building. *(Ranjit Doroszkeiwicz/Alamy)*

***Periodic motion* is motion of an object that regularly returns** to a given position after a fixed time interval. With a little thought, we can identify several types of periodic motion in everyday life. Your car returns to the driveway each afternoon. You return to the dinner table each night to eat. A bumped chandelier swings back and forth, returning to the same position at a regular rate. The Earth returns to the same position in its orbit around the Sun each year, resulting in the variation among the four seasons.

In addition to these everyday examples, numerous other systems exhibit periodic motion. The molecules in a solid oscillate about their equilibrium positions; electromagnetic waves, such as light waves, radar, and radio waves, are characterized by oscillating electric and magnetic field vectors; and in alternating-current electrical circuits, voltage, current, and electric charge vary periodically with time.

A special kind of periodic motion occurs in mechanical systems when the force acting on an object is proportional to the position of the object relative to some equilibrium position. If this force is always directed toward the equilibrium position, the motion is called *simple harmonic motion,* which is the primary focus of this chapter.

15.1 Motion of an Object Attached to a Spring

As a model for simple harmonic motion, consider a block of mass m attached to the end of a spring, with the block free to move on a frictionless, horizontal surface (Active Fig. 15.1). When the spring is neither stretched nor compressed, the block is at rest at the position called the **equilibrium position** of the system, which we identify as $x = 0$ (Active Fig. 15.1b). We know from experience that such a system oscillates back and forth if disturbed from its equilibrium position.

We can understand the oscillating motion of the block in Active Figure 15.1 qualitatively by first recalling that when the block is displaced to a position x, the spring exerts on the block a force that is proportional to the position and given by **Hooke's law** (see Section 7.4):

$$F_s = -kx \quad (15.1)$$

◀ Hooke's law

We call F_s a **restoring force** because it is always directed toward the equilibrium position and therefore *opposite* the displacement of the block from equilibrium. That is, when the block is displaced to the right of $x = 0$ in Active Figure 15.1a, the position is positive and the restoring force is directed to the left. When the block is displaced to the left of $x = 0$ as in Figure 15.1c, the position is negative and the restoring force is directed to the right.

When the block is displaced from the equilibrium point and released, it is a particle under a net force and consequently undergoes an acceleration. Applying Newton's second law to the motion of the block, with Equation 15.1 providing the net force in the x direction, we obtain

$$-kx = ma_x$$

$$a_x = -\frac{k}{m}x \quad (15.2)$$

That is, the acceleration of the block is proportional to its position, and the direction of the acceleration is opposite the direction of the displacement of the block from equilibrium. Systems that behave in this way are said to exhibit **simple harmonic motion.** An object moves with simple harmonic motion whenever its acceleration is proportional to its position and is oppositely directed to the displacement from equilibrium.

If the block in Active Figure 15.1 is displaced to a position $x = A$ and released from rest, its *initial* acceleration is $-kA/m$. When the block passes through the equilibrium position $x = 0$, its acceleration is zero. At this instant, its speed is a maximum because the acceleration changes sign. The block then continues to travel to the left of equilibrium with a positive acceleration and finally reaches $x = -A$, at which time its acceleration is $+kA/m$ and its speed is again zero as discussed in Sections 7.4 and 7.9. The block completes a full cycle of its motion by returning to the

Pitfall Prevention 15.1

The Orientation of the Spring
Active Figure 15.1 shows a *horizontal* spring, with an attached block sliding on a frictionless surface. Another possibility is a block hanging from a *vertical* spring. All the results we discuss for the horizontal spring are the same for the vertical spring with one exception: when the block is placed on the vertical spring, its weight causes the spring to extend. If the resting position of the block is defined as $x = 0$, the results of this chapter also apply to this vertical system.

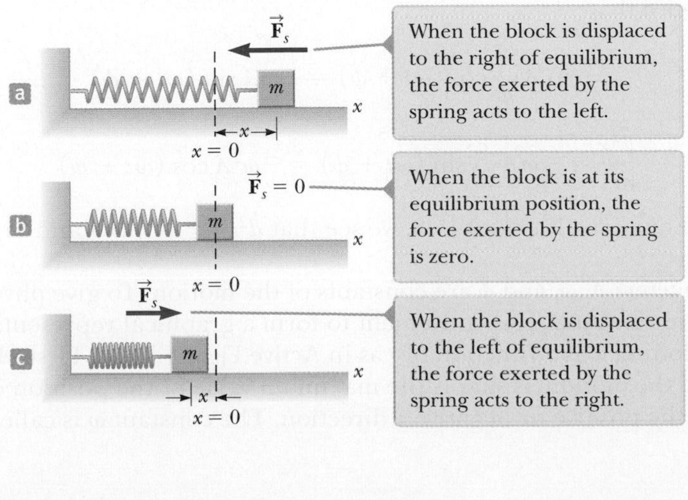

ACTIVE FIGURE 15.1
A block attached to a spring moving on a frictionless surface.

original position, again passing through $x = 0$ with maximum speed. Therefore, the block oscillates between the turning points $x = \pm A$. In the absence of friction, this idealized motion will continue forever because the force exerted by the spring is conservative. Real systems are generally subject to friction, so they do not oscillate forever. We shall explore the details of the situation with friction in Section 15.6.

***Quick Quiz* 15.1** A block on the end of a spring is pulled to position $x = A$ and released from rest. In one full cycle of its motion, through what total distance does it travel? **(a)** $A/2$ **(b)** A **(c)** $2A$ **(d)** $4A$

15.2 Analysis Model: Particle in Simple Harmonic Motion

The motion described in the preceding section occurs so often that we identify the **particle in simple harmonic motion** model to represent such situations. To develop a mathematical representation for this model, we will generally choose x as the axis along which the oscillation occurs; hence, we will drop the subscript-x notation in this discussion. Recall that, by definition, $a = dv/dt = d^2x/dt^2$, so we can express Equation 15.2 as

$$\frac{d^2x}{dt^2} = -\frac{k}{m}x \tag{15.3}$$

Pitfall Prevention 15.2

A Nonconstant Acceleration

The acceleration of a particle in simple harmonic motion is not constant. Equation 15.3 shows that its acceleration varies with position x. Therefore, we *cannot* apply the kinematic equations of Chapter 2 in this situation.

If we denote the ratio k/m with the symbol ω^2 (we choose ω^2 rather than ω so as to make the solution we develop below simpler in form), then

$$\omega^2 = \frac{k}{m} \tag{15.4}$$

and Equation 15.3 can be written in the form

$$\frac{d^2x}{dt^2} = -\omega^2 x \tag{15.5}$$

Let's now find a mathematical solution to Equation 15.5, that is, a function $x(t)$ that satisfies this second-order differential equation and is a mathematical representation of the position of the particle as a function of time. We seek a function whose second derivative is the same as the original function with a negative sign and multiplied by ω^2. The trigonometric functions sine and cosine exhibit this behavior, so we can build a solution around one or both of them. The following cosine function is a solution to the differential equation:

Position versus time for a particle in simple harmonic motion ▶

$$x(t) = A\cos(\omega t + \phi) \tag{15.6}$$

where A, ω, and ϕ are constants. To show explicitly that this solution satisfies Equation 15.5, notice that

$$\frac{dx}{dt} = A\frac{d}{dt}\cos(\omega t + \phi) = -\omega A \sin(\omega t + \phi) \tag{15.7}$$

$$\frac{d^2x}{dt^2} = -\omega A\frac{d}{dt}\sin(\omega t + \phi) = -\omega^2 A\cos(\omega t + \phi) \tag{15.8}$$

Pitfall Prevention 15.3

Where's the Triangle?

Equation 15.6 includes a trigonometric function, a *mathematical function* that can be used whether it refers to a triangle or not. In this case, the cosine function happens to have the correct behavior for representing the position of a particle in simple harmonic motion.

Comparing Equations 15.6 and 15.8, we see that $d^2x/dt^2 = -\omega^2 x$ and Equation 15.5 is satisfied.

The parameters A, ω, and ϕ are constants of the motion. To give physical significance to these constants, it is convenient to form a graphical representation of the motion by plotting x as a function of t as in Active Figure 15.2a. First, A, called the **amplitude** of the motion, is simply the maximum value of the position of the particle in either the positive or negative x direction. The constant ω is called the **angu-**

lar frequency, and it has units[1] of radians per second. It is a measure of how rapidly the oscillations are occurring; the more oscillations per unit time, the higher the value of ω. From Equation 15.4, the angular frequency is

$$\omega = \sqrt{\frac{k}{m}} \tag{15.9}$$

The constant angle ϕ is called the **phase constant** (or initial phase angle) and, along with the amplitude A, is determined uniquely by the position and velocity of the particle at $t = 0$. If the particle is at its maximum position $x = A$ at $t = 0$, the phase constant is $\phi = 0$ and the graphical representation of the motion is as shown in Active Figure 15.2b. The quantity $(\omega t + \phi)$ is called the **phase** of the motion. Notice that the function $x(t)$ is periodic and its value is the same each time ωt increases by 2π radians.

Equations 15.1, 15.5, and 15.6 form the basis of the mathematical representation of the particle in simple harmonic motion model. If you are analyzing a situation and find that the force on an object modeled as a particle is of the mathematical form of Equation 15.1, you know the motion is that of a simple harmonic oscillator and the position of the particle is described by Equation 15.6. If you analyze a system and find that it is described by a differential equation of the form of Equation 15.5, the motion is that of a simple harmonic oscillator. If you analyze a situation and find that the position of a particle is described by Equation 15.6, you know the particle undergoes simple harmonic motion.

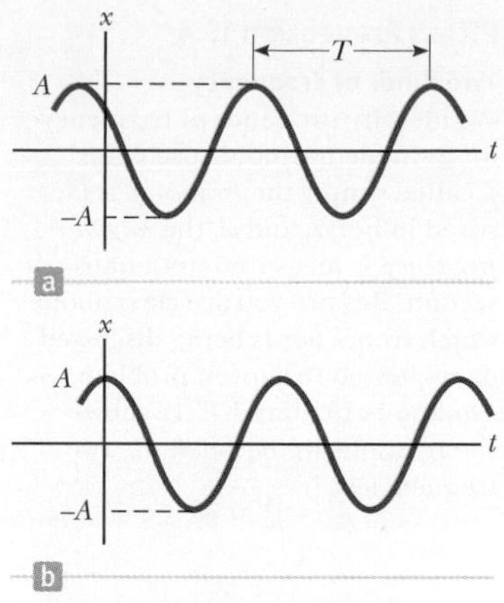

ACTIVE FIGURE 15.2

(a) An x–t graph for a particle undergoing simple harmonic motion. The amplitude of the motion is A, and the period (defined in Eq. 15.10) is T. (b) The x–t graph for the special case in which $x = A$ at $t = 0$ and hence $\phi = 0$.

***Quick Quiz* 15.2** Consider a graphical representation (Fig. 15.3) of simple harmonic motion as described mathematically in Equation 15.6. When the particle is at point Ⓐ on the graph, what can you say about its position and velocity? **(a)** The position and velocity are both positive. **(b)** The position and velocity are both negative. **(c)** The position is positive, and the velocity is zero. **(d)** The position is negative, and the velocity is zero. **(e)** The position is positive, and the velocity is negative. **(f)** The position is negative, and the velocity is positive.

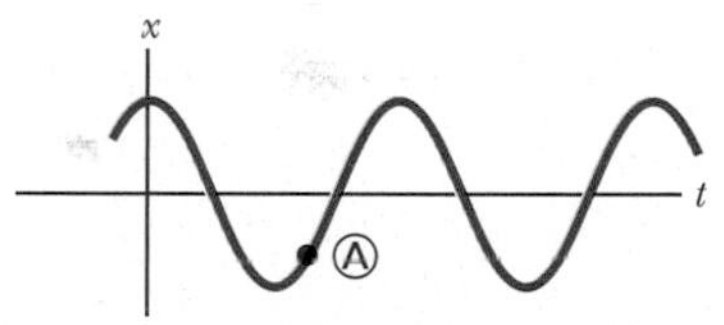

Figure 15.3 (Quick Quiz 15.2) An x–t graph for a particle undergoing simple harmonic motion. At a particular time, the particle's position is indicated by Ⓐ in the graph.

***Quick Quiz* 15.3** Figure 15.4 shows two curves representing particles undergoing simple harmonic motion. The correct description of these two motions is that the simple harmonic motion of particle B is **(a)** of larger angular frequency and larger amplitude than that of particle A, **(b)** of larger angular frequency and smaller amplitude than that of particle A, **(c)** of smaller angular frequency and larger amplitude than that of particle A, or **(d)** of smaller angular frequency and smaller amplitude than that of particle A.

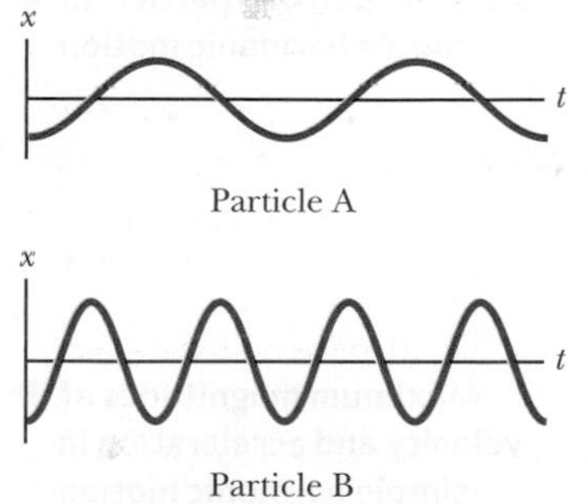

Figure 15.4 (Quick Quiz 15.3) Two x–t graphs for particles undergoing simple harmonic motion. The amplitudes and frequencies are different for the two particles.

Let us investigate further the mathematical description of simple harmonic motion. The **period** T of the motion is the time interval required for the particle to go through one full cycle of its motion (Active Fig. 15.2a). That is, the values of x and v for the particle at time t equal the values of x and v at time $t + T$. Because the phase increases by 2π radians in a time interval of T,

$$[\omega(t + T) + \phi] - (\omega t + \phi) = 2\pi$$

[1]We have seen many examples in earlier chapters in which we evaluate a trigonometric function of an angle. The argument of a trigonometric function, such as sine or cosine, *must* be a pure number. The radian is a pure number because it is a ratio of lengths. Angles in degrees are pure numbers because the degree is an artificial "unit"; it is not related to measurements of lengths. The argument of the trigonometric function in Equation 15.6 must be a pure number. Therefore, ω *must* be expressed in radians per second (and not, for example, in revolutions per second) if t is expressed in seconds. Furthermore, other types of functions such as logarithms and exponential functions require arguments that are pure numbers.

Pitfall Prevention 15.4

Two Kinds of Frequency
We identify two kinds of frequency for a simple harmonic oscillator: f, called simply the *frequency*, is measured in hertz, and ω, the *angular frequency*, is measured in radians per second. Be sure you are clear about which frequency is being discussed or requested in a given problem. Equations 15.11 and 15.12 show the relationship between the two frequencies.

Simplifying this expression gives $\omega T = 2\pi$, or

$$T = \frac{2\pi}{\omega} \tag{15.10}$$

The inverse of the period is called the **frequency** f of the motion. Whereas the period is the time interval per oscillation, the frequency represents the number of oscillations the particle undergoes per unit time interval:

$$f = \frac{1}{T} = \frac{\omega}{2\pi} \tag{15.11}$$

The units of f are cycles per second, or **hertz** (Hz). Rearranging Equation 15.11 gives

$$\omega = 2\pi f = \frac{2\pi}{T} \tag{15.12}$$

Equations 15.9 through 15.11 can be used to express the period and frequency of the motion for the particle in simple harmonic motion in terms of the characteristics m and k of the system as

Period ▶

$$T = \frac{2\pi}{\omega} = 2\pi\sqrt{\frac{m}{k}} \tag{15.13}$$

Frequency ▶

$$f = \frac{1}{T} = \frac{1}{2\pi}\sqrt{\frac{k}{m}} \tag{15.14}$$

That is, the period and frequency depend *only* on the mass of the particle and the force constant of the spring and *not* on the parameters of the motion, such as A or ϕ. As we might expect, the frequency is larger for a stiffer spring (larger value of k) and decreases with increasing mass of the particle.

We can obtain the velocity and acceleration[2] of a particle undergoing simple harmonic motion from Equations 15.7 and 15.8:

Velocity of a particle in simple harmonic motion ▶

$$v = \frac{dx}{dt} = -\omega A \sin(\omega t + \phi) \tag{15.15}$$

Acceleration of a particle in simple harmonic motion ▶

$$a = \frac{d^2x}{dt^2} = -\omega^2 A \cos(\omega t + \phi) \tag{15.16}$$

From Equation 15.15, we see that because the sine and cosine functions oscillate between ± 1, the extreme values of the velocity v are $\pm\omega A$. Likewise, Equation 15.16 shows that the extreme values of the acceleration a are $\pm\omega^2 A$. Therefore, the *maximum* values of the magnitudes of the velocity and acceleration are

Maximum magnitudes of velocity and acceleration in simple harmonic motion ▶

$$v_{\text{max}} = \omega A = \sqrt{\frac{k}{m}}A \tag{15.17}$$

$$a_{\text{max}} = \omega^2 A = \frac{k}{m}A \tag{15.18}$$

Figure 15.5a plots position versus time for an arbitrary value of the phase constant. The associated velocity–time and acceleration–time curves are illustrated in Figures 15.5b and 15.5c, respectively. They show that the phase of the velocity differs from the phase of the position by $\pi/2$ rad, or 90°. That is, when x is a maximum or a minimum, the velocity is zero. Likewise, when x is zero, the speed is a maximum. Furthermore, notice that the phase of the acceleration differs from the phase of the position by π radians, or 180°. For example, when x is a maximum, a has a maximum magnitude in the opposite direction.

[2]Because the motion of a simple harmonic oscillator takes place in one dimension, we denote velocity as v and acceleration as a, with the direction indicated by a positive or negative sign as in Chapter 2.

Quick Quiz **15.4** An object of mass m is hung from a spring and set into oscillation. The period of the oscillation is measured and recorded as T. The object of mass m is removed and replaced with an object of mass $2m$. When this object is set into oscillation, what is the period of the motion? **(a)** $2T$ **(b)** $\sqrt{2}\,T$ **(c)** T **(d)** $T/\sqrt{2}$ **(e)** $T/2$

Equation 15.6 describes simple harmonic motion of a particle in general. Let's now see how to evaluate the constants of the motion. The angular frequency ω is evaluated using Equation 15.9. The constants A and ϕ are evaluated from the initial conditions, that is, the state of the oscillator at $t = 0$.

Suppose a block is set into motion by pulling it from equilibrium by a distance A and releasing it from rest at $t = 0$ as in Active Figure 15.6. We must then require our solutions for $x(t)$ and $v(t)$ (Eqs. 15.6 and 15.15) to obey the initial conditions that $x(0) = A$ and $v(0) = 0$:

$$x(0) = A\cos\phi = A$$

$$v(0) = -\omega A\sin\phi = 0$$

These conditions are met if $\phi = 0$, giving $x = A\cos\omega t$ as our solution. To check this solution, notice that it satisfies the condition that $x(0) = A$ because $\cos 0 = 1$.

The position, velocity, and acceleration of the block versus time are plotted in Figure 15.7a for this special case. The acceleration reaches extreme values of $\mp\omega^2 A$ when the position has extreme values of $\pm A$. Furthermore, the velocity has extreme values of $\pm\omega A$, which both occur at $x = 0$. Hence, the quantitative solution agrees with our qualitative description of this system.

Let's consider another possibility. Suppose the system is oscillating and we define $t = 0$ as the instant the block passes through the unstretched position of the spring while moving to the right (Active Fig. 15.8). In this case, our solutions for $x(t)$ and $v(t)$ must obey the initial conditions that $x(0) = 0$ and $v(0) = v_i$:

$$x(0) = A\cos\phi = 0$$

$$v(0) = -\omega A\sin\phi = v_i$$

The first of these conditions tells us that $\phi = \pm\pi/2$. With these choices for ϕ, the second condition tells us that $A = \mp v_i/\omega$. Because the initial velocity is positive and the amplitude must be positive, we must have $\phi = -\pi/2$. Hence, the solution is

$$x = \frac{v_i}{\omega}\cos\left(\omega t - \frac{\pi}{2}\right)$$

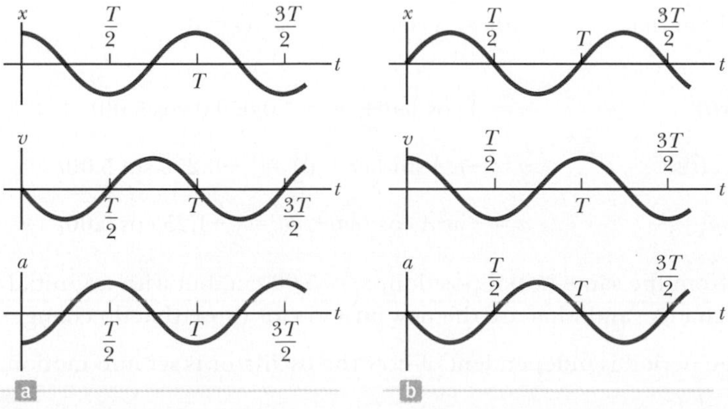

Figure 15.7 (a) Position, velocity, and acceleration versus time for the block in Active Figure 15.6 under the initial conditions that at $t = 0$, $x(0) = A$, and $v(0) = 0$. (b) Position, velocity, and acceleration versus time for the block in Active Figure 15.8 under the initial conditions that at $t = 0$, $x(0) = 0$, and $v(0) = v_i$.

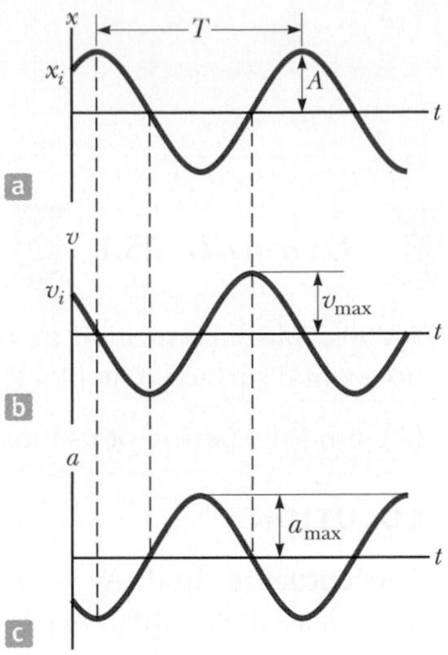

Figure 15.5 Graphical representation of simple harmonic motion. (a) Position versus time. (b) Velocity versus time. (c) Acceleration versus time. Notice that at any specified time the velocity is 90° out of phase with the position and the acceleration is 180° out of phase with the position.

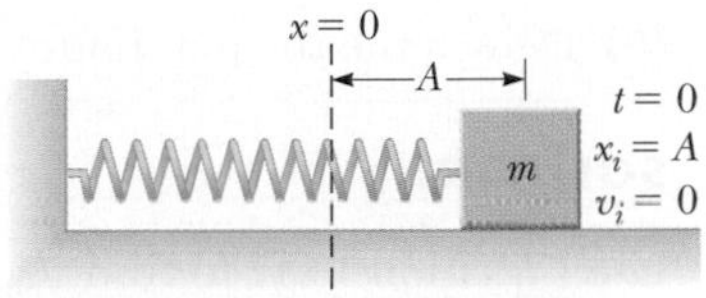

ACTIVE FIGURE 15.6

A block–spring system that begins its motion from rest with the block at $x = A$ at $t = 0$.

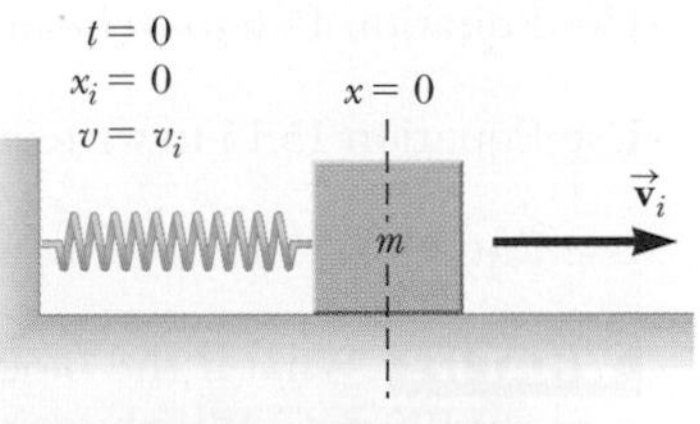

ACTIVE FIGURE 15.8

The block–spring system is undergoing oscillation, and $t = 0$ is defined at an instant when the block passes through the equilibrium position $x = 0$ and is moving to the right with speed v_i.

The graphs of position, velocity, and acceleration versus time for this choice of $t = 0$ are shown in Figure 15.7b. Notice that these curves are the same as those in Figure 15.7a, but shifted to the right by one-fourth of a cycle. This shift is described mathematically by the phase constant $\phi = -\pi/2$, which is one-fourth of a full cycle of 2π.

Example 15.1 A Block–Spring System

A 200-g block connected to a light spring for which the force constant is 5.00 N/m is free to oscillate on a frictionless, horizontal surface. The block is displaced 5.00 cm from equilibrium and released from rest as in Active Figure 15.6.

(A) Find the period of its motion.

SOLUTION

Conceptualize Study Active Figure 15.6 and imagine the block moving back and forth in simple harmonic motion once it is released. Set up an experimental model in the vertical direction by hanging a heavy object such as a stapler from a strong rubber band.

Categorize The block is modeled as a particle in simple harmonic motion. We find values from equations developed in this section for the particle in simple harmonic motion model, so we categorize this example as a substitution problem.

Use Equation 15.9 to find the angular frequency of the block–spring system:

$$\omega = \sqrt{\frac{k}{m}} = \sqrt{\frac{5.00 \text{ N/m}}{200 \times 10^{-3} \text{ kg}}} = 5.00 \text{ rad/s}$$

Use Equation 15.13 to find the period of the system:

$$T = \frac{2\pi}{\omega} = \frac{2\pi}{5.00 \text{ rad/s}} = \boxed{1.26 \text{ s}}$$

(B) Determine the maximum speed of the block.

SOLUTION

Use Equation 15.17 to find v_{max}:

$$v_{\text{max}} = \omega A = (5.00 \text{ rad/s})(5.00 \times 10^{-2} \text{ m}) = \boxed{0.250 \text{ m/s}}$$

(C) What is the maximum acceleration of the block?

SOLUTION

Use Equation 15.18 to find a_{max}:

$$a_{\text{max}} = \omega^2 A = (5.00 \text{ rad/s})^2(5.00 \times 10^{-2} \text{ m}) = \boxed{1.25 \text{ m/s}^2}$$

(D) Express the position, velocity, and acceleration as functions of time in SI units.

SOLUTION

Find the phase constant from the initial condition that $x = A$ at $t = 0$:

$$x(0) = A \cos \phi = A \rightarrow \phi = 0$$

Use Equation 15.6 to write an expression for $x(t)$:

$$x = A \cos (\omega t + \phi) = \boxed{0.050\,0 \cos 5.00t}$$

Use Equation 15.15 to write an expression for $v(t)$:

$$v = -\omega A \sin (\omega t + \phi) = \boxed{-0.250 \sin 5.00t}$$

Use Equation 15.16 to write an expression for $a(t)$:

$$a = -\omega^2 A \cos (\omega t + \phi) = \boxed{-1.25 \cos 5.00t}$$

WHAT IF? What if the block were released from the same initial position, $x_i = 5.00$ cm, but with an initial velocity of $v_i = -0.100$ m/s? Which parts of the solution change, and what are the new answers for those that do change?

Answers Part (A) does not change because the period is independent of how the oscillator is set into motion. Parts (B), (C), and (D) will change.

Write position and velocity expressions for the initial conditions:

$$(1) \quad x(0) = A \cos \phi = x_i$$

$$(2) \quad v(0) = -\omega A \sin \phi = v_i$$

15.1 *cont.*

Divide Equation (2) by Equation (1) to find the phase constant:

$$\frac{-\omega A \sin \phi}{A \cos \phi} = \frac{v_i}{x_i}$$

$$\tan \phi = -\frac{v_i}{\omega x_i} = -\frac{-0.100 \text{ m/s}}{(5.00 \text{ rad/s})(0.050\,0 \text{ m})} = 0.400$$

$$\phi = \tan^{-1}(0.400) = 0.121\pi$$

Use Equation (1) to find A:

$$A = \frac{x_i}{\cos \phi} = \frac{0.050\,0 \text{ m}}{\cos(0.121\pi)} = 0.053\,9 \text{ m}$$

Find the new maximum speed:

$$v_{\max} = \omega A = (5.00 \text{ rad/s})(5.39 \times 10^{-2} \text{ m}) = 0.269 \text{ m/s}$$

Find the new magnitude of the maximum acceleration:

$$a_{\max} = \omega^2 A = (5.00 \text{ rad/s})^2(5.39 \times 10^{-2} \text{ m}) = 1.35 \text{ m/s}^2$$

Find new expressions for position, velocity, and acceleration in SI units:

$$x = 0.053\,9 \cos(5.00t + 0.121\pi)$$

$$v = -0.269 \sin(5.00t + 0.121\pi)$$

$$a = -1.35 \cos(5.00t + 0.121\pi)$$

As we saw in Chapters 7 and 8, many problems are easier to solve using an energy approach rather than one based on variables of motion. This particular What If? is easier to solve from an energy approach. Therefore, we shall investigate the energy of the simple harmonic oscillator in the next section.

Example 15.2 Watch Out for Potholes!

A car with a mass of 1 300 kg is constructed so that its frame is supported by four springs. Each spring has a force constant of 20 000 N/m. Two people riding in the car have a combined mass of 160 kg. Find the frequency of vibration of the car after it is driven over a pothole in the road.

SOLUTION

Conceptualize Think about your experiences with automobiles. When you sit in a car, it moves downward a small distance because your weight is compressing the springs further. If you push down on the front bumper and release it, the front of the car oscillates a few times.

Categorize We imagine the car as being supported by a single spring and model the car as a particle in simple harmonic motion.

Analyze First, let's determine the effective spring constant of the four springs combined. For a given extension x of the springs, the combined force on the car is the sum of the forces from the individual springs.

Find an expression for the total force on the car:

$$F_{\text{total}} = \sum(-kx) = -\left(\sum k\right)x$$

In this expression, x has been factored from the sum because it is the same for all four springs. The effective spring constant for the combined springs is the sum of the individual spring constants.

Evaluate the effective spring constant:

$$k_{\text{eff}} = \sum k = 4 \times 20\,000 \text{ N/m} = 80\,000 \text{ N/m}$$

Use Equation 15.14 to find the frequency of vibration:

$$f = \frac{1}{2\pi}\sqrt{\frac{k_{\text{eff}}}{m}} = \frac{1}{2\pi}\sqrt{\frac{80\,000 \text{ N/m}}{1\,460 \text{ kg}}} = \boxed{1.18 \text{ Hz}}$$

Finalize The mass we used here is that of the car plus the people because that is the total mass that is oscillating. Also notice that we have explored only up-and-down motion of the car. If an oscillation is established in which the car rocks back and forth such that the front end goes up when the back end goes down, the frequency will be different.

continued

15.2 *cont.*

WHAT IF? Suppose the car stops on the side of the road and the two people exit the car. One of them pushes downward on the car and releases it so that it oscillates vertically. Is the frequency of the oscillation the same as the value we just calculated?

Answer The suspension system of the car is the same, but the mass that is oscillating is smaller: it no longer includes the mass of the two people. Therefore, the frequency should be higher. Let's calculate the new frequency, taking the mass to be 1 300 kg:

$$f = \frac{1}{2\pi}\sqrt{\frac{k_{\text{eff}}}{m}} = \frac{1}{2\pi}\sqrt{\frac{80\ 000\ \text{N/m}}{1\ 300\ \text{kg}}} = 1.25\ \text{Hz}$$

As predicted, the new frequency is a bit higher.

15.3 Energy of the Simple Harmonic Oscillator

Let us examine the mechanical energy of a system in which a particle undergoes simple harmonic motion, such as the block–spring system illustrated in Active Figure 15.1. Because the surface is frictionless, the system is isolated and we expect the total mechanical energy of the system to be constant. We assume a massless spring, so the kinetic energy of the system corresponds only to that of the block. We can use Equation 15.15 to express the kinetic energy of the block as

Kinetic energy of a simple harmonic oscillator ▶

$$K = \tfrac{1}{2}mv^2 = \tfrac{1}{2}m\omega^2 A^2 \sin^2(\omega t + \phi) \quad \textbf{(15.19)}$$

The elastic potential energy stored in the spring for any elongation x is given by $\frac{1}{2}kx^2$ (see Eq. 7.22). Using Equation 15.6 gives

Potential energy of a simple harmonic oscillator ▶

$$U = \tfrac{1}{2}kx^2 = \tfrac{1}{2}kA^2 \cos^2(\omega t + \phi) \quad \textbf{(15.20)}$$

We see that K and U are *always* positive quantities or zero. Because $\omega^2 = k/m$, we can express the total mechanical energy of the simple harmonic oscillator as

$$E = K + U = \tfrac{1}{2}kA^2[\sin^2(\omega t + \phi) + \cos^2(\omega t + \phi)]$$

From the identity $\sin^2\theta + \cos^2\theta = 1$, we see that the quantity in square brackets is unity. Therefore, this equation reduces to

Total energy of a simple harmonic oscillator ▶

$$E = \tfrac{1}{2}kA^2 \quad \textbf{(15.21)}$$

That is, the total mechanical energy of a simple harmonic oscillator is a constant of the motion and is proportional to the square of the amplitude. The total mechanical energy is equal to the maximum potential energy stored in the spring when $x = \pm A$ because $v = 0$ at these points and there is no kinetic energy. At the equilibrium position, where $U = 0$ because $x = 0$, the total energy, all in the form of kinetic energy, is again $\frac{1}{2}kA^2$.

Plots of the kinetic and potential energies versus time appear in Active Figure 15.9a, where we have taken $\phi = 0$. At all times, the sum of the kinetic and potential energies is a constant equal to $\frac{1}{2}kA^2$, the total energy of the system.

The variations of K and U with the position x of the block are plotted in Active Figure 15.9b. Energy is continuously being transformed between potential energy stored in the spring and kinetic energy of the block.

Active Figure 15.10 illustrates the position, velocity, acceleration, kinetic energy, and potential energy of the block–spring system for one full period of the motion. Most of the ideas discussed so far are incorporated in this important figure. Study it carefully.

Finally, we can obtain the velocity of the block at an arbitrary position by expressing the total energy of the system at some arbitrary position x as

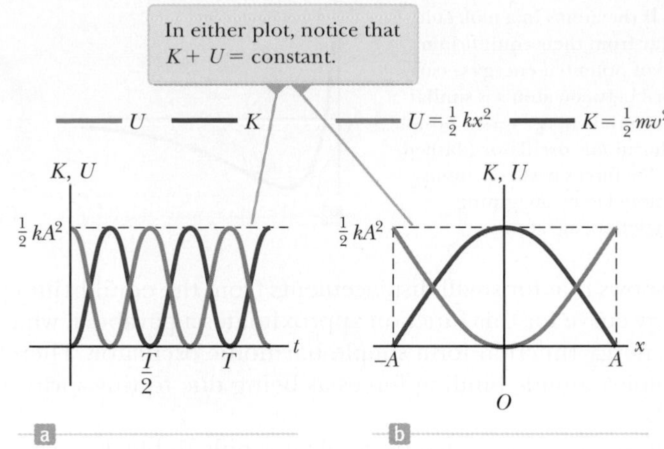

ACTIVE FIGURE 15.9

(a) Kinetic energy and potential energy versus time for a simple harmonic oscillator with $\phi = 0$. (b) Kinetic energy and potential energy versus position for a simple harmonic oscillator.

$$E = K + U = \tfrac{1}{2}mv^2 + \tfrac{1}{2}kx^2 = \tfrac{1}{2}kA^2$$

$$v = \pm\sqrt{\frac{k}{m}(A^2 - x^2)} = \pm\omega\sqrt{A^2 - x^2} \quad (15.22)$$

◀ **Velocity as a function of position for a simple harmonic oscillator**

When you check Equation 15.22 to see whether it agrees with known cases, you find that it verifies that the speed is a maximum at $x = 0$ and is zero at the turning points $x = \pm A$.

You may wonder why we are spending so much time studying simple harmonic oscillators. We do so because they are good models of a wide variety of physical phenomena. For example, recall the Lennard–Jones potential discussed in Example 7.9. This complicated function describes the forces holding atoms together.

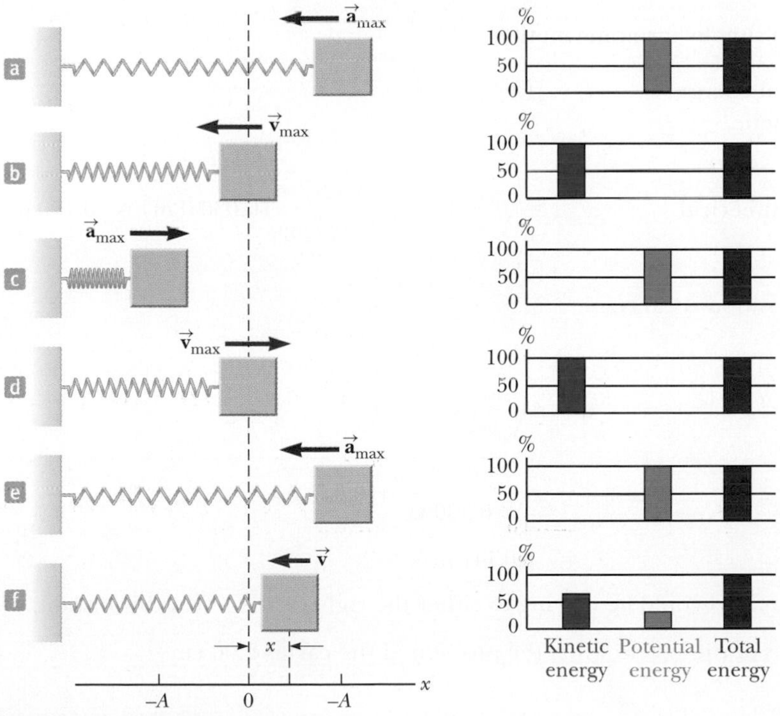

t	x	v	a	K	U
0	A	0	$-\omega^2 A$	0	$\frac{1}{2}kA^2$
$\frac{T}{4}$	0	$-\omega A$	0	$\frac{1}{2}kA^2$	0
$\frac{T}{2}$	$-A$	0	$\omega^2 A$	0	$\frac{1}{2}kA^2$
$\frac{3T}{4}$	0	ωA	0	$\frac{1}{2}kA^2$	0
T	A	0	$-\omega^2 A$	0	$\frac{1}{2}kA^2$
t	x	v	$-\omega^2 x$	$\frac{1}{2}mv^2$	$\frac{1}{2}kx^2$

ACTIVE FIGURE 15.10

(a) through (e) Several instants in the simple harmonic motion for a block–spring system. Energy bar graphs show the distribution of the energy of the system at each instant. The parameters in the table at the right refer to the block–spring system, assuming at $t = 0$, $x = A$; hence, $x = A \cos \omega t$. For these five special instants, one of the types of energy is zero. (f) An arbitrary point in the motion of the oscillator. The system possesses both kinetic energy and potential energy at this instant as shown in the bar graph.

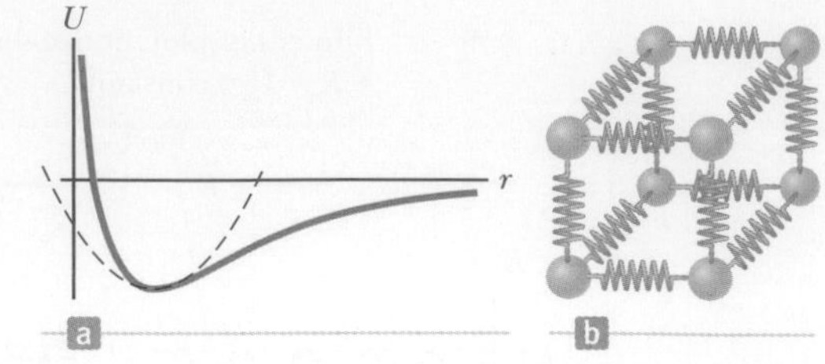

Figure 15.11 (a) If the atoms in a molecule do not move too far from their equilibrium positions, a graph of potential energy versus separation distance between atoms is similar to the graph of potential energy versus position for a simple harmonic oscillator (dashed black curve). (b) The forces between atoms in a solid can be modeled by imagining springs between neighboring atoms.

Figure 15.11a shows that for small displacements from the equilibrium position, the potential energy curve for this function approximates a parabola, which represents the potential energy function for a simple harmonic oscillator. Therefore, we can model the complex atomic binding forces as being due to tiny springs as depicted in Figure 15.11b.

The ideas presented in this chapter apply not only to block–spring systems and atoms, but also to a wide range of situations that include bungee jumping, playing a musical instrument, and viewing the light emitted by a laser. You will see more examples of simple harmonic oscillators as you work through this book.

Example 15.3 Oscillations on a Horizontal Surface

A 0.500-kg cart connected to a light spring for which the force constant is 20.0 N/m oscillates on a frictionless, horizontal air track.

(A) Calculate the maximum speed of the cart if the amplitude of the motion is 3.00 cm.

SOLUTION

Conceptualize The system oscillates in exactly the same way as the block in Active Figure 15.10, so use that figure in your mental image of the motion.

Categorize The cart is modeled as a particle in simple harmonic motion.

Analyze Use Equation 15.21 to express the total energy of the oscillator system and equate it to the kinetic energy of the system when the cart is at $x = 0$:

$$E = \tfrac{1}{2}kA^2 = \tfrac{1}{2}mv_{\text{max}}^2$$

Solve for the maximum speed and substitute numerical values:

$$v_{\text{max}} = \sqrt{\frac{k}{m}}A = \sqrt{\frac{20.0\text{ N/m}}{0.500\text{ kg}}}(0.030\,0\text{ m}) = \boxed{0.190\text{ m/s}}$$

(B) What is the velocity of the cart when the position is 2.00 cm?

SOLUTION

Use Equation 15.22 to evaluate the velocity:

$$v = \pm\sqrt{\frac{k}{m}(A^2 - x^2)}$$

$$= \pm\sqrt{\frac{20.0\text{ N/m}}{0.500\text{ kg}}[(0.030\,0\text{ m})^2 - (0.020\,0\text{ m})^2]}$$

$$= \boxed{\pm 0.141\text{ m/s}}$$

The positive and negative signs indicate that the cart could be moving to either the right or the left at this instant.

(C) Compute the kinetic and potential energies of the system when the position of the cart is 2.00 cm.

SOLUTION

Use the result of part (B) to evaluate the kinetic energy at $x = 0.020\,0$ m:

$$K = \tfrac{1}{2}mv^2 = \tfrac{1}{2}(0.500\text{ kg})(0.141\text{ m/s})^2 = \boxed{5.00 \times 10^{-3}\text{ J}}$$

Evaluate the elastic potential energy at $x = 0.020\,0$ m:

$$U = \tfrac{1}{2}kx^2 = \tfrac{1}{2}(20.0\text{ N/m})(0.0200\text{ m})^2 = \boxed{4.00 \times 10^{-3}\text{ J}}$$

15.3 *cont.*

Finalize The sum of the kinetic and potential energies in part (C) is equal to the total energy, which can be found from Equation 15.21. That must be true for *any* position of the cart.

WHAT IF? The cart in this example could have been set into motion by releasing the cart from rest at $x = 3.00$ cm. What if the cart were released from the same position, but with an initial velocity of $v = -0.100$ m/s? What are the new amplitude and maximum speed of the cart?

Answer This question is of the same type we asked at the end of Example 15.1, but here we apply an energy approach.

First calculate the total energy of the system at $t = 0$:

$$E = \tfrac{1}{2}mv^2 + \tfrac{1}{2}kx^2$$
$$= \tfrac{1}{2}(0.500 \text{ kg})(-0.100 \text{ m/s})^2 + \tfrac{1}{2}(20.0 \text{ N/m})(0.030\,0 \text{ m})^2$$
$$= 1.15 \times 10^{-2} \text{ J}$$

Equate this total energy to the potential energy of the system when the cart is at the endpoint of the motion:

$$E = \tfrac{1}{2}kA^2$$

Solve for the amplitude A:

$$A = \sqrt{\frac{2E}{k}} = \sqrt{\frac{2(1.15 \times 10^{-2} \text{ J})}{20.0 \text{ N/m}}} = 0.033\,9 \text{ m}$$

Equate the total energy to the kinetic energy of the system when the cart is at the equilibrium position:

$$E = \tfrac{1}{2}mv_{\text{max}}^2$$

Solve for the maximum speed:

$$v_{\text{max}} = \sqrt{\frac{2E}{m}} = \sqrt{\frac{2(1.15 \times 10^{-2} \text{ J})}{0.500 \text{ kg}}} = 0.214 \text{ m/s}$$

The amplitude and maximum velocity are larger than the previous values because the cart was given an initial velocity at $t = 0$.

15.4 Comparing Simple Harmonic Motion with Uniform Circular Motion

Some common devices in everyday life exhibit a relationship between oscillatory motion and circular motion. For example, consider the drive mechanism for a non-electric sewing machine in Figure 15.12. The operator of the machine places her feet on the treadle and rocks them back and forth. This oscillatory motion causes the large wheel at the right to undergo circular motion. The red drive belt seen in the photograph transfers this circular motion to the sewing machine mechanism (above the photo) and eventually results in the oscillatory motion of the sewing

Figure 15.12 The bottom of a treadle-style sewing machine from the early twentieth century. The treadle is the wide, flat foot pedal with the metal grillwork.

needle. In this section, we explore this interesting relationship between these two types of motion.

Active Figure 15.13 is a view of an experimental arrangement that shows this relationship. A ball is attached to the rim of a turntable of radius A, which is illuminated from above by a lamp. The ball casts a shadow on a screen. As the turntable rotates with constant angular speed, the shadow of the ball moves back and forth in simple harmonic motion.

Consider a particle located at point P on the circumference of a circle of radius A as in Figure 15.14a, with the line OP making an angle ϕ with the x axis at $t = 0$. We call this circle a *reference circle* for comparing simple harmonic motion with uniform circular motion, and we choose the position of P at $t = 0$ as our reference position. If the particle moves along the circle with constant angular speed ω until OP makes an angle θ with the x axis as in Figure 15.14b, at some time $t > 0$ the angle between OP and the x axis is $\theta = \omega t + \phi$. As the particle moves along the circle, the projection of P on the x axis, labeled point Q, moves back and forth along the x axis between the limits $x = \pm A$.

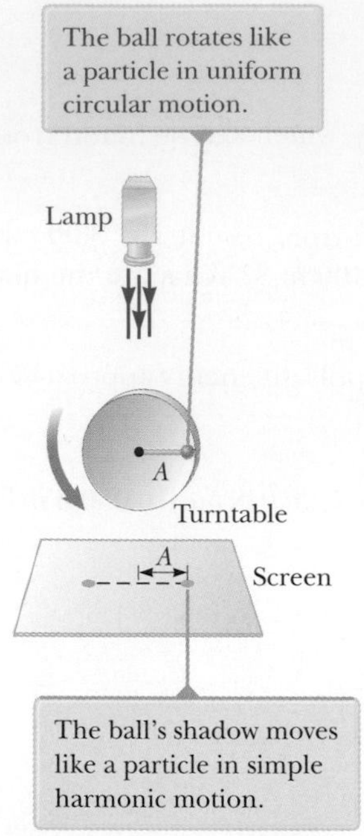

ACTIVE FIGURE 15.13

An experimental setup for demonstrating the connection between simple harmonic motion and uniform circular motion.

Notice that points P and Q always have the same x coordinate. From the right triangle OPQ, we see that this x coordinate is

$$x(t) = A\cos(\omega t + \phi) \tag{15.23}$$

This expression is the same as Equation 15.6 and shows that the point Q moves with simple harmonic motion along the x axis. Therefore, simple harmonic motion along a straight line can be represented by the projection of uniform circular motion along a diameter of a reference circle.

This geometric interpretation shows that the time interval for one complete revolution of the point P on the reference circle is equal to the period of motion T for simple harmonic motion between $x = \pm A$. Therefore, the angular speed ω of P is the same as the angular frequency ω of simple harmonic motion along the x axis (which is why we use the same symbol). The phase constant ϕ for simple harmonic motion corresponds to the initial angle OP makes with the x axis. The radius A of the reference circle equals the amplitude of the simple harmonic motion.

Because the relationship between linear and angular speed for circular motion is $v = r\omega$ (see Eq. 10.10), the particle moving on the reference circle of radius A has a velocity of magnitude ωA. From the geometry in Figure 15.14c, we see that the x component of this velocity is $-\omega A\sin(\omega t + \phi)$. By definition, point Q has a velocity given by dx/dt. Differentiating Equation 15.23 with respect to time, we find that the velocity of Q is the same as the x component of the velocity of P.

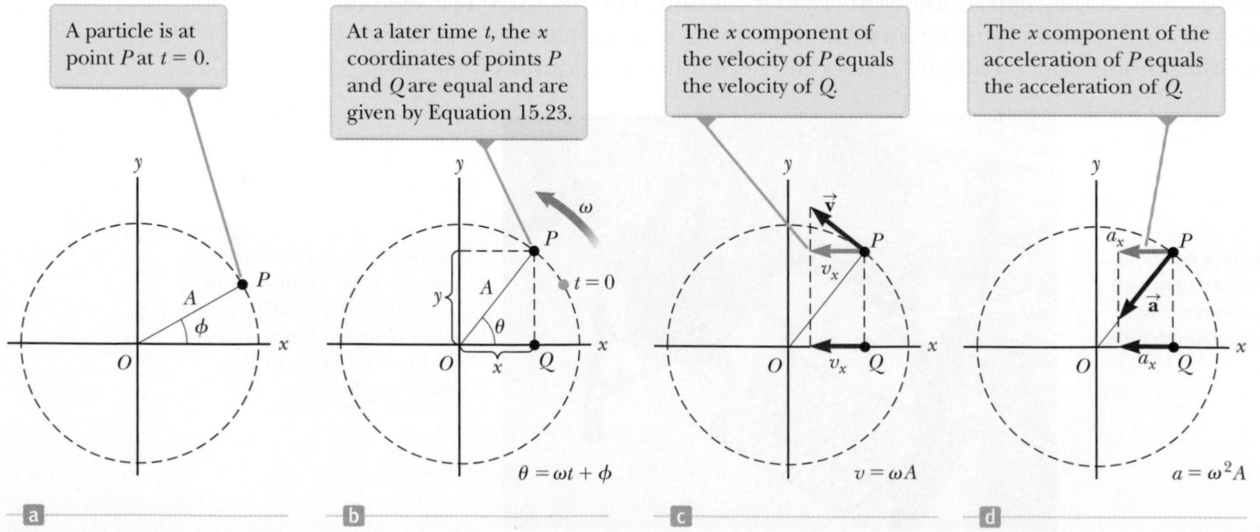

Figure 15.14 Relationship between the uniform circular motion of a point P and the simple harmonic motion of a point Q. A particle at P moves in a circle of radius A with constant angular speed ω.

The acceleration of P on the reference circle is directed radially inward toward O and has a magnitude $v^2/A = \omega^2 A$. From the geometry in Figure 15.14d, we see that the x component of this acceleration is $-\omega^2 A \cos(\omega t + \phi)$. This value is also the acceleration of the projected point Q along the x axis, as you can verify by taking the second derivative of Equation 15.23.

***Quick Quiz* 15.5** Figure 15.15 shows the position of an object in uniform circular motion at $t = 0$. A light shines from above and projects a shadow of the object on a screen below the circular motion. What are the correct values for the *amplitude* and *phase constant* (relative to an x axis to the right) of the simple harmonic motion of the shadow? **(a)** 0.50 m and 0 **(b)** 1.00 m and 0 **(c)** 0.50 m and π **(d)** 1.00 m and π

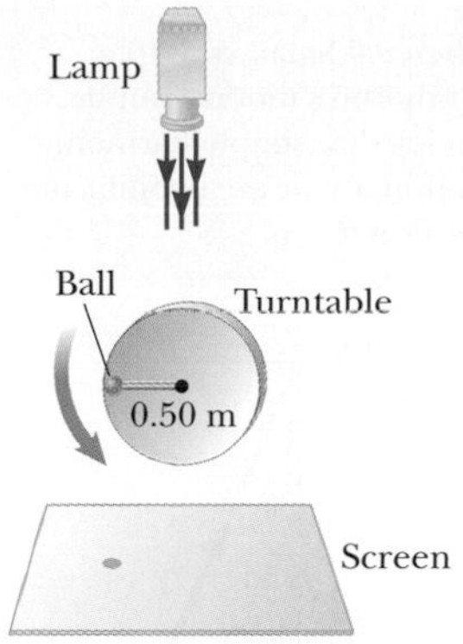

Figure 15.15 (Quick Quiz 15.5) An object moves in circular motion, casting a shadow on the screen below. Its position at an instant of time is shown.

Example 15.4 Circular Motion with Constant Angular Speed

The ball in Active Figure 15.13 rotates counterclockwise in a circle of radius 3.00 m with a constant angular speed of 8.00 rad/s. At $t = 0$, its shadow has an x coordinate of 2.00 m and is moving to the right.

(A) Determine the x coordinate of the shadow as a function of time in SI units.

SOLUTION

Conceptualize Be sure you understand the relationship between circular motion of the ball and simple harmonic motion of its shadow as described in Active Figure 15.13. Notice that the shadow is *not* at is maximum position at $t = 0$.

Categorize The ball on the turntable is a particle in uniform circular motion. The shadow is modeled as a particle in simple harmonic motion.

Analyze Use Equation 15.23 to write an expression for the x coordinate of the rotating ball:

$$x = A\cos(\omega t + \phi)$$

Solve for the phase constant:

$$\phi = \cos^{-1}\left(\frac{x}{A}\right) - \omega t$$

Substitute numerical values for the initial conditions:

$$\phi = \cos^{-1}\left(\frac{2.00\text{ m}}{3.00\text{ m}}\right) - 0 = \pm 48.2° = \pm 0.841\text{ rad}$$

If we were to take $\phi = +0.841$ rad as our answer, the shadow would be moving to the left at $t = 0$. Because the shadow is moving to the right at $t = 0$, we must choose $\phi = -0.841$ rad.

Write the x coordinate as a function of time:

$$x = \boxed{3.00\cos(8.00t - 0.841)}$$

(B) Find the x components of the shadow's velocity and acceleration at any time t.

SOLUTION

Differentiate the x coordinate with respect to time to find the velocity at any time in m/s:

$$v_x = \frac{dx}{dt} = (-3.00\text{ m})(8.00\text{ rad/s})\sin(8.00t - 0.841)$$

$$= \boxed{-24.0\sin(8.00t - 0.841)}$$

Differentiate the velocity with respect to time to find the acceleration at any time in m/s²:

$$a_x = \frac{dv_x}{dt} = (-24.0\text{ m/s})(8.00\text{ rad/s})\cos(8.00t - 0.841)$$

$$= \boxed{-192\cos(8.00t - 0.841)}$$

Finalize These results are equally valid for the ball moving in uniform circular motion and the shadow moving in simple harmonic motion. Notice that the value of the phase constant puts the ball in the fourth quadrant of the xy coordinate system of Figure 15.14, which is consistent with the shadow having a positive value for x and moving toward the right.

When θ is small, a simple pendulum's motion can be modeled as simple harmonic motion about the equilibrium position $\theta = 0$.

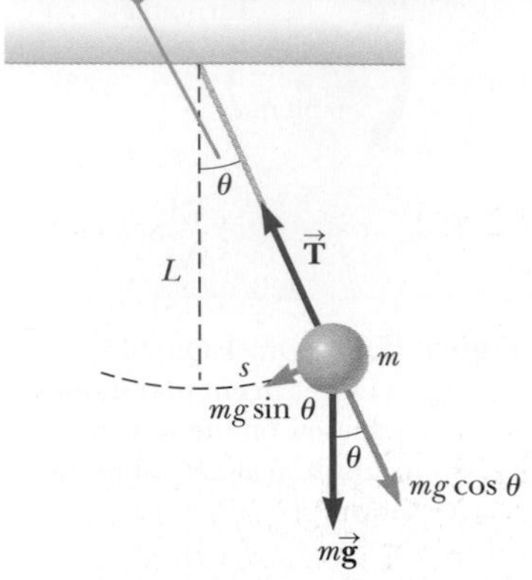

ACTIVE FIGURE 15.16

A simple pendulum.

15.5 The Pendulum

The **simple pendulum** is another mechanical system that exhibits periodic motion. It consists of a particle-like bob of mass m suspended by a light string of length L that is fixed at the upper end as shown in Active Figure 15.16. The motion occurs in the vertical plane and is driven by the gravitational force. We shall show that, provided the angle θ is small (less than about 10°), the motion is very close to that of a simple harmonic oscillator.

The forces acting on the bob are the force $\vec{\mathbf{T}}$ exerted by the string and the gravitational force $m\vec{\mathbf{g}}$. The tangential component $mg \sin \theta$ of the gravitational force always acts toward $\theta = 0$, opposite the displacement of the bob from the lowest position. Therefore, the tangential component is a restoring force, and we can apply Newton's second law for motion in the tangential direction:

$$F_t = ma_t \rightarrow -mg \sin \theta = m\frac{d^2s}{dt^2}$$

where the negative sign indicates that the tangential force acts toward the equilibrium (vertical) position and s is the bob's position measured along the arc. We have expressed the tangential acceleration as the second derivative of the position s. Because $s = L\theta$ (Eq. 10.1a with $r = L$) and L is constant, this equation reduces to

$$\frac{d^2\theta}{dt^2} = -\frac{g}{L}\sin \theta$$

Considering θ as the position, let us compare this equation with Equation 15.3. Does it have the same mathematical form? The right side is proportional to $\sin \theta$ rather than to θ; hence, we would not expect simple harmonic motion because this expression is not of the same mathematical form as Equation 15.3. If we assume θ is *small* (less than about 10° or 0.2 rad), however, we can use the **small angle approximation,** in which $\sin \theta \approx \theta$, where θ is measured in radians. Table 15.1 shows angles in degrees and radians and the sines of these angles. As long as θ is less than approximately 10°, the angle in radians and its sine are the same to within an accuracy of less than 1.0%.

Pitfall Prevention 15.5

Not True Simple Harmonic Motion
The pendulum *does not* exhibit true simple harmonic motion for *any* angle. If the angle is less than about 10°, the motion is close to and can be *modeled* as simple harmonic.

Therefore, for small angles, the equation of motion becomes

$$\frac{d^2\theta}{dt^2} = -\frac{g}{L}\theta \quad \text{(for small values of } \theta) \qquad \textbf{(15.24)}$$

Equation 15.24 has the same mathematical form as Equation 15.3, so we conclude that the motion for small amplitudes of oscillation can be modeled as simple harmonic motion. Therefore, the solution of Equation 15.24 is $\theta = \theta_{max} \cos(\omega t + \phi)$, where θ_{max} is the *maximum angular position* and the angular frequency ω is

Angular frequency for a simple pendulum ▶

$$\omega = \sqrt{\frac{g}{L}} \qquad \textbf{(15.25)}$$

The period of the motion is

Period of a simple pendulum ▶

$$T = \frac{2\pi}{\omega} = 2\pi\sqrt{\frac{L}{g}} \qquad \textbf{(15.26)}$$

In other words, the period and frequency of a simple pendulum depend only on the length of the string and the acceleration due to gravity. Because the period is independent of the mass, we conclude that all simple pendula that are of equal length and are at the same location (so that g is constant) oscillate with the same period.

The simple pendulum can be used as a timekeeper because its period depends only on its length and the local value of g. It is also a convenient device for making precise measurements of the free-fall acceleration. Such measurements are impor-

TABLE 15.1 *Angles and Sines of Angles*

Angle in Degrees	Angle in Radians	Sine of Angle	Percent Difference
0°	0.000 0	0.000 0	0.0%
1°	0.017 5	0.017 5	0.0%
2°	0.034 9	0.034 9	0.0%
3°	0.052 4	0.052 3	0.0%
5°	0.087 3	0.087 2	0.1%
10°	0.174 5	0.173 6	0.5%
15°	0.261 8	0.258 8	1.2%
20°	0.349 1	0.342 0	2.1%
30°	0.523 6	0.500 0	4.7%

tant because variations in local values of g can provide information on the location of oil and other valuable underground resources.

***Quick Quiz* 15.6** A grandfather clock depends on the period of a pendulum to keep correct time. **(i)** Suppose a grandfather clock is calibrated correctly and then a mischievous child slides the bob of the pendulum downward on the oscillating rod. Does the grandfather clock run (a) slow, (b) fast, or (c) correctly? **(ii)** Suppose a grandfather clock is calibrated correctly at sea level and is then taken to the top of a very tall mountain. Does the grandfather clock now run (a) slow, (b) fast, or (c) correctly?

Example 15.5 A Connection Between Length and Time

Christian Huygens (1629–1695), the greatest clockmaker in history, suggested that an international unit of length could be defined as the length of a simple pendulum having a period of exactly 1 s. How much shorter would our length unit be if his suggestion had been followed?

SOLUTION

Conceptualize Imagine a pendulum that swings back and forth in exactly 1 second. Based on your experience in observing swinging objects, can you make an estimate of the required length? Hang a small object from a string and simulate the 1-s pendulum.

Categorize This example involves a simple pendulum, so we categorize it as an application of the concepts introduced in this section.

Analyze Solve Equation 15.26 for the length and substitute the known values:

$$L = \frac{T^2 g}{4\pi^2} = \frac{(1.00 \text{ s})^2(9.80 \text{ m/s}^2)}{4\pi^2} = \boxed{0.248 \text{ m}}$$

Finalize The meter's length would be slightly less than one-fourth of its current length. Also, the number of significant digits depends only on how precisely we know g because the time has been defined to be exactly 1 s.

WHAT IF? What if Huygens had been born on another planet? What would the value for g have to be on that planet such that the meter based on Huygens's pendulum would have the same value as our meter?

Answer Solve Equation 15.26 for g:

$$g = \frac{4\pi^2 L}{T^2} = \frac{4\pi^2(1.00 \text{ m})}{(1.00 \text{ s})^2} = 4\pi^2 \text{ m/s}^2 = 39.5 \text{ m/s}^2$$

No planet in our solar system has an acceleration due to gravity that large.

Physical Pendulum

Suppose you balance a wire coat hanger so that the hook is supported by your extended index finger. When you give the hanger a small angular displacement with your other hand and then release it, it oscillates. If a hanging object oscillates

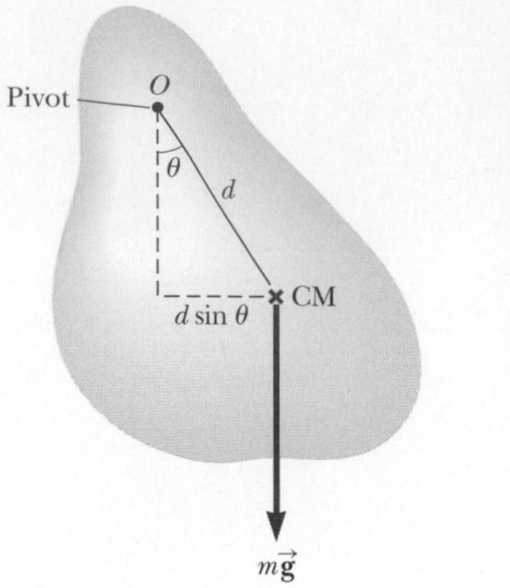

Figure 15.17 A physical pendulum pivoted at O.

about a fixed axis that does not pass through its center of mass and the object cannot be approximated as a point mass, we cannot treat the system as a simple pendulum. In this case, the system is called a **physical pendulum.**

Consider a rigid object pivoted at a point O that is a distance d from the center of mass (Fig. 15.17). The gravitational force provides a torque about an axis through O, and the magnitude of that torque is $mgd \sin \theta$, where θ is as shown in Figure 15.17. We model the object as a rigid object under a net torque and use the rotational form of Newton's second law, $\sum \tau_{\text{ext}} = I\alpha$, where I is the moment of inertia of the object about the axis through O. The result is

$$-mgd \sin \theta = I\frac{d^2\theta}{dt^2}$$

The negative sign indicates that the torque about O tends to decrease θ. That is, the gravitational force produces a restoring torque. If we again assume θ is small, the approximation $\sin \theta \approx \theta$ is valid and the equation of motion reduces to

$$\frac{d^2\theta}{dt^2} = -\left(\frac{mgd}{I}\right)\theta = -\omega^2\theta \tag{15.27}$$

Because this equation is of the same mathematical form as Equation 15.3, its solution is that of the simple harmonic oscillator. That is, the solution of Equation 15.27 is given by $\theta = \theta_{\max} \cos(\omega t + \phi)$, where $\theta_{\max}$ is the maximum angular position and

$$\omega = \sqrt{\frac{mgd}{I}}$$

The period is

Period of a physical pendulum ▶

$$T = \frac{2\pi}{\omega} = 2\pi\sqrt{\frac{I}{mgd}} \tag{15.28}$$

This result can be used to measure the moment of inertia of a flat, rigid object. If the location of the center of mass—and hence the value of d—is known, the moment of inertia can be obtained by measuring the period. Finally, notice that Equation 15.28 reduces to the period of a simple pendulum (Eq. 15.26) when $I = md^2$, that is, when all the mass is concentrated at the center of mass.

Example 15.6 A Swinging Rod

A uniform rod of mass M and length L is pivoted about one end and oscillates in a vertical plane (Fig. 15.18). Find the period of oscillation if the amplitude of the motion is small.

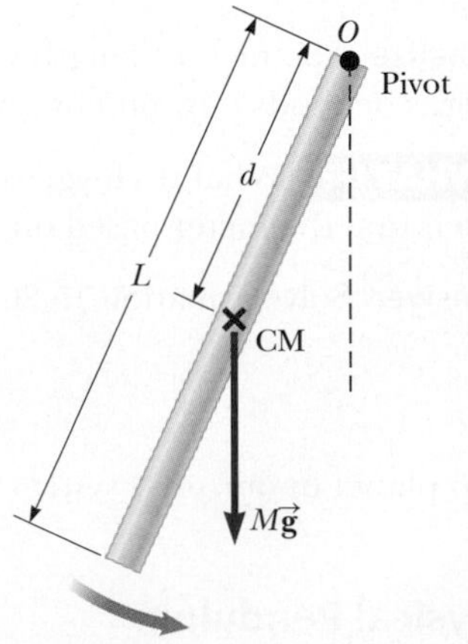

Figure 15.18 (Example 15.6) A rigid rod oscillating about a pivot through one end is a physical pendulum with $d = L/2$.

SOLUTION

Conceptualize Imagine a rod swinging back and forth when pivoted at one end. Try it with a meterstick or a scrap piece of wood.

Categorize Because the rod is not a point particle, we categorize it as a physical pendulum.

Analyze In Chapter 10, we found that the moment of inertia of a uniform rod about an axis through one end is $\frac{1}{3}ML^2$. The distance d from the pivot to the center of mass of the rod is $L/2$.

15.6 *cont.*

Substitute these quantities into Equation 15.28: $T = 2\pi\sqrt{\frac{\frac{1}{3}ML^2}{Mg(L/2)}} = 2\pi\sqrt{\frac{2L}{3g}}$

Finalize In one of the Moon landings, an astronaut walking on the Moon's surface had a belt hanging from his space suit, and the belt oscillated as a physical pendulum. A scientist on the Earth observed this motion on television and used it to estimate the free-fall acceleration on the Moon. How did the scientist make this calculation?

Torsional Pendulum

Figure 15.19 shows a rigid object such as a disk suspended by a wire attached at the top to a fixed support. When the object is twisted through some angle θ, the twisted wire exerts on the object a restoring torque that is proportional to the angular position. That is,

$$\tau = -\kappa\theta$$

where κ (Greek letter kappa) is called the *torsion constant* of the support wire and is a rotational analog to the force constant k for a spring. The value of κ can be obtained by applying a known torque to twist the wire through a measurable angle θ. Applying Newton's second law for rotational motion, we find that

$$\sum \tau = I\alpha \rightarrow -\kappa\theta = I\frac{d^2\theta}{dt^2}$$

$$\frac{d^2\theta}{dt^2} = -\frac{\kappa}{I}\theta \quad \textbf{(15.29)}$$

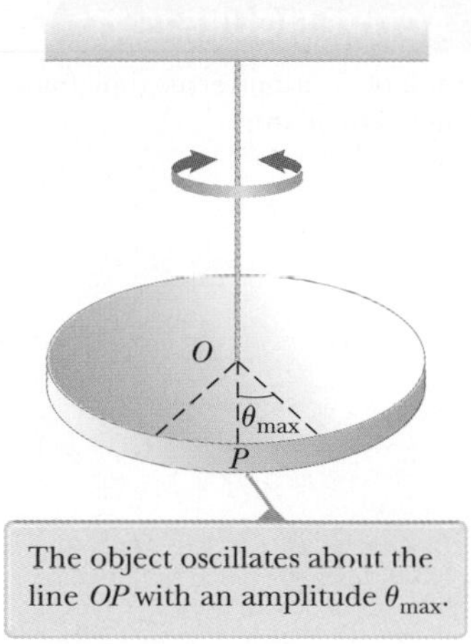

Figure 15.19 A torsional pendulum.

Again, this result is the equation of motion for a simple harmonic oscillator, with $\omega = \sqrt{\kappa/I}$ and a period

$$T = 2\pi\sqrt{\frac{I}{\kappa}} \quad \textbf{(15.30)}$$

◀ **Period of a torsional pendulum**

This system is called a *torsional pendulum.* There is no small-angle restriction in this situation as long as the elastic limit of the wire is not exceeded.

15.6 Damped Oscillations

The oscillatory motions we have considered so far have been for ideal systems, that is, systems that oscillate indefinitely under the action of only one force, a linear restoring force. In many real systems, nonconservative forces such as friction or air resistance retard the motion. Consequently, the mechanical energy of the system diminishes in time, and the motion is said to be *damped.* The lost mechanical energy is transformed into internal energy in the object and the retarding medium. Figure 15.20 depicts one such system: an object attached to a spring and submersed in a viscous liquid. The opening photograph for this chapter depicts damped oscillations in practice. The piston-like devices below the sphere are dampers that transform mechanical energy of the oscillating sphere into internal energy.

One common type of retarding force is that discussed in Section 6.4, where the force is proportional to the speed of the moving object and acts in the direction opposite the velocity of the object with respect to the medium. This retarding force is often observed when an object moves through air, for instance. Because the retarding force can be expressed as $\vec{\mathbf{R}} = -b\vec{\mathbf{v}}$ (where b is a constant called the *damping coefficient*) and the restoring force of the system is $-kx$, we can write Newton's second law as

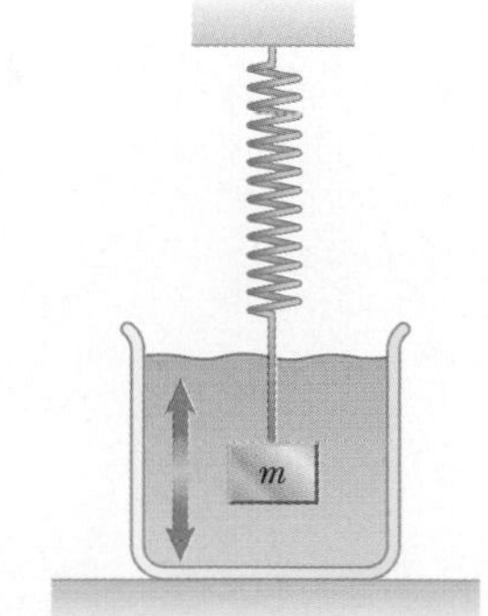

Figure 15.20 One example of a damped oscillator is an object attached to a spring and submersed in a viscous liquid.

$$\sum F_x = -kx - bv_x = ma_x$$

$$-kx - b\frac{dx}{dt} = m\frac{d^2x}{dt^2} \quad \textbf{(15.31)}$$

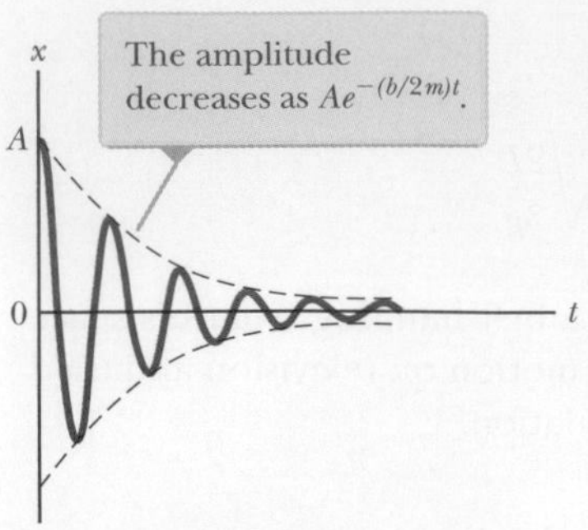

ACTIVE FIGURE 15.21

Graph of position versus time for a damped oscillator.

The solution to this equation requires mathematics that may be unfamiliar to you; we simply state it here without proof. When the retarding force is small compared with the maximum restoring force—that is, when b is small—the solution to Equation 15.31 is

$$x = Ae^{-(b/2m)t}\cos(\omega t + \phi) \tag{15.32}$$

where the angular frequency of oscillation is

$$\omega = \sqrt{\frac{k}{m} - \left(\frac{b}{2m}\right)^2} \tag{15.33}$$

This result can be verified by substituting Equation 15.32 into Equation 15.31. It is convenient to express the angular frequency of a damped oscillator in the form

$$\omega = \sqrt{\omega_0^{\,2} - \left(\frac{b}{2m}\right)^2}$$

where $\omega_0 = \sqrt{k/m}$ represents the angular frequency in the absence of a retarding force (the undamped oscillator) and is called the **natural frequency** of the system.

Active Figure 15.21 shows the position as a function of time for an object oscillating in the presence of a retarding force. When the retarding force is small, the oscillatory character of the motion is preserved but the amplitude decreases exponentially in time, with the result that the motion ultimately becomes undetectable. Any system that behaves in this way is known as a **damped oscillator.** The dashed black lines in Active Figure 15.21, which define the *envelope* of the oscillatory curve, represent the exponential factor in Equation 15.32. This envelope shows that the amplitude decays exponentially with time. For motion with a given spring constant and object mass, the oscillations dampen more rapidly for larger values of the retarding force.

When the magnitude of the retarding force is small such that $b/2m < \omega_0$, the system is said to be **underdamped.** The resulting motion is represented by the blue curve in Figure 15.22. As the value of b increases, the amplitude of the oscillations decreases more and more rapidly. When b reaches a critical value b_c such that $b_c/2m = \omega_0$, the system does not oscillate and is said to be **critically damped.** In this case, the system, once released from rest at some nonequilibrium position, approaches but does not pass through the equilibrium position. The graph of position versus time for this case is the red curve in Figure 15.22.

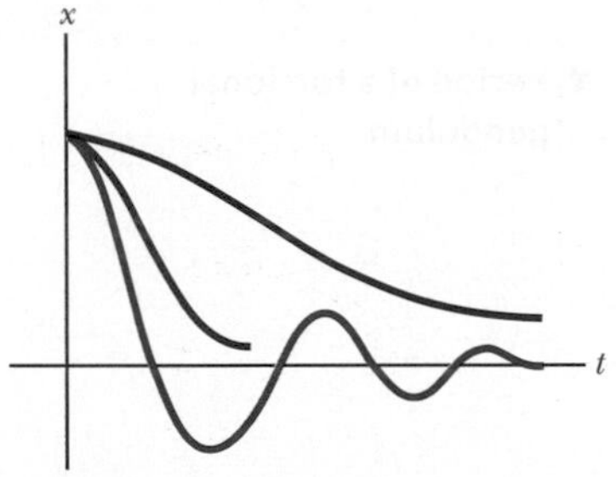

Figure 15.22 Graphs of position versus time for an underdamped oscillator (blue curve), a critically damped oscillator (red curve), and an overdamped oscillator (black curve).

If the medium is so viscous that the retarding force is large compared with the restoring force—that is, if $b/2m > \omega_0$—the system is **overdamped.** Again, the displaced system, when free to move, does not oscillate but rather simply returns to its equilibrium position. As the damping increases, the time interval required for the system to approach equilibrium also increases as indicated by the black curve in Figure 15.22. For critically damped and overdamped systems, there is no angular frequency ω and the solution in Equation 15.32 is not valid.

15.7 Forced Oscillations

We have seen that the mechanical energy of a damped oscillator decreases in time as a result of the resistive force. It is possible to compensate for this energy decrease by applying a periodic external force that does positive work on the system. At any instant, energy can be transferred into the system by an applied force that acts in the direction of motion of the oscillator. For example, a child on a swing can be kept in motion by appropriately timed "pushes." The amplitude of motion remains constant if the energy input per cycle of motion exactly equals the decrease in mechanical energy in each cycle that results from resistive forces.

A common example of a forced oscillator is a damped oscillator driven by an external force that varies periodically, such as $F(t) = F_0 \sin \omega t$, where F_0 is a constant and ω is the angular frequency of the driving force. In general, the frequency ω of the driving force is variable, whereas the natural frequency ω_0 of the oscillator is fixed by the values of k and m. Newton's second law in this situation gives

$$\sum F_x = ma_x \rightarrow F_0 \sin \omega t - b\frac{dx}{dt} - kx = m\frac{d^2x}{dt^2} \quad \textbf{(15.34)}$$

Again, the solution of this equation is rather lengthy and will not be presented. After the driving force on an initially stationary object begins to act, the amplitude of the oscillation will increase. After a sufficiently long period of time, when the energy input per cycle from the driving force equals the amount of mechanical energy transformed to internal energy for each cycle, a steady-state condition is reached in which the oscillations proceed with constant amplitude. In this situation, the solution of Equation 15.34 is

$$x = A \cos (\omega t + \phi) \quad \textbf{(15.35)}$$

where

$$A = \frac{F_0/m}{\sqrt{(\omega^2 - {\omega_0}^2)^2 + \left(\frac{b\omega}{m}\right)^2}} \quad \textbf{(15.36)}$$

◀ **Amplitude of a driven oscillator**

and where $\omega_0 = \sqrt{k/m}$ is the natural frequency of the undamped oscillator ($b = 0$).

Equations 15.35 and 15.36 show that the forced oscillator vibrates at the frequency of the driving force and that the amplitude of the oscillator is constant for a given driving force because it is being driven in steady-state by an external force. For small damping, the amplitude is large when the frequency of the driving force is near the natural frequency of oscillation, or when $\omega \approx \omega_0$. The dramatic increase in amplitude near the natural frequency is called **resonance,** and the natural frequency ω_0 is also called the **resonance frequency** of the system.

The reason for large-amplitude oscillations at the resonance frequency is that energy is being transferred to the system under the most favorable conditions. We can better understand this concept by taking the first time derivative of x in Equation 15.35, which gives an expression for the velocity of the oscillator. We find that v is proportional to $\sin(\omega t + \phi)$, which is the same trigonometric function as that describing the driving force. Therefore, the applied force $\vec{\mathbf{F}}$ is in phase with the velocity. The rate at which work is done on the oscillator by $\vec{\mathbf{F}}$ equals the dot product $\vec{\mathbf{F}} \cdot \vec{\mathbf{v}}$; this rate is the power delivered to the oscillator. Because the product $\vec{\mathbf{F}} \cdot \vec{\mathbf{v}}$ is a maximum when $\vec{\mathbf{F}}$ and $\vec{\mathbf{v}}$ are in phase, we conclude that at resonance, the applied force is in phase with the velocity and the power transferred to the oscillator is a maximum.

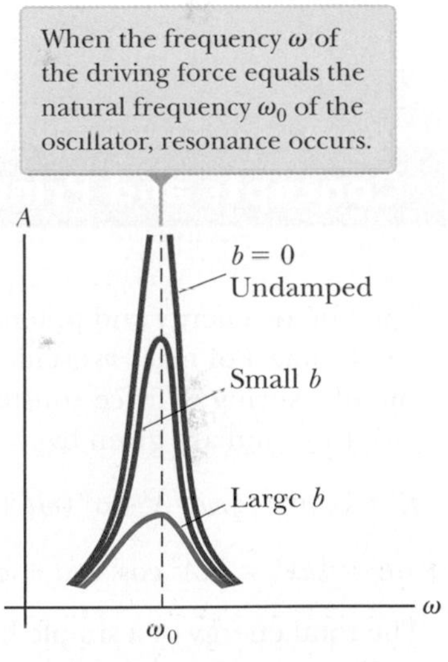

Figure 15.23 Graph of amplitude versus frequency for a damped oscillator when a periodic driving force is present. Notice that the shape of the resonance curve depends on the size of the damping coefficient b.

Figure 15.23 is a graph of amplitude as a function of driving frequency for a forced oscillator with and without damping. Notice that the amplitude increases with decreasing damping ($b \rightarrow 0$) and that the resonance curve broadens as the damping increases. In the absence of a damping force ($b = 0$), we see from Equation 15.36 that the steady-state amplitude approaches infinity as ω approaches ω_0. In other words, if there are no losses in the system and we continue to drive an initially motionless oscillator with a periodic force that is in phase with the velocity, the amplitude of motion builds without limit (see the red-brown curve in Fig. 15.23). This limitless building does not occur in practice because some damping is always present in reality.

Later in this book we shall see that resonance appears in other areas of physics. For example, certain electric circuits have natural frequencies and can be set into strong resonance by a varying voltage applied at a given frequency. A bridge has natural frequencies that can be set into resonance by an appropriate driving force. A dramatic example of such resonance occurred in 1940 when the Tacoma Narrows Bridge in the state of Washington was destroyed by resonant vibrations. Although the winds were not particularly strong on that occasion, the "flapping" of the wind across the roadway (think of the "flapping" of a flag in a strong wind) provided a

Figure 15.24 (a) In 1940, turbulent winds set up torsional vibrations in the Tacoma Narrows Bridge, causing it to oscillate at a frequency near one of the natural frequencies of the bridge structure. (b) Once established, this resonance condition led to the bridge's collapse. (Mathematicians and physicists are currently challenging this interpretation.)

periodic driving force whose frequency matched that of the bridge. The resulting oscillations of the bridge caused it to ultimately collapse (Fig. 15.24) because the bridge design had inadequate built-in safety features.

Many other examples of resonant vibrations can be cited. A resonant vibration you may have experienced is the "singing" of telephone wires in the wind. Machines often break if one vibrating part is in resonance with some other moving part. Soldiers marching in cadence across a bridge have been known to set up resonant vibrations in the structure and thereby cause it to collapse. Whenever any real physical system is driven near its resonance frequency, you can expect oscillations of very large amplitudes.

Summary

Concepts and Principles

The kinetic energy and potential energy for an object of mass m oscillating at the end of a spring of force constant k vary with time and are given by

$$K = \tfrac{1}{2}mv^2 = \tfrac{1}{2}m\omega^2A^2\sin^2(\omega t + \phi) \quad \textbf{(15.19)}$$

$$U = \tfrac{1}{2}kx^2 = \tfrac{1}{2}kA^2\cos^2(\omega t + \phi) \quad \textbf{(15.20)}$$

The total energy of a simple harmonic oscillator is a constant of the motion and is given by

$$E = \tfrac{1}{2}kA^2 \quad \textbf{(15.21)}$$

A **simple pendulum** of length L moves in simple harmonic motion for small angular displacements from the vertical. Its period is

$$T = 2\pi\sqrt{\frac{L}{g}} \quad \textbf{(15.26)}$$

A **physical pendulum** is an extended object that, for small angular displacements, moves in simple harmonic motion about a pivot that does not go through the center of mass. The period of this motion is

$$T = 2\pi\sqrt{\frac{I}{mgd}} \quad \textbf{(15.28)}$$

where I is the moment of inertia of the object about an axis through the pivot and d is the distance from the pivot to the center of mass of the object.

If an oscillator experiences a damping force $\vec{\mathbf{R}} = -b\vec{\mathbf{v}}$, its position for small damping is described by

$$x = Ae^{-(b/2m)t}\cos(\omega t + \phi) \quad \textbf{(15.32)}$$

where

$$\omega = \sqrt{\frac{k}{m} - \left(\frac{b}{2m}\right)^2} \quad \textbf{(15.33)}$$

If an oscillator is subject to a sinusoidal driving force that is described by $F(t) = F_0 \sin \omega t$, it exhibits **resonance,** in which the amplitude is largest when the driving frequency ω matches the natural frequency $\omega_0 = \sqrt{k/m}$ of the oscillator.

Analysis Model for Problem Solving

Particle in Simple Harmonic Motion If a particle is subject to a force of the form of Hooke's law $F = -kx$, the particle exhibits **simple harmonic motion.** Its position is described by

$$x(t) = A\cos(\omega t + \phi) \quad \textbf{(15.6)}$$

where A is the **amplitude** of the motion, ω is the **angular frequency,** and ϕ is the **phase constant.** The value of ϕ depends on the initial position and initial velocity of the oscillator.

The **period** of the oscillation is

$$T = \frac{2\pi}{\omega} = 2\pi\sqrt{\frac{m}{k}} \quad \textbf{(15.13)}$$

and the inverse of the period is the **frequency.**

Objective Questions

☐ denotes answer available in Student Solutions Manual/Study Guide

1. The position of an object moving with simple harmonic motion is given by $x = 4\cos(6\pi t)$, where x is in meters and t is in seconds. What is the period of the oscillating system? (a) 4 s (b) $\frac{1}{6}$ s (c) $\frac{1}{3}$ s (d) 6π s (e) impossible to determine from the information given

2. Which of the following statements is *not* true regarding a mass–spring system that moves with simple harmonic motion in the absence of friction? (a) The total energy of the system remains constant. (b) The energy of the system is continually transformed between kinetic and potential energy. (c) The total energy of the system is proportional to the square of the amplitude. (d) The potential energy stored in the system is greatest when the mass passes through the equilibrium position. (e) The velocity of the oscillating mass has its maximum value when the mass passes through the equilibrium position.

3. A block–spring system vibrating on a frictionless, horizontal surface with an amplitude of 6.0 cm has an energy of 12 J. If the block is replaced by one whose mass is twice the mass of the original block and the amplitude of the motion is again 6.0 cm, what is the energy of the system? (a) 12 J (b) 24 J (c) 6 J (d) 48 J (e) none of those answers

4. If an object of mass m attached to a light spring is replaced by one of mass $9m$, the frequency of the vibrating system changes by what factor? (a) $\frac{1}{9}$ (b) $\frac{1}{3}$ (c) 3.0 (d) 9.0 (e) 6.0

5. An object of mass 0.40 kg, hanging from a spring with a spring constant of 8.0 N/m, is set into an up-and-down simple harmonic motion. What is the magnitude of the acceleration of the object when it is at its maximum displacement of 0.10 m? (a) zero (b) 0.45 m/s^2 (c) 1.0 m/s^2 (d) 2.0 m/s^2 (e) 2.4 m/s^2

6. A runaway railroad car, with mass 3.0×10^5 kg, coasts across a level track at 2.0 m/s when it collides elastically with a spring-loaded bumper at the end of the track. If the spring constant of the bumper is 2.0×10^6 N/m, what is the maximum compression of the spring during the collision? (a) 0.77 m (b) 0.58 m (c) 0.34 m (d) 1.07 m (e) 1.24 m

7. If a simple pendulum oscillates with small amplitude and its length is doubled, what happens to the frequency of its motion? (a) It doubles. (b) It becomes $\sqrt{2}$ times as large. (c) It becomes half as large. (d) It becomes $1/\sqrt{2}$ times as large. (e) It remains the same.

8. An object–spring system moving with simple harmonic motion has an amplitude A. When the kinetic energy of the object equals twice the potential energy stored in the spring, what is the position x of the object? (a) A (b) $\frac{1}{3}A$ (c) $A/\sqrt{3}$ (d) 0 (e) none of those answers

9. A particle on a spring moves in simple harmonic motion along the x axis between turning points at $x_1 = 100$ cm and $x_2 = 140$ cm. **(i)** At which of the following positions does the particle have maximum speed? (a) 100 cm (b) 110 cm (c) 120 cm (d) at none of those positions **(ii)** At which position does it have maximum acceleration? Choose from the same possibilities as in part (i). **(iii)** At which position is the greatest net force exerted on the particle? Choose from the same possibilities as in part (i).

10. A mass–spring system moves with simple harmonic motion along the x axis between turning points at $x_1 = 20$ cm and $x_2 = 60$ cm. For parts (i) through (iii), choose from the same five possibilities. **(i)** At which position does the particle have the greatest magnitude of momentum? (a) 20 cm (b) 30 cm (c) 40 cm (d) some other position (e) The greatest value occurs at multiple points. **(ii)** At which position

does the particle have greatest kinetic energy? **(iii)** At which position does the particle-spring system have the greatest total energy?

11. A block with mass $m = 0.1$ kg oscillates with amplitude $A = 0.1$ m at the end of a spring with force constant $k = 10$ N/m on a frictionless, horizontal surface. Rank the periods of the following situations from greatest to smallest. If any periods are equal, show their equality in your ranking. (a) The system is as described above. (b) The system is as described in situation (a) except the amplitude is 0.2 m. (c) The situation is as described in situation (a) except the mass is 0.2 kg. (d) The situation is as described in situation (a) except the spring has force constant 20 N/m. (e) A small resistive force makes the motion underdamped.

12. For a simple harmonic oscillator, answer yes or no to the following questions. (a) Can the quantities position and velocity have the same sign? (b) Can velocity and acceleration have the same sign? (c) Can position and acceleration have the same sign?

13. The top end of a spring is held fixed. A block is hung on the bottom end as in Figure OQ15.13a, and the frequency f of the oscillation of the system is measured. The block, a second identical block, and the spring are carried up in a space shuttle to Earth orbit. The two blocks are attached to the ends of the spring. The spring is compressed without making adjacent coils touch (Fig. OQ15.13b), and the system is released to oscillate while floating within the shuttle cabin (Fig. OQ15.13c). What is the frequency of oscillation for this system in terms of f? (a) $f/2$ (b) $f/\sqrt{2}$ (c) f (d) $\sqrt{2}f$ (e) $2f$

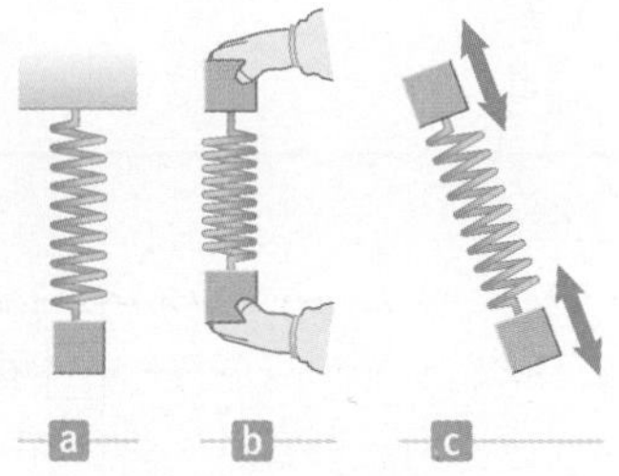

Figure OQ15.13

14. You attach a block to the bottom end of a spring hanging vertically. You slowly let the block move down and find that it hangs at rest with the spring stretched by 15.0 cm. Next, you lift the block back up to the initial position and release it from rest with the spring unstretched. What maximum distance does it move down? (a) 7.5 cm (b) 15.0 cm (c) 30.0 cm (d) 60.0 cm (e) The distance cannot be determined without knowing the mass and spring constant.

15. A simple pendulum has a period of 2.5 s. **(i)** What is its period if its length is made four times larger? (a) 1.25 s (b) 1.77 s (c) 2.5 s (d) 3.54 s (e) 5 s **(ii)** What is its period if the length is held constant at its initial value and the mass of the suspended bob is made four times larger? Choose from the same possibilities.

16. A simple pendulum is suspended from the ceiling of a stationary elevator, and the period is determined. **(i)** When the elevator accelerates upward, is the period (a) greater, (b) smaller, or (c) unchanged? **(ii)** When the elevator has a downward acceleration, is the period (a) greater, (b) smaller, or (c) unchanged? **(iii)** When the elevator moves with constant upward velocity, is the period of the pendulum (a) greater, (b) smaller, or (c) unchanged?

17. You stand on the end of a diving board and bounce to set it into oscillation. You find a maximum response in terms of the amplitude of oscillation of the end of the board when you bounce at frequency f. You now move to the middle of the board and repeat the experiment. Is the resonance frequency for forced oscillations at this point (a) higher, (b) lower, or (c) the same as f?

Conceptual Questions

☐ denotes answer available in Student Solutions Manual/Study Guide

1. Is a bouncing ball an example of simple harmonic motion? Is the daily movement of a student from home to school and back simple harmonic motion? Why or why not?

2. The equations listed in Table 2.2 give position as a function of time, velocity as a function of time, and velocity as a function of position for an object moving in a straight line with constant acceleration. The quantity v_{xi} appears in every equation. (a) Do any of these equations apply to an object moving in a straight line with simple harmonic motion? (b) Using a similar format, make a table of equations describing simple harmonic motion. Include equations giving acceleration as a function of time and acceleration as a function of position. State the equations in such a form that they apply equally to a block–spring system, to a pendulum, and to other vibrating systems. (c) What quantity appears in every equation?

3. (a) If the coordinate of a particle varies as $x = -A \cos \omega t$, what is the phase constant in Equation 15.6? (b) At what position is the particle at $t = 0$?

4. A simple pendulum can be modeled as exhibiting simple harmonic motion when θ is small. Is the motion periodic when θ is large?

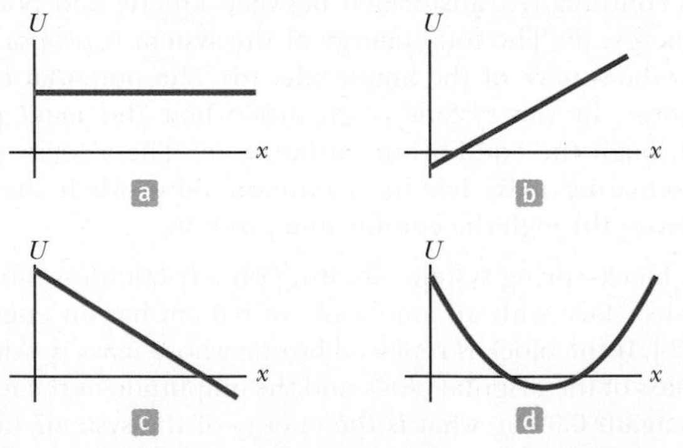

Figure CQ15.5

5. Figure CQ15.5 shows graphs of the potential energy of four different systems versus the position of a particle in each system. Each particle is set into motion with a push at an arbitrarily chosen location. Describe its subsequent motion in each case (a), (b), (c), and (d).

6. A student thinks that any real vibration must be damped. Is the student correct? If so, give convincing reasoning. If not, give an example of a real vibration that keeps constant amplitude forever if the system is isolated.

7. The mechanical energy of an undamped block–spring system is constant as kinetic energy transforms to elastic potential energy and vice versa. For comparison, explain what happens to the energy of a damped oscillator in terms of the mechanical, potential, and kinetic energies.

8. Is it possible to have damped oscillations when a system is at resonance? Explain.

9. Will damped oscillations occur for any values of b and k? Explain.

10. If a pendulum clock keeps perfect time at the base of a mountain, will it also keep perfect time when it is moved to the top of the mountain? Explain.

11. You are looking at a small, leafy tree. You do not notice any breeze, and most of the leaves on the tree are motionless. One leaf, however, is fluttering back and forth wildly. After a while, that leaf stops moving and you notice a different leaf moving much more than all the others. Explain what could cause the large motion of one particular leaf.

12. A pendulum bob is made from a sphere filled with water. What would happen to the frequency of vibration of this pendulum if there were a hole in the sphere that allowed the water to leak out slowly?

13. Consider the simplified single-piston engine in Figure CQ15.13. Assuming the wheel rotates with constant angular speed, explain why the piston rod oscillates in simple harmonic motion.

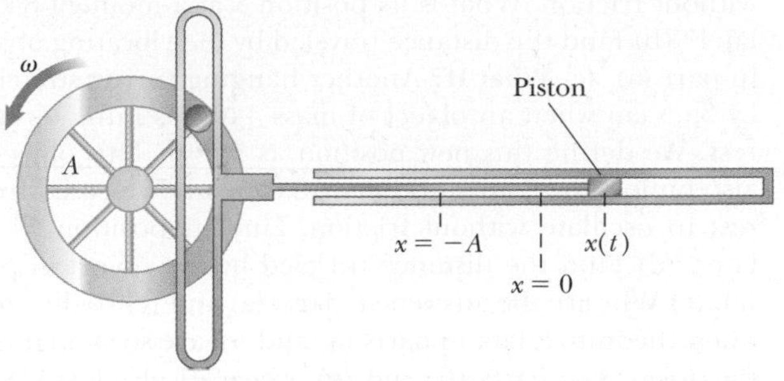

Figure CQ15.13

Problems

WebAssign The problems found in this chapter may be assigned online in Enhanced WebAssign

1. denotes straightforward problem; 2. denotes intermediate problem; 3. denotes challenging problem

1. full solution available in the Student Solutions Manual/Study Guide

1. denotes problems most often assigned in Enhanced WebAssign; these provide students with targeted feedback and either a Master It tutorial or a Watch It solution video.

QC denotes asking for quantitative and conceptual reasoning

S denotes symbolic reasoning problem

M denotes Master It tutorial available in Enhanced WebAssign

GP denotes guided problem

shaded denotes "paired problems" that develop reasoning with symbols and numerical values

Note: Ignore the mass of every spring, except in Problems 64 and 75.

Section 15.1 Motion of an Object Attached to a Spring

Problems 16, 17, 18, 22, and 65 in Chapter 7 can also be assigned with this section.

1. A 0.60-kg block attached to a spring with force constant 130 N/m is free to move on a frictionless, horizontal surface as in Active Figure 15.1. The block is released from rest when the spring is stretched 0.13 m. At the instant the block is released, find (a) the force on the block and (b) its acceleration.

2. When a 4.25-kg object is placed on top of a vertical spring, the spring compresses a distance of 2.62 cm. What is the force constant of the spring?

Section 15.2 Analysis Model: Particle in Simple Harmonic Motion

3. A vertical spring stretches 3.9 cm when a 10-g object is hung from it. The object is replaced with a block of mass 25 g that oscillates up and down in simple harmonic motion. Calculate the period of motion.

4. In an engine, a piston oscillates with simple harmonic motion so that its position varies according to the expression

$$x = 5.00 \cos\left(2t + \frac{\pi}{6}\right)$$

where x is in centimeters and t is in seconds. At $t = 0$, find (a) the position of the particle, (b) its velocity, and (c) its acceleration. Find (d) the period and (e) the amplitude of the motion.

5. The position of a particle is given by the expression $x = 4.00 \cos(3.00\pi t + \pi)$, where x is in meters and t is in seconds. Determine (a) the frequency and (b) period of the motion, (c) the amplitude of the motion, (d) the phase constant, and (e) the position of the particle at $t = 0.250$ s.

6. A piston in a gasoline engine is in simple harmonic motion. The engine is running at the rate of 3 600 rev/min. Taking the extremes of its position relative to its center point as ±5.00 cm, find the magnitudes of the (a) maximum velocity and (b) maximum acceleration of the piston.

7. A 1.00-kg object is attached to a horizontal spring. The spring is initially stretched by 0.100 m, and the object is released from rest there. It proceeds to move without friction. The next time the speed of the object is zero is 0.500 s later. What is the maximum speed of the object?

8. A simple harmonic oscillator takes 12.0 s to undergo five complete vibrations. Find (a) the period of its motion, (b) the frequency in hertz, and (c) the angular frequency in radians per second.

9. A 7.00-kg object is hung from the bottom end of a vertical spring fastened to an overhead beam. The object is set into vertical oscillations having a period of 2.60 s. Find the force constant of the spring.

10. Q|C (a) A hanging spring stretches by 35.0 cm when an object of mass 450 g is hung on it at rest. In this situation, we define its position as $x = 0$. The object is pulled down an additional 18.0 cm and released from rest to oscillate without friction. What is its position x at a moment 84.4 s later? (b) Find the distance traveled by the vibrating object in part (a). (c) **What If?** Another hanging spring stretches by 35.5 cm when an object of mass 440 g is hung on it at rest. We define this new position as $x = 0$. This object is also pulled down an additional 18.0 cm and released from rest to oscillate without friction. Find its position 84.4 s later. (d) Find the distance traveled by the object in part (c). (e) Why are the answers to parts (a) and (c) so different when the initial data in parts (a) and (c) are so similar and the answers to parts (b) and (d) are relatively close? Does this circumstance reveal a fundamental difficulty in calculating the future?

11. **Review.** A particle moves along the x axis. It is initially at the position 0.270 m, moving with velocity 0.140 m/s and acceleration -0.320 m/s^2. Suppose it moves as a particle under constant acceleration for 4.50 s. Find (a) its position and (b) its velocity at the end of this time interval. Next, assume it moves as a particle in simple harmonic motion for 4.50 s and $x = 0$ is its equilibrium position. Find (c) its position and (d) its velocity at the end of this time interval.

12. Q|C A ball dropped from a height of 4.00 m makes an elastic collision with the ground. Assuming no mechanical energy is lost due to air resistance, (a) show that the ensuing motion is periodic and (b) determine the period of the motion. (c) Is the motion simple harmonic? Explain.

13. A particle moving along the x axis in simple harmonic motion starts from its equilibrium position, the origin, at $t = 0$ and moves to the right. The amplitude of its motion is 2.00 cm, and the frequency is 1.50 Hz. (a) Find an expression for the position of the particle as a function of time. Determine (b) the maximum speed of the particle and (c) the earliest time ($t > 0$) at which the particle has this speed. Find (d) the maximum positive acceleration of the particle and (e) the earliest time ($t > 0$) at which the particle has this acceleration. (f) Find the total distance traveled by the particle between $t = 0$ and $t = 1.00$ s.

14. A 1.00-kg glider attached to a spring with a force constant of 25.0 N/m oscillates on a frictionless, horizontal air track. At $t = 0$, the glider is released from rest at $x = -3.00$ cm (that is, the spring is compressed by 3.00 cm). Find (a) the period of the glider's motion, (b) the maximum values of its speed and acceleration, and (c) the position, velocity, and acceleration as functions of time.

15. A 0.500-kg object attached to a spring with a force constant of 8.00 N/m vibrates in simple harmonic motion with an amplitude of 10.0 cm. Calculate the maximum value of its (a) speed and (b) acceleration, (c) the speed and (d) the acceleration when the object is 6.00 cm from the equilibrium position, and (e) the time interval required for the object to move from $x = 0$ to $x = 8.00$ cm.

16. Q|C You attach an object to the bottom end of a hanging vertical spring. It hangs at rest after extending the spring 18.3 cm. You then set the object vibrating. (a) Do you have enough information to find its period? (b) Explain your answer and state whatever you can about its period.

Section 15.3 Energy of the Simple Harmonic Oscillator

17. M To test the resiliency of its bumper during low-speed collisions, a 1 000-kg automobile is driven into a brick wall. The car's bumper behaves like a spring with a force constant 5.00×10^6 N/m and compresses 3.16 cm as the car is brought to rest. What was the speed of the car before impact, assuming no mechanical energy is transformed or transferred away during impact with the wall?

18. A 200-g block is attached to a horizontal spring and executes simple harmonic motion with a period of 0.250 s. The total energy of the system is 2.00 J. Find (a) the force constant of the spring and (b) the amplitude of the motion.

19. A 50.0-g object connected to a spring with a force constant of 35.0 N/m oscillates with an amplitude of 4.00 cm on a frictionless, horizontal surface. Find (a) the total energy of the system and (b) the speed of the object when its position is 1.00 cm. Find (c) the kinetic energy and (d) the potential energy when its position is 3.00 cm.

20. A 2.00-kg object is attached to a spring and placed on a frictionless, horizontal surface. A horizontal force of 20.0 N is required to hold the object at rest when it is pulled 0.200 m from its equilibrium position (the origin of the x axis). The object is now released from rest from this stretched position, and it subsequently undergoes simple harmonic oscillations. Find (a) the force constant of the spring, (b) the frequency of the oscillations, and (c) the maximum speed of the object. (d) Where does this maximum speed occur? (e) Find the maximum acceleration of the object. (f) Where does the maximum acceleration occur? (g) Find the total energy of the oscillating system. Find (h) the speed and (i) the acceleration of the object when its position is equal to one-third the maximum value.

21. Q|C S A simple harmonic oscillator of amplitude A has a total energy E. Determine (a) the kinetic energy and (b) the potential energy when the position is one-third the amplitude. (c) For what values of the position does the kinetic energy equal one-half the potential energy? (d) Are there any values of the position where the kinetic energy is greater than the maximum potential energy? Explain.

22. GP **Review.** A 65.0-kg bungee jumper steps off a bridge with a light bungee cord tied to her body and to the bridge. The unstretched length of the cord is 11.0 m. The jumper reaches the bottom of her motion 36.0 m below the bridge before bouncing back. We wish to find the time interval between her leaving the bridge and her arriving at the bottom of her motion. Her overall motion can be separated into an 11.0-m free fall and a 25.0-m section of simple harmonic oscillation. (a) For the free-fall part, what is the appropriate analysis model to describe her motion? (b) For what time interval is she in free fall? (c) For the simple harmonic oscillation part of the plunge, is the system of the bungee jumper, the spring, and the Earth isolated or non-

isolated? (d) From your response in part (c) find the spring constant of the bungee cord. (e) What is the location of the equilibrium point where the spring force balances the gravitational force exerted on the jumper? (f) What is the angular frequency of the oscillation? (g) What time interval is required for the cord to stretch by 25.0 m? (h) What is the total time interval for the entire 36.0-m drop?

23. **Q|C** **Review.** A 0.250-kg block resting on a frictionless, horizontal surface is attached to a spring whose force constant is 83.8 N/m as in Figure P15.23. A horizontal force $\vec{F}$ causes the spring to stretch a distance of 5.46 cm from its equilibrium position. (a) Find the magnitude of $\vec{F}$. (b) What is the total energy stored in the system when the spring is stretched? (c) Find the magnitude of the acceleration of the block just after the applied force is removed. (d) Find the speed of the block when it first reaches the equilibrium position. (e) If the surface is not frictionless but the block still reaches the equilibrium position, would your answer to part (d) be larger or smaller? (f) What other information would you need to know to find the actual answer to part (d) in this case? (g) What is the largest value of the coefficient of friction that would allow the block to reach the equilibrium position?

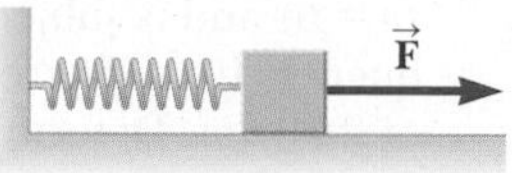

Figure P15.23

24. A 326-g object is attached to a spring and executes simple harmonic motion with a period of 0.250 s. If the total energy of the system is 5.83 J, find (a) the maximum speed of the object, (b) the force constant of the spring, and (c) the amplitude of the motion.

Section 15.4 Comparing Simple Harmonic Motion with Uniform Circular Motion

25. **Q|C** While driving behind a car traveling at 3.00 m/s, you notice that one of the car's tires has a small hemispherical bump on its rim as shown in Figure P15.25. (a) Explain why the bump, from your viewpoint behind the car, executes simple harmonic motion. (b) If the radii of the car's tires are 0.300 m, what is the bump's period of oscillation?

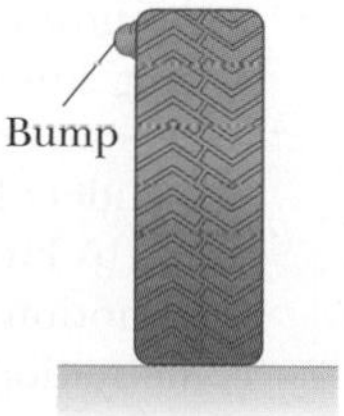

Figure P15.25

Section 15.5 The Pendulum

Problem 62 in Chapter 1 can also be assigned with this section.

26. A "seconds pendulum" is one that moves through its equilibrium position once each second. (The period of the pendulum is precisely 2 s.) The length of a seconds pendulum is 0.992 7 m at Tokyo, Japan, and 0.994 2 m at Cambridge, England. What is the ratio of the free-fall accelerations at these two locations?

27. A simple pendulum makes 120 complete oscillations in 3.00 min at a location where $g = 9.80$ m/s^2. Find (a) the period of the pendulum and (b) its length.

28. **S** A particle of mass m slides without friction inside a hemispherical bowl of radius R. Show that if the particle starts from rest with a small displacement from equilibrium, it moves in simple harmonic motion with an angular frequency equal to that of a simple pendulum of length R. That is, $\omega = \sqrt{g/R}$.

29. A physical pendulum in the form of a planar object moves in simple harmonic motion with a frequency of 0.450 Hz. The pendulum has a mass of 2.20 kg, and the pivot is located 0.350 m from the center of mass. Determine the moment of inertia of the pendulum about the pivot point.

30. **S** A physical pendulum in the form of a planar object moves in simple harmonic motion with a frequency f. The pendulum has a mass m, and the pivot is located a distance d from the center of mass. Determine the moment of inertia of the pendulum about the pivot point.

31. **Q|C** A simple pendulum has a mass of 0.250 kg and a length of 1.00 m. It is displaced through an angle of 15.0° and then released. Using the analysis model of a particle in simple harmonic motion, what are (a) the maximum speed of the bob, (b) its maximum angular acceleration, and (c) the maximum restoring force on the bob? (d) **What If?** Solve parts (a) through (c) again by using analysis models introduced in earlier chapters. (e) Compare the answers.

32. **S** Consider the physical pendulum of Figure 15.17. (a) Represent its moment of inertia about an axis passing through its center of mass and parallel to the axis passing through its pivot point as I_{CM}. Show that its period is

$$T = 2\pi\sqrt{\frac{I_{CM} + md^2}{mgd}}$$

where d is the distance between the pivot point and the center of mass. (b) Show that the period has a minimum value when d satisfies $md^2 = I_{CM}$.

33. **Review.** A simple pendulum is 5.00 m long. What is the period of small oscillations for this pendulum if it is located in an elevator (a) accelerating upward at 5.00 m/s^2? (b) Accelerating downward at 5.00 m/s^2? (c) What is the period of this pendulum if it is placed in a truck that is accelerating horizontally at 5.00 m/s^2?

34. A very light rigid rod of length 0.500 m extends straight out from one end of a meterstick. The combination is suspended from a pivot at the upper end of the rod as shown in Figure P15.34. The combination is then pulled out by a small angle and released. (a) Determine the period of oscillation of the system. (b) By what percentage does the period differ from the period of a simple pendulum 1.00 m long?

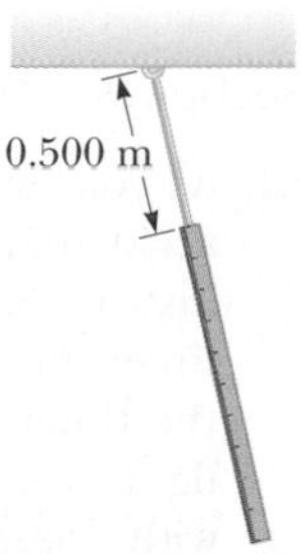

Figure P15.34

35. A watch balance wheel (Fig. P15.35) has a period of oscillation of 0.250 s. The wheel is constructed so that its mass of 20.0 g is concentrated around a rim of radius 0.500 cm. What

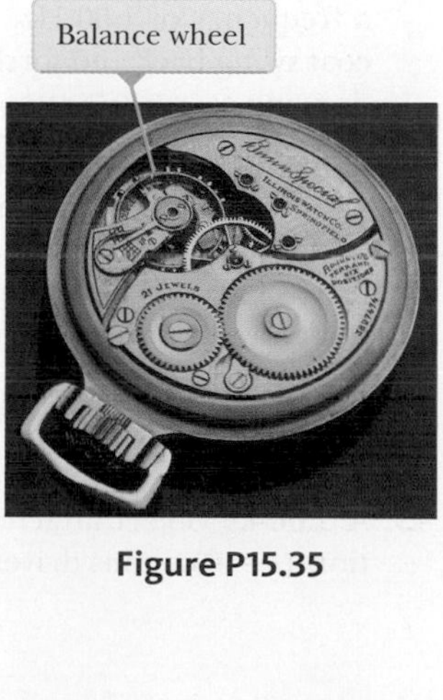

Figure P15.35

are (a) the wheel's moment of inertia and (b) the torsion constant of the attached spring?

36. A small object is attached to the end of a string to form a simple pendulum. The period of its harmonic motion is measured for small angular displacements and three lengths. For lengths of 1.000 m, 0.750 m, and 0.500 m, total time intervals for 50 oscillations of 99.8 s, 86.6 s, and 71.1 s are measured with a stopwatch. (a) Determine the period of motion for each length. (b) Determine the mean value of g obtained from these three independent measurements and compare it with the accepted value. (c) Plot T^2 versus L and obtain a value for g from the slope of your best-fit straight-line graph. (d) Compare the value found in part (c) with that obtained in part (b).

Section 15.6 Damped Oscillations

37. A pendulum with a length of 1.00 m is released from an initial angle of 15.0°. After 1 000 s, its amplitude has been reduced by friction to 5.50°. What is the value of $b/2m$?

38. **S** Show that the time rate of change of mechanical energy for a damped, undriven oscillator is given by $dE/dt = -bv^2$ and hence is always negative. To do so, differentiate the expression for the mechanical energy of an oscillator, $E = \frac{1}{2}mv^2 + \frac{1}{2}kx^2$, and use Equation 15.31.

39. A 10.6-kg object oscillates at the end of a vertical spring that has a spring constant of 2.05×10^4 N/m. The effect of air resistance is represented by the damping coefficient $b = 3.00$ N · s/m. (a) Calculate the frequency of the damped oscillation. (b) By what percentage does the amplitude of the oscillation decrease in each cycle? (c) Find the time interval that elapses while the energy of the system drops to 5.00% of its initial value.

40. **S** Show that Equation 15.32 is a solution of Equation 15.31 provided that $b^2 < 4mk$.

Section 15.7 Forced Oscillations

41. As you enter a fine restaurant, you realize that you have accidentally brought a small electronic timer from home instead of your cell phone. In frustration, you drop the timer into a side pocket of your suit coat, not realizing that the timer is operating. The arm of your chair presses the light cloth of your coat against your body at one spot. Fabric with a length L hangs freely below that spot, with the timer at the bottom. At one point during your dinner, the timer goes off and a buzzer and a vibrator turn on and off with a frequency of 1.50 Hz. It makes the hanging part of your coat swing back and forth with remarkably large amplitude, drawing everyone's attention. Find the value of L.

42. A baby bounces up and down in her crib. Her mass is 12.5 kg, and the crib mattress can be modeled as a light spring with force constant 700 N/m. (a) The baby soon learns to bounce with maximum amplitude and minimum effort by bending her knees at what frequency? (b) If she were to use the mattress as a trampoline—losing contact with it for part of each cycle—what minimum amplitude of oscillation does she require?

43. A 2.00-kg object attached to a spring moves without friction ($b = 0$) and is driven by an external force given by the expression $F = 3.00 \sin(2\pi t)$, where F is in newtons and t is in seconds. The force constant of the spring is 20.0 N/m. Find (a) the resonance angular frequency of the system, (b) the angular frequency of the driven system, and (c) the amplitude of the motion.

44. A block weighing 40.0 N is suspended from a spring that has a force constant of 200 N/m. The system is undamped ($b = 0$) and is subjected to a harmonic driving force of frequency 10.0 Hz, resulting in a forced-motion amplitude of 2.00 cm. Determine the maximum value of the driving force.

45. **M** Damping is negligible for a 0.150-kg object hanging from a light, 6.30-N/m spring. A sinusoidal force with an amplitude of 1.70 N drives the system. At what frequency will the force make the object vibrate with an amplitude of 0.440 m?

46. **S** Considering an undamped, forced oscillator ($b = 0$), show that Equation 15.35 is a solution of Equation 15.34, with an amplitude given by Equation 15.36.

Additional Problems

47. The mass of the deuterium molecule (D_2) is twice that of the hydrogen molecule (H_2). If the vibrational frequency of H_2 is 1.30×10^{14} Hz, what is the vibrational frequency of D_2? Assume the "spring constant" of attracting forces is the same for the two molecules.

48. **Q|C** **Review.** This problem extends the reasoning of Problem 59 in Chapter 9. Two gliders are set in motion on an air track. Glider 1 has mass $m_1 = 0.240$ kg and moves to the right with speed 0.740 m/s. It will have a rear-end collision with glider 2, of mass $m_2 = 0.360$ kg, which initially moves to the right with speed 0.120 m/s. A light spring of force constant 45.0 N/m is attached to the back end of glider 2 as shown in Figure P9.59. When glider 1 touches the spring, superglue instantly and permanently makes it stick to its end of the spring. (a) Find the common speed the two gliders have when the spring is at maximum compression. (b) Find the maximum spring compression distance. The motion after the gliders become attached consists of a combination of (1) the constant-velocity motion of the center of mass of the two-glider system found in part (a) and (2) simple harmonic motion of the gliders relative to the center of mass. (c) Find the energy of the center-of-mass motion. (d) Find the energy of the oscillation.

49. **Q|C** An object of mass m moves in simple harmonic motion with amplitude 12.0 cm on a light spring. Its maximum acceleration is 108 cm/s^2. Regard m as a variable. (a) Find the period T of the object. (b) Find its frequency f. (c) Find the maximum speed v_{max} of the object. (d) Find the total energy E of the object–spring system. (e) Find the force constant k of the spring. (f) Describe the pattern of dependence of each of the quantities T, f, v_{max}, E, and k on m.

50. **Q|C** **Review.** A rock rests on a concrete sidewalk. An earthquake strikes, making the ground move vertically in simple harmonic motion with a constant frequency of 2.40 Hz and with gradually increasing amplitude. (a) With what amplitude does the ground vibrate when the rock begins to lose contact with the sidewalk? Another rock is sitting on the concrete bottom of a swimming pool full of

water. The earthquake produces only vertical motion, so the water does not slosh from side to side. (b) Present a convincing argument that when the ground vibrates with the amplitude found in part (a), the submerged rock also barely loses contact with the floor of the swimming pool.

51. **M** A small ball of mass M is attached to the end of a uniform rod of equal mass M and length L that is pivoted at the top (Fig. P15.51). Determine the tensions in the rod (a) at the pivot and (b) at the point P when the system is stationary. (c) Calculate the period of oscillation for small displacements from equilibrium and (d) determine this period for $L = 2.00$ m.

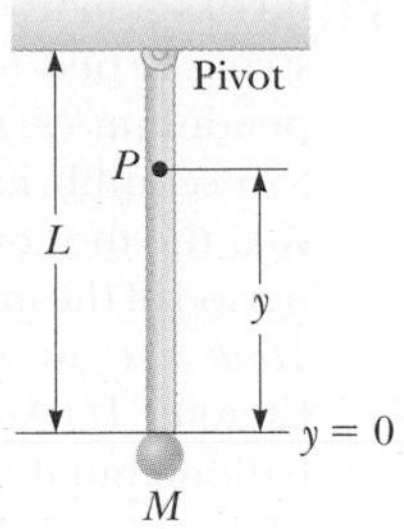

Figure P15.51

52. An object attached to a spring vibrates with simple harmonic motion as described by Figure P15.52. For this motion, find (a) the amplitude, (b) the period, (c) the angular frequency, (d) the maximum speed, (e) the maximum acceleration, and (f) an equation for its position x as a function of time.

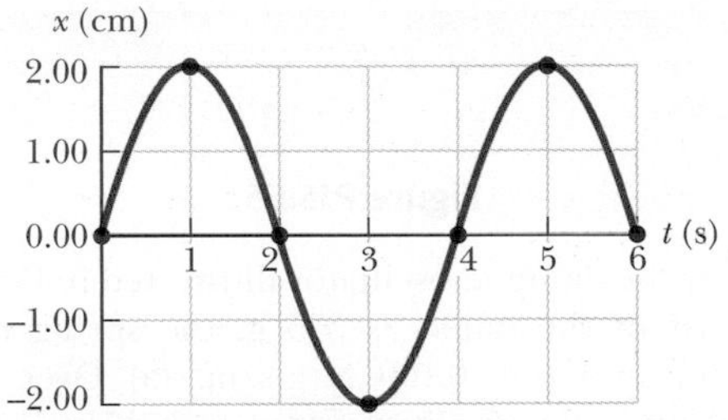

Figure P15.52

53. **Review.** A large block P attached to a light spring executes horizontal, simple harmonic motion as it slides across a frictionless surface with a frequency $f = 1.50$ Hz. Block B rests on it as shown in Figure P15.53, and the coefficient of static friction between the two is $\mu_s = 0.600$. What maximum amplitude of oscillation can the system have if block B is not to slip?

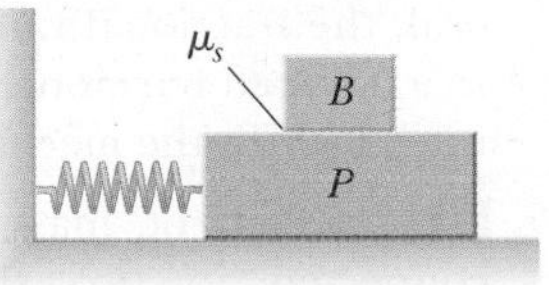

Figure P15.53
Problems 53 and 54.

54. **S** **Review.** A large block P attached to a light spring executes horizontal, simple harmonic motion as it slides across a frictionless surface with a frequency f. Block B rests on it as shown in Figure P15.53, and the coefficient of static friction between the two is μ_s. What maximum amplitude of oscillation can the system have if block B is not to slip?

55. **S** A pendulum of length L and mass M has a spring of force constant k connected to it at a distance h below its point of suspension (Fig. P15.55). Find the frequency of vibration of the system for small values of the amplitude (small θ). Assume the vertical suspension rod of length L is rigid, but ignore its mass.

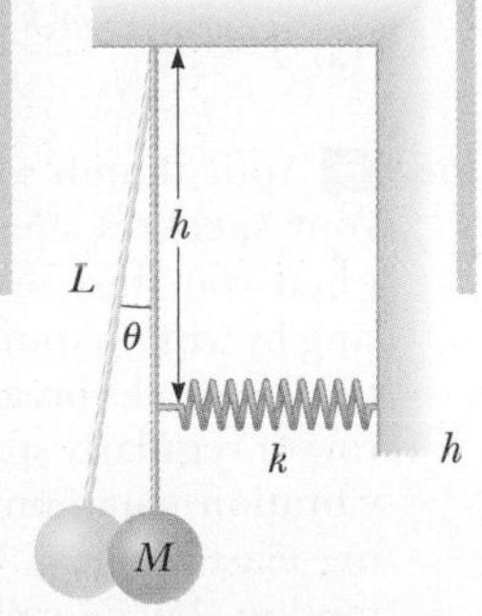

Figure P15.55

56. A particle with a mass of 0.500 kg is attached to a horizontal spring with a force constant of 50.0 N/m. At the moment $t = 0$, the particle has its maximum speed of 20.0 m/s and is moving to the left. (a) Determine the particle's equation of motion, specifying its position as a function of time. (b) Where in the motion is the potential energy three times the kinetic energy? (c) Find the minimum time interval required for the particle to move from $x = 0$ to $x = 1.00$ m. (d) Find the length of a simple pendulum with the same period.

57. A horizontal plank of mass 5.00 kg and length 2.00 m is pivoted at one end. The plank's other end is supported by a spring of force constant 100 N/m (Fig. P15.57). The plank is displaced by a small angle θ from its horizontal equilibrium position and released. Find the angular frequency with which the plank moves with simple harmonic motion.

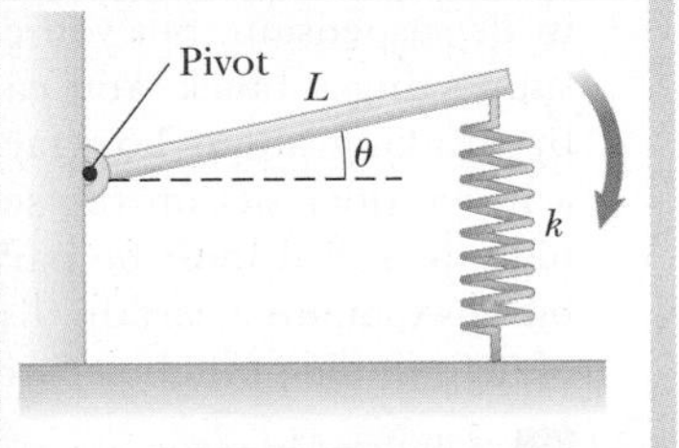

Figure P15.57
Problems 57 and 58.

58. **S** A horizontal plank of mass m and length L is pivoted at one end. The plank's other end is supported by a spring of force constant k (Fig. P15.57). The plank is displaced by a small angle θ from its horizontal equilibrium position and released. Find the angular frequency with which the plank moves with simple harmonic motion.

59. **Q|C** **Review.** A particle of mass 4.00 kg is attached to a spring with a force constant of 100 N/m. It is oscillating on a frictionless, horizontal surface with an amplitude of 2.00 m. A 6.00-kg object is dropped vertically on top of the 4.00-kg object as it passes through its equilibrium point. The two objects stick together. (a) What is the new amplitude of the vibrating system after the collision? (b) By what factor has the period of the system changed? (c) By how much does the energy of the system change as a result of the collision? (d) Account for the change in energy.

60. A simple pendulum with a length of 2.23 m and a mass of 6.74 kg is given an initial speed of 2.06 m/s at its equilibrium position. Assume it undergoes simple harmonic motion. Determine (a) its period, (b) its total energy, and (c) its maximum angular displacement.

61. **Review.** One end of a light spring with force constant $k = 100$ N/m is attached to a vertical wall. A light string is tied to the other end of the horizontal spring. As shown in Figure P15.61, the string changes from horizontal to vertical as it passes over a pulley of mass M in the shape of a solid disk of radius $R = 2.00$ cm. The pulley is free to turn on a fixed, smooth axle. The vertical section of the string supports an object of mass $m = 200$ g.

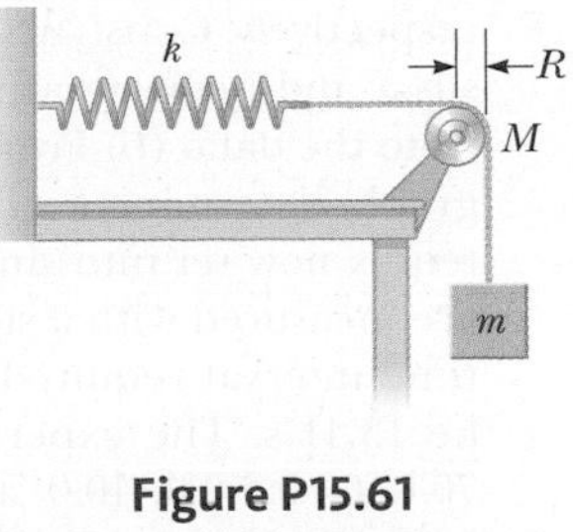

Figure P15.61

The string does not slip at its contact with the pulley. The object is pulled downward a small distance and released. (a) What is the angular frequency ω of oscillation of the object in terms of the mass M? (b) What is the highest possible value of the angular frequency of oscillation of the object? (c) What is the highest possible value of the angular frequency of oscillation of the object if the pulley radius is doubled to $R = 4.00$ cm?

62. **Q|C** People who ride motorcycles and bicycles learn to look out for bumps in the road and especially for *washboarding,* a condition in which many equally spaced ridges are worn into the road. What is so bad about washboarding? A motorcycle has several springs and shock absorbers in its suspension, but you can model it as a single spring supporting a block. You can estimate the force constant by thinking about how far the spring compresses when a heavy rider sits on the seat. A motorcyclist traveling at highway speed must be particularly careful of washboard bumps that are a certain distance apart. What is the order of magnitude of their separation distance?

63. **S** A ball of mass m is connected to two rubber bands of length L, each under tension T as shown in Figure P15.63. The ball is displaced by a small distance y perpendicular to the length of the rubber bands. Assuming the tension does not change, show that (a) the restoring force is $-(2T/L)y$ and (b) the system exhibits simple harmonic motion with an angular frequency $\omega = \sqrt{2T/mL}$.

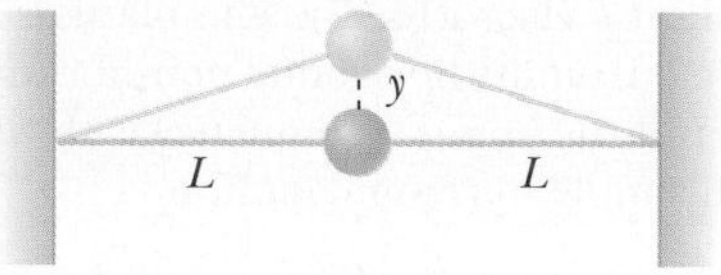

Figure P15.63

64. When a block of mass M, connected to the end of a spring of mass $m_s = 7.40$ g and force constant k, is set into simple harmonic motion, the period of its motion is

$$T = 2\pi\sqrt{\frac{M + (m_s/3)}{k}}$$

A two-part experiment is conducted with the use of blocks of various masses suspended vertically from the spring as shown in Figure P15.64. (a) Static extensions of 17.0, 29.3, 35.3, 41.3, 47.1, and 49.3 cm are measured for M values of 20.0, 40.0, 50.0, 60.0, 70.0, and 80.0 g, respectively. Construct a graph of Mg versus x and perform a linear least-squares fit to the data. (b) From the slope of your graph, determine a value for k for this spring. (c) The system is now set into simple harmonic motion, and periods are measured with a stopwatch. With $M = 80.0$ g, the total time interval required for ten oscillations is measured to be 13.41 s. The experiment is repeated with M values of 70.0, 60.0, 50.0, 40.0, and 20.0 g, with corresponding time intervals for ten oscillations of 12.52, 11.67, 10.67, 9.62, and 7.03 s. (d) Compute the experimental value for T from each of these measurements. (e) Plot a graph of T^2 versus M and (f) determine a value for k from the slope of the linear least-squares fit through the data points. (g) Compare this value of k with that obtained in part (b). (h) Obtain a value for m_s from your graph and compare it with the given value of 7.40 g.

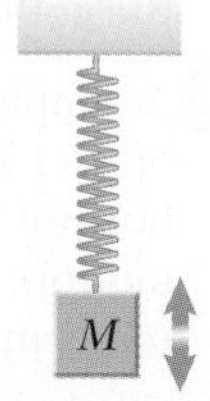

Figure P15.64

65. **Review.** A light balloon filled with helium of density 0.179 kg/m^3 is tied to a light string of length $L = 3.00$ m. The string is tied to the ground forming an "inverted" simple pendulum (Fig. 15.65a). If the balloon is displaced slightly from equilibrium as in Figure P15.65b and released, (a) show that the motion is simple harmonic and (b) determine the period of the motion. Take the density of air to be 1.20 kg/m^3. *Hint:* Use an analogy with the simple pendulum and see Chapter 14. Assume the air applies a buoyant force on the balloon but does not otherwise affect its motion.

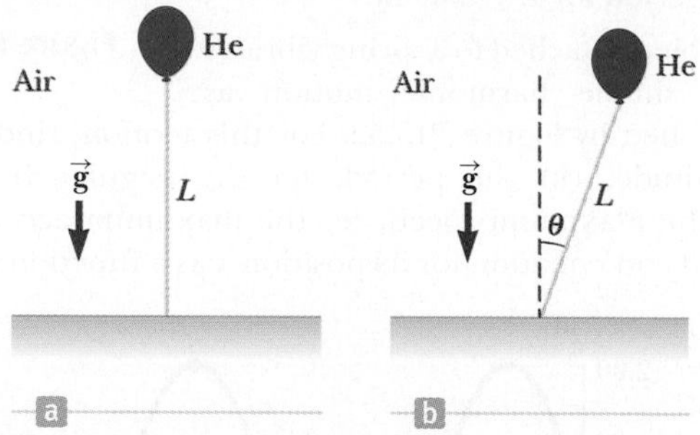

Figure P15.65

66. Consider the damped oscillator illustrated in Figure 15.20. The mass of the object is 375 g, the spring constant is 100 N/m, and $b = 0.100$ N · s/m. (a) Over what time interval does the amplitude drop to half its initial value? (b) **What If?** Over what time interval does the mechanical energy drop to half its initial value? (c) Show that, in general, the fractional rate at which the amplitude decreases in a damped harmonic oscillator is one-half the fractional rate at which the mechanical energy decreases.

67. **S** A block of mass m is connected to two springs of force constants k_1 and k_2 in two ways as shown in Figure P15.67. In both cases, the block moves on a frictionless table after it is displaced from equilibrium and released. Show that in the two cases the block exhibits simple harmonic motion with periods

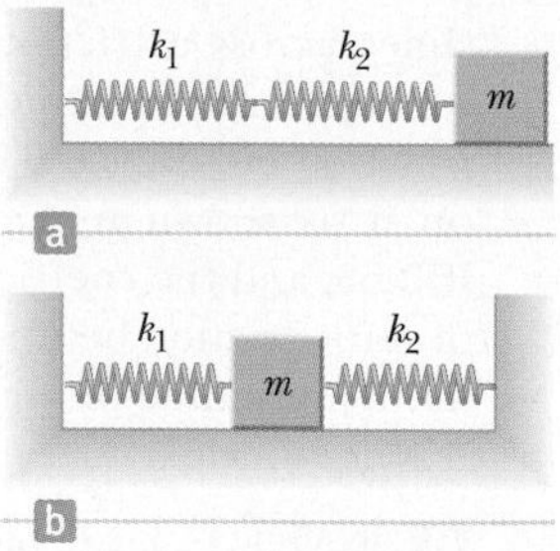

Figure P15.67

$$\text{(a) } T = 2\pi\sqrt{\frac{m(k_1 + k_2)}{k_1 k_2}} \quad \text{and} \quad \text{(b) } T = 2\pi\sqrt{\frac{m}{k_1 + k_2}}$$

68. **S** Your thumb squeaks on a plate you have just washed. Your sneakers squeak on the gym floor. Car tires squeal when you start or stop abruptly. You can make a goblet sing by wiping your moistened finger around its rim. When chalk squeaks on a blackboard, you can see that it makes a row of regularly spaced dashes. As these examples suggest, vibration commonly results when friction acts on a moving elastic object. The oscillation is not simple harmonic motion, but is called *stick-and-slip.* This problem models stick-and-slip motion.

A block of mass m is attached to a fixed support by a horizontal spring with force constant k and negligible mass (Fig. P15.68). Hooke's law describes the spring both in extension and in compression. The block sits on a long horizontal board, with which it has coefficient of static friction μ_s and a smaller coefficient of kinetic friction μ_k. The board moves to the right at constant speed v. Assume the block spends most of its time sticking to the board and moving to the right with it, so the speed v is small in comparison to the average speed the block has as it slips back toward the left. (a) Show that the maximum extension of the spring from its unstressed position is very nearly given by $\mu_s mg/k$. (b) Show that the block oscillates around an equilibrium position at which the spring is stretched by $\mu_k mg/k$. (c) Graph the block's position versus time. (d) Show that the amplitude of the block's motion is

$$A = \frac{(\mu_s - \mu_k)mg}{k}$$

(e) Show that the period of the block's motion is

$$T = \frac{2(\mu_s - \mu_k)mg}{vk} + \pi\sqrt{\frac{m}{k}}$$

It is the excess of static over kinetic friction that is important for the vibration. "The squeaky wheel gets the grease" because even a viscous fluid cannot exert a force of static friction.

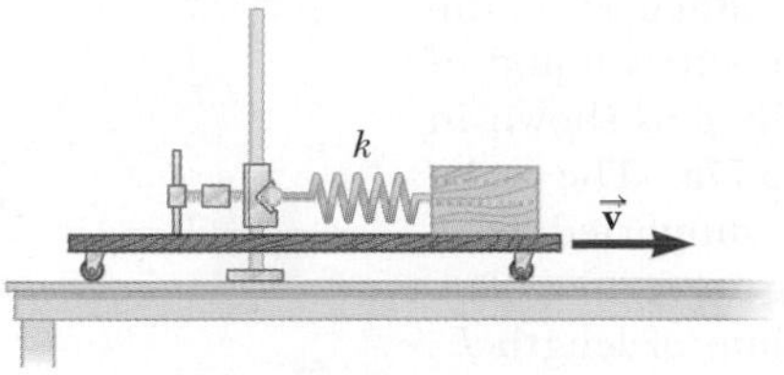

Figure P15.68

69. **S** **Review.** A lobsterman's buoy is a solid wooden cylinder of radius r and mass M. It is weighted at one end so that it floats upright in calm seawater, having density ρ. A passing shark tugs on the slack rope mooring the buoy to a lobster trap, pulling the buoy down a distance x from its equilibrium position and releasing it. (a) Show that the buoy will execute simple harmonic motion if the resistive effects of the water are ignored. (b) Determine the period of the oscillations.

70. *Why is the following situation impossible?* Your job involves building very small damped oscillators. One of your designs involves a spring–object oscillator with a spring of force constant $k = 10.0$ N/m and an object of mass $m = 1.00$ g. Your design objective is that the oscillator undergo many oscillations as its amplitude falls to 25.0% of its initial value in a certain time interval. Measurements on your latest design show that the amplitude falls to the 25.0% value in 23.1 ms. This time interval is too long for what is needed in your project. To shorten the time interval, you double the damping constant b for the oscillator. This doubling allows you to reach your design objective.

71. Two identical steel balls, each of mass 67.4 g, are moving in opposite directions at 5.00 m/s. They collide head-on and bounce apart elastically. By squeezing one of the balls in a vise while precise measurements are made of the resulting amount of compression, you find that Hooke's law is a good model of the ball's elastic behavior. A force of 16.0 kN exerted by each jaw of the vise reduces the diameter by 0.200 mm. Model the motion of each ball, while the balls are in contact, as one-half of a cycle of simple harmonic motion. Compute the time interval for which the balls are in contact. (If you solved Problem 57 in Chapter 7, compare your results from this problem with your results from that one.)

Challenge Problems

72. **S** A smaller disk of radius r and mass m is attached rigidly to the face of a second larger disk of radius R and mass M as shown in Figure P15.72. The center of the small disk is located at the edge of the large disk. The large disk is mounted at its center on a frictionless axle. The assembly is rotated through a small angle θ from its equilibrium position and released. (a) Show that the speed of the center of the small disk as it passes through the equilibrium position is

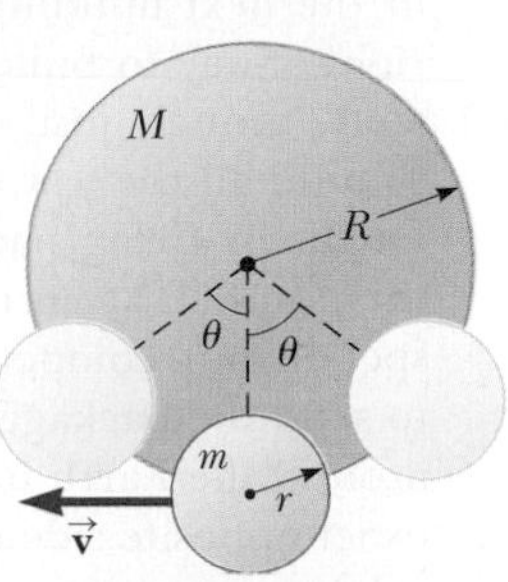

Figure P15.72

$$v = 2\left[\frac{Rg(1 - \cos\theta)}{(M/m) + (r/R)^2 + 2}\right]^{1/2}$$

(b) Show that the period of the motion is

$$T = 2\pi\left[\frac{(M + 2m)R^2 + mr^2}{2mgR}\right]^{1/2}$$

73. An object of mass $m_1 = 9.00$ kg is in equilibrium when connected to a light spring of constant $k = 100$ N/m that is fastened to a wall as shown in Figure P15.73a. A second object, $m_2 = 7.00$ kg, is slowly pushed up against m_1, compressing the spring by the amount $A = 0.200$ m (see Fig. P15.73b).

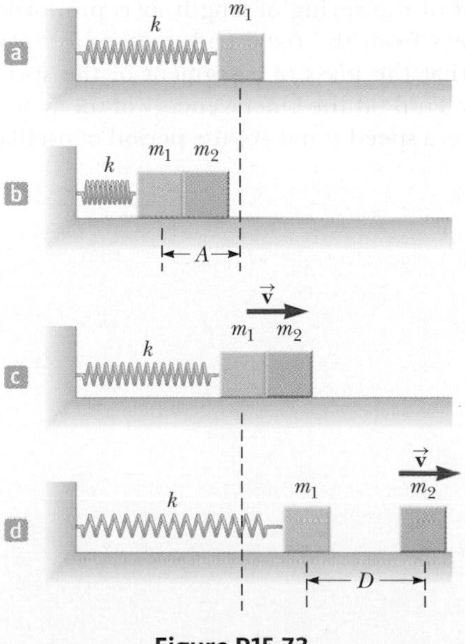

Figure P15.73

The system is then released, and both objects start moving to the right on the frictionless surface. (a) When m_1 reaches the equilibrium point, m_2 loses contact with m_1 (see Fig. P15.73c) and moves to the right with speed v. Determine the value of v. (b) How far apart are the objects when the spring is fully stretched for the first time (the distance D in Fig. P15.48d)?

74. S **Review.** *Why is the following situation impossible?* You are in the high-speed package delivery business. Your competitor in the next building gains the right-of-way to build an evacuated tunnel just above the ground all the way around the Earth. By firing packages into this tunnel at just the right speed, your competitor is able to send the packages into orbit around the Earth in this tunnel so that they arrive on the exact opposite side of the Earth in a very short time interval. You come up with a competing idea. Figuring that the distance *through* the Earth is shorter than the distance *around* the Earth, you obtain permits to build an evacuated tunnel through the center of the Earth (Fig. P15.74). By simply dropping packages into this tunnel, they fall downward and arrive at the other end of your tunnel, which is in a building right next to the other end of your competitor's tunnel. Because your packages arrive on the other side of the Earth in a shorter time interval, you win the competition and your business flourishes. *Note:* An object at a distance r from the center of the Earth is pulled toward the center of the Earth only by the mass within the sphere of radius r (the reddish region in Fig. P15.74). Assume the Earth has uniform density.

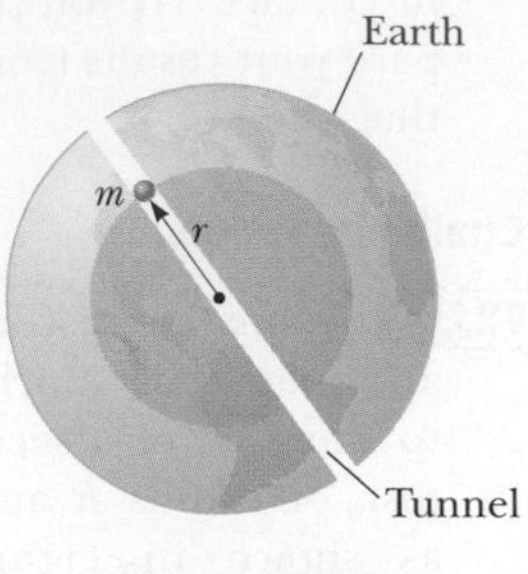

Figure P15.74

75. S A block of mass M is connected to a spring of mass m and oscillates in simple harmonic motion on a frictionless, horizontal track (Fig. P15.75). The force constant of the spring is k, and the equilibrium length is ℓ. Assume all portions of the spring oscillate in phase and the velocity of a segment of the spring of length dx is proportional to the distance x from the fixed end; that is, $v_x = (x/\ell)v$. Also, notice that the mass of a segment of the spring is $dm = (m/\ell)dx$. Find (a) the kinetic energy of the system when the block has a speed v and (b) the period of oscillation.

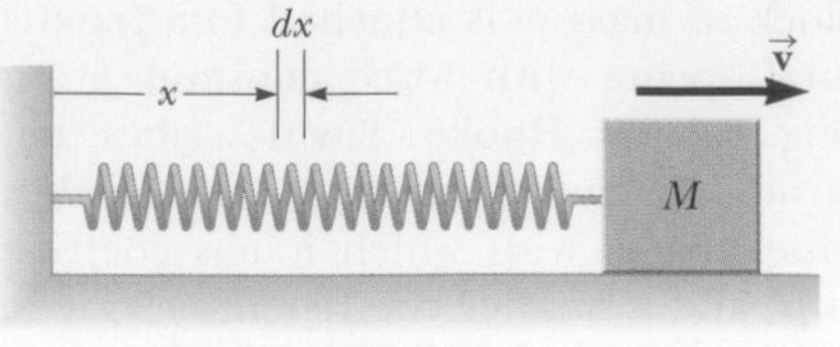

Figure P15.75

76. Review. A system consists of a spring with force constant k = 1 250 N/m, length L = 1.50 m, and an object of mass m = 5.00 kg attached to the end (Fig. P15.76). The object is placed at the level of the point of attachment with the spring unstretched, at position $y_i = L$, and then it is released so that it swings like a pendulum. (a) Find the y position of the object at the lowest point. (b) Will the pendulum's period be greater or less than the period of a simple pendulum with the same mass m and length L? Explain.

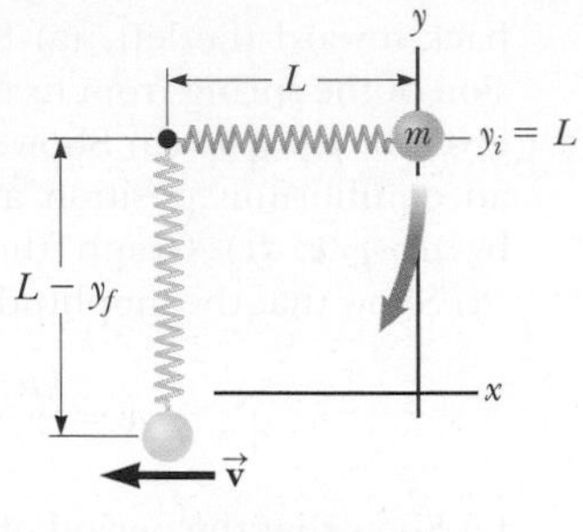

Figure P15.76

77. S A light, cubical container of volume a^3 is initially filled with a liquid of mass density ρ as shown in Figure P15.77a. The cube is initially supported by a light string to form a simple pendulum of length L_i, measured from the center of mass of the filled container, where $L_i >> a$. The liquid is allowed to flow from the bottom of the container at a constant rate (dM/dt). At any time t, the level of the liquid in the container is h and the length of the pendulum is L (measured relative to the instantaneous center of mass) as shown in Figure P15.77b. (a) Find the period of the pendulum as a function of time. (b) What is the period of the pendulum after the liquid completely runs out of the container?

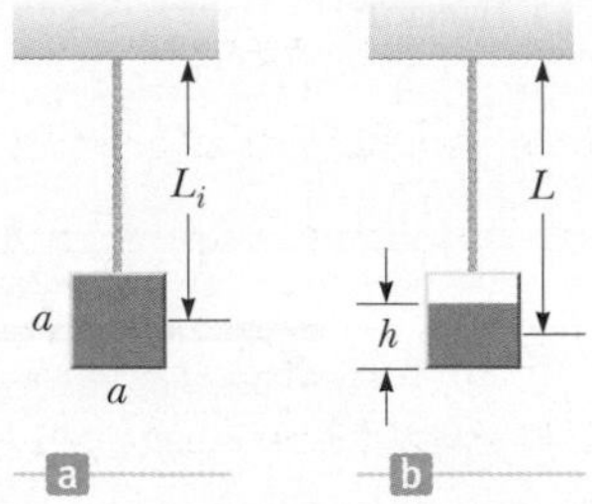

Figure P15.77

chapter 16

Wave Motion

Lifeguards in New South Wales, Australia practice taking their boat over large water waves breaking near the shore. A wave moving over the surface of water is one example of a mechanical wave. *(Travel Ink/Gallo Images/Getty Images)*

Many of us experienced waves as children when we dropped a pebble into a pond. At the point the pebble hits the water's surface, circular waves are created. These waves move outward from the creation point in expanding circles until they reach the shore. If you were to examine carefully the motion of a small object floating on the disturbed water, you would see that the object moves vertically and horizontally about its original position but does not undergo any net displacement away from or toward the point at which the pebble hit the water. The small elements of water in contact with the object, as well as all the other water elements on the pond's surface, behave in the same way. That is, the water *wave* moves from the point of origin to the shore, but the water is not carried with it.

The world is full of waves, the two main types being *mechanical* waves and *electromagnetic* waves. In the case of mechanical waves, some physical medium is being disturbed; in our pebble example, elements of water are disturbed. Electromagnetic waves do not require a medium to propagate; some examples of electromagnetic waves are visible light, radio waves, television signals, and x-rays. Here, in this part of the book, we study only mechanical waves.

Consider again the small object floating on the water. We have caused the object to move at one point in the water by dropping a pebble at another location. The object has gained kinetic energy from our action, so energy must have transferred from the point at which the pebble is dropped to the position of the object. This feature is central to wave motion: *energy* is transferred over a distance, but *matter* is not.

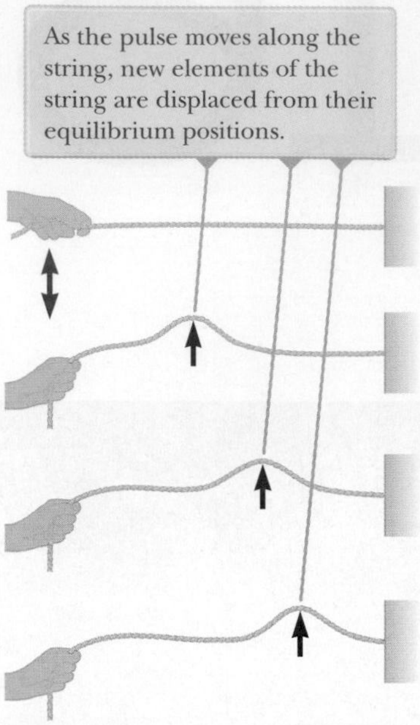

Figure 16.1 A hand moves the end of a stretched string up and down once (red arrow), causing a pulse to travel along the string.

16.1 Propagation of a Disturbance

The introduction to this chapter alluded to the essence of wave motion: the transfer of energy through space without the accompanying transfer of matter. In the list of energy transfer mechanisms in Chapter 8, two mechanisms—mechanical waves and electromagnetic radiation—depend on waves. By contrast, in another mechanism, matter transfer, the energy transfer is accompanied by a movement of matter through space with no wave character in the process.

All mechanical waves require (1) some source of disturbance, (2) a medium containing elements that can be disturbed, and (3) some physical mechanism through which elements of the medium can influence each other. One way to demonstrate wave motion is to flick one end of a long string that is under tension and has its opposite end fixed as shown in Figure 16.1. In this manner, a single bump (called a *pulse*) is formed and travels along the string with a definite speed. Figure 16.1 represents four consecutive "snapshots" of the creation and propagation of the traveling pulse. The hand is the source of the disturbance. The string is the medium through which the pulse travels—individual elements of the string are disturbed from their equilibrium position. Furthermore, the elements are connected together so they influence each other. The pulse has a definite height and a definite speed of propagation along the medium. The shape of the pulse changes very little as it travels along the string.[1]

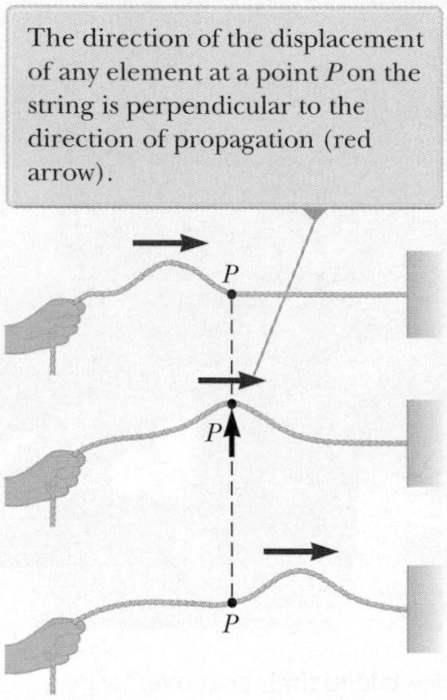

Figure 16.2 The displacement of a particular string element for a transverse pulse traveling on a stretched string.

We shall first focus on a pulse traveling through a medium. Once we have explored the behavior of a pulse, we will then turn our attention to a *wave,* which is a *periodic* disturbance traveling through a medium. We create a pulse on our string by flicking the end of the string once as in Figure 16.1. If we were to move the end of the string up and down repeatedly, we would create a traveling wave, which has characteristics a pulse does not have. We shall explore these characteristics in Section 16.2.

As the pulse in Figure 16.1 travels, each disturbed element of the string moves in a direction *perpendicular* to the direction of propagation. Figure 16.2 illustrates this point for one particular element, labeled *P*. Notice that no part of the string ever moves in the direction of the propagation. A traveling wave or pulse that causes the elements of the disturbed medium to move perpendicular to the direction of propagation is called a **transverse wave.**

Compare this wave with another type of pulse, one moving down a long, stretched spring as shown in Figure 16.3. The left end of the spring is pushed briefly to the

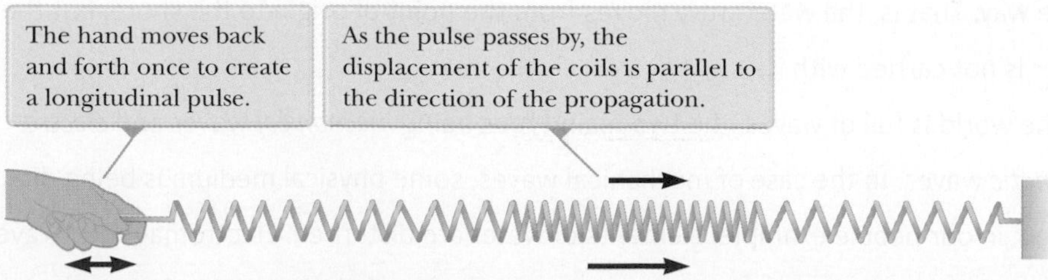

Figure 16.3 A longitudinal pulse along a stretched spring.

[1]In reality, the pulse changes shape and gradually spreads out during the motion. This effect, called *dispersion,* is common to many mechanical waves as well as to electromagnetic waves. We do not consider dispersion in this chapter.

right and then pulled briefly to the left. This movement creates a sudden compression of a region of the coils. The compressed region travels along the spring (to the right in Fig. 16.3). Notice that the direction of the displacement of the coils is *parallel* to the direction of propagation of the compressed region. A traveling wave or pulse that causes the elements of the medium to move parallel to the direction of propagation is called a **longitudinal wave.**

Sound waves, which we shall discuss in Chapter 17, are another example of longitudinal waves. The disturbance in a sound wave is a series of high-pressure and low-pressure regions that travel through air.

Some waves in nature exhibit a combination of transverse and longitudinal displacements. Surface-water waves are a good example. When a water wave travels on the surface of deep water, elements of water at the surface move in nearly circular paths as shown in Active Figure 16.4. The disturbance has both transverse and longitudinal components. The transverse displacements seen in Active Figure 16.4 represent the variations in vertical position of the water elements. The longitudinal displacements represent elements of water moving back and forth in a horizontal direction.

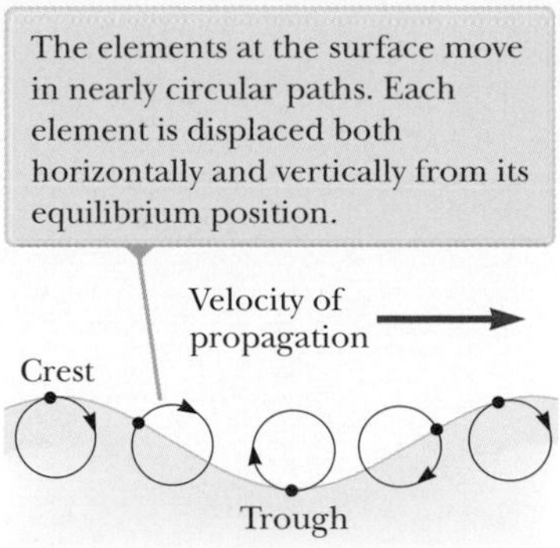

ACTIVE FIGURE 16.4

The motion of water elements on the surface of deep water in which a wave is propagating is a combination of transverse and longitudinal displacements.

The three-dimensional waves that travel out from a point under the Earth's surface at which an earthquake occurs are of both types, transverse and longitudinal. The longitudinal waves are the faster of the two, traveling at speeds in the range of 7 to 8 km/s near the surface. They are called **P waves,** with "P" standing for *primary,* because they travel faster than the transverse waves and arrive first at a seismograph (a device used to detect waves due to earthquakes). The slower transverse waves, called **S waves,** with "S" standing for *secondary,* travel through the Earth at 4 to 5 km/s near the surface. By recording the time interval between the arrivals of these two types of waves at a seismograph, the distance from the seismograph to the point of origin of the waves can be determined. This distance is the radius of an imaginary sphere centered on the seismograph. The origin of the waves is located somewhere on that sphere. The imaginary spheres from three or more monitoring stations located far apart from one another intersect at one region of the Earth, and this region is where the earthquake occurred.

Consider a pulse traveling to the right on a long string as shown in Figure 16.5. Figure 16.5a represents the shape and position of the pulse at time $t = 0$. At this time, the shape of the pulse, whatever it may be, can be represented by some mathematical function that we will write as $y(x, 0) = f(x)$. This function describes the transverse position y of the element of the string located at each value of x at time $t = 0$. Because the speed of the pulse is v, the pulse has traveled to the right a distance vt at the time t (Fig. 16.5b). We assume the shape of the pulse does not change with time. Therefore, at time t, the shape of the pulse is the same as it was at time $t = 0$ as in Figure 16.5a. Consequently, an element of the string at x at this time has the same y position as an element located at $x - vt$ had at time $t = 0$:

$$y(x, t) = y(x - vt, 0)$$

In general, then, we can represent the transverse position y for all positions and times, measured in a stationary frame with the origin at O, as

$$y(x, t) = f(x - vt) \tag{16.1}$$

Similarly, if the pulse travels to the left, the transverse positions of elements of the string are described by

$$y(x, t) = f(x + vt) \tag{16.2}$$

The function y, sometimes called the **wave function,** depends on the two variables x and t. For this reason, it is often written $y(x, t)$, which is read "y as a function of x and t."

It is important to understand the meaning of y. Consider an element of the string at point P in Figure 16.5, identified by a particular value of its x coordinate. As the pulse passes through P, the y coordinate of this element increases, reaches

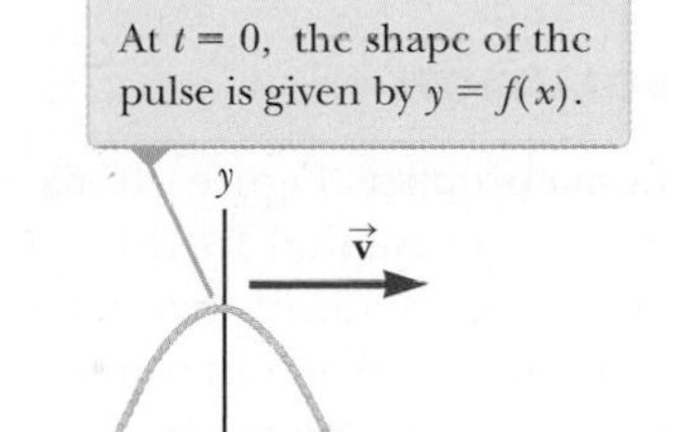

Figure 16.5 A one-dimensional pulse traveling to the right on a string with a speed v.

a maximum, and then decreases to zero. The wave function $y(x, t)$ represents the y coordinate—the transverse position—of any element located at position x at any time t. Furthermore, if t is fixed (as, for example, in the case of taking a snapshot of the pulse), the wave function $y(x)$, sometimes called the **waveform,** defines a curve representing the geometric shape of the pulse at that time.

***Quick Quiz* 16.1** **(i)** In a long line of people waiting to buy tickets, the first person leaves and a pulse of motion occurs as people step forward to fill the gap. As each person steps forward, the gap moves through the line. Is the propagation of this gap (a) transverse or (b) longitudinal? **(ii)** Consider "the wave" at a baseball game: people stand up and raise their arms as the wave arrives at their location, and the resultant pulse moves around the stadium. Is this wave (a) transverse or (b) longitudinal?

Example 16.1 A Pulse Moving to the Right

A pulse moving to the right along the x axis is represented by the wave function

$$y(x, t) = \frac{2}{(x - 3.0t)^2 + 1}$$

where x and y are measured in centimeters and t is measured in seconds. Find expressions for the wave function at $t = 0$, $t = 1.0$ s, and $t = 2.0$ s.

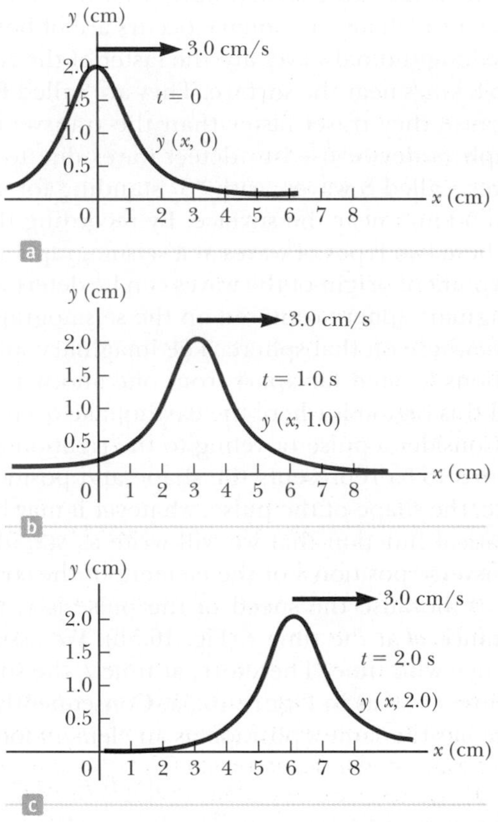

Figure 16.6 (Example 16.1) Graphs of the function $y(x, t) = 2/[(x - 3.0t)^2 + 1]$ at (a) $t = 0$, (b) $t = 1.0$ s, and (c) $t = 2.0$ s.

SOLUTION

Conceptualize Figure 16.6a shows the pulse represented by this wave function at $t = 0$. Imagine this pulse moving to the right and maintaining its shape as suggested by Figures 16.6b and 16.6c.

Categorize We categorize this example as a relatively simple analysis problem in which we interpret the mathematical representation of a pulse.

Analyze The wave function is of the form $y = f(x - vt)$. Inspection of the expression for $y(x, t)$ and comparison to Equation 16.1 reveal that the wave speed is $v = 3.0$ cm/s. Furthermore, by letting $x - 3.0t = 0$, we find that the maximum value of y is given by $A = 2.0$ cm.

Write the wave function expression at $t = 0$:

$$y(x, 0) = \frac{2}{x^2 + 1}$$

Write the wave function expression at $t = 1.0$ s:

$$y(x, 1.0) = \frac{2}{(x - 3.0)^2 + 1}$$

Write the wave function expression at $t = 2.0$ s:

$$y(x, 2.0) = \frac{2}{(x - 6.0)^2 + 1}$$

For each of these expressions, we can substitute various values of x and plot the wave function. This procedure yields the wave functions shown in the three parts of Figure 16.6.

16.1 *cont.*

Finalize These snapshots show that the pulse moves to the right without changing its shape and that it has a constant speed of 3.0 cm/s.

WHAT IF? What if the wave function were

$$y(x, t) = \frac{4}{(x + 3.0t)^2 + 1}$$

How would that change the situation?

Answer One new feature in this expression is the plus sign in the denominator rather than the minus sign. The new expression represents a pulse with a similar shape as that in Figure 16.6, but moving to the left as time progresses. Another new feature here is the numerator of 4 rather than 2. Therefore, the new expression represents a pulse with twice the height of that in Figure 16.6.

16.2 Analysis Model: Traveling Wave

In this section, we introduce an important wave function whose shape is shown in Active Figure 16.7. The wave represented by this curve is called a **sinusoidal wave** because the curve is the same as that of the function sin θ plotted against θ. A sinusoidal wave could be established on the rope in Figure 16.1 by shaking the end of the rope up and down in simple harmonic motion.

The sinusoidal wave is the simplest example of a periodic continuous wave and can be used to build more complex waves (see Section 18.8). The brown curve in Active Figure 16.7 represents a snapshot of a traveling sinusoidal wave at $t = 0$, and the blue curve represents a snapshot of the wave at some later time t. Imagine two types of motion that can occur. First, the entire waveform in Active Figure 16.7 moves to the right so that the brown curve moves toward the right and eventually reaches the position of the blue curve. This movement is the motion of the *wave*. If we focus on one element of the medium, such as the element at $x = 0$, we see that each element moves up and down along the y axis in simple harmonic motion. This movement is the motion of the *elements of the medium*. It is important to differentiate between the motion of the wave and the motion of the elements of the medium.

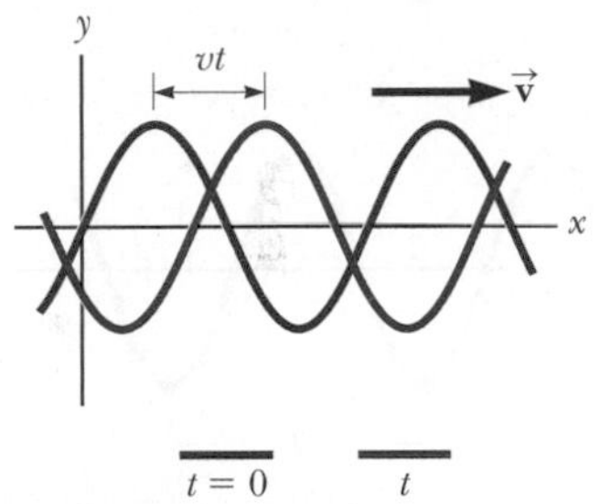

ACTIVE FIGURE 16.7

A one-dimensional sinusoidal wave traveling to the right with a speed v. The brown curve represents a snapshot of the wave at $t = 0$, and the blue curve represents a snapshot at some later time t.

In the early chapters of this book, we developed several analysis models based on three simplification models: the particle, the system, and the rigid object. With our introduction to waves, we can develop a new simplification model, the **wave,** that will allow us to explore more analysis models for solving problems. An ideal particle has zero size. We can build physical objects with nonzero size as combinations of particles. Therefore, the particle can be considered a basic building block. An ideal wave has a single frequency and is infinitely long; that is, the wave exists throughout the Universe. (A wave of finite length must necessarily have a mixture of frequencies.) When this concept is explored in Section 18.8, we will find that ideal waves can be combined to build complex waves, just as we combined particles.

In what follows, we will develop the principal features and mathematical representations of the analysis model of a **traveling wave.** This model is used in situations in which a wave moves through space without interacting with other waves or particles.

Active Figure 16.8a (page 470) shows a snapshot of a wave moving through a medium. Active Figure 16.8b shows a graph of the position of one element of the medium as a function of time. A point in Active Figure 16.8a at which the displacement of the element from its normal position is highest is called the **crest** of the wave. The lowest point is called the **trough.** The distance from one crest to the next is called the **wavelength** λ (Greek letter lambda). More generally, the wavelength is the minimum distance between any two identical points on adjacent waves as shown in Active Figure 16.8a.

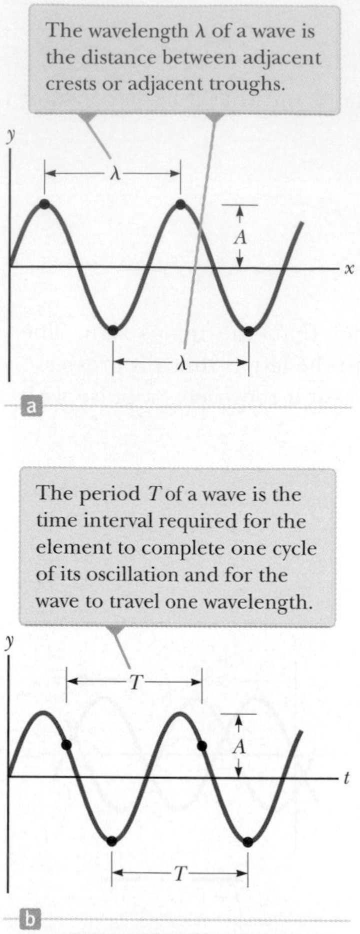

ACTIVE FIGURE 16.8

(a) A snapshot of a sinusoidal wave. (b) The position of one element of the medium as a function of time.

If you count the number of seconds between the arrivals of two adjacent crests at a given point in space, you measure the **period** T of the waves. In general, the period is the time interval required for two identical points of adjacent waves to pass by a point as shown in Active Figure 16.8b. The period of the wave is the same as the period of the simple harmonic oscillation of one element of the medium.

The same information is more often given by the inverse of the period, which is called the **frequency** f. In general, the frequency of a periodic wave is the number of crests (or troughs, or any other point on the wave) that pass a given point in a unit time interval. The frequency of a sinusoidal wave is related to the period by the expression

$$f = \frac{1}{T} \tag{16.3}$$

The frequency of the wave is the same as the frequency of the simple harmonic oscillation of one element of the medium. The most common unit for frequency, as we learned in Chapter 15, is s^{-1}, or **hertz** (Hz). The corresponding unit for T is seconds.

The maximum position of an element of the medium relative to its equilibrium position is called the **amplitude** A of the wave as indicated in Active Figure 16.8.

Waves travel with a specific speed, and this speed depends on the properties of the medium being disturbed. For instance, sound waves travel through room-temperature air with a speed of about 343 m/s (781 mi/h), whereas they travel through most solids with a speed greater than 343 m/s.

Consider the sinusoidal wave in Active Figure 16.8a, which shows the position of the wave at $t = 0$. Because the wave is sinusoidal, we expect the wave function at this instant to be expressed as $y(x, 0) = A \sin ax$, where A is the amplitude and a is a constant to be determined. At $x = 0$, we see that $y(0, 0) = A \sin a(0) = 0$, consistent with Active Figure 16.8a. The next value of x for which y is zero is $x = \lambda/2$. Therefore,

$$y\left(\frac{\lambda}{2}, 0\right) = A \sin\left(a\frac{\lambda}{2}\right) = 0$$

For this equation to be true, we must have $a\lambda/2 = \pi$, or $a = 2\pi/\lambda$. Therefore, the function describing the positions of the elements of the medium through which the sinusoidal wave is traveling can be written

$$y(x, 0) = A \sin\left(\frac{2\pi}{\lambda}x\right) \tag{16.4}$$

where the constant A represents the wave amplitude and the constant λ is the wavelength. Notice that the vertical position of an element of the medium is the same whenever x is increased by an integral multiple of λ. Based on our discussion of Equation 16.1, if the wave moves to the right with a speed v, the wave function at some later time t is

$$y(x, t) = A \sin\left[\frac{2\pi}{\lambda}(x - vt)\right] \tag{16.5}$$

If the wave were traveling to the left, the quantity $x - vt$ would be replaced by $x + vt$ as we learned when we developed Equations 16.1 and 16.2.

By definition, the wave travels through a displacement Δx equal to one wavelength λ in a time interval Δt of one period T. Therefore, the wave speed, wavelength, and period are related by the expression

$$v = \frac{\Delta x}{\Delta t} = \frac{\lambda}{T} \tag{16.6}$$

Substituting this expression for v into Equation 16.5 gives

$$y = A \sin\left[2\pi\left(\frac{x}{\lambda} - \frac{t}{T}\right)\right] \tag{16.7}$$

Pitfall Prevention 16.1

What's the Difference Between Active Figures 16.8a and 16.8b?

Notice the visual similarity between Active Figures 16.8a and 16.8b. The shapes are the same, but (a) is a graph of vertical position versus horizontal position, whereas (b) is vertical position versus time. Active Figure 16.8a is a pictorial representation of the wave *for a series of elements of the medium*; it is what you would see at an instant of time. Active Figure 16.8b is a graphical representation of the position of *one element of the medium* as a function of time. That both figures have the identical shape represents Equation 16.1: a wave is the *same* function of both x and t.

This form of the wave function shows the *periodic* nature of y. Note that we will often use y rather than $y(x, t)$ as a shorthand notation. At any given time t, y has the *same* value at the positions x, $x + \lambda$, $x + 2\lambda$, and so on. Furthermore, at any given position x, the value of y is the same at times t, $t + T$, $t + 2T$, and so on.

We can express the wave function in a convenient form by defining two other quantities, the **angular wave number** k (usually called simply the **wave number**) and the **angular frequency** ω:

$$k \equiv \frac{2\pi}{\lambda} \qquad (16.8)$$

◀ **Angular wave number**

$$\omega \equiv \frac{2\pi}{T} = 2\pi f \qquad (16.9)$$

◀ **Angular frequency**

Using these definitions, Equation 16.7 can be written in the more compact form

$$y = A \sin (kx - \omega t) \qquad (16.10)$$

◀ **Wave function for a sinusoidal wave**

Using Equations 16.3, 16.8, and 16.9, the wave speed v originally given in Equation 16.6 can be expressed in the following alternative forms:

$$v = \frac{\omega}{k} \qquad (16.11)$$

$$v = \lambda f \qquad (16.12)$$

◀ **Speed of a sinusoidal wave**

The wave function given by Equation 16.10 assumes the vertical position y of an element of the medium is zero at $x = 0$ and $t = 0$. That need not be the case. If it is not, we generally express the wave function in the form

$$y = A \sin (kx - \omega t + \phi) \qquad (16.13)$$

◀ **General expression for a sinusoidal wave**

where ϕ is the **phase constant,** just as we learned in our study of periodic motion in Chapter 15. This constant can be determined from the initial conditions. The primary equations in the mathematical representation of the traveling wave analysis model are Equations 16.3, 16.10, and 16.12.

***Quick Quiz* 16.2** A sinusoidal wave of frequency f is traveling along a stretched string. The string is brought to rest, and a second traveling wave of frequency $2f$ is established on the string. **(i)** What is the wave speed of the second wave? (a) twice that of the first wave (b) half that of the first wave (c) the same as that of the first wave (d) impossible to determine **(ii)** From the same choices, describe the wavelength of the second wave. **(iii)** From the same choices, describe the amplitude of the second wave.

Example 16.2 A Traveling Sinusoidal Wave

A sinusoidal wave traveling in the positive x direction has an amplitude of 15.0 cm, a wavelength of 40.0 cm, and a frequency of 8.00 Hz. The vertical position of an element of the medium at $t = 0$ and $x = 0$ is also 15.0 cm as shown in Figure 16.9.

(A) Find the wave number k, period T, angular frequency ω, and speed v of the wave.

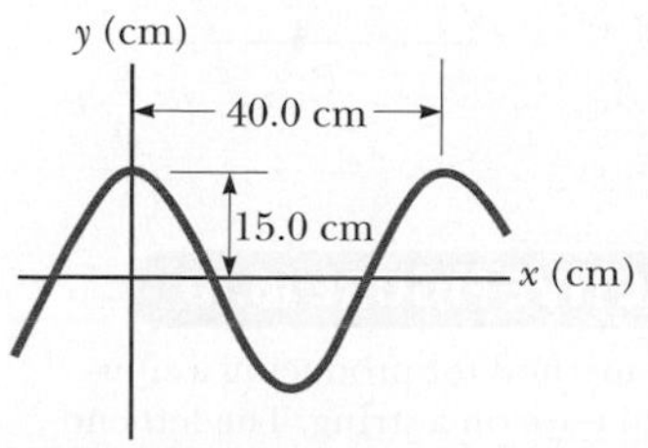

Figure 16.9 (Example 16.2) A sinusoidal wave of wavelength $\lambda = 40.0$ cm and amplitude $A = 15.0$ cm.

SOLUTION

Conceptualize Figure 16.9 shows the wave at $t = 0$. Imagine this wave moving to the right and maintaining its shape.

continued

16.2 *cont.*

Categorize We will evaluate parameters of the wave using equations generated in the preceding discussion, so we categorize this example as a substitution problem.

Evaluate the wave number from Equation 16.8:

$$k = \frac{2\pi}{\lambda} = \frac{2\pi \text{ rad}}{40.0 \text{ cm}} = \boxed{15.7 \text{ rad/m}}$$

Evaluate the period of the wave from Equation 16.3:

$$T = \frac{1}{f} = \frac{1}{8.00 \text{ s}^{-1}} = \boxed{0.125 \text{ s}}$$

Evaluate the angular frequency of the wave from Equation 16.9:

$$\omega = 2\pi f = 2\pi(8.00 \text{ s}^{-1}) = \boxed{50.3 \text{ rad/s}}$$

Evaluate the wave speed from Equation 16.12:

$$v = \lambda f = (40.0 \text{ cm})(8.00 \text{ s}^{-1}) = \boxed{3.20 \text{ m/s}}$$

(B) Determine the phase constant ϕ and write a general expression for the wave function.

SOLUTION

Substitute $A = 15.0$ cm, $y = 15.0$ cm, $x = 0$, and $t = 0$ into Equation 16.13:

$$15.0 = (15.0) \sin \phi \rightarrow \sin \phi = 1 \rightarrow \phi = \frac{\pi}{2} \text{ rad}$$

Write the wave function:

$$y = A \sin\left(kx - \omega t + \frac{\pi}{2}\right) = A \cos(kx - \omega t)$$

Substitute the values for A, k, and ω in SI units into this expression:

$$y = \boxed{0.150 \cos(15.7x - 50.3t)}$$

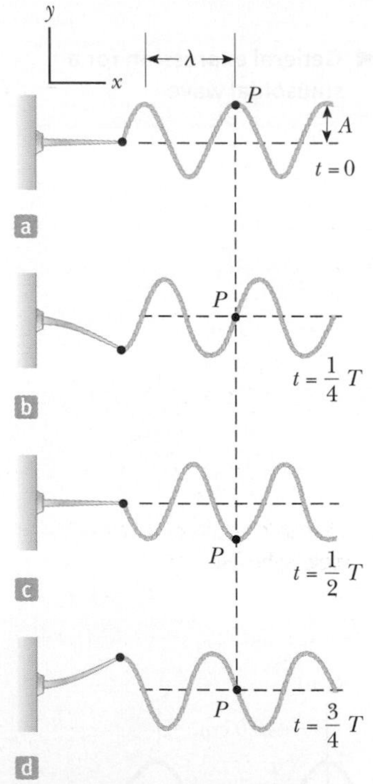

ACTIVE FIGURE 16.10

One method for producing a sinusoidal wave on a string. The left end of the string is connected to a blade that is set into oscillation. Every element of the string, such as that at point P, oscillates with simple harmonic motion in the vertical direction.

Sinusoidal Waves on Strings

In Figure 16.1, we demonstrated how to create a pulse by jerking a taut string up and down once. To create a series of such pulses—a wave—let's replace the hand with an oscillating blade vibrating in simple harmonic motion. Active Figure 16.10 represents snapshots of the wave created in this way at intervals of $T/4$. Because the end of the blade oscillates in simple harmonic motion, each element of the string, such as that at P, also oscillates vertically with simple harmonic motion. Therefore, every element of the string can be treated as a simple harmonic oscillator vibrating with a frequency equal to the frequency of oscillation of the blade.[2] Notice that while each element oscillates in the y direction, the wave travels in the x direction with a speed v. Of course, that is the definition of a transverse wave.

If we define $t = 0$ as the time for which the configuration of the string is as shown in Active Figure 16.10a, the wave function can be written as

$$y = A \sin(kx - \omega t)$$

We can use this expression to describe the motion of any element of the string. An element at point P (or any other element of the string) moves only vertically, and so its x coordinate remains constant. Therefore, the **transverse speed** v_y (not to be confused with the wave speed v) and the **transverse acceleration** a_y of elements of the string are

$$v_y = \left.\frac{dy}{dt}\right]_{x=\text{constant}} = \frac{\partial y}{\partial t} = -\omega A \cos(kx - \omega t) \quad \textbf{(16.14)}$$

$$a_y = \left.\frac{dv_y}{dt}\right]_{x=\text{constant}} = \frac{\partial v_y}{\partial t} = -\omega^2 A \sin(kx - \omega t) \quad \textbf{(16.15)}$$

[2] In this arrangement, we are assuming that a string element always oscillates in a vertical line. The tension in the string would vary if an element were allowed to move sideways. Such motion would make the analysis very complex.

These expressions incorporate partial derivatives because y depends on both x and t. In the operation $\partial y/\partial t$, for example, we take a derivative with respect to t while holding x constant. The maximum magnitudes of the transverse speed and transverse acceleration are simply the absolute values of the coefficients of the cosine and sine functions:

$$v_{y,\max} = \omega A \quad (16.16)$$

$$a_{y,\max} = \omega^2 A \quad (16.17)$$

The transverse speed and transverse acceleration of elements of the string do not reach their maximum values simultaneously. The transverse speed reaches its maximum value (ωA) when $y = 0$, whereas the magnitude of the transverse acceleration reaches its maximum value ($\omega^2 A$) when $y = \pm A$. Finally, Equations 16.16 and 16.17 are identical in mathematical form to the corresponding equations for simple harmonic motion, Equations 15.17 and 15.18.

Pitfall Prevention 16.2

Two Kinds of Speed/Velocity

Do not confuse v, the speed of the wave as it propagates along the string, with v_y, the transverse velocity of a point on the string. The speed v is constant for a uniform medium, whereas v_y varies sinusoidally.

***Quick Quiz* 16.3** The amplitude of a wave is doubled, with no other changes made to the wave. As a result of this doubling, which of the following statements is correct? **(a)** The speed of the wave changes. **(b)** The frequency of the wave changes. **(c)** The maximum transverse speed of an element of the medium changes. **(d)** Statements **(a)** through **(c)** are all true. **(e)** None of statements **(a)** through **(c)** is true.

16.3 The Speed of Waves on Strings

In this section, we determine the speed of a transverse pulse traveling on a taut string. Let's first conceptually predict the parameters that determine the speed. If a string under tension is pulled sideways and then released, the force of tension is responsible for accelerating a particular element of the string back toward its equilibrium position. According to Newton's second law, the acceleration of the element increases with increasing tension. If the element returns to equilibrium more rapidly due to this increased acceleration, we would intuitively argue that the wave speed is greater. Therefore, we expect the wave speed to increase with increasing tension.

Likewise, because it is more difficult to accelerate an element of a massive string than that of a light string, the wave speed should decrease as the mass per unit length of the string increases. If the tension in the string is T and its mass per unit length is μ (Greek letter mu), the wave speed, as we shall show, is

$$v = \sqrt{\frac{T}{\mu}} \quad (16.18)$$

◄ **Speed of a wave on a stretched string**

Let us use a mechanical analysis to derive Equation 16.18. Consider a pulse moving on a taut string to the right with a uniform speed v measured relative to a stationary frame of reference as shown in Figure 16.11a (page 474). Instead of staying in this reference frame, it is more convenient to choose a different inertial reference frame that moves along with the pulse with the same speed as the pulse so that the pulse is at rest within the frame. This change of reference frame is permitted because Newton's laws are valid in either a stationary frame or one that moves with constant velocity. In our new reference frame, shown in the magnified view of Figure 16.11b, all elements of the string move to the left: a given element of the string initially to the right of the pulse moves to the left, rises up and follows the shape of the pulse, and then continues to move to the left. Both parts of Figure 16.11 show such an element at the instant it is located at the top of the pulse.

The small element of the string of length Δs forms an approximate arc of a circle of radius R. In the moving frame of reference (which moves to the right at a speed v along with the pulse), the shaded element moves to the left with a speed v. This

Pitfall Prevention 16.3

Multiple Ts

Do not confuse the T in Equation 16.18 for the tension with the symbol T used in this chapter for the period of a wave. The context of the equation should help you identify which quantity is meant. There simply aren't enough letters in the alphabet to assign a unique letter to each variable!

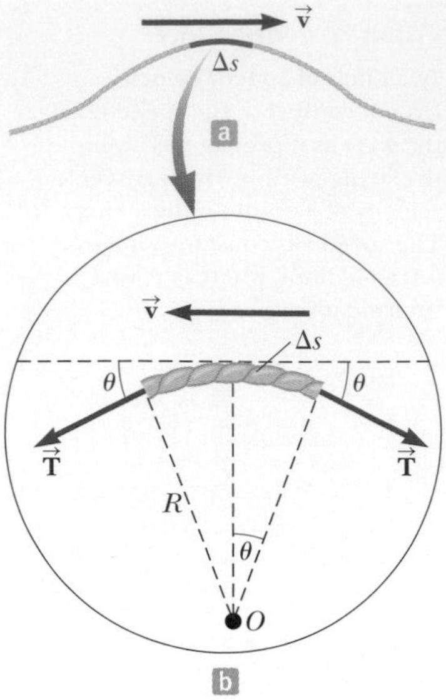

Figure 16.11 (a) In the reference frame of the Earth, a pulse moves to the right on a string with speed v. (b) In a frame of reference moving to the right with the pulse, the small element of length Δs moves to the left with speed v.

element has a centripetal acceleration equal to v^2/R, which is supplied by components of the force $\vec{T}$ whose magnitude is the tension in the string. The force $\vec{T}$ acts on both sides of the element and is tangent to the arc as shown in Figure 16.11b. The horizontal components of $\vec{T}$ cancel, and each vertical component $T \sin \theta$ acts downward. Hence, the total force on the element is $2T \sin \theta$ toward the arc's center. Because the element is small, θ is small, and we can therefore use the small-angle approximation $\sin \theta \approx \theta$. So, the total radial force is

$$F_r = 2T \sin \theta \approx 2T\theta$$

The element has a mass $m = \mu \, \Delta s$. Because the element forms part of a circle and subtends an angle 2θ at the center, $\Delta s = R(2\theta)$, and

$$m = \mu \, \Delta s = 2\mu R\theta$$

Applying Newton's second law to this element in the radial direction gives

$$F_r = ma = \frac{mv^2}{R}$$

$$2T\theta = \frac{2\mu R\theta v^2}{R} \quad \rightarrow \quad v = \sqrt{\frac{T}{\mu}}$$

This expression for v is Equation 16.18.

Notice that this derivation is based on the assumption that the pulse height is small relative to the length of the string. Using this assumption, we were able to use the approximation $\sin \theta \approx \theta$. Furthermore, the model assumes the tension T is not affected by the presence of the pulse; therefore, T is the same at all points on the string. Finally, this proof does *not* assume any particular shape for the pulse. Therefore, a pulse of *any shape* travels along the string with speed $v = \sqrt{T/\mu}$ without any change in pulse shape.

***Quick Quiz* 16.4** Suppose you create a pulse by moving the free end of a taut string up and down once with your hand beginning at $t = 0$. The string is attached at its other end to a distant wall. The pulse reaches the wall at time t. Which of the following actions, taken by itself, decreases the time interval required for the pulse to reach the wall? More than one choice may be correct. **(a)** moving your hand more quickly, but still only up and down once by the same amount **(b)** moving your hand more slowly, but still only up and down once by the same amount **(c)** moving your hand a greater distance up and down in the same amount of time **(d)** moving your hand a lesser distance up and down in the same amount of time **(e)** using a heavier string of the same length and under the same tension **(f)** using a lighter string of the same length and under the same tension **(g)** using a string of the same linear mass density but under decreased tension **(h)** using a string of the same linear mass density but under increased tension

Example 16.3 The Speed of a Pulse on a Cord

A uniform string has a mass of 0.300 kg and a length of 6.00 m (Fig. 16.12). The string passes over a pulley and supports a 2.00-kg object. Find the speed of a pulse traveling along this string.

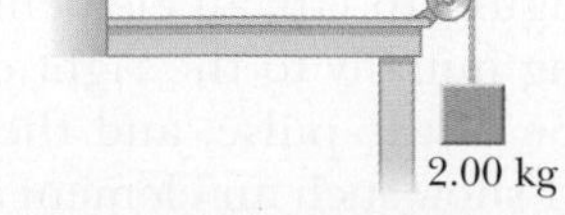

Figure 16.12 (Example 16.3) The tension T in the cord is maintained by the suspended object. The speed of any wave traveling along the cord is given by $v = \sqrt{T/\mu}$.

SOLUTION

Conceptualize In Figure 16.12, the hanging block establishes a tension in the horizontal string. This tension determines the speed with which waves move on the string.

16.3 *cont.*

Categorize To find the tension in the string, we model the hanging block as a particle in equilibrium. Then we use the tension to evaluate the wave speed on the string using Equation 16.18.

Analyze Apply the particle in equilibrium model to the block:

$$\sum F_y = T - m_{\text{block}}g = 0$$

Solve for the tension in the string:

$$T = m_{\text{block}}g$$

Use Equation 16.18 to find the wave speed, using $\mu = m_{\text{string}}/\ell$ for the linear mass density of the string:

$$v = \sqrt{\frac{T}{\mu}} = \sqrt{\frac{m_{\text{block}}g\,\ell}{m_{\text{string}}}}$$

Evaluate the wave speed:

$$v = \sqrt{\frac{(2.00\text{ kg})(9.80\text{ m/s}^2)(6.00\text{ m})}{0.300\text{ kg}}} = \boxed{19.8\text{ m/s}}$$

Finalize The calculation of the tension neglects the small mass of the string. Strictly speaking, the string can never be exactly straight; therefore, the tension is not uniform.

WHAT IF? What if the block were swinging back and forth with respect to the vertical like a pendulum? How would that affect the wave speed on the string?

Answer The swinging block is categorized as a particle under a net force. The magnitude of one of the forces on the block is the tension in the string, which determines the wave speed. As the block swings, the tension changes, so the wave speed changes.

When the block is at the bottom of the swing, the string is vertical and the tension is larger than the weight of the block because the net force must be upward to provide the centripetal acceleration of the block. Therefore, the wave speed must be greater than 19.8 m/s.

When the block is at its highest point at the end of a swing, it is momentarily at rest, so there is no centripetal acceleration at that instant. The block is a particle in equilibrium in the radial direction. The tension is balanced by a component of the gravitational force on the block. Therefore, the tension is smaller than the weight and the wave speed is less than 19.8 m/s. With what frequency does the speed of the wave vary? Is it the same frequency as the pendulum?

Example 16.4 Rescuing the Hiker

An 80.0-kg hiker is trapped on a mountain ledge following a storm. A helicopter rescues the hiker by hovering above him and lowering a cable to him. The mass of the cable is 8.00 kg, and its length is 15.0 m. A sling of mass 70.0 kg is attached to the end of the cable. The hiker attaches himself to the sling, and the helicopter then accelerates upward. Terrified by hanging from the cable in midair, the hiker tries to signal the pilot by sending transverse pulses up the cable. A pulse takes 0.250 s to travel the length of the cable. What is the acceleration of the helicopter? Assume the tension in the cable is uniform.

SOLUTION

Conceptualize Imagine the effect of the acceleration of the helicopter on the cable. The greater the upward acceleration, the larger the tension in the cable. In turn, the larger the tension, the higher the speed of pulses on the cable.

Categorize This problem is a combination of one involving the speed of pulses on a string and one in which the hiker and sling are modeled as a particle under a net force.

Analyze Use the time interval for the pulse to travel from the hiker to the helicopter to find the speed of the pulses on the cable:

$$v = \frac{\Delta x}{\Delta t} = \frac{15.0\text{ m}}{0.250\text{ s}} = 60.0\text{ m/s}$$

Solve Equation 16.18 for the tension in the cable:

$$v = \sqrt{\frac{T}{\mu}} \rightarrow T = \mu v^2$$

continued

16.4 *cont.*

Model the hiker and sling as a particle under a net force, noting that the acceleration of this particle of mass m is the same as the acceleration of the helicopter:

$$\sum F = ma \rightarrow T - mg = ma$$

Solve for the acceleration:

$$a = \frac{T}{m} - g = \frac{\mu v^2}{m} - g = \frac{m_{\text{cable}}\, v^2}{\ell_{\text{cable}} m} - g$$

Substitute numerical values:

$$a = \frac{(8.00 \text{ kg})(60.0 \text{ m/s})^2}{(15.0 \text{ m})(150.0 \text{ kg})} - 9.80 \text{ m/s}^2 = \boxed{3.00 \text{ m/s}^2}$$

Finalize A real cable has stiffness in addition to tension. Stiffness tends to return a wire to its original straight-line shape even when it is not under tension. For example, a piano wire straightens if released from a curved shape; package-wrapping string does not.

Stiffness represents a restoring force in addition to tension and increases the wave speed. Consequently, for a real cable, the speed of 60.0 m/s that we determined is most likely associated with a smaller acceleration of the helicopter.

16.4 Reflection and Transmission

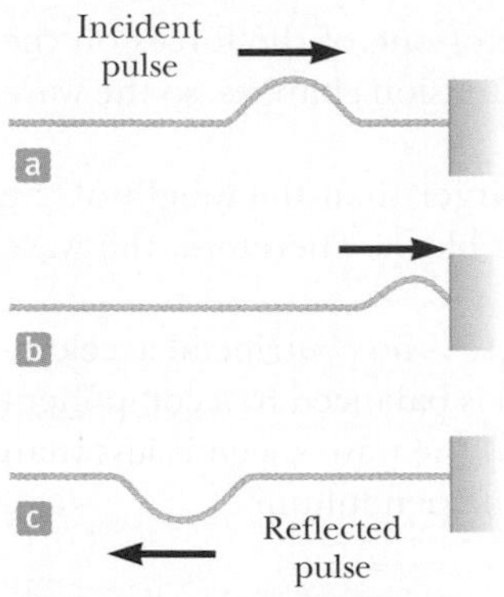

ACTIVE FIGURE 16.13

The reflection of a traveling pulse at the fixed end of a stretched string. The reflected pulse is inverted, but its shape is otherwise unchanged.

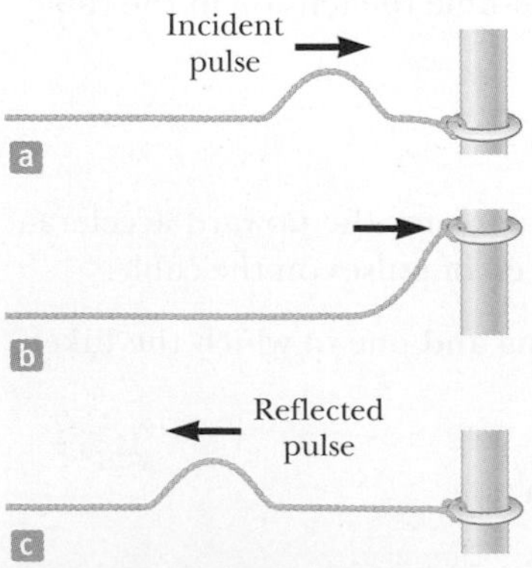

ACTIVE FIGURE 16.14

The reflection of a traveling pulse at the free end of a stretched string. The reflected pulse is not inverted.

The traveling wave model describes waves traveling through a uniform medium without interacting with anything along the way. We now consider how a traveling wave is affected when it encounters a change in the medium. For example, consider a pulse traveling on a string that is rigidly attached to a support at one end as in Active Figure 16.13. When the pulse reaches the support, a severe change in the medium occurs: the string ends. As a result, the pulse undergoes **reflection;** that is, the pulse moves back along the string in the opposite direction.

Notice that the reflected pulse is *inverted.* This inversion can be explained as follows. When the pulse reaches the fixed end of the string, the string produces an upward force on the support. By Newton's third law, the support must exert an equal-magnitude and oppositely directed (downward) reaction force on the string. This downward force causes the pulse to invert upon reflection.

Now consider another case. This time, the pulse arrives at the end of a string that is free to move vertically as in Active Figure 16.14. The tension at the free end is maintained because the string is tied to a ring of negligible mass that is free to slide vertically on a smooth post without friction. Again, the pulse is reflected, but this time it is not inverted. When it reaches the post, the pulse exerts a force on the free end of the string, causing the ring to accelerate upward. The ring rises as high as the incoming pulse, and then the downward component of the tension force pulls the ring back down. This movement of the ring produces a reflected pulse that is not inverted and that has the same amplitude as the incoming pulse.

Finally, consider a situation in which the boundary is intermediate between these two extremes. In this case, part of the energy in the incident pulse is reflected and part undergoes **transmission;** that is, some of the energy passes through the boundary. For instance, suppose a light string is attached to a heavier string as in Active Figure 16.15. When a pulse traveling on the light string reaches the boundary between the two strings, part of the pulse is reflected and inverted and part is transmitted to the heavier string. The reflected pulse is inverted for the same reasons described earlier in the case of the string rigidly attached to a support.

The reflected pulse has a smaller amplitude than the incident pulse. In Section 16.5, we show that the energy carried by a wave is related to its amplitude. According to the principle of the conservation of energy, when the pulse breaks up into a reflected pulse and a transmitted pulse at the boundary, the sum of the energies of these two pulses must equal the energy of the incident pulse. Because the reflected

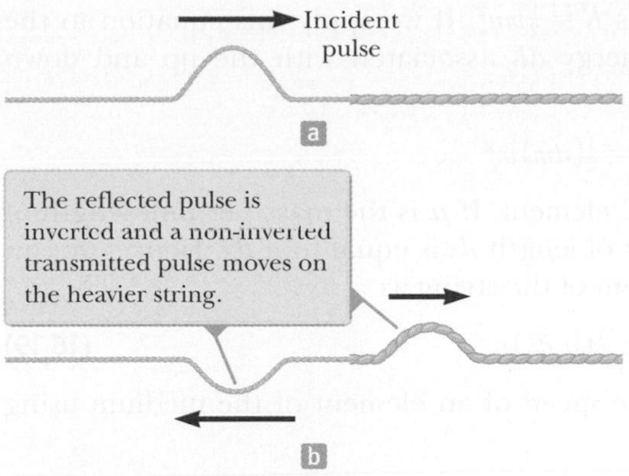

ACTIVE FIGURE 16.15
(a) A pulse traveling to the right on a light string approaches the junction with a heavier string. (b) The situation after the pulse reaches the junction.

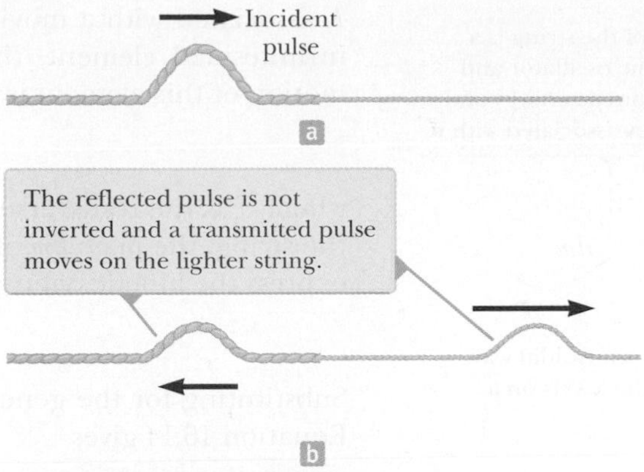

ACTIVE FIGURE 16.16
(a) A pulse traveling to the right on a heavy string approaches the junction with a lighter string. (b) The situation after the pulse reaches the junction.

pulse contains only part of the energy of the incident pulse, its amplitude must be smaller.

When a pulse traveling on a heavy string strikes the boundary between the heavy string and a lighter one as in Active Figure 16.16, again part is reflected and part is transmitted. In this case, the reflected pulse is not inverted.

In either case, the relative heights of the reflected and transmitted pulses depend on the relative densities of the two strings. If the strings are identical, there is no discontinuity at the boundary and no reflection takes place.

According to Equation 16.18, the speed of a wave on a string increases as the mass per unit length of the string decreases. In other words, a wave travels more slowly on a heavy string than on a light string if both are under the same tension. The following general rules apply to reflected waves: When a wave or pulse travels from medium A to medium B and $v_A > v_B$ (that is, when B is denser than A), it is inverted upon reflection. When a wave or pulse travels from medium A to medium B and $v_A < v_B$ (that is, when A is denser than B), it is not inverted upon reflection.

16.5 Rate of Energy Transfer by Sinusoidal Waves on Strings

Waves transport energy through a medium as they propagate. For example, suppose an object is hanging on a stretched string and a pulse is sent down the string as in Figure 16.17a. When the pulse meets the suspended object, the object is momentarily displaced upward as in Figure 16.17b. In the process, energy is transferred to the object and appears as an increase in the gravitational potential energy of the object–Earth system. This section examines the rate at which energy is transported along a string. We shall assume a one-dimensional sinusoidal wave in the calculation of the energy transferred.

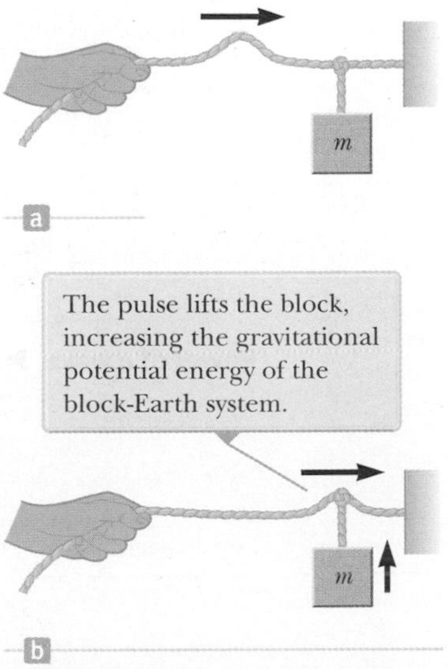

Figure 16.17 (a) A pulse travels to the right on a stretched string, carrying energy with it. (b) The energy of the pulse arrives at the hanging block.

Consider a sinusoidal wave traveling on a string (Fig. 16.18 on page 478). The source of the energy is some external agent at the left end of the string. We can consider the string to be a nonisolated system. As the external agent performs work on the end of the string, moving it up and down, energy enters the system of the string and propagates along its length. Let's focus our attention on an infinitesimal element of the string of length dx and mass dm. Each such element moves vertically with simple harmonic motion. Therefore, we can model each element of the string as a simple harmonic oscillator, with the oscillation in the y direction. All elements have the same angular frequency ω and the same amplitude A. The kinetic energy

Each element of the string is a simple harmonic oscillator and therefore has kinetic energy and potential energy associated with it.

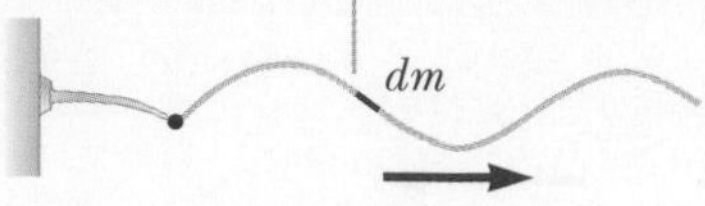

Figure 16.18 A sinusoidal wave traveling along the x axis on a stretched string.

K associated with a moving particle is $K = \frac{1}{2}mv^2$. If we apply this equation to the infinitesimal element, the kinetic energy dK associated with the up and down motion of this element is

$$dK = \tfrac{1}{2}(dm)v_y^2$$

where v_y is the transverse speed of the element. If μ is the mass per unit length of the string, the mass dm of the element of length dx is equal to $\mu\ dx$. Hence, we can express the kinetic energy of an element of the string as

$$dK = \tfrac{1}{2}(\mu\ dx)v_y^2 \tag{16.19}$$

Substituting for the general transverse speed of an element of the medium using Equation 16.14 gives

$$dK = \tfrac{1}{2}\mu[-\omega A\cos(kx - \omega t)]^2\ dx = \tfrac{1}{2}\mu\omega^2A^2\cos^2(kx - \omega t)\ dx$$

If we take a snapshot of the wave at time $t = 0$, the kinetic energy of a given element is

$$dK = \tfrac{1}{2}\mu\omega^2\ A^2\cos^2 kx\ dx$$

Integrating this expression over all the string elements in a wavelength of the wave gives the total kinetic energy K_λ in one wavelength:

$$K_\lambda = \int dK = \int_0^\lambda \tfrac{1}{2}\mu\omega^2A^2\cos^2 kx\ dx = \tfrac{1}{2}\mu\omega^2A^2\int_0^\lambda \cos^2 kx\ dx$$

$$= \tfrac{1}{2}\mu\omega^2A^2\left[\tfrac{1}{2}x + \frac{1}{4k}\sin 2kx\right]_0^\lambda = \tfrac{1}{2}\mu\omega^2A^2\left[\tfrac{1}{2}\lambda\right] = \tfrac{1}{4}\mu\omega^2A^2\lambda$$

In addition to kinetic energy, there is potential energy associated with each element of the string due to its displacement from the equilibrium position and the restoring forces from neighboring elements. A similar analysis to that above for the total potential energy U_λ in one wavelength gives exactly the same result:

$$U_\lambda = \tfrac{1}{4}\mu\omega^2A^2\lambda$$

The total energy in one wavelength of the wave is the sum of the potential and kinetic energies:

$$E_\lambda = U_\lambda + K_\lambda = \tfrac{1}{2}\mu\omega^2A^2\lambda \tag{16.20}$$

As the wave moves along the string, this amount of energy passes by a given point on the string during a time interval of one period of the oscillation. Therefore, the power P, or rate of energy transfer T_{MW} associated with the mechanical wave, is

$$P = \frac{T_{\text{MW}}}{\Delta t} = \frac{E_\lambda}{T} = \frac{\frac{1}{2}\mu\omega^2A^2\lambda}{T} = \tfrac{1}{2}\mu\omega^2A^2\left(\frac{\lambda}{T}\right)$$

Power of a wave ▶

$$P = \tfrac{1}{2}\mu\omega^2A^2v \tag{16.21}$$

Equation 16.21 shows that the rate of energy transfer by a sinusoidal wave on a string is proportional to (a) the square of the frequency, (b) the square of the amplitude, and (c) the wave speed. In fact, the rate of energy transfer in *any* sinusoidal wave is proportional to the square of the angular frequency and to the square of the amplitude.

***Quick Quiz* 16.5** Which of the following, taken by itself, would be most effective in increasing the rate at which energy is transferred by a wave traveling along a string? **(a)** reducing the linear mass density of the string by one half **(b)** doubling the wavelength of the wave **(c)** doubling the tension in the string **(d)** doubling the amplitude of the wave

Example 16.5 Power Supplied to a Vibrating String

A taut string for which $\mu = 5.00 \times 10^{-2}$ kg/m is under a tension of 80.0 N. How much power must be supplied to the string to generate sinusoidal waves at a frequency of 60.0 Hz and an amplitude of 6.00 cm?

SOLUTION

Conceptualize Consider Active Figure 16.10 again and notice that the vibrating blade supplies energy to the string at a certain rate. This energy then propagates to the right along the string.

Categorize We evaluate quantities from equations developed in the chapter, so we categorize this example as a substitution problem.

Use Equation 16.21 to evaluate the power: $P = \frac{1}{2}\mu\omega^2A^2v$

Use Equations 16.9 and 16.18 to substitute for ω and v: $P = \frac{1}{2}\mu(2\pi f)^2A^2\left(\sqrt{\frac{T}{\mu}}\right) = 2\pi^2 f^2A^2\sqrt{\mu T}$

Substitute numerical values: $P = 2\pi^2(60.0 \text{ Hz})^2(0.060\,0 \text{ m})^2\sqrt{(0.050\,0 \text{ kg/m})(80.0 \text{ N})} = \boxed{512 \text{ W}}$

WHAT IF? What if the string is to transfer energy at a rate of 1 000 W? What must be the required amplitude if all other parameters remain the same?

Answer Let us set up a ratio of the new and old power, reflecting only a change in the amplitude:

$$\frac{P_{\text{new}}}{P_{\text{old}}} = \frac{\frac{1}{2}\mu\omega^2A_{\text{new}}^2\,v}{\frac{1}{2}\mu\omega^2A_{\text{old}}^2\,v} = \frac{A_{\text{new}}^2}{A_{\text{old}}^2}$$

Solving for the new amplitude gives

$$A_{\text{new}} = A_{\text{old}}\sqrt{\frac{P_{\text{new}}}{P_{\text{old}}}} = (6.00 \text{ cm})\sqrt{\frac{1\,000 \text{ W}}{512 \text{ W}}} = 8.39 \text{ cm}$$

16.6 The Linear Wave Equation

In Section 16.1, we introduced the concept of the wave function to represent waves traveling on a string. All wave functions $y(x, t)$ represent solutions of an equation called the *linear wave equation.* This equation gives a complete description of the wave motion, and from it one can derive an expression for the wave speed. Furthermore, the linear wave equation is basic to many forms of wave motion. In this section, we derive this equation as applied to waves on strings.

Suppose a traveling wave is propagating along a string that is under a tension T. Let's consider one small string element of length Δx (Fig. 16.19). The ends of the element make small angles θ_A and θ_B with the x axis. The net force acting on the element in the vertical direction is

$$\sum F_y = T\sin\theta_B - T\sin\theta_A = T(\sin\theta_B - \sin\theta_A)$$

Because the angles are small, we can use the approximation $\sin\theta \approx \tan\theta$ to express the net force as

$$\sum F_y \approx T(\tan\theta_B - \tan\theta_A) \quad \textbf{(16.22)}$$

Imagine undergoing an infinitesimal displacement outward from the right end of the rope element in Figure 16.19 along the blue line representing the force $\vec{\mathbf{T}}$. This displacement has infinitesimal x and y components and can be represented by the vector $dx\,\hat{\mathbf{i}} + dy\,\hat{\mathbf{j}}$. The tangent of the angle with respect to the x axis for this displacement is dy/dx. Because we evaluate this tangent at a particular instant of time,

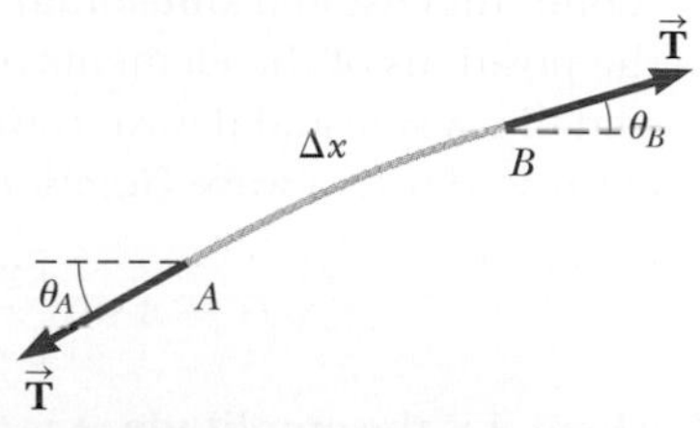

Figure 16.19 An element of a string under tension T.

we must express it in partial form as $\partial y/\partial x$. Substituting for the tangents in Equation 16.22 gives

$$\sum F_y \approx T\left[\left(\frac{\partial y}{\partial x}\right)_B - \left(\frac{\partial y}{\partial x}\right)_A\right] \quad \textbf{(16.23)}$$

Now let's apply Newton's second law to the element, with the mass of the element given by $m = \mu\,\Delta x$:

$$\sum F_y = ma_y = \mu\,\Delta x\left(\frac{\partial^2 y}{\partial t^2}\right) \quad \textbf{(16.24)}$$

Combining Equation 16.23 with Equation 16.24 gives

$$\mu\,\Delta x\left(\frac{\partial^2 y}{\partial t^2}\right) = T\left[\left(\frac{\partial y}{\partial x}\right)_B - \left(\frac{\partial y}{\partial x}\right)_A\right]$$

$$\frac{\mu}{T}\frac{\partial^2 y}{\partial t^2} = \frac{(\partial y/\partial x)_B - (\partial y/dx)_A}{\Delta x} \quad \textbf{(16.25)}$$

The right side of Equation 16.25 can be expressed in a different form if we note that the partial derivative of any function is defined as

$$\frac{\partial f}{\partial x} \equiv \lim_{\Delta x \to 0}\frac{f(x + \Delta x) - f(x)}{\Delta x}$$

Associating $f(x + \Delta x)$ with $(\partial y/\partial x)_B$ and $f(x)$ with $(\partial y/\partial x)_A$, we see that, in the limit $\Delta x \to 0$, Equation 16.25 becomes

Linear wave equation for a string ▶

$$\frac{\mu}{T}\frac{\partial^2 y}{\partial t^2} = \frac{\partial^2 y}{\partial x^2} \quad \textbf{(16.26)}$$

This expression is the linear wave equation as it applies to waves on a string.

The linear wave equation (Eq. 16.26) is often written in the form

Linear wave equation in general ▶

$$\frac{\partial^2 y}{\partial x^2} = \frac{1}{v^2}\frac{\partial^2 y}{\partial t^2} \quad \textbf{(16.27)}$$

Equation 16.27 applies in general to various types of traveling waves. For waves on strings, y represents the vertical position of elements of the string. For sound waves propagating through a gas, y corresponds to longitudinal position of elements of the gas from equilibrium or variations in either the pressure or the density of the gas. In the case of electromagnetic waves, y corresponds to electric or magnetic field components.

We have shown that the sinusoidal wave function (Eq. 16.10) is one solution of the linear wave equation (Eq. 16.27). Although we do not prove it here, the linear wave equation is satisfied by *any* wave function having the form $y = f(x \pm vt)$. Furthermore, we have seen that the linear wave equation is a direct consequence of Newton's second law applied to any element of a string carrying a traveling wave.

Summary

Definitions

A one-dimensional **sinusoidal wave** is one for which the positions of the elements of the medium vary sinusoidally. A sinusoidal wave traveling to the right can be expressed with a **wave function**

$$y(x, t) = A\sin\left[\frac{2\pi}{\lambda}(x - vt)\right] \quad \textbf{(16.5)}$$

where A is the **amplitude**, λ is the **wavelength**, and v is the **wave speed**.

The **angular wave number** k and **angular frequency** ω of a wave are defined as follows:

$$k \equiv \frac{2\pi}{\lambda} \quad \textbf{(16.8)}$$

$$\omega \equiv \frac{2\pi}{T} = 2\pi f \quad \textbf{(16.9)}$$

where T is the **period** of the wave and f is its **frequency**.

A **transverse wave** is one in which the elements of the medium move in a direction *perpendicular* to the direction of propagation.

A **longitudinal wave** is one in which the elements of the medium move in a direction *parallel* to the direction of propagation.

Concepts and Principles

Any one-dimensional wave traveling with a speed v in the x direction can be represented by a wave function of the form

$$y(x, t) = f(x \pm vt) \quad \text{(16.1, 16.2)}$$

where the positive sign applies to a wave traveling in the negative x direction and the negative sign applies to a wave traveling in the positive x direction. The shape of the wave at any instant in time (a snapshot of the wave) is obtained by holding t constant.

The speed of a wave traveling on a taut string of mass per unit length μ and tension T is

$$v = \sqrt{\frac{T}{\mu}} \quad \text{(16.18)}$$

A wave is totally or partially reflected when it reaches the end of the medium in which it propagates or when it reaches a boundary where its speed changes discontinuously. If a wave traveling on a string meets a fixed end, the wave is reflected and inverted. If the wave reaches a free end, it is reflected but not inverted.

The **power** transmitted by a sinusoidal wave on a stretched string is

$$P = \tfrac{1}{2}\mu\omega^2 A^2 v \quad \text{(16.21)}$$

Wave functions are solutions to a differential equation called the **linear wave equation:**

$$\frac{\partial^2 y}{\partial x^2} = \frac{1}{v^2}\frac{\partial^2 y}{\partial t^2} \quad \text{(16.27)}$$

Analysis Model for Problem Solving

Traveling Wave. The wave speed of a sinusoidal wave is

$$v = \frac{\lambda}{T} = \lambda f \quad \text{(16.6, 16.12)}$$

A sinusoidal wave can be expressed as

$$y = A \sin(kx - \omega t) \quad \text{(16.10)}$$

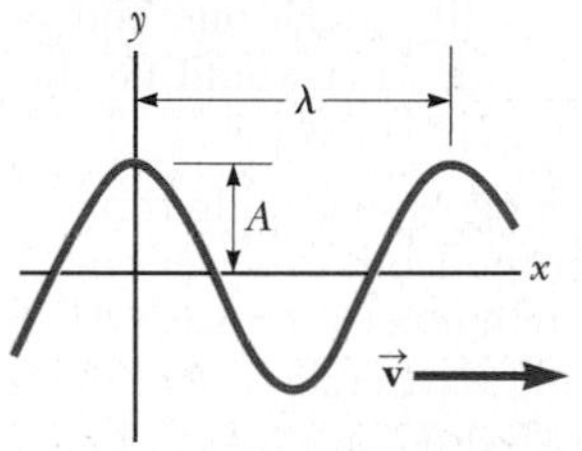

Objective Questions

☐ denotes answer available in Student Solutions Manual/Study Guide

1. The distance between two successive peaks of a sinusoidal wave traveling along a string is 2 m. If the frequency of this wave is 4 Hz, what is the speed of the wave? (a) 4 m/s (b) 1 m/s (c) 8 m/s (d) 2 m/s (e) impossible to answer from the information given

2. Which of the following statements is not necessarily true regarding mechanical waves? (a) They are formed by some source of disturbance. (b) They are sinusoidal in nature. (c) They carry energy. (d) They require a medium through which to propagate. (e) The wave speed depends on the properties of the medium in which they travel.

3. Rank the waves represented by the following functions from the largest to the smallest according to **(i)** their amplitudes, **(ii)** their wavelengths, **(iii)** their frequencies, **(iv)** their periods, and **(v)** their speeds. If the values of a quantity are equal for two waves, show them as having

equal rank. For all functions, x and y are in meters and t is in seconds. (a) $y = 4 \sin (3x - 15t)$ (b) $y = 6 \cos (3x + 15t - 2)$ (c) $y = 8 \sin (2x + 15t)$ (d) $y = 8 \cos (4x + 20t)$ (e) $y = 7 \sin (6x - 24t)$

4. By what factor would you have to multiply the tension in a stretched string so as to double the wave speed? Assume the string does not stretch. (a) a factor of 8 (b) a factor of 4 (c) a factor of 2 (d) a factor of 0.5 (e) You could not change the speed by a predictable factor by changing the tension.

5. When all the strings on a guitar (Fig. OQ16.5) are stretched to the same tension, will the speed of a wave along the most massive bass string be (a) faster, (b) slower, or (c) the same as the speed of a wave on the lighter strings? Alternatively, (d) is the speed on the bass string not necessarily any of these answers?

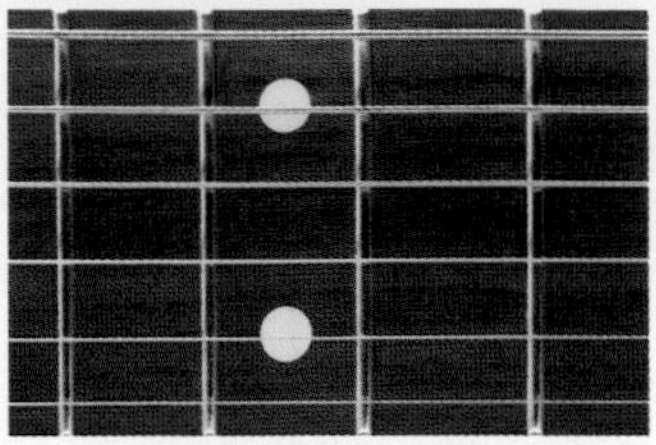
Aaron Graubart/Getty Images

Figure OQ16.5

6. If you stretch a rubber hose and pluck it, you can observe a pulse traveling up and down the hose. **(i)** What happens to the speed of the pulse if you stretch the hose more tightly? (a) It increases. (b) It decreases. (c) It is constant. (d) It changes unpredictably. **(ii)** What happens to the speed if you fill the hose with water? Choose from the same possibilities.

7. (a) Can a wave on a string move with a wave speed that is greater than the maximum transverse speed $v_{y,\max}$ of an element of the string? (b) Can the wave speed be much greater than the maximum element speed? (c) Can the wave speed be equal to the maximum element speed? (d) Can the wave speed be less than $v_{y,\max}$?

8. A source vibrating at constant frequency generates a sinusoidal wave on a string under constant tension. If the power delivered to the string is doubled, by what factor does the amplitude change? (a) a factor of 4 (b) a factor of 2 (c) a factor of $\sqrt{2}$ (d) a factor of 0.707 (e) cannot be predicted

9. If one end of a heavy rope is attached to one end of a lightweight rope, a wave can move from the heavy rope into the lighter one. **(i)** What happens to the speed of the wave? (a) It increases. (b) It decreases. (c) It is constant. (d) It changes unpredictably. **(ii)** What happens to the frequency? Choose from the same possibilities. **(iii)** What happens to the wavelength? Choose from the same possibilities.

Conceptual Questions

☐ denotes answer available in Student Solutions Manual/Study Guide

1. Why is a pulse on a string considered to be transverse?

2. (a) How would you create a longitudinal wave in a stretched spring? (b) Would it be possible to create a transverse wave in a spring?

3. When a pulse travels on a taut string, does it always invert upon reflection? Explain.

4. Does the vertical speed of an element of a horizontal, taut string, through which a wave is traveling, depend on the wave speed? Explain.

5. If you steadily shake one end of a taut rope three times each second, what would be the period of the sinusoidal wave set up in the rope?

6. (a) If a long rope is hung from a ceiling and waves are sent up the rope from its lower end, why does the speed of the waves change as they ascend? (b) Does the speed of the ascending waves increase or decrease? Explain.

7. Why is a solid substance able to transport both longitudinal waves and transverse waves, but a homogeneous fluid is able to transport only longitudinal waves?

8. In mechanics, massless strings are often assumed. Why is that not a good assumption when discussing waves on strings?

9. In an earthquake, both S (transverse) and P (longitudinal) waves propagate from the focus of the earthquake. The focus is in the ground radially below the epicenter on the surface (Fig. CQ16.9). Assume the waves move in straight lines through uniform material. The S waves travel through the Earth more slowly than the P waves (at about 5 km/s versus 8 km/s). By detecting the time of arrival of the waves at a seismograph, (a) how can one determine the distance to the focus of the earthquake? (b) How many detection stations are necessary to locate the focus unambiguously?

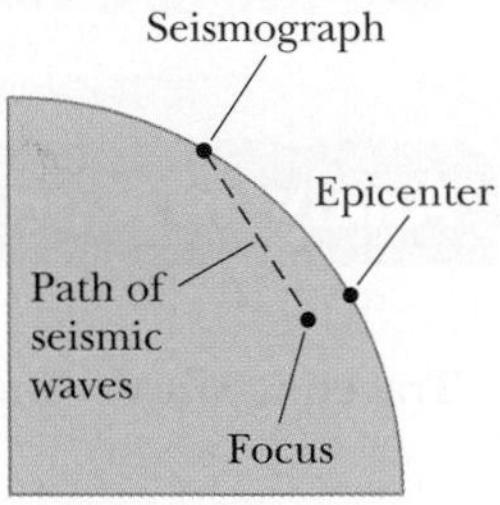

Figure CQ16.9

Problems

WebAssign The problems found in this chapter may be assigned online in Enhanced WebAssign

1. denotes straightforward problem; **2.** denotes intermediate problem; **3.** denotes challenging problem

1. full solution available in the Student Solutions Manual/Study Guide

1. denotes problems most often assigned in Enhanced WebAssign; these provide students with targeted feedback and either a Master It tutorial or a Watch It solution video.

QC denotes asking for quantitative and conceptual reasoning

S denotes symbolic reasoning problem

M denotes Master It tutorial available in Enhanced WebAssign

GP denotes guided problem

shaded denotes "paired problems" that develop reasoning with symbols and numerical values

Section 16.1 Propagation of a Disturbance

1. At $t = 0$, a transverse pulse in a wire is described by the function

$$y = \frac{6.00}{x^2 + 3.00}$$

where x and y are in meters. If the pulse is traveling in the positive x direction with a speed of 4.50 m/s, write the function $y(x, t)$ that describes this pulse.

2. **Q|C** Ocean waves with a crest-to-crest distance of 10.0 m can be described by the wave function

$$y(x, t) = 0.800 \sin [0.628(x - vt)]$$

where x and y are in meters, t is in seconds, and $v =$ 1.20 m/s. (a) Sketch $y(x, t)$ at $t = 0$. (b) Sketch $y(x, t)$ at $t =$ 2.00 s. (c) Compare the graph in part (b) with that for part (a) and explain similarities and differences. (d) How has the wave moved between graph (a) and graph (b)?

3. A seismographic station receives S and P waves from an earthquake, separated in time by 17.3 s. Assume the waves have traveled over the same path at speeds of 4.50 km/s and 7.80 km/s. Find the distance from the seismograph to the focus of the quake.

4. Two points A and B on the surface of the Earth are at the same longitude and 60.0° apart in latitude as shown in Figure P16.4. Suppose an earthquake at point A creates a P wave that reaches point B by traveling straight through the body of the Earth at a constant speed of 7.80 km/s. The earthquake also radiates a Rayleigh wave that travels at 4.50 km/s. In addition to P and S waves, Rayleigh waves are a third type of seismic wave that travels along the *surface* of the Earth rather than through the *bulk* of the Earth. (a) Which of these two seismic waves arrives at B first? (b) What is the time difference between the arrivals of these two waves at B?

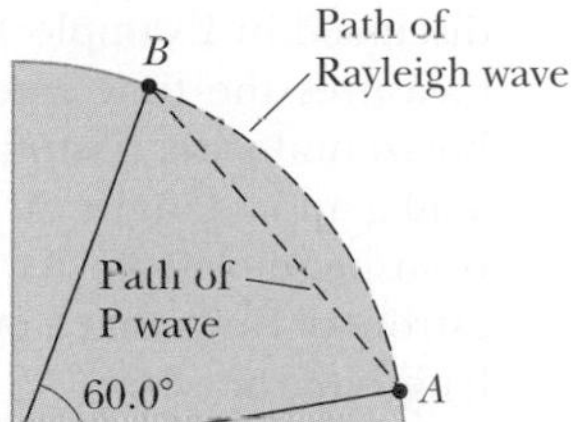

Figure P16.4

Section 16.2 Analysis Model: Traveling Wave

5. **M** The wave function for a traveling wave on a taut string is (in SI units)

$$y(x,t) = 0.350 \sin \left(10\pi t - 3\pi x + \frac{\pi}{4}\right)$$

(a) What are the speed and direction of travel of the wave? (b) What is the vertical position of an element of the string at $t = 0$, $x = 0.100$ m? What are (c) the wavelength and (d) the frequency of the wave? (e) What is the maximum transverse speed of an element of the string?

6. **Q|C** A certain uniform string is held under constant tension. (a) Draw a side-view snapshot of a sinusoidal wave on a string as shown in diagrams in the text. (b) Immediately below diagram (a), draw the same wave at a moment later by one-quarter of the period of the wave. (c) Then, draw a wave with an amplitude 1.5 times larger than the wave in diagram (a). (d) Next, draw a wave differing from the one in your diagram (a) just by having a wavelength 1.5 times larger. (e) Finally, draw a wave differing from that in diagram (a) just by having a frequency 1.5 times larger.

7. A sinusoidal wave is traveling along a rope. The oscillator that generates the wave completes 40.0 vibrations in 30.0 s. A given crest of the wave travels 425 cm along the rope in 10.0 s. What is the wavelength of the wave?

8. For a certain transverse wave, the distance between two successive crests is 1.20 m, and eight crests pass a given point along the direction of travel every 12.0 s. Calculate the wave speed.

9. A wave is described by $y = 0.020\,0 \sin (kx - \omega t)$, where $k =$ 2.11 rad/m, $\omega = 3.62$ rad/s, x and y are in meters, and t is in seconds. Determine (a) the amplitude, (b) the wavelength, (c) the frequency, and (d) the speed of the wave.

10. When a particular wire is vibrating with a frequency of 4.00 Hz, a transverse wave of wavelength 60.0 cm is produced. Determine the speed of waves along the wire.

11. The string shown in Figure P16.11 is driven at a frequency of 5.00 Hz. The amplitude of the motion is $A = 12.0$ cm, and the wave speed is $v = 20.0$ m/s. Furthermore, the wave is such that $y = 0$ at $x = 0$ and $t = 0$. Determine (a) the angular frequency and (b) the wave number for this wave. (c) Write an expression for the wave function. Calculate (d) the maximum transverse speed and (e) the maximum transverse acceleration of an element of the string.

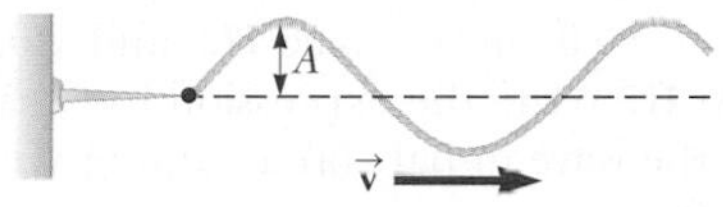

Figure P16.11

12. Consider the sinusoidal wave of Example 16.2 with the wave function

$$y = 0.150 \cos (15.7x - 50.3t)$$

where x and y are in meters and t is in seconds. At a certain instant, let point A be at the origin and point B be the closest point to A along the x axis where the wave is 60.0° out of phase with A. What is the coordinate of B?

13. A sinusoidal wave is described by the wave function $y =$ $0.25 \sin (0.30x - 40t)$ where x and y are in meters and t is in seconds. Determine for this wave (a) the amplitude, (b) the angular frequency, (c) the angular wave number, (d) the wavelength, (e) the wave speed, and (f) the direction of motion.

14. **Q|C** (a) Plot y versus t at $x = 0$ for a sinusoidal wave of the form $y = 0.150 \cos (15.7x - 50.3t)$, where x and y are in meters and t is in seconds. (b) Determine the period of vibration. (c) State how your result compares with the value found in Example 16.2.

15. A transverse wave on a string is described by the wave function

$$y = 0.120 \sin \left(\frac{\pi}{8} x + 4\pi t\right)$$

where x and y are in meters and t is in seconds. Determine (a) the transverse speed and (b) the transverse acceleration at $t = 0.200$ s for an element of the string located at $x = 1.60$ m. What are (c) the wavelength, (d) the period, and (e) the speed of propagation of this wave?

16. A wave on a string is described by the wave function $y = 0.100 \sin(0.50x - 20t)$, where x and y are in meters and t is in seconds. (a) Show that an element of the string at $x = 2.00$ m executes harmonic motion. (b) Determine the frequency of oscillation of this particular element.

17. A sinusoidal wave of wavelength 2.00 m and amplitude 0.100 m travels on a string with a speed of 1.00 m/s to the right. At $t = 0$, the left end of the string is at the origin. For this wave, find (a) the frequency, (b) the angular frequency, (c) the angular wave number, and (d) the wave function in SI units. Determine the equation of motion in SI units for (e) the left end of the string and (f) the point on the string at $x = 1.50$ m to the right of the left end. (g) What is the maximum speed of any element of the string?

18. A transverse sinusoidal wave on a string has a period $T = 25.0$ ms and travels in the negative x direction with a speed of 30.0 m/s. At $t = 0$, an element of the string at $x = 0$ has a transverse position of 2.00 cm and is traveling downward with a speed of 2.00 m/s. (a) What is the amplitude of the wave? (b) What is the initial phase angle? (c) What is the maximum transverse speed of an element of the string? (d) Write the wave function for the wave.

19. (a) Write the expression for y as a function of x and t in SI units for a sinusoidal wave traveling along a rope in the negative x direction with the following characteristics: $A = 8.00$ cm, $\lambda = 80.0$ cm, $f = 3.00$ Hz, and $y(0, t) = 0$ at $t = 0$. (b) **What If?** Write the expression for y as a function of x and t for the wave in part (a) assuming $y(x, 0) = 0$ at the point $x = 10.0$ cm.

20. GP A sinusoidal wave traveling in the negative x direction (to the left) has an amplitude of 20.0 cm, a wavelength of 35.0 cm, and a frequency of 12.0 Hz. The transverse position of an element of the medium at $t = 0$, $x = 0$ is $y = -3.00$ cm, and the element has a positive velocity here. We wish to find an expression for the wave function describing this wave. (a) Sketch the wave at $t = 0$. (b) Find the angular wave number k from the wavelength. (c) Find the period T from the frequency. Find (d) the angular frequency ω and (e) the wave speed v. (f) From the information about $t = 0$, find the phase constant ϕ. (g) Write an expression for the wave function $y(x, t)$.

Section 16.3 The Speed of Waves on Strings

21. An Ethernet cable is 4.00 m long. The cable has a mass of 0.200 kg. A transverse pulse is produced by plucking one end of the taut cable. The pulse makes four trips down and back along the cable in 0.800 s. What is the tension in the cable?

22. A piano string having a mass per unit length equal to 5.00×10^{-3} kg/m is under a tension of 1 350 N. Find the speed with which a wave travels on this string.

23. M Transverse waves travel with a speed of 20.0 m/s on a string under a tension of 6.00 N. What tension is required for a wave speed of 30.0 m/s on the same string?

24. Q|C S A student taking a quiz finds on a reference sheet the two equations

$$f = \frac{1}{T} \quad \text{and} \quad v = \sqrt{\frac{T}{\mu}}$$

She has forgotten what T represents in each equation. (a) Use dimensional analysis to determine the units required for T in each equation. (b) Explain how you can identify the physical quantity each T represents from the units.

25. Review. The elastic limit of a steel wire is 2.70×10^8 Pa. What is the maximum speed at which transverse wave pulses can propagate along this wire without exceeding this stress? (The density of steel is 7.86×10^3 kg/m^3.)

26. A transverse traveling wave on a taut wire has an amplitude of 0.200 mm and a frequency of 500 Hz. It travels with a speed of 196 m/s. (a) Write an equation in SI units of the form $y = A \sin(kx - \omega t)$ for this wave. (b) The mass per unit length of this wire is 4.10 g/m. Find the tension in the wire.

27. Transverse pulses travel with a speed of 200 m/s along a taut copper wire whose diameter is 1.50 mm. What is the tension in the wire? (The density of copper is 8.92 g/cm^3.)

28. *Why is the following situation impossible?* An astronaut on the Moon is studying wave motion using the apparatus discussed in Example 16.3 and shown in Figure 16.12. He measures the time interval for pulses to travel along the horizontal wire. Assume the horizontal wire has a mass of 4.00 g and a length of 1.60 m and assume a 3.00-kg object is suspended from its extension around the pulley. The astronaut finds that a pulse requires 26.1 ms to traverse the length of the wire.

29. Tension is maintained in a string as in Figure P16.29. The observed wave speed is $v = 24.0$ m/s when the suspended mass is $m = 3.00$ kg. (a) What is the mass per unit length of the string? (b) What is the wave speed when the suspended mass is $m = 2.00$ kg?

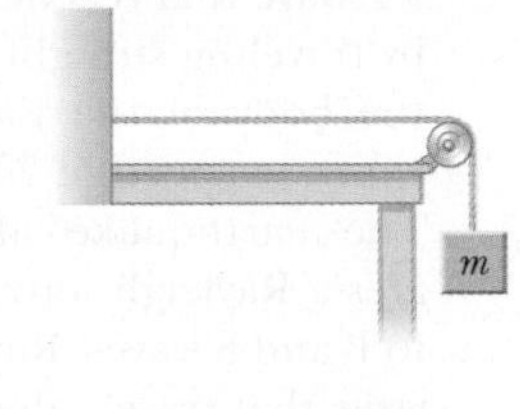

Figure P16.29
Problems 29 and 47.

30. Review. A light string with a mass per unit length of 8.00 g/m has its ends tied to two walls separated by a distance equal to three-fourths the length of the string (Fig. P16.30). An object of mass m is suspended from the center of the string, putting a tension in the string. (a) Find an expression for the transverse wave speed in the string as a function of the mass of the hanging object. (b) What should be the mass of the object suspended from the string if the wave speed is to be 60.0 m/s?

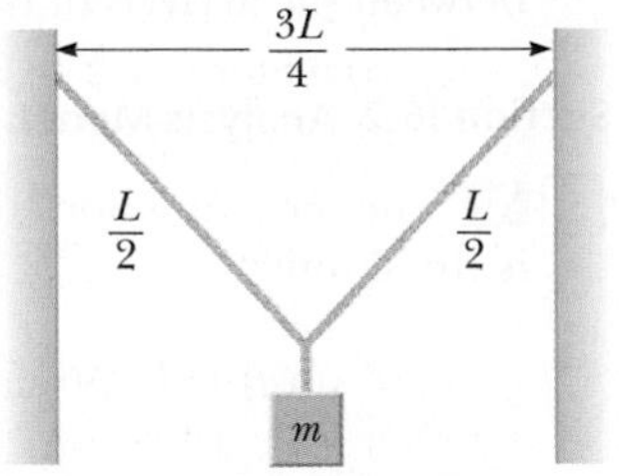

Figure P16.30

31. M A steel wire of length 30.0 m and a copper wire of length 20.0 m, both with 1.00-mm diameters, are connected end to end and stretched to a tension of 150 N. During what time interval will a transverse wave travel the entire length of the two wires?

Section 16.5 Rate of Energy Transfer by Sinusoidal Waves on Strings

32. A taut rope has a mass of 0.180 kg and a length of 3.60 m. What power must be supplied to the rope so as to generate sinusoidal waves having an amplitude of 0.100 m and a wavelength of 0.500 m and traveling with a speed of 30.0 m/s?

33. Transverse waves are being generated on a rope under constant tension. By what factor is the required power increased or decreased if (a) the length of the rope is doubled and the angular frequency remains constant, (b) the amplitude is doubled and the angular frequency is halved, (c) both the wavelength and the amplitude are doubled, and (d) both the length of the rope and the wavelength are halved?

34. M Sinusoidal waves 5.00 cm in amplitude are to be transmitted along a string that has a linear mass density of 4.00×10^{-2} kg/m. The source can deliver a maximum power of 300 W, and the string is under a tension of 100 N. What is the highest frequency f at which the source can operate?

35. M A sinusoidal wave on a string is described by the wave function

$$y = 0.15 \sin (0.80x - 50t)$$

where x and y are in meters and t is in seconds. The mass per unit length of this string is 12.0 g/m. Determine (a) the speed of the wave, (b) the wavelength, (c) the frequency, and (d) the power transmitted by the wave.

36. Q|C In a region far from the epicenter of an earthquake, a seismic wave can be modeled as transporting energy in a single direction without absorption, just as a string wave does. Suppose the seismic wave moves from granite into mudfill with similar density but with a much smaller bulk modulus. Assume the speed of the wave gradually drops by a factor of 25.0, with negligible reflection of the wave. (a) Explain whether the amplitude of the ground shaking will increase or decrease. (b) Does it change by a predictable factor? (This phenomenon led to the collapse of part of the Nimitz Freeway in Oakland, California, during the Loma Prieta earthquake of 1989.)

37. A long string carries a wave; a 6.00-m segment of the string contains four complete wavelengths and has a mass of 180 g. The string vibrates sinusoidally with a frequency of 50.0 Hz and a peak-to-valley displacement of 15.0 cm. (The "peak-to-valley" distance is the vertical distance from the farthest positive position to the farthest negative position.) (a) Write the function that describes this wave traveling in the positive x direction. (b) Determine the power being supplied to the string.

38. S A two-dimensional water wave spreads in circular ripples. Show that the amplitude A at a distance r from the initial disturbance is proportional to $1/\sqrt{r}$. *Suggestion:* Consider the energy carried by one outward-moving ripple.

39. The wave function for a wave on a taut string is

$$y(x, t) = 0.350 \sin \left(10\pi t - 3\pi x + \frac{\pi}{4}\right)$$

where x and y are in meters and t is in seconds. If the linear mass density of the string is 75.0 g/m, (a) what is the average rate at which energy is transmitted along the string? (b) What is the energy contained in each cycle of the wave?

40. S A horizontal string can transmit a maximum power P_0 (without breaking) if a wave with amplitude A and angular frequency ω is traveling along it. To increase this maximum power, a student folds the string and uses this "double string" as a medium. Assuming the tension in the two strands together is the same as the original tension in the single string and the angular frequency of the wave remains the same, determine the maximum power that can be transmitted along the "double string."

Section 16.6 The Linear Wave Equation

41. S Show that the wave function $y = e^{b(x-vt)}$ is a solution of the linear wave equation (Eq. 16.27), where b is a constant.

42. Q|C (a) Evaluate A in the scalar equality $4\ (7 + 3) = A$. (b) Evaluate A, B, and C in the vector equality $700\hat{\mathbf{i}} + 3.00\hat{\mathbf{k}} = A\hat{\mathbf{i}} + B\hat{\mathbf{j}} + C\hat{\mathbf{k}}$. (c) Explain how you arrive at the answers to convince a student who thinks that you cannot solve a single equation for three different unknowns. (d) **What If?** The functional equality or identity

$$A + B \cos (Cx + Dt + E) = 7.00 \cos (3x + 4t + 2)$$

is true for all values of the variables x and t, measured in meters and in seconds, respectively. Evaluate the constants A, B, C, D, and E. (e) Explain how you arrive at your answers to part (d).

43. S Show that the wave function $y = \ln [b(x - vt)]$ is a solution to Equation 16.27, where b is a constant.

44. S (a) Show that the function $y(x, t) = x^2 + v^2t^2$ is a solution to the wave equation. (b) Show that the function in part (a) can be written as $f(x + vt) + g(x - vt)$ and determine the functional forms for f and g. (c) **What If?** Repeat parts (a) and (b) for the function $y(x, t) = \sin (x) \cos (vt)$.

Additional Problems

45. Motion-picture film is projected at a frequency of 24.0 frames per second. Each photograph on the film is the same height of 19.0 mm, just like each oscillation in a wave is the same length. Model the height of a frame as the wavelength of a wave. At what constant speed does the film pass into the projector?

46. "The wave" is a particular type of pulse that can propagate through a large crowd gathered at a sports arena (Fig. P16.46). The elements of the medium are the spectators,

Joe Klamar/AFP/Getty Images

Figure P16.46

with zero position corresponding to their being seated and maximum position corresponding to their standing and raising their arms. When a large fraction of the spectators participates in the wave motion, a somewhat stable pulse shape can develop. The wave speed depends on people's reaction time, which is typically on the order of 0.1 s. Estimate the order of magnitude, in minutes, of the time interval required for such a pulse to make one circuit around a large sports stadium. State the quantities you measure or estimate and their values.

47. A sinusoidal wave in a rope is described by the wave function

$$y = 0.20 \sin (0.75\pi x + 18\pi t)$$

where x and y are in meters and t is in seconds. The rope has a linear mass density of 0.250 kg/m. The tension in the rope is provided by an arrangement like the one illustrated in Figure P16.29. What is the mass of the suspended object?

48. **Q|C** A sinusoidal wave in a string is described by the wave function

$$y = 0.150 \sin (0.800x - 50.0t)$$

where x and y are in meters and t is in seconds. The mass per length of the string is 12.0 g/m. (a) Find the maximum transverse acceleration of an element of this string. (b) Determine the maximum transverse force on a 1.00-cm segment of the string. (c) State how the force found in part (b) compares with the tension in the string.

49. **Review.** A 2.00-kg block hangs from a rubber cord, being supported so that the cord is not stretched. The unstretched length of the cord is 0.500 m, and its mass is 5.00 g. The "spring constant" for the cord is 100 N/m. The block is released and stops momentarily at the lowest point. (a) Determine the tension in the cord when the block is at this lowest point. (b) What is the length of the cord in this "stretched" position? (c) If the block is held in this lowest position, find the speed of a transverse wave in the cord.

50. **S** **Review.** A block of mass M hangs from a rubber cord. The block is supported so that the cord is not stretched. The unstretched length of the cord is L_0, and its mass is m, much less than M. The "spring constant" for the cord is k. The block is released and stops momentarily at the lowest point. (a) Determine the tension in the string when the block is at this lowest point. (b) What is the length of the cord in this "stretched" position? (c) If the block is held in this lowest position, find the speed of a transverse wave in the cord.

51. A transverse wave on a string is described by the wave function

$$y(x, t) = 0.350 \sin (1.25x + 99.6t)$$

where x and y are in meters and t is in seconds. Consider the element of the string at $x = 0$. (a) What is the time interval between the first two instants when this element has a position of $y = 0.175$ m? (b) What distance does the wave travel during the time interval found in part (a)?

52. **Q|C** An undersea earthquake or a landslide can produce an ocean wave of short duration carrying great energy, called a tsunami. When its wavelength is large compared to the ocean depth d, the speed of a water wave is given approximately by $v = \sqrt{gd}$. Assume an earthquake occurs all along a tectonic plate boundary running north to south and produces a straight tsunami wave crest moving everywhere to the west. (a) What physical quantity can you consider to be constant in the motion of any one wave crest? (b) Explain why the amplitude of the wave increases as the wave approaches shore. (c) If the wave has amplitude 1.80 m when its speed is 200 m/s, what will be its amplitude where the water is 9.00 m deep? (d) Explain why the amplitude at the shore should be expected to be still greater, but cannot be meaningfully predicted by your model.

53. **S** **Review.** A block of mass M, supported by a string, rests on a frictionless incline making an angle θ with the horizontal (Fig. P16.53). The length of the string is L, and its mass is $m << M$. Derive an expression for the time interval required for a transverse wave to travel from one end of the string to the other.

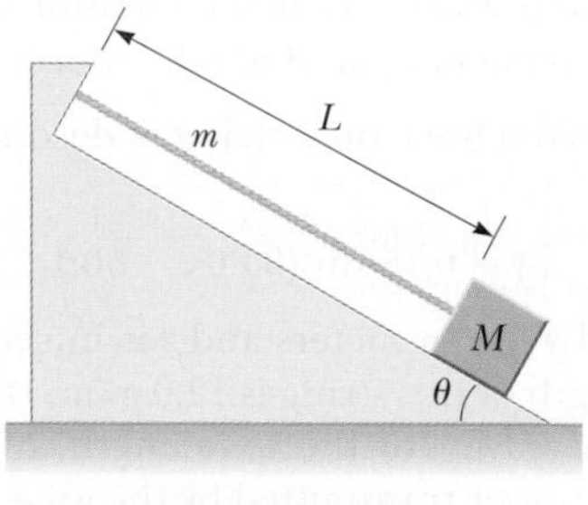

Figure P16.53

54. **Q|C** A string with linear density 0.500 g/m is held under tension 20.0 N. As a transverse sinusoidal wave propagates on the string, elements of the string move with maximum speed $v_{y,\max}$. (a) Determine the power transmitted by the wave as a function of $v_{y,\max}$. (b) State in words the proportionality between power and $v_{y,\max}$. (c) Find the energy contained in a section of string 3.00 m long as a function of $v_{y,\max}$. (d) Express the answer to part (c) in terms of the mass m of this section. (e) Find the energy that the wave carries past a point in 6.00 s.

55. **Review.** A block of mass $M = 0.450$ kg is attached to one end of a cord of mass 0.003 20 kg; the other end of the cord is attached to a fixed point. The block rotates with constant angular speed in a circle on a frictionless, horizontal table as shown in Figure P16.55. Through what angle does the block rotate in the time interval during which a transverse wave travels along the string from the center of the circle to the block?

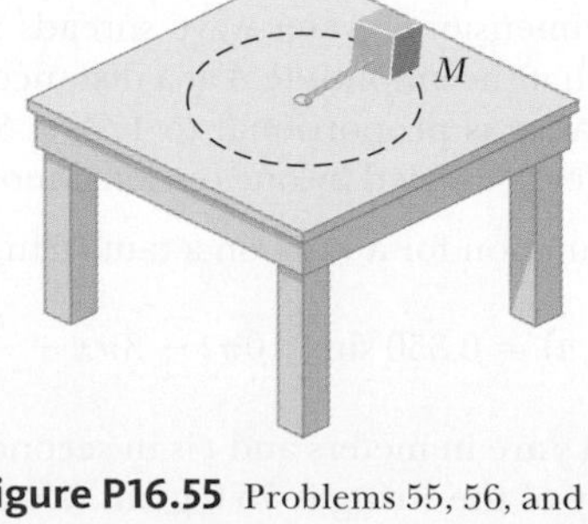

Figure P16.55 Problems 55, 56, and 57.

56. Review. A block of mass $M = 0.450$ kg is attached to one end of a cord of mass $m = 0.003\ 20$ kg; the other end of the cord is attached to a fixed point. The block rotates with constant angular speed $\omega = 10.0$ rad/s in a circle on a frictionless, horizontal table as shown in Figure P16.55. What time interval is required for a transverse wave to travel along the string from the center of the circle to the block?

57. **S** **Review.** A block of mass M is attached to one end of a cord of mass m; the other end of the cord is attached to a fixed point. The block rotates with constant angular speed ω in a circle on a frictionless, horizontal table as shown in Figure P16.55. What time interval is required for a transverse wave to travel along the string from the center of the circle to the block?

58. **S** A rope of total mass m and length L is suspended vertically. Analysis shows that for short transverse pulses, the waves above a short distance from the free end of the rope can be represented to a good approximation by the linear wave equation discussed in Section 16.6. Show that a transverse pulse travels the length of the rope in a time interval that is given approximately by $\Delta t \approx 2\sqrt{L/g}$. *Suggestion:* First find an expression for the wave speed at any point a distance x from the lower end by considering the rope's tension as resulting from the weight of the segment below that point.

59. A wire of density ρ is tapered so that its cross-sectional area varies with x according to

$$A = 1.00 \times 10^{-5}\ x + 1.00 \times 10^{-6}$$

where A is in meters squared and x is in meters. The tension in the wire is T. (a) Derive a relationship for the speed of a wave as a function of position. (b) **What If?** Assume the wire is aluminum and is under a tension $T = 24.0$ N. Determine the wave speed at the origin and at $x = 10.0$ m.

60. **M** **Review.** An aluminum wire is held between two clamps under zero tension at room temperature. Reducing the temperature, which results in a decrease in the wire's equilibrium length, increases the tension in the wire. Taking the cross-sectional area of the wire to be 5.00×10^{-6} m^2, the density to be 2.70×10^3 kg/m^3, and Young's modulus to be 7.00×10^{10} N/m^2, what strain ($\Delta L/L$) results in a transverse wave speed of 100 m/s?

61. **S** A pulse traveling along a string of linear mass density μ is described by the wave function

$$y = [A_0 e^{-bx}] \sin (kx - \omega t)$$

where the factor in brackets is said to be the amplitude. (a) What is the power $P(x)$ carried by this wave at a point x? (b) What is the power $P(0)$ carried by this wave at the origin? (c) Compute the ratio $P(x)/P(0)$.

62. *Why is the following situation impossible?* Tsunamis are ocean surface waves that have enormous wavelengths (100 to 200 km), and the propagation speed for these waves is $v \approx \sqrt{gd_{avg}}$, where d_{avg} is the average depth of the water. An earthquake on the ocean floor in the Gulf of Alaska produces a tsunami that reaches Hilo, Hawaii, 4 450 km away, in a time interval of 5.88 h. (This method was used in 1856 to estimate the average depth of the Pacific Ocean long before soundings were made to give a direct determination.)

Challenge Problems

63. **S** A rope of total mass m and length L is suspended vertically. As shown in Problem 58, a pulse travels from the bottom to the top of the rope in an approximate time interval $\Delta t = 2\sqrt{L/g}$ with a speed that varies with position x measured from the bottom of the rope as $v = \sqrt{gx}$. Assume the linear wave equation in Section 16.6 describes waves at all locations on the rope. (a) Over what time interval does a pulse travel halfway up the rope? Give your answer as a fraction of the quantity $2\sqrt{L/g}$. (b) A pulse starts traveling up the rope. How far has it traveled after a time interval $\sqrt{L/g}$?

64. **S** Assume an object of mass M is suspended from the bottom of the rope of mass m and length L in Problem 58. (a) Show that the time interval for a transverse pulse to travel the length of the rope is

$$\Delta t = 2\sqrt{\frac{L}{mg}}(\sqrt{M+m} - \sqrt{M})$$

(b) **What If?** Show that the expression in part (a) reduces to the result of Problem 58 when $M = 0$. (c) Show that for $m << M$, the expression in part (a) reduces to

$$\Delta t = \sqrt{\frac{mL}{Mg}}$$

65. **S** If a loop of chain is spun at high speed, it can roll along the ground like a circular hoop without collapsing. Consider a chain of uniform linear mass density μ whose center of mass travels to the right at a high speed v_0 as shown in Figure P16.65. (a) Determine the tension in the chain in terms of μ and v_0. Assume the weight of an individual link is negligible compared to the tension. (b) If the loop rolls over a small bump, the resulting deformation of the chain causes two transverse pulses to propagate along the chain, one moving clockwise and one moving counterclockwise. What is the speed of the pulses traveling along the chain? (c) Through what angle does each pulse travel during the time interval over which the loop makes one revolution?

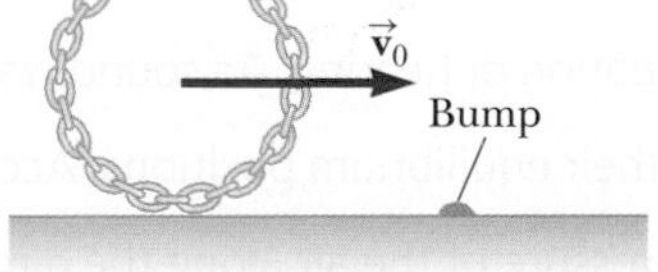

Figure P16.65

66. **S** A string on a musical instrument is held under tension T and extends from the point $x = 0$ to the point $x = L$. The string is overwound with wire in such a way that its mass per unit length $\mu(x)$ increases uniformly from μ_0 at $x = 0$ to μ_L at $x = L$. (a) Find an expression for $\mu(x)$ as a function of x over the range $0 \le x \le L$. (b) Find an expression for the time interval required for a transverse pulse to travel the length of the string.

chapter 17

Sound Waves

Three musicians play the alpenhorn in Valais, Switzerland. In this chapter, we explore the behavior of sound waves such as those coming from these large musical instruments. *(Stefano Cellai/AGE fotostock)*

Most of the waves we studied in Chapter 16 are constrained to move along a one-dimensional medium. For example, the wave in Active Figure 16.7 is a purely mathematical construct moving along the *x* axis. The wave in Active Figure 16.10 is constrained to move along the length of the string. We have also seen waves moving through a two-dimensional medium, such as the ripples on the water surface in the introduction to Part 2 on page 433 and the waves moving over the surface of the ocean in Active Figure 16.4. In this chapter, we investigate mechanical waves that move through three-dimensional bulk media. For example, seismic waves leaving the focus of an earthquake travel through the three-dimensional interior of the Earth.

We will focus our attention on **sound waves,** which travel through any material, but are most commonly experienced as the mechanical waves traveling through air that result in the human perception of hearing. As sound waves travel through air, elements of air are disturbed from their equilibrium positions. Accompanying these movements are changes in density and pressure of the air along the direction of wave motion. If the source of the sound waves vibrates sinusoidally, the density and pressure variations are also sinusoidal. The mathematical description of sinusoidal sound waves is very similar to that of sinusoidal waves on strings, as discussed in Chapter 16.

Sound waves are divided into three categories that cover different frequency ranges. (1) *Audible waves* lie within the range of sensitivity of the human ear. They can be generated in a variety of ways, such as by musical instruments, human voices, or loudspeakers. (2) *Infrasonic waves* have frequencies below the audible range. Elephants can use infrasonic waves to communicate with one another, even when separated by many kilometers. (3) *Ultrasonic waves* have frequencies above the audible range. You may have used a "silent" whistle to retrieve your dog. Dogs easily hear the ultrasonic sound this whistle emits, although humans cannot detect it at all. Ultrasonic waves are also used in medical imaging.

This chapter begins with a discussion of the pressure variations in a sound wave, the speed of sound waves, and wave intensity, which is a function of wave amplitude. We then provide an alternative description of the intensity of sound waves that compresses the wide range of intensities to which the ear is sensitive into a smaller range for convenience. The effects of the motion of sources and listeners on the frequency of a sound are also investigated.

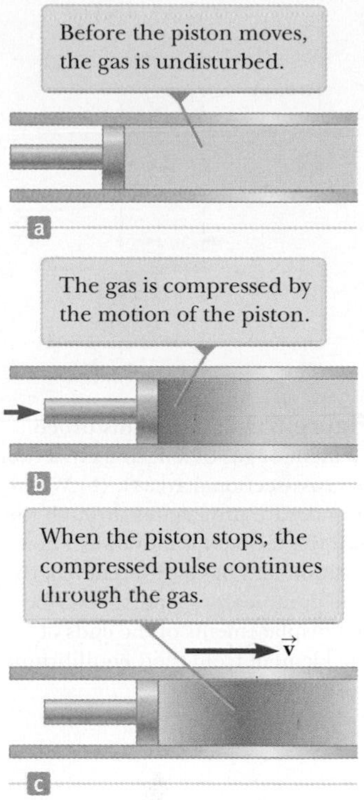

Figure 17.1 Motion of a longitudinal pulse through a compressible gas. The compression (darker region) is produced by the moving piston.

17.1 Pressure Variations in Sound Waves

In Chapter 16, we began our investigation of waves by imagining the creation of a single pulse that traveled down a string (Figure 16.1) or a spring (Figure 16.3). Let's do something similar for sound. We describe pictorially the motion of a one-dimensional longitudinal sound pulse moving through a long tube containing a compressible gas as shown in Figure 17.1. A piston at the left end can be quickly moved to the right to compress the gas and create the pulse. Before the piston is moved, the gas is undisturbed and of uniform density as represented by the uniformly shaded region in Figure 17.1a. When the piston is pushed to the right (Fig. 17.1b), the gas just in front of it is compressed (as represented by the more heavily shaded region); the pressure and density in this region are now higher than they were before the piston moved. When the piston comes to rest (Fig. 17.1c), the compressed region of the gas continues to move to the right, corresponding to a longitudinal pulse traveling through the tube with speed v.

One can produce a one-dimensional *periodic* sound wave in the tube of gas in Figure 17.1 by causing the piston to move in simple harmonic motion. The results are shown in Active Figure 17.2. The darker parts of the colored areas in this figure represent regions in which the gas is compressed and the density and pressure are above their equilibrium values. A compressed region is formed whenever the piston is pushed into the tube. This compressed region, called a **compression,** moves through the tube, continuously compressing the region just in front of itself. When the piston is pulled back, the gas in front of it expands and the pressure and density in this region fall below their equilibrium values (represented by the lighter parts of the colored areas in Active Fig. 17.2). These low-pressure regions, called **rarefactions,** also propagate along the tube, following the compressions. Both regions move at the speed of sound in the medium.

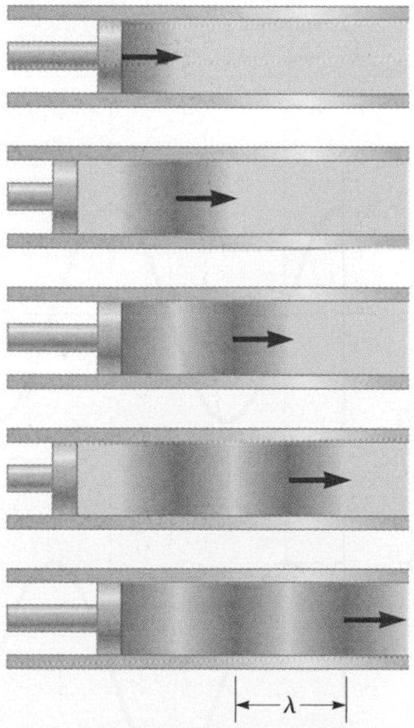

ACTIVE FIGURE 17.2

A longitudinal wave propagating through a gas-filled tube. The source of the wave is an oscillating piston at the left.

As the piston oscillates sinusoidally, regions of compression and rarefaction are continuously set up. The distance between two successive compressions (or two successive rarefactions) equals the wavelength λ of the sound wave. Because the sound wave is longitudinal, as the compressions and rarefactions travel through the tube, any small element of the gas moves with simple harmonic motion parallel to the direction of the wave. If $s(x, t)$ is the position of a small element relative to its equilibrium position,[1] we can express this harmonic position function as

$$s(x, t) = s_{\max} \cos (kx - \omega t) \tag{17.1}$$

[1]We use $s(x, t)$ here instead of $y(x, t)$ because the displacement of elements of the medium is not perpendicular to the x direction.

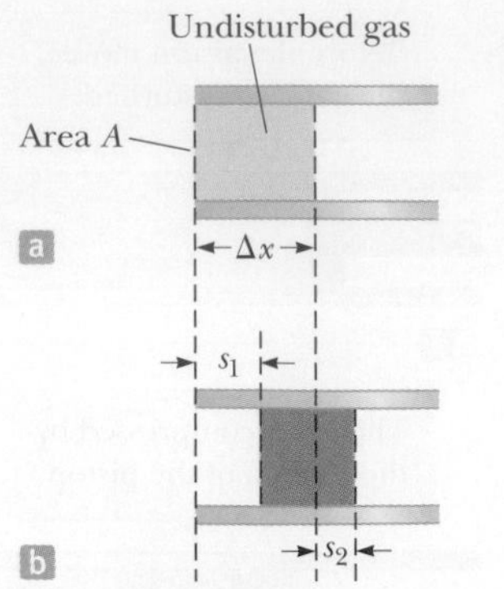

Figure 17.3 (a) An undisturbed element of gas of length Δx in a tube of cross-sectional area A. (b) When a sound wave propagates through the gas, the element is moved to a new position and has a different length. The parameters s_1 and s_2 describe the displacements of the ends of the element from their equilibrium positions.

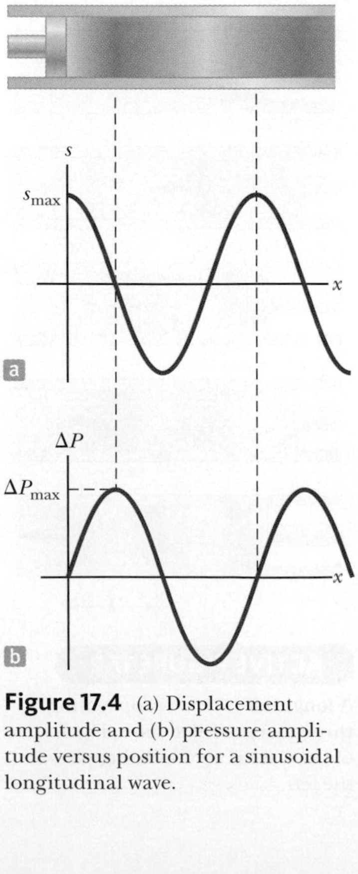

Figure 17.4 (a) Displacement amplitude and (b) pressure amplitude versus position for a sinusoidal longitudinal wave.

where s_{max} is the maximum position of the element relative to equilibrium. This parameter is often called the **displacement amplitude** of the wave. The parameter k is the wave number, and ω is the angular frequency of the wave. Notice that the displacement of the element is along x, in the direction of propagation of the sound wave.

The variation in the gas pressure ΔP measured from the equilibrium value is also periodic with the same wave number and angular frequency as for the displacement in Equation 17.1. Therefore, we can write

$$\Delta P = \Delta P_{max} \sin (kx - \omega t) \tag{17.2}$$

where **the pressure amplitude ΔP_{max}** is the maximum change in pressure from the equilibrium value.

Notice that we have expressed the displacement by means of a cosine function and the pressure by means of a sine function. We will justify this choice in the procedure that follows and relate the pressure amplitude P_{max} to the displacement amplitude s_{max}. Consider the piston–tube arrangement of Figure 17.1 once again. In Figure 17.3a, we focus our attention on a small cylindrical element of undisturbed gas of length Δx and area A. The volume of this element is $V_i = A\,\Delta x$.

Figure 17.3b shows this element of gas after a sound wave has moved it to a new position. The disk's two flat faces move through different distances s_1 and s_2. The change in volume ΔV of the element in the new position is equal to $A\,\Delta s$, where $\Delta s = s_1 - s_2$.

From the definition of bulk modulus (see Eq. 12.8), we express the pressure variation in the element of gas as a function of its change in volume:

$$\Delta P = -B\frac{\Delta V}{V_i}$$

Let's substitute for the initial volume and the change in volume of the element:

$$\Delta P = -B\frac{A\,\Delta s}{A\,\Delta x}$$

Let the thickness Δx of the disk approach zero so that the ratio $\Delta s/\Delta x$ becomes a partial derivative:

$$\Delta P = -B\frac{\partial s}{\partial x} \tag{17.3}$$

Substitute the position function given by Equation 17.1:

$$\Delta P = -B\frac{\partial}{\partial x}[s_{max} \cos (kx - \omega t)] = Bs_{max}k \sin (kx - \omega t)$$

From this result, we see that a displacement described by a cosine function leads to a pressure described by a sine function. We also see that the displacement and pressure amplitudes are related by

$$\Delta P_{max} = Bs_{max}k \tag{17.4}$$

This relationship depends on the bulk modulus of the gas, which is not as readily available as is the density of the gas. Once we determine the speed of sound in a gas in Section 17.2, we will be able to provide an expression that relates ΔP_{max} and s_{max} in terms of the density of the gas.

This discussion shows that a sound wave may be described equally well in terms of either pressure or displacement. A comparison of Equations 17.1 and 17.2 shows that the pressure wave is 90° out of phase with the displacement wave. Graphs of these functions are shown in Figure 17.4. The pressure variation is a maximum when the displacement from equilibrium is zero, and the displacement from equilibrium is a maximum when the pressure variation is zero.

Quick Quiz **17.1** If you blow across the top of an empty soft-drink bottle, a pulse of sound travels down through the air in the bottle. At the moment the pulse reaches the bottom of the bottle, what is the correct description of the displacement of elements of air from their equilibrium positions and the pressure of the air at this point? **(a)** The displacement and pressure are both at a maximum. **(b)** The displacement and pressure are both at a minimum. **(c)** The displacement is zero, and the pressure is a maximum. **(d)** The displacement is zero, and the pressure is a minimum.

17.2 Speed of Sound Waves

We now extend the discussion begun in Section 17.1 to evaluate the speed of sound in a gas. In Figure 17.5a, consider the cylindrical element of gas between the piston and the dashed line. This element of gas is in equilibrium under the influence of forces of equal magnitude, from the piston on the left and from the rest of the gas on the right. The magnitude of these forces is PA, where P is the pressure in the gas and A is the cross-sectional area of the tube.

Figure 17.5b shows the situation after a time interval Δt during which the piston moves to the right at a constant speed v_x due to a force from the left on the piston that has increased in magnitude to $(P + \Delta P)A$. By the end of the time interval Δt, every bit of gas in the element is moving with speed v_x. That will not be true in general for a macroscopic element of gas, but it will become true if we shrink the length of the element to an infinitesimal value.

The length of the undisturbed element of gas is chosen to be $v\,\Delta t$, where v is the speed of sound in the gas and Δt is the time interval between the configurations in Figures 17.5a and 17.5b. Therefore, at the end of the time interval Δt, the sound wave will just reach the right end of the cylindrical element of gas. The gas to the right of the element is undisturbed because the sound wave has not reached it yet.

The element of gas is modeled as a nonisolated system in terms of momentum. The force from the piston has provided an impulse to the element, which in turn exhibits a change in momentum. Therefore, we evaluate both sides of the impulse–momentum theorem:

$$\vec{\mathbf{I}} = \Delta\vec{\mathbf{p}} \tag{17.5}$$

On the left, the impulse is provided by the constant force due to the increased pressure on the piston:

$$\vec{\mathbf{I}} = \sum \vec{\mathbf{F}}\,\Delta t = (A\,\Delta P\,\Delta t)\hat{\mathbf{i}}$$

The pressure change ΔP can be related to the volume change and then to the speeds v and v_x through the bulk modulus:

$$\Delta P = -B\frac{\Delta V}{V_i} = -B\frac{(-v_x A\,\Delta t)}{vA\,\Delta t} = B\frac{v_x}{v}$$

Therefore, the impulse becomes

$$\vec{\mathbf{I}} = \left(AB\frac{v_x}{v}\Delta t\right)\hat{\mathbf{i}} \tag{17.6}$$

On the right-hand side of the impulse–momentum theorem, Equation 17.5, the change in momentum of the element of gas of mass m is as follows:

$$\Delta\vec{\mathbf{p}} = m\,\Delta\vec{\mathbf{v}} = (\rho V_i)(v_x\hat{\mathbf{i}} - 0) = (\rho v v_x A\,\Delta t)\hat{\mathbf{i}} \tag{17.7}$$

Substituting Equations 17.6 and 17.7 into Equation 17.5, we find

$$AB\frac{v_x}{v}\Delta t = \rho v v_x A\,\Delta t$$

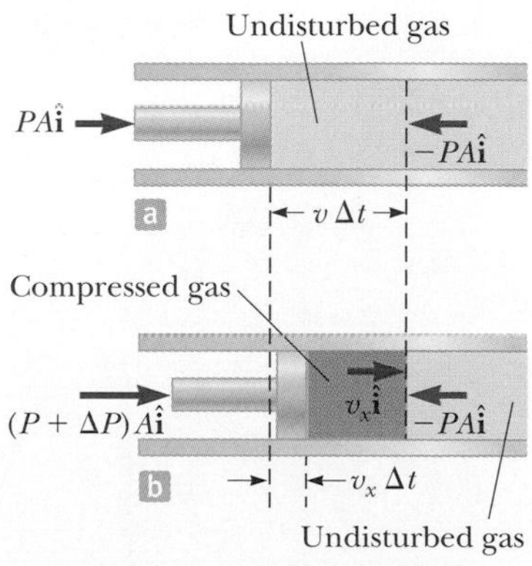

Figure 17.5 (a) An undisturbed element of gas of length $v\,\Delta t$ in a tube of cross-sectional area A. The element is in equilibrium between forces on either end. (b) When the piston moves inward at constant velocity v_x due to an increased force on the left, the element also moves with the same velocity.

which reduces to an expression for the speed of sound in a gas:

$$v = \sqrt{\frac{B}{\rho}} \qquad \textbf{(17.8)}$$

It is interesting to compare this expression with Equation 16.18 for the speed of transverse waves on a string, $v = \sqrt{T/\mu}$. In both cases, the wave speed depends on an elastic property of the medium (bulk modulus B or string tension T) and on an inertial property of the medium (volume density ρ or linear density μ). In fact, the speed of all mechanical waves follows an expression of the general form

$$v = \sqrt{\frac{\text{elastic property}}{\text{inertial property}}}$$

For longitudinal sound waves in a solid rod of material, for example, the speed of sound depends on Young's modulus Y and the density ρ. Table 17.1 provides the speed of sound in several different materials.

The speed of sound also depends on the temperature of the medium. For sound traveling through air, the relationship between wave speed and air temperature is

$$v = 331\sqrt{1 + \frac{T_C}{273}} \qquad \textbf{(17.9)}$$

where v is in meters/second, 331 m/s is the speed of sound in air at 0°C, and T_C is the air temperature in degrees Celsius. Using this equation, one finds that at 20°C, the speed of sound in air is approximately 343 m/s.

This information provides a convenient way to estimate the distance to a thunderstorm. First count the number of seconds between seeing the flash of lightning and hearing the thunder. Dividing this time by 3 gives the approximate distance to the lightning in kilometers because 343 m/s is approximately $\frac{1}{3}$ km/s. Dividing the time in seconds by 5 gives the approximate distance to the lightning in miles because the speed of sound is approximately $\frac{1}{5}$ mi/s.

Having an expression (Eq. 17.8) for the speed of sound, we can now express the relationship between pressure amplitude and displacement amplitude for a sound wave (Eq. 17.4) as

$$\Delta P_{\max} = Bs_{\max}k = (\rho v^2)s_{\max}\left(\frac{\omega}{v}\right) = \rho v \omega s_{\max} \qquad \textbf{(17.10)}$$

This expression is a bit more useful than Equation 17.4 because the density of a gas is more readily available than is the bulk modulus.

TABLE 17.1 *Speed of Sound in Various Media*

Medium	v (m/s)	Medium	v (m/s)	Medium	v (m/s)
Gases		**Liquids at 25°C**		**Solids**[a]	
Hydrogen (0°C)	1 286	Glycerol	1 904	Pyrex glass	5 640
Helium (0°C)	972	Seawater	1 533	Iron	5 950
Air (20°C)	343	Water	1 493	Aluminum	6 420
Air (0°C)	331	Mercury	1 450	Brass	4 700
Oxygen (0°C)	317	Kerosene	1 324	Copper	5 010
		Methyl alcohol	1 143	Gold	3 240
		Carbon tetrachloride	926	Lucite	2 680
				Lead	1 960
				Rubber	1 600

[a]Values given are for propagation of longitudinal waves in bulk media. Speeds for longitudinal waves in thin rods are smaller, and speeds of transverse waves in bulk are smaller yet.

17.3 Intensity of Periodic Sound Waves

In Chapter 16, we showed that a wave traveling on a taut string transports energy, consistent with the notion of energy transfer by mechanical waves in Equation 8.2. Naturally, we would expect sound waves to also represent a transfer of energy. Consider the element of gas acted on by the piston in Figure 17.5. Imagine that the piston is moving back and forth in simple harmonic motion at angular frequency ω. Imagine also that the length of the element becomes very small so that the entire element moves with the same velocity as the piston. Then we can model the element as a particle on which the piston is doing work. The rate at which the piston is doing work on the element at any instant of time is given by Equation 8.19:

$$Power = \vec{\mathbf{F}} \cdot \vec{\mathbf{v}}_x$$

where we have used *Power* rather than *P* so that we don't confuse power *P* with pressure *P*! The force $\vec{\mathbf{F}}$ on the element of gas is related to the pressure and the velocity $\vec{\mathbf{v}}_x$ of the element is the derivative of the displacement function, so we find

$$\begin{aligned} Power &= [\Delta P(x, t)A]\hat{\mathbf{i}} \cdot \frac{\partial}{\partial t}[s(x, t)\hat{\mathbf{i}}] \\ &= [\rho v \omega A s_{\max} \sin(kx - \omega t)]\left\{\frac{\partial}{\partial t}[s_{\max}\cos(kx - \omega t)]\right\} \\ &= \rho v \omega A s_{\max} \sin(kx - \omega t)][\omega s_{\max} \sin(kx - \omega t)] \\ &= \rho v \omega^2 A s_{\max}^2 \sin^2(kx - \omega t) \end{aligned}$$

We now find the time average power over one period of the oscillation. For any given value of x, which we can choose to be $x = 0$, the average value of $\sin^2(kx - \omega t)$ over one period T is

$$\frac{1}{T}\int_0^T \sin^2(0 - \omega t)\,dt = \frac{1}{T}\int_0^T \sin^2 \omega t\,dt = \frac{1}{T}\left(\frac{t}{2} + \frac{\sin 2\omega t}{2\omega}\right)\Bigg|_0^T = \tfrac{1}{2}$$

Therefore,

$$(Power)_{\text{avg}} = \tfrac{1}{2}\rho v \omega^2 A s_{\max}^2$$

We define the **intensity** I of a wave, or the power per unit area, as the rate at which the energy transported by the wave transfers through a unit area A perpendicular to the direction of travel of the wave:

$$I \equiv \frac{(Power)_{\text{avg}}}{A} \tag{17.11}$$

◀ **Intensity of a sound wave**

In this case, the intensity is therefore

$$I = \tfrac{1}{2}\rho v(\omega s_{\max})^2$$

Hence, the intensity of a periodic sound wave is proportional to the square of the displacement amplitude and to the square of the angular frequency. This expression can also be written in terms of the pressure amplitude $\Delta P_{\max}$; in this case, we use Equation 17.10 to obtain

$$I = \frac{(\Delta P_{\max})^2}{2\rho v} \tag{17.12}$$

The string waves we studied in Chapter 16 are constrained to move along the one-dimensional string, as discussed in the introduction to this chapter. The sound waves we have studied with regard to Figures 17.1 through 17.3 and 17.5 are constrained to move in one dimension along the length of the tube. As we mentioned

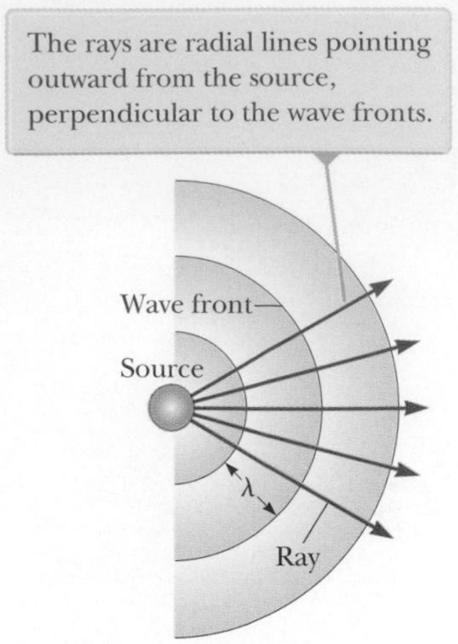

Figure 17.6 Spherical waves emitted by a point source. The circular arcs represent the spherical wave fronts that are concentric with the source.

in the introduction, however, sound waves can move through three-dimensional bulk media, so let's place a sound source in the open air and study the results.

Consider the special case of a point source emitting sound waves equally in all directions. If the air around the source is perfectly uniform, the sound power radiated in all directions is the same, and the speed of sound in all directions is the same. The result in this situation is called a **spherical wave.** Figure 17.6 shows these spherical waves as a series of circular arcs concentric with the source. Each arc represents a surface over which the phase of the wave is constant. We call such a surface of constant phase a **wave front.** The radial distance between adjacent wave fronts that have the same phase is the wavelength λ of the wave. The radial lines pointing outward from the source, representing the direction of propagation of the waves, are called **rays.**

The average power emitted by the source must be distributed uniformly over each spherical wave front of area $4\pi r^2$. Hence, the wave intensity at a distance r from the source is

$$I = \frac{(Power)_{avg}}{A} = \frac{(Power)_{avg}}{4\pi r^2} \quad \textbf{(17.13)}$$

The intensity decreases as the square of the distance from the source. This inverse-square law is reminiscent of the behavior of gravity in Chapter 13.

Quick Quiz **17.2** A vibrating guitar string makes very little sound if it is not mounted on the guitar body. Why does the sound have greater intensity if the string is attached to the guitar body? **(a)** The string vibrates with more energy. **(b)** The energy leaves the guitar at a greater rate. **(c)** The sound power is spread over a larger area at the listener's position. **(d)** The sound power is concentrated over a smaller area at the listener's position. **(e)** The speed of sound is higher in the material of the guitar body. **(f)** None of these answers is correct.

Example 17.1 Hearing Limits

The faintest sounds the human ear can detect at a frequency of 1 000 Hz correspond to an intensity of about 1.00×10^{-12} W/m^2, which is called *threshold of hearing.* The loudest sounds the ear can tolerate at this frequency correspond to an intensity of about 1.00 W/m^2, the *threshold of pain.* Determine the pressure amplitude and displacement amplitude associated with these two limits.

SOLUTION

Conceptualize Think about the quietest environment you have ever experienced. It is likely that the intensity of sound in even this quietest environment is higher than the threshold of hearing.

Categorize Because we are given intensities and asked to calculate pressure and displacement amplitudes, this problem requires the concepts discussed in this section.

Analyze To find the amplitude of the pressure variation at the threshold of hearing, use Equation 17.12, taking the speed of sound waves in air to be $v =$ 343 m/s and the density of air to be $\rho = 1.20$ kg/m^3:

$$\Delta P_{max} = \sqrt{2\rho v I}$$
$$= \sqrt{2(1.20 \text{ kg/m}^3)(343 \text{ m/s})(1.00 \times 10^{-12} \text{ W/m}^2)}$$
$$= 2.87 \times 10^{-5} \text{ N/m}^2$$

Calculate the corresponding displacement amplitude using Equation 17.10, recalling that $\omega = 2\pi f$ (Eq. 16.9):

$$s_{max} = \frac{\Delta P_{max}}{\rho v \omega} = \frac{2.87 \times 10^{-5} \text{ N/m}^2}{(1.20 \text{ kg/m}^3)(343 \text{ m/s})(2\pi \times 1\,000 \text{ Hz})}$$
$$= 1.11 \times 10^{-11} \text{ m}$$

In a similar manner, one finds that the loudest sounds the human ear can tolerate (the threshold of pain) correspond to a pressure amplitude of 28.7 N/m^2 and a displacement amplitude equal to 1.11×10^{-5} m.

17.1 *cont.*

Finalize Because atmospheric pressure is about 10^5 N/m^2, the result for the pressure amplitude tells us that the ear is sensitive to pressure fluctuations as small as 3 parts in 10^{10}! The displacement amplitude is also a remarkably small number! If we compare this result for s_{max} to the size of an atom (about 10^{-10} m), we see that the ear is an extremely sensitive detector of sound waves.

Example 17.2 Intensity Variations of a Point Source

A point source emits sound waves with an average power output of 80.0 W.

(A) Find the intensity 3.00 m from the source.

SOLUTION

Conceptualize Imagine a small loudspeaker sending sound out at an average rate of 80.0 W uniformly in all directions. You are standing 3.00 m away from the speakers. As the sound propagates, the energy of the sound waves is spread out over an ever-expanding sphere.

Categorize We evaluate the intensity from an equation generated in this section, so we categorize this example as a substitution problem.

Because a point source emits energy in the form of spherical waves, use Equation 17.13 to find the intensity:

$$I = \frac{(Power)_{avg}}{4\pi r^2} = \frac{80.0\text{ W}}{4\pi(3.00\text{ m})^2} = 0.707\text{ W/m}^2$$

This intensity is close to the threshold of pain.

(B) Find the distance at which the intensity of the sound is 1.00×10^{-8} W/m^2.

SOLUTION

Solve for r in Equation 17.13 and use the given value for I:

$$r = \sqrt{\frac{(Power)_{avg}}{4\pi I}} = \sqrt{\frac{80.0\text{ W}}{4\pi(1.00 \times 10^{-8}\text{ W/m}^2)}}$$

$$= 2.52 \times 10^4\text{ m}$$

Sound Level in Decibels

Example 17.1 illustrates the wide range of intensities the human ear can detect. Because this range is so wide, it is convenient to use a logarithmic scale, where the **sound level** β (Greek letter beta) is defined by the equation

$$\beta \equiv 10 \log \left(\frac{I}{I_0}\right) \tag{17.14}$$

The constant I_0 is the *reference intensity*, taken to be at the threshold of hearing ($I_0 = 1.00 \times 10^{-12}$ W/m^2), and I is the intensity in watts per square meter to which the sound level β corresponds, where β is measured[2] in **decibels** (dB). On this scale, the threshold of pain ($I = 1.00$ W/m^2) corresponds to a sound level of $\beta = 10 \log [(1\text{ W/m}^2)/(10^{-12}\text{ W/m}^2)] = 10 \log (10^{12}) = 120$ dB, and the threshold of hearing corresponds to $\beta = 10 \log [(10^{-12}\text{ W/m}^2)/(10^{-12}\text{ W/m}^2)] = 0$ dB.

Prolonged exposure to high sound levels may seriously damage the human ear. Ear plugs are recommended whenever sound levels exceed 90 dB. Recent evidence suggests that "noise pollution" may be a contributing factor to high blood pressure, anxiety, and nervousness. Table 17.2 gives some typical sound levels.

TABLE 17.2

Sound Levels

Source of Sound	β (dB)
Nearby jet airplane	150
Jackhammer; machine gun	130
Siren; rock concert	120
Subway; power lawn mower	100
Busy traffic	80
Vacuum cleaner	70
Normal conversation	60
Mosquito buzzing	40
Whisper	30
Rustling leaves	10
Threshold of hearing	0

[2]The unit *bel* is named after the inventor of the telephone, Alexander Graham Bell (1847–1922). The prefix *deci-* is the SI prefix that stands for 10^{-1}.

***Quick Quiz* 17.3** Increasing the intensity of a sound by a factor of 100 causes the sound level to increase by what amount? **(a)** 100 dB **(b)** 20 dB **(c)** 10 dB **(d)** 2 dB

Example 17.3 Sound Levels

Two identical machines are positioned the same distance from a worker. The intensity of sound delivered by each operating machine at the worker's location is $2.0 \times 10^{-7}\ \text{W/m}^2$.

(A) Find the sound level heard by the worker when one machine is operating.

SOLUTION

Conceptualize Imagine a situation in which one source of sound is active and is then joined by a second identical source, such as one person speaking and then a second person speaking at the same time or one musical instrument playing and then being joined by a second instrument.

Categorize Because we are asked for a sound level, we will perform calculations with Equation 17.14.

Analyze Use Equation 17.14 to calculate the sound level at the worker's location with one machine operating:

$$\beta_1 = 10 \log \left(\frac{2.0 \times 10^{-7}\ \text{W/m}^2}{1.00 \times 10^{-12}\ \text{W/m}^2} \right) = 10 \log (2.0 \times 10^5) = \boxed{53\ \text{dB}}$$

(B) Find the sound level heard by the worker when two machines are operating.

SOLUTION

Use Equation 17.14 to calculate the sound level at the worker's location with double the intensity:

$$\beta_2 = 10 \log \left(\frac{4.0 \times 10^{-7}\ \text{W/m}^2}{1.00 \times 10^{-12}\ \text{W/m}^2} \right) = 10 \log (4.0 \times 10^5) = \boxed{56\ \text{dB}}$$

Finalize These results show that when the intensity is doubled, the sound level increases by only 3 dB. This 3-dB increase is independent of the original sound level. (Prove this to yourself!)

WHAT IF? *Loudness* is a psychological response to a sound. It depends on both the intensity and the frequency of the sound. As a rule of thumb, a doubling in loudness is approximately associated with an increase in sound level of 10 dB. (This rule of thumb is relatively inaccurate at very low or very high frequencies.) If the loudness of the machines in this example is to be doubled, how many machines at the same distance from the worker must be running?

Answer Using the rule of thumb, a doubling of loudness corresponds to a sound level increase of 10 dB. Therefore,

$$\beta_2 - \beta_1 = 10\ \text{dB} = 10 \log \left(\frac{I_2}{I_0} \right) - 10 \log \left(\frac{I_1}{I_0} \right) = 10 \log \left(\frac{I_2}{I_1} \right)$$

$$\log \left(\frac{I_2}{I_1} \right) = 1 \rightarrow I_2 = 10 I_1$$

Therefore, ten machines must be operating to double the loudness.

Loudness and Frequency

The discussion of sound level in decibels relates to a *physical* measurement of the strength of a sound. Let us now extend our discussion from the What If? section of Example 17.3 concerning the *psychological* "measurement" of the strength of a sound.

Of course, we don't have instruments in our bodies that can display numerical values of our reactions to stimuli. We have to "calibrate" our reactions somehow by comparing different sounds to a reference sound, but that is not easy to accom-

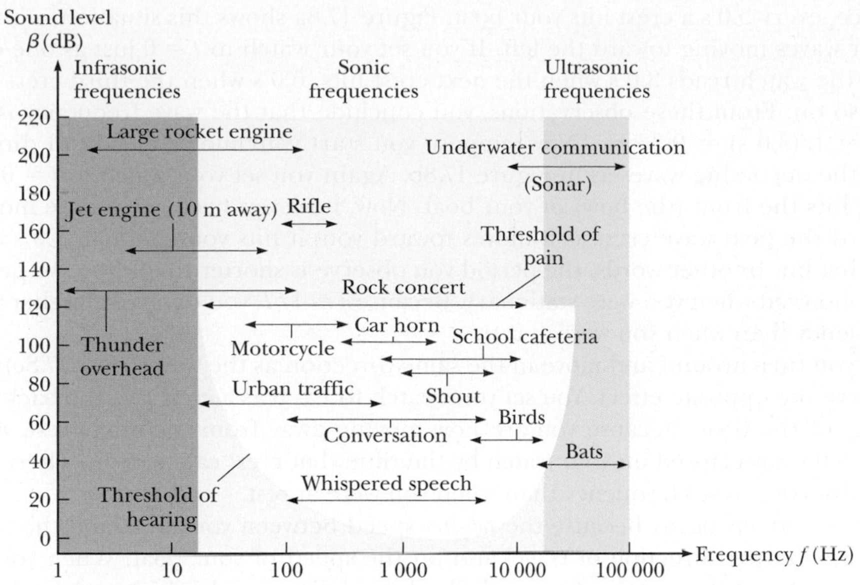

Figure 17.7 Approximate ranges of frequency and sound level of various sources and that of normal human hearing, shown by the white area. (From R. L. Reese, *University Physics*, Pacific Grove, Brooks/Cole, 2000.)

plish. For example, earlier we mentioned that the threshold intensity is 10^{-12} W/m^2, corresponding to an intensity level of 0 dB. In reality, this value is the threshold only for a sound of frequency 1 000 Hz, which is a standard reference frequency in acoustics. If we perform an experiment to measure the threshold intensity at other frequencies, we find a distinct variation of this threshold as a function of frequency. For example, at 100 Hz, a barely audible sound must have an intensity level of about 30 dB! Unfortunately, there is no simple relationship between physical measurements and psychological "measurements." The 100-Hz, 30-dB sound is psychologically "equal" in loudness to the 1 000-Hz, 0-dB sound (both are just barely audible), but they are not physically equal in sound level (30 dB $\neq$ 0 dB).

By using test subjects, the human response to sound has been studied, and the results are shown in the white area of Figure 17.7 along with the approximate frequency and sound-level ranges of other sound sources. The lower curve of the white area corresponds to the threshold of hearing. Its variation with frequency is clear from this diagram. Notice that humans are sensitive to frequencies ranging from about 20 Hz to about 20 000 Hz. The upper bound of the white area is the threshold of pain. Here the boundary of the white area appears straight because the psychological response is relatively independent of frequency at this high sound level.

The most dramatic change with frequency is in the lower left region of the white area, for low frequencies and low intensity levels. Our ears are particularly insensitive in this region. If you are listening to your stereo and the bass (low frequencies) and treble (high frequencies) sound balanced at a high volume, try turning the volume down and listening again. You will probably notice that the bass seems weak, which is due to the insensitivity of the ear to low frequencies at low sound levels as shown in Figure 17.7.

17.4 The Doppler Effect

Perhaps you have noticed how the sound of a vehicle's horn changes as the vehicle moves past you. The frequency of the sound you hear as the vehicle approaches you is higher than the frequency you hear as it moves away from you. This experience is one example of the **Doppler effect.**[3]

To see what causes this apparent frequency change, imagine you are in a boat that is lying at anchor on a gentle sea where the waves have a period of $T = 3.0$ s.

[3]Named after Austrian physicist Christian Johann Doppler (1803–1853), who in 1842 predicted the effect for both sound waves and light waves.

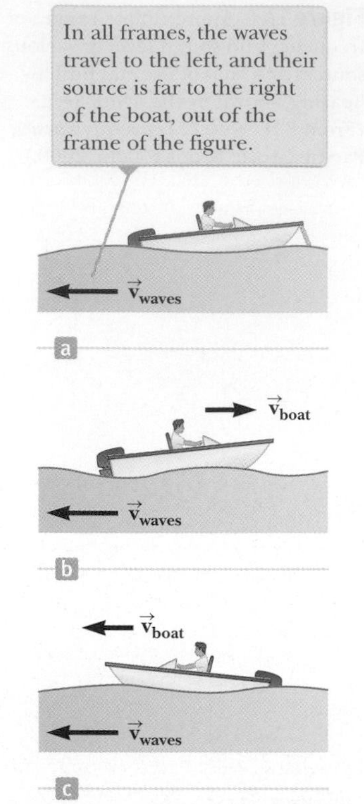

Figure 17.8 (a) Waves moving toward a stationary boat. (b) The boat moving toward the wave source. (c) The boat moving away from the wave source.

Hence, every 3.0 s a crest hits your boat. Figure 17.8a shows this situation, with the water waves moving toward the left. If you set your watch to $t = 0$ just as one crest hits, the watch reads 3.0 s when the next crest hits, 6.0 s when the third crest hits, and so on. From these observations, you conclude that the wave frequency is $f = 1/T = 1/(3.0 \text{ s}) = 0.33$ Hz. Now suppose you start your motor and head directly into the oncoming waves as in Figure 17.8b. Again you set your watch to $t = 0$ as a crest hits the front (the bow) of your boat. Now, however, because you are moving toward the next wave crest as it moves toward you, it hits you less than 3.0 s after the first hit. In other words, the period you observe is shorter than the 3.0-s period you observed when you were stationary. Because $f = 1/T$, you observe a higher wave frequency than when you were at rest.

If you turn around and move in the same direction as the waves (Fig. 17.8c), you observe the opposite effect. You set your watch to $t = 0$ as a crest hits the back (the stern) of the boat. Because you are now moving away from the next crest, more than 3.0 s has elapsed on your watch by the time that crest catches you. Therefore, you observe a lower frequency than when you were at rest.

These effects occur because the *relative* speed between your boat and the waves depends on the direction of travel and on the speed of your boat. When you are moving toward the right in Figure 17.8b, this relative speed is higher than that of the wave speed, which leads to the observation of an increased frequency. When you turn around and move to the left, the relative speed is lower, as is the observed frequency of the water waves.

Let's now examine an analogous situation with sound waves in which the water waves become sound waves, the water becomes the air, and the person on the boat becomes an observer listening to the sound. In this case, an observer O is moving and a sound source S is stationary. For simplicity, we assume the air is also stationary and the observer moves directly toward the source (Active Fig. 17.9). The observer moves with a speed v_O toward a stationary point source ($v_S = 0$), where *stationary* means at rest with respect to the medium, air.

If a point source emits sound waves and the medium is uniform, the waves move at the same speed in all directions radially away from the source; the result is a spherical wave as mentioned in Section 17.3. The distance between adjacent wave fronts equals the wavelength λ. In Active Figure 17.9, the circles are the intersections of these three-dimensional wave fronts with the two-dimensional paper.

We take the frequency of the source in Active Figure 17.9 to be f, the wavelength to be λ, and the speed of sound to be v. If the observer were also stationary, he would detect wave fronts at a frequency f. (That is, when $v_O = 0$ and $v_S = 0$, the observed frequency equals the source frequency.) When the observer moves toward the source, the speed of the waves relative to the observer is $v' = v + v_O$, as in the case of the boat in Figure 17.8, but the wavelength λ is unchanged. Hence, using Equation 16.12, $v = \lambda f$, we can say that the frequency f' heard by the observer is *increased* and is given by

$$f' = \frac{v'}{\lambda} = \frac{v + v_O}{\lambda}$$

Because $\lambda = v/f$, we can express f' as

$$f' = \left(\frac{v + v_O}{v}\right) f \quad \text{(observer moving toward source)} \qquad \textbf{(17.15)}$$

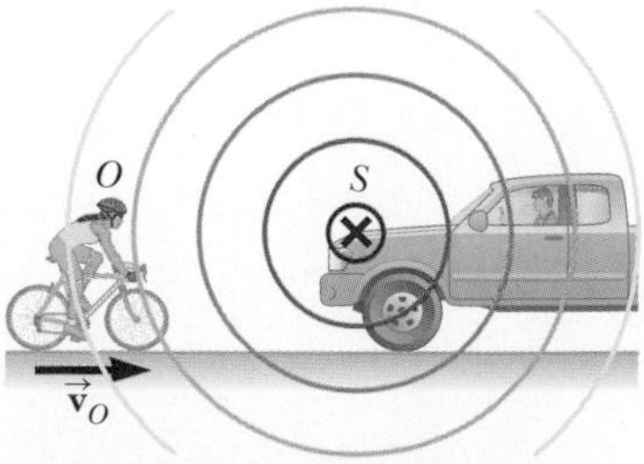

ACTIVE FIGURE 17.9

An observer O (the cyclist) moves with a speed v_O toward a stationary point source S, the horn of a parked truck. The observer hears a frequency f' that is greater than the source frequency.

If the observer is moving away from the source, the speed of the wave relative to the observer is $v' = v - v_O$. The frequency heard by the observer in this case is *decreased* and is given by

$$f' = \left(\frac{v - v_O}{v}\right) f \quad \text{(observer moving away from source)} \qquad \textbf{(17.16)}$$

These last two equations can be reduced to a single equation by adopting a sign convention. Whenever an observer moves with a speed v_O relative to a stationary

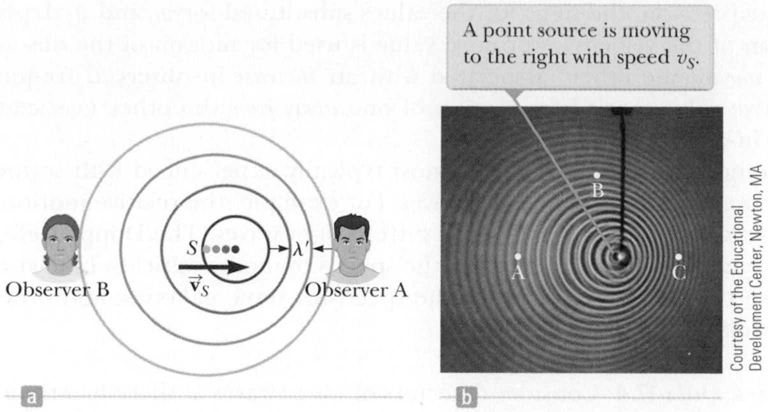

ACTIVE FIGURE 17.10

(a) A source *S* moving with a speed v_S toward a stationary observer A and away from a stationary observer B. Observer A hears an increased frequency, and observer B hears a decreased frequency. (b) The Doppler effect in water, observed in a ripple tank. Letters shown in the photo refer to Quick Quiz 17.4.

source, the frequency heard by the observer is given by Equation 17.15, with v_O interpreted as follows: a positive value is substituted for v_O when the observer moves toward the source, and a negative value is substituted when the observer moves away from the source.

Now suppose the source is in motion and the observer is at rest. If the source moves directly toward observer A in Active Figure 17.10a, each new wave is emitted from a position to the right of the origin of the previous wave. As a result, the wave fronts heard by the observer are closer together than they would be if the source were not moving. (Active Fig. 17.10b shows this effect for waves moving on the surface of water.) As a result, the wavelength λ' measured by observer A is shorter than the wavelength λ of the source. During each vibration, which lasts for a time interval T (the period), the source moves a distance $v_S T = v_S/f$ and the wavelength is *shortened* by this amount. Therefore, the observed wavelength λ' is

$$\lambda' = \lambda - \Delta\lambda = \lambda - \frac{v_S}{f}$$

Because $\lambda = v/f$, the frequency f' heard by observer A is

$$f' = \frac{v}{\lambda'} = \frac{v}{\lambda - (v_S/f)} = \frac{v}{(v/f) - (v_S/f)}$$

$$f' = \left(\frac{v}{v - v_S}\right) f \quad \text{(source moving toward observer)} \qquad \textbf{(17.17)}$$

That is, the observed frequency is *increased* whenever the source is moving toward the observer.

When the source moves away from a stationary observer, as is the case for observer B in Active Figure 17.10a, the observer measures a wavelength λ' that is *greater* than λ and hears a *decreased* frequency:

$$f' = \left(\frac{v}{v + v_S}\right) f \quad \text{(source moving away from observer)} \qquad \textbf{(17.18)}$$

We can express the general relationship for the observed frequency when a source is moving and an observer is at rest as Equation 17.17, with the same sign convention applied to v_S as was applied to v_O: a positive value is substituted for v_S when the source moves toward the observer, and a negative value is substituted when the source moves away from the observer.

Finally, combining Equations 17.15 and 17.17 gives the following general relationship for the observed frequency that includes all four conditions described by Equations 17.15 through 17.18:

$$f' = \left(\frac{v + v_O}{v - v_S}\right) f \qquad \textbf{(17.19)}$$

◀ **General Doppler-shift expression**

Pitfall Prevention 17.1

Doppler Effect Does Not Depend on Distance

Some people think that the Doppler effect depends on the distance between the source and the observer. Although the *intensity* of a sound varies as the distance changes, the apparent *frequency* depends only on the relative speed of source and observer. As you listen to an approaching source, you will detect increasing intensity but constant frequency. As the source passes, you will hear the frequency suddenly drop to a new constant value and the intensity begin to decrease.

In this expression, the signs for the values substituted for v_O and v_S depend on the direction of the velocity. A positive value is used for motion of the observer or the source *toward* the other (associated with an *increase* in observed frequency), and a negative value is used for motion of one *away from* the other (associated with a *decrease* in observed frequency).

Although the Doppler effect is most typically experienced with sound waves, it is a phenomenon common to all waves. For example, the relative motion of source and observer produces a frequency shift in light waves. The Doppler effect is used in police radar systems to measure the speeds of motor vehicles. Likewise, astronomers use the effect to determine the speeds of stars, galaxies, and other celestial objects relative to the Earth.

***Quick Quiz* 17.4** Consider detectors of water waves at three locations A, B, and C in Active Figure 17.10b. Which of the following statements is true? **(a)** The wave speed is highest at location A. **(b)** The wave speed is highest at location C. **(c)** The detected wavelength is largest at location B. **(d)** The detected wavelength is largest at location C. **(e)** The detected frequency is highest at location C. **(f)** The detected frequency is highest at location A.

***Quick Quiz* 17.5** You stand on a platform at a train station and listen to a train approaching the station at a constant velocity. While the train approaches, but before it arrives, what do you hear? **(a)** the intensity and the frequency of the sound both increasing **(b)** the intensity and the frequency of the sound both decreasing **(c)** the intensity increasing and the frequency decreasing **(d)** the intensity decreasing and the frequency increasing **(e)** the intensity increasing and the frequency remaining the same **(f)** the intensity decreasing and the frequency remaining the same

Example 17.4 The Broken Clock Radio

Your clock radio awakens you with a steady and irritating sound of frequency 600 Hz. One morning, it malfunctions and cannot be turned off. In frustration, you drop the clock radio out of your fourth-story dorm window, 15.0 m from the ground. Assume the speed of sound is 343 m/s. As you listen to the falling clock radio, what frequency do you hear just before you hear it striking the ground?

SOLUTION

Conceptualize The speed of the clock radio increases as it falls. Therefore, it is a source of sound moving away from you with an increasing speed so the frequency you hear should be less than 600 Hz.

Categorize We categorize this problem as one in which we combine the particle under constant acceleration model for the falling radio with our understanding of the frequency shift of sound due to the Doppler effect.

Analyze Because the clock radio is modeled as a particle under constant acceleration due to gravity, use Equation 2.13 to express the speed of the source of sound:

$$(1)\quad v_S = v_{yi} + a_y t = 0 - gt = -gt$$

From Equation 2.16, find the time at which the clock radio strikes the ground:

$$y_f = y_i + v_{yi}t - \tfrac{1}{2}gt^2 = 0 + 0 - \tfrac{1}{2}gt^2 \rightarrow t = \sqrt{-\frac{2y_f}{g}}$$

Substitute into Equation (1):

$$v_S = (-g)\sqrt{-\frac{2y_f}{g}} = -\sqrt{-2gy_f}$$

Use Equation 17.19 to determine the Doppler-shifted frequency heard from the falling clock radio:

$$f' = \left[\frac{v + 0}{v - (-\sqrt{-2gy_f})}\right] f = \left(\frac{v}{v + \sqrt{-2gy_f}}\right) f$$

17.4 *cont.*

Substitute numerical values:

$$f' = \left[\frac{343\ \text{m/s}}{343\ \text{m/s} + \sqrt{-2(9.80\ \text{m/s}^2)(-15.0\ \text{m})}}\right](600\ \text{Hz})$$

$$= \boxed{571\ \text{Hz}}$$

Finalize The frequency is lower than the actual frequency of 600 Hz because the clock radio is moving away from you. If it were to fall from a higher floor so that it passes below $y = -15.0$ m, the clock radio would continue to accelerate and the frequency would continue to drop.

Example 17.5 Doppler Submarines

A submarine (sub A) travels through water at a speed of 8.00 m/s, emitting a sonar wave at a frequency of 1 400 Hz. The speed of sound in the water is 1 533 m/s. A second submarine (sub B) is located such that both submarines are traveling directly toward each other. The second submarine is moving at 9.00 m/s.

(A) What frequency is detected by an observer riding on sub B as the subs approach each other?

SOLUTION

Conceptualize Even though the problem involves subs moving in water, there is a Doppler effect just like there is when you are in a moving car and listening to a sound moving through the air from another car.

Categorize Because both subs are moving, we categorize this problem as one involving the Doppler effect for both a moving source and a moving observer.

Analyze Use Equation 17.19 to find the Doppler-shifted frequency heard by the observer in sub B, being careful with the signs assigned to the source and observer speeds:

$$f' = \left(\frac{v + v_O}{v - v_S}\right)f$$

$$f' = \left[\frac{1\,533\ \text{m/s} + (+9.00\ \text{m/s})}{1\,533\ \text{m/s} - (+8.00\ \text{m/s})}\right](1\,400\ \text{Hz}) = \boxed{1\,416\ \text{Hz}}$$

(B) The subs barely miss each other and pass. What frequency is detected by an observer riding on sub B as the subs recede from each other?

SOLUTION

Use Equation 17.19 to find the Doppler-shifted frequency heard by the observer in sub B, again being careful with the signs assigned to the source and observer speeds:

$$f' = \left(\frac{v + v_O}{v - v_S}\right)f$$

$$f' = \left[\frac{1\,533\ \text{m/s} + (-9.00\ \text{m/s})}{1\,533\ \text{m/s} - (-8.00\ \text{m/s})}\right](1\,400\ \text{Hz}) = \boxed{1\,385\ \text{Hz}}$$

Finalize Notice that the frequency drops from 1 416 Hz to 1 385 Hz as the subs pass. This effect is similar to the drop in frequency you hear when a car passes by you while blowing its horn.

WHAT IF? While the subs are approaching each other, some of the sound from sub A reflects from sub B and returns to sub A. If this sound were to be detected by an observer on sub A, what is its frequency?

Answer The sound of apparent frequency 1 416 Hz found in part (A) is reflected from a moving source (sub B) and then detected by a moving observer (sub A). Therefore, the frequency detected by sub A is

$$f'' = \left(\frac{v + v_O}{v - v_S}\right)f'$$

$$= \left[\frac{1\,533\ \text{m/s} + (+8.00\ \text{m/s})}{1\,533\ \text{m/s} - (+9.00\ \text{m/s})}\right](1\,416\ \text{Hz}) = 1\,432\ \text{Hz}$$

This technique is used by police officers to measure the speed of a moving car. Microwaves are emitted from the police car and reflected by the moving car. By detecting the Doppler-shifted frequency of the reflected microwaves, the police officer can determine the speed of the moving car.

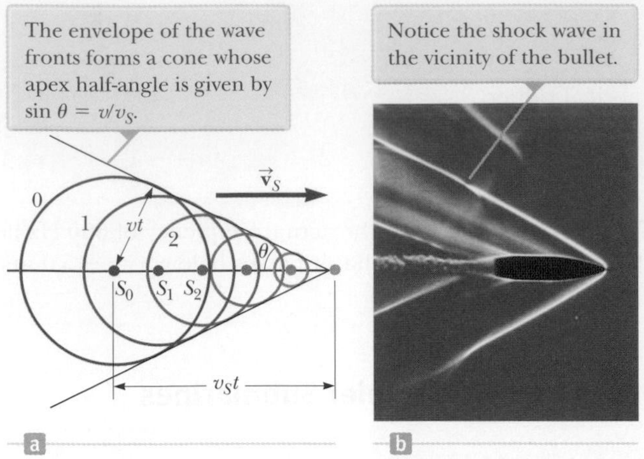

Figure 17.11 (a) A representation of a shock wave produced when a source moves from S_0 to the right with a speed v_S that is greater than the wave speed v in the medium. (b) A stroboscopic photograph of a bullet moving at supersonic speed through the hot air above a candle.

Shock Waves

Figure 17.12 The V-shaped bow wave of a boat is formed because the boat speed is greater than the speed of the water waves it generates. A bow wave is analogous to a shock wave formed by an airplane traveling faster than sound.

Now consider what happens when the speed v_S of a source *exceeds* the wave speed v. This situation is depicted graphically in Figure 17.11a. The circles represent spherical wave fronts emitted by the source at various times during its motion. At $t = 0$, the source is at S_0 and moving toward the right. At later times, the source is at S_1, and then S_2, and so on. At the time t, the wave front centered at S_0 reaches a radius of vt. In this same time interval, the source travels a distance $v_S t$. Notice in Figure 17.11a that a straight line can be drawn tangent to all the wave fronts generated at various times. Therefore, the envelope of these wave fronts is a cone whose apex half-angle θ (the "Mach angle") is given by

$$\sin\theta = \frac{vt}{v_S t} = \frac{v}{v_S}$$

The ratio v_S/v is referred to as the *Mach number,* and the conical wave front produced when $v_S > v$ (supersonic speeds) is known as a *shock wave.* An interesting analogy to shock waves is the V-shaped wave fronts produced by a boat (the bow wave) when the boat's speed exceeds the speed of the surface-water waves (Fig. 17.12).

Jet airplanes traveling at supersonic speeds produce shock waves, which are responsible for the loud "sonic boom" one hears. The shock wave carries a great deal of energy concentrated on the surface of the cone, with correspondingly great pressure variations. Such shock waves are unpleasant to hear and can cause damage to buildings when aircraft fly supersonically at low altitudes. In fact, an airplane flying at supersonic speeds produces a double boom because two shock waves are formed, one from the nose of the plane and one from the tail. People near the path of a space shuttle as it glides toward its landing point often report hearing what sounds like two very closely spaced cracks of thunder.

***Quick Quiz* 17.6** An airplane flying with a constant velocity moves from a cold air mass into a warm air mass. Does the Mach number **(a)** increase, **(b)** decrease, or **(c)** stay the same?

Summary

Definitions

The **intensity** of a periodic sound wave, which is the power per unit area, is

$$I \equiv \frac{(Power)_{avg}}{A} = \frac{(\Delta P_{max})^2}{2\rho v} \qquad \textbf{(17.11, 17.12)}$$

The **sound level** of a sound wave in decibels is

$$\beta \equiv 10 \log\left(\frac{I}{I_0}\right) \qquad \textbf{(17.14)}$$

The constant I_0 is a reference intensity, usually taken to be at the threshold of hearing (1.00×10^{-12} W/m^2), and I is the intensity of the sound wave in watts per square meter.

Concepts and Principles

Sound waves are longitudinal and travel through a compressible medium with a speed that depends on the elastic and inertial properties of that medium. The speed of sound in a gas having a bulk modulus B and density ρ is

$$v = \sqrt{\frac{B}{\rho}} \qquad \textbf{(17.8)}$$

For sinusoidal sound waves, the variation in the position of an element of the medium is

$$s(x, t) = s_{\max} \cos (kx - \omega t) \qquad \textbf{(17.1)}$$

and the variation in pressure from the equilibrium value is

$$\Delta P = \Delta P_{\max} \sin (kx - \omega t) \qquad \textbf{(17.2)}$$

where $\Delta P_{\max}$ is the **pressure amplitude.** The pressure wave is 90° out of phase with the displacement wave. The relationship between $s_{\max}$ and $\Delta P_{\max}$ is

$$\Delta P_{\max} = \rho v \omega s_{\max} \qquad \textbf{(17.10)}$$

The change in frequency heard by an observer whenever there is relative motion between a source of sound waves and the observer is called the **Doppler effect.** The observed frequency is

$$f' = \left(\frac{v + v_O}{v - v_S}\right) f \qquad \textbf{(17.19)}$$

In this expression, the signs for the values substituted for v_O and v_S depend on the direction of the velocity. A positive value for the speed of the observer or source is substituted if the velocity of one is toward the other, whereas a negative value represents a velocity of one away from the other.

Objective Questions

☐ denotes answer available in Student Solutions Manual/Study Guide

1. A sound wave can be characterized as (a) a transverse wave, (b) a longitudinal wave, (c) a transverse wave or a longitudinal wave, depending on the nature of its source, (d) one that carries no energy, or (e) a wave that does not require a medium to be transmitted from one place to the other.

2. What happens to a sound wave as it travels from air into water? (a) Its intensity increases. (b) Its wavelength decreases. (c) Its frequency increases. (d) Its frequency remains the same. (e) Its velocity decreases.

3. As you travel down the highway in your car, an ambulance approaches you from the rear at a high speed (Fig. OQ17.3) sounding its siren at a frequency of 500 Hz. Which statement is correct? (a) You hear a frequency less than 500 Hz. (b) You hear a frequency equal to 500 Hz. (c) You hear a frequency greater than 500 Hz. (d) You hear a frequency greater than 500 Hz, whereas the ambulance driver hears a frequency lower than 500 Hz. (e) You hear a frequency less than 500 Hz, whereas the ambulance driver hears a frequency of 500 Hz.

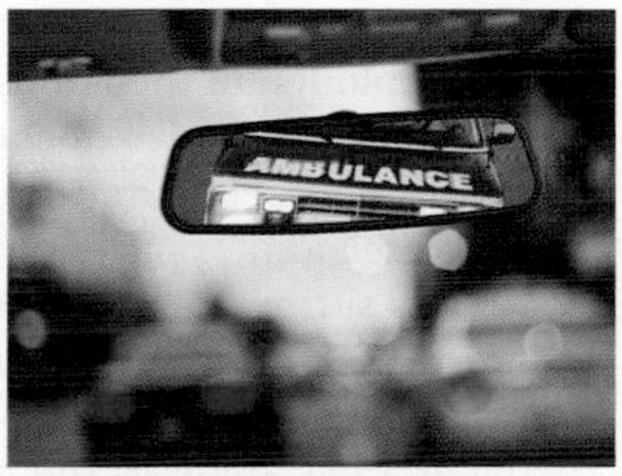

Anthony Redpath/Corbis

Figure OQ17.3

4. Two sirens A and B are sounding so that the frequency from A is twice the frequency from B. Compared with the speed of sound from A, is the speed of sound from B (a) twice as fast, (b) half as fast, (c) four times as fast, (d) one-fourth as fast, or (e) the same?

5. Doubling the power output from a sound source emitting a single frequency will result in what increase in decibel level? (a) 0.50 dB (b) 2.0 dB (c) 3.0 dB (d) 4.0 dB (e) above 20 dB

6. If a 1.00-kHz sound source moves at a speed of 50.0 m/s toward a listener who moves at a speed of 30.0 m/s in a direction away from the source, what is the apparent frequency heard by the listener? (a) 796 Hz (b) 949 Hz (c) 1 000 Hz (d) 1 068 Hz (e) 1 273 Hz

7. Table 17.1 shows the speed of sound is typically an order of magnitude larger in solids than in gases. To what can this higher value be most directly attributed? (a) the difference in density between solids and gases (b) the difference in compressibility between solids and gases (c) the limited size of a solid object compared to a free gas (d) the impossibility of holding a gas under significant tension

8. Assume a change at the source of sound reduces the wavelength of a sound wave in air by a factor of 2. **(i)** What happens to its frequency? (a) It increases by a factor of 4. (b) It increases by a factor of 2. (c) It is unchanged. (d) It decreases by a factor of 2. (e) It changes by an unpredictable factor. **(ii)** What happens to its speed? Choose from the same possibilities as in part (i).

9. A point source broadcasts sound into a uniform medium. If the distance from the source is tripled, how does the intensity change? (a) It becomes one-ninth as large. (b) It becomes one-third as large. (c) It is unchanged. (d) It becomes three times larger. (e) It becomes nine times larger.

10. Of the following sounds, which one is most likely to have a sound level of 60 dB? (a) a rock concert (b) the turning of a page in this textbook (c) dinner-table conversation (d) a cheering crowd at a football game

11. A source of sound vibrates with constant frequency. Rank the frequency of sound observed in the following cases from highest to the lowest. If two frequencies are equal, show their equality in your ranking. All the motions mentioned have the same speed, 25 m/s. (a) The source and observer are stationary. (b) The source is moving toward a stationary observer. (c) The source is moving away from a stationary observer. (d) The observer is moving toward a stationary source. (e) The observer is moving away from a stationary source.

12. With a sensitive sound-level meter, you measure the sound of a running spider as −10 dB. What does the negative sign imply? (a) The spider is moving away from you. (b) The frequency of the sound is too low to be audible to humans. (c) The intensity of the sound is too faint to be audible to humans. (d) You have made a mistake; negative signs do not fit with logarithms.

13. A church bell in a steeple rings once. At 300 m in front of the church, the maximum sound intensity is 2 μW/m^2. At 950 m behind the church, the maximum intensity is 0.2 μW/m^2. What is the main reason for the difference in the intensity? (a) Most of the sound is absorbed by the air before it gets far away from the source. (b) Most of the sound is absorbed by the ground as it travels away from the source. (c) The bell broadcasts the sound mostly toward the front. (d) At a larger distance, the power is spread over a larger area.

14. Suppose an observer and a source of sound are both at rest relative to the ground and a strong wind is blowing away from the source toward the observer. **(i)** What effect does the wind have on the observed frequency? (a) It causes an increase. (b) It causes a decrease. (c) It causes no change. **(ii)** What effect does the wind have on the observed wavelength? Choose from the same possibilities as in part (i). **(iii)** What effect does the wind have on the observed speed of the wave? Choose from the same possibilities as in part (i).

Conceptual Questions

☐ denotes answer available in Student Solutions Manual/Study Guide

1. A sonic ranger is a device that determines the distance to an object by sending out an ultrasonic sound pulse and measuring the time interval required for the wave to return by reflection from the object. Typically, these devices cannot reliably detect an object that is less than half a meter from the sensor. Why is that?

2. How can you determine that the speed of sound is the same for all frequencies by listening to a band or orchestra?

3. A friend sitting in her car far down the road waves to you and beeps her horn at the same moment. How far away must she be for you to calculate the speed of sound to two significant figures by measuring the time interval required for the sound to reach you?

4. Older auto-focus cameras sent out a pulse of sound and measured the time interval required for the pulse to reach an object, reflect off of it, and return to be detected. Can air temperature affect the camera's focus? New cameras use a more reliable infrared system.

5. Explain how the distance to a lightning bolt (Fig. CQ17.5) can be determined by counting the seconds between the flash and the sound of thunder.

© iStockphoto.com/Colin Ortner

Figure CQ17.5

6. You are driving toward a cliff and honk your horn. Is there a Doppler shift of the sound when you hear the echo? If so, is it like a moving source or a moving observer? What if the reflection occurs not from a cliff, but from the forward edge of a huge alien spacecraft moving toward you as you drive?

7. The radar systems used by police to detect speeders are sensitive to the Doppler shift of a pulse of microwaves. Discuss how this sensitivity can be used to measure the speed of a car.

8. *The Tunguska event.* On June 30, 1908, a meteor burned up and exploded in the atmosphere above the Tunguska River valley in Siberia. It knocked down trees over thousands of square kilometers and started a forest fire, but produced no crater and apparently caused no human casualties. A witness sitting on his doorstep outside the zone of falling trees recalled events in the following sequence. He saw a moving light in the sky, brighter than the Sun and descending at a low angle to the horizon. He felt his face become warm. He felt the ground shake. An invisible agent picked him up and immediately dropped him about a meter from where he had been seated. He heard a very loud protracted rumbling. Suggest an explanation for these observations and for the order in which they happened.

9. How can an object move with respect to an observer so that the sound from it is not shifted in frequency?

Problems

WebAssign The problems found in this chapter may be assigned online in Enhanced WebAssign

1. denotes straightforward problem; **2.** denotes intermediate problem; **3.** denotes challenging problem

1. full solution available in the Student Solutions Manual/Study Guide

1. denotes problems most often assigned in Enhanced WebAssign; these provide students with targeted feedback and either a Master It tutorial or a Watch It solution video.

QIC denotes asking for quantitative and conceptual reasoning

S denotes symbolic reasoning problem

M denotes Master It tutorial available in Enhanced WebAssign

GP denotes guided problem

shaded denotes "paired problems" that develop reasoning with symbols and numerical values

Note: Throughout this chapter, pressure variations ΔP are measured relative to atmospheric pressure, 1.013×10^5 Pa.

Section 17.1 Pressure Variations in Sound Waves

1. Write an expression that describes the pressure variation as a function of position and time for a sinusoidal sound wave in air. Assume the speed of sound is 343 m/s, $\lambda = 0.100$ m, and $\Delta P_{max} = 0.200$ Pa.

2. As a certain sound wave travels through the air, it produces pressure variations (above and below atmospheric pressure) given by $\Delta P = 1.27 \sin(\pi x - 340\pi t)$ in SI units. Find (a) the amplitude of the pressure variations, (b) the frequency, (c) the wavelength in air, and (d) the speed of the sound wave.

3. A sinusoidal sound wave moves through a medium and is described by the displacement wave function

$$s(x, t) = 2.00 \cos(15.7x - 858t)$$

where s is in micrometers, x is in meters, and t is in seconds. Find (a) the amplitude, (b) the wavelength, and (c) the speed of this wave. (d) Determine the instantaneous displacement from equilibrium of the elements of the medium at the position $x = 0.050\,0$ m at $t = 3.00$ ms. (e) Determine the maximum speed of the element's oscillatory motion.

Section 17.2 Speed of Sound Waves

Problem 66 in Chapter 2 can also be assigned with this section.

Note: In the rest of this chapter, unless otherwise specified, the equilibrium density of air is $\rho = 1.20$ kg/m^3 and the speed of sound in air is $v = 343$ m/s. Use Table 17.1 to find speeds of sound in other media.

4. An experimenter wishes to generate in air a sound wave that has a displacement amplitude of 5.50×10^{-6} m. The pressure amplitude is to be limited to 0.840 Pa. What is the minimum wavelength the sound wave can have?

5. **QIC** Suppose you hear a clap of thunder 16.2 s after seeing the associated lightning strike. The speed of light in air is 3.00×10^8 m/s. (a) How far are you from the lightning strike? (b) Do you need to know the value of the speed of light to answer? Explain.

6. Earthquakes at fault lines in the Earth's crust create seismic waves, which are longitudinal (P waves) or transverse (S waves). The P waves have a speed of about 7 km/s. Estimate the average bulk modulus of the Earth's crust given that the density of rock is about 2 500 kg/m^3.

7. A dolphin (Fig. P17.7) in seawater at a temperature of 25°C emits a sound wave directed toward the ocean floor 150 m below. How much time passes before it hears an echo?

Stephen Frink/Photographer's Choice/Getty Images

Figure P17.7

8. **QIC** A sound wave propagates in air at 27°C with frequency 4.00 kHz. It passes through a region where the temperature gradually changes and then moves through air at 0°C. Give numerical answers to the following questions to the extent possible and state your reasoning about what happens to the wave physically. (a) What happens to the speed of the wave? (b) What happens to its frequency? (c) What happens to its wavelength?

9. Ultrasound is used in medicine both for diagnostic imaging (Fig. P17.9) and for therapy. For diagnosis, short pulses of ultrasound are passed through the patient's body. An echo reflected from a structure of interest is recorded, and the distance to the structure can be determined from the time delay for the echo's return. To reveal detail, the wavelength of the reflected ultrasound must be small compared to the size of the object reflecting the wave. The speed of ultrasound in human tissue is about 1 500 m/s (nearly the same as the speed of sound in water). (a) What is the wavelength of ultrasound with a frequency of 2.40 MHz? (b) In the whole set of imaging techniques, frequencies in the range 1.00 MHz to 20.0 MHz are used. What is the range of wavelengths corresponding to this range of frequencies?

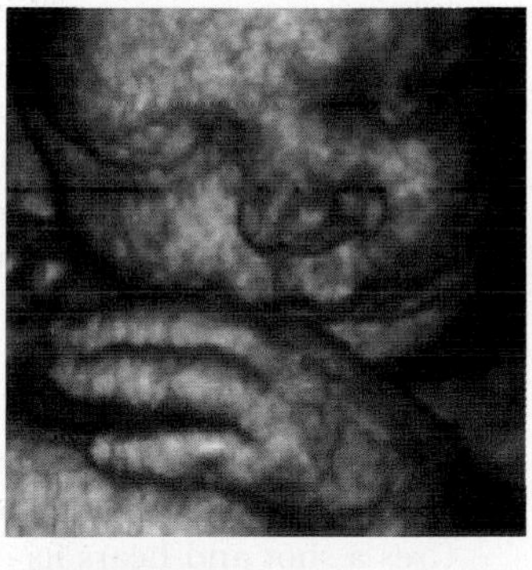

B. Benoit/Photo Researchers, Inc.

Figure P17.9 A view of a fetus in the uterus made with ultrasound imaging.

10. A sound wave in air has a pressure amplitude equal to 4.00×10^{-3} Pa. Calculate the displacement amplitude of the wave at a frequency of 10.0 kHz.

11. A flowerpot is knocked off a window ledge from a height $d = 20.0$ m above the sidewalk as shown in Figure P17.11. It falls toward an unsuspecting man of height $h = 1.75$ m who is standing below. Assume the man requires a time interval of $\Delta t = 0.300$ s to respond to the warning. How close to the sidewalk can the flowerpot fall before it is too late for a warning shouted from the balcony to reach the man in time?

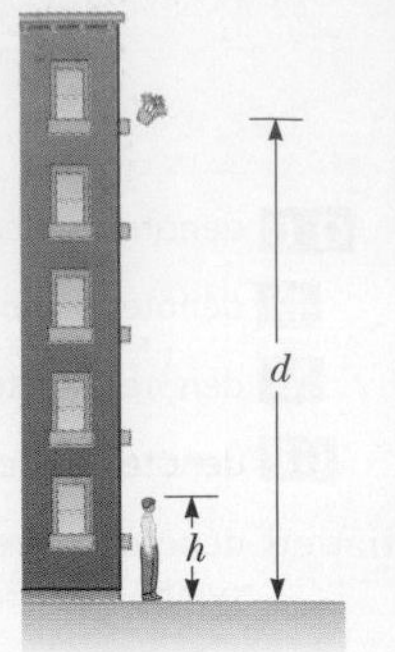

Figure P17.11
Problems 11 and 12.

12. **S** A flowerpot is knocked off a balcony from a height d above the sidewalk as shown in Figure P17.11. It falls toward an unsuspecting man of height h who is standing below. Assume the man requires a time interval of Δt to respond to the warning. How close to the sidewalk can the flowerpot fall before it is too late for a warning shouted from the balcony to reach the man in time? Use the symbol v for the speed of sound.

13. The speed of sound in air (in meters per second) depends on temperature according to the approximate expression

$$v = 331.5 + 0.607T_C$$

where T_C is the Celsius temperature. In dry air, the temperature decreases about 1°C for every 150-m rise in altitude. (a) Assume this change is constant up to an altitude of 9 000 m. What time interval is required for the sound from an airplane flying at 9 000 m to reach the ground on a day when the ground temperature is 30°C? (b) **What If?** Compare your answer with the time interval required if the air were uniformly at 30°C. Which time interval is longer?

14. A rescue plane flies horizontally at a constant speed searching for a disabled boat. When the plane is directly above the boat, the boat's crew blows a loud horn. By the time the plane's sound detector receives the horn's sound, the plane has traveled a distance equal to half its altitude above the ocean. Assuming it takes the sound 2.00 s to reach the plane, determine (a) the speed of the plane and (b) its altitude.

15. A cowboy stands on horizontal ground between two parallel, vertical cliffs. He is not midway between the cliffs. He fires a shot and hears its echoes. The second echo arrives 1.92 s after the first and 1.47 s before the third. Consider only the sound traveling parallel to the ground and reflecting from the cliffs. (a) What is the distance between the cliffs? (b) **What If?** If he can hear a fourth echo, how long after the third echo does it arrive?

16. **S** A sound wave moves down a cylinder as in Active Figure 17.2. Show that the pressure variation of the wave is described by $\Delta P = \pm \rho v \omega \sqrt{s_{max}^2 - s^2}$, where $s = s(x, t)$ is given by Equation 17.1.

17. **Q|C** A hammer strikes one end of a thick iron rail of length 8.50 m. A microphone located at the opposite end of the rail detects two pulses of sound, one that travels through the air and a longitudinal wave that travels through the rail. (a) Which pulse reaches the microphone first? (b) Find the separation in time between the arrivals of the two pulses.

Section 17.3 Intensity of Periodic Sound Waves

18. The area of a typical eardrum is about 5.00×10^{-5} m^2. (a) Calculate the average sound power incident on an eardrum at the threshold of pain, which corresponds to an intensity of 1.00 W/m^2. (b) How much energy is transferred to the eardrum exposed to this sound for 1.00 min?

19. Calculate the sound level (in decibels) of a sound wave that has an intensity of 4.00 μW/m^2.

20. The sound intensity at a distance of 16 m from a noisy generator is measured to be 0.25 W/m^2. What is the sound intensity at a distance of 28 m from the generator?

21. The intensity of a sound wave at a fixed distance from a speaker vibrating at 1.00 kHz is 0.600 W/m^2. (a) Determine the intensity that results if the frequency is increased to 2.50 kHz while a constant displacement amplitude is maintained. (b) Calculate the intensity if the frequency is reduced to 0.500 kHz and the displacement amplitude is doubled.

22. **S** The intensity of a sound wave at a fixed distance from a speaker vibrating at a frequency f is I. (a) Determine the intensity that results if the frequency is increased to f' while a constant displacement amplitude is maintained. (b) Calculate the intensity if the frequency is reduced to $f/2$ and the displacement amplitude is doubled.

23. A person wears a hearing aid that uniformly increases the sound level of all audible frequencies of sound by 30.0 dB. The hearing aid picks up sound having a frequency of 250 Hz at an intensity of 3.0×10^{-11} W/m^2. What is the intensity delivered to the eardrum?

24. A sound wave from a police siren has an intensity of 100.0 W/m^2 at a certain point; a second sound wave from a nearby ambulance has an intensity level that is 10 db greater than the police siren's sound wave at the same point. What is the sound level of the sound wave due to the ambulance?

25. The power output of a certain public-address speaker is 6.00 W. Suppose it broadcasts equally in all directions. (a) Within what distance from the speaker would the sound be painful to the ear? (b) At what distance from the speaker would the sound be barely audible?

26. As the people sing in church, the sound level everywhere inside is 101 dB. No sound is transmitted through the massive walls, but all the windows and doors are open on a summer morning. Their total area is 22.0 m^2. (a) How much sound energy is radiated through the windows and doors in 20.0 min? (b) Suppose the ground is a good reflector and sound radiates from the church uniformly in all horizontal and upward directions. Find the sound level 1.00 km away.

27. The most soaring vocal melody is in Johann Sebastian Bach's Mass in B Minor. In one section, the basses, tenors, altos, and sopranos carry the melody from a low D to a high A. In concert pitch, these notes are now assigned frequencies of 146.8 Hz and 880.0 Hz. Find the wavelengths of (a) the initial note and (b) the final note. Assume the cho-

rus sings the melody with a uniform sound level of 75.0 dB. Find the pressure amplitudes of (c) the initial note and (d) the final note. Find the displacement amplitudes of (e) the initial note and (f) the final note.

28. **S** Show that the difference between decibel levels β_1 and β_2 of a sound is related to the ratio of the distances r_1 and r_2 from the sound source by

$$\beta_2 - \beta_1 = 20 \log\left(\frac{r_1}{r_2}\right)$$

29. **M** A family ice show is held at an enclosed arena. The skaters perform to music with level 80.0 dB. This level is too loud for your baby, who yells at 75.0 dB. (a) What total sound intensity engulfs you? (b) What is the combined sound level?

30. Two small speakers emit sound waves of different frequencies equally in all directions. Speaker A has an output of 1.00 mW, and speaker B has an output of 1.50 mW. Determine the sound level (in decibels) at point C in Figure P17.30 assuming (a) only speaker A emits sound, (b) only speaker B emits sound, and (c) both speakers emit sound.

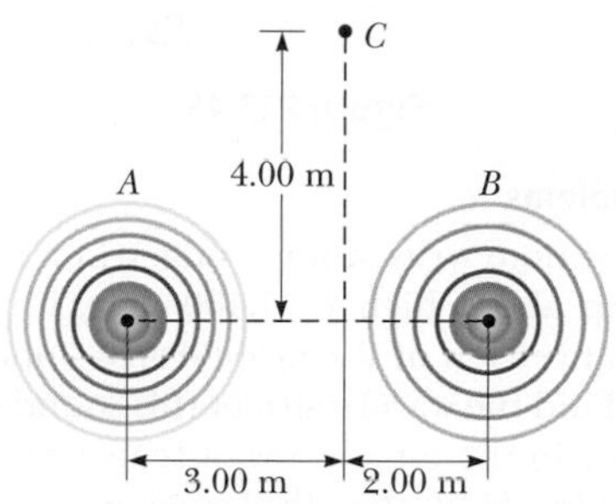

Figure P17.30

31. **M** A firework charge is detonated many meters above the ground. At a distance of $d_1 = 500$ m from the explosion, the acoustic pressure reaches a maximum of $\Delta P_{max} = 10.0$ Pa (Fig. P17.31). Assume the speed of sound is constant at 343 m/s throughout the atmosphere over the region considered, the ground absorbs all the sound falling on it, and the air absorbs sound energy as described by the rate 7.00 dB/km. What is the sound level (in decibels) at a distance of $d_2 = 4.00 \times 10^3$ m from the explosion?

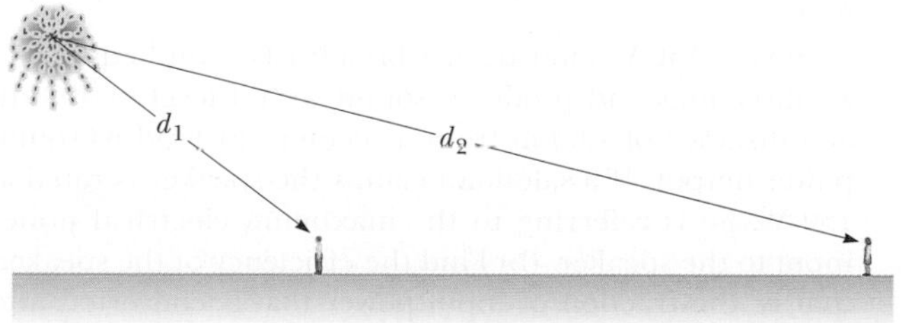

Figure P17.31

32. A fireworks rocket explodes at a height of 100 m above the ground. An observer on the ground directly under the explosion experiences an average sound intensity of 7.00×10^{-2} W/m^2 for 0.200 s. (a) What is the total amount of energy transferred away from the explosion by sound? (b) What is the sound level (in decibels) heard by the observer?

33. The sound level at a distance of 3.00 m from a source is 120 dB. At what distance is the sound level (a) 100 dB and (b) 10.0 dB?

34. *Why is the following situation impossible?* It is early on a Saturday morning, and much to your displeasure your next-door neighbor starts mowing his lawn. As you try to get back to sleep, your next-door neighbor on the other side of your house also begins to mow the lawn with an identical mower the same distance away. This situation annoys you greatly because the total sound now has twice the loudness it had when only one neighbor was mowing.

Section 17.4 The Doppler Effect

35. A driver travels northbound on a highway at a speed of 25.0 m/s. A police car, traveling southbound at a speed of 40.0 m/s, approaches with its siren producing sound at a frequency of 2 500 Hz. (a) What frequency does the driver observe as the police car approaches? (b) What frequency does the driver detect after the police car passes him? (c) Repeat parts (a) and (b) for the case when the police car is behind the driver and travels northbound.

36. **GP** Submarine A travels horizontally at 11.0 m/s through ocean water. It emits a sonar signal of frequency $f = 5.27 \times 10^3$ Hz in the forward direction. Submarine B is in front of submarine A and traveling at 3.00 m/s relative to the water in the same direction as submarine A. A crewman in submarine B uses his equipment to detect the sound waves ("pings") from submarine A. We wish to determine what is heard by the crewman in submarine B. (a) An observer on which submarine detects a frequency f' as described by Equation 17.19? (b) In Equation 17.19, should the sign of v_S be positive or negative? (c) In Equation 17.19, should the sign of v_O be positive or negative? (d) In Equation 17.19, what speed of sound should be used? (e) Find the frequency of the sound detected by the crewman on submarine B.

37. An ambulance moving at 42 m/s sounds its siren whose frequency is 450 Hz. A car is moving in the same direction as the ambulance at 25 m/s. What frequency does a person in the car hear (a) as the ambulance approaches the car? (b) After the ambulance passes the car?

38. When high-energy charged particles move through a transparent medium with a speed greater than the speed of light in that medium, a shock wave, or bow wave, of light is produced. This phenomenon is called the *Cerenkov effect*. When a nuclear reactor is shielded by a large pool of water, Cerenkov radiation can be seen as a blue glow in the vicinity of the reactor core due to high-speed electrons moving through the water (Fig. 17.38). In a particular case, the Cerenkov radiation produces a wave front with an apex half-angle of 53.0°. Calculate the speed of the electrons in the water. The speed of light in water is 2.25×10^8 m/s.

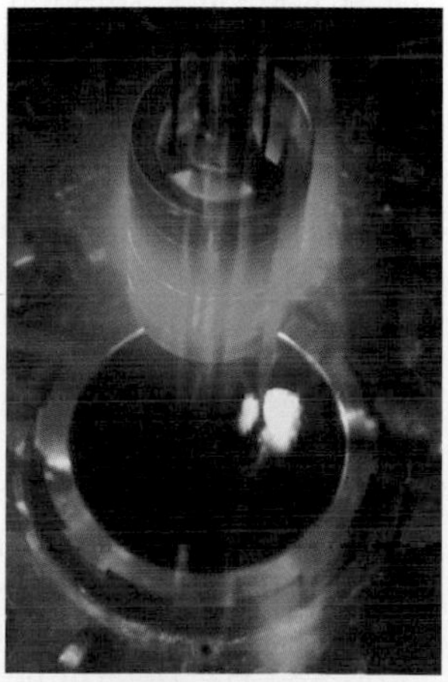

Figure P17.38

39. Review. A block with a speaker bolted to it is connected to a spring having spring constant $k = 20.0$ N/m and oscillates as shown in Figure P17.39. The total mass of the block and speaker is 5.00 kg, and the amplitude of this unit's motion is 0.500 m. The speaker emits sound waves of frequency 440 Hz. Determine (a) the highest and (b) the lowest frequencies heard by the person to the right of the speaker. (c) If the maximum sound level heard by the person is 60.0 dB when the speaker is at its closest distance $d = 1.00$ m from him, what is the minimum sound level heard by the observer?

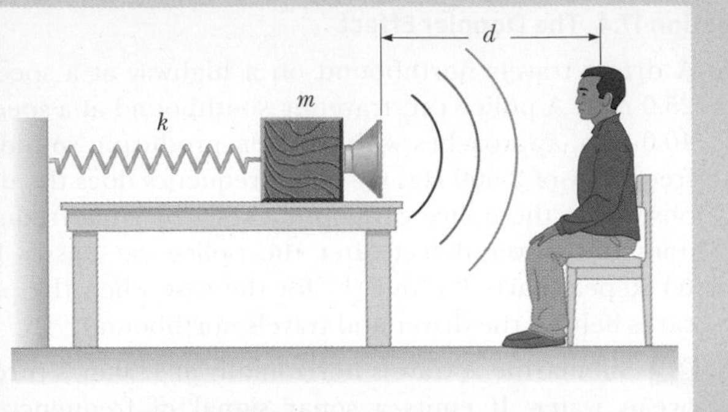

Figure P17.39 Problems 39 and 40.

40. S **Review.** A block with a speaker bolted to it is connected to a spring having spring constant k and oscillates as shown in Figure P17.39. The total mass of the block and speaker is m, and the amplitude of this unit's motion is A. The speaker emits sound waves of frequency f. Determine (a) the highest and (b) the lowest frequencies heard by the person to the right of the speaker. (c) If the maximum sound level heard by the person is β when the speaker is at its closest distance d from him, what is the minimum sound level heard by the observer?

41. Expectant parents are thrilled to hear their unborn baby's heartbeat, revealed by an ultrasonic detector that produces beeps of audible sound in synchronization with the fetal heartbeat. Suppose the fetus's ventricular wall moves in simple harmonic motion with an amplitude of 1.80 mm and a frequency of 115 beats per minute. (a) Find the maximum linear speed of the heart wall. Suppose a source mounted on the detector in contact with the mother's abdomen produces sound at 2 000 000.0 Hz, which travels through tissue at 1.50 km/s. (b) Find the maximum change in frequency between the sound that arrives at the wall of the baby's heart and the sound emitted by the source. (c) Find the maximum change in frequency between the reflected sound received by the detector and that emitted by the source.

42. *Why is the following situation impossible?* At the Summer Olympics, an athlete runs at a constant speed down a straight track while a spectator near the edge of the track blows a note on a horn with a fixed frequency. When the athlete passes the horn, she hears the frequency of the horn fall by the musical interval called a minor third. That is, the frequency she hears drops to five-sixths its original value.

43. Standing at a crosswalk, you hear a frequency of 560 Hz from the siren of an approaching ambulance. After the ambulance passes, the observed frequency of the siren is 480 Hz. Determine the ambulance's speed from these observations.

44. Review. A tuning fork vibrating at 512 Hz falls from rest and accelerates at 9.80 m/s^2. How far below the point of release is the tuning fork when waves of frequency 485 Hz reach the release point?

45. M A supersonic jet traveling at Mach 3.00 at an altitude of $h = 20\,000$ m is directly over a person at time $t = 0$ as shown in Figure P17.45. Assume the average speed of sound in air is 335 m/s over the path of the sound. (a) At what time will the person encounter the shock wave due to the sound emitted at $t = 0$? (b) Where will the plane be when this shock wave is heard?

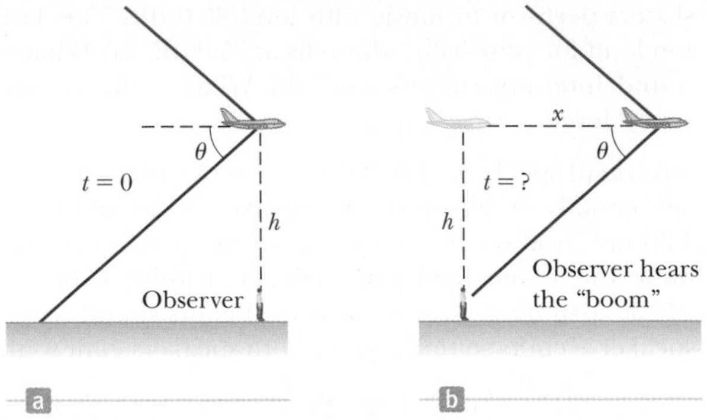

Figure P17.45

Additional Problems

46. The highest note written for a singer in a published score was F-sharp above high C, 1.480 kHz, for Zerbinetta in the original version of Richard Strauss's opera *Ariadne auf Naxos*. (a) Find the wavelength of this sound in air. (b) Suppose people in the fourth row of seats hear this note with level 81.0 dB. Find the displacement amplitude of the sound. (c) **What If?** In response to complaints, Strauss later transposed the note down to F above high C, 1.397 kHz. By what increment did the wavelength change?

47. Trucks carrying garbage to the town dump form a nearly steady procession on a country road, all traveling at 19.7 m/s in the same direction. Two trucks arrive at the dump every 3 min. A bicyclist is also traveling toward the dump, at 4.47 m/s. (a) With what frequency do the trucks pass the cyclist? (b) **What If?** A hill does not slow down the trucks, but makes the out-of-shape cyclist's speed drop to 1.56 m/s. How often do the trucks whiz past the cyclist now?

48. Assume a 150-W loudspeaker broadcasts sound equally in all directions and produces sound with a level of 103 dB at a distance of 1.60 m from its center. (a) Find its sound power output. If a salesman claims the speaker is rated at 150 W, he is referring to the maximum electrical power input to the speaker. (b) Find the efficiency of the speaker, that is, the fraction of input power that is converted into useful output power.

49. An interstate highway has been built through a neighborhood in a city. In the afternoon, the sound level in an apartment in the neighborhood is 80.0 dB as 100 cars pass outside the window every minute. Late at night, the traffic flow is only five cars per minute. What is the average late-night sound level?

50. The tensile stress in a thick copper bar is 99.5% of its elastic breaking point of 13.0×10^{10} N/m^2. If a 500-Hz sound wave is transmitted through the material, (a) what displacement amplitude will cause the bar to break? (b) What is the maximum speed of the elements of copper at this moment? (c) What is the sound intensity in the bar?

51. **QC** **Review.** A 150-g glider moves at $v_1 = 2.30$ m/s on an air track toward an originally stationary 200-g glider as shown in Figure P17.51. The gliders undergo a completely inelastic collision and latch together over a time interval of 7.00 ms. A student suggests roughly half the decrease in mechanical energy of the two-glider system is transferred to the environment by sound. Is this suggestion reasonable? To evaluate the idea, find the implied sound level at a position 0.800 m from the gliders. If the student's idea is unreasonable, suggest a better idea.

Before the collision

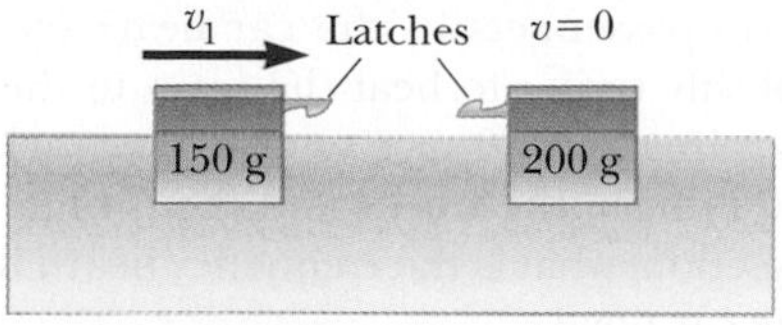

Figure P17.51

52. **QC** Consider the following wave function in SI units:

$$\Delta P(r, t) = \left(\frac{25.0}{r}\right) \sin (1.36r - 2\,030t)$$

Explain how this wave function can apply to a wave radiating from a small source, with r being the radial distance from the center of the source to any point outside the source. Give the most detailed description of the wave that you can. Include answers to such questions as the following and give representative values for any quantities that can be evaluated. (a) Does the wave move more toward the right or the left? (b) As it moves away from the source, what happens to its amplitude? (c) Its speed? (d) Its frequency? (e) Its wavelength? (f) Its power? (g) Its intensity?

53. **Review.** For a certain type of steel, stress is always proportional to strain with Young's modulus 20×10^{10} N/m^2. The steel has density 7.86×10^3 kg/m^3. It will fail by bending permanently if subjected to compressive stress greater than its yield strength $\sigma_y = 400$ MPa. A rod 80.0 cm long, made of this steel, is fired at 12.0 m/s straight at a very hard wall. (a) The speed of a one-dimensional compressional wave moving along the rod is given by $v = \sqrt{Y/\rho}$, where Y is Young's modulus for the rod and ρ is the density. Calculate this speed. (b) After the front end of the rod hits the wall and stops, the back end of the rod keeps moving as described by Newton's first law until it is stopped by excess pressure in a sound wave moving back through the rod. What time interval elapses before the back end of the rod receives the message that it should stop? (c) How far has the back end of the rod moved in this time interval? Find (d) the strain and (e) the stress in the rod. (f) If it is not to fail, what is the maximum impact speed a rod can have in terms of σ_y, Y, and ρ?

54. **QC** A large set of unoccupied football bleachers has solid seats and risers. You stand on the field in front of the bleachers and sharply clap two wooden boards together once. The sound pulse you produce has no definite frequency and no wavelength. The sound you hear reflected from the bleachers has an identifiable frequency and may remind you of a short toot on a trumpet, buzzer, or kazoo. (a) Explain what accounts for this sound. Compute order-of-magnitude estimates for (b) the frequency, (c) the wavelength, and (d) the duration of the sound on the basis of data you specify.

55. **M** To measure her speed, a skydiver carries a buzzer emitting a steady tone at 1 800 Hz. A friend on the ground at the landing site directly below listens to the amplified sound he receives. Assume the air is calm and the speed of sound is independent of altitude. While the skydiver is falling at terminal speed, her friend on the ground receives waves of frequency 2 150 Hz. (a) What is the skydiver's speed of descent? (b) **What If?** Suppose the skydiver can hear the sound of the buzzer reflected from the ground. What frequency does she receive?

56. **QC** Spherical waves of wavelength 45.0 cm propagate outward from a point source. (a) Explain how the intensity at a distance of 240 cm compares with the intensity at a distance of 60.0 cm. (b) Explain how the amplitude at a distance of 240 cm compares with the amplitude at a distance of 60.0 cm. (c) Explain how the phase of the wave at a distance of 240 cm compares with the phase at 60.0 cm at the same moment.

57. **QC** A bat, moving at 5.00 m/s, is chasing a flying insect. If the bat emits a 40.0-kHz chirp and receives back an echo at 40.4 kHz, (a) what is the speed of the insect? (b) Will the bat be able to catch the insect? Explain.

58. **M** Two ships are moving along a line due east (Fig. P17.58). The trailing vessel has a speed relative to a land-based observation point of $v_1 = 64.0$ km/h, and the leading ship has a speed of $v_2 = 45.0$ km/h relative to that point. The two ships are in a region of the ocean where the current is moving uniformly due west at $v_{current} = 10.0$ km/h. The trailing ship transmits a sonar signal at a frequency of 1 200.0 Hz through the water. What frequency is monitored by the leading ship?

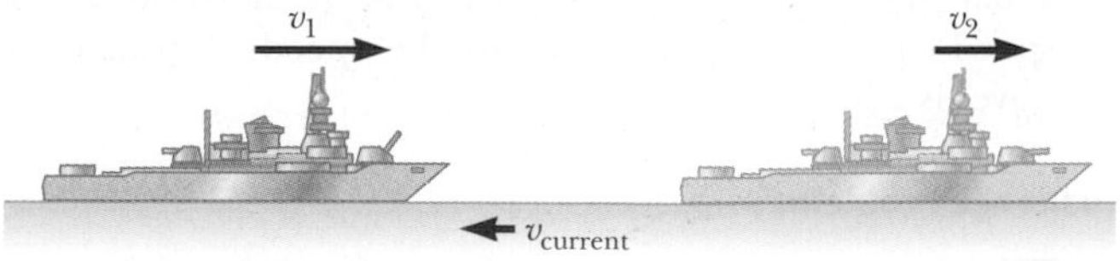

Figure P17.58

59. A police car is traveling east at 40.0 m/s along a straight road, overtaking a car ahead of it moving east at 30.0 m/s. The police car has a malfunctioning siren that is stuck at 1 000 Hz. (a) What would be the wavelength in air of the siren sound if the police car were at rest? (b) What is the wavelength in front of the police car? (c) What is it behind the police car? (d) What is the frequency heard by the driver being chased?

60. **QC** The speed of a one-dimensional compressional wave traveling along a thin copper rod is 3.56 km/s. The rod is given a sharp hammer blow at one end. A listener at the far end of the rod hears the sound twice, transmitted through the metal and through air, with a time interval Δt between the two pulses. (a) Which sound arrives first? (b) Find the length of the rod as a function of Δt. (c) Find the length of the rod if $\Delta t = 127$ ms. (d) Imagine that the copper rod is replaced by another material through which the speed of sound is v_r. What is the length of the rod in terms of t and v_r? (e) Would the answer to part (d) go to a well-defined limit as the speed of sound in the rod goes to infinity? Explain your answer.

61. A large meteoroid enters the Earth's atmosphere at a speed of 20.0 km/s and is not significantly slowed before entering the ocean. (a) What is the Mach angle of the shock wave from the meteoroid in the lower atmosphere? (b) If we assume the meteoroid survives the impact with the ocean surface, what is the (initial) Mach angle of the shock wave the meteoroid produces in the water?

62. Three metal rods are located relative to each other as shown in Figure P17.62, where $L_1 + L_2 = L_3$. The speed of sound in a rod is given by $v = \sqrt{Y/\rho}$, where Y is Young's modulus for the rod and ρ is the density. Values of density and Young's modulus for the three materials are $\rho_1 = 2.70 \times 10^3$ kg/m^3, $Y_1 = 7.00 \times 10^{10}$ N/m^2, $\rho_2 = 11.3 \times 10^3$ kg/m^3, $Y_2 = 1.60 \times 10^{10}$ N/m^2, $\rho_3 = 8.80 \times 10^3$ kg/m^3, $Y_3 = 11.0 \times 10^{10}$ N/m^2. If $L_3 = 1.50$ m, what must the ratio L_1/L_2 be if a sound wave is to travel the length of rods 1 and 2 in the same time interval required for the wave to travel the length of rod 3?

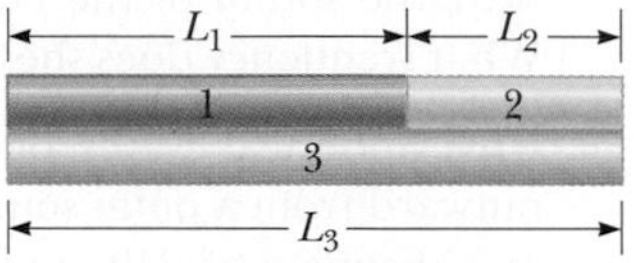

Figure P17.62

63. **M** With particular experimental methods, it is possible to produce and observe in a long, thin rod both a transverse wave whose speed depends primarily on tension in the rod and a longitudinal wave whose speed is determined by Young's modulus and the density of the material according to the expression $v = \sqrt{Y/\rho}$. The transverse wave can be modeled as a wave in a stretched string. A particular metal rod is 150 cm long and has a radius of 0.200 cm and a mass of 50.9 g. Young's modulus for the material is 6.80×10^{10} N/m^2. What must the tension in the rod be if the ratio of the speed of longitudinal waves to the speed of transverse waves is 8.00?

Challenge Problems

64. **S** Equation 17.13 states that at distance r away from a point source with power $(Power)_{avg}$, the wave intensity is

$$I = \frac{(Power)_{avg}}{4\pi r^2}$$

Study Active Figure 17.10 and prove that at distance r straight in front of a point source with power $(Power)_{avg}$ moving with constant speed v_S the wave intensity is

$$I = \frac{(Power)_{avg}}{4\pi r^2}\left(\frac{v - v_S}{v}\right)$$

65. The Doppler equation presented in the text is valid when the motion between the observer and the source occurs on a straight line so that the source and observer are moving either directly toward or directly away from each other. If this restriction is relaxed, one must use the more general Doppler equation

$$f' = \left(\frac{v + v_O \cos\theta_O}{v - v_S \cos\theta_S}\right) f$$

where θ_O and θ_S are defined in Figure P17.65a. Use the preceding equation to solve the following problem. A train moves at a constant speed of $v = 25.0$ m/s toward the intersection shown in Figure P17.65b. A car is stopped near the crossing, 30.0 m from the tracks. The train's horn emits a frequency of 500 Hz when the train is 40.0 m from the intersection. (a) What is the frequency heard by the passengers in the car? (b) If the train emits this sound continuously and the car is stationary at this position long before the train arrives until long after it leaves, what range of frequencies do passengers in the car hear? (c) Suppose the car is foolishly trying to beat the train to the intersection and is traveling at 40.0 m/s toward the tracks. When the car is 30.0 m from the tracks and the train is 40.0 m from the intersection, what is the frequency heard by the passengers in the car now?

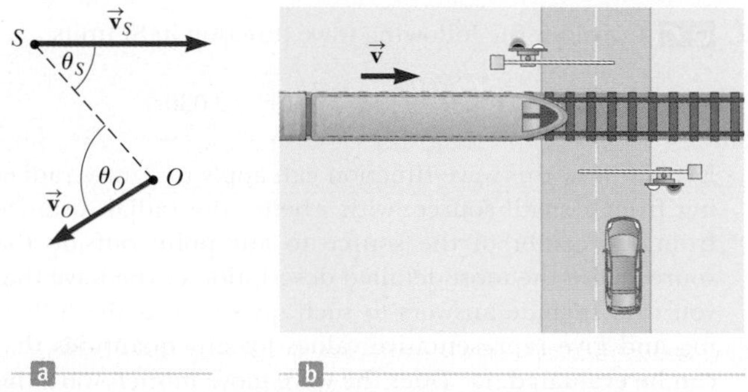

Figure P17.65

66. **S** In Section 17.2, we derived the speed of sound in a gas using the impulse–momentum theorem applied to the cylinder of gas in Figure 17.5. Let us find the speed of sound in a gas using a different approach based on the element of gas in Figure 17.3. Proceed as follows. (a) Draw a force diagram for this element showing the forces exerted on the left and right surfaces due to the pressure of the gas on either side of the element. (b) By applying Newton's second law to the element, show that

$$-\frac{\partial(\Delta P)}{\partial x} A\,\Delta x = \rho A\,\Delta x \frac{\partial^2 s}{\partial t^2}$$

(c) By substituting $\Delta P = -(B\,\partial s/\partial x)$ (Eq. 17.3), derive the following wave equation for sound:

$$\frac{B}{\rho}\frac{\partial^2 s}{\partial x^2} = \frac{\partial^2 s}{\partial t^2}$$

(d) To a mathematical physicist, this equation demonstrates the existence of sound waves and determines their speed. As a physics student, you must take another step or two. Substitute into the wave equation the trial solution $s(x, t) = s_{max} \cos(kx - \omega t)$. Show that this function satisfies the wave equation, provided $\omega/k = v = \sqrt{B/\rho}$.

chapter

18

Superposition and Standing Waves

Blues master B. B. King takes advantage of standing waves on strings. He changes to higher notes on the guitar by pushing the strings against the frets on the fingerboard, shortening the lengths of the portions of the strings that vibrate. *(AP Images)*

The wave model was introduced in the previous two chapters. We have seen that waves are very different from particles. A particle is of zero size, whereas a wave has a characteristic size, its wavelength. Another important difference between waves and particles is that we can explore the possibility of two or more waves combining at one point in the same medium. Particles can be combined to form extended objects, but the particles must be at *different* locations. In contrast, two waves can both be present at the same location. The ramifications of this possibility are explored in this chapter.

When waves are combined in systems with boundary conditions, only certain allowed frequencies can exist and we say the frequencies are *quantized.* Quantization is a notion that is at the heart of quantum mechanics, a subject introduced formally in Chapter 40. There we show that analysis of waves under boundary conditions explains many of the quantum phenomena. In this chapter, we use quantization to understand the behavior of the wide array of musical instruments that are based on strings and air columns.

We also consider the combination of waves having different frequencies. When two sound waves having nearly the same frequency interfere, we hear variations in the loudness called *beats*. Finally, we discuss how any nonsinusoidal periodic wave can be described as a sum of sine and cosine functions.

18.1 Analysis Model: Waves in Interference

Many interesting wave phenomena in nature cannot be described by a single traveling wave. Instead, one must analyze these phenomena in terms of a combination of traveling waves. As noted in the introduction, waves have a remarkable difference from particles in that waves can be combined at the *same* location in space. To analyze such wave combinations, we make use of the **superposition principle:**

Superposition principle ▶

> If two or more traveling waves are moving through a medium, the resultant value of the wave function at any point is the algebraic sum of the values of the wave functions of the individual waves.

Waves that obey this principle are called *linear waves*. In the case of mechanical waves, linear waves are generally characterized by having amplitudes much smaller than their wavelengths. Waves that violate the superposition principle are called *nonlinear waves* and are often characterized by large amplitudes. In this book, we deal only with linear waves.

One consequence of the superposition principle is that two traveling waves can pass through each other without being destroyed or even altered. For instance, when two pebbles are thrown into a pond and hit the surface at different locations, the expanding circular surface waves from the two locations simply pass through each other with no permanent effect. The resulting complex pattern can be viewed as two independent sets of expanding circles.

Active Figure 18.1 is a pictorial representation of the superposition of two pulses. The wave function for the pulse moving to the right is y_1, and the wave function for the pulse moving to the left is y_2. The pulses have the same speed but different shapes, and the displacement of the elements of the medium is in the positive y direction for both pulses. When the waves overlap (Active Fig. 18.1b), the wave function for the resulting complex wave is given by $y_1 + y_2$. When the crests of the pulses coincide (Active Fig. 18.1c), the resulting wave given by $y_1 + y_2$ has a larger amplitude than that of the individual pulses. The two pulses finally separate and continue moving in their original directions (Active Fig. 18.1d). Notice that the pulse shapes remain unchanged after the interaction, as if the two pulses had never met!

Pitfall Prevention 18.1

Do Waves Actually *Interfere*?

In popular usage, the term *interfere* implies that an agent affects a situation in some way so as to preclude something from happening. For example, in American football, *pass interference* means that a defending player has affected the receiver so that the receiver is unable to catch the ball. This usage is very different from its use in physics, where waves pass through each other and interfere, but do not affect each other in any way. In physics, interference is similar to the notion of *combination* as described in this chapter.

The combination of separate waves in the same region of space to produce a resultant wave is called **interference.** For the two pulses shown in Active Figure 18.1, the displacement of the elements of the medium is in the positive y direction for both pulses, and the resultant pulse (created when the individual pulses overlap) exhibits an amplitude greater than that of either individual pulse. Because the displacements caused by the two pulses are in the same direction, we refer to their superposition as **constructive interference.**

Constructive interference ▶

Now consider two pulses traveling in opposite directions on a taut string where one pulse is inverted relative to the other as illustrated in Active Figure 18.2. When these pulses begin to overlap, the resultant pulse is given by $y_1 + y_2$, but the values of the function y_2 are negative. Again, the two pulses pass through each other; because the displacements caused by the two pulses are in opposite directions, however, we refer to their superposition as **destructive interference.**

Destructive interference ▶

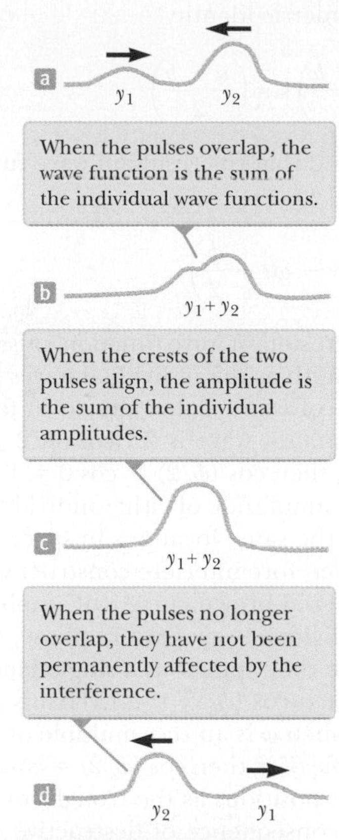

ACTIVE FIGURE 18.1

Constructive interference. Two positive pulses travel on a stretched string in opposite directions and overlap.

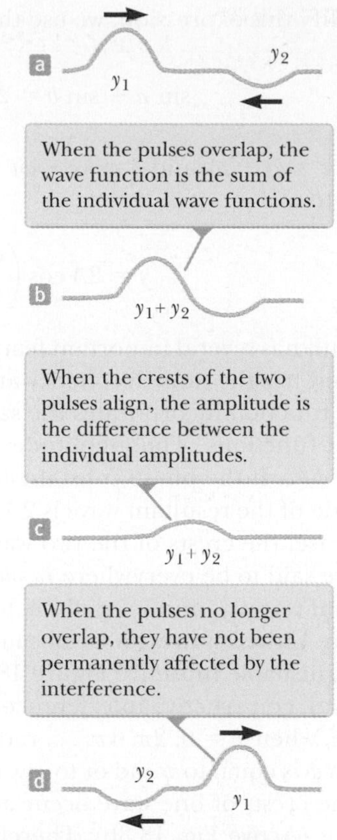

ACTIVE FIGURE 18.2

Destructive interference. Two pulses, one positive and one negative, travel on a stretched string in opposite directions and overlap.

The superposition principle is the centerpiece of the analysis model called **waves in interference.** In many situations, both in acoustics and optics, waves combine according to this principle and exhibit interesting phenomena with practical applications.

***Quick Quiz* 18.1** Two pulses move in opposite directions on a string and are identical in shape except that one has positive displacements of the elements of the string and the other has negative displacements. At the moment the two pulses completely overlap on the string, what happens? **(a)** The energy associated with the pulses has disappeared. **(b)** The string is not moving. **(c)** The string forms a straight line. **(d)** The pulses have vanished and will not reappear.

Superposition of Sinusoidal Waves

Let us now apply the principle of superposition to two sinusoidal waves traveling in the same direction in a linear medium. If the two waves are traveling to the right and have the same frequency, wavelength, and amplitude but differ in phase, we can express their individual wave functions as

$$y_1 = A \sin (kx - \omega t) \qquad y_2 = A \sin (kx - \omega t + \phi)$$

where, as usual, $k = 2\pi/\lambda$, $\omega = 2\pi f$, and ϕ is the phase constant as discussed in Section 16.2. Hence, the resultant wave function y is

$$y = y_1 + y_2 = A\,[\sin (kx - \omega t) + \sin (kx - \omega t + \phi)]$$

To simplify this expression, we use the trigonometric identity

$$\sin a + \sin b = 2\cos\left(\frac{a-b}{2}\right)\sin\left(\frac{a+b}{2}\right)$$

Letting $a = kx - \omega t$ and $b = kx - \omega t + \phi$, we find that the resultant wave function y reduces to

Resultant of two traveling sinusoidal waves ▶

$$y = 2A\cos\left(\frac{\phi}{2}\right)\sin\left(kx - \omega t + \frac{\phi}{2}\right)$$

This result has several important features. The resultant wave function y also is sinusoidal and has the same frequency and wavelength as the individual waves because the sine function incorporates the same values of k and ω that appear in the original wave functions. The amplitude of the resultant wave is $2A\cos(\phi/2)$, and its phase is $\phi/2$. If the phase constant ϕ equals 0, then $\cos(\phi/2) = \cos 0 = 1$ and the amplitude of the resultant wave is $2A$, twice the amplitude of either individual wave. In this case, the crests of the two waves are at the same locations in space and the waves are said to be everywhere *in phase* and therefore interfere constructively. The individual waves y_1 and y_2 combine to form the red-brown curve y of amplitude $2A$ shown in Active Figure 18.3a. Because the individual waves are in phase, they are indistinguishable in Active Figure 18.3a, where they appear as a single blue curve. In general, constructive interference occurs when $\cos(\phi/2) = \pm 1$. That is true, for example, when $\phi = 0, 2\pi, 4\pi, \ldots$ rad, that is, when ϕ is an *even* multiple of π.

When ϕ is equal to π rad or to any *odd* multiple of π, then $\cos(\phi/2) = \cos(\pi/2) = 0$ and the crests of one wave occur at the same positions as the troughs of the second wave (Active Fig. 18.3b). Therefore, as a consequence of destructive interference, the resultant wave has *zero* amplitude everywhere as shown by the straight red-brown line in Active Figure 18.3b. Finally, when the phase constant has an arbitrary value other than 0 or an integer multiple of π rad (Active Fig. 18.3c), the resultant wave has an amplitude whose value is somewhere between 0 and $2A$.

In the more general case in which the waves have the same wavelength but different amplitudes, the results are similar with the following exceptions. In the in-phase case, the amplitude of the resultant wave is not twice that of a single wave, but rather is the sum of the amplitudes of the two waves. When the waves are π rad out of phase, they do not completely cancel as in Active Figure 18.3b. The result is a wave whose amplitude is the difference in the amplitudes of the individual waves.

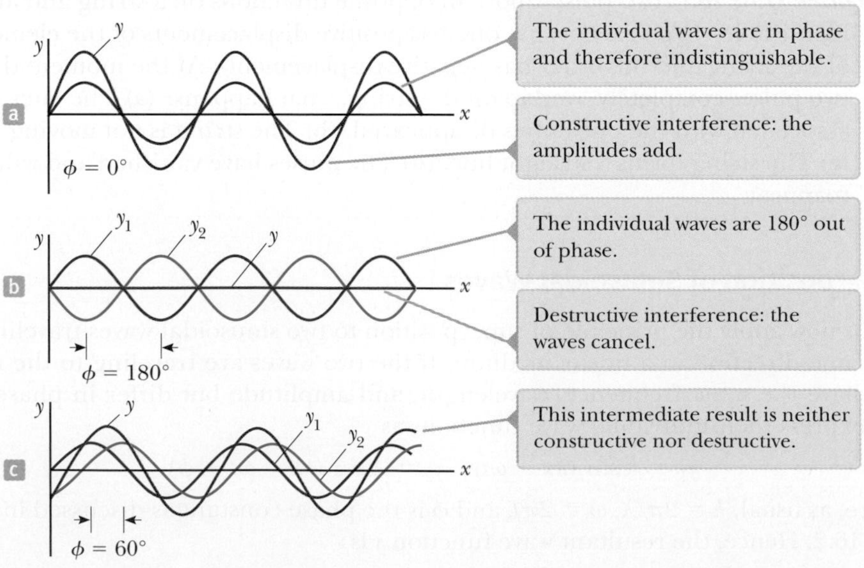

ACTIVE FIGURE 18.3

The superposition of two identical waves y_1 and y_2 (blue and green, respectively) to yield a resultant wave (red-brown).

Interference of Sound Waves

One simple device for demonstrating interference of sound waves is illustrated in Figure 18.4. Sound from a loudspeaker S is sent into a tube at point *P*, where there is a T-shaped junction. Half the sound energy travels in one direction, and half travels in the opposite direction. Therefore, the sound waves that reach the receiver R can travel along either of the two paths. The distance along any path from speaker to receiver is called the **path length** r. The lower path length r_1 is fixed, but the upper path length r_2 can be varied by sliding the U-shaped tube, which is similar to that on a slide trombone. When the difference in the path lengths $\Delta r = |r_2 - r_1|$ is either zero or some integer multiple of the wavelength λ (that is, $\Delta r = n\lambda$, where $n = 0, 1, 2, 3, \ldots$), the two waves reaching the receiver at any instant are in phase and interfere constructively as shown in Active Figure 18.3a. For this case, a maximum in the sound intensity is detected at the receiver. If the path length r_2 is adjusted such that the path difference $\Delta r = \lambda/2, 3\lambda/2, \ldots, n\lambda/2$ (for n odd), the two waves are exactly π rad, or 180°, out of phase at the receiver and hence cancel each other. In this case of destructive interference, no sound is detected at the receiver. This simple experiment demonstrates that a phase difference may arise between two waves generated by the same source when they travel along paths of unequal lengths. This important phenomenon will be indispensable in our investigation of the interference of light waves in Chapter 37.

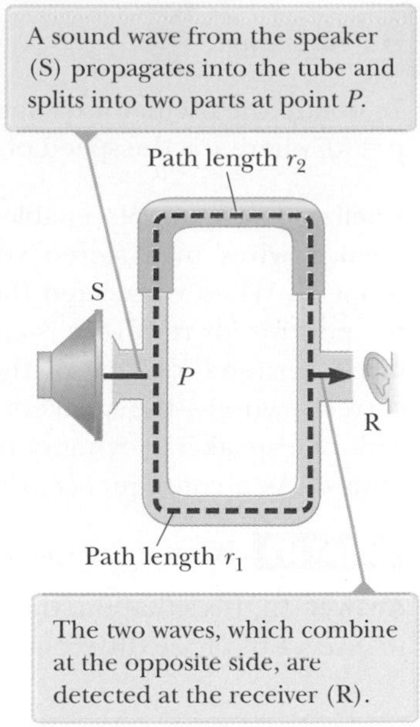

Figure 18.4 An acoustical system for demonstrating interference of sound waves. The upper path length r_2 can be varied by sliding the upper section.

Example 18.1 Two Speakers Driven by the Same Source

Two identical loudspeakers placed 3.00 m apart are driven by the same oscillator (Fig. 18.5). A listener is originally at point *O*, located 8.00 m from the center of the line connecting the two speakers. The listener then moves to point *P*, which is a perpendicular distance 0.350 m from *O*, and she experiences the *first minimum* in sound intensity. What is the frequency of the oscillator?

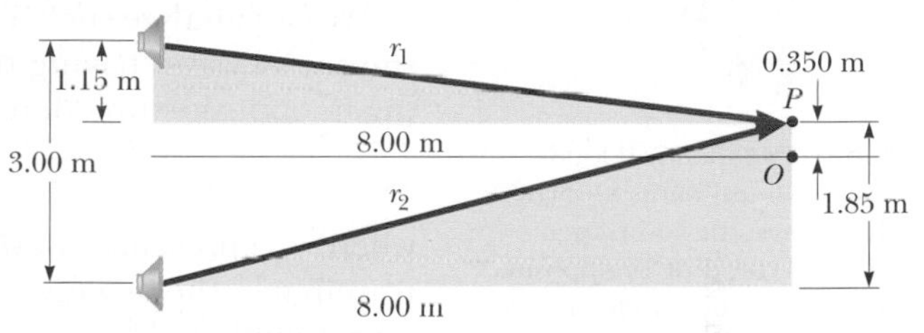

Figure 18.5 (Example 18.1) Two identical loudspeakers emit sound waves to a listener at *P*.

SOLUTION

Conceptualize In Figure 18.4, a sound wave enters a tube and is then *acoustically* split into two different paths before recombining at the other end. In this example, a signal representing the sound is *electrically* split and sent to two different loudspeakers. After leaving the speakers, the sound waves recombine at the position of the listener. Despite the difference in how the splitting occurs, the path difference discussion related to Figure 18.4 can be applied here.

Categorize Because the sound waves from two separate sources combine, we apply the waves in interference analysis model.

Analyze Figure 18.5 shows the physical arrangement of the speakers, along with two shaded right triangles that can be drawn on the basis of the lengths described in the problem. The first minimum occurs when the two waves reaching the listener at point *P* are 180° out of phase, in other words, when their path difference Δr equals $\lambda/2$.

From the shaded triangles, find the path lengths from the speakers to the listener:

$$r_1 = \sqrt{(8.00\text{ m})^2 + (1.15\text{ m})^2} = 8.08\text{ m}$$

$$r_2 = \sqrt{(8.00\text{ m})^2 + (1.85\text{ m})^2} = 8.21\text{ m}$$

Hence, the path difference is $r_2 - r_1 = 0.13$ m. Because this path difference must equal $\lambda/2$ for the first minimum, $\lambda = 0.26$ m.

continued

18.1 *cont.*

To obtain the oscillator frequency, use Equation 16.12, $v = \lambda f$, where v is the speed of sound in air, 343 m/s:

$$f = \frac{v}{\lambda} = \frac{343 \text{ m/s}}{0.26 \text{ m}} = 1.3 \text{ kHz}$$

Finalize This example enables us to understand why the speaker wires in a stereo system should be connected properly. When connected the wrong way—that is, when the positive (or red) wire is connected to the negative (or black) terminal on one of the speakers and the other is correctly wired—the speakers are said to be "out of phase," with one speaker moving outward while the other moves inward. As a consequence, the sound wave coming from one speaker destructively interferes with the wave coming from the other at point O in Figure 18.5. A rarefaction region due to one speaker is superposed on a compression region from the other speaker. Although the two sounds probably do not completely cancel each other (because the left and right stereo signals are usually not identical), a substantial loss of sound quality occurs at point O.

WHAT IF? What if the speakers were connected out of phase? What happens at point P in Figure 18.5?

Answer In this situation, the path difference of $\lambda/2$ combines with a phase difference of $\lambda/2$ due to the incorrect wiring to give a full phase difference of λ. As a result, the waves are in phase and there is a *maximum* intensity at point P.

Figure 18.6 Two identical loudspeakers emit sound waves toward each other. When they overlap, identical waves traveling in opposite directions will combine to form standing waves.

18.2 Standing Waves

The sound waves from the pair of loudspeakers in Example 18.1 leave the speakers in the forward direction, and we considered interference at a point in front of the speakers. Suppose we turn the speakers so that they face each other and then have them emit sound of the same frequency and amplitude. In this situation, two identical waves travel in opposite directions in the same medium as in Figure 18.6. These waves combine in accordance with the waves in interference model.

We can analyze such a situation by considering wave functions for two transverse sinusoidal waves having the same amplitude, frequency, and wavelength but traveling in opposite directions in the same medium:

$$y_1 = A \sin (kx - \omega t) \qquad y_2 = A \sin (kx + \omega t)$$

where y_1 represents a wave traveling in the positive x direction and y_2 represents one traveling in the negative x direction. Adding these two functions gives the resultant wave function y:

$$y = y_1 + y_2 = A \sin (kx - \omega t) + A \sin (kx + \omega t)$$

When we use the trigonometric identity $\sin (a \pm b) = \sin a \cos b \pm \cos a \sin b$, this expression reduces to

$$y = (2A \sin kx) \cos \omega t \qquad \textbf{(18.1)}$$

Pitfall Prevention 18.2

Three Types of Amplitude
We need to distinguish carefully here between the **amplitude of the individual waves,** which is A, and the **amplitude of the simple harmonic motion of the elements of the medium,** which is $2A \sin kx$. A given element in a standing wave vibrates within the constraints of the *envelope* function $2A \sin kx$, where x is that element's position in the medium. Such vibration is in contrast to traveling sinusoidal waves, in which all elements oscillate with the same amplitude and the same frequency and the amplitude A of the wave is the same as the amplitude A of the simple harmonic motion of the elements. Furthermore, we can identify the **amplitude of the standing wave** as $2A$.

Equation 18.1 represents the wave function of a **standing wave.** A standing wave such as the one on a string shown in Figure 18.7 is an oscillation pattern *with a stationary outline* that results from the superposition of two identical waves traveling in opposite directions.

Notice that Equation 18.1 does not contain a function of $kx - \omega t$. Therefore, it is not an expression for a single traveling wave. When you observe a standing wave, there is no sense of motion in the direction of propagation of either original wave. Comparing Equation 18.1 with Equation 15.6, we see that it describes a special kind of simple harmonic motion. Every element of the medium oscillates in simple harmonic motion with the same angular frequency ω (according to the $\cos \omega t$ factor in the equation). The amplitude of the simple harmonic motion of a given element (given by the factor $2A \sin kx$, the coefficient of the cosine function) depends on the location x of the element in the medium, however.

If you can find a noncordless telephone with a coiled cord connecting the handset to the base unit, you can see the difference between a standing wave and a trav-

The amplitude of the vertical oscillation of any element of the string depends on the horizontal position of the element. Each element vibrates within the confines of the envelope function $2A \sin kx$.

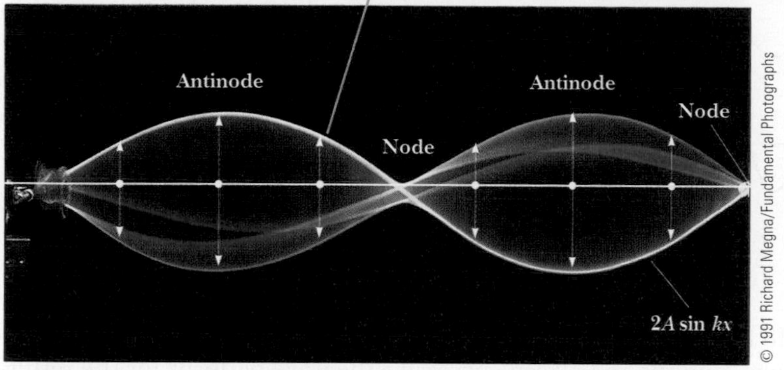

Figure 18.7 Multiflash photograph of a standing wave on a string. The time behavior of the vertical displacement from equilibrium of an individual element of the string is given by $\cos \omega t$. That is, each element vibrates at an angular frequency ω.

eling wave. Stretch the coiled cord out and flick it with a finger. You will see a pulse traveling along the cord. Now shake the handset up and down and adjust your shaking frequency until every coil on the cord is moving up at the same time and then down. That is a standing wave, formed from the combination of waves moving away from your hand and reflected from the base unit toward your hand. Notice that there is no sense of traveling along the cord like there was for the pulse. You only see up-and-down motion of the elements of the cord.

Equation 18.1 shows that the amplitude of the simple harmonic motion of an element of the medium has a minimum value of zero when x satisfies the condition $\sin kx = 0$, that is, when

$$kx = 0, \pi, 2\pi, 3\pi, \ldots$$

Because $k = 2\pi/\lambda$, these values for kx give

$$x = 0, \frac{\lambda}{2}, \lambda, \frac{3\lambda}{2}, \ldots = \frac{n\lambda}{2} \quad n = 0, 1, 2, 3, \ldots \tag{18.2}$$

◀ **Positions of nodes**

These points of zero amplitude are called **nodes.**

The element of the medium with the *greatest* possible displacement from equilibrium has an amplitude of $2A$, which we define as the amplitude of the standing wave. The positions in the medium at which this maximum displacement occurs are called **antinodes.** The antinodes are located at positions for which the coordinate x satisfies the condition $\sin kx = \pm 1$, that is, when

$$kx = \frac{\pi}{2}, \frac{3\pi}{2}, \frac{5\pi}{2}, \ldots$$

Therefore, the positions of the antinodes are given by

$$x = \frac{\lambda}{4}, \frac{3\lambda}{4}, \frac{5\lambda}{4}, \ldots = \frac{n\lambda}{4} \quad n = 1, 3, 5, \ldots \tag{18.3}$$

◀ **Positions of antinodes**

Two nodes and two antinodes are labeled in the standing wave in Figure 18.7. The light blue curve labeled $2A \sin kx$ in Figure 18.7 represents one wavelength of the traveling waves that combine to form the standing wave. Figure 18.7 and Equations 18.2 and 18.3 provide the following important features of the locations of nodes and antinodes:

The distance between adjacent antinodes is equal to $\lambda/2$.
The distance between adjacent nodes is equal to $\lambda/2$.
The distance between a node and an adjacent antinode is $\lambda/4$.

Wave patterns of the elements of the medium produced at various times by two transverse traveling waves moving in opposite directions are shown in Active

ACTIVE FIGURE 18.8

Standing-wave patterns produced at various times by two waves of equal amplitude traveling in opposite directions. For the resultant wave y, the nodes (N) are points of zero displacement and the antinodes (A) are points of maximum displacement.

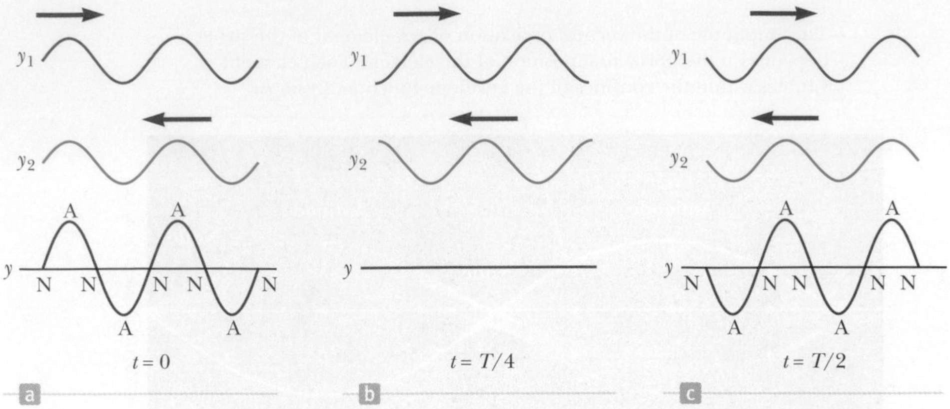

Figure 18.8. The blue and green curves are the wave patterns for the individual traveling waves, and the red-brown curves are the wave patterns for the resultant standing wave. At $t = 0$ (Active Fig. 18.8a), the two traveling waves are in phase, giving a wave pattern in which each element of the medium is at rest and experiencing its maximum displacement from equilibrium. One-quarter of a period later, at $t = T/4$ (Active Fig. 18.8b), the traveling waves have moved one-fourth of a wavelength (one to the right and the other to the left). At this time, the traveling waves are out of phase, and each element of the medium is passing through the equilibrium position in its simple harmonic motion. The result is zero displacement for elements at all values of x; that is, the wave pattern is a straight line. At $t = T/2$ (Active Fig. 18.8c), the traveling waves are again in phase, producing a wave pattern that is inverted relative to the $t = 0$ pattern. In the standing wave, the elements of the medium alternate in time between the extremes shown in Active Figures 18.8a and 18.8c.

***Quick Quiz* 18.2** Consider the waves in Active Figure 18.8 to be waves on a stretched string. Define the velocity of elements of the string as positive if they are moving upward in the figure. **(i)** At the moment the string has the shape shown by the red-brown curve in Active Figure 18.8a, what is the instantaneous velocity of elements along the string? (a) zero for all elements (b) positive for all elements (c) negative for all elements (d) varies with the position of the element **(ii)** From the same choices, at the moment the string has the shape shown by the red-brown curve in Active Figure 18.8b, what is the instantaneous velocity of elements along the string?

Example 18.2 Formation of a Standing Wave

Two waves traveling in opposite directions produce a standing wave. The individual wave functions are

$$y_1 = 4.0 \sin (3.0x - 2.0t)$$

$$y_2 = 4.0 \sin (3.0x + 2.0t)$$

where x and y are measured in centimeters and t is in seconds.

(A) Find the amplitude of the simple harmonic motion of the element of the medium located at $x = 2.3$ cm.

SOLUTION

Conceptualize The waves described by the given equations are identical except for their directions of travel, so they indeed combine to form a standing wave as discussed in this section. We can represent the waves graphically by the blue and green curves in Active Figure 18.8.

Categorize We will substitute values into equations developed in this section, so we categorize this example as a substitution problem.

18.2 *cont.*

From the equations for the waves, we see that $A = 4.0$ cm, $k = 3.0$ rad/cm, and $\omega = 2.0$ rad/s. Use Equation 18.1 to write an expression for the standing wave:

$$y = (2A \sin kx) \cos \omega t = 8.0 \sin 3.0x \cos 2.0t$$

Find the amplitude of the simple harmonic motion of the element at the position $x = 2.3$ cm by evaluating the coefficient of the cosine function at this position:

$$y_{\max} = (8.0 \text{ cm}) \sin 3.0x \,|_{x=2.3}$$
$$= (8.0 \text{ cm}) \sin (6.9 \text{ rad}) = \boxed{4.6 \text{ cm}}$$

(B) Find the positions of the nodes and antinodes if one end of the string is at $x = 0$.

SOLUTION

Find the wavelength of the traveling waves:

$$k = \frac{2\pi}{\lambda} = 3.0 \text{ rad/cm} \rightarrow \lambda = \frac{2\pi}{3.0} \text{ cm}$$

Use Equation 18.2 to find the locations of the nodes:

$$x = n\frac{\lambda}{2} = \boxed{n\left(\frac{\pi}{3.0}\right) \text{ cm} \quad n = 0, 1, 2, 3, \ldots}$$

Use Equation 18.3 to find the locations of the antinodes:

$$x = n\frac{\lambda}{4} = \boxed{n\left(\frac{\pi}{6.0}\right) \text{ cm} \quad n = 1, 3, 5, 7, \ldots}$$

18.3 Analysis Model: Waves Under Boundary Conditions

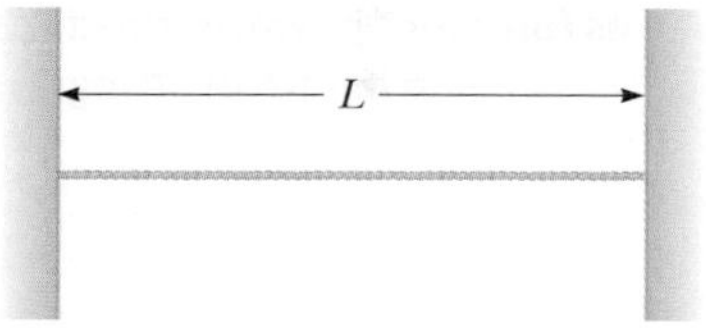

Figure 18.9 A string of length L fixed at both ends.

Consider a string of length L fixed at both ends as shown in Figure 18.9. We will use this system as a model for a guitar string or piano string. Waves can travel in both directions on the string. Therefore, standing waves can be set up in the string by a continuous superposition of waves incident on and reflected from the ends. Notice that there is a *boundary condition* for the waves on the string. Because the ends of the string are fixed, they must necessarily have zero displacement and are therefore nodes by definition. This boundary condition results in the string having a number of discrete natural patterns of oscillation, called **normal modes,** each of which has a characteristic frequency that is easily calculated. This situation in which only certain frequencies of oscillation are allowed is called **quantization.** Quantization is a common occurrence when waves are subject to boundary conditions and is a central feature in our discussions of quantum physics in the extended version of this text. Notice in Active Figure 18.8 that there are no boundary conditions, so standing waves of *any* frequency can be established; there is no quantization without boundary conditions. Because boundary conditions occur so often for waves, we identify an analysis model called **waves under boundary conditions** for the discussion that follows.

The normal modes of oscillation for the string in Figure 18.9 can be described by imposing the boundary conditions that the ends be nodes and that the nodes and antinodes be separated by one-fourth of a wavelength. The first normal mode that is consistent with these requirements, shown in Active Figure 18.10a (page 520), has nodes at its ends and one antinode in the middle. This normal mode is the longest-wavelength mode that is consistent with our boundary conditions. The first normal mode occurs when the wavelength λ_1 is equal to twice the length of the string, or $\lambda_1 = 2L$. The section of a standing wave from one node to the next node is called a *loop*. In the first normal mode, the string is vibrating in one loop. In the second normal mode (see Active Fig. 18.10b), the string vibrates in two loops. In this case, the wavelength λ_2 is equal to the length of the string, as expressed by $\lambda_2 = L$. The third normal mode (see Active Fig. 18.10c) corresponds to the case in which $\lambda_3 = 2L/3$, and our string vibrates in three loops. In general, the wavelengths of the various normal modes for a string of length L fixed at both ends are

$$\lambda_n = \frac{2L}{n} \quad n = 1, 2, 3, \ldots \tag{18.4}$$

◀ Wavelengths of normal modes

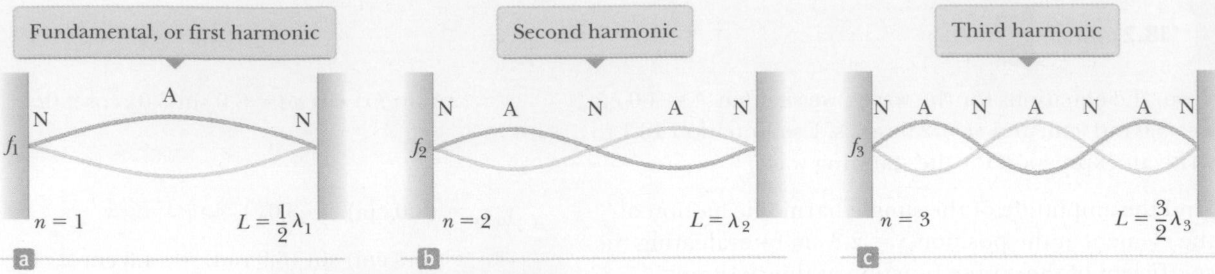

ACTIVE FIGURE 18.10

The normal modes of vibration of the string in Figure 18.9 form a harmonic series. The string vibrates between the extremes shown.

where the index n refers to the nth normal mode of oscillation. These nodes are the *possible* modes of oscillation for the string. The *actual* modes that are excited on a string are discussed shortly.

The natural frequencies associated with the modes of oscillation are obtained from the relationship $f = v/\lambda$, where the wave speed v is the same for all frequencies. Using Equation 18.4, we find that the natural frequencies f_n of the normal modes are

Frequencies of normal modes as functions of wave speed and length of string ▶

$$f_n = \frac{v}{\lambda_n} = n\frac{v}{2L} \quad n = 1, 2, 3, \ldots \tag{18.5}$$

These natural frequencies are also called the *quantized frequencies* associated with the vibrating string fixed at both ends.

Because $v = \sqrt{T/\mu}$ (see Eq. 16.18) for waves on a string, where T is the tension in the string and μ is its linear mass density, we can also express the natural frequencies of a taut string as

Frequencies of normal modes as functions of string tension and linear mass density ▶

$$f_n = \frac{n}{2L}\sqrt{\frac{T}{\mu}} \quad n = 1, 2, 3, \ldots \tag{18.6}$$

The lowest frequency f_1, which corresponds to $n = 1$, is called either the **fundamental** or the **fundamental frequency** and is given by

Fundamental frequency of a taut string ▶

$$f_1 = \frac{1}{2L}\sqrt{\frac{T}{\mu}} \tag{18.7}$$

The frequencies of the remaining normal modes are integer multiples of the fundamental frequency. Frequencies of normal modes that exhibit such an integer-multiple relationship form a **harmonic series,** and the normal modes are called **harmonics.** The fundamental frequency f_1 is the frequency of the first harmonic, the frequency $f_2 = 2f_1$ is the frequency of the second harmonic, and the frequency $f_n = nf_1$ is the frequency of the nth harmonic. Other oscillating systems, such as a drumhead, exhibit normal modes, but the frequencies are not related as integer multiples of a fundamental (see Section 18.6). Therefore, we do not use the term *harmonic* in association with those types of systems.

Let us examine further how the various harmonics are created in a string. To excite only a single harmonic, the string must be distorted into a shape that corresponds to that of the desired harmonic. After being released, the string vibrates at the frequency of that harmonic. This maneuver is difficult to perform, however, and is not how a string of a musical instrument is excited. If the string is distorted such that its shape is not that of just one harmonic, the resulting vibration includes a combination of various harmonics. Such a distortion occurs in musical instruments when the string is plucked (as in a guitar), bowed (as in a cello), or struck (as in a piano). When the string is distorted into a nonsinusoidal shape, only waves that satisfy the boundary conditions can persist on the string. These waves are the harmonics.

The frequency of a string that defines the musical note that it plays is that of the fundamental. The string's frequency can be varied by changing either the string's tension or its length. For example, the tension in guitar and violin strings is varied by a screw adjustment mechanism or by tuning pegs located on the neck of the instrument. As the tension is increased, the frequency of the normal modes increases in accordance with Equation 18.6. Once the instrument is "tuned," players vary the frequency by moving their fingers along the neck, thereby changing the length of the oscillating portion of the string. As the length is shortened, the frequency increases because, as Equation 18.6 specifies, the normal-mode frequencies are inversely proportional to string length.

***Quick Quiz* 18.3** When a standing wave is set up on a string fixed at both ends, which of the following statements is true? **(a)** The number of nodes is equal to the number of antinodes. **(b)** The wavelength is equal to the length of the string divided by an integer. **(c)** The frequency is equal to the number of nodes times the fundamental frequency. **(d)** The shape of the string at any instant shows a symmetry about the midpoint of the string.

Example 18.3 Give Me a C Note!

The middle C string on a piano has a fundamental frequency of 262 Hz, and the string for the first A above middle C has a fundamental frequency of 440 Hz.

(A) Calculate the frequencies of the next two harmonics of the C string.

SOLUTION

Conceptualize Remember that the harmonics of a vibrating string have frequencies that are related by integer multiples of the fundamental.

Categorize This first part of the example is a simple substitution problem.

Knowing that the fundamental frequency is $f_1 = 262$ Hz, find the frequencies of the next harmonics by multiplying by integers:

$$f_2 = 2f_1 = 524 \text{ Hz}$$

$$f_3 = 3f_1 = 786 \text{ Hz}$$

(B) If the A and C strings have the same linear mass density μ and length L, determine the ratio of tensions in the two strings.

SOLUTION

Categorize This part of the example is more of an analysis problem than is part (A).

Analyze Use Equation 18.7 to write expressions for the fundamental frequencies of the two strings:

$$f_{1A} = \frac{1}{2L}\sqrt{\frac{T_A}{\mu}} \quad \text{and} \quad f_{1C} = \frac{1}{2L}\sqrt{\frac{T_C}{\mu}}$$

Divide the first equation by the second and solve for the ratio of tensions:

$$\frac{f_{1A}}{f_{1C}} = \sqrt{\frac{T_A}{T_C}} \rightarrow \frac{T_A}{T_C} = \left(\frac{f_{1A}}{f_{1C}}\right)^2 = \left(\frac{440}{262}\right)^2 = 2.82$$

Finalize If the frequencies of piano strings were determined solely by tension, this result suggests that the ratio of tensions from the lowest string to the highest string on the piano would be enormous. Such large tensions would make it difficult to design a frame to support the strings. In reality, the frequencies of piano strings vary due to additional parameters, including the mass per unit length and the length of the string. The What If? below explores a variation in length.

WHAT IF? If you look inside a real piano, you'll see that the assumption made in part (B) is only partially true. The strings are not likely to have the same length. The string densities for the given notes might be equal, but suppose the length of the A string is only 64% of the length of the C string. What is the ratio of their tensions?

continued

18.3 *cont.*

Answer Using Equation 18.7 again, we set up the ratio of frequencies:

$$\frac{f_{1A}}{f_{1C}} = \frac{L_C}{L_A}\sqrt{\frac{T_A}{T_C}} \rightarrow \frac{T_A}{T_C} = \left(\frac{L_A}{L_C}\right)^2\left(\frac{f_{1A}}{f_{1C}}\right)^2$$

$$\frac{T_A}{T_C} = (0.64)^2\left(\frac{440}{262}\right)^2 = 1.16$$

Notice that this result represents only a 16% increase in tension, compared with the 182% increase in part (B).

Example 18.4 Changing String Vibration with Water

One end of a horizontal string is attached to a vibrating blade, and the other end passes over a pulley as in Figure 18.11a. A sphere of mass 2.00 kg hangs on the end of the string. The string is vibrating in its second harmonic. A container of water is raised under the sphere so that the sphere is completely submerged. In this configuration, the string vibrates in its fifth harmonic as shown in Figure 18.11b. What is the radius of the sphere?

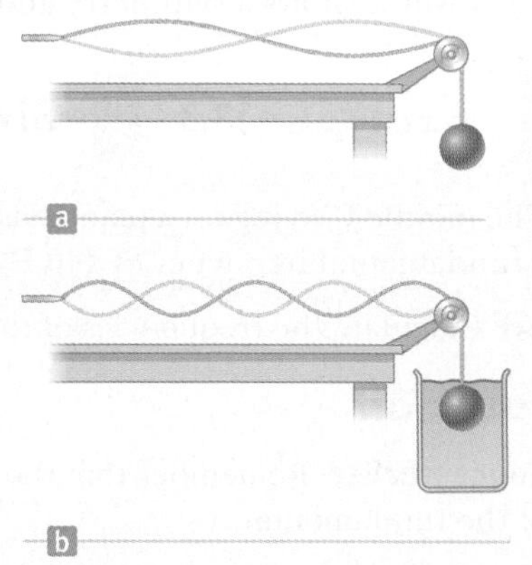

Figure 18.11 (Example 18.4) (a) When the sphere hangs in air, the string vibrates in its second harmonic. (b) When the sphere is immersed in water, the string vibrates in its fifth harmonic.

SOLUTION

Conceptualize Imagine what happens when the sphere is immersed in the water. The buoyant force acts upward on the sphere, reducing the tension in the string. The change in tension causes a change in the speed of waves on the string, which in turn causes a change in the wavelength. This altered wavelength results in the string vibrating in its fifth normal mode rather than the second.

Categorize The hanging sphere is modeled as a particle in equilibrium. One of the forces acting on it is the buoyant force from the water. We also apply the waves under boundary conditions model to the string.

Analyze Apply the particle in equilibrium model to the sphere in Figure 18.11a, identifying T_1 as the tension in the string as the sphere hangs in air:

$$\sum F = T_1 - mg = 0$$

$$T_1 = mg$$

Apply the particle in equilibrium model to the sphere in Figure 18.11b, where T_2 is the tension in the string as the sphere is immersed in water:

$$T_2 + B - mg = 0$$

$$(1)\quad B = mg - T_2$$

The desired quantity, the radius of the sphere, will appear in the expression for the buoyant force B. Before proceeding in this direction, however, we must evaluate T_2 from the information about the standing wave.

Write the equation for the frequency of a standing wave on a string (Eq. 18.6) twice, once before the sphere is immersed and once after. Notice that the frequency f is the same in both cases because it is determined by the vibrating blade. In addition, the linear mass density μ and the length L of the vibrating portion of the string are the same in both cases. Divide the equations:

$$\left.\begin{aligned} f &= \frac{n_1}{2L}\sqrt{\frac{T_1}{\mu}} \\ f &= \frac{n_2}{2L}\sqrt{\frac{T_2}{\mu}} \end{aligned}\right. \rightarrow 1 = \frac{n_1}{n_2}\sqrt{\frac{T_1}{T_2}}$$

Solve for T_2:

$$T_2 = \left(\frac{n_1}{n_2}\right)^2 T_1 = \left(\frac{n_1}{n_2}\right)^2 mg$$

18.4 cont.

Substitute this result into Equation (1):

$$(2)\quad B = mg - \left(\frac{n_1}{n_2}\right)^2 mg = mg\left[1 - \left(\frac{n_1}{n_2}\right)^2\right]$$

Using Equation 14.5, express the buoyant force in terms of the radius of the sphere:

$$B = \rho_{\text{water}} g V_{\text{sphere}} = \rho_{\text{water}} g \left(\tfrac{4}{3}\pi r^3\right)$$

Solve for the radius of the sphere and substitute from Equation (2):

$$r = \left(\frac{3B}{4\pi\rho_{\text{water}} g}\right)^{1/3} = \left\{\frac{3m}{4\pi\rho_{\text{water}}}\left[1 - \left(\frac{n_1}{n_2}\right)^2\right]\right\}^{1/3}$$

Substitute numerical values:

$$r = \left\{\frac{3(2.00\text{ kg})}{4\pi(1\,000\text{ kg/m}^3)}\left[1 - \left(\frac{2}{5}\right)^2\right]\right\}^{1/3}$$

$$= 0.073\,7\text{ m} = 7.37\text{ cm}$$

Finalize Notice that only certain radii of the sphere will result in the string vibrating in a normal mode; the speed of waves on the string must be changed to a value such that the length of the string is an integer multiple of half wavelengths. This limitation is a feature of the *quantization* that was introduced earlier in this chapter: the sphere radii that cause the string to vibrate in a normal mode are *quantized.*

18.4 Resonance

We have seen that a system such as a taut string is capable of oscillating in one or more normal modes of oscillation. Suppose we drive such a string with a vibrating blade as in Figure 18.12. We find that if a periodic force is applied to such a system, the amplitude of the resulting motion of the string is greatest when the frequency of the applied force is equal to one of the natural frequencies of the system. This phenomenon, known as *resonance,* was discussed in Section 15.7. Although a block–spring system or a simple pendulum has only one natural frequency, standing-wave systems have a whole set of natural frequencies, such as that given by Equation 18.6 for a string. Because an oscillating system exhibits a large amplitude when driven at any of its natural frequencies, these frequencies are often referred to as **resonance frequencies.**

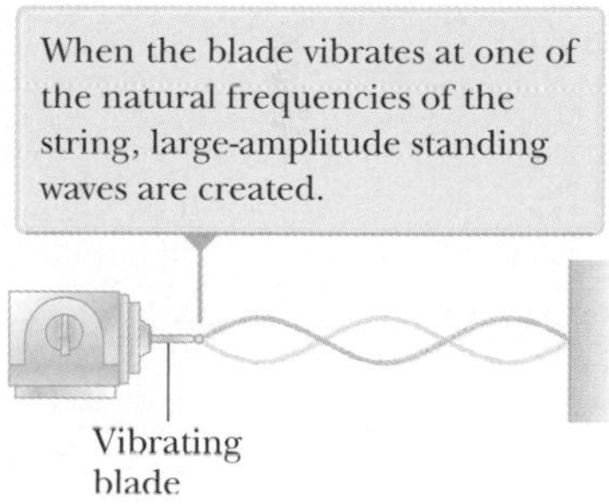

Figure 18.12 Standing waves are set up in a string when one end is connected to a vibrating blade.

Consider the string in Figure 18.12 again. The fixed end is a node, and the end connected to the blade is very nearly a node because the amplitude of the blade's motion is small compared with that of the elements of the string. As the blade oscillates, transverse waves sent down the string are reflected from the fixed end. As we learned in Section 18.3, the string has natural frequencies that are determined by its length, tension, and linear mass density (see Eq. 18.6). When the frequency of the blade equals one of the natural frequencies of the string, standing waves are produced and the string oscillates with a large amplitude. In this resonance case, the wave generated by the oscillating blade is in phase with the reflected wave and the string absorbs energy from the blade. If the string is driven at a frequency that is not one of its natural frequencies, the oscillations are of low amplitude and exhibit no stable pattern.

Resonance is very important in the excitation of musical instruments based on air columns. We shall discuss this application of resonance in Section 18.5.

18.5 Standing Waves in Air Columns

The waves under boundary conditions model can also be applied to sound waves in a column of air such as that inside an organ pipe or a clarinet. Standing waves are the result of interference between longitudinal sound waves traveling in opposite directions.

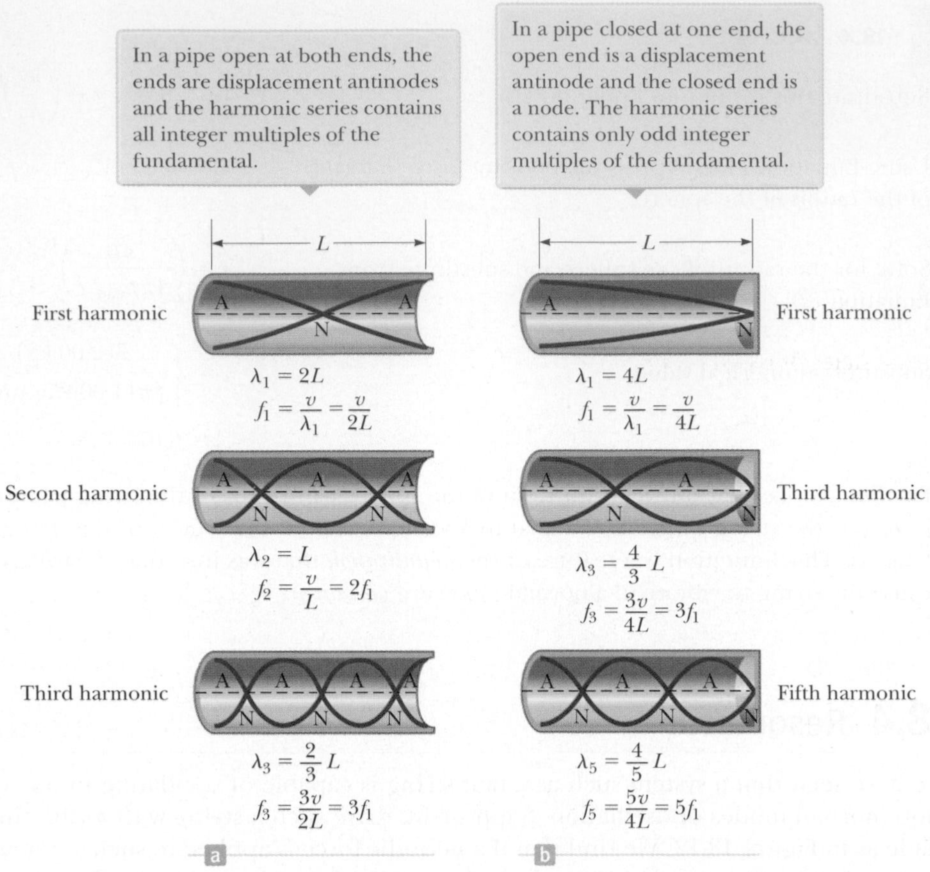

Figure 18.13 Graphical representations of the motion of elements of air in standing longitudinal waves in (a) a column open at both ends and (b) a column closed at one end.

In a pipe closed at one end, the closed end is a **displacement node** because the rigid barrier at this end does not allow longitudinal motion of the air. Because the pressure wave is 90° out of phase with the displacement wave (see Section 17.1), the closed end of an air column corresponds to a **pressure antinode** (that is, a point of maximum pressure variation).

The open end of an air column is approximately a **displacement antinode**[1] and a pressure node. We can understand why no pressure variation occurs at an open end by noting that the end of the air column is open to the atmosphere; therefore, the pressure at this end must remain constant at atmospheric pressure.

You may wonder how a sound wave can reflect from an open end because there may not appear to be a change in the medium at this point: the medium through which the sound wave moves is air both inside and outside the pipe. Sound can be represented as a pressure wave, however, and a compression region of the sound wave is constrained by the sides of the pipe as long as the region is inside the pipe. As the compression region exits at the open end of the pipe, the constraint of the pipe is removed and the compressed air is free to expand into the atmosphere. Therefore, there is a change in the *character* of the medium between the inside of the pipe and the outside even though there is no change in the *material* of the medium. This change in character is sufficient to allow some reflection.

With the boundary conditions of nodes or antinodes at the ends of the air column, we have a set of normal modes of oscillation as is the case for the string fixed at both ends. Therefore, the air column has quantized frequencies.

The first three normal modes of oscillation of a pipe open at both ends are shown in Figure 18.13a. Notice that both ends are displacement antinodes (approx-

Pitfall Prevention 18.3

Sound Waves in Air Are Longitudinal, Not Transverse

The standing longitudinal waves are drawn as transverse waves in Figure 18.13. Because they are in the same direction as the propagation, it is difficult to draw longitudinal displacements. Therefore, it is best to interpret the red-brown curves in Figure 18.13 as a graphical representation of the waves (our diagrams of string waves are pictorial representations), with the vertical axis representing the horizontal displacement $s(x, t)$ of the elements of the medium.

[1]Strictly speaking, the open end of an air column is not exactly a displacement antinode. A compression reaching an open end does not reflect until it passes beyond the end. For a tube of circular cross section, an end correction equal to approximately $0.6R$, where R is the tube's radius, must be added to the length of the air column. Hence, the effective length of the air column is longer than the true length L. We ignore this end correction in this discussion.

imately). In the first normal mode, the standing wave extends between two adjacent antinodes, which is a distance of half a wavelength. Therefore, the wavelength is twice the length of the pipe, and the fundamental frequency is $f_1 = v/2L$. As Figure 18.13a shows, the frequencies of the higher harmonics are $2f_1, 3f_1, \ldots$.

In a pipe open at both ends, the natural frequencies of oscillation form a harmonic series that includes all integral multiples of the fundamental frequency.

Because all harmonics are present and because the fundamental frequency is given by the same expression as that for a string (see Eq. 18.5), we can express the natural frequencies of oscillation as

$$f_n = n\frac{v}{2L} \qquad n = 1, 2, 3, \ldots \tag{18.8}$$

◀ **Natural frequencies of a pipe open at both ends**

Despite the similarity between Equations 18.5 and 18.8, you must remember that v in Equation 18.5 is the speed of waves on the string, whereas v in Equation 18.8 is the speed of sound in air.

If a pipe is closed at one end and open at the other, the closed end is a displacement node (see Fig. 18.13b). In this case, the standing wave for the fundamental mode extends from an antinode to the adjacent node, which is one-fourth of a wavelength. Hence, the wavelength for the first normal mode is $4L$, and the fundamental frequency is $f_1 = v/4L$. As Figure 18.13b shows, the higher-frequency waves that satisfy our conditions are those that have a node at the closed end and an antinode at the open end; hence, the higher harmonics have frequencies $3f_1, 5f_1, \ldots$.

In a pipe closed at one end, the natural frequencies of oscillation form a harmonic series that includes only odd integral multiples of the fundamental frequency.

We express this result mathematically as

$$f_n = n\frac{v}{4L} \qquad n = 1, 3, 5, \ldots \tag{18.9}$$

◀ **Natural frequencies of a pipe closed at one end and open at the other**

It is interesting to investigate what happens to the frequencies of instruments based on air columns and strings during a concert as the temperature rises. The sound emitted by a flute, for example, becomes sharp (increases in frequency) as the flute warms up because the speed of sound increases in the increasingly warmer air inside the flute (consider Eq. 18.8). The sound produced by a violin becomes flat (decreases in frequency) as the strings thermally expand because the expansion causes their tension to decrease (see Eq. 18.6).

Musical instruments based on air columns are generally excited by resonance. The air column is presented with a sound wave that is rich in many frequencies. The air column then responds with a large-amplitude oscillation to the frequencies that match the quantized frequencies in its set of harmonics. In many woodwind instruments, the initial rich sound is provided by a vibrating reed. In brass instruments, this excitation is provided by the sound coming from the vibration of the player's lips. In a flute, the initial excitation comes from blowing over an edge at the mouthpiece of the instrument in a manner similar to blowing across the opening of a bottle with a narrow neck. The sound of the air rushing across the bottle opening has many frequencies, including one that sets the air cavity in the bottle into resonance.

***Quick Quiz* 18.4** A pipe open at both ends resonates at a fundamental frequency f_{open}. When one end is covered and the pipe is again made to resonate, the fundamental frequency is f_{closed}. Which of the following expressions describes how these two resonant frequencies compare? **(a)** $f_{\text{closed}} = f_{\text{open}}$ **(b)** $f_{\text{closed}} = \frac{1}{2}f_{\text{open}}$ **(c)** $f_{\text{closed}} = 2f_{\text{open}}$ **(d)** $f_{\text{closed}} = \frac{3}{2}f_{\text{open}}$

Quick Quiz **18.5** Balboa Park in San Diego has an outdoor organ. When the air temperature increases, the fundamental frequency of one of the organ pipes **(a)** stays the same, **(b)** goes down, **(c)** goes up, or **(d)** is impossible to determine.

Example 18.5 Wind in a Culvert

A section of drainage culvert 1.23 m in length makes a howling noise when the wind blows across its open ends.

(A) Determine the frequencies of the first three harmonics of the culvert if it is cylindrical in shape and open at both ends. Take $v = 343$ m/s as the speed of sound in air.

SOLUTION

Conceptualize The sound of the wind blowing across the end of the pipe contains many frequencies, and the culvert responds to the sound by vibrating at the natural frequencies of the air column.

Categorize This example is a relatively simple substitution problem.

Find the frequency of the first harmonic of the culvert, modeling it as an air column open at both ends:

$$f_1 = \frac{v}{2L} = \frac{343\text{ m/s}}{2(1.23\text{ m})} = \boxed{139\text{ Hz}}$$

Find the next harmonics by multiplying by integers:

$$f_2 = 2f_1 = \boxed{279\text{ Hz}}$$

$$f_3 = 3f_1 = \boxed{418\text{ Hz}}$$

(B) What are the three lowest natural frequencies of the culvert if it is blocked at one end?

SOLUTION

Find the frequency of the first harmonic of the culvert, modeling it as an air column closed at one end:

$$f_1 = \frac{v}{4L} = \frac{343\text{ m/s}}{4(1.23\text{ m})} = \boxed{69.7\text{ Hz}}$$

Find the next two harmonics by multiplying by odd integers:

$$f_3 = 3f_1 = \boxed{209\text{ Hz}}$$

$$f_5 = 5f_1 = \boxed{349\text{ Hz}}$$

Example 18.6 Measuring the Frequency of a Tuning Fork

A simple apparatus for demonstrating resonance in an air column is depicted in Figure 18.14. A vertical pipe open at both ends is partially submerged in water, and a tuning fork vibrating at an unknown frequency is placed near the top of the pipe. The length L of the air column can be adjusted by moving the pipe vertically. The sound waves generated by the fork are reinforced when L corresponds to one of the resonance frequencies of the pipe. For a certain pipe, the smallest value of L for which a peak occurs in the sound intensity is 9.00 cm.

(A) What is the frequency of the tuning fork?

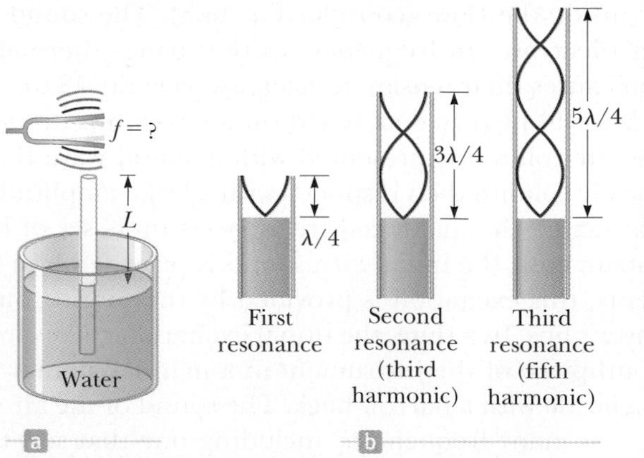

Figure 18.14 (Example 18.6) (a) Apparatus for demonstrating the resonance of sound waves in a pipe closed at one end. The length L of the air column is varied by moving the pipe vertically while it is partially submerged in water. (b) The first three normal modes of the system shown in (a).

SOLUTION

Conceptualize Consider how this problem differs from the preceding example. In the culvert, the length was fixed and the air column was presented with a mixture of very many frequencies. The pipe in this example is presented with one single frequency from the tuning fork, and the length of the pipe is varied until resonance is achieved.

18.6 *cont.*

Categorize This example is a simple substitution problem. Although the pipe is open at its lower end to allow the water to enter, the water's surface acts like a barrier. Therefore, this setup can be modeled as an air column closed at one end.

Use Equation 18.9 to find the fundamental frequency for $L = 0.090\ 0$ m:

$$f_1 = \frac{v}{4L} = \frac{343 \text{ m/s}}{4(0.090\ 0 \text{ m})} = \boxed{953 \text{ Hz}}$$

Because the tuning fork causes the air column to resonate at this frequency, this frequency must also be that of the tuning fork.

(B) What are the values of L for the next two resonance conditions?

SOLUTION

Use Equation 16.12 to find the wavelength of the sound wave from the tuning fork:

$$\lambda = \frac{v}{f} = \frac{343 \text{ m/s}}{953 \text{ Hz}} = 0.360 \text{ m}$$

Notice from Figure 18.14b that the length of the air column for the second resonance is $3\lambda/4$:

$$L = 3\lambda/4 = \boxed{0.270 \text{ m}}$$

Notice from Figure 18.14b that the length of the air column for the third resonance is $5\lambda/4$:

$$L = 5\lambda/4 = \boxed{0.450 \text{ m}}$$

18.6 Standing Waves in Rods and Membranes

Standing waves can also be set up in rods and membranes. A rod clamped in the middle and stroked parallel to the rod at one end oscillates as depicted in Figure 18.15a. The oscillations of the elements of the rod are longitudinal, and so the red-brown curves in Figure 18.15 represent *longitudinal* displacements of various parts of the rod. For clarity, the displacements have been drawn in the transverse direction as they were for air columns. The midpoint is a displacement node because it is fixed by the clamp, whereas the ends are displacement antinodes because they are free to oscillate. The oscillations in this setup are analogous to those in a pipe open at both ends. The red-brown lines in Figure 18.15a represent the first normal mode, for which the wavelength is $2L$ and the frequency is $f = v/2L$, where v is the speed of longitudinal waves in the rod. Other normal modes may be excited by clamping the rod at different points. For example, the second normal mode (Fig. 18.15b) is excited by clamping the rod a distance $L/4$ away from one end.

It is also possible to set up transverse standing waves in rods. Musical instruments that depend on transverse standing waves in rods include triangles, marimbas, xylophones, glockenspiels, chimes, and vibraphones. Other devices that make sounds from vibrating bars include music boxes and wind chimes.

Two-dimensional oscillations can be set up in a flexible membrane stretched over a circular hoop such as that in a drumhead. As the membrane is struck at some point, waves that arrive at the fixed boundary are reflected many times. The resulting sound is not harmonic because the standing waves have frequencies that are *not* related by integer multiples. Without this relationship, the sound may be more correctly described as *noise* rather than as music. The production of noise is in contrast to the situation in wind and stringed instruments, which produce sounds that we describe as musical.

Some possible normal modes of oscillation for a two-dimensional circular membrane are shown in Figure 18.16 on page 528. Whereas nodes are *points* in one-dimensional standing waves on strings and in air columns, a two-dimensional oscillator has *curves* along which there is no displacement of the elements of the medium. The lowest normal mode, which has a frequency f_1, contains only one nodal curve; this curve runs around the outer edge of the membrane. The other

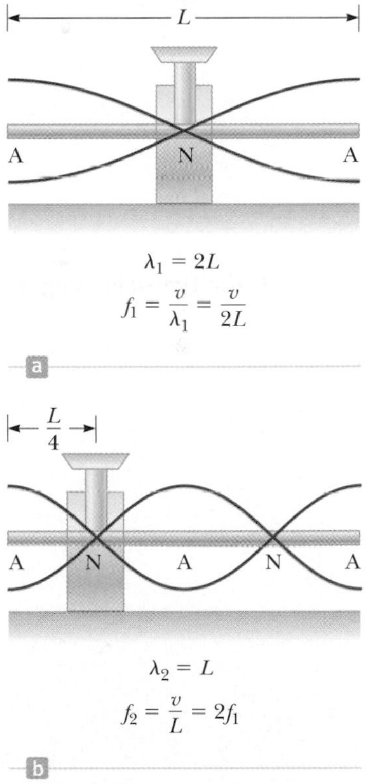

Figure 18.15 Normal-mode longitudinal vibrations of a rod of length L (a) clamped at the middle to produce the first normal mode and (b) clamped at a distance $L/4$ from one end to produce the second normal mode. Notice that the red-brown curves are graphical representations of oscillations parallel to the rod (longitudinal waves).

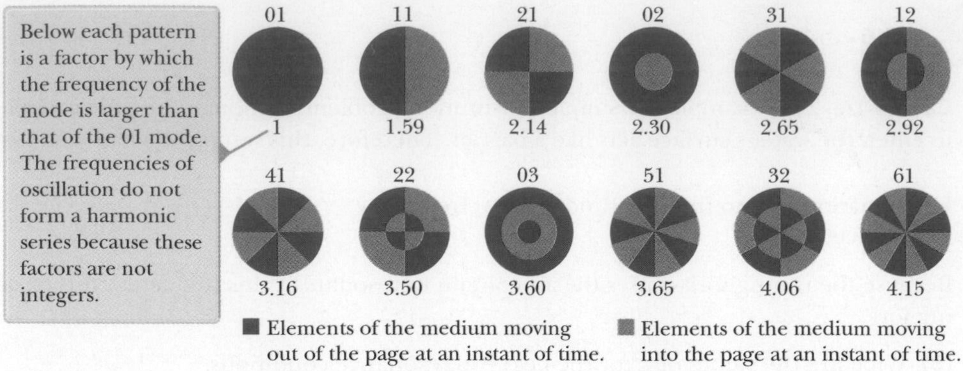

Figure 18.16 Representation of some of the normal modes possible in a circular membrane fixed at its perimeter. The pair of numbers above each pattern corresponds to the number of radial nodes and the number of circular nodes, respectively. In each diagram, elements of the membrane on either side of a nodal line move in opposite directions, as indicated by the colors. *(Adapted from T. D. Rossing,* The Science of Sound, 2nd ed., *Reading, Massachusetts, Addison-Wesley Publishing Co., 1990)*

possible normal modes show additional nodal curves that are circles and straight lines across the diameter of the membrane.

18.7 Beats: Interference in Time

The interference phenomena we have studied so far involve the superposition of two or more waves having the same frequency. Because the amplitude of the oscillation of elements of the medium varies with the position in space of the element in such a wave, we refer to the phenomenon as *spatial interference.* Standing waves in strings and pipes are common examples of spatial interference.

Now let's consider another type of interference, one that results from the superposition of two waves having slightly *different* frequencies. In this case, when the two waves are observed at a point in space, they are periodically in and out of phase. That is, there is a *temporal* (time) alternation between constructive and destructive interference. As a consequence, we refer to this phenomenon as *interference in time* or *temporal interference.* For example, if two tuning forks of slightly different frequencies are struck, one hears a sound of periodically varying amplitude. This phenomenon is called **beating.**

Definition of beating ▶

Beating is the periodic variation in amplitude at a given point due to the superposition of two waves having slightly different frequencies.

The number of amplitude maxima one hears per second, or the *beat frequency,* equals the difference in frequency between the two sources as we shall show below. The maximum beat frequency that the human ear can detect is about 20 beats/s. When the beat frequency exceeds this value, the beats blend indistinguishably with the sounds producing them.

Consider two sound waves of equal amplitude and slightly different frequencies f_1 and f_2 traveling through a medium. We use equations similar to Equation 16.13 to represent the wave functions for these two waves at a point that we identify as $x = 0$. We also choose the phase angle in Equation 16.13 as $\phi = \pi/2$:

$$y_1 = A \sin\left(\frac{\pi}{2} - \omega_1 t\right) = A \cos\,(2\pi f_1 t)$$

$$y_2 = A \sin\left(\frac{\pi}{2} - \omega_2 t\right) = A \cos\,(2\pi f_2 t)$$

Using the superposition principle, we find that the resultant wave function at this point is

$$y = y_1 + y_2 = A\,(\cos 2\pi f_1 t + \cos 2\pi f_2 t)$$

The trigonometric identity

$$\cos a + \cos b = 2 \cos\left(\frac{a-b}{2}\right) \cos\left(\frac{a+b}{2}\right)$$

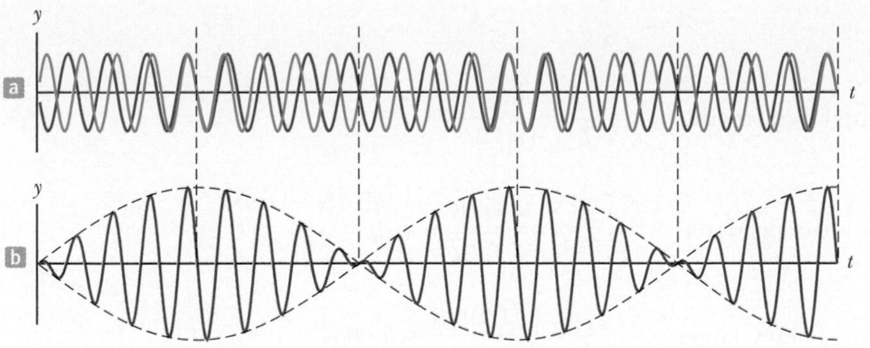

ACTIVE FIGURE 18.17

Beats are formed by the combination of two waves of slightly different frequencies. (a) The individual waves. (b) The combined wave. The envelope wave (dashed line) represents the beating of the combined sounds.

allows us to write the expression for y as

$$y = \left[2A \cos 2\pi\left(\frac{f_1 - f_2}{2}\right)t\right] \cos 2\pi\left(\frac{f_1 + f_2}{2}\right)t \qquad (18.10)$$

◀ **Resultant of two waves of different frequencies but equal amplitude**

Graphs of the individual waves and the resultant wave are shown in Active Figure 18.17. From the factors in Equation 18.10, we see that the resultant wave has an effective frequency equal to the average frequency $(f_1 + f_2)/2$. This wave is multiplied by an envelope wave given by the expression in the square brackets:

$$y_{\text{envelope}} = 2A \cos 2\pi\left(\frac{f_1 - f_2}{2}\right)t \qquad (18.11)$$

That is, the amplitude and therefore the intensity of the resultant sound vary in time. The dashed black line in Active Figure 18.17b is a graphical representation of the envelope wave in Equation 18.11 and is a sine wave varying with frequency $(f_1 - f_2)/2$.

A maximum in the amplitude of the resultant sound wave is detected whenever

$$\cos 2\pi\left(\frac{f_1 - f_2}{2}\right)t = \pm 1$$

Hence, there are *two* maxima in each period of the envelope wave. Because the amplitude varies with frequency as $(f_1 - f_2)/2$, the number of beats per second, or the beat frequency f_{beat}, is twice this value. That is,

$$f_{\text{beat}} = |f_1 - f_2| \qquad (18.12)$$

◀ **Beat frequency**

For instance, if one tuning fork vibrates at 438 Hz and a second one vibrates at 442 Hz, the resultant sound wave of the combination has a frequency of 440 Hz (the musical note A) and a beat frequency of 4 Hz. A listener would hear a 440-Hz sound wave go through an intensity maximum four times every second.

Example 18.7 The Mistuned Piano Strings

Two identical piano strings of length 0.750 m are each tuned exactly to 440 Hz. The tension in one of the strings is then increased by 1.0%. If they are now struck, what is the beat frequency between the fundamentals of the two strings?

SOLUTION

Conceptualize As the tension in one of the strings is changed, its fundamental frequency changes. Therefore, when both strings are played, they will have different frequencies and beats will be heard.

Categorize We must combine our understanding of the waves under boundary conditions model for strings with our new knowledge of beats.

continued

18.7 *cont.*

Analyze Set up a ratio of the fundamental frequencies of the two strings using Equation 18.5:

$$\frac{f_2}{f_1} = \frac{(v_2/2L)}{(v_1/2L)} = \frac{v_2}{v_1}$$

Use Equation 16.18 to substitute for the wave speeds on the strings:

$$\frac{f_2}{f_1} = \frac{\sqrt{T_2/\mu}}{\sqrt{T_1/\mu}} = \sqrt{\frac{T_2}{T_1}}$$

Incorporate that the tension in one string is 1.0% larger than the other; that is, $T_2 = 1.010T_1$:

$$\frac{f_2}{f_1} = \sqrt{\frac{1.010T_1}{T_1}} = 1.005$$

Solve for the frequency of the tightened string:

$$f_2 = 1.005f_1 = 1.005(440\text{ Hz}) = 442\text{ Hz}$$

Find the beat frequency using Equation 18.12:

$$f_{\text{beat}} = 442\text{ Hz} - 440\text{ Hz} = 2\text{ Hz}$$

Finalize Notice that a 1.0% mistuning in tension leads to an easily audible beat frequency of 2 Hz. A piano tuner can use beats to tune a stringed instrument by "beating" a note against a reference tone of known frequency. The tuner can then adjust the string tension until the frequency of the sound it emits equals the frequency of the reference tone. The tuner does so by tightening or loosening the string until the beats produced by it and the reference source become too infrequent to notice.

18.8 Nonsinusoidal Wave Patterns

It is relatively easy to distinguish the sounds coming from a violin and a saxophone even when they are both playing the same note. On the other hand, a person untrained in music may have difficulty distinguishing a note played on a clarinet from the same note played on an oboe. We can use the pattern of the sound waves from various sources to explain these effects.

When frequencies that are integer multiples of a fundamental frequency are combined to make a sound, the result is a *musical* sound. A listener can assign a pitch to the sound based on the fundamental frequency. Pitch is a psychological reaction to a sound that allows the listener to place the sound on a scale from low to high (bass to treble). Combinations of frequencies that are not integer multiples of a fundamental result in a *noise* rather than a musical sound. It is much harder for a listener to assign a pitch to a noise than to a musical sound.

> **Pitfall Prevention 18.4**
>
> **Pitch Versus Frequency**
> Do not confuse the term *pitch* with *frequency*. Frequency is the physical measurement of the number of oscillations per second. Pitch is a psychological reaction to sound that enables a person to place the sound on a scale from high to low or from treble to bass. Therefore, frequency is the stimulus and pitch is the response. Although pitch is related mostly (but not completely) to frequency, they are not the same. A phrase such as "the pitch of the sound" is incorrect because pitch is not a physical property of the sound.

The wave patterns produced by a musical instrument are the result of the superposition of frequencies that are integer multiples of a fundamental. This superposition results in the corresponding richness of musical tones. The human perceptive response associated with various mixtures of harmonics is the *quality* or *timbre* of the sound. For instance, the sound of the trumpet is perceived to have a "brassy" quality (that is, we have learned to associate the adjective *brassy* with that sound); this quality enables us to distinguish the sound of the trumpet from that of the saxophone, whose quality is perceived as "reedy." The clarinet and oboe, however, both contain air columns excited by reeds; because of this similarity, they have similar mixtures of frequencies and it is more difficult for the human ear to distinguish them on the basis of their sound quality.

The sound wave patterns produced by the majority of musical instruments are nonsinusoidal. Characteristic patterns produced by a tuning fork, a flute, and a clarinet, each playing the same note, are shown in Figure 18.18. Each instrument has its own characteristic pattern. Notice, however, that despite the differences in the patterns, each pattern is periodic. This point is important for our analysis of these waves.

The problem of analyzing nonsinusoidal wave patterns appears at first sight to be a formidable task. If the wave pattern is periodic, however, it can be represented as closely as desired by the combination of a sufficiently large number of sinusoidal waves that form a harmonic series. In fact, we can represent any periodic function as a series of sine and cosine terms by using a mathematical technique based on **Fourier's theorem.**[2] The corresponding sum of terms that represents the periodic wave pattern is called a **Fourier series.** Let $y(t)$ be any function that is periodic in time with period T such that $y(t + T) = y(t)$. Fourier's theorem states that this function can be written as

$$y(t) = \sum (A_n \sin 2\pi f_n t + B_n \cos 2\pi f_n t) \quad \textbf{(18.13)}$$

◀ **Fourier's theorem**

where the lowest frequency is $f_1 = 1/T$. The higher frequencies are integer multiples of the fundamental, $f_n = nf_1$, and the coefficients A_n and B_n represent the amplitudes of the various waves. Figure 18.19 represents a harmonic analysis of the wave patterns shown in Figure 18.18. Each bar in the graph represents one of the terms in the series in Equation 18.13 up to $n = 9$. Notice that a struck tuning fork produces only one harmonic (the first), whereas the flute and clarinet produce the first harmonic and many higher ones.

Notice the variation in relative intensity of the various harmonics for the flute and the clarinet. In general, any musical sound consists of a fundamental frequency f plus other frequencies that are integer multiples of f, all having different intensities.

We have discussed the *analysis* of a wave pattern using Fourier's theorem. The analysis involves determining the coefficients of the harmonics in Equation 18.13 from a knowledge of the wave pattern. The reverse process, called *Fourier synthesis,* can also be performed. In this process, the various harmonics are added together to form a resultant wave pattern. As an example of Fourier synthesis, consider the building of a square wave as shown in Active Figure 18.20 (page 532). The symmetry of the square wave results in only odd multiples of the fundamental frequency combining in its synthesis. In Active Figure 18.20a, the blue curve shows the combination of f and $3f$. In Active Figure 18.20b, we have added $5f$ to the combination and obtained the green curve. Notice how the general shape of the square wave is approximated, even though the upper and lower portions are not flat as they should be.

Active Figure 18.20c shows the result of adding odd frequencies up to $9f$. This approximation (red-brown curve) to the square wave is better than the approximations in Active Figures 18.20a and 18.20b. To approximate the square wave as

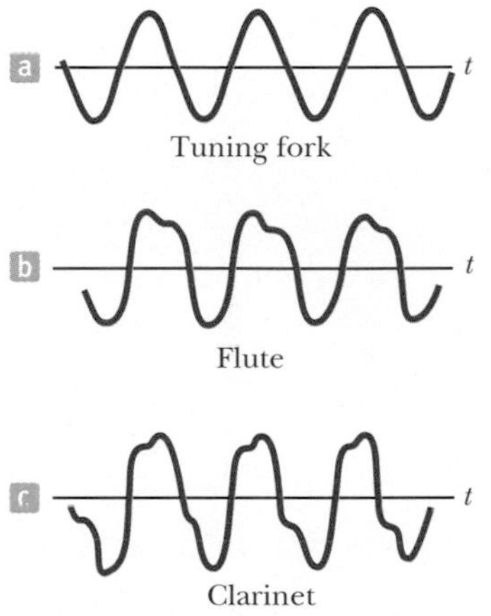

Figure 18.18 Sound wave patterns produced by (a) a tuning fork, (b) a flute, and (c) a clarinet, each at approximately the same frequency.

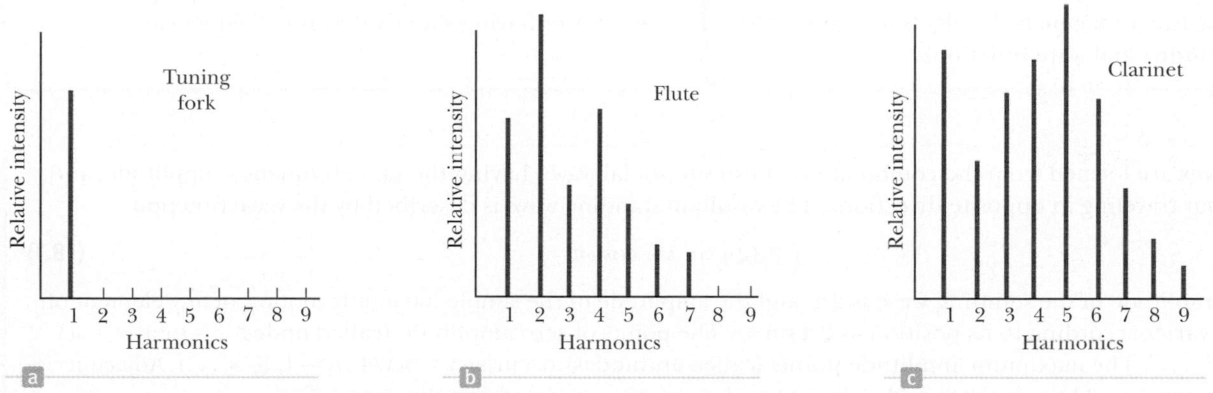

Figure 18.19 Harmonics of the wave patterns shown in Figure 18.18. Notice the variations in intensity of the various harmonics. Parts (a), (b), and (c) correspond to those in Figure 18.18.

[2] Developed by Jean Baptiste Joseph Fourier (1786–1830).

ACTIVE FIGURE 18.20

Fourier synthesis of a square wave, represented by the sum of odd multiples of the first harmonic, which has frequency *f*.

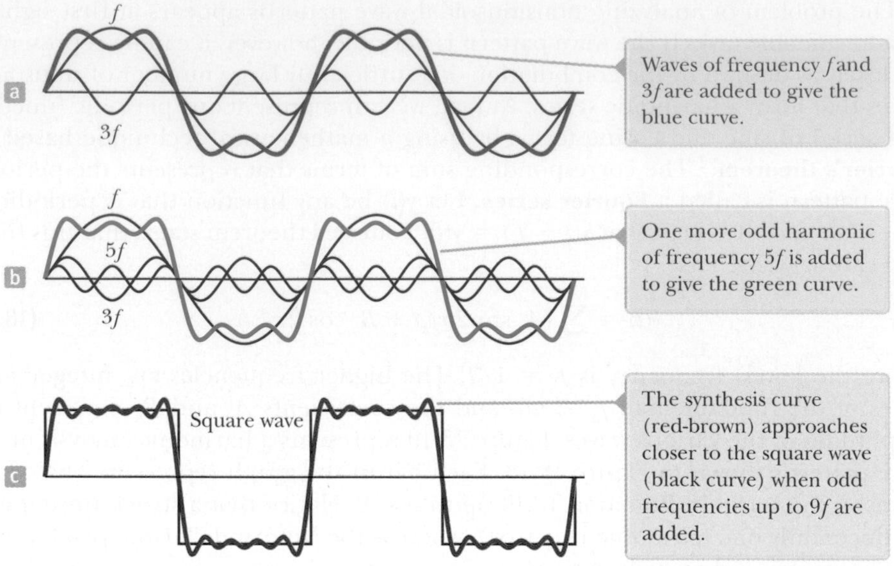

closely as possible, we must add all odd multiples of the fundamental frequency, up to infinite frequency.

Using modern technology, musical sounds can be generated electronically by mixing different amplitudes of any number of harmonics. These widely used electronic music synthesizers are capable of producing an infinite variety of musical tones.

Summary

Concepts and Principles

The **superposition principle** specifies that when two or more waves move through a medium, the value of the resultant wave function equals the algebraic sum of the values of the individual wave functions.

The phenomenon of **beating** is the periodic variation in intensity at a given point due to the superposition of two waves having slightly different frequencies.

Standing waves are formed from the combination of two sinusoidal waves having the same frequency, amplitude, and wavelength but traveling in opposite directions. The resultant standing wave is described by the wave function

$$y = (2A \sin kx) \cos \omega t \qquad \textbf{(18.1)}$$

Hence, the amplitude of the standing wave is $2A$, and the amplitude of the simple harmonic motion of any element of the medium varies according to its position as $2A \sin kx$. The points of zero amplitude (called **nodes**) occur at $x = n\lambda/2$ ($n = 0, 1, 2, 3, \ldots$). The maximum amplitude points (called **antinodes**) occur at $x = n\lambda/4$ ($n = 1, 3, 5, \ldots$). Adjacent antinodes are separated by a distance $\lambda/2$. Adjacent nodes also are separated by a distance $\lambda/2$.

Analysis Models for Problem Solving

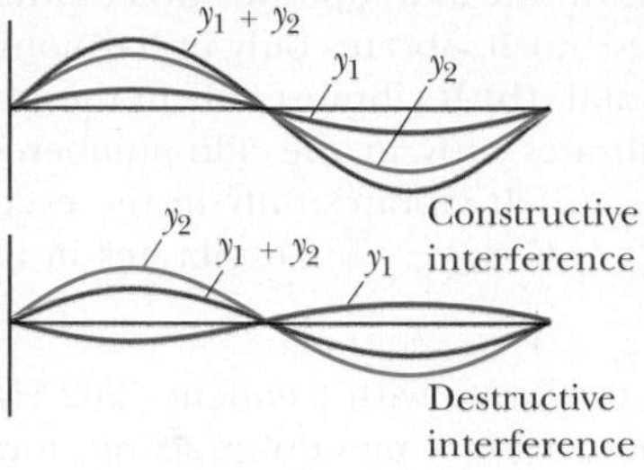

Waves in Interference. When two traveling waves having equal frequencies superimpose, the resultant wave has an amplitude that depends on the phase angle ϕ between the two waves. **Constructive interference** occurs when the two waves are in phase, corresponding to $\phi = 0, 2\pi, 4\pi, \ldots$ rad. **Destructive interference** occurs when the two waves are 180° out of phase, corresponding to $\phi = \pi, 3\pi, 5\pi, \ldots$ rad.

Waves Under Boundary Conditions. When a wave is subject to boundary conditions, only certain natural frequencies are allowed; we say that the frequencies are quantized.

For waves on a string fixed at both ends, the natural frequencies are

$$f_n = \frac{n}{2L}\sqrt{\frac{T}{\mu}} \quad n = 1, 2, 3, \ldots \tag{18.6}$$

where T is the tension in the string and μ is its linear mass density.

For sound waves with speed v in an air column of length L open at both ends, the natural frequencies are

$$f_n = n\frac{v}{2L} \quad n = 1, 2, 3, \ldots \tag{18.8}$$

If an air column is open at one end and closed at the other, only odd harmonics are present and the natural frequencies are

$$f_n = n\frac{v}{4L} \quad n = 1, 3, 5, \ldots \tag{18.9}$$

Objective Questions

☐ denotes answer available in Student Solutions Manual/Study Guide

1. In Figure OQ18.1, a sound wave of wavelength 0.8 m divides into two equal parts that recombine to interfere constructively, with the original difference between their path lengths being $|r_2 - r_1| = 0.8$ m. Rank the following situations according to the intensity of sound at the receiver from the highest to the lowest. Assume the tube walls absorb no sound energy. Give equal ranks to situations in which the intensity is equal. (a) From its original position, the sliding section is moved out by 0.1 m. (b) Next it slides out an additional 0.1 m. (c) It slides out still another 0.1 m. (d) It slides out 0.1 m more.

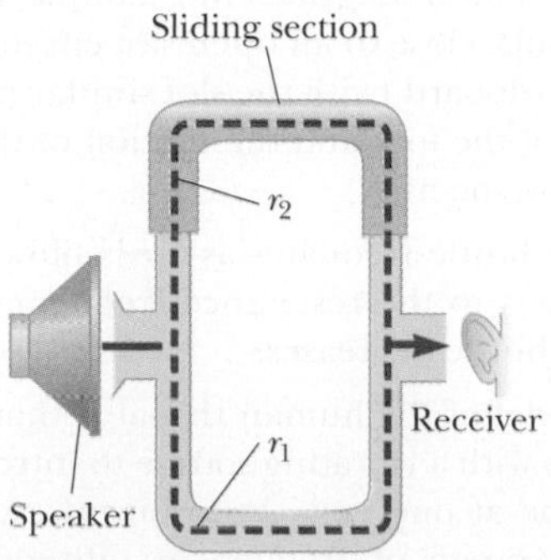

Figure OQ18.1 Objective Question 1 and Problem 6.

2. A series of pulses, each of amplitude 0.1 m, is sent down a string that is attached to a post at one end. The pulses are reflected at the post and travel back along the string without loss of amplitude. **(i)** What is the net displacement at a point on the string where two pulses are crossing? Assume the string is rigidly attached to the post. (a) 0.4 m (b) 0.3 m (c) 0.2 m (d) 0.1 m (e) 0 **(ii)** Next assume the end at which reflection occurs is free to slide up and down. Now what is the net displacement at a point on the string where two pulses are crossing? Choose your answer from the same possibilities as in part (i).

3. In Example 18.1, we investigated an oscillator at 1.3 kHz driving two identical side-by-side speakers. We found that a listener at point O hears sound with maximum intensity, whereas a listener at point P hears a minimum. What is the intensity at P? (a) less than but close to the intensity at O (b) half the intensity at O (c) very low but not zero (d) zero (e) indeterminate

4. A string of length L, mass per unit length μ, and tension T is vibrating at its fundamental frequency. **(i)** If the length

of the string is doubled, with all other factors held constant, what is the effect on the fundamental frequency? (a) It becomes two times larger. (b) It becomes $\sqrt{2}$ times larger. (c) It is unchanged. (d) It becomes $1/\sqrt{2}$ times as large. (e) It becomes one-half as large. **(ii)** If the mass per unit length is doubled, with all other factors held constant, what is the effect on the fundamental frequency? Choose from the same possibilities as in part (i). **(iii)** If the tension is doubled, with all other factors held constant, what is the effect on the fundamental frequency? Choose from the same possibilities as in part (i).

5. Suppose all six equal-length strings of an acoustic guitar are played without fingering, that is, without being pressed down at any frets. What quantities are the same for all six strings? Choose all correct answers. (a) the fundamental frequency (b) the fundamental wavelength of the string wave (c) the fundamental wavelength of the sound emitted (d) the speed of the string wave (e) the speed of the sound emitted

6. Assume two identical sinusoidal waves are moving through the same medium in the same direction. Under what condition will the amplitude of the resultant wave be greater than either of the two original waves? (a) in all cases (b) only if the waves have no difference in phase (c) only if the phase difference is less than 90° (d) only if the phase difference is less than 120° (e) only if the phase difference is less than 180°

7. As oppositely moving pulses of the same shape (one upward, one downward) on a string pass through each other, at one particular instant the string shows no displacement from the equilibrium position at any point. What has happened to the energy carried by the pulses at this instant of time? (a) It was used up in producing the previous motion. (b) It is all potential energy. (c) It is all internal energy. (d) It is all kinetic energy. (e) The positive energy of one pulse adds to zero with the negative energy of the other pulse.

8. An archer shoots an arrow horizontally from the center of the string of a bow held vertically. After the arrow leaves it, the string of the bow will vibrate as a superposition of what standing-wave harmonics? (a) It vibrates only in harmonic number 1, the fundamental. (b) It vibrates only in the second harmonic. (c) It vibrates only in the odd-numbered harmonics 1, 3, 5, 7, ... (d) It vibrates only in the even-numbered harmonics 2, 4, 6, 8, ... (e) It vibrates in all harmonics.

9. A tuning fork is known to vibrate with frequency 262 Hz. When it is sounded along with a mandolin string, four beats are heard every second. Next, a bit of tape is put onto each tine of the tuning fork, and the tuning fork now produces five beats per second with the same mandolin string. What is the frequency of the string? (a) 257 Hz (b) 258 Hz (c) 262 Hz (d) 266 Hz (e) 267 Hz

10. A standing wave having three nodes is set up in a string fixed at both ends. If the frequency of the wave is doubled, how many antinodes will there be? (a) 2 (b) 3 (c) 4 (d) 5 (e) 6

11. A flute has a length of 58.0 cm. If the speed of sound in air is 343 m/s, what is the fundamental frequency of the flute, assuming it is a tube closed at one end and open at the other? (a) 148 Hz (b) 296 Hz (c) 444 Hz (d) 591 Hz (e) none of those answers

12. When two tuning forks are sounded at the same time, a beat frequency of 5 Hz occurs. If one of the tuning forks has a frequency of 245 Hz, what is the frequency of the other tuning fork? (a) 240 Hz (b) 242.5 Hz (c) 247.5 Hz (d) 250 Hz (e) More than one answer could be correct.

Conceptual Questions

☐ denotes answer available in Student Solutions Manual/Study Guide

1. Does the phenomenon of wave interference apply only to sinusoidal waves?

2. When two waves interfere constructively or destructively, is there any gain or loss in energy in the system of the waves? Explain.

3. Explain how a musical instrument such as a piano may be tuned by using the phenomenon of beats.

4. What limits the amplitude of motion of a real vibrating system that is driven at one of its resonant frequencies?

5. Despite a reasonably steady hand, a person often spills his coffee when carrying it to his seat. Discuss resonance as a possible cause of this difficulty and devise a means for preventing the spills.

6. An airplane mechanic notices that the sound from a twin-engine aircraft rapidly varies in loudness when both engines are running. What could be causing this variation from loud to soft?

7. A tuning fork by itself produces a faint sound. Explain how each of the following methods can be used to obtain a louder sound from it. Explain also any effect on the time interval for which the fork vibrates audibly. (a) holding the edge of a sheet of paper against one vibrating tine (b) pressing the handle of the tuning fork against a chalkboard or a tabletop (c) holding the tuning fork above a column of air of properly chosen length as in Example 18.6 (d) holding the tuning fork close to an open slot cut in a sheet of foam plastic or cardboard (with the slot similar in size and shape to one tine of the fork and the motion of the tines perpendicular to the sheet)

8. A soft-drink bottle resonates as air is blown across its top. What happens to the resonance frequency as the level of fluid in the bottle decreases?

9. A crude model of the human throat is that of a pipe open at both ends with a vibrating source to introduce the sound into the pipe at one end. Assuming the vibrating source produces a range of frequencies, discuss the effect of changing the pipe's length.

Problems

ENHANCED WebAssign The problems found in this chapter may be assigned online in Enhanced WebAssign

1. denotes straightforward problem; **2.** denotes intermediate problem; **3.** denotes challenging problem

1 full solution available in the Student Solutions Manual/Study Guide

1. denotes problems most often assigned in Enhanced WebAssign; these provide students with targeted feedback and either a Master It tutorial or a Watch It solution video.

QC denotes asking for quantitative and conceptual reasoning

S denotes symbolic reasoning problem

M denotes Master It tutorial available in Enhanced WebAssign

GP denotes guided problem

shaded denotes "paired problems" that develop reasoning with symbols and numerical values

Note: Unless otherwise specified, assume the speed of sound in air is 343 m/s, its value at an air temperature of 20.0°C. At any other Celsius temperature T_C, the speed of sound in air is described by

$$v = 331\sqrt{1 + \frac{T_C}{273}}$$

where v is in m/s and T is in °C.

Section 18.1 Analysis Model: Waves in Interference

1. Two waves on one string are described by the wave functions

$$y_1 = 3.0 \cos (4.0x - 1.6t) \qquad y_2 = 4.0 \sin (5.0x - 2.0t)$$

where x and y are in centimeters and t is in seconds. Find the superposition of the waves $y_1 + y_2$ at the points (a) $x = 1.00$, $t = 1.00$; (b) $x = 1.00$, $t = 0.500$; and (c) $x = 0.500$, $t = 0$. *Note:* Remember that the arguments of the trigonometric functions are in radians.

2. Two wave pulses A and B are moving in opposite directions, each with a speed $v = 2.00$ cm/s. The amplitude of A is twice the amplitude of B. The pulses are shown in Figure P18.2 at $t = 0$. Sketch the resultant wave at $t = 1.00$ s, 1.50 s, 2.00 s, 2.50 s, and 3.00 s.

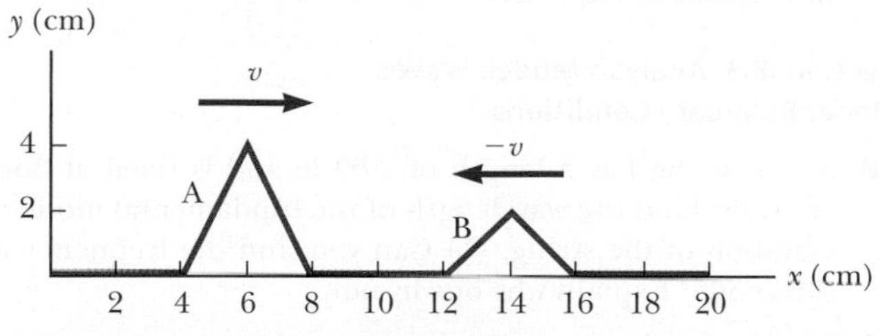

Figure P18.2

3. Two waves are traveling in the same direction along a stretched string. The waves are 90.0° out of phase. Each wave has an amplitude of 4.00 cm. Find the amplitude of the resultant wave.

4. Two pulses of different amplitudes approach each other, each having a speed of $v = 1.00$ m/s. Figure P18.4 shows the positions of the pulses at time $t = 0$. (a) Sketch the resultant wave at $t = 2.00$ s, 4.00 s, 5.00 s, and 6.00 s. (b) **What If?** If the pulse on the right is inverted so that it is upright, how would your sketches of the resultant wave change?

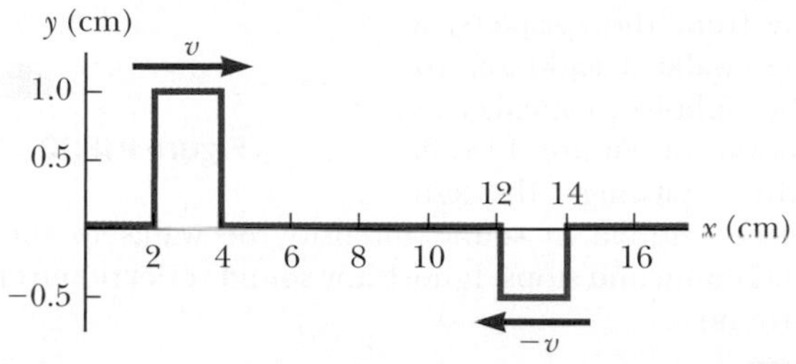

Figure P18.4

5. Two pulses traveling on the same string are described by

$$y_1 = \frac{5}{(3x - 4t)^2 + 2} \qquad y_2 = \frac{-5}{(3x + 4t - 6)^2 + 2}$$

(a) In which direction does each pulse travel? (b) At what instant do the two cancel everywhere? (c) At what point do the two pulses always cancel?

6. The acoustical system shown in Figure OQ18.1 is driven by a speaker emitting sound of frequency 756 Hz. (a) If constructive interference occurs at a particular location of the sliding section, by what minimum amount should the sliding section be moved upward so that destructive interference occurs instead? (b) What minimum distance from the original position of the sliding section will again result in constructive interference?

7. **M** Two traveling sinusoidal waves are described by the wave functions

$$y_1 = 5.00 \sin [\pi(4.00x - 1\,200t)]$$

$$y_2 = 5.00 \sin [\pi(4.00x - 1\,200t - 0.250)]$$

where x, y_1, and y_2 are in meters and t is in seconds. (a) What is the amplitude of the resultant wave function $y_1 + y_2$? (b) What is the frequency of the resultant wave function?

8. Two identical loudspeakers are placed on a wall 2.00 m apart. A listener stands 3.00 m from the wall directly in front of one of the speakers. A single oscillator is driving the speakers at a frequency of 300 Hz. (a) What is the phase difference in radians between the waves from the speakers when they reach the observer? (b) **What If?** What is the frequency closest to 300 Hz to which the oscillator may be adjusted such that the observer hears minimal sound?

9. **M** Two sinusoidal waves in a string are defined by the wave functions

$$y_1 = 2.00 \sin (20.0x - 32.0t) \qquad y_2 = 2.00 \sin (25.0x - 40.0t)$$

where x, y_1, and y_2 are in centimeters and t is in seconds. (a) What is the phase difference between these two waves at the point $x = 5.00$ cm at $t = 2.00$ s? (b) What is the positive x value closest to the origin for which the two phases differ by $\pm\pi$ at $t = 2.00$ s? (At that location, the two waves add to zero.)

10. *Why is the following situation impossible?* Two identical loudspeakers are driven by the same oscillator at frequency 200 Hz. They are located on the ground a distance $d = 4.00$ m from each other. Starting far from the speakers, a man walks straight toward the right-hand speaker as shown in Figure P18.10. After passing through three minima in sound intensity, he walks to the next maximum and stops. Ignore any sound reflection from the ground.

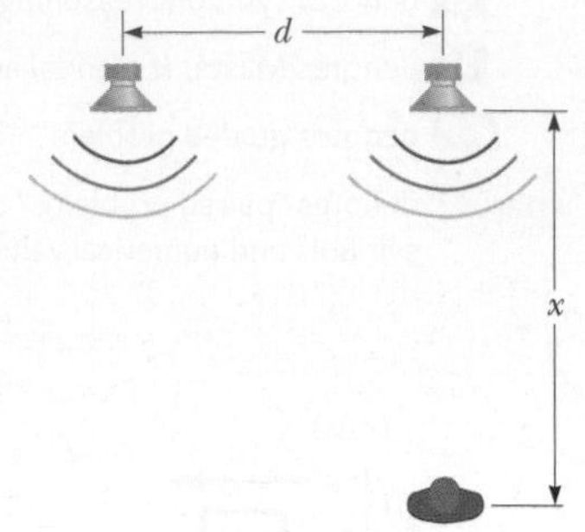

Figure P18.10

11. **Q|C** Two identical loudspeakers 10.0 m apart are driven by the same oscillator with a frequency of $f = 21.5$ Hz (Fig. P18.11) in an area where the speed of sound is 344 m/s. (a) Show that a receiver at point A records a minimum in sound intensity from the two speakers. (b) If the receiver is moved in the plane of the speakers, show that the path it should take so that the intensity remains at a minimum is along the hyperbola $9x^2 - 16y^2 = 144$ (shown in red-brown in Fig. P18.11). (c) Can the receiver remain at a minimum and move very far away from the two sources? If so, determine the limiting form of the path it must take. If not, explain how far it can go.

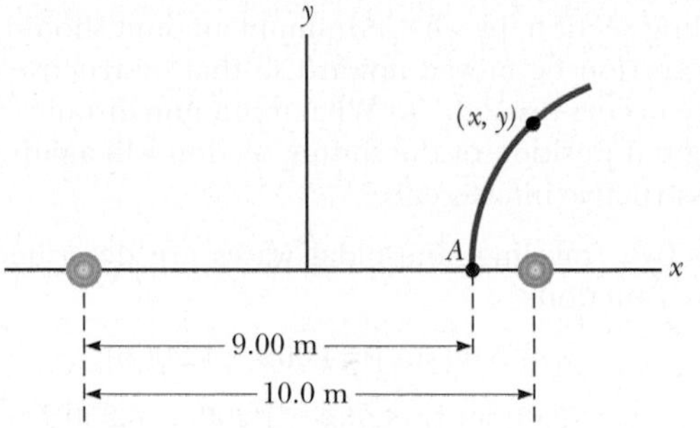

Figure P18.11

Section 18.2 Standing Waves

12. Verify by direct substitution that the wave function for a standing wave given in Equation 18.1,

$$y = (2A \sin kx) \cos \omega t$$

is a solution of the general linear wave equation, Equation 16.27:

$$\frac{\partial^2 y}{\partial x^2} = \frac{1}{v^2}\frac{\partial^2 y}{\partial t^2}$$

13. Two sinusoidal waves traveling in opposite directions interfere to produce a standing wave with the wave function

$$y = 1.50 \sin (0.400x) \cos (200t)$$

where x and y are in meters and t is in seconds. Determine (a) the wavelength, (b) the frequency, and (c) the speed of the interfering waves.

14. **Q|C** Two waves simultaneously present on a long string have a phase difference ϕ between them so that a standing wave formed from their combination is described by

$$y(x, t) = 2A \sin\left(kx + \frac{\phi}{2}\right) \cos\left(\omega t - \frac{\phi}{2}\right)$$

(a) Despite the presence of the phase angle ϕ, is it still true that the nodes are one-half wavelength apart? Explain. (b) Are the nodes different in any way from the way they would be if ϕ were zero? Explain.

15. Two transverse sinusoidal waves combining in a medium are described by the wave functions

$$y_1 = 3.00 \sin \pi(x + 0.600t) \qquad y_2 = 3.00 \sin \pi(x - 0.600t)$$

where x, y_1, and y_2 are in centimeters and t is in seconds. Determine the maximum transverse position of an element of the medium at (a) $x = 0.250$ cm, (b) $x = 0.500$ cm, and (c) $x = 1.50$ cm. (d) Find the three smallest values of x corresponding to antinodes.

16. **Q|C** A standing wave is described by the wave function

$$y = 6 \sin\left(\frac{\pi}{2}x\right) \cos (100\pi t)$$

where x and y are in meters and t is in seconds. (a) Prepare graphs showing y as a function of x for five instants: $t = 0$, 5 ms, 10 ms, 15 ms, and 20 ms. (b) From the graph, identify the wavelength of the wave and explain how to do so. (c) From the graph, identify the frequency of the wave and explain how to do so. (d) From the equation, directly identify the wavelength of the wave and explain how to do so. (e) From the equation, directly identify the frequency and explain how to do so.

17. Two identical loudspeakers are driven in phase by a common oscillator at 800 Hz and face each other at a distance of 1.25 m. Locate the points along the line joining the two speakers where relative minima of sound pressure amplitude would be expected.

Section 18.3 Analysis Model: Waves Under Boundary Conditions

18. A taut string has a length of 2.60 m and is fixed at both ends. (a) Find the wavelength of the fundamental mode of vibration of the string. (b) Can you find the frequency of this mode? Explain why or why not.

19. A certain vibrating string on a piano has a length of 74.0 cm and forms a standing wave having two antinodes. (a) Which harmonic does this wave represent? (b) Determine the wavelength of this wave. (c) How many nodes are there in the wave pattern?

20. A string that is 30.0 cm long and has a mass per unit length of 9.00×10^{-3} kg/m is stretched to a tension of 20.0 N. Find (a) the fundamental frequency and (b) the next three frequencies that could cause standing-wave patterns on the string.

21. A string with a mass $m = 8.00$ g and a length $L = 5.00$ m has one end attached to a wall; the other end is draped over a small, fixed pulley a distance $d = 4.00$ m from the wall and attached to a hanging object with a mass $M = 4.00$ kg as in Figure P18.21. If the horizontal part of the string is plucked, what is the fundamental frequency of its vibration?

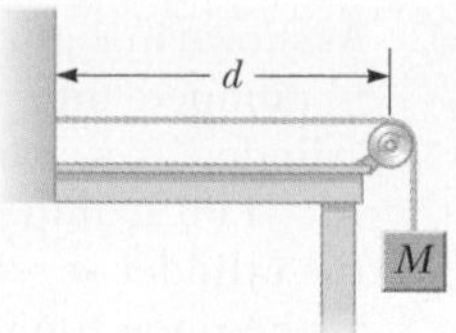

Figure P18.21

22. The 64.0-cm-long string of a guitar has a fundamental frequency of 330 Hz when it vibrates freely along its entire length. A fret is provided for limiting vibration to just the lower two-thirds of the string. (a) If the string is pressed down at this fret and plucked, what is the new fundamental frequency? (b) **What If?** The guitarist can play a "natural harmonic" by gently touching the string at the location of this fret and plucking the string at about one-sixth of the way along its length from the other end. What frequency will be heard then?

23. The A string on a cello vibrates in its first normal mode with a frequency of 220 Hz. The vibrating segment is 70.0 cm long and has a mass of 1.20 g. (a) Find the tension in the string. (b) Determine the frequency of vibration when the string vibrates in three segments.

24. In the arrangement shown in Figure P18.24, an object can be hung from a string (with linear mass density $\mu = 0.002\ 00$ kg/m) that passes over a light pulley. The string is connected to a vibrator (of constant frequency f), and the length of the string between point P and the pulley is $L = 2.00$ m. When the mass m of the object is either 16.0 kg or 25.0 kg, standing waves are observed; no standing waves are observed with any mass between these values, however. (a) What is the frequency of the vibrator? *Note:* The greater the tension in the string, the smaller the number of nodes in the standing wave. (b) What is the largest object mass for which standing waves could be observed?

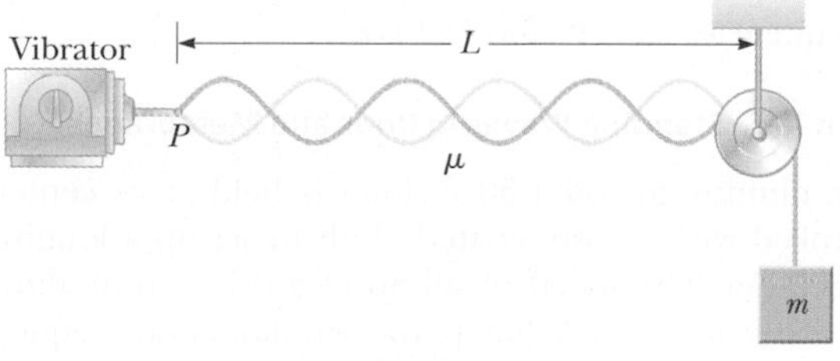

Figure P18.24

25. Review. A sphere of mass $M = 1.00$ kg is supported by a string that passes over a pulley at the end of a horizontal rod of length $L = 0.300$ m (Fig. P18.25). The string makes an angle $\theta = 35.0°$ with the rod. The fundamental frequency of standing waves in the portion of the string above the rod is $f = 60.0$ Hz. Find the mass of the portion of the string above the rod.

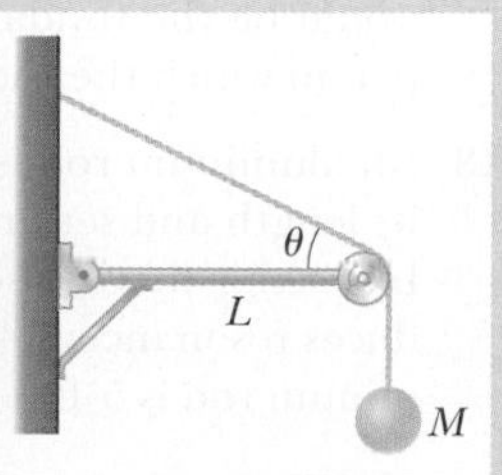

Figure P18.25
Problems 25 and 26.

26. S **Review.** A sphere of mass M is supported by a string that passes over a pulley at the end of a horizontal rod of length L (Fig. P18.25). The string makes an angle θ with the rod. The fundamental frequency of standing waves in the portion of the string above the rod is f. Find the mass of the portion of the string above the rod.

27. A violin string has a length of 0.350 m and is tuned to concert G, with $f_G = 392$ Hz. (a) How far from the end of the string must the violinist place her finger to play concert A, with $f_A = 440$ Hz? (b) If this position is to remain correct to one-half the width of a finger (that is, to within 0.600 cm), what is the maximum allowable percentage change in the string tension?

28. Review. A solid copper object hangs at the bottom of a steel wire of negligible mass. The top end of the wire is fixed. When the wire is struck, it emits sound with a fundamental frequency of 300 Hz. The copper object is then submerged in water so that half its volume is below the water line. Determine the new fundamental frequency.

29. A standing-wave pattern is observed in a thin wire with a length of 3.00 m. The wave function is

$$y = 0.002\ 00 \sin(\pi x) \cos(100\pi t)$$

where x and y are in meters and t is in seconds. (a) How many loops does this pattern exhibit? (b) What is the fundamental frequency of vibration of the wire? (c) **What If?** If the original frequency is held constant and the tension in the wire is increased by a factor of 9, how many loops are present in the new pattern?

Section 18.4 Resonance

30. Q|C The Bay of Fundy, Nova Scotia, has the highest tides in the world. Assume in midocean and at the mouth of the bay the Moon's gravity gradient and the Earth's rotation make the water surface oscillate with an amplitude of a few centimeters and a period of 12 h 24 min. At the head of the bay, the amplitude is several meters. Assume the bay has a length of 210 km and a uniform depth of 36.1 m. The speed of long-wavelength water waves is given by $v = \sqrt{gd}$, where d is the water's depth. Argue for or against the proposition that the tide is magnified by standing-wave resonance.

31. An earthquake can produce a *seiche* in a lake in which the water sloshes back and forth from end to end with remarkably large amplitude and long period. Consider a seiche produced in a farm pond. Suppose the pond is 9.15 m long and assume it has a uniform width and depth. You measure that a pulse produced at one end reaches the other end in 2.50 s. (a) What is the wave speed? (b) What should be the frequency of the ground motion during the earthquake to produce a seiche that is a standing wave with antinodes at each end of the pond and one node at the center?

32. High-frequency sound can be used to produce standing-wave vibrations in a wine glass. A standing-wave vibration in a wine glass is observed to have four nodes and four antinodes equally spaced around the 20.0-cm circumference of the rim of the glass. If transverse waves move around the glass at 900 m/s, an

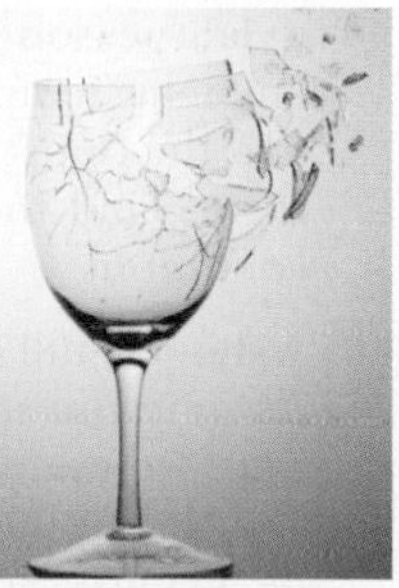

Steve Bronstein/Stone/Getty Images

Figure P18.32

opera singer would have to produce a high harmonic with what frequency to shatter the glass with a resonant vibration as shown in Figure P18.32?

Section 18.5 Standing Waves in Air Columns

33. Calculate the length of a pipe that has a fundamental frequency of 240 Hz assuming the pipe is (a) closed at one end and (b) open at both ends.

34. The overall length of a piccolo is 32.0 cm. The resonating air column is open at both ends. (a) Find the frequency of the lowest note a piccolo can sound. (b) Opening holes in the side of a piccolo effectively shortens the length of the resonant column. Assume the highest note a piccolo can sound is 4 000 Hz. Find the distance between adjacent antinodes for this mode of vibration.

35. The fundamental frequency of an open organ pipe corresponds to middle C (261.6 Hz on the chromatic musical scale). The third resonance of a closed organ pipe has the same frequency. What is the length of (a) the open pipe and (b) the closed pipe?

36. The longest pipe on a certain organ is 4.88 m. What is the fundamental frequency (at 0.00°C) if the pipe is (a) closed at one end and (b) open at each end? (c) What will be the frequencies at 20.0°C?

37. An air column in a glass tube is open at one end and closed at the other by a movable piston. The air in the tube is warmed above room temperature, and a 384-Hz tuning fork is held at the open end. Resonance is heard when the piston is at a distance $d_1 = 22.8$ cm from the open end and again when it is at a distance $d_2 = 68.3$ cm from the open end. (a) What speed of sound is implied by these data? (b) How far from the open end will the piston be when the next resonance is heard?

38. A tuning fork with a frequency of $f = 512$ Hz is placed near the top of the tube shown in Figure P18.38. The water level is lowered so that the length L slowly increases from an initial value of 20.0 cm. Determine the next two values of L that correspond to resonant modes.

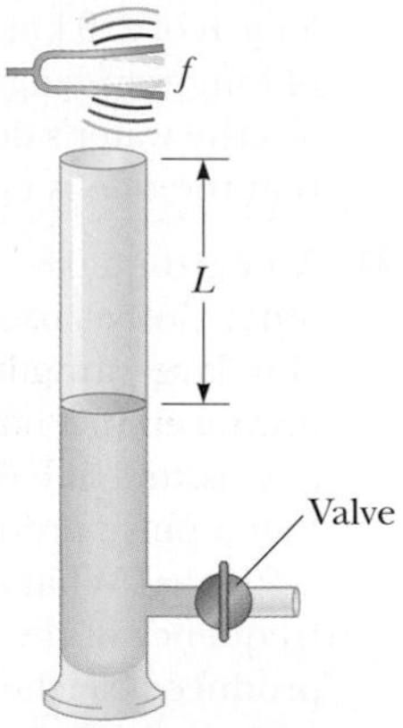

Figure P18.38

39. With a particular fingering, a flute produces a note with frequency 880 Hz at 20.0°C. The flute is open at both ends. (a) Find the air column length. (b) At the beginning of the halftime performance at a late-season football game, the ambient temperature is −5.00°C and the flutist has not had a chance to warm up her instrument. Find the frequency the flute produces under these conditions.

40. A shower stall has dimensions 86.0 cm × 86.0 cm × 210 cm. Assume the stall acts as a pipe closed at both ends, with nodes at opposite sides. Assume singing voices range from 130 Hz to 2 000 Hz and let the speed of sound in the hot air be 355 m/s. For someone singing in this shower, which frequencies would sound the richest (because of resonance)?

41. As shown in Figure P18.41, water is pumped into a tall, vertical cylinder at a volume flow rate $R = 1.00$ L/min. The radius of the cylinder is $r = 5.00$ cm, and at the open top of the cylinder a tuning fork is vibrating with a frequency $f = 512$ Hz. As the water rises, what time interval elapses between successive resonances?

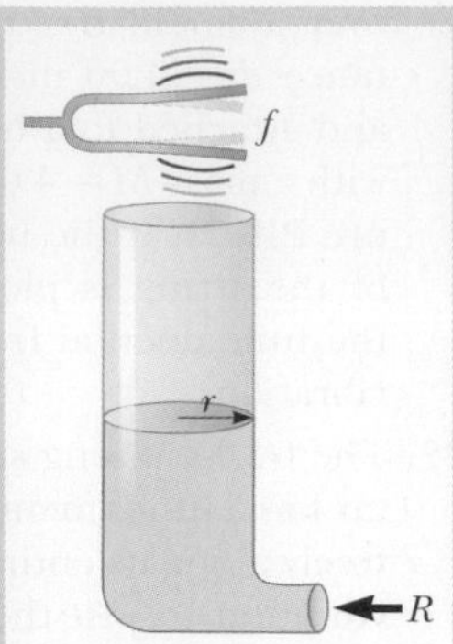

Figure P18.41 Problems 41 and 42.

42. S As shown in Figure P18.41, water is pumped into a tall, vertical cylinder at a volume flow rate R. The radius of the cylinder is r, and at the open top of the cylinder a tuning fork is vibrating with a frequency f. As the water rises, what time interval elapses between successive resonances?

43. M Two adjacent natural frequencies of an organ pipe are determined to be 550 Hz and 650 Hz. Calculate (a) the fundamental frequency and (b) the length of this pipe.

44. Q|C A tunnel under a river is 2.00 km long. (a) At what frequencies can the air in the tunnel resonate? (b) Explain whether it would be good to make a rule against blowing your car horn when you are in the tunnel.

45. A student uses an audio oscillator of adjustable frequency to measure the depth of a water well. The student reports hearing two successive resonances at 51.87 Hz and 59.85 Hz. (a) How deep is the well? (b) How many antinodes are in the standing wave at 51.87 Hz?

46. *Why is the following situation impossible?* A student is listening to the sounds from an air column that is 0.730 m long. He doesn't know if the column is open at both ends or open at only one end. He hears resonance from the air column at frequencies 235 Hz and 587 Hz.

Section 18.6 Standing Waves in Rods and Membranes

47. An aluminum rod 1.60 m long is held at its center. It is stroked with a rosin-coated cloth to set up a longitudinal vibration. The speed of sound in a thin rod of aluminum is 5 100 m/s. (a) What is the fundamental frequency of the waves established in the rod? (b) What harmonics are set up in the rod held in this manner? (c) **What If?** What would be the fundamental frequency if the rod were copper, in which the speed of sound is 3 560 m/s?

48. An aluminum rod is clamped one-fourth of the way along its length and set into longitudinal vibration by a variable-frequency driving source. The lowest frequency that produces resonance is 4 400 Hz. The speed of sound in an aluminum rod is 5 100 m/s. Determine the length of the rod.

Section 18.7 Beats: Interference in Time

49. M In certain ranges of a piano keyboard, more than one string is tuned to the same note to provide extra loudness. For example, the note at 110 Hz has two strings at this frequency. If one string slips from its normal tension of 600 N to 540 N, what beat frequency is heard when the hammer strikes the two strings simultaneously?

50. While attempting to tune the note C at 523 Hz, a piano tuner hears 2.00 beats/s between a reference oscillator and the string. (a) What are the possible frequencies of the string? (b) When she tightens the string slightly, she hears 3.00 beats/s. What is the frequency of the string now? (c) By what percentage should the piano tuner now change the tension in the string to bring it into tune?

51. **M** **Review.** A student holds a tuning fork oscillating at 256 Hz. He walks toward a wall at a constant speed of 1.33 m/s. (a) What beat frequency does he observe between the tuning fork and its echo? (b) How fast must he walk away from the wall to observe a beat frequency of 5.00 Hz?

52. **Review.** Jane waits on a railroad platform while two trains approach from the same direction at equal speeds of 8.00 m/s. Both trains are blowing their whistles (which have the same frequency), and one train is some distance behind the other. After the first train passes Jane but before the second train passes her, she hears beats of frequency 4.00 Hz. What is the frequency of the train whistles?

Section 18.8 Nonsinusoidal Wave Patterns

53. Suppose a flutist plays a 523-Hz C note with first harmonic displacement amplitude $A_1 = 100$ nm. From Figure 18.19b read, by proportion, the displacement amplitudes of harmonics 2 through 7. Take these as the values A_2 through A_7 in the Fourier analysis of the sound and assume $B_1 = B_2 = \cdots = B_7 = 0$. Construct a graph of the waveform of the sound. Your waveform will not look exactly like the flute waveform in Figure 18.18b because you simplify by ignoring cosine terms; nevertheless, it produces the same sensation to human hearing.

Additional Problems

54. Two strings are vibrating at the same frequency of 150 Hz. After the tension in one of the strings is decreased, an observer hears four beats each second when the strings vibrate together. Find the new frequency in the adjusted string.

55. The ship in Figure P18.55 travels along a straight line parallel to the shore and a distance $d = 600$ m from it. The ship's radio receives simultaneous signals of the same frequency from antennas A and B, separated by a distance $L = 800$ m. The signals interfere constructively at point C, which is equidistant from A and B. The signal goes through the first minimum at point D, which is directly outward from the shore from point B. Determine the wavelength of the radio waves.

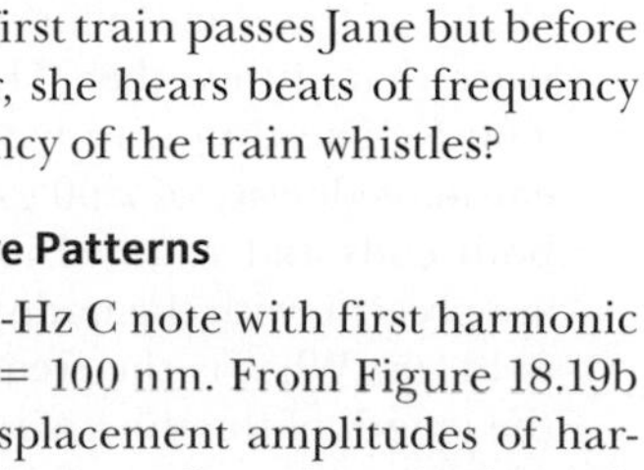

Figure P18.55

56. A 2.00-m-long wire having a mass of 0.100 kg is fixed at both ends. The tension in the wire is maintained at 20.0 N. (a) What are the frequencies of the first three allowed modes of vibration? (b) If a node is observed at a point 0.400 m from one end, in what mode and with what frequency is it vibrating?

57. The fret closest to the bridge on a guitar is 21.4 cm from the bridge as shown in Figure P18.57. When the thinnest string is pressed down at this first fret, the string produces the highest frequency that can be played on that guitar, 2 349 Hz. The next lower note that is produced on the string has frequency 2 217 Hz. How far away from the first fret should the next fret be?

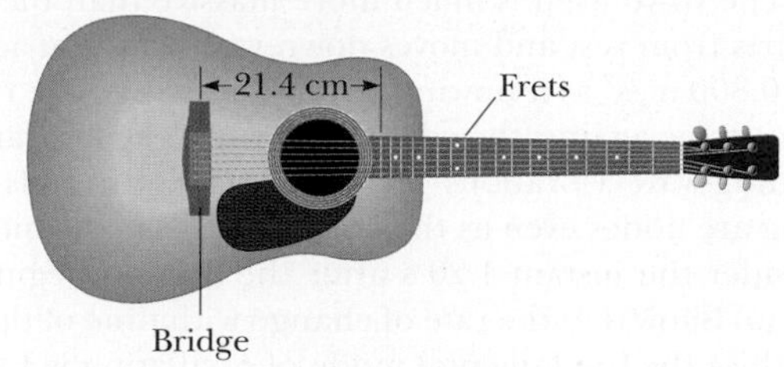

Figure P18.57

58. **QIC** A string fixed at both ends and having a mass of 4.80 g, a length of 2.00 m, and a tension of 48.0 N vibrates in its second ($n = 2$) normal mode. (a) Is the wavelength in air of the sound emitted by this vibrating string larger or smaller than the wavelength of the wave on the string? (b) What is the ratio of the wavelength in air of the sound emitted by this vibrating string and the wavelength of the wave on the string?

59. A quartz watch contains a crystal oscillator in the form of a block of quartz that vibrates by contracting and expanding. An electric circuit feeds in energy to maintain the oscillation and also counts the voltage pulses to keep time. Two opposite faces of the block, 7.05 mm apart, are antinodes, moving alternately toward each other and away from each other. The plane halfway between these two faces is a node of the vibration. The speed of sound in quartz is equal to 3.70×10^3 m/s. Find the frequency of the vibration.

60. **GP** **Review.** For the arrangement shown in Figure P18.60, the inclined plane and the small pulley are frictionless; the string supports the object of mass M at the bottom of the plane; and the string has mass m. The system is in equilibrium, and the vertical part of the string has a length h. We wish to study standing waves set up in the vertical section of the string. (a) What analysis model describes the object of mass M? (b) What analysis model describes the waves on the vertical part of the string? (c) Find the tension in the string. (d) Model the shape of the string as one leg and the hypotenuse of a right triangle. Find the whole length of the string. (e) Find the mass per unit length of the string. (f) Find the speed of waves on the string. (g) Find the lowest frequency for a standing wave on the vertical section of the string. (h) Evaluate this result for $M = 1.50$ kg, $m = 0.750$ g, $h = 0.500$ m, and $\theta = 30.0°$. (i) Find the numerical value for the lowest frequency for a standing wave on the sloped section of the string.

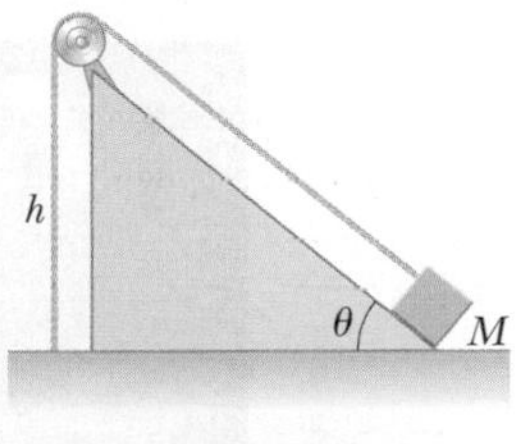

Figure P18.60

61. **QIC** **Review.** Consider the apparatus shown in Figure 18.11 and described in Example 18.4. Suppose the number of antinodes in Figure 18.11b is an arbitrary value n. (a) Find an expression for the radius of the sphere in the water as a function of only n. (b) What is the minimum allowed value of n for a sphere of nonzero size? (c) What is the radius of the largest sphere that will produce a standing wave on the string? (d) What happens if a larger sphere is used?

62. QC **Review.** The top end of a yo-yo string is held stationary. The yo-yo itself is much more massive than the string. It starts from rest and moves down with constant acceleration 0.800 m/s^2 as it unwinds from the string. The rubbing of the string against the edge of the yo-yo excites transverse standing-wave vibrations in the string. Both ends of the string are nodes even as the length of the string increases. Consider the instant 1.20 s after the motion begins from rest. (a) Show that the rate of change with time of the wavelength of the fundamental mode of oscillation is 1.92 m/s. (b) **What if?** Is the rate of change of the wavelength of the second harmonic also 1.92 m/s at this moment? Explain your answer. (c) **What if?** The experiment is repeated after more mass has been added to the yo-yo body. The mass distribution is kept the same so that the yo-yo still moves with downward acceleration 0.800 m/s^2. At the 1.20-s point in this case, is the rate of change of the fundamental wavelength of the string vibration still equal to 1.92 m/s? Explain. (d) Is the rate of change of the second harmonic wavelength the same as in part (b)? Explain.

63. On a marimba (Fig. P18.63), the wooden bar that sounds a tone when struck vibrates in a transverse standing wave having three antinodes and two nodes. The lowest-frequency note is 87.0 Hz, produced by a bar 40.0 cm long. (a) Find the speed of transverse waves on the bar. (b) A resonant pipe suspended vertically below the center of the bar enhances the loudness of the emitted sound. If the pipe is open at the top end only, what length of the pipe is required to resonate with the bar in part (a)?

Figure P18.63

64. A nylon string has mass 5.50 g and length L = 86.0 cm. The lower end is tied to the floor, and the upper end is tied to a small set of wheels through a slot in a track on which the wheels move (Fig. P18.64). The wheels have a mass that is negligible compared with that of the string, and they roll without friction on the track so that the upper end of the string is essentially free. At equilibrium, the string is vertical and motionless. When it is carrying a small-amplitude wave, you may assume the string is always under uniform tension 1.30 N. (a) Find the speed of transverse waves on the string. (b) The string's vibration possibilities are a set of standing-wave states, each with a node at the fixed bottom end and an antinode at the free top end. Find the node–antinode distances for each of the three simplest states. (c) Find the frequency of each of these states.

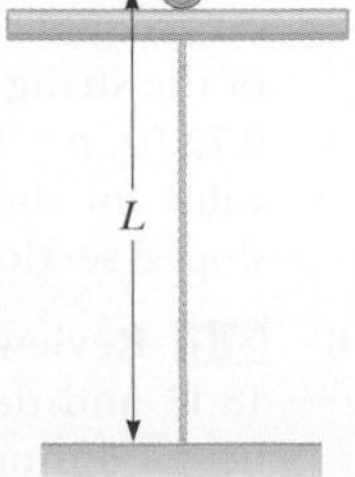

Figure P18.64

65. Two train whistles have identical frequencies of 180 Hz. When one train is at rest in the station and the other is moving nearby, a commuter standing on the station platform hears beats with a frequency of 2.00 beats/s when the whistles operate together. What are the two possible speeds and directions the moving train can have?

66. Two wires are welded together end to end. The wires are made of the same material, but the diameter of one is twice that of the other. They are subjected to a tension of 4.60 N. The thin wire has a length of 40.0 cm and a linear mass density of 2.00 g/m. The combination is fixed at both ends and vibrated in such a way that two antinodes are present, with the node between them being right at the weld. (a) What is the frequency of vibration? (b) What is the length of the thick wire?

67. A string of linear density 1.60 g/m is stretched between clamps 48.0 cm apart. The string does not stretch appreciably as the tension in it is steadily raised from 15.0 N at $t = 0$ to 25.0 N at t = 3.50 s. Therefore, the tension as a function of time is given by the expression $T = 15.0 + 10.0t/3.50$, where T is in newtons and t is in seconds. The string is vibrating in its fundamental mode throughout this process. Find the number of oscillations it completes during the 3.50-s interval.

68. S A standing wave is set up in a string of variable length and tension by a vibrator of variable frequency. Both ends of the string are fixed. When the vibrator has a frequency f, in a string of length L and under tension T, n antinodes are set up in the string. (a) If the length of the string is doubled, by what factor should the frequency be changed so that the same number of antinodes is produced? (b) If the frequency and length are held constant, what tension will produce $n + 1$ antinodes? (c) If the frequency is tripled and the length of the string is halved, by what factor should the tension be changed so that twice as many antinodes are produced?

69. Two waves are described by the wave functions

$$y_1(x, t) = 5.00 \sin (2.00x - 10.0t)$$

$$y_2(x, t) = 10.0 \cos (2.00x - 10.0t)$$

where x, y_1, and y_2 are in meters and t is in seconds. (a) Show that the wave resulting from their superposition can be expressed as a single sine function. (b) Determine the amplitude and phase angle for this sinusoidal wave.

70. A flute is designed so that it produces a frequency of 261.6 Hz, middle C, when all the holes are covered and the temperature is 20.0°C. (a) Consider the flute as a pipe that is open at both ends. Find the length of the flute, assuming middle C is the fundamental. (b) A second player, nearby in a colder room, also attempts to play middle C on an identical flute. A beat frequency of 3.00 Hz is heard when both flutes are playing. What is the temperature of the second room?

71. **Review.** A 12.0-kg object hangs in equilibrium from a string with a total length of $L = 5.00$ m and a linear mass density of $\mu = 0.001\ 00$ kg/m. The string is wrapped around two light, frictionless pulleys that are separated by a distance of $d = 2.00$ m (Fig. P18.71a). (a) Determine the tension in the string. (b) At what frequency must the string between the pulleys vibrate to form the standing-wave pattern shown in Figure P18.71b?

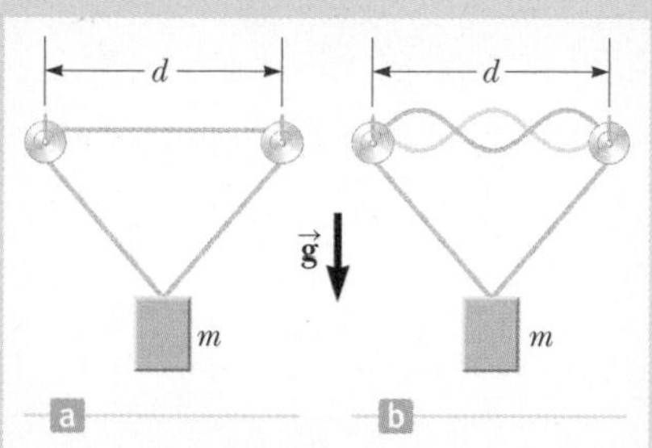

Figure P18.71 Problems 71 and 72.

72. S **Review.** An object of mass m hangs in equilibrium from a string with a total length L and a linear mass density μ. The string is wrapped around two light, frictionless pulleys that are separated by a distance d (Fig. P18.71a). (a) Determine the tension in the string. (b) At what frequency must the string between the pulleys vibrate to form the standing-wave pattern shown in Figure P18.71b?

Challenge Problems

73. S **Review.** Consider the apparatus shown in Figure P18.73a, where the hanging object has mass M and the string is vibrating in its second harmonic. The vibrating blade at the left maintains a constant frequency. The wind begins to blow to the right, applying a constant horizontal force $\vec{F}$ on the hanging object. What is the magnitude of the force the wind must apply to the hanging object so that the string vibrates in its first harmonic as shown in Figure 18.73b?

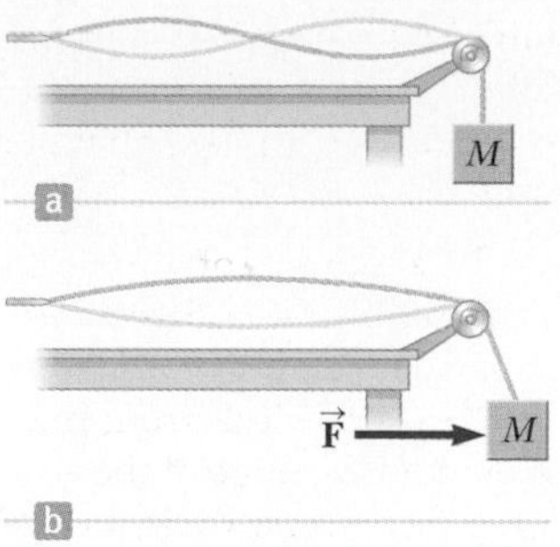

Figure P18.73

74. S In Active Figures 18.20a and 18.20b, notice that the amplitude of the component wave for frequency f is large, that for $3f$ is smaller, and that for $5f$ smaller still. How do we know exactly how much amplitude to assign to each frequency component to build a square wave? This problem helps us find the answer to that question. Let the square wave in Active Figure 18.20c have an amplitude A and let $t = 0$ be at the extreme left of the figure. So, one period T of the square wave is described by

$$y(t) = \begin{cases} A & 0 < t < \dfrac{T}{2} \\ -A & \dfrac{T}{2} < t < T \end{cases}$$

Express Equation 18.13 with angular frequencies:

$$y(t) = \sum_n (A_n \sin n\omega t + B_n \cos n\omega t)$$

Now proceed as follows. (a) Multiply both sides of Equation 18.13 by $\sin m\omega t$ and integrate both sides over one period T. Show that the left-hand side of the resulting equation is equal to 0 if m is even and is equal to $4A/m\omega$ if m is odd. (b) Using trigonometric identities, show that all terms on the right-hand side involving B_n are equal to zero. (c) Using trigonometric identities, show that all terms on the right-hand side involving A_n are equal to zero *except* for the one case of $m = n$. (d) Show that the entire right-hand side of the equation reduces to $\frac{1}{2}A_m T$. (e) Show that the Fourier series expansion for a square wave is

$$y(t) = \sum_n \frac{4A}{n\pi} \sin n\omega t$$

Thermodynamics

part 3

A bubble in one of the many mud pots in Yellowstone National Park is caught just at the moment of popping. A mud pot is a pool of bubbling hot mud that demonstrates the existence of thermodynamic processes below the Earth's surface. *(© David Argyle)*

We now direct our attention to the study of thermodynamics, which involves situations in which the temperature or state (solid, liquid, gas) of a system changes due to energy transfers. As we shall see, thermodynamics is very successful in explaining the bulk properties of matter and the correlation between these properties and the mechanics of atoms and molecules.

Historically, the development of thermodynamics paralleled the development of the atomic theory of matter. By the 1820s, chemical experiments had provided solid evidence for the existence of atoms. At that time, scientists recognized that a connection between thermodynamics and the structure of matter must exist. In 1827, botanist Robert Brown reported that grains of pollen suspended in a liquid move erratically from one place to another as if under constant agitation. In 1905, Albert Einstein used kinetic theory to explain the cause of this erratic motion, known today as *Brownian motion*. Einstein explained this phenomenon by assuming the grains are under constant bombardment by "invisible" molecules in the liquid, which themselves move erratically. This explanation gave scientists insight into the concept of molecular motion and gave credence to the idea that matter is made up of atoms. A connection was thus forged between the everyday world and the tiny, invisible building blocks that make up this world.

Thermodynamics also addresses more practical questions. Have you ever wondered how a refrigerator is able to cool its contents, or what types of transformations occur in a power plant or in the engine of your automobile, or what happens to the kinetic energy of a moving object when the object comes to rest? The laws of thermodynamics can be used to provide explanations for these and other phenomena. ■

chapter 19

Temperature

In our study of mechanics, we carefully defined such concepts as *mass, force,* and *kinetic energy* to facilitate our quantitative approach. Likewise, a quantitative description of thermal phenomena requires careful definitions of such important terms as *temperature, heat,* and *internal energy.* This chapter begins with a discussion of temperature.

Next, we consider the importance when studying thermal phenomena of the particular substance we are investigating. For example, gases expand appreciably when heated, whereas liquids and solids expand only slightly.

This chapter concludes with a study of ideal gases on the macroscopic scale. Here, we are concerned with the relationships among such quantities as pressure, volume, and temperature of a gas. In Chapter 21, we shall examine gases on a microscopic scale, using a model that represents the components of a gas as small particles.

Why would someone designing a pipeline include these strange loops? Pipelines carrying liquids often contain such loops to allow for expansion and contraction as the temperature changes. We will study thermal expansion in this chapter. *(© Lowell Georgia/Encyclopedia/CORBIS)*

19.1 Temperature and the Zeroth Law of Thermodynamics

We often associate the concept of temperature with how hot or cold an object feels when we touch it. In this way, our senses provide us with a qualitative indication of temperature. Our senses, however, are unreliable and often mislead us. For example, if you stand in bare feet with one foot on carpet and the other on an adjacent

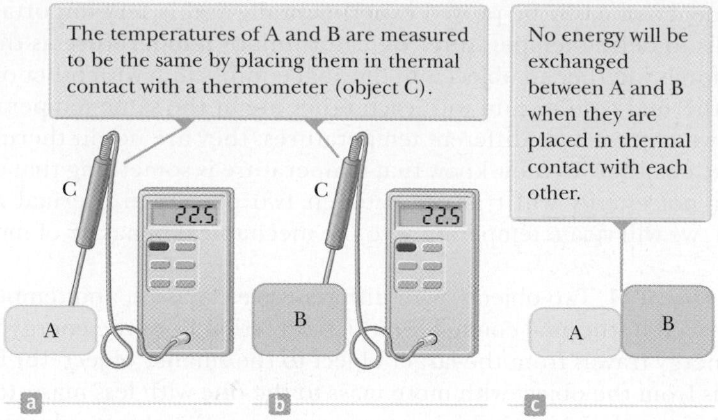

Figure 19.1 The zeroth law of thermodynamics.

tile floor, the tile feels colder than the carpet *even though both are at the same temperature.* The two objects feel different because tile transfers energy by heat at a higher rate than carpet does. Your skin "measures" the rate of energy transfer by heat rather than the actual temperature. What we need is a reliable and reproducible method for measuring the relative hotness or coldness of objects rather than the rate of energy transfer. Scientists have developed a variety of thermometers for making such quantitative measurements.

Two objects at different initial temperatures eventually reach some intermediate temperature when placed in contact with each other. For example, when hot water and cold water are mixed in a bathtub, energy is transferred from the hot water to the cold water and the final temperature of the mixture is somewhere between the initial hot and cold temperatures.

Imagine that two objects are placed in an insulated container such that they interact with each other but not with the environment. If the objects are at different temperatures, energy is transferred between them, even if they are initially not in physical contact with each other. The energy-transfer mechanisms from Chapter 8 that we will focus on are heat and electromagnetic radiation. For purposes of this discussion, let's assume two objects are in **thermal contact** with each other if energy can be exchanged between them by these processes due to a temperature difference. **Thermal equilibrium** is a situation in which two objects would not exchange energy by heat or electromagnetic radiation if they were placed in thermal contact.

Let's consider two objects A and B, which are not in thermal contact, and a third object C, which is our thermometer. We wish to determine whether A and B are in thermal equilibrium with each other. The thermometer (object C) is first placed in thermal contact with object A until thermal equilibrium is reached[1] as shown in Figure 19.1a. From that moment on, the thermometer's reading remains constant and we record this reading. The thermometer is then removed from object A and placed in thermal contact with object B as shown in Figure 19.1b. The reading is again recorded after thermal equilibrium is reached. If the two readings are the same, we can conclude that object A and object B are in thermal equilibrium with each other. If they are placed in contact with each other as in Figure 19.1c, there is no exchange of energy between them.

We can summarize these results in a statement known as the **zeroth law of thermodynamics** (the law of equilibrium):

> If objects A and B are separately in thermal equilibrium with a third object C, then A and B are in thermal equilibrium with each other.

◀ **Zeroth law of thermodynamics**

[1]We assume a negligible amount of energy transfers between the thermometer and object A during the equilibrium process. Without this assumption, which is also made for the thermometer and object B, the measurement of the temperature of an object disturbs the system so that the measured temperature is different from the initial temperature of the object. In practice, whenever you measure a temperature with a thermometer, you measure the disturbed system, not the original system.

This statement can easily be proved experimentally and is very important because it enables us to define temperature. We can think of **temperature** as the property that determines whether an object is in thermal equilibrium with other objects. Two objects in thermal equilibrium with each other are at the same temperature. Conversely, if two objects have different temperatures, they are not in thermal equilibrium with each other. We now know that temperature is something that determines whether or not energy will transfer between two objects in thermal contact. In Chapter 21, we will relate temperature to the mechanical behavior of molecules.

***Quick Quiz* 19.1** Two objects, with different sizes, masses, and temperatures, are placed in thermal contact. In which direction does the energy travel? **(a)** Energy travels from the larger object to the smaller object. **(b)** Energy travels from the object with more mass to the one with less mass. **(c)** Energy travels from the object at higher temperature to the object at lower temperature.

19.2 Thermometers and the Celsius Temperature Scale

Thermometers are devices used to measure the temperature of a system. All thermometers are based on the principle that some physical property of a system changes as the system's temperature changes. Some physical properties that change with temperature are (1) the volume of a liquid, (2) the dimensions of a solid, (3) the pressure of a gas at constant volume, (4) the volume of a gas at constant pressure, (5) the electric resistance of a conductor, and (6) the color of an object.

A common thermometer in everyday use consists of a mass of liquid—usually mercury or alcohol—that expands into a glass capillary tube when heated (Fig. 19.2). In this case, the physical property that changes is the volume of a liquid. Any temperature change in the range of the thermometer can be defined as being proportional to the change in length of the liquid column. The thermometer can be calibrated by placing it in thermal contact with a natural system that remains at constant temperature. One such system is a mixture of water and ice in thermal equilibrium at atmospheric pressure. On the **Celsius temperature scale,** this mixture is defined to have a temperature of zero degrees Celsius, which is written as 0°C; this temperature is called the *ice point* of water. Another commonly used system is a mixture of water and steam in thermal equilibrium at atmospheric pressure; its temperature is defined as 100°C, which is the *steam point* of water. Once the liquid

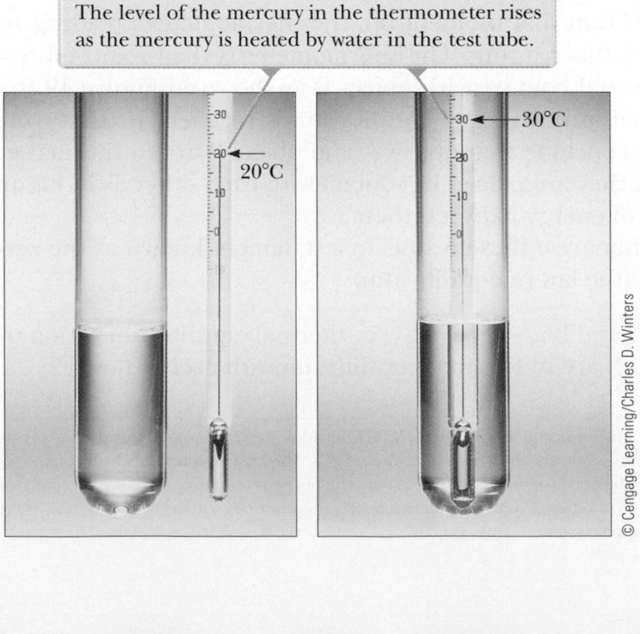

Figure 19.2 A mercury thermometer before and after increasing its temperature.

levels in the thermometer have been established at these two points, the length of the liquid column between the two points is divided into 100 equal segments to create the Celsius scale. Therefore, each segment denotes a change in temperature of one Celsius degree.

Thermometers calibrated in this way present problems when extremely accurate readings are needed. For instance, the readings given by an alcohol thermometer calibrated at the ice and steam points of water might agree with those given by a mercury thermometer only at the calibration points. Because mercury and alcohol have different thermal expansion properties, when one thermometer reads a temperature of, for example, 50°C, the other may indicate a slightly different value. The discrepancies between thermometers are especially large when the temperatures to be measured are far from the calibration points.[2]

An additional practical problem of any thermometer is the limited range of temperatures over which it can be used. A mercury thermometer, for example, cannot be used below the freezing point of mercury, which is −39°C, and an alcohol thermometer is not useful for measuring temperatures above 85°C, the boiling point of alcohol. To surmount this problem, we need a universal thermometer whose readings are independent of the substance used in it. The gas thermometer, discussed in the next section, approaches this requirement.

19.3 The Constant-Volume Gas Thermometer and the Absolute Temperature Scale

One version of a gas thermometer is the constant-volume apparatus shown in Figure 19.3. The physical change exploited in this device is the variation of pressure of a fixed volume of gas with temperature. The flask is immersed in an ice-water bath, and mercury reservoir *B* is raised or lowered until the top of the mercury in column *A* is at the zero point on the scale. The height *h*, the difference between the mercury levels in reservoir *B* and column *A*, indicates the pressure in the flask at 0°C by means of Equation 14.4, $P = P_0 + \rho gh$.

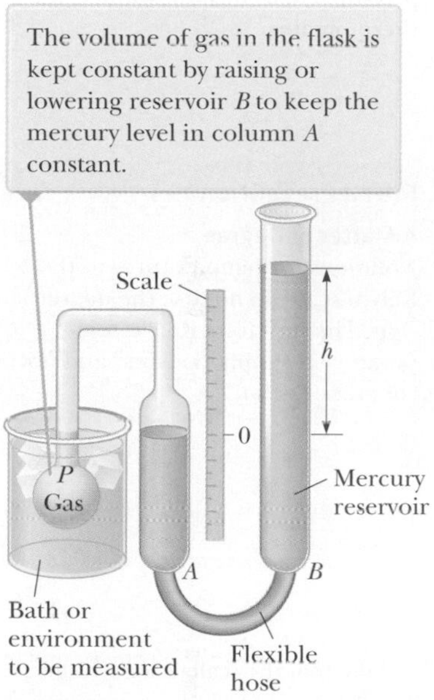

Figure 19.3 A constant-volume gas thermometer measures the pressure of the gas contained in the flask immersed in the bath.

The flask is then immersed in water at the steam point. Reservoir *B* is readjusted until the top of the mercury in column *A* is again at zero on the scale, which ensures that the gas's volume is the same as it was when the flask was in the ice bath (hence the designation "constant volume"). This adjustment of reservoir *B* gives a value for the gas pressure at 100°C. These two pressure and temperature values are then plotted as shown in Figure 19.4. The line connecting the two points serves as a calibration curve for unknown temperatures. (Other experiments show that a linear relationship between pressure and temperature is a very good assumption.) To measure the temperature of a substance, the gas flask of Figure 19.3 is placed in thermal contact with the substance and the height of reservoir *B* is adjusted until the top of the mercury column in *A* is at zero on the scale. The height of the mercury column in *B* indicates the pressure of the gas; knowing the pressure, the temperature of the substance is found using the graph in Figure 19.4.

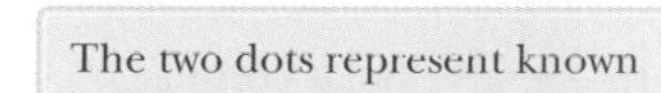

Figure 19.4 A typical graph of pressure versus temperature taken with a constant-volume gas thermometer.

Now suppose temperatures of different gases at different initial pressures are measured with gas thermometers. Experiments show that the thermometer readings are nearly independent of the type of gas used as long as the gas pressure is low and the temperature is well above the point at which the gas liquefies (Fig. 19.5, page 548). The agreement among thermometers using various gases improves as the pressure is reduced.

If we extend the straight lines in Figure 19.5 toward negative temperatures, we find a remarkable result: **in every case, the pressure is zero when the temperature is −273.15°C!** This finding suggests some special role that this particular temperature must play. It is used as the basis for the **absolute temperature scale,** which sets

[2]Two thermometers that use the same liquid may also give different readings, due in part to difficulties in constructing uniform-bore glass capillary tubes.

−273.15°C as its zero point. This temperature is often referred to as **absolute zero.** It is indicated as a zero because at a lower temperature, the pressure of the gas would become negative, which is meaningless. The size of one degree on the absolute temperature scale is chosen to be identical to the size of one degree on the Celsius scale. Therefore, the conversion between these temperatures is

$$T_C = T - 273.15 \qquad (19.1)$$

where T_C is the Celsius temperature and T is the absolute temperature.

Because the ice and steam points are experimentally difficult to duplicate and depend on atmospheric pressure, an absolute temperature scale based on two new fixed points was adopted in 1954 by the International Committee on Weights and Measures. The first point is absolute zero. The second reference temperature for this new scale was chosen as the **triple point of water,** which is the single combination of temperature and pressure at which liquid water, gaseous water, and ice (solid water) coexist in equilibrium. This triple point occurs at a temperature of 0.01°C and a pressure of 4.58 mm of mercury. On the new scale, which uses the unit *kelvin,* the temperature of water at the triple point was set at 273.16 kelvins, abbreviated 273.16 K. This choice was made so that the old absolute temperature scale based on the ice and steam points would agree closely with the new scale based on the triple point. This new absolute temperature scale (also called the **Kelvin scale**) employs the SI unit of absolute temperature, the **kelvin,** which is defined to be 1/273.16 of the difference between absolute zero and the temperature of the triple point of water.

Figure 19.6 gives the absolute temperature for various physical processes and structures. The temperature of absolute zero (0 K) cannot be achieved, although laboratory experiments have come very close, reaching temperatures of less than one nanokelvin.

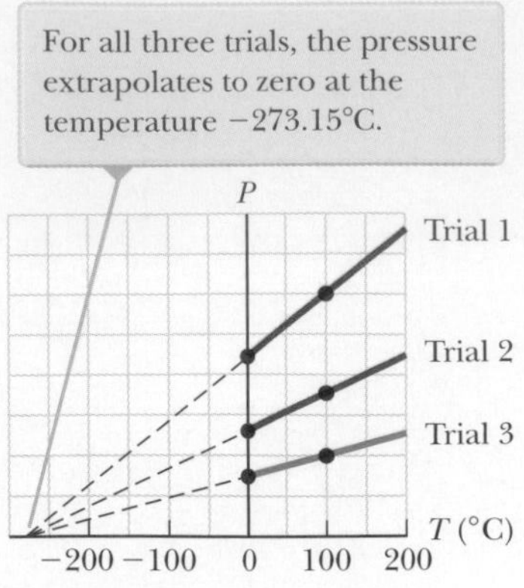

Figure 19.5 Pressure versus temperature for experimental trials in which gases have different pressures in a constant-volume gas thermometer.

Pitfall Prevention 19.1

A Matter of Degree
Notations for temperatures in the Kelvin scale do not use the degree sign. The unit for a Kelvin temperature is simply "kelvins" and not "degrees Kelvin."

The Celsius, Fahrenheit, and Kelvin Temperature Scales[3]

Equation 19.1 shows that the Celsius temperature T_C is shifted from the absolute (Kelvin) temperature T by 273.15°. Because the size of one degree is the same on the two scales, a temperature difference of 5°C is equal to a temperature difference of 5 K. The two scales differ only in the choice of the zero point. Therefore, the ice-point temperature on the Kelvin scale, 273.15 K, corresponds to 0.00°C, and the Kelvin-scale steam point, 373.15 K, is equivalent to 100.00°C.

A common temperature scale in everyday use in the United States is the **Fahrenheit scale.** This scale sets the temperature of the ice point at 32°F and the temperature of the steam point at 212°F. The relationship between the Celsius and Fahrenheit temperature scales is

$$T_F = \tfrac{9}{5}T_C + 32°\text{F} \qquad (19.2)$$

We can use Equations 19.1 and 19.2 to find a relationship between changes in temperature on the Celsius, Kelvin, and Fahrenheit scales:

$$\Delta T_C = \Delta T = \tfrac{5}{9}\Delta T_F \qquad (19.3)$$

Of these three temperature scales, only the Kelvin scale is based on a true zero value of temperature. The Celsius and Fahrenheit scales are based on an arbitrary zero associated with one particular substance, water, on one particular planet, the Earth. Therefore, if you encounter an equation that calls for a temperature T or that involves a ratio of temperatures, you *must* convert all temperatures to kelvins. If the equation contains a change in temperature ΔT, using Celsius temperatures will give you the correct answer, in light of Equation 19.3, but it is always *safest* to convert temperatures to the Kelvin scale.

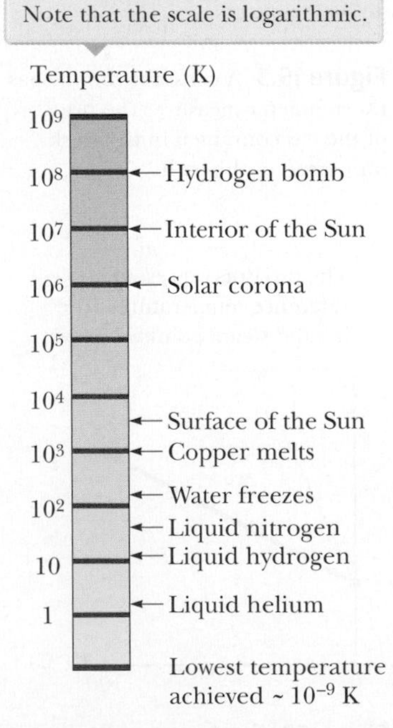

Figure 19.6 Absolute temperatures at which various physical processes occur.

[3]Named after Anders Celsius (1701–1744), Daniel Gabriel Fahrenheit (1686–1736), and William Thomson, Lord Kelvin (1824–1907), respectively.

***Quick Quiz* 19.2** Consider the following pairs of materials. Which pair represents two materials, one of which is twice as hot as the other? **(a)** boiling water at 100°C, a glass of water at 50°C **(b)** boiling water at 100°C, frozen methane at −50°C **(c)** an ice cube at −20°C, flames from a circus fire-eater at 233°C **(d)** none of those pairs

Example 19.1 Converting Temperatures

On a day when the temperature reaches 50°F, what is the temperature in degrees Celsius and in kelvins?

SOLUTION

Conceptualize In the United States, a temperature of 50°F is well understood. In many other parts of the world, however, this temperature might be meaningless because people are familiar with the Celsius temperature scale.

Categorize This example is a simple substitution problem.

Substitute the given temperature into Equation 19.2: $T_C = \frac{5}{9}(T_F - 32) = \frac{5}{9}(50 - 32) = \boxed{10°C}$

Use Equation 19.1 to find the Kelvin temperature: $T = T_C + 273.15 = 10°C + 273.15 = \boxed{283\ K}$

A convenient set of weather-related temperature equivalents to keep in mind is that 0°C is (literally) freezing at 32°F, 10°C is cool at 50°F, 20°C is room temperature, 30°C is warm at 86°F, and 40°C is a hot day at 104°F.

19.4 Thermal Expansion of Solids and Liquids

Our discussion of the liquid thermometer makes use of one of the best-known changes in a substance: as its temperature increases, its volume increases. This phenomenon, known as **thermal expansion,** plays an important role in numerous engineering applications. For example, thermal-expansion joints such as those shown in Figure 19.7 must be included in buildings, concrete highways, railroad tracks, brick walls, and bridges to compensate for dimensional changes that occur as the temperature changes.

Thermal expansion is a consequence of the change in the *average* separation between the atoms in an object. To understand this concept, let's model the atoms as being connected by stiff springs as discussed in Section 15.3 and shown in Figure 15.11b. At ordinary temperatures, the atoms in a solid oscillate about their equilibrium positions with an amplitude of approximately 10^{-11} m and a frequency of approximately 10^{13} Hz. The average spacing between the atoms is about 10^{-10} m.

Without these joints to separate sections of roadway on bridges, the surface would buckle due to thermal expansion on very hot days or crack due to contraction on very cold days.

The long, vertical joint is filled with a soft material that allows the wall to expand and contract as the temperature of the bricks changes.

Figure 19.7 Thermal-expansion joints in (a) bridges and (b) walls.

As the temperature of the solid increases, the atoms oscillate with greater amplitudes; as a result, the average separation between them increases.[4] Consequently, the object expands.

If thermal expansion is sufficiently small relative to an object's initial dimensions, the change in any dimension is, to a good approximation, proportional to the first power of the temperature change. Suppose an object has an initial length L_i along some direction at some temperature and the length increases by an amount ΔL for a change in temperature ΔT. Because it is convenient to consider the fractional change in length per degree of temperature change, we define the **average coefficient of linear expansion** as

$$\alpha \equiv \frac{\Delta L / L_i}{\Delta T}$$

Experiments show that α is constant for small changes in temperature. For purposes of calculation, this equation is usually rewritten as

$$\Delta L = \alpha L_i \, \Delta T \qquad (19.4)$$

or as

$$L_f - L_i = \alpha L_i (T_f - T_i) \qquad (19.5)$$

where L_f is the final length, T_i and T_f are the initial and final temperatures, respectively, and the proportionality constant α is the average coefficient of linear expansion for a given material and has units of $(°C)^{-1}$. Equation 19.4 can be used for both thermal expansion, when the temperature of the material increases, and thermal contraction, when its temperature decreases.

Pitfall Prevention 19.2

Do Holes Become Larger or Smaller?
When an object's temperature is raised, every linear dimension increases in size. That includes any holes in the material, which expand in the same way as if the hole were filled with the material as shown in Active Figure 19.8. Keep in mind the notion of thermal expansion as being similar to a photographic enlargement.

It may be helpful to think of thermal expansion as an effective magnification or as a photographic enlargement of an object. For example, as a metal washer is heated (Active Fig. 19.8), all dimensions, including the radius of the hole, increase according to Equation 19.4. A cavity in a piece of material expands in the same way as if the cavity were filled with the material.

Table 19.1 lists the average coefficients of linear expansion for various materials. For these materials, α is positive, indicating an increase in length with increasing temperature. That is not always the case, however. Some substances—calcite ($CaCO_3$) is one example—expand along one dimension (positive α) and contract along another (negative α) as their temperatures are increased.

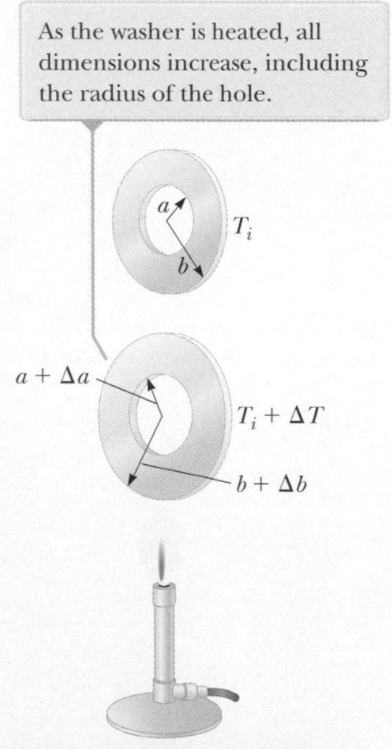

ACTIVE FIGURE 19.8

Thermal expansion of a homogeneous metal washer. (The expansion is exaggerated in this figure.)

Because the linear dimensions of an object change with temperature, it follows that surface area and volume change as well. The change in volume is proportional to the initial volume V_i and to the change in temperature according to the relationship

$$\Delta V = \beta V_i \, \Delta T \qquad (19.6)$$

where β is the **average coefficient of volume expansion.** To find the relationship between β and α, assume the average coefficient of linear expansion of the solid is the same in all directions; that is, assume the material is *isotropic.* Consider a solid box of dimensions ℓ, w, and h. Its volume at some temperature T_i is $V_i = \ell w h$. If the temperature changes to $T_i + \Delta T$, its volume changes to $V_i + \Delta V$, where each dimension changes according to Equation 19.4. Therefore,

$$\begin{aligned} V_i + \Delta V &= (\ell + \Delta \ell)(w + \Delta w)(h + \Delta h) \\ &= (\ell + \alpha \ell \, \Delta T)(w + \alpha w \, \Delta T)(h + \alpha h \, \Delta T) \\ &= \ell w h (1 + \alpha \, \Delta T)^3 \\ &= V_i [1 + 3\alpha \, \Delta T + 3(\alpha \, \Delta T)^2 + (\alpha \, \Delta T)^3] \end{aligned}$$

Dividing both sides by V_i and isolating the term $\Delta V / V_i$, we obtain the fractional change in volume:

[4]More precisely, thermal expansion arises from the *asymmetrical* nature of the potential energy curve for the atoms in a solid as shown in Figure 15.11a. If the oscillators were truly harmonic, the average atomic separations would not change regardless of the amplitude of vibration.

TABLE 19.1 *Average Expansion Coefficients for Some Materials Near Room Temperature*

Material (Solids)	Average Linear Expansion Coefficient $(\alpha)(°C)^{-1}$	Material (Liquids and Gases)	Average Volume Expansion Coefficient $(\beta)(°C)^{-1}$
Aluminum	24×10^{-6}	Acetone	1.5×10^{-4}
Brass and bronze	19×10^{-6}	Alcohol, ethyl	1.12×10^{-4}
Concrete	12×10^{-6}	Benzene	1.24×10^{-4}
Copper	17×10^{-6}	Gasoline	9.6×10^{-4}
Glass (ordinary)	9×10^{-6}	Glycerin	4.85×10^{-4}
Glass (Pyrex)	3.2×10^{-6}	Mercury	1.82×10^{-4}
Invar (Ni–Fe alloy)	0.9×10^{-6}	Turpentine	9.0×10^{-4}
Lead	29×10^{-6}	Air[a] at 0°C	3.67×10^{-3}
Steel	11×10^{-6}	Helium[a]	3.665×10^{-3}

[a]Gases do not have a specific value for the volume expansion coefficient because the amount of expansion depends on the type of process through which the gas is taken. The values given here assume the gas undergoes an expansion at constant pressure.

$$\frac{\Delta V}{V_i} = 3\alpha\,\Delta T + 3(\alpha\,\Delta T)^2 + (\alpha\,\Delta T)^3$$

Because $\alpha\,\Delta T << 1$ for typical values of ΔT ($< \sim 100°C$), we can neglect the terms $3(\alpha\,\Delta T)^2$ and $(\alpha\,\Delta T)^3$. Upon making this approximation, we see that

$$\frac{\Delta V}{V_i} = 3\alpha\,\Delta T \quad \rightarrow \quad \Delta V = (3\alpha)V_i\,\Delta T$$

Comparing this expression to Equation 19.6 shows that

$$\beta = 3\alpha$$

In a similar way, you can show that the change in area of a rectangular plate is given by $\Delta A = 2\alpha A_i\,\Delta T$ (see Problem 49).

A simple mechanism called a *bimetallic strip,* found in practical devices such as thermostats, uses the difference in coefficients of expansion for different materials. It consists of two thin strips of dissimilar metals bonded together. As the temperature of the strip increases, the two metals expand by different amounts and the strip bends as shown in Figure 19.9.

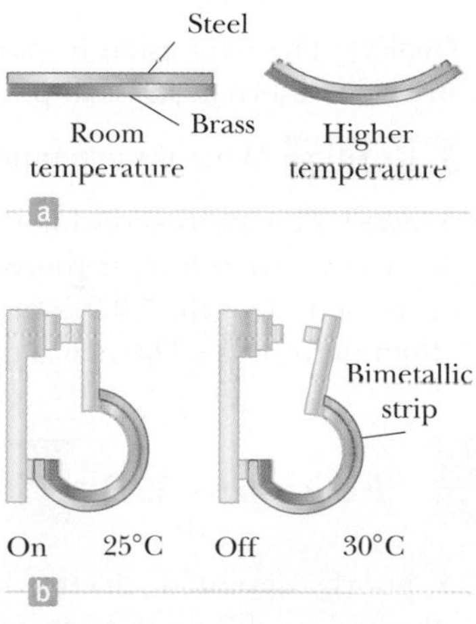

Figure 19.9 (a) A bimetallic strip bends as the temperature changes because the two metals have different expansion coefficients. (b) A bimetallic strip used in a thermostat to break or make electrical contact.

***Quick Quiz* 19.3** If you are asked to make a very sensitive glass thermometer, which of the following working liquids would you choose? **(a)** mercury **(b)** alcohol **(c)** gasoline **(d)** glycerin

***Quick Quiz* 19.4** Two spheres are made of the same metal and have the same radius, but one is hollow and the other is solid. The spheres are taken through the same temperature increase. Which sphere expands more? **(a)** The solid sphere expands more. **(b)** The hollow sphere expands more. **(c)** They expand by the same amount. **(d)** There is not enough information to say.

Example 19.2 Expansion of a Railroad Track

A segment of steel railroad track has a length of 30.000 m when the temperature is 0.0°C.

(A) What is its length when the temperature is 40.0°C?

continued

19.2 *cont.*

SOLUTION

Conceptualize Because the rail is relatively long, we expect to obtain a measurable increase in length for a 40°C temperature increase.

Categorize We will evaluate a length increase using the discussion of this section, so this part of the example is a substitution problem.

Use Equation 19.4 and the value of the coefficient of linear expansion from Table 19.1:

$$\Delta L = \alpha L_i \Delta T = [11 \times 10^{-6} \, (°\text{C})^{-1}](30.000 \text{ m})(40.0°\text{C}) = 0.013 \text{ m}$$

Find the new length of the track:

$$L_f = 30.000 \text{ m} + 0.013 \text{ m} = \boxed{30.013 \text{ m}}$$

(B) Suppose the ends of the rail are rigidly clamped at 0.0°C so that expansion is prevented. What is the thermal stress set up in the rail if its temperature is raised to 40.0°C?

SOLUTION

Categorize This part of the example is an analysis problem because we need to use concepts from another chapter.

Analyze The thermal stress is the same as the tensile stress in the situation in which the rail expands freely and is then compressed with a mechanical force F back to its original length.

Find the tensile stress from Equation 12.6 using Young's modulus for steel from Table 12.1:

$$\text{Tensile stress} = \frac{F}{A} = Y\frac{\Delta L}{L_i}$$

$$\frac{F}{A} = (20 \times 10^{10} \text{ N/m}^2)\left(\frac{0.013 \text{ m}}{30.000 \text{ m}}\right) = \boxed{8.7 \times 10^7 \text{ N/m}^2}$$

Finalize The expansion in part (A) is 1.3 cm. This expansion is indeed measurable as predicted in the Conceptualize step. The thermal stress in part (B) can be avoided by leaving small expansion gaps between the rails.

WHAT IF? What if the temperature drops to −40.0°C? What is the length of the unclamped segment?

Answer The expression for the change in length in Equation 19.4 is the same whether the temperature increases or decreases. Therefore, if there is an increase in length of 0.013 m when the temperature increases by 40°C, there is a decrease in length of 0.013 m when the temperature decreases by 40°C. (We assume α is constant over the entire range of temperatures.) The new length at the colder temperature is 30.000 m − 0.013 m = 29.987 m.

Example 19.3 The Thermal Electrical Short

A poorly designed electronic device has two bolts attached to different parts of the device that almost touch each other in its interior as in Figure 19.10. The steel and brass bolts are at different electric potentials, and if they touch, a short circuit will develop, damaging the device. (We will study electric potential in Chapter 25.) The initial gap between the ends of the bolts is 5.0 μm at 27°C. At what temperature will the bolts touch? Assume the distance between the walls of the device is not affected by the temperature change.

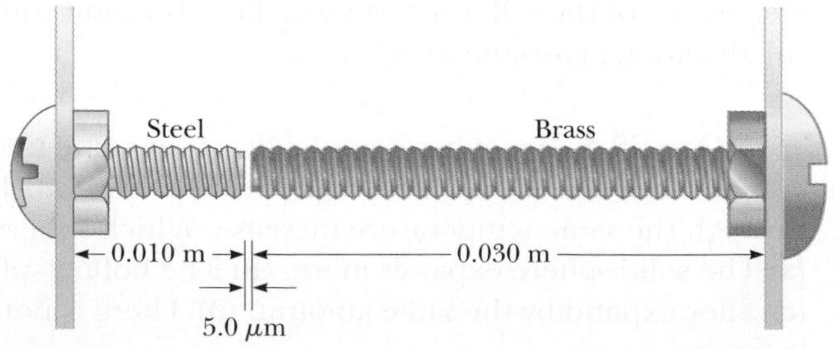

Figure 19.10 (Example 19.3) Two bolts attached to different parts of an electrical device are almost touching when the temperature is 27°C. As the temperature increases, the ends of the bolts move toward each other.

SOLUTION

Conceptualize Imagine the ends of both bolts expanding into the gap between them as the temperature rises.

Categorize We categorize this example as a thermal expansion problem in which the *sum* of the changes in length of the two bolts must equal the length of the initial gap between the ends.

19.3 cont.

Analyze Set the sum of the length changes equal to the width of the gap:

$$\Delta L_{\text{br}} + \Delta L_{\text{st}} = \alpha_{\text{br}} L_{i,\text{br}}\,\Delta T + \alpha_{\text{st}} L_{i,\text{st}}\,\Delta T = 5.0 \times 10^{-6}\text{ m}$$

Solve for ΔT:

$$\Delta T = \frac{5.0 \times 10^{-6}\text{ m}}{\alpha_{\text{br}} L_{i,\text{br}} + \alpha_{\text{st}} L_{i,\text{st}}}$$

$$= \frac{5.0 \times 10^{-6}\text{ m}}{[19 \times 10^{-6}\,(^\circ\text{C})^{-1}](0.030\text{ m}) + [11 \times 10^{-6}\,(^\circ\text{C})^{-1}](0.010\text{ m})} = 7.4^\circ\text{C}$$

Find the temperature at which the bolts touch:

$$T = 27^\circ\text{C} + 7.4^\circ\text{C} = \boxed{34^\circ\text{C}}$$

Finalize This temperature is possible if the air conditioning in the building housing the device fails for a long period on a very hot summer day.

The Unusual Behavior of Water

Liquids generally increase in volume with increasing temperature and have average coefficients of volume expansion about ten times greater than those of solids. Cold water is an exception to this rule as you can see from its density-versus-temperature curve shown in Figure 19.11. As the temperature increases from 0°C to 4°C, water contracts and its density therefore increases. Above 4°C, water expands with increasing temperature and so its density decreases. Therefore, the density of water reaches a maximum value of 1.000 g/cm^3 at 4°C.

We can use this unusual thermal-expansion behavior of water to explain why a pond begins freezing at the surface rather than at the bottom. When the air temperature drops from, for example, 7°C to 6°C, the surface water also cools and consequently decreases in volume. The surface water is denser than the water below it, which has not cooled and decreased in volume. As a result, the surface water sinks, and warmer water from below moves to the surface. When the air temperature is between 4°C and 0°C, however, the surface water expands as it cools, becoming less dense than the water below it. The mixing process stops, and eventually the surface water freezes. As the water freezes, the ice remains on the surface because ice is less dense than water. The ice continues to build up at the surface, while water near the bottom remains at 4°C. If that were not the case, fish and other forms of marine life would not survive.

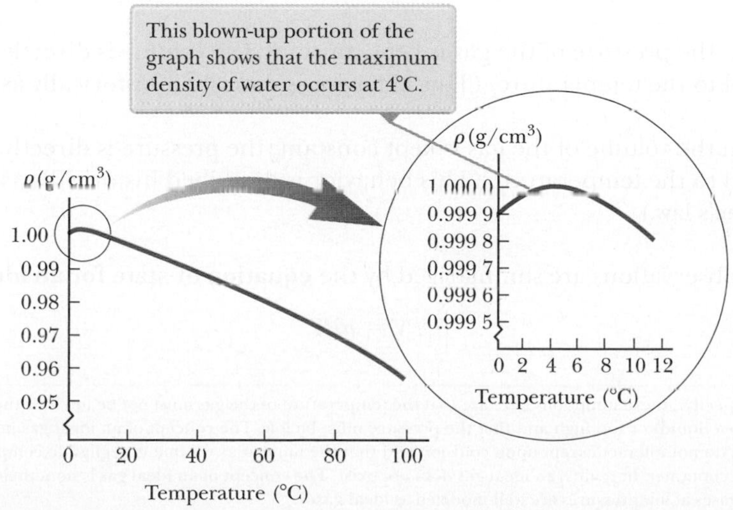

Figure 19.11 The variation in the density of water at atmospheric pressure with temperature.

19.5 Macroscopic Description of an Ideal Gas

The volume expansion equation $\Delta V = \beta V_i\,\Delta T$ is based on the assumption that the material has an initial volume V_i before the temperature change occurs. Such is the case for solids and liquids because they have a fixed volume at a given temperature.

The case for gases is completely different. The interatomic forces within gases are very weak, and, in many cases, we can imagine these forces to be nonexistent and still make very good approximations. Therefore, *there is no equilibrium separation* for the atoms and no "standard" volume at a given temperature; the volume depends on the size of the container. As a result, we cannot express changes in volume ΔV in a process on a gas with Equation 19.6 because we have no defined volume V_i at the beginning of the process. Equations involving gases contain the volume V, rather than a *change* in the volume from an initial value, as a variable.

For a gas, it is useful to know how the quantities volume V, pressure P, and temperature T are related for a sample of gas of mass m. In general, the equation that interrelates these quantities, called the *equation of state,* is very complicated. If the gas is maintained at a very low pressure (or low density), however, the equation of state is quite simple and can be determined from experimental results. Such a low-density gas is commonly referred to as an **ideal gas.**[5] We can use the **ideal gas model** to make predictions that are adequate to describe the behavior of real gases at low pressures.

It is convenient to express the amount of gas in a given volume in terms of the number of moles n. One **mole** of any substance is that amount of the substance that contains **Avogadro's number** $N_A = 6.022 \times 10^{23}$ of constituent particles (atoms or molecules). The number of moles n of a substance is related to its mass m through the expression

$$n = \frac{m}{M} \tag{19.7}$$

where M is the molar mass of the substance. The molar mass of each chemical element is the atomic mass (from the periodic table; see Appendix C) expressed in grams per mole. For example, the mass of one He atom is 4.00 u (atomic mass units), so the molar mass of He is 4.00 g/mol.

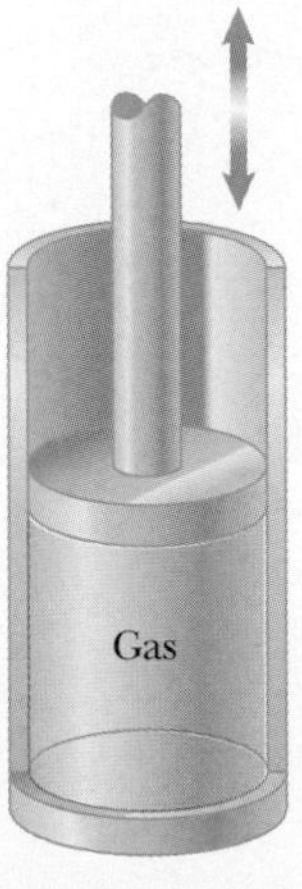

ACTIVE FIGURE 19.12

An ideal gas confined to a cylinder whose volume can be varied by means of a movable piston.

Now suppose an ideal gas is confined to a cylindrical container whose volume can be varied by means of a movable piston as in Active Figure 19.12. If we assume the cylinder does not leak, the mass (or the number of moles) of the gas remains constant. For such a system, experiments provide the following information:

- When the gas is kept at a constant temperature, its pressure is inversely proportional to the volume. (This behavior is described historically as Boyle's law.)
- When the pressure of the gas is kept constant, the volume is directly proportional to the temperature. (This behavior is described historically as Charles's law.)
- When the volume of the gas is kept constant, the pressure is directly proportional to the temperature. (This behavior is described historically as Gay–Lussac's law.)

These observations are summarized by the **equation of state for an ideal gas:**

Equation of state for an ideal gas ▶

$$PV = nRT \tag{19.8}$$

[5]To be more specific, the assumptions here are that the temperature of the gas must not be too low (the gas must not condense into a liquid) or too high and that the pressure must be low. The concept of an ideal gas implies that the gas molecules do not interact except upon collision and that the molecular volume is negligible compared with the volume of the container. In reality, an ideal gas does not exist. The concept of an ideal gas is nonetheless very useful because real gases at low pressures are well-modeled as ideal gases.

In this expression, also known as the **ideal gas law,** n is the number of moles of gas in the sample and R is a constant. Experiments on numerous gases show that as the pressure approaches zero, the quantity PV/nT approaches the same value R for all gases. For this reason, R is called the **universal gas constant.** In SI units, in which pressure is expressed in pascals (1 Pa = 1 N/m^2) and volume in cubic meters, the product PV has units of newton · meters, or joules, and R has the value

$$R = 8.314 \text{ J/mol} \cdot \text{K} \quad (19.9)$$

If the pressure is expressed in atmospheres and the volume in liters (1 L = 10^3 cm^3 = 10^{-3} m^3), then R has the value

$$R = 0.082\,06 \text{ L} \cdot \text{atm/mol} \cdot \text{K}$$

Using this value of R and Equation 19.8 shows that the volume occupied by 1 mol of *any* gas at atmospheric pressure and at 0°C (273 K) is 22.4 L.

The ideal gas law states that if the volume and temperature of a fixed amount of gas do not change, the pressure also remains constant. Consider a bottle of champagne that is shaken and then spews liquid when opened as shown in Figure 19.13. A common misconception is that the pressure inside the bottle is increased when the bottle is shaken. On the contrary, because the temperature of the bottle and its contents remains constant as long as the bottle is sealed, so does the pressure, as can be shown by replacing the cork with a pressure gauge. The correct explanation is as follows. Carbon dioxide gas resides in the volume between the liquid surface and the cork. The pressure of the gas in this volume is set higher than atmospheric pressure in the bottling process. Shaking the bottle displaces some of the carbon dioxide gas into the liquid, where it forms bubbles, and these bubbles become attached to the inside of the bottle. (No new gas is generated by shaking.) When the bottle is opened, the pressure is reduced to atmospheric pressure, which causes the volume of the bubbles to increase suddenly. If the bubbles are attached to the bottle (beneath the liquid surface), their rapid expansion expels liquid from the bottle. If the sides and bottom of the bottle are first tapped until no bubbles remain beneath the surface, however, the drop in pressure does not force liquid from the bottle when the champagne is opened.

The ideal gas law is often expressed in terms of the total number of molecules N. Because the number of moles n equals the ratio of the total number of molecules and Avogadro's number N_A, we can write Equation 19.8 as

$$PV = nRT = \frac{N}{N_A} RT$$

$$PV = N k_B T \quad (19.10)$$

where k_B is **Boltzmann's constant,** which has the value

$$k_B = \frac{R}{N_A} = 1.38 \times 10^{-23} \text{ J/K} \quad (19.11)$$

◀ **Boltzmann's constant**

It is common to call quantities such as P, V, and T the **thermodynamic variables** of an ideal gas. If the equation of state is known, one of the variables can always be expressed as some function of the other two.

***Quick Quiz* 19.5** A common material for cushioning objects in packages is made by trapping bubbles of air between sheets of plastic. Is this material more effective at keeping the contents of the package from moving around inside the package on **(a)** a hot day, **(b)** a cold day, or **(c)** either hot or cold days?

***Quick Quiz* 19.6** On a winter day, you turn on your furnace and the temperature of the air inside your home increases. Assume your home has the normal amount of leakage between inside air and outside air. Is the number of moles of air in your room at the higher temperature **(a)** larger than before, **(b)** smaller than before, or **(c)** the same as before?

Figure 19.13 A bottle of champagne is shaken and opened. Liquid spews out of the opening. A common misconception is that the pressure inside the bottle is increased by the shaking.

Pitfall Prevention 19.3

So Many *k*s

There are a variety of physical quantities for which the letter k is used. Two we have seen previously are the force constant for a spring (Chapter 15) and the wave number for a mechanical wave (Chapter 16). Boltzmann's constant is another k, and we will see k used for thermal conductivity in Chapter 20 and for an electrical constant in Chapter 23. To make some sense of this confusing state of affairs, we use a subscript B for Boltzmann's constant to help us recognize it. In this book, you will see Boltzmann's constant as k_B, but you may see Boltzmann's constant in other resources as simply k.

Example 19.4 Heating a Spray Can

A spray can containing a propellant gas at twice atmospheric pressure (202 kPa) and having a volume of 125.00 cm^3 is at 22°C. It is then tossed into an open fire. (*Warning:* Do not do this experiment; it is very dangerous.) When the temperature of the gas in the can reaches 195°C, what is the pressure inside the can? Assume any change in the volume of the can is negligible.

SOLUTION

Conceptualize Intuitively, you should expect that the pressure of the gas in the container increases because of the increasing temperature.

Categorize We model the gas in the can as ideal and use the ideal gas law to calculate the new pressure.

Analyze Rearrange Equation 19.8:

$$(1)\quad \frac{PV}{T} = nR$$

No air escapes during the compression, so n, and therefore nR, remains constant. Hence, set the initial value of the left side of Equation (1) equal to the final value:

$$(2)\quad \frac{P_iV_i}{T_i} = \frac{P_fV_f}{T_f}$$

Because the initial and final volumes of the gas are assumed to be equal, cancel the volumes:

$$(3)\quad \frac{P_i}{T_i} = \frac{P_f}{T_f}$$

Solve for P_f:

$$P_f = \left(\frac{T_f}{T_i}\right)P_i = \left(\frac{468\text{ K}}{295\text{ K}}\right)(202\text{ kPa}) = \boxed{320\text{ kPa}}$$

Finalize The higher the temperature, the higher the pressure exerted by the trapped gas as expected. If the pressure increases sufficiently, the can may explode. Because of this possibility, you should never dispose of spray cans in a fire.

WHAT IF? Suppose we include a volume change due to thermal expansion of the steel can as the temperature increases. Does that alter our answer for the final pressure significantly?

Answer Because the thermal expansion coefficient of steel is very small, we do not expect much of an effect on our final answer.

Find the change in the volume of the can using Equation 19.6 and the value for α for steel from Table 19.1:

$$\Delta V = \beta V_i \Delta T = 3\alpha V_i \Delta T$$
$$= 3[11 \times 10^{-6}\ (°\text{C})^{-1}](125.00\text{ cm}^3)(173°\text{C}) = 0.71\text{ cm}^3$$

Start from Equation (2) again and find an equation for the final pressure:

$$P_f = \left(\frac{T_f}{T_i}\right)\left(\frac{V_i}{V_f}\right)P_i$$

This result differs from Equation (3) only in the factor V_i/V_f. Evaluate this factor:

$$\frac{V_i}{V_f} = \frac{125.00\text{ cm}^3}{(125.00\text{ cm}^3 + 0.71\text{ cm}^3)} = 0.994 = 99.4\%$$

Therefore, the final pressure will differ by only 0.6% from the value calculated without considering the thermal expansion of the can. Taking 99.4% of the previous final pressure, the final pressure including thermal expansion is 318 kPa.

Summary

Definitions

Two objects are in **thermal equilibrium** with each other if they do not exchange energy when in thermal contact.

Temperature is the property that determines whether an object is in thermal equilibrium with other objects. Two objects in thermal equilibrium with each other are at the same temperature. The SI unit of absolute temperature is the **kelvin,** which is defined to be 1/273.16 of the difference between absolute zero and the temperature of the triple point of water.

Concepts and Principles

The **zeroth law of thermodynamics** states that if objects A and B are separately in thermal equilibrium with a third object C, then objects A and B are in thermal equilibrium with each other.

When the temperature of an object is changed by an amount ΔT, its length changes by an amount ΔL that is proportional to ΔT and to its initial length L_i:

$$\Delta L = \alpha L_i \Delta T \qquad (19.4)$$

where the constant α is the **average coefficient of linear expansion.** The **average coefficient of volume expansion** β for a solid is approximately equal to 3α.

An **ideal gas** is one for which PV/nT is constant. An ideal gas is described by the **equation of state,**

$$PV = nRT \qquad (19.8)$$

where n equals the number of moles of the gas, P is its pressure, V is its volume, R is the universal gas constant (8.314 J/mol · K), and T is the absolute temperature of the gas. A real gas behaves approximately as an ideal gas if it has a low density.

Objective Questions

☐ denotes answer available in Student Solutions Manual/Study Guide

1. A hole is drilled in a metal plate. When the metal is raised to a higher temperature, what happens to the diameter of the hole? (a) It decreases. (b) It increases. (c) It remains the same. (d) The answer depends on the initial temperature of the metal. (e) None of those answers is correct.

2. A cylinder with a piston holds 0.50 m^3 of oxygen at an absolute pressure of 4.0 atm. The piston is pulled outward, increasing the volume of the gas until the pressure drops to 1.0 atm. If the temperature stays constant, what new volume does the gas occupy? (a) 1.0 m^3 (b) 1.5 m^3 (c) 2.0 m^3 (d) 0.12 m^3 (e) 2.5 m^3

3. If the volume of an ideal gas is doubled while its temperature is quadrupled, does the pressure (a) remain the same, (b) decrease by a factor of 2, (c) decrease by a factor of 4, (d) increase by a factor of 2, or (e) increase by a factor of 4

4. On a very cold day in upstate New York, the temperature is −25°C, which is equivalent to what Fahrenheit temperature? (a) −46°F (b) −77°F (c) 18°F (d) −25°F (e) −13°F

5. A temperature of 162°F is equivalent to what temperature in kelvins? (a) 373 K (b) 288 K (c) 345 K (d) 201 K (e) 308 K

6. When a certain gas under a pressure of 5.00×10^6 Pa at 25.0°C is allowed to expand to 3.00 times its original volume, its final pressure is 1.07×10^6 Pa. What is its final temperature? (a) 450 K (b) 233 K (c) 212 K (d) 191 K (e) 115 K

7. What would happen if the glass of a thermometer expanded more on warming than did the liquid in the tube? (a) The thermometer would break. (b) It could be used only for temperatures below room temperature. (c) You would have to hold it with the bulb on top. (d) The scale on the thermometer is reversed so that higher temperature values would be found closer to the bulb. (e) The numbers would not be evenly spaced.

8. Suppose you empty a tray of ice cubes into a bowl partly full of water and cover the bowl. After one-half hour, the contents of the bowl come to thermal equilibrium, with more liquid water and less ice than you started with. Which of the following is true? (a) The temperature of the liquid water is higher than the temperature of the remaining ice. (b) The temperature of the liquid water is the same as that of the ice. (c) The temperature of the liquid water is less than that of the ice. (d) The comparative temperatures of the liquid water and ice depend on the amounts present.

9. The average coefficient of linear expansion of copper is 17×10^{-6} (°C)$^{-1}$. The Statue of Liberty is 93 m tall on a summer morning when the temperature is 25°C. Assume the copper plates covering the statue are mounted edge to edge without expansion joints and do not buckle or bind on the framework supporting them as the day grows hot. What is the order of magnitude of the statue's increase in height? (a) 0.1 mm (b) 1 mm (c) 1 cm (d) 10 cm (e) 1 m

10. ☐ A rubber balloon is filled with 1 L of air at 1 atm and 300 K and is then put into a cryogenic refrigerator at 100 K. The rubber remains flexible as it cools. **(i)** What happens to the volume of the balloon? (a) It decreases to $\frac{1}{3}$ L. (b) It decreases to $1/\sqrt{3}$ L. (c) It is constant. (d) It increases to $\sqrt{3}$ L. (e) It increases to 3 L. **(ii)** What happens to the pressure of the air in the balloon? (a) It decreases to $\frac{1}{3}$ atm. (b) It decreases to $1/\sqrt{3}$ atm. (c) It is constant. (d) It increases to $\sqrt{3}$ atm. (e) It increases to 3 atm.

11. Two cylinders A and B at the same temperature contain the same quantity of the same kind of gas. Cylinder A has three times the volume of cylinder B. What can you conclude about the pressures the gases exert? (a) We can conclude nothing about the pressures. (b) The pressure in A is three times the pressure in B. (c) The pressures must be equal. (d) The pressure in A must be one-third the pressure in B.

12. A cylinder with a piston contains a sample of a thin gas. The kind of gas and the sample size can be changed. The cylinder can be placed in different constant-temperature baths, and the piston can be held in different positions. Rank the following cases according to the pressure of the gas from the highest to the lowest, displaying any cases of equality. (a) A 0.002-mol sample of oxygen is held at 300 K in a 100-cm^3 container. (b) A 0.002-mol sample of oxygen is held at 600 K in a 200-cm^3 container. (c) A 0.002-mol sample of oxygen is held at 600 K in a 300-cm^3 container. (d) A 0.004-mol sample of helium is held at 300 K in a 200-cm^3 container. (e) A 0.004-mol sample of helium is held at 250 K in a 200-cm^3 container.

13. Markings to indicate length are placed on a steel tape in a room that is at a temperature of 22°C. Measurements are then made with the same tape on a day when the temperature is 27°C. Assume the objects you are measuring have a smaller coefficient of linear expansion than steel. Are the measurements (a) too long, (b) too short, or (c) accurate?

14. The pendulum of a certain pendulum clock is made of brass. When the temperature increases, what happens to the period of the clock? (a) It increases. (b) It decreases. (c) It remains the same.

Conceptual Questions

☐ denotes answer available in Student Solutions Manual/Study Guide

1. Is it possible for two objects to be in thermal equilibrium if they are not in contact with each other? Explain.

2. A piece of copper is dropped into a beaker of water. (a) If the water's temperature rises, what happens to the temperature of the copper? (b) Under what conditions are the water and copper in thermal equilibrium?

3. In describing his upcoming trip to the Moon, and as portrayed in the movie *Apollo 13* (Universal, 1995), astronaut Jim Lovell said, "I'll be walking in a place where there's a 400-degree difference between sunlight and shadow." Suppose an astronaut standing on the Moon holds a thermometer in his gloved hand. (a) Is the thermometer reading the temperature of the vacuum at the Moon's surface? (b) Does it read any temperature? If so, what object or substance has that temperature?

4. Use a periodic table of the elements (see Appendix C) to determine the number of grams in one mole of (a) hydrogen, which has diatomic molecules; (b) helium; and (c) carbon monoxide.

5. (a) What does the ideal gas law predict about the volume of a sample of gas at absolute zero? (b) Why is this prediction incorrect?

6. Metal lids on glass jars can often be loosened by running hot water over them. Why does that work?

7. An automobile radiator is filled to the brim with water when the engine is cool. (a) What happens to the water when the engine is running and the water has been raised to a high temperature? (b) What do modern automobiles have in their cooling systems to prevent the loss of coolants?

8. When the metal ring and metal sphere in Figure CQ19.8 are both at room temperature, the sphere can barely be passed through the ring. (a) After the sphere is warmed in a flame, it cannot be passed through the ring. Explain. (b) **What If?** What if the ring is warmed and the sphere is left at room temperature? Does the sphere pass through the ring?

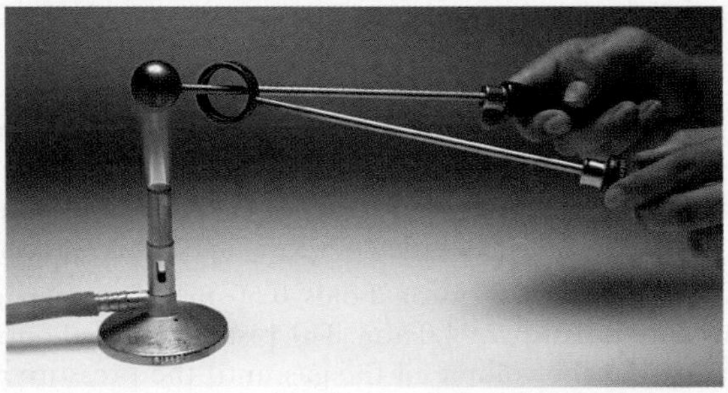

© Cengage Learning/Charles D. Winters

Figure CQ19.8

9. Common thermometers are made of a mercury column in a glass tube. Based on the operation of these thermometers, which has the larger coefficient of linear expansion, glass or mercury? (Don't answer the question by looking in a table.)

10. Some picnickers stop at a convenience store to buy some food, including bags of potato chips. They then drive up into the mountains to their picnic site. When they unload the food, they notice that the bags of chips are puffed up like balloons. Why did that happen?

Problems

ENHANCED WebAssign The problems found in this chapter may be assigned online in Enhanced WebAssign

1. denotes straightforward problem; 2. denotes intermediate problem;
3. denotes challenging problem

1. full solution available in the Student Solutions Manual/Study Guide

1. denotes problems most often assigned in Enhanced WebAssign; these provide students with targeted feedback and either a Master It tutorial or a Watch It solution video.

QC denotes asking for quantitative and conceptual reasoning

S denotes symbolic reasoning problem

M denotes Master It tutorial available in Enhanced WebAssign

GP denotes guided problem

shaded denotes "paired problems" that develop reasoning with symbols and numerical values

Section 19.2 Thermometers and the Celsius Temperature Scale

Section 19.3 The Constant-Volume Gas Thermometer and the Absolute Temperature Scale

1. Convert the following temperatures to their values on the Fahrenheit and Kelvin scales: (a) the sublimation point of dry ice, −78.5°C; (b) human body temperature, 37.0°C.

2. The temperature difference between the inside and the outside of a home on a cold winter day is 57.0°F. Express this difference on (a) the Celsius scale and (b) the Kelvin scale.

3. Q|C A nurse measures the temperature of a patient to be 41.5°C. (a) What is this temperature on the Fahrenheit scale? (b) Do you think the patient is seriously ill? Explain.

4. The boiling point of liquid hydrogen is 20.3 K at atmospheric pressure. What is this temperature on (a) the Celsius scale and (b) the Fahrenheit scale?

5. Liquid nitrogen has a boiling point of −195.81°C at atmospheric pressure. Express this temperature (a) in degrees Fahrenheit and (b) in kelvins.

6. M In a student experiment, a constant-volume gas thermometer is calibrated in dry ice (−78.5°C) and in boiling ethyl alcohol (78.0°C). The separate pressures are 0.900 atm and 1.635 atm. (a) What value of absolute zero in degrees Celsius does the calibration yield? What pressures would be found at (b) the freezing and (c) the boiling points of water? *Hint:* Use the linear relationship $P = A + BT$, where A and B are constants.

Section 19.4 Thermal Expansion of Solids and Liquids

Note: Table 19.1 is available for use in solving problems in this section.

7. A copper telephone wire has essentially no sag between poles 35.0 m apart on a winter day when the temperature is −20.0°C. How much longer is the wire on a summer day when the temperature is 35.0°C?

8. The concrete sections of a certain superhighway are designed to have a length of 25.0 m. The sections are poured and cured at 10.0°C. What minimum spacing should the engineer leave between the sections to eliminate buckling if the concrete is to reach a temperature of 50.0°C?

9. The active element of a certain laser is made of a glass rod 30.0 cm long and 1.50 cm in diameter. Assume the average coefficient of linear expansion of the glass is equal to $9.00 \times 10^{-6}\ (°C)^{-1}$. If the temperature of the rod increases by 65.0°C, what is the increase in (a) its length, (b) its diameter, and (c) its volume?

10. Review. Inside the wall of a house, an L-shaped section of hot-water pipe consists of three parts: a straight, horizontal piece $h = 28.0$ cm long; an elbow; and a straight, vertical piece $\ell =$ 134 cm long (Fig. P19.10). A stud and a second-story floorboard hold the ends of this section of copper pipe stationary. Find the magnitude and direction of the displacement of the pipe elbow when the water flow is turned on, raising the temperature of the pipe from 18.0°C to 46.5°C.

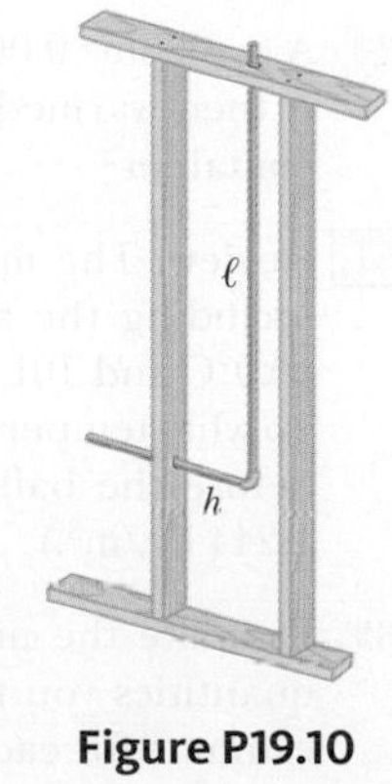

Figure P19.10

11. Q|C At 20.0°C, an aluminum ring has an inner diameter of 5.000 0 cm and a brass rod has a diameter of 5.050 0 cm. (a) If only the ring is warmed, what temperature must it reach so that it will just slip over the rod? (b) **What If?** If both the ring and the rod are warmed together, what temperature must they both reach so that the ring barely slips over the rod? (c) Would this latter process work? Explain. *Hint:* Consult Table 20.2 in the next chapter.

12. *Why is the following situation impossible?* A thin brass ring has an inner diameter 10.00 cm at 20.0°C. A solid aluminum cylinder has diameter 10.02 cm at 20.0°C. Assume the average coefficients of linear expansion of the two metals are constant. Both metals are cooled together to a temperature at which the ring can be slipped over the end of the cylinder.

13. Q|C A volumetric flask made of Pyrex is calibrated at 20.0°C. It is filled to the 100-mL mark with 35.0°C acetone. After the flask is filled, the acetone cools and the flask warms so that the combination of acetone and flask reaches a uniform temperature of 32.0°C. The combination is then cooled back to 20.0°C. (a) What is the volume of the acetone when it cools to 20.0°C? (b) At the temperature of 32.0°C, does the level of acetone lie above or below the 100-mL mark on the flask? Explain.

14. Review. On a day that the temperature is 20.0°C, a concrete walk is poured in such a way that the ends of the walk are unable to move. Take Young's modulus for concrete to be 7.00×10^{9} N/m² and the compressive strength to be 2.00×10^{9} N/m². (a) What is the stress in the cement on a hot day of 50.0°C? (b) Does the concrete fracture?

15. A hollow aluminum cylinder 20.0 cm deep has an internal capacity of 2.000 L at 20.0°C. It is completely filled with turpentine at 20.0°C. The turpentine and the aluminum cylinder are then slowly warmed together to 80.0°C. (a) How much turpentine overflows? (b) What is the volume of turpentine remaining in the cylinder at 80.0°C? (c) If the combination with this amount of turpentine is then cooled back to 20.0°C, how far below the cylinder's rim does the turpentine's surface recede?

16. Q|C **Review.** The Golden Gate Bridge in San Francisco has a main span of length 1.28 km, one of the longest in the world. Imagine that a steel wire with this length and a cross-sectional area of 4.00×10^{-6} m² is laid in a straight line on the bridge deck with its ends attached to the towers of the bridge. On a summer day the temperature of the wire is 35.0°C. (a) When winter arrives, the towers stay the same distance apart and the bridge deck keeps the same shape as its expansion joints open. When the temperature drops to −10.0°C, what is the tension in the wire? Take Young's modulus for steel to be 20.0×10^{10} N/m². (b) Permanent deformation occurs if the stress in the steel exceeds its elastic limit of 3.00×10^{8} N/m². At what temperature would the wire reach its elastic limit? (c) **What**

If? Explain how your answers to parts (a) and (b) would change if the Golden Gate Bridge were twice as long.

17. A sample of lead has a mass of 20.0 kg and a density of 11.3×10^3 kg/m^3 at 0°C. (a) What is the density of lead at 90.0°C? (b) What is the mass of the sample of lead at 90.0°C?

18. **S** A sample of a solid substance has a mass m and a density ρ_0 at a temperature T_0. (a) Find the density of the substance if its temperature is increased by an amount ΔT in terms of the coefficient of volume expansion β. (b) What is the mass of the sample if the temperature is raised by an amount ΔT?

Section 19.5 Macroscopic Description of an Ideal Gas

19. Gas is confined in a tank at a pressure of 11.0 atm and a temperature of 25.0°C. If two-thirds of the gas is withdrawn and the temperature is raised to 75.0°C, what is the pressure of the gas remaining in the tank?

20. A rigid tank contains 1.50 moles of an ideal gas. Determine the number of moles of gas that must be withdrawn from the tank to lower the pressure of the gas from 25.0 atm to 5.00 atm. Assume the volume of the tank and the temperature of the gas remain constant during this operation.

21. Gas is contained in an 8.00-L vessel at a temperature of 20.0°C and a pressure of 9.00 atm. (a) Determine the number of moles of gas in the vessel. (b) How many molecules are in the vessel?

22. **Q|C** Your father and your younger brother are confronted with the same puzzle. Your father's garden sprayer and your brother's water cannon both have tanks with a capacity of 5.00 L (Fig. P19.22). Your father puts a negligible amount of concentrated fertilizer into his tank. They both pour in 4.00 L of water and seal up their tanks, so the tanks also contain air at atmospheric pressure. Next, each uses a hand-operated pump to inject more air until the absolute pressure in the tank reaches 2.40 atm. Now each uses his device to spray out water—not air—until the stream becomes feeble, which it does when the pressure in the tank reaches 1.20 atm. To accomplish spraying out all the water, each finds he must pump up the tank three times. Here is the puzzle: most of the water sprays out after the second pumping. The first and the third pumping-up processes seem just as difficult as the second but result in a much smaller amount of water coming out. Account for this phenomenon.

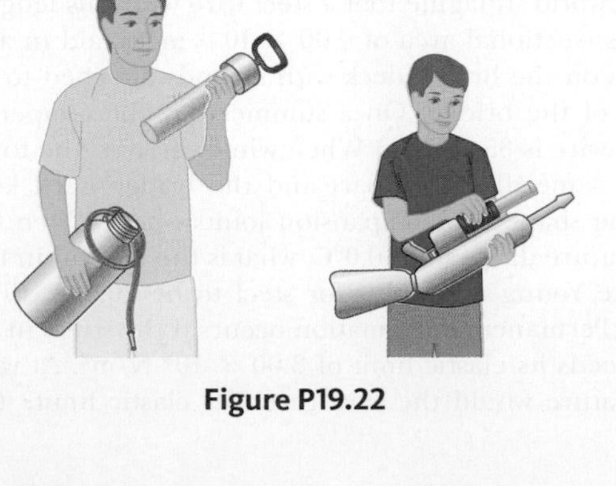

Figure P19.22

23. **M** An auditorium has dimensions 10.0 m × 20.0 m × 30.0 m. How many molecules of air fill the auditorium at 20.0°C and a pressure of 101 kPa (1.00 atm)?

24. **Q|C** A container in the shape of a cube 10.0 cm on each edge contains air (with equivalent molar mass 28.9 g/mol) at atmospheric pressure and temperature 300 K. Find (a) the mass of the gas, (b) the gravitational force exerted on it, and (c) the force it exerts on each face of the cube. (d) Why does such a small sample exert such a great force?

25. **Q|C** (a) Find the number of moles in one cubic meter of an ideal gas at 20.0°C and atmospheric pressure. (b) For air, Avogadro's number of molecules has mass 28.9 g. Calculate the mass of one cubic meter of air. (c) State how this result compares with the tabulated density of air at 20.0°C.

26. **M** The pressure gauge on a tank registers the gauge pressure, which is the difference between the interior pressure and exterior pressure. When the tank is full of oxygen (O_2), it contains 12.0 kg of the gas at a gauge pressure of 40.0 atm. Determine the mass of oxygen that has been withdrawn from the tank when the pressure reading is 25.0 atm. Assume the temperature of the tank remains constant.

27. In state-of-the-art vacuum systems, pressures as low as 1.00×10^{-9} Pa are being attained. Calculate the number of molecules in a 1.00-m^3 vessel at this pressure and a temperature of 27.0°C.

28. **Review.** To measure how far below the ocean surface a bird dives to catch a fish, a scientist uses a method originated by Lord Kelvin. He dusts the interiors of plastic tubes with powdered sugar and then seals one end of each tube. He captures the bird at nighttime in its nest and attaches a tube to its back. He then catches the same bird the next night and removes the tube. In one trial, using a tube 6.50 cm long, water washes away the sugar over a distance of 2.70 cm from the open end of the tube. Find the greatest depth to which the bird dived, assuming the air in the tube stayed at constant temperature.

29. **M** An automobile tire is inflated with air originally at 10.0°C and normal atmospheric pressure. During the process, the air is compressed to 28.0% of its original volume and the temperature is increased to 40.0°C. (a) What is the tire pressure? (b) After the car is driven at high speed, the tire's air temperature rises to 85.0°C and the tire's interior volume increases by 2.00%. What is the new tire pressure (absolute)?

30. A cook puts 9.00 g of water in a 2.00-L pressure cooker that is then warmed to 500°C. What is the pressure inside the container?

31. **Review.** The mass of a hot-air balloon and its cargo (not including the air inside) is 200 kg. The air outside is at 10.0°C and 101 kPa. The volume of the balloon is 400 m^3. To what temperature must the air in the balloon be warmed before the balloon will lift off? (Air density at 10.0°C is 1.244 kg/m^3.)

32. Estimate the mass of the air in your bedroom. State the quantities you take as data and the value you measure or estimate for each.

33. Review. At 25.0 m below the surface of the sea, where the temperature is 5.00°C, a diver exhales an air bubble having a volume of 1.00 cm^3. If the surface temperature of the sea is 20.0°C, what is the volume of the bubble just before it breaks the surface?

34. **S** The pressure gauge on a cylinder of gas registers the gauge pressure, which is the difference between the interior pressure and the exterior pressure P_0. Let's call the gauge pressure P_g. When the cylinder is full, the mass of the gas in it is m_i at a gauge pressure of P_{gi}. Assuming the temperature of the cylinder remains constant, show that the mass of the gas *remaining* in the cylinder when the pressure reading is P_{gf} is given by

$$m_f = m_i\left(\frac{P_{gf} + P_0}{P_{gi} + P_0}\right)$$

Additional Problems

35. A spherical steel ball bearing has a diameter of 2.540 cm at 25.00°C. (a) What is its diameter when its temperature is raised to 100.0°C? (b) What temperature change is required to increase its volume by 1.000%?

36. A steel beam being used in the construction of a skyscraper has a length of 35.000 m when delivered on a cold day at a temperature of 15.000°F. What is the length of the beam when it is being installed later on a warm day when the temperature is 90.000°F?

37. A bicycle tire is inflated to a gauge pressure of 2.50 atm when the temperature is 15.0°C. While a man rides the bicycle, the temperature of the tire rises to 45.0°C. Assuming the volume of the tire does not change, find the gauge pressure in the tire at the higher temperature.

38. *Why is the following situation impossible?* An apparatus is designed so that steam initially at $T = 150°C$, $P = 1.00$ atm, and $V = 0.500$ m^3 in a piston–cylinder apparatus undergoes a process in which (1) the volume remains constant and the pressure drops to 0.870 atm, followed by (2) an expansion in which the pressure remains constant and the volume increases to 1.00 m^3, followed by (3) a return to the initial conditions. It is important that the pressure of the gas never fall below 0.850 atm so that the piston will support a delicate and very expensive part of the apparatus. Without such support, the delicate apparatus can be severely damaged and rendered useless. When the design is turned into a working prototype, it operates perfectly.

39. A student measures the length of a brass rod with a steel tape at 20.0°C. The reading is 95.00 cm. What will the tape indicate for the length of the rod when the rod and the tape are at (a) −15.0°C and (b) 55.0°C?

40. The density of gasoline is 730 kg/m^3 at 0°C. Its average coefficient of volume expansion is 9.60×10^{-4} (°C)$^{-1}$. Assume 1.00 gal of gasoline occupies 0.003 80 m^3. How many extra kilograms of gasoline would you receive if you bought 10.0 gal of gasoline at 0°C rather than at 20.0°C from a pump that is not temperature compensated?

41. **M** A mercury thermometer is constructed as shown in Figure P19.41. The Pyrex glass capillary tube has a diameter of 0.004 00 cm, and the bulb has a diameter of 0.250 cm. Find the change in height of the mercury column that occurs with a temperature change of 30.0°C.

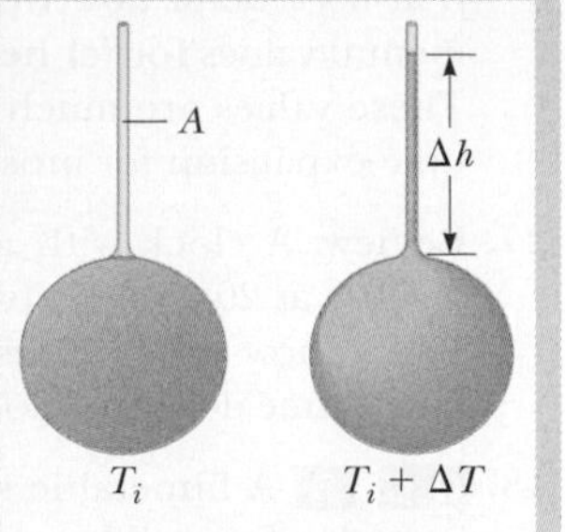

Figure P19.41
Problems 41 and 42.

42. **S** A liquid with a coefficient of volume expansion β just fills a spherical shell of volume V (Fig. P19.41). The shell and the open capillary of area A projecting from the top of the sphere are made of a material with an average coefficient of linear expansion α. The liquid is free to expand into the capillary. Assuming the temperature increases by ΔT, find the distance Δh the liquid rises in the capillary.

43. Review. An aluminum pipe is open at both ends and used as a flute. The pipe is cooled to 5.00°C, at which its length is 0.655 m. As soon as you start to play it, the pipe fills with air at 20.0°C. After that, by how much does its fundamental frequency change as the metal rises in temperature to 20.0°C?

44. **Q|C** Two metal bars are made of invar and a third bar is made of aluminum. At 0°C, each of the three bars is drilled with two holes 40.0 cm apart. Pins are put through the holes to assemble the bars into an equilateral triangle as in Figure P19.44. (a) First ignore the expansion of the invar. Find the angle between the invar bars as a function of Celsius temperature. (b) Is your answer accurate for negative as well as positive temperatures? (c) Is it accurate for 0°C? (d) Solve the problem again, including the expansion of the invar. Aluminum melts at 660°C and invar at 1 427°C. Assume the tabulated expansion coefficients are constant. What are (e) the greatest and (f) the smallest attainable angles between the invar bars?

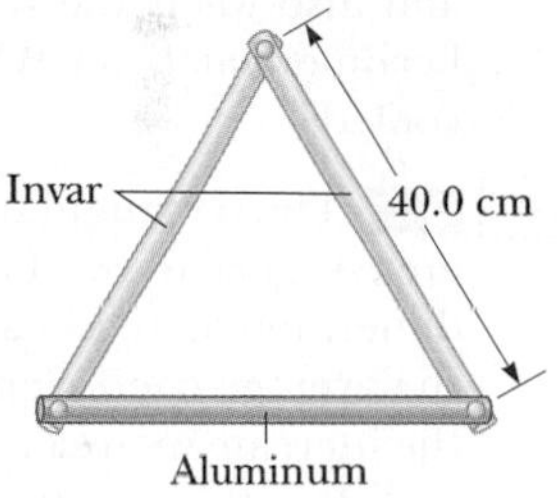

Figure P19.44

45. **Q|C** A liquid has a density ρ. (a) Show that the fractional change in density for a change in temperature ΔT is $\Delta\rho/\rho = -\beta\,\Delta T$. (b) What does the negative sign signify? (c) Fresh water has a maximum density of 1.000 0 g/cm^3 at 4.0°C. At 10.0°C, its density is 0.999 7 g/cm^3. What is β for water over this temperature interval? (d) At 0°C, the density of water is 0.999 9 g/cm^3. What is the value for β over the temperature range 0°C to 4.00°C?

46. **Q|C** (a) Take the definition of the coefficient of volume expansion to be

$$\beta = \frac{1}{V}\frac{dV}{dT}\bigg|_{P=\text{constant}} = \frac{1}{V}\frac{\partial V}{\partial T}$$

Use the equation of state for an ideal gas to show that the coefficient of volume expansion for an ideal gas at constant pressure is given by $\beta = 1/T$, where T is the absolute temperature. (b) What value does this expression predict for

β at 0°C? State how this result compares with the experimental values for (c) helium and (d) air in Table 19.1. *Note:* These values are much larger than the coefficients of volume expansion for most liquids and solids.

47. **Review.** A clock with a brass pendulum has a period of 1.000 s at 20.0°C. If the temperature increases to 30.0°C, (a) by how much does the period change and (b) how much time does the clock gain or lose in one week?

48. Q|C S A bimetallic strip of length L is made of two ribbons of different metals bonded together. (a) First assume the strip is originally straight. As the strip is warmed, the metal with the greater average coefficient of expansion expands more than the other, forcing the strip into an arc with the outer radius having a greater circumference (Fig. P19.48). Derive an expression for the angle of bending θ as a function of the initial length of the strips, their average coefficients of linear expansion, the change in temperature, and the separation of the centers of the strips ($\Delta r = r_2 - r_1$). (b) Show that the angle of bending decreases to zero when ΔT decreases to zero and also when the two average coefficients of expansion become equal. (c) **What If?** What happens if the strip is cooled?

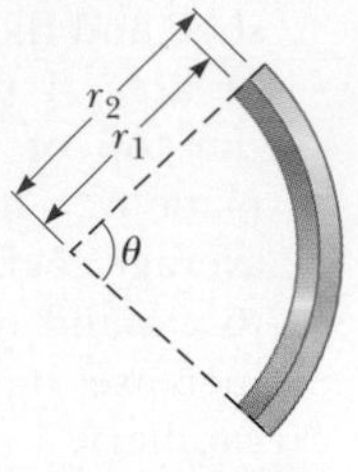

Figure P19.48

49. Q|C The rectangular plate shown in Figure P19.49 has an area A_i equal to ℓw. If the temperature increases by ΔT, each dimension increases according to Equation 19.4, where α is the average coefficient of linear expansion. (a) Show that the increase in area is $\Delta A = 2\alpha A_i \Delta T$. (b) What approximation does this expression assume?

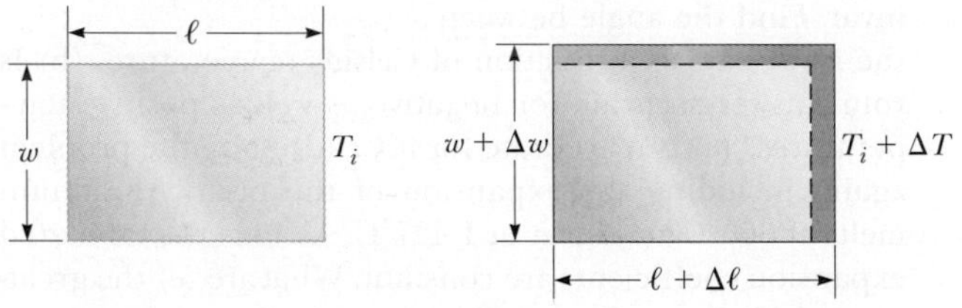

Figure P19.49

50. S The measurement of the average coefficient of volume expansion β for a liquid is complicated because the container also changes size with temperature. Figure P19.50 shows a simple means for measuring β despite the expansion of the container. With this apparatus, one arm of a U-tube is maintained at 0°C in a water–ice bath, and the other arm is maintained at a different temperature T_C in a constant-temperature bath. The connecting tube is horizontal. A difference in the length or diameter of the tube between the two arms of the U-tube has no effect on the pressure balance at the bottom of the tube because the pressure depends only on the depth of the liquid. Derive an expression for β for the liquid in terms of h_0, h_t, and T_C.

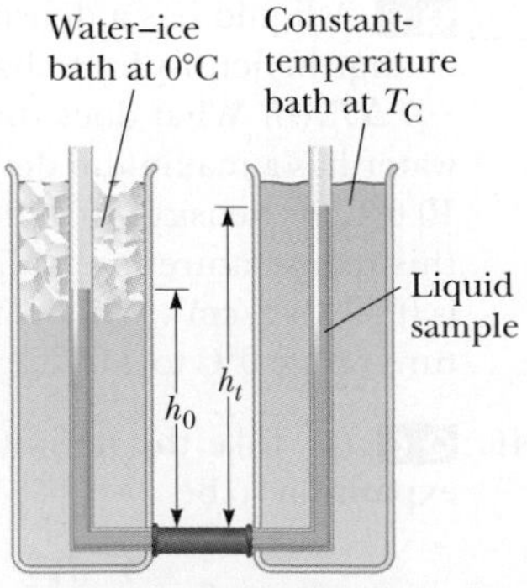

Figure P19.50

51. Q|C A copper rod and a steel rod are different in length by 5.00 cm at 0°C. The rods are warmed and cooled together. (a) Is it possible that the length difference remains constant at all temperatures? Explain. (b) If so, describe the lengths at 0°C as precisely as you can. Can you tell which rod is longer? Can you tell the lengths of the rods?

52. GP S A vertical cylinder of cross-sectional area A is fitted with a tight-fitting, frictionless piston of mass m (Fig. P19.52). The piston is not restricted in its motion in any way and is supported by the gas at pressure P below it. Atmospheric pressure is P_0. We wish to find the height h in Figure P19.52. (a) What analysis model is appropriate to describe the piston? (b) Write an appropriate force equation for the piston from this analysis model in terms of P, P_0, m, A, and g. (c) Suppose n moles of an ideal gas are in the cylinder at a temperature of T. Substitute for P in your answer to part (b) to find the height h of the piston above the bottom of the cylinder.

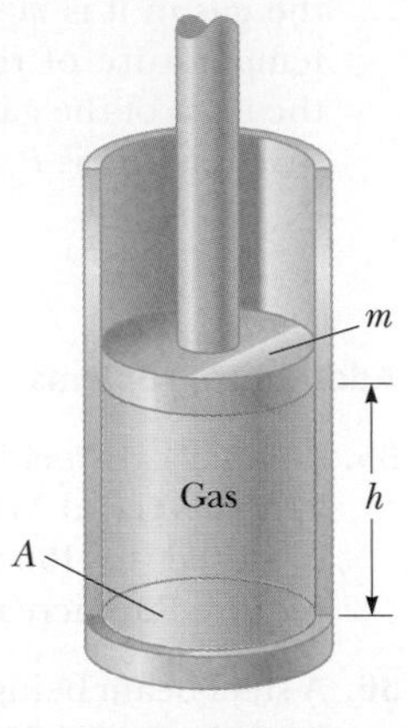

Figure P19.52

53. Q|C **Review.** Consider an object with any one of the shapes displayed in Table 10.2. What is the percentage increase in the moment of inertia of the object when it is warmed from 0°C to 100°C if it is composed of (a) copper or (b) aluminum? Assume the average linear expansion coefficients shown in Table 19.1 do not vary between 0°C and 100°C. (c) Why are the answers for parts (a) and (b) the same for all the shapes?

54. (a) Show that the density of an ideal gas occupying a volume V is given by $\rho = PM/RT$, where M is the molar mass. (b) Determine the density of oxygen gas at atmospheric pressure and 20.0°C.

55. Two concrete spans of a 250-m-long bridge are placed end to end so that no room is allowed for expansion (Fig. P19.55a). If a temperature increase of 20.0°C occurs, what is the height y to which the spans rise when they buckle (Fig. P19.55b)?

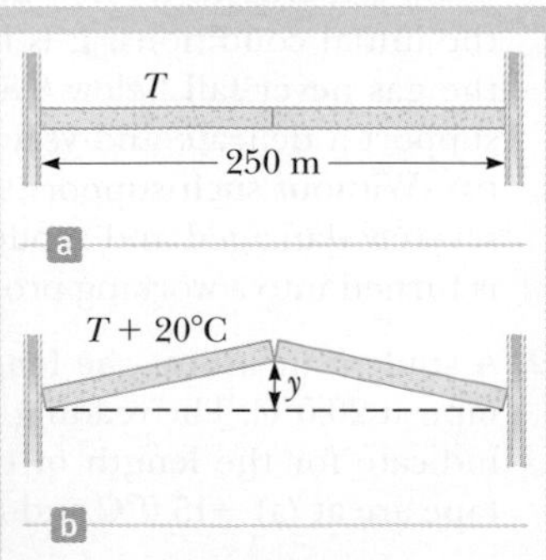

Figure P19.55 Problems 55 and 56.

56. S Two concrete spans that form a bridge of length L are placed end to end so that no room is allowed for expansion (Fig. P19.55a). If a temperature increase of ΔT occurs, what is the height y to which the spans rise when they buckle (Fig. P19.55b)?

57. S **Review.** (a) Derive an expression for the buoyant force on a spherical balloon, submerged in water, as a function

of the depth h below the surface, the volume V_i of the balloon at the surface, the pressure P_0 at the surface, and the density ρ_w of the water. Assume the water temperature does not change with depth. (b) Does the *buoyant* force increase or decrease as the balloon is submerged? (c) At what depth is the buoyant force one-half the surface value?

58. **Q|C** **Review.** Following a collision in outer space, a copper disk at 850°C is rotating about its axis with an angular speed of 25.0 rad/s. As the disk radiates infrared light, its temperature falls to 20.0°C. No external torque acts on the disk. (a) Does the angular speed change as the disk cools? Explain how it changes or why it does not. (b) What is its angular speed at the lower temperature?

59. Starting with Equation 19.10, show that the total pressure P in a container filled with a mixture of several ideal gases is $P = P_1 + P_2 + P_3 + \ldots$, where $P_1, P_2, \ldots$ are the pressures that each gas would exert if it alone filled the container. (These individual pressures are called the *partial pressures* of the respective gases). This result is known as *Dalton's law of partial pressures.*

Challenge Problems

60. A cylinder is closed by a piston connected to a spring of constant 2.00×10^3 N/m (see Fig. P19.60). With the spring relaxed, the cylinder is filled with 5.00 L of gas at a pressure of 1.00 atm and a temperature of 20.0°C. (a) If the piston has a cross-sectional area of 0.010 0 m² and negligible mass, how high will it rise when the temperature is raised to 250°C? (b) What is the pressure of the gas at 250°C?

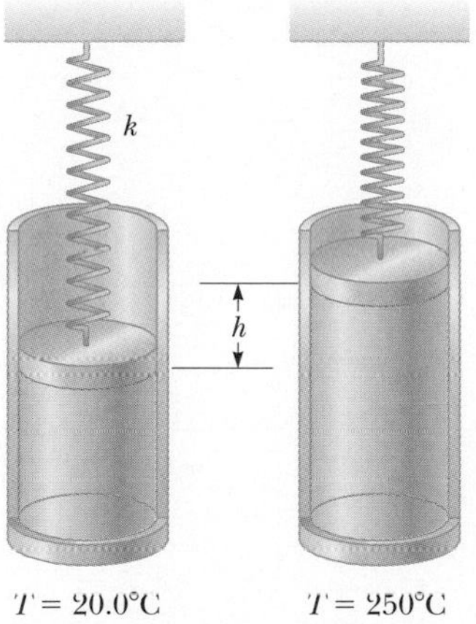

Figure P19.60

61. **Q|C** Helium gas is sold in steel tanks that will rupture if subjected to tensile stress greater than its yield strength of 5×10^8 N/m². If the helium is used to inflate a balloon, could the balloon lift the spherical tank the helium came in? Justify your answer. *Suggestion:* You may consider a spherical steel shell of radius r and thickness t having the density of iron and on the verge of breaking apart into two hemispheres because it contains helium at high pressure.

62. A cylinder that has a 40.0-cm radius and is 50.0 cm deep is filled with air at 20.0°C and 1.00 atm (Fig. P19.62a). A 20.0-kg piston is now lowered into the cylinder, compressing the air trapped inside as it takes equilibrium height h_i (Fig. P19.62b). Finally, a 25.0-kg dog stands on the piston, further compressing the air, which remains at 20°C (Fig. P19.62c). (a) How far down (Δh) does the piston move when the dog steps onto it? (b) To what temperature should the gas be warmed to raise the piston and dog back to h_i?

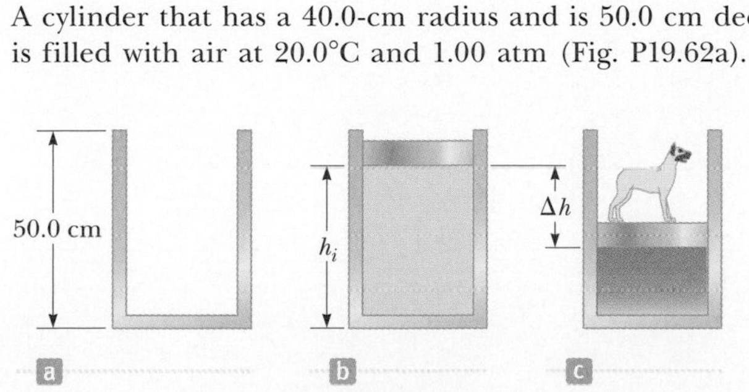

Figure P19.62

63. The relationship $L = L_i + \alpha L_i \Delta T$ is a valid approximation when $\alpha \Delta T$ is small. If $\alpha \Delta T$ is large, one must integrate the relationship $dL = \alpha L\, dT$ to determine the final length. (a) Assuming the coefficient of linear expansion of a material is constant as L varies, determine a general expression for the final length of a rod made of the material. Given a rod of length 1.00 m and a temperature change of 100.0°C, determine the error caused by the approximation when (b) $\alpha = 2.00 \times 10^{-5}$ (°C)$^{-1}$ (a typical value for a metal) and (c) when $\alpha = 0.020\,0$ (°C)$^{-1}$ (an unrealistically large value for comparison). (d) Using the equation from part (a), solve Problem 15 again to find more accurate results.

64. **Review.** A steel wire and a copper wire, each of diameter 2.000 mm, are joined end to end. At 40.0°C, each has an unstretched length of 2.000 m. The wires are connected between two fixed supports 4.000 m apart on a tabletop. The steel wire extends from $x = -2.000$ m to $x = 0$, the copper wire extends from $x = 0$ to $x = 2.000$ m, and the tension is negligible. The temperature is then lowered to 20.0°C. Assume the average coefficient of linear expansion of steel is 11.0×10^{-6} (°C)$^{-1}$ and that of copper is 17.0×10^{-6} (°C)$^{-1}$. Take Young's modulus for steel to be 20.0×10^{10} N/m² and that for copper to be 11.0×10^{10} N/m². At this lower temperature, find (a) the tension in the wire and (b) the x coordinate of the junction between the wires.

65. **Review.** A steel guitar string with a diameter of 1.00 mm is stretched between supports 80.0 cm apart. The temperature is 0.0°C. (a) Find the mass per unit length of this string. (Use the value 7.86×10^3 kg/m³ for the density.) (b) The fundamental frequency of transverse oscillations of the string is 200 Hz. What is the tension in the string? Next, the temperature is raised to 30.0°C. Find the resulting values of (c) the tension and (d) the fundamental frequency. Assume both the Young's modulus of 20.0×10^{10} N/m² and the average coefficient of expansion $\alpha = 11.0 \times 10^{-6}$ (°C)$^{-1}$ have constant values between 0.0°C and 30.0°C.

66. **Q|C** **Review.** A house roof is a perfectly flat plane that makes an angle θ with the horizontal. When its temperature changes, between T_c before dawn each day and T_h in the middle of each afternoon, the roof expands and contracts uniformly with a coefficient of thermal expansion α_1. Resting on the roof is a flat, rectangular metal plate with expansion coefficient α_2, greater than α_1. The length of the plate is L, measured along the slope of the roof. The component of the plate's weight perpendicular to the roof is supported by a normal force uniformly distributed over the area of the plate. The coefficient of kinetic friction between the plate and the roof is μ_k. The plate is always at the same temperature as the roof, so we assume its temperature is continuously changing. Because of the difference in expansion coefficients, each bit of the plate is moving relative to the roof below it, except for points along a certain horizontal line running across the plate called the stationary line.

If the temperature is rising, parts of the plate below the stationary line are moving down relative to the roof and feel a force of kinetic friction acting up the roof. Elements of area above the stationary line are sliding up the roof, and on them kinetic friction acts downward parallel to the roof. The stationary line occupies no area, so we assume no force of static friction acts on the plate while the temperature is changing. The plate as a whole is very nearly in equilibrium, so the net friction force on it must be equal to the component of its weight acting down the incline. (a) Prove that the stationary line is at a distance of

$$\frac{L}{2}\left(1 - \frac{\tan\theta}{\mu_k}\right)$$

below the top edge of the plate. (b) Analyze the forces that act on the plate when the temperature is falling and prove that the stationary line is at that same distance above the bottom edge of the plate. (c) Show that the plate steps down the roof like an inchworm, moving each day by the distance

$$\frac{L}{\mu_k}(\alpha_2 - \alpha_1)(T_h - T_c)\tan\theta$$

(d) Evaluate the distance an aluminum plate moves each day if its length is 1.20 m, the temperature cycles between 4.00°C and 36.0°C, and if the roof has slope 18.5°, coefficient of linear expansion 1.50×10^{-5} (°C)$^{-1}$, and coefficient of friction 0.420 with the plate. (e) **What If?** What if the expansion coefficient of the plate is less than that of the roof? Will the plate creep up the roof?

67. A 1.00-km steel railroad rail is fastened securely at both ends when the temperature is 20.0°C. As the temperature increases, the rail buckles, taking the shape of an arc of a vertical circle. Find the height h of the center of the rail when the temperature is 25.0°C. (You will need to solve a transcendental equation.)

chapter 20

The First Law of Thermodynamics

In this photograph of the Mt. Baker area near Bellingham, Washington, we see evidence of water in all three phases. In the lake is liquid water, and solid water in the form of snow appears on the ground. The clouds in the sky consist of liquid water droplets that have condensed from the gaseous water vapor in the air. Changes of a substance from one phase to another are a result of energy transfer. *(© iStockphoto.com/KingWu)*

Until about 1850, the fields of thermodynamics and mechanics were considered to be two distinct branches of science. The principle of conservation of energy seemed to describe only certain kinds of mechanical systems. Mid-19th-century experiments performed by Englishman James Joule and others, however, showed a strong connection between the transfer of energy by heat in thermal processes and the transfer of energy by work in mechanical processes. Today we know that mechanical energy can be transformed to internal energy, which is formally defined in this chapter. Once the concept of energy was generalized from mechanics to include internal energy, the principle of conservation of energy emerged as a universal law of nature.

This chapter focuses on the concept of internal energy, the first law of thermodynamics, and some important applications of the first law. The first law of thermodynamics describes systems in which the only energy change is that of internal energy and the transfers of energy are by heat and work. A major difference in our discussion of work in this chapter from that in most of the chapters on mechanics is that we will consider work done on *deformable* systems.

20.1 Heat and Internal Energy

At the outset, it is important to make a major distinction between internal energy and heat, terms that are often incorrectly used interchangeably in popular language.

Internal energy is all the energy of a system that is associated with its microscopic components—atoms and molecules—when viewed from a reference frame at rest with respect to the center of mass of the system.

The last part of this sentence ensures that any bulk kinetic energy of the system due to its motion through space is not included in internal energy. Internal energy includes kinetic energy of random translational, rotational, and vibrational motion of molecules; vibrational potential energy associated with forces between atoms in molecules; and electric potential energy associated with forces between molecules. It is useful to relate internal energy to the temperature of an object, but this relationship is limited. We show in Section 20.3 that internal energy changes can also occur in the absence of temperature changes.

Heat is defined as the transfer of energy across the boundary of a system due to a temperature difference between the system and its surroundings.

When you *heat* a substance, you are transferring energy into it by placing it in contact with surroundings that have a higher temperature. Such is the case, for example, when you place a pan of cold water on a stove burner. The burner is at a higher temperature than the water, and so the water gains energy by heat. We shall also use the term *heat* to represent the amount of energy transferred by this method.

Read this definition of heat (Q in Eq. 8.2) very carefully. In particular, notice what heat is *not* in the following common quotes. (1) Heat is *not* energy in a hot substance. For example, "The boiling water has a lot of heat" is incorrect; the boiling water has *internal energy* E_{int}. (2) Heat is *not* radiation. For example, "It was so hot because the sidewalk was radiating heat" is incorrect; energy is leaving the sidewalk by *electromagnetic radiation,* T_{ER} in Equation 8.2. (3) Heat is *not* warmth of an environment. For example, "The heat in the air was so oppressive" is incorrect; on a hot day, the air has a high *temperature* T.

As an analogy to the distinction between heat and internal energy, consider the distinction between work and mechanical energy discussed in Chapter 7. The work done on a system is a measure of the amount of energy transferred to the system from its surroundings, whereas the mechanical energy (kinetic energy plus potential energy) of a system is a consequence of the motion and configuration of the system. Therefore, when a person does work on a system, energy is transferred from the person to the system. It makes no sense to talk about the work *of* a system; one can refer only to the work done *on* or *by* a system when some process has occurred in which energy has been transferred to or from the system. Likewise, it makes no sense to talk about the heat *of* a system; one can refer to heat only when energy has been transferred as a result of a temperature difference. Both heat and work are ways of changing the energy of a system.

Units of Heat

Early studies of heat focused on the resultant increase in temperature of a substance, which was often water. Initial notions of heat were based on a fluid called

Pitfall Prevention 20.1

Internal Energy, Thermal Energy, and Bond Energy

When reading other physics books, you may see terms such as *thermal energy* and *bond energy*. Thermal energy can be interpreted as that part of the internal energy associated with random motion of molecules and therefore related to temperature. Bond energy is the intermolecular potential energy. Therefore,

Internal energy = thermal energy + bond energy

Although this breakdown is presented here for clarification with regard to other books, we will not use these terms because there is no need for them.

Pitfall Prevention 20.2

Heat, Temperature, and Internal Energy Are Different

As you read the newspaper or listen to the radio, be alert for incorrectly used phrases including the word *heat* and think about the proper word to be used in place of *heat*. Incorrect examples include "As the truck braked to a stop, a large amount of heat was generated by friction" and "The heat of a hot summer day . . ."

caloric that flowed from one substance to another and caused changes in temperature. From the name of this mythical fluid came an energy unit related to thermal processes, the **calorie (cal),** which is defined as the amount of energy transfer necessary to raise the temperature of 1 g of water from 14.5°C to 15.5°C.[1] (The "Calorie," written with a capital "C" and used in describing the energy content of foods, is actually a kilocalorie.) The unit of energy in the U.S. customary system is the **British thermal unit (Btu),** which is defined as the amount of energy transfer required to raise the temperature of 1 lb of water from 63°F to 64°F.

Once the relationship between energy in thermal and mechanical processes became clear, there was no need for a separate unit related to thermal processes. The *joule* has already been defined as an energy unit based on mechanical processes. Scientists are increasingly turning away from the calorie and the Btu and are using the joule when describing thermal processes. In this textbook, heat, work, and internal energy are usually measured in joules.

James Prescott Joule
British physicist (1818–1889)
Joule received some formal education in mathematics, philosophy, and chemistry from John Dalton but was in large part self-educated. Joule's research led to the establishment of the principle of conservation of energy. His study of the quantitative relationship among electrical, mechanical, and chemical effects of heat culminated in his announcement in 1843 of the amount of work required to produce a unit of energy, called the mechanical equivalent of heat.

The Mechanical Equivalent of Heat

In Chapters 7 and 8, we found that whenever friction is present in a mechanical system, the mechanical energy in the system decreases; in other words, mechanical energy is not conserved in the presence of nonconservative forces. Various experiments show that this mechanical energy does not simply disappear but is transformed into internal energy. You can perform such an experiment at home by hammering a nail into a scrap piece of wood. What happens to all the kinetic energy of the hammer once you have finished? Some of it is now in the nail as internal energy, as demonstrated by the nail being measurably warmer. Notice that there is *no* transfer of energy by heat in this process. For the nail and board as a nonisolated system, Equation 8.2 becomes $\Delta E_{\text{int}} = W + T_{\text{MW}}$, where W is the work done by the hammer on the nail and T_{MW} is the energy leaving the system by sound waves when the nail is struck. Although this connection between mechanical and internal energy was first suggested by Benjamin Thompson, it was James Prescott Joule who established the equivalence of the decrease in mechanical energy and the increase in internal energy.

A schematic diagram of Joule's most famous experiment is shown in Figure 20.1. The system of interest is the Earth, the two blocks, and the water in a thermally insulated container. Work is done within the system on the water by a rotating paddle wheel, which is driven by heavy blocks falling at a constant speed. If the energy transformed in the bearings and the energy passing through the walls by heat are neglected, the decrease in potential energy of the system as the blocks fall equals the work done by the paddle wheel on the water and, in turn, the increase in internal energy of the water. If the two blocks fall through a distance h, the decrease in potential energy is $2mgh$, where m is the mass of one block; this energy causes the temperature of the water to increase. By varying the conditions of the experiment, Joule found that the decrease in mechanical energy is proportional to the product of the mass of the water and the increase in water temperature. The proportionality constant was found to be approximately 4.18 J/g · °C. Hence, 4.18 J of mechanical energy raises the temperature of 1 g of water by 1°C. More precise measurements taken later demonstrated the proportionality to be 4.186 J/g · °C when the temperature of the water was raised from 14.5°C to 15.5°C. We adopt this "15-degree calorie" value:

$$1 \text{ cal} = 4.186 \text{ J} \tag{20.1}$$

This equality is known, for purely historical reasons, as the **mechanical equivalent of heat.** A more proper name would be *equivalence between mechanical energy and internal energy,* but the historical name is well entrenched in our language, despite the incorrect use of the word *heat.*

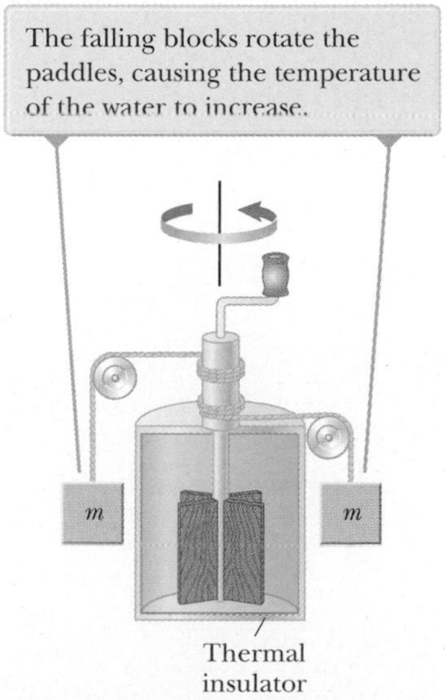

Figure 20.1 Joule's experiment for determining the mechanical equivalent of heat.

[1]Originally, the calorie was defined as the energy transfer necessary to raise the temperature of 1 g of water by 1°C. Careful measurements, however, showed that the amount of energy required to produce a 1°C change depends somewhat on the initial temperature; hence, a more precise definition evolved.

Example 20.1 Losing Weight the Hard Way

A student eats a dinner rated at 2 000 Calories. He wishes to do an equivalent amount of work in the gymnasium by lifting a 50.0-kg barbell. How many times must he raise the barbell to expend this much energy? Assume he raises the barbell 2.00 m each time he lifts it and he regains no energy when he lowers the barbell.

SOLUTION

Conceptualize Imagine the student raising the barbell. He is doing work on the system of the barbell and the Earth, so energy is leaving his body. The total amount of work that the student must do is 2 000 Calories.

Categorize We model the system of the barbell and the Earth as a nonisolated system.

Analyze Reduce the conservation of energy equation, Equation 8.2, to the appropriate expression for the system of the barbell and the Earth:

$$(1)\quad \Delta U_{\text{total}} = W_{\text{total}}$$

Express the change in gravitational potential energy of the system after the barbell is raised once:

$$\Delta U = mgh$$

Express the total amount of energy that must be transferred into the system by work for lifting the barbell n times, assuming energy is not regained when the barbell is lowered:

$$(2)\quad \Delta U_{\text{total}} = nmgh$$

Substitute Equation (2) into Equation (1):

$$nmgh = W_{\text{total}}$$

Solve for n:

$$n = \frac{W_{\text{total}}}{mgh}$$

$$= \frac{(2\,000\text{ Cal})}{(50.0\text{ kg})(9.80\text{ m/s}^2)(2.00\text{ m})}\left(\frac{1.00 \times 10^3\text{ cal}}{\text{Calorie}}\right)\left(\frac{4.186\text{ J}}{1\text{ cal}}\right)$$

$$= 8.54 \times 10^3\text{ times}$$

Finalize If the student is in good shape and lifts the barbell once every 5 s, it will take him about 12 h to perform this feat. Clearly, it is much easier for this student to lose weight by dieting.

In reality, the human body is not 100% efficient. Therefore, not all the energy transformed within the body from the dinner transfers out of the body by work done on the barbell. Some of this energy is used to pump blood and perform other functions within the body. Therefore, the 2 000 Calories can be worked off in less time than 12 h when these other energy processes are included.

20.2 Specific Heat and Calorimetry

When energy is added to a system and there is no change in the kinetic or potential energy of the system, the temperature of the system usually rises. (An exception to this statement is the case in which a system undergoes a change of state—also called a *phase transition*—as discussed in the next section.) If the system consists of a sample of a substance, we find that the quantity of energy required to raise the temperature of a given mass of the substance by some amount varies from one substance to another. For example, the quantity of energy required to raise the temperature of 1 kg of water by 1°C is 4 186 J, but the quantity of energy required to raise the temperature of 1 kg of copper by 1°C is only 387 J. In the discussion that follows, we shall use heat as our example of energy transfer, but keep in mind that the temperature of the system could be changed by means of any method of energy transfer.

The **heat capacity** C of a particular sample is defined as the amount of energy needed to raise the temperature of that sample by 1°C. From this definition, we see that if energy Q produces a change ΔT in the temperature of a sample, then

$$Q = C\,\Delta T \qquad \textbf{(20.2)}$$

TABLE 20.1 *Specific Heats of Some Substances at 25°C and Atmospheric Pressure*

Substance	Specific Heat (J/kg · °C)	Substance	Specific Heat (J/kg · °C)
Elemental solids		*Other solids*	
Aluminum	900	Brass	380
Beryllium	1 830	Glass	837
Cadmium	230	Ice (−5°C)	2 090
Copper	387	Marble	860
Germanium	322	Wood	1 700
Gold	129		
Iron	448	*Liquids*	
Lead	128	Alcohol (ethyl)	2 400
Silicon	703	Mercury	140
Silver	234	Water (15°C)	4 186
		Gas	
		Steam (100°C)	2 010

Note: To convert values to units of cal/g · °C, divide by 4 186.

The **specific heat** c of a substance is the heat capacity per unit mass. Therefore, if energy Q transfers to a sample of a substance with mass m and the temperature of the sample changes by ΔT, the specific heat of the substance is

$$c \equiv \frac{Q}{m\,\Delta T} \quad (20.3)$$

◀ Specific heat

Specific heat is essentially a measure of how thermally insensitive a substance is to the addition of energy. The greater a material's specific heat, the more energy must be added to a given mass of the material to cause a particular temperature change. Table 20.1 lists representative specific heats.

From this definition, we can relate the energy Q transferred between a sample of mass m of a material and its surroundings to a temperature change ΔT as

$$Q = mc\,\Delta T \quad (20.4)$$

For example, the energy required to raise the temperature of 0.500 kg of water by 3.00°C is $Q = (0.500\text{ kg})(4\,186\text{ J/kg}\cdot{}^\circ\text{C})(3.00^\circ\text{C}) = 6.28 \times 10^3\text{ J}$. Notice that when the temperature increases, Q and ΔT are taken to be positive and energy transfers into the system. When the temperature decreases, Q and ΔT are negative and energy transfers out of the system.

We can identify $mc\,\Delta T$ as the change in internal energy of the system if we ignore any thermal expansion or contraction of the system. (Thermal expansion or contraction would result in a very small amount of work being done on the system by the surrounding air.) Then, Equation 20.4 is a reduced form of Equation 8.2: $\Delta E_{\text{int}} = Q$. The internal energy of the system can be changed by transferring energy into the system by any mechanism. For example, if the system is a baked potato in a microwave oven, Equation 8.2 reduces to the following analog to Equation 20.4: $\Delta E_{\text{int}} = T_{\text{ER}} = mc\,\Delta T$, where T_{ER} is the energy transferred to the potato from the microwave oven by electromagnetic radiation. If the system is the air in a bicycle pump, which becomes hot when the pump is operated, Equation 8.2 reduces to the following analog to Equation 20.4: $\Delta E_{\text{int}} = W = mc\,\Delta T$, where W is the work done on the pump by the operator. By identifying $mc\,\Delta T$ as ΔE_{int}, we have taken a step toward a better understanding of temperature: temperature is related to the energy of the molecules of a system. We will learn more details of this relationship in Chapter 21.

Specific heat varies with temperature. If, however, temperature intervals are not too great, the temperature variation can be ignored and c can be treated as a constant.[2] For example, the specific heat of water varies by only about 1% from 0°C

Pitfall Prevention 20.3

An Unfortunate Choice of Terminology

The name *specific heat* is an unfortunate holdover from the days when thermodynamics and mechanics developed separately. A better name would be *specific energy transfer,* but the existing term is too entrenched to be replaced.

Pitfall Prevention 20.4

Energy Can Be Transferred by Any Method

The symbol Q represents the amount of energy transferred, but keep in mind that the energy transfer in Equation 20.4 could be by *any* of the methods introduced in Chapter 8; it does not have to be heat. For example, repeatedly bending a wire coat hanger raises the temperature at the bending point by *work.*

[2]The definition given by Equation 20.4 assumes the specific heat does not vary with temperature over the interval $\Delta T = T_f - T_i$. In general, if c varies with temperature over the interval, the correct expression for Q is $Q = m\int_{T_i}^{T_f} c\,dT$.

to 100°C at atmospheric pressure. Unless stated otherwise, we shall neglect such variations.

***Quick Quiz* 20.1** Imagine you have 1 kg each of iron, glass, and water, and all three samples are at 10°C. **(a)** Rank the samples from highest to lowest temperature after 100 J of energy is added to each sample. **(b)** Rank the samples from greatest to least amount of energy transferred by heat if each sample increases in temperature by 20°C.

Notice from Table 20.1 that water has the highest specific heat of common materials. This high specific heat is in part responsible for the moderate climates found near large bodies of water. As the temperature of a body of water decreases during the winter, energy is transferred from the cooling water to the air by heat, increasing the internal energy of the air. Because of the high specific heat of water, a relatively large amount of energy is transferred to the air for even modest temperature changes of the water. The prevailing winds on the West Coast of the United States are toward the land (eastward). Hence, the energy liberated by the Pacific Ocean as it cools keeps coastal areas much warmer than they would otherwise be. As a result, West Coast states generally have more favorable winter weather than East Coast states, where the prevailing winds do not tend to carry the energy toward land.

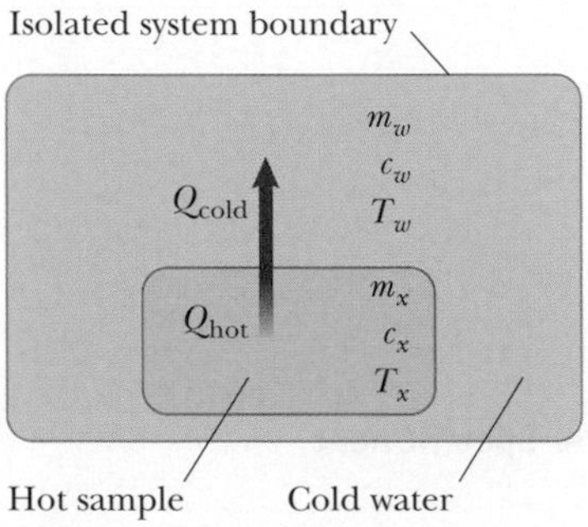

Figure 20.2 In a calorimetry experiment, a hot sample whose specific heat is unknown is placed in cold water in a container that isolates the system from the environment.

Calorimetry

One technique for measuring specific heat involves heating a sample to some known temperature T_x, placing it in a vessel containing water of known mass and temperature $T_w < T_x$, and measuring the temperature of the water after equilibrium has been reached. This technique is called **calorimetry,** and devices in which this energy transfer occurs are called **calorimeters.** Figure 20.2 shows the hot sample in the cold water and the resulting energy transfer by heat from the high-temperature part of the system to the low-temperature part. If the system of the sample and the water is isolated, the principle of conservation of energy requires that the amount of energy Q_{hot} that leaves the sample (of unknown specific heat) equal the amount of energy Q_{cold} that enters the water.[3] Conservation of energy allows us to write the mathematical representation of this energy statement as

$$Q_{cold} = -Q_{hot} \qquad \textbf{(20.5)}$$

Pitfall Prevention 20.5

Remember the Negative Sign

It is *critical* to include the negative sign in Equation 20.5. The negative sign in the equation is necessary for consistency with our sign convention for energy transfer. The energy transfer Q_{hot} has a negative value because energy is leaving the hot substance. The negative sign in the equation ensures that the right side is a positive number, consistent with the left side, which is positive because energy is entering the cold water.

Suppose m_x is the mass of a sample of some substance whose specific heat we wish to determine. Let's call its specific heat c_x and its initial temperature T_x as shown in Figure 20.2. Likewise, let m_w, c_w, and T_w represent corresponding values for the water. If T_f is the final temperature after the system comes to equilibrium, Equation 20.4 shows that the energy transfer for the water is $m_w c_w(T_f - T_w)$, which is positive because $T_f > T_w$, and that the energy transfer for the sample of unknown specific heat is $m_x c_x(T_f - T_x)$, which is negative. Substituting these expressions into Equation 20.5 gives

$$m_w c_w(T_f - T_w) = -m_x c_x(T_f - T_x)$$

This equation can be solved for the unknown specific heat c_x.

Example 20.2 Cooling a Hot Ingot

A 0.050 0-kg ingot of metal is heated to 200.0°C and then dropped into a calorimeter containing 0.400 kg of water initially at 20.0°C. The final equilibrium temperature of the mixed system is 22.4°C. Find the specific heat of the metal.

[3]For precise measurements, the water container should be included in our calculations because it also exchanges energy with the sample. Doing so would require that we know the container's mass and composition, however. If the mass of the water is much greater than that of the container, we can neglect the effects of the container.

20.2 *cont.*

SOLUTION

Conceptualize Imagine the process occurring in the isolated system of Figure 20.2. Energy leaves the hot ingot and goes into the cold water, so the ingot cools off and the water warms up. Once both are at the same temperature, the energy transfer stops.

Categorize We use an equation developed in this section, so we categorize this example as a substitution problem.

Use Equation 20.4 to evaluate each side of Equation 20.5:

$$m_w c_w (T_f - T_w) = -m_x c_x (T_f - T_x)$$

Solve for c_x:

$$c_x = \frac{m_w c_w (T_f - T_w)}{m_x (T_x - T_f)}$$

Substitute numerical values:

$$c_x = \frac{(0.400 \text{ kg})(4\,186 \text{ J/kg}\cdot{}^\circ\text{C})(22.4^\circ\text{C} - 20.0^\circ\text{C})}{(0.050\,0 \text{ kg})(200.0^\circ\text{C} - 22.4^\circ\text{C})}$$

$$= 453 \text{ J/kg}\cdot{}^\circ\text{C}$$

The ingot is most likely iron as you can see by comparing this result with the data given in Table 20.1. The temperature of the ingot is initially above the steam point. Therefore, some of the water may vaporize when the ingot is dropped into the water. We assume the system is sealed and this steam cannot escape. Because the final equilibrium temperature is lower than the steam point, any steam that does result recondenses back into water.

WHAT IF? Suppose you are performing an experiment in the laboratory that uses this technique to determine the specific heat of a sample and you wish to decrease the overall uncertainty in your final result for c_x. Of the data given in this example, changing which value would be most effective in decreasing the uncertainty?

Answer The largest experimental uncertainty is associated with the small difference in temperature of 2.4°C for the water. For example, using the rules for propagation of uncertainty in Appendix Section B.8, an uncertainty of 0.1°C in each of T_f and T_w leads to an 8% uncertainty in their difference. For this temperature difference to be larger experimentally, the most effective change is to *decrease the amount of water.*

Example 20.3 Fun Time for a Cowboy

A cowboy fires a silver bullet with a muzzle speed of 200 m/s into the pine wall of a saloon. Assume all the internal energy generated by the impact remains with the bullet. What is the temperature change of the bullet?

SOLUTION

Conceptualize Imagine similar experiences you may have had in which mechanical energy is transformed to internal energy when a moving object is stopped. For example, as mentioned in Section 20.1, a nail becomes warm after it is hit a few times with a hammer.

Categorize The bullet is modeled as an isolated system. No work is done on the system because the force from the wall moves through no displacement. This example is similar to the skateboarder pushing off a wall in Section 9.7. There, no work is done on the skateboarder by the wall, and potential energy stored in the body from previous meals is transformed to kinetic energy. Here, no work is done by the wall on the bullet, and kinetic energy is transformed to internal energy.

Analyze Reduce the conservation of energy equation, Equation 8.2, to the appropriate expression for the system of the bullet:

$$(1)\quad \Delta K + \Delta E_{\text{int}} = 0$$

The change in the bullet's internal energy is related to its change in temperature:

$$(2)\quad \Delta E_{\text{int}} = mc\,\Delta T$$

Substitute Equation (2) into Equation (1):

$$(0 - \tfrac{1}{2}mv^2) + mc\,\Delta T = 0$$

continued

20.3 *cont.*

Solve for ΔT, using 234 J/kg · °C as the specific heat of silver (see Table 20.1):

$$(3)\ \Delta T = \frac{\frac{1}{2}mv^2}{mc} = \frac{v^2}{2c} = \frac{(200\ \text{m/s})^2}{2(234\ \text{J/kg}\cdot{}^\circ\text{C})} = 85.5^\circ\text{C}$$

Finalize Notice that the result does not depend on the mass of the bullet.

WHAT IF? Suppose the cowboy runs out of silver bullets and fires a lead bullet at the same speed into the wall. Will the temperature change of the bullet be larger or smaller?

Answer Table 20.1 shows that the specific heat of lead is 128 J/kg·°C, which is smaller than that for silver. Therefore, a given amount of energy input or transformation raises lead to a higher temperature than silver and the final temperature of the lead bullet will be larger. In Equation (3), let's substitute the new value for the specific heat:

$$\Delta T = \frac{v^2}{2c} = \frac{(200\ \text{m/s})^2}{2(128\ \text{J/kg}\cdot{}^\circ\text{C})} = 156^\circ\text{C}$$

There is no requirement that the silver and lead bullets have the same mass to determine this change in temperature. The only requirement is that they have the same speed.

20.3 Latent Heat

As we have seen in the preceding section, a substance can undergo a change in temperature when energy is transferred between it and its surroundings. In some situations, however, the transfer of energy does not result in a change in temperature. That is the case whenever the physical characteristics of the substance change from one form to another; such a change is commonly referred to as a **phase change.** Two common phase changes are from solid to liquid (melting) and from liquid to gas (boiling); another is a change in the crystalline structure of a solid. All such phase changes involve a change in the system's internal energy but no change in its temperature. The increase in internal energy in boiling, for example, is represented by the breaking of bonds between molecules in the liquid state; this bond breaking allows the molecules to move farther apart in the gaseous state, with a corresponding increase in intermolecular potential energy.

As you might expect, different substances respond differently to the addition or removal of energy as they change phase because their internal molecular arrangements vary. Also, the amount of energy transferred during a phase change depends on the amount of substance involved. (It takes less energy to melt an ice cube than it does to thaw a frozen lake.) When discussing two phases of a material, we will use the term *higher-phase material* to mean the material existing at the higher temperature. So, for example, if we discuss water and ice, water is the higher-phase material, whereas steam is the higher-phase material in a discussion of steam and water. Consider a system containing a substance in two phases in equilibrium such as water and ice. The initial amount of the higher-phase material, water, in the system is m_i. Now imagine that energy Q enters the system. As a result, the final amount of water is m_f due to the melting of some of the ice. Therefore, the amount of ice that melted, equal to the amount of *new* water, is $\Delta m = m_f - m_i$. We define the **latent heat** for this phase change as

$$L \equiv \frac{Q}{\Delta m} \qquad \textbf{(20.6)}$$

This parameter is called latent heat (literally, the "hidden" heat) because this added or removed energy does not result in a temperature change. The value of L for a substance depends on the nature of the phase change as well as on the properties of the substance. If the entire amount of the lower-phase material undergoes a phase change, the change in mass Δm of the higher-phase material is equal to the

TABLE 20.2 *Latent Heats of Fusion and Vaporization*

Substance	Melting Point (°C)	Latent Heat of Fusion (J/kg)	Boiling Point (°C)	Latent Heat of Vaporization (J/kg)
Helium	−269.65	5.23×10^3	−268.93	2.09×10^4
Oxygen	−218.79	1.38×10^4	−182.97	2.13×10^5
Nitrogen	−209.97	2.55×10^4	−195.81	2.01×10^5
Ethyl alcohol	−114	1.04×10^5	78	8.54×10^5
Water	0.00	3.33×10^5	100.00	2.26×10^6
Sulfur	119	3.81×10^4	444.60	3.26×10^5
Lead	327.3	2.45×10^4	1 750	8.70×10^5
Aluminum	660	3.97×10^5	2 450	1.14×10^7
Silver	960.80	8.82×10^4	2 193	2.33×10^6
Gold	1 063.00	6.44×10^4	2 660	1.58×10^6
Copper	1 083	1.34×10^5	1 187	5.06×10^6

initial mass of the lower-phase material. For example, if an ice cube of mass m on a plate melts completely, the change in mass of the water is $m_f - 0 = m$, which is the mass of new water and is also equal to the initial mass of the ice cube.

From the definition of latent heat, and again choosing heat as our energy transfer mechanism, the energy required to change the phase of a pure substance is

$$Q = L\,\Delta m \qquad (20.7)$$

◀ **Energy transferred to a substance during a phase change**

where Δm is the change in mass of the higher-phase material.

Latent heat of fusion L_f is the term used when the phase change is from solid to liquid (*to fuse* means "to combine by melting"), and **latent heat of vaporization** L_v is the term used when the phase change is from liquid to gas (the liquid "vaporizes").[4] The latent heats of various substances vary considerably as data in Table 20.2 show. When energy enters a system, causing melting or vaporization, the amount of the higher-phase material increases, so Δm is positive and Q is positive, consistent with our sign convention. When energy is extracted from a system, causing freezing or condensation, the amount of the higher-phase material decreases, so Δm is negative and Q is negative, again consistent with our sign convention. Keep in mind that Δm in Equation 20.7 always refers to the higher-phase material.

> **Pitfall Prevention 20.6**
>
> **Signs Are Critical**
>
> Sign errors occur very often when students apply calorimetry equations. For phase changes, remember that Δm in Equation 20.7 is always the change in mass of the higher-phase material. In Equation 20.4, be sure your ΔT is *always* the final temperature minus the initial temperature. In addition, you must *always* include the negative sign on the right side of Equation 20.5.

To understand the role of latent heat in phase changes, consider the energy required to convert a 1.00-g cube of ice at −30.0°C to steam at 120.0°C. Figure 20.3 (page 574) indicates the experimental results obtained when energy is gradually added to the ice. The results are presented as a graph of temperature of the system of the ice cube versus energy added to the system. Let's examine each portion of the red-brown curve, which is divided into parts A through E.

Part A. On this portion of the curve, the temperature of the ice changes from −30.0°C to 0.0°C. Equation 20.4 indicates that the temperature varies linearly with the energy added, so the experimental result is a straight line on the graph. Because the specific heat of ice is 2 090 J/kg · °C, we can calculate the amount of energy added by using Equation 20.4:

$$Q = m_i c_i \,\Delta T = (1.00 \times 10^{-3}\text{ kg})(2\,090\text{ J/kg}\cdot{}^\circ\text{C})(30.0^\circ\text{C}) = 62.7\text{ J}$$

Part B. When the temperature of the ice reaches 0.0°C, the ice–water mixture remains at this temperature—even though energy is being added—until all the ice melts. The energy required to melt 1.00 g of ice at 0.0°C is, from Equation 20.7,

$$Q = L_f\,\Delta m_w = L_f m_i = (3.33 \times 10^5\text{ J/kg})(1.00 \times 10^{-3}\text{ kg}) = 333\text{ J}$$

[4]When a gas cools, it eventually *condenses;* that is, it returns to the liquid phase. The energy given up per unit mass is called the *latent heat of condensation* and is numerically equal to the latent heat of vaporization. Likewise, when a liquid cools, it eventually solidifies, and the *latent heat of solidification* is numerically equal to the latent heat of fusion.

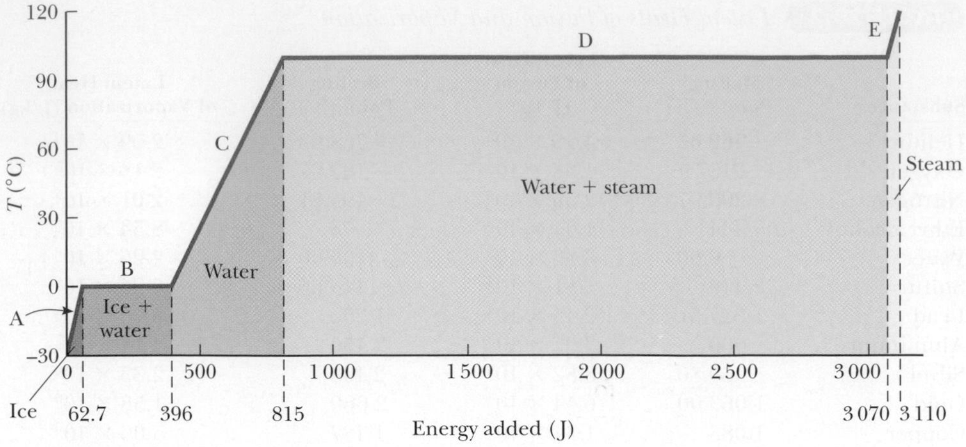

Figure 20.3 A plot of temperature versus energy added when 1.00 g of ice initially at −30.0°C is converted to steam at 120.0°C.

At this point, we have moved to the 396 J (= 62.7 J + 333 J) mark on the energy axis in Figure 20.3.

Part C. Between 0.0°C and 100.0°C, nothing surprising happens. No phase change occurs, and so all energy added to the water is used to increase its temperature. The amount of energy necessary to increase the temperature from 0.0°C to 100.0°C is

$$Q = m_w c_w \,\Delta T = (1.00 \times 10^{-3}\ \text{kg})(4.19 \times 10^3\ \text{J/kg}\cdot{}^\circ\text{C})(100.0^\circ\text{C}) = 419\ \text{J}$$

Part D. At 100.0°C, another phase change occurs as the water changes from water at 100.0°C to steam at 100.0°C. Similar to the ice–water mixture in part B, the water–steam mixture remains at 100.0°C—even though energy is being added—until all the liquid has been converted to steam. The energy required to convert 1.00 g of water to steam at 100.0°C is

$$Q = L_v \,\Delta m_s = L_v m_w = (2.26 \times 10^6\ \text{J/kg})(1.00 \times 10^{-3}\ \text{kg}) = 2.26 \times 10^3\ \text{J}$$

Part E. On this portion of the curve, as in parts A and C, no phase change occurs; therefore, all energy added is used to increase the temperature of the steam. The energy that must be added to raise the temperature of the steam from 100.0°C to 120.0°C is

$$Q = m_s c_s \,\Delta T = (1.00 \times 10^{-3}\ \text{kg})(2.01 \times 10^3\ \text{J/kg}\cdot{}^\circ\text{C})(20.0^\circ\text{C}) = 40.2\ \text{J}$$

The total amount of energy that must be added to change 1 g of ice at −30.0°C to steam at 120.0°C is the sum of the results from all five parts of the curve, which is 3.11×10^3 J. Conversely, to cool 1 g of steam at 120.0°C to ice at −30.0°C, we must remove 3.11×10^3 J of energy.

Notice in Figure 20.3 the relatively large amount of energy that is transferred into the water to vaporize it to steam. Imagine reversing this process, with a large amount of energy transferred out of steam to condense it into water. That is why a burn to your skin from steam at 100°C is much more damaging than exposure of your skin to water at 100°C. A very large amount of energy enters your skin from the steam, and the steam remains at 100°C for a long time while it condenses. Conversely, when your skin makes contact with water at 100°C, the water immediately begins to drop in temperature as energy transfers from the water to your skin.

If liquid water is held perfectly still in a very clean container, it is possible for the water to drop below 0°C without freezing into ice. This phenomenon, called **supercooling,** arises because the water requires a disturbance of some sort for the molecules to move apart and start forming the large, open ice structure that makes the density of ice lower than that of water as discussed in Section 19.4. If supercooled water is disturbed, it suddenly freezes. The system drops into the lower-energy configuration of bound molecules of the ice structure, and the energy released raises the temperature back to 0°C.

Commercial hand warmers consist of liquid sodium acetate in a sealed plastic pouch. The solution in the pouch is in a stable supercooled state. When a disk in the pouch is clicked by your fingers, the liquid solidifies and the temperature increases, just like the supercooled water just mentioned. In this case, however, the freezing point of the liquid is higher than body temperature, so the pouch feels warm to the touch. To reuse the hand warmer, the pouch must be boiled until the solid liquefies. Then, as it cools, it passes below its freezing point into the supercooled state.

It is also possible to create **superheating.** For example, clean water in a very clean cup placed in a microwave oven can sometimes rise in temperature beyond 100°C without boiling because the formation of a bubble of steam in the water requires scratches in the cup or some type of impurity in the water to serve as a nucleation site. When the cup is removed from the microwave oven, the superheated water can become explosive as bubbles form immediately and the hot water is forced upward out of the cup.

***Quick Quiz* 20.2** Suppose the same process of adding energy to the ice cube is performed as discussed above, but instead we graph the internal energy of the system as a function of energy input. What would this graph look like?

Example 20.4 Cooling the Steam

What mass of steam initially at 130°C is needed to warm 200 g of water in a 100-g glass container from 20.0°C to 50.0°C?

SOLUTION

Conceptualize Imagine placing water and steam together in a closed insulated container. The system eventually reaches a uniform state of water with a final temperature of 50.0°C.

Categorize Based on our conceptualization of this situation, we categorize this example as one involving calorimetry in which a phase change occurs.

Analyze Write Equation 20.5 to describe the calorimetry process:

$$(1) \quad Q_{\text{cold}} = -Q_{\text{hot}}$$

The steam undergoes three processes: first a decrease in temperature to 100°C, then condensation into liquid water, and finally a decrease in temperature of the water to 50.0°C. Find the energy transfer in the first process using the unknown mass m_s of the steam:

$$Q_1 = m_s c_s \,\Delta T_s$$

Find the energy transfer in the second process:

$$Q_2 = L_v \,\Delta m_s = L_v(0 - m_s) = -m_s L_v$$

Find the energy transfer in the third process:

$$Q_3 = m_s c_w \,\Delta T_{\text{hot water}}$$

Add the energy transfers in these three stages:

$$(2) \quad Q_{\text{hot}} = Q_1 + Q_2 + Q_3 = m_s(c_s \,\Delta T_s - L_v + c_w \,\Delta T_{\text{hot water}})$$

The 20.0°C water and the glass undergo only one process, an increase in temperature to 50.0°C. Find the energy transfer in this process:

$$(3) \quad Q_{\text{cold}} = m_w c_w \,\Delta T_{\text{cold water}} + m_g c_g \,\Delta T_{\text{glass}}$$

Substitute Equations (2) and (3) into Equation (1):

$$m_w c_w \,\Delta T_{\text{cold water}} + m_g c_g \,\Delta T_{\text{glass}} = -m_s(c_s \,\Delta T_s - L_v + c_w \,\Delta T_{\text{hot water}})$$

Solve for m_s:

$$m_s = -\frac{m_w c_w \,\Delta T_{\text{cold water}} + m_g c_g \,\Delta T_{\text{glass}}}{c_s \,\Delta T_s - L_v + c_w \,\Delta T_{\text{hot water}}}$$

continued

20.4 cont.

Substitute numerical values:

$$m_s = -\frac{(0.200\text{ kg})(4\,186\text{ J/kg}\cdot{}^\circ\text{C})(50.0^\circ\text{C} - 20.0^\circ\text{C}) + (0.100\text{ kg})(837\text{ J/kg}\cdot{}^\circ\text{C})(50.0^\circ\text{C} - 20.0^\circ\text{C})}{(2\,010\text{ J/kg}\cdot{}^\circ\text{C})(100^\circ\text{C} - 130^\circ\text{C}) - (2.26\times10^6\text{ J/kg}) + (4\,186\text{ J/kg}\cdot{}^\circ\text{C})(50.0^\circ\text{C} - 100^\circ\text{C})}$$

$$= 1.09\times10^{-2}\text{ kg} = \boxed{10.9\text{ g}}$$

WHAT IF? What if the final state of the system is water at 100°C? Would we need more steam or less steam? How would the analysis above change?

Answer More steam would be needed to raise the temperature of the water and glass to 100°C instead of 50.0°C. There would be two major changes in the analysis. First, we would not have a term Q_3 for the steam because the water that condenses from the steam does not cool below 100°C. Second, in Q_{cold}, the temperature change would be 80.0°C instead of 30.0°C. For practice, show that the result is a required mass of steam of 31.8 g.

20.4 Work and Heat in Thermodynamic Processes

In thermodynamics, we describe the *state* of a system using such variables as pressure, volume, temperature, and internal energy. As a result, these quantities belong to a category called **state variables.** For any given configuration of the system, we can identify values of the state variables. (For mechanical systems, the state variables include kinetic energy K and potential energy U.) A state of a system can be specified only if the system is in thermal equilibrium internally. In the case of a gas in a container, internal thermal equilibrium requires that every part of the gas be at the same pressure and temperature.

A second category of variables in situations involving energy is **transfer variables.** These variables are those that appear on the right side of the conservation of energy equation, Equation 8.2. Such a variable has a nonzero value if a process occurs in which energy is transferred across the system's boundary. The transfer variable is positive or negative, depending on whether energy is entering or leaving the system. Because a transfer of energy across the boundary represents a change in the system, transfer variables are not associated with a given state of the system, but rather with a *change* in the state of the system.

In the previous sections, we discussed heat as a transfer variable. In this section, we study another important transfer variable for thermodynamic systems, work. Work performed on particles was studied extensively in Chapter 7, and here we investigate the work done on a deformable system, a gas. Consider a gas contained in a cylinder fitted with a movable piston (Fig. 20.4). At equilibrium, the gas occupies a volume V and exerts a uniform pressure P on the cylinder's walls and on the piston. If the piston has a cross-sectional area A, the force exerted by the gas on the piston is $F = PA$. Now let's assume we push the piston inward and compress the gas **quasi-statically,** that is, slowly enough to allow the system to remain essentially in internal thermal equilibrium at all times. As the piston is pushed downward by an external force $\vec{\mathbf{F}} = -F\hat{\mathbf{j}}$ through a displacement of $d\vec{\mathbf{r}} = dy\,\hat{\mathbf{j}}$ (Fig. 20.4b), the work done on the gas is, according to our definition of work in Chapter 7,

$$dW = \vec{\mathbf{F}}\cdot d\vec{\mathbf{r}} = -F\hat{\mathbf{j}}\cdot dy\,\hat{\mathbf{j}} = -F\,dy = -PA\,dy$$

where the magnitude F of the external force is equal to PA because the piston is always in equilibrium between the external force and the force from the gas. The mass of the piston is assumed to be negligible in this discussion. Because $A\,dy$ is the change in volume of the gas dV, we can express the work done on the gas as

$$dW = -P\,dV \qquad \textbf{(20.8)}$$

If the gas is compressed, dV is negative and the work done on the gas is positive. If the gas expands, dV is positive and the work done on the gas is negative. If the volume remains constant, the work done on the gas is zero. The total work done on the gas as its volume changes from V_i to V_f is given by the integral of Equation 20.8:

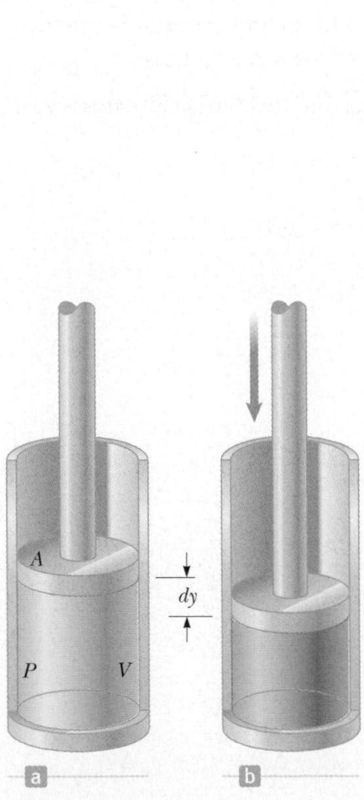

Figure 20.4 Work is done on a gas contained in a cylinder at a pressure P as the piston is pushed downward so that the gas is compressed.

$$W = -\int_{V_i}^{V_f} P\,dV \qquad (20.9)$$

◀ **Work done on a gas**

To evaluate this integral, you must know how the pressure varies with volume during the process.

In general, the pressure is not constant during a process followed by a gas, but depends on the volume and temperature. If the pressure and volume are known at each step of the process, the state of the gas at each step can be plotted on a graphical representation called a ***PV* diagram** as in Active Figure 20.5. This type of diagram allows us to visualize a process through which a gas is progressing. The curve on a *PV* diagram is called the *path* taken between the initial and final states.

Notice that the integral in Equation 20.9 is equal to the area under a curve on a *PV* diagram. Therefore, we can identify an important use for *PV* diagrams:

> The work done on a gas in a quasi-static process that takes the gas from an initial state to a final state is the negative of the area under the curve on a *PV* diagram, evaluated between the initial and final states.

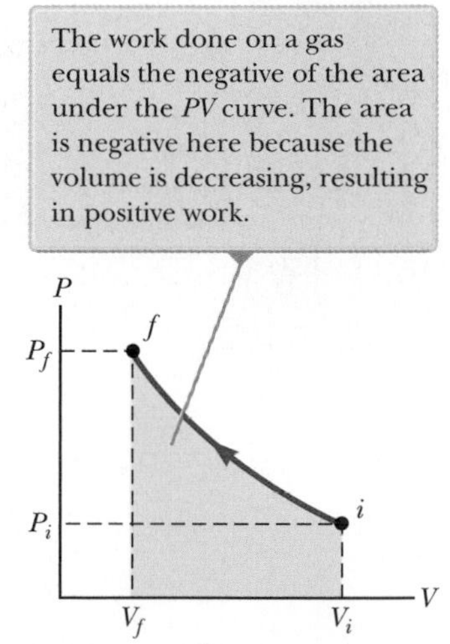

ACTIVE FIGURE 20.5

A gas is compressed quasi-statically (slowly) from state *i* to state *f*. An outside agent must do positive work on the gas to compress it.

For the process of compressing a gas in a cylinder, the work done depends on the particular path taken between the initial and final states as Active Figure 20.5 suggests. To illustrate this important point, consider several different paths connecting *i* and *f* (Active Fig. 20.6). In the process depicted in Active Figure 20.6a, the volume of the gas is first reduced from V_i to V_f at constant pressure P_i and the pressure of the gas then increases from P_i to P_f by heating at constant volume V_f. The work done on the gas along this path is $-P_i(V_f - V_i)$. In Active Figure 20.6b, the pressure of the gas is increased from P_i to P_f at constant volume V_i and then the volume of the gas is reduced from V_i to V_f at constant pressure P_f. The work done on the gas is $-P_f(V_f - V_i)$. This value is greater than that for the process described in Active Figure 20.6a because the piston is moved through the same displacement by a larger force. Finally, for the process described in Active Figure 20.6c, where both *P* and *V* change continuously, the work done on the gas has some value between the values obtained in the first two processes. To evaluate the work in this case, the function $P(V)$ must be known so that we can evaluate the integral in Equation 20.9.

The energy transfer *Q* into or out of a system by heat also depends on the process. Consider the situations depicted in Figure 20.7 (page 578). In each case, the gas has the same initial volume, temperature, and pressure, and is assumed to be ideal. In Figure 20.7a, the gas is thermally insulated from its surroundings except at the bottom of the gas-filled region, where it is in thermal contact with an energy reservoir. An *energy reservoir* is a source of energy that is considered to be so great that a finite transfer of energy to or from the reservoir does not change its temperature. The piston is held at its initial position by an external agent such as a hand. When the force holding the piston is reduced slightly, the piston rises very slowly to its final position shown in Figure 20.7b. Because the piston is moving upward, the

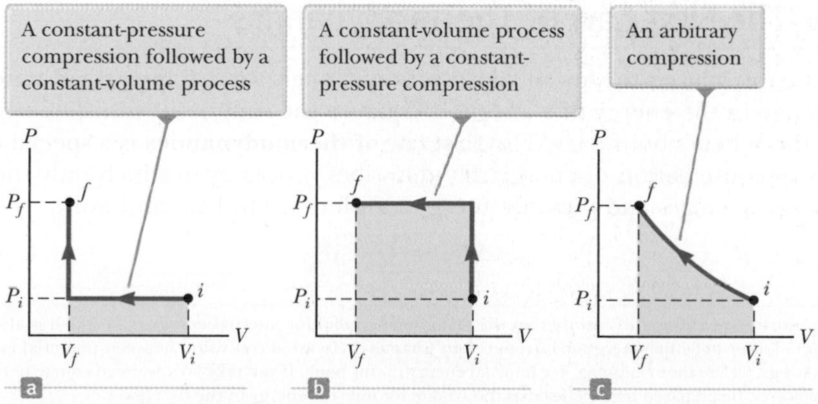

ACTIVE FIGURE 20.6

The work done on a gas as it is taken from an initial state to a final state depends on the path between these states.

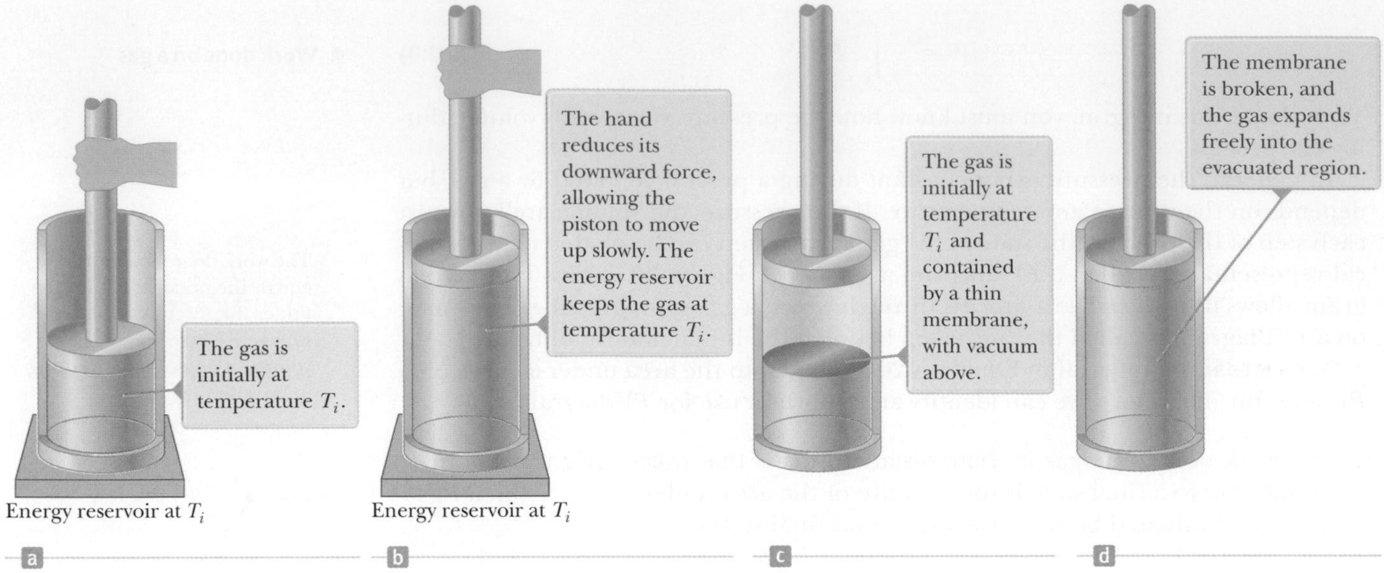

Figure 20.7 Gas in a cylinder. (a) The gas is in contact with an energy reservoir. The walls of the cylinder are perfectly insulating, but the base in contact with the reservoir is conducting. (b) The gas expands slowly to a larger volume. (c) The gas is contained by a membrane in half of a volume, with vacuum in the other half. The entire cylinder is perfectly insulating. (d) The gas expands freely into the larger volume.

gas is doing work on the piston. During this expansion to the final volume V_f, just enough energy is transferred by heat from the reservoir to the gas to maintain a constant temperature T_i.

Now consider the completely thermally insulated system shown in Figure 20.7c. When the membrane is broken, the gas expands rapidly into the vacuum until it occupies a volume V_f and is at a pressure P_f. The final state of the gas is shown in Figure 20.7d. In this case, the gas does no work because it does not apply a force; no force is required to expand into a vacuum. Furthermore, no energy is transferred by heat through the insulating wall.

As we discuss in Section 20.5, experiments show that the temperature of the ideal gas does not change in the process indicated in Figures 20.7c and 20.7d. Therefore, the initial and final states of the ideal gas in Figures 20.7a and 20.7b are identical to the initial and final states in Figures 20.7c and 20.7d, but the paths are different. In the first case, the gas does work on the piston and energy is transferred slowly to the gas by heat. In the second case, no energy is transferred by heat and the value of the work done is zero. Therefore, energy transfer by heat, like work done, depends on the initial, final, and intermediate states of the system. In other words, because heat and work depend on the path, neither quantity is determined solely by the endpoints of a thermodynamic process.

20.5 The First Law of Thermodynamics

When we introduced the law of conservation of energy in Chapter 8, we stated that the change in the energy of a system is equal to the sum of all transfers of energy across the system's boundary. The **first law of thermodynamics** is a special case of the law of conservation of energy that describes processes in which only the internal energy[5] changes and the only energy transfers are by heat and work:

First law of thermodynamics ▶

$$\Delta E_{\text{int}} = Q + W \qquad (20.10)$$

[5]It is an unfortunate accident of history that the traditional symbol for internal energy is U, which is also the traditional symbol for potential energy as introduced in Chapter 7. To avoid confusion between potential energy and internal energy, we use the symbol E_{int} for internal energy in this book. If you take an advanced course in thermodynamics, however, be prepared to see U used as the symbol for internal energy in the first law.

An important consequence of the first law of thermodynamics is that there exists a quantity known as internal energy whose value is determined by the state of the system. The internal energy is therefore a state variable like pressure, volume, and temperature.

Let us investigate some special cases in which the first law can be applied. First, consider an *isolated system,* that is, one that does not interact with its surroundings. In this case, no energy transfer by heat takes place and the work done on the system is zero; hence, the internal energy remains constant. That is, because $Q = W = 0$, it follows that $\Delta E_{\text{int}} = 0$; therefore, $E_{\text{int},i} = E_{\text{int},f}$. We conclude that the internal energy E_{int} of an isolated system remains constant.

Next, consider the case of a system that can exchange energy with its surroundings and is taken through a **cyclic process,** that is, a process that starts and ends at the same state. In this case, the change in the internal energy must again be zero because E_{int} is a state variable; therefore, the energy Q added to the system must equal the negative of the work W done on the system during the cycle. That is, in a cyclic process,

$$\Delta E_{\text{int}} = 0 \quad \text{and} \quad Q = -W \quad \text{(cyclic process)}$$

On a *PV* diagram, a cyclic process appears as a closed curve. (The processes described in Active Figure 20.6 are represented by open curves because the initial and final states differ.) It can be shown that in a cyclic process, the net work done on the system per cycle equals the area enclosed by the path representing the process on a *PV* diagram.

20.6 Some Applications of the First Law of Thermodynamics

In this section, we consider applications of the first law to processes through which a gas is taken. As a model, let's consider the sample of gas contained in the piston–cylinder apparatus in Active Figure 20.8. This figure shows work being done on the gas and energy transferring in by heat, so the internal energy of the gas is rising. In the following discussion of various processes, refer back to this figure and mentally alter the directions of the transfer of energy to reflect what is happening in the process.

Before we apply the first law of thermodynamics to specific systems, it is useful to first define some idealized thermodynamic processes. An **adiabatic process** is one during which no energy enters or leaves the system by heat; that is, $Q = 0$. An adiabatic process can be achieved either by thermally insulating the walls of the system or by performing the process rapidly so that there is negligible time for energy to transfer by heat. Applying the first law of thermodynamics to an adiabatic process gives

$$\Delta E_{\text{int}} = W \quad \text{(adiabatic process)} \tag{20.11}$$

This result shows that if a gas is compressed adiabatically such that W is positive, then ΔE_{int} is positive and the temperature of the gas increases. Conversely, the temperature of a gas decreases when the gas expands adiabatically.

Adiabatic processes are very important in engineering practice. Some common examples are the expansion of hot gases in an internal combustion engine, the liquefaction of gases in a cooling system, and the compression stroke in a diesel engine.

The process described in Figures 20.7c and 20.7d, called an **adiabatic free expansion,** is unique. The process is adiabatic because it takes place in an insulated container. Because the gas expands into a vacuum, it does not apply a force on a piston as does the gas in Figures 20.7a and 20.7b, so no work is done on or by the gas. Therefore, in this adiabatic process, both $Q = 0$ and $W = 0$. As a result, $\Delta E_{\text{int}} = 0$ for this process as can be seen from the first law. That is, the initial and final

Pitfall Prevention 20.7

Dual Sign Conventions

Some physics and engineering books present the first law as $\Delta E_{\text{int}} = Q - W$, with a minus sign between the heat and work. The reason is that work is defined in these treatments as the work done *by* the gas rather than *on* the gas, as in our treatment. The equivalent equation to Equation 20.9 in these treatments defines work as $W = \int_{V_i}^{V_f} P\,dV$. Therefore, if positive work is done by the gas, energy is leaving the system, leading to the negative sign in the first law.

In your studies in other chemistry or engineering courses, or in your reading of other physics books, be sure to note which sign convention is being used for the first law.

Pitfall Prevention 20.8

The First Law

With our approach to energy in this book, the first law of thermodynamics is a special case of Equation 8.2. Some physicists argue that the first law is the general equation for energy conservation, equivalent to Equation 8.2. In this approach, the first law is applied to a closed system (so that there is no matter transfer), heat is interpreted so as to include electromagnetic radiation, and work is interpreted so as to include electrical transmission ("electrical work") and mechanical waves ("molecular work"). Keep that in mind if you run across the first law in your reading of other physics books.

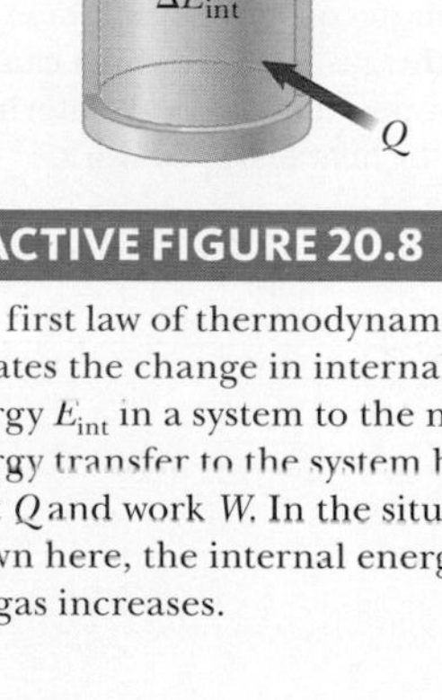

ACTIVE FIGURE 20.8

The first law of thermodynamics equates the change in internal energy E_{int} in a system to the net energy transfer to the system by heat Q and work W. In the situation shown here, the internal energy of the gas increases.

internal energies of a gas are equal in an adiabatic free expansion. As we shall see in Chapter 21, the internal energy of an ideal gas depends only on its temperature. Therefore, we expect no change in temperature during an adiabatic free expansion. This prediction is in accord with the results of experiments performed at low pressures. (Experiments performed at high pressures for real gases show a slight change in temperature after the expansion due to intermolecular interactions, which represent a deviation from the model of an ideal gas.)

A process that occurs at constant pressure is called an **isobaric process.** In Active Figure 20.8, an isobaric process could be established by allowing the piston to move freely so that it is always in equilibrium between the net force from the gas pushing upward and the weight of the piston plus the force due to atmospheric pressure pushing downward. The first process in Active Figure 20.6a and the second process in Active Figure 20.6b are both isobaric.

In such a process, the values of the heat and the work are both usually nonzero. The work done on the gas in an isobaric process is simply

Isobaric process ▶

$$W = -P(V_f - V_i) \quad \text{(isobaric process)} \tag{20.12}$$

where P is the constant pressure of the gas during the process.

A process that takes place at constant volume is called an **isovolumetric process.** In Active Figure 20.8, clamping the piston at a fixed position would ensure an isovolumetric process. The second process in Active Figure 20.6a and the first process in Active Figure 20.6b are both isovolumetric.

Because the volume of the gas does not change in such a process, the work given by Equation 20.9 is zero. Hence, from the first law we see that in an isovolumetric process, because $W = 0$,

Isovolumetric process ▶

$$\Delta E_{\text{int}} = Q \quad \text{(isovolumetric process)} \tag{20.13}$$

This expression specifies that if energy is added by heat to a system kept at constant volume, all the transferred energy remains in the system as an increase in its internal energy. For example, when a can of spray paint is thrown into a fire, energy enters the system (the gas in the can) by heat through the metal walls of the can. Consequently, the temperature, and therefore the pressure, in the can increases until the can possibly explodes.

Isothermal process ▶

A process that occurs at constant temperature is called an **isothermal process.** This process can be established by immersing the cylinder in Active Figure 20.8 in an ice–water bath or by putting the cylinder in contact with some other constant-temperature reservoir. A plot of P versus V at constant temperature for an ideal gas yields a hyperbolic curve called an *isotherm.* The internal energy of an ideal gas is a function of temperature only. Hence, in an isothermal process involving an ideal gas, $\Delta E_{\text{int}} = 0$. For an isothermal process, we conclude from the first law that the energy transfer Q must be equal to the negative of the work done on the gas; that is, $Q = -W$. Any energy that enters the system by heat is transferred out of the system by work; as a result, no change in the internal energy of the system occurs in an isothermal process.

Pitfall Prevention 20.9

$Q \neq 0$ in an Isothermal Process

Do not fall into the common trap of thinking there must be no transfer of energy by heat if the temperature does not change as is the case in an isothermal process. Because the cause of temperature change can be either heat *or* work, the temperature can remain constant even if energy enters the gas by heat, which can only happen if the energy entering the gas by heat leaves by work.

***Quick Quiz* 20.3** In the last three columns of the following table, fill in the boxes with the correct signs (−, +, or 0) for Q, W, and ΔE_{int}. For each situation, the system to be considered is identified.

Situation	System	Q	W	ΔE_{int}
(a) Rapidly pumping up a bicycle tire	Air in the pump			
(b) Pan of room-temperature water sitting on a hot stove	Water in the pan			
(c) Air quickly leaking out of a balloon	Air originally in the balloon			

Isothermal Expansion of an Ideal Gas

Suppose an ideal gas is allowed to expand quasi-statically at constant temperature. This process is described by the PV diagram shown in Figure 20.9. The curve is a hyperbola (see Appendix B, Eq. B.23), and the ideal gas law (Eq. 19.8) with T constant indicates that the equation of this curve is $PV = nRT =$ constant.

Let's calculate the work done on the gas in the expansion from state i to state f. The work done on the gas is given by Equation 20.9. Because the gas is ideal and the process is quasi-static, the ideal gas law is valid for each point on the path. Therefore,

$$W = -\int_{V_i}^{V_f} P\,dV = -\int_{V_i}^{V_f} \frac{nRT}{V}\,dV$$

Because T is constant in this case, it can be removed from the integral along with n and R:

$$W = -nRT\int_{V_i}^{V_f} \frac{dV}{V} = -nRT \ln V\Big|_{V_i}^{V_f}$$

To evaluate the integral, we used $\int(dx/x) = \ln x$. (See Appendix B.) Evaluating the result at the initial and final volumes gives

$$W = nRT \ln\left(\frac{V_i}{V_f}\right) \quad \textbf{(20.14)}$$

Numerically, this work W equals the negative of the shaded area under the PV curve shown in Figure 20.9. Because the gas expands, $V_f > V_i$ and the value for the work done on the gas is negative as we expect. If the gas is compressed, then $V_f < V_i$ and the work done on the gas is positive.

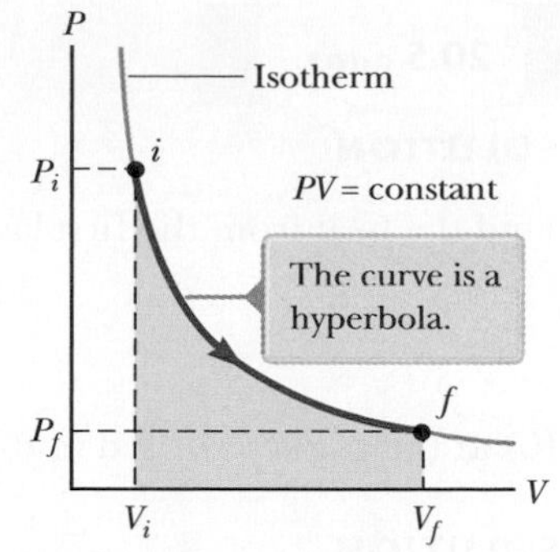

Figure 20.9 The PV diagram for an isothermal expansion of an ideal gas from an initial state to a final state.

***Quick Quiz* 20.4** Characterize the paths in Figure 20.10 as isobaric, isovolumetric, isothermal, or adiabatic. For path B, $Q = 0$. The blue curves are isotherms.

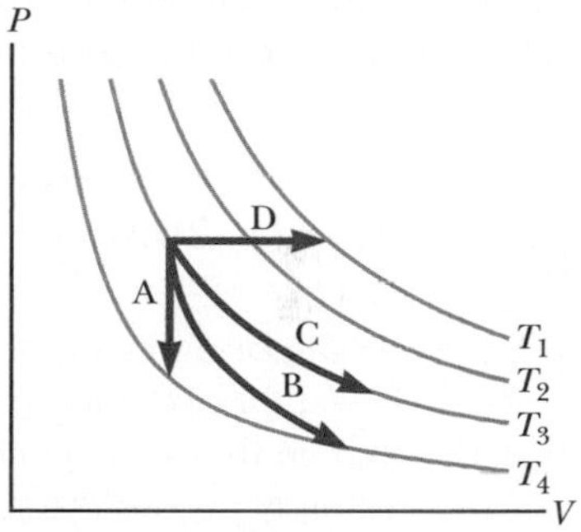

Figure 20.10 (Quick Quiz 20.4) Identify the nature of paths A, B, C, and D.

Example 20.5 An Isothermal Expansion

A 1.0-mol sample of an ideal gas is kept at 0.0°C during an expansion from 3.0 L to 10.0 L.

(A) How much work is done on the gas during the expansion?

SOLUTION

Conceptualize Run the process in your mind: the cylinder in Active Figure 20.8 is immersed in an ice–water bath, and the piston moves outward so that the volume of the gas increases. You can also use the graphical representation in Figure 20.9 to conceptualize the process.

Categorize We will evaluate parameters using equations developed in the preceding sections, so we categorize this example as a substitution problem. Because the temperature of the gas is fixed, the process is isothermal.

Substitute the given values into Equation 20.14:

$$W = nRT \ln\left(\frac{V_i}{V_f}\right)$$

$$= (1.0 \text{ mol})(8.31 \text{ J/mol} \cdot \text{K})(273 \text{ K}) \ln\left(\frac{3.0 \text{ L}}{10.0 \text{ L}}\right)$$

$$= \boxed{-2.7 \times 10^3 \text{ J}}$$

(B) How much energy transfer by heat occurs between the gas and its surroundings in this process?

continued

20.5 *cont.*

SOLUTION

Find the heat from the first law:

$$\Delta E_{int} = Q + W$$
$$0 = Q + W$$
$$Q = -W = \boxed{2.7 \times 10^3\ \text{J}}$$

(C) If the gas is returned to the original volume by means of an isobaric process, how much work is done on the gas?

SOLUTION

Use Equation 20.12. The pressure is not given, so incorporate the ideal gas law:

$$W = -P(V_f - V_i) = -\frac{nRT_i}{V_i}(V_f - V_i)$$
$$= -\frac{(1.0\ \text{mol})(8.31\ \text{J/mol}\cdot\text{K})(273\ \text{K})}{10.0 \times 10^{-3}\ \text{m}^3}(3.0 \times 10^{-3}\ \text{m}^3 - 10.0 \times 10^{-3}\ \text{m}^3)$$
$$= \boxed{1.6 \times 10^3\ \text{J}}$$

We used the initial temperature and volume to calculate the work done because the final temperature was unknown. The work done on the gas is positive because the gas is being compressed.

Example 20.6 Boiling Water

Suppose 1.00 g of water vaporizes isobarically at atmospheric pressure (1.013×10^5 Pa). Its volume in the liquid state is $V_i = V_{liquid} = 1.00\ \text{cm}^3$, and its volume in the vapor state is $V_f = V_{vapor} = 1\,671\ \text{cm}^3$. Find the work done in the expansion and the change in internal energy of the system. Ignore any mixing of the steam and the surrounding air; imagine that the steam simply pushes the surrounding air out of the way.

SOLUTION

Conceptualize Notice that the temperature of the system does not change. There is a phase change occurring as the water evaporates to steam.

Categorize Because the expansion takes place at constant pressure, we categorize the process as isobaric. We will use equations developed in the preceding sections, so we categorize this example as a substitution problem.

Use Equation 20.12 to find the work done on the system as the air is pushed out of the way:

$$W = -P(V_f - V_i)$$
$$= -(1.013 \times 10^5\ \text{Pa})(1\,671 \times 10^{-6}\ \text{m}^3 - 1.00 \times 10^{-6}\ \text{m}^3)$$
$$= \boxed{-169\ \text{J}}$$

Use Equation 20.7 and the latent heat of vaporization for water to find the energy transferred into the system by heat:

$$Q = L_v\,\Delta m_s = m_s L_v = (1.00 \times 10^{-3}\ \text{kg})(2.26 \times 10^6\ \text{J/kg})$$
$$= 2\,260\ \text{J}$$

Use the first law to find the change in internal energy of the system:

$$\Delta E_{int} = Q + W = 2\,260\ \text{J} + (-169\ \text{J}) = \boxed{2.09\ \text{kJ}}$$

The positive value for ΔE_{int} indicates that the internal energy of the system increases. The largest fraction of the energy (2 090 J/ 2260 J = 93%) transferred to the liquid goes into increasing the internal energy of the system. The remaining 7% of the energy transferred leaves the system by work done by the steam on the surrounding atmosphere.

Example 20.7 Heating a Solid

A 1.0-kg bar of copper is heated at atmospheric pressure so that its temperature increases from 20°C to 50°C.

(A) What is the work done on the copper bar by the surrounding atmosphere?

20.7 *cont.*

SOLUTION

Conceptualize This example involves a solid, whereas the preceding two examples involved liquids and gases. For a solid, the change in volume due to thermal expansion is very small.

Categorize Because the expansion takes place at constant atmospheric pressure, we categorize the process as isobaric.

Analyze Find the work done on the copper bar using Equation 20.12:

$$W = -P\Delta V$$

Express the change in volume using Equation 19.6 and that $\beta = 3\alpha$:

$$W = -P(\beta V_i\ \Delta T) = -P(3\alpha V_i\ \Delta T) = -3\alpha P V_i\ \Delta T$$

Substitute for the volume in terms of the mass and density of the copper:

$$W = -3\alpha P\left(\frac{m}{\rho}\right)\Delta T$$

Substitute numerical values:

$$W = -3[1.7 \times 10^{-5}\,(°\text{C})^{-1}](1.013 \times 10^5\ \text{N/m}^2)\left(\frac{1.0\ \text{kg}}{8.92 \times 10^3\ \text{kg/m}^3}\right)(50°\text{C} - 20°\text{C})$$

$$= \boxed{-1.7 \times 10^{-2}\ \text{J}}$$

Because this work is negative, work is done *by* the copper bar on the atmosphere.

(B) How much energy is transferred to the copper bar by heat?

SOLUTION

Use Equation 20.4 and the specific heat of copper from Table 20.1:

$$Q = mc\,\Delta T = (1.0\ \text{kg})(387\ \text{J/kg}\cdot°\text{C})(50°\text{C} - 20°\text{C})$$

$$= \boxed{1.2 \times 10^4\ \text{J}}$$

(C) What is the increase in internal energy of the copper bar?

SOLUTION

Use the first law of thermodynamics:

$$\Delta E_{\text{int}} = Q + W = 1.2 \times 10^4\ \text{J} + (-1.7 \times 10^{-2}\ \text{J})$$

$$= \boxed{1.2 \times 10^4\ \text{J}}$$

Finalize Most of the energy transferred into the system by heat goes into increasing the internal energy of the copper bar. The fraction of energy used to do work on the surrounding atmosphere is only about 10^{-6}. Hence, when the thermal expansion of a solid or a liquid is analyzed, the small amount of work done on or by the system is usually ignored.

20.7 Energy Transfer Mechanisms in Thermal Processes

In Chapter 8, we introduced a global approach to the energy analysis of physical processes through Equation 8.1, $\Delta E_{\text{system}} = \sum T$, where T represents energy transfer, which can occur by several mechanisms. Earlier in this chapter, we discussed two of the terms on the right side of this equation, work W and heat Q. In this section, we explore more details about heat as a means of energy transfer and two other energy transfer methods often related to temperature changes: convection (a form of matter transfer T_{MT}) and electromagnetic radiation T_{ER}.

Thermal Conduction

The process of energy transfer by heat (Q in Eq. 8.2) can also be called **conduction** or **thermal conduction.** In this process, the transfer can be represented on an

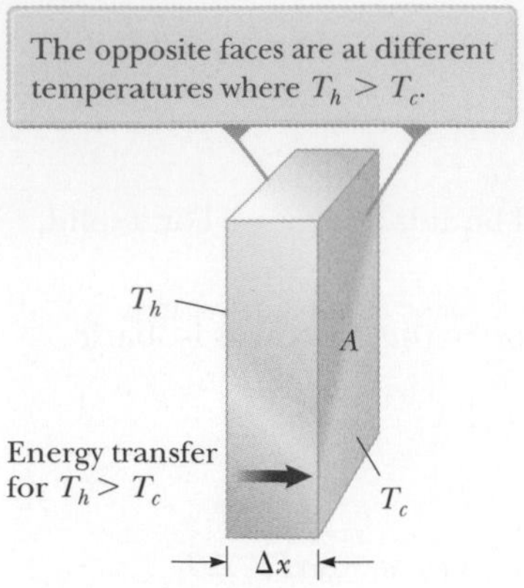

Figure 20.11 Energy transfer through a conducting slab with a cross-sectional area A and a thickness Δx.

atomic scale as an exchange of kinetic energy between microscopic particles—molecules, atoms, and free electrons—in which less-energetic particles gain energy in collisions with more-energetic particles. For example, if you hold one end of a long metal bar and insert the other end into a flame, you will find that the temperature of the metal in your hand soon increases. The energy reaches your hand by means of conduction. Initially, before the rod is inserted into the flame, the microscopic particles in the metal are vibrating about their equilibrium positions. As the flame raises the temperature of the rod, the particles near the flame begin to vibrate with greater and greater amplitudes. These particles, in turn, collide with their neighbors and transfer some of their energy in the collisions. Slowly, the amplitudes of vibration of metal atoms and electrons farther and farther from the flame increase until eventually those in the metal near your hand are affected. This increased vibration is detected by an increase in the temperature of the metal and of your potentially burned hand.

The rate of thermal conduction depends on the properties of the substance being heated. For example, it is possible to hold a piece of asbestos in a flame indefinitely, which implies that very little energy is conducted through the asbestos. In general, metals are good thermal conductors and materials such as asbestos, cork, paper, and fiberglass are poor conductors. Gases also are poor conductors because the separation distance between the particles is so great. Metals are good thermal conductors because they contain large numbers of electrons that are relatively free to move through the metal and so can transport energy over large distances. Therefore, in a good conductor such as copper, conduction takes place by means of both the vibration of atoms and the motion of free electrons.

Conduction occurs only if there is a difference in temperature between two parts of the conducting medium. Consider a slab of material of thickness Δx and cross-sectional area A. One face of the slab is at a temperature T_c, and the other face is at a temperature $T_h > T_c$ (Fig. 20.11). Experimentally, it is found that energy Q transfers in a time interval Δt from the hotter face to the colder one. The rate $P = Q/\Delta t$ at which this energy transfer occurs is found to be proportional to the cross-sectional area and the temperature difference $\Delta T = T_h - T_c$ and inversely proportional to the thickness:

$$P = \frac{Q}{\Delta t} \propto A\frac{\Delta T}{\Delta x}$$

Notice that P has units of watts when Q is in joules and Δt is in seconds. That is not surprising because P is power, the rate of energy transfer by heat. For a slab of infinitesimal thickness dx and temperature difference dT, we can write the **law of thermal conduction** as

$$P = kA\left|\frac{dT}{dx}\right| \qquad \textbf{(20.15)}$$

where the proportionality constant k is the **thermal conductivity** of the material and $|dT/dx|$ is the **temperature gradient** (the rate at which temperature varies with position).

Substances that are good thermal conductors have large thermal conductivity values, whereas good thermal insulators have low thermal conductivity values. Table 20.3 lists thermal conductivities for various substances. Notice that metals are generally better thermal conductors than nonmetals.

Suppose a long, uniform rod of length L is thermally insulated so that energy cannot escape by heat from its surface except at the ends as shown in Figure 20.12. One end is in thermal contact with an energy reservoir at temperature T_c, and the other end is in thermal contact with a reservoir at temperature $T_h > T_c$. When a steady state has been reached, the temperature at each point along the rod is con-

TABLE 20.3

Thermal Conductivities

Substance	Thermal Conductivity (W/m · °C)
Metals (at 25°C)	
Aluminum	238
Copper	397
Gold	314
Iron	79.5
Lead	34.7
Silver	427
Nonmetals (approximate values)	
Asbestos	0.08
Concrete	0.8
Diamond	2 300
Glass	0.8
Ice	2
Rubber	0.2
Water	0.6
Wood	0.08
Gases (at 20°C)	
Air	0.023 4
Helium	0.138
Hydrogen	0.172
Nitrogen	0.023 4
Oxygen	0.023 8

stant in time. In this case, if we assume k is not a function of temperature, the temperature gradient is the same everywhere along the rod and is

$$\left|\frac{dT}{dx}\right| = \frac{T_h - T_c}{L}$$

Therefore, the rate of energy transfer by conduction through the rod is

$$P = kA\left(\frac{T_h - T_c}{L}\right) \qquad (20.16)$$

For a compound slab containing several materials of thicknesses $L_1, L_2, \ldots$ and thermal conductivities $k_1, k_2, \ldots$, the rate of energy transfer through the slab at steady state is

$$P = \frac{A(T_h - T_c)}{\sum_i (L_i/k_i)} \qquad (20.17)$$

where T_h and T_c are the temperatures of the outer surfaces (which are held constant) and the summation is over all slabs. Example 20.8 shows how Equation 20.17 results from a consideration of two thicknesses of materials.

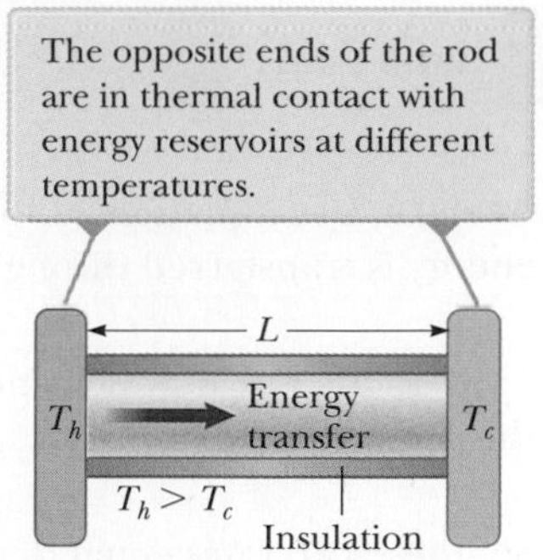

Figure 20.12 Conduction of energy through a uniform, insulated rod of length L.

Quick Quiz **20.5** You have two rods of the same length and diameter, but they are formed from different materials. The rods are used to connect two regions at different temperatures so that energy transfers through the rods by heat. They can be connected in series as in Figure 20.13a or in parallel as in Figure 20.13b. In which case is the rate of energy transfer by heat larger? **(a)** The rate is larger when the rods are in series. **(b)** The rate is larger when the rods are in parallel. **(c)** The rate is the same in both cases.

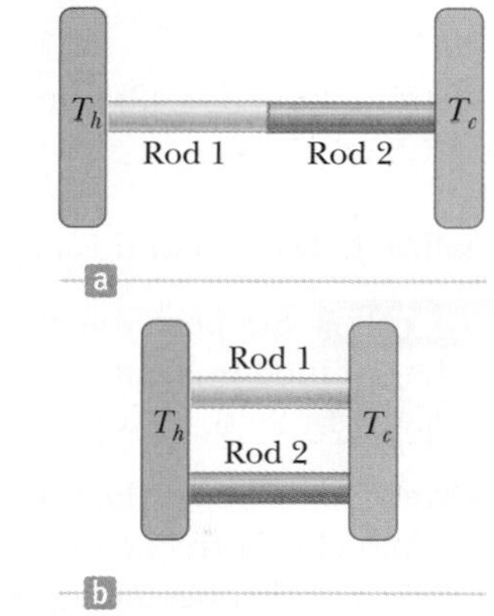

Figure 20.13 (Quick Quiz 20.5) In which case is the rate of energy transfer larger?

Example 20.8 Energy Transfer Through Two Slabs

Two slabs of thickness L_1 and L_2 and thermal conductivities k_1 and k_2 are in thermal contact with each other as shown in Figure 20.14. The temperatures of their outer surfaces are T_c and T_h, respectively, and $T_h > T_c$. Determine the temperature at the interface and the rate of energy transfer by conduction through an area A of the slabs in the steady-state condition.

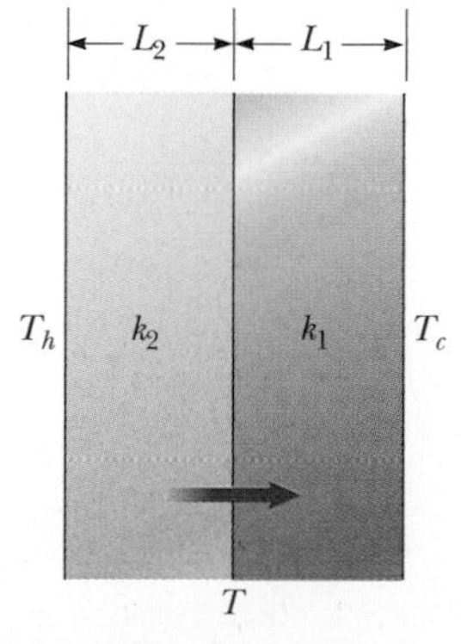

Figure 20.14 (Example 20.8) Energy transfer by conduction through two slabs in thermal contact with each other. At steady state, the rate of energy transfer through slab 1 equals the rate of energy transfer through slab 2.

SOLUTION

Conceptualize Notice the phrase "in the steady-state condition." We interpret this phrase to mean that energy transfers through the compound slab at the same rate at all points. Otherwise, energy would be building up or disappearing at some point. Furthermore, the temperature varies with position in the two slabs, most likely at different rates in each part of the compound slab. When the system is in steady state, the interface is at some fixed temperature T.

Categorize We categorize this example as a thermal conduction problem and impose the condition that the power is the same in both slabs of material.

continued

20.8 cont.

Analyze Use Equation 20.16 to express the rate at which energy is transferred through an area A of slab 1:

$$(1)\quad P_1 = k_1 A\left(\frac{T - T_c}{L_1}\right)$$

Express the rate at which energy is transferred through the same area of slab 2:

$$(2)\quad P_2 = k_2 A\left(\frac{T_h - T}{L_2}\right)$$

Set these two rates equal to represent the steady-state situation:

$$k_1 A\left(\frac{T - T_c}{L_1}\right) = k_2 A\left(\frac{T_h - T}{L_2}\right)$$

Solve for T:

$$(3)\quad T = \frac{k_1 L_2 T_c + k_2 L_1 T_h}{k_1 L_2 + k_2 L_1}$$

Substitute Equation (3) into either Equation (1) or Equation (2):

$$(4)\quad P = \frac{A(T_h - T_c)}{(L_1/k_1) + (L_2/k_2)}$$

Finalize Extension of this procedure to several slabs of materials leads to Equation 20.17.

WHAT IF? Suppose you are building an insulated container with two layers of insulation and the rate of energy transfer determined by Equation (4) turns out to be too high. You have enough room to increase the thickness of one of the two layers by 20%. How would you decide which layer to choose?

Answer To decrease the power as much as possible, you must increase the denominator in Equation (4) as much as possible. Whichever thickness you choose to increase, L_1 or L_2, you increase the corresponding term L/k in the denominator by 20%. For this percentage change to represent the largest absolute change, you want to take 20% of the larger term. Therefore, you should increase the thickness of the layer that has the larger value of L/k.

Home Insulation

In engineering practice, the term L/k for a particular substance is referred to as the ***R*-value** of the material. Therefore, Equation 20.17 reduces to

$$P = \frac{A(T_h - T_c)}{\sum_i R_i} \qquad \textbf{(20.18)}$$

where $R_i = L_i/k_i$. The R-values for a few common building materials are given in Table 20.4. In the United States, the insulating properties of materials used in

TABLE 20.4 R-*Values for Some Common Building Materials*

Material	R-value (ft^2 · °F · h/Btu)
Hardwood siding (1 in. thick)	0.91
Wood shingles (lapped)	0.87
Brick (4 in. thick)	4.00
Concrete block (filled cores)	1.93
Fiberglass insulation (3.5 in. thick)	10.90
Fiberglass insulation (6 in. thick)	18.80
Fiberglass board (1 in. thick)	4.35
Cellulose fiber (1 in. thick)	3.70
Flat glass (0.125 in. thick)	0.89
Insulating glass (0.25-in. space)	1.54
Air space (3.5 in. thick)	1.01
Stagnant air layer	0.17
Drywall (0.5 in. thick)	0.45
Sheathing (0.5 in. thick)	1.32

buildings are usually expressed in U.S. customary units, not SI units. Therefore, in Table 20.4, *R*-values are given as a combination of British thermal units, feet, hours, and degrees Fahrenheit.

At any vertical surface open to the air, a very thin stagnant layer of air adheres to the surface. One must consider this layer when determining the *R*-value for a wall. The thickness of this stagnant layer on an outside wall depends on the speed of the wind. Energy transfer through the walls of a house on a windy day is greater than that on a day when the air is calm. A representative *R*-value for this stagnant layer of air is given in Table 20.4.

Example 20.9 The *R*-Value of a Typical Wall

Calculate the total *R*-value for a wall constructed as shown in Figure 20.15a. Starting outside the house (toward the front in the figure) and moving inward, the wall consists of 4 in. of brick, 0.5 in. of sheathing, an air space 3.5 in. thick, and 0.5 in. of drywall.

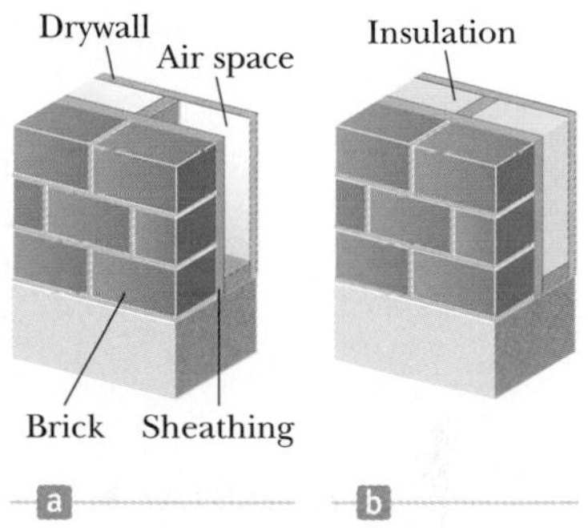

Figure 20.15 (Example 20.9) An exterior house wall containing (a) an air space and (b) insulation.

SOLUTION

Conceptualize Use Figure 20.15 to help conceptualize the structure of the wall. Do not forget the stagnant air layers inside and outside the house.

Categorize We will use specific equations developed in this section on home insulation, so we categorize this example as a substitution problem.

Use Table 20.4 to find the *R*-value of each layer:

$$R_1 \text{ (outside stagnant air layer)} = 0.17 \text{ ft}^2 \cdot {}^\circ\text{F} \cdot \text{h/Btu}$$
$$R_2 \text{ (brick)} = 4.00 \text{ ft}^2 \cdot {}^\circ\text{F} \cdot \text{h/Btu}$$
$$R_3 \text{ (sheathing)} = 1.32 \text{ ft}^2 \cdot {}^\circ\text{F} \cdot \text{h/Btu}$$
$$R_4 \text{ (air space)} = 1.01 \text{ ft}^2 \cdot {}^\circ\text{F} \cdot \text{h/Btu}$$
$$R_5 \text{ (drywall)} = 0.45 \text{ ft}^2 \cdot {}^\circ\text{F} \cdot \text{h/Btu}$$
$$R_6 \text{ (inside stagnant air layer)} = 0.17 \text{ ft}^2 \cdot {}^\circ\text{F} \cdot \text{h/Btu}$$

Add the *R*-values to obtain the total *R*-value for the wall:

$$R_{\text{total}} = R_1 + R_2 + R_3 + R_4 + R_5 + R_6 = \boxed{7.12 \text{ ft}^2 \cdot {}^\circ\text{F} \cdot \text{h/Btu}}$$

WHAT IF? Suppose you are not happy with this total *R*-value for the wall. You cannot change the overall structure, but you can fill the air space as in Figure 20.15b. To *maximize* the total *R*-value, what material should you choose to fill the air space?

Answer Looking at Table 20.4, we see that 3.5 in. of fiberglass insulation is more than ten times as effective as 3.5 in. of air. Therefore, we should fill the air space with fiberglass insulation. The result is that we add 10.90 ft^2 · °F · h/Btu of *R*-value, and we lose 1.01 ft^2 · °F · h/Btu due to the air space we have replaced. The new total *R*-value is equal to 7.12 ft^2 · °F · h/Btu + 9.89 ft^2 · °F · h/Btu = 17.01 ft^2 · °F · h/Btu.

Convection

At one time or another, you probably have warmed your hands by holding them over an open flame. In this situation, the air directly above the flame is heated and expands. As a result, the density of this air decreases and the air rises. This hot air warms your hands as it flows by. Energy transferred by the movement of a warm substance is said to have been transferred by **convection,** which is a form of matter transfer, T_{MT} in Equation 8.2. When resulting from differences in density, as with air around a fire, the process is referred to as *natural convection.* Airflow at a beach is an example of natural convection, as is the mixing that occurs as surface water in a lake cools and sinks (see Section 19.4). When the heated substance is forced to move by a fan or pump, as in some hot-air and hot-water heating systems, the process is called *forced convection.*

If it were not for convection currents, it would be very difficult to boil water. As water is heated in a teakettle, the lower layers are warmed first. This water expands and rises to the top because its density is lowered. At the same time, the denser, cool water at the surface sinks to the bottom of the kettle and is heated.

The same process occurs when a room is heated by a radiator. The hot radiator warms the air in the lower regions of the room. The warm air expands and rises to the ceiling because of its lower density. The denser, cooler air from above sinks, and the continuous air current pattern shown in Figure 20.16 is established.

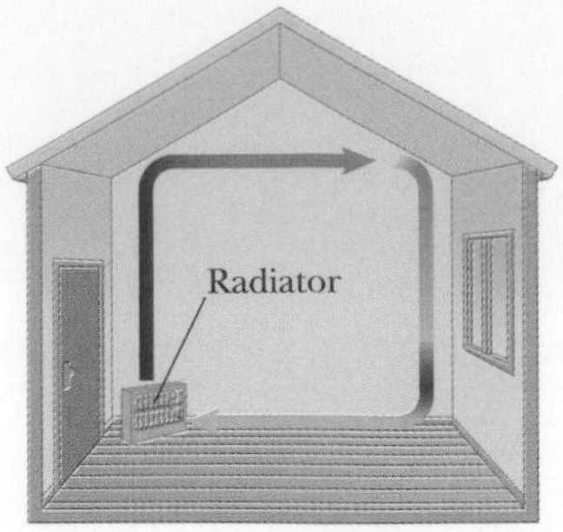

Figure 20.16 Convection currents are set up in a room warmed by a radiator.

Radiation

The third means of energy transfer we shall discuss is **thermal radiation,** T_{ER} in Equation 8.2. All objects radiate energy continuously in the form of electromagnetic waves (see Chapter 34) produced by thermal vibrations of the molecules. You are likely familiar with electromagnetic radiation in the form of the orange glow from an electric stove burner, an electric space heater, or the coils of a toaster.

The rate at which an object radiates energy is proportional to the fourth power of its absolute temperature. Known as **Stefan's law,** this behavior is expressed in equation form as

Stefan's law ▶

$$P = \sigma A e T^4 \tag{20.19}$$

where P is the power in watts of electromagnetic waves radiated from the surface of the object, σ is a constant equal to $5.669\,6 \times 10^{-8}\ \text{W/m}^2 \cdot \text{K}^4$, A is the surface area of the object in square meters, e is the **emissivity,** and T is the surface temperature in kelvins. The value of e can vary between zero and unity depending on the properties of the surface of the object. The emissivity is equal to the **absorptivity,** which is the fraction of the incoming radiation that the surface absorbs. A mirror has very low absorptivity because it reflects almost all incident light. Therefore, a mirror surface also has a very low emissivity. At the other extreme, a black surface has high absorptivity and high emissivity. An **ideal absorber** is defined as an object that absorbs all the energy incident on it, and for such an object, $e = 1$. An object for which $e = 1$ is often referred to as a **black body.** We shall investigate experimental and theoretical approaches to radiation from a black body in Chapter 40.

Every second, approximately 1 370 J of electromagnetic radiation from the Sun passes perpendicularly through each 1 m^2 at the top of the Earth's atmosphere. This radiation is primarily visible and infrared light accompanied by a significant amount of ultraviolet radiation. We shall study these types of radiation in detail in Chapter 34. Enough energy arrives at the surface of the Earth each day to supply all our energy needs on this planet hundreds of times over, if only it could be captured and used efficiently. The growth in the number of solar energy–powered houses built in the United States reflects the increasing efforts being made to use this abundant energy.

What happens to the atmospheric temperature at night is another example of the effects of energy transfer by radiation. If there is a cloud cover above the Earth, the water vapor in the clouds absorbs part of the infrared radiation emitted by the Earth and re-emits it back to the surface. Consequently, temperature levels at the surface remain moderate. In the absence of this cloud cover, there is less in the way to prevent this radiation from escaping into space; therefore, the temperature decreases more on a clear night than on a cloudy one.

As an object radiates energy at a rate given by Equation 20.19, it also absorbs electromagnetic radiation from the surroundings, which consist of other objects that radiate energy. If the latter process did not occur, an object would eventually radiate all its energy and its temperature would reach absolute zero. If an object is at a temperature T and its surroundings are at an average temperature T_0, the net rate of energy gained or lost by the object as a result of radiation is

$$P_{\text{net}} = \sigma A e(T^4 - T_0^4) \tag{20.20}$$

When an object is in equilibrium with its surroundings, it radiates and absorbs energy at the same rate and its temperature remains constant. When an object is hotter than its surroundings, it radiates more energy than it absorbs and its temperature decreases.

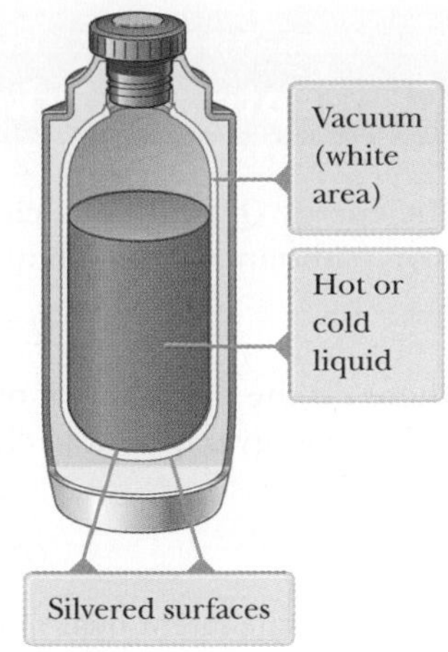

Figure 20.17 A cross-sectional view of a Dewar flask, which is used to store hot or cold substances.

The Dewar Flask

The *Dewar flask*[6] is a container designed to minimize energy transfers by conduction, convection, and radiation. Such a container is used to store cold or hot liquids for long periods of time. (An insulated bottle, such as a Thermos, is a common household equivalent of a Dewar flask.) The standard construction (Fig. 20.17) consists of a double-walled Pyrex glass vessel with silvered walls. The space between the walls is evacuated to minimize energy transfer by conduction and convection. The silvered surfaces minimize energy transfer by radiation because silver is a very good reflector and has very low emissivity. A further reduction in energy loss is obtained by reducing the size of the neck. Dewar flasks are commonly used to store liquid nitrogen (boiling point 77 K) and liquid oxygen (boiling point 90 K).

To confine liquid helium (boiling point 4.2 K), which has a very low heat of vaporization, it is often necessary to use a double Dewar system in which the Dewar flask containing the liquid is surrounded by a second Dewar flask. The space between the two flasks is filled with liquid nitrogen.

Newer designs of storage containers use "superinsulation" that consists of many layers of reflecting material separated by fiberglass. All this material is in a vacuum, and no liquid nitrogen is needed with this design.

[6]Invented by Sir James Dewar (1842–1923).

Summary

Definitions

Internal energy is all a system's energy that is associated with the system's microscopic components. Internal energy includes kinetic energy of random translation, rotation, and vibration of molecules; vibrational potential energy within molecules; and potential energy between molecules.

Heat is the transfer of energy across the boundary of a system resulting from a temperature difference between the system and its surroundings. The symbol Q represents the amount of energy transferred by this process.

A **calorie** is the amount of energy necessary to raise the temperature of 1 g of water from 14.5°C to 15.5°C.

The **heat capacity** C of any sample is the amount of energy needed to raise the temperature of the sample by 1°C.

The **specific heat** c of a substance is the heat capacity per unit mass:

$$c \equiv \frac{Q}{m\,\Delta T} \tag{20.3}$$

The **latent heat** of a substance is defined as the ratio of the energy input to a substance to the change in mass of the higher-phase material:

$$L \equiv \frac{Q}{\Delta m} \tag{20.6}$$

continued

Concepts and Principles

The energy Q required to change the temperature of a mass m of a substance by an amount ΔT is

$$Q = mc\,\Delta T \qquad (20.4)$$

where c is the specific heat of the substance.

The energy required to change the phase of a pure substance is

$$Q = L\,\Delta m \qquad (20.7)$$

where L is the latent heat of the substance, which depends on the nature of the phase change and the substance, and Δm is the change in mass of the higher-phase material.

The **work** done on a gas as its volume changes from some initial value V_i to some final value V_f is

$$W = -\int_{V_i}^{V_f} P\,dV \qquad (20.9)$$

where P is the pressure of the gas, which may vary during the process. To evaluate W, the process must be fully specified; that is, P and V must be known during each step. The work done depends on the path taken between the initial and final states.

The **first law of thermodynamics** states that when a system undergoes a change from one state to another, the change in its internal energy is

$$\Delta E_{\text{int}} = Q + W \qquad (20.10)$$

where Q is the energy transferred into the system by heat and W is the work done on the system. Although Q and W both depend on the path taken from the initial state to the final state, the quantity ΔE_{int} does not depend on the path.

In a **cyclic process** (one that originates and terminates at the same state), $\Delta E_{\text{int}} = 0$ and therefore $Q = -W$. That is, the energy transferred into the system by heat equals the negative of the work done on the system during the process.

In an **adiabatic process,** no energy is transferred by heat between the system and its surroundings ($Q = 0$). In this case, the first law gives $\Delta E_{\text{int}} = W$. In the **adiabatic free expansion** of a gas, $Q = 0$ and $W = 0$, so $\Delta E_{\text{int}} = 0$. That is, the internal energy of the gas does not change in such a process.

An **isobaric process** is one that occurs at constant pressure. The work done on a gas in such a process is $W = -P(V_f - V_i)$.

An **isovolumetric process** is one that occurs at constant volume. No work is done in such a process, so $\Delta E_{\text{int}} = Q$.

An **isothermal process** is one that occurs at constant temperature. The work done on an ideal gas during an isothermal process is

$$W = nRT\ln\left(\frac{V_i}{V_f}\right) \qquad (20.14)$$

Conduction can be viewed as an exchange of kinetic energy between colliding molecules or electrons. The rate of energy transfer by conduction through a slab of area A is

$$P = kA\left|\frac{dT}{dx}\right| \qquad (20.15)$$

where k is the **thermal conductivity** of the material from which the slab is made and $|dT/dx|$ is the **temperature gradient.**

In **convection,** a warm substance transfers energy from one location to another.

All objects emit **thermal radiation** in the form of electromagnetic waves at the rate

$$P = \sigma A e T^4 \qquad (20.19)$$

Objective Questions

☐ denotes answer available in Student Solutions Manual/Study Guide

1. How long would it take a 1 000 W heater to melt 1.00 kg of ice at −20.0°C, assuming all the energy from the heater is absorbed by the ice? (a) 4.18 s (b) 41.8 s (c) 5.55 min (d) 6.25 min (e) 38.4 min

2. A 100-g piece of copper, initially at 95.0°C, is dropped into 200 g of water contained in a 280-g aluminum can; the water and can are initially at 15.0°C. What is the final temperature of the system? (Specific heats of copper and

aluminum are 0.092 and 0.215 cal/g · °C, respectively.) (a) 16°C (b) 18°C (c) 24°C (d) 26°C (e) none of those answers

3. The specific heat of substance A is greater than that of substance B. Both A and B are at the same initial temperature when equal amounts of energy are added to them. Assuming no melting or vaporization occurs, which of the following can be concluded about the final temperature T_A of substance A and the final temperature T_B of substance B? (a) $T_A > T_B$ (b) $T_A < T_B$ (c) $T_A = T_B$ (d) More information is needed.

4. An amount of energy is added to ice, raising its temperature from −10°C to −5°C. A larger amount of energy is added to the same mass of water, raising its temperature from 15°C to 20°C. From these results, what would you conclude? (a) Overcoming the latent heat of fusion of ice requires an input of energy. (b) The latent heat of fusion of ice delivers some energy to the system. (c) The specific heat of ice is less than that of water. (d) The specific heat of ice is greater than that of water. (e) More information is needed to draw any conclusion.

5. How much energy is required to raise the temperature of 5.00 kg of lead from 20.0°C to its melting point of 327°C? The specific heat of lead is 128 J/kg · °C. (a) 4.04×10^5 J (b) 1.07×10^5 J (c) 8.15×10^4 J (d) 2.13×10^4 J (e) 1.96×10^5 J

6. Ethyl alcohol has about one-half the specific heat of water. Assume equal amounts of energy are transferred by heat into equal-mass liquid samples of alcohol and water in separate insulated containers. The water rises in temperature by 25°C. How much will the alcohol rise in temperature? (a) It will rise by 12°C. (b) It will rise by 25°C. (c) It will rise by 50°C. (d) It depends on the rate of energy transfer. (e) It will not rise in temperature.

7. Assume you are measuring the specific heat of a sample of originally hot metal by using a calorimeter containing water. Because your calorimeter is not perfectly insulating, energy can transfer by heat between the contents of the calorimeter and the room. To obtain the most accurate result for the specific heat of the metal, you should use water with which initial temperature? (a) slightly lower than room temperature (b) the same as room temperature (c) slightly higher than room temperature (d) whatever you like because the initial temperature makes no difference

8. Beryllium has roughly one-half the specific heat of water (H_2O). Rank the quantities of energy input required to produce the following changes from the largest to the smallest. In your ranking, note any cases of equality. (a) raising the temperature of 1 kg of H_2O from 20°C to 26°C (b) raising the temperature of 2 kg of H_2O from 20°C to 23°C (c) raising the temperature of 2 kg of H_2O from 1°C to 4°C (d) raising the temperature of 2 kg of beryllium from −1°C to 2°C (e) raising the temperature of 2 kg of H_2O from −1°C to 2°C

9. A person shakes a sealed insulated bottle containing hot coffee for a few minutes. **(i)** What is the change in the temperature of the coffee? (a) a large decrease (b) a slight decrease (c) no change (d) a slight increase (e) a large increase **(ii)** What is the change in the internal energy of the coffee? Choose from the same possibilities.

10. A poker is a stiff, nonflammable rod used to push burning logs around in a fireplace. For safety and comfort of use, should the poker be made from a material with (a) high specific heat and high thermal conductivity, (b) low specific heat and low thermal conductivity, (c) low specific heat and high thermal conductivity, or (d) high specific heat and low thermal conductivity?

11. Star *A* has twice the radius and twice the absolute surface temperature of star *B*. The emissivity of both stars can be assumed to be 1. What is the ratio of the power output of star *A* to that of star *B*? (a) 4 (b) 8 (c) 16 (d) 32 (e) 64

12. If a gas is compressed isothermally, which of the following statements is true? (a) Energy is transferred into the gas by heat. (b) No work is done on the gas. (c) The temperature of the gas increases. (d) The internal energy of the gas remains constant. (e) None of those statements is true.

13. When a gas undergoes an adiabatic expansion, which of the following statements is true? (a) The temperature of the gas does not change. (b) No work is done by the gas. (c) No energy is transferred to the gas by heat. (d) The internal energy of the gas does not change. (e) The pressure increases.

14. If a gas undergoes an isobaric process, which of the following statements is true? (a) The temperature of the gas doesn't change. (b) Work is done on or by the gas. (c) No energy is transferred by heat to or from the gas. (d) The volume of the gas remains the same. (e) The pressure of the gas decreases uniformly.

15. An ideal gas is compressed to half its initial volume by means of several possible processes. Which of the following processes results in the most work done on the gas? (a) isothermal (b) adiabatic (c) isobaric (d) The work done is independent of the process.

Conceptual Questions

☐ denotes answer available in Student Solutions Manual/Study Guide

1. Pioneers stored fruits and vegetables in underground cellars. In winter, why did the pioneers place an open barrel of water alongside their produce?

2. In usually warm climates that experience a hard freeze, fruit growers will spray the fruit trees with water, hoping that a layer of ice will form on the fruit. Why would such a layer be advantageous?

3. What is wrong with the following statement: "Given any two bodies, the one with the higher temperature contains more heat."

4. Why is a person able to remove a piece of dry aluminum foil from a hot oven with bare fingers, whereas a burn results if there is moisture on the foil?

5. Using the first law of thermodynamics, explain why the *total* energy of an isolated system is always constant.

6. Is it possible to convert internal energy to mechanical energy? Explain with examples.

7. It is the morning of a day that will become hot. You just purchased drinks for a picnic and are loading them, with ice, into a chest in the back of your car. (a) You wrap a wool blanket around the chest. Does doing so help to keep the beverages cool, or should you expect the wool blanket to warm them up? Explain your answer. (b) Your younger sister suggests you wrap her up in another wool blanket to keep her cool on the hot day like the ice chest. Explain your response to her.

8. You need to pick up a very hot cooking pot in your kitchen. You have a pair of cotton oven mitts. To pick up the pot most comfortably, should you soak them in cold water or keep them dry?

9. Suppose you pour hot coffee for your guests, and one of them wants it with cream. He wants the coffee to be as warm as possible several minutes later when he drinks it. To have the warmest coffee, should the person add the cream just after the coffee is poured or just before drinking? Explain.

10. When camping in a canyon on a still night, a camper notices that as soon as the sun strikes the surrounding peaks, a breeze begins to stir. What causes the breeze?

11. Rub the palm of your hand on a metal surface for about 30 seconds. Place the palm of your other hand on an unrubbed portion of the surface and then on the rubbed portion. The rubbed portion will feel warmer. Now repeat this process on a wood surface. Why does the temperature difference between the rubbed and unrubbed portions of the wood surface seem larger than for the metal surface?

12. In 1801, Humphry Davy rubbed together pieces of ice inside an icehouse. He made sure that nothing in the environment was at a higher temperature than the rubbed pieces. He observed the production of drops of liquid water. Make a table listing this and other experiments or processes to illustrate each of the following situations. (a) A system can absorb energy by heat, increase in internal energy, and increase in temperature. (b) A system can absorb energy by heat and increase in internal energy without an increase in temperature. (c) A system can absorb energy by heat without increasing in temperature or in internal energy. (d) A system can increase in internal energy and in temperature without absorbing energy by heat. (e) A system can increase in internal energy without absorbing energy by heat or increasing in temperature.

Problems

ENHANCED WebAssign The problems found in this chapter may be assigned online in Enhanced WebAssign

1. denotes straightforward problem; **2.** denotes intermediate problem; **3.** denotes challenging problem

[1.] full solution available in the Student Solutions Manual/Study Guide

1. (shaded) denotes problems most often assigned in Enhanced WebAssign; these provide students with targeted feedback and either a Master It tutorial or a Watch It solution video.

Q|C denotes asking for quantitative and conceptual reasoning

S denotes symbolic reasoning problem

M denotes Master It tutorial available in Enhanced WebAssign

GP denotes guided problem

shaded denotes "paired problems" that develop reasoning with symbols and numerical values

Section 20.1 Heat and Internal Energy

1. A 55.0-kg woman cheats on her diet and eats a 540 Calorie (540 kcal) jelly doughnut for breakfast. (a) How many joules of energy are the equivalent of one jelly doughnut? (b) How many steps must the woman climb on a very tall stairway to change the gravitational potential energy of the woman–Earth system by a value equivalent to the food energy in one jelly doughnut? Assume the height of a single stair is 15.0 cm. (c) If the human body is only 25.0% efficient in converting chemical potential energy to mechanical energy, how many steps must the woman climb to work off her breakfast?

Section 20.2 Specific Heat and Calorimetry

[2.] The highest waterfall in the world is the Salto Angel Falls in Venezuela. Its longest single falls has a height of 807 m. If water at the top of the falls is at 15.0°C, what is the maximum temperature of the water at the bottom of the falls? Assume all the kinetic energy of the water as it reaches the bottom goes into raising its temperature.

3. A combination of 0.250 kg of water at 20.0°C, 0.400 kg of aluminum at 26.0°C, and 0.100 kg of copper at 100°C is mixed in an insulated container and allowed to come to thermal equilibrium. Ignore any energy transfer to or from the container. What is the final temperature of the mixture?

4. Consider Joule's apparatus described in Figure 20.1. The mass of each of the two blocks is 1.50 kg, and the insulated tank is filled with 200 g of water. What is the increase in the water's temperature after the blocks fall through a distance of 3.00 m?

5. What mass of water at 25.0°C must be allowed to come to thermal equilibrium with a 1.85-kg cube of aluminum initially at 150°C to lower the temperature of the aluminum to 65.0°C? Assume any water turned to steam subsequently condenses.

[6.] The temperature of a silver bar rises by 10.0°C when it absorbs 1.23 kJ of energy by heat. The mass of the bar is 525 g. Determine the specific heat of silver from these data.

7. In cold climates, including the northern United States, a house can be built with very large windows facing south to take advantage of solar heating. Sunlight shining in during the daytime is absorbed by the floor, interior walls, and objects in the room, raising their temperature to 38.0°C. If the house is well insulated, you may model it as losing energy by heat steadily at the rate 6 000 W on a day in April when the average exterior temperature is 4°C and when the conventional heating system is not used at all. During the period between 5:00 p.m. and 7:00 a.m., the temperature of the house drops and a sufficiently large "thermal mass" is required to keep it from dropping too far. The thermal mass can be a large quantity of stone (with specific heat 850 J/kg · °C) in the floor and the interior walls exposed to sunlight. What mass of stone is required if the temperature is not to drop below 18.0°C overnight?

8. An aluminum cup of mass 200 g contains 800 g of water in thermal equilibrium at 80.0°C. The combination of cup and water is cooled uniformly so that the temperature decreases by 1.50°C per minute. At what rate is energy being removed by heat? Express your answer in watts.

9. A 1.50-kg iron horseshoe initially at 600°C is dropped into a bucket containing 20.0 kg of water at 25.0°C. What is the final temperature of the water–horseshoe system? Ignore the heat capacity of the container and assume a negligible amount of water boils away.

10. **Q|C** An electric drill with a steel drill bit of mass m = 27.0 g and diameter 0.635 cm is used to drill into a cubical steel block of mass M = 240 g. Assume steel has the same properties as iron. The cutting process can be modeled as happening at one point on the circumference of the bit. This point moves in a helix at constant tangential speed 40.0 m/s and exerts a force of constant magnitude 3.20 N on the block. As shown in Figure P20.10, a groove in the bit carries the chips up to the top of the block, where they form a pile around the hole. The drill is turned on and drills into the block for a time interval of 15.0 s. Let's assume this time interval is long enough for conduction within the steel to bring it all to a uniform temperature. Furthermore, assume the steel objects lose a negligible amount of energy by conduction, convection, and radiation into their environment. (a) Suppose the drill bit cuts three-quarters of the way through the block during 15.0 s. Find the temperature change of the whole quantity of steel. (b) **What If?** Now suppose the drill bit is dull and cuts only one eighth of the way through the block in 15.0 s. Identify the temperature change of the whole quantity of steel in this case. (c) What pieces of data, if any, are unnecessary for the solution? Explain.

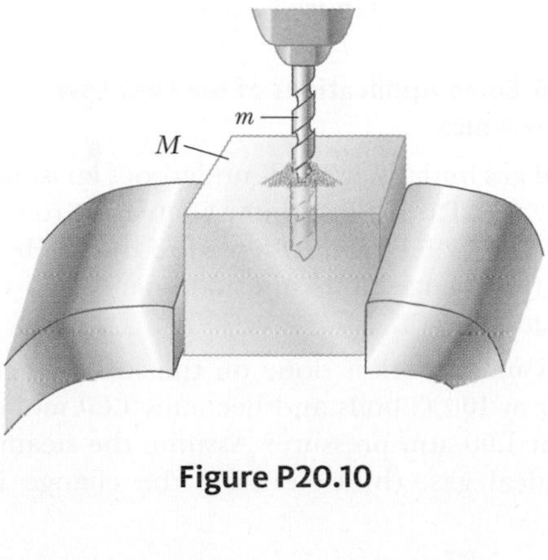

Figure P20.10

11. **Q|C** An aluminum calorimeter with a mass of 100 g contains 250 g of water. The calorimeter and water are in thermal equilibrium at 10.0°C. Two metallic blocks are placed into the water. One is a 50.0-g piece of copper at 80.0°C. The other has a mass of 70.0 g and is originally at a temperature of 100°C. The entire system stabilizes at a final temperature of 20.0°C. (a) Determine the specific heat of the unknown sample. (b) Using the data in Table 20.1, can you make a positive identification of the unknown material? Can you identify a possible material? (c) Explain your answers for part (b).

12. **Q|C** A 3.00-g copper coin at 25.0°C drops 50.0 m to the ground. (a) Assuming 60.0% of the change in gravitational potential energy of the coin–Earth system goes into increasing the internal energy of the coin, determine the coin's final temperature. (b) **What If?** Does the result depend on the mass of the coin? Explain.

13. Two thermally insulated vessels are connected by a narrow tube fitted with a valve that is initially closed as shown in Figure P20.13. One vessel of volume 16.8 L contains oxygen at a temperature of 300 K and a pressure of 1.75 atm. The other vessel of volume 22.4 L contains oxygen at a temperature of 450 K and a pressure of 2.25 atm. When the valve is opened, the gases in the two vessels mix and the temperature and pressure become uniform throughout. (a) What is the final temperature? (b) What is the final pressure?

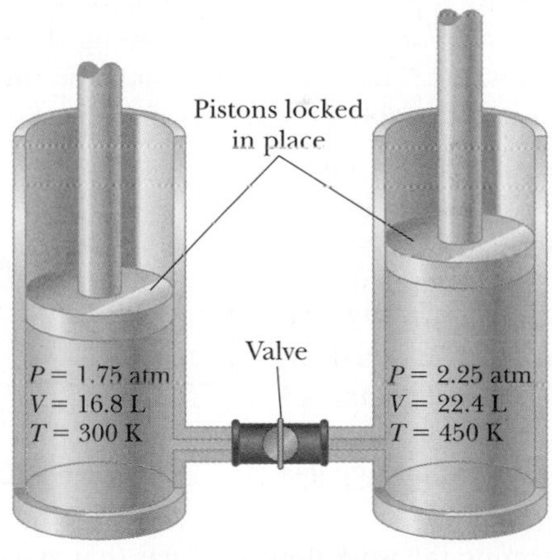

Figure P20.13

Section 20.3 Latent Heat

14. How much energy is required to change a 40.0-g ice cube from ice at −10.0°C to steam at 110°C?

15. A 75.0-g ice cube at 0°C is placed in 825 g of water at 25.0°C. What is the final temperature of the mixture?

16. A 3.00-g lead bullet at 30.0°C is fired at a speed of 240 m/s into a large block of ice at 0°C, in which it becomes embedded. What quantity of ice melts?

17. Steam at 100°C is added to ice at 0°C. (a) Find the amount of ice melted and the final temperature when the mass of

steam is 10.0 g and the mass of ice is 50.0 g. (b) **What If?** Repeat when the mass of steam is 1.00 g and the mass of ice is 50.0 g.

18. A 1.00-kg block of copper at 20.0°C is dropped into a large vessel of liquid nitrogen at 77.3 K. How many kilograms of nitrogen boil away by the time the copper reaches 77.3 K? (The specific heat of copper is 0.092 0 cal/g · °C, and the latent heat of vaporization of nitrogen is 48.0 cal/g.)

19. In an insulated vessel, 250 g of ice at 0°C is added to 600 g of water at 18.0°C. (a) What is the final temperature of the system? (b) How much ice remains when the system reaches equilibrium?

20. **Q|C** An automobile has a mass of 1 500 kg, and its aluminum brakes have an overall mass of 6.00 kg. (a) Assume all the mechanical energy that transforms into internal energy when the car stops is deposited in the brakes and no energy is transferred out of the brakes by heat. The brakes are originally at 20.0°C. How many times can the car be stopped from 25.0 m/s before the brakes start to melt? (b) Identify some effects ignored in part (a) that are important in a more realistic assessment of the warming of the brakes.

Section 20.4 Work and Heat in Thermodynamic Processes

21. An ideal gas is enclosed in a cylinder with a movable piston on top of it. The piston has a mass of 8 000 g and an area of 5.00 cm^2 and is free to slide up and down, keeping the pressure of the gas constant. How much work is done on the gas as the temperature of 0.200 mol of the gas is raised from 20.0°C to 300°C?

22. **S** An ideal gas is enclosed in a cylinder that has a movable piston on top. The piston has a mass m and an area A and is free to slide up and down, keeping the pressure of the gas constant. How much work is done on the gas as the temperature of n mol of the gas is raised from T_1 to T_2?

23. An ideal gas is taken through a quasi-static process described by $P = \alpha V^2$, with $\alpha = 5.00$ atm/m^6, as shown in Figure P20.23. The gas is expanded to twice its original volume of 1.00 m^3. How much work is done on the expanding gas in this process?

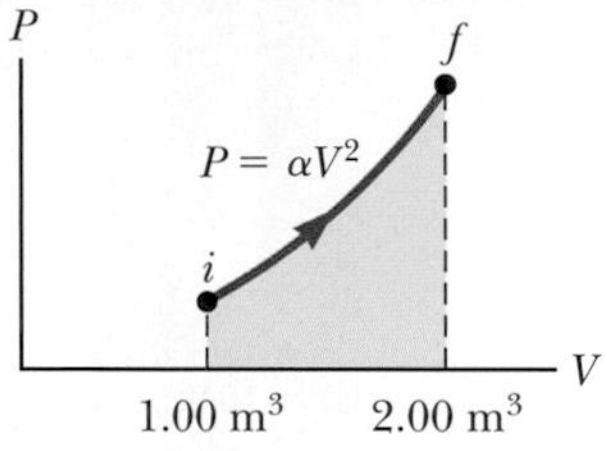

Figure P20.23

24. (a) Determine the work done on a gas that expands from i to f as indicated in Figure P20.24. (b) **What If?** How much work is done on the gas if it is compressed from f to i along the same path?

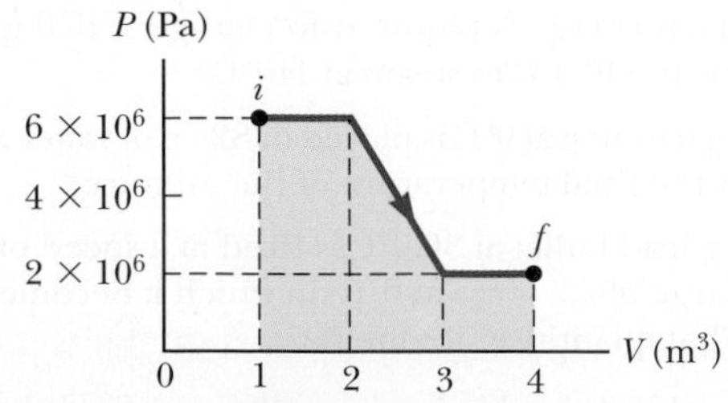

Figure P20.24

25. **Q|C** **S** One mole of an ideal gas is warmed slowly so that it goes from the PV state (P_i, V_i) to ($3P_i$, $3V_i$) in such a way that the pressure of the gas is directly proportional to the volume. (a) How much work is done on the gas in the process? (b) How is the temperature of the gas related to its volume during this process?

Section 20.5 The First Law of Thermodynamics

26. A gas is taken through the cyclic process described in Figure P20.26. (a) Find the net energy transferred to the system by heat during one complete cycle. (b) **What If?** If the cycle is reversed—that is, the process follows the path $ACBA$—what is the net energy input per cycle by heat?

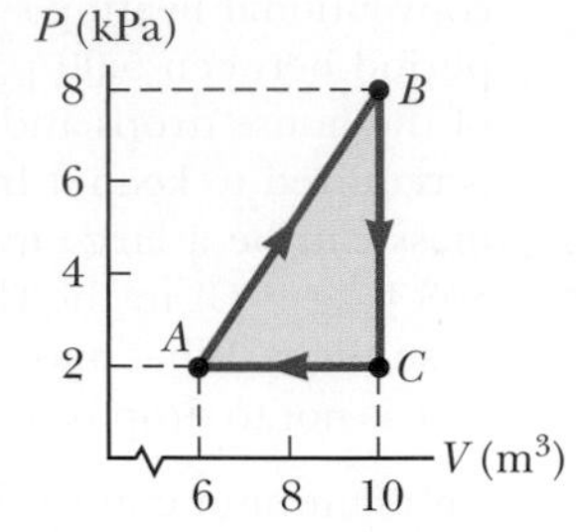

Figure P20.26
Problems 26 and 27.

27. Consider the cyclic process depicted in Figure P20.26. If Q is negative for the process BC and ΔE_{int} is negative for the process CA, what are the signs of Q, W, and ΔE_{int} that are associated with each of the three processes?

28. *Why is the following situation impossible?* An ideal gas undergoes a process with the following parameters: $Q = 10.0$ J, $W = 12.0$ J, and $\Delta T = -2.00$°C.

29. A thermodynamic system undergoes a process in which its internal energy decreases by 500 J. Over the same time interval, 220 J of work is done on the system. Find the energy transferred from it by heat.

30. A sample of an ideal gas goes through the process shown in Figure P20.30. From A to B, the process is adiabatic; from B to C, it is isobaric with 100 kJ of energy entering the system by heat; from C to D, the process is isothermal; and from D to A, it is isobaric with 150 kJ of energy leaving the system by heat. Determine the difference in internal energy $E_{int,B} - E_{int,A}$.

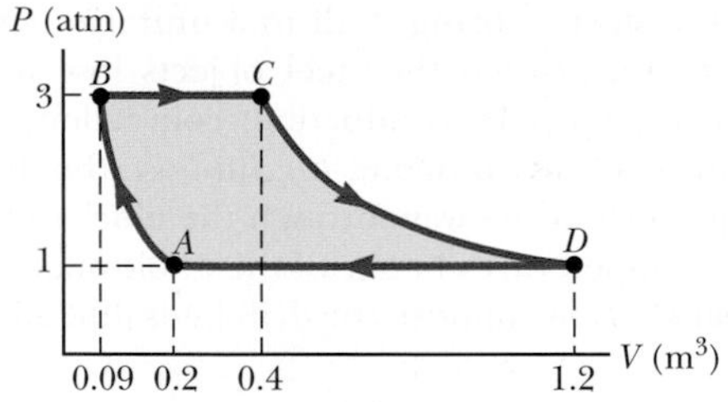

Figure P20.30

Section 20.6 Some Applications of the First Law of Thermodynamics

31. An ideal gas initially at 300 K undergoes an isobaric expansion at 2.50 kPa. If the volume increases from 1.00 m^3 to 3.00 m^3 and 12.5 kJ is transferred to the gas by heat, what are (a) the change in its internal energy and (b) its final temperature?

32. (a) How much work is done on the steam when 1.00 mol of water at 100°C boils and becomes 1.00 mol of steam at 100°C at 1.00 atm pressure? Assume the steam to behave as an ideal gas. (b) Determine the change in internal

energy of the system of the water and steam as the water vaporizes.

33. M A 2.00-mol sample of helium gas initially at 300 K, and 0.400 atm is compressed isothermally to 1.20 atm. Noting that the helium behaves as an ideal gas, find (a) the final volume of the gas, (b) the work done on the gas, and (c) the energy transferred by heat.

34. One mole of an ideal gas does 3 000 J of work on its surroundings as it expands isothermally to a final pressure of 1.00 atm and volume of 25.0 L. Determine (a) the initial volume and (b) the temperature of the gas.

35. An ideal gas initially at P_i, V_i, and T_i is taken through a cycle as shown in Figure P20.35. (a) Find the net work done on the gas per cycle for 1.00 mol of gas initially at 0°C. (b) What is the net energy added by heat to the gas per cycle?

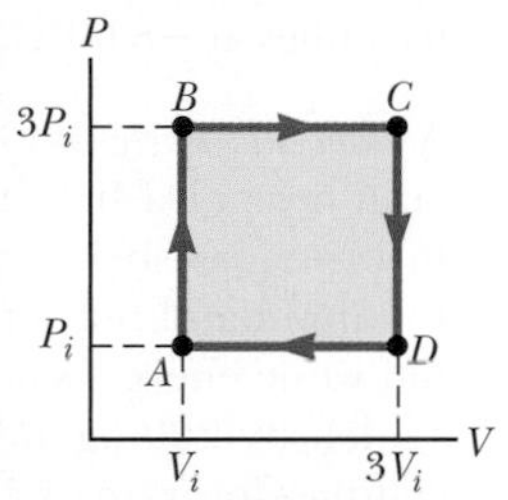

Figure P20.35
Problems 35 and 36.

36. S An ideal gas initially at P_i, V_i, and T_i is taken through a cycle as shown in Figure P20.35. (a) Find the net work done on the gas per cycle. (b) What is the net energy added by heat to the system per cycle?

37. A 1.00-kg block of aluminum is warmed at atmospheric pressure so that its temperature increases from 22.0°C to 40.0°C. Find (a) the work done on the aluminum, (b) the energy added to it by heat, and (c) the change in its internal energy.

38. In Figure P20.38, the change in internal energy of a gas that is taken from *A* to *C* along the blue path is +800 J. The work done on the gas along the red path *ABC* is −500 J. (a) How much energy must be added to the system by heat as it goes from *A* through *B* to *C*? (b) If the pressure at point *A* is five times that of point *C*, what is the work done on the system in going from *C* to *D*? (c) What is the energy exchanged with the surroundings by heat as the gas goes from *C* to *A* along the green path? (d) If the change in internal energy in going from point *D* to point *A* is +500 J, how much energy must be added to the system by heat as it goes from point *C* to point *D*?

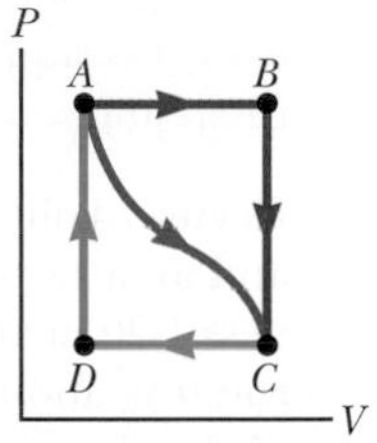

Figure P20.38

Section 20.7 Energy Transfer Mechanisms in Thermal Processes

39. A glass windowpane in a home is 0.620 cm thick and has dimensions of 1.00 m × 2.00 m. On a certain day, the temperature of the interior surface of the glass is 25.0°C and the exterior surface temperature is 0°C. (a) What is the rate at which energy is transferred by heat through the glass? (b) How much energy is transferred through the window in one day, assuming the temperatures on the surfaces remain constant?

40. A concrete slab is 12.0 cm thick and has an area of 5.00 m^2. Electric heating coils are installed under the slab to melt the ice on the surface in the winter months. What minimum power must be supplied to the coils to maintain a temperature difference of 20.0°C between the bottom of the slab and its surface? Assume all the energy transferred is through the slab.

41. A student is trying to decide what to wear. His bedroom is at 20.0°C. His skin temperature is 35.0°C. The area of his exposed skin is 1.50 m^2. People all over the world have skin that is dark in the infrared, with emissivity about 0.900. Find the net energy transfer from his body by radiation in 10.0 min.

42. The surface of the Sun has a temperature of about 5 800 K. The radius of the Sun is 6.96×10^8 m. Calculate the total energy radiated by the Sun each second. Assume the emissivity of the Sun is 0.986.

43. The tungsten filament of a certain 100-W lightbulb radiates 2.00 W of light. (The other 98 W is carried away by convection and conduction.) The filament has a surface area of 0.250 mm^2 and an emissivity of 0.950. Find the filament's temperature. (The melting point of tungsten is 3 683 K.)

44. At high noon, the Sun delivers 1 000 W to each square meter of a blacktop road. If the hot asphalt transfers energy only by radiation, what is its steady-state temperature?

45. Two lightbulbs have cylindrical filaments much greater in length than in diameter. The evacuated bulbs are identical except that one operates at a filament temperature of 2 100°C and the other operates at 2 000°C. (a) Find the ratio of the power emitted by the hotter lightbulb to that emitted by the cooler lightbulb. (b) With the bulbs operating at the same respective temperatures, the cooler lightbulb is to be altered by making its filament thicker so that it emits the same power as the hotter one. By what factor should the radius of this filament be increased?

46. Q|C At our distance from the Sun, the intensity of solar radiation is 1 370 W/m^2. The temperature of the Earth is affected by the *greenhouse effect* of the atmosphere. This phenomenon describes the effect of absorption of infrared light emitted by the surface so as to make the surface temperature of the Earth higher than if it were airless. For comparison, consider a spherical object of radius r with no atmosphere at the same distance from the Sun as the Earth. Assume its emissivity is the same for all kinds of electromagnetic waves and its temperature is uniform over its surface. (a) Explain why the projected area over which it absorbs sunlight is πr^2 and the surface area over which it radiates is $4\pi r^2$. (b) Compute its steady-state temperature. Is it chilly?

47. (a) Calculate the *R*-value of a thermal window made of two single panes of glass each 0.125 in. thick and separated by a 0.250-in. air space. (b) By what factor is the transfer of energy by heat through the window reduced by using the thermal window instead of the single-pane window? Include the contributions of inside and outside stagnant air layers.

48. Q|C For bacteriological testing of water supplies and in medical clinics, samples must routinely be incubated for 24 h at 37°C. Peace Corps volunteer and MIT engineer Amy Smith invented a low-cost, low-maintenance incubator. The incubator consists of a foam-insulated box containing a waxy material that melts at 37.0°C interspersed among tubes, dishes, or bottles containing the test samples and growth medium (bacteria food). Outside the box, the waxy material is first melted by a stove or solar energy collector. Then the waxy material is put into the box to keep the test samples warm as the material solidifies. The heat of fusion of the phase-change material is 205 kJ/kg. Model the insulation as a panel with surface area 0.490 m^2, thickness 4.50 cm, and conductivity 0.012 0 W/m · °C. Assume the exterior temperature is 23.0°C for 12.0 h and 16.0°C for 12.0 h. (a) What mass of the waxy material is required to conduct the bacteriological test? (b) Explain why your calculation can be done without knowing the mass of the test samples or of the insulation.

49. A bar of gold (Au) is in thermal contact with a bar of silver (Ag) of the same length and area (Fig. P20.49). One end of the compound bar is maintained at 80.0°C, and the opposite end is at 30.0°C. When the energy transfer reaches steady state, what is the temperature at the junction?

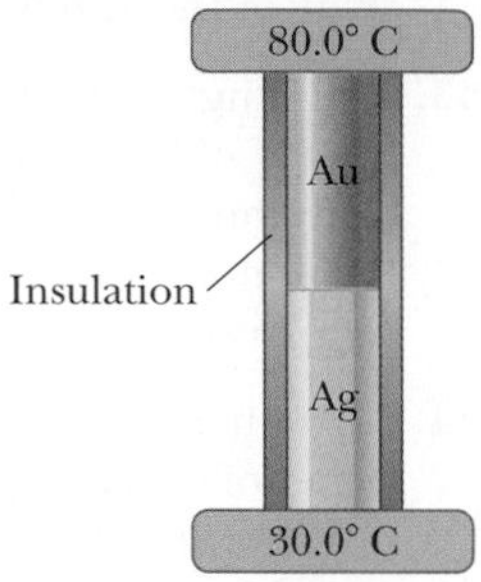

Figure P20.49

50. A large, hot pizza floats in outer space after being jettisoned as refuse from a spacecraft. What is the order of magnitude (a) of its rate of energy loss and (b) of its rate of temperature change? List the quantities you estimate and the value you estimate for each.

Additional Problems

51. Liquid nitrogen has a boiling point of 77.3 K and a latent heat of vaporization of 2.01×10^5 J/kg. A 25.0-W electric heating element is immersed in an insulated vessel containing 25.0 L of liquid nitrogen at its boiling point. How many kilograms of nitrogen are boiled away in a period of 4.00 h?

52. GP **Review.** Two speeding lead bullets, one of mass 12.0 g moving to the right at 300 m/s and one of mass 8.00 g moving to the left at 400 m/s, collide head-on, and all the material sticks together. Both bullets are originally at temperature 30.0°C. Assume the change in kinetic energy of the system appears entirely as increased internal energy. We would like to determine the temperature and phase of the bullets after the collision. (a) What two analysis models are appropriate for the system of two bullets for the time interval from before to after the collision? (b) From one of these models, what is the speed of the combined bullets after the collision? (c) How much of the initial kinetic energy has transformed to internal energy in the system after the collision? (d) Does all the lead melt due to the collision? (e) What is the temperature of the combined bullets after the collision? (f) What is the phase of the combined bullets after the collision?

53. M An aluminum rod 0.500 m in length and with a cross-sectional area of 2.50 cm^2 is inserted into a thermally insulated vessel containing liquid helium at 4.20 K. The rod is initially at 300 K. (a) If one-half of the rod is inserted into the helium, how many liters of helium boil off by the time the inserted half cools to 4.20 K? Assume the upper half does not yet cool. (b) If the circular surface of the upper end of the rod is maintained at 300 K, what is the approximate boil-off rate of liquid helium in liters per second after the lower half has reached 4.20 K? (Aluminum has thermal conductivity of 3 100 W/m · K at 4.20 K; ignore its temperature variation. The density of liquid helium is 125 kg/m^3.)

54. Q|C An ice-cube tray is filled with 75.0 g of water. After the filled tray reaches an equilibrium temperature of 20.0°C, it is placed in a freezer set at −8.00°C to make ice cubes. (a) Describe the processes that occur as energy is being removed from the water to make ice. (b) Calculate the energy that must be removed from the water to make ice cubes at −8.00°C.

55. A *flow calorimeter* is an apparatus used to measure the specific heat of a liquid. The technique of flow calorimetry involves measuring the temperature difference between the input and output points of a flowing stream of the liquid while energy is added by heat at a known rate. A liquid of density 900 kg/m^3 flows through the calorimeter with volume flow rate of 2.00 L/min. At steady state, a temperature difference 3.50°C is established between the input and output points when energy is supplied at the rate of 200 W. What is the specific heat of the liquid?

56. S A *flow calorimeter* is an apparatus used to measure the specific heat of a liquid. The technique of flow calorimetry involves measuring the temperature difference between the input and output points of a flowing stream of the liquid while energy is added by heat at a known rate. A liquid of density ρ flows through the calorimeter with volume flow rate R. At steady state, a temperature difference ΔT is established between the input and output points when energy is supplied at the rate P. What is the specific heat of the liquid?

57. **Review.** Following a collision between a large spacecraft and an asteroid, a copper disk of radius 28.0 m and thickness 1.20 m at a temperature of 850°C is floating in space, rotating about its symmetry axis with an angular speed of 25.0 rad/s. As the disk radiates infrared light, its temperature falls to 20.0°C. No external torque acts on the disk. (a) Find the change in kinetic energy of the disk. (b) Find the change in internal energy of the disk. (c) Find the amount of energy it radiates.

58. Q|C S One mole of an ideal gas is contained in a cylinder with a movable piston. The initial pressure, volume, and temperature are P_i, V_i, and T_i, respectively. Find the work done on the gas in the following processes. In operational terms, describe how to carry out each process and show each process on a *PV* diagram. (a) an isobaric compression in which the final volume is one-half the initial volume (b) an isothermal compression in which the final pressure is four times the initial pressure (c) an isovolumetric process in which the final pressure is three times the initial pressure

59. **Review.** A 670-kg meteoroid happens to be composed of aluminum. When it is far from the Earth, its temperature is −15.0°C and it moves at 14.0 km/s relative to the planet. As it crashes into the Earth, assume the internal energy transformed from the mechanical energy of the meteoroid–

Earth system is shared equally between the meteoroid and the Earth and all the material of the meteoroid rises momentarily to the same final temperature. Find this temperature. Assume the specific heat of liquid and of gaseous aluminum is 1 170 J/kg · °C.

60. *Why is the following situation impossible?* A group of campers arises at 8:30 a.m. and uses a solar cooker, which consists of a curved, reflecting surface that concentrates sunlight onto the object to be warmed (Fig. P20.60). During the day, the maximum solar intensity reaching the Earth's surface at the cooker's location is I = 600 W/m^2. The cooker faces the Sun and has a face diameter of d = 0.600 m. Assume a fraction f of 40.0% of the incident energy is transferred to 1.50 L of water in an open container, initially at 20.0°C. The water comes to a boil, and the campers enjoy hot coffee for breakfast before hiking ten miles and returning by noon for lunch.

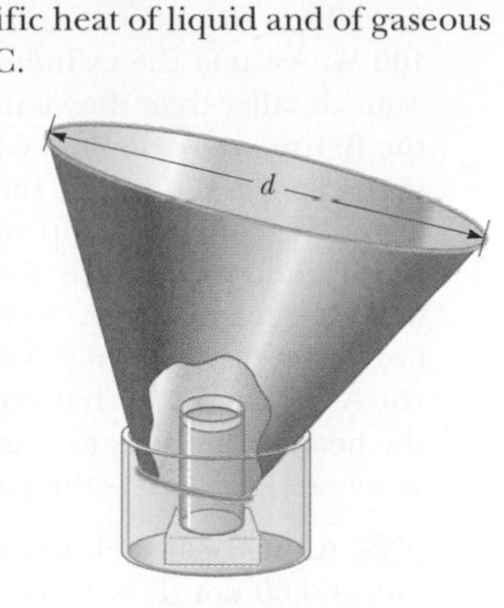

Figure P20.60

61. **M** Water in an electric teakettle is boiling. The power absorbed by the water is 1.00 kW. Assuming the pressure of vapor in the kettle equals atmospheric pressure, determine the speed of effusion of vapor from the kettle's spout if the spout has a cross-sectional area of 2.00 cm^2. Model the steam as an ideal gas.

62. (a) In air at 0°C, a 1.60-kg copper block at 0°C is set sliding at 2.50 m/s over a sheet of ice at 0°C. Friction brings the block to rest. Find the mass of the ice that melts. (b) As the block slows down, identify its energy input Q, its change in internal energy ΔE_{int}, and the change in mechanical energy for the block–ice system. (c) For the ice as a system, identify its energy input Q and its change in internal energy ΔE_{int}. (d) A 1.60-kg block of ice at 0°C is set sliding at 2.50 m/s over a sheet of copper at 0°C. Friction brings the block to rest. Find the mass of the ice that melts. (e) Evaluate Q and ΔE_{int} for the block of ice as a system and ΔE_{mech} for the block–ice system. (f) Evaluate Q and ΔE_{int} for the metal sheet as a system. (g) A thin, 1.60-kg slab of copper at 20°C is set sliding at 2.50 m/s over an identical stationary slab at the same temperature. Friction quickly stops the motion. Assuming no energy is transferred to the environment by heat, find the change in temperature of both objects. (h) Evaluate Q and ΔE_{int} for the sliding slab and ΔE_{mech} for the two-slab system. (i) Evaluate Q and ΔE_{int} for the stationary slab.

63. A cooking vessel on a slow burner contains 10.0 kg of water and an unknown mass of ice in equilibrium at 0°C at time t = 0. The temperature of the mixture is measured at various times, and the result is plotted in Figure P20.63. During the first 50.0 min, the mixture remains at 0°C. From 50.0 min to 60.0 min, the temperature increases to 2.00°C. Ignoring the heat capacity of the vessel, determine the initial mass of the ice.

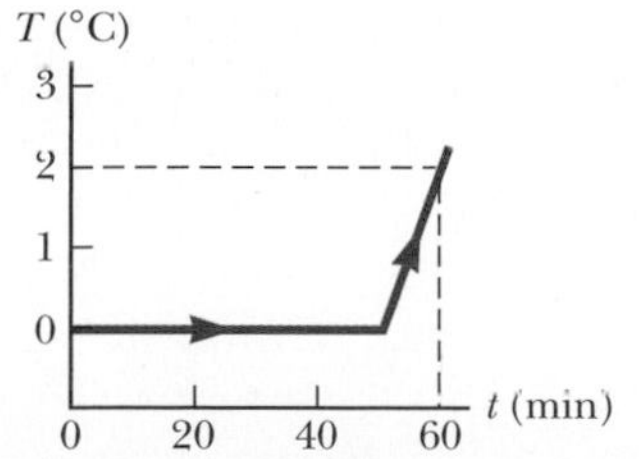

Figure P20.63

64. The average thermal conductivity of the walls (including the windows) and roof of the house depicted in Figure P20.64 is 0.480 W/m · °C, and their average thickness is 21.0 cm. The house is kept warm with natural gas having a heat of combustion (that is, the energy provided per cubic meter of gas burned) of 9 300 kcal/m^3. How many cubic meters of gas must be burned each day to maintain an inside temperature of 25.0°C if the outside temperature is 0.0°C? Disregard radiation and the energy transferred by heat through the ground.

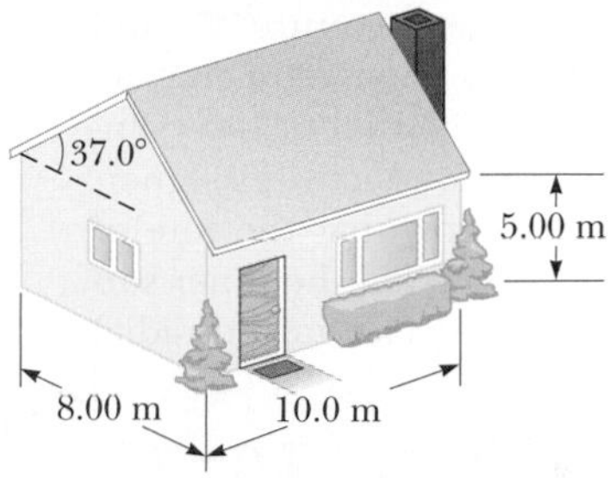

Figure P20.64

65. **Q|C** During periods of high activity, the Sun has more sunspots than usual. Sunspots are cooler than the rest of the luminous layer of the Sun's atmosphere (the photosphere). Paradoxically, the total power output of the active Sun is not lower than average but is the same or slightly higher than average. Work out the details of the following crude model of this phenomenon. Consider a patch of the photosphere with an area of 5.10 × 10^{14} m^2. Its emissivity is 0.965. (a) Find the power it radiates if its temperature is uniformly 5 800 K, corresponding to the quiet Sun. (b) To represent a sunspot, assume 10.0% of the patch area is at 4 800 K and the other 90.0% is at 5 890 K. Find the power output of the patch. (c) State how the answer to part (b) compares with the answer to part (a). (d) Find the average temperature of the patch. Note that this cooler temperature results in a higher power output. (The next sunspot maximum is expected around the year 2012.)

66. **Q|C** A student measures the following data in a calorimetry experiment designed to determine the specific heat of aluminum:

Initial temperature of water and calorimeter:	70.0°C
Mass of water:	0.400 kg
Mass of calorimeter:	0.040 kg
Specific heat of calorimeter:	0.63 kJ/kg · °C
Initial temperature of aluminum:	27.0°C
Mass of aluminum:	0.200 kg
Final temperature of mixture:	66.3°C

(a) Use these data to determine the specific heat of aluminum. (b) Explain whether your result is within 15% of the value listed in Table 20.1.

Challenge Problems

67. A pond of water at 0°C is covered with a layer of ice 4.00 cm thick. If the air temperature stays constant at −10.0°C, what time interval is required for the ice thickness to increase to 8.00 cm? *Suggestion:* Use Equation 20.16 in the form

$$\frac{dQ}{dt} = kA\frac{\Delta T}{x}$$

and note that the incremental energy dQ extracted from the water through the thickness x of ice is the amount required to freeze a thickness dx of ice. That is, $dQ = L_f \rho A\, dx$, where ρ is the density of the ice, A is the area, and L_f is the latent heat of fusion.

68. (a) The inside of a hollow cylinder is maintained at a temperature T_a, and the outside is at a lower temperature, T_b (Fig. P20.68). The wall of the cylinder has a thermal conductivity k. Ignoring end effects, show that the rate of energy conduction from the inner surface to the outer surface in the radial direction is

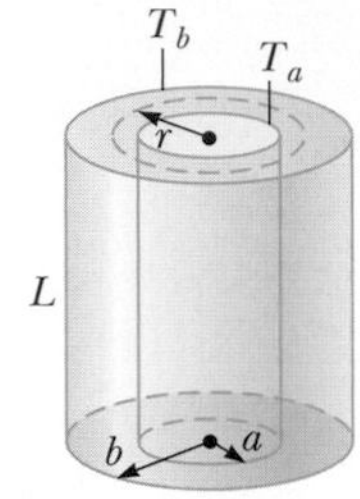

Figure P20.68

$$\frac{dQ}{dt} = 2\pi Lk\left[\frac{T_a - T_b}{\ln(b/a)}\right]$$

Suggestions: The temperature gradient is dT/dr. A radial energy current passes through a concentric cylinder of area $2\pi rL$. (b) The passenger section of a jet airliner is in the shape of a cylindrical tube with a length of 35.0 m and an inner radius of 2.50 m. Its walls are lined with an insulating material 6.00 cm in thickness and having a thermal conductivity of 4.00×10^{-5} cal/s · cm · °C. A heater must maintain the interior temperature at 25.0°C while the outside temperature is −35.0°C. What power must be supplied to the heater?

69. Consider the piston–cylinder apparatus shown in Figure P20.69. The bottom of the cylinder contains 2.00 kg of water at just under 100.0°C. The cylinder has a radius of $r = 7.50$ cm. The piston of mass $m = 3.00$ kg sits on the surface of the water. An electric heater in the cylinder base transfers energy into the water at a rate of 100 W. Assume the cylinder is much taller than shown in the figure, so we don't need to be concerned about the piston reaching the top of the cylinder. (a) Once the water begins boiling, how fast is the piston rising? Model the steam as an ideal gas. (b) After the water has completely turned to steam and the heater continues to transfer energy to the steam at the same rate, how fast is the piston rising?

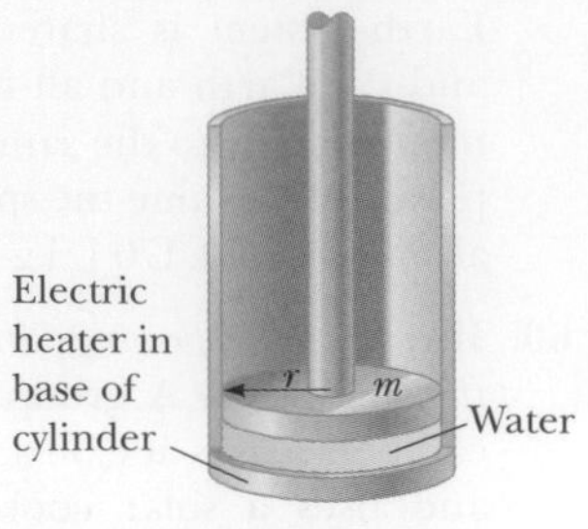

Figure P20.69

70. QIC A spherical shell has inner radius 3.00 cm and outer radius 7.00 cm. It is made of material with thermal conductivity $k = 0.800$ W/m · °C. The interior is maintained at temperature 5°C and the exterior at 40°C. After an interval of time, the shell reaches a steady state with the temperature at each point within it remaining constant in time. (a) Explain why the rate of energy transfer P must be the same through each spherical surface, of radius r, within the shell and must satisfy

$$\frac{dT}{dr} = \frac{P}{4\pi kr^2}$$

(b) Next, prove that

$$\int_5^{40} dT = \frac{P}{4\pi k}\int_{0.03}^{0.07} r^{-2}\, dr$$

where T is in degrees Celsius and r is in meters. (c) Find the rate of energy transfer through the shell. (d) Prove that

$$\int_5^{T} dT = 1.84\int_{0.03}^{r} r^{-2}\, dr$$

where T is in degrees Celsius and r is in meters. (e) Find the temperature within the shell as a function of radius. (f) Find the temperature at $r = 5.00$ cm, halfway through the shell.

chapter

21

The Kinetic Theory of Gases

A boy inflates his bicycle tire with a hand-operated pump. Kinetic theory helps to describe the details of the air in the pump. (*© Cengage Learning/George Semple*)

In Chapter 19, we discussed the properties of an ideal gas by using such macroscopic variables as pressure, volume, and temperature. Such large-scale properties can be related to a description on a microscopic scale, where matter is treated as a collection of molecules. Applying Newton's laws of motion in a statistical manner to a collection of particles provides a reasonable description of thermodynamic processes. To keep the mathematics relatively simple, we shall consider primarily the behavior of gases because in gases the interactions between molecules are much weaker than they are in liquids or solids.

We shall begin by relating pressure and temperature directly to the details of molecular motion in a sample of gas. Based on these results, we will make predictions of molar specific heats of gases. Some of these predictions will be correct and some will not. We will extend our model to explain those values that are not predicted correctly by the simpler model. Finally, we discuss the distribution of molecular speeds in a gas.

21.1 Molecular Model of an Ideal Gas

We begin this chapter by developing a microscopic model of an ideal gas, called **kinetic theory.** In developing this model, we make the following assumptions:

Assumptions of the microscopic model of an ideal gas ▶

1. **The number of molecules in the gas is large, and the average separation between them is large compared with their dimensions.** In other words, the molecules occupy a negligible volume in the container. That is consistent with the ideal gas model, in which we model the molecules as particles.
2. **The molecules obey Newton's laws of motion, but as a whole they move randomly.** By "randomly," we mean that any molecule can move in any direction with any speed.
3. **The molecules interact only by short-range forces during elastic collisions.** That is consistent with the ideal gas model, in which the molecules exert no long-range forces on each other.
4. **The molecules make elastic collisions with the walls.** These collisions lead to the macroscopic pressure on the walls of the container.
5. **The gas under consideration is a pure substance; that is, all molecules are identical.**

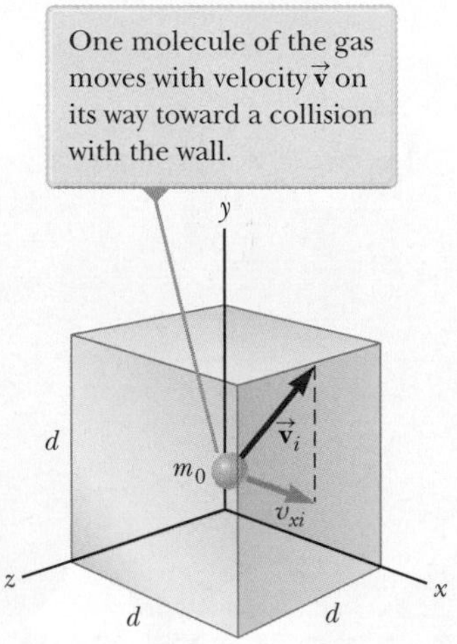

Figure 21.1 A cubical box with sides of length d containing an ideal gas.

Although we often picture an ideal gas as consisting of single atoms, the behavior of molecular gases approximates that of ideal gases rather well at low pressures. Usually, molecular rotations or vibrations have no effect on the motions considered here.

For our first application of kinetic theory, let us relate the macroscope variable of pressure P to microscopic quantities. Consider a collection of N molecules of an ideal gas in a container of volume V. The container is a cube with edges of length d (Fig. 21.1). We shall first focus our attention on one of these molecules of mass m_0 and assume it is moving so that its component of velocity in the x direction is v_{xi} as in Active Figure 21.2. (The subscript i here refers to the ith molecule in the collection, not to an initial value. We will combine the effects of all the molecules shortly.) As the molecule collides elastically with any wall (assumption 4), its velocity component perpendicular to the wall is reversed because the mass of the wall is far greater than the mass of the molecule. The molecule is modeled as a nonisolated system for which the impulse from the wall causes a change in the molecule's momentum. Because the momentum component p_{xi} of the molecule is $m_0 v_{xi}$ before the collision and $-m_0 v_{xi}$ after the collision, the change in the x component of the momentum of the molecule is

$$\Delta p_{xi} = -m_0 v_{xi} - (m_0 v_{xi}) = -2m_0 v_{xi}$$

Because the molecules obey Newton's laws (assumption 2), we can apply the impulse-momentum theorem (Eq. 9.10) to the molecule to give

$$\overline{F}_{i,\text{on molecule}}\, \Delta t_{\text{collision}} = \Delta p_{xi} = -2m_0 v_{xi}$$

where $\overline{F}_{i,\text{on molecule}}$ is the x component of the average force[1] the wall exerts on the molecule during the collision and $\Delta t_{\text{collision}}$ is the duration of the collision. For the molecule to make another collision with the same wall after this first collision, it must travel a distance of $2d$ in the x direction (across the container and back). Therefore, the time interval between two collisions with the same wall is

$$\Delta t = \frac{2d}{v_{xi}}$$

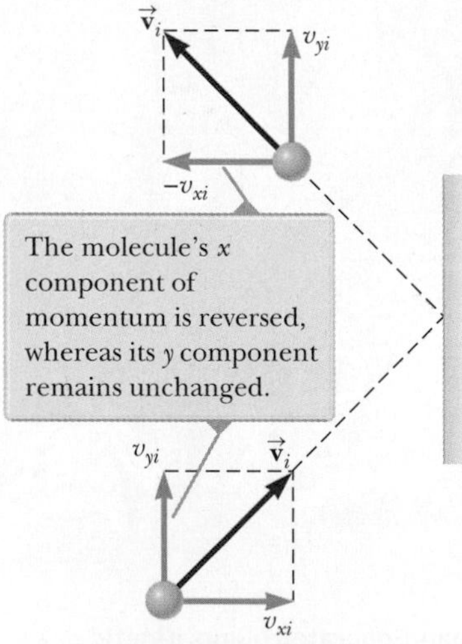

ACTIVE FIGURE 21.2

A molecule makes an elastic collision with the wall of the container. In this construction, we assume the molecule moves in the xy plane.

The force that causes the change in momentum of the molecule in the collision with the wall occurs only during the collision. We can, however, average the force

[1]For this discussion, we use a bar over a variable to represent the average value of the variable, such as $\overline{F}$ for the average force, rather than the subscript "avg" that we have used before. This notation is to save confusion because we already have a number of subscripts on variables.

over the time interval for the molecule to move across the cube and back. Sometime during this time interval the collision occurs, so the change in momentum for this time interval is the same as that for the short duration of the collision. Therefore, we can rewrite the impulse-momentum theorem as

$$\overline{F}_i \, \Delta t = -2m_0 v_{xi}$$

where $\overline{F}_i$ is the average force component over the time interval for the molecule to move across the cube and back. Because exactly one collision occurs for each such time interval, this result is also the long-term average force on the molecule over long time intervals containing any number of multiples of Δt.

This equation and the preceding one enable us to express the x component of the long-term average force exerted by the wall on the molecule as

$$\overline{F}_i = -\frac{2m_0 v_{xi}}{\Delta t} = -\frac{2m_0 v_{xi}^{\,2}}{2d} = -\frac{m_0 v_{xi}^{\,2}}{d}$$

Now, by Newton's third law, the x component of the long-term average force exerted by the *molecule* on the *wall* is equal in magnitude and opposite in direction:

$$\overline{F}_{i,\text{on wall}} = -\overline{F}_i = -\left(-\frac{m_0 v_{xi}^{\,2}}{d}\right) = \frac{m_0 v_{xi}^{\,2}}{d}$$

The total average force $\overline{F}$ exerted by the gas on the wall is found by adding the average forces exerted by the individual molecules. Adding terms such as that above for all molecules gives

$$\overline{F} = \sum_{i=1}^{N} \frac{m_0 v_{xi}^{\,2}}{d} = \frac{m_0}{d} \sum_{i=1}^{N} v_{xi}^{\,2}$$

where we have factored out the length of the box and the mass m_0 because assumption 5 tells us that all the molecules are the same. We now impose assumption 1, that the number of molecules is large. For a small number of molecules, the actual force on the wall would vary with time. It would be nonzero during the short interval of a collision of a molecule with the wall and zero when no molecule happens to be hitting the wall. For a very large number of molecules such as Avogadro's number, however, these variations in force are smoothed out so that the average force given above is the same over *any* time interval. Therefore, the *constant* force F on the wall due to the molecular collisions is

$$F = \frac{m_0}{d} \sum_{i=1}^{N} v_{xi}^{\,2}$$

To proceed further, let's consider how to express the average value of the square of the x component of the velocity for N molecules. The traditional average of a set of values is the sum of the values over the number of values:

$$\overline{v_x^{\,2}} = \frac{\sum_{i=1}^{N} v_{xi}^{\,2}}{N}$$

The numerator of this expression is contained in the right side of the preceding equation. Therefore, by combining the two expressions the total force on the wall can be written

$$F = \frac{m_0}{d} N \overline{v_x^{\,2}} \qquad \textbf{(21.1)}$$

Now let's focus again on one molecule with velocity components v_{xi}, v_{yi}, and v_{zi}. The Pythagorean theorem relates the square of the speed of the molecule to the squares of the velocity components:

$$v_i^2 = v_{xi}^{\,2} + v_{yi}^{\,2} + v_{zi}^{\,2}$$

Hence, the average value of v^2 for all the molecules in the container is related to the average values of v_x^2, v_y^2, and v_z^2 according to the expression

$$\overline{v^2} = \overline{v_x^2} + \overline{v_y^2} + \overline{v_z^2}$$

Because the motion is completely random (assumption 2), the average values $\overline{v_x^2}$, $\overline{v_y^2}$, and $\overline{v_z^2}$ are equal to one another. Using this fact and the preceding equation, we find that

$$\overline{v^2} = 3\overline{v_x^2}$$

Therefore, from Equation 21.1, the total force exerted on the wall is

$$F = \tfrac{1}{3}N\frac{m_0\overline{v^2}}{d}$$

Using this expression, we can find the total pressure exerted on the wall:

$$P = \frac{F}{A} = \frac{F}{d^2} = \tfrac{1}{3}N\frac{m_0\overline{v^2}}{d^3} = \tfrac{1}{3}\left(\frac{N}{V}\right)m_0\overline{v^2}$$

Relationship between pressure and molecular kinetic energy ▶

$$P = \tfrac{2}{3}\left(\frac{N}{V}\right)\left(\tfrac{1}{2}m_0\overline{v^2}\right) \quad \textbf{(21.2)}$$

This result indicates that the pressure of a gas is proportional to (1) the number of molecules per unit volume and (2) the average translational kinetic energy of the molecules, $\frac{1}{2}m_0\overline{v^2}$. In analyzing this simplified model of an ideal gas, we obtain an important result that relates the macroscopic quantity of pressure to a microscopic quantity, the average value of the square of the molecular speed. Therefore, a key link between the molecular world and the large-scale world has been established.

Notice that Equation 21.2 verifies some features of pressure with which you are probably familiar. One way to increase the pressure inside a container is to increase the number of molecules per unit volume N/V in the container. That is what you do when you add air to a tire. The pressure in the tire can also be raised by increasing the average translational kinetic energy of the air molecules in the tire. That can be accomplished by increasing the temperature of that air, which is why the pressure inside a tire increases as the tire warms up during long road trips. The continuous flexing of the tire as it moves along the road surface results in work done on the rubber as parts of the tire distort, causing an increase in internal energy of the rubber. The increased temperature of the rubber results in the transfer of energy by heat into the air inside the tire. This transfer increases the air's temperature, and this increase in temperature in turn produces an increase in pressure.

Molecular Interpretation of Temperature

Let's now consider another macroscopic variable, the temperature T of the gas. We can gain some insight into the meaning of temperature by first writing Equation 21.2 in the form

$$PV = \tfrac{2}{3}N\left(\tfrac{1}{2}m_0\overline{v^2}\right)$$

Let's now compare this expression with the equation of state for an ideal gas (Eq. 19.10):

$$PV = Nk_BT$$

Recall that the equation of state is based on experimental facts concerning the macroscopic behavior of gases. Equating the right sides of these expressions gives

Relationship between temperature and molecular kinetic energy ▶

$$T = \frac{2}{3k_B}\left(\tfrac{1}{2}m_0\overline{v^2}\right) \quad \textbf{(21.3)}$$

This result tells us that temperature is a direct measure of average molecular kinetic energy. By rearranging Equation 21.3, we can relate the translational molecular kinetic energy to the temperature:

$$\tfrac{1}{2}m_0\overline{v^2} = \tfrac{3}{2}k_BT \qquad (21.4)$$

◀ **Average kinetic energy per molecule**

That is, the average translational kinetic energy per molecule is $\frac{3}{2}k_BT$. Because $\overline{v_x^2} = \frac{1}{3}\overline{v^2}$, it follows that

$$\tfrac{1}{2}m_0\overline{v_x^2} = \tfrac{1}{2}k_BT \qquad (21.5)$$

In a similar manner, for the *y* and *z* directions,

$$\tfrac{1}{2}m_0\overline{v_y^2} = \tfrac{1}{2}k_BT \text{ and } \tfrac{1}{2}m_0\overline{v_z^2} = \tfrac{1}{2}k_BT$$

Therefore, each translational degree of freedom contributes an equal amount of energy, $\frac{1}{2}k_BT$, to the gas. (In general, a "degree of freedom" refers to an independent means by which a molecule can possess energy.) A generalization of this result, known as the **theorem of equipartition of energy,** is as follows:

> Each degree of freedom contributes $\frac{1}{2}k_BT$ to the energy of a system, where possible degrees of freedom are those associated with translation, rotation, and vibration of molecules.

◀ **Theorem of equipartition of energy**

The total translational kinetic energy of *N* molecules of gas is simply *N* times the average energy per molecule, which is given by Equation 21.4:

$$K_{\text{tot trans}} = N\left(\tfrac{1}{2}m_0\overline{v^2}\right) = \tfrac{3}{2}Nk_BT = \tfrac{3}{2}nRT \qquad (21.6)$$

◀ **Total translational kinetic energy of *N* molecules**

where we have used $k_B = R/N_A$ for Boltzmann's constant and $n = N/N_A$ for the number of moles of gas. If the gas molecules possess only translational kinetic energy, Equation 21.6 represents the internal energy of the gas. This result implies that the internal energy of an ideal gas depends *only* on the temperature. We will follow up on this point in Section 21.2.

The square root of $\overline{v^2}$ is called the **root-mean-square (rms) speed** of the molecules. From Equation 21.4, we find that the rms speed is

$$v_{\text{rms}} = \sqrt{\overline{v^2}} = \sqrt{\frac{3k_BT}{m_0}} = \sqrt{\frac{3RT}{M}} \qquad (21.7)$$

◀ **Root-mean-square speed**

where *M* is the molar mass in kilograms per mole and is equal to m_0N_A. This expression shows that, at a given temperature, lighter molecules move faster, on the average, than do heavier molecules. For example, at a given temperature, hydrogen molecules, whose molar mass is 2.02×10^{-3} kg/mol, have an average speed approximately four times that of oxygen molecules, whose molar mass is 32.0×10^{-3} kg/mol. Table 21.1 lists the rms speeds for various molecules at 20°C.

Pitfall Prevention 21.1

The Square Root of the Square? Taking the square root of $\overline{v^2}$ does not "undo" the square because we have taken an average *between* squaring and taking the square root. Although the square root of $(\overline{v})^2$ is $\overline{v} = v_{\text{avg}}$ because the squaring is done after the averaging, the square root of $\overline{v^2}$ is *not* v_{avg}, but rather v_{rms}.

***Quick Quiz* 21.1** Two containers hold an ideal gas at the same temperature and pressure. Both containers hold the same type of gas, but container B has twice the volume of container A. **(i)** What is the average translational kinetic energy per molecule in container B? (a) twice that of container A (b) the same as that of container A (c) half that of container A (d) impossible to determine **(ii)** From the same choices, describe the internal energy of the gas in container B.

TABLE 21.1 *Some Root-Mean-Square (rms) Speeds*

Gas	Molar Mass (g/mol)	v_{rms} at 20°C (m/s)	Gas	Molar Mass (g/mol)	v_{rms} at 20°C (m/s)
H_2	2.02	1902	NO	30.0	494
He	4.00	1352	O_2	32.0	478
H_2O	18.0	637	CO_2	44.0	408
Ne	20.2	602	SO_2	64.1	338
N_2 or CO	28.0	511			

Example 21.1 A Tank of Helium

A tank used for filling helium balloons has a volume of 0.300 m^3 and contains 2.00 mol of helium gas at 20.0°C. Assume the helium behaves like an ideal gas.

(A) What is the total translational kinetic energy of the gas molecules?

SOLUTION

Conceptualize Imagine a microscopic model of a gas in which you can watch the molecules move about the container more rapidly as the temperature increases.

Categorize We evaluate parameters with equations developed in the preceding discussion, so this example is a substitution problem.

Use Equation 21.6 with $n = 2.00$ mol and $T = 293$ K:

$$K_{\text{tot trans}} = \tfrac{3}{2}nRT = \tfrac{3}{2}(2.00 \text{ mol})(8.31 \text{ J/mol} \cdot \text{K})(293 \text{ K})$$
$$= \boxed{7.30 \times 10^3 \text{ J}}$$

(B) What is the average kinetic energy per molecule?

SOLUTION

Use Equation 21.4:

$$\tfrac{1}{2}m_0 v^2 = \tfrac{3}{2}k_B T = \tfrac{3}{2}(1.38 \times 10^{-23} \text{ J/K})(293 \text{ K})$$
$$= \boxed{6.07 \times 10^{-21} \text{ J}}$$

WHAT IF? What if the temperature is raised from 20.0°C to 40.0°C? Because 40.0 is twice as large as 20.0, is the total translational energy of the molecules of the gas twice as large at the higher temperature?

Answer The expression for the total translational energy depends on the temperature, and the value for the temperature must be expressed in kelvins, not in degrees Celsius. Therefore, the ratio of 40.0 to 20.0 is *not* the appropriate ratio. Converting the Celsius temperatures to kelvins, 20.0°C is 293 K and 40.0°C is 313 K. Therefore, the total translational energy increases by a factor of only 313 K/293 K = 1.07.

21.2 Molar Specific Heat of an Ideal Gas

Consider an ideal gas undergoing several processes such that the change in temperature is $\Delta T = T_f - T_i$ for all processes. The temperature change can be achieved by taking a variety of paths from one isotherm to another as shown in Figure 21.3. Because ΔT is the same for each path, the change in internal energy ΔE_{int} is the same for all paths. The work W done on the gas (the negative of the area under the curves) is different for each path. Therefore, from the first law of thermodynamics, the heat associated with a given change in temperature does *not* have a unique value as discussed in Section 20.4.

We can address this difficulty by defining specific heats for two special processes: isovolumetric and isobaric. Because the number of moles n is a convenient measure of the amount of gas, we define the **molar specific heats** associated with these processes as follows:

$$Q = nC_V \Delta T \quad \text{(constant volume)} \tag{21.8}$$

$$Q = nC_P \Delta T \quad \text{(constant pressure)} \tag{21.9}$$

where C_V is the **molar specific heat at constant volume** and C_P is the **molar specific heat at constant pressure.** When energy is added to a gas by heat at constant pressure, not only does the internal energy of the gas increase, but (negative) work is done on the gas because of the change in volume required to keep the pressure constant. Therefore, the heat Q in Equation 21.9 must account for both the increase in internal energy and the transfer of energy out of the system by work.

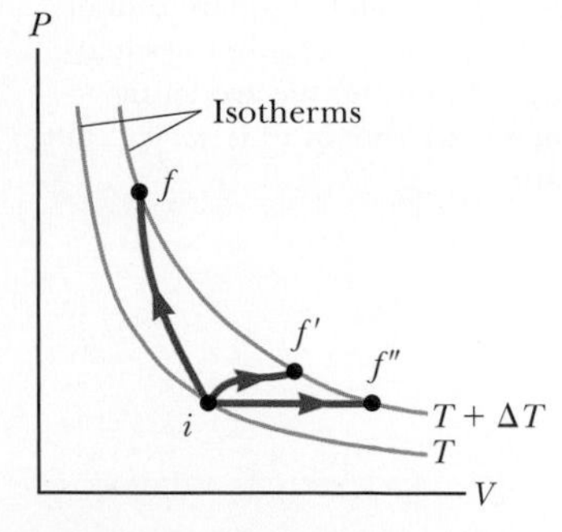

Figure 21.3 An ideal gas is taken from one isotherm at temperature T to another at temperature $T + \Delta T$ along three different paths.

For this reason, Q is greater in Equation 21.9 than in Equation 21.8 for given values of n and ΔT. Therefore, C_P is greater than C_V.

In the previous section, we found that the temperature of a gas is a measure of the average translational kinetic energy of the gas molecules. This kinetic energy is associated with the motion of the center of mass of each molecule. It does not include the energy associated with the internal motion of the molecule, namely, vibrations and rotations about the center of mass. That should not be surprising because the simple kinetic theory model assumes a structureless molecule.

So, let's first consider the simplest case of an ideal monatomic gas, that is, a gas containing one atom per molecule such as helium, neon, or argon. When energy is added to a monatomic gas in a container of fixed volume, all the added energy goes into increasing the translational kinetic energy of the atoms. There is no other way to store the energy in a monatomic gas. Therefore, from Equation 21.6, we see that the internal energy E_{int} of N molecules (or n mol) of an ideal monatomic gas is

$$E_{\text{int}} = K_{\text{tot trans}} = \tfrac{3}{2}Nk_BT = \tfrac{3}{2}nRT \qquad (21.10)$$

◀ **Internal energy of an ideal monatomic gas**

For a monatomic ideal gas, E_{int} is a function of T only and the functional relationship is given by Equation 21.10. In general, the internal energy of any ideal gas is a function of T only and the exact relationship depends on the type of gas.

If energy is transferred by heat to a system at constant volume, no work is done on the system. That is, $W = -\int P\,dV = 0$ for a constant-volume process. Hence, from the first law of thermodynamics,

$$Q = \Delta E_{\text{int}} \qquad (21.11)$$

In other words, all the energy transferred by heat goes into increasing the internal energy of the system. A constant-volume process from i to f for an ideal gas is described in Active Figure 21.4, where ΔT is the temperature difference between the two isotherms. Substituting the expression for Q given by Equation 21.8 into Equation 21.11, we obtain

$$\Delta E_{\text{int}} = nC_V\,\Delta T \qquad (21.12)$$

This equation applies to all ideal gases, those gases having more than one atom per molecule as well as monatomic ideal gases. In the limit of infinitesimal changes, we can use Equation 21.12 to express the molar specific heat at constant volume as

$$C_V = \frac{1}{n}\frac{dE_{\text{int}}}{dT} \qquad (21.13)$$

Let's now apply the results of this discussion to a monatomic gas. Substituting the internal energy from Equation 21.10 into Equation 21.13 gives

$$C_V = \tfrac{3}{2}R \qquad (21.14)$$

This expression predicts a value of $C_V = \frac{3}{2}R = 12.5\ \text{J/mol}\cdot\text{K}$ for *all* monatomic gases. This prediction is in excellent agreement with measured values of molar specific heats for such gases as helium, neon, argon, and xenon over a wide range of temperatures (Table 21.2 on page 606). Small variations in Table 21.2 from the predicted values are because real gases are not ideal gases. In real gases, weak intermolecular interactions occur, which are not addressed in our ideal gas model.

Now suppose the gas is taken along the constant-pressure path $i \rightarrow f'$ shown in Active Figure 21.4. Along this path, the temperature again increases by ΔT. The energy that must be transferred by heat to the gas in this process is $Q = nC_P\,\Delta T$. Because the volume changes in this process, the work done on the gas is $W = -P\,\Delta V$, where P is the constant pressure at which the process occurs. Applying the first law of thermodynamics to this process, we have

$$\Delta E_{\text{int}} = Q + W = nC_P\,\Delta T + (-P\,\Delta V) \qquad (21.15)$$

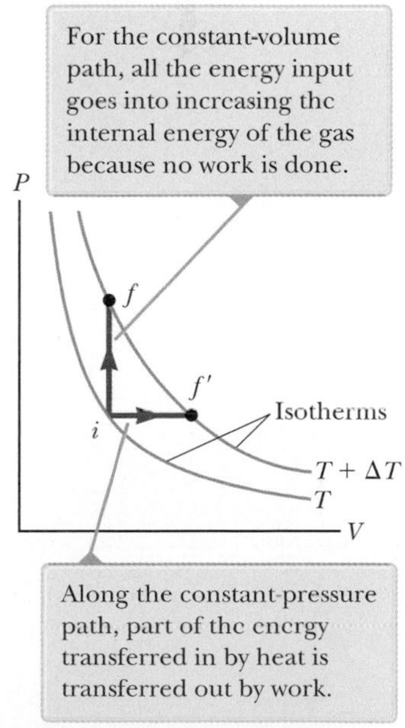

ACTIVE FIGURE 21.4

Energy is transferred by heat to an ideal gas in two ways.

In this case, the energy added to the gas by heat is channeled as follows. Part of it leaves the system by work (that is, the gas moves a piston through a displacement), and the remainder appears as an increase in the internal energy of the gas. The change in internal energy for the process $i \rightarrow f'$, however, is equal to that for the process $i \rightarrow f$ because E_{int} depends only on temperature for an ideal gas and ΔT is the same for both processes. In addition, because $PV = nRT$, note that for a constant-pressure process, $P\,\Delta V = nR\,\Delta T$. Substituting this value for $P\,\Delta V$ into Equation 21.15 with $\Delta E_{\text{int}} = nC_V\,\Delta T$ (Eq. 21.12) gives

$$nC_V\,\Delta T = nC_P\,\Delta T - nR\,\Delta T$$

$$C_P - C_V = R \tag{21.16}$$

This expression applies to *any* ideal gas. It predicts that the molar specific heat of an ideal gas at constant pressure is greater than the molar specific heat at constant volume by an amount R, the universal gas constant (which has the value 8.31 J/mol · K). This expression is applicable to real gases as the data in Table 21.2 show.

Because $C_V = \frac{3}{2}R$ for a monatomic ideal gas, Equation 21.16 predicts a value $C_P = \frac{5}{2}R = 20.8$ J/mol · K for the molar specific heat of a monatomic gas at constant pressure. The ratio of these molar specific heats is a dimensionless quantity γ (Greek letter gamma):

Ratio of molar specific heats for a monatomic ideal gas ▶

$$\gamma = \frac{C_P}{C_V} = \frac{5R/2}{3R/2} = \frac{5}{3} = 1.67 \tag{21.17}$$

Theoretical values of C_V, C_P, and γ are in excellent agreement with experimental values obtained for monatomic gases, but they are in serious disagreement with the values for the more complex gases (see Table 21.2). That is not surprising; the value $C_V = \frac{3}{2}R$ was derived for a monatomic ideal gas, and we expect some additional contribution to the molar specific heat from the internal structure of the more complex molecules. In Section 21.4, we describe the effect of molecular structure on the molar specific heat of a gas. The internal energy—and hence the molar specific heat—of a complex gas must include contributions from the rotational and the vibrational motions of the molecule.

In the case of solids and liquids heated at constant pressure, very little work is done because the thermal expansion is small. Consequently, C_P and C_V are approximately equal for solids and liquids.

TABLE 21.2 *Molar Specific Heats of Various Gases*

	Molar Specific Heat (J/mol · K)[a]			
Gas	C_P	C_V	$C_P - C_V$	$\gamma = C_P/C_V$
Monatomic gases				
He	20.8	12.5	8.33	1.67
Ar	20.8	12.5	8.33	1.67
Ne	20.8	12.7	8.12	1.64
Kr	20.8	12.3	8.49	1.69
Diatomic gases				
H_2	28.8	20.4	8.33	1.41
N_2	29.1	20.8	8.33	1.40
O_2	29.4	21.1	8.33	1.40
CO	29.3	21.0	8.33	1.40
Cl_2	34.7	25.7	8.96	1.35
Polyatomic gases				
CO_2	37.0	28.5	8.50	1.30
SO_2	40.4	31.4	9.00	1.29
H_2O	35.4	27.0	8.37	1.30
CH_4	35.5	27.1	8.41	1.31

[a] All values except that for water were obtained at 300 K.

Quick Quiz **21.2** **(i)** How does the internal energy of an ideal gas change as it follows path $i \rightarrow f$ in Active Figure 21.4? (a) E_{int} increases. (b) E_{int} decreases. (c) E_{int} stays the same. (d) There is not enough information to determine how E_{int} changes. **(ii)** From the same choices, how does the internal energy of an ideal gas change as it follows path $f \rightarrow f'$ along the isotherm labeled $T + \Delta T$ in Active Figure 21.4?

Example 21.2 Heating a Cylinder of Helium

A cylinder contains 3.00 mol of helium gas at a temperature of 300 K.

(A) If the gas is heated at constant volume, how much energy must be transferred by heat to the gas for its temperature to increase to 500 K?

SOLUTION

Conceptualize Run the process in your mind with the help of the piston–cylinder arrangement in Active Figure 19.12. Imagine that the piston is clamped in position to maintain the constant volume of the gas.

Categorize We evaluate parameters with equations developed in the preceding discussion, so this example is a substitution problem.

Use Equation 21.8 to find the energy transfer: $Q_1 = nC_V\,\Delta T$

Substitute the given values: $Q_1 = (3.00\text{ mol})(12.5\text{ J/mol}\cdot\text{K})(500\text{ K} - 300\text{ K})$

$= 7.50 \times 10^3\text{ J}$

(B) How much energy must be transferred by heat to the gas at constant pressure to raise the temperature to 500 K?

SOLUTION

Use Equation 21.9 to find the energy transfer: $Q_2 = nC_P\,\Delta T$

Substitute the given values: $Q_2 = (3.00\text{ mol})(20.8\text{ J/mol}\cdot\text{K})(500\text{ K} - 300\text{ K})$

$= 12.5 \times 10^3\text{ J}$

This value is larger than Q_1 because of the transfer of energy out of the gas by work to raise the piston in the constant pressure process.

21.3 Adiabatic Processes for an Ideal Gas

As noted in Section 20.6, an **adiabatic process** is one in which no energy is transferred by heat between a system and its surroundings. For example, if a gas is compressed (or expanded) rapidly, very little energy is transferred out of (or into) the system by heat, so the process is nearly adiabatic. Such processes occur in the cycle of a gasoline engine, which is discussed in detail in Chapter 22. Another example of an adiabatic process is the slow expansion of a gas that is thermally insulated from its surroundings. All three variables in the ideal gas law—P, V, and T—change during an adiabatic process.

Let's imagine an adiabatic gas process involving an infinitesimal change in volume dV and an accompanying infinitesimal change in temperature dT. The work done on the gas is $-P\,dV$. Because the internal energy of an ideal gas depends only on temperature, the change in the internal energy in an adiabatic process is the same as that for an isovolumetric process between the same temperatures,

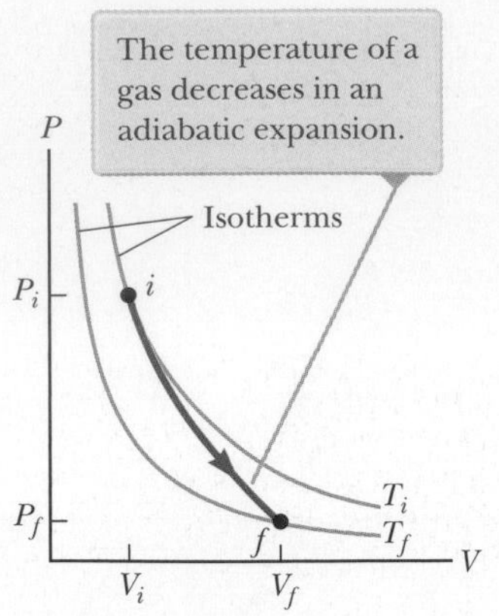

Figure 21.5 The *PV* diagram for an adiabatic expansion of an ideal gas.

$dE_{\text{int}} = nC_V\, dT$ (Eq. 21.12). Hence, the first law of thermodynamics, $\Delta E_{\text{int}} = Q + W$, with $Q = 0$, becomes the infinitesimal form

$$dE_{\text{int}} = nC_V\, dT = -P\, dV$$

Taking the total differential of the equation of state of an ideal gas, $PV = nRT$, gives

$$P\, dV + V\, dP = nR\, dT$$

Eliminating dT from these two equations, we find that

$$P\, dV + V\, dP = -\frac{R}{C_V} P\, dV$$

Substituting $R = C_P - C_V$ and dividing by PV gives

$$\frac{dV}{V} + \frac{dP}{P} = -\left(\frac{C_P - C_V}{C_V}\right)\frac{dV}{V} = (1 - \gamma)\frac{dV}{V}$$

$$\frac{dP}{P} + \gamma \frac{dV}{V} = 0$$

Integrating this expression, we have

$$\ln P + \gamma \ln V = \text{constant}$$

which is equivalent to

Relationship between *P* and *V* for an adiabatic process involving an ideal gas ▶

$$PV^{\gamma} = \text{constant} \tag{21.18}$$

The *PV* diagram for an adiabatic expansion is shown in Figure 21.5. Because $\gamma > 1$, the *PV* curve is steeper than it would be for an isothermal expansion. By the definition of an adiabatic process, no energy is transferred by heat into or out of the system. Hence, from the first law, we see that ΔE_{int} is negative (work is done *by* the gas, so its internal energy decreases) and so ΔT also is negative. Therefore, the temperature of the gas decreases ($T_f < T_i$) during an adiabatic expansion.[2] Conversely, the temperature increases if the gas is compressed adiabatically. Applying Equation 21.18 to the initial and final states, we see that

$$P_i V_i^{\gamma} = P_f V_f^{\gamma} \tag{21.19}$$

Using the ideal gas law, we can express Equation 21.18 as

Relationship between *T* and *V* for an adiabatic process involving an ideal gas ▶

$$TV^{\gamma - 1} = \text{constant} \tag{21.20}$$

Example 21.3 A Diesel Engine Cylinder

Air at 20.0°C in the cylinder of a diesel engine is compressed from an initial pressure of 1.00 atm and volume of 800.0 cm^3 to a volume of 60.0 cm^3. Assume air behaves as an ideal gas with $\gamma = 1.40$ and the compression is adiabatic. Find the final pressure and temperature of the air.

SOLUTION

Conceptualize Imagine what happens if a gas is compressed into a smaller volume. Our discussion above and Figure 21.5 tell us that the pressure and temperature both increase.

Categorize We categorize this example as a problem involving an adiabatic process.

[2]In the adiabatic free expansion discussed in Section 20.6, the temperature remains constant. In this unique process, no work is done because the gas expands into a vacuum. In general, the temperature decreases in an adiabatic expansion in which work is done.

21.3 *cont.*

Analyze Use Equation 21.19 to find the final pressure:

$$P_f = P_i\left(\frac{V_i}{V_f}\right)^\gamma = (1.00 \text{ atm})\left(\frac{800.0 \text{ cm}^3}{60.0 \text{ cm}^3}\right)^{1.40}$$

$$= \boxed{37.6 \text{ atm}}$$

Use the ideal gas law to find the final temperature:

$$\frac{P_iV_i}{T_i} = \frac{P_fV_f}{T_f}$$

$$T_f = \frac{P_fV_f}{P_iV_i}T_i = \frac{(37.6 \text{ atm})(60.0 \text{ cm}^3)}{(1.00 \text{ atm})(800.0 \text{ cm}^3)}(293 \text{ K})$$

$$= 826 \text{ K} = \boxed{553°\text{C}}$$

Finalize The temperature of the gas increases by a factor of 826 K/293 K = 2.82. The high compression in a diesel engine raises the temperature of the gas enough to cause the combustion of fuel without the use of spark plugs.

21.4 The Equipartition of Energy

Predictions based on our model for molar specific heat agree quite well with the behavior of monatomic gases, but not with the behavior of complex gases (see Table 21.2). The value predicted by the model for the quantity $C_P - C_V = R$, however, is the same for all gases. This similarity is not surprising because this difference is the result of the work done on the gas, which is independent of its molecular structure.

To clarify the variations in C_V and C_P in gases more complex than monatomic gases, let's explore further the origin of molar specific heat. So far, we have assumed the sole contribution to the internal energy of a gas is the translational kinetic energy of the molecules. The internal energy of a gas, however, includes contributions from the translational, vibrational, and rotational motion of the molecules. The rotational and vibrational motions of molecules can be activated by collisions and therefore are "coupled" to the translational motion of the molecules. The branch of physics known as *statistical mechanics* has shown that, for a large number of particles obeying the laws of Newtonian mechanics, the available energy is, on average, shared equally by each independent degree of freedom. Recall from Section 21.1 that the equipartition theorem states that, at equilibrium, each degree of freedom contributes $\frac{1}{2}k_BT$ of energy per molecule.

Let's consider a diatomic gas whose molecules have the shape of a dumbbell (Fig. 21.6). In this model, the center of mass of the molecule can translate in the x, y, and z directions (Fig. 21.6a). In addition, the molecule can rotate about three mutually perpendicular axes (Fig. 21.6b). The rotation about the y axis can be neglected because the molecule's moment of inertia I_y and its rotational energy $\frac{1}{2}I_y\omega^2$ about this axis are negligible compared with those associated with the x and z axes. (If the two atoms are modeled as particles, then I_y is identically zero.) Therefore, there are five degrees of freedom for translation and rotation: three associated with the translational motion and two associated with the rotational motion. Because each degree of freedom contributes, on average, $\frac{1}{2}k_BT$ of energy per molecule, the internal energy for a system of N molecules, ignoring vibration for now, is

$$E_{\text{int}} = 3N(\tfrac{1}{2}k_BT) + 2N(\tfrac{1}{2}k_BT) = \tfrac{5}{2}Nk_BT = \tfrac{5}{2}nRT$$

We can use this result and Equation 21.13 to find the molar specific heat at constant volume:

$$C_V = \frac{1}{n}\frac{dE_{\text{int}}}{dT} = \frac{1}{n}\frac{d}{dT}(\tfrac{5}{2}nRT) = \tfrac{5}{2}R \qquad \textbf{(21.21)}$$

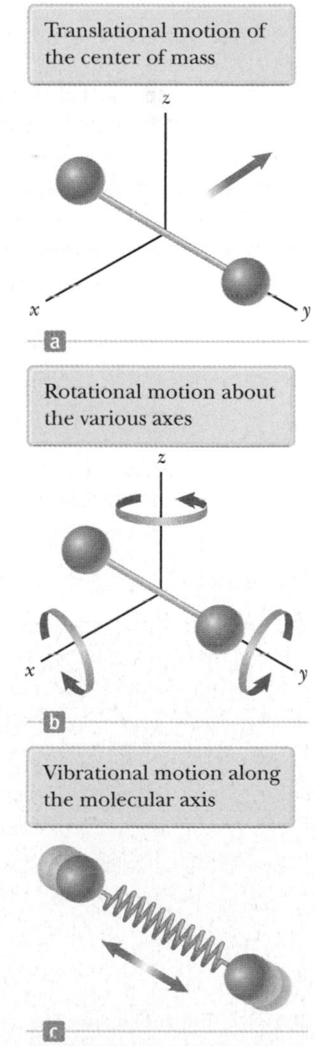

Figure 21.6 Possible motions of a diatomic molecule.

From Equations 21.16 and 21.17, we find that

$$C_P = C_V + R = \tfrac{7}{2}R$$

$$\gamma = \frac{C_P}{C_V} = \frac{\frac{7}{2}R}{\frac{5}{2}R} = \frac{7}{5} = 1.40$$

These results agree quite well with most of the data for diatomic molecules given in Table 21.2. That is rather surprising because we have not yet accounted for the possible vibrations of the molecule.

In the model for vibration, the two atoms are joined by an imaginary spring (see Fig. 21.6c). The vibrational motion adds two more degrees of freedom, which correspond to the kinetic energy and the potential energy associated with vibrations along the length of the molecule. Hence, a model that includes all three types of motion predicts a total internal energy of

$$E_{\text{int}} = 3N(\tfrac{1}{2}k_B T) + 2N(\tfrac{1}{2}k_B T) + 2N(\tfrac{1}{2}k_B T) = \tfrac{7}{2}Nk_B T = \tfrac{7}{2}nRT$$

and a molar specific heat at constant volume of

$$C_V = \frac{1}{n}\frac{dE_{\text{int}}}{dT} = \frac{1}{n}\frac{d}{dT}(\tfrac{7}{2}nRT) = \tfrac{7}{2}R \tag{21.22}$$

This value is inconsistent with experimental data for molecules such as H_2 and N_2 (see Table 21.2) and suggests a breakdown of our model based on classical physics.

It might seem that our model is a failure for predicting molar specific heats for diatomic gases. We can claim some success for our model, however, if measurements of molar specific heat are made over a wide temperature range rather than at the single temperature that gives us the values in Table 21.2. Figure 21.7 shows the molar specific heat of hydrogen as a function of temperature. The remarkable feature about the three plateaus in the graph's curve is that they are at the values of the molar specific heat predicted by Equations 21.14, 21.21, and 21.22! For low temperatures, the diatomic hydrogen gas behaves like a monatomic gas. As the temperature rises to room temperature, its molar specific heat rises to a value for a diatomic gas, consistent with the inclusion of rotation but not vibration. For high temperatures, the molar specific heat is consistent with a model including all types of motion.

Before addressing the reason for this mysterious behavior, let's make some brief remarks about polyatomic gases. For molecules with more than two atoms, the vibrations are more complex than for diatomic molecules and the number of degrees of freedom is even larger. The result is an even higher predicted molar specific heat,

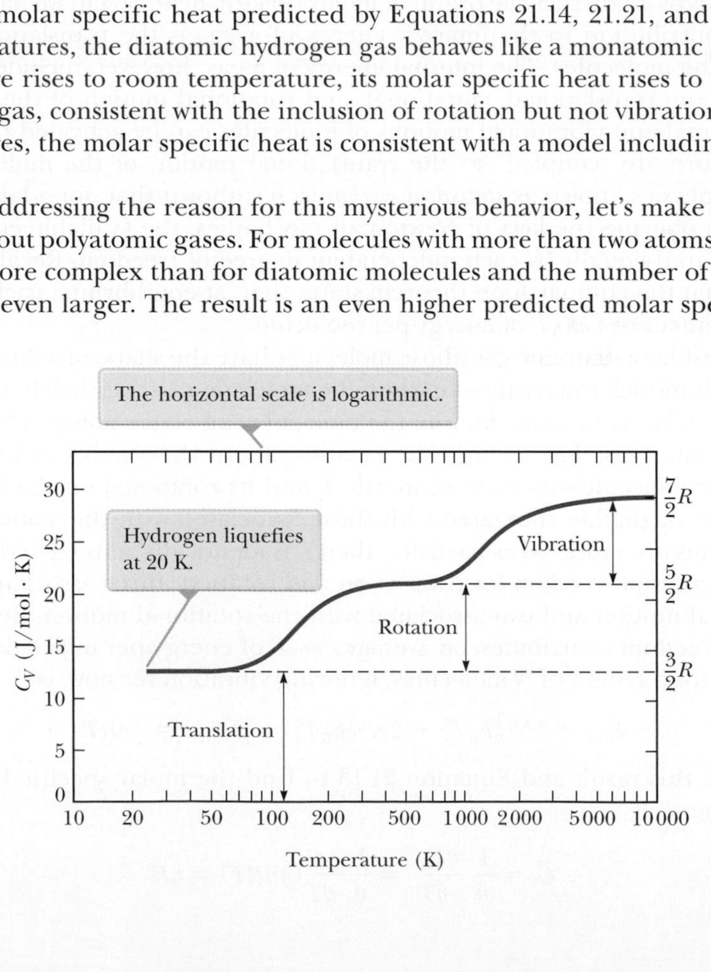

Figure 21.7 The molar specific heat of hydrogen as a function of temperature.

which is in qualitative agreement with experiment. The molar specific heats for the polyatomic gases in Table 21.2 are higher than those for diatomic gases. The more degrees of freedom available to a molecule, the more "ways" there are to store energy, resulting in a higher molar specific heat.

A Hint of Energy Quantization

Our model for molar specific heats has been based so far on purely classical notions. It predicts a value of the specific heat for a diatomic gas that, according to Figure 21.7, only agrees with experimental measurements made at high temperatures. To explain why this value is only true at high temperatures and why the plateaus in Figure 21.7 exist, we must go beyond classical physics and introduce some quantum physics into the model. In Chapter 18, we discussed quantization of frequency for vibrating strings and air columns; only certain frequencies of standing waves can exist. That is a natural result whenever waves are subject to boundary conditions.

Quantum physics (Chapters 40 through 43) shows that atoms and molecules can be described by the physics of waves under boundary conditions. Consequently, these waves have quantized frequencies. Furthermore, in quantum physics, the energy of a system is proportional to the frequency of the wave representing the system. Hence, **the energies of atoms and molecules are quantized.**

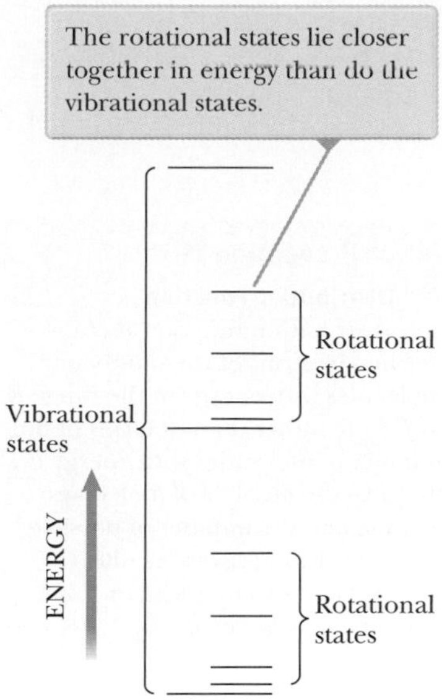

Figure 21.8 An energy-level diagram for vibrational and rotational states of a diatomic molecule.

For a molecule, quantum physics tells us that the rotational and vibrational energies are quantized. Figure 21.8 shows an **energy-level diagram** for the rotational and vibrational quantum states of a diatomic molecule. The lowest allowed state is called the **ground state.** Notice that vibrational states are separated by larger energy gaps than are rotational states.

At low temperatures, the energy a molecule gains in collisions with its neighbors is generally not large enough to raise it to the first excited state of either rotation or vibration. Therefore, even though rotation and vibration are allowed according to classical physics, they do not occur in reality at low temperatures. All molecules are in the ground state for rotation and vibration. The only contribution to the molecules' average energy is from translation, and the specific heat is that predicted by Equation 21.14.

As the temperature is raised, the average energy of the molecules increases. In some collisions, a molecule may have enough energy transferred to it from another molecule to excite the first rotational state. As the temperature is raised further, more molecules can be excited to this state. The result is that rotation begins to contribute to the internal energy, and the molar specific heat rises. At about room temperature in Figure 21.7, the second plateau has been reached and rotation contributes fully to the molar specific heat. The molar specific heat is now equal to the value predicted by Equation 21.21.

There is no contribution at room temperature from vibration because the molecules are still in the ground vibrational state. The temperature must be raised even further to excite the first vibrational state, which happens in Figure 21.7 between 1 000 K and 10 000 K. At 10 000 K on the right side of the figure, vibration is contributing fully to the internal energy and the molar specific heat has the value predicted by Equation 21.22.

The predictions of this model are supportive of the theorem of equipartition of energy. In addition, the inclusion in the model of energy quantization from quantum physics allows a full understanding of Figure 21.7.

***Quick Quiz* 21.3** The molar specific heat of a diatomic gas is measured at constant volume and found to be 29.1 J/mol · K. What are the types of energy that are contributing to the molar specific heat? **(a)** translation only **(b)** translation and rotation only **(c)** translation and vibration only **(d)** translation, rotation, and vibration

Quick Quiz **21.4** The molar specific heat of a gas is measured at constant volume and found to be $11R/2$. Is the gas most likely to be **(a)** monatomic, **(b)** diatomic, or **(c)** polyatomic?

Pitfall Prevention 21.2

The Distribution Function

The distribution function $n_V(E)$ is defined in terms of the number of molecules with energy in the range E to $E + dE$ rather than in terms of the number of molecules with energy E. Because the number of molecules is finite and the number of possible values of the energy is infinite, the number of molecules with an *exact* energy E may be zero.

21.5 Distribution of Molecular Speeds

Thus far, we have considered only average values of the energies of molecules in a gas and have not addressed the distribution of energies among molecules. In reality, the motion of the molecules is extremely chaotic. Any individual molecule collides with others at an enormous rate, typically a billion times per second. Each collision results in a change in the speed and direction of motion of each of the participant molecules. Equation 21.7 shows that rms molecular speeds increase with increasing temperature. What is the relative number of molecules that possess some characteristic such as energy within a certain range?

We shall address this question by considering the **number density** $n_V(E)$. This quantity, called a *distribution function,* is defined so that $n_V(E)\ dE$ is the number of molecules per unit volume with energy between E and $E + dE$. (The ratio of the number of molecules that have the desired characteristic to the total number of molecules is the probability that a particular molecule has that characteristic.) In general, the number density is found from statistical mechanics to be

Boltzmann distribution law ▶

$$n_V(E) = n_0 e^{-E/k_BT} \qquad \textbf{(21.23)}$$

where n_0 is defined such that $n_0\ dE$ is the number of molecules per unit volume having energy between $E = 0$ and $E = dE$. This equation, known as the **Boltzmann distribution law,** is important in describing the statistical mechanics of a large number of molecules. It states that the probability of finding the molecules in a particular energy state varies exponentially as the negative of the energy divided by k_BT. All the molecules would fall into the lowest energy level if the thermal agitation at a temperature T did not excite the molecules to higher energy levels.

Example 21.4 Thermal Excitation of Atomic Energy Levels

As discussed in Section 21.4, atoms can occupy only certain discrete energy levels. Consider a gas at a temperature of 2 500 K whose atoms can occupy only two energy levels separated by 1.50 eV, where 1 eV (electron volt) is an energy unit equal to 1.60×10^{-19} J (Fig. 21.9). Determine the ratio of the number of atoms in the higher energy level to the number in the lower energy level.

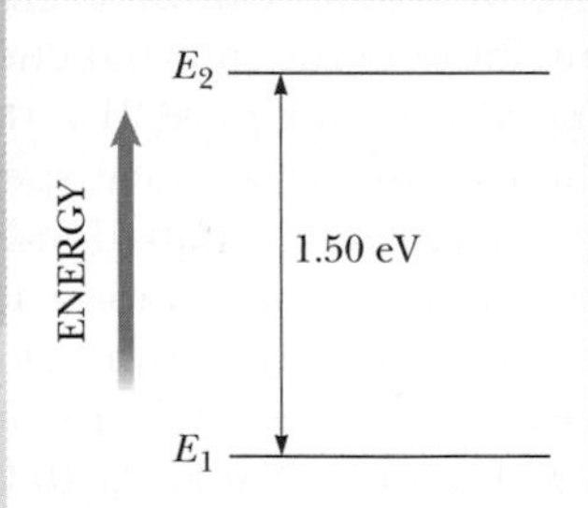

Figure 21.9 (Example 21.4) Energy-level diagram for a gas whose atoms can occupy two energy states.

SOLUTION

Conceptualize In your mental representation of this example, remember that only two possible states are allowed for the system of the atom. Figure 21.9 helps you visualize the two states on an energy-level diagram. In this case, the atom has two possible energies, E_1 and E_2, where $E_1 < E_2$.

Categorize We categorize this example as one in which we apply the Boltzmann distribution law to a quantized system.

Analyze Set up the ratio of the number of atoms in the higher energy level to the number in the lower energy level and use Equation 21.23 to express each number:

$$(1) \quad \frac{n_V(E_2)}{n_V(E_1)} = \frac{n_0 e^{-E_2/k_BT}}{n_0 e^{-E_1/k_BT}} = e^{-(E_2 - E_1)/k_BT}$$

21.4 *cont.*

Evaluate k_BT in the exponent: $$k_BT = (1.38 \times 10^{-23}\ \text{J/K})(2\,500\ \text{K})\left(\frac{1\ \text{eV}}{1.60 \times 10^{-19}\ \text{J}}\right) = 0.216\ \text{eV}$$

Substitute this value into Equation (1): $$\frac{n_V(E_2)}{n_V(E_1)} = e^{-1.50\ \text{eV}/0.216\ \text{eV}} = e^{-6.96} = 9.52 \times 10^{-4}$$

Finalize This result indicates that at $T = 2\,500$ K, only a small fraction of the atoms are in the higher energy level. In fact, for every atom in the higher energy level, there are about 1 000 atoms in the lower level. The number of atoms in the higher level increases at even higher temperatures, but the distribution law specifies that at equilibrium there are always more atoms in the lower level than in the higher level.

WHAT IF? What if the energy levels in Figure 21.9 were closer together in energy? Would that increase or decrease the fraction of the atoms in the upper energy level?

Answer If the excited level is lower in energy than that in Figure 21.9, it would be easier for thermal agitation to excite atoms to this level and the fraction of atoms in this energy level would be larger, which we can see mathematically by expressing Equation (1) as

$$r_2 = e^{-(E_2 - E_1)/k_BT}$$

where r_2 is the ratio of atoms having energy E_2 to those with energy E_1. Differentiating with respect to E_2, we find

$$\frac{dr_2}{dE_2} = \frac{d}{dE_2}\left[e^{-(E_2 - E_1)/k_BT}\right] = -\frac{1}{k_BT}e^{-(E_2 - E_1)/k_BT} < 0$$

Because the derivative has a negative value, as E_2 decreases, r_2 increases.

Now that we have discussed the distribution of energies, let's think about the distribution of molecular speeds. In 1860, James Clerk Maxwell (1831–1879) derived an expression that describes the distribution of molecular speeds in a very definite manner. His work and subsequent developments by other scientists were highly controversial because direct detection of molecules could not be achieved experimentally at that time. About 60 years later, however, experiments were devised that confirmed Maxwell's predictions.

Let's consider a container of gas whose molecules have some distribution of speeds. Suppose we want to determine how many gas molecules have a speed in the range from, for example, 400 to 401 m/s. Intuitively, we expect the speed distribution to depend on temperature. Furthermore, we expect the distribution to peak in the vicinity of v_{rms}. That is, few molecules are expected to have speeds much less than or much greater than v_{rms} because these extreme speeds result only from an unlikely chain of collisions.

The observed speed distribution of gas molecules in thermal equilibrium is shown in Active Figure 21.10 (page 614). The quantity N_v, called the **Maxwell–Boltzmann speed distribution function,** is defined as follows. If N is the total number of molecules, the number of molecules with speeds between v and $v + dv$ is $dN = N_v\, dv$. This number is also equal to the area of the shaded rectangle in Active Figure 21.10. Furthermore, the fraction of molecules with speeds between v and $v + dv$ is $(N_v\, dv)/N$. This fraction is also equal to the probability that a molecule has a speed in the range v to $v + dv$.

The fundamental expression that describes the distribution of speeds of N gas molecules is

$$N_v = 4\pi N\left(\frac{m_0}{2\pi k_BT}\right)^{3/2} v^2 e^{-m_0v^2/2k_BT} \qquad \textbf{(21.24)}$$

University of Vienna, courtesy AIP Emilio Segre Visual Archives

Ludwig Boltzmann
Austrian physicist (1844–1906)
Boltzmann made many important contributions to the development of the kinetic theory of gases, electromagnetism, and thermodynamics. His pioneering work in the field of kinetic theory led to the branch of physics known as statistical mechanics.

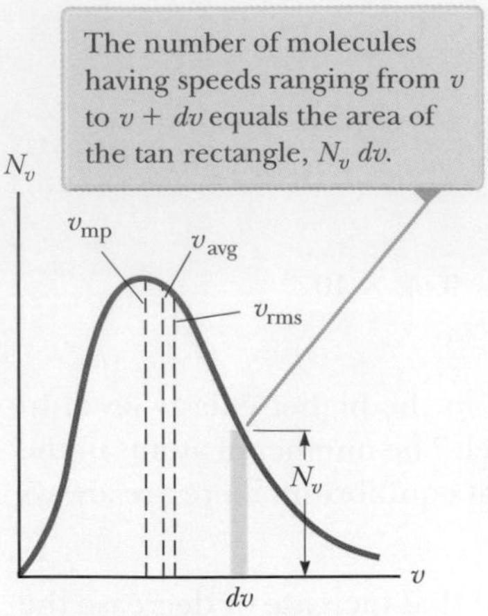

ACTIVE FIGURE 21.10

The speed distribution of gas molecules at some temperature. The function N_v approaches zero as v approaches infinity.

where m_0 is the mass of a gas molecule, k_B is Boltzmann's constant, and T is the absolute temperature.[3] Observe the appearance of the Boltzmann factor e^{-E/k_BT} with $E = \frac{1}{2}m_0v^2$.

As indicated in Active Figure 21.10, the average speed is somewhat lower than the rms speed. The *most probable speed* v_{mp} is the speed at which the distribution curve reaches a peak. Using Equation 21.24, we find that

$$v_{rms} = \sqrt{\overline{v^2}} = \sqrt{\frac{3k_BT}{m_0}} = 1.73\sqrt{\frac{k_BT}{m_0}} \quad \textbf{(21.25)}$$

$$v_{avg} = \sqrt{\frac{8k_BT}{\pi m_0}} = 1.60\sqrt{\frac{k_BT}{m_0}} \quad \textbf{(21.26)}$$

$$v_{mp} = \sqrt{\frac{2k_BT}{m_0}} = 1.41\sqrt{\frac{k_BT}{m_0}} \quad \textbf{(21.27)}$$

Equation 21.25 has previously appeared as Equation 21.7. The details of the derivations of these equations from Equation 21.24 are left for the end-of-chapter problems (see Problems 38 and 65). From these equations, we see that

$$v_{rms} > v_{avg} > v_{mp}$$

Active Figure 21.11 represents speed distribution curves for nitrogen, N_2. The curves were obtained by using Equation 21.24 to evaluate the distribution function at various speeds and at two temperatures. Notice that the peak in each curve shifts to the right as T increases, indicating that the average speed increases with increasing temperature, as expected. Because the lowest speed possible is zero and the upper classical limit of the speed is infinity, the curves are asymmetrical. (In Chapter 39, we show that the actual upper limit is the speed of light.)

Equation 21.24 shows that the distribution of molecular speeds in a gas depends both on mass and on temperature. At a given temperature, the fraction of molecules with speeds exceeding a fixed value increases as the mass decreases. Hence, lighter molecules such as H_2 and He escape into space more readily from the

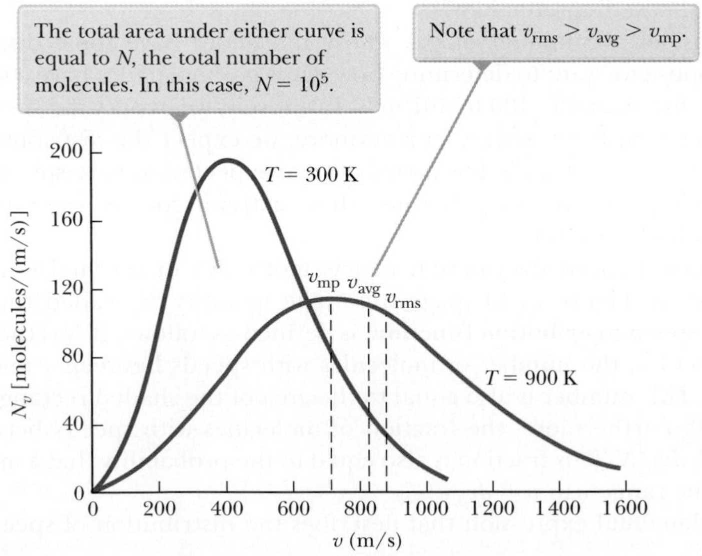

ACTIVE FIGURE 21.11

The speed distribution function for 10^5 nitrogen molecules at 300 K and 900 K.

[3] For the derivation of this expression, see an advanced textbook on thermodynamics.

Earth's atmosphere than do heavier molecules such as N_2 and O_2. (See the discussion of escape speed in Chapter 13. Gas molecules escape even more readily from the Moon's surface than from the Earth's because the escape speed on the Moon is lower than that on the Earth.)

The speed distribution curves for molecules in a liquid are similar to those shown in Active Figure 21.11. We can understand the phenomenon of evaporation of a liquid from this distribution in speeds, given that some molecules in the liquid are more energetic than others. Some of the faster-moving molecules in the liquid penetrate the surface and even leave the liquid at temperatures well below the boiling point. The molecules that escape the liquid by evaporation are those that have sufficient energy to overcome the attractive forces of the molecules in the liquid phase. Consequently, the molecules left behind in the liquid phase have a lower average kinetic energy; as a result, the temperature of the liquid decreases. Hence, evaporation is a cooling process. For example, an alcohol-soaked cloth can be placed on a feverish head to cool and comfort a patient.

Example 21.5 A System of Nine Particles

Nine particles have speeds of 5.00, 8.00, 12.0, 12.0, 12.0, 14.0, 14.0, 17.0, and 20.0 m/s.

(A) Find the particles' average speed.

SOLUTION

Conceptualize Imagine a small number of particles moving in random directions with the few speeds listed.

Categorize Because we are dealing with a small number of particles, we can calculate the average speed directly.

Analyze Find the average speed of the particles by dividing the sum of the speeds by the total number of particles:

$$v_{\text{avg}} = \frac{(5.00 + 8.00 + 12.0 + 12.0 + 12.0 + 14.0 + 14.0 + 17.0 + 20.0)\ \text{m/s}}{9}$$

$$= \boxed{12.7\ \text{m/s}}$$

(B) What is the rms speed of the particles?

SOLUTION

Find the average speed squared of the particles by dividing the sum of the speeds squared by the total number of particles:

$$\overline{v^2} = \frac{(5.00^2 + 8.00^2 + 12.0^2 + 12.0^2 + 12.0^2 + 14.0^2 + 14.0^2 + 17.0^2 + 20.0^2)\ \text{m}^2/\text{s}^2}{9}$$

$$= 178\ \text{m}^2/\text{s}^2$$

Find the rms speed of the particles by taking the square root:

$$v_{\text{rms}} = \sqrt{\overline{v^2}} = \sqrt{178\ \text{m}^2/\text{s}^2} = \boxed{13.3\ \text{m/s}}$$

(C) What is the most probable speed of the particles?

SOLUTION

Three of the particles have a speed of 12.0 m/s, two have a speed of 14.0 m/s, and the remaining four have different speeds. Hence, the most probable speed v_{mp} is 12.0 m/s.

Finalize Compare this example, in which the number of particles is small and we know the individual particle speeds, with the next example.

Example 21.6 Molecular Speeds in a Hydrogen Gas

A 0.500-mol sample of hydrogen gas is at 300 K.

(A) Find the average speed, the rms speed, and the most probable speed of the hydrogen molecules.

SOLUTION

Conceptualize Imagine a huge number of particles in a real gas, all moving in random directions with different speeds.

Categorize We cannot calculate the averages as was done in Example 21.5 because the individual speeds of the particles are not known. We are dealing with a very large number of particles, however, so we can use the Maxwell-Boltzmann speed distribution function.

Analyze Use Equation 21.26 to find the average speed:

$$v_{\text{avg}} = 1.60\sqrt{\frac{k_B T}{m_0}} = 1.60\sqrt{\frac{(1.38 \times 10^{-23}\text{ J/K})(300\text{ K})}{2(1.67 \times 10^{-27}\text{ kg})}}$$

$$= \boxed{1.78 \times 10^3\text{ m/s}}$$

Use Equation 21.25 to find the rms speed:

$$v_{\text{rms}} = 1.73\sqrt{\frac{k_B T}{m_0}} = 1.73\sqrt{\frac{(1.38 \times 10^{-23}\text{ J/K})(300\text{ K})}{2(1.67 \times 10^{-27}\text{ kg})}}$$

$$= \boxed{1.93 \times 10^3\text{ m/s}}$$

Use Equation 21.27 to find the most probable speed:

$$v_{\text{mp}} = 1.41\sqrt{\frac{k_B T}{m_0}} = 1.41\sqrt{\frac{(1.38 \times 10^{-23}\text{ J/K})(300\text{ K})}{2(1.67 \times 10^{-27}\text{ kg})}}$$

$$= \boxed{1.57 \times 10^3\text{ m/s}}$$

(B) Find the number of molecules with speeds between 400 m/s and 401 m/s.

SOLUTION

Use Equation 21.24 to evaluate the number of molecules in a narrow speed range between v and $v + dv$:

$$(1) \quad N_v\, dv = 4\pi N\left(\frac{m_0}{2\pi k_B T}\right)^{3/2} v^2 e^{-m_0 v^2/2k_B T}\, dv$$

Evaluate the constant in front of v^2:

$$4\pi N\left(\frac{m_0}{2\pi k_B T}\right)^{3/2} = 4\pi n N_A\left(\frac{m_0}{2\pi k_B T}\right)^{3/2}$$

$$= 4\pi(0.500\text{ mol})(6.02 \times 10^{23}\text{ mol}^{-1})\left[\frac{2(1.67 \times 10^{-27}\text{ kg})}{2\pi(1.38 \times 10^{-23}\text{ J/K})(300\text{ K})}\right]^{3/2}$$

$$= 1.74 \times 10^{14}\text{ s}^3/\text{m}^3$$

Evaluate the exponent of e that appears in Equation (1):

$$-\frac{m_0 v^2}{2k_B T} = -\frac{2(1.67 \times 10^{-27}\text{ kg})(400\text{ m/s})^2}{2(1.38 \times 10^{-23}\text{ J/K})(300\text{ K})} = -0.064\,5$$

Evaluate $N_v\, dv$ using Equation (1):

$$N_v\, dv = (1.74 \times 10^{14}\text{ s}^3/\text{m}^3)(400\text{ m/s})^2 e^{-0.064\,5}(1\text{ m/s})$$

$$= \boxed{2.61 \times 10^{19}\text{ molecules}}$$

Finalize In this evaluation, we could calculate the result without integration because $dv = 1$ m/s is much smaller than $v = 400$ m/s. Had we sought the number of particles between, say, 400 m/s and 500 m/s, we would need to integrate Equation (1) between these speed limits.

Summary

Concepts and Principles

The pressure of N molecules of an ideal gas contained in a volume V is

$$P = \frac{2}{3}\left(\frac{N}{V}\right)\left(\frac{1}{2}m_0\overline{v^2}\right) \quad (21.2)$$

The average translational kinetic energy per molecule of a gas, $\frac{1}{2}m_0\overline{v^2}$, is related to the temperature T of the gas through the expression

$$\frac{1}{2}m_0\overline{v^2} = \frac{3}{2}k_B T \quad (21.4)$$

where k_B is Boltzmann's constant. Each translational degree of freedom (x, y, or z) has $\frac{1}{2}k_B T$ of energy associated with it.

The internal energy of N molecules (or n mol) of an ideal monatomic gas is

$$E_{int} = \frac{3}{2}Nk_B T = \frac{3}{2}nRT \quad (21.10)$$

The change in internal energy for n mol of any ideal gas that undergoes a change in temperature ΔT is

$$\Delta E_{int} = nC_V\,\Delta T \quad (21.12)$$

where C_V is the **molar specific heat at constant volume.**

The molar specific heat of an ideal monatomic gas at constant volume is $C_V = \frac{3}{2}R$; the molar specific heat at constant pressure is $C_P = \frac{5}{2}R$. The ratio of specific heats is given by $\gamma = C_P/C_V = \frac{5}{3}$.

If an ideal gas undergoes an adiabatic expansion or compression, the first law of thermodynamics, together with the equation of state, shows that

$$PV^{\gamma} = \text{constant} \quad (21.18)$$

The **Boltzmann distribution law** describes the distribution of particles among available energy states. The relative number of particles having energy between E and $E + dE$ is $n_V(E)\,dE$, where

$$n_V(E) = n_0 e^{-E/k_B T} \quad (21.23)$$

The **Maxwell–Boltzmann speed distribution function** describes the distribution of speeds of molecules in a gas:

$$N_v = 4\pi N\left(\frac{m_0}{2\pi k_B T}\right)^{3/2} v^2 e^{-m_0 v^2/2k_B T} \quad (21.24)$$

Equation 21.24 enables us to calculate the **root-mean-square speed,** the **average speed,** and the **most probable speed** of molecules in a gas:

$$v_{rms} = \sqrt{\overline{v^2}} = \sqrt{\frac{3k_B T}{m_0}} = 1.73\sqrt{\frac{k_B T}{m_0}} \quad (21.25)$$

$$v_{avg} = \sqrt{\frac{8k_B T}{\pi m_0}} = 1.60\sqrt{\frac{k_B T}{m_0}} \quad (21.26)$$

$$v_{mp} = \sqrt{\frac{2k_B T}{m_0}} = 1.41\sqrt{\frac{k_B T}{m_0}} \quad (21.27)$$

Objective Questions

☐ denotes answer available in Student Solutions Manual/Study Guide

1. Which of the assumptions below is *not* made in the kinetic theory of gases? (a) The number of molecules is very large. (b) The molecules obey Newton's laws of motion. (c) The forces between molecules are long range. (d) The gas is a pure substance. (e) The average separation between molecules is large compared to their dimensions.
2. An ideal gas is maintained at constant pressure. If the temperature of the gas is increased from 200 K to 600 K, what happens to the rms speed of the molecules? (a) It increases by a factor of 3. (b) It remains the same. (c) It is one-third the original speed. (d) It is $\sqrt{3}$ times the original speed. (e) It increases by a factor of 6.
3. A sample of gas with a thermometer immersed in the gas is held over a hot plate. A student is asked to give a step-by-step account of what makes our observation of the temperature of the gas increase. His response includes the following steps. (a) The molecules speed up. (b) Then the molecules collide with one another more often. (c) Internal friction makes the collisions inelastic. (d) Heat is produced in the collisions. (e) The molecules of the gas transfer more energy to the thermometer when they strike it, so we observe that the temperature has gone up. (f) The same process can take place without the use of a hot plate if you quickly push in the piston in an insulated cylinder containing the

gas. **(i)** Which of the parts (a) through (f) of this account are correct statements necessary for a clear and complete explanation? **(ii)** Which are correct statements that are not necessary to account for the higher thermometer reading? **(iii)** Which are incorrect statements?

4. A helium-filled latex balloon initially at room temperature is placed in a freezer. The latex remains flexible. **(i)** Does the balloon's volume (a) increase, (b) decrease, or (c) remain the same? **(ii)** Does the pressure of the helium gas (a) increase significantly, (b) decrease significantly, or (c) remain approximately the same?

5. A gas is at 200 K. If we wish to double the rms speed of the molecules of the gas, to what value must we raise its temperature? (a) 283 K (b) 400 K (c) 566 K (d) 800 K (e) 1130 K

6. Rank the following from largest to smallest, noting any cases of equality. (a) the average speed of molecules in a particular sample of ideal gas (b) the most probable speed (c) the root-mean-square speed (d) the average vector velocity of the molecules

7. Two samples of the same ideal gas have the same pressure and density. Sample B has twice the volume of sample A. What is the rms speed of the molecules in sample B? (a) twice that in sample A (b) equal to that in sample A (c) half that in sample A (d) impossible to determine

8. An ideal gas is contained in a vessel at 300 K. The temperature of the gas is then increased to 900 K. **(i)** By what factor does the average kinetic energy of the molecules change, (a) a factor of 9, (b) a factor of 3, (c) a factor of $\sqrt{3}$, (d) a factor of 1, or (e) a factor of $\frac{1}{3}$? Using the same choices as in part (i), by what factor does each of the following change: **(ii)** the rms molecular speed of the molecules, **(iii)** the average momentum change that one molecule undergoes in a collision with one particular wall, **(iv)** the rate of collisions of molecules with walls, and **(v)** the pressure of the gas.

9. Cylinder A contains oxygen (O_2) gas, and cylinder B contains nitrogen (N_2) gas. If the molecules in the two cylinders have the same rms speeds, which of the following statements is *false*? (a) The two gases have different temperatures. (b) The temperature of cylinder B is less than the temperature of cylinder A. (c) The temperature of cylinder B is greater than the temperature of cylinder A. (d) The average kinetic energy of the nitrogen molecules is less than the average kinetic energy of the oxygen molecules.

Conceptual Questions

☐ denotes answer available in Student Solutions Manual/Study Guide

1. Dalton's law of partial pressures states that the total pressure of a mixture of gases is equal to the sum of the pressures that each gas in the mixture would exert if it were alone in the container. Give a convincing argument for this law based on the kinetic theory of gases.

2. One container is filled with helium gas and another with argon gas. Both containers are at the same temperature. Which molecules have the higher rms speed? Explain.

3. When alcohol is rubbed on your body, it lowers your skin temperature. Explain this effect.

4. What happens to a helium-filled latex balloon released into the air? Does it expand or contract? Does it stop rising at some height?

5. Which is denser, dry air or air saturated with water vapor? Explain.

6. Why does a diatomic gas have a greater energy content per mole than a monatomic gas at the same temperature?

7. Hot air rises, so why does it generally become cooler as you climb a mountain? *Note:* Air has low thermal conductivity.

Problems

ENHANCED WebAssign The problems found in this chapter may be assigned online in Enhanced WebAssign

1. denotes straightforward problem; **2.** denotes intermediate problem; **3.** denotes challenging problem

1. full solution available in the Student Solutions Manual/Study Guide

1. denotes problems most often assigned in Enhanced WebAssign; these provide students with targeted feedback and either a Master It tutorial or a Watch It solution video.

Q|C denotes asking for quantitative and conceptual reasoning

S denotes symbolic reasoning problem

M denotes Master It tutorial available in Enhanced WebAssign

GP denotes guided problem

shaded denotes "paired problems" that develop reasoning with symbols and numerical values

Section 21.1 Molecular Model of an Ideal Gas

Note: Problem 24 in Chapter 19 can be assigned with this section.

1. Calculate the mass of an atom of (a) helium, (b) iron, and (c) lead. Give your answers in kilograms. The atomic masses of these atoms are 4.00 u, 55.9 u, and 207 u, respectively.

2. **M** A cylinder contains a mixture of helium and argon gas in equilibrium at 150°C. (a) What is the average kinetic energy for each type of gas molecule? (b) What is the rms speed of each type of molecule?

3. In a 30.0-s interval, 500 hailstones strike a glass window of area 0.600 m^2 at an angle of 45.0° to the window surface. Each hailstone has a mass of 5.00 g and a speed of 8.00 m/s. Assuming the collisions are elastic, find (a) the average force and (b) the average pressure on the window during this interval.

4. In an ultrahigh vacuum system (with typical pressures lower than 10^{-7} pascal), the pressure is measured to be 1.00×10^{-10} torr (where 1 torr = 133 Pa). Assuming the temperature is 300 K, find the number of molecules in a volume of 1.00 m^3.

5. A spherical balloon of volume 4.00×10^3 cm^3 contains helium at a pressure of 1.20×10^5 Pa. How many moles of helium are in the balloon if the average kinetic energy of the helium atoms is 3.60×10^{-22} J?

6. **S** A spherical balloon of volume V contains helium at a pressure P. How many moles of helium are in the balloon if the average kinetic energy of the helium atoms is $\overline{K}$?

7. A 2.00-mol sample of oxygen gas is confined to a 5.00-L vessel at a pressure of 8.00 atm. Find the average translational kinetic energy of the oxygen molecules under these conditions.

8. Oxygen, modeled as an ideal gas, is in a container and has a temperature of 77.0°C. What is the rms-average magnitude of the momentum of the gas molecules in the container?

9. (a) How many atoms of helium gas fill a spherical balloon of diameter 30.0 cm at 20.0°C and 1.00 atm? (b) What is the average kinetic energy of the helium atoms? (c) What is the rms speed of the helium atoms?

10. The rms speed of an oxygen molecule (O_2) in a container of oxygen gas is 625 m/s. What is the temperature of the gas?

11. In a period of 1.00 s, 5.00×10^{23} nitrogen molecules strike a wall with an area of 8.00 cm^2. Assume the molecules move with a speed of 300 m/s and strike the wall head-on in elastic collisions. What is the pressure exerted on the wall? *Note:* The mass of one N_2 molecule is 4.65×10^{-26} kg.

12. **Q|C** A 7.00-L vessel contains 3.50 moles of gas at a pressure of 1.60×10^6 Pa. Find (a) the temperature of the gas and (b) the average kinetic energy of the gas molecules in the vessel. (c) What additional information would you need if you were asked to find the average speed of the gas molecules?

Section 21.2 Molar Specific Heat of an Ideal Gas

Note: You may use data in Table 21.2 about particular gases. Here we define a "monatomic ideal gas" to have molar specific heats $C_V = \frac{3}{2}R$ and $C_P = \frac{5}{2}R$, and a "diatomic ideal gas" to have $C_V = \frac{5}{2}R$ and $C_P = \frac{7}{2}R$.

13. **S** A sample of a diatomic ideal gas has pressure P and volume V. When the gas is warmed, its pressure triples and its volume doubles. This warming process includes two steps, the first at constant pressure and the second at constant volume. Determine the amount of energy transferred to the gas by heat.

14. In a constant-volume process, 209 J of energy is transferred by heat to 1.00 mol of an ideal monatomic gas initially at 300 K. Find (a) the work done on the gas, (b) the increase in internal energy of the gas, and (c) its final temperature.

15. A 1.00-mol sample of hydrogen gas is heated at constant pressure from 300 K to 420 K. Calculate (a) the energy transferred to the gas by heat, (b) the increase in its internal energy, and (c) the work done on the gas.

16. **Review.** A house has well-insulated walls. It contains a volume of 100 m^3 of air at 300 K. (a) Calculate the energy required to increase the temperature of this diatomic ideal gas by 1.00°C. (b) **What If?** If all this energy could be used to lift an object of mass m through a height of 2.00 m, what is the value of m?

17. A vertical cylinder with a heavy piston contains air at 300 K. The initial pressure is 2.00×10^5 Pa, and the initial volume is 0.350 m^3. Take the molar mass of air as 28.9 g/mol and assume $C_V = \frac{5}{2}R$. (a) Find the specific heat of air at constant volume in units of J/kg · °C. (b) Calculate the mass of the air in the cylinder. (c) Suppose the piston is held fixed. Find the energy input required to raise the temperature of the air to 700 K. (d) **What If?** Assume again the conditions of the initial state and assume the heavy piston is free to move. Find the energy input required to raise the temperature to 700 K.

18. A 1.00-L insulated bottle is full of tea at 90.0°C. You pour out one cup of tea and immediately screw the stopper back on the bottle. Make an order-of-magnitude estimate of the change in temperature of the tea remaining in the bottle that results from the admission of air at room temperature. State the quantities you take as data and the values you measure or estimate for them.

Section 21.3 Adiabatic Processes for an Ideal Gas

19. During the compression stroke of a certain gasoline engine, the pressure increases from 1.00 atm to 20.0 atm. If the process is adiabatic and the air–fuel mixture behaves as a diatomic ideal gas, (a) by what factor does the volume change and (b) by what factor does the temperature change? Assuming the compression starts with 0.016 0 mol of gas at 27.0°C, find the values of (c) Q, (d) ΔE_{int}, and (e) W that characterize the process.

20. A 2.00-mol sample of a diatomic ideal gas expands slowly and adiabatically from a pressure of 5.00 atm and a volume of 12.0 L to a final volume of 30.0 L. (a) What is the final pressure of the gas? (b) What are the initial and final temperatures? Find (c) Q, (d) ΔE_{int}, and (e) W for the gas during this process.

21. **M** Air in a thundercloud expands as it rises. If its initial temperature is 300 K and no energy is lost by thermal conduction on expansion, what is its temperature when the initial volume has doubled?

22. How much work is required to compress 5.00 mol of air at 20.0°C and 1.00 atm to one-tenth of the original volume (a) by an isothermal process? (b) **What If?** How much work is required to produce the same compression in an adiabatic process? (c) What is the final pressure in part (a)? (d) What is the final pressure in part (b)?

23. During the power stroke in a four-stroke automobile engine, the piston is forced down as the mixture of combustion products and air undergoes an adiabatic expansion. Assume (1) the engine is running at 2 500 cycles/min; (2) the gauge pressure immediately before the expansion is 20.0 atm; (3) the volumes of the mixture immediately before and after the expansion are 50.0 cm^3 and 400 cm^3, respectively (Fig. P21.23); (4) the time interval for the expansion is one-fourth that of the total cycle; and (5) the mixture behaves like an ideal gas with specific heat ratio 1.40. Find the average power generated during the power stroke.

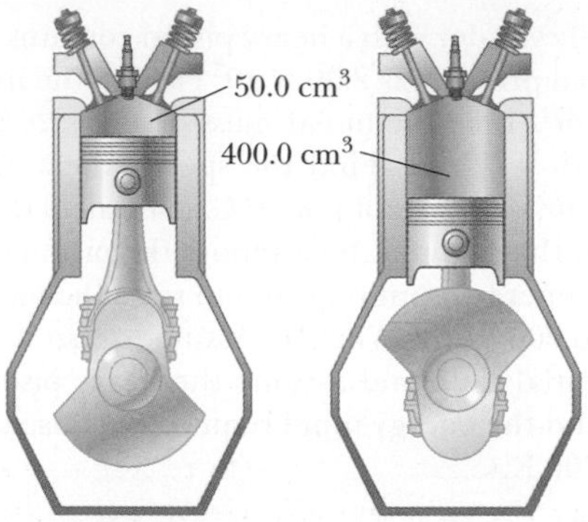

Figure P21.23

24. *Why is the following situation impossible?* A new diesel engine that increases fuel economy over previous models is designed. Automobiles fitted with this design become incredible best sellers. Two design features are responsible for the increased fuel economy: (1) the engine is made entirely of aluminum to reduce the weight of the automobile, and (2) the exhaust of the engine is used to prewarm the air to 50°C before it enters the cylinder to increase the final temperature of the compressed gas. The engine has a *compression ratio*—that is, the ratio of the initial volume of the air to its final volume after compression—of 14.5. The compression process is adiabatic, and the air behaves as a diatomic ideal gas with $\gamma = 1.40$.

25. A 4.00-L sample of a diatomic ideal gas with specific heat ratio 1.40, confined to a cylinder, is carried through a closed cycle. The gas is initially at 1.00 atm and 300 K. First, its pressure is tripled under constant volume. Then, it expands adiabatically to its original pressure. Finally, the gas is compressed isobarically to its original volume. (a) Draw a *PV* diagram of this cycle. (b) Determine the volume of the gas at the end of the adiabatic expansion. (c) Find the temperature of the gas at the start of the adiabatic expansion. (d) Find the temperature at the end of the cycle. (e) What was the net work done on the gas for this cycle?

26. **S** An ideal gas with specific heat ratio γ confined to a cylinder is put through a closed cycle. Initially, the gas is at P_i, V_i, and T_i. First, its pressure is tripled under constant volume. It then expands adiabatically to its original pressure and finally is compressed isobarically to its original volume. (a) Draw a *PV* diagram of this cycle. (b) Determine the volume at the end of the adiabatic expansion. Find (c) the temperature of the gas at the start of the adiabatic expansion and (d) the temperature at the end of the cycle. (e) What was the net work done on the gas for this cycle?

27. **GP** Air (a diatomic ideal gas) at 27.0°C and atmospheric pressure is drawn into a bicycle pump (see the chapter-opening photo on page 599) that has a cylinder with an inner diameter of 2.50 cm and length 50.0 cm. The downstroke adiabatically compresses the air, which reaches a gauge pressure of 8.00×10^5 Pa before entering the tire. We wish to investigate the temperature increase of the pump. (a) What is the initial volume of the air in the pump? (b) What is the number of moles of air in the pump? (c) What is the absolute pressure of the compressed air? (d) What is the volume of the compressed air? (e) What is the temperature of the compressed air? (f) What is the increase in internal energy of the gas during the compression? **What If?** The pump is made of steel that is 2.00 mm thick. Assume 4.00 cm of the cylinder's length is allowed to come to thermal equilibrium with the air. (g) What is the volume of steel in this 4.00-cm length? (h) What is the mass of steel in this 4.00-cm length? (i) Assume the pump is compressed once. After the adiabatic expansion, conduction results in the energy increase in part (f) being shared between the gas and the 4.00-cm length of steel. What will be the increase in temperature of the steel after one compression?

Section 21.4 The Equipartition of Energy

28. **S** A certain molecule has f degrees of freedom. Show that an ideal gas consisting of such molecules has the following properties: (a) its total internal energy is $fnRT/2$, (b) its molar specific heat at constant volume is $fR/2$, (c) its molar specific heat at constant pressure is $(f + 2)R/2$, and (d) its specific heat ratio is $\gamma = C_P/C_V = (f + 2)/f$.

29. In a crude model (Fig. P21.29) of a rotating diatomic chlorine molecule (Cl_2), the two Cl atoms are 2.00×10^{-10} m apart and rotate about their center of mass with angular speed $\omega = 2.00 \times 10^{12}$ rad/s. What is the rotational kinetic energy of one molecule of Cl_2, which has a molar mass of 70.0 g/mol?

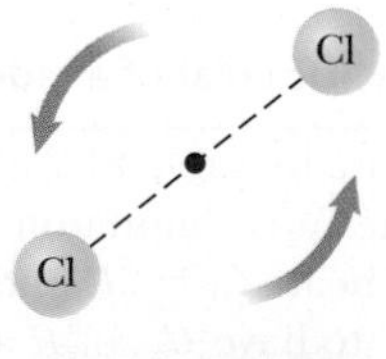

Figure P21.29

30. *Why is the following situation impossible?* A team of researchers discovers a new gas, which has a value of $\gamma = C_P/C_V$ of 1.75.

31. **M** The relationship between the heat capacity of a sample and the specific heat of the sample material is discussed in Section 20.2. Consider a sample containing 2.00 mol of an ideal diatomic gas. Assuming the molecules rotate but do

not vibrate, find (a) the total heat capacity of the sample at constant volume and (b) the total heat capacity at constant pressure. (c) **What If?** Repeat parts (a) and (b), assuming the molecules both rotate and vibrate.

Section 21.5 Distribution of Molecular Speeds

32. **M** Fifteen identical particles have various speeds: one has a speed of 2.00 m/s, two have speeds of 3.00 m/s, three have speeds of 5.00 m/s, four have speeds of 7.00 m/s, three have speeds of 9.00 m/s, and two have speeds of 12.0 m/s. Find (a) the average speed, (b) the rms speed, and (c) the most probable speed of these particles.

33. One cubic meter of atomic hydrogen at 0°C at atmospheric pressure contains approximately 2.70×10^{25} atoms. The first excited state of the hydrogen atom has an energy of 10.2 eV above that of the lowest state, called the ground state. Use the Boltzmann factor to find the number of atoms in the first excited state (a) at 0°C and at (b) (1.00×10^4)°C.

34. Two gases in a mixture diffuse through a filter at rates proportional to their rms speeds. (a) Find the ratio of speeds for the two isotopes of chlorine, ^{35}Cl and ^{37}Cl, as they diffuse through the air. (b) Which isotope moves faster?

35. **Review.** At what temperature would the average speed of helium atoms equal (a) the escape speed from the Earth, 1.12×10^4 m/s, and (b) the escape speed from the Moon, 2.37×10^3 m/s? *Note:* The mass of a helium atom is 6.64×10^{-27} kg.

36. **Q|C** Consider a container of nitrogen gas molecules at 900 K. Calculate (a) the most probable speed, (b) the average speed, and (c) the rms speed for the molecules. (d) State how your results compare with the values displayed in Active Figure 21.11.

37. Assume the Earth's atmosphere has a uniform temperature of 20.0°C and uniform composition, with an effective molar mass of 28.9 g/mol. (a) Show that the number density of molecules depends on height y above sea level according to

$$n_V(y) = n_0 e^{-m_0 g y / k_B T}$$

where n_0 is the number density at sea level (where $y = 0$). This result is called the *law of atmospheres.* (b) Commercial jetliners typically cruise at an altitude of 11.0 km. Find the ratio of the atmospheric density there to the density at sea level.

38. **S** From the Maxwell–Boltzmann speed distribution, show that the most probable speed of a gas molecule is given by Equation 21.27. *Note:* The most probable speed corresponds to the point at which the slope of the speed distribution curve dN_v/dv is zero.

39. The law of atmospheres states that the number density of molecules in the atmosphere depends on height y above sea level according to

$$n_V(y) = n_0 e^{-m_0 g y / k_B T}$$

where n_0 is the number density at sea level (where $y = 0$). The average height of a molecule in the Earth's atmosphere is given by

$$y_{avg} = \frac{\int_0^\infty y n_V(y)\, dy}{\int_0^\infty n_V(y)\, dy} = \frac{\int_0^\infty y e^{-m_0 g y / k_B T}\, dy}{\int_0^\infty e^{-m_0 g y / k_B T}\, dy}$$

(a) Prove that this average height is equal to $k_B T / m_0 g$. (b) Evaluate the average height, assuming the temperature is 10.0°C and the molecular mass is 28.9 u, both uniform throughout the atmosphere.

Additional Problems

40. Eight molecules have speeds of 3.00 km/s, 4.00 km/s, 5.80 km/s, 2.50 km/s, 3.60 km/s, 1.90 km/s, 3.80 km/s, and 6.60 km/s. Find (a) the average speed of the molecules and (b) the rms speed of the molecules.

41. A small oxygen tank at a gauge pressure of 125 atm has a volume of 6.88 L at 21.0°C. (a) If an athlete breathes oxygen from this tank at the rate of 8.50 L/min when measured at atmospheric pressure and the temperature remains at 21.0°C, how long will the tank last before it is empty? (b) At a particular moment during this process, what is the ratio of the rms speed of the molecules remaining in the tank to the rms speed of those being released at atmospheric pressure?

42. **Q|C** The dimensions of a classroom are 4.20 m × 3.00 m × 2.50 m. (a) Find the number of molecules of air in the classroom at atmospheric pressure and 20.0°C. (b) Find the mass of this air, assuming the air consists of diatomic molecules with molar mass 28.9 g/mol. (c) Find the average kinetic energy of the molecules. (d) Find the rms molecular speed. (e) **What If?** Assume the molar specific heat of the air is independent of temperature. Find the change in internal energy of the air in the room as the temperature is raised to 25.0°C. (f) Explain how you could convince a fellow student that your answer to part (e) is correct, even though it sounds surprising.

43. **Q|C** The Earth's atmosphere consists primarily of oxygen (21%) and nitrogen (78%). The rms speed of oxygen molecules (O_2) in the atmosphere at a certain location is 535 m/s. (a) What is the temperature of the atmosphere at this location? (b) Would the rms speed of nitrogen molecules (N_2) at this location be higher, equal to, or lower than 535 m/s? Explain. (c) Determine the rms speed of N_2 at his location.

44. **Q|C** The *mean free path* ℓ of a molecule is the average distance that a molecule travels before colliding with another molecule. It is given by

$$\ell = \frac{1}{\sqrt{2}\pi d^2 N_V}$$

where d is the diameter of the molecule and N_V is the number of molecules per unit volume. The number of collisions that a molecule makes with other molecules per unit time, or *collision frequency* f, is given by

$$f = \frac{v_{avg}}{\ell}$$

(a) If the diameter of an oxygen molecule is 2.00×10^{-10} m, find the mean free path of the molecules in a scuba tank that has a volume of 12.0 L and is filled with oxygen at a gauge pressure of 100 atm at a temperature of 25.0°C. (b) What is the average time interval between molecular collisions for a molecule of this gas?

45. A certain ideal gas has a molar specific heat of $C_V = \frac{7}{2}R$. A 2.00-mol sample of the gas always starts at pressure 1.00×10^5 Pa and temperature 300 K. For each of the following processes, determine (a) the final pressure, (b) the final volume, (c) the final temperature, (d) the change in internal energy of the gas, (e) the energy added to the gas by heat, and (f) the work done on the gas. **(i)** The gas is heated at constant pressure to 400 K. **(ii)** The gas is heated at constant volume to 400 K. **(iii)** The gas is compressed at constant temperature to 1.20×10^5 Pa. **(iv)** The gas is compressed adiabatically to 1.20×10^5 Pa.

46. **Q|C** In a sample of a solid metal, each atom is free to vibrate about some equilibrium position. The atom's energy consists of kinetic energy for motion in the x, y, and z directions plus elastic potential energy associated with the Hooke's law forces exerted by neighboring atoms on it in the x, y, and z directions. According to the theorem of equipartition of energy, assume the average energy of each atom is $\frac{1}{2}k_BT$ for each degree of freedom. (a) Prove that the molar specific heat of the solid is $3R$. The *Dulong–Petit law* states that this result generally describes pure solids at sufficiently high temperatures. (You may ignore the difference between the specific heat at constant pressure and the specific heat at constant volume.) (b) Evaluate the specific heat c of iron. Explain how it compares with the value listed in Table 20.1. (c) Repeat the evaluation and comparison for gold.

47. **S** Twenty particles, each of mass m_0 and confined to a volume V, have various speeds: two have speed v, three have speed $2v$, five have speed $3v$, four have speed $4v$, three have speed $5v$, two have speed $6v$, and one has speed $7v$. Find (a) the average speed, (b) the rms speed, (c) the most probable speed, (d) the average pressure the particles exert on the walls of the vessel, and (e) the average kinetic energy per particle.

48. **Q|C** **S** In a cylinder, a sample of an ideal gas with number of moles n undergoes an adiabatic process. (a) Starting with the expression $W = -\int P\,dV$ and using the condition PV^γ = constant, show that the work done on the gas is

$$W = \left(\frac{1}{\gamma - 1}\right)(P_fV_f - P_iV_i)$$

(b) Starting with the first law of thermodynamics, show that the work done on the gas is equal to $nC_V(T_f - T_i)$. (c) Are these two results consistent with each other? Explain.

49. As a 1.00-mol sample of a monatomic ideal gas expands adiabatically, the work done on it is -2.50×10^3 J. The initial temperature and pressure of the gas are 500 K and 3.60 atm. Calculate (a) the final temperature and (b) the final pressure.

50. **S** A sample consists of an amount n in moles of a monatomic ideal gas. The gas expands adiabatically, with work W done on it. (Work W is a negative number.) The initial temperature and pressure of the gas are T_i and P_i. Calculate (a) the final temperature and (b) the final pressure.

51. An air rifle shoots a lead pellet by allowing high-pressure air to expand, propelling the pellet down the rifle barrel. Because this process happens very quickly, no appreciable thermal conduction occurs and the expansion is essentially adiabatic. Suppose the rifle starts with 12.0 cm^3 of compressed air, which behaves as an ideal gas with $\gamma = 1.40$. The expanding air pushes a 1.10-g pellet as a piston with cross-sectional area 0.030 0 cm^2 along the 50.0-cm-long gun barrel. What initial pressure is required to eject the pellet with a muzzle speed of 120 m/s? Ignore the effects of the air in front of the bullet and friction with the inside walls of the barrel.

52. The compressibility κ of a substance is defined as the fractional change in volume of that substance for a given change in pressure:

$$\kappa = -\frac{1}{V}\frac{dV}{dP}$$

(a) Explain why the negative sign in this expression ensures κ is always positive. (b) Show that if an ideal gas is compressed isothermally, its compressibility is given by $\kappa_1 = 1/P$. (c) **What If?** Show that if an ideal gas is compressed adiabatically, its compressibility is given by $\kappa_2 = 1/(\gamma P)$. Determine values for (d) κ_1 and (e) κ_2 for a monatomic ideal gas at a pressure of 2.00 atm.

53. **Review.** Oxygen at pressures much greater than 1 atm is toxic to lung cells. Assume a deep-sea diver breathes a mixture of oxygen (O_2) and helium (He). By weight, what ratio of helium to oxygen must be used if the diver is at an ocean depth of 50.0 m?

54. **Q|C** Examine the data for polyatomic gases in Table 21.2 and give a reason why sulfur dioxide has a higher specific heat at constant volume than the other polyatomic gases at 300 K.

55. Model air as a diatomic ideal gas with M = 28.9 g/mol. A cylinder with a piston contains 1.20 kg of air at 25.0°C and 2.00×10^5 Pa. Energy is transferred by heat into the system as it is permitted to expand, with the pressure rising to 4.00×10^5 Pa. Throughout the expansion, the relationship between pressure and volume is given by

$$P = CV^{1/2}$$

where C is a constant. Find (a) the initial volume, (b) the final volume, (c) the final temperature, (d) the work done on the air, and (e) the energy transferred by heat.

56. **Q|C** **Review.** As a sound wave passes through a gas, the compressions are either so rapid or so far apart that thermal conduction is prevented by a negligible time interval or by effective thickness of insulation. The compressions and rarefactions are adiabatic. (a) Show that the speed of sound in an ideal gas is

$$v = \sqrt{\frac{\gamma RT}{M}}$$

where M is the molar mass. The speed of sound in a liquid is given by Equation 17.8; use that equation and the definition of the bulk modulus from Section 12.4. (b) Compute the theoretical speed of sound in air at 20.0°C and state how it compares with the value in Table 17.1. Take M = 28.9 g/mol. (c) Show that the speed of sound in an ideal gas is

$$v = \sqrt{\frac{\gamma k_B T}{m_0}}$$

where m_0 is the mass of one molecule. (d) State how the result in part (c) compares with the most probable, average, and rms molecular speeds.

57. A pitcher throws a 0.142-kg baseball at 47.2 m/s. As it travels 16.8 m to home plate, the ball slows down to 42.5 m/s because of air resistance. Find the change in temperature of the air through which it passes. To find the greatest possible temperature change, you may make the following assumptions. Air has a molar specific heat of $C_P = \frac{7}{2}R$ and an equivalent molar mass of 28.9 g/mol. The process is so rapid that the cover of the baseball acts as thermal insulation and the temperature of the ball itself does not change. A change in temperature happens initially only for the air in a cylinder 16.8 m in length and 3.70 cm in radius. This air is initially at 20.0°C.

58. **Q|C** The latent heat of vaporization for water at room temperature is 2 430 J/g. Consider one particular molecule at the surface of a glass of liquid water, moving upward with sufficiently high speed that it will be the next molecule to join the vapor. (a) Find its translational kinetic energy. (b) Find its speed. Now consider a thin gas made only of molecules like that one. (c) What is its temperature? (d) Why are you not burned by water evaporating from a vessel at room temperature?

59. For a Maxwellian gas, use a computer or programmable calculator to find the numerical value of the ratio $N_v(v)/N_v(v_{mp})$ for the following values of v: (a) $v = (v_{mp}/50.0)$, (b) $(v_{mp}/10.0)$, (c) $(v_{mp}/2.00)$, (d) v_{mp}, (e) $2.00v_{mp}$, (f) $10.0v_{mp}$, and (g) $50.0v_{mp}$. Give your results to three significant figures.

60. **Q|C** A triatomic molecule can have a linear configuration, as does CO_2 (Fig. P21.60a), or it can be nonlinear, like H_2O (Fig. P21.60b). Suppose the temperature of a gas of triatomic molecules is sufficiently low that vibrational motion is negligible. What is the molar specific heat at constant volume, expressed as a multiple of the universal gas constant, (a) if the molecules are linear and (b) if the molecules are nonlinear? At high temperatures, a triatomic molecule has two modes of vibration, and each contributes $\frac{1}{2}R$ to the molar specific heat for its kinetic energy and another $\frac{1}{2}R$ for its potential energy. Identify the high-temperature molar specific heat at constant volume for a triatomic ideal gas of (c) linear molecules and (d) nonlinear molecules. (e) Explain how specific heat data can be used to determine whether a triatomic molecule is linear or nonlinear. Are the data in Table 21.2 sufficient to make this determination?

61. When a small particle is suspended in a fluid, bombardment by molecules makes the particle jitter about at random. Robert Brown discovered this motion in 1827 while studying plant fertilization, and the motion has become known as *Brownian motion.* The particle's average kinetic energy can be taken as $\frac{3}{2}k_BT$, the same as that of a molecule in an ideal gas. Consider a spherical particle of density 1.00×10^3 kg/m^3 in water at 20.0°C. (a) For a particle of diameter d, evaluate the rms speed. (b) The particle's actual motion is a random walk, but imagine that it moves with constant velocity equal in magnitude to its rms speed. In what time interval would it move by a distance equal to its own diameter? (c) Evaluate the rms speed and the time interval for a particle of diameter 3.00 μm. (d) Evaluate the rms speed and the time interval for a sphere of mass 70.0 kg, modeling your own body.

62. A vessel contains 1.00×10^4 oxygen molecules at 500 K. (a) Make an accurate graph of the Maxwell speed distribution function versus speed with points at speed intervals of 100 m/s. (b) Determine the most probable speed from this graph. (c) Calculate the average and rms speeds for the molecules and label these points on your graph. (d) From the graph, estimate the fraction of molecules with speeds in the range 300 m/s to 600 m/s.

63. **Q|C** A sample of a monatomic ideal gas occupies 5.00 L at atmospheric pressure and 300 K (point A in Fig. P21.63). It is warmed at constant volume to 3.00 atm (point B). Then it is allowed to expand isothermally to 1.00 atm (point C) and at last compressed isobarically to its original state. (a) Find the number of moles in the sample. Find (b) the temperature at point B, (c) the temperature at point C, and (d) the volume at point C. (e) Now consider the processes $A \to B$, $B \to C$, and $C \to A$. Describe how to carry out each

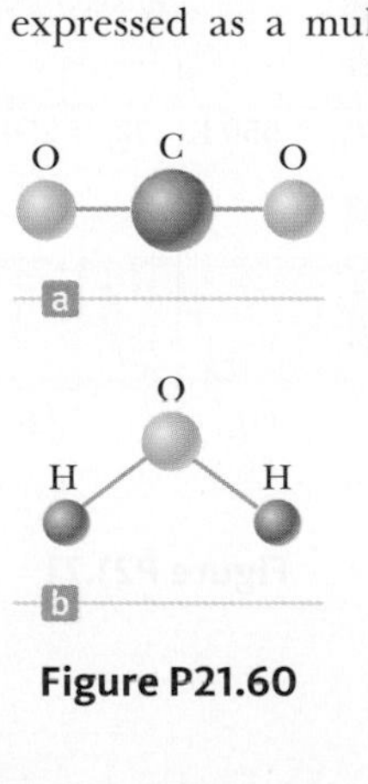

Figure P21.60

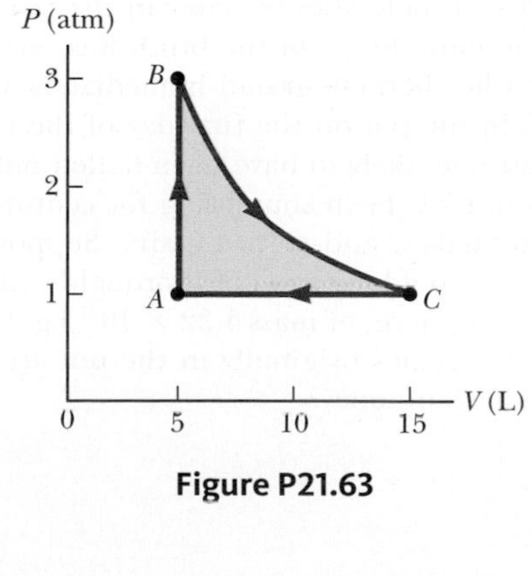

Figure P21.63

process experimentally. (f) Find Q, W, and ΔE_{int} for each of the processes. (g) For the whole cycle $A \rightarrow B \rightarrow C \rightarrow A$, find Q, W, and ΔE_{int}.

64. QC S Consider the particles in a gas centrifuge, a device used to separate particles of different mass by whirling them in a circular path of radius r at angular speed ω. The force acting on a gas molecule toward the center of the centrifuge is $m_0\omega^2 r$. (a) Discuss how a gas centrifuge can be used to separate particles of different mass. (b) Suppose the centrifuge contains a gas of particles of identical mass. Show that the density of the particles as a function of r is

$$n(r) = n_0 e^{m_0 r^2 \omega^2 / 2k_B T}$$

65. S Using the Maxwell–Boltzmann speed distribution function, verify Equations 21.25 and 21.26 for (a) the rms speed and (b) the average speed of the molecules of a gas at a temperature T. The average value of v^n is

$$\overline{v^n} = \frac{1}{N}\int_0^\infty v^n N_v \, dv$$

Use the table of integrals B.6 in Appendix B.

66. On the PV diagram for an ideal gas, one isothermal curve and one adiabatic curve pass through each point as shown in Figure P21.66. Prove that the slope of the adiabatic curve is steeper than the slope of the isotherm at that point by the factor γ.

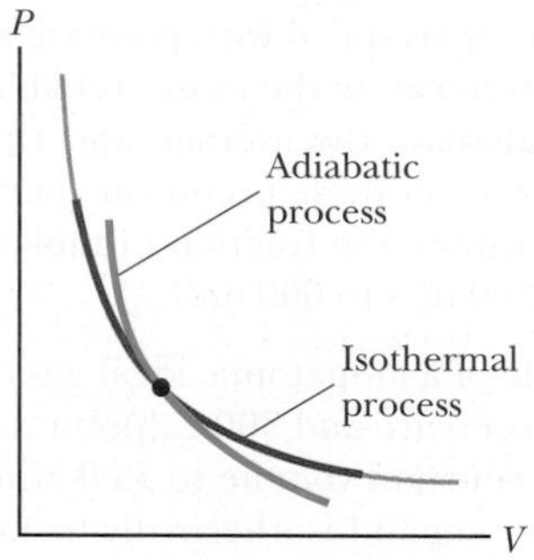

Figure P21.66

67. In Beijing, a restaurant keeps a pot of chicken broth simmering continuously. Every morning, it is topped up to contain 10.0 L of water along with a fresh chicken, vegetables, and spices. The molar mass of water is 18.0 g/mol. (a) Find the number of molecules of water in the pot. (b) During a certain month, 90.0% of the broth was served each day to people who then emigrated immediately. Of the water molecules in the pot on the first day of the month, when was the last one likely to have been ladled out of the pot? (c) The broth has been simmering for centuries, through wars, earthquakes, and stove repairs. Suppose the water that was in the pot long ago has thoroughly mixed into the Earth's hydrosphere, of mass 1.32×10^{21} kg. How many of the water molecules originally in the pot are likely to be present in it again today?

68. S **Review.** (a) If it has enough kinetic energy, a molecule at the surface of the Earth can "escape the Earth's gravitation" in the sense that it can continue to move away from the Earth forever as discussed in Section 13.6. Using the principle of conservation of energy, show that the minimum kinetic energy needed for "escape" is $m_0 g R_E$, where m_0 is the mass of the molecule, g is the free-fall acceleration at the surface, and R_E is the radius of the Earth. (b) Calculate the temperature for which the minimum escape kinetic energy is ten times the average kinetic energy of an oxygen molecule.

69. Using multiple laser beams, physicists have been able to cool and trap sodium atoms in a small region. In one experiment, the temperature of the atoms was reduced to 0.240 mK. (a) Determine the rms speed of the sodium atoms at this temperature. The atoms can be trapped for about 1.00 s. The trap has a linear dimension of roughly 1.00 cm. (b) Over what approximate time interval would an atom wander out of the trap region if there were no trapping action?

Challenge Problems

70. QC S Equations 21.25 and 21.26 show that $v_{\text{rms}} > v_{\text{avg}}$ for a collection of gas particles, which turns out to be true whenever the particles have a distribution of speeds. Let us explore this inequality for a two-particle gas. Let the speed of one particle be $v_1 = av_{\text{avg}}$ and the other particle have speed $v_2 = (2 - a)v_{\text{avg}}$. (a) Show that the average of these two speeds is v_{avg}. (b) Show that

$$v_{\text{rms}}^2 = v_{\text{avg}}^2 (2 - 2a + a^2)$$

(c) Argue that the equation in part (b) proves that, in general, $v_{\text{rms}} > v_{\text{avg}}$. (d) Under what special condition will $v_{\text{rms}} = v_{\text{avg}}$ for the two-particle gas?

71. A cylinder is closed at both ends and has insulating walls. It is divided into two compartments by an insulating piston that is perpendicular to the axis of the cylinder as shown in Figure P21.71a. Each compartment contains 1.00 mol of oxygen that behaves as an ideal gas with $\gamma = 1.40$. Initially, the two compartments have equal volumes and their temperatures are 550 K and 250 K. The piston is then allowed to move slowly parallel to the axis of the cylinder until it comes to rest at an equilibrium position (Fig. P21.71b). Find the final temperatures in the two compartments.

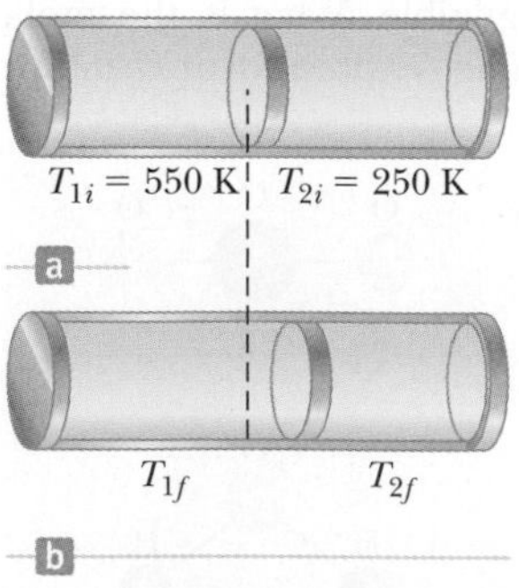

Figure P21.71

chapter 22

Heat Engines, Entropy, and the Second Law of Thermodynamics

A Stirling engine from the early nineteenth century. Air is heated in the lower cylinder using an external source. As this happens, the air expands and pushes against a piston, causing it to move. The air is then cooled, allowing the cycle to begin again. This is one example of a heat engine, which we study in this chapter. *(© SSPL/The Image Works)*

The first law of thermodynamics, which we studied in Chapter 20, is a statement of conservation of energy and is a special-case reduction of Equation 8.2. This law states that a change in internal energy in a system can occur as a result of energy transfer by heat, by work, or by both. Although the first law of thermodynamics is very important, it makes no distinction between processes that occur spontaneously and those that do not. Only certain types of energy conversion and energy transfer processes actually take place in nature, however. The *second law of thermodynamics*, the major topic in this chapter, establishes which processes do and do not occur. The following are examples of

J-L Charmet/Science Photo Library/Photo Researchers, Inc.

Lord Kelvin
British physicist and mathematician (1824–1907)
Born William Thomson in Belfast, Kelvin was the first to propose the use of an absolute scale of temperature. The Kelvin temperature scale is named in his honor. Kelvin's work in thermodynamics led to the idea that energy cannot pass spontaneously from a colder object to a hotter object.

processes that do not violate the first law of thermodynamics if they proceed in either direction, but are observed in reality to proceed in only one direction:

- When two objects at different temperatures are placed in thermal contact with each other, the net transfer of energy by heat is always from the warmer object to the cooler object, never from the cooler to the warmer.
- A rubber ball dropped to the ground bounces several times and eventually comes to rest, but a ball lying on the ground never gathers internal energy from the ground and begins bouncing on its own.
- An oscillating pendulum eventually comes to rest because of collisions with air molecules and friction at the point of suspension. The mechanical energy of the system is converted to internal energy in the air, the pendulum, and the suspension; the reverse conversion of energy never occurs.

All these processes are *irreversible;* that is, they are processes that occur naturally in one direction only. No irreversible process has ever been observed to run backward. If it were to do so, it would violate the second law of thermodynamics.[1]

22.1 Heat Engines and the Second Law of Thermodynamics

A **heat engine** is a device that takes in energy by heat[2] and, operating in a cyclic process, expels a fraction of that energy by means of work. For instance, in a typical process by which a power plant produces electricity, a fuel such as coal is burned and the high-temperature gases produced are used to convert liquid water to steam. This steam is directed at the blades of a turbine, setting it into rotation. The mechanical energy associated with this rotation is used to drive an electric generator. Another device that can be modeled as a heat engine is the internal combustion engine in an automobile. This device uses energy from a burning fuel to perform work on pistons that results in the motion of the automobile.

A heat engine carries some working substance through a cyclic process during which (1) the working substance absorbs energy by heat from a high-temperature energy reservoir, (2) work is done by the engine, and (3) energy is expelled by heat to a lower-temperature reservoir. As an example, consider the operation of a steam engine (Fig. 22.1), which uses water as the working substance. The water in a boiler absorbs energy from burning fuel and evaporates to steam, which then does work by expanding against a piston. After the steam cools and condenses, the liquid water produced returns to the boiler and the cycle repeats.

© Andy Moore/Photolibrary/Jupiterimages

Figure 22.1 A steam-driven locomotive obtains its energy by burning wood or coal. The generated energy vaporizes water into steam, which powers the locomotive. Modern locomotives use diesel fuel instead of wood or coal. Whether old-fashioned or modern, such locomotives can be modeled as heat engines, which extract energy from a burning fuel and convert a fraction of it to mechanical energy.

It is useful to represent a heat engine schematically as in Active Figure 22.2. The engine absorbs a quantity of energy $|Q_h|$ from the hot reservoir. For the mathematical discussion of heat engines, we use absolute values to make all energy transfers

[1]Although a process occurring in the time-reversed sense has never been *observed,* it is *possible* for it to occur. As we shall see later in this chapter, however, the probability of such a process occurring is infinitesimally small. From this viewpoint, processes occur with a vastly greater probability in one direction than in the opposite direction.

[2]We use heat as our model for energy transfer into a heat engine. Other methods of energy transfer are possible in the model of a heat engine, however. For example, the Earth's atmosphere can be modeled as a heat engine in which the input energy transfer is by means of electromagnetic radiation from the Sun. The output of the atmospheric heat engine causes the wind structure in the atmosphere.

by heat positive, and the direction of transfer is indicated with an explicit positive or negative sign. The engine does work W_{eng} (so that *negative* work $W = -W_{eng}$ is done *on* the engine) and then gives up a quantity of energy $|Q_c|$ to the cold reservoir. Because the working substance goes through a cycle, its initial and final internal energies are equal: $\Delta E_{int} = 0$. Hence, from the first law of thermodynamics, $\Delta E_{int} = Q + W = Q - W_{eng} = 0$, and the net work W_{eng} done by a heat engine is equal to the net energy Q_{net} transferred to it. As you can see from Active Figure 22.2, $Q_{net} = |Q_h| - |Q_c|$; therefore,

$$W_{eng} = |Q_h| - |Q_c| \quad \textbf{(22.1)}$$

The **thermal efficiency** e of a heat engine is defined as the ratio of the net work done by the engine during one cycle to the energy input at the higher temperature during the cycle:

$$e \equiv \frac{W_{eng}}{|Q_h|} = \frac{|Q_h| - |Q_c|}{|Q_h|} = 1 - \frac{|Q_c|}{|Q_h|} \quad \textbf{(22.2)}$$

◀ **Thermal efficiency of a heat engine**

You can think of the efficiency as the ratio of what you gain (work) to what you give (energy transfer at the higher temperature). In practice, all heat engines expel only a fraction of the input energy Q_h by mechanical work; consequently, their efficiency is always less than 100%. For example, a good automobile engine has an efficiency of about 20%, and diesel engines have efficiencies ranging from 35% to 40%.

Equation 22.2 shows that a heat engine has 100% efficiency ($e = 1$) only if $|Q_c| = 0$, that is, if no energy is expelled to the cold reservoir. In other words, a heat engine with perfect efficiency would have to expel all the input energy by work. Because efficiencies of real engines are well below 100%, the **Kelvin–Planck form of the second law of thermodynamics** states the following:

> It is impossible to construct a heat engine that, operating in a cycle, produces no effect other than the input of energy by heat from a reservoir and the performance of an equal amount of work.

This statement of the second law means that during the operation of a heat engine, W_{eng} can never be equal to $|Q_h|$ or, alternatively, that some energy $|Q_c|$ must be rejected to the environment. Every heat engine *must* have some energy exhaust. Figure 22.3 is a schematic diagram of the impossible "perfect" heat engine.

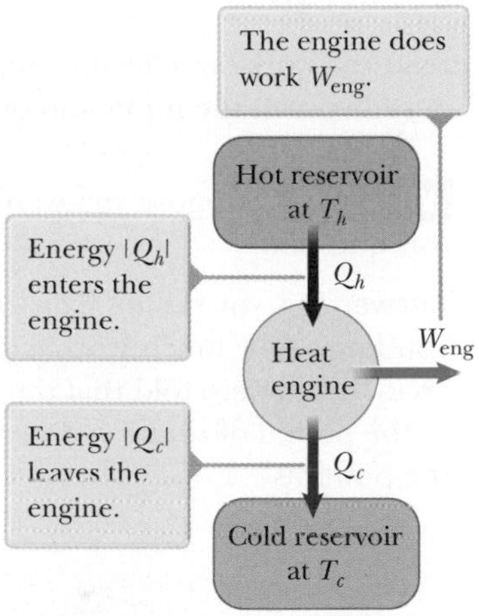

ACTIVE FIGURE 22.2

Schematic representation of a heat engine.

***Quick Quiz* 22.1** The energy input to an engine is 3.00 times greater than the work it performs. **(i)** What is its thermal efficiency? (a) 3.00 (b) 1.00 (c) 0.333 (d) impossible to determine **(ii)** What fraction of the energy input is expelled to the cold reservoir? (a) 0.333 (b) 0.667 (c) 1.00 (d) impossible to determine

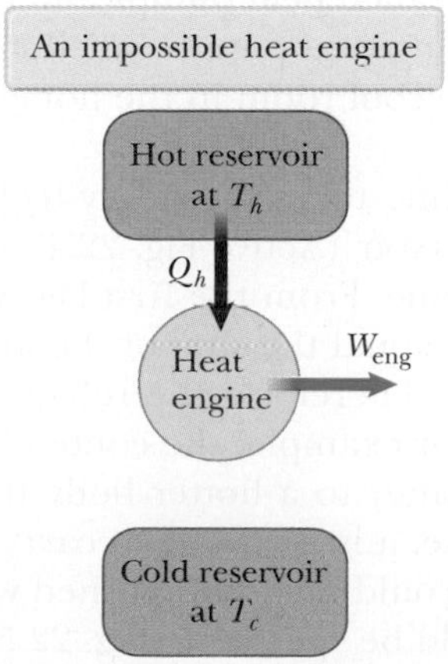

Figure 22.3 Schematic diagram of a heat engine that takes in energy from a hot reservoir and does an equivalent amount of work. It is impossible to construct such a perfect engine.

Pitfall Prevention 22.1

The First and Second Laws

Notice the distinction between the first and second laws of thermodynamics. If a gas undergoes a *one-time isothermal process*, then $\Delta E_{int} = Q + W = 0$ and $W = -Q$. Therefore, the first law allows *all* energy input by heat to be expelled by work. In a heat engine, however, in which a substance undergoes a *cyclic* process, only a *portion* of the energy input by heat can be expelled by work according to the second law.

Example 22.1 The Efficiency of an Engine

An engine transfers 2.00×10^3 J of energy from a hot reservoir during a cycle and transfers 1.50×10^3 J as exhaust to a cold reservoir.

(A) Find the efficiency of the engine.

SOLUTION

Conceptualize Review Active Figure 22.2; think about energy going into the engine from the hot reservoir and splitting, with part coming out by work and part by heat into the cold reservoir.

Categorize This example involves evaluation of quantities from the equations introduced in this section, so we categorize it as a substitution problem.

Find the efficiency of the engine from Equation 22.2:

$$e = 1 - \frac{|Q_c|}{|Q_h|} = 1 - \frac{1.50 \times 10^3 \text{ J}}{2.00 \times 10^3 \text{ J}} = \boxed{0.250, \text{ or } 25.0\%}$$

(B) How much work does this engine do in one cycle?

SOLUTION

Find the work done by the engine by taking the difference between the input and output energies:

$$W_{\text{eng}} = |Q_h| - |Q_c| = 2.00 \times 10^3 \text{ J} - 1.50 \times 10^3 \text{ J}$$

$$= \boxed{5.0 \times 10^2 \text{ J}}$$

WHAT IF? Suppose you were asked for the power output of this engine. Do you have sufficient information to answer this question?

Answer No, you do not have enough information. The power of an engine is the *rate* at which work is done by the engine. You know how much work is done per cycle, but you have no information about the time interval associated with one cycle. If you were told that the engine operates at 2 000 rpm (revolutions per minute), however, you could relate this rate to the period of rotation T of the mechanism of the engine. Assuming there is one thermodynamic cycle per revolution, the power is

$$P = \frac{W_{\text{eng}}}{T} = \frac{5.0 \times 10^2 \text{ J}}{\left(\frac{1}{2\,000} \text{ min}\right)}\left(\frac{1 \text{ min}}{60 \text{ s}}\right) = 1.7 \times 10^4 \text{ W}$$

22.2 Heat Pumps and Refrigerators

In a heat engine, the direction of energy transfer is from the hot reservoir to the cold reservoir, which is the natural direction. The role of the heat engine is to process the energy from the hot reservoir so as to do useful work. What if we wanted to transfer energy from the cold reservoir to the hot reservoir? Because that is not the natural direction of energy transfer, we must put some energy into a device to be successful. Devices that perform this task are called **heat pumps** and **refrigerators.** For example, homes in summer are cooled using heat pumps called *air conditioners.* The air conditioner transfers energy from the cool room in the home to the warm air outside.

In a refrigerator or a heat pump, the engine takes in energy $|Q_c|$ from a cold reservoir and expels energy $|Q_h|$ to a hot reservoir (Active Fig. 22.4), which can be accomplished only if work is done *on* the engine. From the first law, we know that the energy given up to the hot reservoir must equal the sum of the work done and the energy taken in from the cold reservoir. Therefore, the refrigerator or heat pump transfers energy from a colder body (for example, the contents of a kitchen refrigerator or the winter air outside a building) to a hotter body (the air in the kitchen or a room in the building). In practice, it is desirable to carry out this process with a minimum of work. If the process could be accomplished without doing any work, the refrigerator or heat pump would be "perfect" (Fig. 22.5). Again, the

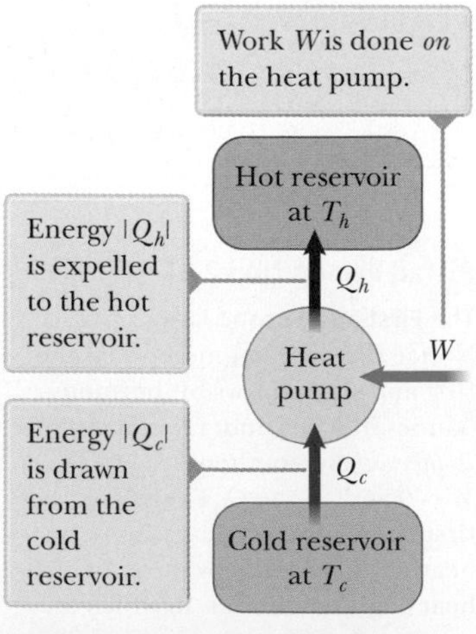

ACTIVE FIGURE 22.4

Schematic representation of a heat pump.

existence of such a device would be in violation of the second law of thermodynamics, which in the form of the **Clausius statement**[3] states:

> It is impossible to construct a cyclical machine whose sole effect is to transfer energy continuously by heat from one object to another object at a higher temperature without the input of energy by work.

In simpler terms, energy does not transfer spontaneously by heat from a cold object to a hot object. Work input is required to run a refrigerator.

The Clausius and Kelvin–Planck statements of the second law of thermodynamics appear at first sight to be unrelated, but in fact they are equivalent in all respects. Although we do not prove so here, if either statement is false, so is the other.[4]

In practice, a heat pump includes a circulating fluid that passes through two sets of metal coils that can exchange energy with the surroundings. The fluid is cold and at low pressure when it is in the coils located in a cool environment, where it absorbs energy by heat. The resulting warm fluid is then compressed and enters the other coils as a hot, high-pressure fluid. There it releases its stored energy to the warm surroundings. In an air conditioner, energy is absorbed into the fluid in coils located in a building's interior; after the fluid is compressed, energy leaves the fluid through coils located outdoors. In a refrigerator, the external coils are behind or underneath the unit (Fig. 22.6). The internal coils are in the walls of the refrigerator and absorb energy from the food.

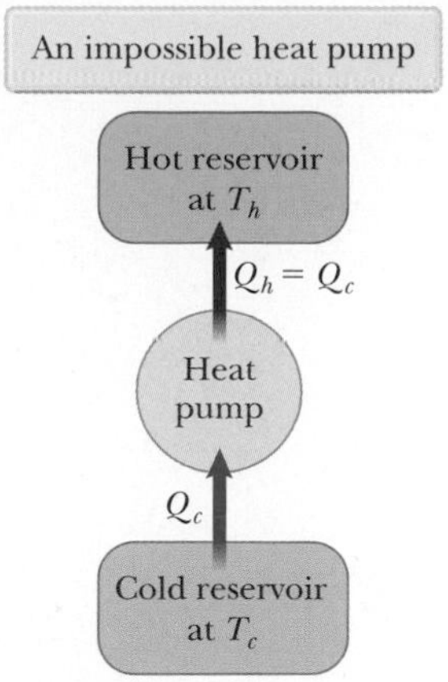

Figure 22.5 Schematic diagram of an impossible heat pump or refrigerator, that is, one that takes in energy from a cold reservoir and expels an equivalent amount of energy to a hot reservoir without the input of energy by work.

The effectiveness of a heat pump is described in terms of a number called the **coefficient of performance** (COP). The COP is similar to the thermal efficiency for a heat engine in that it is a ratio of what you gain (energy transferred to or from a reservoir) to what you give (work input). For a heat pump operating in the cooling mode, "what you gain" is energy removed from the cold reservoir. The most effective refrigerator or air conditioner is one that removes the greatest amount of energy from the cold reservoir in exchange for the least amount of work. Therefore, for these devices operating in the cooling mode, we define the COP in terms of $|Q_c|$:

$$\text{COP (cooling mode)} = \frac{\text{energy transferred at low temperature}}{\text{work done on heat pump}} = \frac{|Q_c|}{W} \qquad \textbf{(22.3)}$$

A good refrigerator should have a high COP, typically 5 or 6.

In addition to cooling applications, heat pumps are becoming increasingly popular for heating purposes. The energy-absorbing coils for a heat pump are located outside a building, in contact with the air or buried in the ground. The other set of coils are in the building's interior. The circulating fluid flowing through the coils absorbs energy from the outside and releases it to the interior of the building from the interior coils.

In the heating mode, the COP of a heat pump is defined as the ratio of the energy transferred to the hot reservoir to the work required to transfer that energy:

$$\text{COP (heating mode)} = \frac{\text{energy transferred at high temperature}}{\text{work done on heat pump}} = \frac{|Q_h|}{W} \qquad \textbf{(22.4)}$$

If the outside temperature is 25°F (−4°C) or higher, a typical value of the COP for a heat pump is about 4. That is, the amount of energy transferred to the building is about four times greater than the work done by the motor in the heat pump. As the outside temperature decreases, however, it becomes more difficult for the heat pump to extract sufficient energy from the air and so the COP decreases. Therefore, the use of heat pumps that extract energy from the air, although satisfactory in

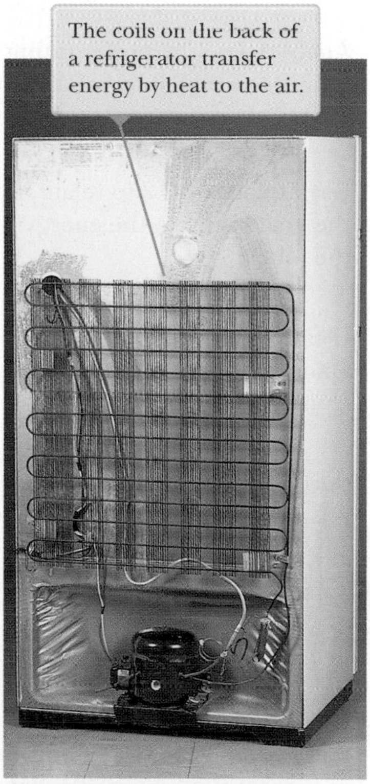

Figure 22.6 The back of a household refrigerator. The air surrounding the coils is the hot reservoir.

[3]First expressed by Rudolf Clausius (1822–1888).

[4]See an advanced textbook on thermodynamics for this proof.

moderate climates, is not appropriate in areas where winter temperatures are very low. It is possible to use heat pumps in colder areas by burying the external coils deep in the ground. In that case, the energy is extracted from the ground, which tends to be warmer than the air in the winter.

***Quick Quiz* 22.2** The energy entering an electric heater by electrical transmission can be converted to internal energy with an efficiency of 100%. By what factor does the cost of heating your home change when you replace your electric heating system with an electric heat pump that has a COP of 4.00? Assume the motor running the heat pump is 100% efficient. **(a)** 4.00 **(b)** 2.00 **(c)** 0.500 **(d)** 0.250

Example 22.2 Freezing Water

A certain refrigerator has a COP of 5.00. When the refrigerator is running, its power input is 500 W. A sample of water of mass 500 g and temperature 20.0°C is placed in the freezer compartment. How long does it take to freeze the water to ice at 0°C? Assume all other parts of the refrigerator stay at the same temperature and there is no leakage of energy from the exterior, so the operation of the refrigerator results only in energy being extracted from the water.

SOLUTION

Conceptualize Energy leaves the water, reducing its temperature and then freezing it into ice. The time interval required for this entire process is related to the rate at which energy is withdrawn from the water, which, in turn, is related to the power input of the refrigerator.

Categorize We categorize this example as one that combines our understanding of temperature changes and phase changes from Chapter 20 and our understanding of heat pumps from this chapter.

Analyze Use the power rating of the refrigerator to find the time interval Δt required for the freezing process to occur:

$$P = \frac{W}{\Delta t} \quad \rightarrow \quad \Delta t = \frac{W}{P}$$

Use Equation 22.3 to relate the work W done on the heat pump to the energy $|Q_c|$ extracted from the water:

$$\Delta t = \frac{|Q_c|}{P(\text{COP})}$$

Use Equations 20.4 and 20.7 to substitute the amount of energy $|Q_c|$ that must be extracted from the water of mass m:

$$\Delta t = \frac{|mc\,\Delta T + L_f\,\Delta m|}{P(\text{COP})}$$

Recognize that the amount of water that freezes is $\Delta m = -m$ because all the water freezes:

$$\Delta t = \frac{|m(c\,\Delta T - L_f)|}{P(\text{COP})}$$

Subsitute numerical values:

$$\Delta t = \frac{|(0.500\text{ kg})[(4\,186\text{ J/kg}\cdot{}^\circ\text{C})(-20.0^\circ\text{C}) - 3.33\times 10^5\text{ J/kg}]|}{(500\text{ W})(5.00)}$$

$$= \boxed{83.3\text{ s}}$$

Finalize In reality, the time interval for the water to freeze in a refrigerator is much longer than 83.3 s, which suggests that the assumptions of our model are not valid. Only a small part of the energy extracted from the refrigerator interior in a given time interval comes from the water. Energy must also be extracted from the container in which the water is placed, and energy that continuously leaks into the interior from the exterior must be extracted.

22.3 Reversible and Irreversible Processes

In the next section, we will discuss a theoretical heat engine that is the most efficient possible. To understand its nature, we must first examine the meaning of reversible and irreversible processes. In a **reversible** process, the system undergoing the process can be returned to its initial conditions along the same path on a *PV* diagram, and every point along this path is an equilibrium state. A process that does not satisfy these requirements is **irreversible.**

Pitfall Prevention 22.2

All Real Processes Are Irreversible
The reversible process is an idealization; all real processes on the Earth are irreversible.

All natural processes are known to be irreversible. Let's examine the adiabatic free expansion of a gas, which was already discussed in Section 20.6, and show that it cannot be reversible. Consider a gas in a thermally insulated container as shown in Figure 22.7. A membrane separates the gas from a vacuum. When the membrane is punctured, the gas expands freely into the vacuum. As a result of the puncture, the system has changed because it occupies a greater volume after the expansion. Because the gas does not exert a force through a displacement, it does no work on the surroundings as it expands. In addition, no energy is transferred to or from the gas by heat because the container is insulated from its surroundings. Therefore, in this adiabatic process, the system has changed but the surroundings have not.

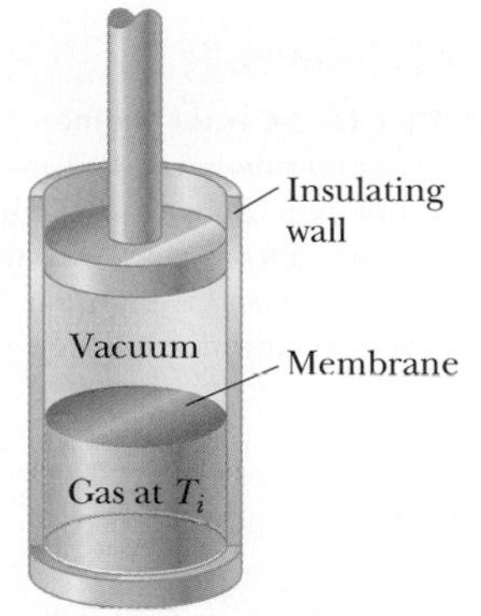

Figure 22.7 Adiabatic free expansion of a gas.

For this process to be reversible, we must return the gas to its original volume and temperature without changing the surroundings. Imagine trying to reverse the process by compressing the gas to its original volume. To do so, we fit the container with a piston and use an engine to force the piston inward. During this process, the surroundings change because work is being done by an outside agent on the system. In addition, the system changes because the compression increases the temperature of the gas. The temperature of the gas can be lowered by allowing it to come into contact with an external energy reservoir. Although this step returns the gas to its original conditions, the surroundings are again affected because energy is being added to the surroundings from the gas. If this energy could be used to drive the engine that compressed the gas, the net energy transfer to the surroundings would be zero. In this way, the system and its surroundings could be returned to their initial conditions and we could identify the process as reversible. The Kelvin–Planck statement of the second law, however, specifies that the energy removed from the gas to return the temperature to its original value cannot be completely converted to mechanical energy in the form of the work done by the engine in compressing the gas. Therefore, we must conclude that the process is irreversible.

We could also argue that the adiabatic free expansion is irreversible by relying on the portion of the definition of a reversible process that refers to equilibrium states. For example, during the sudden expansion, significant variations in pressure occur throughout the gas. Therefore, there is no well-defined value of the pressure for the entire system at any time between the initial and final states. In fact, the process cannot even be represented as a path on a *PV* diagram. The *PV* diagram for an adiabatic free expansion would show the initial and final conditions as points, but these points would not be connected by a path. Therefore, because the intermediate conditions between the initial and final states are not equilibrium states, the process is irreversible.

Figure 22.8 A method for compressing a gas in a reversible isothermal process.

Although all real processes are irreversible, some are almost reversible. If a real process occurs very slowly such that the system is always very nearly in an equilibrium state, the process can be approximated as being reversible. Suppose a gas is compressed isothermally in a piston–cylinder arrangement in which the gas is in thermal contact with an energy reservoir and we continuously transfer just enough energy from the gas to the reservoir to keep the temperature constant. For example, imagine that the gas is compressed very slowly by dropping grains of sand onto a frictionless piston as shown in Figure 22.8. As each grain lands on the piston and compresses the gas a small amount, the system deviates from an equilibrium state, but it is so close to one that it achieves a new equilibrium state in a relatively short time interval. Each grain added represents a change to a new equilibrium state, but the differences between states are so small that the entire process can be approximated as occurring through continuous equilibrium states. The process can be reversed by slowly removing grains from the piston.

A general characteristic of a reversible process is that no dissipative effects (such as turbulence or friction) that convert mechanical energy to internal energy can be present. Such effects can be impossible to eliminate completely. Hence, it is not surprising that real processes in nature are irreversible.

22.4 The Carnot Engine

Pitfall Prevention 22.3

Don't Shop for a Carnot Engine

The Carnot engine is an idealization; do not expect a Carnot engine to be developed for commercial use. We explore the Carnot engine only for theoretical considerations.

In 1824, a French engineer named Sadi Carnot described a theoretical engine, now called a **Carnot engine,** that is of great importance from both practical and theoretical viewpoints. He showed that a heat engine operating in an ideal, reversible cycle—called a **Carnot cycle**—between two energy reservoirs is the most efficient engine possible. Such an ideal engine establishes an upper limit on the efficiencies of all other engines. That is, the net work done by a working substance taken through the Carnot cycle is the greatest amount of work possible for a given amount of energy supplied to the substance at the higher temperature. **Carnot's theorem** can be stated as follows:

> No real heat engine operating between two energy reservoirs can be more efficient than a Carnot engine operating between the same two reservoirs.

To prove the validity of this theorem, imagine two heat engines operating between the *same* energy reservoirs. One is a Carnot engine with efficiency e_C, and the other is an engine with efficiency e, where we assume $e > e_C$. Because the cycle in the Carnot engine is reversible, the engine can operate in reverse as a refrigerator. The more efficient engine is used to drive the Carnot engine as a Carnot refrigerator. The output by work of the more efficient engine is matched to the input by work of the Carnot refrigerator. For the *combination* of the engine and refrigerator, no exchange by work with the surroundings occurs. Because we have assumed the engine is more efficient than the refrigerator, the net result of the combination is a transfer of energy from the cold to the hot reservoir without work being done on the combination. According to the Clausius statement of the second law, this process is impossible. Hence, the assumption that $e > e_C$ must be false. All real engines are less efficient than the Carnot engine because they do not operate through a reversible cycle. The efficiency of a real engine is further reduced by such practical difficulties as friction and energy losses by conduction.

To describe the Carnot cycle taking place between temperatures T_c and T_h, let's assume the working substance is an ideal gas contained in a cylinder fitted with a movable piston at one end. The cylinder's walls and the piston are thermally nonconducting. Four stages of the Carnot cycle are shown in Active Figure 22.9, and the *PV* diagram for the cycle is shown in Active Figure 22.10. The Carnot cycle consists of two adiabatic processes and two isothermal processes, all reversible:

1. Process $A \rightarrow B$ (Active Fig. 22.9a) is an isothermal expansion at temperature T_h. The gas is placed in thermal contact with an energy reservoir at temperature T_h. During the expansion, the gas absorbs energy $|Q_h|$ from the reservoir through the base of the cylinder and does work W_{AB} in raising the piston.
2. In process $B \rightarrow C$ (Active Fig. 22.9b), the base of the cylinder is replaced by a thermally nonconducting wall and the gas expands adiabatically; that is, no energy enters or leaves the system by heat. During the expansion, the temperature of the gas decreases from T_h to T_c and the gas does work W_{BC} in raising the piston.

J.-L. Charmet/Science Photo Library/Photo Researchers, Inc.

Sadi Carnot
French engineer (1796–1832)
Carnot was the first to show the quantitative relationship between work and heat. In 1824, he published his only work, *Reflections on the Motive Power of Heat,* which reviewed the industrial, political, and economic importance of the steam engine. In it, he defined work as "weight lifted through a height."

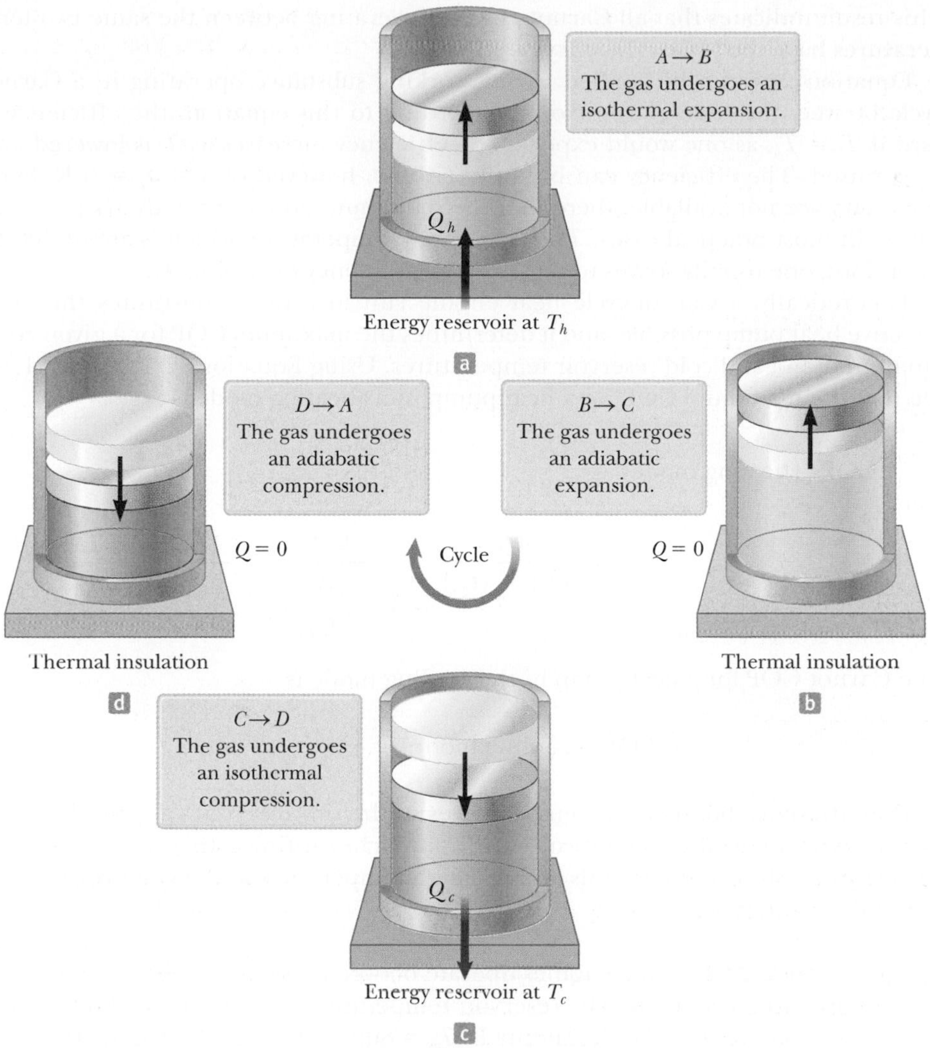

ACTIVE FIGURE 22.9

The Carnot cycle. The letters *A*, *B*, *C*, and *D* refer to the states of the gas shown in Active Figure 22.10. The arrows on the piston indicate the direction of its motion during each process.

3. In process $C \to D$ (Active Fig. 22.9c), the gas is placed in thermal contact with an energy reservoir at temperature T_c and is compressed isothermally at temperature T_c. During this time, the gas expels energy $|Q_c|$ to the reservoir and the work done by the piston on the gas is W_{CD}.
4. In the final process $D \to A$ (Active Fig. 22.9d), the base of the cylinder is replaced by a nonconducting wall and the gas is compressed adiabatically. The temperature of the gas increases to T_h, and the work done by the piston on the gas is W_{DA}.

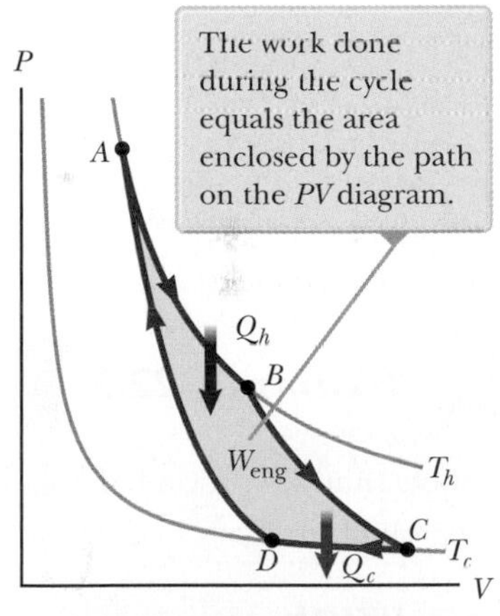

ACTIVE FIGURE 22.10

PV diagram for the Carnot cycle. The net work done W_{eng} equals the net energy transferred into the Carnot engine in one cycle, $|Q_h| - |Q_c|$.

The thermal efficiency of the engine is given by Equation 22.2:

$$e = 1 - \frac{|Q_c|}{|Q_h|}$$

In Example 22.3, we show that for a Carnot cycle,

$$\frac{|Q_c|}{|Q_h|} = \frac{T_c}{T_h} \quad (22.5)$$

Hence, the thermal efficiency of a Carnot engine is

$$e_C = 1 - \frac{T_c}{T_h} \quad (22.6)$$

◀ **Efficiency of a Carnot engine**

This result indicates that all Carnot engines operating between the same two temperatures have the same efficiency.[5]

Equation 22.6 can be applied to any working substance operating in a Carnot cycle between two energy reservoirs. According to this equation, the efficiency is zero if $T_c = T_h$, as one would expect. The efficiency increases as T_c is lowered and T_h is raised. The efficiency can be unity (100%), however, only if $T_c = 0$ K. Such reservoirs are not available; therefore, the maximum efficiency is always less than 100%. In most practical cases, T_c is near room temperature, which is about 300 K. Therefore, one usually strives to increase the efficiency by raising T_h.

Theoretically, a Carnot-cycle heat engine run in reverse constitutes the most effective heat pump possible, and it determines the maximum COP for a given combination of hot and cold reservoir temperatures. Using Equations 22.1 and 22.4, we see that the maximum COP for a heat pump in its heating mode is

$$\text{COP}_\text{C}\,(\text{heating mode}) = \frac{|Q_h|}{W}$$

$$= \frac{|Q_h|}{|Q_h| - |Q_c|} = \frac{1}{1 - \dfrac{|Q_c|}{|Q_h|}} = \frac{1}{1 - \dfrac{T_c}{T_h}} = \frac{T_h}{T_h - T_c}$$

The Carnot COP for a heat pump in the cooling mode is

$$\text{COP}_\text{C}\,(\text{cooling mode}) = \frac{T_c}{T_h - T_c}$$

As the difference between the temperatures of the two reservoirs approaches zero in this expression, the theoretical COP approaches infinity. In practice, the low temperature of the cooling coils and the high temperature at the compressor limit the COP to values below 10.

***Quick Quiz* 22.3** Three engines operate between reservoirs separated in temperature by 300 K. The reservoir temperatures are as follows: Engine A: $T_h = 1\,000$ K, $T_c = 700$ K; Engine B: $T_h = 800$ K, $T_c = 500$ K; Engine C: $T_h = 600$ K, $T_c = 300$ K. Rank the engines in order of theoretically possible efficiency from highest to lowest.

Example 22.3 Efficiency of the Carnot Engine

Show that the ratio of energy transfers by heat in a Carnot engine is equal to the ratio of reservoir temperatures, as given by Equation 22.5.

SOLUTION

Conceptualize Make use of Active Figures 22.9 and 22.10 to help you visualize the processes in the Carnot cycle.

Categorize Because of our understanding of the Carnot cycle, we can categorize the processes in the cycle as isothermal and adiabatic.

[5]For the processes in the Carnot cycle to be reversible, they must be carried out infinitesimally slowly. Therefore, although the Carnot engine is the most efficient engine possible, it has zero power output because it takes an infinite time interval to complete one cycle! For a real engine, the short time interval for each cycle results in the working substance reaching a high temperature lower than that of the hot reservoir and a low temperature higher than that of the cold reservoir. An engine undergoing a Carnot cycle between this narrower temperature range was analyzed by F. L. Curzon and B. Ahlborn ("Efficiency of a Carnot engine at maximum power output," *Am. J. Phys.* **43**(1), 22, 1975), who found that the efficiency at maximum power output depends only on the reservoir temperatures T_c and T_h and is given by $e_{\text{C-A}} = 1 - (T_c/T_h)^{1/2}$. The Curzon–Ahlborn efficiency $e_{\text{C-A}}$ provides a closer approximation to the efficiencies of real engines than does the Carnot efficiency.

22.3 *cont.*

Analyze For the isothermal expansion (process $A \to B$ in Active Fig. 22.9), find the energy transfer by heat from the hot reservoir using Equation 20.14 and the first law of thermodynamics:

$$|Q_h| = |\Delta E_{\text{int}} - W_{AB}| = |0 - W_{AB}| = nRT_h \ln \frac{V_B}{V_A}$$

In a similar manner, find the energy transfer to the cold reservoir during the isothermal compression $C \to D$:

$$|Q_c| = |\Delta E_{\text{int}} - W_{CD}| = |0 - W_{CD}| = nRT_c \ln \frac{V_C}{V_D}$$

Divide the second expression by the first:

$$(1) \quad \frac{|Q_c|}{|Q_h|} = \frac{T_c}{T_h} \frac{\ln (V_C/V_D)}{\ln (V_B/V_A)}$$

Apply Equation 21.20 to the adiabatic processes $B \to C$ and $D \to A$:

$$T_h V_B^{\gamma-1} = T_c V_C^{\gamma-1}$$
$$T_h V_A^{\gamma-1} = T_c V_D^{\gamma-1}$$

Divide the first equation by the second:

$$\left(\frac{V_B}{V_A}\right)^{\gamma-1} = \left(\frac{V_C}{V_D}\right)^{\gamma-1}$$

$$(2) \quad \frac{V_B}{V_A} = \frac{V_C}{V_D}$$

Substitute Equation (2) into Equation (1):

$$\frac{|Q_c|}{|Q_h|} = \frac{T_c}{T_h} \frac{\ln (V_C/V_D)}{\ln (V_B/V_A)} = \frac{T_c}{T_h} \frac{\ln (V_C/V_D)}{\ln (V_C/V_D)} = \frac{T_c}{T_h}$$

Finalize This last equation is Equation 22.5, the one we set out to prove.

Example 22.4 The Steam Engine

A steam engine has a boiler that operates at 500 K. The energy from the burning fuel changes water to steam, and this steam then drives a piston. The cold reservoir's temperature is that of the outside air, approximately 300 K. What is the maximum thermal efficiency of this steam engine?

SOLUTION

Conceptualize In a steam engine, the gas pushing on the piston in Active Figure 22.9 is steam. A real steam engine does not operate in a Carnot cycle, but, to find the maximum possible efficiency, imagine a Carnot steam engine.

Categorize We calculate an efficiency using Equation 22.6, so we categorize this example as a substitution problem.

Substitute the reservoir temperatures into Equation 22.6:

$$e_C = 1 - \frac{T_c}{T_h} = 1 - \frac{300 \text{ K}}{500 \text{ K}} = \boxed{0.400} \quad \text{or} \quad \boxed{40.0\%}$$

This result is the highest *theoretical* efficiency of the engine. In practice, the efficiency is considerably lower.

WHAT IF? Suppose we wished to increase the theoretical efficiency of this engine. This increase can be achieved by raising T_h by ΔT or by decreasing T_c by the same ΔT. Which would be more effective?

Answer A given ΔT would have a larger fractional effect on a smaller temperature, so you would expect a larger change in efficiency if you alter T_c by ΔT. Let's test that numerically. Raising T_h by 50 K, corresponding to $T_h = 550$ K, would give a maximum efficiency of

$$e_C = 1 - \frac{T_c}{T_h} = 1 - \frac{300 \text{ K}}{550 \text{ K}} = 0.455$$

continued

22.4 cont.

Decreasing T_c by 50 K, corresponding to $T_c = 250$ K, would give a maximum efficiency of

$$e_C = 1 - \frac{T_c}{T_h} = 1 - \frac{250\text{ K}}{500\text{ K}} = 0.500$$

Although changing T_c is *mathematically* more effective, often changing T_h is *practically* more feasible.

22.5 Gasoline and Diesel Engines

In a gasoline engine, six processes occur in each cycle; they are illustrated in Active Figure 22.11. In this discussion, let's consider the interior of the cylinder above the piston to be the system that is taken through repeated cycles in the engine's operation. For a given cycle, the piston moves up and down twice, which represents a four-stroke cycle consisting of two upstrokes and two downstrokes. The processes in the cycle can be approximated by the **Otto cycle** shown in the *PV* diagram in Active Figure 22.12. In the following discussion, refer to Active Figure 22.11 for the pictorial representation of the strokes and Active Figure 22.12 for the significance on the *PV* diagram of the letter designations below:

1. During the *intake stroke* (Active Fig. 22.11a and $O \rightarrow A$ in Active Figure 22.12), the piston moves downward and a gaseous mixture of air and fuel is drawn into the cylinder at atmospheric pressure. That is the energy input part of the cycle: energy enters the system (the interior of the cylinder) by matter transfer as potential energy stored in the fuel. In this process, the volume increases from V_2 to V_1. This apparent backward numbering is

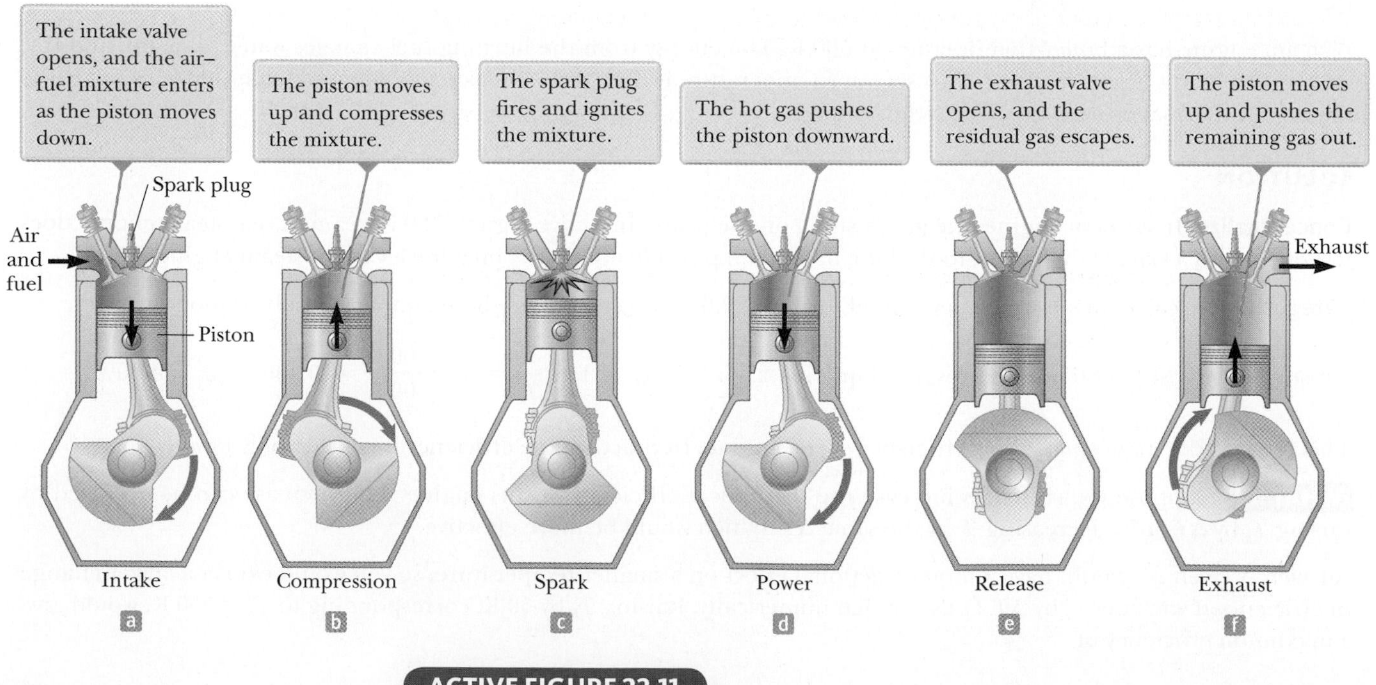

ACTIVE FIGURE 22.11

The four-stroke cycle of a conventional gasoline engine. The arrows on the piston indicate the direction of its motion during each process.

based on the compression stroke (process 2 below), in which the air–fuel mixture is compressed from V_1 to V_2.

2. During the *compression stroke* (Active Fig. 22.11b and $A \rightarrow B$ in Active Fig. 22.12), the piston moves upward, the air–fuel mixture is compressed adiabatically from volume V_1 to volume V_2, and the temperature increases from T_A to T_B. The work done on the gas is positive, and its value is equal to the negative of the area under the curve AB in Active Figure 22.12.
3. Combustion occurs when the spark plug fires (Active Fig. 22.11c and $B \rightarrow C$ in Active Fig. 22.12). That is not one of the strokes of the cycle because it occurs in a very short time interval while the piston is at its highest position. The combustion represents a rapid energy transformation from potential energy stored in chemical bonds in the fuel to internal energy associated with molecular motion, which is related to temperature. During this time interval, the mixture's pressure and temperature increase rapidly, with the temperature rising from T_B to T_C. The volume, however, remains approximately constant because of the short time interval. As a result, approximately no work is done on or by the gas. We can model this process in the PV diagram (Active Fig. 22.12) as that process in which the energy $|Q_h|$ enters the system. (In reality, however, this process is a *conversion* of energy already in the cylinder from process $O \rightarrow A$.)
4. In the *power stroke* (Active Fig. 22.11d and $C \rightarrow D$ in Active Fig. 22.12), the gas expands adiabatically from V_2 to V_1. This expansion causes the temperature to drop from T_C to T_D. Work is done by the gas in pushing the piston downward, and the value of this work is equal to the area under the curve CD.
5. Release of the residual gases occurs when an exhaust valve is opened (Active Fig. 22.11e and $D \rightarrow A$ in Active Fig. 22.12). The pressure suddenly drops for a short time interval. During this time interval, the piston is almost stationary and the volume is approximately constant. Energy is expelled from the interior of the cylinder and continues to be expelled during the next process.
6. In the final process, the *exhaust stroke* (Active Fig. 22.11e and $A \rightarrow O$ in Active Fig. 22.12), the piston moves upward while the exhaust valve remains open. Residual gases are exhausted at atmospheric pressure, and the volume decreases from V_1 to V_2. The cycle then repeats.

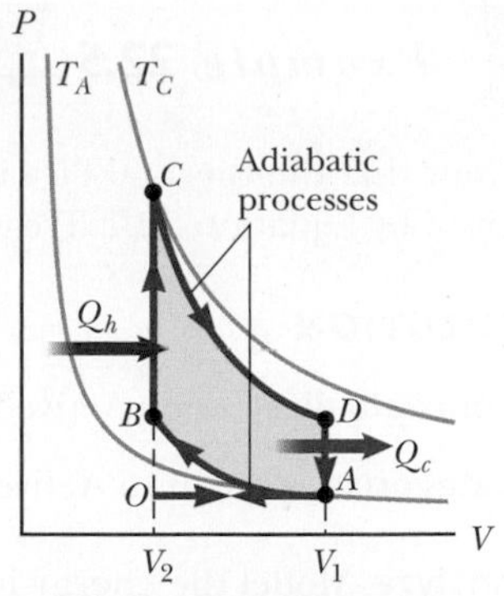

ACTIVE FIGURE 22.12

PV diagram for the Otto cycle, which approximately represents the processes occurring in an internal combustion engine.

If the air–fuel mixture is assumed to be an ideal gas, the efficiency of the Otto cycle is

$$e = 1 - \frac{1}{(V_1/V_2)^{\gamma-1}} \quad \text{(Otto cycle)} \qquad \textbf{(22.7)}$$

where V_1/V_2 is the **compression ratio** and γ is the ratio of the molar specific heats C_P/C_V for the air–fuel mixture. Equation 22.7, which is derived in Example 22.5, shows that the efficiency increases as the compression ratio increases. For a typical compression ratio of 8 and with $\gamma = 1.4$, Equation 22.7 predicts a theoretical efficiency of 56% for an engine operating in the idealized Otto cycle. This value is much greater than that achieved in real engines (15% to 20%) because of such effects as friction, energy transfer by conduction through the cylinder walls, and incomplete combustion of the air–fuel mixture.

Diesel engines operate on a cycle similar to the Otto cycle, but they do not employ a spark plug. The compression ratio for a diesel engine is much greater than that for a gasoline engine. Air in the cylinder is compressed to a very small volume, and, as a consequence, the cylinder temperature at the end of the compression stroke is very high. At this point, fuel is injected into the cylinder. The temperature is high enough for the air–fuel mixture to ignite without the assistance of a spark plug. Diesel engines are more efficient than gasoline engines because of their greater compression ratios and resulting higher combustion temperatures.

Example 22.5 Efficiency of the Otto Cycle

Show that the thermal efficiency of an engine operating in an idealized Otto cycle (see Active Figs. 22.11 and 22.12) is given by Equation 22.7. Treat the working substance as an ideal gas.

SOLUTION

Conceptualize Study Active Figures 22.11 and 22.12 to make sure you understand the working of the Otto cycle.

Categorize As seen in Active Figure 22.12, we categorize the processes in the Otto cycle as isovolumetric and adiabatic.

Analyze Model the energy input and output as occurring by heat in processes $B \to C$ and $D \to A$. (In reality, most of the energy enters and leaves by matter transfer as the air–fuel mixture enters and leaves the cylinder.) Use Equation 21.8 to find the energy transfers by heat for these processes, which take place at constant volume:

$$B \to C \quad |Q_h| = nC_V(T_C - T_B)$$

$$D \to A \quad |Q_c| = nC_V(T_D - T_A)$$

Substitute these expressions into Equation 22.2:

$$(1) \quad e = 1 - \frac{|Q_c|}{|Q_h|} = 1 - \frac{T_D - T_A}{T_C - T_B}$$

Apply Equation 21.20 to the adiabatic processes $A \to B$ and $C \to D$:

$$A \to B \quad T_A V_A^{\gamma-1} = T_B V_B^{\gamma-1}$$

$$C \to D \quad T_C V_C^{\gamma-1} = T_D V_D^{\gamma-1}$$

Solve these equations for the temperatures T_A and T_D, noting that $V_A = V_D = V_1$ and $V_B = V_C = V_2$:

$$(2) \quad T_A = T_B\left(\frac{V_B}{V_A}\right)^{\gamma-1} = T_B\left(\frac{V_2}{V_1}\right)^{\gamma-1}$$

$$(3) \quad T_D = T_C\left(\frac{V_C}{V_D}\right)^{\gamma-1} = T_C\left(\frac{V_2}{V_1}\right)^{\gamma-1}$$

Subtract Equation (2) from Equation (3) and rearrange:

$$(4) \quad \frac{T_D - T_A}{T_C - T_B} = \left(\frac{V_2}{V_1}\right)^{\gamma-1}$$

Substitute Equation (4) into Equation (1):

$$e = 1 - \frac{1}{(V_1/V_2)^{\gamma-1}}$$

Finalize This final expression is Equation 22.7.

Pitfall Prevention 22.4

Entropy Is Abstract
Entropy is one of the most abstract notions in physics, so follow the discussion in this and the subsequent sections very carefully. Do not confuse energy with entropy. Even though the names sound similar, they are very different concepts.

22.6 Entropy

The zeroth law of thermodynamics involves the concept of temperature, and the first law involves the concept of internal energy. Temperature and internal energy are both state variables; that is, the value of each depends only on the thermodynamic state of a system, not on the process that brought it to that state. Another state variable—this one related to the second law of thermodynamics—is *entropy S*. In this section, we define entropy on a macroscopic scale as it was first expressed by Clausius in 1865.

Entropy was originally formulated as a useful concept in thermodynamics. Its importance grew, however, as the field of statistical mechanics developed because the analytical techniques of statistical mechanics provide an alternative means of interpreting entropy and a more global significance to the concept. In statistical mechanics, the behavior of a substance is described in terms of the statistical behavior of its atoms and molecules. An important finding in these studies is that isolated systems tend toward disorder, and **entropy** is a measure of this disorder. For example, consider the molecules of a gas in the air in your room. If half the gas molecules had velocity vectors of equal magnitude directed toward the left and the other half had velocity vectors of the same magnitude directed toward the right, the situation would be very ordered. Such a situation is extremely unlikely, however. If you could view the molecules, you would see that they move haphazardly in all directions, bumping into one another, changing speed upon collision, some going fast and others going slowly. This situation is highly disordered.

The cause of the tendency of an isolated system toward disorder is easily explained. To do so, let's distinguish between *microstates* and *macrostates* of a system. A **microstate** is a particular configuration of the individual constituents of the system. For example, the description of the ordered velocity vectors of the air molecules in your room refers to a particular microstate, and the more likely haphazard motion is another microstate. A **macrostate** is a description of the system's conditions from a macroscopic point of view. For a thermodynamic system, macrostates are described by macroscopic variables such as pressure, density, and temperature.

For any given macrostate of the system, a number of microstates are possible. Let's first consider some nonthermodynamic systems for simplicity. For example, the macrostate of a 4 on a pair of dice can be formed from the possible microstates 1–3, 2–2, and 3–1. The macrostate of 2 has only one microstate, 1–1. It is assumed all microstates are equally probable. When all possible macrostates are examined, however, it is found that macrostates associated with disorder have far more possible microstates than those associated with order. Therefore, 4 is a more disordered macrostate for two dice than 2 because there are three microstates for a 4 and only one for a 2.

There is only one microstate associated with the macrostate of a royal flush in a poker hand of five spades, laid out in order from ten to ace (Fig. 22.13a). Figure 22.13b shows another poker hand. The macrostate here is "worthless hand." The *particular* hand (the microstate) in Figure 22.13b is as equally probable as the hand in Figure 22.13a. There are, however, *many* other hands similar to that in Figure 22.13b; that is, there are many microstates that also qualify as worthless hands. The more microstates that belong to a particular macrostate, the higher the probability that macrostate will occur. The macrostate of a royal flush in spades is ordered, of low probability, and of high value in poker. The macrostate of a worthless hand is disordered, of high probability, and of low poker value.

a

b

Figure 22.13 (a) A royal flush has low probability of occurring. (b) A worthless poker hand, one of many.

***Quick Quiz* 22.4** **(a)** Suppose you select four cards at random from a standard deck of playing cards and end up with a macrostate of four deuces. How many microstates are associated with this macrostate? **(b)** Suppose you pick up two cards and end up with a macrostate of two aces. How many microstates are associated with this macrostate?

We can also imagine ordered macrostates and disordered macrostates in physical processes, not just in games of dice and poker. The result of a dice throw or a poker hand stays fixed once the dice are thrown or the cards are dealt. Physical systems, on the other hand, are in a constant state of flux, changing from moment to moment from one microstate to another. Based on the relationship between the probability of a macrostate and the number of associated microstates, we therefore see that the probability of a system moving in time from an ordered macrostate to a disordered macrostate is far greater than the probability of the reverse because there are more microstates in a disordered macrostate.

The original formulation of entropy in thermodynamics involves the transfer of energy by heat during a reversible process. Consider any infinitesimal process in which a system changes from one equilibrium state to another. If dQ_r is the amount of energy transferred by heat when the system follows a reversible path between the states, the change in entropy dS is equal to this amount of energy for the reversible process divided by the absolute temperature of the system:

Change in entropy for an infinitesimal process

$$dS = \frac{dQ_r}{T} \tag{22.8}$$

We have assumed the temperature is constant because the process is infinitesimal. Because entropy is a state variable, the change in entropy during a process depends only on the endpoints and therefore is independent of the actual path followed. Consequently, the entropy change for an irreversible process can be determined by calculating the entropy change for a *reversible* process that connects the same initial and final states.

The subscript r on the quantity dQ_r is a reminder that the transferred energy is to be measured along a reversible path even though the system may actually have followed some irreversible path. When energy is absorbed by the system, dQ_r is positive and the entropy of the system increases. When energy is expelled by the system, dQ_r is negative and the entropy of the system decreases. Notice that Equation 22.8 does not define entropy but rather the *change* in entropy. Hence, the meaningful quantity in describing a process is the *change* in entropy.

To calculate the change in entropy for a *finite* process, first recognize that T is generally not constant during the process. Therefore, we must integrate Equation 22.8:

Change in entropy for a finite process

$$\Delta S = \int_i^f dS = \int_i^f \frac{dQ_r}{T} \tag{22.9}$$

As with an infinitesimal process, the change in entropy ΔS of a system going from one state to another has the same value for *all* paths connecting the two states. That is, the finite change in entropy ΔS of a system depends only on the properties of the initial and final equilibrium states. Therefore, we are free to choose a particular reversible path over which to evaluate the entropy in place of the actual path as long as the initial and final states are the same for both paths. This point is explored further in Section 22.7.

***Quick Quiz* 22.5** An ideal gas is taken from an initial temperature T_i to a higher final temperature T_f along two different reversible paths. Path A is at constant pressure, and path B is at constant volume. What is the relation between the entropy changes of the gas for these paths? **(a)** $\Delta S_A > \Delta S_B$ **(b)** $\Delta S_A = \Delta S_B$ **(c)** $\Delta S_A < \Delta S_B$

Example 22.6 Change in Entropy: Melting

A solid that has a latent heat of fusion L_f melts at a temperature T_m. Calculate the change in entropy of this substance when a mass m of the substance melts.

SOLUTION

Conceptualize Imagine placing the substance in a warm environment so that energy enters the substance by heat. The process can be reversed by placing the substance in a cool environment so that energy leaves the substance by heat. The mass m of the substance that melts is equal to Δm, the change in mass of the higher-phase (liquid) substance.

Categorize Because the melting takes place at a fixed temperature, we categorize the process as isothermal.

Analyze Use Equation 20.7 in Equation 22.9, noting that the temperature remains fixed:

$$\Delta S = \int \frac{dQ_r}{T} = \frac{1}{T_m}\int dQ_r = \frac{Q_r}{T_m} = \frac{L_f \Delta m}{T_m} = \frac{L_f m}{T_m}$$

Finalize Notice that Δm is positive so that ΔS is positive, representing that energy is added to the ice cube.

22.6 *cont.*

WHAT IF? Suppose you did not have Equation 22.9 available to calculate an entropy change. How could you argue from the statistical description of entropy that the changes in entropy should be positive?

Answer When a solid melts, its entropy increases because the molecules are much more disordered in the liquid state than they are in the solid state. The positive value for ΔS also means that the substance in its liquid state does not spontaneously transfer energy from itself to the warm surroundings and freeze because to do so would involve a spontaneous increase in order and a decrease in entropy.

Let's consider the changes in entropy that occur in a Carnot heat engine that operates between the temperatures T_c and T_h. In one cycle, the engine takes in energy $|Q_h|$ from the hot reservoir and expels energy $|Q_c|$ to the cold reservoir. These energy transfers occur only during the isothermal portions of the Carnot cycle; therefore, the constant temperature can be brought out in front of the integral sign in Equation 22.9. The integral then simply has the value of the total amount of energy transferred by heat. Therefore, the total change in entropy for one cycle is

$$\Delta S = \frac{|Q_h|}{T_h} - \frac{|Q_c|}{T_c}$$

where the minus sign represents that energy is leaving the engine. In Example 22.3, we showed that for a Carnot engine,

$$\frac{|Q_c|}{|Q_h|} = \frac{T_c}{T_h}$$

Using this result in the previous expression for ΔS, we find that the total change in entropy for a Carnot engine operating in a cycle is *zero:*

$$\Delta S = 0$$

Now consider a system taken through an arbitrary (non-Carnot) reversible cycle. Because entropy is a state variable—and hence depends only on the properties of a given equilibrium state—we conclude that $\Delta S = 0$ for *any* reversible cycle. In general, we can write this condition as

$$\oint \frac{dQ_r}{T} = 0 \quad \text{(reversible cycle)} \tag{22.10}$$

where the symbol $\oint$ indicates that the integration is over a closed path.

22.7 Entropy and the Second Law

By definition, a calculation of the change in entropy for a system requires information about a reversible path connecting the initial and final equilibrium states. To calculate changes in entropy for real (irreversible) processes, remember that entropy (like internal energy) depends only on the *state* of the system. That is, entropy is a state variable, and the change in entropy depends only on the initial and final states.

You can calculate the entropy change in some irreversible process between two equilibrium states by devising a reversible process (or series of reversible processes) between the same two states and computing $\Delta S = \int dQ_r/T$ for the reversible process. In irreversible processes, it is important to distinguish between Q, the actual energy transfer in the process, and Q_r, the energy that would have been transferred by heat along a reversible path. Only Q_r is the correct value to be used in calculating the entropy change.

If we consider a system and its surroundings to include the entire Universe, the Universe is always moving toward a higher-probability macrostate, corresponding to greater disorder. Because entropy is a measure of disorder, an alternative way of stating this behavior is as follows:

Entropy statement of the second law of thermodynamics ▶

The entropy of the Universe increases in all real processes.

This statement is yet another wording of the second law of thermodynamics that can be shown to be equivalent to the Kelvin-Planck and Clausius statements.

When dealing with a system that is not isolated from its surroundings, remember that the increase in entropy described in the second law is that of the system *and* its surroundings. When a system and its surroundings interact in an irreversible process, the increase in entropy of one is greater than the decrease in entropy of the other. Hence, the change in entropy of the Universe must be greater than zero for an irreversible process and equal to zero for a reversible process. Ultimately, because real processes are irreversible, the entropy of the Universe should increase steadily and eventually reach a maximum value. At this value, the Universe will be in a state of uniform temperature and density. All physical, chemical, and biological processes will have ceased at this time because a state of perfect disorder implies that no energy is available for doing work. This gloomy state of affairs is sometimes referred to as the *heat death* of the Universe.

***Quick Quiz* 22.6** True or False: The entropy change in an adiabatic process must be zero because $Q = 0$.

Entropy Change in Thermal Conduction

Let's now consider a system consisting of a hot reservoir and a cold reservoir that are in thermal contact with each other and isolated from the rest of the Universe. A process occurs during which energy Q is transferred by heat from the hot reservoir at temperature T_h to the cold reservoir at temperature T_c. The process as described is irreversible (energy would not spontaneously flow from cold to hot), so we must find an equivalent reversible process. Because the temperature of a reservoir does not change during the process, we can replace the real process for each reservoir with a reversible, isothermal process in which the same amount of energy is transferred by heat. Consequently, for a reservoir, the entropy change does not depend on whether the process is reversible or irreversible.

Because the cold reservoir absorbs energy Q, its entropy increases by Q/T_c. At the same time, the hot reservoir loses energy Q, so its entropy change is $-Q/T_h$. Because $T_h > T_c$, the increase in entropy of the cold reservoir is greater than the decrease in entropy of the hot reservoir. Therefore, the change in entropy of the system (and of the Universe) is greater than zero:

$$\Delta S_U = \frac{Q}{T_c} + \frac{-Q}{T_h} > 0$$

Suppose energy were to transfer spontaneously from a cold object to a hot object, in violation of the second law. This impossible energy transfer can be described in terms of disorder. Before the transfer, a certain degree of order is associated with the different temperatures of the objects. The hot object's molecules have a higher average energy than the cold object's molecules. If energy spontaneously transfers from the cold object to the hot object, the cold object becomes colder over a time interval and the hot object becomes hotter. The difference in average molecular energy becomes even greater, which would represent an increase in order for the system and a violation of the second law.

In comparison, the process that does occur naturally is the transfer of energy from the hot object to the cold object. In this process, the difference in average molecular energy decreases, which represents a more random distribution of energy and an increase in disorder.

Entropy Change in a Free Expansion

Let's again consider the adiabatic free expansion of a gas occupying an initial volume V_i (Fig. 22.14). In this situation, a membrane separating the gas from an evacuated region is broken and the gas expands to a volume V_f. This process is irreversible; the gas would not spontaneously crowd into half the volume after filling the entire volume. What are the changes in entropy of the gas and of the Universe during this process? The process is neither reversible nor quasi-static. As shown in Section 20.6, the initial and final temperatures of the gas are the same.

To apply Equation 22.9, we cannot take $Q = 0$, the value for the irreversible process, but must instead find Q_r; that is, we must find an equivalent reversible path that shares the same initial and final states. A simple choice is an isothermal, reversible expansion in which the gas pushes slowly against a piston while energy enters the gas by heat from a reservoir to hold the temperature constant. Because T is constant in this process, Equation 22.9 gives

$$\Delta S = \int_i^f \frac{dQ_r}{T} = \frac{1}{T}\int_i^f dQ_r$$

For an isothermal process, the first law of thermodynamics specifies that $\int_i^f dQ_r$ is equal to the negative of the work done on the gas during the expansion from V_i to V_f, which is given by Equation 20.14. Using this result, we find that the entropy change for the gas is

$$\Delta S = nR\ln\left(\frac{V_f}{V_i}\right) \qquad \textbf{(22.11)}$$

Because $V_f > V_i$, we conclude that ΔS is positive. This positive result indicates that both the entropy and the disorder of the gas *increase* as a result of the irreversible, adiabatic expansion.

It is easy to see that the gas is more disordered after the expansion. Instead of being concentrated in a relatively small space, the molecules are scattered over a larger region.

Because the free expansion takes place in an insulated container, no energy is transferred by heat from the surroundings. (Remember that the isothermal, reversible expansion is only a *replacement* process used to calculate the entropy change for the gas; it is not the *actual* process.) Therefore, the free expansion has no effect on the surroundings, and the entropy change of the surroundings is zero.

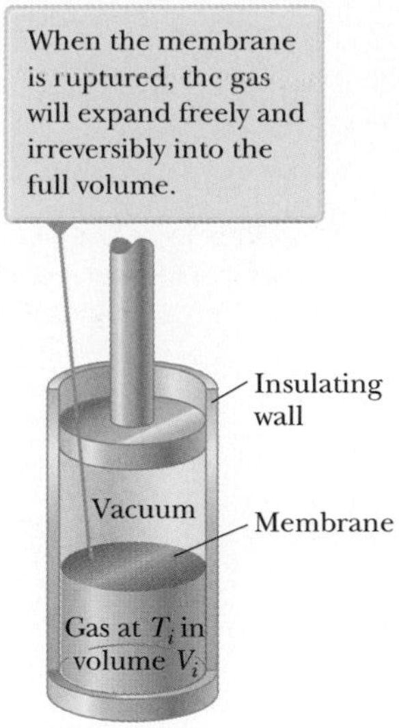

Figure 22.14 Adiabatic free expansion of a gas. The container is thermally insulated from its surroundings; therefore, $Q = 0$.

22.8 Entropy on a Microscopic Scale

As we have seen, entropy can be approached by relying on macroscopic concepts. Entropy can also be treated from a microscopic viewpoint through statistical analysis of molecular motions. Let's use a microscopic model to investigate once again the free expansion of an ideal gas, which was discussed from a macroscopic point of view in Section 22.7.

In the kinetic theory of gases, gas molecules are represented as particles moving randomly. Suppose the gas is initially confined to the volume V_i shown in Figure 22.14. When the membrane is removed, the molecules eventually are distributed throughout the greater volume V_f of the entire container. For a given uniform distribution of gas in the volume, there are a large number of equivalent microstates, and the entropy of the gas can be related to the number of microstates corresponding to a given macrostate.

Let's count the number of microstates by considering the variety of molecular locations available to the molecules. Let's assume each molecule occupies some microscopic volume V_m. The total number of possible locations of a single molecule in a macroscopic initial volume V_i is the ratio $w_i = V_i/V_m$, which is a huge number. We use w_i here to represent either the number of *ways* the molecule can be placed in the initial volume or the number of microstates, which is equivalent to the number of available locations. We assume the probabilities of a molecule occupying any of these locations are equal.

As more molecules are added to the system, the number of possible ways the molecules can be positioned in the volume multiplies. For example, if you consider two molecules, for every possible placement of the first, all possible placements of the second are available. Therefore, there are w_i ways of locating the first molecule, and for each way, there are w_i ways of locating the second molecule. The total number of ways of locating the two molecules is $w_i w_i = w_i^2$.

Neglecting the very small probability of having two molecules occupy the same location, each molecule may go into any of the V_i/V_m locations, and so the number of ways of locating N molecules in the volume becomes $W_i = w_i^N = (V_i/V_m)^N$. ($W_i$ is not to be confused with work.) Similarly, when the volume is increased to V_f, the number of ways of locating N molecules increases to $W_f = w_f^N = (V_f/V_m)^N$. The ratio of the number of ways of placing the molecules in the volume for the initial and final configurations is

$$\frac{W_f}{W_i} = \frac{(V_f/V_m)^N}{(V_i/V_m)^N} = \left(\frac{V_f}{V_i}\right)^N$$

Taking the natural logarithm of this equation and multiplying by Boltzmann's constant gives

$$k_B \ln\left(\frac{W_f}{W_i}\right) = k_B \ln\left(\frac{V_f}{V_i}\right)^N = nN_A\, k_B \ln\left(\frac{V_f}{V_i}\right)$$

where we have used the equality $N = nN_A$. We know from Equation 19.11 that $N_A k_B$ is the universal gas constant R; therefore, we can write this equation as

$$k_B \ln W_f - k_B \ln W_i = nR \ln\left(\frac{V_f}{V_i}\right) \qquad \textbf{(22.12)}$$

From Equation 22.11, we know that when a gas undergoes a free expansion from V_i to V_f, the change in entropy is

$$S_f - S_i = nR \ln\left(\frac{V_f}{V_i}\right) \qquad \textbf{(22.13)}$$

Notice that the right sides of Equations 22.12 and 22.13 are identical. Therefore, from the left sides, we make the following important connection between entropy and the number of microstates for a given macrostate:

Entropy (microscopic definition) ▶

$$S \equiv k_B \ln W \qquad \textbf{(22.14)}$$

The more microstates there are that correspond to a given macrostate, the greater the entropy of that macrostate. As discussed previously, there are many more microstates associated with disordered macrostates than with ordered macrostates. Therefore, Equation 22.14 indicates mathematically our earlier statement that entropy is a measure of disorder. Although our discussion used the specific example of the free expansion of an ideal gas, a more rigorous development of the statistical interpretation of entropy would lead us to the same conclusion.

We have stated that individual microstates are equally probable. Because there are far more microstates associated with a disordered macrostate than with an ordered macrostate, however, a disordered macrostate is much more probable than an ordered one.

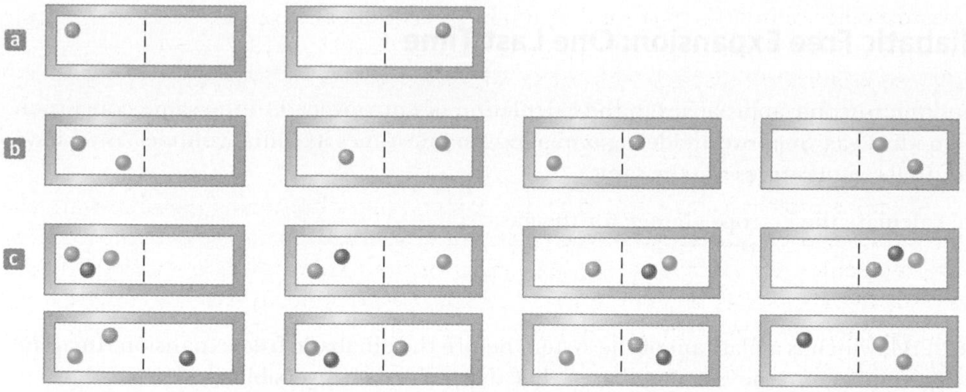

ACTIVE FIGURE 22.15

(a) One molecule in a container has a 1-in-2 chance of being on the left side. (b) Two molecules have a 1-in-4 chance of being on the left side at the same time. (c) Three molecules have a 1-in-8 chance of being on the left side at the same time.

Let's explore this concept by considering 100 molecules in a container. At any given moment, the probability of one molecule being in the left part of the container shown in Active Figure 22.15a as a result of random motion is $\frac{1}{2}$. If there are two molecules as shown in Active Figure 22.15b, the probability of both being in the left part is $(\frac{1}{2})^2$, or 1 in 4. If there are three molecules (Active Fig. 22.15c), the probability of them all being in the left portion at the same moment is $(\frac{1}{2})^3$, or 1 in 8. For 100 independently moving molecules, the probability that the 50 fastest ones will be found in the left part at any moment is $(\frac{1}{2})^{50}$. Likewise, the probability that the remaining 50 slower molecules will be found in the right part at any moment is $(\frac{1}{2})^{50}$. Therefore, the probability of finding this fast–slow separation as a result of random motion is the product $(\frac{1}{2})^{50}(\frac{1}{2})^{50} = (\frac{1}{2})^{100}$, which corresponds to about 1 in 10^{30}. When this calculation is extrapolated from 100 molecules to the number in 1 mol of gas (6.02×10^{23}), the ordered arrangement is found to be *extremely* improbable!

Conceptual Example 22.7 Let's Play Marbles!

Suppose you have a bag of 100 marbles of which 50 are red and 50 are green. You are allowed to draw four marbles from the bag according to the following rules. Draw one marble, record its color, and return it to the bag. Shake the bag and then draw another marble. Continue this process until you have drawn and returned four marbles. What are the possible macrostates for this set of events? What is the most likely macrostate? What is the least likely macrostate?

SOLUTION

Because each marble is returned to the bag before the next one is drawn and the bag is then shaken, the probability of drawing a red marble is always the same as the probability of drawing a green one. All the possible microstates and macrostates are shown in Table 22.1. As this table indicates, there is only one way to draw a macrostate of four red marbles, so there is only one microstate for that macrostate. There are, however, four possible microstates that correspond to the macrostate of one green marble and three red marbles, six microstates that correspond to two green marbles and two red marbles, four microstates that correspond to three green marbles and one red marble, and one microstate that corresponds to four green marbles. The most likely, and most disordered, macrostate—two red marbles and two green marbles—corresponds to the largest number of microstates. The least likely, most ordered macrostates—four red marbles or four green marbles—correspond to the smallest number of microstates.

TABLE 22.1 *Possible Results of Drawing Four Marbles from a Bag*

Macrostate	Possible Microstates	Total Number of Microstates
All R	RRRR	1
1G, 3R	RRRG, RRGR, RGRR, GRRR	4
2G, 2R	RRGG, RGRG, GRRG, RGGR, GRGR, GGRR	6
3G, 1R	GGGR, GGRG, GRGG, RGGG	4
All G	GGGG	1

Example 22.8 Adiabatic Free Expansion: One Last Time

Let's verify that the macroscopic and microscopic approaches to the calculation of entropy lead to the same conclusion for the adiabatic free expansion of an ideal gas. Suppose an ideal gas expands to four times its initial volume. As we have seen for this process, the initial and final temperatures are the same.

(A) Using a macroscopic approach, calculate the entropy change for the gas.

SOLUTION

Conceptualize Look back at Figure 22.14, which is a diagram of the system before the adiabatic free expansion. Imagine breaking the membrane so that the gas moves into the evacuated area. The expansion is irreversible.

Categorize We can replace the irreversible process with a reversible isothermal process between the same initial and final states. This approach is macroscopic, so we use a thermodynamic variable, in particular, the volume V.

Analyze Use Equation 22.11 to evaluate the entropy change:

$$\Delta S = nR\ln\left(\frac{V_f}{V_i}\right) = nR\ln\left(\frac{4V_i}{V_i}\right) = \boxed{nR\ln 4}$$

(B) Using statistical considerations, calculate the change in entropy for the gas and show that it agrees with the answer you obtained in part (A).

SOLUTION

Categorize This approach is microscopic, so we use variables related to the individual molecules.

Analyze The number of microstates available to a single molecule in the initial volume V_i is $w_i = V_i/V_m$. Use this number to find the number of available microstates for N molecules:

$$W_i = w_i^N = \left(\frac{V_i}{V_m}\right)^N$$

Find the number of available microstates for N molecules in the final volume $V_f = 4V_i$:

$$W_f = \left(\frac{V_f}{V_m}\right)^N = \left(\frac{4V_i}{V_m}\right)^N$$

Use Equation 22.14 to find the entropy change:

$$\Delta S = k_B\ln W_f - k_B\ln W_i = k_B\ln\left(\frac{W_f}{W_i}\right)$$

$$= k_B\ln\left(\frac{4V_i}{V_i}\right)^N = k_B\ln(4^N) = Nk_B\ln 4 = \boxed{nR\ln 4}$$

Finalize The answer is the same as that for part (A), which dealt with macroscopic parameters.

WHAT IF? In part (A), we used Equation 22.11, which was based on a reversible isothermal process connecting the initial and final states. Would you arrive at the same result if you chose a different reversible process?

Answer You *must* arrive at the same result because entropy is a state variable. For example, consider the two-step process in Figure 22.16: a reversible adiabatic expansion from V_i to $4V_i$ ($A \rightarrow B$) during which the temperature drops from T_1 to T_2 and a reversible isovolumetric process ($B \rightarrow C$) that takes the gas back to the initial temperature T_1. During the reversible adiabatic process, $\Delta S = 0$ because $Q_r = 0$.

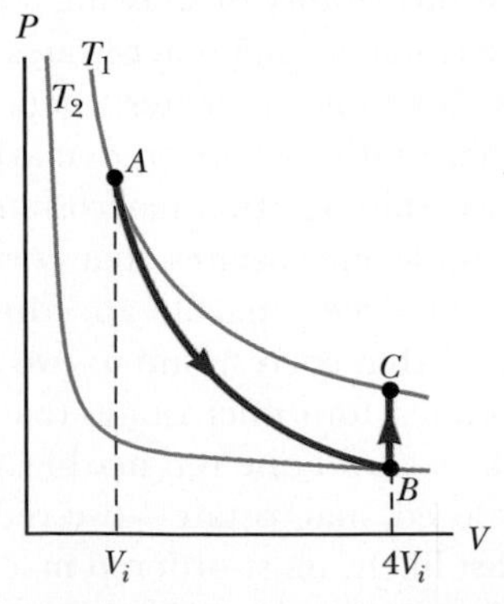

Figure 22.16 (Example 22.8) A gas expands to four times its initial volume and back to the initial temperature by means of a two-step process.

22.8 cont.

For the reversible isovolumetric process ($B \rightarrow C$), use Equation 22.9:

$$\Delta S = \int_i^f \frac{dQ_r}{T} = \int_{T_2}^{T_1} \frac{nC_V dT}{T} = nC_V \ln\left(\frac{T_1}{T_2}\right)$$

Find the ratio of temperature T_1 to T_2 from Equation 21.20 for the adiabatic process:

$$\frac{T_1}{T_2} = \left(\frac{4V_i}{V_i}\right)^{\gamma-1} = (4)^{\gamma-1}$$

Substitute to find ΔS:

$$\Delta S = nC_V \ln (4)^{\gamma-1} = nC_V(\gamma - 1) \ln 4$$

$$= nC_V\left(\frac{C_P}{C_V} - 1\right) \ln 4 = n(C_P - C_V) \ln 4 = nR \ln 4$$

and you do indeed obtain the exact same result for the entropy change.

Summary

Definitions

The **thermal efficiency** e of a heat engine is

$$e \equiv \frac{W_{\text{eng}}}{|Q_h|} = \frac{|Q_h| - |Q_c|}{|Q_h|} = 1 - \frac{|Q_c|}{|Q_h|} \qquad \textbf{(22.2)}$$

From a microscopic viewpoint, the **entropy** of a given macrostate is defined as

$$S \equiv k_B \ln W \qquad \textbf{(22.14)}$$

where k_B is Boltzmann's constant and W is the number of microstates of the system corresponding to the macrostate.

In a **reversible** process, the system can be returned to its initial conditions along the same path on a *PV* diagram, and every point along this path is an equilibrium state. A process that does not satisfy these requirements is **irreversible.**

Concepts and Principles

A **heat engine** is a device that takes in energy by heat and, operating in a cyclic process, expels a fraction of that energy by means of work. The net work done by a heat engine in carrying a working substance through a cyclic process ($\Delta E_{\text{int}} = 0$) is

$$W_{\text{eng}} = |Q_h| - |Q_c| \qquad \textbf{(22.1)}$$

where $|Q_h|$ is the energy taken in from a hot reservoir and $|Q_c|$ is the energy expelled to a cold reservoir.

Two ways the **second law of thermodynamics** can be stated are as follows:

- It is impossible to construct a heat engine that, operating in a cycle, produces no effect other than the input of energy by heat from a reservoir and the performance of an equal amount of work (the Kelvin–Planck statement).
- It is impossible to construct a cyclical machine whose sole effect is to transfer energy continuously by heat from one object to another object at a higher temperature without the input of energy by work (the Clausius statement).

continued

Carnot's theorem states that no real heat engine operating (irreversibly) between the temperatures T_c and T_h can be more efficient than an engine operating reversibly in a Carnot cycle between the same two temperatures.

The thermal efficiency of a heat engine operating in the Carnot cycle is

$$e_C = 1 - \frac{T_c}{T_h} \quad \textbf{(22.6)}$$

The second law of thermodynamics states that when real (irreversible) processes occur, the degree of disorder in the system plus the surroundings increases. When a process occurs in an isolated system, the state of the system becomes more disordered. The measure of disorder in a system is called **entropy** S. Therefore, yet another way the second law can be stated is as follows:

- The entropy of the Universe increases in all real processes.

The **change in entropy** dS of a system during a process between two infinitesimally separated equilibrium states is

$$dS = \frac{dQ_r}{T} \quad \textbf{(22.8)}$$

where dQ_r is the energy transfer by heat for the system for a reversible process that connects the initial and final states.

The change in entropy of a system during an arbitrary process between an initial state and a final state is

$$\Delta S = \int_i^f \frac{dQ_r}{T} \quad \textbf{(22.9)}$$

The value of ΔS for the system is the same for all paths connecting the initial and final states. The change in entropy for a system undergoing any reversible, cyclic process is zero, and when such a process occurs, the entropy of the Universe remains constant.

Objective Questions

☐ denotes answer available in *Student Solutions Manual/Study Guide*

1. A steam turbine operates at a boiler temperature of 450 K and an exhaust temperature of 300 K. What is the maximum theoretical efficiency of this system? (a) 0.240 (b) 0.500 (c) 0.333 (d) 0.667 (e) 0.150

2. An engine does 15.0 kJ of work while exhausting 37.0 kJ to a cold reservoir. What is the efficiency of the engine? (a) 0.150 (b) 0.288 (c) 0.333 (d) 0.450 (e) 1.20

3. A refrigerator has 18.0 kJ of work done on it while 115 kJ of energy is transferred from inside its interior. What is its coefficient of performance? (a) 3.40 (b) 2.80 (c) 8.90 (d) 6.40 (e) 5.20

4. Of the following, which is *not* a statement of the second law of thermodynamics? (a) No heat engine operating in a cycle can absorb energy from a reservoir and use it entirely to do work. (b) No real engine operating between two energy reservoirs can be more efficient than a Carnot engine operating between the same two reservoirs. (c) When a system undergoes a change in state, the change in the internal energy of the system is the sum of the energy transferred to the system by heat and the work done on the system. (d) The entropy of the Universe increases in all natural processes. (e) Energy will not spontaneously transfer by heat from a cold object to a hot object.

5. Consider cyclic processes completely characterized by each of the following net energy inputs and outputs. In each case, the energy transfers listed are the *only* ones occurring. Classify each process as (a) possible, (b) impossible according to the first law of thermodynamics, (c) impossible according to the second law of thermodynamics, or (d) impossible according to both the first and second laws. **(i)** Input is 5 J of work, and output is 4 J of work. **(ii)** Input is 5 J of work, and output is 5 J of energy transferred by heat. **(iii)** Input is 5 J of energy transferred by electrical transmission, and output is 6 J of work. **(iv)** Input is 5 J of energy transferred by heat, and output is 5 J of energy transferred by heat. **(v)** Input is 5 J of energy transferred by heat, and output is 5 J of work. **(vi)** Input is 5 J of energy transferred by heat, and output is 3 J of work plus 2 J of energy transferred by heat.

6. A compact air-conditioning unit is placed on a table inside a well-insulated apartment and is plugged in and turned on. What happens to the average temperature of the apartment? (a) It increases. (b) It decreases. (c) It remains constant. (d) It increases until the unit warms up and then decreases. (e) The answer depends on the initial temperature of the apartment.

7. The second law of thermodynamics implies that the coefficient of performance of a refrigerator must be what? (a) less than 1 (b) less than or equal to 1 (c) greater than or equal to 1 (d) finite (e) greater than 0

8. A thermodynamic process occurs in which the entropy of a system changes by -8 J/K. According to the second law of thermodynamics, what can you conclude about the entropy change of the environment? (a) It must be $+8$ J/K or less. (b) It must be between $+8$ J/K and 0. (c) It must be equal to $+8$ J/K. (d) It must be $+8$ J/K or more. (e) It must be zero.

9. A sample of a monatomic ideal gas is contained in a cylinder with a piston. Its state is represented by the dot in the *PV* diagram shown in Figure OQ22.9. Arrows *A* through *E* represent isobaric, isothermal, adiabatic, and isovolumetric processes that the sample can undergo. In each process except *D*, the volume changes by a factor of 2. All five processes are reversible. Rank the processes according to the change in entropy of the gas from the largest positive value to the largest-magnitude negative value. In your rankings, display any cases of equality.

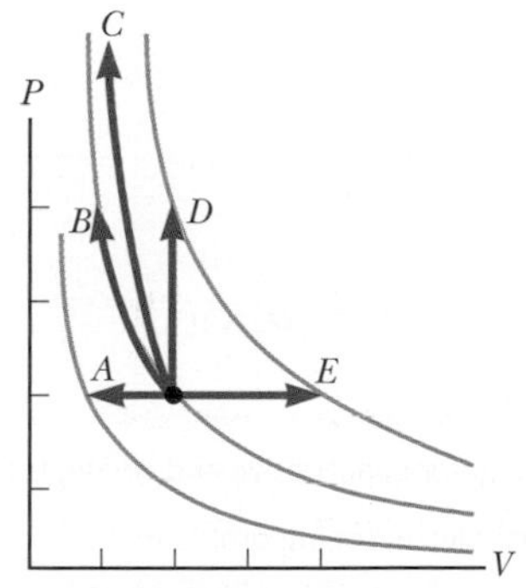

Figure OQ22.9

10. Assume a sample of an ideal gas is at room temperature. What action will *necessarily* make the entropy of the sample increase? (a) Transfer energy into it by heat. (b) Transfer energy into it irreversibly by heat. (c) Do work on it. (d) Increase either its temperature or its volume, without letting the other variable decrease. (e) None of those choices is correct.

11. The arrow *OA* in the *PV* diagram shown in Figure OQ22.11 represents a reversible adiabatic expansion of an ideal gas. The same sample of gas, starting from the same state *O*, now undergoes an adiabatic free expansion to the same final volume. What point on the diagram could represent the final state of the gas? (a) the same point *A* as for the reversible expansion (b) point *B* (c) point *C* (d) any of those choices (e) none of those choices

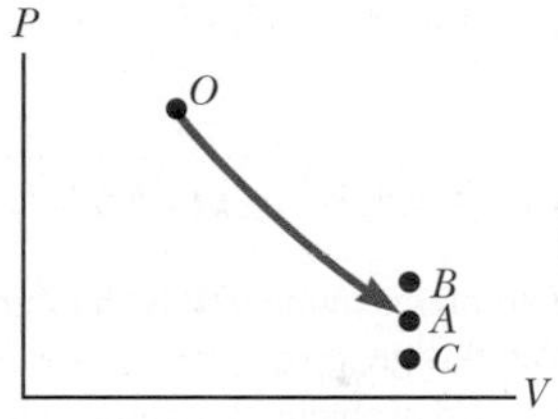

Figure OQ22.11

Conceptual Questions

☐ denotes answer available in Student Solutions Manual/Study Guide

1. What are some factors that affect the efficiency of automobile engines?

2. A steam-driven turbine is one major component of an electric power plant. Why is it advantageous to have the temperature of the steam as high as possible?

3. Does the second law of thermodynamics contradict or correct the first law? Argue for your answer.

4. "The first law of thermodynamics says you can't really win, and the second law says you can't even break even." Explain how this statement applies to a particular device or process; alternatively, argue against the statement.

5. Is it possible to construct a heat engine that creates no thermal pollution? Explain.

6. (a) Give an example of an irreversible process that occurs in nature. (b) Give an example of a process in nature that is nearly reversible.

7. The device shown in Figure CQ22.7, called a thermoelectric converter, uses a series of semiconductor cells to transform internal energy to electric potential energy, which we will study in Chapter 25. In the photograph on the left, both legs of the device are at the same temperature and no electric potential energy is produced. When one leg is at a higher temperature than the other as shown in the photograph on the right, however, electric potential energy is produced as the device extracts energy from the hot reservoir and drives a small electric motor. (a) Why is the difference in temperature necessary to produce electric potential energy in this demonstration? (b) In what sense does this intriguing experiment demonstrate the second law of thermodynamics?

Courtesy of PASCO Scientific Company

Figure CQ22.7

8. Discuss three different common examples of natural processes that involve an increase in entropy. Be sure to account for all parts of each system under consideration.

9. Discuss the change in entropy of a gas that expands (a) at constant temperature and (b) adiabatically.

10. Suppose your roommate cleans and tidies up your messy room after a big party. Because she is creating more order, does this process represent a violation of the second law of thermodynamics?

11. "Energy is the mistress of the Universe, and entropy is her shadow." Writing for an audience of general readers, argue for this statement with at least two examples. Alternatively, argue for the view that entropy is like an executive who instantly determines what will happen, whereas energy is like a bookkeeper telling us how little we can afford. (Arnold Sommerfeld suggested the idea for this question.)

12. (a) If you shake a jar full of jelly beans of different sizes, the larger beans tend to appear near the top and the smaller ones tend to fall to the bottom. Why? (b) Does this process violate the second law of thermodynamics?

13. The energy exhaust from a certain coal-fired electric generating station is carried by "cooling water" into Lake Ontario. The water is warm from the viewpoint of living things in the lake. Some of them congregate around the outlet port and can impede the water flow. (a) Use the theory of heat engines to explain why this action can reduce the electric power output of the station. (b) An engineer says that the electric output is reduced because of "higher back pressure on the turbine blades." Comment on the accuracy of this statement.

Problems

ENHANCED WebAssign The problems found in this chapter may be assigned online in Enhanced WebAssign

1. denotes straightforward problem; **2.** denotes intermediate problem; **3.** denotes challenging problem

1. full solution available in the Student Solutions Manual/Study Guide

1. denotes problems most often assigned in Enhanced WebAssign; these provide students with targeted feedback and either a Master It tutorial or a Watch It solution video.

QC denotes asking for quantitative and conceptual reasoning

S denotes symbolic reasoning problem

M denotes Master It tutorial available in Enhanced WebAssign

GP denotes guided problem

shaded denotes "paired problems" that develop reasoning with symbols and numerical values

Section 22.1 Heat Engines and the Second Law of Thermodynamics

1. An engine absorbs 1.70 kJ from a hot reservoir at 277°C and expels 1.20 kJ to a cold reservoir at 27°C in each cycle. (a) What is the engine's efficiency? (b) How much work is done by the engine in each cycle? (c) What is the power output of the engine if each cycle lasts 0.300 s?

2. The work done by an engine equals one-fourth the energy it absorbs from a reservoir. (a) What is its thermal efficiency? (b) What fraction of the energy absorbed is expelled to the cold reservoir?

3. A heat engine takes in 360 J of energy from a hot reservoir and performs 25.0 J of work in each cycle. Find (a) the efficiency of the engine and (b) the energy expelled to the cold reservoir in each cycle.

4. A gun is a heat engine. In particular, it is an internal combustion piston engine that does not operate in a cycle, but comes apart during its adiabatic expansion process. A certain gun consists of 1.80 kg of iron. It fires one 2.40-g bullet at 320 m/s with an energy efficiency of 1.10%. Assume the body of the gun absorbs all the energy exhaust—the other 98.9%—and increases uniformly in temperature for a short time interval before it loses any energy by heat into the environment. Find its temperature increase.

5. A particular heat engine has a mechanical power output of 5.00 kW and an efficiency of 25.0%. The engine expels 8.00×10^3 J of exhaust energy in each cycle. Find (a) the energy taken in during each cycle and (b) the time interval for each cycle.

6. A multicylinder gasoline engine in an airplane, operating at 2.50×10^3 rev/min, takes in energy 7.89×10^3 J and exhausts 4.58×10^3 J for each revolution of the crankshaft. (a) How many liters of fuel does it consume in 1.00 h of operation if the heat of combustion of the fuel is equal to 4.03×10^7 J/L? (b) What is the mechanical power output of the engine? Ignore friction and express the answer in horsepower. (c) What is the torque exerted by the crankshaft on the load? (d) What power must the exhaust and cooling system transfer out of the engine?

7. Suppose a heat engine is connected to two energy reservoirs, one a pool of molten aluminum (660°C) and the other a block of solid mercury (−38.9°C). The engine runs

by freezing 1.00 g of aluminum and melting 15.0 g of mercury during each cycle. The heat of fusion of aluminum is 3.97×10^5 J/kg; the heat of fusion of mercury is 1.18×10^4 J/kg. What is the efficiency of this engine?

Section 22.2 Heat Pumps and Refrigerators

8. A refrigerator has a coefficient of performance equal to 5.00. The refrigerator takes in 120 J of energy from a cold reservoir in each cycle. Find (a) the work required in each cycle and (b) the energy expelled to the hot reservoir.

9. During each cycle, a refrigerator ejects 625 kJ of energy to a high-temperature reservoir and takes in 550 kJ of energy from a low-temperature reservoir. Determine (a) the work done on the refrigerant in each cycle and (b) the coefficient of performance of the refrigerator.

10. A heat pump has a coefficient of performance of 3.80 and operates with a power consumption of 7.03×10^3 W. (a) How much energy does it deliver into a home during 8.00 h of continuous operation? (b) How much energy does it extract from the outside air?

11. A freezer has a coefficient of performance of 6.30. It is advertised as using electricity at a rate of 457 kWh/yr. (a) On average, how much energy does it use in a single day? (b) On average, how much energy does it remove from the refrigerator in a single day? (c) What maximum mass of water at 20.0°C could the freezer freeze in a single day? *Note:* One kilowatt-hour (kWh) is an amount of energy equal to running a 1-kW appliance for one hour.

12. A heat pump has a coefficient of performance equal to 4.20 and requires a power of 1.75 kW to operate. (a) How much energy does the heat pump add to a home in one hour? (b) If the heat pump is reversed so that it acts as an air conditioner in the summer, what would be its coefficient of performance?

Section 22.3 Reversible and Irreversible Processes

Section 22.4 The Carnot Engine

13. One of the most efficient heat engines ever built is a coal-fired steam turbine in the Ohio River valley, operating between 1 870°C and 430°C. (a) What is its maximum theoretical efficiency? (b) The actual efficiency of the engine is 42.0%. How much mechanical power does the engine deliver if it absorbs 1.40×10^5 J of energy each second from its hot reservoir?

14. A heat engine operates between a reservoir at 25.0°C and one at 375°C. What is the maximum efficiency possible for this engine?

15. A Carnot engine has a power output of 150 kW. The engine operates between two reservoirs at 20.0°C and 500°C. (a) How much energy enters the engine by heat per hour? (b) How much energy is exhausted by heat per hour?

16. **S** A Carnot engine has a power output P. The engine operates between two reservoirs at temperature T_c and T_h. (a) How much energy enters the engine by heat in a time interval Δt? (b) How much energy is exhausted by heat in the time interval Δt?

17. What is the coefficient of performance of a refrigerator that operates with Carnot efficiency between temperatures −3.00°C and +27.0°C?

18. *Why is the following situation impossible?* An inventor comes to a patent office with the claim that her heat engine, which employs water as a working substance, has a thermodynamic efficiency of 0.110. Although this efficiency is low compared with typical automobile engines, she explains that her engine operates between an energy reservoir at room temperature and a water–ice mixture at atmospheric pressure and therefore requires no fuel other than that to make the ice. The patent is approved, and working prototypes of the engine prove the inventor's efficiency claim.

19. **Q|C** A heat engine is being designed to have a Carnot efficiency of 65.0% when operating between two energy reservoirs. (a) If the temperature of the cold reservoir is 20.0°C, what must be the temperature of the hot reservoir? (b) Can the actual efficiency of the engine be equal to 65.0%? Explain.

20. **S** An ideal refrigerator or ideal heat pump is equivalent to a Carnot engine running in reverse. That is, energy $|Q_c|$ is taken in from a cold reservoir and energy $|Q_h|$ is rejected to a hot reservoir. (a) Show that the work that must be supplied to run the refrigerator or heat pump is

$$W = \frac{T_h - T_c}{T_c}|Q_c|$$

(b) Show that the coefficient of performance (COP) of the ideal refrigerator is

$$\text{COP} = \frac{T_c}{T_h - T_c}$$

21. What is the maximum possible coefficient of performance of a heat pump that brings energy from outdoors at −3.00°C into a 22.0°C house? *Note:* The work done to run the heat pump is also available to warm the house.

22. **M** How much work does an ideal Carnot refrigerator require to remove 1.00 J of energy from liquid helium at 4.00 K and expel this energy to a room-temperature (293-K) environment?

23. If a 35.0%-efficient Carnot heat engine (Active Fig. 22.2) is run in reverse so as to form a refrigerator (Active Fig. 22.4), what would be this refrigerator's coefficient of performance?

24. **Q|C** A Carnot heat engine operates between temperatures T_h and T_c. (a) If $T_h = 500$ K and $T_c = 350$ K, what is the efficiency of the engine? (b) What is the change in its efficiency for each degree of increase in T_h above 500 K? (c) What is the change in its efficiency for each degree of change in T_c? (d) Does the answer to part (c) depend on T_c? Explain.

25. An ideal gas is taken through a Carnot cycle. The isothermal expansion occurs at 250°C, and the isothermal compression takes place at 50.0°C. The gas takes in 1.20×10^3 J of energy from the hot reservoir during the isothermal expansion. Find (a) the energy expelled to the cold reservoir in each cycle and (b) the net work done by the gas in each cycle.

26. **Q|C** An electric power plant that would make use of the temperature gradient in the ocean has been proposed. The system is to operate between 20.0°C (surface-water temperature) and 5.00°C (water temperature at a depth of about 1 km). (a) What is the maximum efficiency of such a system? (b) If the electric power output of the plant is 75.0 MW, how much energy is taken in from the warm reservoir per hour? (c) In view of your answer to part (a), explain whether you think such a system is worthwhile. Note that the "fuel" is free.

27. Argon enters a turbine at a rate of 80.0 kg/min, a temperature of 800°C, and a pressure of 1.50 MPa. It expands adiabatically as it pushes on the turbine blades and exits at pressure 300 kPa. (a) Calculate its temperature at exit. (b) Calculate the (maximum) power output of the turning turbine. (c) The turbine is one component of a model closed-cycle gas turbine engine. Calculate the maximum efficiency of the engine.

28. **Q|C** **S** Suppose you build a two-engine device with the exhaust energy output from one heat engine supplying the input energy for a second heat engine. We say that the two engines are running *in series*. Let e_1 and e_2 represent the efficiencies of the two engines. (a) The overall efficiency of the two-engine device is defined as the total work output divided by the energy put into the first engine by heat. Show that the overall efficiency e is given by

$$e = e_1 + e_2 - e_1 e_2$$

What If? For parts (b) through (e) that follow, assume the two engines are Carnot engines. Engine 1 operates between temperatures T_h and T_i. The gas in engine 2 varies in temperature between T_i and T_c. In terms of the temperatures, (b) what is the efficiency of the combination engine? (c) Does an improvement in net efficiency result from the use of two engines instead of one? (d) What value of the intermediate temperature T_i results in equal work being done by each of the two engines in series? (e) What value of T_i results in each of the two engines in series having the same efficiency?

29. **Q|C** An electric generating station is designed to have an electric output power of 1.40 MW using a turbine with two-thirds the efficiency of a Carnot engine. The exhaust energy is transferred by heat into a cooling tower at 110°C. (a) Find the rate at which the station exhausts energy by heat as a function of the fuel combustion temperature T_h. (b) If the firebox is modified to run hotter by using more advanced combustion technology, how does the amount of energy exhaust change? (c) Find the exhaust power for T_h = 800°C. (d) Find the value of T_h for which the exhaust power would be only half as large as in part (c). (e) Find the value of T_h for which the exhaust power would be one-fourth as large as in part (c).

30. At point A in a Carnot cycle, 2.34 mol of a monatomic ideal gas has a pressure of 1 400 kPa, a volume of 10.0 L, and a temperature of 720 K. The gas expands isothermally to point B and then expands adiabatically to point C, where its volume is 24.0 L. An isothermal compression brings it to point D, where its volume is 15.0 L. An adiabatic process returns the gas to point A. (a) Determine all the unknown pressures, volumes, and temperatures as you fill in the following table:

	P	V	T
A	1 400 kPa	10.0 L	720 K
B			
C		24.0 L	
D		15.0 L	

(b) Find the energy added by heat, the work done by the engine, and the change in internal energy for each of the steps $A \rightarrow B$, $B \rightarrow C$, $C \rightarrow D$, and $D \rightarrow A$. (c) Calculate the efficiency $W_{net}/|Q_h|$. (d) Show that the efficiency is equal to $1 - T_C/T_A$, the Carnot efficiency.

31. A heat pump used for heating shown in Figure P22.31 is essentially an air conditioner installed backward. It extracts energy from colder air outside and deposits it in a warmer room. Suppose the ratio of the actual energy entering the room to the work done by the device's motor is 10.0% of the theoretical maximum ratio. Determine the energy entering the room per joule of work done by the motor given that the inside temperature is 20.0°C and the outside temperature is −5.00°C.

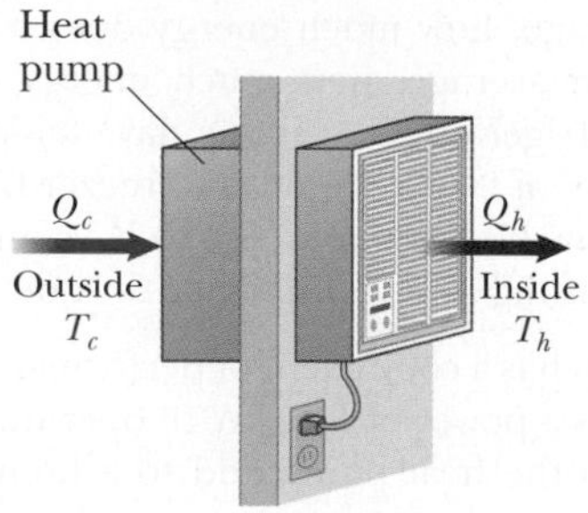

Figure P22.31

32. An ideal (Carnot) freezer in a kitchen has a constant temperature of 260 K, whereas the air in the kitchen has a constant temperature of 300 K. Suppose the insulation for the freezer is not perfect but rather conducts energy into the freezer at a rate of 0.150 W. Determine the average power required for the freezer's motor to maintain the constant temperature in the freezer.

Section 22.5 Gasoline and Diesel Engines

Note: For problems in this section, assume the gas in the engine is diatomic with γ = 1.40.

33. **M** In a cylinder of an automobile engine, immediately after combustion the gas is confined to a volume of 50.0 cm^3 and has an initial pressure of 3.00 × 10^6 Pa. The piston moves outward to a final volume of 300 cm^3, and the gas expands without energy transfer by heat. (a) What is the final pressure of the gas? (b) How much work is done by the gas in expanding?

34. A gasoline engine has a compression ratio of 6.00. (a) What is the efficiency of the engine if it operates in an idealized Otto cycle? (b) **What If?** If the actual efficiency is 15.0%, what fraction of the fuel is wasted as a result of friction and energy transfers by heat that could be avoided in a reversible engine? Assume complete combustion of the air–fuel mixture.

35. **S** An idealized diesel engine operates in a cycle known as the *air-standard diesel cycle* shown in Figure P22.35. Fuel is sprayed into the cylinder at the point of maximum compression, B. Combustion occurs during the expansion $B \rightarrow C$, which is modeled as an isobaric process. Show that the efficiency of an engine operating in this idealized diesel cycle is

$$e = 1 - \frac{1}{\gamma}\left(\frac{T_D - T_A}{T_C - T_B}\right)$$

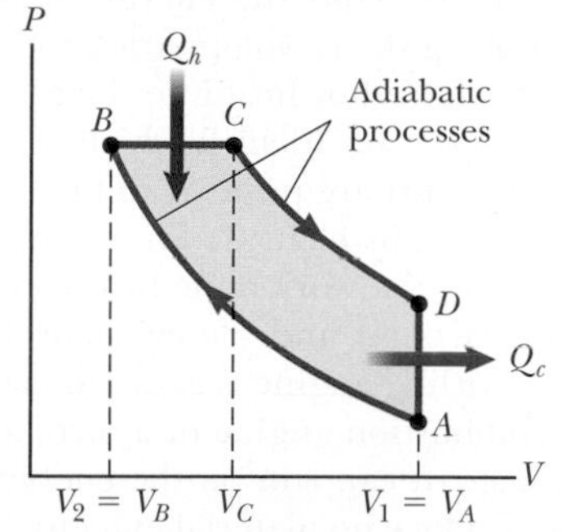

Figure P22.35

Section 22.6 Entropy

Section 22.7 Entropy and the Second Law

36. An ice tray contains 500 g of liquid water at 0°C. Calculate the change in entropy of the water as it freezes slowly and completely at 0°C.

37. A Styrofoam cup holding 125 g of hot water at 100°C cools to room temperature, 20.0°C. What is the change in entropy of the room? Neglect the specific heat of the cup and any change in temperature of the room.

38. Two 2.00×10^3-kg cars both traveling at 20.0 m/s undergo a head-on collision and stick together. Find the change in entropy of the surrounding air resulting from the collision if the air temperature is 23.0°C. Ignore the energy carried away from the collision by sound.

39. A 70.0-kg log falls from a height of 25.0 m into a lake. If the log, the lake, and the air are all at 300 K, find the change in entropy of the air during this process.

40. A 1.00-mol sample of H_2 gas is contained in the left side of the container shown in Figure P22.40, which has equal volumes on the left and right. The right side is evacuated. When the valve is opened, the gas streams into the right side. (a) What is the entropy change of the gas? (b) Does the temperature of the gas change? Assume the container is so large that the hydrogen behaves as an ideal gas.

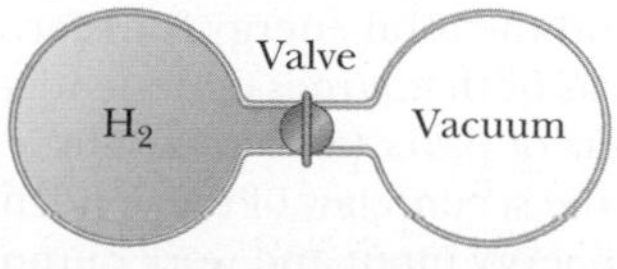

Figure P22.40

41. A 2.00-L container has a center partition that divides it into two equal parts as shown in Figure P22.41. The left side contains 0.044 0 mol of H_2 gas, and the right side contains 0.044 0 mol of O_2 gas. Both gases are at room temperature and at atmospheric pressure. The partition is removed, and the gases are allowed to mix. What is the entropy increase of the system?

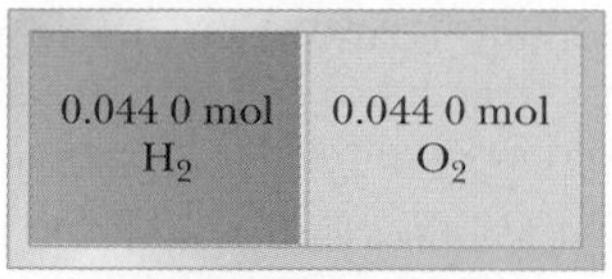

Figure P22.41

42. How fast are you personally making the entropy of the Universe increase right now? Compute an order-of-magnitude estimate, stating what quantities you take as data and the values you measure or estimate for them.

43. When an aluminum bar is connected between a hot reservoir at 725 K and a cold reservoir at 310 K, 2.50 kJ of energy is transferred by heat from the hot reservoir to the cold reservoir. In this irreversible process, calculate the change in entropy of (a) the hot reservoir, (b) the cold reservoir, and (c) the Universe, neglecting any change in entropy of the aluminum rod.

44. **S** When a metal bar is connected between a hot reservoir at T_h and a cold reservoir at T_c, the energy transferred by heat from the hot reservoir to the cold reservoir is Q. In this irreversible process, find expressions for the change in entropy of (a) the hot reservoir, (b) the cold reservoir, and (c) the Universe, neglecting any change in entropy of the metal rod.

45. The temperature at the surface of the Sun is approximately 5 800 K, and the temperature at the surface of the Earth is approximately 290 K. What entropy change of the Universe occurs when 1.00×10^3 J of energy is transferred by radiation from the Sun to the Earth?

Section 22.8 Entropy on a Microscopic Scale

46. If you roll two dice, what is the total number of ways in which you can obtain (a) a 12 and (b) a 7?

47. Prepare a table like Table 22.1 by using the same procedure (a) for the case in which you draw three marbles from your bag rather than four and (b) for the case in which you draw five marbles rather than four.

48. (a) Prepare a table like Table 22.1 for the following occurrence. You toss four coins into the air simultaneously and then record the results of your tosses in terms of the numbers of heads (H) and tails (T) that result. For example, HHTH and HTHH are two possible ways in which three heads and one tail can be achieved. (b) On the basis of your table, what is the most probable result recorded for a toss? In terms of entropy, (c) what is the most ordered macrostate, and (d) what is the most disordered?

Additional Problems

49. The energy absorbed by an engine is three times greater than the work it performs. (a) What is its thermal efficiency?

(b) What fraction of the energy absorbed is expelled to the cold reservoir?

50. A steam engine is operated in a cold climate where the exhaust temperature is 0°C. (a) Calculate the theoretical maximum efficiency of the engine using an intake steam temperature of 100°C. (b) If, instead, superheated steam at 200°C is used, find the maximum possible efficiency.

51. Find the maximum (Carnot) efficiency of an engine that absorbs energy from a hot reservoir at 545°C and exhausts energy to a cold reservoir at 185°C.

52. Every second at Niagara Falls, some 5.00×10^3 m^3 of water falls a distance of 50.0 m. What is the increase in entropy of the Universe per second due to the falling water? Assume the mass of the surroundings is so great that its temperature and that of the water stay nearly constant at 20.0°C. Also assume a negligible amount of water evaporates.

53. M Energy transfers by heat through the exterior walls and roof of a house at a rate of 5.00×10^3 J/s = 5.00 kW when the interior temperature is 22.0°C and the outside temperature is −5.00°C. (a) Calculate the electric power required to maintain the interior temperature at 22.0°C if the power is used in electric resistance heaters that convert all the energy transferred in by electrical transmission into internal energy. (b) **What If?** Calculate the electric power required to maintain the interior temperature at 22.0°C if the power is used to drive an electric motor that operates the compressor of a heat pump that has a coefficient of performance equal to 60.0% of the Carnot-cycle value.

54. Q|C In 1993, the U.S. government instituted a requirement that all room air conditioners sold in the United States must have an energy efficiency ratio (EER) of 10 or higher. The EER is defined as the ratio of the cooling capacity of the air conditioner, measured in British thermal units per hour, or Btu/h, to its electrical power requirement in watts. (a) Convert the EER of 10.0 to dimensionless form, using the conversion 1 Btu = 1 055 J. (b) What is the appropriate name for this dimensionless quantity? (c) In the 1970s, it was common to find room air conditioners with EERs of 5 or lower. State how the operating costs compare for 10 000-Btu/h air conditioners with EERs of 5.00 and 10.0. Assume each air conditioner operates for 1 500 h during the summer in a city where electricity costs 17.0¢ per kWh.

55. An airtight freezer holds n moles of air at 25.0°C and 1.00 atm. The air is then cooled to −18.0°C. (a) What is the change in entropy of the air if the volume is held constant? (b) What would the entropy change be if the pressure were maintained at 1.00 atm during the cooling?

56. Q|C Suppose an ideal (Carnot) heat pump could be constructed for use as an air conditioner. (a) Obtain an expression for the coefficient of performance (COP) for such an air conditioner in terms of T_h and T_c. (b) Would such an air conditioner operate on a smaller energy input if the difference in the operating temperatures were greater or smaller? (c) Compute the COP for such an air conditioner if the indoor temperature is 20.0°C and the outdoor temperature is 40.0°C.

57. GP S In 1816, Robert Stirling, a Scottish clergyman, patented the *Stirling engine,* which has found a wide variety of applications ever since, including the solar power application discussed on the cover of this textbook. Fuel is burned externally to warm one of the engine's two cylinders. A fixed quantity of inert gas moves cyclically between the cylinders, expanding in the hot one and contracting in the cold one. Figure P22.57 represents a model for its thermodynamic cycle. Consider n moles of an ideal monatomic gas being taken once through the cycle, consisting of two isothermal processes at temperatures $3T_i$ and T_i and two constant-volume processes. Let us find the efficiency of this engine. (a) Find the energy transferred by heat into the gas during the isovolumetric process AB. (b) Find the energy transferred by heat into the gas during the isothermal process BC. (c) Find the energy transferred by heat into the gas during the isovolumetric process CD. (d) Find the energy transferred by heat into the gas during the isothermal process DA. (e) Identify which of the results from parts (a) through (d) are positive and evaluate the energy input to the engine by heat. (f) From the first law of thermodynamics, find the work done by the engine. (g) From the results of parts (e) and (f), evaluate the efficiency of the engine. A Stirling engine is easier to manufacture than an internal combustion engine or a turbine. It can run on burning garbage. It can run on the energy transferred by sunlight and produce no material exhaust. Stirling engines are not currently used in automobiles due to long startup times and poor acceleration response.

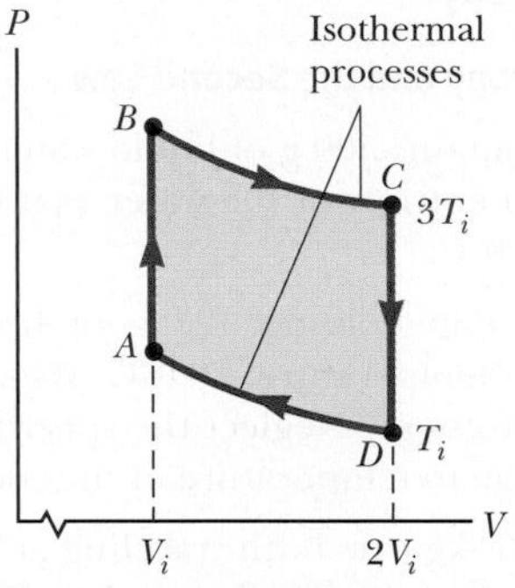

Figure P22.57

58. Q|C A firebox is at 750 K, and the ambient temperature is 300 K. The efficiency of a Carnot engine doing 150 J of work as it transports energy between these constant-temperature baths is 60.0%. The Carnot engine must take in energy 150 J/0.600 = 250 J from the hot reservoir and must put out 100 J of energy by heat into the environment. To follow Carnot's reasoning, suppose some other heat engine S could have an efficiency of 70.0%. (a) Find the energy input and exhaust energy output of engine S as it does 150 J of work. (b) Let engine S operate as in part (a) and run the Carnot engine in reverse between the same reservoirs. The output work of engine S is the input work for the Carnot refrigerator. Find the total energy transferred to or from the firebox and the total energy transferred to or from the environment as both engines operate together. (c) Explain how the results of parts (a) and (b) show that the Clausius statement of the second law of thermodynamics is violated. (d) Find the energy input and work output of engine S as it puts out exhaust energy of 100 J. Let engine S operate as in part (c) and contribute 150 J of its work output to running the Carnot engine in reverse. Find (e) the total energy the firebox puts out as both engines operate together, (f) the total work output, and (g) the total energy transferred to the environment. (h) Explain how the results show that the Kelvin–Planck statement of the second law is violated. Therefore, our assumption about the efficiency of engine S must be false. (i) Let the engines operate together through one cycle as in part (d). Find the change in entropy of the Universe. (j) Explain how the result of part (i) shows that the entropy statement of the second law is violated.

59. **Review.** This problem complements Problem 84 in Chapter 10. In the operation of a single-cylinder internal combustion piston engine, one charge of fuel explodes to drive the piston outward in the *power stroke.* Part of its energy output is stored in a turning flywheel. This energy is then used to push the piston inward to compress the next charge of fuel and air. In this compression process, assume an original volume of 0.120 L of a diatomic ideal gas at atmospheric pressure is compressed adiabatically to one-eighth of its original volume. (a) Find the work input required to compress the gas. (b) Assume the flywheel is a solid disk of mass 5.10 kg and radius 8.50 cm, turning freely without friction between the power stroke and the compression stroke. How fast must the flywheel turn immediately after the power stroke? This situation represents the minimum angular speed at which the engine can operate without stalling. (c) When the engine's operation is well above the point of stalling, assume the flywheel puts 5.00% of its maximum energy into compressing the next charge of fuel and air. Find its maximum angular speed in this case.

60. A biology laboratory is maintained at a constant temperature of 7.00°C by an air conditioner, which is vented to the air outside. On a typical hot summer day, the outside temperature is 27.0°C and the air-conditioning unit emits energy to the outside at a rate of 10.0 kW. Model the unit as having a coefficient of performance (COP) equal to 40.0% of the COP of an ideal Carnot device. (a) At what rate does the air conditioner remove energy from the laboratory? (b) Calculate the power required for the work input. (c) Find the change in entropy of the Universe produced by the air conditioner in 1.00 h. (d) **What If?** The outside temperature increases to 32.0°C. Find the fractional change in the COP of the air conditioner.

61. A heat engine operates between two reservoirs at T_2 = 600 K and T_1 = 350 K. It takes in 1.00×10^3 J of energy from the higher-temperature reservoir and performs 250 J of work. Find (a) the entropy change of the Universe ΔS_U for this process and (b) the work W that could have been done by an ideal Carnot engine operating between these two reservoirs. (c) Show that the difference between the amounts of work done in parts (a) and (b) is $T_1 \Delta S_U$.

62. **Q|C** **S** A 1.00-mol sample of a monatomic ideal gas is taken through the cycle shown in Figure P22.62. At point A, the pressure, volume, and temperature are P_i, V_i, and T_i, respectively. In terms of R and T_i, find (a) the total energy entering the system by heat per cycle, (b) the total energy leaving the system by heat per cycle, and (c) the efficiency of an engine operating in this cycle. (d) Explain how the efficiency compares with that of an engine operating in a Carnot cycle between the same temperature extremes.

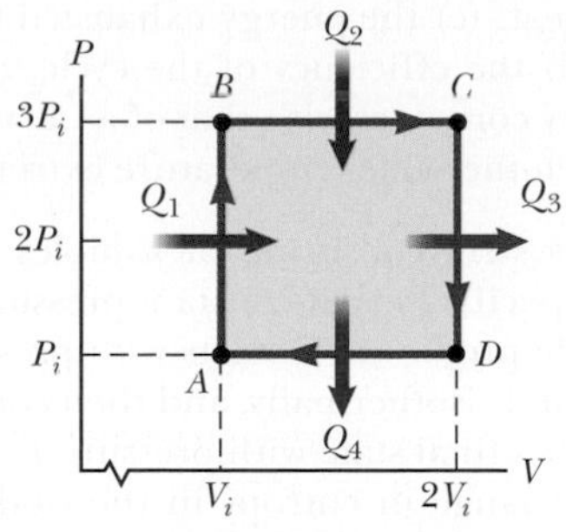

Figure P22.62

63. A power plant, having a Carnot efficiency, produces 1.00 GW of electrical power from turbines that take in steam at 500 K and reject water at 300 K into a flowing river. The water downstream is 6.00 K warmer due to the output of the power plant. Determine the flow rate of the river.

64. **S** A power plant, having a Carnot efficiency, produces electric power P from turbines that take in energy from steam at temperature T_h and discharge energy at temperature T_c through a heat exchanger into a flowing river. The water downstream is warmer by ΔT due to the output of the power plant. Determine the flow rate of the river.

65. **S** A sample consisting of n moles of an ideal gas undergoes a reversible isobaric expansion from volume V_i to volume $3V_i$. Find the change in entropy of the gas by calculating $\int_i^f dQ/T$, where $dQ = nC_P\,dT$.

66. **Q|C** An athlete whose mass is 70.0 kg drinks 16.0 ounces (454 g) of refrigerated water. The water is at a temperature of 35.0°F. (a) Ignoring the temperature change of the body that results from the water intake (so that the body is regarded as a reservoir always at 98.6°F), find the entropy increase of the entire system. (b) **What If?** Assume the entire body is cooled by the drink and the average specific heat of a person is equal to the specific heat of liquid water. Ignoring any other energy transfers by heat and any metabolic energy release, find the athlete's temperature after she drinks the cold water given an initial body temperature of 98.6°F. (c) Under these assumptions, what is the entropy increase of the entire system? (d) State how this result compares with the one you obtained in part (a).

67. **Q|C** A 1.00-mol sample of an ideal monatomic gas is taken through the cycle shown in Figure P22.67. The process $A \to B$ is a reversible isothermal expansion. Calculate (a) the net work done by the gas, (b) the energy added to

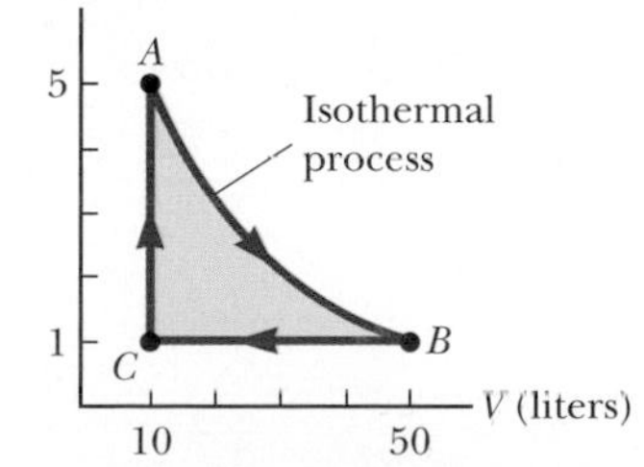

Figure P22.67

the gas by heat, (c) the energy exhausted from the gas by heat, and (d) the efficiency of the cycle. (e) Explain how the efficiency compares with that of a Carnot engine operating between the same temperature extremes.

68. **QC** **S** A system consisting of n moles of an ideal gas with molar specific heat at constant pressure C_P undergoes two reversible processes. It starts with pressure P_i and volume V_i, expands isothermally, and then contracts adiabatically to reach a final state with pressure P_i and volume $3V_i$. (a) Find its change in entropy in the isothermal process. (The entropy does not change in the adiabatic process.) (b) **What If?** Explain why the answer to part (a) must be the same as the answer to Problem 65. (You do not need to solve Problem 65 to answer this question.)

69. **QC** **S** A sample of an ideal gas expands isothermally, doubling in volume. (a) Show that the work done on the gas in expanding is $W = -nRT\ln 2$. (b) Because the internal energy E_{int} of an ideal gas depends solely on its temperature, the change in internal energy is zero during the expansion. It follows from the first law that the energy input to the gas by heat during the expansion is equal to the energy output by work. Does this process have 100% efficiency in converting energy input by heat into work output? (c) Does this conversion violate the second law? Explain.

70. *Why is the following situation impossible?* Two samples of water are mixed at constant pressure inside an insulated container: 1.00 kg of water at 10.0°C and 1.00 kg of water at 30.0°C. Because the container is insulated, there is no exchange of energy by heat between the water and the environment. Furthermore, the amount of energy that leaves the warm water by heat is equal to the amount that enters the cool water by heat. Therefore, the entropy change of the Universe is zero for this process.

Challenge Problems

71. A 1.00-mol sample of an ideal gas ($\gamma = 1.40$) is carried through the Carnot cycle described in Active Figure 22.10. At point A, the pressure is 25.0 atm and the temperature is 600 K. At point C, the pressure is 1.00 atm and the temperature is 400 K. (a) Determine the pressures and volumes at points A, B, C, and D. (b) Calculate the net work done per cycle.

72. The compression ratio of an Otto cycle as shown in Active Figure 22.12 is $V_A/V_B = 8.00$. At the beginning A of the compression process, 500 cm^3 of gas is at 100 kPa and 20.0°C. At the beginning of the adiabatic expansion, the temperature is $T_C = 750$°C. Model the working fluid as an ideal gas with $\gamma = 1.40$. (a) Fill in this table to follow the states of the gas:

	T (K)	P (kPa)	V (cm^3)
A	293	100	500
B			
C	1 023		
D			

(b) Fill in this table to follow the processes:

	Q	W	ΔE_{int}
$A \rightarrow B$			
$B \rightarrow C$			
$C \rightarrow D$			
$D \rightarrow A$			
$ABCDA$			

(c) Identify the energy input $|Q_h|$, (d) the energy exhaust $|Q_c|$, and (e) the net output work W_{eng}. (f) Calculate the thermal efficiency. (g) Find the number of crankshaft revolutions per minute required for a one-cylinder engine to have an output power of 1.00 kW = 1.34 hp. *Note:* The thermodynamic cycle involves four piston strokes.

appendix A
Tables

TABLE A.1 *Conversion Factors*

Length

	m	cm	km	in.	ft	mi
1 meter	1	10^2	10^{-3}	39.37	3.281	6.214×10^{-4}
1 centimeter	10^{-2}	1	10^{-5}	0.393 7	3.281×10^{-2}	6.214×10^{-6}
1 kilometer	10^3	10^5	1	3.937×10^4	3.281×10^3	0.621 4
1 inch	2.540×10^{-2}	2.540	2.540×10^{-5}	1	8.333×10^{-2}	1.578×10^{-5}
1 foot	0.304 8	30.48	3.048×10^{-4}	12	1	1.894×10^{-4}
1 mile	1 609	1.609×10^5	1.609	6.336×10^4	5 280	1

Mass

	kg	g	slug	u
1 kilogram	1	10^3	6.852×10^{-2}	6.024×10^{26}
1 gram	10^{-3}	1	6.852×10^{-5}	6.024×10^{23}
1 slug	14.59	1.459×10^4	1	8.789×10^{27}
1 atomic mass unit	1.660×10^{-27}	1.660×10^{-24}	1.137×10^{-28}	1

Note: 1 metric ton = 1 000 kg.

Time

	s	min	h	day	yr
1 second	1	1.667×10^{-2}	2.778×10^{-4}	1.157×10^{-5}	3.169×10^{-8}
1 minute	60	1	1.667×10^{-2}	6.994×10^{-4}	1.901×10^{-6}
1 hour	3 600	60	1	4.167×10^{-2}	1.141×10^{-4}
1 day	8.640×10^4	1 440	24	1	2.738×10^{-5}
1 year	3.156×10^7	5.259×10^5	8.766×10^3	365.2	1

Speed

	m/s	cm/s	ft/s	mi/h
1 meter per second	1	10^2	3.281	2.237
1 centimeter per second	10^{-2}	1	3.281×10^{-2}	2.237×10^{-2}
1 foot per second	0.304 8	30.48	1	0.681 8
1 mile per hour	0.447 0	44.70	1.467	1

Note: 1 mi/min = 60 mi/h = 88 ft/s.

Force

	N	lb
1 newton	1	0.224 8
1 pound	4.448	1

(Continued)

TABLE A.1 *Conversion Factors (Continued)*

Energy, Energy Transfer

	J	ft · lb	eV
1 joule	1	0.737 6	6.242×10^{18}
1 foot-pound	1.356	1	8.464×10^{18}
1 electron volt	1.602×10^{-19}	1.182×10^{-19}	1
1 calorie	4.186	3.087	2.613×10^{19}
1 British thermal unit	1.055×10^{3}	7.779×10^{2}	6.585×10^{21}
1 kilowatt-hour	3.600×10^{6}	2.655×10^{6}	2.247×10^{25}

	cal	Btu	kWh
1 joule	0.238 9	9.481×10^{-4}	2.778×10^{-7}
1 foot-pound	0.323 9	1.285×10^{-3}	3.766×10^{-7}
1 electron volt	3.827×10^{-20}	1.519×10^{-22}	4.450×10^{-26}
1 calorie	1	3.968×10^{-3}	1.163×10^{-6}
1 British thermal unit	2.520×10^{2}	1	2.930×10^{-4}
1 kilowatt-hour	8.601×10^{5}	3.413×10^{2}	1

Pressure

	Pa	atm
1 pascal	1	9.869×10^{-6}
1 atmosphere	1.013×10^{5}	1
1 centimeter mercury[a]	1.333×10^{3}	1.316×10^{-2}
1 pound per square inch	6.895×10^{3}	6.805×10^{-2}
1 pound per square foot	47.88	4.725×10^{-4}

	cm Hg	lb/in.^{2}	lb/ft^2
1 pascal	7.501×10^{-4}	1.450×10^{-4}	2.089×10^{-2}
1 atmosphere	76	14.70	2.116×10^{3}
1 centimeter mercury[a]	1	0.194 3	27.85
1 pound per square inch	5.171	1	144
1 pound per square foot	3.591×10^{-2}	6.944×10^{-3}	1

[a]At 0°C and at a location where the free-fall acceleration has its "standard" value, 9.806 65 m/s^2.

TABLE A.2 *Symbols, Dimensions, and Units of Physical Quantities*

Quantity	Common Symbol	Unit[a]	Dimensions[b]	Unit in Terms of Base SI Units
Acceleration	$\vec{\mathbf{a}}$	m/s^2	L/T^2	m/s^2
Amount of substance	n	MOLE		mol
Angle	θ, ϕ	radian (rad)	1	
Angular acceleration	$\vec{\boldsymbol{\alpha}}$	rad/s^2	T^{-2}	s^{-2}
Angular frequency	ω	rad/s	T^{-1}	s^{-1}
Angular momentum	$\vec{\mathbf{L}}$	kg · m^2/s	ML^2/T	kg · m^2/s
Angular velocity	$\vec{\boldsymbol{\omega}}$	rad/s	T^{-1}	s^{-1}
Area	A	m^2	L^2	m^2
Atomic number	Z			
Capacitance	C	farad (F)	Q^{2}T^2/ML^2	A^2 · s^4/kg · m^2
Charge	q, Q, e	coulomb (C)	Q	A · s

(*Continued*)

TABLE A.2 *Symbols, Dimensions, and Units of Physical Quantities* (Continued)

Quantity	Common Symbol	Unit[a]	Dimensions[b]	Unit in Terms of Base SI Units
Charge density				
Line	λ	C/m	Q/L	A · s/m
Surface	σ	C/m^2	Q/L^2	A · s/m^2
Volume	ρ	C/m^3	Q/L^3	A · s/m^3
Conductivity	σ	1/Ω · m	Q^{2}T/ML^3	A^2 · s^3/kg · m^3
Current	I	AMPERE	Q/T	A
Current density	J	A/m^2	Q/TL^2	A/m^2
Density	ρ	kg/m^3	M/L^3	kg/m^3
Dielectric constant	κ			
Electric dipole moment	$\vec{\mathbf{p}}$	C · m	QL	A · s · m
Electric field	$\vec{\mathbf{E}}$	V/m	ML/QT^2	kg · m/A · s^3
Electric flux	Φ_E	V · m	ML^3/QT^2	kg · m^3/A · s^3
Electromotive force	$\mathcal{E}$	volt (V)	ML^2/QT^2	kg · m^2/A · s^3
Energy	E, U, K	joule (J)	ML^2/T^2	kg · m^2/s^2
Entropy	S	J/K	ML^2/T^{2}K	kg · m^2/s^2 · K
Force	$\vec{\mathbf{F}}$	newton (N)	ML/T^2	kg · m/s^2
Frequency	f	hertz (Hz)	T^{-1}	s^{-1}
Heat	Q	joule (J)	ML^2/T^2	kg · m^2/s^2
Inductance	L	henry (H)	ML^2/Q^2	kg · m^2/A^2 · s^2
Length	ℓ, L	METER	L	m
Displacement	$\Delta x, \Delta\vec{\mathbf{r}}$			
Distance	d, h			
Position	$x, y, z, \vec{\mathbf{r}}$			
Magnetic dipole moment	$\vec{\boldsymbol{\mu}}$	N · m/T	QL^2/T	A · m^2
Magnetic field	$\vec{\mathbf{B}}$	tesla (T) (= Wb/m^2)	M/QT	kg/A · s^2
Magnetic flux	Φ_B	weber (Wb)	ML^2/QT	kg · m^2/A · s^2
Mass	m, M	KILOGRAM	M	kg
Molar specific heat	C	J/mol · K		kg · m^2/s^2 · mol · K
Moment of inertia	I	kg · m^2	ML^2	kg · m^2
Momentum	$\vec{\mathbf{p}}$	kg · m/s	ML/T	kg · m/s
Period	T	s	T	s
Permeability of free space	μ_0	N/A^2 (= H/m)	ML/Q^2	kg · m/A^2 · s^2
Permittivity of free space	ϵ_0	C^2/N · m^2 (= F/m)	Q^{2}T^2/ML^3	A^2 · s^4/kg · m^3
Potential	V	volt (V)(= J/C)	ML^2/QT^2	kg · m^2/A · s^3
Power	P	watt (W)(= J/s)	ML^2/T^3	kg · m^2/s^3
Pressure	P	pascal (Pa)(= N/m^2)	M/LT^2	kg/m · s^2
Resistance	R	ohm (Ω)(= V/A)	ML^2/Q^{2}T	kg · m^2/A^2 · s^3
Specific heat	c	J/kg · K	L^2/T^{2}K	m^2/s^2 · K
Speed	v	m/s	L/T	m/s
Temperature	T	KELVIN	K	K
Time	t	SECOND	T	s
Torque	$\vec{\boldsymbol{\tau}}$	N · m	ML^2/T^2	kg · m^2/s^2
Velocity	$\vec{\mathbf{v}}$	m/s	L/T	m/s
Volume	V	m^3	L^3	m^3
Wavelength	λ	m	L	m
Work	W	joule (J)(= N · m)	ML^2/T^2	kg · m^2/s^2

[a]The base SI units are given in uppercase letters.

[b]The symbols M, L, T, K, and Q denote mass, length, time, temperature, and charge, respectively.

appendix **B**

Mathematics Review

This appendix in mathematics is intended as a brief review of operations and methods. Early in this course, you should be totally familiar with basic algebraic techniques, analytic geometry, and trigonometry. The sections on differential and integral calculus are more detailed and are intended for students who have difficulty applying calculus concepts to physical situations.

B.1 Scientific Notation

Many quantities used by scientists often have very large or very small values. The speed of light, for example, is about 300 000 000 m/s, and the ink required to make the dot over an *i* in this textbook has a mass of about 0.000 000 001 kg. Obviously, it is very cumbersome to read, write, and keep track of such numbers. We avoid this problem by using a method incorporating powers of the number 10:

$$10^0 = 1$$
$$10^1 = 10$$
$$10^2 = 10 \times 10 = 100$$
$$10^3 = 10 \times 10 \times 10 = 1\,000$$
$$10^4 = 10 \times 10 \times 10 \times 10 = 10\,000$$
$$10^5 = 10 \times 10 \times 10 \times 10 \times 10 = 100\,000$$

and so on. The number of zeros corresponds to the power to which ten is raised, called the **exponent** of ten. For example, the speed of light, 300 000 000 m/s, can be expressed as 3.00×10^8 m/s.

In this method, some representative numbers smaller than unity are the following:

$$10^{-1} = \frac{1}{10} = 0.1$$

$$10^{-2} = \frac{1}{10 \times 10} = 0.01$$

$$10^{-3} = \frac{1}{10 \times 10 \times 10} = 0.001$$

$$10^{-4} = \frac{1}{10 \times 10 \times 10 \times 10} = 0.000\,1$$

$$10^{-5} = \frac{1}{10 \times 10 \times 10 \times 10 \times 10} = 0.000\,01$$

In these cases, the number of places the decimal point is to the left of the digit 1 equals the value of the (negative) exponent. Numbers expressed as some power of ten multiplied by another number between one and ten are said to be in **scientific notation.** For example, the scientific notation for 5 943 000 000 is 5.943×10^9 and that for 0.000 083 2 is 8.32×10^{-5}.

When numbers expressed in scientific notation are being multiplied, the following general rule is very useful:

$$10^n \times 10^m = 10^{n+m} \qquad \textbf{(B.1)}$$

where n and m can be *any* numbers (not necessarily integers). For example, $10^2 \times 10^5 = 10^7$. The rule also applies if one of the exponents is negative: $10^3 \times 10^{-8} = 10^{-5}$.

When dividing numbers expressed in scientific notation, note that

$$\frac{10^n}{10^m} = 10^n \times 10^{-m} = 10^{n-m} \quad \textbf{(B.2)}$$

Exercises

With help from the preceding rules, verify the answers to the following equations:

1. $86\,400 = 8.64 \times 10^4$
2. $9\,816\,762.5 = 9.816\,762\,5 \times 10^6$
3. $0.000\,000\,039\,8 = 3.98 \times 10^{-8}$
4. $(4.0 \times 10^8)(9.0 \times 10^9) = 3.6 \times 10^{18}$
5. $(3.0 \times 10^7)(6.0 \times 10^{-12}) = 1.8 \times 10^{-4}$
6. $\dfrac{75 \times 10^{-11}}{5.0 \times 10^{-3}} = 1.5 \times 10^{-7}$
7. $\dfrac{(3 \times 10^6)(8 \times 10^{-2})}{(2 \times 10^{17})(6 \times 10^5)} = 2 \times 10^{-18}$

B.2 Algebra

Some Basic Rules

When algebraic operations are performed, the laws of arithmetic apply. Symbols such as x, y, and z are usually used to represent unspecified quantities, called the **unknowns.**

First, consider the equation

$$8x = 32$$

If we wish to solve for x, we can divide (or multiply) each side of the equation by the same factor without destroying the equality. In this case, if we divide both sides by 8, we have

$$\frac{8x}{8} = \frac{32}{8}$$

$$x = 4$$

Next consider the equation

$$x + 2 = 8$$

In this type of expression, we can add or subtract the same quantity from each side. If we subtract 2 from each side, we have

$$x + 2 - 2 = 8 - 2$$

$$x = 6$$

In general, if $x + a = b$, then $x = b - a$.

Now consider the equation

$$\frac{x}{5} = 9$$

If we multiply each side by 5, we are left with x on the left by itself and 45 on the right:

$$\left(\frac{x}{5}\right)(5) = 9 \times 5$$

$$x = 45$$

In all cases, *whatever operation is performed on the left side of the equality must also be performed on the right side.*

The following rules for multiplying, dividing, adding, and subtracting fractions should be recalled, where a, b, c, and d are four numbers:

	Rule	Example
Multiplying	$\left(\frac{a}{b}\right)\left(\frac{c}{d}\right) = \frac{ac}{bd}$	$\left(\frac{2}{3}\right)\left(\frac{4}{5}\right) = \frac{8}{15}$
Dividing	$\frac{(a/b)}{(c/d)} = \frac{ad}{bc}$	$\frac{2/3}{4/5} = \frac{(2)(5)}{(4)(3)} = \frac{10}{12}$
Adding	$\frac{a}{b} \pm \frac{c}{d} = \frac{ad \pm bc}{bd}$	$\frac{2}{3} - \frac{4}{5} = \frac{(2)(5) - (4)(3)}{(3)(5)} = -\frac{2}{15}$

Exercises

In the following exercises, solve for x.

	Answers
1. $a = \frac{1}{1 + x}$	$x = \frac{1 - a}{a}$
2. $3x - 5 = 13$	$x = 6$
3. $ax - 5 = bx + 2$	$x = \frac{7}{a - b}$
4. $\frac{5}{2x + 6} = \frac{3}{4x + 8}$	$x = -\frac{11}{7}$

Powers

When powers of a given quantity x are multiplied, the following rule applies:

$$x^n x^m = x^{n+m} \quad \textbf{(B.3)}$$

For example, $x^2x^4 = x^{2+4} = x^6$.

When dividing the powers of a given quantity, the rule is

$$\frac{x^n}{x^m} = x^{n-m} \quad \textbf{(B.4)}$$

For example, $x^8/x^2 = x^{8-2} = x^6$.

A power that is a fraction, such as $\frac{1}{3}$, corresponds to a root as follows:

$$x^{1/n} = \sqrt[n]{x} \quad \textbf{(B.5)}$$

For example, $4^{1/3} = \sqrt[3]{4} = 1.587\,4$. (A scientific calculator is useful for such calculations.)

Finally, any quantity x^n raised to the mth power is

$$(x^n)^m = x^{nm} \quad \textbf{(B.6)}$$

Table B.1 summarizes the rules of exponents.

TABLE B.1

Rules of Exponents

$x^0 = 1$

$x^1 = x$

$x^n x^m = x^{n+m}$

$x^n/x^m = x^{n-m}$

$x^{1/n} = \sqrt[n]{x}$

$(x^n)^m = x^{nm}$

Exercises

Verify the following equations:

1. $3^2 \times 3^3 = 243$

2. $x^5x^{-8} = x^{-3}$

3. $x^{10}/x^{-5} = x^{15}$
4. $5^{1/3} = 1.709\ 976$ (Use your calculator.)
5. $60^{1/4} = 2.783\ 158$ (Use your calculator.)
6. $(x^4)^3 = x^{12}$

Factoring

Some useful formulas for factoring an equation are the following:

$ax + ay + az = a(x + y + z)$ common factor

$a^2 + 2ab + b^2 = (a + b)^2$ perfect square

$a^2 - b^2 = (a + b)(a - b)$ differences of squares

Quadratic Equations

The general form of a quadratic equation is

$$ax^2 + bx + c = 0 \tag{B.7}$$

where x is the unknown quantity and a, b, and c are numerical factors referred to as **coefficients** of the equation. This equation has two roots, given by

$$x = \frac{-b \pm \sqrt{b^2 - 4ac}}{2a} \tag{B.8}$$

If $b^2 \geq 4ac$, the roots are real.

Example B.1

The equation $x^2 + 5x + 4 = 0$ has the following roots corresponding to the two signs of the square-root term:

$$x = \frac{-5 \pm \sqrt{5^2 - (4)(1)(4)}}{2(1)} = \frac{-5 \pm \sqrt{9}}{2} = \frac{-5 \pm 3}{2}$$

$$x_+ = \frac{-5 + 3}{2} = \boxed{-1} \qquad x_- = \frac{-5 - 3}{2} = \boxed{-4}$$

where x_+ refers to the root corresponding to the positive sign and x_- refers to the root corresponding to the negative sign.

Exercises

Solve the following quadratic equations:

	Answers	
1. $x^2 + 2x - 3 = 0$	$x_+ = 1$	$x_- = -3$
2. $2x^2 - 5x + 2 = 0$	$x_+ = 2$	$x_- = \frac{1}{2}$
3. $2x^2 - 4x - 9 = 0$	$x_+ = 1 + \sqrt{22}/2$	$x_- = 1 - \sqrt{22}/2$

Linear Equations

A linear equation has the general form

$$y = mx + b \tag{B.9}$$

where m and b are constants. This equation is referred to as linear because the graph of y versus x is a straight line as shown in Figure B.1. The constant b, called the **y-intercept**, represents the value of y at which the straight line intersects the y

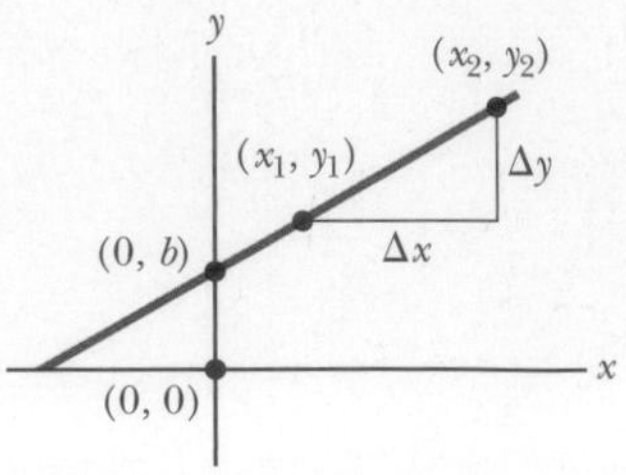

Figure B.1 A straight line graphed on an xy coordinate system. The slope of the line is the ratio of Δy to Δx.

axis. The constant m is equal to the **slope** of the straight line. If any two points on the straight line are specified by the coordinates (x_1, y_1) and (x_2, y_2) as in Figure B.1, the slope of the straight line can be expressed as

$$\text{Slope} = \frac{y_2 - y_1}{x_2 - x_1} = \frac{\Delta y}{\Delta x} \qquad \textbf{(B.10)}$$

Note that m and b can have either positive or negative values. If $m > 0$, the straight line has a *positive* slope as in Figure B.1. If $m < 0$, the straight line has a *negative* slope. In Figure B.1, both m and b are positive. Three other possible situations are shown in Figure B.2.

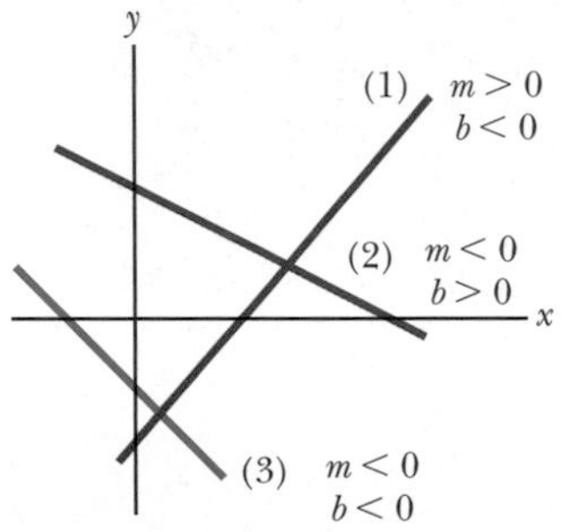

Figure B.2 The brown line has a positive slope and a negative y-intercept. The blue line has a negative slope and a positive y-intercept. The green line has a negative slope and a negative y-intercept.

Exercises

1. Draw graphs of the following straight lines: (a) $y = 5x + 3$ (b) $y = -2x + 4$ (c) $y = -3x - 6$
2. Find the slopes of the straight lines described in Exercise 1.

Answers (a) 5 (b) -2 (c) -3

3. Find the slopes of the straight lines that pass through the following sets of points: (a) $(0, -4)$ and $(4, 2)$ (b) $(0, 0)$ and $(2, -5)$ (c) $(-5, 2)$ and $(4, -2)$

Answers (a) $\frac{3}{2}$ (b) $-\frac{5}{2}$ (c) $-\frac{4}{9}$

Solving Simultaneous Linear Equations

Consider the equation $3x + 5y = 15$, which has two unknowns, x and y. Such an equation does not have a unique solution. For example, $(x = 0, y = 3)$, $(x = 5, y = 0)$, and $(x = 2, y = \frac{9}{5})$ are all solutions to this equation.

If a problem has two unknowns, a unique solution is possible only if we have *two* pieces of information. In most common cases, those two pieces of information are equations. In general, if a problem has n unknowns, its solution requires n equations. To solve two simultaneous equations involving two unknowns, x and y, we solve one of the equations for x in terms of y and substitute this expression into the other equation.

In some cases, the two pieces of information may be (1) one equation and (2) a condition on the solutions. For example, suppose we have the equation $m = 3n$ and the condition that m and n must be the smallest positive nonzero integers possible. Then, the single equation does not allow a unique solution, but the addition of the condition gives us that $n = 1$ and $m = 3$.

Example B.2

Solve the two simultaneous equations

$$(1) \quad 5x + y = -8$$
$$(2) \quad 2x - 2y = 4$$

Solution From Equation (2), $x = y + 2$. Substitution of this equation into Equation (1) gives

$$5(y + 2) + y = -8$$
$$6y = -18$$
$$y = -3$$
$$x = y + 2 = -1$$

B.2 *cont.*

Alternative Solution Multiply each term in Equation (1) by the factor 2 and add the result to Equation (2):

$$\begin{aligned} 10x + 2y &= -16 \\ 2x - 2y &= 4 \\ \hline 12x \quad &= -12 \\ x &= -1 \\ y = x - 2 &= -3 \end{aligned}$$

Two linear equations containing two unknowns can also be solved by a graphical method. If the straight lines corresponding to the two equations are plotted in a conventional coordinate system, the intersection of the two lines represents the solution. For example, consider the two equations

$$x - y = 2$$

$$x - 2y = -1$$

These equations are plotted in Figure B.3. The intersection of the two lines has the coordinates $x = 5$ and $y = 3$, which represents the solution to the equations. You should check this solution by the analytical technique discussed earlier.

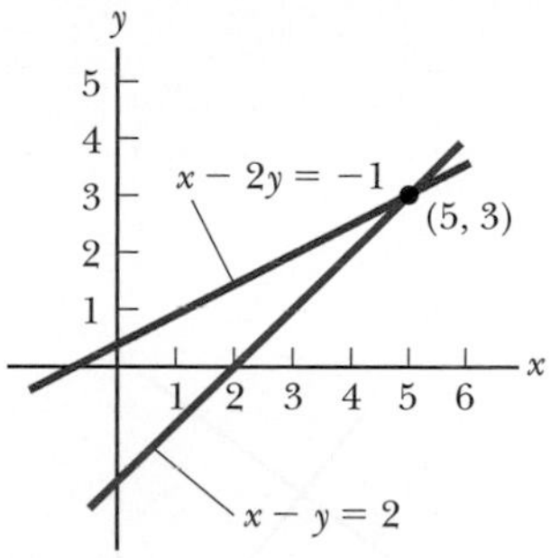

Figure B.3 A graphical solution for two linear equations.

Exercises

Solve the following pairs of simultaneous equations involving two unknowns:

	Answers
1. $x + y = 8$ $x - y = 2$	$x = 5,\ y = 3$
2. $98 - T = 10a$ $T - 49 = 5a$	$T = 65,\ a = 3.27$
3. $6x + 2y = 6$ $8x - 4y = 28$	$x = 2,\ y = -3$

Logarithms

Suppose a quantity x is expressed as a power of some quantity a:

$$x = a^y \tag{B.11}$$

The number a is called the **base** number. The **logarithm** of x with respect to the base a is equal to the exponent to which the base must be raised to satisfy the expression $x = a^y$:

$$y = \log_a x \tag{B.12}$$

Conversely, the **antilogarithm** of y is the number x:

$$x = \text{antilog}_a\, y \tag{B.13}$$

In practice, the two bases most often used are base 10, called the *common* logarithm base, and base $e = 2.718\,282$, called Euler's constant or the *natural* logarithm base. When common logarithms are used,

$$y = \log_{10} x \quad (\text{or } x = 10^y) \tag{B.14}$$

When natural logarithms are used,

$$y = \ln x \quad (\text{or } x = e^y) \tag{B.15}$$

For example, $\log_{10} 52 = 1.716$, so $\text{antilog}_{10}\, 1.716 = 10^{1.716} = 52$. Likewise, $\ln 52 = 3.951$, so $\text{antiln}\, 3.951 = e^{3.951} = 52$.

In general, note you can convert between base 10 and base e with the equality

$$\ln x = (2.302\,585) \log_{10} x \tag{B.16}$$

Finally, some useful properties of logarithms are the following:

$$\left.\begin{aligned} \log(ab) &= \log a + \log b \\ \log(a/b) &= \log a - \log b \\ \log(a^n) &= n \log a \end{aligned}\right\} \text{any base}$$

$$\ln e = 1$$

$$\ln e^a = a$$

$$\ln\left(\frac{1}{a}\right) = -\ln a$$

B.3 Geometry

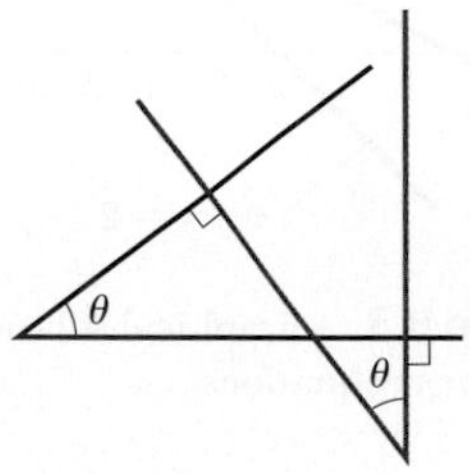

Figure B.4 The angles are equal because their sides are perpendicular.

The **distance** d between two points having coordinates (x_1, y_1) and (x_2, y_2) is

$$d = \sqrt{(x_2 - x_1)^2 + (y_2 - y_1)^2} \tag{B.17}$$

Two angles are equal if their sides are perpendicular, right side to right side and left side to left side. For example, the two angles marked θ in Figure B.4 are the same because of the perpendicularity of the sides of the angles. To distinguish the left and right sides of an angle, imagine standing at the angle's apex and facing into the angle.

Radian measure: The arc length s of a circular arc (Fig. B.5) is proportional to the radius r for a fixed value of θ (in radians):

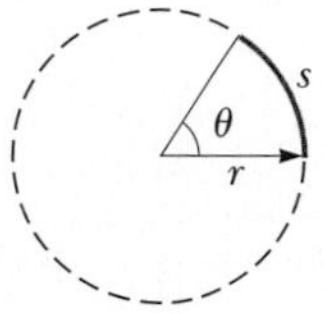

Figure B.5 The angle θ in radians is the ratio of the arc length s to the radius r of the circle.

$$s = r\theta \qquad \theta = \frac{s}{r} \tag{B.18}$$

Table B.2 gives the **areas** and **volumes** for several geometric shapes used throughout this text.

TABLE B.2 *Useful Information for Geometry*

Shape	Area or Volume	Shape	Area or Volume
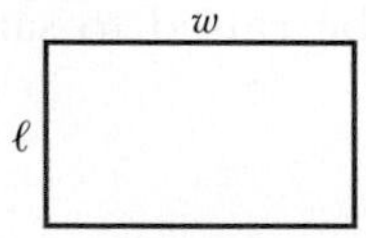 Rectangle	Area $= \ell w$	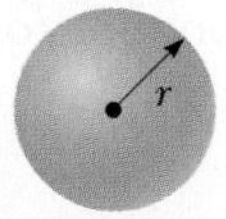 Sphere	Surface area $= 4\pi r^2$ Volume $= \frac{4\pi r^3}{3}$
 Circle	Area $= \pi r^2$ Circumference $= 2\pi r$	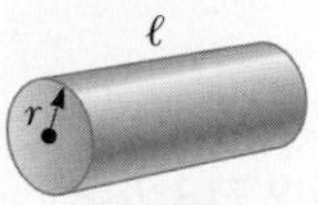 Cylinder	Lateral surface area $= 2\pi r\ell$ Volume $= \pi r^2 \ell$
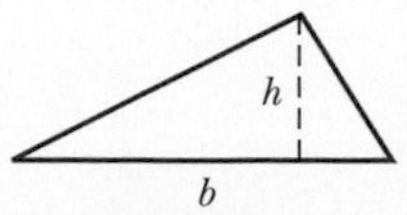 Triangle	Area $= \frac{1}{2}bh$	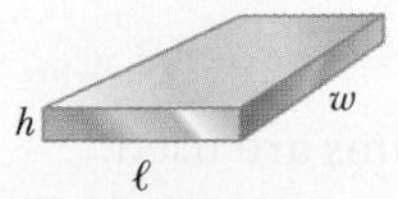 Rectangular box	Surface area $= 2(\ell h + \ell w + hw)$ Volume $= \ell wh$

The equation of a **straight line** (Fig. B.6) is

$$y = mx + b \tag{B.19}$$

where b is the y-intercept and m is the slope of the line.

The equation of a **circle** of radius R centered at the origin is

$$x^2 + y^2 = R^2 \tag{B.20}$$

The equation of an **ellipse** having the origin at its center (Fig. B.7) is

$$\frac{x^2}{a^2} + \frac{y^2}{b^2} = 1 \tag{B.21}$$

where a is the length of the semimajor axis (the longer one) and b is the length of the semiminor axis (the shorter one).

The equation of a **parabola** the vertex of which is at $y = b$ (Fig. B.8) is

$$y = ax^2 + b \tag{B.22}$$

The equation of a **rectangular hyperbola** (Fig. B.9) is

$$xy = \text{constant} \tag{B.23}$$

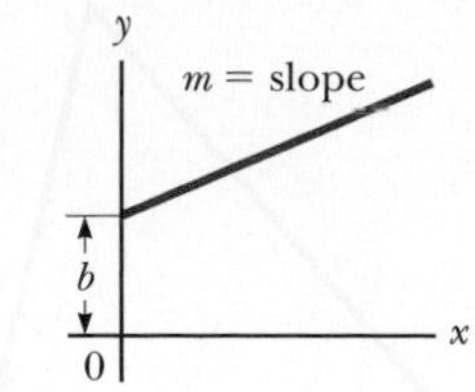

Figure B.6 A straight line with a slope of m and a y-intercept of b.

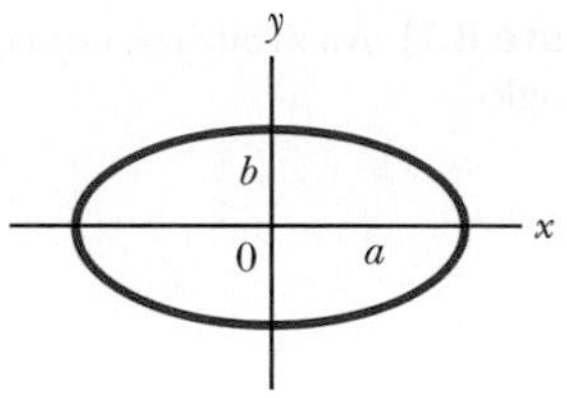

Figure B.7 An ellipse with semi major axis a and semiminor axis b.

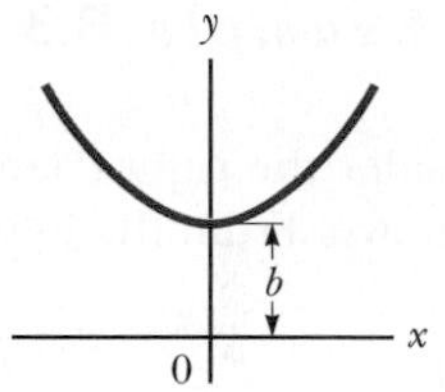

Figure B.8 A parabola with its vertex at $y = b$.

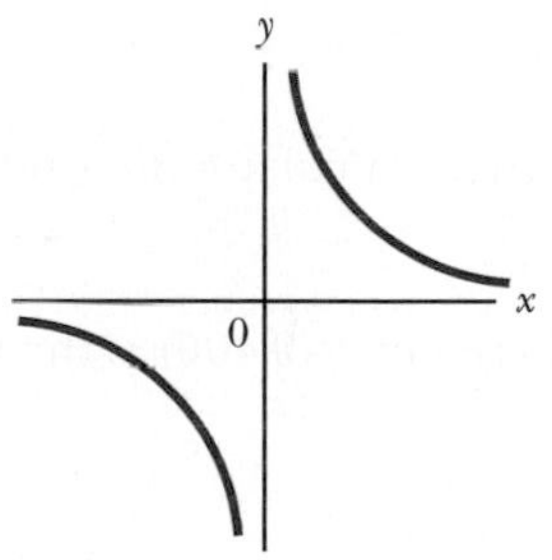

Figure B.9 A hyperbola.

B.4 Trigonometry

That portion of mathematics based on the special properties of the right triangle is called trigonometry. By definition, a right triangle is a triangle containing a 90° angle. Consider the right triangle shown in Figure B.10, where side a is opposite the angle θ, side b is adjacent to the angle θ, and side c is the hypotenuse of the triangle. The three basic trigonometric functions defined by such a triangle are the sine (sin), cosine (cos), and tangent (tan). In terms of the angle θ, these functions are defined as follows:

$$\sin\theta = \frac{\text{side opposite } \theta}{\text{hypotenuse}} = \frac{a}{c} \tag{B.24}$$

$$\cos\theta = \frac{\text{side adjacent to } \theta}{\text{hypotenuse}} = \frac{b}{c} \tag{B.25}$$

$$\tan\theta = \frac{\text{side opposite } \theta}{\text{side adjacent to } \theta} = \frac{a}{b} \tag{B.26}$$

The Pythagorean theorem provides the following relationship among the sides of a right triangle:

$$c^2 = a^2 + b^2 \tag{B.27}$$

From the preceding definitions and the Pythagorean theorem, it follows that

$$\sin^2\theta + \cos^2\theta = 1$$

$$\tan\theta = \frac{\sin\theta}{\cos\theta}$$

The cosecant, secant, and cotangent functions are defined by

$$\csc\theta = \frac{1}{\sin\theta} \qquad \sec\theta = \frac{1}{\cos\theta} \qquad \cot\theta = \frac{1}{\tan\theta}$$

The following relationships are derived directly from the right triangle shown in Figure B.10:

$$\sin\theta = \cos(90° - \theta)$$

$$\cos\theta = \sin(90° - \theta)$$

$$\cot\theta = \tan(90° - \theta)$$

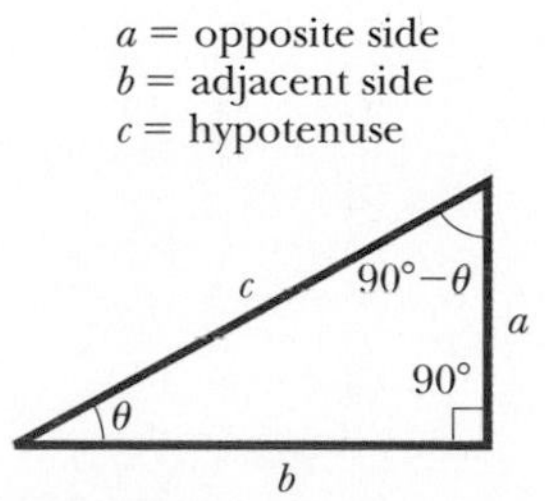

Figure B.10 A right triangle, used to define the basic functions of trigonometry.

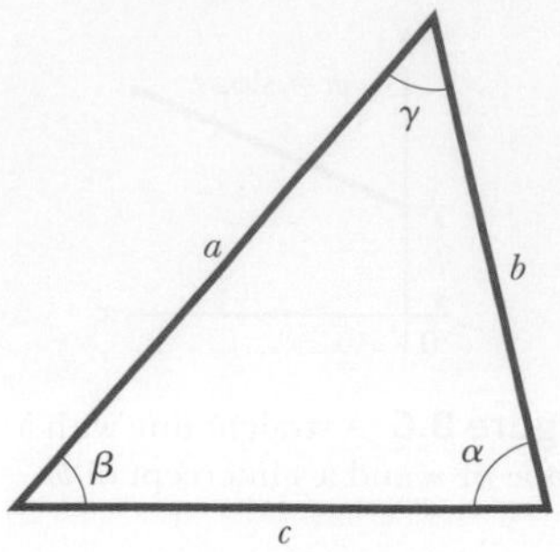

Figure B.11 An arbitrary, nonright triangle.

Some properties of trigonometric functions are the following:

$$\sin(-\theta) = -\sin\theta$$
$$\cos(-\theta) = \cos\theta$$
$$\tan(-\theta) = -\tan\theta$$

The following relationships apply to *any* triangle as shown in Figure B.11:

$$\alpha + \beta + \gamma = 180°$$

$$\text{Law of cosines}\quad \begin{cases} a^2 = b^2 + c^2 - 2bc\cos\alpha \\ b^2 = a^2 + c^2 - 2ac\cos\beta \\ c^2 = a^2 + b^2 - 2ab\cos\gamma \end{cases}$$

$$\text{Law of sines}\quad \frac{a}{\sin\alpha} = \frac{b}{\sin\beta} = \frac{c}{\sin\gamma}$$

Table B.3 lists a number of useful trigonometric identities.

Example B.3

Consider the right triangle in Figure B.12 in which $a = 2.00$, $b = 5.00$, and c is unknown. From the Pythagorean theorem, we have

$$c^2 = a^2 + b^2 = 2.00^2 + 5.00^2 = 4.00 + 25.0 = 29.0$$

$$c = \sqrt{29.0} = 5.39$$

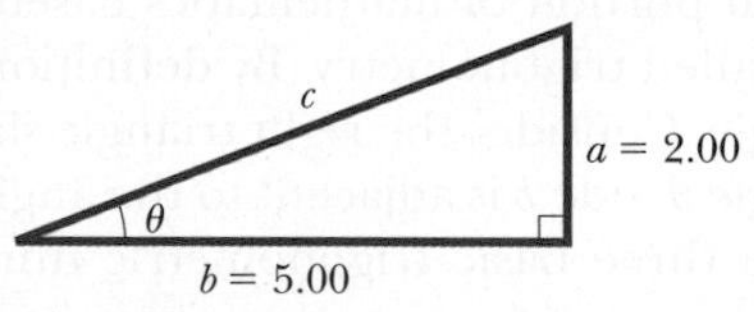

Figure B.12 (Example B.3)

To find the angle θ, note that

$$\tan\theta = \frac{a}{b} = \frac{2.00}{5.00} = 0.400$$

Using a calculator, we find that

$$\theta = \tan^{-1}(0.400) = 21.8°$$

where $\tan^{-1}(0.400)$ is the notation for "angle whose tangent is 0.400," sometimes written as arctan (0.400).

TABLE B.3 *Some Trigonometric Identities*

$\sin^2\theta + \cos^2\theta = 1$	$\csc^2\theta = 1 + \cot^2\theta$
$\sec^2\theta = 1 + \tan^2\theta$	$\sin^2\frac{\theta}{2} = \frac{1}{2}(1 - \cos\theta)$
$\sin 2\theta = 2\sin\theta\cos\theta$	$\cos^2\frac{\theta}{2} = \frac{1}{2}(1 + \cos\theta)$
$\cos 2\theta = \cos^2\theta - \sin^2\theta$	$1 - \cos\theta = 2\sin^2\frac{\theta}{2}$
$\tan 2\theta = \frac{2\tan\theta}{1 - \tan^2\theta}$	$\tan\frac{\theta}{2} = \sqrt{\frac{1 - \cos\theta}{1 + \cos\theta}}$

$\sin(A \pm B) = \sin A\cos B \pm \cos A\sin B$

$\cos(A \pm B) = \cos A\cos B \mp \sin A\sin B$

$\sin A \pm \sin B = 2\sin[\frac{1}{2}(A \pm B)]\cos[\frac{1}{2}(A \mp B)]$

$\cos A + \cos B = 2\cos[\frac{1}{2}(A + B)]\cos[\frac{1}{2}(A - B)]$

$\cos A - \cos B = 2\sin[\frac{1}{2}(A + B)]\sin[\frac{1}{2}(B - A)]$

Exercises

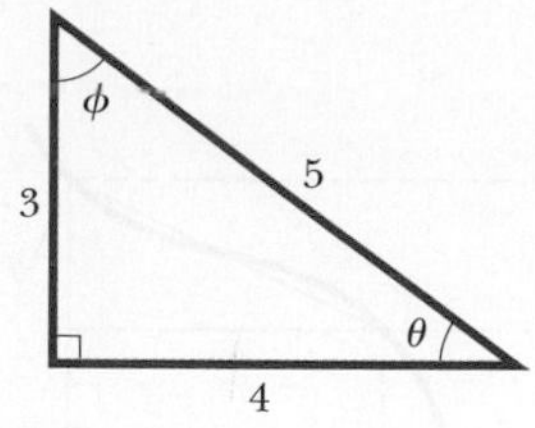

Figure B.13 (Exercise 1)

1. In Figure B.13, identify (a) the side opposite θ (b) the side adjacent to ϕ and then find (c) $\cos\theta$, (d) $\sin\phi$, and (e) $\tan\phi$.

Answers (a) 3 (b) 3 (c) $\frac{4}{5}$ (d) $\frac{4}{5}$ (e) $\frac{4}{3}$

2. In a certain right triangle, the two sides that are perpendicular to each other are 5.00 m and 7.00 m long. What is the length of the third side?

Answer 8.60 m

3. A right triangle has a hypotenuse of length 3.0 m, and one of its angles is 30°. (a) What is the length of the side opposite the 30° angle? (b) What is the side adjacent to the 30° angle?

Answers (a) 1.5 m (b) 2.6 m

B.5 Series Expansions

$$(a+b)^n = a^n + \frac{n}{1!}a^{n-1}b + \frac{n(n-1)}{2!}a^{n-2}b^2 + \cdots$$

$$(1+x)^n = 1 + nx + \frac{n(n-1)}{2!}x^2 + \cdots$$

$$e^x = 1 + x + \frac{x^2}{2!} + \frac{x^3}{3!} + \cdots$$

$$\ln(1 \pm x) = \pm x - \tfrac{1}{2}x^2 \pm \tfrac{1}{3}x^3 - \cdots$$

$$\left.\begin{aligned} \sin x &= x - \frac{x^3}{3!} + \frac{x^5}{5!} - \cdots \\ \cos x &= 1 - \frac{x^2}{2!} + \frac{x^4}{4!} - \cdots \\ \tan x &= x + \frac{x^3}{3} + \frac{2x^5}{15} + \cdots \quad |x| < \frac{\pi}{2} \end{aligned}\right\} x \text{ in radians}$$

For $x << 1$, the following approximations can be used:[1]

$$(1+x)^n \approx 1 + nx \qquad \sin x \approx x$$
$$e^x \approx 1 + x \qquad \cos x \approx 1$$
$$\ln(1 \pm x) \approx \pm x \qquad \tan x \approx x$$

B.6 Differential Calculus

In various branches of science, it is sometimes necessary to use the basic tools of calculus, invented by Newton, to describe physical phenomena. The use of calculus is fundamental in the treatment of various problems in Newtonian mechanics, electricity, and magnetism. In this section, we simply state some basic properties and "rules of thumb" that should be a useful review to the student.

First, a **function** must be specified that relates one variable to another (e.g., a coordinate as a function of time). Suppose one of the variables is called y (the dependent variable), and the other x (the independent variable). We might have a function relationship such as

$$y(x) = ax^3 + bx^2 + cx + d$$

If a, b, c, and d are specified constants, y can be calculated for any value of x. We usually deal with continuous functions, that is, those for which y varies "smoothly" with x.

[1]The approximations for the functions $\sin x$, $\cos x$, and $\tan x$ are for $x \le 0.1$ rad.

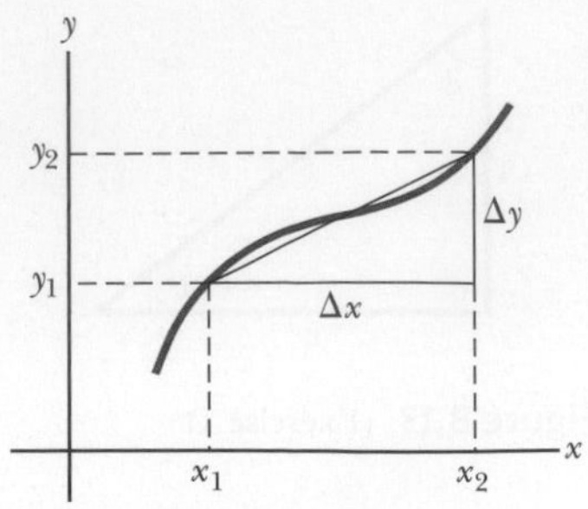

Figure B.14 The lengths Δx and Δy are used to define the derivative of this function at a point.

The **derivative** of y with respect to x is defined as the limit as Δx approaches zero of the slopes of chords drawn between two points on the y versus x curve. Mathematically, we write this definition as

$$\frac{dy}{dx} = \lim_{\Delta x \to 0} \frac{\Delta y}{\Delta x} = \lim_{\Delta x \to 0} \frac{y(x + \Delta x) - y(x)}{\Delta x} \quad \textbf{(B.28)}$$

where Δy and Δx are defined as $\Delta x = x_2 - x_1$ and $\Delta y = y_2 - y_1$ (Fig. B.14). Note that dy/dx *does not* mean dy divided by dx, but rather is simply a notation of the limiting process of the derivative as defined by Equation B.28.

A useful expression to remember when $y(x) = ax^n$, where a is a *constant* and n is *any* positive or negative number (integer or fraction), is

$$\frac{dy}{dx} = nax^{n-1} \quad \textbf{(B.29)}$$

If $y(x)$ is a polynomial or algebraic function of x, we apply Equation B.29 to *each* term in the polynomial and take $d[\text{constant}]/dx = 0$. In Examples B.4 through B.7, we evaluate the derivatives of several functions.

TABLE B.4

Derivative for Several Functions

$\frac{d}{dx}(a) = 0$
$\frac{d}{dx}(ax^n) = nax^{n-1}$
$\frac{d}{dx}(e^{ax}) = ae^{ax}$
$\frac{d}{dx}(\sin ax) = a \cos ax$
$\frac{d}{dx}(\cos ax) = -a \sin ax$
$\frac{d}{dx}(\tan ax) = a \sec^2 ax$
$\frac{d}{dx}(\cot ax) = -a \csc^2 ax$
$\frac{d}{dx}(\sec x) = \tan x \sec x$
$\frac{d}{dx}(\csc x) = -\cot x \csc x$
$\frac{d}{dx}(\ln ax) = \frac{1}{x}$
$\frac{d}{dx}(\sin^{-1} ax) = \frac{a}{\sqrt{1 - a^2x^2}}$
$\frac{d}{dx}(\cos^{-1} ax) = \frac{-a}{\sqrt{1 - a^2x^2}}$
$\frac{d}{dx}(\tan^{-1} ax) = \frac{a}{1 + a^2x^2}$

Note: The symbols a and n represent constants.

Special Properties of the Derivative

A. Derivative of the product of two functions If a function $f(x)$ is given by the product of two functions—say, $g(x)$ and $h(x)$—the derivative of $f(x)$ is defined as

$$\frac{d}{dx} f(x) = \frac{d}{dx}[g(x)h(x)] = g\frac{dh}{dx} + h\frac{dg}{dx} \quad \textbf{(B.30)}$$

B. Derivative of the sum of two functions If a function $f(x)$ is equal to the sum of two functions, the derivative of the sum is equal to the sum of the derivatives:

$$\frac{d}{dx} f(x) = \frac{d}{dx}[g(x) + h(x)] = \frac{dg}{dx} + \frac{dh}{dx} \quad \textbf{(B.31)}$$

C. Chain rule of differential calculus If $y = f(x)$ and $x = g(z)$, then dy/dz can be written as the product of two derivatives:

$$\frac{dy}{dz} = \frac{dy}{dx}\frac{dx}{dz} \quad \textbf{(B.32)}$$

D. The second derivative The second derivative of y with respect to x is defined as the derivative of the function dy/dx (the derivative of the derivative). It is usually written as

$$\frac{d^2y}{dx^2} = \frac{d}{dx}\left(\frac{dy}{dx}\right) \quad \textbf{(B.33)}$$

Some of the more commonly used derivatives of functions are listed in Table B.4.

Example B.4

Suppose $y(x)$ (that is, y as a function of x) is given by

$$y(x) = ax^3 + bx + c$$

where a and b are constants. It follows that

$$y(x + \Delta x) = a(x + \Delta x)^3 + b(x + \Delta x) + c$$
$$= a(x^3 + 3x^2\,\Delta x + 3x\,\Delta x^2 + \Delta x^3) + b(x + \Delta x) + c$$

B.4 *cont.*

so

$$\Delta y = y(x + \Delta x) - y(x) = a(3x^2\,\Delta x + 3x\,\Delta x^2 + \Delta x^3) + b\Delta x$$

Substituting this into Equation B.28 gives

$$\frac{dy}{dx} = \lim_{\Delta x \to 0} \frac{\Delta y}{\Delta x} = \lim_{\Delta x \to 0} [3ax^2 + 3ax\,\Delta x + a\,\Delta x^2] + b$$

$$\frac{dy}{dx} = \boxed{3ax^2 + b}$$

Example B.5

Find the derivative of

$$y(x) = 8x^5 + 4x^3 + 2x + 7$$

Solution Applying Equation B.29 to each term independently and remembering that d/dx (constant) = 0, we have

$$\frac{dy}{dx} = 8(5)x^4 + 4(3)x^2 + 2(1)x^0 + 0$$

$$\frac{dy}{dx} = \boxed{40x^4 + 12x^2 + 2}$$

Example B.6

Find the derivative of $y(x) = x^3/(x + 1)^2$ with respect to x.

Solution We can rewrite this function as $y(x) = x^3(x + 1)^{-2}$ and apply Equation B.30:

$$\frac{dy}{dx} = (x + 1)^{-2}\frac{d}{dx}(x^3) + x^3\frac{d}{dx}(x + 1)^{-2}$$

$$= (x + 1)^{-2}\,3x^2 + x^3\,(-2)(x + 1)^{-3}$$

$$\frac{dy}{dx} = \boxed{\frac{3x^2}{(x + 1)^2} - \frac{2x^3}{(x + 1)^3}} = \boxed{\frac{x^2(x + 3)}{(x + 1)^3}}$$

Example B.7

A useful formula that follows from Equation B.30 is the derivative of the quotient of two functions. Show that

$$\frac{d}{dx}\left[\frac{g(x)}{h(x)}\right] = \frac{h\dfrac{dg}{dx} - g\dfrac{dh}{dx}}{h^2}$$

Solution We can write the quotient as gh^{-1} and then apply Equations B.29 and B.30:

$$\frac{d}{dx}\left(\frac{g}{h}\right) = \frac{d}{dx}(gh^{-1}) = g\frac{d}{dx}(h^{-1}) + h^{-1}\frac{d}{dx}(g)$$

$$= -gh^{-2}\frac{dh}{dx} + h^{-1}\frac{dg}{dx}$$

$$= \frac{h\dfrac{dg}{dx} - g\dfrac{dh}{dx}}{h^2}$$

B.7 Integral Calculus

We think of integration as the inverse of differentiation. As an example, consider the expression

$$f(x) = \frac{dy}{dx} = 3ax^2 + b \qquad \textbf{(B.34)}$$

which was the result of differentiating the function

$$y(x) = ax^3 + bx + c$$

in Example B.4. We can write Equation B.34 as $dy = f(x)\,dx = (3ax^2 + b)\,dx$ and obtain $y(x)$ by "summing" over all values of x. Mathematically, we write this inverse operation as

$$y(x) = \int f(x)\,dx$$

For the function $f(x)$ given by Equation B.34, we have

$$y(x) = \int (3ax^2 + b)\,dx = ax^3 + bx + c$$

where c is a constant of the integration. This type of integral is called an *indefinite integral* because its value depends on the choice of c.

A general **indefinite integral** $I(x)$ is defined as

$$I(x) = \int f(x)\,dx \qquad \textbf{(B.35)}$$

where $f(x)$ is called the *integrand* and $f(x) = dI(x)/dx$.

For a *general continuous* function $f(x)$, the integral can be described as the area under the curve bounded by $f(x)$ and the x axis, between two specified values of x, say, x_1 and x_2, as in Figure B.15.

The area of the blue element in Figure B.15 is approximately $f(x_i)\,\Delta x_i$. If we sum all these area elements between x_1 and x_2 and take the limit of this sum as $\Delta x_i \rightarrow 0$, we obtain the *true* area under the curve bounded by $f(x)$ and the x axis, between the limits x_1 and x_2:

$$\text{Area} = \lim_{\Delta x_i \rightarrow 0} \sum_i f(x_i)\Delta x_i = \int_{x_1}^{x_2} f(x)\,dx \qquad \textbf{(B.36)}$$

Integrals of the type defined by Equation B.36 are called **definite integrals.**

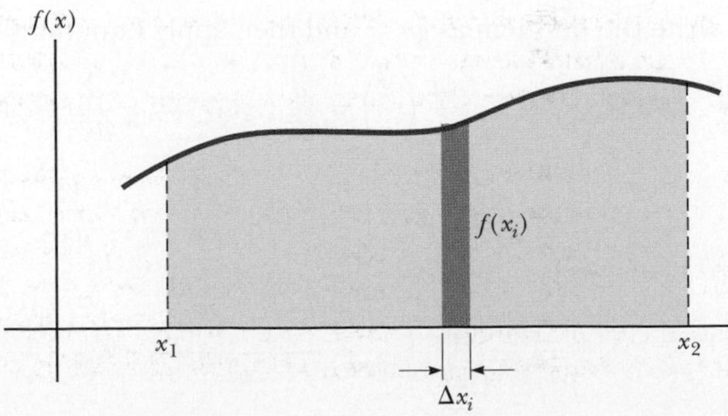

Figure B.15 The definite integral of a function is the area under the curve of the function between the limits x_1 and x_2.

One common integral that arises in practical situations has the form

$$\int x^n \, dx = \frac{x^{n+1}}{n+1} + c \quad (n \neq -1) \tag{B.37}$$

This result is obvious, being that differentiation of the right-hand side with respect to x gives $f(x) = x^n$ directly. If the limits of the integration are known, this integral becomes a *definite integral* and is written

$$\int_{x_1}^{x_2} x^n \, dx = \frac{x^{n+1}}{n+1}\bigg|_{x_1}^{x_2} = \frac{x_2^{\,n+1} - x_1^{\,n+1}}{n+1} \quad (n \neq -1) \tag{B.38}$$

Examples

1. $\displaystyle\int_0^a x^2 \, dx = \frac{x^3}{3}\bigg]_0^a = \frac{a^3}{3}$

2. $\displaystyle\int_0^b x^{3/2} \, dx = \frac{x^{5/2}}{5/2}\bigg]_0^b = \tfrac{2}{5} b^{5/2}$

3. $\displaystyle\int_3^5 x \, dx = \frac{x^2}{2}\bigg]_3^5 = \frac{5^2 - 3^2}{2} = 8$

Partial Integration

Sometimes it is useful to apply the method of *partial integration* (also called "integrating by parts") to evaluate certain integrals. This method uses the property

$$\int u \, dv = uv - \int v \, du \tag{B.39}$$

where u and v are *carefully* chosen so as to reduce a complex integral to a simpler one. In many cases, several reductions have to be made. Consider the function

$$I(x) = \int x^2 e^x \, dx$$

which can be evaluated by integrating by parts twice. First, if we choose $u = x^2$, $v = e^x$, we obtain

$$\int x^2 e^x \, dx = \int x^2 \, d(e^x) = x^2 e^x - 2\int e^x x \, dx + c_1$$

Now, in the second term, choose $u = x$, $v = e^x$, which gives

$$\int x^2 e^x \, dx = x^2 e^x - 2x e^x + 2\int e^x \, dx + c_1$$

or

$$\int x^2 e^x \, dx = x^2 e^x - 2xe^x + 2e^x + c_2$$

The Perfect Differential

Another useful method to remember is that of the *perfect differential,* in which we look for a change of variable such that the differential of the function is the

TABLE B.5 *Some Indefinite Integrals (An arbitrary constant should be added to each of these integrals.)*

$$\int x^n\,dx = \frac{x^{n+1}}{n+1} \text{ (provided } n \neq 1)$$

$$\int \frac{dx}{x} = \int x^{-1}\,dx = \ln x$$

$$\int \frac{dx}{a+bx} = \frac{1}{b}\ln(a+bx)$$

$$\int \frac{x\,dx}{a+bx} = \frac{x}{b} - \frac{a}{b^2}\ln(a+bx)$$

$$\int \frac{dx}{x(x+a)} = -\frac{1}{a}\ln\frac{x+a}{x}$$

$$\int \frac{dx}{(a+bx)^2} = -\frac{1}{b(a+bx)}$$

$$\int \frac{dx}{a^2+x^2} = \frac{1}{a}\tan^{-1}\frac{x}{a}$$

$$\int \frac{dx}{a^2-x^2} = \frac{1}{2a}\ln\frac{a+x}{a-x} \quad (a^2 - x^2 > 0)$$

$$\int \frac{dx}{x^2-a^2} = \frac{1}{2a}\ln\frac{x-a}{x+a} \quad (x^2 - a^2 > 0)$$

$$\int \frac{x\,dx}{a^2 \pm x^2} = \pm\tfrac{1}{2}\ln(a^2 \pm x^2)$$

$$\int \frac{dx}{\sqrt{a^2-x^2}} = \sin^{-1}\frac{x}{a} = -\cos^{-1}\frac{x}{a} \quad (a^2 - x^2 > 0)$$

$$\int \frac{dx}{\sqrt{x^2 \pm a^2}} = \ln(x + \sqrt{x^2 \pm a^2})$$

$$\int \frac{x\,dx}{\sqrt{a^2-x^2}} = -\sqrt{a^2-x^2}$$

$$\int \frac{x\,dx}{\sqrt{x^2 \pm a^2}} = \sqrt{x^2 \pm a^2}$$

$$\int \sqrt{a^2-x^2}\,dx = \tfrac{1}{2}\left(x\sqrt{a^2-x^2} + a^2\sin^{-1}\frac{x}{|a|}\right)$$

$$\int x\sqrt{a^2-x^2}\,dx = -\tfrac{1}{3}(a^2-x^2)^{3/2}$$

$$\int \sqrt{x^2 \pm a^2}\,dx = \tfrac{1}{2}\left[x\sqrt{x^2 \pm a^2} \pm a^2\ln(x + \sqrt{x^2 \pm a^2})\right]$$

$$\int x(\sqrt{x^2 \pm a^2})\,dx = \tfrac{1}{3}(x^2 \pm a^2)^{3/2}$$

$$\int e^{ax}\,dx = \frac{1}{a}e^{ax}$$

$$\int \ln ax\,dx = (x\ln ax) - x$$

$$\int xe^{ax}\,dx = \frac{e^{ax}}{a^2}(ax-1)$$

$$\int \frac{dx}{a+be^{cx}} = \frac{x}{a} - \frac{1}{ac}\ln(a+be^{cx})$$

$$\int \sin ax\,dx = -\frac{1}{a}\cos ax$$

$$\int \cos ax\,dx = \frac{1}{a}\sin ax$$

$$\int \tan ax\,dx = -\frac{1}{a}\ln(\cos ax) = \frac{1}{a}\ln(\sec ax)$$

$$\int \cot ax\,dx = \frac{1}{a}\ln(\sin ax)$$

$$\int \sec ax\,dx = \frac{1}{a}\ln(\sec ax + \tan ax) = \frac{1}{a}\ln\left[\tan\left(\frac{ax}{2} + \frac{\pi}{4}\right)\right]$$

$$\int \csc ax\,dx = \frac{1}{a}\ln(\csc ax - \cot ax) = \frac{1}{a}\ln\left(\tan\frac{ax}{2}\right)$$

$$\int \sin^2 ax\,dx = \frac{x}{2} - \frac{\sin 2ax}{4a}$$

$$\int \cos^2 ax\,dx = \frac{x}{2} + \frac{\sin 2ax}{4a}$$

$$\int \frac{dx}{\sin^2 ax} = -\frac{1}{a}\cot ax$$

$$\int \frac{dx}{\cos^2 ax} = \frac{1}{a}\tan ax$$

$$\int \tan^2 ax\,dx = \frac{1}{a}(\tan ax) - x$$

$$\int \cot^2 ax\,dx = -\frac{1}{a}(\cot ax) - x$$

$$\int \sin^{-1} ax\,dx = x(\sin^{-1} ax) + \frac{\sqrt{1-a^2x^2}}{a}$$

$$\int \cos^{-1} ax\,dx = x(\cos^{-1} ax) - \frac{\sqrt{1-a^2x^2}}{a}$$

$$\int \frac{dx}{(x^2+a^2)^{3/2}} = \frac{x}{a^2\sqrt{x^2+a^2}}$$

$$\int \frac{x\,dx}{(x^2+a^2)^{3/2}} = -\frac{1}{\sqrt{x^2+a^2}}$$

TABLE B.6 *Gauss's Probability Integral and Other Definite Integrals*

$$\int_0^\infty x^n e^{-ax}\, dx = \frac{n!}{a^{n+1}}$$

$$I_0 = \int_0^\infty e^{-ax^2}\, dx = \frac{1}{2}\sqrt{\frac{\pi}{a}} \quad \text{(Gauss's probability integral)}$$

$$I_1 = \int_0^\infty x e^{-ax^2}\, dx = \frac{1}{2a}$$

$$I_2 = \int_0^\infty x^2 e^{-ax^2}\, dx = -\frac{dI_0}{da} = \frac{1}{4}\sqrt{\frac{\pi}{a^3}}$$

$$I_3 = \int_0^\infty x^3 e^{-ax^2}\, dx = -\frac{dI_1}{da} = \frac{1}{2a^2}$$

$$I_4 = \int_0^\infty x^4 e^{-ax^2}\, dx = \frac{d^2 I_0}{da^2} = \frac{3}{8}\sqrt{\frac{\pi}{a^5}}$$

$$I_5 = \int_0^\infty x^5 e^{-ax^2}\, dx = \frac{d^2 I_1}{da^2} = \frac{1}{a^3}$$

$$\vdots$$

$$I_{2n} = (-1)^n \frac{d^n}{da^n} I_0$$

$$I_{2n+1} = (-1)^n \frac{d^n}{da^n} I_1$$

differential of the independent variable appearing in the integrand. For example, consider the integral

$$I(x) = \int \cos^2 x \sin x\, dx$$

This integral becomes easy to evaluate if we rewrite the differential as $d(\cos x) = -\sin x\, dx$. The integral then becomes

$$\int \cos^2 x \sin x\, dx = -\int \cos^2 x\, d(\cos x)$$

If we now change variables, letting $y = \cos x$, we obtain

$$\int \cos^2 x \sin x\, dx = -\int y^2\, dy = -\frac{y^3}{3} + c = -\frac{\cos^3 x}{3} + c$$

Table B.5 lists some useful indefinite integrals. Table B.6 gives Gauss's probability integral and other definite integrals. A more complete list can be found in various handbooks, such as *The Handbook of Chemistry and Physics* (Boca Raton, FL: CRC Press, published annually).

B.8 Propagation of Uncertainty

In laboratory experiments, a common activity is to take measurements that act as raw data. These measurements are of several types—length, time interval, temperature, voltage, and so on—and are taken by a variety of instruments. Regardless of the measurement and the quality of the instrumentation, **there is always uncertainty associated with a physical measurement.** This uncertainty is a combination of that associated with the instrument and that related to the system being measured. An example of the former is the inability to exactly determine the position of a length measurement between the lines on a meterstick. An example of uncertainty related to the system being measured is the variation of temperature within a sample of water so that a single temperature for the sample is difficult to determine.

Uncertainties can be expressed in two ways. **Absolute uncertainty** refers to an uncertainty expressed in the same units as the measurement. Therefore, the length of a computer disk label might be expressed as (5.5 ± 0.1) cm. The uncertainty of ± 0.1 cm by itself is not descriptive enough for some purposes, however. This uncertainty is large if the measurement is 1.0 cm, but it is small if the measurement is 100 m. To give a more descriptive account of the uncertainty, **fractional uncertainty** or **percent uncertainty** is used. In this type of description, the uncertainty is divided by the actual measurement. Therefore, the length of the computer disk label could be expressed as

$$\ell = 5.5 \text{ cm} \pm \frac{0.1 \text{ cm}}{5.5 \text{ cm}} = 5.5 \text{ cm} \pm 0.018 \quad \text{(fractional uncertainty)}$$

or as

$$\ell = 5.5 \text{ cm} \pm 1.8\% \quad \text{(percent uncertainty)}$$

When combining measurements in a calculation, the percent uncertainty in the final result is generally larger than the uncertainty in the individual measurements. This is called **propagation of uncertainty** and is one of the challenges of experimental physics.

Some simple rules can provide a reasonable estimate of the uncertainty in a calculated result:

Multiplication and division: When measurements with uncertainties are multiplied or divided, add the *percent uncertainties* to obtain the percent uncertainty in the result.

Example: The Area of a Rectangular Plate

$$A = \ell w = (5.5 \text{ cm} \pm 1.8\%) \times (6.4 \text{ cm} \pm 1.6\%) = 35 \text{ cm}^2 \pm 3.4\%$$
$$= (35 \pm 1) \text{ cm}^2$$

Addition and subtraction: When measurements with uncertainties are added or subtracted, add the *absolute uncertainties* to obtain the absolute uncertainty in the result.

Example: A Change in Temperature

$$\Delta T = T_2 - T_1 = (99.2 \pm 1.5)°\text{C} - (27.6 \pm 1.5)°\text{C} = (71.6 \pm 3.0)°\text{C}$$
$$= 71.6°\text{C} \pm 4.2\%$$

Powers: If a measurement is taken to a power, the percent uncertainty is multiplied by that power to obtain the percent uncertainty in the result.

Example: The Volume of a Sphere

$$V = \tfrac{4}{3}\pi r^3 = \tfrac{4}{3}\pi(6.20 \text{ cm} \pm 2.0\%)^3 = 998 \text{ cm}^3 \pm 6.0\%$$
$$= (998 \pm 60) \text{ cm}^3$$

For complicated calculations, many uncertainties are added together, which can cause the uncertainty in the final result to be undesirably large. Experiments should be designed such that calculations are as simple as possible.

Notice that uncertainties in a calculation always add. As a result, an experiment involving a subtraction should be avoided if possible, especially if the measurements being subtracted are close together. The result of such a calculation is a small difference in the measurements and uncertainties that add together. It is possible that the uncertainty in the result could be larger than the result itself!

appendix **C**

Periodic Table of the Elements

Symbol — **Ca** 20 — Atomic number
Atomic mass† — 40.078
$4s^2$ — Electron configuration

Group I	Group II	Transition elements						
H 1 1.007 9 $1s$								
Li 3 6.941 $2s^1$	**Be** 4 9.0122 $2s^2$							
Na 11 22.990 $3s^1$	**Mg** 12 24.305 $3s^2$							
K 19 39.098 $4s^1$	**Ca** 20 40.078 $4s^2$	**Sc** 21 44.956 $3d^14s^2$	**Ti** 22 47.867 $3d^24s^2$	**V** 23 50.942 $3d^34s^2$	**Cr** 24 51.996 $3d^54s^1$	**Mn** 25 54.938 $3d^54s^2$	**Fe** 26 55.845 $3d^64s^2$	**Co** 27 58.933 $3d^74s^2$
Rb 37 85.468 $5s^1$	**Sr** 38 87.62 $5s^2$	**Y** 39 88.906 $4d^15s^2$	**Zr** 40 91.224 $4d^25s^2$	**Nb** 41 92.906 $4d^45s^1$	**Mo** 42 95.94 $4d^55s^1$	**Tc** 43 (98) $4d^55s^2$	**Ru** 44 101.07 $4d^75s^1$	**Rh** 45 102.91 $4d^85s^1$
Cs 55 132.91 $6s^1$	**Ba** 56 137.33 $6s^2$	57–71*	**Hf** 72 178.49 $5d^26s^2$	**Ta** 73 180.95 $5d^36s^2$	**W** 74 183.84 $5d^46s^2$	**Re** 75 186.21 $5d^56s^2$	**Os** 76 190.23 $5d^66s^2$	**Ir** 77 192.2 $5d^76s^2$
Fr 87 (223) $7s^1$	**Ra** 88 (226) $7s^2$	89–103**	**Rf** 104 (261) $6d^27s^2$	**Db** 105 (262) $6d^37s^2$	**Sg** 106 (266)	**Bh** 107 (264)	**Hs** 108 (277)	**Mt** 109 (268)

*Lanthanide series	**La** 57 138.91 $5d^16s^2$	**Ce** 58 140.12 $5d^14f^16s^2$	**Pr** 59 140.91 $4f^36s^2$	**Nd** 60 144.24 $4f^46s^2$	**Pm** 61 (145) $4f^56s^2$	**Sm** 62 150.36 $4f^66s^2$
**Actinide series	**Ac** 89 (227) $6d^17s^2$	**Th** 90 232.04 $6d^27s^2$	**Pa** 91 231.04 $5f^26d^17s^2$	**U** 92 238.03 $5f^36d^17s^2$	**Np** 93 (237) $5f^46d^17s^2$	**Pu** 94 (244) $5f^66d^07s^2$

Note: Atomic mass values given are averaged over isotopes in the percentages in which they exist in nature.
†For an unstable element, mass number of the most stable known isotope is given in parentheses.
††Elements 112, 114, and 116 have not yet been officially named.
†††For a description of the atomic data, visit *physics.nist.gov/PhysRefData/Elements/per_text.html*

			Group III	Group IV	Group V	Group VI	Group VII	Group 0
							H 1 1.007 9 $1s^1$	**He** 2 4.002 6 $1s^2$
			B 5 10.811 $2p^1$	**C** 6 12.011 $2p^2$	**N** 7 14.007 $2p^3$	**O** 8 15.999 $2p^4$	**F** 9 18.998 $2p^5$	**Ne** 10 20.180 $2p^6$
			Al 13 26.982 $3p^1$	**Si** 14 28.086 $3p^2$	**P** 15 30.974 $3p^3$	**S** 16 32.066 $3p^4$	**Cl** 17 35.453 $3p^5$	**Ar** 18 39.948 $3p^6$
Ni 28 58.693 $3d^84s^2$	**Cu** 29 63.546 $3d^{10}4s^1$	**Zn** 30 65.41 $3d^{10}4s^2$	**Ga** 31 69.723 $4p^1$	**Ge** 32 72.64 $4p^2$	**As** 33 74.922 $4p^3$	**Se** 34 78.96 $4p^4$	**Br** 35 79.904 $4p^5$	**Kr** 36 83.80 $4p^6$
Pd 46 106.42 $4d^{10}$	**Ag** 47 107.87 $4d^{10}5s^1$	**Cd** 48 112.41 $4d^{10}5s^2$	**In** 49 114.82 $5p^1$	**Sn** 50 118.71 $5p^2$	**Sb** 51 121.76 $5p^3$	**Te** 52 127.60 $5p^4$	**I** 53 126.90 $5p^5$	**Xe** 54 131.29 $5p^6$
Pt 78 195.08 $5d^96s^1$	**Au** 79 196.97 $5d^{10}6s^1$	**Hg** 80 200.59 $5d^{10}6s^2$	**Tl** 81 204.38 $6p^1$	**Pb** 82 207.2 $6p^2$	**Bi** 83 208.98 $6p^3$	**Po** 84 (209) $6p^4$	**At** 85 (210) $6p^5$	**Rn** 86 (222) $6p^6$
Ds 110 (271)	**Rg** 111 (272)	112†† (285)		114†† (289)		116†† (292)		

Eu 63 151.96 $4f^76s^2$	**Gd** 64 157.25 $4f^75d^16s^2$	**Tb** 65 158.93 $4f^85d^16s^2$	**Dy** 66 162.50 $4f^{10}6s^2$	**Ho** 67 164.93 $4f^{11}6s^2$	**Er** 68 167.26 $4f^{12}6s^2$	**Tm** 69 168.93 $4f^{13}6s^2$	**Yb** 70 173.04 $4f^{14}6s^2$	**Lu** 71 174.97 $4f^{14}5d^16s^2$
Am 95 (243) $5f^77s^2$	**Cm** 96 (247) $5f^76d^17s^2$	**Bk** 97 (247) $5f^86d^17s^2$	**Cf** 98 (251) $5f^{10}7s^2$	**Es** 99 (252) $5f^{11}7s^2$	**Fm** 100 (257) $5f^{12}7s^2$	**Md** 101 (258) $5f^{13}7s^2$	**No** 102 (259) $5f^{14}7s^2$	**Lr** 103 (262) $6d^15f^{14}7s^2$

appendix D

SI Units

TABLE D.1 *SI Units*

	SI Base Unit	
Base Quantity	**Name**	**Symbol**
Length	meter	m
Mass	kilogram	kg
Time	second	s
Electric current	ampere	A
Temperature	kelvin	K
Amount of substance	mole	mol
Luminous intensity	candela	cd

TABLE D.2 *Some Derived SI Units*

Quantity	**Name**	**Symbol**	**Expression in Terms of Base Units**	**Expression in Terms of Other SI Units**
Plane angle	radian	rad	m/m	
Frequency	hertz	Hz	s^{-1}	
Force	newton	N	$kg \cdot m/s^2$	J/m
Pressure	pascal	Pa	$kg/m \cdot s^2$	N/m^2
Energy	joule	J	$kg \cdot m^2/s^2$	$N \cdot m$
Power	watt	W	$kg \cdot m^2/s^3$	J/s
Electric charge	coulomb	C	$A \cdot s$	
Electric potential	volt	V	$kg \cdot m^2/A \cdot s^3$	W/A
Capacitance	farad	F	$A^2 \cdot s^4/kg \cdot m^2$	C/V
Electric resistance	ohm	Ω	$kg \cdot m^2/A^2 \cdot s^3$	V/A
Magnetic flux	weber	Wb	$kg \cdot m^2/A \cdot s^2$	$V \cdot s$
Magnetic field	tesla	T	$kg/A \cdot s^2$	
Inductance	henry	H	$kg \cdot m^2/A^2 \cdot s^2$	$T \cdot m^2/A$

Answers to Quick Quizzes and Odd-Numbered Problems

Chapter 1

Answers to Quick Quizzes

1.1 (a) **1.2** False **1.3** (b)

Answers to Odd-Numbered Problems

1. (a) 5.52×10^3 kg/m^3 (b) It is between the density of aluminum and that of iron and is greater than the densities of typical surface rocks.

3. 23.0 kg **5.** 7.69 cm **7.** 0.141 nm

9. (b) only

11. (a) kg · m/s (b) N · s

13. 871 m^2 **15.** 11.4×10^3 kg/m^3

17. By measuring the pages, we find that each page has area 0.277 m × 0.217 m = 0.060 m^2. The room has wall area 37 m^2, requiring 616 sheets that would be counted as 1 232 pages. Volume 1 of this textbook contains only 736 pages.

19. 1.00×10^{10} lb **21.** 2.86 cm **23.** 151 μm

25. (a) 317 years (b) 2.48×10^9 bills

27. $\sim 10^6$ balls in a room 4 m by 4 m by 3 m

29. $\sim 10^2$ piano tuners **31.** 31 556 926.0 s

33. (209 ± 4) cm^2 **35.** 8.80% **37.** 19

39. 63 **43.** 48.6 kg

45. Answers may vary somewhat due to variation in reading precise numbers off the graph. (a) 0.015 g (b) 8% (c) 5.2 g/m^2 (d) For shapes cut from this copy paper, the mass of the cutout is proportional to its area. The proportionality constant is 5.2 g/m^2 ± 8%, where the uncertainty is estimated. (e) This result is to be expected if the paper has thickness and density that are uniform within the experimental uncertainty. (f) The slope is the areal density of the paper, its mass per unit area.

47. We could write $(1.61 \pm 0.17) \times 10^3$ kg/m^3 or $(1.6 \pm 0.2) \times 10^3$ kg/m^3.

49. 316 m **51.** 5.0 m **53.** 3.41 m

55. (a) aluminum, 2.75 g/cm^3; copper, 9.36 g/cm^3; brass, 8.91 g/cm^3; tin, 7.68 g/cm^3; iron, 7.88 g/cm^3 (b) The tabulated values are smaller by 2% for aluminum, by 5% for copper, by 6% for brass, by 5% for tin, and by 0.3% for iron.

57. 1×10^{10} gal/yr

59. Answers may vary. (a) $\sim 10^{29}$ prokaryotes (b) $\sim 10^{14}$ kg

61. (a) $B = 2.70$ g/cm^3, $C = 1.19$ g/cm^4 (b) 1.39 kg

63. $V = 0.579t + (1.19 \times 10^{-9})t^2$, where V is in cubic feet and t is in seconds

65. (a) 0.529 cm/s (b) 11.5 cm/s

67. (a) 12.1 m (b) 135° (c) 25.2° (d) 135°

Chapter 2

Answers to Quick Quizzes

1. (c) **2.** (b)

3. False. Your graph should look something like the one shown at the top of the right column. This v_x–t graph shows that the maximum speed is about 5.0 m/s, which is 18 km/h (= 11 mi/h), so the driver was not speeding.

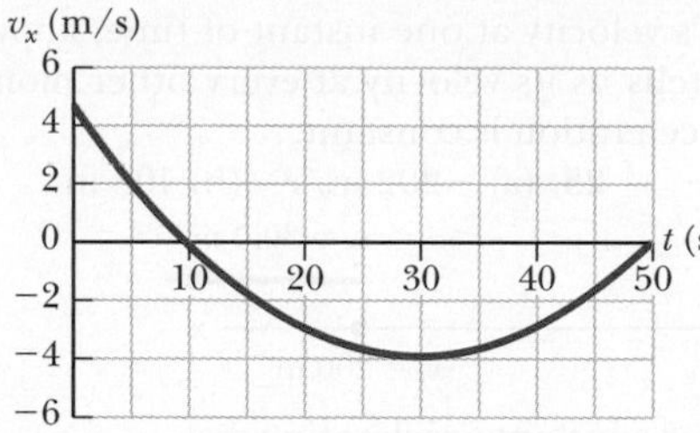

4. (b) **5.** (c) **6.** (a)–(e), (b)–(d), (c)–(f)

7. (i) (e) (ii) (d)

Answers to Odd-Numbered Problems

1. (a) 5 m/s (b) 1.2 m/s (c) −2.5 m/s (d) −3.3 m/s (e) 0

3. (a) 3.75 m/s (b) 0

5. (a) −2.4 m/s (b) −3.8 m/s (c) 4.0 s

7. (a) L/t_1 (b) $-L/t_2$ (c) 0 (d) $2L/(t_1 + t_2)$

9. 1.9×10^8 years ago

11. (a) 5.00 m (b) 4.88×10^3 s

13. (a)

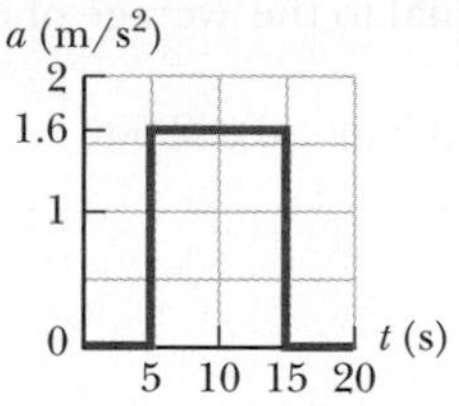

(b) 1.60 m/s^2 (c) 0.800 m/s^2

15. (a) 1.3 m/s^2 (b) $t = 3$ s, $a = 2$ m/s^2 (c) $t = 6$ s, $t > 10$ s (d) $a = -1.5$ m/s^2, $t = 8$ s

17. (a) 2.00 m (b) −3.00 m/s (c) −2.00 m/s^2

19.

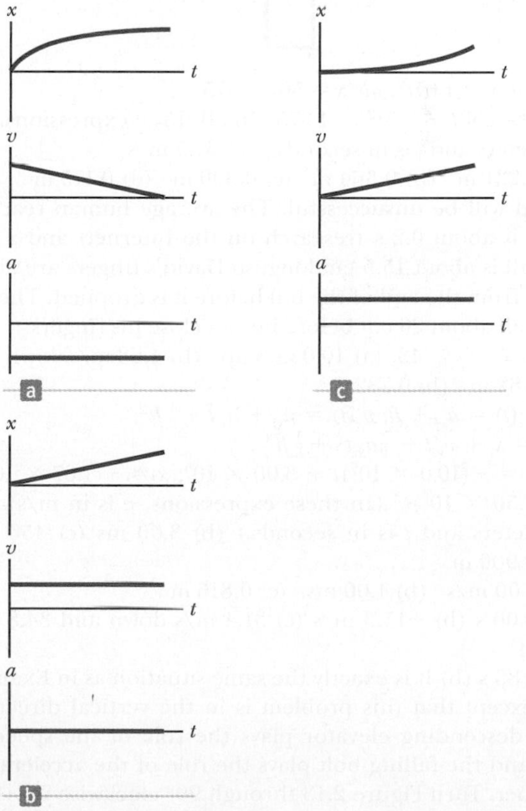

21. (a) 9.00 m/s (b) −3.00 m/s (c) 17.0 m/s (d) The graph of velocity versus time is a straight line passing through 13 m/s at 10:05 a.m. and sloping downward, decreasing by 4 m/s for each second thereafter. (e) If and only if we know the object's velocity at one instant of time, knowing its acceleration tells us its velocity at every other moment as long as the acceleration is constant.

23. -16.0 cm/s^2 **25.** (a) -202 m/s^2 (b) 198 m

27. (a)

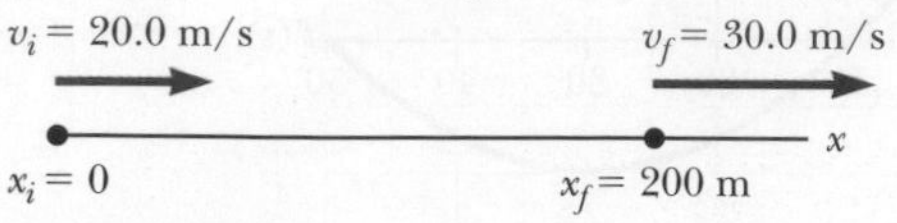

(b) Particle under constant acceleration
(c) $v_f^2 = v_i^2 + 2a(x_f - x_i)$ (Equation 2.17)
(d) $a = \dfrac{v_f^2 - v_i^2}{2(x_f - x_i)}$ (e) 1.25 m/s^2 (f) 8.00 s

29. 3.10 m/s

31. (a) The idea is false unless the acceleration is zero. We define constant acceleration to mean that the velocity is changing steadily in time. So, the velocity cannot be changing steadily in space. (b) This idea is true. Because the velocity is changing steadily in time, the velocity halfway through an interval is equal to the average of its initial and final values.

33. (a) 13.5 m (b) 13.5 m (c) 13.5 m (d) 22.5 m

35. (a) 1.88 km (b) 1.46 km

(c)

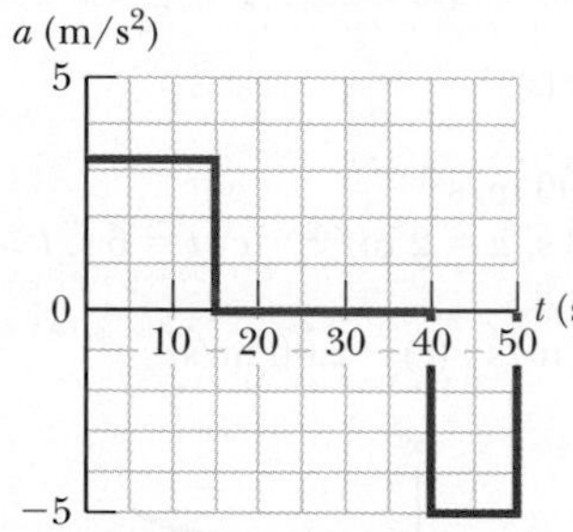

(d) $0a$: $x = 1.67t^2$; ab: $x = 50t - 375$;
bc: $x = 250t - 2.5t^2 - 4\,375$ (In all three expressions, x is in meters and t is in seconds.) (e) 37.5 m/s

37. (a) 0.231 m (b) 0.364 m (c) 0.399 m (d) 0.175 m

39. David will be unsuccessful. The average human reaction time is about 0.2 s (research on the Internet) and a dollar bill is about 15.5 cm long, so David's fingers are about 8 cm from the end of the bill before it is dropped. The bill will fall about 20 cm before he can close his fingers.

41. 1.79 s **43.** (a) 10.0 m/s up (b) 4.68 m/s down

45. (a) 7.82 m (b) 0.782 s

47. (a) $a_x(t) = a_{xi} + Jt$; $v_x(t) = v_{xi} + a_{xi}t + \frac{1}{2}Jt^2$; $x(t) = x_i + v_{xi}t + \frac{1}{2}a_{xi}t^2 + \frac{1}{6}Jt^3$

49. (a) $a = -(10.0 \times 10^7)t + 3.00 \times 10^5$; $x = -(1.67 \times 10^7)t^3 + (1.50 \times 10^5)t^2$ (In these expressions, a is in m/s², x is in meters and t is in seconds.) (b) 3.00 ms (c) 450 m/s (d) 0.900 m

51. (a) 4.00 m/s (b) 1.00 ms (c) 0.816 m

53. (a) 3.00 s (b) −15.3 m/s (c) 31.4 m/s down and 34.8 m/s down

55. (a) 2.83 s (b) It is exactly the same situation as in Example 2.8 except that this problem is in the vertical direction. The descending elevator plays the role of the speeding car, and the falling bolt plays the role of the accelerating trooper. Turn Figure 2.13 through 90° clockwise to visualize the elevator–bolt problem! (c) If each floor is 3 m high, the highest floor that can be reached is the 13th floor.

57. (a) From the graph, we see that the Acela is cruising at a constant positive velocity in the positive x direction from about −50 s to 50 s. From 50 s to 200 s, the Acela accelerates in the positive x direction reaching a top speed of about 170 mi/h. Around 200 s, the engineer applies the brakes, and the train, still traveling in the positive x direction, slows down and then stops at 350 s. Just after 350 s, the train reverses direction (v becomes negative) and steadily gains speed in the negative x direction. (b) approximately 2.2 mi/h/s (c) approximately 6.7 mi

59. (a) Here, v_1 must be greater than v_2 and the distance between the leading athlete and the finish line must be great enough so that the trailing athlete has time to catch up.
(b) $t = \dfrac{d_1}{v_1 - v_2}$ (c) $d_2 = \dfrac{v_2 d_1}{v_1 - v_2}$

61. (a) 5.46 s (b) 73.0 m
(c) $v_{\text{Stan}} = 22.6$ m/s, $v_{\text{Kathy}} = 26.7$ m/s

63. (a) $v_B = (1/\tan\theta)v$ (b) The velocity v_B starts off larger than v for small angles θ and then decreases, approaching zero as θ approaches 90°.

65. (a) 5.32 m/s² for Laura and 3.75 m/s² for Healan
(b) 10.6 m/s for Laura and 11.2 m/s for Healan
(c) Laura, by 2.63 m (d) 4.47 m at $t = 2.84$ s

67. (a) 35.9 m (b) 4.04 s (c) 45.8 m (d) 22.6 m/s

69. (b) $\Delta t = \sqrt{\dfrac{2D}{g}}$ (c) $\Delta t = \sqrt{\dfrac{2L}{g\sin\theta}}$

Chapter 3

Answers to Quick Quizzes

1. vectors: (b), (c); scalars: (a), (d), (e)
2. (c) **4.** (b)
3. (b) and (c) **5.** (c)

Answers to Odd-Numbered Problems

1. (−2.75, −4.76) m **3.** (a) 2.24 m (b) 2.24 m at 26.6°

5. (a) (−3.56 cm, −2.40 cm) (b) (r = 4.30 cm, θ = 326°) (c) (r = 8.60 cm, θ = 34.0°) (d) (r = 12.9 cm, θ = 146°)

7. 70.0 m

9. This situation can *never* be true because the distance is an arc of a circle between two points, whereas the magnitude of the displacement vector is a straight-line chord of the circle between the same points.

11. (a) 5.2 m at 60° (b) 3.0 m at 330° (c) 3.0 m at 150° (d) 5.2 m at 300°

13. approximately 420 ft at −3° **15.** 47.2 units at 122°

17. (a) yes (b) The speed of the camper should be 28.3 m/s or more to satisfy this requirement.

19. (a) $(-11.1\hat{\mathbf{i}} + 6.40\hat{\mathbf{j}})$ m (b) $(1.65\hat{\mathbf{i}} + 2.86\hat{\mathbf{j}})$ cm (c) $(-18.0\hat{\mathbf{i}} - 12.6\hat{\mathbf{j}})$ in.

21. 358 m at 2.00° S of E

23. (a) $2.00\hat{\mathbf{i}} - 6.00\hat{\mathbf{j}}$ (b) $4.00\hat{\mathbf{i}} + 2.00\hat{\mathbf{j}}$ (c) 6.32 (d) 4.47 (e) 288°; 26.6°

25. 9.48 m at 166° **27.** 4.64 m at 78.6° N of E

29. (a) 185 N at 77.8° from the positive x axis (b) $(-39.3\hat{\mathbf{i}} - 181\hat{\mathbf{j}})$ N

31. (a) 2.83 m at θ = 315° (b) 13.4 m at θ = 117°

33. (a) $8.00\hat{\mathbf{i}} + 12.0\hat{\mathbf{j}} - 4.00\hat{\mathbf{k}}$ (b) $2.00\hat{\mathbf{i}} + 3.00\hat{\mathbf{j}} - 1.00\hat{\mathbf{k}}$ (c) $-24.0\hat{\mathbf{i}} - 36.0\hat{\mathbf{j}} + 12.0\hat{\mathbf{k}}$

35. (a) $-3.00\hat{\mathbf{i}} + 2.00\hat{\mathbf{j}}$ (b) 3.61 at 146° (c) $3.00\hat{\mathbf{i}} - 6.00\hat{\mathbf{j}}$

37. (a) $a = 5.00$ and $b = 7.00$ (b) For vectors to be equal, all their components must be equal. A vector equation contains more information than a scalar equation.

39. 196 cm at 345°

41. (a) $(-20.5\hat{\mathbf{i}} + 35.5\hat{\mathbf{j}})$ m/s (b) $25.0\hat{\mathbf{j}}$ m/s (c) $(-61.5\hat{\mathbf{i}} + 107\hat{\mathbf{j}})$ m (d) $37.5\hat{\mathbf{j}}$ m (e) 157 km

43. 1.43×10^4 m at 32.2° above the horizontal

45. (a) $\vec{\mathbf{E}} = (15.1\hat{\mathbf{i}} + 7.72\hat{\mathbf{j}})$ cm (b) $\vec{\mathbf{F}} = (-7.72\hat{\mathbf{i}} + 15.1\hat{\mathbf{j}})$ cm (c) $\vec{\mathbf{G}} = (-7.72\hat{\mathbf{i}} - 15.1\hat{\mathbf{j}})$ cm

47. 240 m at 237°

49. (a)

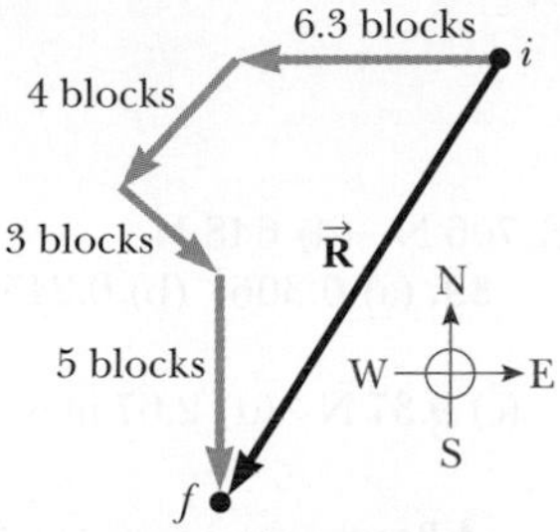

(b) 18.3 b (c) 12.4 b at 233° counterclockwise from east

51. (a) 25.4 s (b) 15.0 km/h

53. (a) The *x*, *y*, and *z* components are, respectively, 2.00, 1.00, and 3.00. (b) 3.74 (c) $\theta_x = 57.7°$, $\theta_y = 74.5°$, $\theta_z = 36.7°$

55. (a) 0.079 8 N (b) 57.9° (c) 32.1° **57.** 1.15°

59. (a) $(10\,000 - 9\,600 \sin\theta)^{1/2}$ cm (b) 270°; 140 cm (c) 90°; 20.0 cm (d) They do make sense. The maximum value is attained when $\vec{\mathbf{A}}$ and $\vec{\mathbf{B}}$ are in the same direction, and it is 60 cm + 80 cm. The minimum value is attained when $\vec{\mathbf{A}}$ and $\vec{\mathbf{B}}$ are in opposite directions, and it is 80 cm − 60 cm.

61. (a) $-2.00\hat{\mathbf{k}}$ m/s (b) its velocity vector

63. (a) $\vec{\mathbf{R}}_1 = a\hat{\mathbf{i}} + b\hat{\mathbf{j}}$ (b) $R_1 = (a^2 + b^2)^{1/2}$ (c) $\vec{\mathbf{R}}_2 = a\hat{\mathbf{i}} + b\hat{\mathbf{j}} + c\hat{\mathbf{k}}$

Chapter 4

Answers to Quick Quizzes

1. (a)

2. (i) (b) **(ii)** (a)

3. 15°, 30°, 45°, 60°, 75°

4. **(i)** (d) **(ii)** (b)

5. **(i)** (b) **(ii)** (d)

Answers to Odd-Numbered Problems

1. (a) 4.87 km at 209° from E (b) 23.3 m/s (c) 13.5 m/s at 209°

3. (a) $\vec{\mathbf{r}} = 18.0t\hat{\mathbf{i}} + (4.00t - 4.90t^2)\hat{\mathbf{j}}$, where $\vec{\mathbf{r}}$ is in meters and *t* is in seconds
(b) $\vec{\mathbf{v}} = 18.0\hat{\mathbf{i}} + (4.00 - 9.80t)\hat{\mathbf{j}}$, where $\vec{\mathbf{v}}$ is in meters per second and *t* is in seconds
(c) $\vec{\mathbf{a}} = -9.80\hat{\mathbf{j}}$ m/s²
(d) $\vec{\mathbf{r}} = (54.0\hat{\mathbf{i}} - 32.1\hat{\mathbf{j}})$ m; $\vec{\mathbf{v}} = (18.0\hat{\mathbf{i}} - 25.4\hat{\mathbf{j}})$ m/s; $\vec{\mathbf{a}} = -9.80\hat{\mathbf{j}}$ m/s²

5. (a) $\vec{\mathbf{v}} = -12.0t\hat{\mathbf{j}}$, where $\vec{\mathbf{v}}$ is in meters per second and *t* is in seconds (b) $\vec{\mathbf{a}} = -12.0\hat{\mathbf{j}}$ m/s (c) $\vec{\mathbf{r}} = (3.00\hat{\mathbf{i}} - 6.00\hat{\mathbf{j}})$ m; $\vec{\mathbf{v}} = -12.0\hat{\mathbf{j}}$ m/s

7. (a) $(0.800\hat{\mathbf{i}} - 0.300\hat{\mathbf{j}})$ m/s² (b) 339° (c) $(360\hat{\mathbf{i}} - 72.7\hat{\mathbf{j}})$ m, −15.2°

9. (a) 2.81 m/s horizontal (b) 60.2° below the horizontal

11. 53.1° **13.** (a) 3.96 m/s horizontally forward (b) 9.6%

15. $d \tan\theta_i - \dfrac{gd^2}{2v_i^{\,2}\cos^2\theta_i}$

17. (a) The ball clears by 0.89 m. (b) while descending

19. (a) 18.1 m/s (b) 1.13 m (c) 2.79 m **21.** 9.91 m/s

23. (a) (0, 50.0 m) (b) $v_{xi} = 18.0$ m/s; $v_{yi} = 0$ (c) Particle under constant acceleration (d) Particle under constant velocity (e) $v_{xf} = v_{xi}$; $v_{yf} = -gt$ (f) $x_f = v_{xi}t$; $y_f = y_i - \frac{1}{2}gt^2$ (g) 3.19 s (h) 36.1 m/s, −60.1°

25. 1.92 s **27.** 377 m/s² **29.** 0.749 rev/s

31. 1.48 m/s² inward and 29.9° backward

33. (a) Yes. The particle can be either speeding up or slowing down, with a tangential component of acceleration of magnitude $\sqrt{6^2 - 4.5^2} = 3.97$ m/s². (b) No. The magnitude of the acceleration cannot be less than $v^2/r = 4.5$ m/s².

35. (a) 15.0 km/h east (b) 15.0 km/h west (c) 0.016 7 h = 60.0 s

37. (a) 1.26 h (b) 1.13 h (c) 1.19 h

39. (a) 9.80 m/s² down and 2.50 m/s² south (b) 9.80 m/s² down (c) The bolt moves on a parabola with its axis downward and tilting to the south. It lands south of the point directly below its starting point. (d) The bolt moves on a parabola with a vertical axis.

41. (a) $\dfrac{2d/c}{1 - v^2/c^2}$ (b) $\dfrac{2d}{c}$
(c) The trip in flowing water takes a longer time interval. The swimmer travels at the low upstream speed for a longer time interval, so his average speed is reduced below *c*. Mathematically, $1/(1 - v^2/c^2)$ is always greater than 1. In the extreme, as $v \to c$, the time interval becomes infinite. In that case, the student can never return to the starting point because he cannot swim fast enough to overcome the river current.

43. 15.3 m

45. The relationship between the height *h* and the walking speed is $h = (4.16 \times 10^{-3})v_x^{\,2}$, where *h* is in meters and v_x is in meters per second. At a typical walking speed of 4 to 5 km/h, the ball would have to be dropped from a height of about 1 cm, clearly much too low for a person's hand. Even at Olympic-record speed for the 100-m run (confirm on the Internet), this situation would only occur if the ball is dropped from about 0.4 m, which is also below the hand of a normally proportioned person.

47. (a) 101 m/s (b) 3.27×10^4 ft (c) 20.6 s

49. (a) 26.9 m/s (b) 67.3 m (c) $(2.00\hat{\mathbf{i}} - 5.00\hat{\mathbf{j}})$ m/s²

51. (a) $(7.62\hat{\mathbf{i}} - 6.48\hat{\mathbf{j}})$ cm (b) $(10.0\hat{\mathbf{i}} - 7.05\hat{\mathbf{j}})$ cm

53. The initial height of the ball when struck is 3.94 m, which is too high for the batter to hit the ball.

55. (a) 1.69 km/s (b) 1.80 h

57. (a) $x = v_i(0.164\,3 + 0.002\,299v_i^{\,2})^{1/2} + 0.047\,94v_i^{\,2}$, where *x* is in meters and v_i is in meters per second (b) 0.041 0 m (c) 961 m (d) $x \approx 0.405v_i$ (e) $x \approx 0.095\,9v_i^{\,2}$ (f) The graph of *x* versus v_i starts from the origin as a straight line with slope 0.405 s. Then it curves upward above this tangent line, getting closer and closer to the parabola $x = 0.095\,9v_i^{\,2}$, where *x* is in meters and v_i is in meters per second.

59. (a) 6.80 km (b) 3.00 km vertically above the impact point (c) 66.2°

61. (a) 46.5 m/s (b) −77.6° (c) 6.34 s

63. (a) 20.0 m/s (b) 5.00 s (c) $(16.0\hat{\mathbf{i}} - 27.1\hat{\mathbf{j}})$ m/s (d) 6.53 s (e) $24.5\hat{\mathbf{i}}$ m

65. (a) 4.00 km/h (b) 4.00 km/h

67. (a) 43.2 m (b) $(9.66\hat{\mathbf{i}} - 25.6\hat{\mathbf{j}})$ m/s (c) Air resistance would ordinarily make the jump distance smaller and the final horizontal and vertical velocity components both somewhat smaller. If a skilled jumper makes her body into an airfoil, however, she can deflect downward the air through which she passes so that it deflects her upward, giving her more time in the air and a longer jump.

69. 33.5° below the horizontal

71. Safe distances are less than 270 m or greater than 3.48×10^3 m from the western shore.

Chapter 5

Answers to Quick Quizzes

1. (d) **4.** (b)
2. (a) **5.** (i) (c) (ii) (a)
3. (d) **6.** (b)
7. (b) Pulling up on the rope decreases the normal force, which, in turn, decreases the force of kinetic friction.

Answers to Odd-Numbered Problems

1. (a) $(6.00\hat{\mathbf{i}} + 15.0\hat{\mathbf{j}})$ N (b) 16.2 N

3. (a) $(2.50\hat{\mathbf{i}} + 5.00\hat{\mathbf{j}})$ N (b) 5.59 N

5. (a) 1.53 m (b) 24.0 N forward and upward at 5.29° with the horizontal

7. (a) 3.64×10^{-18} N (b) 8.93×10^{-30} N is 408 billion times smaller

9. (a) force exerted by spring on hand, to the left; force exerted by spring on wall, to the right (b) force exerted by wagon on handle, downward to the left; force exerted by wagon on planet, upward; force exerted by wagon on ground, downward (c) force exerted by football on player, downward to the right; force exerted by football on planet, upward (d) force exerted by small-mass object on large-mass object, to the left (e) force exerted by negative charge on positive charge, to the left (f) force exerted by iron on magnet, to the left

11. (a) $t = \sqrt{\dfrac{2h}{g}}$ (b) $a_x = \dfrac{F}{m}$ (c) $x = \dfrac{Fh}{mg}$ (d) $a = \sqrt{(F/m)^2 + g^2}$

13. (a) 5.00 m/s² at 36.9° (b) 6.08 m/s² at 25.3°

15. (a) 15.0 lb up (b) 5.00 lb up (c) 0

17. (a) 3.43 kN (b) 0.967 m/s horizontally forward

19. (a) $P\cos 40° - n = 0$ and $P\sin 40°$ 220 N = 0; $P = 342$ N and $n = 262$ N (b) $P - n\cos 40° - (220 \text{ N})\sin 40° = 0$ and $n \sin 40 - (220 \text{ N})\cos 40° = 0$; $n = 262$ N and $P = 342$ N (c) The results agree. The methods are of the same level of difficulty. Each involves one equation in one unknown and one equation in two unknowns. If we are interested in finding n without finding P, method (b) is simpler.

21. (a)

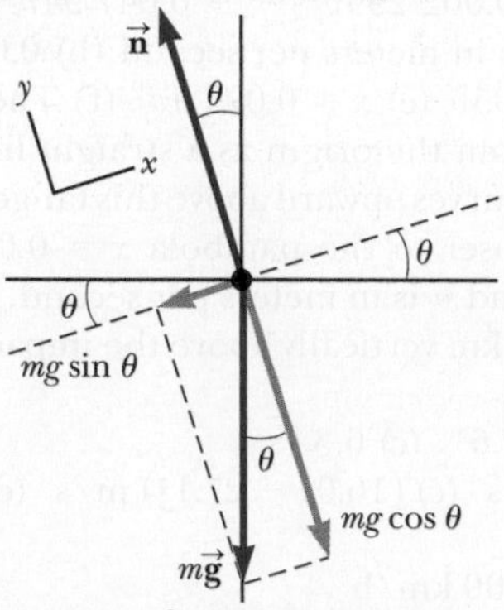

(b) 2.54 m/s² (c) 3.19 m/s

23. (a)

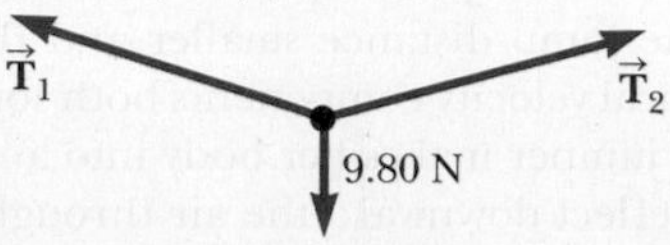

(b) 613 N

27. 8.66 N east

29. (a) 646 N up (b) 646 N up (c) 627 N up (d) 589 N up

31. (a) $T_1 = 79.8$ N, $T_2 = 39.9$ N (b) 2.34 m/s²

33. (a) $F_x > 19.6$ N (b) $F_x \leq -78.4$ N (c)

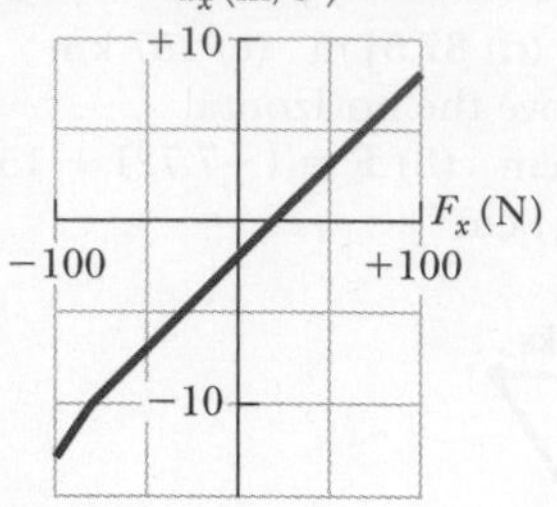

35. (a) 706 N (b) 814 N (c) 706 N (d) 648 N

37. 1.76 kN to the left

39. (a) 0.306 (b) 0.245

41. (a) 1.11 s (b) 0.875 s

43. (a) 1.78 m/s² (b) 0.368 (c) 9.37 N (d) 2.67 m/s

45. 37.8 N

47. (a)

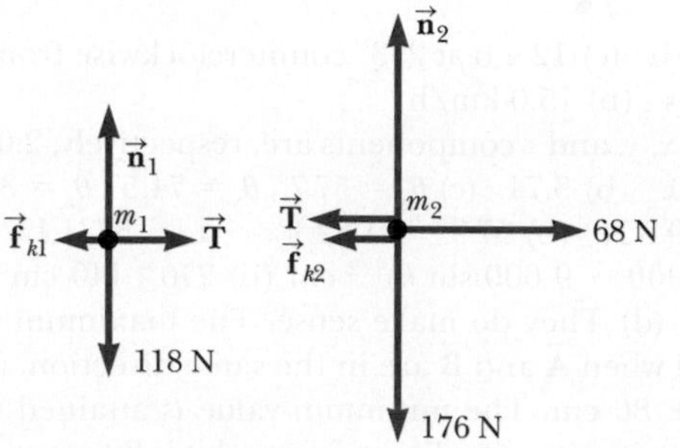

(b) 1.29 m/s² to the right (c) 27.2 N

49. 6.84 m

51. 0.060 0 m

53. (a) $a = -\mu_k g$ (b) $d = \dfrac{v_i^{\,2}}{2\mu_k g}$

55. (a)

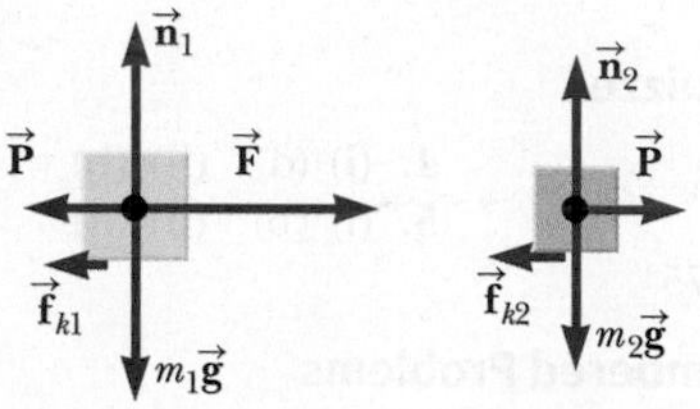

(b) F (c) $F - P$ (d) P (e) m_1: $F - P = m_1 a$; m_2: $P = m_2 a$

(f) $a = \dfrac{F - \mu_1 m_1 g - \mu_2 m_2 g}{m_1 + m_2}$,

(g) $P = \dfrac{m_2}{m_1 + m_2}[F + m_1(\mu_2 - \mu_1)g]$

57. (a)

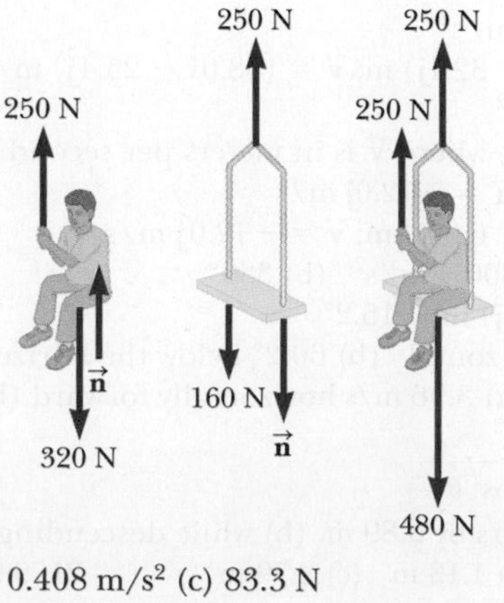

(b) 0.408 m/s² (c) 83.3 N

59. (a)

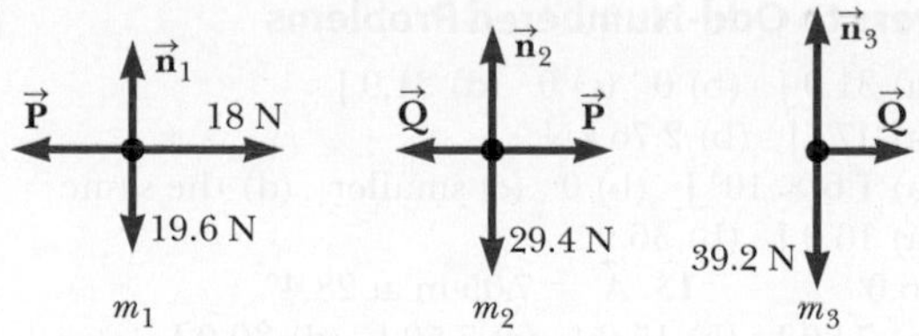

(b) 2.00 m/s² to the right (c) 4.00 N on m_1, 6.00 N right on m_2, 8.00 N right on m_3 (d) 14.0 N between m_1 and m_2, 8.00 N between m_2 and m_3 (e) The m_2 block models the heavy block of wood. The contact force on your back is modeled by the force between the m_2 and the m_3 blocks, which is much less than the force F. The difference between F and this contact force is the net force causing the acceleration of the 5-kg pair of objects. The acceleration is real and nonzero, but it lasts for so short a time that it is never associated with a large velocity. The frame of the building and your legs exert forces, small in magnitude relative to the hammer blow, to bring the partition, block, and you to rest again over a time large relative to the hammer blow.

61. (a) *Upper pulley:* *Lower pulley:*

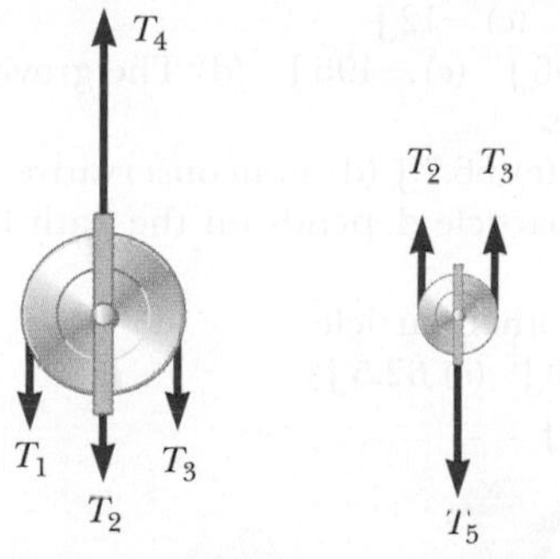

(b) $Mg/2$, $Mg/2$, $Mg/2$, $3Mg/2$, Mg (c) $Mg/2$

63. (b) If θ is greater than $\tan^{-1}(1/\mu_s)$, motion is impossible.

65. (a) The net force on the cushion is in a fixed direction, downward and forward making angle $\tan^{-1}(F/mg)$ with the vertical. Starting from rest, it will move along this line with (b) increasing speed. Its velocity changes in magnitude. (c) 1.63 m (d) It will move along a parabola. The axis of the parabola is parallel to the line described in part (a). If the cushion is thrown in a direction above this line, its path will be concave downward, making its velocity become more and more nearly parallel to the line over time. If the cushion is thrown down more steeply, its path will be concave upward, again making its velocity turn toward the fixed direction of its acceleration.

67. $(M + m_1 + m_2)(m_1 g/m_2)$

69. (a) 30.7° (b) 0.843 N **71.** 72.0 N

73. (a) 0.931 m/s² (b) From a value of 0.625 m/s² for large x, the acceleration gradually increases, passes through a maximum, and then drops more rapidly, becoming negative and reaching −2.10 m/s² at $x = 0$. (c) 0.976 m/s² at $x =$ 25.0 cm (d) 6.10 cm

75. (a) 4.90 m/s² (b) 3.13 m/s at 30.0° below the horizontal (c) 1.35 m (d) 1.14 s (e) The mass of the block makes no difference.

Chapter 6

Answers to Quick Quizzes

1. **(i)** (a) **(ii)** (b)

2. **(i)** Because the speed is constant, the only direction the force can have is that of the centripetal acceleration. The force is larger at Ⓒ than at Ⓐ because the radius at Ⓒ is smaller. There is no force at Ⓑ because the wire is straight. **(ii)** In addition to the forces in the centripetal direction in part (a), there are now tangential forces to provide the tangential acceleration. The tangential force is the same at all three points because the tangential acceleration is constant.

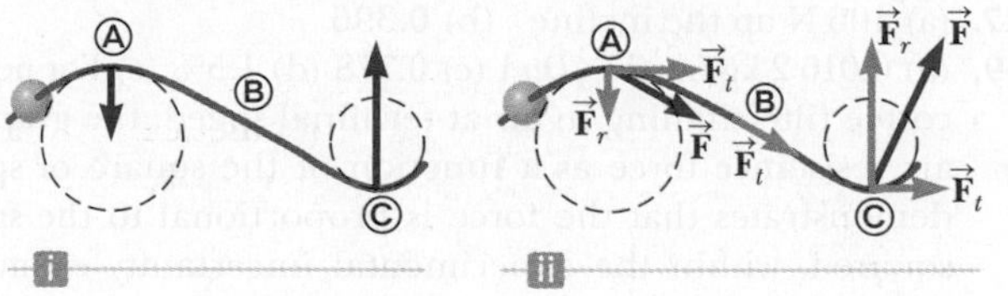

3. (c) **4.** (a)

Answers to Odd-Numbered Problems

1. any speed up to 8.08 m/s

3. (a) 8.32×10^{-8} N toward the nucleus (b) 9.13×10^{22} m/s² inward

5. 6.22×10^{-12} N **7.** 2.14 rev/min **9.** $v \leq 14.3$ m/s

11. (a) static friction (b) 0.085 0

13. (a) 1.33 m/s² (b) 1.79 m/s² at 48.0° inward from the direction of the velocity

15. (a) $v = \sqrt{R\left(\frac{2T}{m} - g\right)}$ (b) $2T$

17. No. The archeologist needs a vine of tensile strength equal to or greater than 1.38 kN to make it across.

19. (a) 8.62 m (b) Mg, downward (c) 8.45 m/s² (d) Calculation of the normal force shows it to be negative, which is impossible. We interpret it to mean that the normal force goes to zero at some point and the passengers will fall out of their seats near the top of the ride if they are not restrained in some way. We could arrive at this same result without calculating the normal force by noting that the acceleration in part (c) is smaller than that due to gravity. The teardrop shape has the advantage of a larger acceleration of the riders at the top of the arc for a path having the same height as the circular path, so the passengers stay in the cars.

21. (a) 17.0° (b) 5.12 N

23. (a) 491 N (b) 50.1 kg (c) 2.00 m/s² **25.** 0.527°

27. 0.212 m/s², opposite the velocity vector **29.** 3.01 N up

31. (a) 1.47 N · s/m (b) 2.04×10^{-3} s (c) 2.94×10^{-2} N

35. (a) 0.034 7 s⁻¹ (b) 2.50 m/s (c) $a = -cv$

37. (a) At Ⓐ, the velocity is east and the acceleration is south. (b) At Ⓑ, the velocity is south and the acceleration is west.

39. 781 N

41. (a) $mg - \dfrac{mv^2}{R}$ (b) $\sqrt{gR}$

43. (a) $v = v_i\, e^{-bt/m}$ (b)

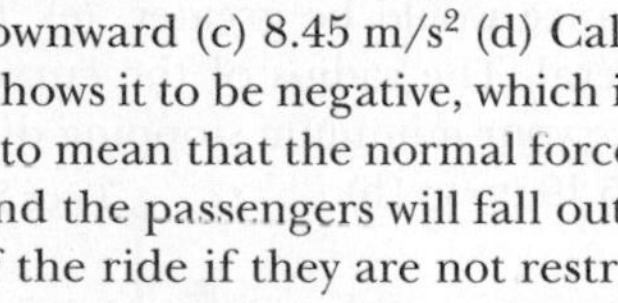

(c) In this model, the object keeps moving forever. (d) It travels a finite distance in an infinite time interval.

45. (a) the downward gravitational force and the tension force in the string, always directed toward the center of the path

(b)

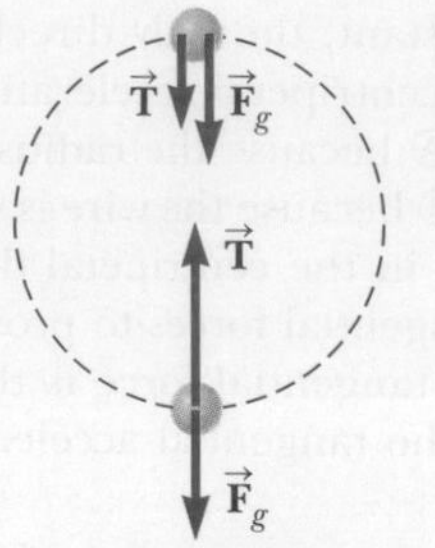

(c) 6.05 N (d) 7.82 m/s

47. (a) 106 N up the incline (b) 0.396

49. (a) 0.016 2 kg/m (b) $\frac{1}{2}D\rho A$ (c) 0.778 (d) 1.5% (e) For nested coffee filters falling in air at terminal speed, the graph of air resistance force as a function of the square of speed demonstrates that the force is proportional to the speed squared, within the experimental uncertainty estimated as 2%. This proportionality agrees with the theoretical model of air resistance at high speeds. The drag coefficient of a coffee filter is $D = 0.78 \pm 2\%$.

51. $g(\cos\phi\tan\theta - \sin\phi)$

53. (a) 735 N (b) 732 N (c) The gravitational force is larger. The normal force is smaller, just like it is when going over the top of a Ferris wheel.

55. (a) The only horizontal force on the car is the force of friction, with a maximum value determined by the surface roughness (described by the coefficient of static friction) and the normal force (here equal to the gravitational force on the car). (b) 34.3 m (c) 68.6 m (d) Braking is better. You should not turn the wheel. If you used any of the available friction force to change the direction of the car, it would be unavailable to slow the car and the stopping distance would be greater. (e) The conclusion is true in general. The radius of the curve you can barely make is twice your minimum stopping distance.

57. (a) 5.19 m/s (b)

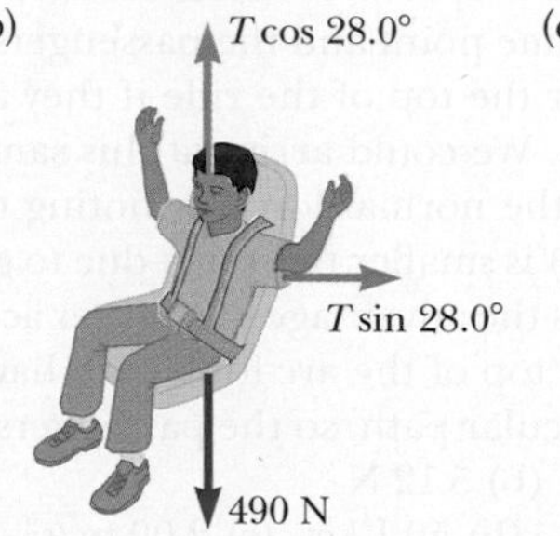

(c) 555 N

59. (b) The gravitational and friction forces remain constant, the normal force increases, and the person remains in motion with the wall. (c) The gravitational force remains constant, the normal and friction forces decrease, and the person slides relative to the wall and downward into the pit.

61. (a) $v_{\min} = \sqrt{\dfrac{Rg(\tan\theta - \mu_s)}{1 + \mu_s\tan\theta}}$, $v_{\max} = \sqrt{\dfrac{Rg(\tan\theta + \mu_s)}{1 - \mu_s\tan\theta}}$
(b) $\mu_s = \tan\theta$

63. 12.8 N **65.** (a) 78.3 m/s (b) 11.1 s (c) 121 m

67. (a) 8.04 s (b) 379 m/s (c) 1.19×10^{-2} m/s (d) 9.55 cm

69. (a) 0.013 2 m/s (b) 1.03 m/s (c) 6.87 m/s

Chapter 7

Answers to Quick Quizzes

1. (a)
2. (c), (a), (d), (b)
3. (d)
4. (a)
5. (b)
6. (c)
7. (i) (c) (ii) (a)
8. (d)

Answers to Odd-Numbered Problems

1. (a) 31.9 J (b) 0 (c) 0 (d) 31.9 J
3. (a) 472 J (b) 2.76 kN
5. (a) 1.6×10^3 J (b) 0 (c) smaller (d) the same
9. (a) 16.0 J (b) 36.9°
11. 16.0 **13.** $\vec{A} = 7.05$ m at 28.4°
15. (a) 7.50 J (b) 15.0 J (c) 7.50 J (d) 30.0 J
17. (a) 0.938 cm (b) 1.25 J **19.** 7.37 N/m
21. (a) $x = mg\left(\dfrac{1}{k_1} + \dfrac{1}{k_2}\right)$ (b) $k = \left(\dfrac{1}{k_1} + \dfrac{1}{k_2}\right)^{-1}$
23. (a) Design the spring constant so that the weight of one tray removed from the pile causes an extension of the springs equal to the thickness of one tray. (b) 316 N/m (c) We do not need to know the length and width of the tray.
25. 50.0 J **27.** 0.299 m/s **29.** (b) mgR
31. (a) 1.20 J (b) 5.00 m/s (c) 6.30 J
33. (a) 60.0 J (b) 60.0 J **35.** 878 kN up
37. (a) 97.8 J (b) $(-4.31\hat{i} + 31.6\hat{j})$ N (c) 8.73 m/s
39. (a) 4.56 kJ (b) 4.56 kJ (c) 6.34 kN (d) 422 km/s^2 (e) 6.34 kN (f) The two theories agree.
41. (a) 2.5 J (b) −9.8 J (c) −12 J
43. (a) −196 J (b) −196 J (c) −196 J (d) The gravitational force is conservative.
45. (a) 125 J (b) 50.0 J (c) 66.7 J (d) nonconservative (e) The work done on the particle depends on the path followed by the particle.
47. A/r^2 away from the other particle
49. (a) 40.0 J (b) −40.0 J (c) 62.5 J
51. $(7 - 9x^2y)\hat{i} - (3x^3)\hat{j}$
53.

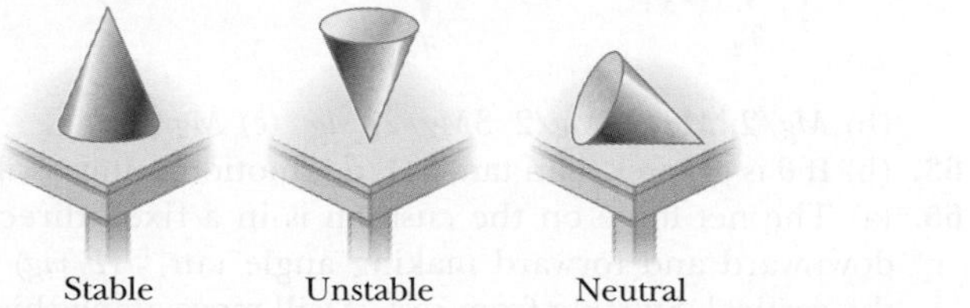

55. 90.0 J
57. (a) 8×10^7 N/m (b) It lasts for a time interval. If the interaction occupied no time interval, the force exerted by each ball on the other would be infinite, and that cannot happen. (c) 0.8 J (d) 0.15 mm (e) 10^{-4} s
59. (a) $U(x) = 1 + 4e^{-2x}$ (b) The force must be conservative because the work the force does on the particle on which it acts depends only on the original and final positions of the particle, not on the path between them.
61. (a) $\vec{F}_1 = (20.5\hat{i} + 14.3\hat{j})$ N, $\vec{F}_2 = (-36.4\hat{i} + 21.0\hat{j})$ N
(b) $\sum\vec{F} = (-15.9\hat{i} + 35.3\hat{j})$ N
(c) $\vec{a} = (-3.18\hat{i} + 7.07\hat{j})$ m/s^2
(d) $\vec{v} = (-5.54\hat{i} + 23.7\hat{j})$ m/s
(e) $\vec{r} = (-2.30\hat{i} + 39.3\hat{j})$ m (f) 1.48 kJ (g) 1.48 kJ
(h) The work–kinetic energy theorem is consistent with Newton's second law.
63. 0.131 m
65. (a)

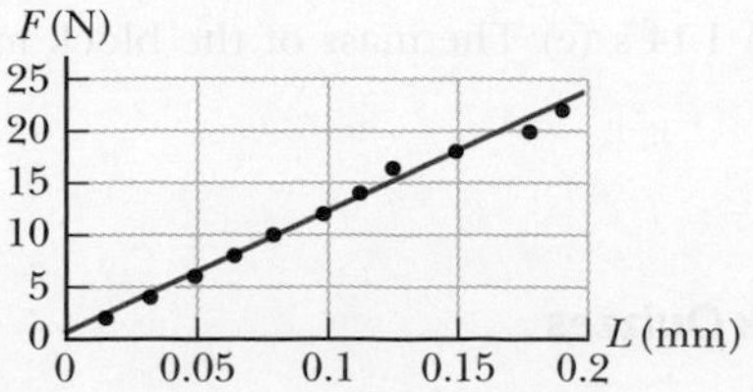

(b) The slope of the line is 116 N/m. (c) We use all the points listed and also the origin. There is no visible evidence for a bend in the graph or nonlinearity near either end. (d) 116 N/m (e) 12.7 N

67. (a) $x = 3.62m/(4.30 - 23.4m)$, where x is in meters and m is in kilograms (b) 0.095 1 m (c) 0.492 m (d) 6.85 m (e) The situation is impossible. (f) The extension is directly proportional to m when m is only a few grams. Then it grows faster and faster, diverging to infinity for $m = 0.184$ kg

Chapter 8

Answers to Quick Quizzes

1. (a) For the television set, energy enters by electrical transmission (through the power cord). Energy leaves by heat (from hot surfaces into the air), mechanical waves (sound from the speaker), and electromagnetic radiation (from the screen). (b) For the gasoline-powered lawn mower, energy enters by matter transfer (gasoline). Energy leaves by work (on the blades of grass), mechanical waves (sound), and heat (from hot surfaces into the air). (c) For the hand-cranked pencil sharpener, energy enters by work (from your hand turning the crank). Energy leaves by work (done on the pencil), mechanical waves (sound), and heat due to the temperature increase from friction.

2. (i) (b) (ii) (b) (iii) (a) **4.** $v_1 = v_2 = v_3$

3. (a) **5.** (c)

Answers to Odd-Numbered Problems

1. (a) $\Delta E_{\text{int}} = Q + T_{\text{ET}} + T_{\text{ER}}$
(b) $\Delta K + \Delta U + \Delta E_{\text{int}} = W + Q + T_{\text{MW}} + T_{\text{MT}}$
(c) $\Delta U = Q + T_{\text{MT}}$ (d) $0 = Q + T_{\text{MT}} + T_{\text{ET}} + T_{\text{ER}}$

3. 10.2 m **5.** (a) $v = (3gR)^{1/2}$ (b) 0.098 0 N down

7. (a) 4.43 m/s (b) 5.00 m

9. 5.49 m/s **11.** $\sqrt{\dfrac{8gh}{15}}$ **13.** $\dfrac{v^2}{2\mu_k g}$

15. (a) 0.791 m/s (b) 0.531 m/s

17. (a) 5.60 J (b) 2.29 rev **19.** (a) 168 J

21. (a) 4.12 m (b) 3.35 m

23. (a) −160 J (b) 73.5 J (c) 28.8 N (d) 0.679

25. (a) 1.40 m/s (b) 4.60 cm after release (c) 1.79 m/s

27. (a) Isolated. The only external influence on the system is the normal force from the slide, but this force is always perpendicular to its displacement so it performs no work on the system. (b) No, the slide is frictionless. (c) $E_{\text{system}} = mgh$ (d) $E_{\text{system}} = \frac{1}{5}mgh + \frac{1}{2}mv_i^2$ (e) $E_{\text{system}} = mgy_{\text{max}} + \frac{1}{2}mv_{xi}^2$ (f) $v_i = \sqrt{\dfrac{8gh}{5}}$ (g) $y_{\text{max}} = h(1 - \frac{4}{5}\cos^2\theta)$ (h) If friction is present, mechanical energy of the system would *not* be conserved, so the child's kinetic energy at all points after leaving the top of the waterslide would be reduced when compared with the frictionless case. Consequently, her launch speed and maximum height would be reduced as well.

29. 1.23 kW **31.** ~ 10^4 W **33.** \$145

35. 4.5×10^3 N **37.** 236 s or 3.93 min

39. (a) 10.2 kW (b) 10.6 kW (c) 5.82 MJ

41. (a) 0.588 J (b) 0.588 J (c) 2.42 m/s (d) $K = 0.196$ J, $U = 0.392$ J

43. $H = h + \dfrac{d^2}{4h}$

45. (a) 11.1 m/s (b) 1.00×10^3 J (c) 1.35 m

47. (a) $K = 2 + 24t^2 + 72t^4$, where t is in seconds and K is in joules (b) $a = 12t$ and $F = 48t$, where t is in seconds, a is in m/s², and F is in newtons (c) $P = 48t + 288t^3$, where t is in seconds and P is in watts (d) 1.25×10^3 J

49. (a) -6.08×10^3 J (b) -4.59×10^3 J (c) 4.59×10^3 J

51. (a) 1.38×10^4 J (b) 5.51×10^3 W (c) The value in part (b) represents only energy that leaves the engine and is transformed to kinetic energy of the car. Additional energy leaves the energy by sound and heat. More energy leaves the engine to do work against friction forces and air resistance.

53. (a) 1.53 J at $x = 6.00$ cm, 0 J at $x = 0$ (b) 1.75 m/s (c) 1.51 m/s (d) The answer to part (c) is not half the answer to part (b) because the equation for the speed of an oscillator is not linear in position

55. (a) 2.17 kW (b) 58.6 kW

57. (a) $x = -4.0$ mm (b) −1.0 cm **59.** 33.4 kW

61. (a) 100 J (b) 0.410 m (c) 2.84 m/s (d) −9.80 mm (e) 2.85 m/s

63. 0.328

65. (a) 0.400 m (b) 4.10 m/s (c) The block stays on the track.

71. (b) 0.342

73. (a) 14.1 m/s (b) 800 N (c) 771 N (d) 1.57 kN up

75. (a) $-\mu_k gx/L$ (b) $(\mu_k gL)^{1/2}$

77. (a) 6.15 m/s (b) 9.87 m/s

79. (a) 25.8 m (b) 27.1 m/s²

Chapter 9

Answers to Quick Quizzes

1. (d)

2. (b), (c), (a)

3. (i) (c), (e) (ii) (b), (d)

4. (a) All three are the same. (b) dashboard, seat belt, air bag

5. (a) **7.** (b)

6. (b) **8.** (i) (a) (ii) (b)

Answers to Odd-Numbered Problems

1. (b) $p = \sqrt{2mK}$ **3.** 7.00 N

5. (a) She moves at 4.71 m/s east. (b) 717 J (c) System momentum is conserved with the value zero. (d) The forces on the two siblings are internal forces, which cannot change the momentum of the system; the system is isolated. (e) Even though there is motion afterward, the final momenta are of equal magnitude in opposite directions, so the final momentum of the system is still zero.

7. (a) $\vec{\mathbf{v}}_{pi} = -\left(\dfrac{m_g}{m_g + m_p}\right)v_{gp}\hat{\mathbf{i}}$ (b) $\vec{\mathbf{v}}_{gi} = \left(\dfrac{m_p}{m_g + m_p}\right)v_{gp}\hat{\mathbf{i}}$

9. (a) $-6.00\hat{\mathbf{i}}$ m/s (b) 8.40 J (c) The original energy is in the spring. (d) A force had to be exerted over a displacement to compress the spring, transferring energy into it by work. The cord exerts force, but over no displacement. (e) System momentum is conserved with the value zero. (f) The forces on the two blocks are internal forces, which cannot change the momentum of the system; the system is isolated. (g) Even though there is motion afterward, the final momenta are of equal magnitude in opposite directions, so the final momentum of the system is still zero.

11. (a) 13.5 N · s (b) 9.00 kN **13.** (c) no difference

15. (a) $12.0\hat{\mathbf{i}}$ N · s (b) $4.80\hat{\mathbf{i}}$ m/s (c) $2.80\hat{\mathbf{i}}$ m/s (d) $2.40\hat{\mathbf{i}}$ N

17. 16.5 N **19.** 301 m/s
21. (a) 0.284 (b) 1.15×10^{-13} J and 4.54×10^{-14} J
23. 91.2 m/s
25. (a) 2.24 m/s (b) Coupling order makes no difference.
27. 0.556 m **29.** $(3.00\hat{\mathbf{i}} - 1.20\hat{\mathbf{j}})$ m/s
31. $v_O = v_i \cos\theta$, $v_Y = v_i \sin\theta$ **33.** 2.50 m/s at $-60.0°$
35. (a) $(-9.33\hat{\mathbf{i}} - 8.33\hat{\mathbf{j}})$ Mm/s (b) 439 fJ
37. $\vec{\mathbf{r}}_{CM} = (0\hat{\mathbf{i}} + 1.00\hat{\mathbf{j}})$ m **39.** 3.57×10^8 J
41. (a) $(1.40\hat{\mathbf{i}} + 2.40\hat{\mathbf{j}})$ m/s (b) $(7.00\hat{\mathbf{i}} + 12.0\hat{\mathbf{j}})$ kg · m/s
43. 0.700 m
45. (a) Yes; $18.0\hat{\mathbf{i}}$ kg · m/s (b) No; the floor does zero work. (c) Yes; we could say that the final momentum of the cart came from the floor or from the Earth through the floor. (d) No; the kinetic energy came from the original gravitational potential energy of the Earth-elevated load system, in the amount 27.0 J. (e) Yes; the acceleration is caused by the static friction force exerted by the floor that prevents the wheels from slipping backward.
47. (b) The bumper continues to exert a force to the left until the particle has swung down to its lowest point.
49. (a) 787 m/s (b) 138 m/s
51. (a) 442 metric tons (b) 19.2 metric tons (c) It is much less than the suggested value of 442/2.50. Mathematically, the logarithm in the rocket propulsion equation is not a linear function. Physically, a higher exhaust speed has an extra-large cumulative effect on the rocket body's final speed by counting again and again in the speed the body attains second after second during its burn.
53. (a) zero (b) $\dfrac{mv_i}{\sqrt{2}}$ upward **55.** 260 N normal to the wall
57. (a) Momentum of the bullet–block system is conserved in the collision, so you can relate the speed of the block and bullet immediately after the collision to the initial speed of the bullet. Then, you can use conservation of mechanical energy for the bullet–block–Earth system to relate the speed after the collision to the maximum height. (b) 521 m/s upward
59. (a) $\dfrac{m_1 v_1 + m_2 v_2}{m_1 + m_2}$ (b) $(v_1 - v_2)\sqrt{\dfrac{m_1 m_2}{k(m_1 + m_2)}}$
(c) $v_{1f} = \dfrac{(m_1 - m_2)v_1 + 2m_2 v_2}{m_1 + m_2}$, $v_{2f} = \dfrac{2m_1 v_1 + (m_2 - m_1)v_2}{m_1 + m_2}$
61. m_1: 13.9 m m_2: 0.556 m **63.** 0.960 m
65. $\left(\dfrac{M + m}{m}\right)\sqrt{\dfrac{gd^2}{2h}}$
67. (a) 0; inelastic (b) $(-0.250\hat{\mathbf{i}} + 0.75\hat{\mathbf{j}} - 2.00\hat{\mathbf{k}})$ m/s; perfectly inelastic (c) either $a = -6.74$ with $\vec{\mathbf{v}} = -0.419\hat{\mathbf{k}}$ m/s or $a = 2.74$ with $\vec{\mathbf{v}} = -3.58\hat{\mathbf{k}}$ m/s
69. 0.403
71. (a) $-0.256\hat{\mathbf{i}}$ m/s and $0.128\hat{\mathbf{i}}$ m/s (b) $-0.064\,2\hat{\mathbf{i}}$ m/s and 0 (c) 0 and 0
73. (a) 100 m/s (b) 374 J
75. (a) particle of mass m: $\sqrt{2}v_i$; particle of mass $3m$: $\sqrt{\frac{2}{3}}v_i$ (b) 35.3°
77. (a) $v_{CM} = \sqrt{\dfrac{F}{2m}(x_1 + x_2)}$
(b) $\theta = \cos^{-1}\left[1 - \dfrac{F}{2mgL}(x_1 - x_2)\right]$
79. $\vec{\mathbf{F}} = \left(\dfrac{3Mgx}{L}\right)\hat{\mathbf{j}}$

Chapter 10

Answers to Quick Quizzes

1. (i) (c) (ii) (b)
2. (b)
3. (i) (b) (ii) (a)
4. (a)
5. (i) (b) (ii) (a)
6. (b)
7. (b)

Answers to Odd-Numbered Problems

1. (a) 7.27×10^{-5} rad/s (b) Because of its angular speed, the Earth bulges at the equator.
3. (a) 5.00 rad, 10.0 rad/s, 4.00 rad/s² (b) 53.0 rad, 22.0 rad/s, 4.00 rad/s²
5. (a) 4.00 rad/s² (b) 18.0 rad
7. (a) 5.24 s (b) 27.4 rad **9.** 13.7 rad/s²
11. (a) $\omega(2h^3/g)^{1/2}$ (b) 0.011 6 m (c) Yes; the deflection is only 0.02% of the original height. (d) It would decrease. Because the displacement is proportional to angular speed and the angular acceleration is constant, the displacement decreases linearly in time.
13. (a) 0.180 rad/s (b) 8.10 m/s² radially inward
15. (a) 25.0 rad/s (b) 39.8 rad/s² (c) 0.628 s
17. (a) 8.00 rad/s (b) 8.00 m/s (c) 64.1 m/s² at an angle 3.58° from the radial line to point P (d) 9.00 rad
19. (a) 126 rad/s (b) 3.77 m/s (c) 1.26 km/s² (d) 20.1 m
21. (a) 3.47 rad/s (b) 1.74 m/s (c) 2.78 s (d) 1.02 rotations
23. 0.572
25. (a) 143 kg · m² (b) 2.57 kJ
27. (a) 24.5 m/s (b) no (c) no (d) no (e) no (f) yes
29. (a) 5.80 kg · m² (b) Yes, knowing the height of the door is unnecessary.
31. 1.28 kg · m² **33.** $\frac{11}{12}mL^2$
35. −3.55 N · m **37.** 21.5 N
39. (a) 24.0 N · m (b) 0.035 6 rad/s² (c) 1.07 m/s²
41. (a) 21.6 kg · m² (b) 3.60 N · m (c) 52.5 rev
43. 0.312 **45.** 1.03×10^{-3} J **47.** 149 rad/s
49. (a) 1.59 m/s (b) 53.1 rad/s
51. (a) 11.4 N (b) 7.57 m/s² (c) 9.53 m/s (d) 9.53 m/s
53. (a) $2(Rg/3)^{1/2}$ (b) $4(Rg/3)^{1/2}$ (c) $(Rg)^{1/2}$
55. (a) 500 J (b) 250 J (c) 750 J
57. (a) $\frac{2}{3}g\sin\theta$ (b) The acceleration of $\frac{1}{2}g\sin\theta$ for the hoop is smaller than that for the disk. (c) $\frac{1}{3}\tan\theta$
59. (a) The disk (b) disk: $\sqrt{\frac{4}{3}gh}$; hoop: $\sqrt{gh}$
61. (a) 1.21×10^{-4} kg · m² (b) Knowing the height of the can is unnecessary. (c) The mass is not uniformly distributed; the density of the metal can is larger than that of the soup.
63. (a) 4.00 J (b) 1.60 s (c) 0.80 m
65. (a) 12.5 rad/s (b) 128 rad
67. (a) $(3g/L)^{1/2}$ (b) $3g/2L$ (c) $-\frac{3}{2}g\hat{\mathbf{i}} - \frac{3}{4}g\hat{\mathbf{j}}$ (d) $-\frac{3}{2}Mg\hat{\mathbf{i}} + \frac{1}{4}Mg\hat{\mathbf{j}}$
69. $\dfrac{g(h_2 - h_1)}{2\pi R^2}$
71. (a) Particle under a net force (b) Rigid object under a net torque (c) 118 N (d) 156 N (e) $\dfrac{r^2}{a}(T_2 - T_1)$ (f) 1.17 kg · m²
75. $\sqrt{\dfrac{10}{7}\left[\dfrac{g(R - r)(1 - \cos\theta)}{r^2}\right]}$
77. (a) 2.70R (b) $F_x = -20mg/7$, $F_y = -mg$

79. (a) $\sqrt{\frac{3}{4}gh}$ (b) $\sqrt{\frac{3}{4}gh}$

83. (a) $\sigma = 0.060\,2\ \text{s}^{-1}$, $\omega_0 = 3.50$ rad/s (b) $\alpha = -0.176$ rad/s^2 (c) 1.29 rev (d) 9.26 rev

85. (b) to the left **87.** (a) 2.88 s (b) 12.8 s

Chapter 11

Answers to Quick Quizzes

1. (d) **3.** (b)
2. **(i)** (a) **(ii)** (c) **4.** (a)

Answers to Odd-Numbered Problems

1. $\hat{\mathbf{i}} + 8.00\hat{\mathbf{j}} + 22.0\hat{\mathbf{k}}$ **3.** (a) $7.00\hat{\mathbf{k}}$ (b) 60.3°

5. (a) 30 N · m (counterclockwise) (b) 36 N · m (counterclockwise)

7. 45.0° **9.** (a) $F_3 = F_1 + F_2$ (b) no

11. $17.5\hat{\mathbf{k}}$ kg · m^2/s **13.** $m(xv_y - yv_x)\hat{\mathbf{k}}$

15. $60.0\hat{\mathbf{k}}$ kg · m^2/s **17.** $mvR[\cos(vt/R) + 1]\hat{\mathbf{k}}$

19. (a) zero (b) $(-mv_i^3 \sin^2\theta \cos\theta/2g)\hat{\mathbf{k}}$ (c) $(-2mv_i^3 \sin^2\theta \cos\theta/g)\hat{\mathbf{k}}$ (d) The downward gravitational force exerts a torque on the projectile in the negative z direction.

21. (a) $-m\ell gt\cos\theta\hat{\mathbf{k}}$ (b) The Earth exerts a gravitational torque on the ball. (c) $-mg\ell\cos\theta\hat{\mathbf{k}}$

23. 1.20 kg · m^2/s

25. (a) 0.360 kg · m^2/s (b) 0.540 kg · m^2/s

27. (a) 0.433 kg · m^2/s (b) 1.73 kg · m^2/s

29. (a) 1.57×10^8 kg · m^2/s (b) 6.26×10^3 s = 1.74 h

31. 7.14 rev/min

33. (a) 11.1 rad/s counterclockwise (b) No; 507 J is transformed into internal energy. (c) No; the turntable bearing promptly imparts impulse 44.9 kg · m/s north into the turntable–clay system and thereafter keeps changing the system momentum.

35. (a) The mechanical energy of the system is not constant. Some chemical energy is converted into mechanical energy. (b) The momentum of the system is not constant. The turntable bearing exerts an external northward force on the axle. (c) The angular momentum of the system is constant. (d) 0.360 rad/s counterclockwise (e) 99.9 J

37. (a) $mv\ell$ down (b) $M/(M + m)$

39. (a) $\omega = 2mv_id/[M + 2m]R^2$ (b) No; some mechanical energy of the system changes into internal energy. (c) The momentum of the system is not constant. The axle exerts a backward force on the cylinder when the clay strikes.

41. (a) yes (b) 4.50 kg · m^2/s (c) No. In the perfectly inelastic collision, kinetic energy is transformed to internal energy. (d) 0.749 rad/s (e) $K_i = 2.50 \times 10^3$ J; $K_f = 1.69$ J. Most of the initial kinetic energy is transformed to internal energy in the collision.

43. 5.46×10^{22} N · m **45.** 0.910 km/s **47.** 7.50×10^{-11} s

49. (a) $7md^2/3$ (b) $mgd\hat{\mathbf{k}}$ (c) $3g/7d$ counterclockwise (d) $2g/7$ upward (e) mgd (f) $\sqrt{6g/7d}$ (g) $m\sqrt{14gd^3/3}$ (h) $\sqrt{2gd/21}$

51. (a) isolated system (angular momentum) (b) $mv_id/2$ (c) $(\frac{1}{12}M + \frac{1}{4}m)d^2$ (d) $(\frac{1}{12}M + \frac{1}{4}m)d^2\omega$ (e) $\dfrac{6mv_i}{(M + 3m)d}$ (f) $\frac{1}{2}mv_i^2$ (g) $\dfrac{3m^2v_i^2}{2(M + 3m)}$ (h) $-\dfrac{M}{M + 3m}$

53. (a) v_ir_i/r (b) $(mv_i^2r_i^2)r^{-3}$ (c) $\frac{1}{2}mv_i^2(r_i^2/r^2 - 1)$

55. (a) 3 750 kg · m^2/s (b) 1.88 kJ (c) 3 750 kg · m^2/s (d) 10.0 m/s (e) 7.50 kJ (f) 5.62 kJ

57. (a) $2mv_0$ (b) $2v_0/3$ (c) $4m\ell v_0/3$ (d) $4v_0/9\ell$ (e) mv_0^2 (f) $26mv_0^2/27$ (g) No horizontal forces act on the bola from outside after release, so the horizontal momentum stays constant. Its center of mass moves steadily with the horizontal velocity it had at release. No torques about its axis of rotation act on the bola, so the angular momentum stays constant. Internal forces cannot affect momentum conservation and angular momentum conservation, but they can affect mechanical energy.

59. an increase of 6.368×10^{-4} % or 0.550 s, which is not significant

61. (a) $\frac{1}{3}\omega_i$ (b) $-\frac{2}{3}$ (c) $\dfrac{R\omega_i}{3\mu g}$ (d) $\dfrac{R^2\omega_i^2}{18\mu g}$

63. $4\sqrt{\frac{1}{3}ga(\sqrt{2} - 1)}$

Chapter 12

Answers to Quick Quizzes

1. (a) **3.** (b)
2. (b) **4.** **(i)** (b) **(ii)** (a) **(iii)** (c)

Answers to Odd-Numbered Problems

1. $F_x - R_x = 0$, $F_y + R_y - F_g = 0$, $F_y\cos\theta - F_x\sin\theta - 0.5F_g\cos\theta = 0$

3. (3.85 cm, 6.85 cm) **5.** 0.750 m

7. (2.54 m, 4.75 m) **9.** 177 kg

11. Sam exerts an upward force of 176 N, and Joe exerts an upward force of 274 N.

13. (a) $f_s = 268$ N, $n = 1\,300$ N (b) 0.324

15. (a) 1.04 kN at 60.0° upward and to the right (b) $(370\hat{\mathbf{i}} + 910\hat{\mathbf{j}})$ N

17. (a) 29.9 N (b) 22.2 N

19. (a) 27.7 kN (b) 11.5 kN (c) 4.19 kN

21. (a) 859 N (b) 1.04 kN at 36.9° to the left and upward

23. 2.81 m **25.** $T_1 = 501$ N, $T_2 = 672$ N, $T_3 = 384$ N

27. 4.90 mm **29.** 23.8 μm

31. (a) 3.14×10^4 N (b) 6.28×10^4 N

33. (a) $-0.053\,8$ m^3 (b) 1.09×10^3 kg/m^3 (c) With only a 5% change in volume in this extreme case, liquid water is indeed nearly incompressible in biological and student laboratory situations.

35. 0.029 2 mm **37.** $n_A = 5.98 \times 10^5$ N, $n_B = 4.80 \times 10^5$ N

39. 0.896 m **41.** $F_t = 724$ N, $F_s = 716$ N

43. (a)

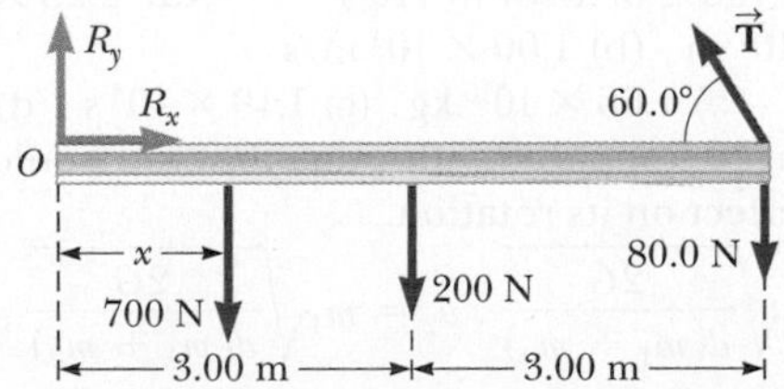

(b) $T = 343$ N, $R_x = 171$ N to the right, $R_y = 683$ N up (c) 5.14 m

45. (a) $T = F_g(L + d)/[\sin\theta\,(2L + d)]$ (b) $R_x = F_g(L + d)\cot\theta/(2L + d)$; $R_y = F_gL/(2L + d)$

47. $\vec{\mathbf{F}}_A = (-6.47 \times 10^5\hat{\mathbf{i}} + 1.27 \times 10^5\hat{\mathbf{j}})$ N, $\vec{\mathbf{F}}_B = 6.47 \times 10^5\hat{\mathbf{i}}$ N

49. (a) 5.08 kN (b) 4.77 kN (c) 8.26 kN

51. (a) $\frac{1}{2}m\left(\dfrac{2\mu_s\sin\theta - \cos\theta}{\cos\theta - \mu_s\sin\theta}\right)$ (b) $(m + M)g\sqrt{1 + \mu_s^2}$

(c) $g\sqrt{M^2 + \mu_s^2(m + M)^2}$

53. (a) 9.28 kN (b) The moment arm of the force $\vec{\mathbf{F}}_h$ is no longer 70 cm from the shoulder joint but only 49.5 cm, therefore reducing $\vec{\mathbf{F}}_m$ to 6.56 kN.

55. (a) 66.7 N (b) increasing at 0.125 N/s

57. (a) $\frac{1}{\sqrt{15}}\frac{mgd}{\ell}$ (b) $n_A = mg\left(\frac{2\ell - d}{2\ell}\right)$, $n_B = \frac{mgd}{2\ell}$

(c) $R_x = \frac{1}{\sqrt{15}}\frac{mgd}{\ell}$, $R_y = \frac{mgd}{2\ell}$ (to the right and downward on the right half of the ladder)

59. (a) $P_1 = P_3 = 1.67$ N, $P_2 = 3.33$ N (b) 2.36 N

61. 5.73 rad/s **63.** 4.90 cm **65.** 9.00 ft **67.** $3F_g/8$

Chapter 13

Answers to Quick Quizzes

1. (e)
2. (c)
3. (a)
4. (a) Perihelion (b) Aphelion (c) Perihelion (d) All points

Answers to Odd-Numbered Problems

1. 2.67×10^{-7} m/s^2

3. (a) 2.50×10^{-7} N toward the 500-kg object (b) between the objects and 2.45 m from the 500-kg object

5. 7.41×10^{-10} N **7.** 2.00 kg and 3.00 kg **9.** 30.6 nm

11. (a) 7.61 cm/s^2 (b) 363 s (c) 3.08 km (d) 28.9 m/s at 72.9° below the horizontal

13. 1.50 h or 90.0 min

15. (a) 0.71 yr (b) The departure must be timed so that the spacecraft arrives at the aphelion when the target planet is there.

17. 1.26×10^{32} kg **19.** 4.99 days **21.** 8.92×10^7 m

23. (a) yes (b) 3.93 yr

25. $\frac{GM}{\ell^2}(\frac{1}{2} + \sqrt{2})$ at 45° to the positive x axis

27. (a) 1.31×10^{17} N (b) 2.62×10^{12} N/kg **29.** 2.82×10^9 J

31. (a) 1.84×10^9 kg/m^3 (b) 3.27×10^6 m/s^2 (c) -2.08×10^{13} J

33. 2.52×10^7 m

35. (a) 4.69×10^8 J (b) -4.69×10^8 J (c) 9.38×10^8 J

37. (a) 0.980 (b) 127 yr (c) -2.13×10^{17} J

39. (a) 5.30×10^3 s (b) 7.79 km/s (c) 6.43×10^9 J

41. (a) same size force (b) 15.6 km/s

43. $\omega = 0.057\,2$ rad/s or 1 rev in 110 s **45.** 2.25×10^{-7}

47. (a) 1.00×10^7 m (b) 1.00×10^4 m/s

49. (a) 15.3 km (b) 1.66×10^{16} kg (c) 1.13×10^4 s (d) No; its mass is so large compared with yours that you would have a negligible effect on its rotation.

51. (a) $v_1 = m_2\sqrt{\frac{2G}{d(m_1 + m_2)}}$, $v_2 = m_1\sqrt{\frac{2G}{d(m_1 + m_2)}}$,

$v_{\text{rel}} = \sqrt{\frac{2G(m_1 + m_2)}{d}}$ (b) 1.07×10^{32} J and 2.67×10^{31} J

53. (a) -7.04×10^4 J (b) -1.57×10^5 J (c) 13.2 m/s

55. 7.79×10^{14} kg

57. (a) 2×10^8 yr (b) ~ 10^{41} kg (c) 10^{11}

59. (a) 2.93×10^4 m/s (b) $K = 2.74 \times 10^{33}$ J, $U = -5.39 \times 10^{33}$ J (c) $K = 2.56 \times 10^{33}$ J, $U = -5.21 \times 10^{33}$ J (d) Yes; $E = -2.65 \times 10^{33}$ J at both aphelion and perihelion.

61. 119 km **63.** $\sqrt{\frac{GM}{4R_E}}$

65. $(800 + 1.73 \times 10^{-4})\hat{\mathbf{i}}$ m/s and $(800 - 1.73 \times 10^{-4})\hat{\mathbf{i}}$ m/s

67. 18.2 ms

69. (a) -3.67×10^7 J (b) 9.24×10^{10} kg · m^2/s (c) $v = 5.58$ km/s, $r = 1.04 \times 10^7$ m (d) 8.69×10^6 m (e) 134 min

Chapter 14

Answers to Quick Quizzes

1. (a)
2. (a)
3. (c)
4. (b) or (c)
5. (a)

Answers to Odd-Numbered Problems

1. 24.8 kg

3. (a) 6.24 MPa (b) Yes; this pressure could puncture the vinyl flooring.

5. 2.96×10^6 Pa **7.** 8.46 m **9.** 7.74×10^{-3} m^2

11. (a) 3.71×10^5 Pa (b) 3.57×10^4 N

13. (a) 2.94×10^4 N (b) 1.63×10^4 N · m

15. 0.072 1 mm **17.** 98.6 kPa

19. (a) 116 kPa (b) 52.0 Pa

21. (a) 20.0 cm (b) 0.490 cm **23.** 0.258 N down

25. (a) 4.9 N down, 16.7 N up (b) 86.2 N (c) By either method of evaluation, the buoyant force is 11.8 N up.

27. (a) 7.00 cm (b) 2.80 kg

29. (a) 1 250 kg/m^3 (b) 500 kg/m^3

31. (a) 408 kg/m^3 (b) When m is less than 0.310 kg, the wooden block will be only partially submerged in the water. (c) When m is greater than 0.310 kg, the wooden block and steel object will sink.

33. (a) 3.82×10^3 N (b) 1.04×10^3 N; the balloon rises because the net force is positive: the upward buoyant force is greater than the downward gravitational force. (c) 106 kg

35. (a) 11.6 cm (b) 0.963 g/cm^3 (c) No; the density ρ is not linear in h.

37. (a) 17.7 m/s (b) 1.73 mm

39. (a) 0.638 m/s (b) 2.55 m/s (c) 1.80×10^{-3} m^3/s

41. 0.247 cm

43. (a) 2.28 N toward Holland (b) 1.74×10^6 s

45. (a) 15.1 MPa (b) 2.95 m/s

47. (a) 1.91 m/s (b) 8.65×10^{-4} m^3/s

49. 347 m/s **51.** 12.6 m/s

53. (a) 1.02×10^7 Pa (b) 6.61×10^5 N

55. (a) 6.70 cm (b) 5.74 cm

57. 2.25 m **59.** 455 kPa **61.** 0.556 m

63. (a) 8.01 km (b) yes

65. upper scale: 17.3 N; lower scale: 31.7 N

67. 91.64% **69.** 758 Pa **71.** 4.43 m/s

73. (a) 1.25 cm (b) 14.3 m/s

77. (a) 18.3 mm (b) 14.3 mm (c) 8.56 mm

Chapter 15

Answers to Quick Quizzes

1. (d)
2. (f)
3. (a)
4. (b)
5. (c)
6. **(i)** (a) **(ii)** (a)

Answers to Odd-Numbered Problems

1. (a) 17 N to the left (b) 28 m/s^2 to the left

3. 0.63 s
5. (a) 1.50 Hz (b) 0.667 s (c) 4.00 m (d) π rad (e) 2.83 m
7. 0.628 m/s **9.** 40.9 N/m
11. (a) −2.34 m (b) −1.30 m/s (c) −0.076 3 m (d) 0.315 m/s
13. (a) $x = 2.00 \cos (3.00\pi t - 90°)$ or $x = 2.00 \sin (3.00\pi t)$ where x is in centimeters and t is in seconds (b) 18.8 cm/s (c) 0.333 s (d) 178 cm/s^2 (e) 0.500 s (f) 12.0 cm
15. (a) 40.0 cm/s (b) 160 cm/s^2 (c) 32.0 cm/s (d) −96.0 cm/s^2 (e) 0.232 s
17. 2.23 m/s
19. (a) 28.0 mJ (b) 1.02 m/s (c) 12.2 mJ (d) 15.8 mJ
21. (a) $\frac{8}{9}E$ (b) $\frac{1}{9}E$ (c) $x = \pm\sqrt{\frac{2}{3}}A$ (d) No; the maximum potential energy is equal to the total energy of the system. Because the total energy must remain constant, the kinetic energy can never be greater than the maximum potential energy.
23. (a) 4.58 N (b) 0.125 J (c) 18.3 m/s^2 (d) 1.00 m/s (e) smaller (f) the coefficient of kinetic friction between the block and surface (g) 0.934
25. (b) 0.628 s
27. (a) 1.50 s (b) 0.559 m **29.** 0.944 kg · m^2
31. (a) 0.820 m/s (b) 2.57 rad/s^2 (c) 0.641 N (d) $v_{\max} =$ 0.817 m/s, $\alpha_{\max} =$ 2.54 rad/s^2, $F_{\max} =$ 0.634 N (e) The answers are close but not exactly the same. The answers computed from conservation of energy and from Newton's second law are more precise.
33. (a) 3.65 s (b) 6.41 s (c) 4.24 s
35. (a) 5.00×10^{-7} kg · m^2 (b) 3.16×10^{-4} N · m/rad
37. 1.00×10^{-3} s^{-1}
39. (a) 7.00 Hz (b) 2.00% (c) 10.6 s **41.** 11.0 cm
43. (a) 3.16 s^{-1} (b) 6.28 s^{-1} (c) 5.09 cm
45. 0.641 Hz or 1.31 Hz **47.** 9.19×10^{13} Hz
49. (a) 2.09 s (b) 0.477 Hz (c) 36.0 cm/s (d) $E = 0.064\,8m$, where E is in joules and m is in kilograms (e) $k = 9.00m$, where k is in newtons/meter and m is in kilograms (f) Period, frequency, and maximum speed are all independent of mass in this situation. The energy and the force constant are directly proportional to mass.
51. (a) $2Mg$ (b) $Mg\left(1 + \frac{y}{L}\right)$ (c) $\frac{4\pi}{3}\sqrt{\frac{2L}{g}}$ (d) 2.68 s
53. 6.62 cm **55.** $\frac{1}{2\pi L}\sqrt{gL + \frac{kh^2}{M}}$ **57.** 7.75 s^{-1}
59. (a) 1.26 m (b) 1.58 (c) The energy decreases by 120 J. (d) Mechanical energy is transformed into internal energy in the perfectly inelastic collision.
61. (a) $\omega = \sqrt{\frac{200}{0.400 + M}}$, where ω is in s^{-1} and M is in kilograms (b) 22.4 s^{-1} (c) 22.4 s^{-1}
65. (b) 1.46 s
69. (b) $T = \frac{2}{r}\sqrt{\frac{\pi M}{\rho g}}$ **71.** 9.12×10^{-5} s
73. (a) 0.500 m/s (b) 8.56 cm
75. (a) $\frac{1}{2}(M + \frac{1}{3}m)v^2$ (b) $2\pi\sqrt{\frac{M + \frac{1}{3}m}{k}}$
77. (a) $\frac{2\pi}{\sqrt{g}}\sqrt{L_i + \frac{1}{2\rho a^2}\left(\frac{dM}{dt}\right)t}$ (b) $2\pi\sqrt{\frac{L_i}{g}}$

Chapter 16

Answers to Quick Quizzes

1. (i) (b) (ii) (a) **4.** (f) and (h)
2. (i) (c) (ii) (b) (iii) (d) **5.** (d)
3. (c)

Answers to Odd-Numbered Problems

1. $y = \frac{6.00}{(x - 4.50t)^2 + 3.00}$ where x and y are in meters and t is in seconds
3. 184 km
5. (a) $3.33\hat{\mathbf{i}}$ m/s (b) −5.48 cm (c) 0.667 m (d) 5.00 Hz (e) 11.0 m/s
7. 0.319 m
9. (a) 2.00 cm (b) 2.98 m (c) 0.576 Hz (d) 1.72 m/s
11. (a) 31.4 rad/s (b) 1.57 rad/m (c) $y = 0.120 \sin (1.57x - 31.4t)$, where x and y are in meters and t is in seconds (d) 3.77 m/s (e) 118 m/s^2
13. (a) 0.250 m (b) 40.0 rad/s (c) 0.300 rad/m (d) 20.9 m (e) 133 m/s (f) positive x direction
15. (a) −1.51 m/s (b) 0 (c) 16.0 m (d) 0.500 s (e) 32.0 m/s
17. (a) 0.500 Hz (b) 3.14 rad/s (c) 3.14 rad/m (d) $0.100 \sin (\pi x - \pi t)$ (e) $0.100 \sin (-\pi t)$ (f) $0.100 \sin (4.71 - \pi t)$ (g) 0.314 m/s
19. (a) $y = 0.080\,0 \sin (2.5\pi x + 6\pi t)$ (b) $y = 0.080\,0 \sin (2.5\pi x + 6\pi t - 0.25\pi)$
21. 80.0 N **23.** 13.5 N **25.** 185 m/s **27.** 631 N
29. (a) 0.051 0 kg/m (b) 19.6 m/s **31.** 0.329 s
33. (a) 1 (b) 1 (c) 1 (d) increased by a factor of 4
35. (a) 62.5 m/s (b) 7.85 m (c) 7.96 Hz (d) 21.1 W
37. (a) $y = 0.075 \sin (4.19x - 314t)$ (b) 625 W
39. (a) 15.1 W (b) 3.02 J
45. 0.456 m/s **47.** 14.7 kg
49. (a) 39.2 N (b) 0.892 m (c) 83.6 m/s
51. (a) 21.0 ms (b) 1.68 m
53. $\sqrt{\frac{mL}{Mg \sin \theta}}$ **55.** 0.084 3 rad **57.** $\frac{1}{\omega}\sqrt{\frac{m}{M}}$
59. (a) $v = \sqrt{\frac{T}{\rho(1.00 \times 10^{-5}\,x + 1.00 \times 10^{-6})}}$, where v is in meters per second, T is in newtons, and x is in meters (b) $v(0) = 94.3$ m/s, $v(10.0 \text{ m}) = 9.38$ m/s
61. (a) $\frac{\mu\omega^3}{2k}A_0{}^2e^{-2bx}$ (b) $\frac{\mu\omega^3}{2k}A_0{}^2$ (c) e^{-2bx}
63. (a) $(0.707)(2\sqrt{L/g})$ (b) $L/4$
65. (a) $\mu v_0{}^2$ (b) v_0 (c) clockwise: 4π; counterclockwise: 0

Chapter 17

Answers to Quick Quizzes

1. (c) **4.** (e)
2. (b) **5.** (e)
3. (b) **6.** (b)

Answers to Odd-Numbered Problems

1. $\Delta P = 0.200 \sin (20\pi x - 6\,860\pi t)$ where ΔP is in pascals, x is in meters, and t is in seconds
3. (a) 2.00 μm (b) 40.0 cm (c) 54.6 m/s (d) −0.433 μm (e) 1.72 mm/s
5. (a) 5.56 km (b) No. The speed of light is much greater than the speed of sound, so the time interval required for

the light to reach you is negligible compared to the time interval for the sound.

7. 0.196 s

9. (a) 0.625 mm (b) 1.50 mm to 75.0 μm **11.** 7.82 m

13. (a) 27.2 s (b) 25.7 s; the time interval in part (a) is longer.

15. (a) 833 m (b) 1.47 s

17. (a) the pulse that travels through the rail (b) 23.4 ms

19. 66.0 dB **21.** (a) 3.75 W/m^2 (b) 0.600 W/m^2

23. 3.0×10^{-8} W/m^2 **25.** (a) 0.691 m (b) 691 km

27. (a) 2.34 m (b) 0.390 m (c) 0.161 Pa (d) 0.161 Pa (e) 4.25×10^{-7} m (f) 7.09×10^{-8} m

29. (a) 1.32×10^{-4} W/m^2 (b) 81.2 dB

31. 68.3 dB **33.** (a) 30.0 m (b) 9.49×10^5 m

35. (a) 3.04 kHz (b) 2.08 kHz (c) 2.62 kHz; 2.40 kHz

37. (a) 475 Hz (b) 430 Hz

39. (a) 441 Hz (b) 439 Hz (c) 54.0 dB

41. (a) 0.021 7 m/s (b) 28.9 Hz (c) 57.8 Hz

43. 26.4 m/s **45.** (a) 56.3 s (b) 56.6 km farther along

47. (a) 0.515 trucks per minute (b) 0.614 trucks per minute

49. 67.0 dB

51. It is unreasonable, implying a sound level of 123 dB. Nearly all the decrease in mechanical energy becomes internal energy in the latch.

53. (a) 5.04×10^3 m/s (b) 1.59×10^{-4} s (c) 1.90×10^{-3} m (d) 2.38×10^{-3} (e) 4.76×10^8 N/m^2 (f) $\dfrac{\sigma_y}{\sqrt{\rho Y}}$

55. (a) 55.8 m/s (b) 2 500 Hz

57. (a) 3.29 m/s (b) The bat will be able to catch the insect because the bat is traveling at a higher speed in the same direction as the insect.

59. (a) 0.343 m (b) 0.303 m (c) 0.383 m (d) 1.03 kHz

61. (a) 0.983° (b) 4.40° **63.** 1.34×10^4 N

65. (a) 531 Hz (b) 466 Hz to 539 Hz (c) 568 Hz

Chapter 18

Answers to Quick Quizzes

1. (c) **4.** (b)
2. (i) (a) (ii) (d) **5.** (c)
3. (d)

Answers to Odd-Numbered Problems

1. (a) −1.65 cm (b) −6.02 cm (c) 1.15 cm

3. 5.66 cm

5. (a) y_1: positive x direction; y_2: negative x direction (b) 0.750 s (c) 1.00 m

7. (a) 9.24 m (b) 600 Hz **9.** (a) 156° (b) 0.058 4 cm

11. (c) Yes; the limiting form of the path is two straight lines through the origin with slope ±0.75.

13. (a) 15.7 m (b) 31.8 Hz (c) 500 m/s

15. (a) 4.24 cm (b) 6.00 cm (c) 6.00 cm (d) 0.500 cm, 1.50 cm, 2.50 cm

17. at 0.089 1 m, 0.303 m, 0.518 m, 0.732 m, 0.947 m, and 1.16 m from one speaker

19. (a) second harmonic (b) 74.0 cm (c) 3

21. 19.6 Hz **23.** (a) 163 N (b) 660 Hz

25. 1.86 g **27.** (a) 3.8 cm (b) 3.85%

29. (a) three loops (b) 16.7 Hz (c) one loop

31. (a) 3.66 m/s (b) 0.200 Hz

33. (a) 0.357 m (b) 0.715 m

35. (a) 0.656 m (b) 1.64 m

37. (a) 349 m/s (b) 1.14 m

39. (a) 0.195 m (b) 841 Hz **41.** 158 s

43. (a) 50.0 Hz (b) 1.72 m **45.** (a) 21.5 m (b) seven

47. (a) 1.59 kHz (b) odd-numbered harmonics (c) 1.11 kHz

49. 5.64 beats/s **51.** (a) 1.99 beats/s (b) 3.38 m/s

53. The following coefficients are approximate: $A_1 = 100$, $A_2 = 156$, $A_3 = 62$, $A_4 = 104$, $A_5 = 52$, $A_6 = 29$, $A_7 = 25$.

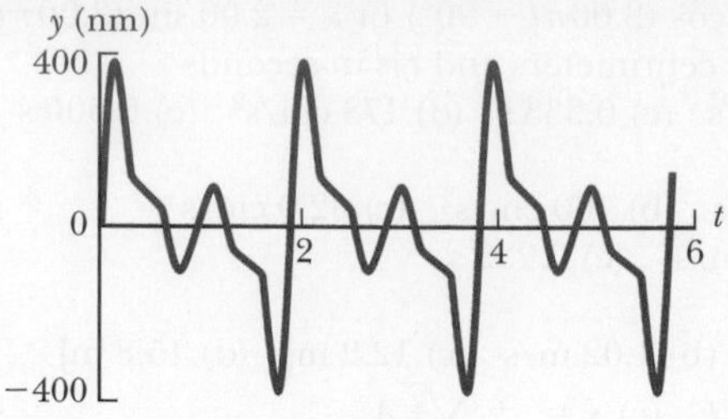

55. 800 m **57.** 1.27 cm **59.** 262 kHz

61. (a) $r = 0.078\,2\left(1 - \dfrac{4}{n^2}\right)^{1/3}$ (b) 3 (c) 0.078 2 m (d) The sphere floats on the water.

63. (a) 34.8 m/s (b) 0.986 m

65. 3.85 m/s away from the station or 3.77 m/s toward the station

67. 407 cycles **69.** (b) $A = 11.2$ m, $\phi = 63.4°$

71. (a) 78.9 N (b) 211 Hz **73.** $\sqrt{15Mg}$

Chapter 19

Answers to Quick Quizzes

1. (c) **4.** (c)
2. (c) **5.** (a)
3. (c) **6.** (b)

Answers to Odd-Numbered Problems

1. (a) −109°F, 195 K (b) 98.6°F, 310 K

3. (a) 106.7°F (b) Yes; the normal body temperature is 98.6°F, so the patient has a high fever that needs immediate attention.

5. (a) −320°F (b) 77.3 K **7.** 3.27 cm

9. (a) 0.176 mm (b) 8.78 μm (c) 0.093 0 cm^3

11. (a) 437°C (b) 2.1×10^3 °C (c) No; aluminum melts at 660°C (Table 20.2). Also, although it is not in Table 20.2, Internet research shows that brass (an alloy of copper and zinc) melts at about 900°C.

13. (a) 99.8 mL (b) It lies below the mark. The acetone has reduced in volume, and the flask has increased in volume.

15. (a) 99.4 mL (b) 2.01 L (c) 0.998 cm

17. (a) 11.2×10^3 kg/m^3 (b) 20.0 kg **19.** 4.28 atm

21. (a) 2.99 mol (b) 1.80×10^{24} molecules

23. 1.50×10^{29} molecules

25. (a) 41.6 mol (b) 1.20 kg (c) This value is in agreement with the tabulated density.

27. 2.42×10^{11} molecules

29. (a) 3.95 atm = 400 kPa (b) 4.43 atm = 449 kPa

31. 473 K **33.** 3.68 cm^3

35. (a) 2.542 cm (b) $\Delta T = 300$°C **37.** 2.86 atm

39. (a) 94.97 cm (b) 95.03 cm

41. 3.37 cm **43.** −0.094 2 Hz

45. (b) As the temperature increases, the density decreases (assuming β is positive). (c) 5×10^{-5} (°C)$^{-1}$ (d) -2.5×10^{-5} (°C)$^{-1}$

47. (a) 9.5×10^{-5} s (b) It loses 57.5 s.

49. (b) It assumes $\alpha\,\Delta T$ is much less than 1.

51. (a) yes, as long as the coefficients of expansion remain constant (b) The lengths L_C and L_S at 0°C need to satisfy $17L_C = 11L_S$. Then the steel rod must be longer. With $L_S - L_C = 5.00$ cm, the only possibility is $L_S = 14.2$ cm and $L_C = 9.17$ cm.
53. (a) 0.34% (b) 0.48% (c) All the moments of inertia have the same mathematical form: the product of a constant, the mass, and a length squared.
55. 2.74 m
57. (a) $\frac{\rho_w g P_0 V_i}{P_0 + \rho_w g h}$ (b) decrease (c) $h = \frac{P_0}{\rho_w g} = 10.3$ m
61. No; steel would need to be 2.30 times stronger.
63. (a) $L_f = L_i e^{\alpha \Delta T}$ (b) $(2.00 \times 10^{-4})\%$ (c) 59.4% (d) With this approach, 102 mL of turpentine spills, 2.01 L remains in the cylinder at 80.0°C, and the turpentine level at 20.0°C is 0.969 cm below the cylinder's rim.
65. (a) 6.17×10^{-3} kg/m (b) 632 N (c) 580 N (d) 192 Hz
67. 4.54 m

Chapter 20

Answers to Quick Quizzes

1. (i) iron, glass, water (ii) water, glass, iron
2. The figure shows a graphical representation of the internal energy of the ice as a function of energy added. Notice that this graph looks quite different from Figure 20.3 in that it doesn't have the flat portions during the phase changes. Regardless of how the temperature is varying in Figure 20.3, the internal energy of the system simply increases linearly with energy input.

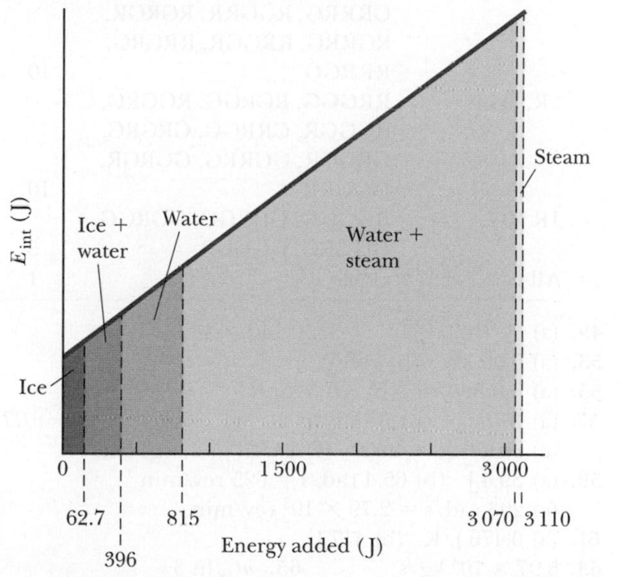

3.

Situation	System	Q	W	ΔE_{int}
(a) Rapidly pumping up a bicycle tire	Air in the pump	0	+	+
(b) Pan of room-temperature water sitting on a hot stove	Water in the pan	+	0	+
(c) Air quickly leaking out of a balloon	Air originally in the balloon	0	−	−

4. Path A is isovolumetric, path B is adiabatic, path C is isothermal, and path D is isobaric.
5. (b)

Answers to Odd-Numbered Problems

1. (a) 2.26×10^6 J (b) 2.80×10^4 steps (c) 6.99×10^3 steps
3. 23.6°C **5.** 0.845 kg **7.** 1.78×10^4 kg **9.** 29.6°C
11. (a) 1 822 J/kg · °C (b) We cannot make a definite identification. It might be beryllium. (c) The material might be an unknown alloy or a material not listed in the table.
13. (a) 380 K (b) 2.04 atm **15.** 16.3°C
17. (a) 10.0 g of ice melts, $T_f = 40.4$°C (b) 8.04 g of ice melts, $T_f = 0$°C
19. (a) 0°C (b) 114 g **21.** −466 J **23.** −1.18 MJ
25. (a) $-4P_iV_i$ (b) According to $T = (P_i/nRV_i)V^2$, it is proportional to the square of the volume.
27.

Process	Q	W	ΔE_{int}
BC	−	0	−
CA	−	+	−
AB	+	−	+

29. 720 J **31.** (a) 7.50 kJ (b) 900 K
33. (a) 0.041 0 m³ (b) +5.48 kJ (c) −5.48 kJ
35. (a) −9.08 kJ (b) 9.08 kJ
37. (a) −0.048 6 J (b) 16.2 kJ (c) 16.2 kJ
39. (a) 6.45×10^3 W (b) 5.57×10^8 J
41. 74.8 kJ **43.** 3.49×10^3 K
45. (a) 1.19 (b) a factor of 1.19
47. (a) 1.85 ft² · °F · h/Btu (b) a factor of 1.78
49. 51.2°C **51.** 1.79 kg
53. (a) 17.2 L (b) 0.351 L/s **55.** 1.90×10^3 J/kg · °C
57. (a) 9.31×10^{10} J (b) -8.47×10^{12} J (c) 8.38×10^{12} J
59. (5.87×10^4)°C **61.** 3.76 m/s **63.** 1.44 kg
65. (a) 3.16×10^{22} W (b) 3.17×10^{22} W (c) It is 0.408% larger. (d) 5.78×10^3 K
67. 3.66×10^4 s = 10.2 h **69.** (a) 4.19 mm/s (b) 12.6 mm/s

Chapter 21

Answers to Quick Quizzes

1. (i) (b) (ii) (a) **3.** (d)
2. (i) (a) (ii) (c) **4.** (c)

Answers to Odd-Numbered Problems

1. (a) 4.00 u = 6.64×10^{-27} kg (b) 55.9 u = 9.28×10^{-26} kg (c) 207 u = 3.44×10^{-25} kg
3. (a) 0.943 N (b) 1.57 Pa
5. 3.32 mol **7.** 5.05×10^{-21} J
9. (a) 3.54×10^{23} atoms (b) 6.07×10^{-21} J (c) 1.35 km/s
11. 17.4 kPa **13.** 13.5PV
15. (a) 3.46 kJ (b) 2.45 kJ (c) −1.01 kJ
17. (a) 719 J/kg · °C (b) 0.811 kg (c) 233 kJ (d) 327 kJ
19. (a) a factor of 0.118 (b) a factor of 2.35 (c) 0 (d) 135 J (e) 135 J
21. 227 K **23.** 25.0 kW
25. (a)

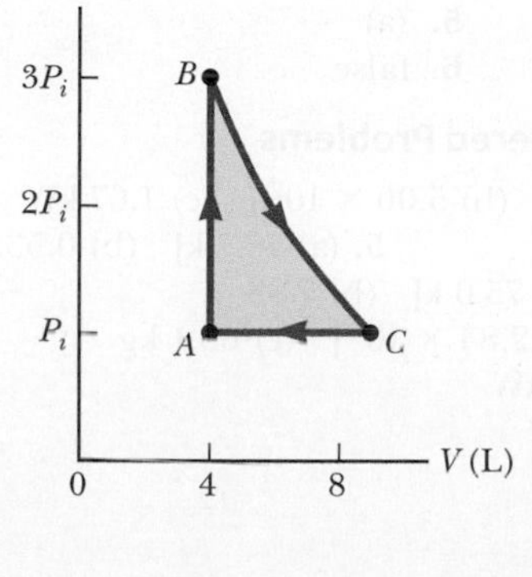

(b) 8.77 L (c) 900 K (d) 300 K (e) −336 J

27. (a) 2.45×10^{-4} m^3 (b) 9.97×10^{-3} mol (c) 9.01×10^5 Pa (d) 5.15×10^{-5} m^3 (e) 560 K (f) 53.9 J (g) 6.79×10^{-6} m^3 (h) 53.3 g (i) 2.24 K

29. 2.32×10^{-21} J

31. (a) 41.6 J/K (b) 58.2 J/K (c) 58.2 J/K, 74.8 J/K

33. (a) $2.00 \times 10^{-163} \rightarrow 0$ atoms (b) 2.70×10^{20} atoms

35. (a) 2.37×10^4 K (b) 1.06×10^3 K

37. (b) 0.278 **39.** (b) 8.31 km **41.** (a) 1.70 h (b) 1.00

43. (a) 367 K (b) The rms speed of nitrogen would be higher because the molar mass of nitrogen is less than that of oxygen. (c) 572 m/s

45. **(i)** (a) 100 kPa (b) 66.5 L (c) 400 K (d) +5.82 kJ (e) +7.48 kJ (f) −1.66 kJ; **(ii)** (a) 133 kPa (b) 49.9 L (c) 400 K (d) +5.82 kJ (e) +5.82 kJ (f) 0; **(iii)** (a) 120 kPa (b) 41.6 L (c) 300 K (d) 0 (e) −909 J (f) +909 J; **(iv)** (a) 120 kPa (b) 43.3 L (c) 312 K (d) +722 J (e) 0 (f) +722 J

47. (a) $3.65v$ (b) $3.99v$ (c) $3.00v$ (d) $\dfrac{106 m_0 v^2}{V}$ (e) $7.98 m_0 v^2$

49. (a) 300 K (b) 1.00 atm

51. 5.74×10^6 Pa = 56.6 atm **53.** 0.623

55. (a) 0.514 m^3 (b) 2.06 m^3 (c) 2.38×10^3 K (d) −480 kJ (e) 2.28 MJ

57. 0.480°C

59. (a) 1.09×10^{-3} (b) 2.69×10^{-2} (c) 0.529 (d) 1.00 (e) 0.199 (f) 1.01×10^{-41} (g) $1.25 \times 10^{-1\,082}$

61. (a) $v_{\text{rms}} = (18 k_B T/\pi \rho d^3)^{1/2} = (4.81 \times 10^{-12}) d^{-3/2}$, where v_{rms} is in meters per second and d is in meters (b) $t = (2.08 \times 10^{11}) d^{5/2}$, where t is in seconds and d is in meters (c) 0.926 mm/s and 3.24 ms (d) 1.32×10^{-11} m/s and 3.88×10^{10} s

63. (a) 0.203 mol (b) 900 K (c) 900 K (d) 15.0 L (e) $A \rightarrow B$: Lock the piston in place and put the cylinder into an oven at 900 K. $B \rightarrow C$: Keep the gas in the oven while gradually letting the gas expand to lift the piston as far as it can. $C \rightarrow A$: Move the cylinder from the oven back to the 300-K room and let the gas cool and contract.
(f, g)

	Q, kJ	W, kJ	ΔE_{int}, kJ
AB	1.52	0	1.52
BC	1.67	−1.67	0
CA	−2.53	+1.01	−1.52
ABCA	0.656	−0.656	0

67. (a) 3.34×10^{26} molecules (b) during the 27th day (c) 2.53×10^6

69. (a) 0.510 m/s (b) 20 ms **71.** 510 K and 290 K

Chapter 22

Answers to Quick Quizzes

1. **(i)** (c) **(ii)** (b)
2. (d)
3. C, B, A
4. (a) one (b) six
5. (a)
6. false

Answers to Odd-Numbered Problems

1. (a) 0.294 (or 29.4%) (b) 5.00×10^2 J (c) 1.67 kW

3. (a) 6.94% (b) 335 J **5.** (a) 10.7 kJ (b) 0.533 s

7. 55.4% **9.** (a) 75.0 kJ (b) 7.33

11. (a) 4.51×10^6 J (b) 2.84×10^7 J (c) 68.1 kg

13. (a) 67.2% (b) 58.8 kW

15. (a) 8.70×10^8 J (b) 3.30×10^8 J **17.** 9.00

19. (a) 564°C (b) No; a real engine will always have an efficiency *less* than the Carnot efficiency because it operates in an irreversible manner.

21. 11.8 **23.** 1.86

25. (a) 741 J (b) 459 J

27. (a) 564 K (b) 212 kW (c) 47.5%

29. (a) $\dfrac{Q_c}{\Delta t} = 1.40\left(\dfrac{0.5T_h + 383}{T_h - 383}\right)$, where $Q_c/\Delta t$ is in megawatts and T is in kelvins (b) The exhaust power decreases as the firebox temperature increases. (c) 1.87 MW (d) 3.84×10^3 K (e) No answer exists. The energy exhaust cannot be that small.

31. 1.17 **33.** (a) 244 kPa (b) 192 J

37. 143 J/K **39.** 57.2 J/K **41.** 0.507 J/K

43. (a) −3.45 J/K (b) +8.06 J/K (c) +4.62 J/K

45. 3.28 J/K

47. (a)

Macrostate	Microstates	Number of ways to draw
All R	RRR	1
2 R, 1 G	GRR, RGR, RRG	3
1 R, 2 G	GGR, GRG, RGG	3
All G	GGG	1

(b)

Macrostate	Microstates	Number of ways to draw
All R	RRRR	1
4R, 1G	GRRRR, RGRRR, RRGRR, RRRGR, RRRRG	5
3R, 2G	GGRRR, GRGRR, GRRGR, GRRRG, RGGRR, RGRGR, RGRRG, RRGGR, RRGRG, RRRGG	10
2R, 3G	RRGGG, RGRGG, RGGRG, RGGGR, GRRGG, GRGRG, GRGGR, GGRRG, GGRGR, GGGRR	10
1R, 4G	RGGGG, GRGGG, GGRGG, GGGRG, GGGGR	5
All G	GGGGG	1

49. (a) $\frac{1}{3}$ (b) $\frac{2}{3}$ **51.** 0.440 = 44.0%

53. (a) 5.00 kW (b) 763 W

55. (a) $-0.390nR$ (b) $-0.545nR$

57. (a) $3nRT_i$ (b) $3nRT_i \ln 2$ (c) $-3nRT_i$ (d) $-nRT_i \ln 2$ (e) $3nRT_i(1 + \ln 2)$ (f) $2nRT_i \ln 2$ (g) 0.273

59. (a) 39.4 J (b) 65.4 rad/s = 625 rev/min (c) 293 rad/s = 2.79×10^3 rev/min

61. (a) 0.476 J/K (b) 417 J

63. 5.97×10^4 kg/s **65.** $nC_P \ln 3$

67. (a) 4.10×10^3 J (b) 1.42×10^4 J (c) 1.01×10^4 J (d) 28.8% (e) Because $e_C = 80.0\%$, the efficiency of the cycle is much lower than that of a Carnot engine operating between the same temperature extremes.

69. (b) yes (c) No; the second law refers to an engine operating in a cycle, whereas this problem involves only a single process.

71. (a) $P_A = 25.0$ atm, $V_A = 1.97 \times 10^{-3}$ m^3; $P_B = 4.13$ atm, $V_B = 1.19 \times 10^{-2}$ m^3; $P_C = 1.00$ atm, $V_C = 3.28 \times 10^{-2}$ m^3; $P_D = 6.05$ atm, $V_D = 5.43 \times 10^{-3}$ m^3 (b) 2.99×10^3 J

index

Locator note: **boldface** indicates a definition; *italics* indicates a figure; *t* indicates a table

Standard Abbreviations and Symbols for Units

Symbol	Unit	Symbol	Unit
A	ampere	K	kelvin
u	atomic mass unit	kg	kilogram
atm	atmosphere	kmol	kilomole
Btu	British thermal unit	L	liter
C	coulomb	lb	pound
°C	degree Celsius	ly	lightyear
cal	calorie	m	meter
d	day	min	minute
eV	electron volt	mol	mole
°F	degree Fahrenheit	N	newton
F	farad	Pa	pascal
ft	foot	rad	radian
G	gauss	rev	revolution
g	gram	s	second
H	henry	T	tesla
h	hour	V	volt
hp	horsepower	W	watt
Hz	hertz	Wb	weber
in.	inch	yr	year
J	joule	Ω	ohm

Mathematical Symbols Used in the Text and Their Meaning

Symbol	Meaning
$=$	is equal to
$\equiv$	is defined as
$\neq$	is not equal to
$\propto$	is proportional to
$\sim$	is on the order of
$>$	is greater than
$<$	is less than
$\gg(\ll)$	is much greater (less) than
$\approx$	is approximately equal to
Δx	the change in x
$\sum_{i=1}^{N} x_i$	the sum of all quantities x_i from $i = 1$ to $i = N$
$\|x\|$	the absolute value of x (always a nonnegative quantity)
$\Delta x \to 0$	Δx approaches zero
$\frac{dx}{dt}$	the derivative of x with respect to t
$\frac{\partial x}{\partial t}$	the partial derivative of x with respect to t
$\int$	integral